COMPREHENSIVE MEDICINAL CHEMISTRY II

COMPREHENSIVE MEDICINAL CHEMISTRY II

Editors-in-Chief

Dr John B Taylor

Former Senior Vice-President for Drug Discovery, Rhône-Poulenc Rorer, Worldwide, UK

Professor David J Triggle

State University of New York, Buffalo, NY, USA

Volume 1

GLOBAL PERSPECTIVE

Volume Editor

Dr Peter D Kennewell

Swindon, UK

ELSEVIER

AMSTERDAM BOSTON HEIDELBERG LONDON NEW YORK OXFORD
PARIS SAN DIEGO SAN FRANCISCO SINGAPORE SYDNEY TOKYO

Elsevier Ltd.
The Boulevard, Langford Lane, Kidlington, Oxford OX5 1GB, UK

First edition 2007

British Library Cataloguing in Publication Data
A catalogue record for this book is available from the British Library

Library of Congress Catalog Number: 2006936669

ISBN-13: 978-0-08-044513-7
ISBN-10: 0-08-044513-6

Printed and bound in Spain

06 07 08 09 10 10 9 8 7 6 5 4 3 2 1

Disclaimers

Contents

Sources of New Drugs

Animal Experimentation

The Role of Industry

Drug Discovery: Revolution, Decline?

Healthcare in the Social Context

Ethical Issues

Funding and Regulation of Research

Intellectual Property

Contents of all Volumes

Volume 5 ADME-Tox Approaches

Preface

The first edition of *Comprehensive Medicinal Chemistry* was published in 1990 and was intended to present an integrated and comprehensive overview of the then rapidly developing science of medicinal chemistry from its origins in organic chemistry. In the last two decades, the field has grown to embrace not only all the sophisticated synthetic and technological advances in organic chemistry but also major advances in the biological sciences. The mapping of the human genome has resulted in the provision of a multitude of new biological targets for the medicinal chemist with the prospect of more rational drug design (CADD). In addition, the development of sophisticated in silico technologies for structure–property relationships (ADMET) enables a much better understanding of the fate of potential new drugs in the body with the subsequent development of better new medicines.

It was our ambitious aim for this second edition, published 16 years after the first edition, to provide both scientists and research managers in all relevant fields with a comprehensive treatise covering all aspects of current medicinal chemistry, a science that has been transformed in the twenty-first century. The second edition is a complete reference source, published in eight volumes, encompassing all aspects of modern drug discovery from its mechanistic basis, through the underlying general principles and exemplified with comprehensive therapeutic applications. The broad scope and coverage of *Comprehensive Medicinal Chemistry II* would not have been possible without our panel of authoritative Volume Editors whose international recognition in their respective fields has been of paramount importance in the enlistment of the world-class scientists who have provided their individual 'state of the science' contributions. Their collective contributions have been invaluable.

Volume 1 (edited by Peter D Kennewell) overviews the general socioeconomic and political factors influencing modern R&D in both the developed and developing worlds. Volume 2 (edited by Walter H Moos) addresses the various strategic and organizational aspects of modern R&D. Volume 3 (edited by Hugo Kubinyi) critically reviews the multitude of modern technologies that underpin current discovery and development activities. Volume 4 (edited by Jonathan S Mason) highlights the historical progress, current status, and future potential in the field of computer-assisted drug design (CADD). Volume 5 (edited by Bernard Testa and Han van de Waterbeemd) reviews the fate of drugs in the body (ADMET), including the most recent progress in the application of 'in silico' tools. Volume 6 (edited by Michael Williams) and Volume 7 (edited by Jacob J Plattner and Manoj C Desai) cover the pivotal roles undertaken by the medicinal chemist and pharmacologist in integrating all the preceding scientific input into the design and synthesis of viable new medicines. Volume 8 (edited by John B Taylor and David J Triggle) illustrates the evolution of modern medicinal chemistry with a selection of personal accounts by eminent scientists describing their lifetime experiences in the field, together with some illustrative case histories of successful drug discovery and development.

We believe that this major work will serve as the single most authoritative reference source for all aspects of medicinal chemistry for the next decade and it is intended to maintain its ongoing value by systematic electronic upgrades. We hope that the material provided here will serve to fulfill the words of Antoine de Saint-Exupery (1900–44) and allow future generations of medicinal chemists to discover the future.

'As for the future, your task is not to foresee it but to enable it'
Citadelle (1948)

John B Taylor and David J Triggle

Preface to Volume 1

The objective of this volume is to offer an educative and authoritative background discussion about the drug discovery and development process. By giving a wide contextual element to the subject, the intention is to be of interest to two general groups: the specialist scientific community wishing to gain insights into the socioeconomic factors influencing drug development and the nonscientific professionals in the industry wishing to know more about the background to current products and the trends which influence the discovery of new agents.

In the twenty-first century it is difficult to envisage any scientific activity that does not have a social dimension, but this is particularly true for all aspects of medical research including the discovery and development of drugs. However, there are many problems. We all want to live healthier and longer lives, and we have, in all probability, either experienced or known someone who has experienced the trauma that severe illness can bring to an individual. At least in the developed world, life expectancy continues to improve and conditions that only a short while ago would have proved severely debilitating or even fatal can now be controlled. However, all such societies are struggling with the costs of providing treatments and in meeting the increasing demand for medical interventions and with the ethical dilemmas thrown up by such developments. The situation in the developing world is, of course, very different with too many of our fellow humans having, at best, access to only rudimentary health care and often to none at all. Too many diseases that could be readily controlled in developed countries are fatal and many diseases are not even the subject of intensive research because of the fear that patients would not be able to pay for the treatment. The pharmaceutical industry, despite being responsible for the introduction of virtually all important drugs to the market, has a poor public image. It receives scant praise for the discovery of major new medicines but plenty of criticisms when products are withdrawn or when it appears to price its products too high. There appears to be a general perception that drugs are made for profit rather than medical necessity.

This volume does not claim to resolve these issues but does attempt to discuss some of them in a balanced way, thus giving a view of the world in which drug discovery operates.

The volume is arranged into ten sections and opens with a personal reminiscence from one of the pioneers of medicinal chemistry research, Ralph Hirschmann, who describes his experiences during a long and highly successful career with Merck. Two chapters describe the evolution of medicinal chemistry/drug discovery and update the list of drugs introduced since the first edition of this work. In the next section, three chapters deal with the impact of the revolution in understanding of gene function, much of which has arisen since the first edition of this work was published. Thus, epigenetics and the promise of individually tailored medicines are described before the chemical challenges of producing gene therapy-derived treatments are discussed. This is a field that is changing so rapidly that it is probably impossible to predict what the situation will be at the time when the next edition of this work is prepared.

The next section comprises five chapters describing different possible sources of new drugs, including one on the growth and potential of biological macromolecules. Then there is a chapter discussing ongoing efforts to reduce the number of animals used in essential testing of drug candidates. Such efforts will never convince the hard core of activists totally opposed to the use of any animals, but they do show how it is possible to reduce the numbers involved to the lowest possible level. Attention then moves from the scientific to the social dimension with six chapters, in two sections, dealing with different aspects of the pharmaceutical industry.

Since the first edition of the volume, there has been further amalgamation of big companies along with an explosive growth of new small companies. There also appears to have been a significant fall in the efficiency of drug discovery and this is discussed.

The seventh section discusses the provision of health care in developed and developing countries and the roles of orphan and generic drugs. Ethical issues, both as they affect the industry and the wider community, are discussed in the eighth section while the ninth discusses the role of governments in sponsoring medical research and in regulating products already released to the market. Finally, the patenting of drug discoveries is the subject of the last chapter.

I would like to thank all the authors for their efforts in producing this work and hope that both groups of readers described above will find it interesting.

Peter D Kennewell

Editors-in-Chief

John B Taylor, DSc, was formerly Senior Vice President for Drug Discovery at Rhône-Poulenc Rorer. He obtained his BSc in chemistry from the University of Nottingham in 1956 and his PhD in organic chemistry at the Imperial College of Science and Technology with Nobel Laureate Professor Sir Derek Barton in 1962. He subsequently undertook postdoctoral research fellowships at the Research Institute for Medicine and Chemistry in Cambridge (US) with Sir Derek and at the University of Liverpool (UK), before entering the pharmaceutical industry.

During his career in the pharmaceutical industry Dr Taylor spent more than 30 years covering all aspects of research and development in an international environment. From 1970 to 1985 he held a number of positions in the Hoechst Roussel organization, ultimately as research director for Roussel Uclaf (France). In 1985 he joined Rhône-Poulenc Rorer holding various management positions in the research groups worldwide before becoming Senior Vice President for Drug Discovery in Rhône-Poulenc Rorer.

Dr Taylor is the co-author of two books on medicinal chemistry and has more than 50 publications and patents in medicinal chemistry. He was joint executive editor for the first edition of Comprehensive Medicinal Chemistry, a visiting professor for medicinal chemistry at the City University (London) from 1974 to 1984 and was awarded a DSc in medicinal chemistry from the University of London in 1991.

David J Triggle, PhD, is the University Professor and a Distinguished Professor in the School of Pharmacy and Pharmaceutical Sciences at the State University of New York at Buffalo. Professor Triggle received his education in the UK with a BSc degree in chemistry at the University of Southampton and a PhD degree in chemistry at the University of Hull working with Professor Norman Chapman. Following postdoctoral fellowships at the University of Ottawa (Canada) with Bernard Belleau and the University of London (UK) with Peter de la Mare he assumed a position in the School of Pharmacy at the University at Buffalo. He served as Chairman of the Department of Biochemical Pharmacology from 1971 to 1985 and as Dean of the School of Pharmacy from 1985 to 1995. From 1996 to 2001 he served as Dean of the Graduate School and from 1999 to 2001 was also the University Provost. He is currently the University Professor, in which capacity he teaches bioethics and science policy, and is President of the Center for Inquiry Institute, a secular think tank located in Amherst, New York.

Professor Triggle is the author of three books dealing with the autonomic nervous system and drug–receptor interactions, the editor of a further dozen books, some 280 papers, some 150 chapters and reviews, and has presented over 1000 invited lectures worldwide. The Institute for Scientific Information lists him as one of the 100 most highly cited scientists in the field of pharmacology. His principal research interests have been in the areas of drug–receptor interactions, the chemical pharmacology of drugs active at ion channels, and issues of graduate education and scientific research policy.

Editor of Volume 1

Peter D Kennewell studied Natural Sciences at Churchill College, Cambridge, and then moved to the University of East Anglia to complete a PhD with Professor Alan Katritzky. This was followed by a postdoctoral fellowship with Professor Ned Heindel at Lehigh University, Bethlehem, Pennsylvania, working on the synthesis of novel antimalarial agents. On his return to the United Kingdom, he joined Roussel Laboratories where he remained in a variety of managerial positions until the Swindon research unit was closed following the merger which created Hoechst Marion Roussel. He then joined the Biotechnology and Biological Sciences Research Council, and worked for the Biomolecular Sciences Committee and the Biochemistry and Cell Biology Committees. His particular interests were in the encouragement of the application of chemistry to biological problems. He finally retired from this organization in June 2005.

Dr Kennewell acts as a course monitor and occasional lecturer on a number of online distance learning courses (Medicinal and Pharmaceutical Chemistry, Heterocyclic Chemistry) run by Lehigh University. He is the author or co-author of 65 publications, 13 patents and, with John Taylor, co-authored *Introductory Medicinal Chemistry* and *Modern Medicinal Chemistry*. He was a member of the Editorial Board and Editor of Volume 1 of *Comprehensive Medicinal Chemistry*.

Contributors to Volume 1

R Barker
Association of the British Pharmaceutical Industry (ABPI), London, UK

D J Bower
University of Dundee, Dundee, UK

D Brown
Alchemy Biomedical Consulting, Cambridge, UK

L A Cabanilla
Tufts University, Boston, MA, USA

D Callahan
The Hastings Center, Garrison, NY, USA

J Cockbain
Frank B Dehn & Co, Oxford, UK

R D Combes
FRAME, Nottingham, UK

G M Cragg
NCI, Frederick, MD, USA

L Da Ros
GlaxoSmithKline, Verona, Italy

M Darnbrough
London, UK

Y T Das
Rutgers – The State University of New Jersey, New Brunswick, NJ, USA

M L Eaton
Stanford University, Graduate School of Business, Stanford, CA, USA

P W Erhardt
University of Toledo, Toledo, OH, USA

Q C Franco
StratEdge, Washington, DC, USA

D Gurwitz
Tel-Aviv University, Tel-Aviv, Israel

R F Hirschmann
University of Pennsylvania, Philadelphia, PA, USA

M M Hopkins
University of Sussex, Brighton, UK

J R Howard,
University of Pennsylvania, Philadelphia, PA, USA

P D Kennewell
Swindon, UK

A Kraft
University of Sussex, Brighton, UK

P A Lachance
Rutgers – The State University of New Jersey, New Brunswick, NJ, USA

A Li Wan Po
Centre for Evidence-Based Pharmacotherapy, Nottingham, UK

S Mahdi,
University of Sussex, Brighton, UK

V G Manolopoulos
Democritus University of Thrace, Alexandroupolis, Greece

P A Martin
Nottingham University, Nottingham, UK

A D Miller
Imperial College London, London, UK

C-P Milne
Tufts University, Boston, MA, USA

L A Mitscher
University of Kansas, Lawrence, KS, USA

D J Newman
NCI, Frederick, MD, USA

C G Newton
BioFocus DRI, Saffron Walden, UK

P Nightingale
University of Sussex, Brighton, UK

J K Osbourn
Cambridge Antibody Technology, Cambridge, UK

J K Ozawa
SRI International, Arlington, VA, USA

J R Proudfoot
Boehringer Ingelheim Inc., Ridgefield, CT, USA

E Ratti
GlaxoSmithKline, Verona, Italy

M Sahl
MJSahl Consulting, Philadelphia, PA, USA

M Skingle
GlaxoSmithKline, Stevenage, UK

J L Sturchio
Merck & Co., Inc., White House Station, NJ, USA

M Thanou
Imperial College London, London, UK

D Trist
GlaxoSmithKline, Verona, Italy

S Waddington
University College London, London, UK

A A Wasunna
The Hastings Center, Garrison, NY, USA

W W Weber
University of Michigan, Ann Arbor, MI, USA

P R Wolpe
University of Pennsylvania, Philadelphia, PA, USA

1.01 Reflections of a Medicinal Chemist: Formative Years through Thirty-Seven Years Service in the Pharmaceutical Industry

R F Hirschmann, University of Pennsylvania, Philadelphia, PA, USA
J L Sturchio, Merck & Co., Inc., Whitehouse Station, NJ, USA

1.01.1 Introduction

It has been my good fortune to have been trained and influenced throughout my career by outstanding mentors and collaborators, both biologists and chemists. Their collective impact is immeasurable. They differed from each other in their fields of expertise, their research philosophy, and their 'Weltanschauung.' Taken together, the thoughts contained in this chapter reflect – to paraphrase Ralph Waldo Emerson[1]– the amassed thoughts and experiences of innumerable minds.

1.01.2 The Training of a Medicinal Chemist

1.01.2.1 Graduate School

Organic chemistry is the foundation of medicinal chemistry. I was very fortunate that Professor William S. Johnson of the University of Wisconsin, Madison, accepted me as a graduate student in 1946. My PhD thesis involved natural product total synthesis, and the target was a steroid. My knowledge of natural product total synthesis made me an attractive candidate for the pharmaceutical industry, for reasons that have remained a tradition widely accepted by big Pharma, but not by biotech companies. The attraction of a natural product as a synthetic target lies in part in the fact that the target was set by Nature, which gives it an aura of legitimacy. The most challenging part in the synthesis of a complex natural product often concerns the retrosynthetic analysis of its synthesis, which is generally determined by the major professor, not by the student or postdoctoral fellow. Successive heads of medicinal chemistry departments in big Pharma were trained in natural product total synthesis and they, in turn, tended to prefer candidates with a similar background when making job offers. This preference tends to become a self-fulfilling proposition.

Nature has evolved expertise in the use of reactions such as aldol condensations and others, and they are used repeatedly. Partly as a result thereof, chemists can rely on volumes of literature dealing with these reactions. To be sure there will be one or several steps in any natural product synthesis that will require creativity, which will ultimately make the difference between success and failure.

The situation is very different in the synthesis of unnatural products, which are today generally designed to display predetermined physical, chemical, or biological properties. There may be little or no prior art to guide the synthesis. Let me give one example: my colleague at the University of Pennsylvania, Professor Amos B. Smith, III, an acknowledged leader in natural product total synthesis, and I initiated a program in 1988 to design and synthesize inhibitors of proteolytic enzymes using pyrrolinone-based mimics of amino acids. Interestingly, chiral pyrrolinone-based amino acid mimetics had not been described. The endeavor proved to be rewarding, in that it led to an inhibitor of HIV-1 protease, which has better pharmacokinetic properties than its peptidal precursor, both because the pyrroline bonds are stable to proteases and because transport across membranes involves a lower desolvation penalty.[2] The x-ray crystal structure of the enzyme-inhibitor complex has also been reported.[2] In a recent publication, Smith *et al.* reported the 'total synthesis' of an unnatural tetrapyrrolinone mimicking the β-turn of the peptide hormone somatostatin. By current standards, and especially by Smith's own standards, it is not an imposing structure. However, the synthesis, including the preparation of the required unnatural building blocks, entailed some 53 steps, and required 46 person months. The synthesis of each of the four amino acid mimicking building blocks required about 11 steps, even though only one of the pyrrolinones contained a functional side chain. Thus, the total synthesis of unnatural products can be challenging. Moreover, the total synthesis of an unnatural product may not be publishable unless it possesses the predetermined physical and/or biological properties. Finally, in natural product total synthesis, very small amounts of material suffice to establish the identity with the natural product. In the case of nonnatural products, considerable amounts of material may be required to determine whether the substance possesses the desired properties. For these reasons I believe that natural product total synthesis should no longer be regarded as the only appropriate training for a potential employee by big Pharma.

My initial synthetic target as a graduate student in Madison was estrone. Because Professors William S. Johnson and Alfred L. Wilds were recognized leaders in steroid synthesis, they were quickly informed of the dramatic results obtained by Dr Philip Hench at the Mayo Clinic in the first clinical trial of cortisone in April 1949. Professor Johnson, therefore, shifted me from the total synthesis of estrone to that of cortisone. I was told only that cortisone "did something" in the clinic. Looking back, I am surprised that I did not ask any further questions of my mentor. Professor Johnson remains for me the perfect major professor, because of his ethical and scientific standards, his dedication to research and to his students, and for his creativity. As a graduate of Oberlin College I had taken relatively few courses in chemistry. As a first year graduate student in Madison I was, therefore, concerned whether I could compete with fellow students from larger universities, who had taken many more hours of chemistry. This concern did not, however, prove to be a problem, suggesting perhaps that there is only so much chemistry that one can absorb as an undergraduate. Moreover, Dr Johnson was tolerant toward anyone willing to be in the laboratory in the evenings and over weekends. I was also very fortunate that my laboratory accommodated students of both Professor Johnson and Professor Alfred L. Wilds. Professor Wilds kindly took an interest in my training, and contributed significantly to my education. It will be difficult for today's reader of this textbook to comprehend that in 1950, the year I received the doctorate degree from the University of Wisconsin, nuclear magnetic resonance (NMR) spectroscopy was unknown and infrared spectral capability was not available in Madison. It has always amazed me that in the early twentieth century the structures of steroids, both the scaffold and the substituents, were elucidated through very hard and brilliant work, using such unremarkable reagents as potassium permanganate and acetic anhydride. Reading the history of early steroid research thus has always had a very humbling effect on me. I believe that it is sad that modern textbooks, and often classroom lectures in organic chemistry, have generally become dehumanized. It is argued that there is so much to be taught that there is neither time nor space to name the chemists on whose shoulders we stand. Perhaps so, but in my opinion, we pay a high price for this exclusive concern with 'the facts,' at the expense of our invaluable scientific heritage.

In the 1940s ultraviolet spectroscopy was well established as a tool in spectroscopy, but its value to organic chemists was questioned. A notable exception was Professor Wilds. Because he was very generous with his time, UV spectroscopy contributed significantly to my PhD thesis. For example, I was able to use this training to discover that a rearrangement of a double bond had taken place during a saponification step in Professor Johnson's total synthesis of equilenin, which caused a double bond to migrate from conjugation with an ester/acid into conjugation with a ketone. I was so excited by this discovery, that I expected the world to stand still once the news was out. This did not turn out to be the case! Moreover a physical chemist on my oral exam committee was amused that a synthetic organic chemist would take UV spectra seriously. I was to realize only much later, after I had joined the faculty at the University of Pennsylvania, that few, even well-trained organic chemists, make optimal use of UV spectroscopy. To give just one example: few medicinal chemists use UV spectroscopy to determine the precise concentration of a test compound in an aqueous solution prior to a biological assay. I am also amused when graduate students sign up for NMR time, when the UV spectrum would have provided the desired information right away.

In addition to Professors Johnson and Wilds, I also benefited greatly from a course taught by Professor Samuel McElvain on the role of electronic concepts in organic chemistry as developed by Lowery, Kossel, Remick, Ingold, and others. In combination with the research by H.C. Brown on steric effects, the two pillars of twentieth century medicinal chemistry were in place. Finally, it was a privilege to take Professor Homer Adkins' course in advanced organic chemistry. He was the only professor I have encountered who taught a philosophy of science, stressing skepticism as epitomized by his observation that "logic is the organized way of going wrong with confidence."

1.01.2.2 My 37 Years At Merck & Co., Inc.

1.01.2.2.1 The cortisol era

1.01.2.2.1.1 My years in process research

In the fall of 1949, Dr Max Tishler, then Director of Process Research at Merck & Co., Inc., and Dr Karl Folkers, Director of Fundamental Research, came to Madison on a recruiting trip. Before his arrival I realized that Professor Johnson had a healthy respect for Dr Tishler, because my mentor told me not to be too specific about our strategy for the total synthesis of cortisone "lest Dr Tishler figure out exactly what we are trying to do." The interviews were a success, and for the first time, I saw Dr Tishler's eyes flash with excitement. To an extraordinary degree he was both a passionate scientist and a hard-nosed boss. I was offered a position by both Dr Folkers and Dr Tishler. I should not have been surprised, because at that time Merck was interested in chemists with prior experience in steroid chemistry. I accepted Dr Tishler's offer, but I learned that I would be working on a total synthesis of folic acid, a somewhat unexpected assignment.

Several years later Dr Tishler was elected a member of the National Academy of Sciences (NAS) upon nomination by Professor Robert Woodward. It was extraordinary for a Process Research Chemist to be elected to the NAS, and it greatly strengthened his position with corporate management. Dr Tishler also hired a Bryn Mawr graduate trained by Professor Marshall Gates, Ms Lucy Aliminosa, who became my wife in 1951. The Tishlers were good friends of Lucy, and Dr Tishler therefore took an interest in me, I believe.

Thus, in the spring of 1950 I started to work as a Process Research Chemist reporting to Dr Norman Wendler, a kind person and an outstanding organic chemist. My assignment had changed, however. Instead of working on a synthesis of folic acid, I was to develop a new method to convert cholic acid into desoxycholic acid, the first step in the partial synthesis of cortisone. Dr Wendler was to be the first of the 13 sequential superiors to whom I reported at Merck. I learned an important lesson from having so many different mentors. To succeed in the art of directing research, there is not much point in acquiring a supervisor's style. It is best to be one's natural self – for better or worse – because one's subordinates are too astute to be fooled by acting.

I also learned that communication is a challenging art. When one is really excited about an idea, one must be prepared to devote time and energy to convey one's enthusiasm to colleagues. Surprisingly, at the end of a detailed discussion, the two participants often leave a conversation with different impressions of the conclusion that was reached.

It is worth mentioning that the environment at Merck in the 1950s was different from that which prevails throughout the industry today. Dr Tishler was a no nonsense, demanding research director, and I always made sure that I acted promptly on all of his instructions and suggestions. This is not surprising. What is noteworthy is that I was nevertheless free to pursue my own ideas. Some of that work was done in the evenings and over weekends, but some of it was carried out on 'company time.' Importantly, when an unauthorized project produced a useful result, my supervisors were pleased. I have the impression that the environment has changed, because today bench chemists and their supervisors are expected to devote all of their energy to the officially assigned tasks. Clearly the assigned tasks must be the first priority, but when leadership becomes excessively autocratic, and when creative initiative is stifled, something invaluable is lost. I believe that the insightful manager knows which of his associates should be closely supervised, and who should be left some measure of freedom. 'Management by objectives' is a sound concept, but when carried too far, i.e., when it is assumed that all good ideas come from the top or from a committee, productivity is likely to suffer.

Although I was not aware of it, I started my industrial career during one of the most remarkable periods in the history of the pharmaceutical industry. In the 1940s drug discovery at Merck, as at other major pharmaceutical companies, was 'largely based on blind, empirical screening of myriad chemical entities or on extracting compounds from microbial broths derived from soil samples.'[3] The age of 'rational drug design' was yet to come. Arguably the cortisone era served as a bridge between the two. Merck's decision to go forward was based on the faith, intuition, and insightfulness that cortisol, the major constituents of the adrenal gland, must play a major role in physiology, and that cortisol and cortisone might have great potential also in therapy. In addition, there was the false rumor that the Luftwaffe could fly at higher altitudes than our airforce because they were supplied with cortisone. Actually 'rational' analysis of the biological properties of the then known close analogs of cortisone was definitely not encouraging. Cortisone has two hydroxyl groups, at C17α and at C21, and three ketones at C3, C11, and C20. Kendall's compound A, lacking only the 17α hydroxyl of cortisone, and Reichstein's substance S, lacking only the 11-ketone, were devoid of any interesting biological properties even though between them they possessed all of the oxygen functions of cortisone. A rational drug designer should be forgiven had he or she decided that the synthesis of cortisone from cholic acid was not worth the effort of a synthesis of some 36 steps. Yet Professor Homer Adkins' quote (see above) would have been relevant. Fortunately Dr Tishler and Dr Lewis H. Sarett undertook the enormous challenge of making cortisone available by synthesis to make a clinical trial possible. Even though Kendall's compound A and Reichstein's substance S lacked any useful biological properties, the synthesis of cortisone was initiated at Merck in March of 1946.[4] Arguably, Merck and Co., Inc., played the role of a venture capitalist at that time. Nowadays, venture capitalists who have sought my opinion about investing in a given new technology want to be all but assured that success is certain. Merck had no such assurance in 1946.

As reported by Fieser and Fieser,[4] three Merck scientists and a technician processed 577 kg of desoxycholic acid and Dr Edward C. Kendall sent some partially purified desoxycholic acid to Rahway, NJ, for further purification. Cortisone prepared from desoxycholic acid became available just 2 years later, in April of 1948, and was sent to Hench and Kendall at the Mayo Clinic for initiation of the clinical trial the following month. Many persons are under the impression that natural cortisone was used in the successful clinical trial that resulted in the massive effort at Merck. This was not the case. The huge effort had to precede the clinical trial, which fully validated Dr Hench's intuition that cortisone might prove to be an anti-inflammatory chemical entity, based on his astute observation that the condition of women suffering from arthritis improved when they became pregnant. The enthusiasm following the first clinical trial involving a bedridden 29-year-old woman declined somewhat when several side effects associated with steroid therapy became apparent. These include adrenal atrophy, negative nitrogen balance, osteoporosis, 'moon face,' and others. Ultimately these side effects were to lead to the search for and discovery of nonsteroidal anti-inflammatory medications known as nonsteroidal anti-inflammatory drugs (NSAIDS).

Prior to the synthesis by Dr T. Shen *et al.* of indomethacin,[5] the first important NSAID in 1963, attempts were made to discover analogs of cortisol with fewer side effects than those of cortisone or its 11-dihydro analog, the endogenous hormone cortisol. These two steroids are interconverted in vivo. The partial synthesis of cortisol, possessing the critical 11-β-hydroxyl group, was accomplished by Dr Wendler and Dr Robert Graber in Rahway. The decade that followed became my introduction to medicinal chemistry. The goal was to discover an analog of cortisone/cortisol devoid of the side effects of the endogenous compounds. Along with all other medicinal chemists, it took me nearly a decade to appreciate that the strategy to find a safer drug by generating more potent compounds, permitting a reduction in dosage, was doomed to failure because the desired biological effects and the side effects of cortisol were mediated via a common receptor. This is probably the most important lesson in medicinal chemistry that I learned during my first 10 years in industry. More recent research has, however, opened a new window,[6] via novel structures that are devoid of the classic steroid scaffold.[7,8] With hindsight then, it is no surprise that the only side effect of cortisol that was successfully abolished via analog synthesis is salt retention, because it is mediated via a separate receptor, for which aldosterone is the endogenous ligand. That cortisol, an endogenous steroid hormone, can produce significant side effects, reflects the difference between physiologic and pharmacologic doses.

1.01.2.2.1.1.1 Toward more potent and safer analogs of cortisol The search for alternative synthetic routes to cortisol and other research led to an understanding of the structural factors affecting relative glucocorticoid and mineralocorticoid activities, i.e. the desired anti-inflammatory activity and the unwanted salt retention.

Dr Hans Hirschmann, my brother, and his coauthors at Case Western Reserve University found unexpectedly that 16α-hydroxysteroids do not display mineralocorticoid activity.[9] Therefore, Dr Bernard Ellis *et al.* at the British Drug House[10] and Dr Allen Bernstein at Lederle[11] synthesized 16α-hydroxycortisol. These chemists discovered between 1955 and 1956 that the compound displayed considerable glucocorticoid activity, but did not cause sodium retention. Thus, the 16α-hydroxy-steroids represented an advance in steroid therapy.

The search for a route to cortisol via an 11α-hydroxylated steroid led in 1954 to the interesting discovery by Dr Josef Fried and Ms Emily F. Sabo at Squibb, that the 9α-fluoro analog of cortisol showed a ten-fold increase in potency over the endogenous cortisol.[12] The discovery was made possible by the intelligent pursuit of fortuitous observations and their insightful interpretation by Fried and Sabo. The issue was not trivial. On the one hand, there was the desire to remain focused, and on the other, the danger of neglecting unexpected results that might prove to be more significant than the original objective. 9α-Fluorocortisol was of no clinical interest because of its unacceptably high salt retention properties, even in the 1-dehydro series.[13] This problem was later overcome at Schering and at Merck by the additional introduction of 16α- or 16β-methyl substituents, which like the 16α-hydroxyl, blocked the interaction of the resulting steroids with the aldosterone receptor but enhanced affinity at the desired glucocorticoid receptor. This led ultimately to the synthesis of the medically important C-16 diastereomeric 16-methyl-9α-fluoroprednisolone analogs.

The conversion of the relevant 9β,11β-epoxides into the 9α-fluoro-11β-hydroxy-steroids on treatment with anhydrous hydrogen fluoride (HF) in alcohol-free chloroform afforded only a 50% yield. At about that time I was transferred from Dr Wendler's group to that of Dr Robert E. Jones to become more involved in typical Process Research. The yield for the HF reaction was an issue of sufficient importance to the company that Corporate Management was kept informed on a daily basis. I found within a day that the reaction actually afforded much better yields in straight alcohol and in other organic bases, as long as the concentration of anhydrous HF was high enough to permit acid-catalyzed opening of the epoxide.[14] We concluded that the yields were low in alcohol-free chloroform because the concentration of anhydrous HF, and therefore of 'fluoride ion,' was too low to favor fluorohydrin formation over dehydration. Dr Karl Pfister, the inventor of methyldopa, and Head of Process Research at that time, graciously told me that I had taken a big load off his shoulders and that he had wanted me to report directly to him, but since this was not possible for organizational reasons, he transferred me to Dr John Chemerda, the leader of the largest group in Process Research.

At about the same time (1955) the group of Dr John Hogg and coauthors at the Upjohn Company were the first to prepare alkylated analogs of the glucocorticoid hormone. The first,[15] 2α-methylcortisone, was inactive, presumably because the 2α-methyl substituent blocks the in vivo reduction of the 11-ketone to the 11β-hydroxyl. On the other hand, 2α-methylcortisol was a more potent glucocorticoid than cortisol, but it also showed enhanced unwanted mineralocorticoid activity and was therefore not developed. Fortunately, the Upjohn scientists persevered, and the 6α-methylcortisol was found to be more potent than cortisol as a glucocorticoid, and pleasingly it displayed negligible salt retention. Later, the 6α-methyl derivative of prednisolone[13] was marketed by The Upjohn Company.

Given that the glucocorticoidal and mineralocorticoidal receptors represent different proteins, it was not surprising that the first significant clinical advance over cortisol was the discovery of the 1,2-dehydro analogs by Dr Emanuel B. Hershberg and coauthors at Schering in 1955.[13] These congeners, like the 9α-fluoro cortisol, were more potent than the hormone, but unlike the latter, they displayed reduced salt retention. The new compounds, named prednisone and prednisolone, were obtained by microbial dehydrogenation of the hormones. The use of microorganisms including both bacteria and fungi to achieve chemical transformation was not new and was employed also at Squibb, Upjohn, Merck, and by Wettstein and coauthors. Today it seems amusing that some synthetic organic chemists initially took a dim view of the use of microorganisms in a synthetic endeavor, which they regarded as 'cheating.' The introduction of the 1,2 double bond via the use of microorganisms became routine thereafter.

1.01.2.2.1.1.2 Opthalmic application The ophthalmic application of anti-inflammatory steroids as eye drops had been accomplished via the administration of a suspension of steroidal 21-acetates. Upjohn then developed the 'more elegant' water-soluble 21-succinate ester. I was therefore given the assignment of preparing the 'more physiologic' 21-phosphate. In those days organic chemists dreaded the assignment of a phosphate as a synthetic target because we were used to discarding the aqueous layers in the isolation of our reaction products and we were uncomfortable when the aqueous layer contained the desired product as was the case with phosphates. Dr George I. Poos, a member of my group, developed a practical synthesis of the 21-phosphate of cortisol.[16] The marketed product was attractive also for intravenous administration, but lost its glamour for the ophthalmic market when the patent of the equally effective, albeit less elegant, acetate expired. Not all seemingly 'important' research objectives are truly worth pursuing, especially if their principal appeal relates to marketing objectives rather than medical need.

1.01.2.2.1.1.3 Alternative starting materials Under Dr Tishler's superb leadership the overall yield for the conversion of bile acids into cortisone had become so high that they were hard to believe, even for insiders. Merck, however, kept this achievement a secret and talked only about the complexity of the process. In retrospect this may have been a mistake, because it encouraged the competition, especially at Syntex, to explore starting materials other than bile acids. Indeed, such research was also underway at Merck,[17] typified by the synthesis of allopregnan-3β ol-11,20-dione acetate from ergosterol, stigmasterol, and diosgenin. This research led to—what turned out to be—an interesting in-house competition. Starting with the 12-keto steroid hecogenin (please see below) Mr. C. Stewart Snoddy, Dr Wendler, and I synthesized the corresponding $\Delta^{9,11}$ olefin which was converted into 22-isoallospirostane-3β,9α,11α-triol with osmium tetroxide.[18] At the same time Dr John Chemerda and his collaborators[19] studied the conversion of $\Delta^{7,9(11)}$-steroids into 11-oxygenated congeners. In the critical step Ms Lucy Aliminosa brought the Δ^7-9α,11α epoxide into contact with acid-washed alumina during chromatographic purification; in the process the oxide ring is opened by acid catalyzed hydrolysis to afford the corresponding Δ^8-unsaturated 7α,11α-diol. Two Merck laboratories had thus achieved the desired 11-hydroxylation. Dr Tishler expressed little joy about my process. He informed me that he hated osmium tetroxide. By that time I knew him to be a good friend, albeit a demanding boss. On the other hand, Lucy Aliminosa was the heroine of the day. As Dr Tishler informed me, "we will do it Lucy's way." I was not unhappy about the outcome.

1.01.2.2.1.1.3.1 The first example of a reaction recognized to be under 'stereoelectronic control' My own 18 years in Process Research were scientifically stimulating and personally satisfying in every way. I enjoyed working in the laboratory with a group of three or four chemists. I was privileged to work with remarkably devoted and gifted collaborators under a research-oriented corporate management. I can recall only one disappointment. Early in my career in 1952 I made what was to become a significant contribution to synthetic organic chemistry. I reduced the 12-ketone of the sapogenin hecogenin and separated the resulting diols. Stereochemical assignments were readily made by then well-established methods and the separated epimers were converted into the corresponding mesylates. Contrary to expectations, solvolysis of the equatorial 12β- mesylate afforded an olefin in high yield whereas the axial 12α-mesylate was recovered largely unchanged. The olefin possessed unexpectedly strong bands as revealed by infrared spectral absorbances at 6.04 and 11.24 μm. Hydroxylation of the presumed 3-hydroxy-11,12-dehydro sapogenin with osmium tetroxide afforded a triol that was quantitatively converted into a diacetate, revealing an OH band in the infrared at 2.8 μm. Saponification of this diacetate quantitatively regenerated the triol, and the latter on cleavage with periodic acid produced formaldehyde in 60–65% yield as determined by chromatropic acid titration and isolation of its dimedone derivatives together with a quantitative yield of a nor-ketone having a single carbonyl bond in the infrared at 5.84 μm and only end absorption in the UV spectrum. We were able to explain the unexpected results and the infrared spectral properties of the olefin by showing that the elimination reaction in the 12β mesylate series resulted in a rearrangement of the steroidal C/D rings affording a C-nor-D-homo steroid. My colleagues Mr. Stewart C. Snoddy, Jr., Dr N. L. Wendler, and I reported in 1952[20] that the formation of the olefin from the equatorial 12β mesylate "represents a rearrangement path wherein the stereoelectronic requirements are fulfilled only in the case of the natural C_{12}-β-configuration" of the equatorial hydroxyl. We pointed out that "the significance of this geometrical factor is reflected in the extraordinary ease with which this rearrangement occurs." We also noted that 'stereoelectronic control'[20] considerations provide an attractive mechanism to explain the biosynthesis of the alkaloids jervine and veratramine, the structures of which had just been elucidated by Wintersteiner and Fried at Squibb. The concept of stereolectronic analysis built on the prior work of Derek Barton, who had introduced the concept of axial and equatorial substituents for the A, B, and C rings of the steroid. Barton had thereby rationalized the chemical reactivity of epimeric steroidal alcohols and esters almost overnight. To my knowledge but contrary to the commonly held belief, ours was the first reported use of the term 'stereoelectronic' to describe the mechanistic underpinning of a molecular rearrangement.

1.01.2.2.1.2 Transfer to fundamental research

It had become apparent to Dr Sarett, who had responsibility for new drug discovery, that Ralph Hirschmann, a member of the Process Research team, was 'inappropriately' interested in the synthesis of new compounds for biological testing, the domain of Fundamental Research. In 1958 I was, therefore, transferred from Dr Tishler's Process Research Department to Fundamental Research as one of Dr Sarett's five group leaders for steroid research. The transfer proved to be a high-pressure assignment. Dr Sarett met with the steroid group leaders early every Monday morning. We were expected not only to report on the progress of our laboratories, but also to present new research proposals. This assignment did not make for relaxing weekends. Nevertheless, I have always regarded my internship with Dr Sarett as exceptionally valuable. Dr Sarett's approach was distinct from those of other successful medicinal chemists whom I had observed because it was very systematic. He had realized early on that in medicinal chemistry, all effects are either steric or electronic. Sarett also introduced me to the concept of 'minimum systematic variation' in lead development. If, for example, the biological effects of the introduction of either a chloro or a methyl substituent alpha to a ketone are the same, the effects are steric; if the effects are opposite, they are electronic, etc. Further, if introduction of a methyl substituent decreases potency, there is little point in giving an ethyl group a 'try'.

Dr Sarett stressed that when selecting the next synthetic target, it is important first to write down all possible candidates, and then make a careful selection to ensure that the maximum of information is obtained from the smallest synthetic effort. He taught us never to decide on one's next synthetic target on the spur of the moment.

1.01.2.2.1.2.1 An early example of prodrug design based on a biochemical rationale (MK-700)

Of all the steroids that my group synthesized in the general search for a less toxic cortisol, the most original was the 2-acetamido-2-deoxy-β-D-glucopyranoside of prednisolone (MK-700).[21] To my knowledge, it was the first glucocorticoid that was 'designed' to have a better therapeutic index than prednisolone. The design was based on the assumption that the 21-glucoside, like the 21-methyl ethers, would be biologically inactive and that its conversion to the free steroid (prednisolone) would occur primarily in the synovial fluid of inflamed joints that we and others had shown to have strikingly higher concentrations of *N*-acetylglucosaminidase than normal joints or plasma. This prodrug did indeed display an improved therapeutic index in the rat granuloma assay but lacked oral bioavailability. At least in hospitalized patients, parenteral administration of medicines in solution would not have been a problem. However, the potential of the compound as a parenteral drug was not explored. Its therapeutic potential, therefore, remains unknown. On the other hand, the application of the underlying concept was to become routine. Arguably, making oral bioavailability a *conditio sine qua non*, without considering each case on its merits, is unwise.

On 27 and 28 August 1991, an International Symposium on the History of Steroid Chemistry was held in New York City. Participants included[22] Pedro Lehmann, Arthur Birch, Gilbert Stork, Jeffrey Sturchio, George Rosenkranz, E. R. H. Jones, Carl Djerassi, Leon Gortler, Josef Fried, Ralph Hirschmann, Alejandro Zaffaroni, Seymour Bernstein, Konrad Bloch, Frank Colton, John Hogg, and Koji Nakanishi.[22]

I provided a synopsis of contributions of the Merck laboratories, not mentioned herein, at that meeting.[23]

I have not mentioned the important work of Rosenkranz and others at Syntex, because as a whole it did not significantly relate to my research at Merck. Marker's important work during the early 1940s clarified the structures of sapogenins. His work led to the formation of the Mexican Steroid Industry. I do recall a conversation with Russel E. Marker in my office about the chemistry of the sapogenin hecogenin that he had isolated, and which became an important starting material in my research.[20]

1.01.2.2.1.2.2 The discovery of the steroidal preg-4-eno [3,2-] pyrazoles

The synthesis of steroidal preg-4-eno [3,2-] pyrazoles[24] emerged from my tenure in Dr Sarett's department. I had started the project on my own. Treatment of a 3-keto-Δ^4 -steroid with ethyl formate followed by hydrazine or phenylhydrazine, respectively, led to the desired heterocycles in high yield. It became an official endeavor only later, and then because of Dr Tishler's support. The pyrazoles, especially the *p*-fluorophenyl pyrazoles, proved to be the most potent activity-enhancing modifications discovered for steroidal anti-inflammatory agents. One of these pyrazoles was marketed in Europe. The pyrazoles also disproved the heretofore held belief that an α,β-unsaturated ketone in the A-ring is required for cortisol-like biological activity.

1.01.2.2.2 Transfer to exploratory research

1.01.2.2.2.1 The total synthesis of an enzyme in solution

When Dr Robert G. Denkewalter, the Vice President for Process Development informed me in 1963 that he was about to assume the newly created position of Vice President for Exploratory Research to undertake basic research in nucleic acid and peptide chemistry, he invited me to head a group of three chemists to establish peptide chemistry at Merck. We started from ground zero. When Smyth, Stein, and Moore published the complete primary structure of ribonuclease A[25] that same year, it became clear just a few months later that my assignment was nothing less than the total synthesis of ribonuclease.

Although in recent years steroid chemistry has regained considerable momentum, in 1963, steroid research was on the decline. This had become obvious to me and, therefore, I gave an immediate and positive response to Dr Denkewalter who had urged me not to make a hasty decision. I was excited about the opportunity to learn about a field of natural product chemistry that was completely new to me. I suspect that abandoning one's 20-year experience in one field of organic chemistry to enter another field may entail a significant risk in academia, but this was not the case in industry.

1.01.2.2.2.1.1 The use of *N*-carboxyanhydrides Denkewalter's reading of the peptide literature led him to a 1950 paper by J.L. Bailey,[26] which reported the controlled synthesis of small peptide esters in anhydrous medium at low temperatures, using amino acid *N*-carboxyanhydrides. These anhydrides were prepared by allowing amino acids to react with phosgene. *N*-carboxyanhydrides had also been successfully employed in polymerization reactions. Bailey's procedure was impractical primarily because of solubility problems in organic solvents at low temperatures. Although we were unaware of the fact at the time, Professor Paul Bartlett[27] at Harvard had explored the use of *N*-carboxyanhydrides in aqueous medium in controlled peptide synthesis without isolation of intermediates, but had abandoned the project because of side reactions that he was not able to control. The appeal of the use of *N*-carboxyanhydrides in peptide synthesis in aqueous solution lies in the fact that the desired coupling reaction proceeds very rapidly. Treatment of the potassium salt of an amino acid with a slight excess of an *N*-carboxyanhydride generates the dipeptide, the new amino group of which is protected as the carbamate potassium salt. Acidification results in the formation of the carbamic acid, which loses CO_2 spontaneously. It takes only about 5 min to generate the desired dipeptide. The cycle can then be repeated with retention of enantiomeric integrity. While we were able to synthesize a crude octapeptide quickly and without isolation of intermediates, purifying only the final product, the method generally failed to afford multiple step products of the desired purity[28] without chromatographic purification.

1.01.2.2.2.2 The state of the art in 1963

In 1906, in a much-quoted lecture, Emil Fischer expressed the view that the daunting challenges of an enzyme synthesis notwithstanding, chemists should at least give it a try. In 1907 he had synthesized peptides containing 20 amino acids, using his own acid chloride method and the azide procedure of Curtius. Nevertheless, little real progress was made in peptide synthesis until Bergman and Zervas[29] introduced the benzyloxycarbonyl protecting group in 1932, which, unlike an acyl protecting group, permitted the introduction of an activated amino acid without loss of enantiomeric purity and which could be removed by catalytic hydrogenation. In combination with other protecting groups and new coupling reactions, du Vigneaud achieved the synthesis of the first compound of biological interest, oxytocin, in 1953.[30] This was a remarkable achievement because this nonapeptide amide contains five functional amino acid side chains including a cystine bridge. It would be a mistake to conclude that with the synthesis of a hormone in hand, the synthesis of any peptide in solution had become routine. Each polypeptide, like any other organic molecule, has its own personality and synthesis of 'model compounds' has gone out of style for just that reason. In 1968, the synthesis of such biologically active peptides as angiotensin, adrenocorticotropic hormone (ACTH), gastrin, insulin, secretin, glucagon, and calcitonin had been accomplished.

The unique role of enzymes was described well by F. H. C. Crick in 1958,[31] who stated that the:

> "nearest rivals [of enzymes] are the nucleic acids.... The most significant thing about proteins is that they can do almost anything...[and] the main function of proteins is to act as enzymes.... Once the central and unique role of proteins is admitted, there seems to be little point in genes doing anything else [but protein synthesis]."

1.01.2.2.2.3 The Merck synthesis of an enzyme

The Merck group undertook the synthesis of the 104 membered chain called ribonuclease S- protein, so named because Richards had shown that ribonuclease A can be split by the protease subtilisin into two fragments: (1) S-protein the C-terminal fragment, which contains 104 amino acids including the 4 cystine bridges, i.e., 8 cysteine residues; and (2) the N-terminal 20-membered fragment, named S-peptide. Neither fragment displayed any enzymatic activity, but when the two fragments were mixed in equimolar proportion in aqueous solution, the resulting mixture, named ribonuclease S^1, was active, with an enzymatic activity equal to that of ribonuclease A. Hofmann and associates had synthesized S-peptide in 1966; therefore, the synthesis of S-protein would complete the total synthesis of an enzyme. Importantly, Haber and Anfinsen had shown that fully reduced S-protein can be oxidized in the presence of S-peptide to afford enzymatically active material in 35% yield. Had this not been established, we would not have undertaken the synthesis of S-protein.

It is important to note that at the time we initiated the synthesis of an enzyme, there were only three other laboratories who thought it reasonable to contemplate that objective. These teams were the laboratories of Professor Bruce Merrifield at Rockefeller University, Professor Klaus Hofmann in Pittsburgh, and Dr Christian B. Anfinsen at the NIH. Most of our colleagues, however, thought it likely that our objective could not possibly be achieved for the following reason: ribonuclease contains eight cysteine residues. The final step of any synthesis includes the oxidation of these eight residues to generate four disulfide bridges. In theory, there are 105 different ways in which eight cysteine residues can be combined to form four cystine bridges, only one of which corresponds to ribonuclease. It was generally believed at that time that a template was required to ensure proper folding, i.e., to provide the information required to favor the one isomer out of 105 that is the natural product. This argument is not without merit as evidenced by the subsequent discovery of chaperones that serve that purpose.[32] On the other hand, in 1958 F. H. Crick wrote the following:[31]

> "It is conventional at the moment to consider separately the synthesis of the polypeptide chains and their folding. It is of course possible that there is a special mechanism for folding up the chain, but the more likely hypothesis is that the folding is simply a function of the order of the amino acids.... I think myself that this latter idea may well be correct...."

Indeed, White and Anfinsen[33] provided support for Crick's insightful suggestion. They showed that the enzyme's activity, which is lost when the four cystine bridges are reduced under denaturing conditions, is completely restored following dialysis and subsequent reoxidation. The significance of this experiment has been questioned, because the denaturation may have been incomplete. The Anfinsen hypothesis was validated by the total syntheses of ribonuclease.

Clearly the careful design of the strategy for the synthesis was all important. In addition to the question related to the proper foldings of the protein, discussed above, the choice of the protecting groups was of paramount importance. The presence of several sulfur-containing amino acid residues in the proteins argued against the use of a strategy that entailed catalytic-hydrogenation for the removal of any of the protecting groups. Furthermore, protection of the cysteine residue with the widely employed S-benzyl group was unattractive because its removal requires sodium-liquid ammonia, a system not attractive for proline-containing peptides. We had therefore invented the acetamidomethyl group for the protection of the eight cysteine residues.[34] It met our requirements of being stable to trifluoroacetic acid at 25 degrees, to anhydrous HF at zero degrees, and to hydrazine. It can, however, be selectively removed with Hg(II) under mild conditions. The use of the *N*-carboxyanhydrides of unprotected arginine, and unprotected aspartic and glutamic acids as well as the related 2,5 thiazolidinediones (NTAs)[35] for the introduction of unprotected histidine permitted the development of a strategy in which the third functionality needed to be protected only for lysine and cysteine. Thus, rearrangements involving the esterified β-carboxyl of aspartic acid and the γ-carboxyl of glutamic acid were avoided. The formation of the four disulfide bridges subsequent to the liberation of the ε-amino groups of the eight lysine residues was considered to be a desirable feature. For the ε-amino group of lysine we employed the benzyloxycarbonyl protecting group, which can be removed with anhydrous HF. Importantly, S-protein is stable in this solvent at zero degree. This strategy allowed us to use the butyloxycarbonyl group as the acid-labile temporary blocking group for the introduction of amino acids via Anderson's hydroxysuccinimide esters (please see below). Finally, this combination of protecting groups enabled us to remove all the *N*-blocking groups of the tetrahectapeptide with liquid HF, while leaving the cysteines protected.

To synthesize S-protein we relied on the fragment condensation method. A total of 19 fragments was prepared using NCA's, NTA's, and the Boc-hydroxysuccinimide esters of G. W. Anderson.[36] To permit the use of unprotected ω-carboxy groups of aspartic and glutamic acids, we utilized the azide method for fragment condensation as mentioned above. The plan devised to accomplish the synthesis of ribonuclease S-protein is described in the first[37] of five consecutive Communications to the Editor of the *Journal of the American Chemical Society*. As we pointed out,[38] it is one of the characteristics of peptide chemistry that the success or failure of a given synthetic approach depends to an unusual degree both on the precise experimental conditions, and on the judicious selection of protecting groups and coupling reactions employed. For example, it had been claimed that the hydrazinolysis of peptide esters is not an attractive procedure with larger peptides and that the azide coupling procedure is attended to a large degree by the Curtius rearrangement. Our hydrazinolysis reactions, which allowed us to convert C- terminal ester fragments to azides via hydrazides, were successful because: (1) the β-carboxy group of aspartic acid and the γ-carboxy group of glutamic acid were unprotected; and (2) we developed protocols that avoided side-reactions such as the conversion of arginine to ornithine or of asparagine and glutamine to the corresponding hydrazides. By leaving the β-carboxy group of aspartic acid unprotected, we avoided such side reactions as succinimide formation, which can either survive subsequent reactions unchanged or rearrange to a mixture of α- and β-carboxy linked aspartyl residues.

To study the removal of the eight sulfhydryl acetamidomethyl blocking groups from our synthetic protein we attempted to acetamidomethylate reduced natural S-protein. Using the aqueous conditions that afford the acetamidomethylated cysteine itself, did not yield an intermediate capable of regenerating enzymatically active protein. S-Alkylation in anhydrous HF, however, proved satisfactory. Cleavage of the sulfhydryl blocking groups with Hg (II) in acetic acid afforded the reduced S-protein that could be converted to enzymatically active material. The regeneration of enzyme activity by air oxidation as described by Haber and Anfinsen proceeded in low yield. We were able to improve the protocol for the oxidation step to a nearly quantitative yield. The final coupling reaction was carried out by Dr Ruth Nutt, who had only 1.6 mg of the precious hexacontapeptide at her disposal! The unprotected, oxidized S-protein revealed optimal enzymatic activity only when the oxidation step was carried out in the presence of S-peptide. All other control experiments also gave the expected results.

The successful completion of the endeavor was due above all to the experimental skill of the bench chemists, the leadership of three group leaders, Dr Daniel Veber, the late Dr Fred Holly, and Dr Erwin Schoenewaldt, as well as the spirit of collaboration within the entire team. I am particularly indebted to Mrs Susan R. Jenkins and Dr Ruth Nutt for carrying out the most challenging coupling reactions involving the large fragments. It is also a pleasure to acknowledge the skill of Tom Beesley in using Sephadex gel filtration to purify our growing fragments and Carl Homnick who provided amino acid analyses within a deviation of 3%. To my knowledge such accuracy is no longer available today. These analyses were invaluable in demonstrating purity. An advantage of the fragment condensation strategy is the fact that we were generally confronted with impurities that differed significantly in molecular weight from the desired product. The remarkable spirit of collaboration, which is a great credit to the entire group of chemists, deserves further comment. Because we had chosen fragment condensation as our underlying strategy, a chemist A might have synthesized fragment 1 and a chemist B, fragment 2. To couple these two building blocks, one of the two chemists would have to turn over all of his or her precious material to the other. The further the synthesis progressed, the more precious the fragment that was handed over. To the best of my knowledge, these arbitrary assignments of the fragment coupling reactions did not become a major issue.

I suspect that chemists outside the peptide field may not fully appreciate the role that purification techniques available to us and to Professor Merrifield played in achieving success. In the chemical synthesis of peptides it can become very challenging to detect impurities that may arise, such as diastereomers, or asparagines that rearranged to succinimides, or to β-linked aspartates. More often than not, such impurities do not change the desired properties of the synthetic proteins, but they are impurities nevertheless. Therefore, the more cavalier characterization of synthetic proteins now widely employed leaves something to be desired. In our synthesis of RNase S[1] we went to great lengths to characterize our intermediates as fully as was possible at the time when mass spectrometry was not available. We made extensive use of generating and fully examining enzymatic digests, which permitted us to show that asparagines and glutamines had not been hydrolyzed to the corresponding dibasic acids. Enzyme digests also demonstrated chiral purity.

While reporting to Dr Denkewalter I had one and only one responsibility: to ensure the successful total synthesis of our enzyme. I had no time-consuming administrative tasks. During that period I spent hours discussing the synthesis with Dr Daniel Veber. I made an interesting observation during those months. Dr Veber and I would discuss a particular chemical issue and reach what we thought to be a logical conclusion. Because we had no other tasks, we would consider the issue further only to realize half-an-hour later that there was a better conclusion than the one we had reached earlier. It is not a reassuring thought, because I had come to realize that when I 'make rounds,' the recommendations I make are often not as good as those I could have made, had we continued the dialog longer.

The chemical syntheses of an enzyme were announced at a joint press conference held at Rockefeller University in January 1969 by two groups, one by Professor Bruce Merrifield and Dr Gutte, and the other by our group at Merck. Brief remarks were made by Drs Merrifield and Gutte, by Dr Denkewalter and by me. One sentence of mine was quoted in the *New York Times* the following weekend: "We all build on the work of those who came before us and we never know what the future will bring."

1.01.2.2.2.4 The Merrifield 'solid phase' synthesis

The Merrifield solid phase synthesis announced in 1962 has revolutionized not only peptide chemistry but also nonpeptide synthetic organic chemistry. Its distinguishing feature, known to every chemist today, is the fact that the C-terminal amino acid is permanently attached to a resin throughout the elongation of the peptide chain. The peptide is cleaved from the resin in the final step and is then purified. Remarkably, Merrifield was able to achieve sufficiently high yields to complete the synthesis of biologically active material without purification of intermediates. Such operations are now completely automated. Unfortunately, the pharmaceutical industry widely embraced combinational chemistry for lead discovery in the late twentieth century before that concept had been validated. As pointed out elsewhere,[39] only Pharmacopeia appreciated early on that libraries, to be of potential value, need to be sufficiently complex to generate structures of potential interest; to ensure adequate chemical purity, the chemistry has to be studied first.

Over 30 years have passed since the completion of the synthesis of ribonuclease A and ribonuclease S. I have been told that the Merrifield/Merck enzyme syntheses were one of only three chemical achievements that were reported on the front page of the *New York Times*.

The impact the Merrifield solid phase approach was to have on synthetic organic chemistry was not anticipated in January of 1969. This is also true of the fact that these syntheses stimulated the pharmaceutical industry to take a fresh look at the potential for synthetic peptides not only in support of basic research, but also for the discovery of marketable products. Thus, the enzyme synthesis stimulated synthetic organic chemists in academia and in industry to take a greater interest in peptides generally, resulting in the development of new technologies, the discovery of invaluable peptidal medicines and ultimately of peptidomimetics.[40,41]

From a purely personal perspective, it has always been a source of satisfaction to me that the relationships between Professor Merrifield and the Merck group, which were competing for the first total synthesis of an enzyme, were always cordial. I remember driving to the shore many years later, when I heard the announcement on the radio that Professor Merrifield had been named recipient of the Nobel Prize. I stopped my car and dispatched a congratulatory telegram to which Bruce responded most graciously.

As pointed out above, we undertook the synthesis of an enzyme because we thought that Dr Anfinsen had shown that the reduced, denatured protein can be oxidized to regenerate enzymatic activity. The validity of this experiment is now questioned because of the possibility that some of the desired tertiary structure is retained even after denaturation. As a result, the most profound significance of the two total syntheses may lie in the fact that only a total synthesis can prove unequivocally that the amino acid sequence does indeed encode the tertiary structure of RNase.

To put the advances made in peptide research during the twentieth century into perspective, it is worth recalling that prior to Dr Sumner's 1926 isolation of urease, no one could have predicted that enzymes would prove to be proteins.[42]

Last but not least, it is remarkable that Merck & Co. supported this enterprise on the recommendation of Dr Max Tishler and with the approval of Mr. Henry Gadsden, Chairman of the Merck Board of Directors. I am deeply grateful to Dr Denkewalter, who initiated this endeavor, for giving me the opportunity to participate in this work. I am equally indebted to all of my collaborators whose expertise as organic chemists and whose experimental skills and, importantly, whose remarkable teamwork assured the success of this very significant endeavor. My next assignment came in 1972, and it came as another major surprise.

1.01.3 The Merger Between Merck & Co., Inc. and Sharp & Dohme

In 1953 Merck & Co., Inc. merged with Sharp & Dohme in Philadelphia resulting in the formation of Merck Sharp & Dohme. The cultures of the two companies could scarcely have been more dissimilar. As pointed out by Drs Lou Galambos and Jeffrey Sturchio,[43] Merck "valued innovative science and high quality products, had experience in industrial sales, but very little experience or capability in marketing. The most important players at Merck were the medicinal and fermentation chemists, the firm's acknowledged stars." On the other hand, "Sharp and Dohme valued aggressive salesmanship, conducted largely on a face-to-face basis. The business employed marketing techniques that were frequently more effective than informative. "The two cultures were thus sustained by two contrasting status systems."

To this day I am fond of saying facetiously "that the merger between the two companies is proceeding on schedule."

I should like to stress, however, that the first post-merger product was the discovery by the Sharp and Dohme arm of the diuretic chlorothiazide, a major breakthrough. The compound was synthesized by one of Dr James Sprague's chemists, Dr Novello. The biology was headed by Karl Beyer, MD, PhD, a renowned pharmacologist. Chlorothiazide was truly a breakthrough diuretic. Its marketing potential was greatly underestimated for two reasons: chlorothiazide was the first diuretic which, unlike the prior mercurials, was a drug, not a poison, and, secondly, it proved to be not only a safe diuretic but, importantly, also a very effective antihypertensive.

The difference between Rahway and Philadelphia/West Point went beyond the differences identified above. West Point Research was dominated more by pharmacology than chemistry whereas in Rahway at that time chemistry dominated the scene and biology centered on biochemistry and microbiology rather than pharmacology. Research in Rahway referred to West Point as a country club, and West Point Research referred to Rahway as 'Emerald City,' where a constant state of agitation is confused with productivity.

Be this as it may, in 1972 Dr Sprague was approaching retirement. He had built the West Point Medicinal Chemistry Department into a first rate organization, he was the unquestioned boss, and the only department head his chemists had ever known. In 1972 I had set foot on West Point soil only once, that being in 1969 as a result of Dr Denkewalter's hint to West Point management that his peptide chemists would be pleased to give lectures about the synthesis of RNase. Drs Holly and Veber, as well as Dr Denkewalter and I drove to Pennsylvania where Drs Veber and Holly reported on our work.

1.01.3.1 Transfer to West Point, PA

None of this was on my mind in 1972. I didn't even know that Dr Sprague was approaching retirement. Fortunately, Dr Tishler forewarned me and, therefore, when Dr Beyer came to Rahway to invite me to head medicinal chemistry at West Point, I remained fairly calm. At that time, our daughter, Carla, was a senior in high school who obviously did not want to move to Pennsylvania. Therefore, during the first year, I spent my weekends in Scotch Plains, NJ, and the week at the Holiday Inn in Kulpsville, PA. It was actually a good arrangement. I had no knowledge of pharmacology in general, and no information about any of the projects at West Point. Thus, I spent all my evenings trying to learn about the ongoing West Point programs. I immediately recruited Professor Samuel Danishefsky as a consultant in organic chemistry as well as two biochemists, Professors Jeremy Knowles and John Law. Professor Knowles was about to move from Oxford to Harvard, and he alleged that his consultantship income from Merck significantly enhanced his standard of living in the Cambridge, MA area. Professors Jeremy Knowles and John Law brought the all-important discipline of biochemistry to the Medicinal Chemistry Department.

I believe that Dr Beyer offered me the job because he expected, correctly, that peptides would become increasingly prominent in drug discovery. Fortunately, I was one of the individuals on Dr Sarett's list of 'acceptable candidates.' As part of the reorganization, 12 of the 13 members of the peptide group agreed to move to West Point with me, at least in part because housing was cheaper. I shall never forget my first day at West Point. I held a departmental meeting, where I faced a very nervous audience. I wanted very much to reassure my new team and I like to think that the meeting went reasonably well. I said in essence that I had tremendous respect for Dr Sprague and all he had accomplished, but I also said that I was not Dr Sprague, and that I would do some things differently. My main purpose, however, was to assure the group that it was not my intent to see how much chaos and fear I could create during my first week as Department Head. I will always be most grateful to Drs Edward T. Cragoe and Edward Engelhardt, my chief lieutenants, for helping me succeed in my new assignment.

I now reported to Dr Karl Beyer, Jr, who headed West Point Research. The most senior biologist at West Point under Dr Beyer was Dr Clement A. Stone, and my principal biological contact was the pharmacologist Dr Alexander Scriabine, with whom I met at least once a week. It was a marvelous learning opportunity for me.

In 1971, after the synthesis of RNase and subsequent to the retirements of Drs Pfister and Denkewalter, I reported to Dr Sarett, as Senior Director of the newly formed 'New Lead Discovery' group in Rahway. We synthesized, inter alia, the so-called 'renin substrate' as the first step toward the establishment of a renin inhibition program. Dr Veber and his colleagues designed and synthesized cyclic penta- and decapeptides, which Dr Charles Sweet found to be renin inhibitors in vitro and to behave like renin inhibitors in the spontaneously hypertensive (SH) rat. These compounds also induced hemorrhaging. It is not known to what extent the latter effect contributed to the lowering of blood pressure. Clearly these compounds were not attractive leads, but they attest to the long-standing interest in renin inhibitors by the pharmacologists at West Point and the peptide chemists in Rahway. In 1973 the peptide group had been scheduled to join me in West Point. That had been part of Dr Beyer's desire to bring peptide research to West Point. I was therefore somewhat unprepared when Mr John Horan, who was to succeed Mr Henry Gadsden as Chairman of the Board, and CEO, called me to his office to inquire why the company should move me and my group of 12 chemists from Rahway to West Point. I do not recall what I said during that unexpected interrogation. I must have said something about trying to bring the two sites closer together, because I do remember Mr Horan asking whether the move, then, was to be a 'social experiment.' That was the low point in the discussion, but in the end, Mr Horan gave his approval. I believe, with hindsight, that this was his expectation from the beginning. Nevertheless, I was told subsequently that Mr Gadsden had remarked that "it cost me $1 million to move one man from New Jersey to Pennsylvania." I hasten to add that both Mr. Gadsden and Mr. Horan remained very supportive of me thereafter, and I felt much affection for them.

1.01.3.1.1 **Compounds possessing two symbiotic biological activities**

1.01.3.1.1.1 Uricosuric diuretics

Dr Sprague had initiated a program to discover potential drugs possessing two different biological properties that would be useful for the treatment of a given medical problem, e.g., diuretics, which also induce the excretion of uric acid. Dr Sprague had identified a lead compound at the time of his retirement that was a uricosuric diuretic. Subsequently, we discovered a compound good enough to be nominated as a candidate for safety assessment. Unfortunately, it failed to pass that critical step.

1.01.3.1.1.2 Vasodilators with β_2 adrenergic blocking activity

We tried a different approach to discover compounds with two symbiotic biological activities. Instead of screening for leads possessing two desired symbiotic properties, we sought to discover them by design. Since the structural requirements for beta adrenergic blockade are reasonably simple, we started with vasodilators and we tried to incorporate β_2 adrenergic blockade by design. In fact, in a 1979 publication, the Medicinal Chemistry and Pharmacology Department at West Point, PA reported the design and synthesis of an antihypertensive beta adrenergic blocking agent that also acted as a vasodilator. We thought that the vasodilating property of this compound was an intrinsic property of the pharmacologic profile and that the vasodilating component was not due to β_2 adrenergic agonism.[44] This molecule was therefore thought to represent an example of the "symbiotic approach to drug design." Later on we concluded, however, that in fact the vasodilating properties of the compounds in question were due to β_2 adrenergic agonism after all. In a subsequent paper[45] we sought to incorporate β_2 adrenergic properties into a dihydrolutidine-type vasodilator. We concluded again, that "the development of a useful bivalent agent cannot be achieved reliably simply by combining pharmacophoric elements," that is to say "the incorporation of an aminohydroxypropoxy moiety into an aryl ring of a vasodilator does not guarantee the introduction of significant beta adrenoreceptor antagonism."

I subsequently abandoned the search for symbiotic medicines because I had concluded that the chances that one could identify a chemical entity where the potencies for the two biological activities are in good balance, are not very good. Further, the differences in metabolism between animals and humans might further complicate the matter. While Merck's physicians liked the concept, i.e., having to prescribe only one pill rather than two, we concluded that it represents a very long shot.

1.01.3.1.2 **Methyldopa progenitors**

A second significant assignment related to one of Merck's most important products at that time, methyldopa the antihypertensive discovered by Drs Pfister and Stein. Methyldopa had one significant liability: its oral bioavailability varied from patient to patient. Our task was, therefore, to discover an oral prodrug with high and predictable bioavailability. Given the medical importance of this antihypertensive medicine, I made an unusual decision: I put every organic chemist in the department on the problem and launched a crash program. As I had expected, the well-disciplined West Point medicinal chemists responded beautifully. The project, which was termed "ester progenitors of methyldopa" was the subject of a 1978[46] publication by Walfred S. Saari and his collaborators. The program led to the discovery that the pivaloyloxyethyl and the succinimidoethyl esters of methyldopa were more potent antihypertensive agents than methyldopa after oral administration in the SH rat. All esters that were found to be more potent antihypertensives in the SH rat were also hydrolyzed to α-methyldopa at a relatively rapid rate, but Dr W. Saari *et al.* were able to show that the chemical rate of hydrolysis cannot be the sole determinant of antihypertensive potency.

The work demonstrated that in animals the two above-mentioned progenitors were indeed better absorbed and yielded higher plasma and brain levels than the amino acid. In human subjects, however, the pivaloyloxyethyl ester appeared to be more potent than the succinimidoethyl ester; importantly, the latter did not produce the increased potency expected from animal studies. Neither compound became an approved drug.

1.01.3.1.3 The somatostatin program

In 1973 Brazeau, *et al.* reported the isolation and chemical and biological characterization of somatotropin-release inhibiting factor (SRIF-14, somatostatin). At one of the regular meetings of all the department heads, both chemists and biologists, Dr Sarett asked whether there were any ideas that could serve as the basis for a novel antidiabetic program. I was aware of the studies by Dr Luft and his team at the Karolinska Institute that growth hormone (GH) may play a permissive role in the development of retinopathy in diabetes. In addition, Dr Unger[47] pointed out that while glucagon levels are within the normal range in diabetics, they are inappropriately high if one considers the plasma levels of glucose in these patients. Because SRIF-14 suppresses both GH and glucagon release, I had become interested in finding an analog of SRIF-14 with a half-life sufficiently long to test the concept that lowering both GH and glucagon levels in diabetics would be beneficial. Since SRIF-14 also inhibits insulin release, I thought of juvenile diabetics, who lack insulin, as the initial target patient population. Dr Daniel Veber and his colleagues initiated a spectacularly successful program,[48] building on Dr Rivier's alanine and D-amino acid scans. The research at West Point led first to the elucidation of the bioactive conformations of SRIF-14, and then to the design and synthesis of a cyclic hexapeptide (MK-678), which, pleasingly, was more potent and which, as expected, was completely stable to proteases. In addition it displayed some oral activity – a significant advance. Oral bioavailability was, however, less than 5%, presumably because of a high desolvation penalty. Dr E. M. Scolnick, who had succeeded Dr Vagelos as head of Research, like many others, had little or no hope for peptides as drugs, precisely because of their low oral bioavailability. The compound was therefore tested in a cavalier clinical trial and then dropped. Interestingly, however, Sandoz subsequently discovered and successfully developed sandostatin as a parenteral SRIF-14 mimetic, albeit for clinical targets other than juvenile diabetics, showing again that parenteral administration is not inconsistent with commercial success.

1.01.3.1.4 The protease inhibitor program

Two major developments toward a mechanism-based design of protease inhibitors were the identification of protease inhibitors typified by pepstatin[49] by Dr H. Umezawa at the Institute of Microbial Chemistry in Tokyo in 1970 and by a publication of equally profound importance by Professor R. Wolfenden[50] of the Department of Biochemistry of the University of North Carolina, in Chapel Hill, in which he provides his interpretations of Pauling's transition state analog hypothesis. The work of Professor Wolfenden did not receive the recognition that it deserved, in my opinion.

In 1972 Miller and Poper[51] at Eli Lilly reported the exciting observation that pepstatin, a nonspecific inhibitor of renin and other aspartate proteases, lowered blood pressure in rats. Professor Dan Rich of the University of Wisconsin, Madison, immediately recognized the implications of the work at Lilly in terms of the earlier reports by Umezawa and Wolfenden, i.e., that pepstatin might in fact be a transition analog inhibitor. Dr Rich reasoned further that if his speculation proved to be correct, it should be possible to alter the peptide backbone in the regions contiguous to the critical secondary alcohol, thereby creating a new class of protease inhibitors that would be specific for the protease of interest. Such compounds would have great potential for drug discovery. These concepts were, however, sufficiently unconventional that Dr Rich's NIH proposals did not fare well. In 1978, I became the second member of the Study Section with a background in peptide research, and therefore one of the two assigned principal reviewers of the proposal. I believe that my enthusiastic review played a role in the decision by the Study Section to recommend approval and funding of the proposal. Dr Rich thus pioneered what has become a seminal, novel approach to protease inhibitor design. It may be worth noting that the discovery of enzyme inhibitors incorporating transition state analogs exemplifies that nature is one of a medicinal chemist's best teachers.

I first met Dr Dan Rich in 1969 while he served as a postdoctoral fellow with Professor W. S. Johnson at Stanford University. We went for a long walk that became the beginning of a lasting friendship. I invited Dr Rich to consider joining Merck. My colleague Dr Patchett, who had an opening, offered Dr Rich a position at Merck Rahway, an offer that Dr Rich declined, because he had decided to pursue an academic career.

Independently, Dr Veber, like Dr Rich, had considered that pepstatin might represent a transition state analog, which led him also to consider incorporating its statine residue generally into protease inhibitors. In the meantime, Dr Rich had written identical letters to twenty different industrial medicinal chemistry departments, asking for financial support. Nineteen replies were very polite expressions of "no interest," but I was very interested and offered to support a postdoctoral fellow to work on the design of transition state analog-based inhibitors of renin under Dr Rich's direction in Madison. Later, the project was expanded to involve aspartate protease inhibitors generally. Equally important, these developments led to Dr Rich's appointment as a consultant and to a highly productive collaboration between "the two Dans," Drs Rich and Veber. Although bioavailable renin inhibitors are now being tested in the clinic, renin inhibitors discovered in the 1970s and 1980s were not sufficiently bioavailable after oral administration because of their inappropriately high molecular weight and/or large number of rotatable bonds. Renin inhibitors, unlike angiotensin I-converting enzyme inhibitors, will not block the metabolism of bradykinin and may thus be free of one side effect: coughing induced by converting enzyme inhibitors.

Dr Rich, supported by both the Wisconsin Heart Association and Merck, also provided experimental support for the transition state analog concept. The hydroxyl group in the central statine residue contributes four orders of magnitude of binding affinity of the statine residue, and he and his group showed it to be a tight binding inhibitor; competitive with substrate. The Merck-Wisconsin collaboration developed potent and highly selective inhibitors. Particularly noteworthy is the elegant paper by Dr Joshua Boger (now CEO of the Vertex Corporation) and his Merck associates, published jointly with Drs Rich and Bopari in 1983.[52] This paper reports that the net gain in binding energy through the incorporation of the transition state analogs is greater than 4–5 kcal. This observation was followed 2 years later by a second paper[53] by Drs Boger and Payne and their collaborators, which reported incorporation of a novel analog of statine and afforded a subnanomolar renin inhibitor. In a 1990 paper Dr Veber and his coworkers reviewed the design of long-acting renin inhibitors.[54]

Dr Rich is in the process of writing a Perspectives Article for the *Journal of Medicinal Chemistry* which will also be the subject of his Smissman Award Lecture this year. He will point out that in 1970 only five crystal structures of peptides had been solved, none of them aspartic acid proteases, and that around 1980 the approaches to discover protease inhibitors generally followed the antimetabolite strategy of Drs Hitching and Elion, but with little success. Further, as I pointed out elsewhere[39] "in the 1980s the recognition that 'rational design' of enzyme inhibitors is a fruitful approach to drug discovery, led to the belief that knowing the tertiary structure of active sites of such enzyme targets would greatly facilitate the discovery process" [of enzyme inhibitors]. "Significant time and effort was invested in this approach by several companies before it was recognized that the x-ray structure of uninhibited enzymes is likely" [to be of little value].[39] I believe that this effort illustrates that the broad acceptance by the scientific community of an unvalidated concept can actually slow down the discovery process. The same mistake was made again 10 years later, when the industry put its reliance for the discovery of new leads prematurely on an unsophisticated use of combinational chemistry.[39]

1.01.4 P. Roy Vagelos, MD, Successor to Dr L. H. Sarett

In 1974, P. Roy Vagelos, MD accepted an offer to move from Washington University in St. Louis, MO, to assume the position of Senior Vice President for Basic Research with the Merck Research Laboratories. Dr Vagelos had studied with the renowned biochemist Dr Earl Stadtman at the NIH and thus became an expert in lipid metabolism. Dr Vagelos' office was located in Rahway, but his initial responsibilities as Senior Vice President were at West Point with both Dr Clement Stone and I reporting to him. It had also been agreed that after 1 year Dr Vagelos would succeed Dr Sarett as the Head of Research. These developments were to have an important impact on my career. At that time some reorganizations seemed called for in Rahway at the levels below those of Drs Sarett and Vagelos.

I received a phone call in my office at West Point from Dr Vagelos in which he discussed this latter problem quite openly with me, and he concluded the conversation by asking me to return to Rahway as Vice President for Basic Research. I attributed this promotion to two unrelated factors: one was the synthesis of RNase, which gave me broader visibility than the synthesis of some more conventional natural product, i.e., the glory belonged to the enzyme, not to me. The other was the fact that Dr Vagelos considered it to be an important responsibility of any department head to have identified individuals in the department, who could serve as one's replacement, should the need arise. I was most fortunate that under the two Senior Directors reporting to me at West Point, I had four outstanding back-ups as potential heads of the Medicinal Chemistry Department. They were Drs Paul Anderson, John J. Baldwin, Robert L. Smith, and Daniel F. Veber. All were excellent and experienced medicinal chemists. Dr Anderson had exceptional administrative skills. Indeed, he succeeded me as the Head of the Department and he was elected President of the American Chemical Society a few years later. I have long considered Dr John Baldwin to be the most knowledgeable medicinal chemist I have met. Dr Smith, whose recent untimely death was a great loss to his family, friends and to the company, brought tremendous creativity and enthusiasm to all of his projects and Dr Daniel Veber was even then recognized as one of the world's most respected peptide chemists. Taken together, I suspect that the enzyme synthesis, and the stature of my West Point lieutenants played a role in my being offered the position of Vice President of Basic Research in 1976.

1.01.4.1 A New Assignment: Vice President of Basic Research

As Vice President I was to retain responsibility for the West Point Medicinal Chemistry Department. Since my wife and I had only recently moved into a new house in Blue Bell, PA, in December of 1972, and Lucy had started to build beautiful gardens, moving back to New Jersey in 1976 was not attractive. Dr Vagelos gave me the option of remaining in Blue Bell and commuting by limo to Rahway on a daily basis. He made it clear, however, that my responsibilities would be the same, even if I commuted. I accepted this proposal, generally spending one night a week in New Jersey. I remember using the 90 min ride on the Pennsylvania and New Jersey Turnpikes from home to Rahway to read the latest journals, and promptly falling asleep on the way home after a typical day in Rahway! There was never a dull moment.

In my new position two Rahway basic research departments in chemistry were to report to me as well as microbiology under Dr Jerome Birnbaum. My immediate assignments were to restore calm among the chemists and to recruit heads of biochemistry and immunology. Dr Eugene H. Cordes, the chair of the chemistry department at Indiana University, Bloomington, IN, and Dr Alan Rosenthal, then a research fellow at the National Institute of Allergy and Infectious Diseases at the NIH, who had clarified the role of the macrophage in immune response, agreed to join Basic Research in Rahway.

I also appointed Dr Burton G. Christensen, a long-time associate and a renowned expert in beta lactam chemistry, to the position of head of the Department of Synthetic Organic Chemistry. Dr Arthur A. Patchett continued as Head of New Lead Discovery. New responsibilities related to program reviews in Terlings Park, UK and Montreal, Canada, which were also attended by Dr Stone.

In the pharmaceutical industry some companies promote scientists based on their administrative skills. However, this was not the case at Merck, where scientific potential was considered more important. Thus, at my senior staff meetings, the emphasis was definitely on scientific issues. There was one exception: I had realized that the heads of chemistry were more demanding than their biology counterparts when they evaluated their subordinates. I made the assumption that chemists and biologists as a group were equally competent at Merck, and I instituted a rating system based on that premise, which Dr Vagelos eventually adopted for all of research.

1.01.4.2 A Breakthrough at the Squibb Institute of Medical Research

The first acute challenge to face Basic Research after the reorganization was the breakthrough discovery of captopril, the first potent, orally bioavailable, competitive inhibitor of the angiotensin-converting enzyme by Drs Ondetti and Cushman and their collaborators at the Squibb Institute of Medical Research in Princeton, NJ, in 1977.[55] The discovery of an inhibitor for this carboxydipeptidase represented the first successful blockade of the renin-angiotensin system, thus all but validating also renin as a target for enzyme inhibitor design. Captopril proved to be an important new medicine for the treatment of hypertension.

When the Squibb discovery of captopril was announced, Dr Arthur A. Patchett had already started an angiotensin-converting enzyme inhibitor program of his own, which, like that at Squibb, was based on Wolfenden's work with carboxypeptidase A.

I mentioned earlier that of my many mentors, Dr Sarett taught me the most about medicinal chemistry. I believe that Dr Vagelos was the mentor from whom I learned the most in terms of basic strategic concepts. One of these was to appreciate the importance of recognizing significant breakthroughs made by our competitors and to respond promptly and with sufficient manpower to be effective. In the case at hand, the discovery of captopril led to an increase in manpower in Dr Patchett's Department working on the angiotensin-converting enzyme inhibitor program from two chemists to about 20 almost overnight, reminiscent of the methyldopa crash program at West Point referred to above.

The angiotensin-converting enzyme is a zinc-containing exopeptidase. In the design of captopril, Drs Ondetti, Cushman, and their collaborators, built on the presumed similarities of this enzyme and carboxypeptidase A. As pointed out by Dr Cushman *et al.*[56] angiotensin-converting enzyme may play a role in blood pressure regulation both because it 'converts' the biologically inactive angiotensin I to angiotensin II, which raises blood pressure, and because it metabolizes, i.e., inactivates, the antihypertensive bradykinin. As mentioned above, and as clearly acknowledged by the Squibb team, the observation by Drs Byers and Wolfenden[57,58] that D-2- benzylsuccinic acid is a potent competitive inhibitor of carboxypeptidase A was the point of departure for the converting enzyme inhibitor program at Squibb. The Squibb workers designed and synthesized the first angiotensin-converting enzyme inhibitors, including captopril, wherein the sulfhydryl group powerfully bound the enzyme's zinc atom. The one fairly common side effect of captopril, loss of taste, was attributed to the free sulfhydryl group. This side effect provided us with an opportunity to discover an improved second generation chemical entity.

1.01.4.3 Improved Second-Generation Converting Enzyme Inhibitors

By having now some 20 chemists on the angiotensin-converting enzyme inhibitors program, Dr Patchett was able to pursue simultaneously several independent approaches toward an improved second-generation product. Dr Patchett has commented on these in detail. From my perspective, the reason why Dr Patchett succeeded where others failed is that their unsuccessful approaches looked for a replacement of the sulfhydryl by a functionality able to bind the zinc atom of the enzyme as tightly as the sulfyldryl. No such replacement has been found. In designing enalapril maleate, Dr Patchett accepted a carboxyl group as a much weaker ligand for the zinc atom, but compensated for the resulting loss of binding affinity by introducing the phenethyl function, which increased the binding affinity of the enzyme inhibitor.[59] Dr Patchett also designed and synthesized lisinopril, which incorporates the same phenethyl substituent.

After my retirement from Merck, when consulting for various biotechnology companies and also for big Pharma, I would encounter situations where a breakthrough had been achieved either by those companies themselves, or by their competitors. When I urged them to act on these events more expeditiously, I was sometimes told that switching manpower would be bad for the morale of the scientists being transferred. I do not believe that this assessment is generally valid, because the company's interests may demand such action and also because medicinal chemists prefer to be on a 'hot project' rather than a lukewarm one. Both of these considerations guided the decision at Merck to increase manpower dramatically on the angiotensin-converting enzyme inhibitor program. Patchett's success paid off handsomely for many hypertensive patients and, in the process, also for Merck & Co., Inc.

Taken together, enalapril and lisinopril strikingly remind us of the wisdom of one of the giants of the pharmaceutical industry, Mr George W. Merck, who founded Merck & Co., Inc. during the depth of the depression and whose famous passage from a speech given at the Medical College of Virginia (now part of Virginia Commonwealth University) in Richmond, Virginia is quoted below. He gave that speech on 1 December, 1950, less than a year after I started working for Merck. It made a deep impression on me at that time.

> We try to remember that medicine is for the patient. We try never to forget that medicine is for the people. It is not for the profits. The profits follow, and if we have remembered that, they have never failed to appear. The better we have remembered it, the larger they have been.
>
> Nor is medicine for the politicians, except in so far as they are statesmen. I could add that medicine also is not for the professions, unless it is for the patient, first and last! How can we bring the best of medicine to each and every person? It won't be solved by wrangling with words, and it won't be settled by slogans and by calling names. We will fall into gross error with fatal consequences unless we find the answer – how to get the best of all medicine to all the people. It is up to us in research work, in industries, and in colleges and other institutions, to help keep the problem in focus. We cannot step aside and say that we have achieved our goal by inventing a new drug or a new way by which to treat presently incurable diseases, a new way to help those who suffer from malnutrition, or the creation of ideal balanced diets on a worldwide scale. We cannot rest till the way has been found, with our help, to bring our finest achievement to everyone.

1.01.4.4 The Concept of 'The Champion' in Drug Discovery: Benign Prostatic Hypertrophy and the Inhibition of 5α-Reductase

From time to time, a person in a non-managerial position will have a marked impact on the research organization by single handedly becoming the champion for a specific research project. To those who were privileged to interact with him, Dr Glen Arth served as a splendid example of a champion. Dr Arth, an outstanding experimentalist, had been one of the important players in Dr Sarett's total synthesis of cortisone, which was the only practical rather than 'formal' synthesis of this hormone. Later, Dr Arth championed the search for a treatment of a disease that burdens many elderly men, known as benign prostatic hypertrophy (BPH). In the 1950s, Merck started to take an interest in investigating the role of androgens in several disorders linked to male sex hormones such as prostate disease, acne, and 'male pattern baldness.' Critical to the eventual success of the program was the recognition in the late 1960s that the enzyme 5α-reductase converts the male hormone testosterone into the more potent androgen dihydrotestosterone (DTH) in which the steroid A/B rings are *trans*-fused. Dr Arth and his Merck collaborators (biochemists, chemists, and biologists, as well as consultants) appreciated the potential advantages of 5α-reductase inhibitors over the more toxic anti-androgens. It was, however, a report in 1974 by Dr Julianne Imperato-McGinley and Dr Ralph Perterson at Cornell Medical College and another by Drs Patrick Walsh and Jean Wilson at the University of Texas Southwestern Medical School that put

Dr Arth and his associates into high gear, because these studies demonstrated clinically that DHT and testosterone played distinct roles in the sexual differentiation of a fetus and in the development of males during puberty. As a result, the Merck Research Coordinating Committee (RCC) agreed to start screening for 5α-reductase inhibitors. Sadly, Dr Arth died unexpectedly in 1976. His colleague, Dr Gary Rasmusson, referred to him rightly as the "sole personality through the late 1960s and early 70s [who drove the BPH project.]." Dr Rassmusson and Glenn Reynolds eventually synthesized MK-906, a steroid 5α-reductase inhibitor devoid of any intrinsic hormonal activity. Dr Eugene Cordes (please see above), the head of the Biochemistry Department, shepherded MK-906 through the critical remaining preclinical programs. Finally in 1992, MK-906 was approved by the FDA. It is marketed as finasteride. As is always the case, FDA approval was the result of dedicated interdisciplinary research, but to my knowledge, no one has questioned that the BPH project owes its beginning and early successes to the insightfulness and enthusiasm of Dr Glenn Arth.

1.01.4.5 Hypercholesterolemia: A Challenge for the Pharmaceutical Industry

To be a contributor to the discovery of a new medicine that reduces morbidity and mortality and thereby enables a large percentage of our population to lead productive lives is surely highly satisfying. Lovastatin, a cholesterol biosynthesis inhibitor, is such a medicine. It was approved by the FDA on 31 August 1987. The events that culminated in that FDA approval go back nearly 200 years. Louis F. Fieser and Mary Fieser in their 1959 edition of 'Steroids'[60] attribute the "discovery of cholesterol to Michel Eugène Chevreul, who in 1812 first differentiated between saponifiable and nonsaponifiable animal lipids."

Cholesterol has long been known to be the dominant sterol of all higher vertebrates. Mammalian sterol, except for the gastrointestinal tract, is nearly pure cholesterol. The nervous tissue, especially brain, is rich in cholesterol. But it has also been known for over 100 years that cholesterol plays an important role also in pathophysiology. My former colleague, Dr Jonathan A. Tobert of the Medical Affairs Department at Merck pointed out[58] that more than 100 years ago, the German pathologist Dr Virchow observed that patients who had died of diseases such as heart attacks had arteries that were often thickened by deposits of a yellowish fatty substance now known as cholesterol, a condition termed atheroma. Although it was known that feeding cholesterol to rabbits rapidly produced the equivalent of atheroma, the idea that there is a cause-and-effect relationship between dietary cholesterol and coronary heart disease (CHD) was not readily accepted by the medical community. However, the well-known Framingham study in the 1950s established a correlation between plasma cholesterol levels and CHD, especially in the US and in northern Europe. This cause-and-effect relationship was attributed mainly to low-density lipoprotein (LDL) cholesterol, which thus became known as the bad cholesterol, in contrast to high-density cholesterol, which correlated inversely with CHD mortality. This led directly to the proposition that reducing LDL cholesterol will reduce the incidence of myocardial infarction. This hypothesis was, nevertheless, not immediately accepted by the medical community although "a good case could be made that lowering cholesterol reduced the risk of coronary events."[61] In 1984, some 70 years after the demonstration that feeding cholesterol to rabbits rapidly induced the equivalent of atheroma, an NIH Consensus Conference concluded that lowering elevated LDL cholesterol levels with diet and drugs would reduce CHD.[62]

The Medical Community, notably the National Institutes of Health, had done its part. It remained for the pharmaceutical industry to take the lead.

1.01.4.5.1 A breakthrough at Sankyo: the discovery of compactin

As Dr Jonathan Tobert pointed out, "early attempts to reduce cholesterol biosynthesis were disastrous."[61] In the 1960s triparenol, also known as MER-29, entered clinical trial as an inhibitor of cholesterol biosynthesis. The complex biosynthetic pathway leading to cholesterol is well understood. It is a process involving more than 30 enzyme-catalyzed steps. Unfortunately, triparenol inhibits that process at a late step and thus leads to the irreversible accumulation of desmosterol which is more harmful than cholesterol. It thus became clear that any drug that would safely inhibit cholesterol biosynthesis should block an early step in the pathway and one that does not result in the build-up of an intermediate. The enzyme β-hydroxy-β-methylglutaryl-CoA (HMG-CoA) reductase catalyzes the rate-limiting step in the conversion of HMG-CoA into mevalonate. Therefore, it appeared an attractive target for inhibition, since its inhibitors would be expected to be devoid of any mechanism-based toxicity such as is exhibited by triparanol.

In 1976, the Japanese microbiologist Akina Endo and Masao Kuroda of the Fermentation Research Laboratories at Sankyo Co Ltd., Japan reported that citrinin, which had originally been isolated as an antibiotic from *Penicillium citrinum* in 1914 by Raistrack and Smith,[63] inhibits cholesterol synthesis from ^{14}C-acetate in rat liver.[64] The Japanese workers emphasized that unlike clofibrate, citrinin did not cause an increase in liver weight. Later that same year Endo *et al.* reported in the same journal[65] that subsequent work in search of cholesterol biosynthesis inhibitors produced by microorganisms led to the isolation and chemical characterization of three chemically closely related metabolites named ML-236 A, ML-236 B, and ML-236 C, respectively. ML-236 B, also known as CS 500 or compactin, was the major metabolite and also the best inhibitor. They reported that its effect at a dose of $20\,\mathrm{mg\,kg^{-1}}$ in the rat was to lower cholesterol levels by 30% and that the effect lasted for 18 h.

Earlier Drs Z. H. Beg and P. J. Lupien, at the University Loval in Quebec had reported that 3-hydroxy-3-methylglutaric acid (HMG) inhibits a specific step in cholesterol biosynthesis that is mediated by HMG-CoA reductase.[66] As mentioned in an excellent review article by Mr Albert W. Alberts,[67] the discovery of ML-236 B represented a breakthrough in the search for a selective, competitive inhibitor of HMG CoA reductase, precisely because it leads to the accumulation of HMG CoA, a water soluble intermediate "capable of being readily metabolized to simpler molecules."[66]

1.01.4.5.2 The discovery of the first approved β-hydroxy-β-methylglutaryl-CoA reductase inhibitor, lovastatin

Fortunately, Merck was in an excellent position to respond quickly and effectively to the reports from Sankyo that compactin is effective in lowering LDL-cholesterol levels in humans. This was due in part to the fact that, as mentioned above, the head of Research at the time, Dr P. Roy Vagelos, an authority in lipid metabolism and Mr Alfred W. Alberts, a long-time associate of Dr Vagelos, had both moved from Washington University in St. Louis to Rahway, providing critically important expertise. Equally important proved to be the fact that Dr Arthur A. Patchett had initiated a fermentation product for screening project (FERPS) at Merck in 1974 to supply microbial extracts for both in vitro and in vivo screens.[68] Indeed, in 1978 that program achieved a major breakthrough very quickly, leading to isolation of lovastatin (then called mevinolin). Initial concerns that Merck's HMG-CoA reductase inhibitor might be Endo's compactin were laid to rest when it was shown to differ from Sankyo's ML236B (later called mevastatin) by the presence of an additional methyl substituent. Interestingly, the isolation of lovastatin was guided by Dr Carl Hoffman, who had isolated,[69] characterized,[70] and synthesized[71] mevalonic acid in 1956 and 1957. The development of lovastatin was halted when Sankyo announced that it had discontinued clinical studies with compactin. Fortunately, Nobel Prize winners Dr Michael Brown and Dr Joseph Goldstein of the University of Texas Health Science Center in Dallas were able to persuade Dr Vagelos to reinitiate the development of lovastatin, pointing out that it is in fact life saving in patients with uncontrolled very high levels of plasma cholesterol. Even more importantly, based on safety studies in animals and in humans, the formal clinical studies were resumed, leading ultimately to regulatory approval.

1.01.4.5.3 The discovery of simvastatin

As mentioned above, compactin and lovastatin differ from each other in the presence of an additional methyl substituent in the latter in the naphthalenyl ring. The two enzyme inhibitors have in common a chiral 2-methylbutanote ester side chain that is required for enzyme inhibition, but which is lost in vivo by ester hydrolysis. Interestingly, the potency of the inhibitor is the same irrespective of the enantiomer of the butanoate that is incorporated. This observation led the Medicinal Chemistry Department at West Point, under the leadership of Dr Paul S. Anderson and the late Dr Robert L. Smith, to replace the 2-methylbutanoate side chain of lovastatin by the gem dimethyl analog[72] to afford simvastatin, which does not contain a chiral carbon in the side chain. Importantly, it is more resistant to inactivating ester hydrolysis for steric reasons and is twice as potent.

1.01.4.5.4 Lipitor

Today, Parke-Davis' 'Lipitor,' also marketed by Pfizer, is the most widely prescribed HMG-CoA reductase inhibitor.

1.01.4.5.5 A new mechanism to lower plasma cholesterol levels in humans: the discovery of ezetimibe at Schering

More recently ezetimibe was discovered at Schering. The drug lowers cholesterol levels in humans by a mechanism that is complementary to that of the statin: it blocks intestinal absorption of cholesterol and related phytosterols. The fixed combination of ezetimibe/simvastatin, which is marketed as Vytorin, represents a valuable new approach toward optimizing lipid-lowering treatments.

It is of interest to note that combining antihypertensive medicines that lower blood pressure via diverse mechanisms represents a widely accepted regimen. Indeed, a combination of two or more different modes of action are required in some patients to achieve adequate control of blood pressure. The fixed combination ezetimibe/simvastatin is another example that shows that combining drugs that approach a medical problem via two different biochemical/pharmacologic mechanisms can be beneficial. The concept is not new and is used extensively also in the treatment of cancer, asthma, and congestive heart failure.

This fixed combination approach raises the question whether patients (and therefore the industry) would benefit if a greater effort were made to find novel biochemical and/or pharmacologic pathways to treat diseases for which we currently have only one modality of treatment.

1.01.5 The Discovery of Imipenem/Cilastatin, a Life-Saving Fixed Combination Antibiotic

1.01.5.1 Introduction

The discovery of HMG-CoA reductase inhibitors as antihypercholesterolemics, as described in the preceding section, may be said to have been reasonably straightforward from the initial discovery of the desired biological activity in Japan to approval of lovastatin in the US.

In contrast, the road from the initial discovery of the exciting antibacterial profile of thienamycin to FDA approval of imipenem/cilastatin proved to be an unexpected obstacle course. The fixed combination (Primaxin) has been described in an in-house publication[73] as "one of the most difficult, most costly, most frustrating, and most rewarding in the history of Merck research."

The discovery of benzylpenicillin was a breakthrough in the history of antibacterial agents. For a long time, different beta lactam antibiotics had a common scaffold characterized chemically by the fusion of a four- and a five-membered ring. Cephalosporin C, which is produced with penicillin *N* by a *Cephalosporium* sp., cultivated from sea water on the coast of Sardinia near a sewage outlet, has a broader spectrum, i.e., it is active against a larger number of bacteria. It differs chemically from penicillins in its scaffold. Both penicillins and cephalosporins interfere with the synthesis of the bacterial cell wall, an attractive mechanism, since human cells lack that structural feature. One therefore expected correctly, that beta lactams would not prove to be toxic to humans. Chemical changes in the structures of beta lactams resulted in antibiotics differing in their antibacterial profile. The discovery of the cephamycins provided protection against Gram-negative bacteria, but at the cost of a trade-off, since they lacked efficacy against Gram-positive pathogens.

Further chemical modifications of the cephamycins afforded cefoxitin, the first beta lactam antibiotic that was effective against anaerobic pathogens.

1.01.5.2 The Discovery of Thienamycin, an Unstable Antibiotic with a Remarkably Broad Profile as an Inhibitor of Bacterial Cell Wall Synthesis

In the late 1970s, a soil sample that had been collected in New Jersey, was evaluated by Dr Sebastian Hernandez and his associates in the Merck Laboratories in Spain. The culture that contained the antibiotic thienamycin attracted the attention of Dr Hernandez's staff because it had an unusual lavender blue pigmentation. Further screening in Spain revealed the presence of inhibitors of bacterial cell wall synthesis. The organism was sent to Dr Edward Stapley, Executive Director of Basic Microbiology in Rahway, NJ. Dr Sheldon Zimmerman, Associate Director of Analytical Microbiology, concluded that the culture contained not only two known antibiotics, but also one new chemical entity. In another Rahway laboratory, Frederick and Jean Kahan were searching for cell wall synthesis inhibitors in Gram-positive bacteria, such as *Streptococcus* and *Staphylococcus*. The Kahans encountered what became the first major challenge of the thienamycin problem: its chemical instability. At the same time that the antibacterial profile of the new antibiotic aroused great interest in its chemical structure, the stability problem made it clear that the chemical challenges would be formidable. Dr Helmut Kropp, a Senior Research fellow, made the exciting discovery that partially purified thienamycin protected mice that had been infected by *Pseudomonas*. At about the same time (1974) Dr Jerome Birnbaum was appointed Vice President for Microbiology and Agricultural Research and thus assumed overall responsibility for the biology of the program. The Kahans skills as biochemists matched their tremendous dedication to the thienemycin project; working around the clock, they were able to generate a small amount of purified material, which they made available to Dr Georg Albers-Schönberg and Dr Byron Arison for structural studies. The latter concluded that thienamycin is the first member of a family of antibiotics possessing a des-thiacarbapenem nucleus that has a thioethylamino side chain attached to the unsaturated 5-membered ring. Dr Arison's colleagues were reluctant to accept his structure for the antibiotic. Nevertheless, Dr Karst Hoogsteen and Mr. Jordan Hirschfield confirmed the Arison structure by x-ray crystallography.

The optimal stability of the new antibiotic was found to be in the pH range of 6–7. Above that pH the unprotected primary amino group attacks the beta lactam functionality of another molecule of the antibiotic, resulting in the loss of the antibacterial activity.

1.01.5.3 The Discovery of *N*-Formimidoyl Thienamycin (Imipenem), a Stable Molecule with a Superior Profile as an Inhibitor of Bacterial Cell Wall Synthesis

Given the exciting antibacterial profile of thienamycin on the one hand, and its unacceptably poor stability on the other, it became the task of the medicinal chemists under the direction of Dr Burton Christensen, Vice President of Basic Chemistry, to design and synthesize a more stable analog of thienamycin, which would retain the antibacterial properties of that beta lactam. The task proved to be a formidable one. It should be kept in mind that thienamycin, the starting material, was in very short supply and that it was also very unstable. Nevertheless, over 300 derivatives were synthesized over a 3-year period. These derivatization experiments were carried out on a 1 mg scale. Mr Ken Wildonger was the first to achieve the desired goal with the synthesis of *N*-formimidoyl thienamycin (imipenem), which was crystallized by Mr Thomas Miller. The antibacterial properties of imipenem were actually superior to those of thienamycin. With a possible product candidate in hand, the next challenge facing Dr Christensen and his associates was to generate adequate amounts to supply Safety Assessment and Clinical Research, both located at West Point. Supplies of imipenem were initially transported from Rahway by chartered plane!

At this juncture another totally unexpected problem arose. It was discovered that imipenem and thienamycin are metabolized rapidly in vivo in several mammalian species. The metabolic degradation was found to occur in the kidney. Scientists working with beta lactam antibiotics expect to find metabolism by bacterial enzymes. The Merck team was now confronted by the ironic fact that an antibiotic stable to bacterial beta lactamases is rapidly inactivated by a mammalian enzyme bound to kidney membranes, known as dehydropeptidase I. Going forward with imipenem would mean that the new Merck antibiotic would not be useful in the treatment of urinary infections.

Thus, two major problems remained: (1) how to deal with the susceptibility of imipenem to degradation by the dehydropeptidase I in the kidney; and (2) the supply problem of the starting material, thienamycin. As it turned out, both of these problems were solved by the organic chemists in the Basic Chemistry Department.

It was expected that the yields of the fermentation process that provides thienamycin would increase steadily until an adequate supply was assured. Atypically this turned out not to be the case. It was finally recognized that thienamycin would have to be supplied by total synthesis, and the process would have to be economically practical. The Merck team, including the Process Research chemists led by Dr Seemon Pines (in particular, Drs Len Weinstock and Victor Grenda and their colleagues) were up to the task. The total synthesis of thienamycin was first achieved by Dr Christensen, Dr Salzmann, and their associates[74] who developed a stereocontrolled, enantiomerically specific total synthesis. The synthesis from L-aspartic acid entailed novel chemistry. This was required because of the unique structure of thienamycin, which differs from the lactam structure of the penicillins and the cephalosporins by its highly strained ring system, which lacks a sulfur atom. In addition, the ring substituents are also unlike those of the conventional beta lactams.

The second problem, susceptibility of the imipenem to degradation in the kidney required that the antibiotic be combined with an enzyme inhibitor to prevent the renal metabolism in animals and in humans by dehydropeptidase-1. This led to the design and synthesis of cilastatin, a highly substituted heptenoic acid derivative inhibitor. Importantly, the pharmacokinetic properties of cilastatin match those of the imipenem. Pleasingly, cilastatin provided a fringe benefit: it excludes the entry of imipenem into the proximal tubular epithalium of the kidney where it might cause tubular necrosis. The fixed combination of imipenem – cilastatin completes the tour de force, made possible by the splendid interdisciplinary research of the industrial chemists and biologists. An excellent paper[75] gives a detailed account of the development of imipenem – cilastatin, which overcame all of the many shortcomings of thienamycin, an antibiotic with an unusually broad spectrum including a high order of bactericidal activity against *Pseudomonas aeruginosa*, *Serratia*, *Bacteroides fragilis*, *Enterococci*, and other species resistant to other antibiotics. Because imipenem is resistant to hydrolysis by bacterial beta lactamases and because of is extraordinary spectrum, the new fixed combination represents to this day a remarkable contribution to therapy.

1.01.6 Ivermectin: from the Discovery of an Animal Health Anthelmintic to A Wonder Drug for the Developing World

1.01.6.1 Introduction

The discovery of an animal health medication faces an obstacle not generally encountered in the search for a new medicine for humans: the issue of the cost of producing the drug. Generally speaking, if a new drug for humans is superior to heretofore available therapy, it will be produced by industry and prescribed by physicians even if it is more expensive than the older treatment. This is not necessarily the case for animal health products. A farmer will carefully assess the cost/benefit ratio before switching to a superior but costlier medication. This is an important distinction and probably the principal reason why few pharmaceutical companies enter the animal health arena.

As has been pointed out by Dr William C. Campbell of the Merck Institute for Therapeutic Research in Rahway, NJ,[76,77] the anthelmintic program at the Merck Sharp Dohme Research Laboratories (MSDRL) began in 1953. From it emerged thiabendazole, the first of the benzimidazole anthelmintics, as well as cambendazole, rafoxanide, and clorsulon.[76,77] Building on the experience gained from human health directed antibiotics, Merck Research subsequently decided to explore fermentation broths as a source of new anthelmintics. Wisely, the animal health team chose to use an in vitro screen in mice as the primary assay for fermentation products.

1.01.6.2 The Collaboration with the Kitasato Institute

Another major innovation was the decision to enter into a research agreement with the Kitasato Institute in Japan. It was reached in 1974, and involved the shipment of bacterial cultures from Japan to the MSDRL, where they were inoculated in microbiological medium to permit fermentation. The resultant broths were then tested in a so-called tandem coccidiosis-helminthiasis assay. Mice fed on a diet incorporating one of the broths was found to be free of worms. It was established fairly quickly that an anthelmintic had indeed been present in the broth. As Dr Campbell was careful to emphasize, "had a particular mouse not been examined properly [by a technician]" a major new drug might have been missed. This serves as a reminder that drug discovery requires, of course, an idea generated by a sophisticated scientist, but it will lead to a drug only if those 'down the line' perform perfectly. In this case that task was the careful examination of a mouse for worms.

1.01.6.3 The Broth from Culture OS3153

Subsequent testing of broths obtained from culture OS3153 confirmed the presence of an anthelmintic. The next tasks were to isolate and to characterize chemically the active components in the culture now named C-076. Mass spectrometry revealed that there were four biologically active components, each of which consisted of a major and a minor component. Using mass spectrometric separations and then NMR spectroscopic techniques, the active principles of C-076 were found to be glycosidic derivatives of pentacyclic 16-membered lactones, similar in structure to the milbemycins. C-076 differed from the latter primarily in that the former lacked two glycosidic attachments. The new anthelmintic was named avermectin.

1.01.6.4 Avermectin

Avermectin fulfilled its promise as a cost-effective animal health anthelmintic. Like every other biologically active compound, it causes side effects. Had avermectin been a drug for humans, one could have launched a major effort to improve on the drug's therapeutic index, regardless of the cost. I remember discussing that issue with the late Dr Michael H. Fisher, an outstanding, softly spoken scientist who was responsible for the chemistry of animal health research at Merck. Given that avermectin has five double bonds, I was not very optimistic that we would be able to increase safety in a cost-effective way. Dr Fisher's more optimistic outlook was fully vindicated by subsequent developments. The medicinal chemists succeeded in selectively reducing the C 22–23 double bond in good yield. The resulting dihydroavermectin (ivermectin) was actually slightly less potent than avermectin B_1, but, to our delight, it displayed a therapeutic index superior to that of its natural product precursor.

1.01.6.5 Ivermectin

The extraordinary potency of the new anthelmintics against *N. dubius* led Drs Campbell and Fisher to explore the potential of the avermectins against a new host of parasitic nematodes. The results proved to be exciting. C-076 was effective against parasitic nematodes occupying all the major segments of the gastrointestinal tract, including nematode strains resistant to benzimidazole anthelminitics.

The Merck chemists under the leadership of Drs Mike Fisher and Helmut Mrozik synthesized more than 1000 analogs, all of which were tested by the parasitologists. None were significantly more potent than the natural product.

It was also exciting to discover that avermectin and ivermectin are not only broad-spectrum anthelmintics, but also insecticides and acaricides. As Dr Campbell pointed out[76] the activity "against arthropods changed the scientific and commercial prospects dramatically. When ivermectin was eventually launched as a product for cattle, it was offered not as an anthelmintic, but as an antiparasitic agent for the control of endoparasites and ectoparasites."

As mentioned above, the thienamycin project presented one challenge after another.[73] The avermectin program began with some interesting results and thereafter kept getting better and better. Some of this was good luck. For example, as Dr Campbell pointed out,[76] avermectin and ivermectin were fortunately active against the microfilariae of dog heartworm, but, again fortunately, they were inactive against the adult stage. "The result was not only a successful commercial product for heartworm prevention, but also a ...dramatic example of stage specificity in chemotherapy."[74] To be sure it was the excellence of the scientists involved that ensured that all developing opportunities were recognized and then implemented.

What the antibiotic and the anthelmintic programs had in common was the excellence of the research, from the top on down, which ensured ultimate success. Top management cannot ensure success, it can only help set the course and provide the optimal environment; conversely, it can also create an atmosphere that is not conducive to creative research.

The next major development came in April 1978, when the parasitologists made an important observation and fully recognized its potential.[73] To quote from Dr Campbell' s review paper:[76]

> Another observation that had far-reaching consequences was again a matter of the filarial group of worms. In April, 1978 ivermectin was about to be tested against the gastrointestinal nematodes in horses. At the last minute a decision was made (prompted by a suggestion from L. S. Blair nee Slayton) to examine pieces of skin from the treated and untreated horses in order to detect a possible effect on the microfilariae of *Onchocerca cervicalis* – should those horses happen to be infected with that parasite. The horses did happen to be infected, and an effect on the microfilariae was clearly evident (Egerton *et al.* 1981a [as cited in [74]]). The significance of the observation lay not in the importance of the parasite in horses, for it is an obscure parasite that was then considered of no importance whatever, but lay rather in the relationship of the parasite to an important pathogen of man. To those of us who were actively interested in human parasitology, the potential utility of the finding was clear. In the summer of 1978 I sent the drug (and relayed our results) to an investigator in Australia who was conducting anti-filarial drug tests under the sponsorship of the World Health Organization. The investigator, Dr Bruce Copeman, quickly showed that ivermectin was active also against the microfilariae of *Onchocerca spp* in cattle, The information from horses and cattle, together with the toxicological data that had by then been gathered, served as the basis for recommending the testing of ivermectin in man.

1.01.6.6 The Human Formulation of Ivermectin (Mectizan)

The microfilariae of *Onchocerca volvulus* are the pathogens that are responsible for onchocerciasis in humans. These microfilariae cause progressive skin lesions and frequently induce ocular lesions that lead to blindness. This disease affects people in 35 countries in Africa, America, and the Arabian Peninsula,[77,78] and is known as river blindness. Suramin was effective in killing the adult worm, but proved to be very toxic.[78] Diethylcarbamazine citrate similarly proved to be a very unattractive therapy.[78] Ivermectin turned out to be a miracle drug for the treatment of onchocerciasis, not only because of its effectiveness and safety, but also because it only needs to be given to the patient once a year, a matter of great importance in countries where the health delivery infrastructure is fragile. In their assessment of the impact of ivermectin on illness and disability associated with onchocerciasis, Drs James M. Tielsch and Arleyne Beeche[79] concluded that "regular distribution to populations living in endemic areas has demonstrated significant reduction in blinding ocular complications, transmission and disability caused by onchocercal skin disease," not to mention the impact on intestinal helminth infection, lymphatic filariasis, and human scabies infection.[79]

During my tenure as Vice President for Basic Research my colleagues made many significant contributions to human and animal health. None of these is more pleasing to me than the Mectizan Project.

The Merck Mectizan Donation Program has been a source of enormous satisfaction to all Merck employees and retirees since its inception on 21 October 1987, under the leadership of its then Chairman and CEO, Dr P. Roy Vagelos. It was reaffirmed by his successor, Mr Raymond V. Gilmartin. In essence, Merck & Co., Inc. announced its intention to donate Mectizan for as long as it might be needed and whenever it is needed. "This unprecedented decision came twelve years after the discovery of avermectin by Merck scientists and nearly seven years after the first human clinical trials in Dakar, Senegal."[80]

A priori patient compliance would be expected to be a major hurdle. As mentioned above, almost miraculously, Mectizan is effective when administered once a year. The treatment programs are organized as community-directed interventions – entire villages line up for the treatments annually. The program now reaches some 45 million people in more than 88 000 villages in 34 countries. All of the gods of classical antiquity are smiling on this project.

In the chapters and volumes of this book that follow, the state of the art is brought up to date. This textbook should give a sense of pride and satisfaction to every person who has made medicinal chemistry his or her career. It goes without saying that all of the collaborators of medicinal chemists should share in this sense of satisfaction.

1.01.7 Epilogue

This chapter was written by two coauthors who have been associated with the pharmaceutical industry, although one of us (RH) has spent the past eighteen and a half years in academe. It is fitting and proper to devote part of this epilogue to the evolving interactions of academe and industry. Medicinal chemistry has always stood with one foot in the academic world and the other in the industrial world, as many of the examples noted above illustrate. The roles of these two sectors need to be profoundly different if their combined efforts are to benefit the patient. It is the responsibility and function of academe and government to undertake fundamental research, and it is the task of industry to perform the developmental research required if the knowledge generated by the former research is to lead to medicines and vaccines that are available to all of us in our hospitals and drugstores.

Let us be very clear. We believe that our academic and governmental institutions have no monopoly on outstanding research, and is it not unknown for mundane results to come from university or public institutions. This is equally true of industry. What has worked well in the past and will succeed in the foreseeable future is to understand that the perspectives and priorities of academic and industrial research are complementary. To maximize the opportunity for successful innovation, it is critical to ensure that the best fundamental research in academia and government is followed by the best applied research in the industry. Neither endeavor is nobler, or more challenging, than the other. Everyone would agree that it is an enormous achievement for a scientist in academe or government to generate a truly new idea. It may be more difficult for anyone outside the industry to appreciate the equally creative ideas required in successful drug discovery and development. The above discussion illustrates this point. Consider Section 1.01.2.2.1 on 'The Cortisol Era'. The basic research was done by Drs Reichstein, Hench, and Kendall, all academics, but their fundamental research would not have benefited a single patient without the industrial research that followed, which posed equally challenging intellectual puzzles, and also sparked new lines of academic investigation in the process.

We suspect that the easier problems may have been solved first, which may explain in part why the drug company pipelines – though still remarkably productive – seem dryer than they used to be. There may also be other reasons why it is becoming more difficult to generate a new approved drug. One is that for diseases such as diabetes, Alzheimer's disease, and others, there are no good animal models. As a result we have to go from an in vitro experiment directly to a clinical trial without the benefit of an in vivo experiment in animals. Also, the enormous cost of clinical trials has made the development of later entries in a therapeutic class more uncertain. But second and third generation drugs can also lead to significant advances, and it is also true that individuals vary in their response to different drugs that supposedly work via a common mechanism. Since second, but not third generation drugs have generally represented significant advances, the disappearance of 'me-too' medicines may not be that great a loss. It is nevertheless also true that individuals vary in their response to different drugs that supposedly work via a common mechanism.

Sadly, public opinion of the industry is presently at a low point. To a certain degree the industry itself has become a political football, a matter of concern cited by Mr Merck already in 1950. We suspect, however, that even the industry's most severe critics would not wish their families to be deprived of the medications discovered by the industry during the past 15 years. Recent advances for the treatments of rheumatoid arthritis, cancer, asthma, and diabetes, which are the result of excellent research in academia, government, and industry, are a case in point. We have every confidence that the combined efforts of governmental, academic, and industrial research will lead in the years ahead to advances in medicinal chemistry that today we can only imagine.

References

1. Emmerson, R.W. *Letters and Social Aims*, 1875.
2. Smith, A. B., III; Hirschmann, R.; Pasternak, A.; Yao, W.; Sprengeler, P. A.; Holloway, M. K.; Kuo, L. C.; Chen, Z.; Darke, P. L.; Schleif, W. A. *J. Med. Chem.* **1997**, *40*, 2440–2444.
3. (*C&E New*, 2005, 7 February, p. 32, quoted from book reviews by D.M. Kiefer.)
4. Fieser L.; Fieser, M. *Steroids*; Reinhold Publishing Corp.: New York; Chapman & Hall, Ltd: London, UK, **1950**, p 648.
5. Shen, T. Y.; Windholz, T. B.; Rosegay, A.; Witzel, B. E.; Wilson, A. N.; Willett, J. D.; Holtz, W. J.; Ellis, R. L.; Matzuk, A. R.; Lucas, S.; Stammer, C. H.; Holly, F. W.; Sarett, L. H.; Risley, E. A.; Nuss, G. W.; Winter, C. A. *J. Am. Chem. Soc.* **1963**, *85*, 488.
6. Kym, P. R.; Kort, M. E.; Coghlan, M. J.; Moore, J. L.; Tang, R.; Ratajczyk, J. D.; Larson, D. P.; Elmore, S. W.; Pratt, J. K.; Stashko, M. A. et al. *J. Med. Chem.* **2003**, *46*, 1016–1030.
7. Schäcke, H.; Schottelius, A.; Döcke, W.-D.; Strehlke, P.; Jaroch, S.; Schmees, N.; Rehwinkel, H.; Hennekes, H.; Asadullah, K. *Proc. Natl. Acad. Sci. USA* **2004**, *101*, 227–232.
8. Ali, A.; Thompson, C. F.; Balkovec, J. M.; Graham, D. W.; Hammond, M. L.; Quraishi, N.; Tata, J. R.; Einstein, M.; Ge, L.; Harris, G. et al. *J. Med. Chem.* **2004**, *47*, 2441–2452.
9. Hirschmann, H.; Hirschmann, F. B.; Farrell, G. L. *J. Am. Chem. Soc.* **1953**, *75*, 4862–4863.
10. Ellis, B.; Hartley, F.; Petrow, V.; and Wedlake, D. *J. Chem. Soc.* **1955**, 4383–4388.

11. Allen, W. S.; Bernstein, S. *J. Am. Chem. Soc.* **1955**, 77, 1028–1032; 1956, 78, 1909–1913.
12. Fried, J.; Sabo, E. F. *J. Am. Chem. Soc.* **1954**, *76*, 1455–1456.
13. Herzog, H. L.; Nobile, A.; Tolksdorf, S.; Charney, W.; Hershberg, E. B.; Perlman, P. L. *Science* **1955**, *121*, 3136–3176.
14. Hirschmann, R.F.; Miller, R.; Wood, J.; Jones, R. E. *J. Am. Chem. Soc.* **1956**, *78*, 4956–4959.
15. Hogg, J.A.; Lincoln, F.H.; Jackson, R.W.; Schneider, W. P. *J. Am. Chem. Soc.* **1955**, 77, 6401 – 6402.
16. Poos, G. I.; Hirschmann, R.; Bailey, G. A.; Cutler, F. A., Jr.; Sarett, L. H.; Chemerda, J. C. *Chem. Ind.* **1958**, *39*, 1260–1261.
17. Chamberlin, E. M.; Ruyle, W. V.; Erickson, A. E.; Chemerda, J. M.; Aliminosa, L. M.; Erickson, R. L.; Sita, G. E.; Tishler, M. *J. Am. Chem. Soc.* **1951**, *73*, 2396–2397.
18. Hirschmann, R.; Snoddy, C. S., Jr.; Wendler, N. L. *J. Am. Chem. Soc.* **1953**, *75*, 3252–3255.
19. Chamberlin, E. M.; Ruyle, W. V.; Erickson, A. E.; Chemerda, J. M.; Aliminosa, L. M.; Erickson, R. L.; Sita, G. E.; Tischler, M. *J. Am. Chem. Soc.* **1953**, *75*, 3477–3483.
20. Hirschmann, R.; Snoddy, C. S., Jr.; Wendler, N. L. *J. Am. Chem. Soc.* **1952**, *74*, 2693.
21. Hirschmann, R.; Strachan, R. G.; Buchschacher, P.; Sarett, L. H.; Steelman, S. L.; Silber, R. *J. Am. Chem. Soc.* **1964**, *86*, 3903.
22. Gortler, L.; Sturchio, J. L. *Steroids* **1992**, *57*, 355–356.
23. Hirschmann, R. *Steroids* **1992**, *57*, 579–592.
24. Hirschmann, R.; Buchschacher, P.; Steinberg, N. G.; Fried, J. H.; Ellis, R.; Kent, G. J.; Tishler, M. *J. Am. Chem. Soc.* **1964**, *86*, 1520–1527; Hirschmann, R.; Steinberg, N. G.; Schoenewaldt, E. F.; Paleveda, W. J.; Tishler, M. *J. Med. Chem.* **1964**, 7, 352–355.
25. Smyth, D. G.; Stein, W. H.; Moore, S. *J. Biol. Chem.* **1963**, *238*, 227.
26. Bailey, J. L. *J. Chem. Soc.* **1950**, 3461–3466.
27. Bartlett, P. D.; Jones, E. R. H. *J. Am. Chem. Soc.* **1957**, *79*, 2153; Bartlett, P. D.; Dittmer, D. C., ibid, **1957**, *79*, 2159.
28. Hirschmann, R.; Strachan, R. G.; Schwam, H.; Schoenewaldt, E. F.; Joshua, H.; Barkemeyer, B.; Veber, D. F.; Paleveda, W. J., Jr.; Jacob, T. A.; Beesley, T. E.; Denkewalter, R. G. *J. Org. Chem.* **1967**, *32*, 3415–3425.
29. Bergman, M.; Zervas, L. *Berichte* **1932**, *65*, 1192–1201.
30. du Vigneaud, V.; Ressler, C.; Swan, J. M.; Roberts, C. W.; Katsoyannis, P. G.; Gordon, S. *J. Am. Chem. Soc.* **1953**, *75*, 4879–4880.
31. Crick, F. H. C. *Symp. Soc. Exp. Biol.* **1958**, 138.
32. For a recent example, see Hundley, H.A., Walter, W., Bairstow, S., Craig, E.A. Published online, *Science Express Reports* March 31, **2005**.
33. White, F. H.; Anfinsen, C. B. *Ann. NY Acad. Sci.* **1959**, *81*, 515–523.
34. Veber, D. F.; Milkowski, J. D.; Denkewalter, R. G.; Hirschmann, R. *Tetrahedron Lett.* **1968**, *26*, 3057–3058; Veber, D. F.; Milkowski, J. D.; Varga, S. L.; Denkewalter, R. G.; Hirschmann, R. *J. Am. Chem. Soc.* **1972**, *94*, 5456–5461.
35. Dewey, R. S.; Schoenewaldt, E. F.; Joshua, H.; Paleveda, W. J.; Schwain, J. H.; Barkemeyer, H.; Arison, B. H.; Veber, D. F.; Denkewalter, R. G.; Hirschmann, R. *J. Am. Chem. Soc.* **1968**, *90*, 3254.
36. Anderson, G. W.; Zimmerman, J. E.; Callahan, F. M. *J. Am. Chem. Soc.* **1964**, *86*, 1839.
37. Denkewalter, R. G.; Veber, D. F.; Holly, F. W.; Hirschmann, R. *J. Am. Chem. Soc.* **1969**, *91*, 502.
38. Hirschmann, R.; Denkewalter, R. G. *Naturwissenschaften* **1970**, *57*, 145–151.
39. Hirschmann, R. In *The Practice of Medicinal Chemistry*; Hansch, C., Sammes, P. G., Taylor, J. B., Eds.; Academic Press/Elsevier Ltd.: Oxford, UK, 1990.
40. Prasad, V.; Birzin, E. T.; McVaugh, C. T.; van Rijn, R. D.; Rohrer, S. P.; Chicchi, G.; Underwood, D. J.; Thornton, E. R.; Smith, A. B., III; Hirschmann, R. *J. Med. Chem.* **2003**, *46*, 1858–1869.
41. Smith, A. B., III; Charnley, A. K.; Mesaros, O. K.; Wang, W.; Benowitz, A.; Chu, C.-L.; Feng, J.-J.; Chen, K.-H.; Lin, A.; Cheng, F.-C.; Taylor, L.; Hirschmann, R. *Org. Lett.* **2005**, 7, 399–402.
42. Sumner, J. B. *J. Biol. Chem.* **1926**, *69*, 435; Hirschmann, R. *Chem. Biol. Pept.* **1972**, 721–728.
43. Galambos, L.; Sturchio, J. L. "Sustaining Innovation: Critical Transitions at Merck & Co., Inc.," Presented at Conference on "Understanding Innovation" The Johns Hopkins University, Baltimore, Maryland, June **1997**, p 17.
44. Baldwin, J. J.; Lumma, W. C.; Lundell, G. F.; Ponticello, G. S.; Raab, A. W.; Engelhardt, E. L.; Hirschmann, R. *J. Med. Chem.* **1979**, *22*, 1284–1290.
45. Baldwin, J. J.; Hirschmann, R.; Engelhardt, E. L.; Ponticello, G. S. *J. Med. Chem.* **1981**, *24*, 628–631.
46. Saari, W. S.; Freedman, M. B.; Hartman, R. D.; King, S. W.; Raab, A. W.; Randall, W. C.; Engelhardt, E. L.; Hirschmann, R. *J. Med. Chem.* **1978**, *21*, 746–753.
47. Unger, R. H. *Diabetes* **1976**, *25*, 36.
48. Veber, F.; Holly, F. W.; Paleveda, W. J.; Nutt, R. F.; Bergstrand, S. J.; Torchiana, M.; Glitzer, M. S.; Saperstein, R.; Hirschmann, R. *Proc. Natl. Acad. Sci. USA* **1978**, *75*, 2636–2640.
49. Umezawa, H.; Aoyagi, T.; Morishima, H.; Matsuzaki, M.; Hamada, M.; Takeuchi, T. *J. Antibiot.* **1970**, *23*, 259–262.
50. Wolfenden, R. *Acc. Chem. Res.* **1972**, *5*, 10–18.
51. Miller, R. P.; Poper, C. J.; Wilson, C. W.; DeVito, E. *Dep. Biochem.* **1972**, *21*, 1294–2941.
52. Boger, J.; Lohr, N. W.; Ulm, E. H.; Poe, M.; Blaine, E. H.; Fanelli, G. M.; Lin, T. Y.; Payne, L. S.; Schorn, T. W.; LaMont, T. W. et al. *Nature* **1983**, *303*, 81–84.
53. Boger, J.; Payne, L. S.; Perlow, D. S.; Lohr, N. S.; Poe, M.; Blaine, E. H.; Ulm, E. H.; Schorn, T. W.; LaMont, B. I.; Lin, T.-Y. et al. *J. Med. Chem.* **1985**, *28*, 1779–1790.
54. Veber, D. F.; Payne, L. S.; Williams, P. D.; Perlow, D. S.; Lundell, G. F.; Gould, N. P.; Siegel, P. K. S.; Sweet, C. S.; Freidinger, R. M. *Biochem. Soc. Trans.* **1990**, *18*, 1291–1294.
55. Ondetti, M. A.; Rubin, B.; Cushman, D. W. *Science* **1977**, *196*, 441–444.
56. Cushman, D. W.; Cheung, H. S.; Sabo, E. F.; Ondetti, M. A. *Biochemistry* **1977**, *16*, 5484–5491.
57. Byers, L. D.; Wolfenden, R. *J. Biol. Chem.* **1972**, *247*, 606–608.
58. Byers, L. D.; Wolfenden, R. *Biochemistry* **1973**, *12*, 2070–2078.
59. Patchett, A. A.; Harris, E.; Tristram, E. W.; Wyvratt, M. J.; Wu, M. T.; Taub, D.; Peterson, E. R.; Ikeler, T. J.; ten Broeke, J.; Payne, L. G. et al. *Nature* **1980**, *288*, 280–283.
60. Fieser, L.; Fieser, M. *Steroids*; Reinhold Publishing Corp.: New York, 1959; p 3.
61. Tobert, J. A. *Nat. Rev. Drug Disc.* **2003**, *2*, 517–526.
62. NIH Consensus Development Conference Statement. *Nutr. Rev.* **1985**. 43, 283 – 291.
63. Raistrack, H.; Smith, G. *Chem. Ind.* **1941**, 828–830.

64. Endo, A.; Kuroda, M. *J. Antibiot.* **1976**, *29*, 841–843.
65. Endo, A.; Kuroda, M.; Tsujita, Y. *J. Antibiot.* **1976**, *29*, 1346–1348.
66. Beg, Z.H.; Lupien, P. J. *Biochim. Biophys. Acta* **1972**, 439 – 448.
67. Alberts, A. W. *Am. J. Cardiol.* **1988**, *62*, 10J–15J.
68. Patchett, A. A. *J. Med. Chem.* **2002**, *45*, 5609–5616.
69. Skeggs, H. R.; Wright, L. D.; Cresson, E. L.; MacRae, G. D. E.; Hoffman, C. H.; Wolf, D. E.; Folkers, K. *J. Bacteriol.* **1956**, *72*, 519–524.
70. Wolf, D. E.; Hoffman, C. H.; Aldrich, P. E.; Skeggs, H. R.; Wright, L. D.; Folkers, K. *J. Am. Chem. Soc.* **1957**, *79*, 1486–1487.
71. Hoffman, C. H.; Wagner, A. F.; Wilson, A. N.; Walton, E.; Shunk, C. H.; Wolf, D. E.; Holly, F. W.; Folkers, K. *J. Am. Chem. Soc.* **1957**, *79*, 2316–2318.
72. Hoffman, W. F.; Alberts, A. W.; Anderson, P. S.; Chen, J. S.; Smith, R. L.; Willard, A. K. *J. Med. Chem.* **1986**, *29*, 849–852.
73. *Merck World* 1986, 7, 4–17.
74. Salzmann, T. N.; Retcliffe, R. W.; Bouffard, F. A.; Christensen, B. G. *Philos. Trans. R. Soc. Lond. Ser. B* **1980**, *289*, 191–195.
75. Kahan, F. M.; Kropp, H.; Sundelof, J. G.; Birnbaumm, J. *J. Antimicrob. Chemother.* **1983**, *12*, 1–35.
76. Campbell, W. C. The genesis of the antiparasitic drug ivermectin. In *Inventive Minds: Creativity in Technology*; Weber, R. J., Perkins, D. W., Eds.; Oxford University Press: New York, 1992, pp 000–000.
77. Campbell, W. C. *Med. Res. Rev.* **1963**, *13*, 61–79.
78. Thylefors, B. *Trop. Med. Int. Health* **2004**, *9*, A1–A3.
79. Tielsch, J. M.; Beeche, A. *Trop. Med. Int. Health* **2004**, *9*, A45–A56.
80. Sturchio, J. L.; Colatrella, B. D. Successful public-private partnerships in global health: Lessons from the MECTIZAN Donation Program. In *The Economics of Essential Medicines*; Granville, B., Ed.; Royal Institute of International Affairs: London, 2002, pp 255–274.

Biographies

Ralph F Hirschmann born in Bavaria, Germany, came to the US in his teens. He graduated from Oberlin College and then served in the US Army in the Pacific Theater during World War II. He resumed his education at the University of Wisconsin (Madison) as the Sterling Winthrop Fellow with W. S. Johnson as mentor (PhD 1950). He joined Merck & Co., Inc. that year, retiring at 65 in 1987, as Senior Vice President for Basic Research when he joined the Faculty at the University of Pennsylvania as the Makineni Professor. At Merck, his team discovered Mevacor, Vasotec, Prinivil, Primaxin, Proscar, and Ivermectin. In 1969 Robert G Denkewalter, Hirschmann, and their collaborators reported the first total synthesis of an enzyme in solution.

He is a member of the American Academy of Arts and Sciences and of the National Academy of Sciences, and its Institute of Medicine. He received the Cope Medal and the Willard Gibbs Medal. In 2000, he was awarded the National Medal of Science from President Clinton.

Jeffrey L Sturchio is Vice President, External Affairs, Human Health – Europe, Middle East & Africa at Merck & Co., Inc., in Whitehouse Station, New Jersey. He is responsible for the development, coordination, and implementation of a range of health policy and communications initiatives for the region. He has been centrally involved in Merck's participation in the UN/Industry Accelerating Access Initiative to help improve HIV/AIDS care and treatment in the developing world. He is also a member of the private sector delegation to the Board of the Global Fund to Fight AIDS, TB, and Malaria.

Dr Sturchio received an AB in history (1973) from Princeton University and a PhD in the history and sociology of science from the University of Pennsylvania (1981). His previous positions include the AT&T Archives, the Beckman Center for the History of Chemistry at the University of Pennsylvania, Rutgers University, and the New Jersey Institute of Technology. He has also been a postdoctoral fellow and senior fellow at the Smithsonian Institution's National Museum of American History (NMAH). In 2004, he was appointed a visiting fellow of LSE Health and Social Care at the London School of Economics and elected a fellow of the American Association for the Advancement of Science. He joined Merck in June 1989 as the Company's first Corporate Archivist.

His publications include *Chemistry in America, 1876–1976: Historical Indicators* (Reidel, 1985; paperback edition, 1988), written with A Thackray, P T Carroll, and R F Bud; *Values & Visions: A Merck Century* (Merck & Co., Inc., 1991); 'Pharmaceutical firms and the transition to biotechnology: a study in strategic innovation' (with L Galambos), *Business History Review* 72 (Summer 1998): 250–278; 'Against: Direct to consumer advertising is medicalising normal human experience' (with S Bonaccorso), *British Medical Journal* 324 (13 April 2002): 910–911; 'Successful public-private partnerships in global health: lessons from the MECTIZAN Donation Program,' (with B Colatrella), in *The Economics of Essential Medicines*, ed. by B Granville (London: Royal Institute of International Affairs, 2002); and 'Partnership for action: the experience of the Accelerating Access Initiative, 2000–04, and lessons learned,' in *Delivering Essential Medicines*, ed. by A Attaran and B Granville (London: Royal Institute of International Affairs, 2004).

Comprehensive Medicinal Chemistry II
ISBN (set): 0-08-044513-6

ISBN (Volume 1) 0-08-044514-4; pp. 1–27

1.02 Drug Discovery: Historical Perspective, Current Status, and Outlook

P W Erhardt, University of Toledo, OH, USA
J R Proudfoot, Boehringer Ingelheim Inc., Ridgefield, CT, USA

1.02.1 Introduction

Amidst the wide array of drug research activities now being undertaken by numerous life science-related disciplines, it becomes useful to devise a working definition for the practice of medicinal chemistry that allows for its distinction and, in so doing, also establishes the focus sought for this chapter. Thus, we will begin by stating that medicinal chemistry uses physical organic principles to understand the interaction of small molecular displays with biological surfaces.[1] Within this context, physical organic principles should be thought of as encompassing overall conformational considerations and molecular electrostatic potentials, as well as selected chemical properties such as distinct stereochemical, lipophilic/hydrophilic, electronic, and steric parameters. Understanding such interactions provides fundamental knowledge that may be applied in either a general or specific manner toward designing a new drug or enhancing the overall profile of a given compound. Small molecular displays should be thought of in terms of low molecular weight structures that are typically of a xenobiotic origin, and thus not in terms of biotechnology-derived polymers. While the latter are being enthusiastically pursued by other disciplines, it should be additionally appreciated that whenever the consideration of chemical details associated with a specific molecular region of a given biomolecule is undertaken, then by our definition these other disciplines have also entered into the purview of medicinal chemistry's stated interest. Finally, biological surfaces should be thought of very broadly, namely to encompass the complete span of tissues and endogenous molecules associated with a drug's absorption, distribution, metabolism, excretion, and toxicity (the so-called ADMET processes), as well as with the more traditional span of biological surfaces that might be exploited for some type of therapeutically efficacious interaction.

By intent, the numerous technologies that can be deployed as tools to study these types of interactions at medicinal chemistry's fundamental level of understanding, have been completely dissociated from this definition. In addition to the long-standing analytical, synthetic and computational chemistry approaches that have been traditionally conducted by medicinal chemists in a systematic manner on either one or the other of the interacting partners in order to explore structure–activity relationships (SARs), such tools also now include biotechnology-related methods such as site-directed mutagenesis, and combinatorial chemistry methods provided the latter are coupled with chemistry knowledge-capturing bioassay and structural databases. While this definition emphasizes the basic science nature of medicinal chemistry, medicinal chemistry's deployment as an applied research activity is equally important. **Figure 1** reflects the synergistic relationship between medicinal chemistry's basic and applied research activities wherein one dimension becomes multiplied by the other to create the actual working space. The importance of this relationship should become clear as the present chapter, dedicated to medicinal chemistry's historical, current and future roles within the realm of drug discovery, unfolds.

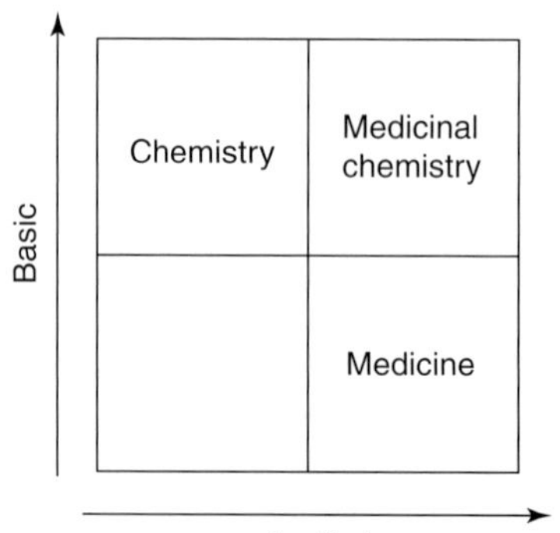

Figure 1 Inherent relationship of medicinal chemistry to both basic and applied research. Adapted from a figure provided by F.A. Cotton[2] as part of his summary and commentary about a book entitled *Pasteur's Quadrant*.[3] Surveys suggest that pharmaceutical companies spend about 5%, 37%, and 54% of their research dollars on basic, applied, and developmental aspects of drug discovery-related research, respectively.[4]

It should be noted that considerable portions of the historical section, as well as portions of the sections thereafter, have been taken from a related publication.[1] Thus, readers may want to look at this earlier document and its use of selected case studies that were specifically dedicated to medicinal chemistry's entry into the present millennium. Interestingly, some of the latter's forecasts already appear to be coming into play while others, as for most attempts at prediction, already appear to be in need of revision. Further references to this prior work will not be repetitiously cited herein.

1.02.2 Historical Perspective

By considering how medicinal chemistry has been practiced across jumps of about 25-year increments, this section provides a short discourse about medicinal chemistry's emergence as a formalized discipline and then considers several key events that occurred during its early development. This section does not include a chronological list of medicinal chemistry's many contributions, nor does it attempt to highlight the many accomplishments of its notable investigators. Both of the latter can be found elsewhere as part of more traditional, historical treatments.[5] Focusing on drug research, old and new drug discovery paradigms are compared, with the latter also being critiqued as part of a lead-in to the next section which then continues by considering the current status of medicinal chemistry relative to the continually evolving process of drug discovery.

1.02.2.1 Emergence of Medicinal Chemistry as an Academic Discipline and Its Full-Scale Adoption by the Industry

Medicinal chemistry's roots can be found in the fertile mix of ancient folk medicine and early natural-product chemistry, hence its name. That this solid foundation drawing from both applied and basic research has not only sustained itself, but continues to be multiplied in a synergistic manner today, is captured nicely within **Figure 1**. Similarly, that medicinal chemistry's connection with natural products immediately sets the stage for its later adoption by the drug industry, is reflected by what some consider to be the likely origin of the word 'drug' as we use it today; namely from the Old English 'dryge,' in turn hypothetically reconstructed from the Primitive Germanic series 'dreug-, draug-, and drug-' which all historically meant 'dry' and therefore can be construed to be conveying how herbs were commonly treated so as to produce dried powders for use as early medicinals.

Responding to a growing appreciation for the links between chemical structure and observed biological activity, medicinal chemistry emerged from this mix about 150 years ago as a distinct discipline intending to explore these relationships via chemical modification and structural mimicry of nature's materials, particularly with an eye toward enhancing the efficacy of substances thought to be of therapeutic value.[6] In the US, medicinal chemistry became formally recognized as a graduate-level discipline about 75 years ago within the academic framework of pharmacy education. From this setting, overviews of medicinal chemistry's subject matter have been offered to both undergraduate and graduate pharmacy students for many years.[7,8]

Understanding structure[biological]–activity relationships (SARs) at the level of inherent physical organic properties such as lipophilic, electronic, and steric parameters, coupled with consideration of molecular conformation, soon became the hallmark of medicinal chemistry research, a development that further distinguished it from its natural-products predecessor which had begun to focus more upon understanding plant phytochemistry. It follows that because these fundamental SARs could be useful during the design of new drugs, applications toward drug discovery became the principal domain for the still sprouting, basic science component of this discipline. Perhaps somewhat prematurely, medicinal chemistry's drug design role became especially important in the private sector where its practice quickly spread and grew rampantly among the rich fields being staked out across the acres of patents and intellectual property that were of particular interest to the pharmaceutical industry. This full-scale adoption into the industry provided a second home for the continued growth of medicinal chemistry wherein critical influences important to the private sector would also serve to further shape its maturation.

1.02.2.2 An Adolescent Heyday of Rational Drug Design

As a more comprehensive understanding about the links between observed activity and pharmacological mechanisms began to develop about 50 years ago and then also proceeded to grow rapidly in biochemical sophistication, medicinal chemistry, in turn, entered into what can now be considered to be an adolescent phase. Confidently instilled with new knowledge about what was happening at the biomolecular level and wanting to respond favorably to the pressing needs of its home within the private sector, the ensuing period was characterized by the high hope of being able to independently design new drugs in a rational, ab initio manner, rather than by relying solely upon Mother Nature's templates and guidance for such. Unfortunately, while this adolescent 'heyday of rational drug design'[9] can be credited with having spurred significant advances in the methods that are deployed for considering molecular conformation, the rate of actually delivering useful therapeutic entities having novel chemical structures, referred to today as new chemical entities (NCEs), was not significantly improved for most pharmacological targets unless the latter's relevant biomolecules also happened to lend themselves to rigorous analysis, such as obtainment of an x-ray diffraction pattern for a crystallized enyzyme's active site with or without a bound ligand. One of the main reasons why rational drug design fell short of its promise was because without experimental data like that afforded by x-ray views of the drug's target site, medicinal chemistry's hypothetical SAR-derived models often reflected rather speculative notions that were far easier to conceive than were the actual syntheses required to produce the molecular probes needed to assess a given model's associated hypotheses. Thus, with only a small number of clinical success stories to relay, medicinal chemistry's 'preconceived notions about what a new drug ought to look like' began to take on negative, rather than positive, connotations, particularly when being hand-waved within the results-oriented setting of a private-sector drug discovery program.[10]

Compounding this disappointment, from a practical point of view, the pharmaceutical industry, by and large, soon concluded that it was more advantageous to employ synthetic organic chemists and have them learn some pharmacology, than to employ formally trained medicinal chemists and have them rectify any shortcomings in synthetic chemistry experience that they might have due to their time spent on a much broader range of nonchemical subject matter during graduate school. Indeed, given the propensity for like to hire like, the vast majority of today's investigators who practice medicinal chemistry within industry have academic backgrounds from organic chemistry rather than from formalized programs of medicinal chemistry.

Two points from this historical segment are worth noting. The first has now become a familiar theme in the life sciences arena whenever a new clinical treatment or technological approach arrives on the horizon, such as that of the former discourse about rational drug design. Invariably, such developments are greeted by such a high level of expectation and hype that, upon falling short, they become subject to a rebound reaction that then overly critiques what might have otherwise been regarded as a positive development. Only after several reiterations does a realistic view finally become equilibrated so that a given technology is matched with appropriate expectation and practical usage. Thus, looking back from today's vantage point, one should not be too concerned about the early disappointment with medicinal chemistry's initial foray into rational drug design. The second point worth noting provides useful insights for a later discussion about the formalized training of future medicinal chemists, namely the importance of obtaining a solid background in synthetic organic chemistry. Having emphasized the significance of the latter, further references to medicinal chemistry throughout the remainder of this chapter should be taken to mean its actual practice as a discipline, regardless of how a given investigator may have become trained to do so. Thus, importantly and key to this historical juncture, no matter how its practitioners were being derived or selected for employment within the pharmaceutical industry, and even though the design of drugs in a rational manner had fallen short of the hyped delivery, medicinal chemistry continued to thrive quite nicely during this period and certainly progressed in its maturation as both a recognized discipline and a central player within the academic and private sectors, respectively.

1.02.2.3 High-Throughput Screening, Combinatorial Chemistry, and a Marriage Made on Earth, if Not in Heaven

Arriving at the next historical segment, however, one finds that medicinal chemistry's previous inability to accelerate the production of NCEs, became greatly exacerbated when the biotechnology rainfall began to hover over the field of drug discovery just about 25 years ago.[11] With this development, not only did the number of interesting biological targets begin to rise rapidly, but also the ability to assay targets in a high-throughput manner suddenly prompted the screening of huge numbers of compounds in very quick timeframes. Ultimately, the need to satisfy the immense appetite of high-throughput screening (HTS) for compounds was addressed not by any of natural-product, synthetic medicinal, or organic chemistry, but instead by further developments within what had quickly become a flood level, continuous downpour of biotechnology-related breakthroughs.[12–14] Starting as gene-cassette-directed, biosynthetic peptide libraries and quickly moving into solid-phase, randomly generated peptide and nucleotide libraries,[15–19] this novel technique soon spread across other disciplines to also spawn the new field of small-molecule, combinatorial chemistry. Today, using equipment and platforms available from a variety of suppliers, huge libraries of small molecules can be readily produced in either a random fashion, or in a directed manner based upon a structural template afforded by a suspected lead compound.[20–27] Thus, the marriage of HTS with combinatorial chemistry would soon propagate a new approach toward the discovery of drugs.

1.02.2.4 Birth of a New Drug Discovery Paradigm

As the marriage of compound library technologies to HTS became consummated, this pairing produced what quickly came to be regarded as a new paradigm for the discovery of NCEs across the entire big pharma segment of the pharmaceutical enterprise.[28–30] **Figures 2** and **3** provide a side-by-side comparison of the old (classical) and new drug discovery paradigms, respectively.

Interestingly, this marriage also led to a situation where identifying initial lead compounds was no longer considered to be a bottleneck for the overall process of drug discovery. Indeed, many of the programs within pharmaceutical companies soon began to suffer from 'compound overload'[38] with far too many initial leads to effectively follow up. Shifting the rate-limiting step away from discovery into development, created a new situation wherein there then became a need to move much more quickly toward characterizing a given lead compound's pharmacokinetic (PK) profile, namely its absorption, distribution, metabolism, and excretion (ADME) behaviors that would be anticipated upon administration to humans. When the latter are also combined with concerns about potential toxicity, the entire process of ADMET assessment thus became the new bottleneck for the overall process of drug discovery, along with the traditionally sluggish clinical and associated regulatory steps as shown near the end of **Figures 2** and **3**.

Responding to this pressure, an emphasis was also placed upon moving ADMET-related assessments into more of a HTS format that could be undertaken at earlier decision points associated with defining a lead compound. Thus, even though efficacy-related HTS and combinatorial chemistry reflect very significant incorporations of new methodologies, from a strategic point of view the most striking feature of the new drug discovery and development paradigm shown in **Figure 3** compared to the classical approach depicted in **Figure 2** actually becomes the movement of ADMET-related assays closer to the beginning of the overall process by also deploying HTS methods. Certainly, with the plethora of biologically based therapeutic concepts continuing to rise even further, and the identification of compounds capable of interacting with those concepts that become selected as targets now being much quicker because of the marriage of HTS to compound library and combinatorial chemistry technologies, a much more efficient handling of ADMET-related concerns clearly represents one of the most significant challenges facing today's drug discovery and development enterprise.

1.02.2.5 Site-Directed Mutagenesis

While the drug discovery process has been influenced by biotechnology in numerous ways (e.g., **Table 1**), one development deserves to be especially noted as this account of medicinal chemistry's history is brought to a close. This development has made a major impact directly upon the process of uncovering SAR relevant to small-molecule drug design.[39] The method utilizes site-directed mutagenesis (SDM). Within the context of medicinal chemistry, SDM involves making systematic point mutations directed toward selected sites on genes that translate to specific amino acid residues within proteins associated with enzyme active sites or ligand receptor systems, such that the targeted changes can be used to study SAR details while holding one or more active site/receptor ligands constant during the assessment. This approach is depicted in **Figure 4**. Numerous investigators are now utilizing this reverse SAR technique to explore both enzyme–ligand and receptor–ligand interactions.

Once an enzyme's peptide sequence is in hand, it is not uncommon that transformation of a bacterial host with an appropriate gene copy plasmid can allow at least one point mutated and overexpressed protein to be examined per month in an associated functional biochemical assay. This clearly demonstrates that SDM should no longer be considered to be lengthy and tedious compared to classical SAR studies undertaken by rational synthetic modifications of an enzyme's substrates or inhibitors.

1.02.3 Current Status

1.02.3.1 Solving an Identity Crisis and Moving On to Play a Key Role as the Central Interpreter of Drug Discovery Information

The deployment of these new approaches toward identifying lead compounds and for studying SAR represent exciting developments that have clearly had an invigorating effect upon the practice of medicinal chemistry in many areas. Alternatively, that the SAR hallmark and drug design intellectual domains of medicinal chemistry, along with chemically servicing the experimental approaches toward identifying novel lead structures, have all been overrun by technologies initially derived from other disciplines, also has had somewhat of an unnerving impact upon medicinal chemistry. This is because it had previously been the nearly exclusive deliverance of these roles from a chemical orientation that had served to distinguish medicinal chemistry as a distinct discipline for so many years. Thus, as a field, medicinal

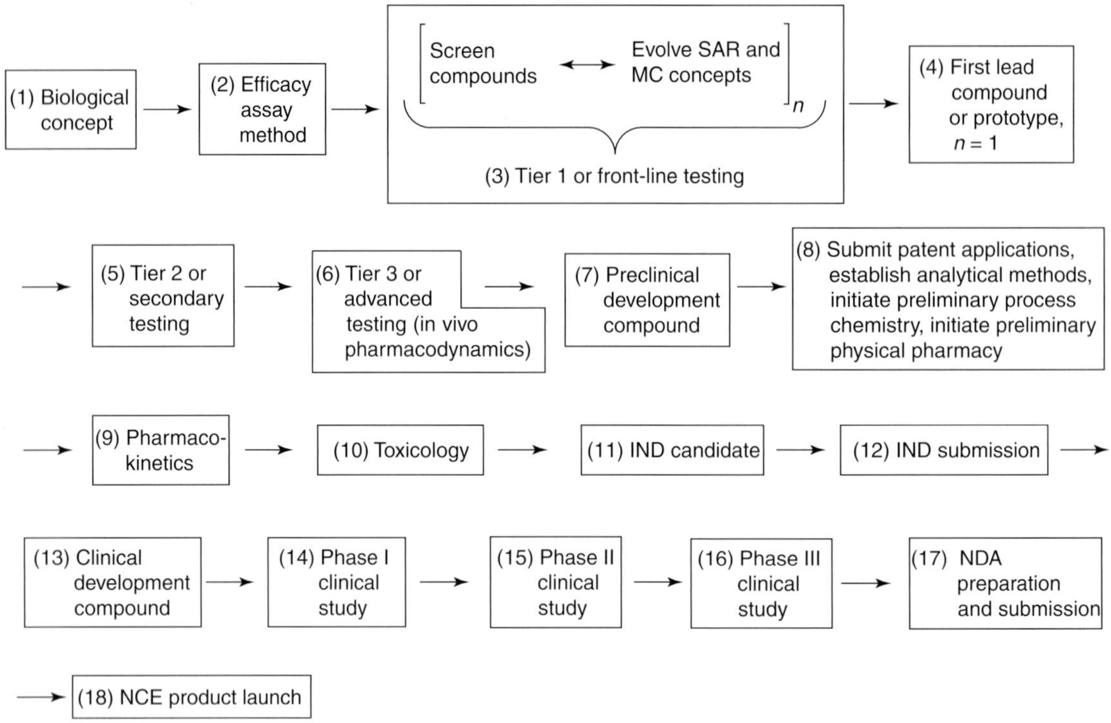

Figure 2 Classical drug discovery and development paradigm.[28,31–36] This model portrays interactions with US regulatory agencies and uses terms related to those interactions for steps 11 to 17. All of the other terms typify generic phrases that have been commonly used by the international pharmaceutical community. While some of the noted activities can be conducted in parallel or in an overlapping manner, the stepwise, sequential nature of this paradigm's overall process is striking. Furthermore, whenever a progressing compound fails to meet criteria set at an advanced step, the process returns or draws again from step 3 for another reiteration. Numerous reiterations eventually identify a compound that is able to traverse the entire process. A successful passage through the entire process to produce just one product compound has been estimated to require about 15 years at a total cost of about $500 million. While the largest share of these time and cost requirements occur during the latter steps, the identification of a promising preclinical development compound, step 7, can be estimated to take about 4 years from the time of initiating a therapeutic concept. Step 1 is typically associated with some type of physiological or pharmacological notion that intends to amplify or attenuate a specific biological mechanism so as to return some pathophysiology to an overall state of homeostasis. Step 2 typically involves one or two biochemical level assay(s) for the interaction of compounds intending to amplify or attenuate the concept-related mechanism. As discussed in the text, steps 3 and 4 reflect key contributions from medicinal chemistry and typically use all sources of available information to provide for compound efficacy hits, e.g., everything from natural product surveys to rational approaches based upon x-ray diffractions of the biological target. Step 5 generally involves larger in vitro models (e.g., tissue level rather than biochemical level) for efficacy and efficacy-related selectivity. Step 6 generally involves in vivo testing and utilizes a pharmacodynamic (observable pharmacologic effect) approach toward compound availability and duration of action. Step 7 typically derives from a formal review conducted by an interdisciplinary team upon examination of a formalized compilation of all data obtained to that point. Step 8 specifies parallel activities that are typically initiated at this juncture by distinct disciplines within a given organization. Step 9 begins more refined pharmacokinetic evaluations by utilizing analytical methods for the drug itself to address in vivo availability and duration of action. Step 10 represents short-term (e.g., 2-week) dose-ranging studies to initially identify toxic markers within one or more small animal populations. Expanded toxicology studies typically progress while overlapping with steps 11 through 14. Steps 11 to 13 represent formalized reviews undertaken by both the sponsoring organization and the U.S. Food and Drug Administration (FDA). Step 14 is typically a dose-ranging study conducted in healthy humans. Steps 15 and 16 reflect efficacy testing in sick patients, possible drug-interactions, etc. Step 17 again reflects formalized reviews undertaken by both the sponsoring organization and the FDA. The FDA's 'fast-track' review of this information is now being said to have been reduced to an average of about 18 months. It is estimated that it costs a company about $150 000 for each day that a compound spends in development. Finally, step 18 represents the delivery of an NCE to the marketplace.

chemistry moved from its disappointing adolescence only to find itself in the middle of an identity crisis. This crisis was occurring among both its private sector practitioners and its academicians. Indeed, the trend established during this period and still in place today, that places a major emphasis upon genomics and proteomics within the public sector funding arena, is still causing many academic medicinal chemistry and basic chemistry investigators to turn their

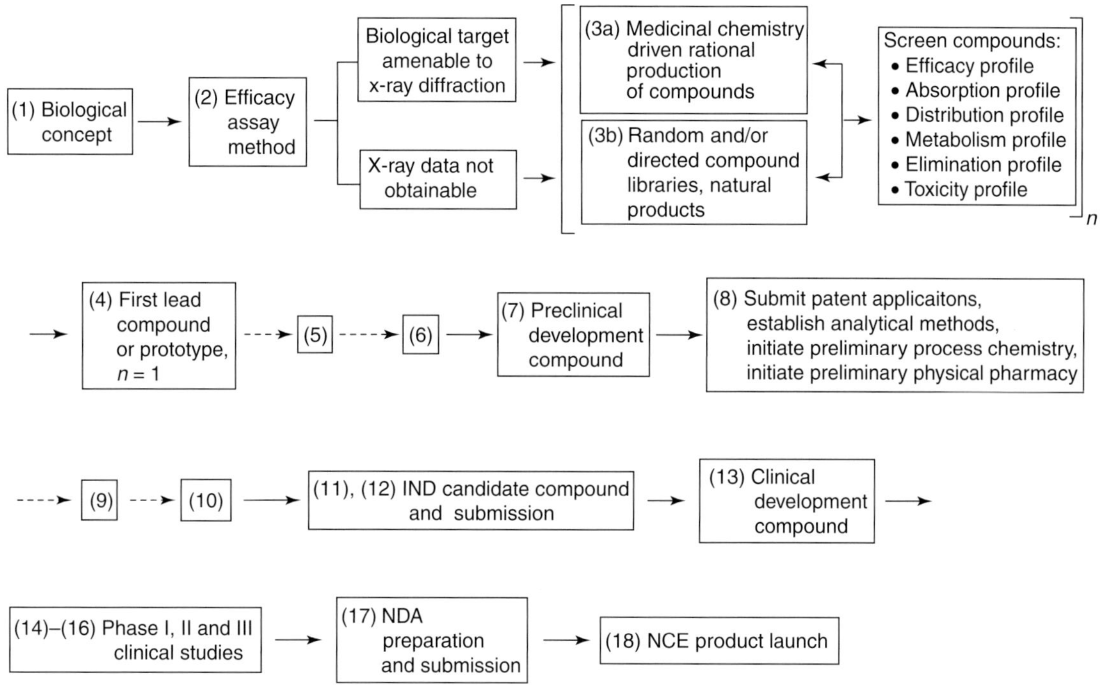

Figure 3 New drug discovery and development paradigm.[28,31–36] This model portrays interactions with US regulatory agencies and uses terms related to those interactions for steps 11 to 17. All of the other terms typify generic phrases that are commonly used by the international pharmaceutical community. The battery of profiling included in step 3 represents a striking contrast to the classical drug discovery paradigm (**Figure 2**). Furthermore, all of these screens are/will be of a high-throughput nature such that huge numbers of compounds can be simultaneously tested in extremely short time periods. As the predictive value of the resulting profiles improve, selected compounds will have higher and higher propensities to successfully proceed through steps 5, 6, 9, and 10 (**Figure 2**) to the point that these assays may become more of a confirmatory nature or may even be able to be completely omitted (hence their dotted lines in this figure).[37] The efficiency of successfully traversing the various clinical testing steps 14 to 16 will also be improved but their complete removal from the overall process is highly unlikely. After the initial investments to upgrade step 3 and enough time has passed to allow for the generation of knowledge from step 3's raw data results (see text for discussion), the overall timeframe and cost for a single NCE to traverse the new paradigm should be considerably improved from the estimates provided in **Figure 2**. Step 1 is likely to be associated with some type of genomic and/or proteomic derived notion that intends to amplify or attenuate a specific biological mechanism so as to return some pathophysiology to an overall state of homeostasis. Step 2 will be a high-throughput assay derived from using molecular biology and bioengineering techniques. Because step 3a exploits actual pictures of what type of structural arrangements are needed to interact with the biological targets, this approach toward identifying new compound hits will continue to operate with high efficiency. However, because of this same efficiency, the biological targets that lend themselves to such experimental depiction (by affording crystals suitable for x-ray diffraction) could become depleted much fastser than the range of more difficult targets. Step 3b represents various combinatorial chemistry-derived libraries, natural product collections, and elicited natural product libraries (see text for discussion of this topic). Steps 4 to 18 are similar to the descriptions noted in **Figure 2**.

intellectual pursuits further and further toward biochemistry, if not to molecular biology.[41] Likewise, the undergraduate instruction of pharmacy students, which had for so long represented medicinal chemistry's academic bread and butter, has shifted its emphasis away from the basic sciences toward more of a clinically oriented curriculum. A final development that contributed to medicinal chemistry's identity crisis is the fact that the new drug discovery paradigm supplants medicinal chemistry's long-standing position wherein its practitioners have typically been regarded as the primary inventors of the composition of matter specifications associated with NCE patent applications.

Although classical medicinal chemistry rationale, as alluded to earlier and as now practiced in tandem with SDM studies, can always be counted on to effectively identify and fine-tune lead compounds for systems whose biomolecules lend themselves to x-ray diffraction and/or nuclear magnetic resonance (NMR) analyses, i.e., structure-based drug design,[42,43] medicinal chemists had to prepare to face the possibility that the complement of such amenable pharmacological targets could become quickly exploited, perhaps even exhausted. Such a scenario, in turn, suggests that this last stronghold for the practice of rational medicinal chemistry could also be lost as a bastion against what could then potentially become an even more serious identity crisis in the future.

Table 1 The impact of biotechnology on small-molecule drug discovery and development[a]

Activity	*Impact*
Genomics and proteomics	Plethora of new and better-defined mechanisms to pursue as therapeutic targets
High-throughput efficacy assays	Screen huge numbers of therapeutic candidates in short timeframes using low compound quantities
High-throughput ADMET assays	An evolving development: once validated and coupled with efficacy assays, should eventually allow for selection or drug design/synthesis of clinical candidate compounds rather than lead compounds that still require considerable preclinical testing and additional chemical tailoring
Peptide and oligonucleotide compound libraries	Provide huge numbers of compounds for screening (spawned the field of combinatorial chemistry as now applied to small organic compounds); can be used as SAR probes and, pending further developments in formulation and delivery, may also become useful as drug candidates
Site-directed mutagenesis	Allows 'reverse' SAR explorations
Transgenic species	Novel in vivo models of pathophysiology that allow 'pharmacological proof of principle' in animal models that mimic the human situation; and animal models modified to have human metabolism genes so as to provide more accurate PK data and risk assessment[40]
Peptide version of pharmacological prototype	Developed to the IND Phase as an intravenous agent can allow for 'clinical proof of principle' or concept
Pharmacogenetics	An evolving development: should soon refine clinical studies, market indications/contraindications, and allow for subgrouping of populations to optimize therapeutic regimens; eventually should allow for classification of prophylactic treatment subgroups

[a] This listing is not intended to highlight the numerous activities associated with the development of specific 'biotech' or large-molecule therapeutics. The arrangement of activities follows the order conveyed in **Figures 2** and **3**, rather than being alphabetical.

Early trends contributing to medicinal chemistry's present status thus initially suggested the following path for the immediate future of small-molecule drug discovery (**Figure 3**):

1. Genomics and proteomics would continue to uncover numerous new pharmacological targets, i.e., to the extent that choosing the most appropriate and validating such targets among the many therapeutic possibilities would also continue to rise as a growing challenge in itself.
2. Biotechnology would, in turn, continue to quickly respond by generating ligand-identification assays for all new targets chosen to be pursued, namely by deploying HTS protocols.[44]
3. Targets that lend themselves to x-ray diffraction and structure-based drug design would likely be exploited quickly.
4. Ligands for HTS would be supplied by existing and new combinatorial-derived compound libraries, as well as from wild[45] and biotechnology-elicited natural sources.
5. Assessment of ADMET parameters, considered to be the new bottleneck for the overall drug discovery process (**Table 2**), would continue to move toward HTS modes that can be placed at earlier and earlier points within the decision trees utilized to select lead compounds for further development as drugs.

It is important to emphasize at this juncture, that in order to place confidence in the predictive value of ADMET HTS surveys, these particular screens must become validated relative to actual clinical-related outcomes. Today, however, this situation is best likened to a deep, dark chasm that the still evolving ADMET HTS surveys need to traverse if they are to ultimately become successful. Fortunately, while many items of (1) to (5) listed above appear to have already come into play, it seems that the most negatively boding possibility has not, namely that of potentially exhausting those targets that readily lend themselves to x-ray and structure-based drug design. Instead, x-ray methods and the associated proteomic-related technologies needed to obtain adequate quality crystals, seem to be rising in very admirable fashion to meet this challenge.[52–58]

Most importantly, however, applying our recited definition for medicinal chemistry across the present drug discovery paradigm reveals that there is an even more critical activity that medicinal chemistry needs to become immediately involved with. Indeed, as HTS results are amassed into mountains not just for efficacy data, but for each of the ADMET parameters as well, it should ultimately become medicinal chemistry's role, by definition, to attempt to

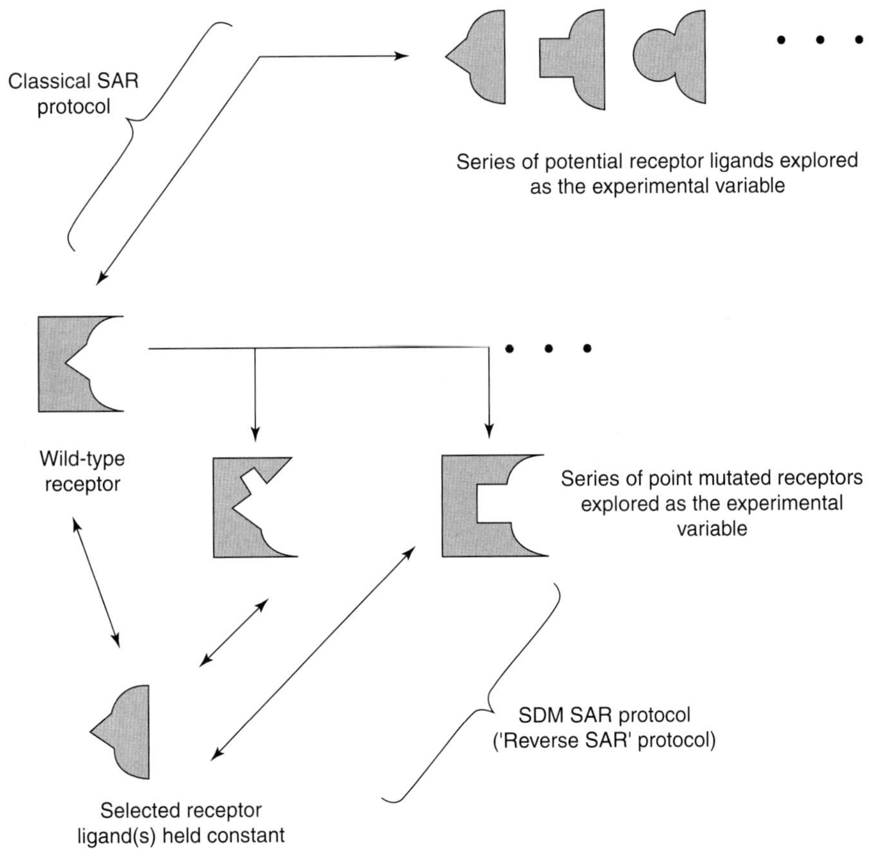

Figure 4 Classical versus site-directed mutagenesis (SDM) exploration of structure–activity relationships (SARs).

understand and codify these awesome, criss-crossing ranges of biological data if such data is to be effectively merged and used to either select or design the most promising preclinical development compounds. For example, while medicinal chemistry's principles and logic may not be needed to identify hits or leads from a single HTS efficacy survey across a library of potential ligands, or perhaps not even needed for two or three of such consecutive surveys involving a few additional ADMET-related parameters, present experience with the new drug discovery paradigm is already demonstrating that the same series of compounds identified from within an initial library as a hit subset or as further generated within a directed library based upon the initial hit subset, are typically not able to sustain themselves as the most preferred leads upon continued HTS parameter surveys. In other words, the identification of the optimal end product structure, namely the best preclinical candidate compound, is unlikely to be derivable from an experimental process that does not represent a 'knowledge'[59,60] generating system that then allows for rational medicinal chemistry-based assessments and adjustments, or even complete revamping, to be interspersed at several points along the way. In this sense, the move toward 'focused libraries'[61,62] and 'smarter,' presorted relational databases actually represents more than just the often-touted desire to 'be more efficient.'[63] Indeed, this may be the only way for the new drug discovery paradigm, now well into its own adolescent phase, to eventually work as it continues to mature and take on more of the ADMET-related considerations in a HTS format. Taken together, it can now be emphasized that the common denominator required to correlate the HTS data from one pharmacological setting to that of another ultimately resides in the precise chemical structure language that medicinal chemistry has been evolving since its emergence as a distinct field, i.e., SAR principles defined in terms of physical organic properties displayed in three-dimensional (3D) space. This, in turn, suggests that it should now be medicinal chemistry that rises to become the central interpreter and distinct facilitator that will eventually allow the entire new drug discovery paradigm to become successful. Indeed, this central role for medicinal chemistry may become even more critical longer term. Speculating that the new drug discovery paradigm will mature within the next 25 years into a synergistic merger of efficacy and thorough ADMET HTS systems that allows for an effective multiparameter survey to be conducted at the onset of the discovery process,

Table 2 Assessment of drug discovery and development bottlenecks[46–51]

Activity	*Estimated timeframe*	*Percentage successfully traversing associated criteria*
Biological conception	A plethora of new characterizations lies waiting to be exploited; this situation is expected to prevail well into the future	The challenge lies in prioritizing which of the numerous mechanisms might be best to pursue (see next entry)
Proof of therapeutic principle	Ultimately requires reaching phase II clinical testing; Biotech derived humanized and/or transgenic disease state models may be able to be substituted at an earlier time point depending upon the confidence associated with their validation	Generally high although there are some distinct therapeutic categories that continue to have low success rates or lack definitive validation such as the attempted treatments of septic shock or the pursuit of endothelin modulators
Identification of lead compound based upon efficacy screen	Using HTS, thousands of compounds can be tested in a matter of days or less (10–100 times more with UHTS); companies are beginning to have more lead compounds than they can move forward in any given program	One compound out of 5000 from random libraries/1 out of 10 from directed libraries; despite the low efficiency, this is not regarded as a bottleneck because HTS can be done so quickly; much higher percentages can be obtained during ligand- and structure-based drug design but synthesis is then correspondingly slower
Progression to preclinical development compound[a]	Approximately 2 years	About 1 out of 50 wherein all can be examined during the indicated time frame
Progression to clinical development compound[a]	Approximately 2 years	About 1 out of 10 wherein all can be examined during the indicated timeframe
Phase I study[b]	Approximately 1 year[c]	About 1 out of 2
Phase II study[b]	Approximately 2 year[c]	About 1 out of 2
Phase III study	Approximately 3 year[c]	About 1 out of 1.5 and often with modified labeling details
Product launch	Approximately 2 years (NDA submission/approval)	About one out of 1.5[d]

[a] Presently regarded as the bottleneck for the overall process. These are the points where ADMET properties have been historically assessed. Approximately 40% are rejected because of poor pharmacokinetics and about another 20% because they show toxicity in animals. In the new drug discovery paradigm (**Figure 3**), the ADMET assessments are being moved to an earlier point in the overall process and are being conducted in a HTS mode. However, in most cases, validation of the new methods relative to clinical success still needs to be accomplished.

[b] While these clinical studies may not be able to be accomplished any quicker, they may be able to be done more efficiently (e.g., smaller numbers and focused phenotypes within selected patient populations) and with a greater success rate based upon making the same improvements in the ADMET assessment area noted in footnote *a*.

[c] Timing includes generation and submission of formal reports.

[d] About 1 out of 5 compounds entering into clinical trials becomes approved. The overall process to obtain one marketed drug takes about 12–15 years at a cost of about $500 million.

the accompanying and by then validated, predictive data that will have been generated over this initial period should be statistically adequate to actually realize today's dream of 'virtual' or 'in silico screening'[64–71] through virtual compound and virtual informational libraries not just for identifying potential efficacy leads to synthesize, as is already being attempted, but across the entire preclinical portion of the new paradigm wherein the best overall preclinical candidate compound is selected with high precision for synthesis at the outset of a new therapeutic program. This futuristic drug discovery paradigm is depicted in **Figure 5**.

This evolution, however, depends upon the entire maturation process being able to proceed in a knowledge-generating manner. Central to the latter remains medicinal chemistry as the common denominator. For example, with

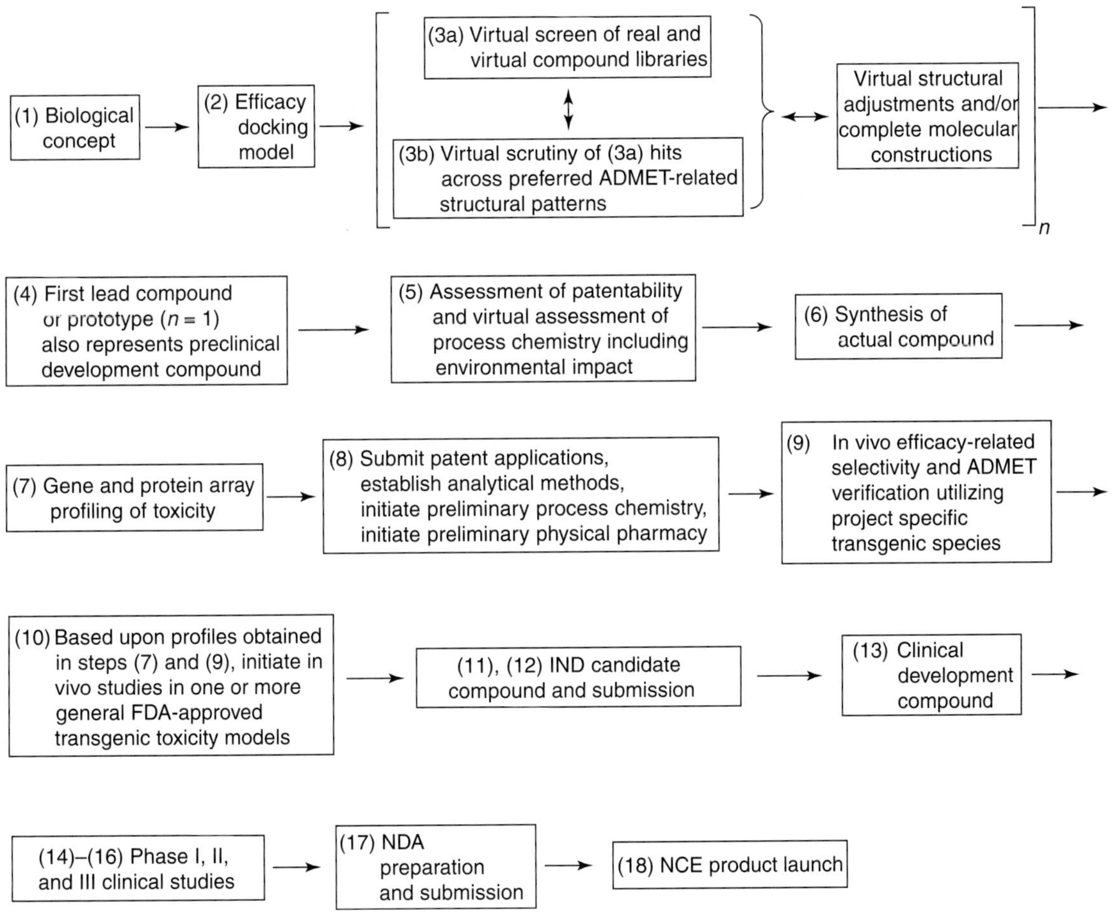

Figure 5 Future drug discovery and development paradigm. This model portrays interactions with US regulatory agencies and uses terms related to those interactions for steps 11 to 17. 'Future' implies about another 50 to 75 years. The most striking feature of this paradigm compared to **Figures 2** and **3** is the considerable number of decisions that will be made from virtual constructs rather than from experimental results. Confidence in the virtual decisions will be directly proportional to the level of knowledge that is learned from the huge amounts of drug screening data being amassed during the next 50 years coupled with the overall ability to predict clinical outcomes. Step 1 is likely to be associated with some type of genomic and/or proteomic derived notion that intends to amplify or attenuate one or more specific biological mechanisms so as to either return some pathophysiology to an overall state of homeostasis, or to modify some system in a manner that prevents or provides prophylaxis toward an otherwise anticipated pathophysiological development. A growing emphasis of treatments will be directed toward prevention. Step 2 may be based upon an actual x-ray diffraction version of the biological target or upon a computationally constructed version derived from similar known systems that have been catalogued for such extrapolations. In either case, docking studies will be conducted in a virtual mode. Steps 3, 4, and 5 will be conducted in a virtual mode. Steps 6 and 7 represent the first laboratory-based activities. After submission of patents, it is proposed that in vivo testing involving steps 9 and 10 will be able to take advantage of project specific and FDA approved generic toxicity model transgenic species. Steps 11 to 18 are similar to those in **Figures 2** and **3** except that the likelihood for a compound to fall short of the desired criteria will be significantly reduced. Subject inclusion/exclusion criteria will also be much more refined based upon advances in the field of pharmacogenetics.

time it can be expected that just as various pharmacophores and toxicophores have already been identified for specific structures associated with efficacious or toxic endpoints, respectively, distinguishing molecular properties and structural features will, likewise, become associated with each of the ADME behaviors. Indeed, work toward such characterizations is already progressing in all of these areas. Understanding the pharmacophores, metabophores,[9] toxicophores, etc. in terms of subtle differences in molecular electrostatic potentials (from which medicinal chemistry's physical organic properties of interest are derived) as well as in terms of simple chemical structural patterns, will eventually allow for identifying optimal composites of all of these parameters across virtual compound libraries as long as the latter databases have also been constructed in terms of both accurate 3D molecular electrostatic potentials and gross structural properties. Thus, in the very least, medicinal chemistry's recent identity crises should now be considered to have been more than solved by the very same factors that served to prompt such feelings in the first place. Indeed, implementation of the new drug discovery paradigm continues to be accompanied by a desperate call for medicinal chemistry's hallmark SAR-related logic and rationale to now make it all come together so as to become a success.

As an initial follow-up to this call, the present status of assessing molecular conformation will be reviewed first since this is such an integral part of practicing medicinal chemistry along any venue. In particular, we will look most closely at the handling of chemical structures and chemical-related information within database settings, an additionally challenging activity now considered to be a field of its own, namely that of chemoinformatics.[72–75]

1.02.3.2 Chemoinformatics

Given the exponential proliferation of technical data and our increasing ability to rapidly disseminate it through electronic networks, it is no wonder that new systems capable of 'managing and integrating information'[76] have been regarded among 'the most important of the emerging technologies for future growth and economic development across the globe.'[76,77] That 'information technology' (IT), in turn, is now receiving high priority in all sectors is quite clear,[78–85] particularly with regard to systems directed toward integrating bioinformatics-related information as promoted via the world wide web.[86] As mentioned, medicinal chemistry's contributions toward this assessment of the future importance of IT primarily reside in the area of handling chemical structures and chemical information. In this regard, the increasing use of databases to link chemical structures with biological properties has already been alluded to in terms of both real experimental data sets and virtual compilations. Although serious strides have been taken in this area, however, there is still a significant need for improvement in the handling of chemical structures beyond what is suggested by what now appears to be occurring within today's database assemblies. For example, that "better correlations are sometimes obtained by using 2D displays of a database's chemical structures than by using 3D displays" only testifies to the fact that we are still not doing a very good job at developing the latter.[87] How medicinal chemistry must step up and rise to the challenges already posed by this situation in order to fulfill the key roles delineated above for its near and longer-term future are addressed within the next several paragraphs of this section.

Assessment of molecular conformation, particularly with regard to database-housed structures, represents a critical aspect of chemoinformatics. While new proteins of interest can be addressed reasonably well by examining long-standing databases such as the Protein Data Bank,[88] and other Web-based resources[89] for either explicit or similar structural motifs and then deploying x-ray (pending a suitable crystal), NMR, and molecular dynamic/simulation computational studies[90–95] as appropriate, the handling of small molecules and of highly flexible molecular systems in general, remains controversial.[96] For example, the only clear consensus is that treatments of small molecules for use within database collections "have, to date, been extremely inadequate."[97] Certainly, a variety of automated, 3D chemical structure drawing programs are available that can start from simple 2D representations by using Dreiding molecular mechanics or other, user-friendly automated molecular mechanics-based algorithms, as well as when data is expressed by a connection table or linear string.[98] Some programs are able to derive 3D structure "from more than 20 different types of import formats."[99] Furthermore, several of these programs can be directly integrated with the latest versions of more sophisticated quantum mechanics packages such as Gaussian 98, MOPAC (with MNDO/d), and extended Huckel.[98] Thus, electronic handling of chemical structures, and to a certain extent comparing them, in 3D formats has already become reasonably well worked out.[72–75,98–101]

Nevertheless, the fundamental problem still remains as to how the 3D structure is initially derived in terms of its chemical correctness based upon what assumptions might have been made during the process. Further, there are still challenges associated with how readily 3D structure information can be linked with other, nonchemical types of informational fields. As has been pointed out by others, the reason that such mingling of data fields often does not afford good fits is because "each was initially designed to optimize some aspect of its own process and the data relationships and structures are not consistent."[72] At this point, inexpensive Web-based tools that can integrate chemical structure data with other types of information from a variety of sources, including genomic data, have already begun to emerge.[102] While this trend is likely to continue to pick up, user-friendly solutions that cut across all disciplines again beg for the central interpretation that can be afforded by medicinal chemistry's hallmark rationale of SAR-derived principles and unifying concepts.

Unfortunately, the quick assignment of chemically correct 3D structures may not be as readily solvable. Recalling from Section 1.02.2, medicinal chemistry has been concerning itself with this task for quite some time. Medicinal chemistry's interest in small-molecule chemical structure is further complicated, however, by the additional need to also understand how a given drug molecule's conformational family behaves during its interactions with each of the biological environments of interest. For example, as a drug embarks on its 'random walk'[103] through the biological realm (**Figure 6**), the ensuing series of interactions have unique effects upon each other's conformations[104,105] at each step of the journey and not with just the step that finally consummates the drug's encounter and meaningful relationship that is struck with its desired efficacy-related receptor/active site.

In order to track such behavior in a comprehensive manner, it becomes necessary to consider a drug's multiple conformational behaviors by engaging as many different types of conformational assessment technologies as possible, while initially taking an approach that is unbiased by any knowledge that may be available from a specific interacting environment. For example, the three common approaches depicted in **Figure 7** include: (1) x-ray (itself prone to bias from solid-state interactions within the crystal lattice); (2) solution spectroscopic methods, namely NMR, which can

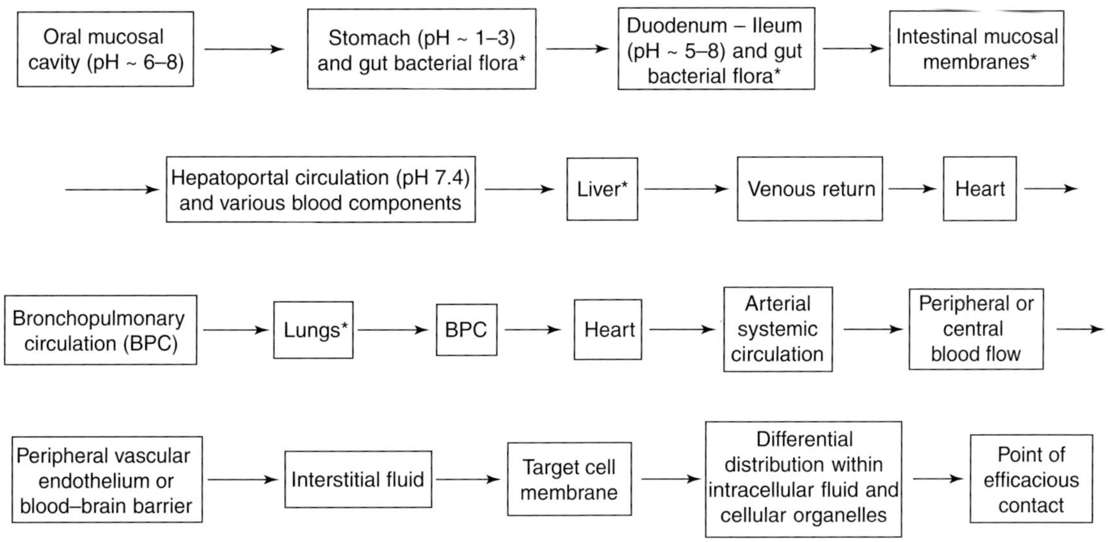

Figure 6 Random walk[103] taken by an oral drug on route to its point of efficacious interaction within a human target cell. This continuum of interactions between a drug and various biological surfaces within the human biological realm is typically divided into categories associated with ADMET and efficacy. Biological milieus marked with an asterisk represent compartments having particularly high metabolic capabilities. Blood is notably high in esterase capability. In the future, medicinal chemists will utilize knowledge about ADMET-related SARs to more effectively identify the best drug leads and to further enhance the therapeutic profiles of selected compounds.

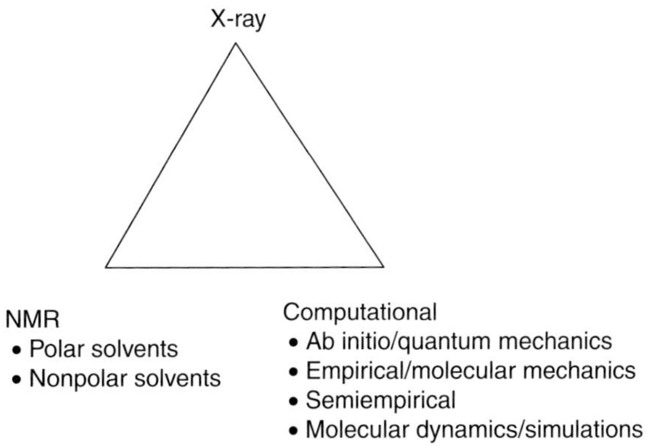

Figure 7 Techniques employed to assess conformational detail. X-ray diffraction requires a suitable crystal and its results are subject to solid-state interactions. Computational paradigms are most accurate when done at the highest levels of calculation but these types of calculations become prohibitively computer-time intensive. NMR requires that the molecule be soluble in the chosen solvent and that adequate compound supplies be available. Mass spectrometry is also becoming an important tool for larger molecules although it provides smaller amounts of descriptive conformational data. A composite of all approaches provides for the best possible assessment of molecular conformation.

often be done in both polar and nonpolar media (this technique, however, being more limited by the amount of descriptive data that it can generate); and (3) computational approaches that can be done with various levels of solvation and heightened energy content (limited, however, by the assumptions and approximations that need to be taken in order to simplify the mathematical rigor so as to allow solutions to be derived in practical computational timeframes). Analogous to the simple, drawing program starting points, programs are also available for converting x-ray and NMR data into 3D structures.[106]

While the importance of taking such a three-pronged approach is not new, it is emphasized herein because today's medicinal chemistry literature suggests that investigators sometimes still fall into the single technique trap from which further extrapolations of data are then often made with great conviction. This may be because it is difficult to obtain an acceptable crystal for x-ray analysis, have adequate solubility for highfield conformational analysis by NMR, or perhaps, to become aligned with appropriate computational expertise and computing power. At any rate, advances in all three of these areas can be expected to alleviate such implementation-related shortcomings so that medicinal chemists of just the near-term will be more readily able to consider structures from at least a three-pronged starting point either independently or through collaboration with other specialists and experts dedicated to each of these areas.

As mentioned above, after taking an unbiased structural starting point, medicinal chemistry needs to especially consider structures (and the energies thereof) by ascertaining what their relevant conformations might be during interactions within various biological milieus. It can be imagined that at least for the immediate future, a useful range of such media to be considered will include: aqueous solutions at acidic and neutral pH, namely at about 2 (stomach) and 7.4 (physiological), respectively; one or more lipophilic settings, such as might be encountered during passive transport through membranes; and finally, specific biological receptors and/or enzyme active site settings that are of particular interest. Importantly, with time this list can then be expected to further grow so as to also include: several distinct biological surface models deemed to be representative for interaction with various transportophore relationships; several distinct biological surface models deemed to be relevant for interaction with specific metabophore relationships such as within the active site of a specific cytochrome P450 metabolizing enzyme; and finally, several distinct biological surface models deemed to be relevant for interaction with specific toxicophore relationships. It should also be appreciated that the interaction of even just one ligand within just one of the various biological settings could still involve a wide range of conformational relationships wherein the biological surface may also exist as an equilibrium mixture of various conformational family members. If x-ray, NMR, etc. can be further deployed to assess any one or combination of these types of interactions, then a composite approach that deploys as many as possible of these techniques will again represent the most ideal way to approach future conformational considerations within the variously biased settings. Advances toward experimentally studying the nature of complexes where compounds are docked into real and model biological environments are proceeding rapidly in all of these areas, with mass spectrometery (MS)[107–109] and microcalorimetry[110,111] also now adding themselves alongside x-ray and NMR[112,113] as extremely useful experimental techniques for the study of such SARs. Besides the experimental approaches, computational schemes will likely always be deployed because they can provide the relative energies associated with all of the different species. Furthermore, computational methods can be used to derive energy paths to get from the first set of unbiased structures to a second set of environmentally accommodated conformations in both aqueous media and at mutually molded biological surfaces. Importantly, these paths and their energy differences can then be compared along with the direct comparison of structures per se, while attempting to uncover and define correlations between chemical structure and some other informational field within or between various databases.

Finally, by using computational paradigms, these same types of comparisons (i.e., among and between distinct families of conformationally related members) can also be done for additional sets of conformational family members that become accessible at appropriately increased energy levels (i.e., at one or more 5 kcal mol^{-1} increments of energy) to thus address the beneficial losses of energy that might be obtained during favorable binding with receptors or active sites.[114] These types of altered conformations can also become candidates for structural comparisons between databases. The latter represents another important refinement that could become utilized as part of SAR queries that will need to be undertaken across the new efficacy and ADMET-related parameters of the future. With time, each structural family might be ultimately addressed by treating the 3D displays in terms of coordinate point schemes or graph theory matrices.[115] This is because these types of methods lend themselves to the latest thoughts pertaining to utilizing intentionally 'fuzzy coordinates,'[116,117] e.g., $x \pm x'$, $y \pm y'$, and $z \pm z'$ (rather than just x, y, and z plots), for each atomic point within a molecular matrix wherein the specified variations might be intelligently derived from the composite of aforementioned, energy-biased and energy-raised computational and experimental approaches. Likewise, the fuzzy coordinate strategy might become better deployed during the searching routines, or perhaps both knowledgeably fuzzy data entry and knowledgeably fuzzy data searching engines handled, in turn, by fuzzy hardware,[118] will ultimately best identify the correlations which are being sought in any given search paradigm of the future.

It should be noted, however, that for the fuzzy types of structural treatments, queries will be most effective when the database has become large enough to statistically rid itself of the additional noise that such fuzziness will inherently create.

It can be noted that it is probably already feasible to place most of the clinically used drugs into a structural database that could at least begin to approach the low to mid-tier levels of sophistication discussed above because considerable portions of such data and detail are typically already available within the literature for each drug even if it is presently spread across a variety of technical journals. On the other hand, it should also be clear that an alternate strategy will be needed to handle the mountains of research compounds associated with just a single HTS parameter survey. Unfortunately, it appears that some of the large compound surveys being conducted today do not even have systematically treated 2D structural representations. Indeed, while the present status of handling chemical structure and data associated with HTS is wisely being directed toward controlling the size of the haystack,[119] the dire status of handling conformational detail is reflected by attempts that try to grossly distinguish between druglike and nondruglike molecules[120] in a 2D manner or, at best, to identify certain 'privileged structures'[121,122] while using 3D constructs derived from less than completely rigorous experimental and computational assessments. Furthermore, in certain companies, notions about druglike patterns (or actually the lack thereof) are already being set up as the first in silico filter to be deployed against a given compound library's members while the latter are still en route to a HTS efficacy screen. Unfortunately, this scenario can detract from the definition of an initial efficacy pharmacophore along structural motifs that might, alternatively, be able to readily take advantage of neutral areas by making straightforward chemical modifications that then serve to avoid the nondruglike features. At present, and for probably much of the near term as well, strategies that use nondruglike parameters to limit the number of compounds that can otherwise contribute toward the definition of a given efficacy-related structural space would appear to be premature. In the very least, such strategies are counter to the need to continue to accumulate greater knowledge in the overall ADMET arena, let alone in the specific handling of 3D chemical structure at this particular time. Finally, when it is additionally appreciated that in most cases the connection of HTS ADMET data with actual clinical outcomes still remains to be much more securely validated, the strategy to deploy notions about nondruglike structural hurdles as decision steps prior to efficacy screening becomes reminiscent of medicinal chemistry's own adolescent phase wherein medicinal chemistry's efforts to rationally design drugs without the benefit of the additional knowledge afforded by an x-ray of the actual target site, ultimately did not enhance either the production of NCEs or the image of medicinal chemistry.

With regard to chemical structure, the present situation thus indicates that we have a long way to go toward achieving the aforementioned tiers of conformational treatments when dealing with large databases and applying them toward the process of drug discovery. Nevertheless, because of the importance of chemoinformatics toward understanding, fully appreciating, and, ultimately, actually implementing bioinformatics along the practical avenues of new drug discovery, it can be imagined that future structural fields within databases, including those associated with HTS, may be handled according to the following scenario, as summarized from the ongoing discussion within this section and as also conveyed within **Figure 8**.

For optimal use in the future, it is suggested that several levels of sophistication will be built into database architectures so that a simple 2D format can be input immediately. Accompanying the simple 2D structure field would be a field for experimentally obtained or calculated physicochemical properties (the latter data also to be upgraded as structures are matured). While this simple starting point would lend itself to some types of rudimentary structure-related searching paradigms, the same compound would then gradually progress by further conformational study through a series of more sophisticated chemical structure displays. As mentioned earlier, x-ray, NMR, and computational approaches toward considering molecular conformation will be deployed for real compounds given that it is also likely that advances in all of these areas will allow them to be more readily applied in each case. Obviously, virtual compound libraries and databases will have to rely solely upon computational approaches and upon knowledgeable extrapolation from experimental data derivable by analogy to structures within overlapping similarity space. Eventually, structures would be manipulated to a top tier of chemical conformational information. This top tier might portray the population ratios within a conformational family for a given structure entry expressed as both distinct member and averaged electrostatic surface potentials wherein the latter can be further expanded so as to display their atomic orientations by fuzzy graph theory or fuzzy 3D coordinate systems. Thus, at this point, one might speculate that an intelligently fuzzy coordinate system could eventually represent the highest level of development for the 3D quantitative SAR (3D-QSAR)[123,124] based searching paradigms seemingly rising to the forefront of today's trends in the form of comparative molecular field analysis (CoMFA).[125,126] Furthermore, one can imagine that this tier might actually be developed in triplicate for each compound, that is: one informational field for the environmentally unbiased structural entries; another involving several subsets associated with known or suspected interactions with the biological realm; and a third for tracking conformational families when raised by about 5 and 10 kcal mol^{-1} in energy. Finally, as

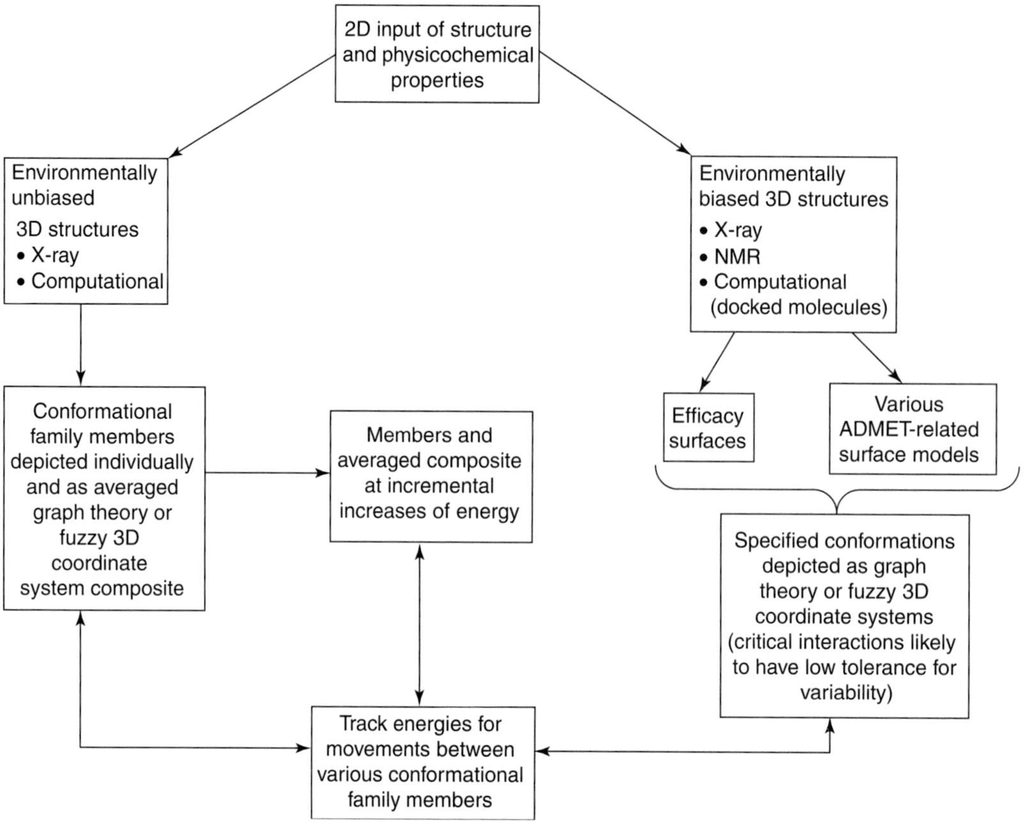

Figure 8 Handling chemical structures within databases now and into the future. This figure depicts the quick entry and gradual maturation of structures. Search engines, in turn, would also provide for a variety of flexible paradigms involving physical properties with both full and partial (sub)structure searching capabilities using pattern overlap/recognition, similarity/dissimilarity, comparative molecular field analysis (CoMFA), etc. Structure entry would be initiated by a simple 2D depiction that is gradually matured in conformational sophistication via experimental and computational studies. Note that structures would be evolved in both an unbiased and in several environmentally biased formats. The highest structural tier would represent tracking/searching the energies required for various conformational movements that members would take when going from one family to another.

chemoinformatics continues to churn its computational technologies forward, conformational and energetic considerations pertaining to a compound's movement between its various displays can also be expected to be further refined so as to ultimately allow future characterization and searching of the dynamic chemical events that occur at the drug–biological interface, e.g., modes and energies of docking trajectories and their associated molecular motions relative to both ligand and receptor/active site. That this top tier is extremely valuable for understanding the interactions of interest to medicinal chemistry is apparent from the large amount of effort already going on today in this area,[127–132] particularly when such studies are able to take advantage of an x-ray-derived starting point.

By the same token, chemical structure search engines of the future will probably be set up so that they can also be undertaken at several tiers of sophistication, the more sophisticated requiring more expert-based enquiries and longer search times for the attempted correlations to be assessed. A reasonable hierarchy for search capability relative to the structural portion of any query might become: (1) simple 2D structure with and without physicochemical properties; (2) 3D structure at incremented levels of refinement; (3) 2D and 3D substructures; (4) molecular similarity/dissimilarity indices; (5) fuzzy coordinate matrices; (6) docked systems from either the drug's or the receptor/active site's view at various levels of specifiable precision; and finally, in the more distant future, (7) energy paths for a drug's movement across various biological milieus including the trajectories and molecular motions associated with drug receptor/active site docking scenarios. Emphasizing informatics flexibility, this type of approach where data entry can occur rapidly for starting structure displays and then be gradually matured to more sophisticated displays as conformational details are accurately accrued, coupled with the ability to query at different levels of chemical complexity and visual displays[133] at any point during database maturation, should allow for chemically creative database mining strategies to be effected in the near term, as well as into the more distant future.

What this section points to is that, ultimately, structural databases of the future will probably have several 'tiers'[134] of organized chemical and conformational information available which can be distinctly mined according to the specified needs of a directed (biased) searching scheme while still being able to be completely mixed within an overall relational architecture such that undirected (unbiased), 'knowledge-generating mining paradigms' can also be undertaken.[135–140] Certainly, simple physicochemical data will need to be included among the parameters for chemical structure storage. Likewise, searching engines will need to allow for discrete substructure queries as well as for assessing overall patterns of similarity and dissimilarity[141–148] across entire electronic surfaces.

1.02.3.3 Impact of Genomics and Proteomics (Target Identification and Validation)

As mentioned in Section 1.02.2, advances in biochemistry and molecular biology have had a huge impact on medicinal chemistry and the process of drug discovery over the course of the last 25 years (e.g., **Table 1**). Whereas drug structures were previously optimized through animal or tissue-based testing, most drug discovery campaigns now deploy assays for selected molecular targets in discrete biochemical systems or in well-defined cell lines. The ability to produce large quantities of active proteins and to generate cell lines overexpressing particular targets has enabled HTS to become a keystone of the modern drug discovery process (**Figure 3**). Access to large amounts of pure protein or protein subdomains, often engineered to ease handling and purification, has also facilitated structure determination by x-ray crystallography and NMR methods to the extent that structure-based drug design (**Table 3**) is now a routine a part of the lead optimization process for many target classes.

In addition to such influences on medicinal chemistry-related lead identification and structural optimization, advances in biochemistry and molecular biology have proliferated the fields of genomics, proteomics, and pharmacogenomics – areas that impact directly upon the initial selection and validation of disease targets. In the past, the combination of animal and tissue testing along with the exploitation of clinically validated competitor compounds led to the development of many successful 'drug analogs.' With the tools afforded by the modern drug discovery paradigm (**Figure 3**), there has been a shift in focus toward first-in-class molecules acting at first-in-class targets that address unmet medical needs. However, these targets usually come with a relatively decreased level of clinical validation such that genomic and proteomic approaches are usually employed to provide a link between the target and the disease to be treated. From a technology perspective, the introduction of DNA microarray methods has enabled an unprecedented exploration of the transcriptional state of cell populations.[149] Examination of the transcriptional state of normal and disease modified cells can shed light on potential targets involved in the disease process. Furthermore, transcriptional response to treatment with pharmacological agents can also be used to assess drug efficacy and toxicity at the molecular level.

Modern genomic technologies allow the selective deletion of target genes in the mouse and other species. Evaluation of the phenotype for these "knockout animals" can provide information about the role and importance of

Table 3 Medicinal chemistry-related drug discovery approaches[a]

Method	*Key features*
Structure-based drug design	An experimentally derived (e.g., by x-ray, NMR) model of the target site is available (sometimes with one or more bound ligands as well); often is coupled with molecular modeling studies and with SDM studies
Analog-based drug design	A special case of ligand-based drug design wherein the parent ligand is a successfully marketed drug
Ligand-based drug design	At least one, weakly binding ligand for the target is known and it serves as a molecular template for the design of further compounds or exploration of related structural space
Library-based drug discovery	A library of compounds is assessed (typically) via a HTS in search of a hit or lead compound; with the latter in hand, this approach combines with ligand-based drug design to form a directed library
Natural product drug discovery	Matrices or isolates from natural sources are (typically) assessed in various biological assays, often times in a HTS mode

[a] These approaches are meant to focus upon just chemistry activities such that within this context, HTS and 'target-based' technologies for example, are not listed as approaches toward drug discovery. Today, rational drug design can be thought of as encompassing the first three entries and, since all entries are listed in an approximate ranking for their decreasing efficiency toward producing an active compound, medicinal chemistry has thus turned to a new page in history compared to its first experiences with rational drug design as conveyed in Section 1.02.2.

the gene products.[150] Target gene deletion, in an ideal setting, corresponds to the effects of a perfect antagonist compound. Recent publications[151,152] indicate that target knockout phenotypes recapitulate the corresponding drug effects in up to 85% of the cases, either in the context of markers of drug efficacy or drug side effects. However, there are also instances where the gene deletion phenotype can mislead. In this regard it should be appreciated that many of the drug/target combinations assessed historically related to drugs developed using rodent models of disease to evaluate efficacy, and in those cases one might expect a good correlation of target role between mouse and human for this particular selection of therapeutics. Alternatively, there are circumstances where knockout experiments may not be able to provide target validation, i.e., in instances where the target functions differently in mouse compared to human, or the target gene deletion may be embryonically or neonatally lethal precluding the assessment of target abrogation in the adult. Likewise, unanticipated compensatory mechanisms during development can diminish the effects of gene deletion. In these cases, conditional knockouts, in which the target gene is abrogated in a particular tissue or at a defined time, may provide target validation. However, to date, the latter have proven more difficult to implement on as large a scale.[153] Finally, even when gene modification in the mouse does provide robust target validation, the overall process is typically expensive and time consuming. One technique that has been widely adopted since the demonstration of its utility in mammalian cells[154] is that of gene silencing by small interfering RNA (siRNA).[155] Several reports of target identification using this technique have recently appeared.[156–158] It has been applied in a high-throughput fashion for target identification[159] and validation,[160] and it is likely that, in the future, siRNA methods in combination with expression analysis will become a mainstay of the target discovery/validation process.

Proteomics, in turn, probes the protein content of the cell (the proteome) in terms of function and interactions. In this regard, it has also been used to identify potentially useful disease targets. Since proteins represent the biological surfaces that serve as mediators for most small-molecule drug actions, proteomics can also be used to identify specific protein targets for useful drug actions. From affinity-based ligand approaches, exemplified by the identification of FK506 binding protein (FKBP) as a target for FK506,[161] methods for protein target identification have evolved to more sophisticated yeast, three-hybrid systems.[162] The combination of proteomics and genomics has allowed the construction of detailed protein interaction maps for entire signal transduction pathways, e.g., tumor necrosis factor alpha (TNF-α)/nuclear factor kappa B (NFκB) pathway[163,164] and even at the level of entire organisms, e.g., *Caenorhabditis elegans*.[165] These pathways can give a clearer picture for the potential desirable effects and unwanted side effects of molecules interacting with a given target in a particular pathway.

Pharmacogenomics attempts to relate variable drug response to variation in particular genes and gene products. Thus, pharmacogenomics encompasses an analysis of the genome-wide contributions of numerous genes and gene products to drug response and toxicity.[166] This field has also benefited from the development of high-throughput gene array technology. As mentioned above, gene expression profiling is proving to be very useful in the validation of drug efficacy in the discovery stage and it is likely that, in the near future, expression profiling will also prove useful in the evaluation of drug response[167] and toxicology screening.[168] Ultimately, these sciences hold the promise of personalized medicine based on the identification of genes related to disease susceptibility or drug susceptibility. Pharmacogenomic analysis can potentially identify disease susceptibility genes as well as drug susceptibility genes: target effects, off-target effects, new targets. Perhaps the most advanced example of the impact to date is the identification of chemotactic chemokine receptor 5 (CCR5) as a co-receptor for human immunodeficiency virus 1 (HIV-1) and a potential target for the treatment of acquired immune deficiency syndrome (AIDS).[169] From the first clinical validation disclosed in 1996, it is remarkable that only 9 years later there are now drug molecules in phase II and phase III clinical trials that target this receptor for the treatment of AIDS. Although it is not yet possible to point to examples for the discovery of marketed drugs influenced by one of these applications, genomics, proteomics, and pharmacogenomics clearly have already impacted upon the discovery process in the areas of target identification and validation, as well as further contributing toward the biotechnology-derived tools that have inspired the still evolving, new drug discovery paradigm.

1.02.3.4 High Attrition during Drug Development and an Actual Decline in New Chemical Entities

When these new ways to identify novel therapeutic targets are coupled with the new paradigm for the identification of potential drug molecules, the entire process of drug discovery can certainly be said to be operating in an exhilarating period. For today's practitioner, however, it becomes immediately disappointing to note that the statistics associated with the production of new molecular entities (NMEs) over the last few years indicate that NMEs are actually undergoing a decline.[170] This disturbing development is depicted in **Figure 9**. Since the NMEs reflect the sum of the NCEs that are closely associated with medicinal chemistry efforts, plus new biological entities (NBEs) that are more

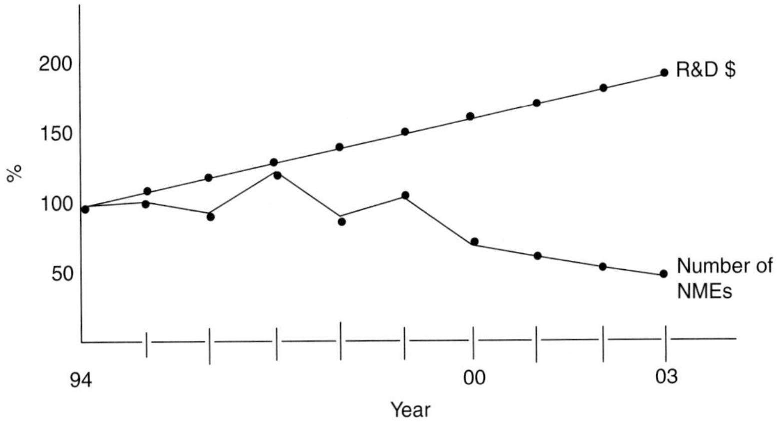

Figure 9 Diminishing returns.[170] Downward trend continues in the number of new drugs entering the marketplace while R&D expenditures continue to climb. NME = new molecular entity = ΣNCE + NBE (new chemical entity plus new biological entity).[171]

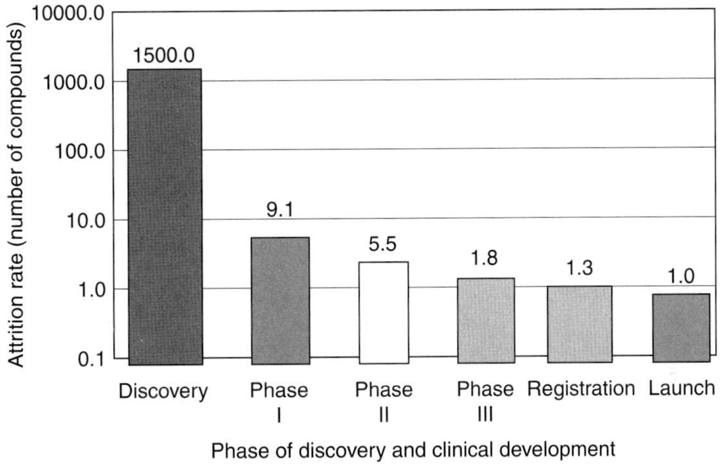

Figure 10 Drug candidate attrition: the numbers.[172] Bar graph numbers represent the numbers of compounds that typically are tested in order to advance to the next stage of the overall drug discovery process in order to achieve the successful market launch of one new product as the last step.

closely associated with biotechnology-related disciplines, one might wonder if just one of these two subcategories is largely responsible for the overall decline. However, this is not the case and, if anything, it is the NCEs that appear to be taking the biggest dive.[171] Further analysis of this situation reveals that, on a percentage basis, there has been little improvement in the overall attrition rate for lead compounds moving from late preclinical development through the early phases of clinical testing.[172] The numbers of compounds typically needed to be tested at various stages in order to ultimately launch just a single compound into the marketplace, along with the reasons for such high attrition are depicted in **Figures 10** and **11**, respectively.

It might be acceptable to fail at the same percentage level when overall throughput moves at a higher rate because the absolute number of drugs ultimately delivered on an annual basis would still increase. However, even this less than ideal scenario, in terms of efficicency, does not appear to be the case since the absolute number of NCEs is, instead, going down. Thus, with the lead identification stage typically moving at a faster pace and with attrition and throughput essentially unchanged during the subsequent preclinical and clinical development stages, one may want to turn toward the initial target selection stage and consider its need for a closer examination. Such questions about 'target-based drug discovery' do indeed appear to now be arising within the literature.[173] As suggested in the preceding section, however, today's pursuit of highly novel and seemingly more sophisticated mechanistic targets is certainly meritorious. Furthermore, whether the novelty and complexity of today's targets are to blame or are not to blame, it should be apparent that there is still good reason to pursue the possibility that the development stages might be able to be made

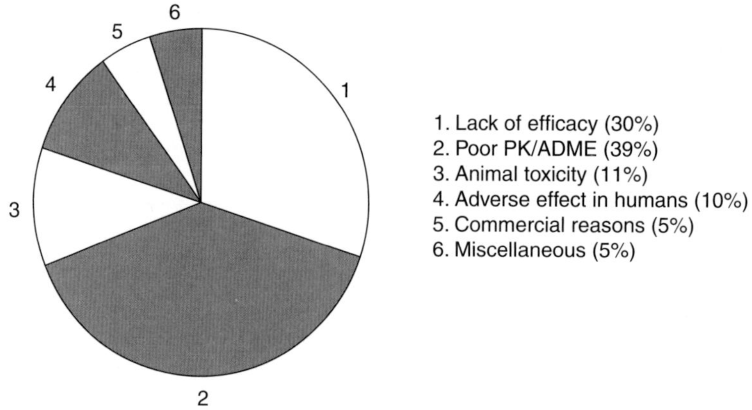

Figure 11 Drug candidate attrition: the reasons.[172] Several reasons for the high attrition rate are compared via their percentage of contribution. Note that while the lack of human efficacy (30%) is certainly high, problems associated with the combined ADMET properties double (60%) this figure. Note that these recent percentages, reflecting the new paradigm in drug discovery (**Figure 3**), remain similar to those previously obtained from the classical drug discovery process (**Figure 2**).

more efficient in terms of attrition and/or more rapid in terms of throughput. Both of these goals are closely linked with making improvements in the preclinical evaluation/clinical prediction and molecular adjustment/enhancement of a given compound's PK-related properties. Since the latter, reflected as a composite of the various ADME parameters, are best addressed during initial drug design and subsequent molecular modification, the knowledge-generating SAR activities that represent the hallmark of medicinal chemistry, are again clearly being called for.

1.02.3.5 Drug Design-Related Multitasking

In order to effectively play this key role as a central interpreter of the wide variety of drug discovery-related SAR information, today's medicinal chemists will have to remain well versed in physical organic principles and conformational considerations while becoming just as adept at applying them toward each of the ADMET areas as they have been when previously applying them toward the singular pursuit of efficacy-related biochemical scenarios. The required interplay of activities associated with assessing such varied types of data coupled with that for the various methods of assessing and modifying chemical structure are depicted in **Figure 12**.

Responding to this bombardment of data, the various approaches that can be deployed by today's medicinal chemists to discover new drugs are listed in **Table 3** along with a brief description of each method. These methods are oftentimes used in combination or in a parallel and complementary manner. They have been arranged according to how much information is known about the biological surface, the latter then generally being directly proportional to a medicinal chemist's success at having a proposed ligand actually hit the desired target. While most of the methods have been adopted to pursue small-molecule SAR, it is unfortunate that it is precisely this vital aspect that is most glaringly at risk when conducting a rapid survey across a compound library using a HTS strategy. This particular irony is further discussed below.

As has been alluded to earlier, HTS efficacy hits, per se, can certainly be pursued without the aid of medicinal chemistry. Indeed, one can imagine that with one or more compound libraries already in hand from an automated synthesis,[174] and the areas of robotics[52,175–176] and laboratory information management or LIMS[177] also continuing to rapidly evolve, HTS in the brute force mode may be able to essentially proceed without any significant human intervention. However, as has been repeatedly emphasized, if the new drug discovery paradigm is to ultimately become successful, this type of screening will need to be accompanied by structure-associated knowledge generation and assessment, with the latter being conducted using the rationale and logic that can only be interjected by human intervention. In this regard, structural-related decision-making quickly falls right into the middle of the domain of medicinal chemistry and its distinct area of small-molecule SAR expertise.

Given this responsibility, it becomes important to briefly review some critical aspects about SAR that would be worthwhile to include within the database assemblies that are currently being drawn up to handle the mountains of data already arriving from today's HTS programs. One can predict that once ADMET profiling by HTS is validated in the future, it will become extremely valuable for a knowledge-generating paradigm to be able to discern not just the active hits within an efficacy database and to be able to compare their structural patterns to those in another database,

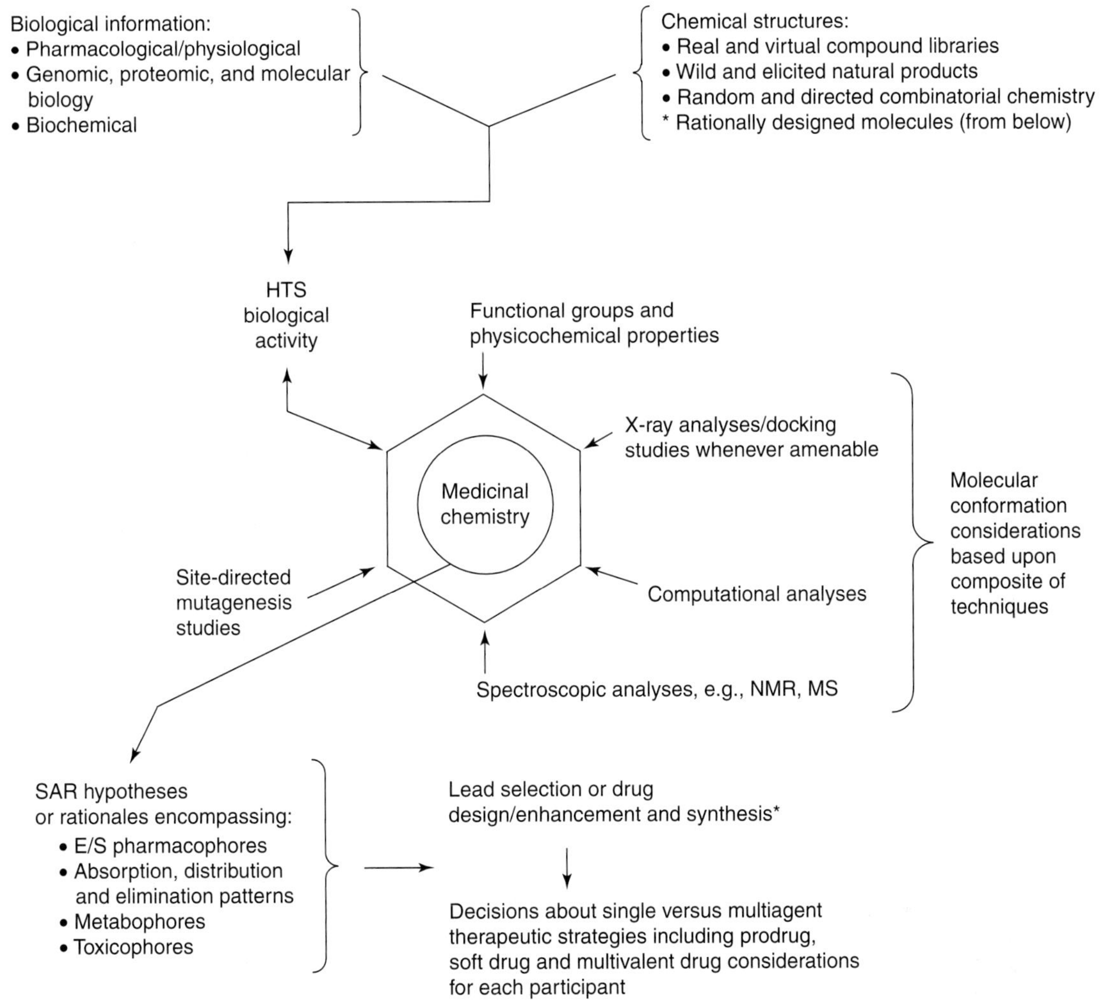

Figure 12 Today's practice of medicinal chemistry: Drug design-related multi-tasking. The most striking differences from the long-standing practice of medicinal chemistry are: (1) data reduction of huge amounts of rapidly derived HTS biological results; (2) greater emphasis upon multitechnique chemical structure considerations; and, most importantly, (3) the simultaneous attention given to all of the ADMET-related parameters along with efficacy and efficacy-related selectivity (E/S) during lead compound selection and further design or enhancement coupled with an expanding knowledge base. In the future, the latter should also offer the possibility for achieving synergistic benefits by taking advantage of various combinations of multiagent, prodrug, soft drug and/or multivalent drug strategies.

but to also be able to demarcate the regions on compounds that can be altered with little effect upon the desired biological activity as well as those areas that are intolerant toward structural modification. The neutral areas, in particular, represent ideal points for seamless merging of one set of a database's hits with that of another regardless of the degree of pattern overlap, or for further chemical manipulation of a hit so as to adjust it to the structural requirements defined by another data set that may be so distant in structural similarity space that attempted overlap or pattern recognition routines are otherwise futile. The regions that are intolerant of modification represent areas to be avoided during knowledge-based tailoring of an efficacy lead. Alternatively, the intolerant regions represent areas that can be exploited when attempting to negate a particular action, e.g., metabolism or toxicity. Actual examples of utilizing both neutral and negative SARs to advantage are provided in Section 1.02.4 to further illustrate how these types of data sets might also be simultaneously deployed by future medicinal chemists with significantly stepped up complexity as more and more parameters become added to the process of early lead identification/optimization.

As a summary to this section, **Table 4** lists several points that are relevant to chemical compound categories associated with modern drug discovery. **Table 4** is important because it addresses a noteworthy concern, namely that today's trend to aggressively filter nondruglike compounds from the initial drug discovery HTS process may work

Table 4 Molecular attributes of discovery libraries, compound hits, lead structures, and final compounds

Discovery libraries

- 'Size' is less important than 'diversity' (with allowance for structurally redundant series, e.g., Me, Et, etc.)
- 'Diversity' is also much more important than 'druglike properties,' e.g., presence of nondruglike members can be extremely useful toward initially probing overall structural space. Alternatively, 'assay likable properties' are mandatory, i.e., compound members must be able to be delivered (e.g., solubilized, etc.) according to demands of a given assay

Compound hits

- 'Druglike properties' are less important than 'efficacy tolerance,' i.e., flexibility for altering structure without altering efficacy (access to 'neutral regions')

Initial lead structures

- 'Individual structures' are less important than a detailed description of the 'pharmacophore' in terms of electrostatic surface potentials plus knowledge about structural space that is neutral or intolerable toward modifications relative to the measured biological parameter, e.g., efficacy
- 'Nondruglike' features that may be present within certain members contributing to the overall pharmacophore should be 'red-flagged' but not necessarily ruled out as potential building-blocks while the overall process of merging structural space across all parameters is continued, e.g., efficacy plus ADMET

Final lead compounds

- Optimal blend of efficacy and druglike properties (nondruglike features now completely removed or adequately modified according to experimentally ascertained criteria that have been validated for their correlation to the clinical response)
- At least one neutral region or prodrug/soft drug option remains such that unanticipated hurdles presenting themselves during further development might still be addressed by additional chemical modification
- One or more back-up compounds having distinctly different molecular scaffolds while still fulfilling the overall ensemble of pharmacophore and druglike patterns

Notice that while emphasis is placed upon defining a given pharmacophore to the maximum possible detail by de-emphasizing the use of 'druglike property' parameters as an early filtering mechanism, components of the pharmacophore that are presumed to be undesirable should still be 'red-flagged' as such. ADMET SAR can then be superimposed within the distinct molecular contexts of each identified pharmacophore so as to be more efficiently deployed as a filter and, importantly, in a proactive manner while initial lead structures move toward final lead compounds. This two-step approach will allow for continued knowledge-building within all of the key parameters relative to various therapeutic areas.

against the accumulation of knowledge that will be useful toward continually improving the overall process, let alone being useful toward fully characterizing the aforementioned structural space subtleties associated with an initial hit or later lead compound. In other words, while it can be argued that a certain efficiency in the production of NCEs might already be obtainable at this juncture by engaging in this type of negative strategy, an overemphasis upon this approach runs the risk of having the drug discovery and development process becoming forever locked into just this point of evolution. Alternatively, **Table 4** conveys how some of today's trends that pertain to the molecular attributes of discovery libraries, compound hits, and lead compounds, might be best deployed so as to allow future developments to be derived from a truly solid base of accumulating knowledge.

1.02.3.6 Drug Development-Related Multitasking

Although the seamless nature of today's drug discovery paradigm is a natural consequence of moving what used to be later considerations into earlier time-points of the overall process, the distinct types of chemical-related activities that need to be addressed when a lead compound progresses to advanced stages of development are worth mentioning within their own category. Like the close interplay of medicinal chemistry and the new era of biological assessments discussed to this point, most of the chemical-related activities conveyed by **Figure 13** are also best undertaken in a simultaneous and close interdisciplinary manner.

Briefly, of foremost importance is the preparation and complete characterization of an analytical standard for 'the' final lead compound. All process chemistry-related activities and advanced biological testing activities, particularly those associated with toxicological assessments, will rely on the analytical standard for verification of substance and solution integrity. As part of the characterization of the high purity standard, chemical precursors along with deliberately synthesized reaction side products and potential chemical breakdown products, are typically utilized to establish that they can be unambiguously distinguished, for example by high-performance liquid chromatography (HPLC), from the desired material. Since this same type of characterization is typically mandated by government regulatory bodies to validate an assay for a new drug substance, the complementary relationship between synthetic medicinal chemistry and analytical chemistry becomes quite clear at this particular juncture of the overall drug

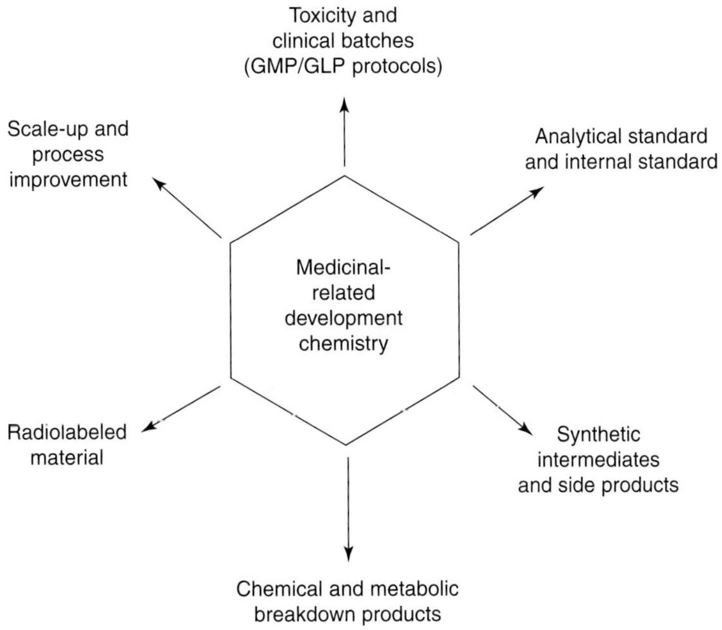

Figure 13 Drug development-related multi tasking. While big pharma typically has specialized teams of chemists dedicated to various of these activities, it is not uncommon in small pharma for drug discovery medicinal chemists to become actively involved with several (or all) of these tasks relative to each final lead compound that may progress to such status. Note that the final toxicity and clinical batches must be prepared according to Good Manufacturing Practices (GMP) accompanied by Good Laboratory Practices (GLP) analytical support.

discovery paradigm. Similarly, the need for syntheses of both anticipated metabolites and experimentally derived metabolite possibilities serves to maintain the complementary relationship between medicinal chemistry and the biologically oriented disciplines working in the field of drug metabolism. This is an especially valuable relationship when the latter do not have a full appreciation for the nuances of chemical structure or when the lead compound itself prompts an analog program directed toward enhancement of drug metabolism-related PK properties. Finally, process chemistry becomes important at this point not just for longer-range thinking in terms of an eventual product cost, but because several hundred grams will be needed immediately for preliminary formulation and toxicology studies. The need for several kilograms will soon follow if the compound is to then undergo phase I clinical testing which is usually also accompanied by extended toxicity testing. Four key issues pertaining to the interplay of medicinal chemistry with process chemistry relative to current developments are mentioned below. Like the trend for the ADMET-related assessments, all of these issues, except for that of intellectual property (IP) which has always been at the forefront, are moving toward earlier consideration points within the overall process of drug discovery.

First, issues pertaining to IP involve both the structural novelty of biologically interesting compositions of matter and the latter's synthetic accessibility via practical process chemistry routes that do not infringe patent-protected methodologies. Today's fixation upon emphasizing compound library members that have druglike properties and the future possibility for further instillation of additionally preselected molecular arrangements that can endow preferred metabolic profiles in some ways become contradictory to enhancing the future molecular diversity, and thereby patentability, of the overall drug discovery structural landscape. This seeming paradox is taken up again for further discussion within the final sections of this chapter. Second, the eventual production cost for a new therapeutic agent is much more important today than it has been in the past. This is because pharmaceutical companies must now garner their profits from a marketplace that has, right or wrong, become sensitized about the cost of ethical pharmaceutical agents. The days of simply raising the price of such products in parallel to increasing costs associated with discovering and developing them have been over for quite some time.[178] In this regard, the cost-effectiveness of small-molecule drugs will probably maintain an edge over biotechnology-derived therapeutic agents for at least the near term. A third point to be mentioned pertains to the impact of the 'green chemistry' movement.[179–182] This movement has prompted pharmaceutical companies to insure that their productions of drugs are friendly toward the environment in terms of all materials and methods that may be deployed in the process. Finally, the USFDA's initiative to have all stereoisomers present within a drug defined both chemically and biologically has prompted industry's pursuit of drugs that either do not contain asymmetric centers or are enantiomerically pure.[183] This, in turn, has prompted the need for better stereochemically controlled processes during production. Stereocontrol has always represented an extremely interesting area for synthetic chemistry exploration and now for biotechnology-derived chemistry and reagent research as well, e.g., exploitation of enzymes at the chemical manufacturing scale. Considerable progress is being made toward developing such methods on many fronts including enzymatic[184–186] and microarray technologies.[187] Oftentimes, however, the new laboratory techniques do not readily lend themselves toward inexpensive, scale-up/manufacturing type of green chemistry. Thus, of the various challenges facing medicinal chemists also involved in drug development, devising practical routes to single stereoisomer compounds that are both cost-effective and conducive toward green chemistry, probably represents the largest.

1.02.3.7 Absorption, Distribution, Metabolism, Excretion, and Toxicity

Given the pivotal nature of ADMET assessments within the overall scheme of drug discovery and their present separation from truly reliable clinical prediction by a dark chasm that cries for the type of bridging knowledge that can be derived from medicinal chemistry's SAR forte, the ADMET area deserves to be specifically reconsidered as part of this account of medicinal chemistry's current status. Because of this critical importance, each of the ADMET parameters is separately discussed below.

1.02.3.7.1 Assuring absorption

In addition to conducting in vivo bioavailability studies on selected compounds at a later stage of development, early in vitro assessments of structural information that might be useful toward assuring absorption after a drug's oral administration are now being conducted on a routine basis.[188–190] Similar studies are being directed toward assessing penetration across the blood–brain barrier.[191,192] Determination of the pK_a values for ionizable groups, determination of partition coefficients (e.g., using various types of log P calculations and measurements), and measurement of passage across models of biological membranes (e.g., Caco cell lines) represent data that has now been shifted toward HTS experimental and purely computational modes.[193–204] These types of studies can be thought of as absorption high-throughput screening (AHTS). Since recent results suggest that the biological transporter systems are extremely important factors in this area,[205–207] their study is also becoming part of AHTS (e.g., passage of drugs across Caco cell layers from both directions[208,209]). This trend toward increasing sophistication within AHTS can be expected to continue. That genomics and proteomics will help to identify and initially define absorption-related systems biochemically should be clear. Alternatively, that biotechnology working with bioengineering and nanotechnology might also be directed toward instilling passageways or specific pores for drugs across the human gastrointestinal endothelial system[210] is more speculative, as are purely chemical[211–213] and nanotechnology[214–218] approaches toward prompting or constructing passageways, respectively. Likewise, that advances in formulation and alternate delivery technologies[219–228] could eventually obviate the need for oral administration is also speculative. Nevertheless, all of these possibilities need to be mentioned because, taken together, they make the point that significant advances in any of the ADMET areas, regardless of their technological source, have the potential to eliminate the need for assaying certain of their presently related parameters, perhaps even reversing the initial portions of the present drug discovery paradigm (**Figure 3**) back to where it originated, i.e., to being concerned primarily with just efficacy and selectivity during front-line testing (**Figure 2**).

As for deciphering selective efficacy-related SARs, medicinal chemistry's role should be directed toward making sense out of the AHTS data mountains looming ahead using molecular structure information as the common code, in this instance by relating the latter to structure–absorption relationships (SAbRs). Such efforts might eventually culminate in affording molecular blueprints for affecting absorption-related structural modifications that are correlated with certain structural themes and absorption characteristics for which efficacy hits may be able to be categorized using structural similarity/dissimilarity indices. Notable advances have already been made toward defining useful SAbR in terms of database and virtual compound profiling, e.g., the so-called 'rule of five.'[229] The latter should be recognized as just a first, gross step in this direction that can be expected to continue in a more sophisticated manner in the future, e.g., along the lines of 3D structural considerations relevant to the transporter systems, as well as more refined parameterization of physicochemical properties.[230]

1.02.3.7.2 Directing distribution

The same types of studies mentioned above, along with a panel of assays specific for certain depot tissues such as red blood cells, plasma protein binding factors, adipose tissue, etc.[231–234] will be additionally mobilized toward directing distribution of a xenobiotic. Thus, as the handling of chemical structure improves and more sophisticated correlations begin to unfold in the future, AHTS can be thought of as A/DHTS that provides both SAbR and SDR. Simultaneous collection of such data can allow investigators to reflect upon drug absorption and distribution as a continuum of drug events that can be effectively incorporated together at an earlier point of the overall lead decision process. Furthermore, in the case of directing distribution it can be anticipated that genomics and proteomics will become instrumental toward identifying numerous key factors that are overexpressed in various pathophysiological states. For example, cancer cells are already known to overexpress a variety of specified factors.[235–239] Ligands designed to interact with such factors residing on cell surfaces can then be coupled with diagnostic and therapeutic agents so as to be delivered at higher concentrations to these locales. Such strategies can be thought of as placing both an 'address' and a 'message' within a molecular construct[240,241] that may involve an overlap of two small-molecule-related SAR patterns, or perhaps a small molecule conjugated to a bioengineered biomolecule wherein the latter typically serves as the address system. Indeed, the bioconjugate or immunoconjugate strategy has been around for quite a while[242] and it appears to be benefiting from a renewed interest[243] in that chemotherapeutic 'smart bombs'[244] are now being added to our older arsenals of single 'arrows' and 'combinations' of small molecule 'magic bullets.'[245] A later case study involving paclitaxel (PAC) is also used to further emphasize this theme wherein the chemical knowledge in the area of PAC protection and coupling reactions was additionally used to construct compounds that would be directed toward some of the factors that are overexpressed on certain human cancer cells so as to enhance 'selective toxicity,'[246] particularly since there is some precedent in this case that this might be feasible by combining two small molecules. One can imagine that as data continues to be amassed for these types of factors, the most promising ones will be quickly pursued according to both of the aforementioned scenarios, paired small-molecule SARs and small molecule–bioconjugate pairs. Whether undertaken in a rational manner or via the merger of two HTS-generated databases (i.e., one for an efficacious message and one for determining a selective address) these types of pursuits fall into the general category of tailoring a lead. Therefore, it can be expected that the expertise afforded by medicinal chemistry will again be an integral component of such activities.

Before turning to those parameters that might be considered to be associated with ending a drug's random walk through the biological realm (e.g., metabolism and excretion), it is necessary to discuss a practical limitation to where this overall discourse is leading. Clearly, there will be ceilings for how many molecular adjustments can be stacked into a single compound no matter how knowledgeable we become about the various ADMET-related structural parameters and how they might be merged so as to best take advantage of molecular overlaps. This will be the case even when 'prodrug' strategies are adopted[247] (**Figure 14**), wherein certain addresses or messages that have been added to deal with one or more aspects of ADMET, become programmatically jettisoned along the way while simultaneously activating the efficacy payload that is to be delivered to only the desired locale as the final statement.

Thus, this situation prompts the prediction that in order to interact optimally with the entire gamut of efficacy and ADMET-related parameters during a given course of drug therapy, the latter may need to be delivered not as a single agent but as a distinct set of multiple agents wherein each individual component or player makes a specified contribution toward optimizing one or more of the efficacy and ADMET parameters relative to the overall drug team's therapeutic game plan.

1.02.3.7.3 Modulating metabolism

There is no doubt that one of the largest challenges facing industry-based medicinal chemists today is that of modifying the metabolism of a lead structure to enhance the latter's overall PK profile, usually within the context of trying to prevent or attenuate a given metabolic event so as to prolong biological half-life. Numerous texts,[250] monographs,[251,252] and reviews[253,254] are available for readers interested in taking up this particular call for medicinal chemistry input. Focusing upon phase 1 metabolic pathways, the most aggressive and thus also the most frequently encountered biotransformation reactions are depicted in **Figure 15** relative to chemical functionality typically present within a drug molecule. Short of removing the susceptible functionality altogether, the most reliable approach that can be taken to avoid these pathways is that of introducing steric hindrance either directly into that site or into as close a neighboring position as possible. Although numerous exceptions can be cited, no other physicochemical property comes close to being even a distant second in terms of its successful manipulation in this regard. Thus, given today's fondness within the drug discovery community for simple numerical-related theorems, the aforementioned steric-hindrance strategy can be thought of as 'drug metabolism's rule of one.'

Focusing upon phase 2 metabolic events, two pathways take on prominence, i.e., when such functionality is present, these particular events have a high likelihood of occurring. These two pathways are also depicted in **Figure 15** and, in perfect accord with 'drug metabolism's rule of one' can again almost always be counted on to be highly susceptible to steric hindrance. The glutathione detoxification pathway is also a rapid phase 2 biotransformation reaction but since it

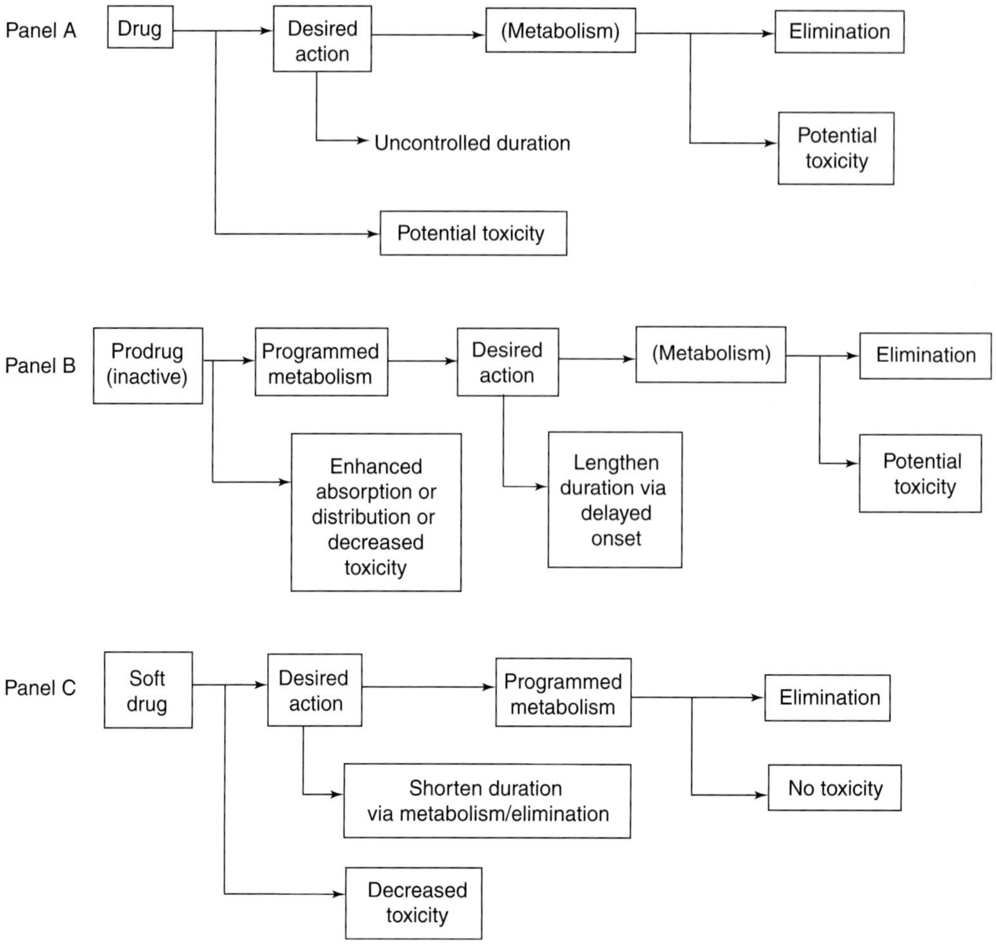

Figure 14 Contrasting dispositions of a drug, prodrug and soft drug.[247–249] Panel A depicts a generalized version of a standard drug's pattern of observed activities. Panel B depicts how a prodrug approach can be used to modify the entry-side portion of a given drug's overall profile of actions. Panel C depicts how a soft drug approach can be used to modify the elimination-side portion of a given drug's overall profile of action. Both prodrugs and soft drugs can be used to decrease toxicity.

is more typically encountered by toxicants and reactive metabolites rather than by parent drug molecules, it has not been displayed here.

Although these select examples represent only a very much abbreviated highlight of just a few of the many common biotransformations to which drug molecules are subject, their display serves to exemplify the challenges posed to medicinal chemists when trying to fully appreciate, and thereby better modulate, this particular aspect of the ADMET profile. For example, it should be immediately apparent that a variety of metabolic possibilities compete for reaction with a given drug. Two practical complications result from this situation. First, while each type of biotransformation is subject to classical organic chemical reactivity variables associated with the local electronic and steric properties near the substrate's reaction site, when it comes to predicting drug metabolism, exposure itself becomes an ill-defined variable that is equally important to inherent chemical reactivity. Thus, even a poor substrate for a particular enzyme can still be highly subject to that enzymatic event if the substrate achieves a significant concentration and spends considerable time in the compartment where that enzyme happens to predominate and, vice versa, an excellent substrate for a particular enzyme, e.g., as determined by a purified biochemical assay, will not be subject to that pathway if the substrate has limited access to the enzyme. Because of these possibilities, it is imperative to assess potential drug metabolism events not only in isolated enzyme systems such as a battery of isolated or overexpressed cytochrome P450 (CYPs) so as to ascertain inherent substrate suitability, but also in systems that begin to further address drug distribution factors relative to both gross physicochemical properties and to the more subtle nuances of chemical structure such as the importance of interactions with the various transporter proteins. Indeed, such aspects are so critical that the present status of our efforts to predict a drug's half-life (i.e., metabolic stability coupled with

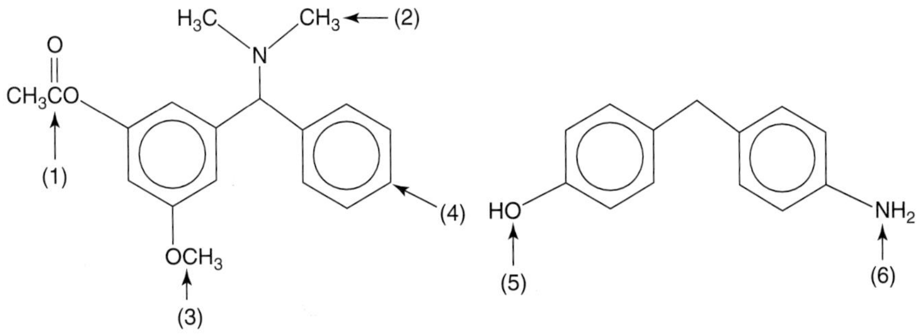

Figure 15 Metabolic possibilities for model compounds having representative functionality. Selected phase 1 reactions: (1) hydrolysis of various types of esters, in this case mediated by a carboxyesterase; (2) *N*-dealkylation mediated by certain of the cytochrome P450 (CYP) enzymes; (3) *O*-dealkylation mediated by certain of the CYPs; and (4) aromatic hydroxylation also mediated by certain of the CYPs. Depending upon the subtleties of their electronic and steric environments, the relative competitive biotransformation rates for these processes will generally be: (1) ≫ (2) > (3) ∼ (4). Selected phase 2 reactions: (5) formation of a glucuronic acid conjugate (or in some cases a sulfate conjugate); and (6) *N*-acetylation. In terms of relative biotransformation rates in general: (5) ≫ (6).

Table 5 Four major themes link drug metabolism to drug design

1. Metabolism itself is the mechanistic concept that is to be manipulated for therapeutic gain
   - Inhibition
   - Induction
2. Avoiding metabolism may enhance the overall therapeutic profile
   - Improve observed potency and duration
   - Prevent toxicity
3. Prodrugs
   - Improve formulation properties (solubility, taste)
   - Enhance absorption
   - Prolong duration (depot version)
   - Direct distribution
4. Soft drugs
   - Program duration
   - Avoid toxicity
   - Pair with administration (depot release) to obtain 'steady state' concentrations
   - Limit distribution (topicals and other selected compartments)

avoidance of excretion) in humans is probably better accomplished by comparing in vivo data from another species altogether (such as rat) than when attempted by comparing biochemical or in vitro data from the same species, i.e., purified human or humanized enzyme systems.[255,256]

A second complication of the pathway multiplicity impacts directly upon medicinal chemists' attempts to enhance metabolic stability even after the culprit pathway has been clearly established. In this case, when the latter is attenuated, e.g., by the effective incorporation of steric hindrance, it is not uncommon for an alternate pathway to then become problematic, a situation which in this context is sometimes referred to as 'metabolic switching' (and thus not to be confused with older definitions for the latter associated with the fields of biochemistry or nutrition).

Aside from dedicated texts, more complete listings of metabolism events are available as drug metabolism databases and expert systems which can also be used toward metabolic prediction. A few of the longer-standing of these products include: (1) Metabol Expert[257]; (2) META[258]; (3) Metabolite[259]; (4) Synopsis[260]; and (5) Meteor.[261] Some of these commercial products have attempted to assign relative rates to each competing pathway based upon substrate suitability and, to a certain extent, the abundance and distribution of the various enzymes along with the anticipated distribution of the drug. However, it is still far more appropriate to consider these types of products as conveying all of the metabolic possibilities, rather than reliable metabolic probabilities, for any given case of a new drug molecule.[251,262]

In addition to modulating metabolism to impact upon PK profile, three other major drug discovery-related themes link metabolism to drug design (**Table 5**). The first of these is when metabolism itself is the mechanistic concept that is to be directly manipulated for some type of therapeutic gain. A successful example of this theme within the area of cancer treatment would be the now well-established aromatase inhibitors.[263–267] Another theme involves prodrugs. The use of prodrug strategies, which typically rely upon metabolism to liberate the active drug molecule (**Figure 14**) was alluded to in the preceding section, and numerous successful examples can likewise be cited for this. On the opposite side of a drug's action, there is a final area pertaining to controlling metabolism that falls so specifically into medicinal chemistry's domain of lead tailoring that it also merits a brief discussion. This topic involves exploiting what has come to be called 'soft drug technology' (**Figure 14**)[248,249] where a metabophore is placed within an established drug or lead compound's structure in order to program a specified course of metabolic inactivation of the resulting hybrid molecule. A successful example of the soft drug approach has been placed within Section 1.02.4.

Analogous to the importance of merging SAbR and SDR with efficacy and selectivity-related SAR, structure–metabolism relationships (SMRs) and numerous metabophore patterns can be expected to be gradually discerned and put to extensive use by medicinal chemistry in the future to either attenuate or enhance a candidate drug's metabolism. Modulation of the latter will be used along all of the themes conveyed in **Table 5**, as well as toward enhancing the overall therapeutic profile and reducing unwanted toxicity of drug candidates.

1.02.3.7.4 Engaging excretion

SAbR, SDR, and SMR can all be applied toward engaging the excretion pattern of administered drugs in a proactive manner so as to optimize this aspect of a drug's overall profile, the soft drug example applying here in particular. Analogous to the distribution area, genomics and proteomics can be expected to soon delineate important systems, such as specialized transporters within tissues like the kidney and liver, that are especially responsible for the elimination of xenobiotic drugs and their metabolites. The concept of negative SAR will be critical toward avoiding these transporters and there is at least a possibility that steric factors might again be used to advantage in this area. Medicinal chemistry's involvement toward uncovering structure–excretion relationships (SERs) and deriving generally useful structural patterns that might be used to rationally tailor lead compounds or for merging of different types of databases while attempting to select lead compounds, again falls into the central theme for medicinal chemistry's future as being elaborated within this chapter. As for the other ADMET areas, more speculative notions in this area can provide some interesting alternatives for the natural excretion of xenobiotics. Although SERs are likely to also encompass various endogenous materials and their catabolic fragments, one might still imagine that just like the futuristic examples sited for absorption, in the more distant future, biotechnology, chemical and nanotechnology approaches might all be successfully applied toward engineering specific drug excretion passages through selected tissues.

1.02.3.7.5 Tending to toxicity

That toxicology has now become a protagonist through its participation in the design or early selection of drug leads, already represents a remarkable turnaround from its historical, antagonist role as a gatekeeper or policeman standing at an advanced stage of drug development with an eye toward halting the progression of potentially toxic compounds on route to the clinic.[268] Nevertheless, it may very well be that the most profound effect that genomics and proteomics are going to have within the ADMET arena will further pertain to advances in avoiding toxicity. Like the field of drug metabolism, toxicology has been collecting its data within databases for quite some time (**Table 6**). In fact, some of the structural patterns that have come to be associated with distinct toxicities (toxicophores) are probably on much firmer ground than are the metabophore relationships. On the other hand, drug metabolism derives from a finite number of genetic constructs that translate into metabolic activity (proteinaceous enzymes albeit notorious for their seeming molecular promiscuities) such that with enough time the entire set of metabolic options should eventually become well characterized. Toxic endpoints, alternatively, have no such limitation associated with their possible origins. In other words, to show that a drug and its known or anticipated metabolites are completely nontoxic is comparable to trying to prove the null hypothesis, even when a limited concentration range is specified. Nevertheless, genomics, proteomics, and biotechnology do, indeed, appear to already be producing some promising technologies that can be directed toward this difficult area. For example, array technologies, as mentioned in the prior section on genomics, are already becoming available to assess the influence of a drug on enormous numbers of genes and proteins in HTS fashion.[269–276]

Once enough standard data of this type are produced by taking known agents up in dose until their toxicity becomes fingerprinted via distinctive patterns of hot spots, array patterns may be used to cross-check against the profiles obtained in the same HTS mode for new lead compounds. Given the quick rate that these important trends are being further developed and are likely to eventually become validated, medicinal chemistry could certainly become overwhelmed trying to keep up with its complementary role to identify the corresponding structure–toxicity relationships (STR) for each array hot spot.

In the case of toxicity then, medicinal chemistry will probably need to approach STR in a different manner, e.g., initially from just the exogenous compound side of the equation for a given toxicity relative to the observed hot spot

Table 6 Long-standing toxicology databases and related organizations

Database or organization	*Description*
Centers for Health Research (formerly Chemical Industry Institute of Toxicology or CIIT)	Industry consortium sponsored collection/dissemination of toxicology data; also conduct research and training in toxicology [277]
American Chemistry Council Long-Range Research Initiative	Industry consortium sponsored initiative to advance knowledge about the health, safety and environmental effects of products and processes[278]
LHASA Ltd. (UK-based, nonprofit segment)	Facilitates collaborations in which companies share information to establish rules for knowledge bases associated with toxicology[279]
International Toxicology Information Center (ITIC)[a]	Pilot program to share data in order to eventually be able to predict the toxicology of small molecules, thus lessoning the expense of in vitro and in vivo testing[279]
US Environmental Protection Agency (EPA) High Volume Chemicals (HVP) Screening Information Data Set (SIDS)	US user-friendly version that will also be submitted to the Organization for Economic Cooperation and Development (OECD) and its tie-in with IUCLID[a279]
SNP Consortium (Nonprofit: makes its information available to public)	Addresses phenotypic aspects relative to individual responses to xenobiotics, e.g., metabolic phenotype and toxicity
Tox Express/Gene Express database offered by Gene Logic[b] (commercial)	Offers a gene-expression approach toward toxicity assessment[279]
National Institute for Environmental Health Science (NIEHS)[c]	Compiling a database of results from toxicogenomic studies in order to divide chemicals into various classes of toxicity based on which genes they stimulate or repress[280]
International Program on Chemical Safety/Organization for Economic Cooperation and Development (IPCS/OECD)	Risk assessment terminology standardization and harmonization[281,282]
MULTICASE (commercial)	Prediction of carcinogenicity and other potential toxicities[283]
MDL Toxicity Database (commercial)	Allows structure-based searches of more than 145 000 (January 2001) toxic chemical substances, drugs, and drug-development compounds[284]
DEREK and STAR (LHASA commercial segments)	Prediction of toxicity[285]
SciVision's TOXSYS (commercial)	General toxicity database to be developed in collaboration with the US FDA[286]
Phase-1's Molecular Toxicology Platform gene expression microarays (commercial)[b]	Allows detection of gene expression changes in many toxicologic pathways[287]

[a] Includes cooperative efforts with the European Union and the European Chemicals Bureau (ECB) in using the International Uniform Chemical Database (IUCLID) and its relationship to high-volume chemicals (HVP).
[b] This company's product is representative of several of such technologies that are also being made available by a variety of other vendors.
[c] Includes cooperative efforts with the Environmental Protection Agency (EPA) and the Information Division at the National Institute for Occupational Safety and Health (NIOSH).

patterns (unless proteomics and biotechnology also quickly step in to additionally define the biochemical nature of the actual endogenous partners that are involved in a given toxic event). Taking a chemically oriented starting point, however, should serve reasonably well in that there will likely become a finite number of chemical reactivity patterns that can be associated with toxicity. Medicinal chemistry can be expected to elaborate these reactivities into general STR and to then use them toward defining the liabilities in new compounds. The notion that there should be a finite number of structurally identifiable toxiclike patterns is analogous to the notion that there should be a finite number of amenable druglike patterns that reside within a given structural database having a certain degree of molecular diversity. Indeed, the case for toxicity is certainly on firmer ground at this particular point in time since there will likely be little added to the area of fundamental chemical reactivity as opposed to proteomic's anticipated revelation of numerous new biochemistries that will, in turn, provide numerous new pharmacologic targets wherein many can be expected to have their own distinct pharmacophore (and potentially new druglike patterns). Finally, since the precise locales where the toxicity hot spots may ultimately occur are endless, the latter will perhaps be better addressed by directing a second set of database queries toward the ADME profile and intracellular localization patterns that a given drug may exhibit. In the end, after array technologies are producing useful toxicology-related knowledge, the interplay of all of the ADME parameters with STR should become just as important as they are for efficacy in terms of what type of toxicity may ultimately be observed within the clinic.

1.02.3.8 An Intriguing Opportunity to Go Back to the Future by Revisiting One's Roots

The concern that to optimize all of the efficacy and ADMET parameters within a single molecule might be difficult, if not impossible, to achieve on a routine basis even when prodrug strategies are relied upon to address absorption and distribution properties was raised within the prior discussion about drug distribution. Alternatively, the use of two or more molecules that might work together as a team also presents itself as a possibility toward potentially achieving such a multiparameter optimized profile. That some herbal remedies appear to demonstrate greater benefits as their mixtures than as their isolated components adds to this intriguing possibility and thus prompts a brief consideration of this topic. Although today's trend in the US to self-administer herbal remedies and preventatives is admittedly driven more by consumerism than by solid science,[288,289] this potential reconnection of basic science to medicinal chemistry's historical roots also becomes noteworthy. Aside from their poorly defined analytical characterization and notoriously inadequate quality control, one of the major, basic science questions about herbals that do possess validated pharmacological properties is why their natural forms are sometimes superior to the more purified versions of their active constituents, even when the latter are adjusted to reflect varying concentration ratios thought to coincide with the natural relative abundances. Given their incomplete analytical characterizations, it should be apparent relative to the present discourse that, in addition to simple prodrug possibilities, numerous unidentified, nonefficacious, and otherwise silent constituents within any given herb could have an interaction with one or more of the efficaciously active constituents at any one or more levels of the latter's ADMET steps. When these interactions are favorable, the resulting overall pharmacological profile becomes altered in a seemingly synergistic manner that is obtainable from the more natural forms of the mixture but lost upon purification to matrices containing only the efficaciously active components.[290] In fact, there is already experimental precedent for this scenario relative to efficaciously silent components improving the absorption,[291] enhancing the distribution,[292,293] and favorably altering the metabolism[294] of their active herbal counterparts, as well as more classical synergies involving direct interactions that occur at the sites involved with efficacy.[295] Therapeutic enhancements derived directly from multiple interactions at efficacy sites have been pursued for many years, with multivalent, single drug entities reflecting the latest trend in this direction.[296] What should be truly remarkable in the following discourse, which at this point can be likened to going "back to the future," is that the rapidly evolving process of drug discovery will undoubtedly continue to add the sophistication of the entire ADMET profile into such multi-action-directed considerations.[297–301]

Optimization of the overall pharmacological profile is precisely what is being striven for when selecting and/or chemically tailoring an NCE lead according to either the old or new paradigm of drug discovery. Restating, however, that it may be expecting too much even upon extending the new paradigm into the future as a knowledge-generating process, to obtain complete optimization within a single, multiparameterized molecule, perhaps it will again be Mother Nature that will once more lend her hand by revealing some of the modes of ADMET synergy that she, long ago, has already instilled into some of her herbal productions. At the very least, medicinal chemistry should take care to not forget its roots in natural product chemistry as it marches forward with biotechnology just behind genomics and proteomics further into the new millennium. For example, efforts can be directed toward uncovering efficacy and ADMET-related synergies that may be present among the constituents of herbs purported to have anticancer or cancer-preventative properties by taking advantage of the common cell culture panels already in place to assess anticancer activity along with various transporter system interactions via HTS format. However, because anticancer/cancer-preventative synergy could derive from favorable interactions across a wide variety of ADMET processes relative to any combination of one or more efficacy-related endpoints, several mechanism-based assays associated with several key possibilities for efficacy will also need to be deployed as part of such a program. One can only imagine how sophisticated this type of pursuit will become in the future when such highly interdisciplinary, efficacy networks are further coupled to an even wider network of ADMET parameter experimental protocols.

A more classical approach toward the interactions of multicomponent systems would be to utilize clinical investigations to study the interactions, either positive or negative, that herbals may have with drugs when both are administered to humans. Importantly, for all of these herb-related studies, it becomes imperative that extensive chemical constituent fingerprinting is also undertaken so that the observed effects, particularly those suggestive of synergy, can be correlated with overall chemical composition patterns and not with just the distinct concentrations of preselected components already known to possess established activity.[302]

In contrast to both of the aforementioned types of studies that can be considered to represent systematic examinations of 'herbal-directed, small libraries' and specified herb–drug clinical combinations, respectively, it becomes interesting to speculate how a truly random, brute force approach toward identifying synergy might proceed as a HTS survey of a random compound library in pursuit of optimal pair or even triple compound teams rather than as an attempt to identify a single, 'blue-chip' drug that can do it all. In this regard, however, it must first be recognized that the present trend to test mixtures of several compounds within a given well does not even begin to address synergy. This is because based upon considerable experience with various chemotherapeutic agents,[303] synergy is most likely to be observed at very select ratios within very distinct concentrations of the involved players. In other words, looking at the most simple case of assessing the potential synergy between just two molecules, A and B, requires testing A in the presence of B across a range of molar ratios presented across a range of absolute concentrations.[304–306] This situation is depicted in **Figure 16**.

Considering the brute force approach from a purely mathematical viewpoint and in a minimally elaborated pharmacological format, suppose the possibility for A plus B synergy relating to just a single, efficacy or ADMET-related HTS parameter is examined across a compound library having only 100 members wherein paired combinations are tested at just three relative molar ratios (e.g., A/B at 0.5/1, 1/1, and 1/0.5) at only three total concentrations of both members (e.g., 0.1, 1.0, and 10 μM), then a total of 44 850 drug tests plus numerous control runs would be required for an $n = 1$ pass through the library.[307] Perhaps because of these rather impressive numbers, brute force HTS will undoubtedly relish such pursuits. Once the HTS forces become mobilized in this area, such testing could set up an interesting 'John Henry' competition with more directed investigations, such as those that have been elaborated above that seek to systematically identify the specific synergies seemingly present within certain herbals. Ultimately, no matter how the identification of such favorable drug–drug partnering possibilities are uncovered and are able to better deal with the various ADMET parameters of tomorrow, as well as for the classical efficacy relationships of today, they will certainly prove to be invaluable toward alleviating the situation of trying to establish all of the most desired behaviors for a given therapeutic target within the context of a single molecular framework. Furthermore, it can be anticipated that this type of information will become extremely useful when it becomes further elaborated by medicinal chemistry into general structural motifs that have potential synergistic utilities and applications beyond what was initially uncovered by the specific mixtures of defined compounds.

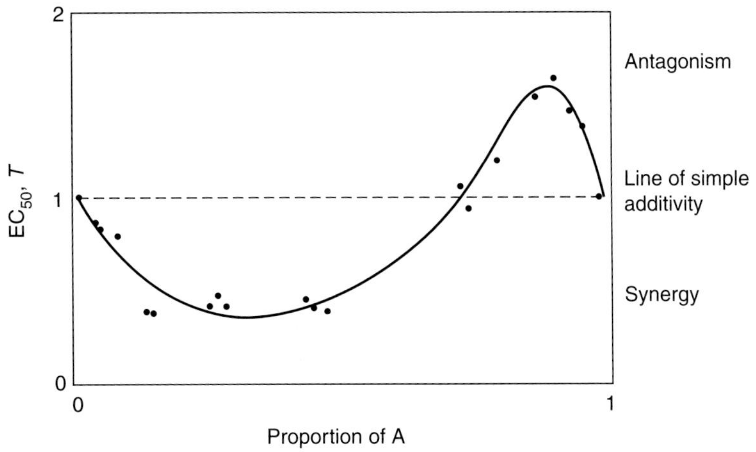

Figure 16 Drug–drug interaction plot for two drugs A and B.[303] EC_{50}, T is the total concentration of the combined drugs which gives 50% of the maximum possible effect. The EC_{50}, T is shown as a function of the fraction of drug A (drug B's fraction is one minus the fraction shown). Rescaling of drug concentrations to units of their EC_{50} allows simple additivity to be set at unity such that deviations below or above this line indicate synergism or antagonism, respectively. The dots are actual experimental results obtained for two anticancer agents, wherein the observed EC_{50}, T values reflect 20 rays of fixed drug fractions as estimated from the data along that ray alone. The fitted curve was generated by a global model for the entire data set and indicates the complicated nature of interaction relationships within even a well-controlled cell culture environment. That synergism can be accompanied not only by simple additivity but also by ratio-dependent antagonistic relationships is apparent.

1.02.4 Case Studies

This section intends to provide a few examples of drug discovery-related medicinal chemistry research and development activities taken from the academic and private sectors, respectively. The first set of studies involve efforts being undertaken by the Center for Drug Design and Development (CD3) within the University of Toledo College of Pharmacy. Four projects are briefly described wherein the first reflects a novel approach toward creating new compound libraries, the second and third address the challenges of handing 3D structures within a database setting, and the fourth demonstrates the importance of utilizing neutral and negative SAR data, as well as positive SAR. The second set of studies involve efforts undertaken by the drug discovery community with industry playing the important role of finally delivering fully developed products to the market place. Two projects are described wherein the first was accomplished in its entirety several years ago by American Critical Care, a small pharma company responsible for the successful design and development of what is now regarded as the prototypical soft drug. The second series of projects represent a continuum of research activities conducted more recently within both academic and industry settings in an effort to abate the growing AIDS epidemic. A focus is placed upon several of the larger companies that have successfully launched an NCE in this area and, as a result, these last examples demonstrate many of the new approaches that are presently being undertaken by big pharma toward discovering novel chemotherapeutic agents in general.

1.02.4.1 Academeme

1.02.4.1.1 Combinatorial phytochemistry

This example relates to the chemical library side of the new drug discovery paradigm. In particular, this project seeks to enhance molecular diversity[308,309] along phytochemical structural themes that have shown promising activity during initial efficacy screening such that having related compound libraries would be highly desirable.[310] It has a certain appeal for our laboratories due to its ultimate practicality since it seeks to produce directed molecular diversity within common plants indigenous to the midwestern USA rather than by relying upon a 'medicine man' to retrieve exotic samples from faraway lands. This possibility is being explored by simultaneously exposing plants to both an elicitor (botany's designation for an inducing agent) and selected biochemical feedstocks. For example, the biochemical pathways leading to the anticancer phytoalexins from soybeans (**Scheme 1**) may be able to be elicited by soybean cyst nematode infections to produce a more diverse family of active principles when grown in environments containing biochemically biased nutrients.[311,312]

Toward this end, it has been established that the statistical reproducibility of HPLC-derived phytochemical constituent fingerprints from soybean controls is adequate to discern real fluctuations in these types of natural products.[302] Work is now progressing toward ascertaining the differences that result upon exposures of soybeans to various stimuli and feedstocks. Interesting results will be followed up by studying the genetic control of the involved pathways. In this regard it should also be noted that a reverse approach that leads to similar 'combinatorial biosynthesis'[316] endpoints is also being undertaken by various other groups, particularly with an interest toward the production of proteins and peptide families from plant systems.[317] In those studies, directed modification of the genetic regions controlling one or more established phytochemical pathways is first effected, and then these types of biotechnology interventions are followed up by characterization of the altered biocombinatorial expression products.

1.02.4.1.2 Human drug metabolism database

This example serves to further illustrate how database structures might be quickly entered in two dimensions and gradually matured into three dimensions using a computational approach. Experimental details will also be entered based upon x-ray and NMR data even though not elaborated as part of the example. The specific research problem pertains to the consideration of structures to be placed within a human drug metabolism database.[318]

Structures are initially considered as closed-shell molecules in their electronic and vibrational ground states with protonated and unprotonated forms, as appropriate, also being entered. If a structure possesses tautomeric options or if there is evidence for the involvement of internal hydrogen bonding, then the tautomeric forms and the hydrogen-bonded forms are additionally considered from the onset. Determination of 3D structure is carried out in two steps. Preliminary geometry optimization is affected by using a molecular mechanics method. For example, in this case the gas-phase structure is determined by applying the MacroModel 6.5 modeling package running on a Silicon Graphics Indigo 2 workstation with modified (and extended) AMBER parameters also being applied from this package. Multiconformational assessment using systematic rotations about several predefined chemical bonds with selected rotational angles is then conducted to define the low-energy conformers and conformationally flexible regions for each starting structure. In the second step, the initial family of entry structures are subjected to ab initio geometry

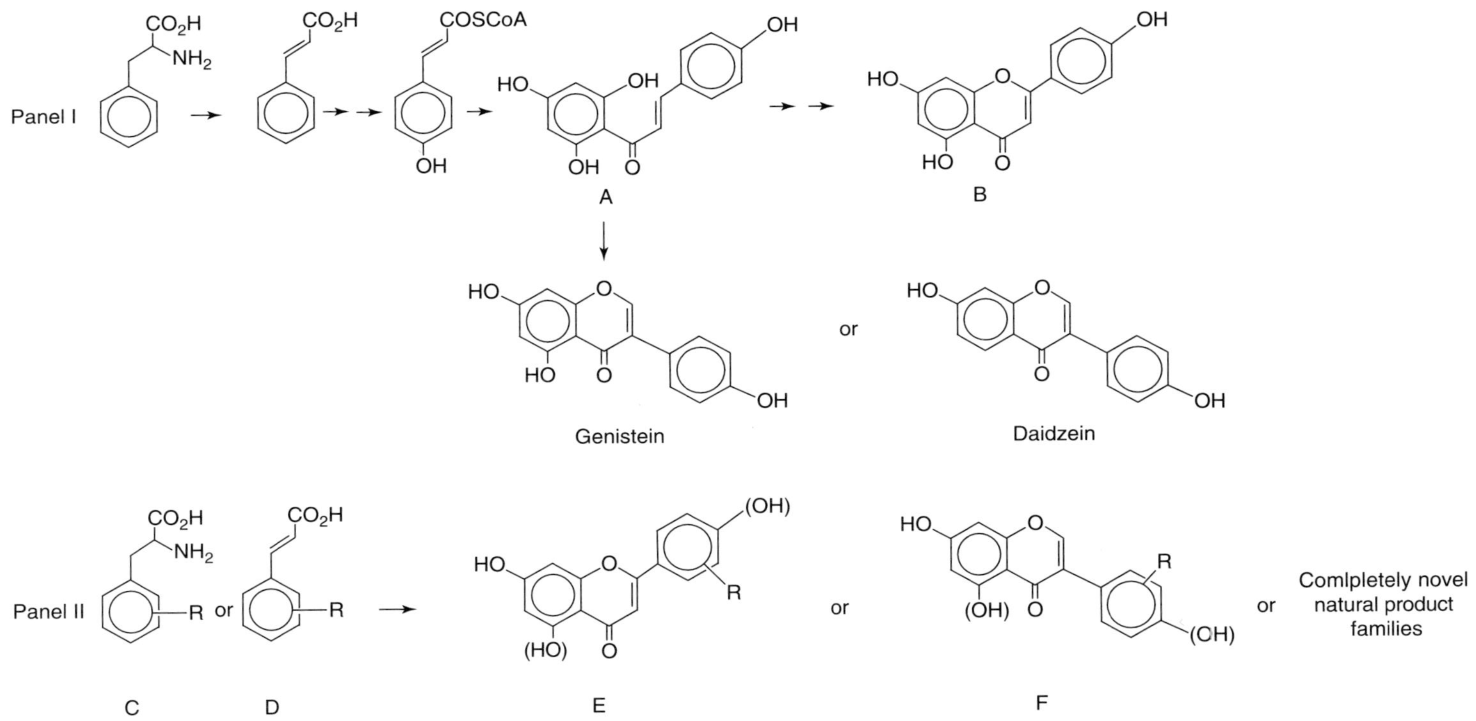

Scheme 1 Elicitation of directed and novel natural product families from soybean. Panel I: Normal biochemical pathways leading to the flavones[8] and to the key anticancer isoflavones genistein and daidzein wherein phenylalanine is first converted to cinnamic acid, *p*-coumaroyl CoA, and finally to key intermediate A, naringenin chalcone. Panel II: Feeding unnatural starting materials such as the aryl-substituted phenylalanine and cinnamic acid derivatives shown as C and D, respectively, under circumstances where this pathway is also being elicited by external stimuli (e.g., soybean cyst nematode infections), could be expected to produce new flavone derivatives E, isoflavone derivatives F, or completely novel natural product families. This diagram of the phytochemical pathway leading to flavones and isoflavones represents a composite of several references.[313–315] An array of inexpensive analogs related to C and D is available from commercial sources.

optimizations, which in this case use a Gaussian 98 package running on a T90 machine housed in a State-level US supercomputer center resource. Depending upon the size of the molecule, 3–21G* or 6–31G* basis sets[319] are used for conformational and tautomeric assessments. Density functional theory using the B3LYP functional[320] is applied for the consideration of exchange correlation energy while keeping the required computer time at reasonable levels. The highest-level structure determination is performed at the B3LYP/6–31G* level. To ascertain the local energy minimum character of an optimized structure, vibrational frequency analysis is carried out using the harmonic oscillator approximation. Determination of vibrational frequencies also allows for obtaining thermal corrections to the energy calculated at 0 K. Free energies are then calculated at 310 K (human body temperature). The latter values become particularly important for cases where structural (conformational or tautomeric) equilibria occur.

From the calculated relative free energies, the gas-phase equilibrium constant and the composition of the equilibrium mixture can be directly determined. Although these values may not be relevant in an aqueous or blood compartment, the calculated conformational distribution is relevant for nonpolar environments such as may be encountered when a drug passively traverses membranes or enters the cavity of a nonhydrated receptor/enzyme active site just prior to binding. Repetition of this computational scheme from biased starting structures based upon actual knowledge about the interacting biological systems or from x-ray or NMR studies (particularly when the latter have been conducted in polar media), followed by studies of how the various sets of conformations become interchanged and how they additionally behave when further raised in energy, complete the computational analysis for each structure being adopted into the human drug metabolism database.

1.02.4.1.3 In pursuit of an oxetane

This example attempts to demonstrate how the dynamic energy relationships between a compound's different conformations might be addressed within a database setting. Eventually, structures within the aforementioned human drug metabolism database will also be subjected to this type of analysis. This example also demonstrates the interplay of deploying NMR, x-ray, and computational approaches toward a more complete understanding of compound structure. The example draws from the CD3's ongoing anticancer chemotherapeutic program where an early effort was being directed toward replacing the complex scaffold of PAC (see **Scheme 3**) with a very simple molecular format that still displays PAC's key pharmacophoric groups in the appropriate 3D orientations purported at the time to be preferable for activity.[321–323] Toward this end, initial interest involved defining the role of the β-acetoxyoxetane system, particularly when the latter is immediately adjacent to planar structural motifs. Since such systems are rather unique among natural products[324] as well as across the synthetic literature,[325–328] it became useful to first study their formation within model systems relevant for this project. 2-Phenylglycerol was synthesized[329] and deployed as a model to study the molecule's conformation by x-ray, NMR, and computational techniques (**Figure 7**) as a prelude to affecting its cyclization.[330] The energy differences that result as the molecule is reoriented so as to be lined up for the cyclization were also calculated. Finally, once properly oriented in 3D space, the energy required to actually traverse the S_n2 reaction trajectory between the 1- and 3-positions was calculated (one of which positions utilizes its oxygen substituent for the attack while the other relinquishes its oxygen as part of a leaving group). The synthesis of 2-phenylglycerol and the pathway and energies associated with the intermediate species and cyclization process to form the oxetane are summarized in **Scheme 2**.[329,330]

Not surprisingly, given the strained-ring nature of this system, the energy needed to effect the ring closure from the lowest of three closely related local minima conformations similarly belonging to a family common to the independent x-ray and computationally derived starting points was about 28 kcal mol^{-1}. What becomes interesting, however, is that within this particular system, nearly half of this energy requirement results from the need to disrupt hydrogen bonds in order to initially reorient the molecule into a conformation appropriate for the reaction. The actual movement of the relevant atoms along the reaction trajectory (**Scheme 2**, dotted line), despite the resulting strain that becomes placed upon the overall system's bond angles, then accounts for only slightly more than one-half of the total reaction energy. Therefore, from a synthetic point of view the results suggest that it should be beneficial to employ a hydrogen bond acceptor solvent that has a high boiling point, the first property assisting in disruption of the internal hydrogen bonds that need to be broken for conformational reorientation and the second property for allowing enough thermal energy to be conveniently added so as to prompt progression across the reaction trajectory. That such favorable conformational perturbations could indeed be achieved by simply deploying these types of solvents was then confirmed by reexamination of the independent results obtained from our initial NMR studies conducted in polar, protic media.

Alternatively, from a medicinal chemistry point of view, these results serve as a reminder about the longstanding arguments pertaining to the importance and energetics of drug desolvation prior to receptor/active site interaction and, alternatively, the roles that stoichiometric water molecules can play within such sites. That such concerns can be

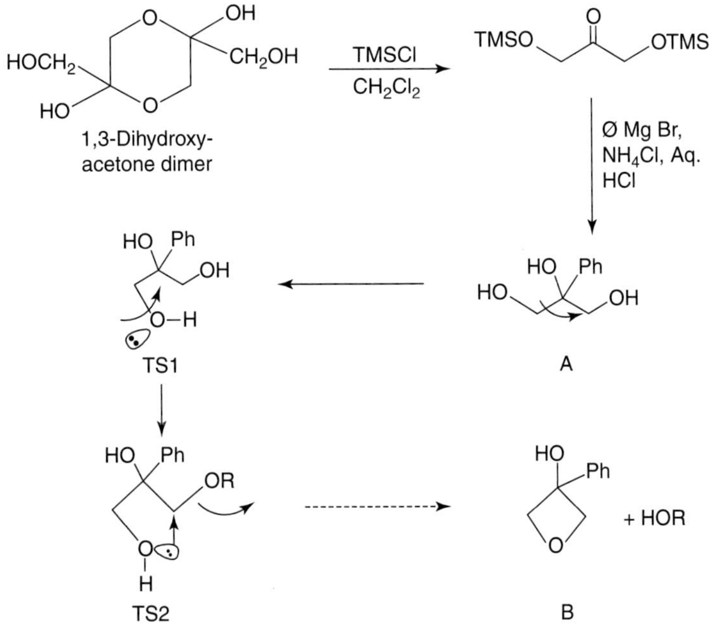

Scheme 2 Synthesis of 2-phenylglycerol and computational investigation of its conversion to 3-hydroxy-3-phenyloxetane.[329–331] 2-Phenylglycerol, A, is depicted so as to convey the lowest energy structure of the three close local minima observed during ab initio calculations performed at the HF/6-31G* level. TS1 and TS2 represent transition conformers obtained after the indicated bond rotations, while B represents the desired 3,3-disubstituted oxetane system. The respective relative energies in kcal mol^{-1} for A, TS1, and TS2 along with the product oxetane, B, are as follows: 0.00, 13.6, 12.4, and 28.2.

addressed in a deliberate manner by using multidisciplinary approaches similar to the example cited herein seems clear. Indeed, a quick survey of the present medical chemistry literature suggests that consideration of the dynamic nature of conformational perturbations associated with efficacious events is already beginning to take hold.[127–132] However, as has been emphasized earlier, it is predicted that it will be even more critical in the future to correlate SARs from one database to another according to the dynamic energy differences between the various molecule's conformational family members when several ADMET-related interactions are additionally factored into the overall behavior of a molecule being contemplated for further development. In other words, simple comparisons of static structures, even when rigorously assigned in 3D, will probably not be adequate to address a molecule's behavior across all of the efficacy and ADMET-related biological surfaces that become of interest as part of the molecule's optimization during future, new drug design and development paradigms.

1.02.4.1.4 Targeting cancer chemotherapy

The investigations to be exemplified in this final case study from academe have been directed toward studying the SAR associated with biological transporter systems with the hope of eventually establishing a database of transportophore relationships that might be generally applicable toward enhancing the selection and/or development of efficacy leads from another data set. To provide immediate relevance to this long-term project, it has initially focused upon the P-glycoprotein pump (P-gp)[332,333] that is associated, in part,[334,335] with the development of multidrug resistance (MDR)[336,337] during cancer chemotherapy. P-gp is a 170 kDa transmembrane glycoprotein belonging to the ABC class of transporters that serves as an energy-dependent, unidirectional efflux pump having broad substrate specificity. In humans it is encoded by the MDR gene, MDR1, whose classical phenotype is characterized by a reduced ability to accumulate drugs intracellularly, and thus the deleterious impact of increased P-gp activity upon cancer chemotherapy.[338–342] By way of practical example, the cytotoxicity of PAC is decreased by nearly three orders of magnitude when breast cancer cell lines become subject to MDR, largely via a P-gp mechanism.[343] In order to explore a series of probes that will systematically span a specified range of physicochemical properties when coupled to the PAC framework, one needs to first identify a region on PAC that is tolerant toward such modification in terms of PAC's inherent efficacy via its ability to overstabilize microtubules and prompt apoptosis. Thus, neutral SAR was sought where changes are known to not significantly alter PAC's cytotoxicity toward nonresistant breast cancer cells.

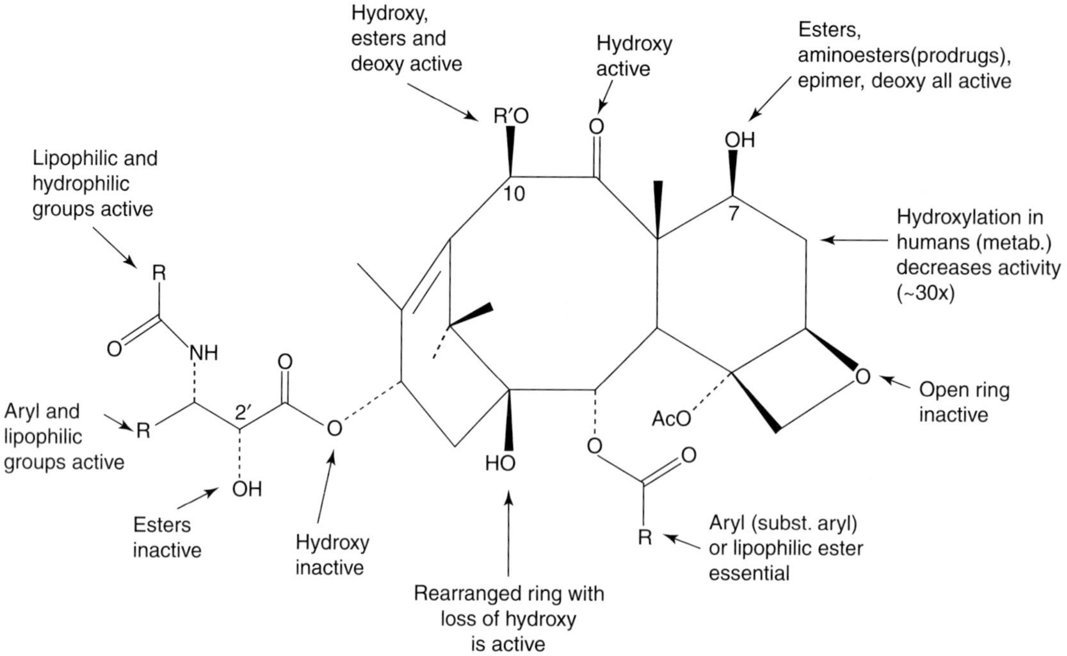

Scheme 3 Overall SAR profile for paclitaxel-related compounds. This summary represents a consolidation of SAR information contained in several review articles.[344–348] Note the tolerance for structural modification along the northern hemisphere of the taxane ring system. Paclitaxel has R=Ø and R′=CH_3CO. Positions 7, 10 and 2′ (on the side chain) are also noted.

Scheme 3 provides a summary of the accumulated SAR data obtained from the PAC-related review literature,[344–348] wherein it becomes clear that several positions along the northern edge represent neutral areas that can lend themselves toward such an exploration.

Inhibitors of P-gp have already been identified by several different investigators, and these types of compounds belong to a class of agents referred to as chemosensitizer drugs for which there are a variety of mechanisms.[349,350] While P-gp inhibitors can be coadministered with a cytotoxic agent in order to negate MDR toward the latter when studied in cell culture, historically these types of chemosensitizers have not fared well clinically.[351] One of the reasons that the inhibitors have not fared well is that they must compete with the accompanying cytotoxic agent for access to the P-gp MDR receptors. Thus, it can be imagined that if an SAR can be identified that is unfavorable for binding with P-gp MDR receptors and, furthermore, that if such a negative transportophore could be incorporated onto the original cytotoxic agent in a neutral position, then the cytotoxic agent might itself avoid MDR or at the very least become better equipped to do so in the presence of a coadministered P-gp MDR inhibitor.[352]

Toward this end, initial studies have been directed toward exploring the possibility that it may be feasible to identify negative SAR that is undesirable to the P-gp system within the specific chemical context of PAC by manipulating the latter at neutral positions that do not significantly affect PAC's inherent efficacy. To ascertain the generality toward potentially being able to place such a negative transportophore onto other established chemotherapeutic agents and onto lead compounds being contemplated for preclinical development, a similar series of negative SAR probes is being examined within the context of a completely different molecular scaffold, namely that of the camptothecin (CPT) family of natural products for which topotecan represents a clinically useful anticancer drug.[353] CPT, accompanied by a summary account of its SAR-related literature,[344,353–376] is depicted in **Scheme 4**, wherein it becomes clear that the northwest corner represents the key neutral region in CPT that might be manipulated analogously to those in PAC. Since these two compounds have very different molecular templates and owe their cytotoxicities to two distinctly different mechanisms (i.e., PAC largely, but not exclusively[377–379] to overstabilization of microtubules[380,381] and CPT largely, but not exclusively,[382] to 'poisoning' of topoisomerase I[383]), and because topotecan, a clinically deployed CPT analog (**Scheme 4**), is at the lower end of the spectrum in terms of being subject to P-gp-related MDR (it loses about one order of magnitude from its initial potency [343]), taken together these two molecules represent an excellent pair to examine the generality of the transportophore-related SAR findings, i.e., different chemical structures with different efficacy-related mechanisms and with differing sensitivity

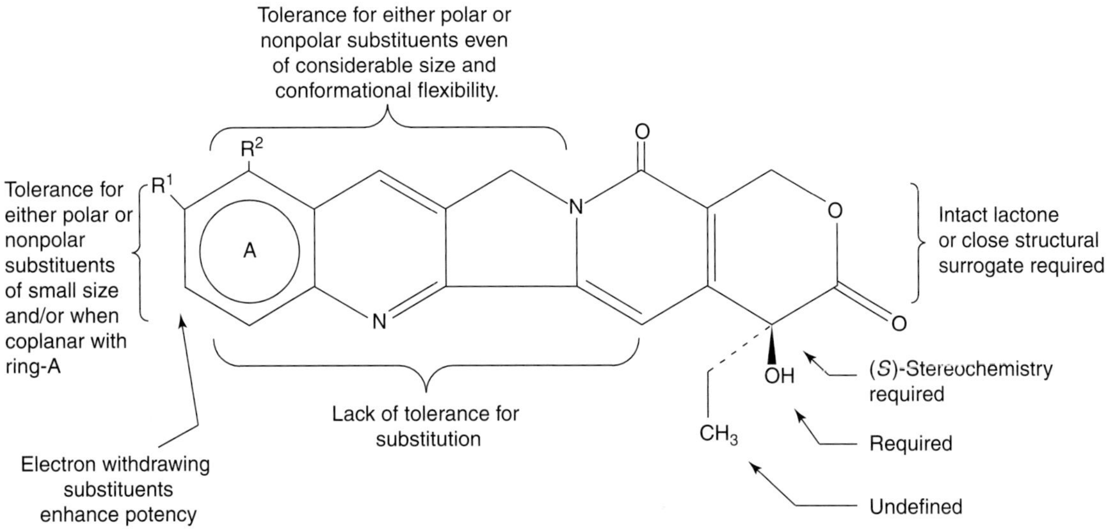

Scheme 4 Overall SAR profile for camptothecin-related compounds. This summary represents a consolidation of SAR information contained in several primary references.[344,353–376] Note the tolerance for structural modification along the northern hemisphere of the overall molecule, especially when approaching the western edge. Camptothecin, the parent natural product, has $R^1 = R^2 = H$. Topotecan, used clinically, has $R^1 = OH$ and $R^2 = CH_2N(CH_3)_2$.

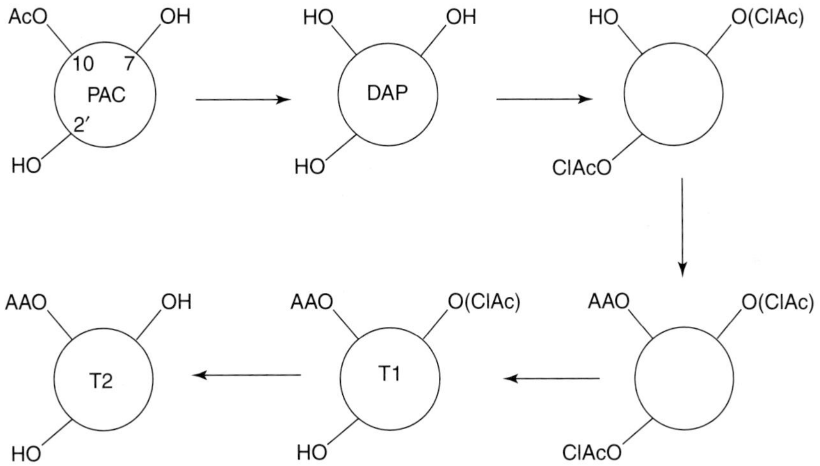

Scheme 5 Schematic overview for synthesis of paclitaxel analogs. PAC = paclitaxel with its 2′ and 7-OH groups displayed along with its 10-acetoxy group (see **Scheme 3** for complete structure). DAP = 10-deacetylpaclitaxel; ClAc, chloroacetyl protecting group; AA, (N-BOC) Asp (α-Benzyl Ester) wherein the two Asp protecting groups were also selectively removed to provide four compounds in each of the target families T1 and T2: (1) fully protected Asp, (2) amine-exposed Asp, (3) α-carboxy-exposed Asp, and (4) fully deprotected Asp.

toward P-gp-related MDR. Other molecular scaffold systems and biological testing models can also be imagined so as to extend such P-gp investigations into the areas of drug absorption, uptake into hepatic tissue (drug metabolism), and passage across the blood-brain barrier (e.g., for either enhancing or attenuating drug penetration into the CNS). Likewise, additional transporters within the ABC class can be systematically explored by using a similar approach.

After establishing chemical methods to conveniently protect/deprotect the reactive 2′-position of PAC,[384–386] two series of 10-position-modified 10-deacetyl-PAC (DAP) analogs were prepared in which the 7-position was either liberated as PAC's parent hydroxyl functionality (**Scheme 3**) or was protected with a chloroacetyl ester-group. For our academic laboratory, the cost of starting materials was a factor, so a series of analogs was devised to maximize the SAR that could be gained while minimizing the reactions and starting material requirements for PAC. Our synthetic medicinal chemistry strategy is depicted in **Scheme 5**.

From this approach, the diprotected Asp derivatives (N-BOC, benzyl ester) provided readouts about the tolerance for steric bulk at the microtubule binding site and at the P-gp transporter binding site, both of which proved to be very tolerant in this regard. These derivatives also provided similar readouts about lipophilicity, both of which impacted rather inconsequentially upon inherent efficacy but actually appeared to favor binding and cellular export by P-gp. Likewise, exposure of the basic amine functionality significantly worsened the P-gp susceptibility (increased binding and export) while having only modest effects on inherent efficacy. Alternatively, exposure of the acidic carboxyl moiety, although decreasing inherent efficacy by about 10-fold, was seemingly able to decrease the susceptibility toward P-gp by about one-third. Finally, exposure of both the amine and carboxylic acid functionalities essentially restored the inherent efficacy-related potency while either not effecting PAC's P-gp liability or reducing it by about one-half. Encouraged by the results from this initial series of P-gp probes, we decided to prepare a second series of probes. In this regard we noted that our most promising negative P-gp SAR functionality, namely a full-blown carboxylate anion, can also be used by virtue of its sheer polarity, to enhance PAC's notorious low aqueous solubility. Less obvious, however, we also discerned that this same structural space had a certain degree of similarity overlap with 'address' systems (*see* Section 1.02.3.7.2) which could hone to certain types of cancer cells such as those associated with the folate transporter which contains a Glu moiety attached to pteroic acid via an amide linkage. This situation is depicted in **Figure 17** in a general, strategic design manner.

Ongoing efforts are thus exploring the possibility that from the SAR-neutral, northern edge of PAC, we will be able to construct a single, permanent appendage (and not a prodrug-related cargo) that will: (1) enhance PAC's aqueous solubility by virtue of its simple physicochemical properties; (2) improve its selectivity for cancer cells compared to rapidly dividing healthy cells by virtue of its folate transportophore positive SAR; and (3) have a significantly reduced liability toward MDR by virtue of its P-gp transportophore negative SAR associated with the presence of a free carboxylate anion.

Although this particular example reflects a rational SAR strategy, the same types of informational endpoints can certainly be pursued via a coupled HTS/combinatorial chemistry approach providing that the chemical structure components within the resulting databases are initially constructed with architectures flexible enough to allow for such queries. Likewise, while this example reflects a simple query between two different biological behaviors, the same types of queries can be conducted across multiple databases for multiple parameters. That the next 25 years of medicinal chemistry will involve a considerable amount of making sense out of such multiple parameter correlations based upon experimentally derived data is quite clear. That the next 50 to 75 years might then be able to be fruitfully spent in more of a virtual correlations mode is certainly more speculative but is, at least, probably reasonable providing that we can build our knowledge base and fundamental understanding of how the various parameters, as assessed in isolation according to the above HTS scenarios, interact simultaneously within the whole system.

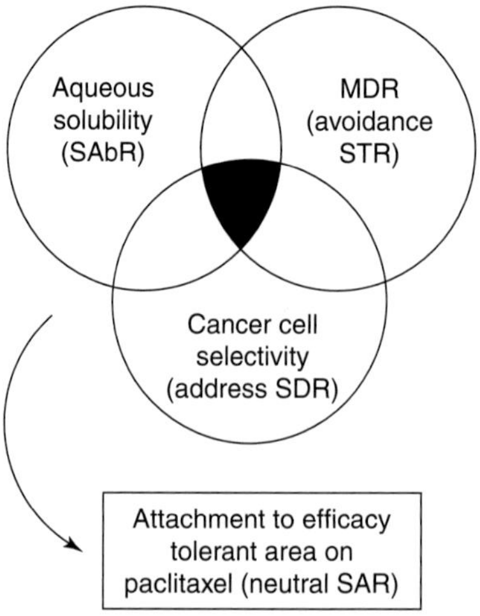

Figure 17 Design of optimized drug candidate based upon simultaneous consideration of several fields of paclitaxel-related SAR. Studies indicate that there is a distinct region of structural space that is simultaneously overlapped by SAR pertaining to enhanced aqueous solubility, avoidance of multidrug resistance, and the propensity to selectively associate with cancer cells compared to healthy cells. Coupling of this distinct structural space (depicted as the shaded, bulls-eye region) onto an area of paclitaxel that can accommodate structural modification without losing efficacy provides a hybridized drug candidate.

1.02.4.2 Industry

1.02.4.2.1 A soft drug from small pharma: esmolol stat

Soft drugs (**Figure 14**) are compounds that have been specifically modified so as to program a selected route and rate for their metabolism. While nature has provided numerous templates for the design of soft drugs, esmolol (**Scheme 6**) has come to be regarded as the prototypical soft drug that was obtained via rational drug design.[387] In this case, a methyl propionate was appended to the classical aryloxypropanolamine template associated with beta-adrenergic receptor blockade (**Scheme 6**) in order to program the latter's metabolism along the ubiquitous esterase pathways such that the resulting beta-blocker would possess an ultrashort duration of action.[388–390] Thus, a methyl 3-arylpropionate system (bolded atoms within **Scheme 6**) represents a useful metabophore already having clinical proof of principle within the molecular context of an aryloxypropanolamine template. This metabophore can be used to program human drug metabolism by esterases.

It can be noted that the rational design of esmolol simultaneously drew upon several of medicinal chemistry's basic science principles mentioned thus far: (1) negative SAR, wherein it was envisioned that only lipophilic or, at most, moderately polar groups could be deployed in the aryl portion of the aryloxypropanolamine pharmacophore if activity was to be retained; (2) electronic physicochemical properties operative within a biological matrix, wherein it was imagined that while an ester would be permissible in the aryl portion (neutral SAR), a carboxylic acid moiety placed within the same aryl portion would become too foreign to be recognized by beta-adrenergic receptors upon ionization of the carboxylic acid at physiological pH; (3) general structure–metabolism relationships (SMRs), wherein it was appreciated that an ester linkage might be relied upon to program a quick metabolism; (4) steric physicochemical properties, wherein it was imagined that the metabolic hydrolysis rate could be quickened by extending the initial ester linkages away from the bulky aryl group such that the sterically unhindered methyl 3-arylpropionate metabophore was finally identified; and (5) appreciation for the physiologic, drug SER, where there is a general propensity to excrete low-molecular-weight acids, probably via the anionic transporters (thus itself a useful transportophore relationship for excretion). These are all fundamental physical organic principles applied in a straightforward manner within very specific contexts of the biological realm. Thus, this particular case study serves four purposes. The first is to again emphasize that beyond activity hits per se, neutral and negative SAR should also be tracked so as to be readily retrievable from the databases associated with a given parameter survey of the future. The second is to again emphasize that medicinal chemistry will need to become an active participant in the merging of various HTS parameter surveys by using chemical structure as a common denominator, especially when such activities become considerably more complicated than the esmolol case. Third, the esmolol case demonstrates that even when problems can be reduced to what appears to be a rather simple set of factors, it will still be medicinal chemistry's unique desire to systematically characterize the complete pattern of chemical structural relationships that is likely to be called upon to finalize what other disciplines might consider to then be rather subtle or even mundane details. In other words, who, besides a medicinal chemist, can be expected to enthusiastically pursue methyl-, ethyl-, propyl-, etc. relationships either synthetically or by tediously purveying huge databases of the future, just to look for those SAR 'Goldilocks' situations[387] that could become relevant toward addressing a problem within another structural setting while attempting to merge the two data sets within a common chemical context? Indeed, as shown in **Figure 18**, the latter was precisely the case for the esmolol-related metabophore upon comparison of methyl benzoate, methyl α-phenylacetate, methyl 3-phenylpropionate and methyl 4-phenylbutyrate, wherein the observed half-lives for these systems when incorporated into the molecular context of a beta-blocker pharmacophore became about 40, 20, 10, and 60 min, respectively.

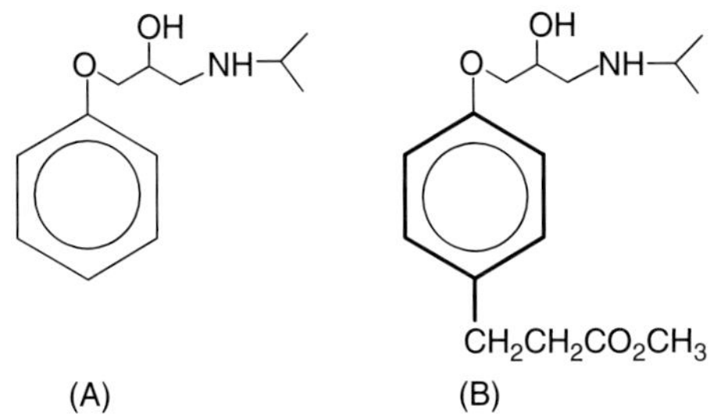

Scheme 6 Esmolol as the prototypical soft drug.[387] Compound A represents the classical aryloxypropanolamine pharmacophore associated with blockade of beta-adrenergic receptors. Compound B is esmolol, a soft drug version of A that has been programmed to have an ultrashort duration of action due to hydrolysis of the methyl ester by the ubiquitous esterases. The methyl 3-arylpropionate (bolded within B) thus represents a useful metabophore for the associated human esterases.

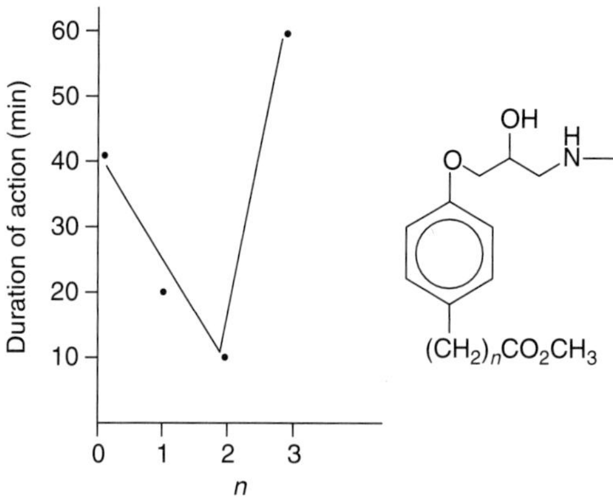

Figure 18 Relationship between methylene-extended esters and duration of action within a series of esmolol analogs.[387–390] The 'Goldilocks' nature of the ethylene extension relative to the desired 10 min duration of action is apparent.

In addition to the well-established, critical care uses of esmolol within the emergency room (ER) setting, e.g., 'esmolol stat' situations, numerous other uses of the soft drug technology are likewise being pursued.[391,392] For example, the esterase capability in newborns was recently compared to that of adults, and it was found that esmolol's half-life in cord blood (baby side) is about twice as long as that in adult blood. Furthermore, individual variation is significantly more pronounced within the newborns.[393,394] These findings, in turn, have prompted an exploration of the generality of deploying the esmolol metabophore within the chemical contexts of several other types of therapeutic agents, namely those that are commonly used to treat the neonatal population in critical care settings. Thus, this fourth and last aspect of the esmolol example clearly demonstrates the potential impact that such classical medicinal chemistry studies can have upon the new field of pharmacogenetics as the latter is surely to become further evolved. It is also interesting to note how this applied research problem initially pertaining to a practical desire to shorten the half-life for a beta-blocker drug, was pursued in a basic research manner directed toward uncovering fundamental SAR about the interactions between esterases and their substrates which, in turn, has produced knowledge that can now be applied in a general manner within the context of various other structural settings or parent molecules. Thus, this case study serves as true testimony to the benefits that can be derived in a synergistic (multiplicative) manner by virtue of medicinal chemistry's entwined basic and applied research nature (**Figure 1**).

1.02.4.2.2 Anti-acquired immune deficiency syndrome drugs

Despite the noted decrease in recent NCEs and its implications for the pharmaceutical industry overall, one area where implementation of modern technology has been enormously successful has been in the discovery of drugs to treat AIDS. Many of the new drugs in this area have come from the application of HTS and structure-based drug design approaches, with advances in molecular biology having enabled the molecular assays used to drive many of the medicinal chemistry efforts. Interestingly, since drugs for the treatment of AIDS have typically been approved more rapidly than is usual, it is possible that the success in this particular area presages what we may also see in the coming years from newer technologies applied to the discovery process.

Although the earliest cases of AIDS occurred in the 1950s, the first documented cases came to the attention of the US Center for Disease Control in 1981. By 1985 it was discovered that AIDS is caused by a retrovirus, subsequently named human immunodeficiency virus type 1 (HIV-1). In that year there were 12 000 new cases and 7000 deaths from AIDS in the USA. In 2004, the World Health Organization estimated that, worldwide, over 30 million people were infected with HIV-1 and that 20 million people had died from AIDS since the start of the epidemic. With the discovery of HIV-1 as the cause of AIDS, major drug discovery efforts were launched to identify effective antiviral agents to combat the disease. Arising mostly within big pharma, this drug discovery effort coincided with dramatic changes in the technologies and science available for medicinal chemistry such that an overview of this area becomes particularly instructive with regard to the impact of new technologies on modern drug discovery.

HIV-1 invades and destroys the $CD4^+$ cells of the immune system. The virus life cycle requires the incorporation of viral genetic material into the host genome. Subsequent transcription and translation events generate virus progeny, and the release of the new virus particles is fatal to the host (e.g., $CD4^+$) cell. The gradual depletion of immune cells occurs over a period of years and compromises the immune response to such an extent that infections that would normally be relatively benign can become life-threatening and deadly.

Drug discovery efforts have focused on interfering with the various processes in the virus lifecycle. As a retrovirus, the genetic material of HIV-1 is RNA. Two enzymes not found in human cells, reverse transcriptase (RT) and integrase, are essential for incorporating the information from this genetic material into the host genome. RT has multiple catalytic activities – it can function as an RNA-dependent DNA polymerase, a DNA-dependent DNA polymerase, and an RNA hydrolase. It transforms the virus RNA into double-stranded DNA. Subsequently, this viral DNA is incorporated into the host genome by the integrase enzyme. After transcription provides new viral RNA and the template for the synthesis of new virus proteins, the activity of one additional virus enzyme, HIV protease, is essential for the production of new, infectious viral progeny. To date, the reverse transcriptase enzyme and the protease enzyme have proven to be the most druggable targets in the virus life cycle and have yielded most of the drugs currently used to combat the virus and treat the disease. More recently, effective inhibitors of the integrase enzyme have been discovered and some have proceeded to clinical evaluation. There has also been success in targeting the processes required for virus to recognize and enter host cells with one approved polypeptide drug acting by this mechanism and several small molecule clinical candidates under evaluation.

Initial efforts to identify anti-HIV agents involved assaying molecules for the prevention of virus replication in cell culture. Zidovudine (AZT) (**Scheme 7**), the first antiviral drug approved for the treatment of AIDS, was discovered by this approach. The discovery process has been disclosed in detail,[395] and it provides an example of knowledge and experience-based drug discovery. Analogs of the natural nucleosides had been used successfully for many years in the clinic as antiviral agents (e.g., idoxuridine, acyclovir: **Scheme 7**). AZT, first synthesized in the mid-1960s,[396] was known to possess anticancer and antiviral properties, and in 1985 was demonstrated to have activity against HIV-1 replication in cell culture.[397] AZT is an analog of the nucleoside deoxythymidine and is converted by endogenous kinases into the active triphosphate derivative which is recognized by RT and incorporated into the growing strand of viral DNA. However, because AZT lacks the 3′-OH group necessary for the attachment of additional nucleotides, further elongation of the DNA is impossible and the viral replication cycle is disrupted. AZT proved effective in the clinic in suppressing the virus and prolonging the life of AIDS patients and was approved in 1987.

The discovery of AZT as an effective therapeutic agent validated RT as a target in the virus lifecycle, and several additional nucleoside analogs that act in an analogous manner have since been approved (also shown in **Scheme 7**). They are analogs of the endogenous nucleosides, deoxythymidine, deoxycytosine, deoxyadenosine, or deoxyguanosine and all lack the 3′-OH group of the natural nucleosides that is necessary for DNA strand elongation. AZT, stavudine,[398] zalcitabine,[399] and didanosine were previously known and were disclosed as effective anti-HIV-1 agents around the same time as AZT. All are readily recognizable as nucleoside analogs and this structural similarity to the natural substrates is likely required since they must be recognized not only by RT but also by the endogenous kinases that transform them into the active triphosphates.

Since the first phosphorylation step that produces the nucleoside monophosphates is the slow step in the activation of nucleoside analogs, the monophosphate derivatives could be more effective therapeutic agents. However, phosphatase as drug substances have two major potential liabilities: metabolic instability due to the action of phosphates enzymes; and, poor cell permeability due to the two negative charges at physiological pH. The design of tenofovir disoproxil provided a creative solution to both of these issues. Susceptibility to phosphatase activity was solved by incorporation of a noncleavable phosphonate moiety (appended to a modified sugar ring system precedented by acyclovir) to give tenofovir,[400] which provided proof of concept in the clinic but which was not orally bioavailable and could only be delivered by the intravenous route. To achieve oral bioavailability and cellular penetration, the phosphonate group was modified to the double ester prodrug (**Figure 14**) which is cleaved by intracellular esterase enzymes.[401,402]

After its approval in 1987 it became apparent that prolonged use of AZT (and indeed any other anti-AIDS drug) as a monotherapeutic agent engenders resistant virus.[403] RT lacks the transcriptional fidelity of mammalian DNA polymerases and no correction/editing features are involved in virus replication. Transcriptional errors occur at a rate as high as 1 per 2000 bases transcribed. Several mutations may be introduced during each replication cycle and the virus can be described as a quasispecies with many variants present at any time.[404] Given the facility with which mutations may be generated, it was clear that agents from a single therapeutic class would probably not be sufficient to suppress the virus indefinitely and prevent disease progression. Additionally, since nucleoside inhibitors are analogs of natural cellular components, they have the potential to interact with endogenous DNA polymerase enzymes, and some of the toxicities seen with AZT and other nucleoside analogs are thought to arise from such lack of selectivity.[405] Many efforts were therefore initiated to identify structurally and mechanistically novel RT inhibitors. Whereas the discovery efforts in the area of nucleoside inhibitors can be characterized as a knowledge-based analog approach to drug design, the identification of nonnucleoside reverse transcriptase inhibitors (NNRTIs) hinged on the success of HTS to generate lead structures.

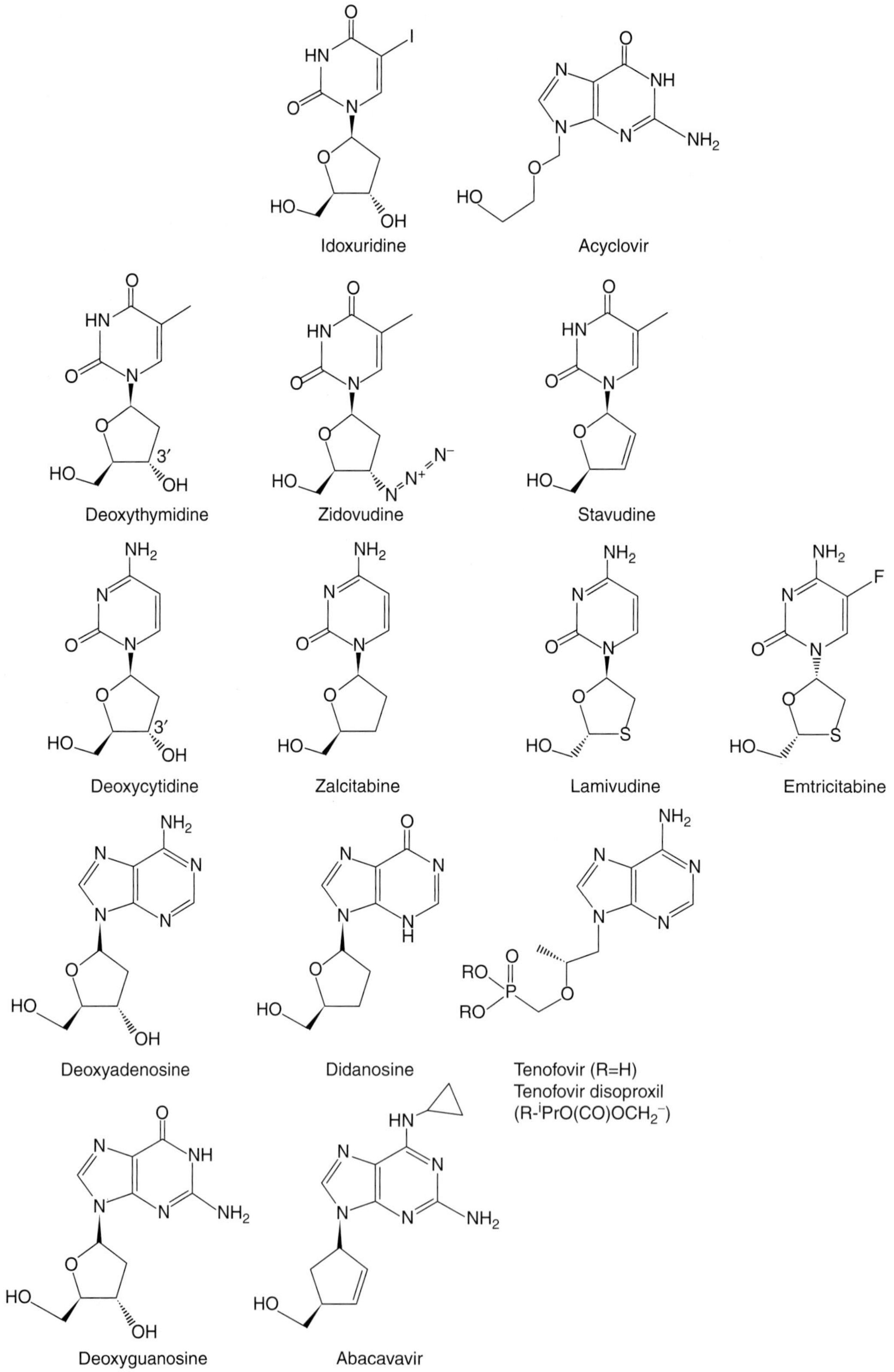

Scheme 7 Zidovudine (AZT) and related nucleoside analogs. The endogenous nucleoside analogs are shown to the left of each series of related drugs. Idoxuridine (approved in 1963) and acyclovir (approved in 1981) represent early antiviral agents that paved the way for AZT and stavudine which were approved in 1987 and 1994, respectively.

As mentioned in the previous sections of this chapter, over the past decade HTS technology has advanced to the extent that hundred of thousands of compounds can be screened in a short period of time. The drugs below came from earlier screening campaigns involving the evaluation of hundreds or perhaps thousands of compounds per month, and as such, they constitute early examples of the impact of HTS on the drug discovery process. Three successfully launched NNRTI drugs, (nevirapine,[406] delavirdine,[407] and efavirenz[408]: **Scheme 8**) and numerous additional clinical candidates (representative examples: **Scheme 9**) arose from these screening campaigns.

The three campaigns that resulted in successfully launched drugs all dealt with comparable issues during lead optimization.[409–411] As is expected for lead optimization campaigns beginning with HTS-derived structures, improvement in potency was a common goal. In contrast to the nucleoside inhibitors above, the nonnucleoside inhibitors were generally first optimized for inhibition of RT in a molecular assay and then tested for antiviral activity in cell culture assays. It is interesting to note, in the context of the recent observations on the changes in druglike properties generally seen in progressing from lead to drug,[412] that optimization of these lead molecules did not always require the purported expectation for an increase in molecular weight. In addition to improving potency and in line with the earlier discussions within this chapter that relate to the challenges of addressing inadequate PK and ADMET-related properties, attaining metabolic stability (and also chemical stability in the case of the efavirenz lead structure) proved to be critical. Also worthy of note is that the optimization of the efavirenz lead structure required the successful replacement of a metabolically labile thiourea motif. The first NNRTI inhibitor class described in the literature (tivirapine, R28913: **Scheme 9**) contained a similar thiourea motif which could not be successfully replaced and although these molecules proceeded to the clinic, they failed to progress beyond phase I.

Additional examples of molecules from the NNRTI class that were optimized from leads discovered in cellular or molecular screening campaigns and which progressed to clinical trials are also shown in **Scheme 9**. Again, for these early HTS leads, chemical and metabolic stability as well as potency were the common issues addressed during lead

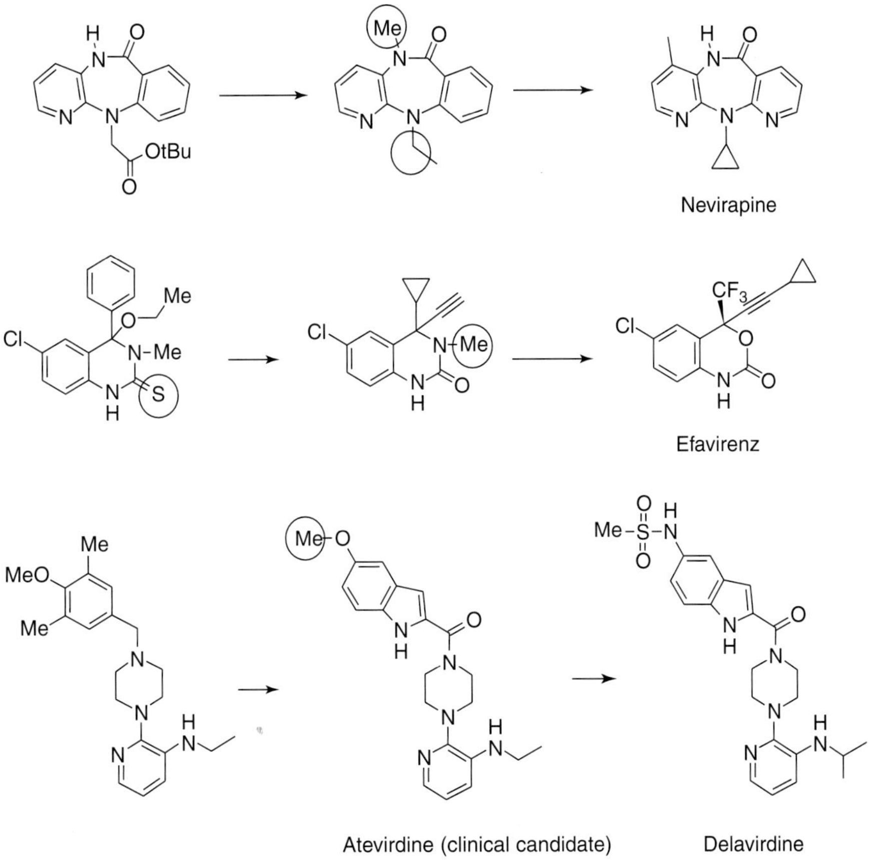

Scheme 8 Nevirapine, efavirenz, and delavirdine. Circled portions of structures represent areas having problematic metabolism that were addressed during lead optimization.

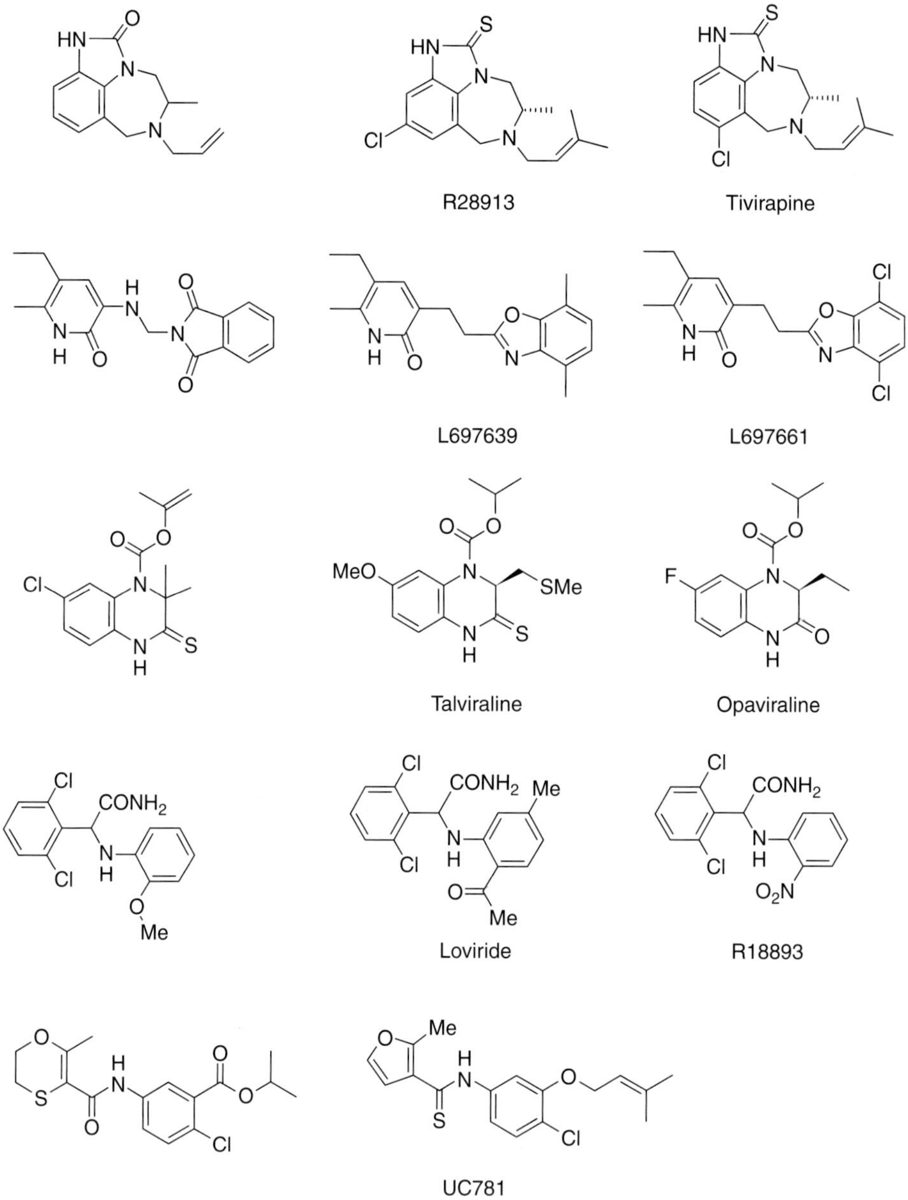

Scheme 9 Additional NNRTI clinical candidates obtained from HTS. Lead compounds are shown in the far-left column with their corresponding clinical candidates shown in the middle and far-right columns.

optimization. As above, it appears that for this drug class, progression from lead to drug does not necessarily require substantial increases in either molecular weight or complexity.

One remarkable feature of the NNRTIs when compared to their nucleoside counterparts is the selectivity they exhibit for HIV-1 RT compared to HIV-2 RT. They are typically inactive against HIV-2 RT whereas nucleoside inhibitors (as their triphosphates) are usually equally effective against both enzymes. Efforts to understand this phenomenon involved biophysical and structural studies to identify the binding site occupied by NNRTIs. Biochemical studies showed that the NNRTIs are noncompetitive inhibitors of RT, thus indicating that they do not compete with substrates at the enzyme active site.[413] Photoaffinity labeling experiments identified two tyrosine residues, tyrosines 181 and 188, as components of the NNRTI binding site. In sequence, these tyrosine residues are close to aspartic acid residues 185 and 186 which constitute part of the enzyme active site.[414] Subsequently, co-crystal structures of RT with nevirapine[415] and other inhibitors[416,417] provided a more detailed structural perspective on the interactions of the

NNRTIs and RT. The NNRTIs all bind to the same lipophilic pocket on the enzyme and moderate the activity of RT through allosteric inhibition. A comparison of the HIV-1 and HIV-2 protein sequences in the region of the binding pocket helped to rationalize the different efficacy of NNRTIs against HIV-1 and HIV-2. Sequence information showed positions 181 and 188 which contain tyrosine residues in HIV-1 RT, are occupied by isoleucine and leucine residues in HIV-2 RT. SDM studies confirmed that mutation of even the single amino acid Y188 in HIV-1 RT to L188 abrogated the effects of the drug nevirapine.[418] These mutagenesis studies were the precursor to cell passaging studies of the wild-type virus in the presence of nevirapine that generated resistant virus carrying the mutation Y181C in HIV-1 RT.[419] This ready generation of resistant virus was confirmed for nevirapine in the clinic and meant that the drug could not be used as a monotherapeutic agent. However, subsequent clinical studies showed that nevirapine, and other NNRTIs, are effective antiviral agents when used in combination with nucleoside[420] or protease inhibitors.[421,422] Although structural information on RT was available at an early stage, it was not a driving force in the design for the first-generation NNRTI compounds. It did, however, allow for a better understanding of resistance at the molecular level. **Figure 19** is derived from the structure of nevirapine bound to RT. Even though mutations conferring resistance may be far apart in the primary sequence of the protein sequence, when the mutations are mapped on to the 3D structure of the enzyme, they all cluster around either the NNRTI binding pocket or regions important for substrate binding.

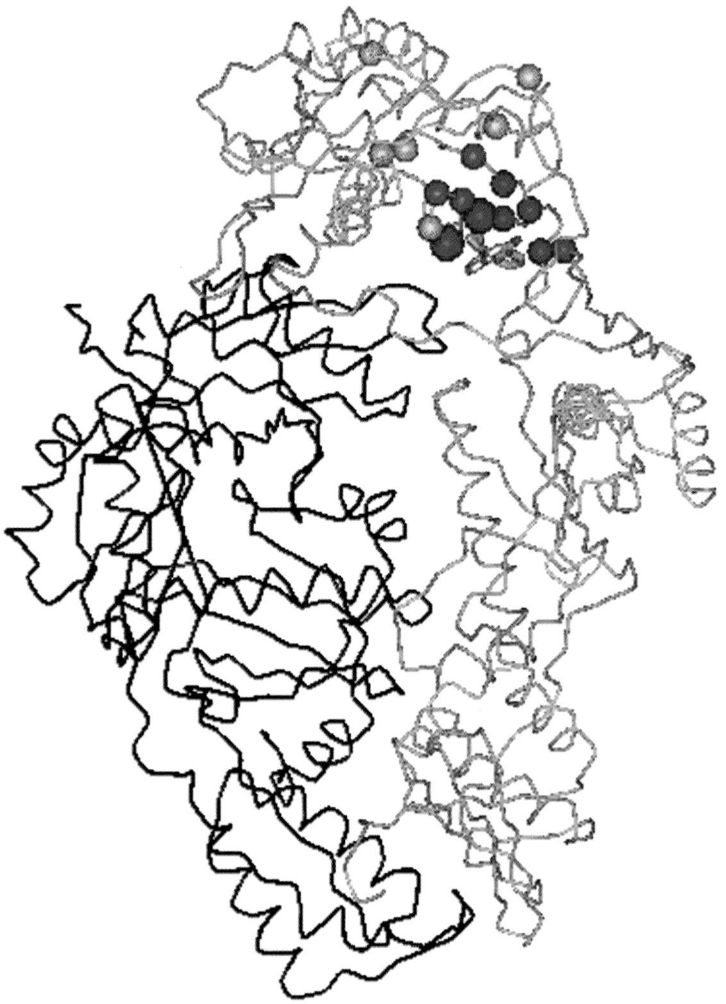

Figure 19 NNRTI mutations/NRTI mutations. The alpha-carbons of the key aspartic acid residues of the active site are represented by red spheres. The blue spheres represent sites of mutation that confer resistance to nevirapine and other NNRTI drugs (the larger blue spheres represent the positions of tyrosines 181 and 188). The orange spheres represent sites that confer resistance to nucleoside inhibitors.

The ready emergence of resistance to NNRTIs is a consequence of the relative inefficiency of RT mentioned above and the fact that the NNRTI binding pocket can accommodate many different amino acid side chain changes and still yield a functional enzyme. Efforts to design second generation, more effective NNRTIs have focused on attaining efficacy against many of the resistant mutant RT enzymes along with the wild-type enzyme. The ability to generate and provide mutant enzymes for molecular assays and for structural studies has been an enabling force for compound optimization and would not have been possible without the techniques of modern molecular biology. Again, requirements for second-generation PK properties conducive to once-a-day dosing and compatibility of compounds with multidrug regimens become important issues. Indeed, the latter two features are an important part of all current anti-AIDS drug discovery efforts. Since patients often take many drugs as part of a therapeutic regimen, any drug features that accommodate patient compliance lessen the likelihood for emergence of resistant virus. Some of the drugs currently under evaluation as second-generation NRTIs are shown in **Scheme 10**. Calanolide A[423] was discovered through bioassay directed fractionation of a tropical rainforest tree organic extract. It is the only unmodified natural product currently progressing in the clinic as a treatment of AIDS. The antiviral activity demonstrated initially in cell culture was subsequently found to be due to RT inhibition.[424] Etravirine evolved from the loviride structure (**Scheme 9**) via the thiourea derivative. GW5634 derived from the screening lead shown. The discovery process for capravirine has not yet been described, but it is likely that one of the early NNRTI compounds, HEPT,[425] provided a structural lead. These second-generation NRTIs show advantages over the currently approved drugs, generally in the area of activity against resistant virus carrying various RT mutations.[426]

In contrast to these successes in the RT inhibitor area which were enabled through analog-based discovery approaches and HTS, success in the area of HIV-1 protease inhibitors has been achieved mostly through rational,

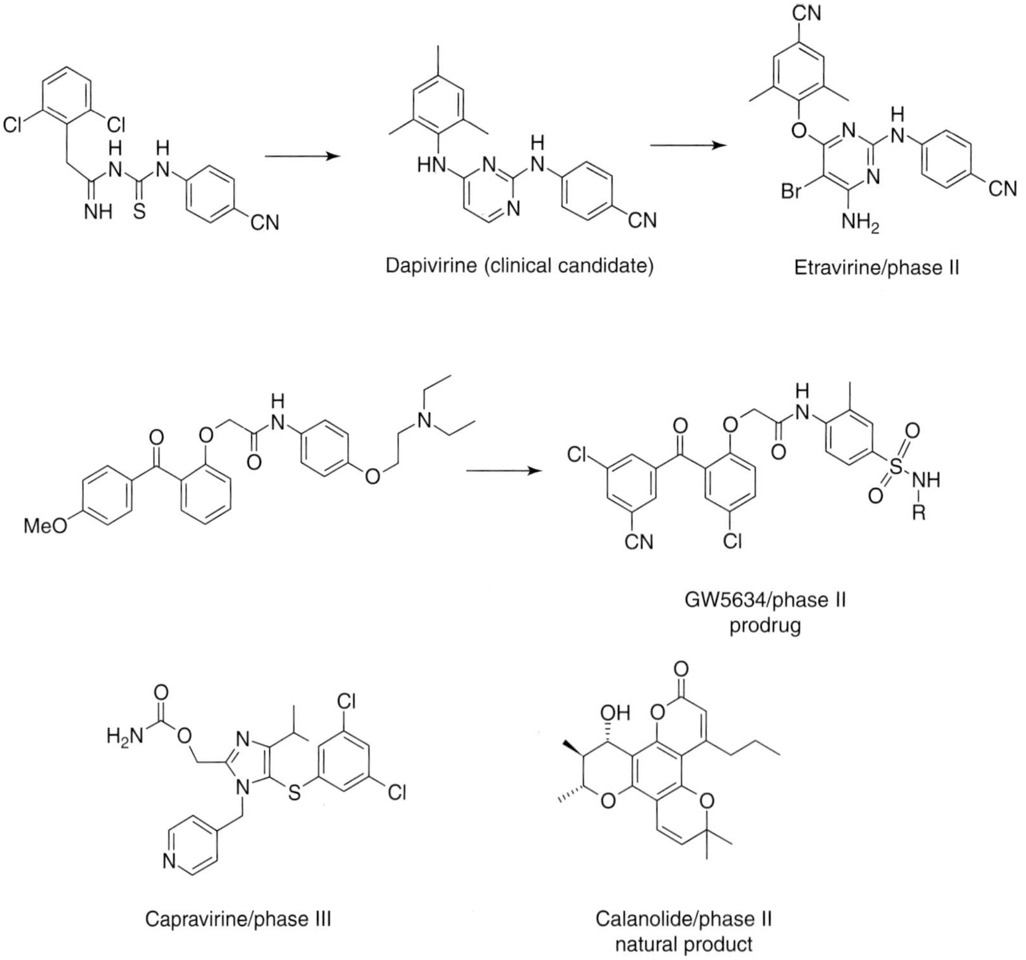

Scheme 10 NNRTI agents undergoing clinical study.

structure-based drug design (**Table 3**). Early confusion about the mechanism of action of HIV-1 protease was clarified when it was recognized that the enzyme is an aspartic protease and functions as a dimer. The first x-ray crystal structures were published in 1987 (one enabled by chemically synthesized protein).[427] This information, in combination with an experience base in the drug discovery community targeting other proteases (particularly the aspartic protease renin), laid the foundation for success in the area of HIV-1 protease inhibitor design. Alternatively, enzyme structural information was not a factor in the design of the first protease inhibitor to reach the market, saquinavir.[428] This drug was derived from a ligand-based approach where peptidic substrates were modified to incorporate noncleavable peptide bond surrogates in order to attain potent enzyme inhibition. As shown in **Scheme 11**,

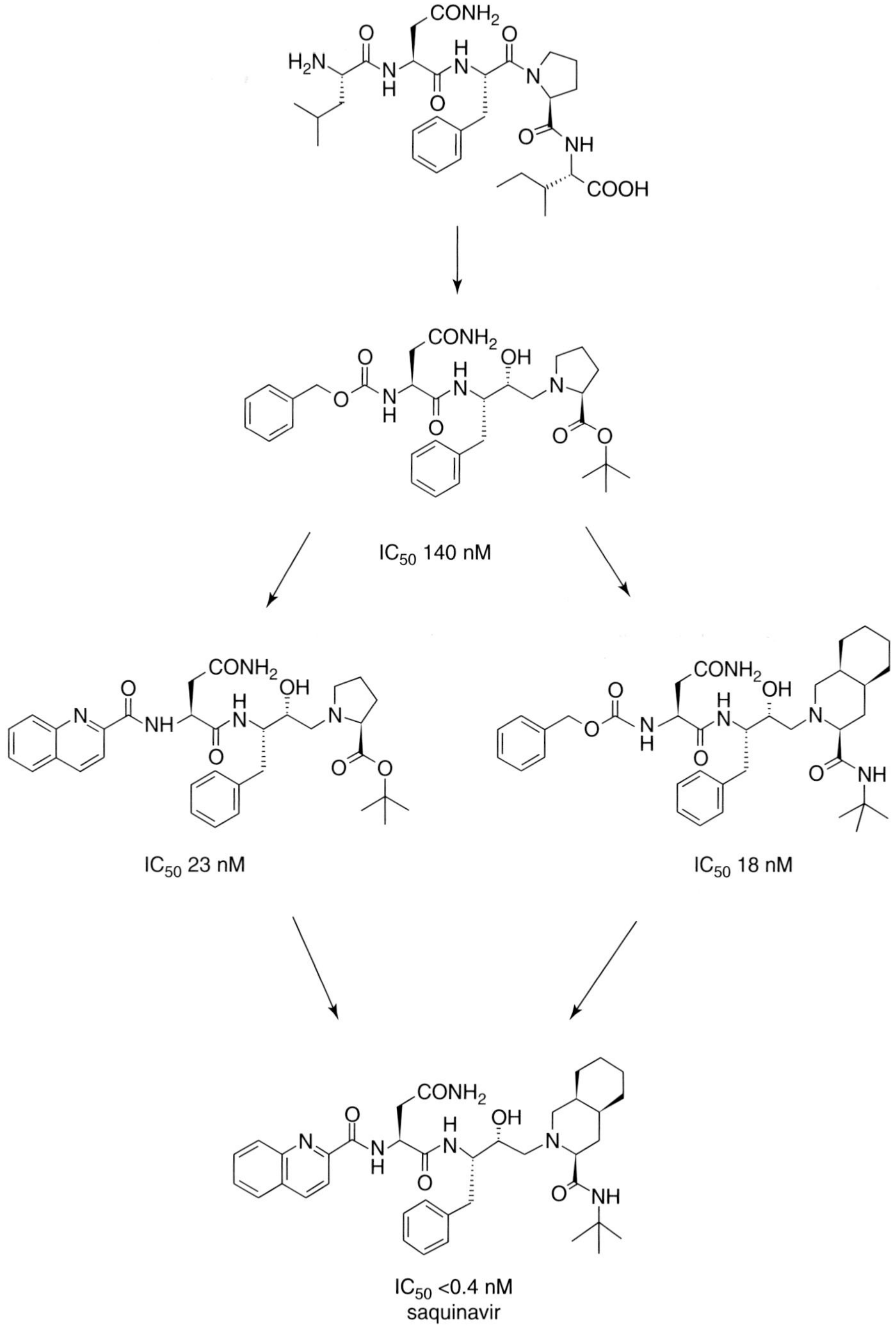

Scheme 11 Saquinavir. Some of the immediate precursor compounds and their potencies are also shown.

incorporation of a hydroxyethylene peptide bond surrogate into the substrate analog gave the lead structure with moderate potency against the protease. Stepwise modification of the left- and right-hand sides of the lead structure led to significant improvements in potency and, ultimately, combination of the preferred modifications gave saquinavir. Additional marketed protease inhibitors, such as nelfinavir,[429] amprenavir,[430] and atazanavir[431] (**Scheme 12**) and the additional clinical candidate TMC-114[432] also incorporate the hydroxyethylene isostere. Amprenavir has also been developed as a phosphate prodrug that has improved solubility. The phosphate group is readily cleaved in the gut to release amprenavir for absorption.

In an alternative approach to lead generation, researchers at Abbott hypothesized that since the protease enzyme functions as a dimer possessing C2 symmetry, inhibitors could also be designed based partially on symmetry considerations.[433] Modification of the symmetrical lead compound (**Scheme 13**) ultimately led, via two clinical candidates, to ritonavir[434] (approved in 1996). Efforts to generate a second-generation inhibitor based on ritonavir focused on addressing two issues: (1) achieving potency against resistant proteases where Val 82 in the active site is mutated; and, (2) attenuating the significant CYP inhibition seen with ritonavir. Analysis of structural information indicated that the resistance of the mutant enzymes was due to interactions with the isopropyl group on the thiazole ring and modification of the right-hand side region of ritonavir to incorporate a smaller heterocyclic system gave activity against the mutant enzymes. **Figure 20** is derived from the crystal structure of ritonavir bound to HIV protease. Somewhat interestingly, the CYP inhibition properties of ritonavir may make it a useful partner in various cocktail regimens of AIDS drugs. For drugs that are metabolized by CYP co-administration with ritonavir can provide a significant boosting of plasma levels and efficacy. Lopinavir, for example, has been approved as a formulation with ritonavir (Kaletra).

Two screening-based approaches have also been successful in generating effective protease inhibitors. Screening of a collection of rennin inhibitors at Merck provided a peptidic lead structure[437] (**Scheme 14**) which was modified to produce indinavir[438] in a process which involved the use of crystal structure information and competitor structural information as part of the optimization process. In comparison, screening at Upjohn provided warfarin as a low-molecular-weight lead.[439] Evolution of this structure proceeded through two clinical candidates with increasing molecular complexity to tipranavir, which is currently in preregistration. This design effort was also guided significantly by the availability of structural information. Ultimately, it is interesting to note that the drug structures themselves are quite similar in molecular weight despite the huge differences in the starting points.

In addition to the marketed inhibitors of RT and HIV protease, researchers have recently been successful in generating inhibitors of the integrase enzyme. This enzyme is responsible for incorporating the double-stranded DNA virus genetic material into the host genome and has always been a promising target for anti-AIDS drug discovery. Since combined inhibition of RT and HIV protease is the mainstay of current therapies, and the drug combinations are

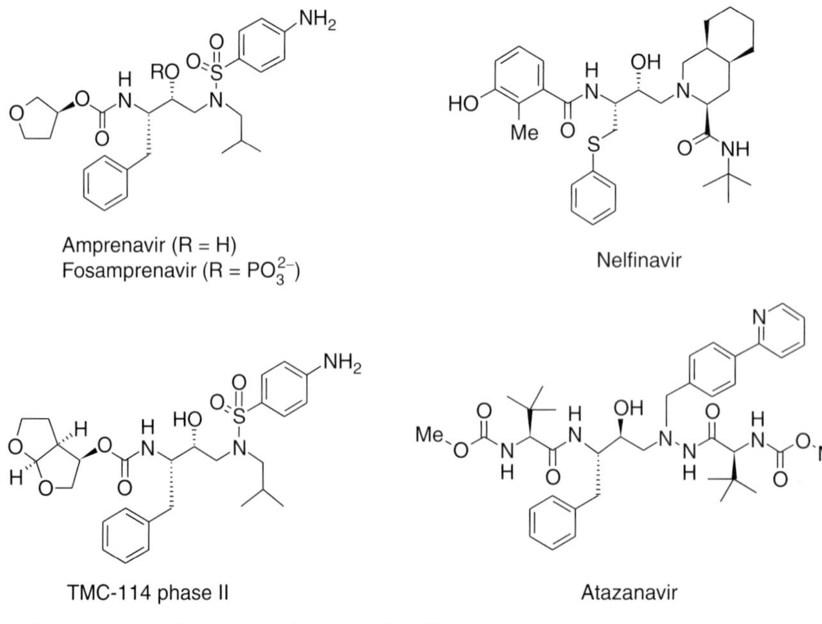

Scheme 12 Additional hydroxyethylene-based protease inhibitors.

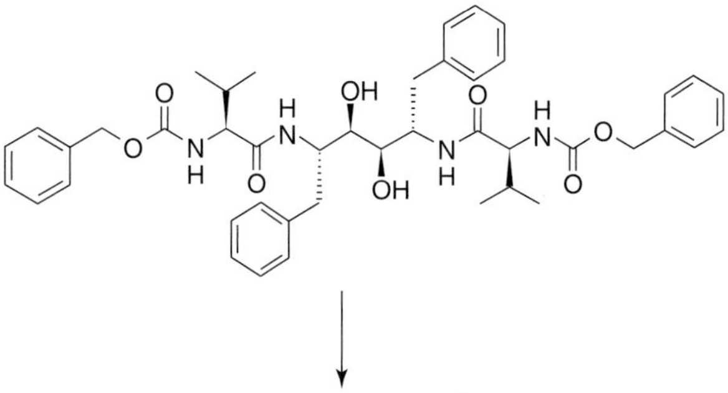

Lead structure

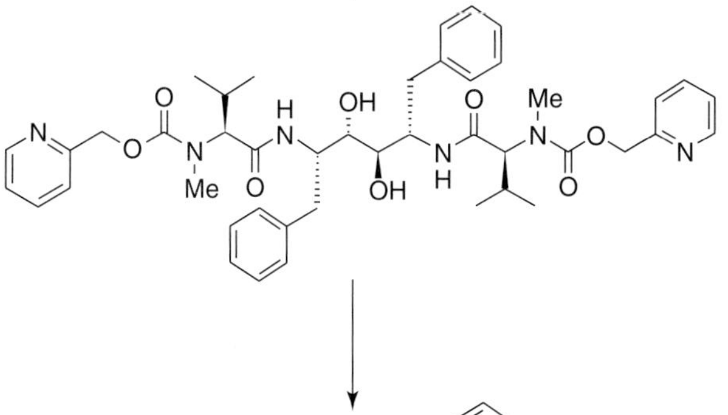

Clinical candidate

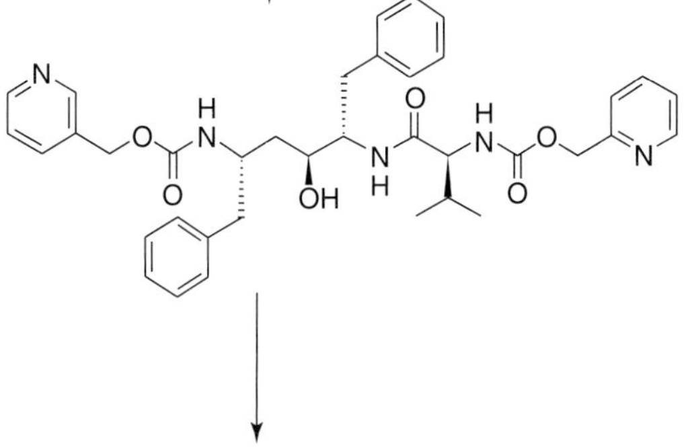

Clinical candidate:
better physical properties, good oral bioavailability, low $t_{1/2}$ due to metabolism

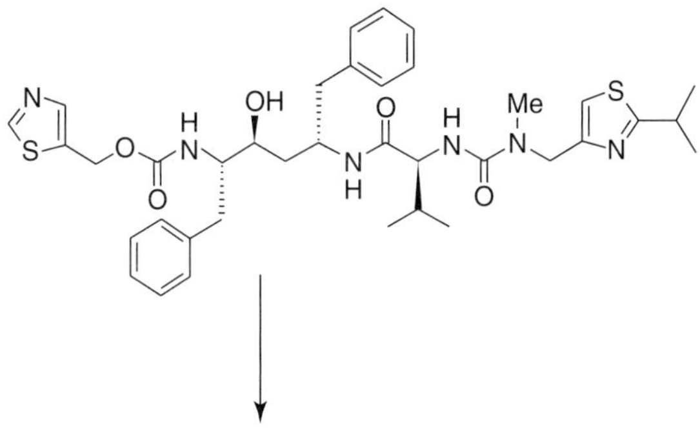

Ritonavir

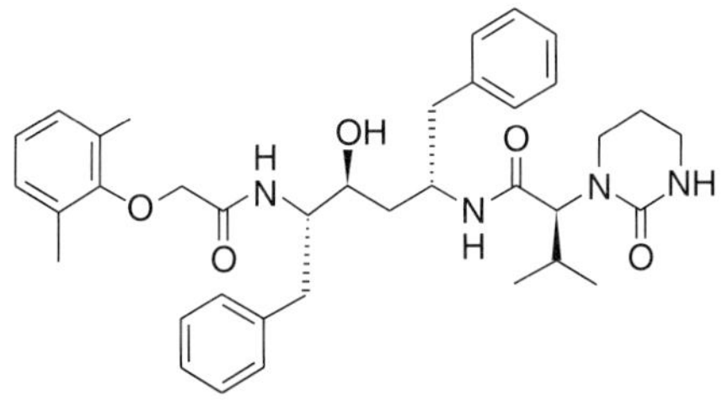

Lopinavir

Scheme 13 Ritonavir and lopinavir. Note that in this case several clinical candidates were traversed in order to eventually obtain the successful agent ritonavir.

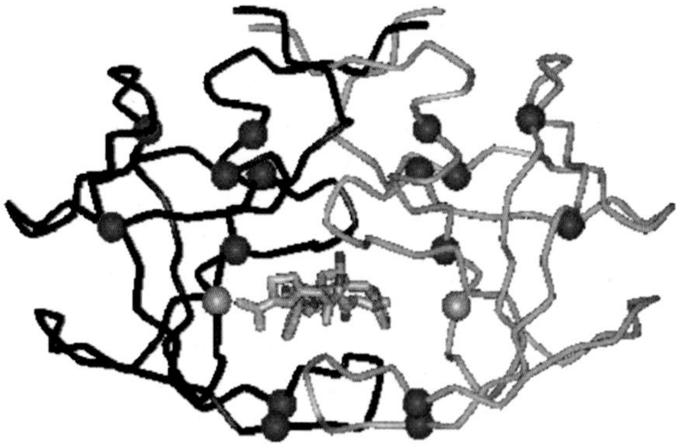

Figure 20 Ritonavir bound to HIV protease with resistance mutation sites also noted. The orange spheres represent the alpha-carbons of Val 82 and the blue spheres represent additional mutations in the protease enzyme that confer resistance to other protease inhibitors.[435] The CYP issue was localized to the left-hand side thiazole ring. Modification of this ring system to the dimethylphenyl ether gave lopinavir.[436]

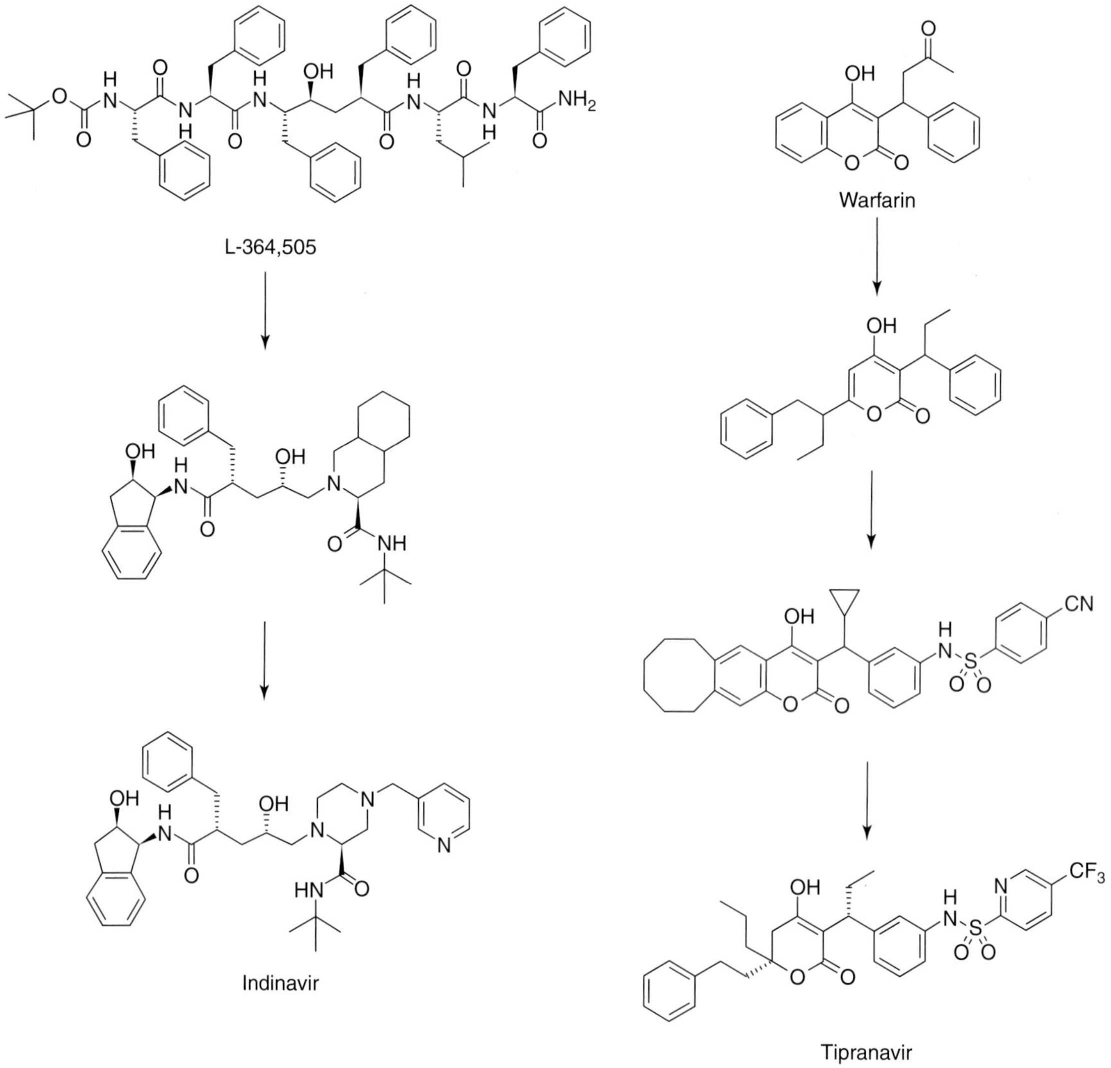

Scheme 14 Indinavir and tipranavir. A few early compounds are also shown.

superior to the actions of individual drugs alone, an effective inhibitor of other key targets in the life cycle could provide substantial therapeutic benefit. Like the RT enzyme, integrase has no host counterpart and offers the potential for a selective and safe therapeutic mechanism. After many years of effort, and the report of several unattractive structures as inhibitors, a series of ketoacid derivatives, discovered through HTS, was recently disclosed to effectively inhibit the enzyme[440] (**Scheme 15**). Optimization of this lead structure, particularly with regard to the metabolically unstable ketoacid moiety, has yielded very effective in vitro inhibitors of the enzyme.[441] A molecule with similar structural features, S-1360, has progressed to clinical trials.[442] **Figure 21** is derived from the crystal structure of an analog of S-1360 bound to the integrase enzyme.[443] As with the other classes of AIDS therapeutics, there is already evidence that therapeutic agents targeting integrase will also be subject to resistance.[444]

One additional area of anti-AIDS drug discovery that has shown progress recently has been the field of viral entry inhibitors. The prototype proof of concept molecule for this class, enfuvirtide, was approved for marketing. Enfuvirtide is a polypeptide derived from the C-terminal domain of the HIV-1 fusion protein, and prevents fusion of the virus with the cell and subsequent cell entry. The drug is somewhat remarkable due to the fact that its synthesis requires 106 steps (99 on solid phase).[445] Although enfuvirtide did not come from a classical medicinal chemistry design effort, it provided evidence that prevention of virus entry constituted a valid approach to combating HIV infection. More recently, additional opportunities for preventing virus entry were identified with the recognition that HIV-1 makes use of the chemokine co-receptors to enter cells. It is remarkable that in the relatively short period from this disclosure, chemokine receptor antagonists have already progressed from screening to clinical evaluation. SCH 417690, for example, evolved from the HTS hit shown in **Scheme 16**.[446] During the optimization process, chemists had to solve issues involving metabolic stability and selectivity away from muscarinic receptors (negative SAR). One newer issue which had to be addressed during this optimization effort was activity against the hERG (human ether-a-go-go-related gene) channel, considered to be an indicator of QT (interval between Q and T waves) prolongation and potential serious cardiac side effects.[447]

In the 20 years since the recognition that AIDS is caused by human immunodeficiency virus, the drug discovery community has provided some 20 antiviral agents targeting various components of the virus life cycle. These drug discovery efforts have made use of historical knowledge, HTS, and structure-based drug design. That the efforts have also been facilitated by advances in molecular biology should be quite clear.

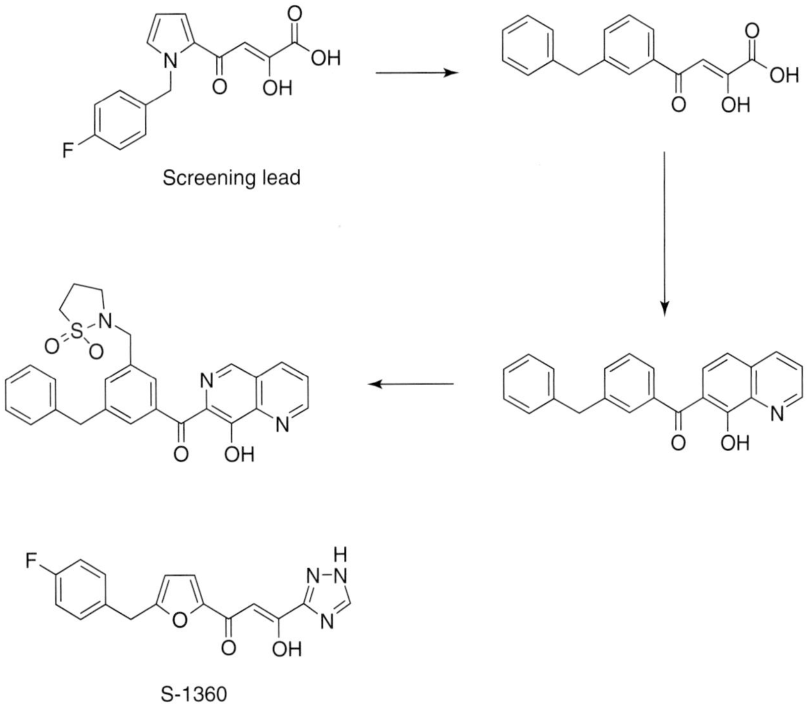

Scheme 15 Integrase inhibitors.

Figure 21 Inhibitor bound to integrase with resistance mutation sites also noted. Only that portion of the protein is shown so as to depict how the inhibitors bind close to three carboxylates of the active site. In addition, the sites of resistant mutations that have arisen in vitro are illustrated by the blue spheres. This structural information should provide additional guidance in the optimization of integrase inhibitors.

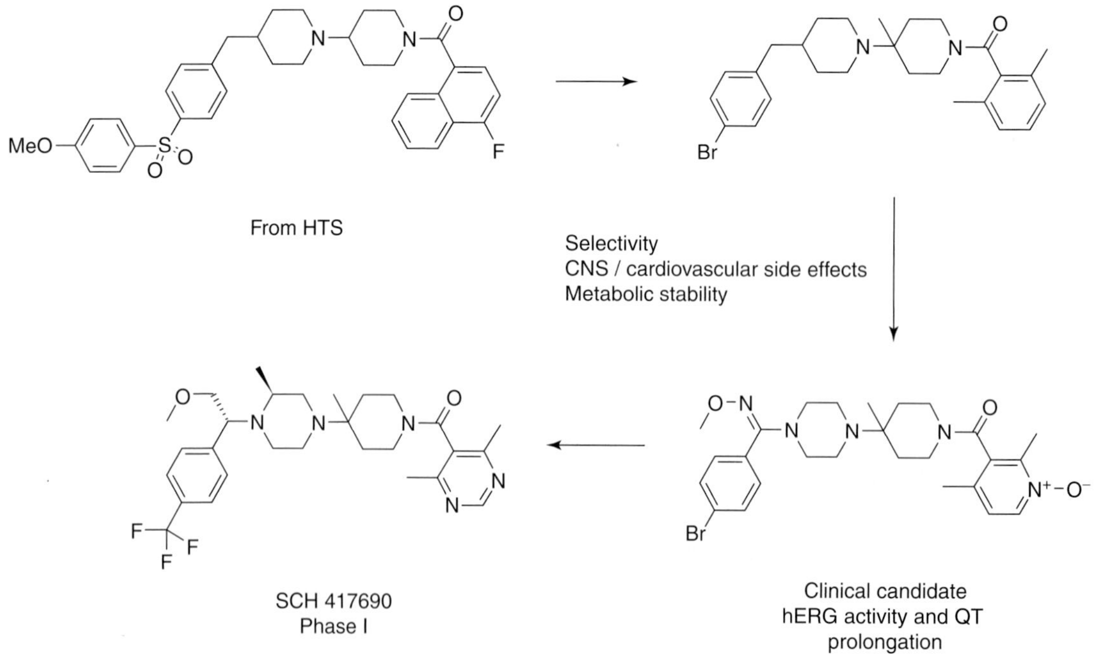

Scheme 16 Viral entry inhibitors.

1.02.5 Outlook

1.02.5.1 Drug Discovery's Future

Drawing from the foregoing discussions and consideration of selected case studies from academe and the private sector, the future of drug discovery is further contemplated below. As a first step, several points are listed in bullet fashion. Their sequential progression can be thought of as being driven by an ever-increasing accumulation of knowledge within the life sciences arena.

- Genomics and especially proteomics, will continue to unveil myriad new biomechanisms applicable to modification for therapeutic benefit and toward gaining a better understanding of the processes associated with ADMET. Assessing the value of these mechanisms as worthwhile targets will continue to grow as a practical challenge.
- The systems associated with meritorious targets that also lend themselves to crystallization followed by x-ray diffraction, will be readily exploited by structure-based drug design.
- Biotech will continue to evolve HTS methodologies to assay therapeutic mechanisms and to assess ADMET properties as a front-line initiative during drug discovery.
- For cases where structure-based drug design is not possible, collected compound libraries and combinatorial libraries, along with wild and genetically altered or elicited collections of "natural" materials, will provide initial hits which will then be followed up by directed libraries and by ligand-based drug design.
- Accumulated data in all areas will become well managed via a variety of databases including, in particular, an evolved level of 3D structural sophistication within their chemoinformatic treatments. The latter will ultimately provide the common language for the integration of information across all databases with medicinal chemists playing a lead role in this regard.
- As ADMET information accumulates from preclinical models, which will eventually become validated relative to the human case, distinct structural motifs will arise that can be used with statistically derived confidence limits in a predictive manner during drug design.
- HTS of manmade libraries, directed analyses of nature's mixtures, and importantly, rational exploitation of the newly elaborated and fully understood ADMET structural motifs will reveal synergistic combinations that involve more than one compound wherein at least one of the components is, itself, efficaciously silent.
- For each case of new drug design, a summation of the accumulated knowledge from efficacy testing, ADMET, and potential synergistic relationships will guide final tailoring of the clinical candidate drug or multidrug combination. This situation is further depicted in **Figure 22**.

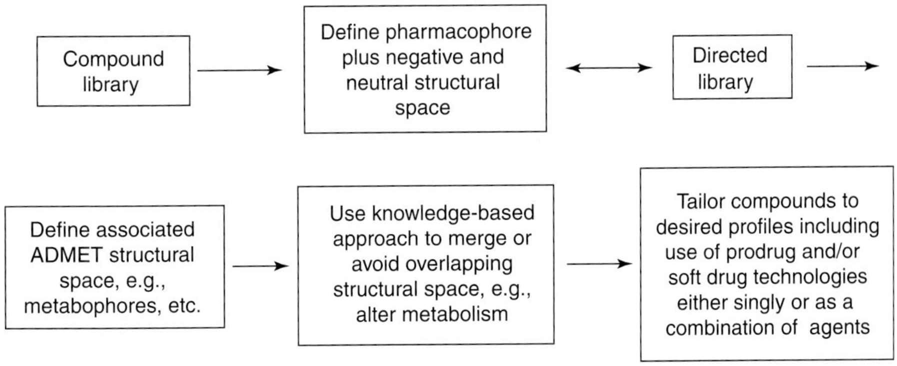

Figure 22 Future drug design and development strategy. Note that this strategy emphasizes a complete definition of the efficacy-related pharmacophore as the central theme such that it can be effectively merged with simultaneously generated ADMET-related pharmacophores via a knowledge-driven, proactive process that will ultimately produce the optimized clinical candidate or combination of agents to be deployed for therapy or prophylaxis. It should be noted that this 'knowledge-based' decision tree contrasts some of the futuristic schemes that have been suggested by others wherein the various ADMET issues are simply used as consecutive or simultaneous filters to eliminate compounds being selected from huge compound libraries or virtual databases. As elaborated within the text, it is these authors' opinion that by medicinal chemistry input, HTS data of the future will be able to take drug discovery investigations to significantly greater heights of knowledge such that the latter can then be used for proactively assembling the positive type of enhanced-property molecular constructs mentioned above. Indeed, if this scenario does not unfold, then the overall new and future drug discovery processes will be forever locked into a negative mode that simply keeps eliminating compounds failing to meet certain criteria placed at each parameter.

- Whereas a reductionist approach toward achieving the most simple, single-molecular system possible will remain appealing, it will no longer dictate the status quo for drug development. Thus, any combination of one or more agents wherein each member can independently use prodrug, soft drug or multivalent strategies, will also be considered within the context of deploying whatever is envisioned to ultimately work the best for the clinical indication at hand. This strategy is further depicted in **Figure 23**.
- 'A better understanding of [biochemical] data [coupled with a better appreciation for systems biology] within the context of [human] disease mechanisms will enable smaller, more targeted clinical trials with a higher success rate.'[448]
- As clinical successes unfold, and the aforementioned dark chasm presently associated with ADMET prediction is bridged, a gradual move from experimental to virtual screening will occur. While this is already being done for initial lead finding by docking 'druglike' virtual compound libraries into pockets defined by x-ray, virtual methods should eventually be able to essentially replace the entire preclinical, front-line testing paradigm.
- Palliative and curative targets will remain, the latter particularly for combating the ever-evolving populations of microorganisms and viruses, and the former for retaining the quality of life as the human lifespan continues to elongate. However, preventative and prophylactic treatment paradigms will gradually take on more significance and will eventually reside at the highest priority of life sciences research.
- For all treatments, and especially for preventative and prophylactic paradigms, pharmacogenomics will define population subgroups that will then receive treatment protocols optimized for their individuality relative to the particular treatment that is being rendered. Testing to ascertain an individual's pharmacogenetic profile will begin at birth and continue at periodic intervals throughout one's life, all relative to the optimal deployment of preventative protocols. Using array technologies, such testing will also be conducted immediately prior to and during any treatment of a detected pathophysiology.
- In the more distant future, the combination of nanotechnologies with bioengineering and biotechnology may allow instillation of devices not only for immediate diagnoses of any deviations from homeostasis, but also of programmable

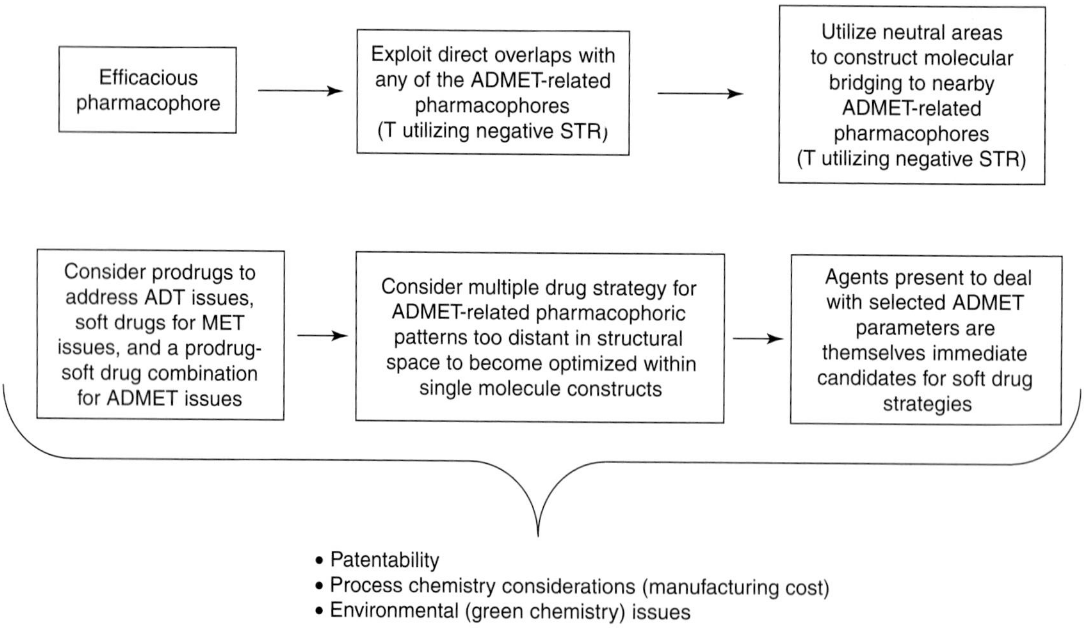

Figure 23 Lead selection and drug design decision flowchart. Decisions and design strategies are based upon efficacy and ADMET-related pharmacophoric parameters. This flowchart has been set up to represent the case where a single-molecule construct having the lowest level of complexity/sophistication is initially sought. However, in the future it is also likely that well-established templates that optimize a certain parameter will be able to be effectively paired with the efficacy-related agent at an early point in the overall design process. For example, a compound that inhibits a transporter system responsible for a given lead's poor passage through the gastrointestinal endothelium might be ideally coadministered as a soft drug version (**Figure 14**). In this way the partner compound would solve the oral absorption problem and then be quickly metabolized and eliminated without doing much of anything else. Structural manipulation of the efficacy construct could then be directed toward enhancing other DMET-related profiles.

portals for user-friendly drug administration other than via the oral route. Likewise, this same mix of technologies could eventually allow for instillation of programmable exits for the controlled elimination of drugs with or without the need for metabolism. Interestingly, such developments would ultimately cause the entire drug discovery process to return primarily to the pursuit of only efficacy and toxicity issues, i.e., a return to something similar to the classical drug discovery and development paradigm shown in **Figure 2**.

- Somewhere during gene therapy's complete eradication of defective gene-based disease and bolstering of gene-linked defenses, along with the nanotechnology/bioengineering effort to reconstruct humans relative to improving health, public policies and opinions dealing with what other attributes might be manipulated to enhance the 'quality of life' will need to be clarified by significant input from the nonbasic science disciplines. Thus, the social sciences, humanities, and philosophy fields, along with religion and the lay public at large, should look forward to providing what will soon become desperately needed input into the continued directions that life sciences research is likely to take well before the end of the next century, let alone before the conclusion of the present millennium.

1.02.5.2 Medicinal Chemistry's Niche within a Highly Interdisciplinary Process

In addition to applying the various drug discovery-related approaches listed in **Table 3**, medicinal chemistry's activities reside within a matrix of several other disciplines that have now also become intimately involved with the overall process of drug discovery. Indeed, it is the biotechnology-inspired trends that remain at the forefront of moving the drug discovery process forward. HTS has already led to the situation where there are now mountains of in vitro data available for input toward drug-related considerations. Within just the near term, the entire gamut of ADMET parameters can be expected to join efficacy surveys being conducted by HTS. Importantly, during this period the latter's output will also become validated in terms of predicting clinical correlates. The common link needed to bridge all these different informational data sets will be molecular structure knowledge as afforded by the probe compounds or compound library members that become deployed during a given assay. Molecular structure can be best appreciated by the precise language that medicinal chemistry has been learning since its formalization as a distinct discipline about 75 years ago. Thus, medicinal chemistry is obliged to also step to the forefront and assist in understanding and translating what the mountains of new data mean so that they might be optimally applied toward the development of new therapeutic agents.

That the aforementioned situation constitutes a rather unique niche for medicinal chemistry should be clear. Furthermore, that medicinal chemistry must approach this role via close collaborations with numerous other disciplines should also be very clear. For example, recognizing that we have not been very effective to date, the appropriate handling of 3D chemical structure within large databases represents a significant challenge that needs to be resolved by a cooperative effort between medicinal chemists and computational chemists along with both bioinformatic and chemoinformatic database experts as quickly as possible. That medicinal chemists have a good appreciation for the biological nature of the data within one mountain versus that of another is an equally challenging interdisciplinary problem that will need to be addressed by cooperative efforts between medicinal chemists and investigators from all of the biochemical and biological-related sciences. Resolving both of these challenges will eventually allow the in vitro data sets to be intermeshed so as to provide knowledge-generating assemblies that accurately predict the results that are eventually obtained in vivo and, ultimately, within the clinic.

1.02.5.3 Training Future Medicinal Chemists

Faced with these immediate and critical roles for medicinal chemistry within drug discovery research, how should academia be preparing doctoral-level investigators to contribute as medicinal chemists of the future? The first and foremost aim should be to retain medicinal chemistry's emphasis upon the physical organic principles that define chemical behavior in a given biological setting. This is fundamental to being a medicinal chemist. Such principles cannot be learned well by relying only upon textbook/e-instruction or even by predesigned laboratory outcome exposures. Thus, a laboratory-based thesis project that involves physical organic principles as the underlying variables of its scientific enquiries is mandated. While several types of chemical problems might be envisioned to provide such a learning experience, the laboratory practices of synthetic and physical organic chemistry represent extremely useful tools to permanently drive home the principles associated with intra- and intermolecular behavior and chemical reactivity. Likewise, with regard to synthesis/compound production, it is also extremely important to first learn how to isolate and characterize pure materials. Combinatorial mixtures and biochemical manipulations that rely upon chemical kinetics and the process of natural selection to dictate their concentrations can then be better appreciated if approached at a later point in time. Finally, while a multistep synthesis of a complex natural product can instill fundamental chemical principles, it may be more effective for a budding medicinal chemist to prepare one or more series of probes wherein most of the members in the series are novel in structure but are (seemingly) obtainable via reasonably close literature precedent according to short synthetic sequences, e.g., five or six steps to each template that is to then be further derivatized. The latter positions the student closer to eventually appreciating structural trends and patterns that may reside within databases.

While this solid foundation in organic synthetic chemistry is being derived from experimental laboratory work, graduate-level exposures to various other fields and aspects of life science research will, instead, have to rely upon available courses, seminars, or independent reading. Merging a student's chemical learning base with a specific biological area being targeted by the student's molecular probes, however, ought to be additionally feasible via actual experimentation without jeopardizing either subject's rigor. In the end, however, postgraduate, continuing education is probably the only way that an investigator intending to practice medicinal chemistry will be able to traverse the explosion of information occurring in all of the areas relevant toward assuming the roles needed to resolve the aforementioned challenges of the new drug discovery paradigm. That a practitioner may be able to have a head start along this learning path by initially pursuing a formalized medicinal chemistry curriculum rather than an organic chemistry curriculum, has been suggested by others.[449]

It should be emphasized that the broader exposure to the life sciences is a critical component for a medicinal chemist's continuing education not because medicinal chemists should eventually attempt to independently pursue each of such endeavors, but because these exposures will allow them to be able to interact and collaborate more meaningfully with dedicated experts in each of the numerous other fields. Thus, the ability of medicinal chemists to participate in interdisciplinary research while serving as scientific 'scholars' during their attempts to integrate knowledge across broad sets of data and scientific fields,[450] are key behaviors that also need to be instilled early in the overall, graduate-level educational process. By their very nature, medicinal chemistry experiments often prompt fundamental questions or hurdles that may be related to a variety of other disciplines. Thus, the interplay of the subject matter from various biological disciplines during medicinal chemistry research is as inherent to the broader medicinal chemistry intellectual process and notion of scholarship as is the practical requisite for a solid-based knowledge of fundamental physical organic principles accompanied by the ability to conduct chemical syntheses.

In the future, increasing numbers of formalized short programs pertaining to a given biological area are likely to be offered to practicing medicinal chemists at technical meetings, at academic centers, at corporate cites, and via e-instruction.[451] Given the interdisciplinary nature of the problems already at hand, along with the proposition that they will become significantly more complex as we progress further into the future, it is likely that companies that encourage such interdisciplinary types of continuing education will also eventually become the leaders that are able to most effectively implement the new paradigm of drug discovery, i.e., not just toward generating more data faster while working on smaller scale, but toward producing interdisciplinary knowledge bundles that actually lead to better NCEs at a quicker pace while spending less money.

1.02.5.4 Intellectual Property Considerations

As an appropriate part of closing this chapter, it becomes important to consider how all of the aforementioned technical and operational possibilities could impact upon where drug discovery and medicinal chemistry may be headed in terms of pharmaceutical intellectual property (IP).[452,453] Comments in this area will be directed only toward small-molecule compounds and not toward biomolecules despite the noted turmoil that was initially created in the gene-related IP arena.[454,455] As indicated earlier, the highly interdisciplinary nature of today's life science research endeavors, coupled with the new paradigm in drug discovery, indicates that the small-molecule, composition-of-matter-related IP arena is no longer the exclusive domain of medicinal chemistry. Nevertheless, even though the appropriate list of inventors for any given case that has utilized HTS and combinatorial chemistry could become quite complex, with patience these situations should all be reconcilable. Likewise, that there may already be reasonable recommendations for addressing the increasing costs in obtaining patents seems clear.[456] Alternatively, there are some other issues that are also beginning to hit the IP arena for which answers and precedented procedural models may not be as clear. Given that the desirable goal of enhancing world trade has prompted the need to recognize (if not to completely harmonize) patents on a global basis,[457,458] it is likely that the unique position held by the USA with regard to acknowledging notebook entries as the earliest dates of an invention's conception will ultimately give way to the more practical European process that simply acknowledges the first to file. However, this move will further encourage multinational-based companies to file patents on technologies that are less mature. For example, casting this possibility within the trends elaborated throughout this chapter, companies will need to resist the urge to file on complete compound libraries and instead focus upon claims that protect only a reasonable subfamily of leads for which several members have indeed been identified as being meritorious by both efficacy and selectivity and at least preliminary ADMET HTS, i.e., experimentally ascertained privileged structures for the indications (utilities) actually in hand. Unfortunately, an even worse scenario has already begun to unfold, namely in that applications appear to be pending and arguments are being directed toward the validity of patenting huge virtual libraries considered to be druglike in their make-up.[459] Emphasizing the notion that an actual reduction to practice should remain paramount for a patent, this chapter's authors presently stand in opposition to such attempts to garner protection of virtual libraries. Along this same line, the authors feel compelled to further note that the current requests for assignment of CAS numbers to virtual compounds also represents a step in the wrong direction.[459] Finally, while patent protection of an existing scaffold that has experimentally demonstrated its utility in one or more therapeutic areas is certainly meritorious, in the future, companies will still need to refrain from overelaborating these same scaffolds in an attempt to generate NCEs across several other therapeutic areas based solely upon already having secured IP protection within the context of compositions of matter. In other words, force-fitting a given scaffold via its array of appendage options, rather than by conducting a HTS survey of other structural systems across the complete profile of selective efficacy and ADMET parameters, could easily be taken as a step backwards in terms of both the molecular diversity and therapeutic quality that is ultimately being delivered to the marketplace.

1.02.5.5 A Knowledge versus Diversity Paradox

In this same regard, however, a seeming paradox will be created by the insertion of greater knowledge bundles into the new drug discovery paradigm. Since medicinal chemistry will seek to define SAR in terms of 3D electrostatic potentials that become predictive of preferred ADMET and efficacy and selectivity behaviors so that their various assemblages can lead to privileged drug structural motifs (or to ensembles of privileged structural motifs that are deployed as drug teams), once this process begins to become effective, it will also play against molecular diversity. While this situation is not nearly as limiting as the situation conveyed in the preceding section and while it will always be subject to an expansion of diversity based upon the uniqueness of a given efficacy pharmacophoric component, enhanced ADMET knowledge in particular, will indeed work in a direction away from overall diversity. Hopefully, however, the saving factor in this evolution will remain the pursuit of therapeutically preferred arrangements and not the overutilization of a particular motif just because it has been able to garner an exceptionally favorable ADMET profile, perhaps accompanied by a strong patent position as well (see above). Finally, while enhanced knowledge inherently leads to more credible and useful predictions, it is the overextrapolation, extra weight, or zeal that is sometimes placed upon a given prediction versus other options including that of having no prediction, that can become problematic. Thus, even when all of the challenges cited in this chapter appear to be resolved, the various disciplines caught up in drug discovery, including that of medicinal chemistry, should all remain cognizant of the earlier days of "preconceived notions" while also recalling the old adage that "a little bit of knowledge can sometimes be dangerous" such that when the ideal drugs/drug ensembles of the near-term future are constructed from experimental data, and those of the more distant future from virtual data, the subsequent, laboratory-based preclinical and clinical investigations, will still remain open to the possibility that, at any point along the way, anything might still be able to happen. Casting this last sentiment in a favorable direction, medicinal chemists of the future, no matter how knowledgeably and guided by wisdom the overall process of drug discovery may seem to have become, should always remain on the alert for serendipity.

1.02.6 Conclusions

Medicinal chemistry has been defined as both an applied and a basic science that considers fundamental physical organic principles to understand the interactions of small molecular displays with biological surfaces. Regardless of discipline or background, when investigators seek this level of understanding, they are embarking upon basic medicinal chemistry research. Using a variety of input data, medicinal chemistry's applications, in turn, become that of designing or selecting new drug candidates as well as providing molecular blueprints for improving the therapeutic profiles of existing drugs or of new pharmaceutical agent lead structures. Historically, medicinal chemistry has also been heavily involved with generating input data by designing and synthesizing probe molecules that can systematically test the roles being played by the various physical organic principles during a given interaction. More recently, medicinal chemistry has begun to utilize site-directed mutagenesis as an additional tool to understand these same types of interactions.

Having already spawned a new paradigm for discovering new drugs, biotechnology and its ever-growing list of spin-off disciplines continue to have a major impact upon drug discovery and medicinal chemistry. Although the initial steps in drug discovery appear to have become accelerated by such new technologies, the most recent trend analyses for the endpoint production of NCEs suggest that the latter have actually decreased. While many of these 'biotech' developments are still new on the horizon and should probably be considered to be in their own adolescent phase, the authors suggest that unless the overall process of drug discovery deliberately strives to constantly incorporate mechanisms for capturing chemical associated data, relationships and knowledge, the trend in NCE production may not improve significantly. In this regard, the authors have also emphasized that the role of central interpreter for bridging biological data with chemical data can be best played by no other discipline than medicinal chemistry. Thus, while it may indeed be true that "the next 100 years are going to become the century of biology within pharmaceutical discovery,"[460] it will be medicinal chemistry that ultimately makes sense out of all of the data so as to be able to definitively put it to some practical use in a systematic manner involving SARs that then become directly coupled to actual drug discovery within the context of producing new molecules that are endowed with therapeutic activity while also exhibiting an overall ADMET profile that is well suited for use in humans.

References

1. Erhardt, P. W. *Pure Appl. Chem.* **2002**, *74*, 703–785.
2. Cotton, F. A. *Chem. Eng. News* **2000**, *Dec 4*, 5.
3. Stokes, D. E. *Pasteur's Quadrant: Basic Science and Technological Innovation*; Brookings Press: Washington, DC, 1997.
4. Reisch, M. S. *Chem. Eng. News* **2001**, *Feb 12*, 15–17.
5. *The Pharmaceutical Century*; Jacobs, M., Newman, A., Ryan, J., Eds.; American Chemical Society: Washington, DC, 2000.
6. *Medicinal Chemistry Part 1*, 3rd ed.; Burger, A., Ed.; Wiley-Interscience: New York, 1970 (first published in 1951).
7. *Textbook of Organic Medicinal and Pharmaceutical Chemistry*; 6th ed.; Wilson, C. O., Gisvold, O., Doerge, R. F., Eds.; Lippincott: Philadelphia, PA, 1971 (first published in 1949).
8. *Inorganic Medicinal and Pharmaceutical Chemistry*; Block, J. H., Roche, E. B., Soine, T. O., Wilson, C. O., Eds.; Lee and Febiger: Philadelphia, PA, 1974.
9. Erhardt, P. W. Drug Metabolism Data Past Present and Future Considerations. In *Drug Metabolism: Databases and High-Throughput Testing During Drug Design and Development*; Erhardt, P. W., Ed.; IUPAC/Blackwell: Boston, MA, 1999, pp 2–15.
10. Erhardt, P. W. In *Chronicles of Drug Discovery*; Lednicer, D., Ed.; American Chemical Society: Washington, DC, 1993, pp 191–206.
11. Venuti, M. C. *Ann. Rep. Med. Chem.* **1990**, *25*, 289–298.
12. Oldenburg, K. R. *Ann. Rep. Med. Chem.* **1998**, *33*, 301–311.
13. Ausman, D. J. *Modern Drug Disc.* **2001**, *Jan*, 18–23.
14. Felton, M. J. *Modern Drug Disc.* **2001**, *Jan*, 24–28.
15. Hylands, P. J.; Nisbet, L. J. *Ann. Rep. Med. Chem.* **1991**, *26*, 259–269.
16. Dower, W. J.; Fodor, S. P. *Ann. Rep. Med. Chem.* **1991**, *26*, 271–280.
17. Moos, W. H.; Green, G. D.; Pavia, M. R. *Ann. Rep. Med. Chem.* **1993**, *28*, 315–324.
18. Gallop, M. A.; Barrett, R. W.; Dower, W. J.; Fodor, S.; Gordon, E. M. *J. Med. Chem.* **1994**, *37*, 1234–1251.
19. Gordon, E. M.; Barrett, R. W.; Dower, W. J.; Fodor, S.; Gallop, M. A. *J. Med. Chem.* **1994**, *37*, 1385–1401.
20. Terrett, N. K. *Combinatorial Chemistry*; American Chemical Society: New York, 1998.
21. Trainor, G. L. *Ann. Rep. Med. Chem.* **1999**, *34*, 267–286.
22. Kyranos, J. N.; Hogan, J. C., Jr. *Modern Drug Disc.* **1999**, *July/Aug*, 73–82.
23. Brown, R. K. *Modern Drug Disc.* **1999**, *July/Aug*, 63–71.
24. Borman, S. *Chem. Eng. News* **2000**, *May 15*, 53–65.
25. LeProust, E.; Pellois, J. P.; Yu, P.; Zhang, H.; Gao, X. *J. Comb. Chem.* **2000**, *2*, 349–354.
26. Scharn, D.; Wenschuh, H.; Reineke, U.; Schneider-Mergener, J.; Germeroth, L. *J. Comb. Chem.* **2000**, *2*, 361–369.
27. Dolle, R. E. *Mod. Drug Disc.* **2001**, *Feb*, 43–48 (the complete survey article which appeared in *J. Comb. Chem.* **2000**, *2*, 383–433 and includes 290 references, 10 tables, and 26 figures, is also available on the web at http://pubs.acs.org/hotarticl/jcchff/2000/cc000055x_rev.html).
28. Boguslavsky, J. *Drug Disc. Dev.* **1999**, *Sept*, 71.
29. Karet, G. *Drug Disc. Dev.* **2000**, *Mar*, 28–32.
30. U'Prichard, D. C. *Drug Disc. World* **2000**, *Fall*, 12–22.
31. *From Bench to Market*; Cabri, W., Di Fabio, R., Eds.; Oxford University Press: New York, 2000.
32. Knapman, K. *Modern Drug Disc.* **2000**, *June*, 75–77.
33. Hume, S. *Innov. Pharm. Tech.* **2000**, *6*, 108–111.
34. Court, J.; Oughton, N. *Innov. Pharm. Tech.* **2000**, *6*, 113–117.
35. M. D. Lemonick, *Time* **2001**, *Jan 15*, 58–67.
36. Zall, M. *Modern Drug Disc.* **2001**, *Mar*, 37–42.
37. Brennan, M. B. *Chem. Eng. News* **2000**, *Feb 14*, 81–84.
38. Paul Erhardt personal communications while visiting several big pharma during 2000–2004.
39. Strader, C. D. *J. Med. Chem.* **1996**, *39*, 1.
40. Hanson, D. *Chem. Eng. News* **2000**, *June 19*, 37.
41. Jacobs, M. *Chem. Eng. News* **2000**, *Oct 30*, 43–49.
42. Kubinyi, H. *Curr. Opin. Drug Disc. Dev.* **1998**, *1*, 4–15.
43. Murcko, M. A.; Caron, P. R.; Charifson, P. S. *Annu. Rep. Med. Chem.* **1999**, *34*, 297–306.
44. Rubenstein, K. *Drug and Market Development* **2000**, Report 1993, 1–200.
45. Josephson, J. *Modern Drug Disc.* **2000**, *May*, 45–50.
46. Karet, G. *Drug Disc. Dev.* **1999**, *Nov/Dec*, 71–74.
47. Lesney, M. S. *Modern Drug Disc.* **2000**, *May*, 31–32.

48. Brennan, M. B. *Chem. Eng. News* **2000**, *June 5*, 63–73.
49. F. Golden, *Time* **2001**, *Jan 15*, 74–76.
50. Thayer, A. *Chem. Eng. News* **2001**, *Feb 5*, 18.
51. Clinical trial information available on the Internet: Royal Coll. Physicians of Edinburgh at http://www.rcpe.ac.uk/cochrane/intro.html; Pharmaceutical Information Network at http://pharminfo.com/conference/clintrial/fda_1.html; PhRMA (association of 100 major pharmaceutical companies) at http://www.pharma.org/charts/approval.html; Univ. Pittsburgh IRB manual at http://www.ofreshs.upmc.edu/irb/preface.htm; Center Watch (a comprehensive venue and listing for clinical trials) at http://www.centerwatch.com; FDA Center for Drug Evaluation and Research (CDER) at http://www.fda.gov/cder; eOrange Book (approved drug products with therapeutic equivalence evaluations) at http://www.fda.gov/cder/ob.; FDA CDER Handbook at http://www.fda.gov/cder/handbook/index.htm; FDA CDER Fact Book 1997 at http://www.fda.gov/cder/reports/cderfact.pdf; FDA Consumer Magazine at http://www.fda.gov/fdac/fdacindex.html; FDA Guidelines on content and format for INDs at http://www.fda.gov/cder/guidance/ clin2.pdf (accessed Aug 2006).
52. Karet, G. *Drug Disc. Dev.* **2000**, *June/July*, 34–40.
53. Moran, S.; Stewart, L. *Drug Disc. World* **2000**, *Fall*, 41–46.
54. Wu, Z. P. Accelerating Lead and Solid Form Selection in Drug Discovery and Development Using Non-Ambient X-Ray Diffraction; New Chemical Technologies Satellite Program, 221st National ACS Meeting, March 2001.
55. Pearlman, D. A.; Charifson, P. S. *J. Med. Chem.* **2001**, *44*, 502–511.
56. Schoenlein, R. W.; Chattopadhyay, S.; Chong, H. H. W.; Glover, T. E.; Heimann, P. A.; Shank, C. V.; Zholents, A. A.; Zolotorev, M. S. *Science* **2000**, *287*, 2237–2240.
57. Wilson, E. *Chem. Eng. News* **2000**, *Mar 27*, 7–8.
58. Yarnell, A. *Chem. Eng. News* **2005**, *Feb 8*, 13.
59. Flores, T.; Garcia, D.; McKenzie, T.; Pettigrew, J. Planning for Knowledge-Led R&D. In *Innovations in Pharmaceutical Technology*; Barnacal, P. A., Ed.; Chancery Media: London, 2000, pp 18–22.
60. Studt, T. *Drug Disc. Dev.* **2000**, *Mar*, 7.
61. Hodgkin, E.; Andrews-Cramer, K. *Modern Drug Disc.* **2000**, *Jan/Feb*, 55–60.
62. Sleep, N. *Modern Drug Disc.* **2000**, *July/Aug*, 37–42.
63. Karet, G. *Drug Disc. Dev.* **1999**, *Nov/Dec*, 39–43.
64. Boguslavsky, J. *Drug Disc. Dev.* **2000**, *Mar*, 67.
65. Young, S.; Li, J. *Drug Disc. Dev.* **2000**, *Apr*, 34–37.
66. Brennan, M. B. *Chem. Eng. New.* **2000**, *June 5*, 63–73.
67. Estrada, E.; Uriarte, E.; Montero, A.; Teijeira, M.; Santana, L.; De Clercq, E. *J. Med. Chem.* **2000**, *43*, 1975–1985.
68. Boehm, H.-J.; Boehringer, M.; Bur, D.; Gmuender, H.; Huber, W.; Klaus, W.; Kostrewa, D.; Kuehne, H.; Luebbers, T.; Meunier-Keller, N. et al. *J. Med. Chem.* **2000**, *43*, 2664–2674.
69. Bissantz, C.; Folkers, G.; Rognan, D. *J. Med. Chem.* **2000**, *43*, 4759–4767.
70. Waterbeemd, H. vande.; Gifford, E. *Nat. Rev.* **2003**, *2*, 192–204.
71. Lombardo, F.; Gifford, E.; Shalaeva, M. Y. *Mini Rev. Med. Chem.* **2003**, *3*, 861–875.
72. Brown, F. K. *Annu. Rep. Med. Chem.* **1998**, *33*, 375–384.
73. Hann, M.; Green, R. *Curr. Opin. Chem. Biol.* **1999**, *3*, 379–383.
74. Ertl, P.; Miltz, W.; Rohde, B.; Selzer, P. *Drug Disc. World* **2000**, *1*, 45–50.
75. ChemFinder at http://www.chemfinder.com; Chem Navigator at http://www.chemnavigator.com; Info Chem GmbH at http://www.info-chem.com; Institute for Scientific Info. at http://www.isinet.com; MDL Information Systems at http://www.mdli.com; MSI at http://www.msi.com; Oxford Molecular at http://www.oxmol.com; Pharmacopeia at http://www.pharmacopeia.com; QsarIS at http://www.scivision.com/qsaris.html; Synopsys Scientific Systems at http://www.synopsys.co.uk; Tripos at www.tripos.com (accessed Aug 2006).
76. Interactive video conference the role of higher education in economic development. *Old Dominion Univ. Academ. Television Serv.* **1997**, *Jan. 31*, http://www.utoledo.edu (accessed Aug 2006).
77. Erhardt, P. W. Preface. In *Drug Metabolism:- Databases and High-Throughput Testing during Drug Design and Development*; Erhardt, P. W., Ed.; IUPAC/Blackwell Science: London, 1999, pp viii–ix.
78. Ausman, D. *Molecular Connection (MDL 's Newsletter)* **1999**, *18*, 3.
79. Recupero, A. J. *Drug Disc. Dev.* **1999**, *Nov/Dec*, 59–62.
80. Agres, T. *Drug Disc. Dev.* **2000**, *Jan/Feb*, 11–12.
81. Studt, T. *Drug Disc. Dev.* **2000**, *Jan/Feb*, 68–69.
82. Zarrabian, S. *Chem. Eng. News* **2000**, *Feb* 7.
83. Thayer, A. M. *Chem. Eng. News* **2000**, *Feb* 7, 19–32.
84. Tye, S.; Karet, G. *Drug Disc. Dev.* **2000**, *Mar*, 35.
85. Schulz, W. *Chem. Eng. News* **2001**, *Feb 5*, 25.
86. DARPA at http://www.darpa.mil; FAO at http://www.fao.org/GENINFO/partner/defautl.htm; Gen Bank at http://www.ncbi.nlm.nih.gov/entrez/nucleotide.html; DDBJ at http://www.ddbj.nig.ac.jp/; EMBL at http://www.ebi.ac.uk/ebi_docs/embl_db/ebi/topembl.html; Human Genome Sequencing at http://www.ncbi.nlm.nih.gov/genome/seq; KEEGG 12.0 at http://www.genome.ad.jp/kegg/; NASA at http://www.nasa.gov; Nat. Acad. Sci. STEP at http://www4.nas.edu/pd/step.nsf; Nat. Ctr. Biotech. Info. at http://www.ncbi.nlm.nih.gov; Nat. Human Genome Res. Inst. at http://www.nhgri.nih.gov/index.html; Nat. Sci. Fndn. Math/Phys. Sci. at http://www.nsf.gov/home/mps/start.htm; Net Genics 'SYNERGY' at http://www.netgenics.com/SYNERGY.html; Pfam 4.3 at pfam.wustl.edu/; PROSITE at http://www3.oup.co.uk/nar/Volume_27/Issue_01/html/gkc073_gml.html#hdO; Sanger Ctr. At http://www.sanger.ac.uk/; SRS at http://www.embl-heidelberg-de/srs51; TIGER Database at http://www.tigr.org/tdb/tdb.html; UNESCO at http://www.unesco.org/general/eng./programmes/science/life/index.htm; Univ. Cal. Berkeley at http://www.Berkeley.edu (accessed Aug 2006).
87. Paul Erhardt personal communications while attending several recent workshops directed toward expediting drug development.
88. Protein Data Bank. http://www.rcsb.org/bdb/ (accessed Aug 2006).
89. Biotech Validation Suite For Protein Structure at http://biotech.embl-heidelberg.de:8400/; BLAST at http://www.ncbi.nlm.nih.gov/BLAST/; Build Proteins From Scratch at http://csb.stanford.edu/levitt and http://www.chem.cornell.edu/department/Faculty/Scheraga/Scheraga.html; CATH at http://www.biochem.ucl.ac.uk/bsm/cath/; FASTA at http://www2.ebi.ac.uk/fasta3/; MODELLER at http://guitar.rockfeller.edu; PDB at http://www.pdb.bnl.gov/; Pedant at MIPS at http://pedant.mips.biochem.mpg.del; PIR at http://www.nbrf.georgetown.edu/pir/searchdb.html; Predict Protein at http://www.embl-heidelber.de/predictprotein/predictprotein.html; PSIPRED at http://insulin.brunel.ac.uk/

psipred; Rosetta at http://depts.Washington.edu/bakerpg; SAM-98 at http://www.cse.ucsc.edu/research/compbio; SCOP at http://scop.mrc-lmb.cam.ac.uk/sopl; Structure at http://www.ncbi.nlm.nih.gov/Structure/RESEARCH/res.shtml; Swiss-Model at http://expasy.hcuge.ch/swissmod/SWISSMODEL.html; SWISS-PROT at http://expasy.hcuge.ch/sprot/sprottop.html; Threader at http://insulin.brunel.ac.uk/threader/threader.html (accessed Aug 2006).

90. Schulz, G. E.; Schirmer, R. H. Prediction of Secondary Structure from the Amino Acid Sequence. In *Principles of Protein Structure*; Cantor, C. R., Ed.; Springer-Verlag: New York, 1999, pp 108–130.
91. Wilson, E. K. *Chem. Eng. News* **2000**, *Sept 25*, 41–44.
92. Holliman, L. *Modern Drug Disc.* **2000**, *Nov/Dec*, 41–46.
93. Felton, M. J. *Modern Drug Disc.* **2000**, *Nov/Dec*, 49–54.
94. Vieth, M.; Cummins, D. J. *J. Med. Chem.* **2000**, *43*, 3020–3032.
95. Kastenholz, M. A.; Pastor, M.; Cruciani, G.; Haaksma, E.; Fox, T. *J. Med. Chem.* **2000**, *43*, 3033–3044.
96. Karet, G. *Drug Disc. Dev.* **2001**, *Jan*, 28–32.
97. Paul Erhardt personal communications while attending several recent scientific conferences in the area of medicinal chemistry.
98. ChemPen 3D at http://home.ici.net/~hfevans/chempen.htm; CS Chem Draw and CS Chem Draw 3D at http://www.camsoft.com; Corina at http://www.mol-net.de (accessed Aug 2006).
99. Nezlin, A. *Chem. News Com.* **1999**, *9*, 24–25.
100. Endres, M. *Today's Chem. at Work* **1992**, *Oct*, 30–44.
101. Karet, G. *Drug Disc. Dev.* **2000**, *Aug/Sept*, 42–48.
102. Rogers, R. S. *Chem. Eng. News* **1999**, *Apr 26*, 17–18.
103. Hansch, C.; Fujita, T. *J. Am. Chem. Soc.* **1964**, *86*, 1616–1626.
104. Koshland, D. E., Jr.; Neet, K. E. *Annu. Rev. Biochem.* **1968**, *37*, 359–410.
105. Belleau, B. *Adv. Drug Res.* **1965**, *2*, 89–126.
106. QUANTA (system for determining 3D protein structure from x-ray data) and FELIX (system for determining 3D protein structure from NMR data) available from MSI-Pharmacopeia. http://www.msi.com (accessed Aug 2006).
107. Hajduk, P. J.; Boyd, S.; Nettesheim, D.; Nienaber, V.; Severin, J.; Smith, R.; Davidson, D.; Rockway, T.; Fesik, S. W. *J. Med. Chem.* **2000**, *43*, 3862–3866.
108. Tang, K.; Fu, D.-J.; Julien, D.; Braun, A.; Cantor, C. R.; Koster, H. *Proc. Natl. Acad. Sci. USA* **1999**, *96*, 10016–10020.
109. Griffey, R. H.; Hofstadler, S. A.; Sannes-Lowery, K. A.; Ecker, D. J.; Crooke, S. T. *Proc. Natl. Acad. Sci. USA* **1999**, *96*, 10129–10133.
110. Karet, G. *Drug Disc. Dev.* **2000**, *Oct*, 38–40.
111. Montanari, M. L. C.; Beezer, A. E.; Montanari, C. A.; Pilo-Veloso, D. *J. Med. Chem.* **2000**, *43*, 3448–3452.
112. Medek, A.; Hajduk, P. J.; Mack, J.; Fesik, S. W. *J. Am. Chem. Soc.* **2000**, *122*, 1241–1242.
113. Rouhi, M. *Chem. Eng. News* **2000**, *Feb 21*, 30–31.
114. Oostenbrink, B. C.; Pitera, J. W.; van Lipzig, M.; Meerman, J. H.; van Gunsteren, W. F. *J. Med. Chem.* **2000**, *43*, 4594–4605; Wu, N.; Mo, Y.; Gao, J.; Pai, E. F. *Proc. Natl. Acad. Sci USA* **2000**, *97*, 2017–2022; Warshel, A.; Strajbl, M.; Villa, J.; Florian, J. *Biochem.* **2000**, *39*, 14728–14738.
115. *Topological Indices and Related Descriptors in QSAR and QSPR;* Devillers J., Balaban, A.T., Eds.; Gordon and Breach: Amsterdam, 1999.
116. Wrotnowski, C. *Modern Drug Disc.* **1999**, *Nov/Dec*, 46–55.
117. Henry, C. M. *Chem. Eng. News* **2000**, *Nov 27*, 22–26.
118. Wilson, E. K. *Chem. Eng. News* **2000**, *Nov 6*, 35–39.
119. Boyd, D. B. *Modern Drug Disc.* **1998**, *Nov/Dec*, 41–48.
120. Paul Erhardt personal communications while visiting several big pharma and while attending recent scientific conferences in the area of medicinal chemistry.
121. Evans, B. E.; Rittle, K. E.; Bock, M. G.; DiPardo, R. M.; Freidinger, R. M.; Whitter, W. L.; Lundell, G. F.; Veber, D. F.; Anderson, P. S.; Chang, R. S. L. et al. *J. Med. Chem.* **1988**, *31*, 2235–2246.
122. Patchett, A. A.; Nargund, R. P. *Annu. Rep. Med. Chem.* **2000**, *35*, 289–298.
123. Pastor, M.; Cruciani, G.; McLay, I; Pickett, S.; Clementi, S. *J. Med. Chem.* **2000**, *43*, 3233–3243.
124. Gnerre, C.; Catto, M.; Leonetti, F.; Weber, P.; Carrupt, P.-A.; Altomare, C.; Carotti, A.; Testa, B. *J. Med. Chem.* **2000**, *43*, 4747–4758.
125. Wilcox, R. E.; Huang, W. H.; Brusniak, M. Y.; Wilcox, D. M.; Pearlman, R. S.; Teeter, M. M.; DuRand, C. J.; Wiens, B. L.; Neve, K. A. *J. Med. Chem.* **2000**, *43*, 3005–3019.
126. Jayatilleke, P. R. N.; Nair, A. C.; Zauhar, R.; Welsh, W. J. *J. Med. Chem.* **2000**, *43*, 4446–4451.
127. Carlson, H. A.; Masukawa, K. M.; Rubins, K.; Bushman, F. D.; Jorgensen, W. L.; Lins, R. D.; Briggs, J. M.; McCammon, J. A. *J. Med. Chem.* **2000**, *43*, 2100–2114.
128. Bradley, E. K.; Beroza, P.; Penzotti, J. E.; Grootenhuis, P. D.; Spellmeyer, D. C.; Miller, J. L. *J. Med. Chem.* **2000**, *43*, 2770–2774.
129. Graffner-Nordberg, M.; Marelius, J.; Ohlsson, S.; Persson, A.; Swedberg, G.; Andersson, P.; Andersson, S. E.; Aqvist, J.; Hallberg, A. *J. Med. Chem.* **2000**, *43*, 3852–3861.
130. Garcia-Nieto, R.; Manzanares, I.; Cuevas, C.; Gago, F. *J. Med. Chem.* **2000**, *43*, 4367–4369.
131. Vedani, A.; Briem, H.; Dobler, M.; Dollinger, H.; McMasters, D. R. *J. Med. Chem.* **2000**, *43*, 4416–4427.
132. Lopez-Rodriguez, M. L.; Morcillo, M. J.; Fernandez, M.; Rosada, M. L.; Pardo, L.; Schaper, K.-J. *J. Med. Chem.* **2001**, *44*, 198–207.
133. Wedin, R. *Modern Drug Disc.* **1999**, *Sept/Oct*, 39–47.
134. Ladd, B. *Modern Drug Disc.* **2000**, *Jan/Feb*, 46–52.
135. Venkatsubramanian, V. Computer-Aided Molecular Design Using Neural Networks and Genetic Algorithms. In *Genetic Algorithms in Molecular Modeling*; Devillers, J., Ed.; Academic Press: New York, 1996, pp 35–70.
136. Globus, A. *Nanotechnology* **1999**, *10*, 290–299.
137. Studt, T. *Drug Disc. Dev.* **2000**, *Aug/Sept*, 30–36.
138. Resnick, R. *Drug Disc. Dev.* **2000**, *Oct*, 51–52.
139. Jaen-Oltra, J.; Salabert-Salvador, T.; Garcia-March, F. J.; Perez-Gimenez, F.; Tohmas-Vert, F. *J. Med. Chem.* **2000**, *43*, 1143–1148.
140. Bio Reason at http://www.bioreason.com; Columbus Molecular Software (Lead Scope) at http://www.columbus-molecular.com; Daylight Chemical Information Systems at http://www.daylight.com; IBM (Intelligent Miner) at http://www.ibm.com; Incyte (Life Tools and Life Prot) at http://www.incyte.com; Molecular Applications (Gene Mine 3.5.1) at http://www.mag.com; Molecular Simulations at http://www.msi.com; Oxford Molecular (DIVA 1.1) at http://www.oxmol.com; Pangea Systems (Gene World 3.5) at http://www.pangeasystems.com; Pharsight at http://www.pharsight.com; SAS Institute (Enterprise Miner) at http://www.sas.com; SGI (Mine Set 3.0) at http://www.sgi.com;

Spotfire (Spotfire Pro 4.0 and Leads Discover) at http://www.spotfire.com; SPSS (Clementine) at http://www.spss.com; Tripos at http://www.tripos.com (accessed Aug 2006).
141. Johnson, M. A.; Gifford, E.; Tsai, C.-C. In *Concepts and Applications of Molecular Similarity*; Johnson, M. A., Ed.; John Wiley: New York, 1990, pp 289–320.
142. Gifford, E.; Johnson, M.; Tsai, C.-C. *J. Comput.-Aided Mol. Des.* **1991**, *5*, 303–322.
143. Gifford, E. M.; Johnson, M. A.; Kaiser, D. G.; Tsai, C.-C. *J. Chem. Inf. Comput. Sci.* **1992**, *32*, 591–599.
144. Gifford, E. M.; Johnson, M. A.; Kaiser, D. G.; Tsai, C.-C. *Xenobiotica* **1995**, *25*, 125–146.
145. Gifford, E.M. Applications of Molecular Similarity Methods to Visualize Xenobiotic Metabolism Structure–Reactivity Relationships. Ph.D. Thesis, University of Toledo, Toledo, OH, USA, 1996.
146. Andrews, K. M.; Cramer, R. D. *J. Med. Chem.* **2000**, *43*, 1723–1740.
147. Borowski, T.; Krol, M.; Broclawik, E.; Baranowski, T. C.; Strekowski, L.; Mokrosz, M. J. *J. Med. Chem.* **2000**, *43*, 1901–1909.
148. Wintner, E. A.; Moallemi, C. C. *J. Med. Chem.* **2000**, *43*, 1933–2006.
149. Stoughton, R. B.; Friend, S. H. *Nat. Rev. Drug Disc.* **2005**, *4*, 345–350.
150. Prosser, H.; Rastan, S. *Trends Biotechnol.* **2003**, *3*, 224–232.
151. Zambrowicz, B. P.; Sands, A. T. *Nat. Rev. Drug Disc.* **2003**, *2*, 38–51.
152. Harris, S. *Drug Disc. Today* **2001**, *6*, 628–636.
153. Zambrowicz, B. P.; Sands, A. *Drug Disc. Today: Targets* **2004**, 198–207.
154. Elbashir, S. M.; Harborth, J.; Lendeckel, W.; Yalcin, A.; Weber, K.; Tuschl, T. *Nature* **2001**, *4*, 494–498.
155. Dorsett, Y.; Tuschl, T. *Nat. Rev. Drug Disc.* **2004**, *3*, 318–329.
156. Lindvall, J. M.; Blomberg, K.; Emelie, M.; Vaeliaho, J.; Vargas, L.; Heinonen, J. E.; Bergloef, A; Mohamed, A. J.; Nore, B. F.; Vihinen, M. et al. *Immunol. Rev.* **2005**, *203*, 200–215.
157. Lapteva, N.; Yang, A.-G.; Sanders, D. E.; Strube, R. W.; Chen, S.-Y. *Cancer Gene Ther.* **2005**, *12*, 84–89.
158. Li, S.-L.; Dwarakanath, R. S.; Cai, Q.; Lanting, L.; Natarajan, R. *J. Lipid Res.* **2005**, *46*, 220–229.
159. Aza-Blanc, P.; Cooper, C. L.; Wagner, K.; Batalov, S.; Deveraux, Q. L.; Cooke, M. P. *Mol. Cell* **2003**, *12*, 627–-637.
160. Xin, H.; Bernal, A.; Amato, F. A.; Pinhasov, A.; Kauffman, J.; Brenneman, D. E.; Derian, C. K.; Andrade-Gordon, P.; Plata-Salaman, C. R.; Ilyin, S. E. *J. Biomol. Screen.* **2004**, *9*, 286–293.
161. Harding, M. W.; Galat, A.; Uehling, D. E.; Schreiber, S. L. *Nature* **1989**, *341*, 758–760.
162. Becker, F.; Murthi, K.; Smith, C.; Come, J.; Costa-Roldan, N.; Kaufmann, C.; Hanke, U.; Degenhart, C.; Baumann, S.; Wallner, W. et al. *Chem. Biol.* **2004**, *11*, 211–223.
163. Bouwmeester, T.; Bauch, A.; Ruffner, H.; Angrand, P.-O.; Bergamini, G.; Croughton, K.; Cruciat, C.; Eberhard, D.; Gagneur, J.; Ghidelli, S. et al. *Nat. Cell Biol.* **2004**, *6*, 97–105.
164. Colland, F.; Jacq, X.; Trouplin, V.; Mougin, C.; Groizeleau, C.; Hamburger, A.; Meil, A.; Wojcik, J.; Legrain, P.; Gauthier, J.-M. *Genome Res.* **2004**, *14*, 1324–1332.
165. Li, S.; Armstrong, C. M.; Bertin, N.; Ge, H.; Milstein, S.; Boxem, M.; Vidalain, P. O.; Han, J.-D. J.; Chesneau, A.; Hao, T. et al. *Science* **2004**, *303*, 540–544.
166. Mancinelli, R.; Cronin, M.; Sadee, W. *Pharma. Sci.* **2000**, as an electronic reference.
167. Robert, J.; Vekris, A.; Pourquier, P.; Bonnet, J. *Crit. Rev. Oncol. Hematol.* **2004**, *51*, 205–227.
168. Bugelski, P. J. *Curr. Opin. Drug Disc.* **2002**, *5*, 79–89.
169. Samson, M.; Libert, F.; Doranz, B. J.; Rucker, J.; Liesnard, C.; Farber, C.-M.; Saragosti, S.; Lapouméroulie, C.; Cognaux, J.; Forceille, C. et al. *Nature* **1996**, *382*, 722.
170. Mullin, R. *Chem. Eng. News* **2005**, *Feb 28*, 29–39 (Based upon original data provided by the Center for Medicines Research International, IMS Health).
171. Hegde, S.; Carter, J. *Annu. Rep. Med. Chem.* **2004**, *39*, 337–368.
172. Handen, J. S. *Drug Disc. Today* **2002**, 7, 83–85; Van de Waterbeemd, H. *Curr. Opin. Drug Disc. Dev.* **2002**, *5*, 33–43; Chanda, S. K.; Cal dwell, J. S. *Drug Disc. Today* **2003**, *8*, 168–174; Drews, J. *Drug Disc. Today* **2003**, *8*, 411–420; FDA. *FDA White Paper* **2004**; Van den Haak, E. A. *CMR Report* **2004**, *04-234R*.
173. Sams-Dodd, F. *Drug Disc. Today* **2005**, *10*, 139–147.
174. Harness, J. In *Innovations in Pharmaceutical Technology*; Barnacal, P. A., Ed.; Chancery Media: London, 2000, pp 37–45.
175. Boguslavsky, J. *Drug Disc. Dev.* **2000**, *Aug/Sept*, 51–54.
176. Boguslavsky, J. *Drug Disc. Dev.* **2000**, *Oct*, 54–58.
177. Gibson, J.; Karet, G. *Drug Disc. Dev.* **2000**, *June/July*, 61–64.
178. Paul Erhardt personal communications during the 175th Anniversary USP Convention held in Washington, DC: March, 1995.
179. *Green Chemistry: Frontiers in Benign Chemical Syntheses and Processes*; Anastas, P., Williamson, T. C., Eds.; Oxford University Press: Cary, NC, 1998.
180. Anastas, P.; Warner, J. *Green Chemistry: Theory and Practice*; Oxford University Press: Cary, NC, 1998.
181. *Green Chemical Syntheses and Process* (Part 1) and *Green Engineering* (Part 2); Anastas, P. T., Heine, L. G., Williamson, T. C., Eds.; Oxford University Press: Cary, NC, 2000.
182. *Green Chemistry: Challenging Perspectives*; Tundo, P., Anastas, P., Eds.; Oxford University Press: Cary, NC, 2000.
183. Stinson, S. C. *Chem. Eng. News* **2000**, *Oct 23*, 55–78.
184. Auclair, K.; Sutherland, A.; Kennedy, J.; Witter, D. J.; Van den Heever, J. P.; Hutchinson, C. R.; Vederas, J. C. *J. Am. Chem. Soc.* **2000**, *122*, 11519–11520.
185. Watanabe, K.; Oikawa, H.; Yagi, K.; Ohashi, S.; Mie, T.; Ichihara, A.; Honma, M. *J. Biochem.* **2000**, *127*, 467–473.
186. Watanabe, K.; Mie, T.; Ichihara, A.; Oikawa, H.; Honma, M. *J. Biol. Chem.* **2000**, *275*, 38393–38401.
187. Borman, S. *Chem. Eng. News* **2001**, *Jan 15*, 9.
188. Yalkowski, S. H.; Valvani, S. C. *J. Pharm. Sci.* **1980**, *69*, 912–922.
189. Mason, R. P.; Rhodes, D. G.; Herbette, L. G. *J. Med. Chem.* **1991**, *34*, 869–877.
190. Herbette, L. G.; Rhodes, D. G.; Mason, R. P. *Drug Des. Deliv.* **1991**, 7, 75–118.
191. Abraham, M. H.; Chada, S.; Mitchell, R. *J. Pharm. Sci.* **1994**, *83*, 1257–1268.
192. Basak, S. C.; Gute, B. D.; Drewes, L. R. *Pharm. Res.* **1996**, *13*, 775–778.
193. Kansy, M.; Senner, F.; Gubernator, K. *J. Med. Chem.* **1998**, *41*, 1007–1010.
194. Palm, K.; Luthman, K.; Ungell, A.-L.; Strandlund, G.; Beigi, F.; Lundahl, P.; Artursson, P. *J. Med. Chem.* **1998**, *41*, 5382–5392.

195. Smith, D. A.; Van de Waterbeemd, H. *Curr. Opin. Chem. Biol.* **1999**, *3*, 373–378.
196. Stewart, B. H.; Wang, Y.; Surendran, N. *Annu. Rep. Med. Chem.* **2000**, *35*, 299–307.
197. Crivori, P.; Cruciani, G.; Carrupt, P.-A.; Testa, B. *J. Med. Chem.* **2000**, *43*, 2204–2216.
198. Yoshida, F.; Topliss, J. G. *J. Med. Chem.* **2000**, *43*, 2575–2585.
199. Lombardo, F.; Shalaeva, M. Y.; Tupper, K. A.; Gao, F.; Abraham, M. H. *J. Med. Chem.* **2000**, *43*, 2922–2928.
200. Ertl, P.; Rohde, B.; Selzer, P. *J. Med. Chem.* **2000**, *43*, 3714–3717.
201. Egan, W. J.; Merz, K. M., Jr.; Baldwin, J. J. *J. Med. Chem.* **2000**, *43*, 3867–3877.
202. Chait, A. *Drug Disc. Dev.* **2000**, *Apr*, 63.
203. Pickering, L. *Drug Disc. Dev.* **2001**, *Jan*, 34–38.
204. Absorption Systems 'BCS Biowaivers' at http://www.absorption.com; ArQule 'Pilot' at http://www.argule.com; Camitro Corp. 'ADME/Tox' at http://www.camitro.com; Cerep SA 'BioPrints' at http://www.cerep.fr; Exon Hit Therapeutics SA 'Genetic Makeup/Drug Toxicity' at http://www.exonhit.com; LION Bioscience AG 'i-Biology at SRS databases' at http://www.lionbioscience.com; Quintiles 'Early ADMET Screening' at qkan.busdev@quintiles.com; Schrodinger, Inc. 'QikProp' at http://www.schrodiner.com; Simulations Plus 'GastroPlus' at http://www.simulationsplus.com; Trega Biosciences 'In vitro ADME' at http://www.trega.com (accessed Aug 2006).
205. *Transport Processes in Pharmaceutical Systems;* Amidon, G. L., Lee, P. I., Topp, E. M., Eds.; Marcel Dekker: New York, 1999.
206. *Membrane Structure in Disease and Drug Therapy;* Zimmer, G., Ed.; Marcel Dekker: New York, 2000.
207. *Oral Drug Absorption: Prediction and Assessment;* Dressman, J. B., Lennernas, H., Eds.; Marcel Dekker: New York, 2000.
208. Taylor, E. W.; Gibbons, J. A.; Braeckman, R. A. *Pharm. Res.* **1997**, *14*, 572–577.
209. Gibbons, J. A.; Taylor, E. W.; Dietz, C. M.; Luo, Z.-P.; Luo, H.; Braeckman, R. A. High-Throughput Screening for Gastrointestinal Absorption in the Generation of SAR Databases. In *Drug Metabolism: Databases and High Throughput Testing during Drug Design and Development*; Erhardt, P. W., Ed.; IUPAC/Blackwell Science: Oxford, 1999, pp 71–78.
210. Rouhi, M. *Chem. Eng. News* **2001**, *Jan 29*, 10.
211. Janout, V.; DiGiorgio, C.; Regen, S. L. *J. Am. Chem. Soc.* **2000**, *122*, 2671–2672.
212. Wender, P. A.; Mitchell, D. J.; Pattabiraman, K.; Pelkey, E. T.; Steinman, L.; Rothbard, J. B. *Proc. Natl. Acad. Sci. USA* **2000**, *97*, 13003–13008.
213. Rothbard, J. B.; Garlington, S.; Lin, Q.; Kirschberg, T.; Kreides, E.; McGrove, P. L.; Wendy, P. A.; Kharari, P. A. *Nat. Med.* **2000**, *6*, 1253–1257.
214. Henry, C. M. *Chem. Eng. News* **2000**, *Oct 23*, 85–100.
215. Rouhi, M. *Chem. Eng. News* **2000**, *Sept 4*, 43.
216. Dagani, R. *Chem. Eng. News* **2000**, *Oct 16*, 27–32.
217. Schulz, W. *Chem. Eng. News* **2000**, *Oct 16*, 39–42.
218. Zurer, P. *Chem. Eng. News* **2001**, *Mar 12*, 12.
219. *Controlled Drug Delivery;* Park, K., Mrsny, R. J., Eds.; American Chemical Society: Washington, DC, 2000.
220. Mort, M. *Med. Drug Disc.* **2000**, *Apr*, 30–34.
221. Henry, C. M. *Chem. Eng. News* **2000**, *Sept 18*, 49–65.
222. Bracht, S. *Innov. Pharm. Tech.* **2000**, *5*, 92–98.
223. Levy, A. *Innov. Pharm. Tech.* **2000**, *5*, 100–109.
224. Southall, J.; Ellis, C. *Innov. Pharm. Tech.* **2000**, *5*, 110–115.
225. Winnips, C.; Keller, M. *Innov. Pharm. Tech.* **2000**, *6*, 70–75.
226. Winnips, C. *Drug Disc. World* **2000/2001**, *2*, 62–66.
227. Watkins, K. J. *Chem. Eng. News* **2000**, *Jan 8*, 11–15.
228. Jacoby, M. *Chem. Eng. News* **2001**, *Feb 5*, 30–35.
229. Lipinski, C. A.; Lombardo, F.; Dominy, B. W.; Feeney, P. J. *Adv. Drug Deliv. Rev.* **1997**, *23*, 3–25.
230. Erhardt, P. W. Metabolism Prediction. In *High-Throughput ADMETox Estimation: In vitro and In Silico Approaches*; Darvas, F., Dorman, G., Eds.; BioTechniques Press/Easton Publication: Westborough, MA, 2002, pp 41–48.
231. Frostell-Karlsson, A.; Remaeus, A.; Roos, H.; Andersson, K.; Borg, P.; Hamalainen, M.; Karlsson, R. *J. Med. Chem.* **2000**, *43*, 1986–1992.
232. Diaz, N.; Suarez, D.; Sordo, T. L.; Merz, K. M., Jr. *J. Med. Chem.* **2001**, *44*, 250–260.
233. Hajduk, P. J.; Bures, M.; Praestgaard, J.; Fesik, S. W. *J. Med. Chem.* **2000**, *43*, 3443–3447.
234. Czech, M. P. *Chem. Ind. Inst. Tox. Activities* **2000**, *May/June*, 2–4.
235. Dechantsreiter, M. A.; Planker, E.; Matha, B.; Lohof, E.; Holzemann, G.; Jonczyk, A.; Goodman, S. L.; Kessler, H. *J. Med. Chem.* **1999**, *42*, 3033–3040.
236. Pasqualini, R.; Koivunen, E.; Kain, R.; Lahdenranta, J.; Sakamoto, M.; Stryhn, A.; Ashmun, R. A.; Shapiro, L. H.; Arap, W.; Ruoslahti, E. *Cancer Res.* **2000**, *60*, 722–727.
237. Eberhard, A.; Kahlert, S.; Goede, V.; Hemmerlein, B.; Plate, K. H.; Augustin, H. G. *Cancer Res.* **2000**, *60*, 1388–1393.
238. Wyder, L.; Vitaliti, A.; Schneider, H.; Hebbard, L. W.; Moritz, D. R.; Wittmer, M.; Ajmo, M.; Klemenz, R. *Cancer Res.* **2000**, *60*, 4682–4688.
239. Matsuda, M.; Nishimura, S.-I.; Nakajima, F.; Nishimura, T. *J. Med. Chem.* **2001**, *44*, 715–724.
240. Metzger, T. G.; Paterlini, M. G.; Portoghese, P. S.; Ferguson, D. M. *Neurochem. Res.* **1996**, *21*, 1287–1294.
241. Larson, D. L.; Jones, R. M.; Hjorth, S. A.; Schwartz, T. W.; Portoghese, P. S. *J. Med. Chem.* **2000**, *43*, 1573–1576.
242. Marx, J. L. *Science* **1982**, *216*, 283–285.
243. Baeuerle, P. A.; Wolf, E. *Modern Drug Disc.* **2000**, *Apr*, 37–42.
244. Geelhoed, G. W. *S. Afr. J. Surg.* **1988**, *26*, 1–3.
245. Ehrlich, P. *Lancet* **1913**, 445–451.
246. Albert, A. *Selective Toxicity*; John Wiley: New York, 1960.
247. (a) *Prodrugs as Novel Drug Delivery Systems;* Higuchi, T., Stella, V., Eds.; American Chemical Society: Washington, DC, 1975; (b) *Design of Prodrugs;* Bundgaard, H., Ed.; Elsevier: Amsterdam, 1985.
248. Bodor, N. In *Strategy in Drug Research*; Buisman, J. A. K., Ed.; Elsevier: Amsterdam, the Netherlands, 1982, pp 137–164.
249. Bodor, N.; Buchwald, P. *Med. Res. Rev.* **2000**, *20*, 58–101.
250. Rozman, K. K.; Klaassen, C. D.; Parkinson, A.; Medinsky, M. A.; Valentine, J. L. Dispositon of Toxicants. In *Casarett & Doull's Toxicology: The Basic Science of Poisons*, 6th ed.; Klaassen, C. D., Ed.; McGraw-Hill: New York, 2001, pp 105–237.
251. *Drug Metabolism: Databases and High-Throughput Screening during Drug Design and Development*; Erhardt, P., Ed.; IUPAC/Blackwell Science: London, 1999.
252. *Handbook of Drug Metabolism*; Woolf, T., Ed.; Marcel Dekker: New York, 1999.

253. Guengerich, F. P. Cytochrome P450 Enzymes. In *Biotransformation, Vol. 3, Comprehensive Toxicology*; Guengerich, F. P., Ed.; Elsevier Science: Oxford, 1997, pp 37–68.
254. Satoh, T.; Hosokawa, M. *Annu. Rev. Pharmacol. Toxicol.* **1998**, *38*, 257–288.
255. Sarver, J.; White, D.; Erhardt, P.; Bachmann, K. *Envir. Health Perspect.* **1997**, *105*, 1204–1209.
256. Ward, K. W.; Erhardt, P.; Bachmann, K. *J. Pharmacol. Toxicol. Methods* **2005**, *51*, 57–64.
257. Darvas, F. Metabol Expert: An Expert System for Predicting the Metabolism of Substances. In *QSAR in Environmental Toxicology II*; Kaiser, C., Ed.; Reidel: Dordrecht, 1987, pp 71–81.
258. Klopman, G.; Dimayuga, M.; Talafous, J. *J. Chem. Inf. Comput. Sci.* **1994**, *34*, 1320–1325.
259. Gifford, E.; Johnson, M.; Tsai, C. *Molecular Connection: The MDL Newsletter for Communicating with Customers* **1994**, *13*, 12–13.
260. *Biotransformations, 1–7*; Hawkins, D., Ed.; Royal Society of Chemistry: Cambridge, 1997 (first database release).
261. Greene, N. Knowledge Based Expert Systems for Toxicity and Metabolism Prediction. In *Drug Metabolism: Databases and High Throughput Testing during Drug Design and Development*; Erhardt, P., Ed.; IUPAC Blackwell Science: London, 1999, pp 289–296.
262. Erhardt, P. *J. Curr. Drug Metab.* **2003**, *4*, 411–422.
263. Thompson, E. A.; Siiteri, P. K. *J. Biol. Chem.* **1974**, *249*, 5373.
264. Fishman, J.; Goto, J. *J. Biol. Chem.* **1981**, *256*, 4466.
265. Metcalf, B. W. *J. Am. Chem. Soc.* **1981**, *103*, 3221.
266. Marcotte, P.; Robinson, C. H. *Biochemistry* **1982**, *21*, 2773.
267. Covey, D. V. *J. Biol. Chem.* **1983**, *256*, 1076.
268. Erhardt, P. Abstract Soc. Tox. Canada 32nd Annual Symposium, Montreal, 1999.
269. Sina, J. F. *Annu. Rep. Med. Chem.* **1998**, *33*, 283–291.
270. Conolly, R. B. *Chem. Industry Inst. Tox. Activities* **1999**, *19*, 6–8.
271. Jacobs, M. *Chem. Eng. News* **2000**, *Mar 13*, 29.
272. Boguslavsky, J. *Drug Disc. Dev.* **2000**, *June/July*, 43–46.
273. Eggers, M. *Innov. Pharm. Tech.* **2000**, *6*, 36–44.
274. Boguslavsky, J. *Drug Disc. Dev.* **2000**, *Oct*, 30–36.
275. Neft, R. E.; Farr, S. *Drug Disc. World* **2000**, *1* (2), 33–34.
276. For general toxicology: Cellomics, Inc. at http://www.cellomics.com; Celera, Inc. at http://www.celera.com; Chemical Industry Inst. Tox. at http://www.ciit.org; Inpharmatica at http://www.inpharmatica.co.uk; Nat. Ctr. Bioinformatics at http://www.ncbi.nlm.nih.gov/Web/Genebank/index.html and at http://www.ncbi.nlm.nih.gov/Entvez/Genonome/org.html; Rosetta Inpharmatics at http://www.vii.com; SciVision (ToxSYS) at http://www.scivision.com. For drug–drug interactions: Georgetown Univ. at http://dml.Georgetown.edu/depts/pharmacology/clinlist.html; FDA at http://www.fda.gov/cder/drug/advisory/stjwort.htm; Public Citizen Health Research Group at http://www.citizen.org/hrg/newsletters/pillnews.htm (accessed Aug 2006).
277. Kuypu, B. J. *Chem. Ind. Inst. Tox. Activities* **2000**, *20*, 1–2.
278. Henry, C. J.; Bus, J. S. *Chem. Ind. Inst. Tox. Activities* **2000**, *20*, 1–5.
279. Felton, M. J. *Modern Drug Disc.* **2000**, *Oct*, 81–83.
280. Hogue, C. *Chem. Eng. News* **2001**, *Mar 19*, 33–34.
281. Lewalle, P. *Terminol. Stand. Harmon.* **1999**, *11*, 1–28.
282. Duffus, J. H. *Chem. Int. (IUPAC News Magazine)* **2001**, *23*, 34–39.
283. Klopman, G. *Quant. Struct.-Act. Relat.* **1992**, *11*, 176–184.
284. Willis, R.; Felton, M. J. *Modern Drug Disc.* **2001**, *Jan*, 31–36.
285. Sanderson, D.; Earnshaw, C.; Judson, P. *Hum. Expt. Tox.* **1991**, *10*, 261–273.
286. Gad, S. C. *SciVision Update* **2001**, *Jan*, 1–2.
287. Neft, R. E.; Farr, S. *Drug Disc. World* **2000**, *Feb*, 33–34.
288. Wood, F. Natural Cures and Gentle Medicines that Work Better than Dangerous Drugs or Risky Surgery. Newspaper advertisement, *The Blade* (Toledo, OH) 2000, Jan 25, p 3A.
289. Howe, K. Scientists Rip Herbal Remedies Fatal Health Problems Tied to Lack of Regulation. *San Francisco Chronicle*, 2000, Jan 21, p 5.
290. Barnes, J. A *Inpharma* **1999**, *1185*, 1.
291. Keung, W.-M.; Lazo, O.; Kunze, L.; Vallee, B. *Proc. Natl. Acad. Sci. USA* **1996**, *93*, 4284–4289.
292. Stermitz, F.; Lorenz, P.; Tawara, J.; Zenewicz, L.; Lewis, K. *Proc. Natl. Acad. Sci. USA* **2000**, *97*, 1433–1438.
293. Guz, N. R.; Stermitz, F. R. *J. Nat. Prod.* **2000**, *63*, 1140–1145.
294. Keung, W. M.; Lazo, O.; Kunze, L.; Vallee, B. L. *Proc. Natl. Acad. Sci. USA* **1995**, *92*, 8990–8993.
295. Verma, S.; Salamone, E.; Goldin, B. *Biochem. Biophys. Res. Commun.* **1997**, *233*, 692–696.
296. Mammen, M.; Choi, S. K.; Whitesides, G. M. *Angew. Chem. Int. Ed. Engl.* **1998**, *10*, 2754–2794.
297. Kiessling, L. L.; Strong, L. E.; Gestwicki, J. E. *Annu. Rep. Med. Chem.* **2000**, *35*, 321–330.
298. Annoura, H.; Nakanishi, K.; Toba, T.; Takemoto, N.; Imajo, S.; Miyajima, A.; Tamura-Horikawa, Y.; Tamura, S. *J. Med. Chem.* **2000**, *43*, 3372–3376.
299. Gangjee, A.; Yu, J.; McGuire, J. J.; Cody, V.; Galitsky, N.; Kisliuk, R. L.; Queener, S. F. *J. Med. Chem.* **2000**, *43*, 3837–3851.
300. Borman, S. *Chem. Eng. News* **2000**, *Oct 9*, 48–53.
301. Sucheck, S. J.; Wong, A. L.; Koeller, K. M.; Boehr, D. D.; Draker, K.; Sears, P.; Wright, G. D.; Wong, C.-H. *J. Am. Chem. Soc.* **2000**, *122*, 5230–5231.
302. Faghihi, J.; Jiang, X.; Vierling, R.; Goldman, S.; Sharfstein, S.; Sarver, J.; Erhardt, P. *J. Chromatogr., A* **2001**, *915*, 61–74.
303. Brun, Y.; Wrzossek, C.; Parsons, J.; Slocum, H.; White, D.; Greco, W. *Am. Assoc. Cancer Res. Proc.* **2000**, *41*, 102 (abstract).
304. White, D.; James, L. *J. Clin. Epidemiol.* **1996**, *49*, 419–429.
305. Greco, W.; Bravo, G.; Parsons, J. *Pharmacol. Rev.* **1995**, *47*, 331–385.
306. Faessel, H. M.; Slocum, H. K.; Rustum, Y. M.; Greco, W. R. *Biochem. Pharmacol.* **1999**, *57*, 567–577.
307. Paired agent combinations = 100!/(2! 98!) = 4950 agent pairs. Test each pair at 3 concentrations for each agent = 9 combinations per pair. Number of pair tests = 9 (4950) = 44 550. Number of single agent tests at 3 concentrations per agent = 3(100) = 300. Total tests for an N = 1 pass through library = 44 550 + 300 = 44 850.
308. Moos, W.; Green, G. D.; Pavia, M. R. *Annu. Rep. Med. Chem.* **1993**, *28*, 315–324.
309. Spellmeyer, D. C.; Grootenhuis, P. D. J. *Annu. Rep. Med. Chem.* **1999**, *34*, 287–296.

310. Dorfman, A. *Time* **2001**, *July 1*, 97–99.
311. Huang, J.-S.; Barker, K. *Plant Physiol.* **1991**, *96*, 1302–1307.
312. Cheong, J. J.; Alba, R.; Cote, F.; Enkerli, J.; Hahn, M. G. *Plant Physiol.* **1993**, *103*, 1173–1182.
313. *Organic Chemistry of Secondary Plant Metabolism;* Geissman, T., Crout, D., Eds.; Freeman, Cooper and Co.: San Francisco, CA, 1969.
314. *The Biosynthesis of Secondary Metabolites;* Herbert, R., Ed.; Chapman and Hall: New York, 1989.
315. Murashige, T.; Skoog, F. *Physiol. Plant* **1962**, *15*, 473–497.
316. Borman, S. *Chem. Eng. News* **2000**, *Oct 30*, 35–36.
317. Gruber, V.; Theisen, M. *Annu. Rep. Med. Chem.* **2000**, *35*, 357–364.
318. Erhardt, P. W. Epilogue. In *Drug Metabolism: Databases and High-Throughput Testing During Drug Design and Development*; Erhardt, P. W., Ed.; IUPAC/Blackwell Science: London, 1999, p 320.
319. Hehre, W. J.; Radom, L.; Schleyer, P. van R.; Pople, J. A. *Ab Initio Molecular Orbital Theory*; John Wiley: New York, 1986.
320. Becke, A. D. *J. Chem. Phys.* **1993**, *98*, 5648–5652.
321. Hu, Z. M.S. Thesis, University of Toledo, Toledo, OH, USA, 1997.
322. Hu, Z.; Erhardt, P. W. *Org. Proc. Res. Dev.* **1997**, *1*, 387–390.
323. Hu, Z.; Hardie, M. J.; Burckel, P.; Pinkerton, A. A.; Erhardt, P. W. *J. Chem. Crystallogr.* **1999**, *29*, 185–191.
324. Marchand, A. P.; Wang, Y.; Ren, C.-T.; Vidyasagar, V.; Wang, D. *Tetrahedron* **1996**, *52*, 6063–6072.
325. Pomerantz, M.; Hartman, P. H. *Tetrahedron Lett.* **1968**, *10*, 991–993.
326. Turro, N. J.; Wriede, P. A. *J. Am. Chem. Soc.* **1970**, *92*, 320–329.
327. Lewis, F. D.; Turro, N. J. *J. Am. Chem. Soc.* **1970**, *92*, 311–320.
328. Friedrich, L. E.; Lam, P. Y.-S. *J. Org. Chem.* **1981**, *46*, 306–311.
329. Klis, W. A.; Erhardt, P. W. *Synthet. Commun.* **2000**, *30*, 4027–4038.
330. Erhardt, P. W.; Klis, W. A.; Nagy, P. I.; Kirschbaum, K.; Wu, N.; Martin, A.; Pinkerton, A. A. *J. Chem. Crystallogr.* **2000**, *30*, 83–90.
331. Erhardt, P.; Klis, W. US Patent 6, 255,540, 2001.
332. Endicott, J. A.; Ling, V. *Annu. Rev. Biochem.* **1989**, *58*, 137–171.
333. Gottesman, M. M.; Pastan, I. *Annu. Rev. Biochem.* **1993**, *62*, 385–427.
334. Scheffer, G. L.; Kool, M.; Heijn, M.; deHaas, M.; Pijnenborg, A.; Wijnholds, J.; van Helvoort, A.; de Jong, M. C.; Hooijberg, J. H.; Mol, C. et al. *Cancer Res.* **2000**, *60*, 5269–5277.
335. Allen, J. D.; Brinkhuis, R. F.; van Deemter, L.; Wijnholds, J; Schinkel, A. *Cancer Res.* **2000**, *60*, 5761–5766.
336. *Multidrug Resistance in Cancer Cells: Molecular, Biochemical, Physiological and Biological Aspects*; Gupta, S., Tsuruo, T., Eds.; John Wiley: Chichester, UK, 1996.
337. Moscow, J. A.; Schneider, E.; Ivy, S. P.; Cowan, K. H. *Cancer Chemother. Biol. Resp. Modif.* **1997**, *17*, 139–177.
338. Gottesman, M. M.; Pastan, I.; Ambudkar, S. V. *Curr. Opin. Genet. Dev.* **1996**, *6*, 610–617.
339. Gros, P.; Croop, J.; Housman, D. *Cell* **1986**, *47*, 371–380.
340. Cole, S. P.; Bhardwaj, G.; Gerlach, J. H.; Mackie, J. E.; Grant, C. E.; Almquist, K. C.; Stewart, A. J.; Kurz, E. U; Duncan, A. M.; Deeley, R. G. *Science* **1992**, *258*, 1650–1654.
341. Doyle, L. A.; Yang, W.; Abruzzo, L. V.; Krogmann, T.; Gao, Y.; Rishi, A. K.; Ross, D. D. *Proc. Natl. Acad. Sci. USA* **1998**, *95*, 15665–15670.
342. Ramachandran, C.; Melnick, S. J. *Mol. Diagn.* **1999**, *4*, 81–94.
343. Values reported in the NIH/NCI database at http://dtp.nci.nih.gov/docs/cancer/searches/standard_mechanism_list.html#topo1. Similar values are obtained in our labs using MCF-7 versus NCI/ADR-RES cell lines (unreported data).
344. Wall, M. E. Camptothecin and Taxol. In *Chronicles of Drug Discovery*; Lednicer, D., Ed.; American Chemical Society: Washington, DC, 1993, pp 327–348.
345. Suffness, M. *Annu. Rep. Med. Chem.* **1993**, *28*, 305–314.
346. George, G.; Ali, S. M.; Zygmunt, J.; Jayasinghe, L. R. *Exp. Opin. Ther. Patents* **1994**, *4*, 109–120.
347. Nicolaou, K. C.; Dai, W.-M.; Guy, R. K. *Angew. Chem. Int. Ed. Engl.* **1994**, *33*, 15–44.
348. *Taxane Anticancer Agents: Basic Science and Current Status;* Georg, G. I., Chen, T., Ojima, I., Vyas, D. M., Eds.; American Chemical Society: Washington, DC, 1995.
349. Sonneveld, P.; Wiemer, E. *Cur. Opin. Oncol.* **1997**, *9*, 543–548.
350. Ling, V. *Cancer Chemother. Pharmacol.* **1997**, *40*, S3–S8.
351. Fisher, G. A.; Sikic, B. I. *Hematol. Oncol. Clin. N. Am.* **1995**, *9*, 363–382.
352. Sarver, J.; Erhardt, P.; Klis, W.; Byers, J. Abstract, SGK 2000 Reaching for the Cure/Making a Difference Mission Conference 2000.
353. Kingsbury, W. D.; Boehm, J. C.; Jakas, D. R.; Holden, K. G.; Hecht, S. M.; Gallagher, G.; Caranfa, M. J.; McCabe, F. L.; Faucette, L. F.; Johnson, R. K. et al. *J. Med. Chem.* **1991**, *34*, 98–107.
354. Liu, L. *Adv. Pharmacol.* **1994**, *29B*, 1–298.
355. Sinha, B. *Drugs* **1995**, *49*, 11–19.
356. Pantazis, P.; Giovanella, B.; Rothenberg, M. *Ann. NY Acad. Sci.* **1996**, *803*, 1–328.
357. Jaxel, C.; Kohn, K.; Wani, M.; Wall, M.; Pommier, Y. *Cancer Res.* **1989**, *49*, 1465–1469.
358. Wall, M.; Wani, M.; Nicholas, A.; Manikumar, G.; Tele, C.; Moore, L.; Truesdale, A.; Leitner, P.; Besterman, J. *J. Med. Chem.* **1993**, *36*, 2689–2700.
359. Pommier, Y. *Semin. Oncol.* **1996**, *23*, 3–10.
360. Wall, M.; Wani, M. *Ann. NY Acad. Sci.* **1996**, *803*, 1–12.
361. Uehling, D.; Nanthakumar, S.; Croom, D.; Emerson, D.; Leitner, P.; Luzzio, M.; McIntyre, G.; Morton, B.; Profeta, S.; Sisco, J. et al. *J. Med. Chem.* **1995**, *38*, 1106–1118.
362. Sawada, S.; Matsuoka, S.; Nokata, K.; Nagata, H.; Furuta, T.; Yokokura, T.; Miyasaka, T. *Chem. Pharm. Bull.* **1991**, *39*, 3183–3188.
363. Sugimori, M.; Ejima, A.; Ohsuki, S.; Uoto, K.; Mitsui, I.; Matsumoto, K.; Kawato, Y.; Yasuoka, M.; Sato, K.; Tagawa, H. et al. *J. Med. Chem.* **1994**, *37*, 3033–3039.
364. Giovanella, B.; Stehlin, J.; Wall, M.; Wani, M.; Nicholas, A.; Liu, L.; Silber, R.; Potmesil, M. *Science* **1989**, *246*, 1046–1048.
365. Hinz, H.; Harris, N.; Natelson, E.; Giovanella, B. *Cancer Res.* **1994**, *54*, 3096–3100.
366. Wang, H.; Liu, S.; Hwang, K.; Taylor, G.; Lee, K.-H. *Bioorg. Med. Chem.* **1994**, *2*, 1397–1402.
367. Luzzio, M.; Besterman, J.; Emerson, D.; Evans, M.; Lackey, K.; Leitner, P.; McIntyre, G.; Morton, B.; Myers, P.; Peel, M. et al. *J. Med. Chem.* **1995**, *38*, 395–401.

368. Lackey, K.; Besterman, J.; Fletcher, W.; Leitner, P.; Morton, B.; Sternbach, D. *J. Med. Chem.* **1995**, *38*, 906–911.
369. Lackey, K.; Sternbach, D.; Croom, D.; Emerson, D.; Evans, M.; Leitner, P.; Luzzio, M.; McIntyre, G.; Vuong, A.; Yates, J. et al. *J. Med. Chem.* **1996**, *39*, 713–719.
370. Nicholas, A.; Wani, M.; Manikumar, G.; Wall, M.; Kohn, K.; Pommier, Y. *J. Med. Chem.* **1990**, *33*, 972–978.
371. Crow, R.; Crothers, D. *J. Med. Chem.* **1992**, *35*, 4160–4164.
372. Jaxel, C.; Kohn, K. W.; Wani, M. C.; Wall, M. E.; Pommier, Y. *Cancer Res.* **1989**, *49*, 1465–1469.
373. Pommier, Y.; Jaxel, C.; Heise, C. Structure–Activity Relationship of Topoisomerase I Inhibition by Camptothecin Derivatives: Evidence for the Existence of a Ternary Complex. In *DNA Topoisomerases in Cancer*; Potmesil, M., Kohn, K., Eds.; Oxford University Press: New York, 1991, pp 121–132.
374. Lackey, K.; Besterman, J.; Fletcher, W.; Leitner, P.; Morton, B.; Sternbach, D. *J. Med. Chem.* **1995**, *38*, 906–911.
375. Sugimori, M.; Ejima, A.; Ohsuki, S.; Uoto, K.; Mitsui, I.; Matsumoto, K.; Kawato, Y.; Yasuoka, M.; Sato, K.; Tagawa, H. et al. *J. Med. Chem.* **1994**, *37*, 3033–3039.
376. Carrigan, S.; Fox, P.; Wall, M.; Wani, M.; Bowen, J. *J. Comput.-Aided Mol. Des.* **1997**, *11*, 71–78.
377. Rodi, D. J.; Janes, R. W.; Sanganee, H. J.; Holton, R. A.; Wallace, B. A.; Makowski, L. *J. Mol. Biol.* **1999**, *285*, 197–203.
378. Yang, C.-P.; Horwitz, S. B. *Cancer Res.* **2000**, *60*, 5171–5178.
379. Andre, N.; Braguer, D.; Brasseur, G.; Goncalves, A.; Lemesle-Meunier, D.; Guise, S.; Jordan, M. A.; Briand, D. *Cancer Res.* **2000**, *60*, 5349–5353.
380. Schiff, P. B.; Fant, J.; Horwitz, S. B. *Nature* **1979**, *277*, 665–667.
381. Erhardt, P. W. *Taxane J.* **1997**, *3*, 36–42.
382. Nakashio, A.; Fujita, N.; Rokudai, S.; Sato, S.; Tsuruo, T. *Cancer Res.* **2000**, *60*, 5303–5309.
383. Froelich-Ammon, S.; Osheroff, N. *J. Biol. Chem.* **1995**, *270*, 21429–21432.
384. Klis, W.; Sarver, J.; Erhardt, P. *Synthet. Commun.* **2002**, *32*, 2711–2718.
385. Klis, W.; Sarver, J.; Erhardt, P. *Tetrahedron Lett.* **2001**, *42*, 7747–7750.
386. Klis, W.; Sarver, J.; Erhardt, P. US Patent, 6,846,937, B2, 2005.
387. Erhardt, P. *Chron. Drug Disc.* **1993**, *3*, 191–206.
388. Erhardt, P. W.; Woo, C. M.; Gorczynski, R. J.; Anderson, W. G. *J. Med. Chem.* **1982**, *25*, 1402–1407.
389. Erhardt, P. W.; Woo, C. M.; Anderson, W. G.; Gorczynski, R. J. *J. Med. Chem.* **1982**, *25*, 1408–1412.
390. Erhardt, P. W.; Woo, C. M.; Matier, W. L.; Gorczynski, R. J.; Anderson, W. G. *J. Med. Chem.* **1983**, *26*, 1109–1112.
391. Erhardt, P. W. US Patent, 6,750,238 B1, 2004.
392. Erhardt, P. W. US Patent, 6,756,047 B2, 2004.
393. Liao, J.; Golding, K.; Dubischar, K.; Sarver, J.; Erhardt, P.; Kuch, H.; Aouthmany, M. Abstract, 27th National ACS Medicinal Chemistry Symposium, 2000.
394. Liao, J. M.S. Thesis, University of Toledo, Toledo, OH, USA, 2000.
395. Pattisal, K. H. Discovery and Development of Zidovudine. In *The Search for Antiviral Drugs*; Adams, J. A., Merluzzi, J., Eds.; Birkhauser: Boston, MA, 1993, pp 23–43.
396. Horwitz, J. P.; Chua, J.; Noel, M. *J. Org. Chem.* **1964**, *29*, 2076–2078.
397. Mitsuya, H.; Weinhold, K. J.; Furman, P. A.; St. Clair, M. H.; Nusinoff Lehrman, S.; Gallo, R. C.; Bolognesi, D.; Barry, D. W.; Broder, S. *Proc. Natl. Acad. Sci. USA* **1985**, *82*, 7096–7100.
398. Horwitz, J. P.; Chua, J.; Da, M.; Rooge, A.; Noel, M. *Tetrahedron Lett.* **1964**, *5*, 2725–2727; *J. Med. Chem.* **1987**, *30*, 440–444; Lin, T. S.; Schinazi, R. F.; Prusoff, W. H. *Biochem. Pharmacol.* **1987**, *36*, 2713–2718.
399. Horwitz, J. P.; Chua, J.; Noel, M.; Donatti, J. T. *J. Org. Chem.* **1967**, *32*, 817–818.
400. Thormar, H.; Balzarini, J.; Holy, A.; Jindrich, J.; Rosenberg, I.; Debyser, Z.; Desmyter, J.; De Clercq, E. *Antimicrob. Agents Chemother.* **1993**, *37*, 2540–2544.
401. Robbins, B. L.; Srinivas, R. V.; Kim, C.; Bischofberger, N.; Fridland, A. *Antimicrob. Agents Chemother.* **1998**, *42*, 612–617.
402. Tuske, S.; Sarafianos, S. G.; Clark, A. D.; Ding, J.; Naeger, L. K.; White, K. L.; Miller, M. D.; Gibbs, C. S.; Boyer, P. L.; Clark, P. et al. *Nat. Struct. Mol. Biol.* **2004**, *11*, 469–474.
403. Larder, B. A.; Kemp, S. D. *Science* **1989**, *246*, 1155–1158.
404. Roberts, J. D.; Bebenek, K.; Kunkel, T. A. *Science* **1988**, *242*, 1171–1173; Preston, B. D.; Poiesz, B. J.; Loeb, L. A. *Science* **1988**, *242*, 1168–1171.
405. Lewis, W. *Progr. Cardiovasc. Dis.* **2003**, *45*, 305–318.
406. Merluzzi, V. J.; Hargrave, K. D.; Labadia, M.; Grozinger, K.; Skoog, M.; Wu, J. C.; Shih, C. K.; Eckner, K.; Hattox, S.; Adams, J. et al. *Science* **1990**, *250*, 1411–1413.
407. Dueweke, T. J.; Poppe, S. M.; Romero, D. L.; Swaney, S. M.; So, A. G.; Downey, K. M.; Althaus, I. W.; Reusser, F.; Busso, M.; Resnick, L. *Antimicrob. Agents Chemother.* **1993**, *37*, 1127–1131.
408. Young, S. D.; Britcher, S. F.; Tran, L. O.; Payne, L. S.; Lumma, W. C.; Lyle, T. A.; Huff, J. R.; Anderson, P. S.; Olsen, D. B.; Carroll, S. S. *Antimicrob. Agents Chemother.* **1995**, *39*, 2602–2605.
409. Hargrave, K. D.; Proudfoot, J. R.; Grozinger, K. G.; Cullen, E.; Kapadia, S. R.; Patel, U. R.; Fuchs, V. U.; Mauldin, S. C.; Vitous, J.; Behnke, M. L. et al. *J. Med. Chem.* **1991**, *34*, 2231–2241.
410. Romero, D. L.; Morge, R. A.; Genin, M. J.; Biles, C.; Busso, M.; Resnick, L.; Althaus, I. W.; Reusser, F.; Thomas, R. C.; Tarpley, W. G. *J. Med. Chem.* **1993**, *36*, 1505–1508.
411. Tucker, T. J.; Lyle, T. A.; Wiscount, C. M.; Britcher, S. F.; Young, S. D.; Sanders, W. M.; Lumma, W. C.; Goldman, M. E.; O'Brien, J. A.; Ball, R. G. et al. *J. Med. Chem.* **1994**, *37*, 2437–2444.
412. Oprea, T. I.; Davis, A. M.; Teague, S. J.; Leeson, P. D. *J. Chem. Inf. Comput. Sci.* **2001**, *41*, 1308–1315.
413. Wu, J. C.; Warren, T. C.; Adams, J.; Proudfoot, J.; Skiles, J.; Raghavan, P.; Perry, C.; Potocki, I.; Farina, P. R.; Grob, P. M. *Biochemistry* **1991**, *30*, 2022–2026.
414. Cohen, K. A.; Hopkins, J.; Ingraham, R. H.; Pargellis, C.; Wu, J. C.; Palladino, D. E.; Kinkade, P.; Warren, T. C.; Rogers, S.; Adams, J. et al. *J. Biol. Chem.* **1991**, *266*, 14670–14674.
415. Kohlstaedt, L. A.; Wang, J.; Friedman, J. M.; Rice, P. A.; Steitz, T. A. *Science* **1992**, *256*, 1783–1790.
416. Ren, J.; Esnouf, R.; Garman, E.; Somers, D.; Ross, C.; Kirby, I.; Keeling, J.; Darby, G.; Jones, Y.; Stuart, D. et al. *Nat. Struct. Biol.* **1995**, *2*, 293–302.
417. Sarafianos, S. G.; Das, K.; Hughes, S. H.; Arnold, E. *Curr. Opin. Struct. Biol.* **2004**, *14*, 716–730.
418. Shih, C.-K.; Rose, J. M.; Hansen, G. L.; Wu, J. C.; Bacolla, A.; Griffin, J. A. *Proc. Natl. Acad. Sci. USA* **1991**, *88*, 9878–9882.

419. Richman, D.; Shih, C.-K.; Lowy, I.; Rose, J.; Prodanovich, P.; Goff, S.; Griffin, J. *Proc. Natl. Acad. Sci. USA* **1991**, *88*, 11241–11245; Richman, D. D.; Havlir, D.; Corbeil, J.; Looney, D.; Ignacio, C.; Spector, S. A.; Sullivan, J.; Cheeseman, S.; Barringer, K.; Pauletti, D. *J. Virol.* **1994**, *68*, 1660–1666.
420. Montaner, J. S. G.; Reiss, P.; Cooper, D.; Vella, S.; Harris, M.; Conway, B.; Wainberg, M. A.; Smith, D.; Robinson, P.; Hall, D. et al. *JAMA* **1998**, *279*, 930–937.
421. Albrecht, M. A.; Bosch, R. J.; Hammer, S. M.; Liou, S. H.; Kessler, H.; Para, M. F.; Eron, J.; Valdez, H.; Dehlinger, M.; Katzenstein, D. A. *N. Engl. J. Med.* **2001**, *345*, 398–407.
422. Harris, M. *J. HIV Ther.* **1999**, *4*, 37–40.
423. Kashman, Y.; Gustafson, K. R.; Fuller, R.; Cardellina, H.; McMahon, J. B.; Currens, M. J.; Buckheit, R. W., Jr.; Hughes, S. H.; Cragg, G. M.; Boyd, M. R. *J. Med. Chem.* **1992**, *35*, 2735–2743.
424. Currens, M. J.; Gulakowski, R. J.; Mariner, J. M.; Moran, R. A.; Buckheit, R. W., Jr.; Gustafson, K. R.; McMahon, J. B.; Boyd, M. R. *J. Pharmacol. Exp. Ther.* **1996**, *279*, 645–651.
425. Baba, M.; Tanaka, H.; De Clercq, E.; Pauwels, R.; Balzarini, J.; Schols, D.; Nakashima, H.; Perno, C. F.; Walker, R. T.; Miyasaka, T. *Biochem. Biophys. Res. Commun.* **1989**, *165*, 1375–1381.
426. Quan, Y.; Motakis, D.; Buckheit, R., Jr.; Xu, Z. Q.; Flavin, M. T.; Parniak, M. A.; Wainberg, M. A. *Antivir. Ther.* **1999**, *4*, 203–209.
427. Wlodawer, A.; Miller, M.; Jaskolski, M.; Sathyanarayana, B. K.; Baldwin, E.; Weber, I. T.; Selk, L. M.; Clawson, L.; Schneider, J.; Kent, S. B. H. *Science* **1989**, *245*, 616; Navia, M. A.; Fitzgerald, P. M. D.; McKeever, B. M.; Leu, C.-T.; Heimbach, W. C.; Herber, W. K.; Sigal, I. S.; Darke, P. L.; Springer, J. P. *Nature* **1989**, *337*, 615–620.
428. Roberts, N. A.; Martin, J. A.; Kinchington, D.; Broadhurst, A. V.; Craig, J. C.; Duncan, I. B.; Galpin, S. A.; Handa, B. K.; Kay, J.; Krohn, A. et al. *Science* **1990**, *248*, 358–361.
429. Kaldor, S. W.; Kalish, V. J.; Davies, J. F.; Shetty, B. V.; Fritz, J. E.; Appelt, K.; Burgess, J. A.; Campanale, K. M.; Chirgadze, N. Y.; Clawson, D. K. et al. *J. Med. Chem* **1997**, *40*, 3979–3985.
430. Kim, E. E.; Baker, C. T.; Dwyer, M. D.; Murcko, M. A.; Rao, B. G.; Tung, R. D.; Navia, M. A. *J. Am. Chem. Soc.* **1995**, *117*, 1181–1182.
431. Colonno, R. J.; Thiry, A.; Limoli, K.; Parkin, N. *Antimicrob. Agents Chemother.* **2003**, *47*, 1324–1333.
432. Surleraux, D. L. N. G.; Tahri, A.; Verschueren, W. G.; Pille, G. M. E.; de Kock, H. A.; Jonckers, T. H. M.; Peeters, A.; De Meyer, S.; Azijn, H.; Pauwels, R. et al. *J. Med. Chem.* **2005**, *48*, 1813–1822.
433. Erickson, J.; Neidhart, D. J.; VanDrie, J.; Kempf, D. J.; Wang, X. C.; Norbeck, D. W.; Plattner, J. J.; Rittenhouse, J. W.; Turon, M.; Wideburg, N. et al. *Science* **1990**, *249*, 527–533.
434. Kempf, D. J.; Norbeck, D. W.; Codacovi, L.; Wang, X. C.; Kohlbrenner, W. E.; Wideburg, N. E.; Paul, D. A.; Knigge, M. F.; Vasavanonda, S.; Craig-Kennard, A. et al. *J. Med. Chem.* **1990**, *33*, 2687–2689.
435. Clavel, F.; Hance, A. J. *N. Engl. J. Med.* **2004**, *350*, 1023–1035.
436. Sham, H. L.; Kempf, D. J.; Molla, A.; Marsh, K. C.; Kumar, G. N.; Chen, C.-M.; Kati, W.; Stewart, K.; Lal, R.; Hsu, A. et al. *Antimicrob. Agents Chemother.* **1998**, *42*, 3218–3224.
437. Vacca, J. P.; Guare, J. P.; deSolms, S. J.; Sanders, W. M.; Giuliani, E. A.; Young, S. D.; Darke, P. L.; Zugay, J.; Sigal, I. S.; Schleif, W. A. *J. Med. Chem.* **1991**, *34*, 1225–1228.
438. Dorsey, B. D.; Levin, R. B.; McDaniel, S. L.; Vacca, J. P.; Guare, J. P.; Darke, P. L.; Zugay, J. A.; Emini, E. A.; Schleif, W. A.; Quintero, J. C. et al. *J. Med. Chem* **1994**, *37*, 3443–3451.
439. Thaisrivongs, S.; Tomich, P. K.; Watenpaugh, K. D.; Chong, K.-T.; Howe, W. J.; Yang, C.-P; Strohbach, J. W.; Turner, S. R.; McGrath, J. P.; Bohanon, M. J. et al. *J. Med. Chem.* **1994**, *37*, 3200–3204.
440. Hazuda, D. J.; Felock, P.; Witmer, M.; Wolfe, A.; Stillmock, K.; Grobler, J. A.; Espeseth, A.; Gabryelski, L.; Schleif, W.; Blau, C. et al. *Science* **2000**, *6*, 646–650.
441. Zhuang, L.; Wai, J. S.; Embrey, M. W.; Fisher, T. E.; Egbertson, M. S.; Payne, L. S.; Guare, J. P.; Vacca, J. P.; Hazuda, D. J.; Felock, P. J. et al. *J. Med. Chem.* **2003**, *46*, 453–456.
442. Billich, A. *Curr. Opin. Investig. Drugs* **2003**, *4*, 206–209.
443. Goldgur, Y.; Craigie, R.; Cohen, G. H.; Fujiwara, T.; Yoshinaga, T.; Fujishita, T.; Sugimoto, H.; Endo, T.; Murai, H.; Davies, D. R. *Proc. Natl. Acad. Sci. USA* **1999**, *96*, 13040–13043.
444. Fikkert, V.; Hombrouck, A.; van Remoortel, B.; de Maeyer, M.; Pannecouque, C.; de Clercq, E.; Debyser, Z.; Witvrouw, M. *AIDS* **2004**, *18*, 2019–2028.
445. Bray, B. L. *Nat. Rev. Drug Disc.* **2003**, *2*, 587–593.
446. Tagat, J. R.; McCombie, S. W.; Nazareno, D.; Labroli, M. A.; Xiao, Y.; Steensma, R. W.; Strizki, J. M.; Baroudy, B. M.; Cox, K.; Lachowicz, J. et al. *J. Med. Chem.* **2004**, *47*, 2405–2408.
447. Roden, D. M. *N. Engl. J. Med.* **2004**, *350*, 1013–1022.
448. Mullin, R. *Chem. Eng. News* **2005**, *Feb 28*, 29–39.
449. Ganellin, C. R. *Chem. Int. (IUPAC News Magazine)* **2001**, *23*, 43–45.
450. Brent, J. University of Minnesota CLA Today *(1998–99 Annual Report)* **1999**, *Fall*, 10–13.
451. For example, respectively: the long-standing Pharmacology for Medicinal Chemistry course often offered by the ACS as a satellite program to their national scientific meetings; the Residential School on Medicinal Chemistry offered at Drew University. (http://www.depts.drew.edu/resmed, accessed Aug 2006); Drug Metabolism for Medicinal Chemists short course that P. Erhardt offers on-site; and, a long-distance Continuing Learning Program in Medicinal Chemistry offered by the University of Nottingham, UK. (http://www.nottingham.ac.uk/, accessed Aug 2006).
452. Wright, E. G. *Modern Drug Disc.* **2000**, *Oct*, 69–70.
453. Ledbetter, E. *Modern Drug Disc.* **2000**, *Apr/May*, 25–28 and 81–84.
454. Jacobs, M. *Chem. Eng. News* **2000**, *Apr 10*, 39–44.
455. Jacobs, M. *Chem. Eng. News* **2001**, *Jan 15*, 33.
456. US National Academies' National Research Council (NRC) report entitled *A Patent System for the 21st Century*, 2004 available at http://books.nap.edu/catalog/10976.html (accessed Aug 2006).
457. Hanson, D. J. *Chem. Eng. News* **2000**, *Sept 11*, 19–20.
458. Kaminski, M. D. *Modern Drug Disc.* **2001**, *Jan*, 36–37.
459. Erhardt, P. W. *Chem. Int* (*IUPAC Newsletter*) **2002**, *24*, 16.
460. Elliston, K. O. (as quoted in) *Chem. Eng. News* **2005**, *Feb 14*, 54–55.

Biographies

Paul W Erhardt received a PhD in medicinal chemistry from the University of Minnesota in 1974 and undertook a 1-year postdoctoral study in the area of bioanalytical chemistry and drug metabolism at the University of Texas at Austin. His early career involved bench-level research as a synthetic medicinal chemist within the pharmaceutical industry. He was with American Critical Care in Chicago for about 10 years as a Research Scientist, Senior Research Scientist, and Group Leader. During this period he was directly responsible for the chemical design, synthesis, and entire chemical-related preclinical/phase I development of esmolol, a drug presently marketed as Brevibloc. He then joined Berlex Laboratories in New Jersey as a Section Head where over the course of 10 years he became the Assistant Director of Medicinal Chemistry in charge of drug discovery and, finally, the Assistant Director across all pharmaceutical research and development activities. When the medicinal chemistry research operation of Berlex was merged with the biotechnology operations of two new corporate purchases located on the west coast, he became the medicinal chemist representative on a key business task force that evaluated external technologies for the purpose of maintaining the company's drug development pipeline. During this period he became a US PTO-Certified Patent Agent in order to better deal with the patent issues that often accompany external technology and its in-licensing. He also led the development of a unified technology Beschluss (decision-making) document which harmonized the optimal use of R&D resources across the Berlex/SAG corporate triad (Europe, USA, and Japan) relative to the progression of all internal and in-licensed technologies from concept to market.

With a lingering desire to be closer to the day-to-day experimental practice of bench-level medicinal chemistry, Dr Erhardt returned to academia about 10 years ago when he joined The University of Toledo College of Pharmacy as a tenured Professor and Director of the Center for Drug Design and Development (CD3). During this latest period he has been awarded: the College's Outstanding Teaching Faculty Award and the College's Outstanding Research Faculty Award; been nominated for the University's Outstanding Researcher Award; and has stepped-in for 1-year as an Acting Assistant Dean so as to directly participate in the College's formal academic accreditation process. Dr Erhardt has also become active in the IUPAC where he has edited a book about using drug metabolism considerations during drug design and development and where he has recently been voted President for the IUPAC Division of Chemistry and Human Health (Division VII). His present research focuses on medicinal chemistry considerations pertaining to oncology, drug metabolism and soft drug technologies, ADMET-related SAR and synergy, and chiral auxiliary synthetic reagents amenable to practical, drug-related process chemistry. His annual research budget garnered from extramural sources has recently achieved the level of $1 million.

John R Proudfoot received his BSc degree in chemistry from University College Dublin in Ireland in 1978. He proceeded to graduate studies, also at University College Dublin, exploring biomimetic alkaloid syntheses with Professor Dervilla M X Donnelly and received his doctorate in 1982. Dr Proudfoot then moved to Professor Carl Djerassi's group at Stanford University for postdoctoral studies on the identification, synthesis, and biosynthesis of marine sterols. After an additional postdoctoral experience at the university of California–San Francisco with Professor John Cashman, exploring drug metabolism by the flavin-containing monooxygenase, Dr Proudfoot joined the medicinal chemistry department at Boehringer Ingelheim Pharmaceuticals, in Ridgefield, CT, in 1987 and where he has remained to date. He has contributed to numerous drug discovery programs, and was a member of the team that discovered the currently marketed anti-AIDS drug nevirapine.

Comprehensive Medicinal Chemistry II
ISBN (set): 0-08-044513-6

ISBN (Volume 1) 0-08-044514-4; pp. 29–96

1.03 Major Drug Introductions

P D Kennewell, Swindon, UK

1.03.1 Introduction

Volume 6 of Comprehensive Medicinal Chemistry contains a list of over 5500 compounds that have been studied as medicinal agents in man.[1] This chapter is less ambitious in its scope and only covers new drug introductions over the 10 years 1993–2002 as listed in *Annual Reports of Medicinal Chemistry*.[2–11] Compounds included are new chemical entities (NCEs) and a selection of new biological entities (NBEs). The author apologizes most sincerely to anyone whose compound has been missed from this compilation.

During the period 1993–2003, 339 compounds were introduced; an average of 34 per year (**Table 1**). Other chapters in this volume discuss in some detail whether or not the productivity of the pharmaceutical industry has been falling so it will be sufficient here to merely note that the number of new agents does indeed seem to show a downward trend over this time period. **Table 2** lists the countries of origin of the new medicaments along with countries in which the drug was first marketed. The dominance of the US in both drug discovery and initial marketing is apparent from this

Table 1 Number of new compounds introduced between 1994 and 2003

Year	*Number of new compounds*
1994	44
1995	36
1996	38
1997	39
1998	27
1999	35
2000	35
2001	25
2002	33
2003	27
Total	339

table. Previous studies had indicated that Japan held this position but it now seems to have very definitely slipped into second place, with the UK, Germany, and Switzerland being the other major players. One other feature of note is the relatively high number of drugs initially marketed in Sweden.

For the convenience of the reader, it has been decided to group the agents into the categories shown in **Table 3**.

Within these 339 drugs are a number that represent scientific and/or commercial breakthroughs. For example, the continued growth of the use of statins in cardiovascular disease, which resulted in atorvastatin calcium (Lipitor) becoming the first drug to achieve sales of US$1 billion in its first year; the introduction of the COX-2 inhibitors, which

Table 2 Countries of origin of the new medicaments and countries in which first marketed between 1994 and 2003[a]

Country	*Number of compounds discovered*	*Number of compounds first marketed*
USA	122	136
Japan	73	75
UK	39	32
Germany	27	21
Switzerland	22	10
France	15	7
Netherlands	6	7
Italy	5	6
Spain	5	5
Sweden	4	11
Denmark	4	5
Australia	2	4

[a] In addition the following countries had 3 or less compounds in both these categories: India, South Africa, Canada, Czech Republic, Finland, Belgium, Austria, South Korea, Mexico, New Zealand, Israel, Kazakhstan, Russia, and Norway.

Table 3 Number of new agents discovered in each category

Category	*Number of new agents discovered*
Anti-infectives	59
Cancer related	46
Cardiovascular	54
Nervous system	55
Gastrointestinal tract related	14
Lung related	21
Joints and bones	31
Immunology (including vaccines)	7
Hormone related	13
Reproduction and fertility	8
Skin related	7
Eye related	7
Miscellaneous	17

offered the possibility of long-term control of pain and inflammation in osteoarthritis without the gastrointestinal effects of NSAIDs, although rofecoxib (Vioxx) has been withdrawn because of cardiovascular effects and its fate is in the hands of lawyers and unsympathetic jurors; and the introduction of sildenafil (Viagra) as a totally novel therapy for erectile dysfunction, which is now well known to anyone with an email address. Perhaps less well known have been treatments for a number of less common, but very serious diseases. Thus, dornase alfa (Pulmozyme) was the first new treatment for 30 years for cystic fibrosis; two recombinant forms of the enzyme α-galactosidase A were introduced for the treatment of Fabry's disease; imiglucerase (Cerezyme) was introduced as a recombinant enzyme replacement therapy for type I Gaucher's disease; defeiprone (Kelfer) is the first oral iron chelator that provides life-saving benefits for patients with thalassemia; and interferon β-1a (Avonex) and glatiramer acetate (Copaxone) were marketed for multiple sclerosis. The introduction of two new antimalarials, arteether (Artemotil) and bulaquine (Aablaquin), appear to be the only examples of drugs for use predominantly in the less developed countries.

In the anti-infective domain, 30 new antiviral agents were produced, mainly in response to the continuing threat posed by HIV-AIDS. In view of the fact that the disease has only been recognized within the last 40 years, the extent of our knowledge of the life cycle and structure of the virus and our ability to control it is remarkable. This period has seen the introduction of protease inhibitors, second-generation non-nucleoside reverse transcriptase inhibitors, and the first agent to block the uptake of the virus into CD4 T cells. Other compounds targeted at viruses included: two agents against hepatitis B, one of which was the first organogermanium compound to be marked as a pharmaceutical; two neuramidase inhibitors for a wide variety of influenza strains; and the first agent based on antisense technology, which was launched for the treatment of cytomegalovirus infections. Among the new antibiotics introduced were a number with improved spectra of activity against Gram-positive bacteria responsible for life-threatening infections. These included five new quinolones and three new classes of agents; the oxazolidin-2-ones, the cyclic lipopeptides, and the ketolides. In addition, a new treatment for severe sepsis was introduced. Whether or not these will be sufficient to counter the threat posed by drug-resistant bacteria only time will tell. Finally, the six new antifungals included two representatives of the echinocandins, a new class of such compounds.

In the oncology field, perhaps the most remarkable compound is imatinib mesylate (Gleevec), introduced for the treatment of chronic myelogenous leukemia (CML). Its mode of action involves inhibiting the Brc-Abl oncoprotein, a tyrosine kinase that causes CML. Also introduced were inhibitors of topoisomerase I, third-generation orally active aromatase inhibitors for breast cancer, and proteasome inhibitors for multiple myelomas. Biological agents for cancer therapy included recombinant tumor necrosis factor (TNF) for soft tissue sarcomas and malignant melanoma, two radiolabeled antibodies for non-Hodgkin's lymphoma, antibody-targeted antineoplastics for acute myeloid leukemia, a humanized monoclonal antibody for B-cell chronic lymphocytic leukemia, and a monoclonal antibody against epidermal growth factor receptor for colorectal cancer. In addition, new antiemetics appeared to aid patients' tolerance to chemotherapy.

Agents for various forms of cardiovascular disease figured highly over this time period. The role of the statins in controlling blood cholesterol levels was well established and Lipitor made a spectacular appearance in the marketplace followed by the even more potent Crestor. No fewer than three different approaches to the treatment of acute heart failure appeared, namely, colforsin daropate (Adele), an activator of adenylate cyclase, levosemendan (Simdax), a Ca^{2+} sensitizer, and nesiritide (Natrecor), a recombinant form of human B-type natriuretic peptide. Other newcomers included: losartan potassium (Cozaar), the first selective nonpeptide angiotensin II antagonist for hypertension; tirilazad (Freedox), a peroxidation inhibitor for the reduction of tissue damage following subarachnoid hemorrhage in men; anagrelide hydrochloride (Agrylin), the first therapy for raised platelet counts with essential thrombocytopenia; fondaparinux sodium (Arixtra), an agent against deep vein thrombosis; and bosentan (Tracleer), the first endothelin receptor antagonist for pulmonary arterial hypertension.

Within the central nervous system (CNS) domain a number of agents have been introduced with novel mechanisms of action. For example, venlafaxine (Effexor) is an antidepressant with dual serotonin/norepinephrine reuptake inhibitory activity, topiramate (Topomax) belongs to a new class of antiepileptics, while tolcapone (Tasmar) (subsequently withdrawn) and entacopone (Comtess) are the first examples of catechol *O*-methyl transferase inhibitors acting as anti-Parkinsonian agents. Two agents, interferon β-1a (Avonex) and glatiramer acetate (Copaxone), were introduced for the treatment of relapsing multiple sclerosis. Lomerizine (Teranas, Migsis) is the first dual sodium and calcium channel blocker to be launched for the treatment of migraine, while zaleplon (Sonata) represents a new chemical class for the treatment of insomnia. Taltirelin (Ceredist) is an orally active drug against ataxia due to cerebrospinal degeneration, and edaravone (Radicut) is a lipophilic antioxidant for the improvement of neurologic impairment following acute brain infarction.

The number of new agents affecting the gastrointestinal tract was relatively modest over this time period. These induced included ranitidine bismuth citrate (Pylorid), the first therapy specifically tailored to attack *Helicobacter pylori*, and orlistat, which prevents the absorption of dietary fat and is used for the treatment of obesity. Two agents, alosetron

(Lotronex) and tegaserod maleate (Zelmac), were launched for the treatment of irritable bowel syndrome, although the former was rapidly withdrawn in response to the occurence of ischemic colitis.

Interesting novel antiasthmatics included seratrodast (Bronica), the first thromboxane A_2 antagonist; zafirlukast (Accolate), the first LTD_4 antagonist; zileuton (Zyflo), a reversible, orally active direct inhibitor of 5-lipoxygenase; and omalizumab (Xolair), a humanized anti-IgE antibody. Two lung-active biologically derived agents are dornase alfa (Pulmozyme) for cystic fibrosis and pumactant (ALEC), a lung surfactant for the treatment of repiratory distress syndrome in premature infants.

Introductions in the field of osteoarthritis treatments have been dominated by the appearance of the COX-2 inhibitors, celecoxib (Celebrex), rofecoxib (Vioxx), etoricoxib (Arcoxia), valdecoxib (Bextra), and parecoxib (Dynastat). Treatments for joint decay in rheumatoid arthritis include leflunomide (Arava), anakinra (Kineret), a human interleukin-1 (IL1) receptor antagonist, and adalimumab (Humira), a humanized monoclonal antibody that binds human TNF-α. Another biological agent of great interest is OP-1 (Novos), a combination of human recombinant osteogenic protein and a bovine bone-derived collagen carrier that is used to promote bone growth in long bone nonunion fractures.

In the immunology field, a vaccine against Lyme disease and a nasally administered influenza vaccine was introduced.

New treatments for both type 1 and type 2 diabetes have appeared on the market with insulin lispro (Humalog) being claimed to mimic the normal human response in type 1 diabetes. Lanreotide (Somatuline) and pegvisomant (Somavert) have been introduced as treatments for acromegaly (overproduction of growth hormone), while somatomedin-1 (Igef) and somatotropin (Nutropin) are now available for the treatment of growth hormone insensitivity and failure to grow.

In the field of deomatology, two interesting new biological agents are alefacept (Amevine) and efalizumab (Raptiva), which are used in the treatment of plaque psoriasis.

Finally, a number of new treatments for glaucoma have been introduced including brimonidine (Alphagan), a selective α_{2a}-adrenergic agonist; the prostaglandin analogs bimatoprost (Lumigan), latanoprost (Xalatan), travoprost (Travatan), and unoprostone (Rescula); and the carbonic anhydrase inhibitors, brinzolamide (Azopt) and dorzolamide (Trusopt).

1.03.2 Anti-Infectives

This category includes 21 new antibiotics (including one specific for pneumonia), 29 antivirals, 6 antifungals, 2 antimalarials, and 1 agent acting against sepsis.

1.03.2.1 Antibiotics

1.03.2.1.1 Balofloxacin

Trade name	Q-Roxin
Manufacturer	Choongwae Pharma Corporation
Country of origin	Japan
Year of introduction	2002
Country in which first launched	South Korea
CAS registry number	127294-70-6
Structure	

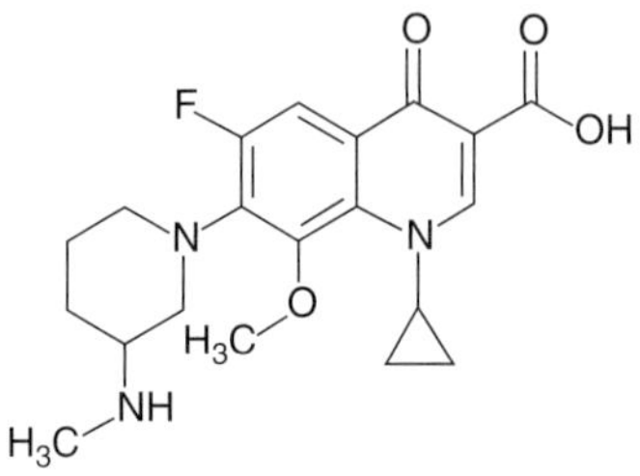

Balofloxacin is an orally active fluoroquinolone antibiotic introduced for the treatment of urinary tract infections. In vitro antibacterial activity against Gram-positive bacteria (*Staphylococcus aureus* including methicillin-resistant *S. aureus* (MRSA), *Staphylococcus epidermis*, *Streptococcus pyrogenes*, and *Streptococcus pneumonia*) was almost equal to that of sparfloxacin or tosufloxacin, while activity against Gram-negative bacteria was atleast twofold lower. Clinically, it is well tolerated and shows activity against urinary tract infections similar to that of ofloxacin. Following oral administration it is well absorbed and eliminated in the urine unchanged with a half-life of approximately 8 h.

1.03.2.1.2 Biapenem

Trade name	Omegacin
Manufacturer	Meiji Seika
Country of origin	USA
Year of introduction	2002
Country in which first launched	Japan
CAS registry number	120410-24-4
Structure	

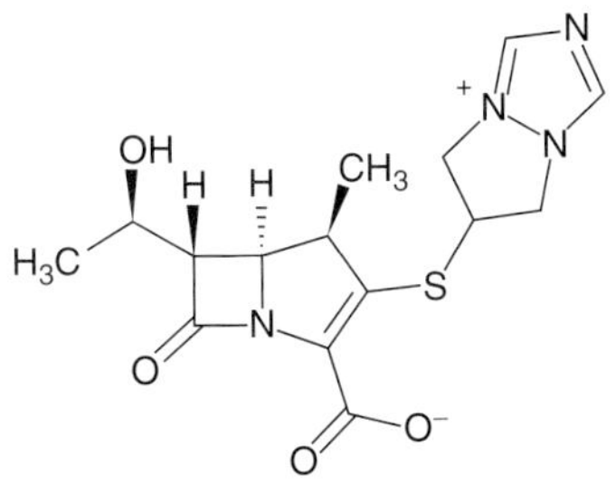

Biapenem is a bacterial cell wall synthesis inhibitor with a broad spectrum of antibiotic activity in vitro. It is stable to hydrolysis by human renal dihydropeptidase I and showed good clinical and microbiological efficacy in the treatment of patients with intra-abdominal, lower respiratory tract, and complicated urinary tract infections. After intravenous administration it is widely distributed, has linear pharmacokinetics, and is mainly eliminated in the urine with a half-life of approximately 1 h. Biapenem is well tolerated with the most common adverse side effects being skin eruptions/rashes, nausea, and diarrhea.

1.03.2.1.3 Cefcapene pivoxil

Trade name	Flomox
Manufacturer	Shionogi
Country of origin	Japan
Year of introduction	1997
Country in which first launched	Japan
CAS registry number	105889-45-0
Structure	

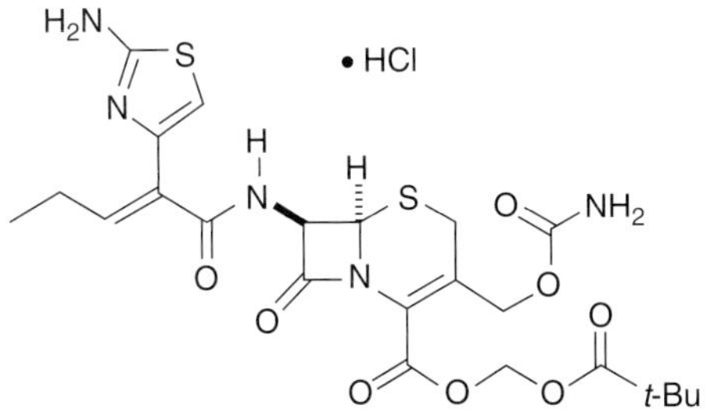

Cefcapene pivoxil functions as a cell wall synthesis inhibitor and is highly active against a wide variety of Gram-positive and Gram-negative bacteria. It is not effective against strains such as *Pseudomonas aeruginosa* and enteroccoci. It was launched as an orally active cephalosporin for respiratory and urinary tract infections, skin/soft tissue infections, and for use in gynecology, dentistry, and oral surgery. This is a prodrug with loss of the pivaloyloxymethyl group giving rise to the active acid.

1.03.2.1.4 **Cefditoren pivoxil**

Trade name	Meiact
Manufacturer	Meiji Seika
Country of origin	Japan
Year of introduction	1994
Country in which first launched	Japan
CAS registry number	117467-28-4
Structure	

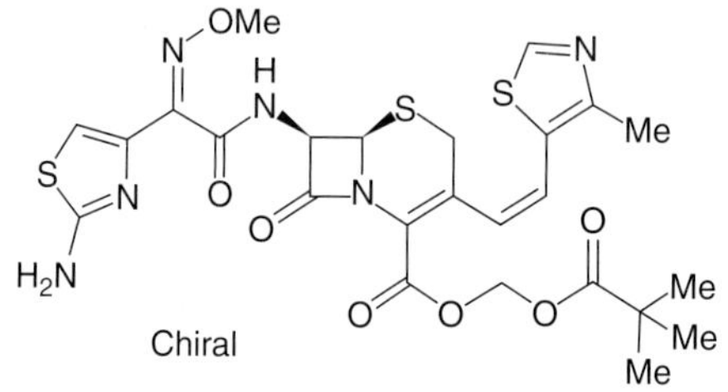

Cefditoren pivoxil is an orally active third-generation cephalosporin that is reported to have a broad spectrum of activity against both Gram-positive and Gram-negative organisms. It was introduced for the treatment of a broad range of bacterial infections including dermatological and other community-acquired infections. Its potency is reported to be greater than many existing agents in this class. It is a prodrug and the free acid, cefditoren, is produced in vivo.

1.03.2.1.5 **Cefoselis**

Trade name	Wincef
Manufacturer	Fujisawa
Country of origin	Japan
Year of introduction	1998
Country in which first launched	Japan
CAS registry number	122841-12-7
Structure	

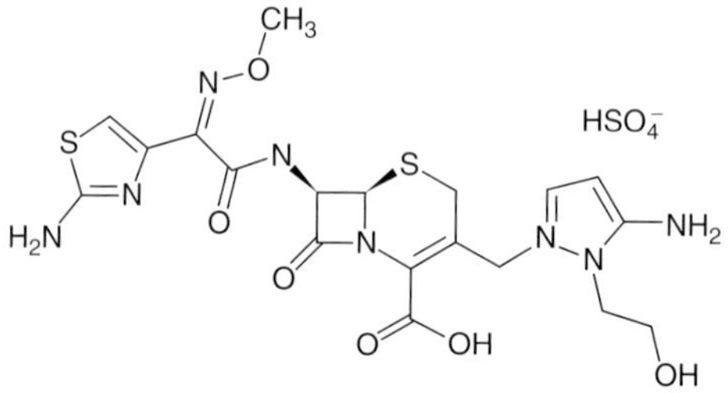

Cefoselis is a fourth-generation cephalosporin, which was launched as a parenteral antibiotic against a variety of infections including methicillin-resistant MRSA and *P. aeruginosa*.

1.03.2.1.6 Cefozopran hydrochloride

Trade name	Firstcin
Manufacturer	Takeda
Country of origin	Japan
Year of introduction	1995
Country in which first launched	Japan
CAS registry number	125905-00-2
Structure	

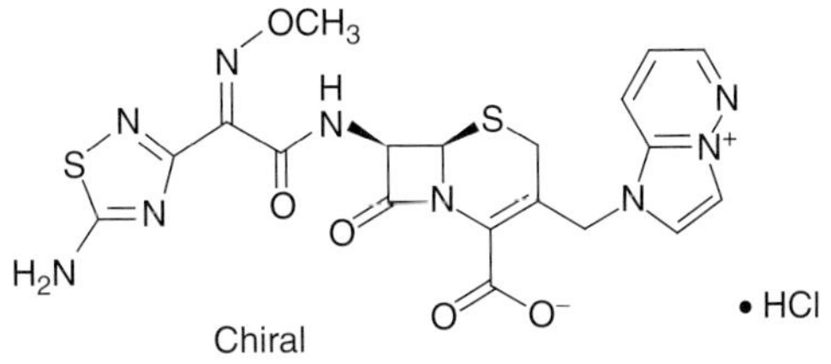

Cefozopran hydrochloride is a third-generation cephalosporin that was launched for the treatment of severe infections in immunocompromised patients caused by staphylococci and enterococci. While it shows a very broad antibacterial spectrum against Gram-positive and Gram-negative organisms, it is particularly potent against *S. aureus*, *Enterococcus faecalis*, *P. aeruginosa*, and *Citrobacter freundii*. It is resistant to hydrolysis by most chromosomal and plasmid mediated β-lactamases and is reported to be active against respiratory, urinary tract, obstetrical, gynecological, soft tissue, and surgical infections.

1.03.2.1.7 Dalfopristin/quinopristin

Trade name	Synercid
Manufacturer	Rhone-Poulenc-Rorer
Country of origin	France
Year of introduction	1999
Country in which first launched	UK
CAS registry number	126602-98-9
Structure	

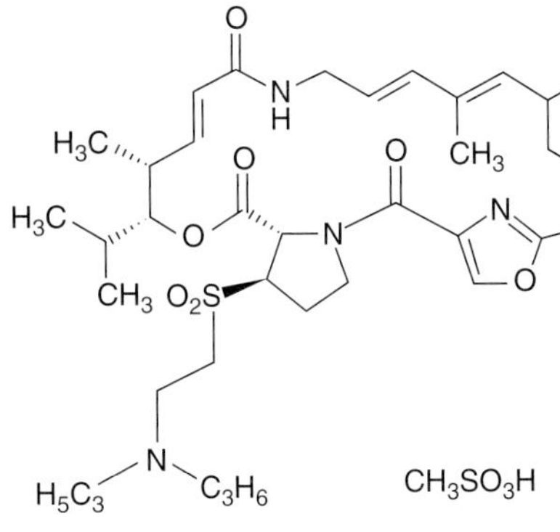

Dalfopristin (methanesulfonate)

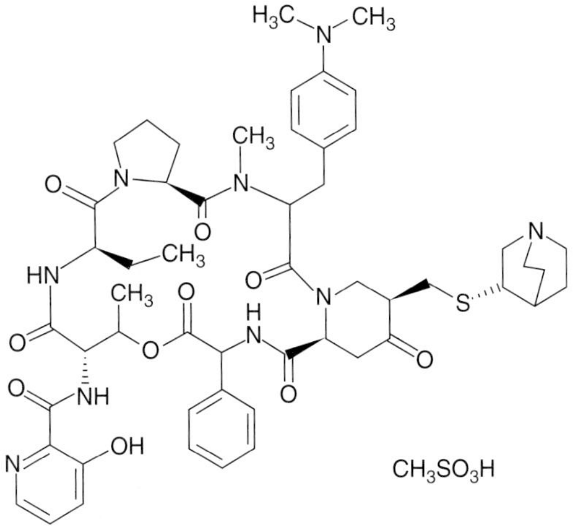

Quinupristine (methanesulfonate)

Dalfopristin and quinupristine are two well-defined semisynthetic antibacterials and are combined in the ratio of 70:30 to produce an injectable antibiotic. The combination of the two antibiotics, which are bacteriostatic in their own right, acts at the ribosomal level to inhibit protein synthesis. This is the first antibiotic in its class to reach the market and appears to be effective in the treatment of severe or life-threatening infections such as Gram-positive nosocomial sepsis, including those caused by vancomycin-resistant *E. faecium* or MRSA.

1.03.2.1.8 Daptomycin

Trade name	Cubicin
Manufacturer	Lilly
Country of origin	USA
Year of introduction	2003
Country in which first launched	USA
CAS registry number	103060-53-2
Structure	

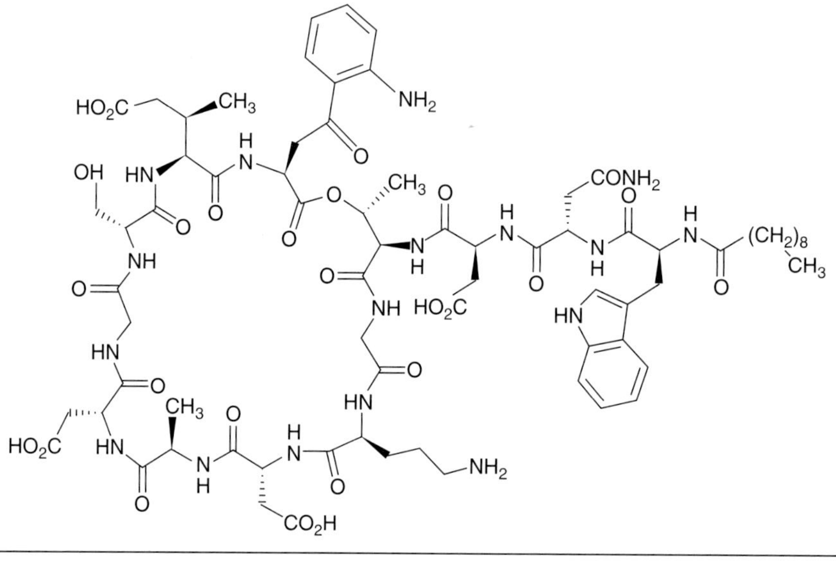

Daptomycin is the first example of a new class of lipopeptide antibiotics that work by disrupting bacterial membrane function at a number of points, i.e., disruption of membrane potential and amino acid transport, inhibition of lipoteichoic acid synthesis, and inhibition of peptidoglycan synthesis. It is indicated for the treatment of complicated skin and skin structure infections caused by a range of Gram-positive bacteria and, due to its novel mode of action, cross-resistance with other antibiotics has not been noted. Dosing is once per day ($4\,mg\,kg^{-1}\,day^{-1}$) by intravenous infusion; the drug has a half-life of 8.1 h. As it is excreted renally, dosing adjustments are required for those with severe renal insufficiency. Clinical trials indicate a more rapid activity in skin and soft tissue infections than was found for vancomycin or semisynthetic penicillins such as cloxacillin, oxacillin, or flucloxacillin.

1.03.2.1.9 Ertapenem sodium

Trade name	Invanz
Manufacturer	AstraZeneca
Country of origin	UK
Year of introduction	2002
Country in which first launched	USA
CAS registry number	153773-38-3

Structure

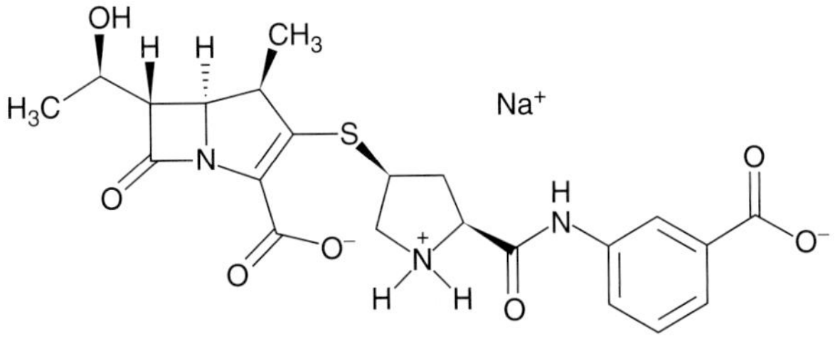

Ertapenem is a bacterial cell wall biosynthesis inhibitor with activity against a wide range of Gram-negative and Gram-positive bacteria and resistance to a broad and extended spectrum of β-lactamases. It showed clinical activity in the treatment of obstetric and gynecological infections, skin and soft tissue infections, community-acquired pneumonia, urinary tract infections, and in intra-abdominal infections. The pharmacokinetics show extended serum half-lives compared to currently available carbapenems and cephalosporins.

1.03.2.1.10 **Flurithromycin ethylsuccinate**

Trade name	Ritro
Manufacturer	Pharmacia & Upjohn
Country of origin	UK
Year of introduction	1997
Country in which first launched	Italy
CAS registry number	82730-23-2
Structure	

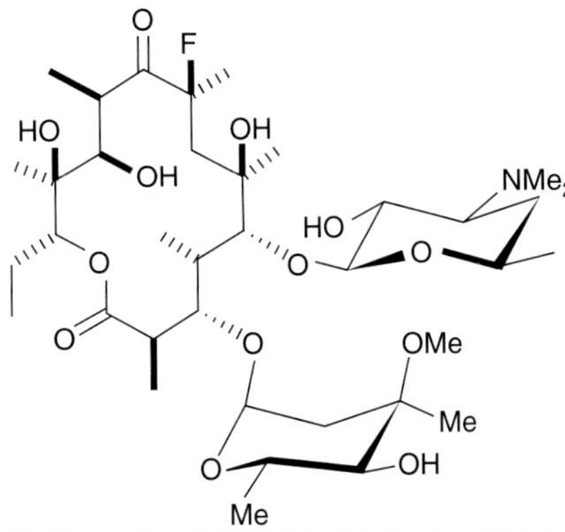

Flurithromycin is used for the treatment of serious nosocomial respiratory infections. The presence of the fluorine atom gives it improved acid stability, prolonged serum half-life, higher tissue penetration, and better bioavailability over erythromycin.

1.03.2.1.11 **Fropenam**

Trade name	Farom
Manufacturer	Suntory
Country of origin	Japan
Year of introduction	1997
Country in which first launched	Japan
CAS registry number	106560-14-9

Structure

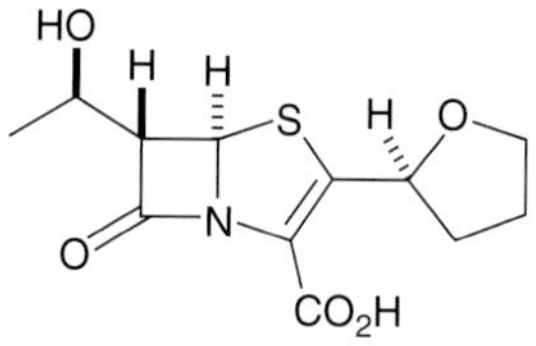

Fropenam is a β-lactamase stable, broad-spectrum oral penem antibiotic with activity against anaerobes, Gram-positive, and Gram-negative bacteria, and the Enterobacteriaceae. It was launched for the treatment of common respiratory tract infections.

1.03.2.1.12 Gatifloxacin

Trade name	Tequin
Manufacturer	Kyori
Country of origin	Japan
Year of introduction	1999
Country in which first launched	USA
CAS registry number	160738-57-8
Structure	

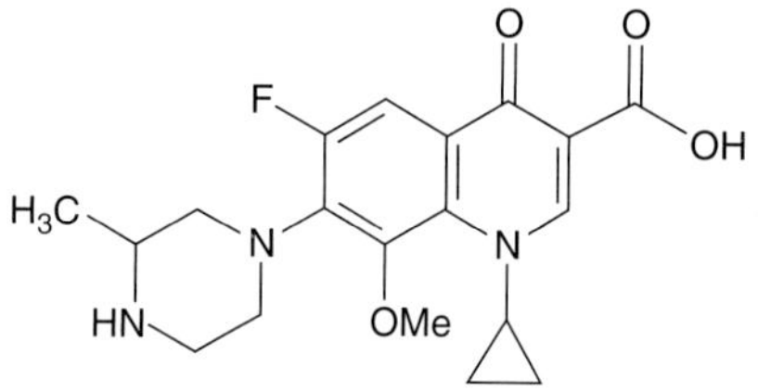

Gatifloxacin is a novel, orally active antibiotic that shows good activity in the treatment of respiratory tract and urinary infections, particularly community-acquired infections including bronchitis, pneumonia, and the common sexually transmitted diseases.

1.03.2.1.13 Linezolid

Trade name	Zyvox
Manufacturer	Pharmacia Corp.
Country of origin	USA
Year of introduction	2000
Country in which first launched	USA
CAS registry number	165800-03-3
Structure	

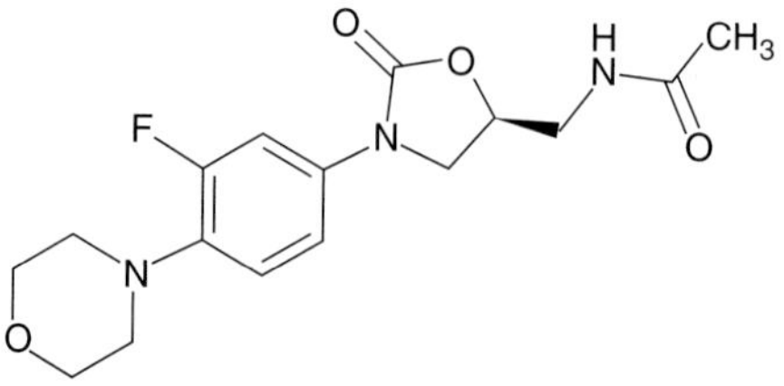

Linezolid can be considered as the first of a new class of antibacterials, the oxazolidinones, which act by inhibiting early ribosomal protein synthesis without directly inhibiting DNA or RNA synthesis. In vitro, linezolid is as potent as vancomycin against staphylococcal, streptococcal, and pneumococcal infections, and enterococcal species. It was launched for the treatment of patients with infections caused by serious Gram-positive pathogens, particularly skin and soft tissue infections, community-acquired pneumonia, and vancomycin-resistant enterococcal infections.

1.03.2.1.14 **Meropenem**

Trade name	Merrem
Manufacturer	Sumitomo/Zeneca
Country of origin	Japan
Year of introduction	1994
Country in which first launched	Italy
CAS registry number	96036-03-2
Structure	

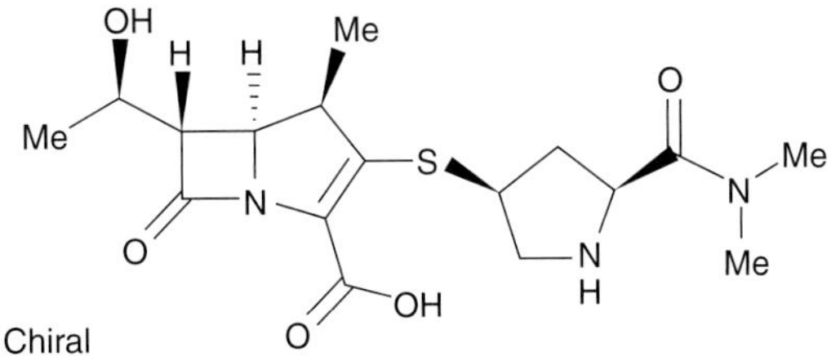

Meropenem has a broad spectrum of antibacterial activity against most clinically important Gram-positive and Gram-negative aerobic and anaerobic bacteria with especially high potency against multiresistant members of the Enterobacteriaceae and *P. aeruginosa*. It is dehydropeptidase 1 stable and is proposed for the intravenous treatment of hospital infections such as lower respiratory tract, urinary tract, intra-abdominal, gynecological, and polymicrobial infections.

1.03.2.1.15 **Moxifloxacin hydrochloride**

Trade name	Avelox
Manufacturer	Bayer
Country of origin	Germany
Year of introduction	1999
Country in which first launched	Germany
CAS registry number	186826-86-8
Structure	

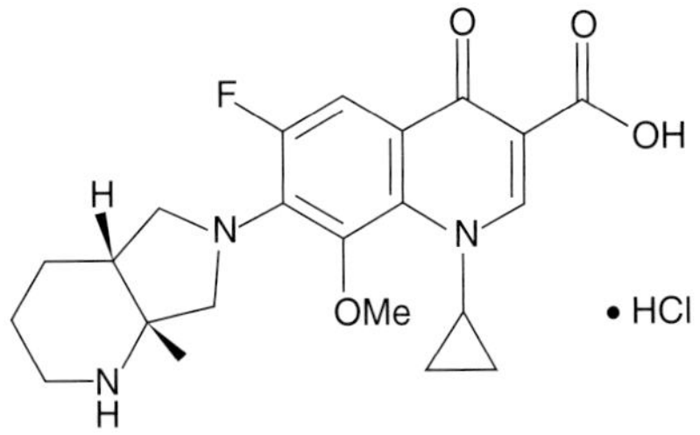

Moxifloxin is a fluoroquinolonecarboxylic acid-derived antibiotic that was introduced for the treatment of respiratory tract infections such as community acquired pneumonia, acute exacerbations of bronchitis, or acute sinusitis. It shows a favorable pharmacokinetic profile with good tissue penetration and plasma concentrations above minimum inhibitory concentrations (MICs), and a lack of phototoxicity.

1.03.2.1.16 **Panipenem/betamipron**

Trade name	Carbenin
Manufacturer	Sankyo
Country of origin	Japan
Year of introduction	1994
Country in which first launched	Japan
CAS registry number	138240-65-0
Structure	

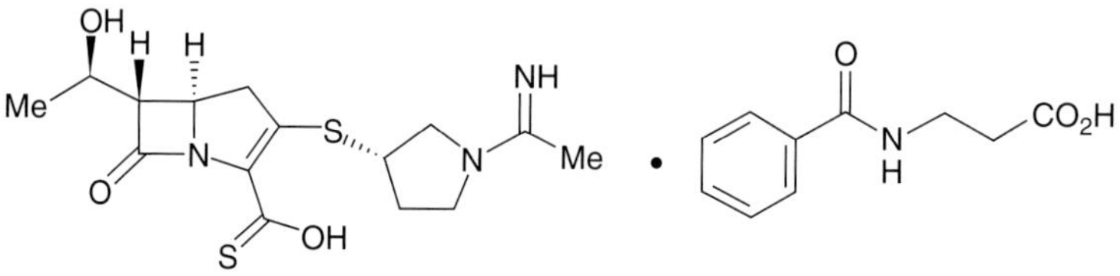

Panipenem is a semisynthetic carbapenem antibiotic that is efficacious in patients with severe infections complicating hematological disorders and in children where other antibiotics have been ineffectual. Betamipron, when mixed with panipenem in a 1:1 ratio, blocks the incorporation of panipenem into renal tubules, thus preventing renal dysfunction.

1.03.2.1.17 **Pazufloxacin**

Trade name	Pasil, Pazucross
Manufacturer	Toyama/Mitsubishi
Country of origin	Japan
Year of introduction	2002
Country in which first launched	Japan
CAS registry number	127045-41-4
Structure	

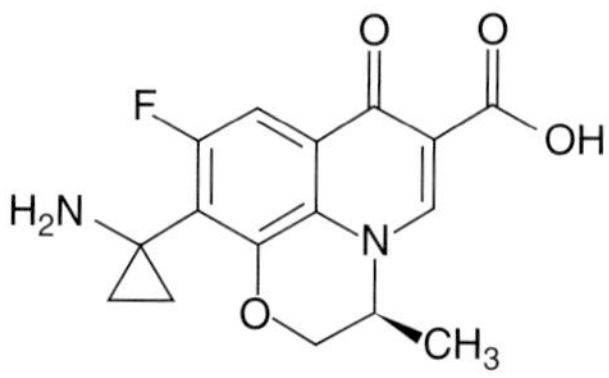

Pazufloxacin displays broad-spectrum activity against Gram-positive and Gram-negative bacteria, although it is less active than ciprofloxacin against pneumococci and is not active against ciprofloxacin-resistant bacteria. Good clinical responses have been seen in patients with urinary tract infections and, to a lesser extent, with respiratory tract infections. It has a short half-life of 2–2.5 h with a phototoxicity equal to that of ciprofloxacin.

1.03.2.1.18 **Prulifloxacin**

Trade name	Sword
Manufacturer	Nippon Shinyaku
Country of origin	Japan
Year of introduction	2002
Country in which first launched	Japan
CAS registry number	123447-62-1
Structure	

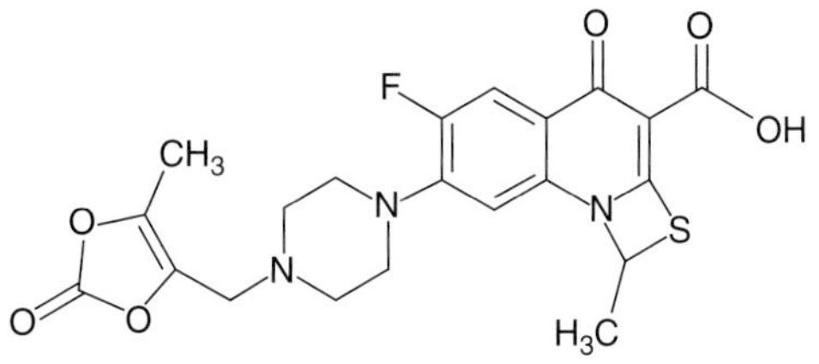

Prulifloxacin is another fluoroquinolonecarboxylic acid-derived antibiotic, which was introduced for the oral treatment of urinary tract infections, respiratory tract infections, and bacterial pneumoniae. It is a prodrug that is rapidly hydrolyzed by paraoxonase-type enzymes in the blood and liver to give the DNA gyrase inhibitor NM 394. This metabolite accounts for all the antibiotic activity. Overall, activity against Gram-positive bacteria is similar to ciprofloxacin but greater in the case of Gram-negative bacteria. Plasma half-life is around 8 h and any adverse effects are similar to those of other fluoroquinolones.

1.03.2.1.19 **Telithromycin**

Trade name	Ketek
Manufacturer	Aventis
Country of origin	France
Year of introduction	2001
Country in which first launched	Germany
CAS registry number	191114-48-4
Structure	

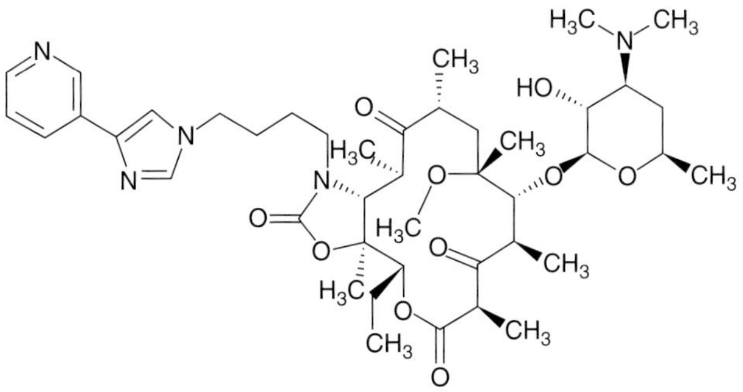

Telithromycin was introduced as a once daily oral treatment for respiratory infections including community-acquired pneumonia, acute bacterial exacerbations of chronic bronchitis, acute sinusitis, and tonsillitis/pharangitis. It is a semisynthetic derivative of the macrolide erythromycin and acts by preventing bacterial protein synthesis by binding to two domains of the 50S subunit bacterial ribosomes. It does not form a stable inhibitory cytochrome P450 Fe^{2+}–nitrosoalkene metabolite complex and therefore should not show hepatotoxity.

1.03.2.1.20 Trimetrexate glucuronate

Trade name	NeuTrexin
Manufacturer	Warner-Lambert/ US Bioscience
Country of origin	USA
Year of introduction	1994
Country in which first launched	USA, Canada
CAS registry number	82952-64-5
Structure	

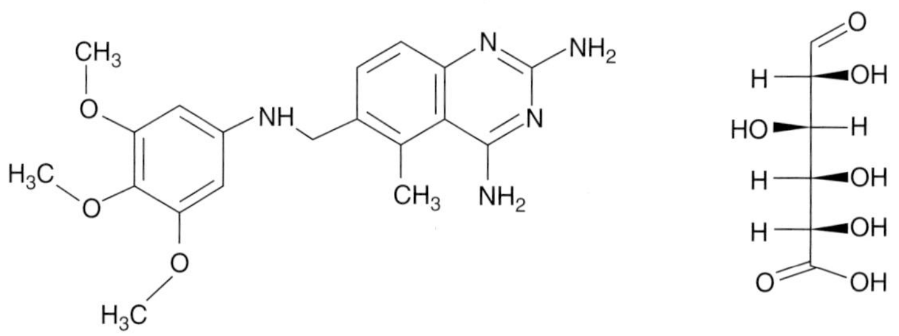

Trimetrexate glucuronate is a nonclassical antifolate that inhibits dihydrofolate reductase and was introduced for the treatment of *Pneumonocystis carinii* pneumonia in patients with AIDS. It is also in clinical trials against a number of cancers.

1.03.2.1.21 Trovafloxacin mesylate

Trade name	Trovan
Manufacturer	Pfizer
Country of origin	Switzerland
Year of introduction	1998
Country in which first launched	US/Switzerland
CAS registry number	157605-25-9
Structure	

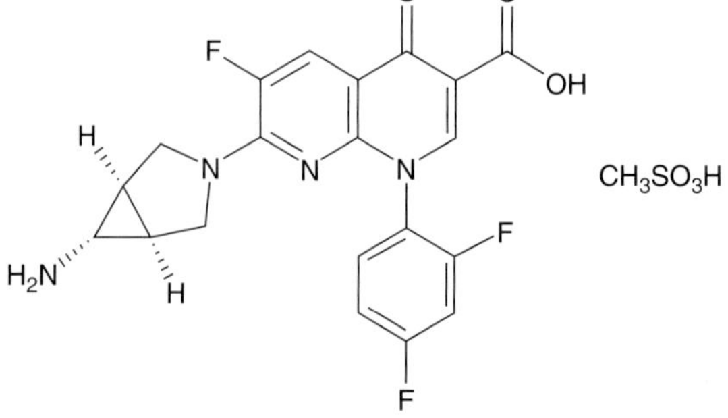

Trovafloxacin mesylate is a fluoroquinolone-derived antibiotic that is given once a day for the treatment of diverse acute bacterial infections, particularly community-acquired respiratory infections. Doses of 200 mg day^{-1} demonstrated advantages over amoxicillin, cephalosporins, and clarithromycin.

1.03.2.2 Antivirals

1.03.2.2.1 Abacavir sulfate

Trade name	Ziagen
Manufacturer	Glaxo Wellcome
Country of origin	UK
Year of introduction	1999
Country in which first launched	USA
CAS registry number	188062-50-2
Structure	

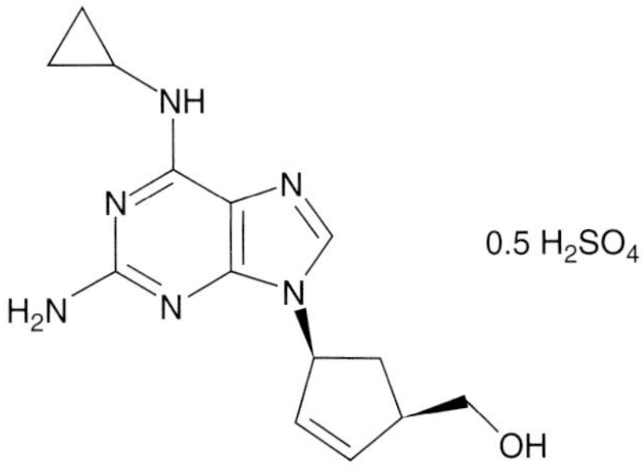

Abacavir is a carbocyclic nucleoside reverse transcriptase inhibitor and is a potent and selective inhibitor of HIV-1 and HIV-2 replication. Resistance to abacavir appears to develop only slowly and, when used in combination with other antiretroviral drugs such as amprenavir, produces durable suppression of viral loads.

1.03.2.2.2 Adefovir dipivoxil

Trade name	Hepsera
Manufacturer	Institute of Organic Chemistry and Biochemistry of the Academy of Sciences in the Czech Republic and the REGA Stichting Research Institute/Gilead
Country of origin	Czech Republic and Belgium
Year of introduction	2002
Country in which first launched	USA
CAS registry number	142340-99-6
Structure	

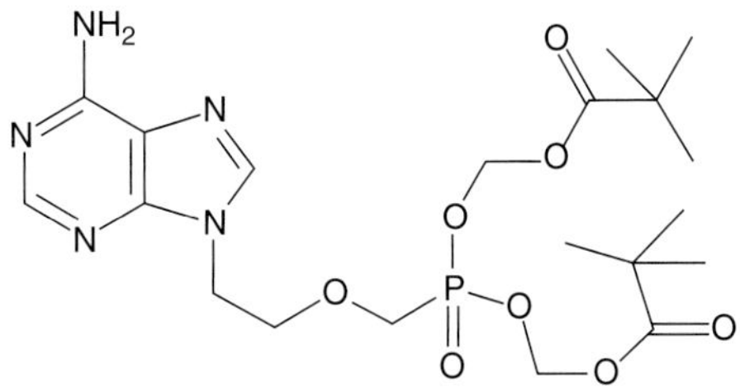

Adefovir dipivoxil was launched as a treatment for hepatitis B viral infections. It is also a prodrug of adefovir, which is phosphorylated twice to give the active species in vivo. Its relatively long half-life enables once daily dosing but nephrotoxicity restricts the possible level of dosing.

1.03.2.2.3 Amprenavir

Trade name	Agenerase
Manufacturer	Vertex
Country of origin	USA
Year of introduction	1999
Country in which first launched	USA
CAS registry number	161814-49-9
Structure	

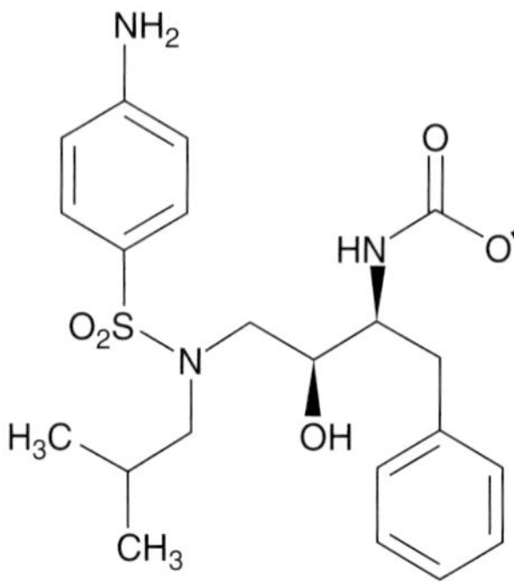

Amprenavir was the fifth nonpeptidic inhibitor of HIV-1 protease to be marketed and was intended for the treatment of AIDS patients in combination with approved antiretroviral nucleoside analogs. It potently inhibits HIV-1 aspartyl protease and displays a good oral bioavailability in humans with penetration into the CNS. In combination it considerably decreases viral load and restores $CD4^+$ T-cell counts in patients with HIV infection.

1.03.2.2.4 Atazanavir

Trade name	Reyataz
Manufacturer	Novartis
Country of origin	USA
Year of introduction	2003
Country in which first launched	USA
CAS registry number	198904-31-3
Structure	

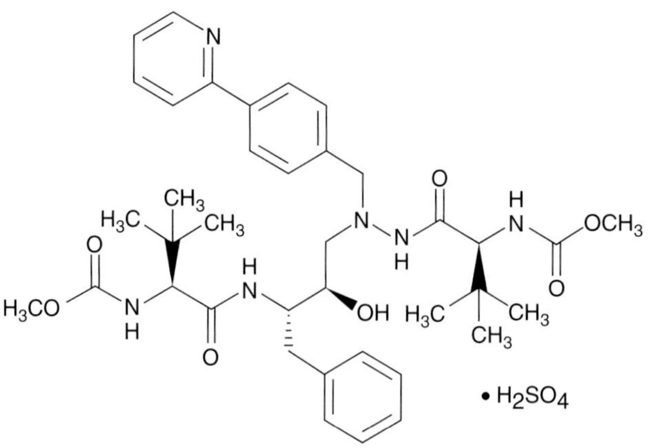

Atazanavir is an inhibitor of HIV-1 protease and is potently active against indanavir- and saquinavir-resistant strains. Recommended dosage is 400 mg once daily and it appears to be well tolerated. It showed marked activity in reducing HIV RNA levels in patients who had not responded to previous treatments.

1.03.2.2.5 Cidofovir

Trade name	Vistide
Manufacturer	Gilead Sciences
Country of origin	USA
Year of introduction	1996
Country in which first launched	USA
CAS registry number	113852-37-2
Structure	

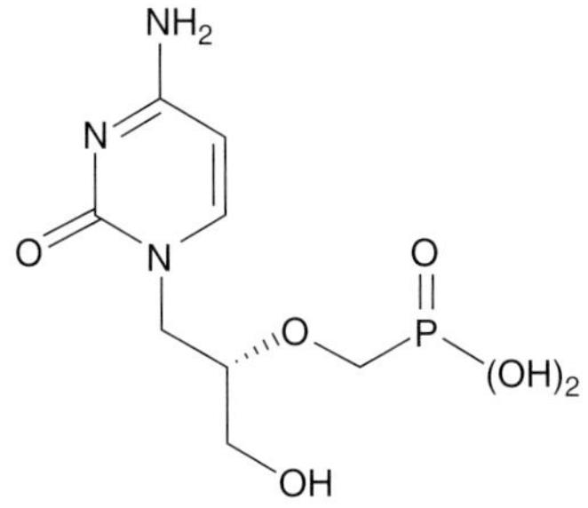

Cidofovir is rapidly converted to its active diphosphate metabolite, which inhibits viral DNA polymerase. It was launched for the treatment of cytomegalovirus retinitis in AIDS patients and is sufficiently long lived to require dosing only once every 1–2 weeks.

1.03.2.2.6 Delaviridine mesylate

Trade name	Rescriptor
Manufacturer	Pharmacia & Upjohn
Country of origin	USA
Year of introduction	1997
Country in which first launched	USA
CAS registry number	147221-93-0
Structure	

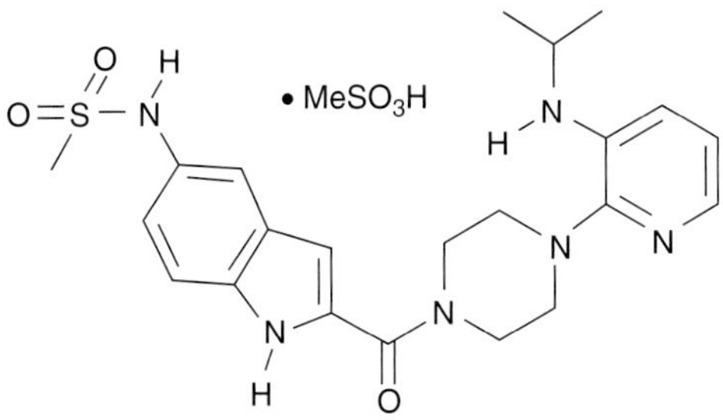

Delaviridine mesylate is a second-generation non-nucleoside HIV-1 reverse transcriptase inhibitor, which acts as an allosteric mixed inhibitor of both RNA- and DNA-directed polymerase domains of reverse transcriptase and binds more strongly to the enzyme–substrate complex than to the free enzyme.

1.03.2.2.7 Efavirenz

Trade name	Sustiva
Manufacturer	Merck
Country of origin	USA
Year of introduction	1998
Country in which first launched	USA
CAS registry number	154598-52-4
Structure	

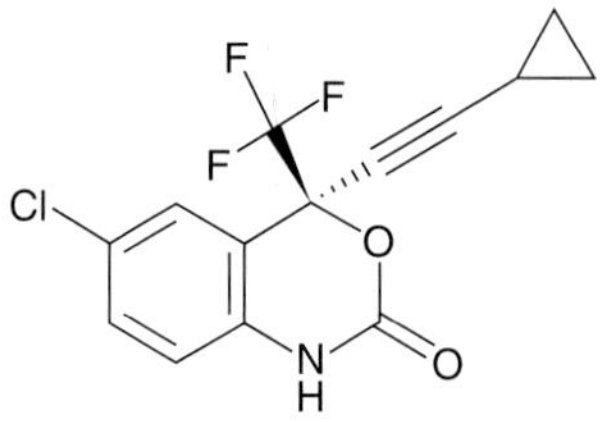

Efavirenz is a non-nucleoside reverse transcriptase inhibitor for the treatment of HIV infections in combination with other antiretroviral agents. It readily crosses the blood–brain barrier resulting in an increased concentration in the cerebrospinal fluid. The pharmacokinetic profile appears to be better than other drugs in this class.

1.03.2.2.8 Emtricitabine

Trade name	Emtriva
Manufacturer	Emory University/Gilead
Country of origin	USA
Year of introduction	2003
Country in which first launched	USA
CAS registry number	143491-57-0
Structure	

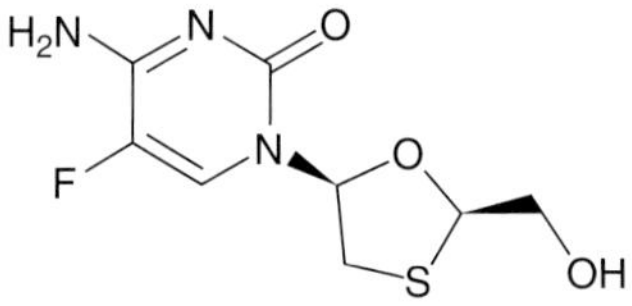

Emtricitabine is a synthetic nucleoside inhibitor of HIV-1 reverse transcriptase. Recommended treatment is 200 mg day^{-1}; there are mild to moderate side effects at this dose. Cross-resistance to lamivudine and zalcitabine is seen.

1.03.2.2.9 Enfuvirtide

Trade name	Fuzeon
Manufacturer	Duke University/Roche/Trimeris
Country of origin	USA
Year of introduction	2003
Country in which first launched	USA
CAS registry number	159519-65-0

Enfuvirtide is the first of a new class of HIV therapeutics that interferes with the entry of HIV-1 by inhibiting fusion of viral and cellular membranes. Because of this unique mechanism it does not show cross-resistance to known agents. The dose is 90 mg twice daily by subcutaneous injection, and it is used in a cocktail with other anti-HIV agents.

1.03.2.2.10 Famciclovir

Trade name	Famvir
Manufacturer	SmithKlineBeecham
Country of origin	UK
Year of introduction	1994
Country in which first launched	UK, USA
CAS registry number	104227-87-4
Structure	

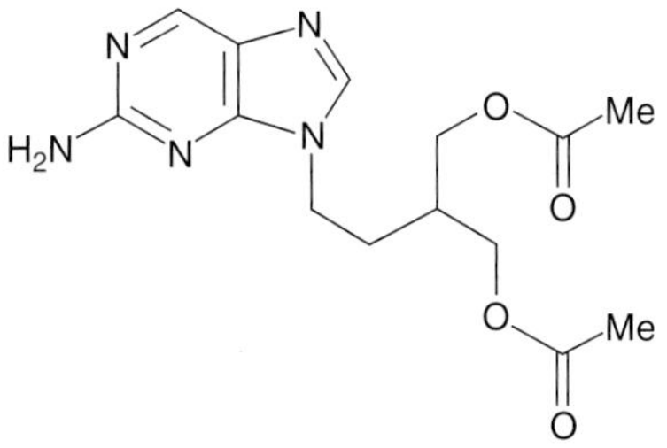

Famciclovir is an orally active prodrug of penciclovir and was launched for the treatment of herpes zoster. Selectivity is achieved by virtue of the fact that penciclovir is only phosphorylated, and hence activated in herpesvirus-infected cells. The metabolite is a strong inhibitor of herpesvirus DNA polymerases and viral DNA synthesis.

1.03.2.2.11 Formivirsen sodium

Trade name	Vitravene
Manufacturer	Isis Pharmaceuticals
Country of origin	USA
Year of introduction	1998
Country in which first launched	USA
CAS registry number	160369-77-7

Vitravene is indicated for the treatment of cytomegalovirus-induced retinitis in patients with AIDS. It acts by an antisense mechanism and targets an intermediate early mRNA of cytomegalovirus, thus inhibiting the expression of two regulatory proteins. Vitravene was thus the first agent based on antisense technology to reach the market.

1.03.2.2.12 Fosamprenavir

Trade name	Lexiva
Manufacturer	Vertex/GlaxoSmithKline
Country of origin	USA
Year of introduction	2003
Country in which first launched	USA
CAS registry number	226700-81-8

Structure

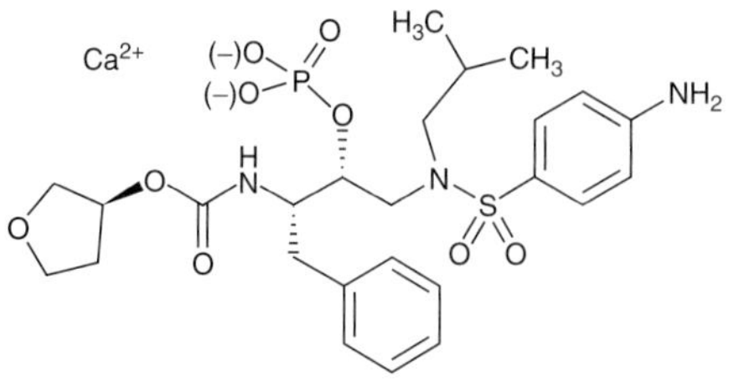

Fosamprenavir is a prodrug of the HIV protease inhibitor amprenavir that is highly active but its low water solubility reduces its utility. Fosamprenavir is rapidly hydrolyzed by phosphatases in the gut epithelium to give good levels of amprenavir. In the clinic, 700–1400 mg are administered daily in conjunction with ritonavir.

1.03.2.2.13 **Imiquinod**

Trade name	Aldara
Manufacturer	3 M Pharmaceuticals
Country of origin	USA
Year of introduction	1997
Country in which first launched	USA
CAS registry number	99011-02-6
Structure	

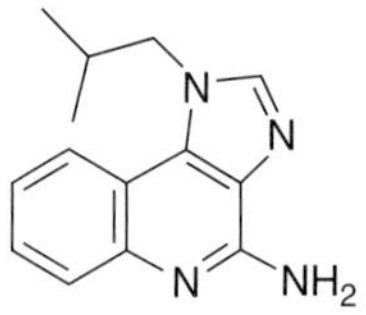

Imiquinod was launched for the topical treatment of genital warts caused by human papillomavirus. It works by the induction of cytokines, in particular interferon alpha, possibly by interaction with a kinase modulating the transduction pathway leading to cytokine genes.

1.03.2.2.14 **Indinavir sulfate**

Trade name	Crixivan
Manufacturer	Merck
Country of origin	USA
Year of introduction	1996
Country in which first launched	USA
CAS registry number	150378-17-9
Structure	

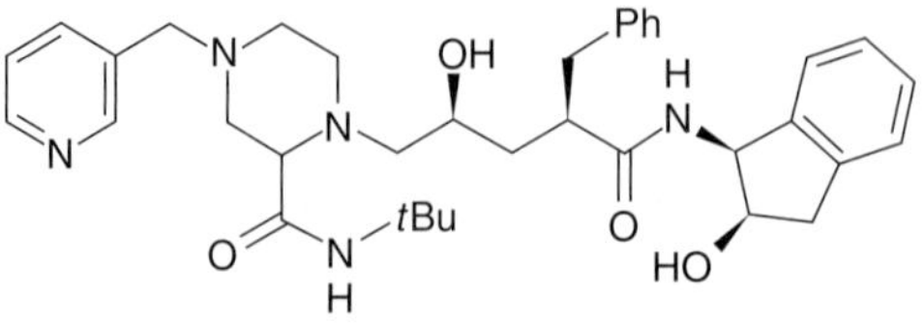

Indinavir sulfate was launched as an orally bioavailable HIV-1 protease inhibitor. It appears to be relatively safe and well tolerated.

1.03.2.2.15 **Interferon alfacon-1**

Trade name	Infergen
Manufacturer	Amgen
Country of origin	USA
Year of introduction	1997
Country in which first launched	USA
CAS registry number	74899-72-2
Structure – protein of molecular weight 19 600	

Infergen was introduced for the treatment of chronic hepatitis C. It is more effective than other interferons in its antiproliferative effects, in activating natural killer (NK) cells, and in immunomodulation.

1.03.2.2.16 **Lamivudine**

Trade name	Epivir
Manufacturer	Biochem Pharma
Country of origin	Canada
Year of introduction	1995
Country in which first launched	USA
CAS registry number	134678-17-4
Structure	

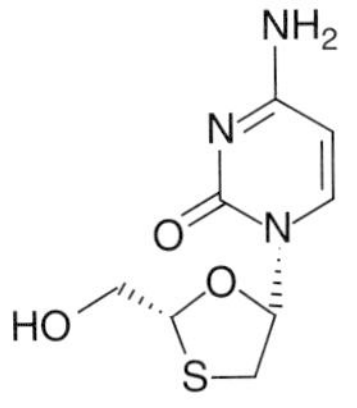

Lamivudine is a new generation orally active nucleoside analog that was launched for use in combination with AZT as a first-line therapy for patients infected with HIV. Its phosphorylated metabolite is an inhibitor of reverse transcriptase. It is less toxic than AZT and appears also to have activity against hepatitis B virus.

1.03.2.2.17 **Lopinavir**

Trade name	Kaletra
Manufacturer	Abbott
Country of origin	USA
Year of introduction	2000
Country in which first launched	USA
CAS registry number	192725-17-0
Structure	

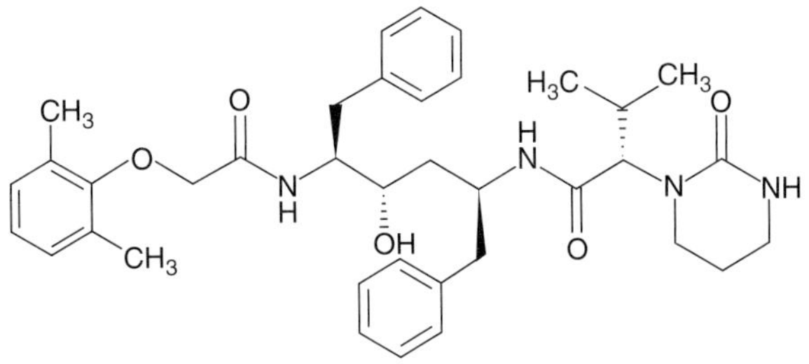

Lopinavir was launched as a joint formulation with ritonavir, an established HIV protease inhibitor. Ritonavir inhibits the cytochrome P450 isoenzyme CYP3A4 and thus the microsomal metabolism of lopinavir resulting in increased plasma levels and duration of action of lopinavir. In patients with AIDS, the plasma HIV RNA level was reduced and the CD^{4+} T-cell counts increased after administration of lopinavir with relatively small doses of ritonavir.

1.03.2.2.18 Neflinavir mesylate

Trade name	Viracept
Manufacturer	Agouron
Country of origin	USA
Year of introduction	1997
Country in which first launched	USA
CAS registry number	159989-65-8
Structure	

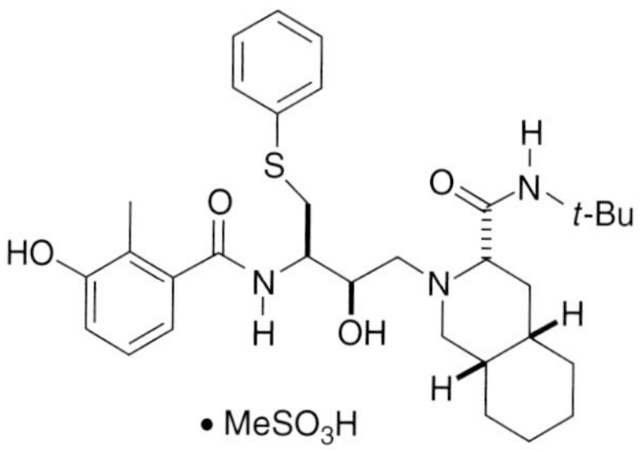

Viracept is an orally available, nonpeptidic HIV protease inhibitor with low toxicity and a resistance profile different from other protease inhibitors. In combination with ziduvidine and lamivudine, it generated a 98% mean reduction from baseline in viral load after 24 weeks compared to ziduvidine and lamivudine alone.

1.03.2.2.19 Nevirapine

Trade name	Viramune
Manufacturer	Boehringer Ingelheim
Country of origin	Germany
Year of introduction	1996
Country in which first launched	USA
CAS registry number	71125-38-7
Structure	

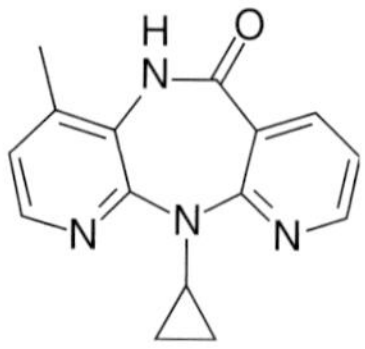

Nevirapine was launched for use in combination with nucleoside analogs to treat HIV-infected patients who had experienced clinical and/or immunologic deterioration. It is a noncompetitive non-nucleoside inhibitor of HIV-1 reverse transcriptase.

1.03.2.2.20 **Oseltamivir phosphate**

Trade name	Tamiflu
Manufacturer	Gilead
Country of origin	USA
Year of introduction	1999
Country in which first launched	Switzerland
CAS registry number	204255-11-8
Structure	

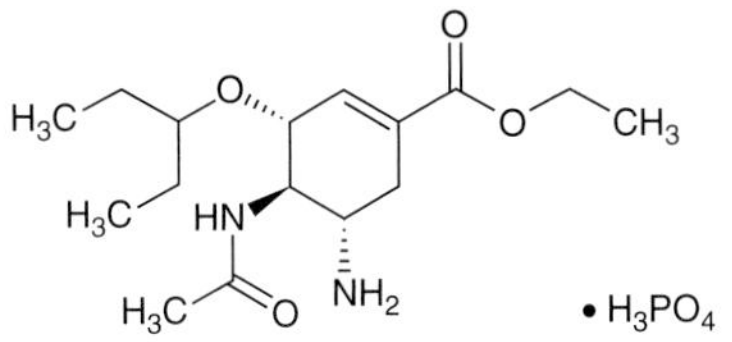

Oseltamivir phosphate was launched for the treatment of influenza infections by all common strain viruses. It is a prodrug of the acid GS-4071, which is a potent inhibitor of both influenza A and B virus neuramidase isoenzymes. Forming the ester greatly improves the oral bioavailability and clinical trials show that it significantly reduces the severity and duration of clinical symptoms including fever, cough, and general malaise.

1.03.2.2.21 **Penciclovir**

Trade name	Vectavir
Manufacture	SmithKlineBeecham
Country of origin	UK
Year of introduction	1996
Country in which first launched	UK
CAS registry number	39809-25-1
Structure	

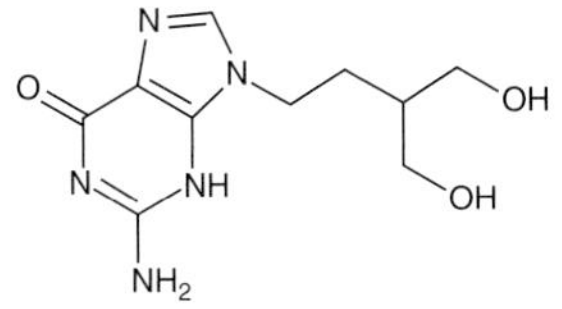

Penciclovir was launched for the treatment of herpes labialis. It acts as a competitive inhibitor of DNA polymerase and is relatively long acting.

1.03.2.2.22 **Propagermanium**

Trade name	Serosion
Manufacturer	Sanwa Kagaku
Country of origin	Japan
Year of introduction	1994
Country in which first launched	Japan
CAS registry number	12758-40-6

Structure

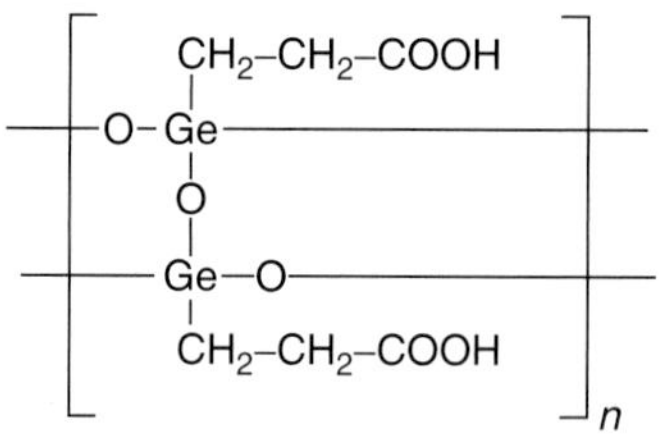

This is the first organogermanium compound to be marketed as a pharmaceutical; it was launched for the treatment of chronic hepatitis B. Its activity is thought to be the result of an enhanced immune response due to the production of interferon α/β.

1.03.2.2.23 **Ritonavir**

Trade name	Norvir
Manufacturer	Abbott
Country of origin	USA
Year of introduction	1996
Country in which first launched	USA
CAS registry number	155213-67-5
Structure	

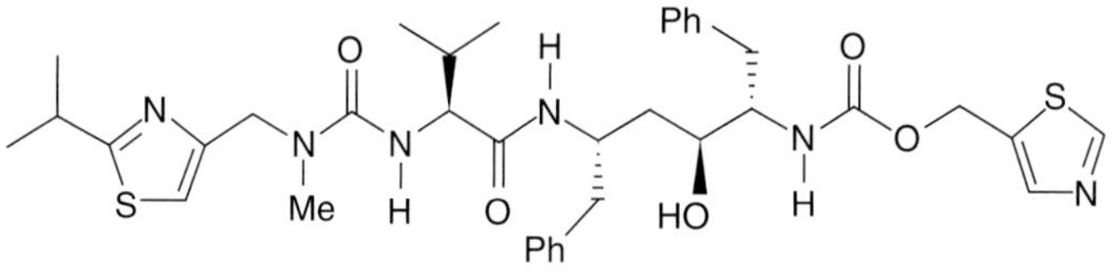

Ritonavir is an inhibitor of HIV aspartic protease. It was launched for the treatment of advanced HIV in combination with antiretroviral nucleoside analogs. Its selectivity for the viral enzyme over the human form is greater than 500-fold, and it has good oral bioavailability.

1.03.2.2.24 **Saquinavir mesylate**

Trade name	Invirase
Manufacturer	Roche
Country of origin	Switzerland
Year of introduction	1995
Country in which first launched	USA
CAS registry number	127779-20-8
Structure	

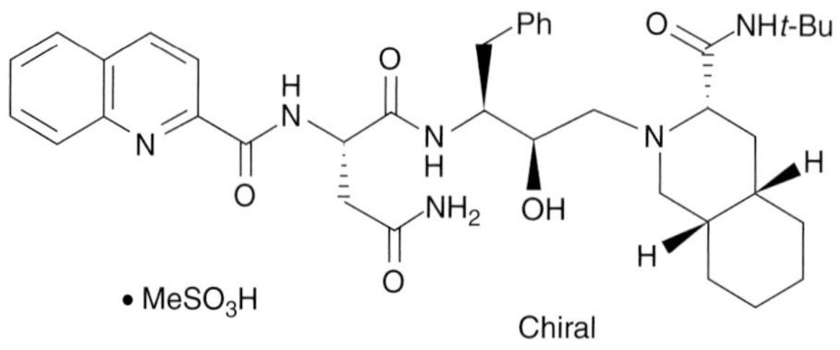

Saquinavir mesylate was the first HIV protease inhibitor to reach the market. Initial use was with approved nucleoside analogs for the treatment of advanced HIV infections. It is well tolerated alone and in combination with AZT.

1.03.2.2.25 **Stavudine**

Trade name	Zerit
Manufacturer	Bristol-Meyers Squibb
Country of origin	USA
Year of introduction	1994
Country in which first launched	USA
CAS registry number	3056-17-5
Structure	

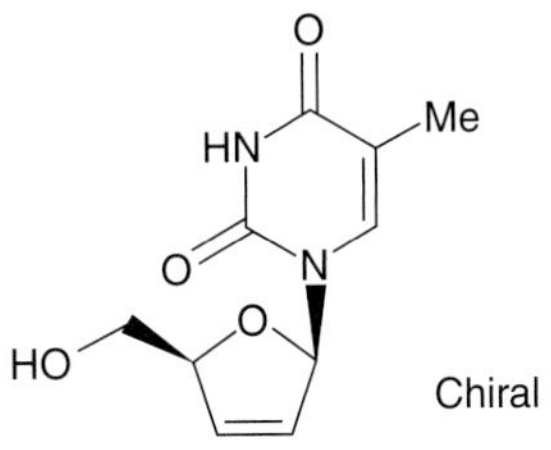

Stavudine is orally active with good bioavailability and its 5′-triphosphate is highly active against reverse transcriptase. It was introduced for the treatment of late-stage patients with AIDS refractory to other treatments.

1.03.2.2.26 **Tenofovir disproxil fumarate**

Trade name	Viread
Manufacturer	Gilead Sciences
Country of origin	USA
Year of introduction	2001
Country in which first launched	USA
CAS registry number	201238-50-9
Structure	

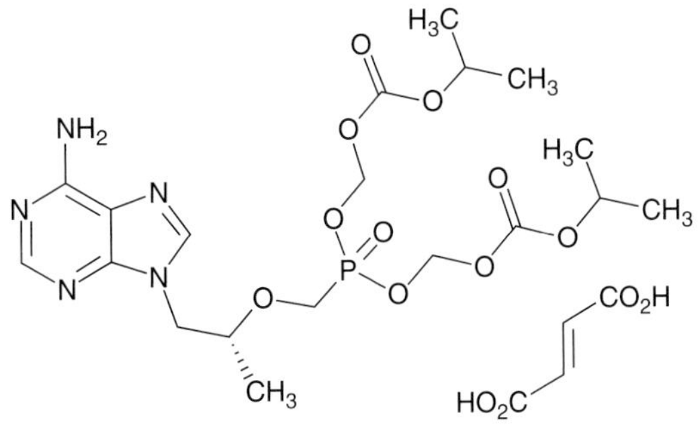

Tenofovir disproxil fumarate was the first nucleotide analog reverse transcriptase inhibitor to be launched in the US as an oral treatment for HIV infection. It is a prodrug, which rapidly hydrolyzes to form the free tenofovir. This undergoes two phosphorylation steps to give the active metabolite, a potent inhibitor of reverse transcriptase. Because of rapid cellular uptake, the prodrug is more active than tenofovir reducing the level of HIV in the blood for up to 48 weeks when added to existing antiretroviral regimens. This activity has been seen in patients bearing resistant strains of HIV. It is eliminated by the kidneys, is not metabolized by the liver, and the half-life is such that once daily dosing is possible.

1.03.2.2.27 **Valaciclovir hydrochloride**

Trade name	Valtrex
Manufacturer	Glaxo Wellcome
Country of origin	UK
Year of introduction	1995
Country in which first launched	UK
CAS registry number	124832-27-5
Structure	

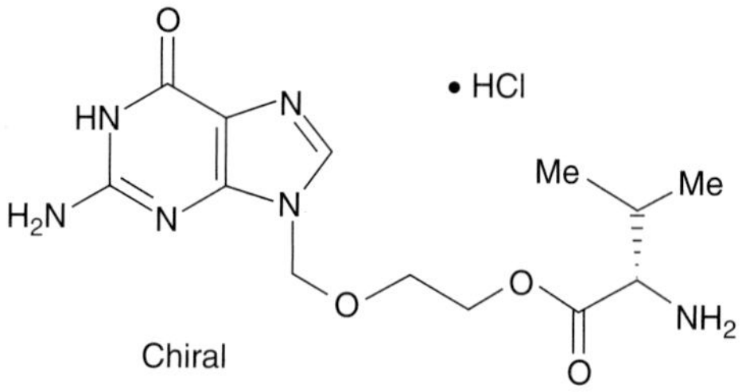

Valaciclovir is an orally active L-valyl ester of the known antiviral aciclovir. It was launched for the treatment of herpes simplex infections of skin and mucous membranes. It is readily absorbed following oral treatment and hydrolyzed in the first pass effect to aciclovir. The phosphorylated metabolite is generated in virus-infected cells and inhibits viral DNA polymerase.

1.03.2.2.28 **Valganciclovir hydrochloride**

Trade name	Valcyte
Manufacturer	Roche
Country of origin	Switzerland
Year of introduction	2001
Country in which first launched	USA
CAS registry number	175865-59-5
Structure	

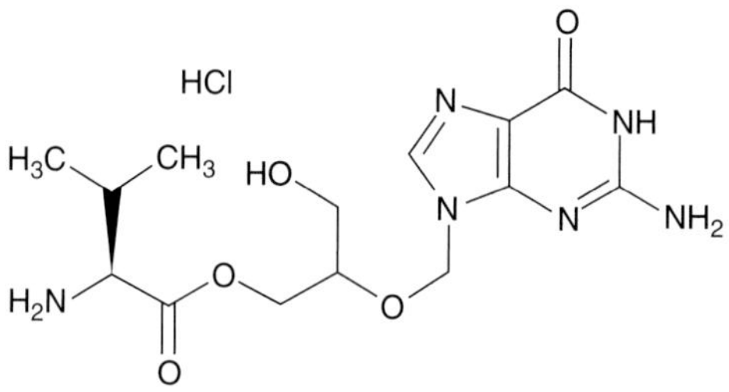

Valganciclovir hydrochloride is a prodrug of the antiviral ganciclovir. It had been marketed for the treatment of cytomegalovirus retinitis. It is rapidly hydrolyzed by intracellular esterases in the intestinal mucosal cells and by hepatic esterases. Oral treatment at 900 mg twice daily for 3 weeks followed by 900 mg once a day for a week was as effective as intravenous ganciclovir (5 $mg\,kg^{-1}$ twice a day for 3 weeks followed by 5 $mg\,kg^{-1}$ once a day for a week).

1.03.2.2.29 **Zanamivir**

Trade name	Relenza
Manufacturer	Biota Scientific
Country of origin	Australia
Year of introduction	1999
Country in which first launched	Australia
CAS registry number	139110-80-8

Structure

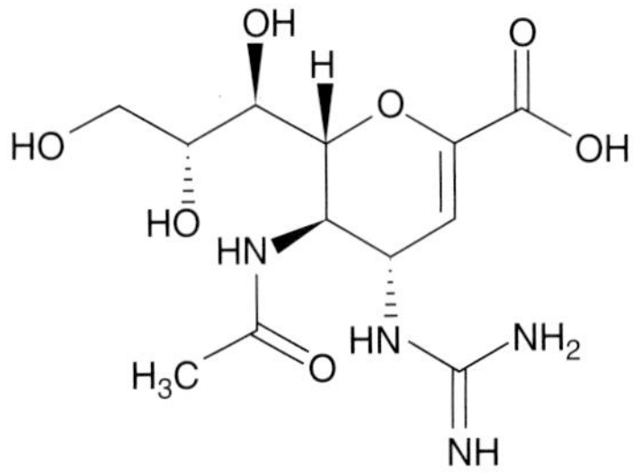

Zanamivir is a potent and specific inhibitor of neuraminidase, a key viral surface glycohydrolase essential for viral replication and disease progression and was launched for the treatment of human influenza A and B virus infections.

1.03.2.3 Antifungals

1.03.2.3.1 Caspofungin acetate

Trade name	Cancidas
Manufacturer	Merck
Country of origin	USA
Year of introduction	2001
Country in which first launched	USA
CAS registry number	179463-17-3
Structure	

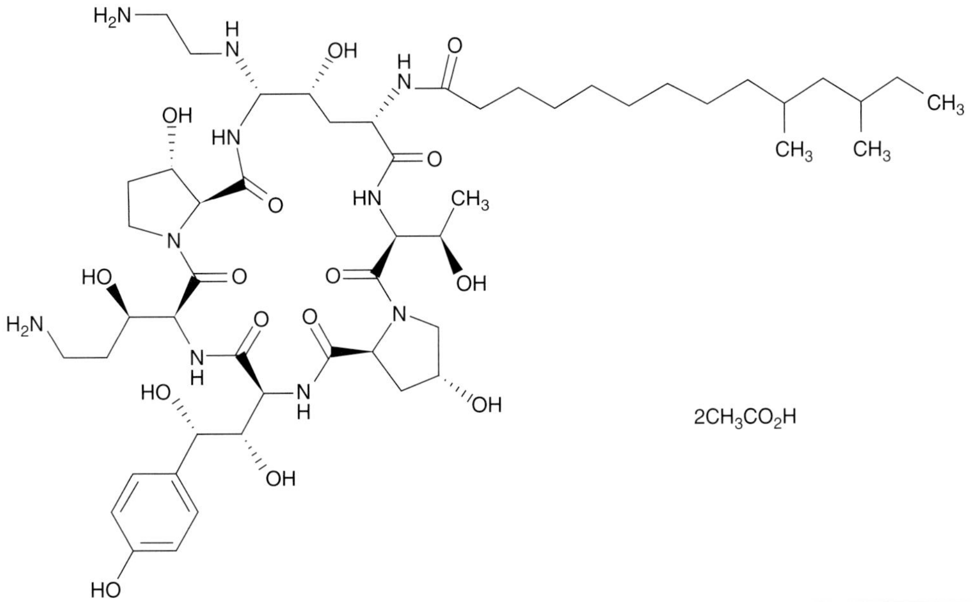

Caspofungin acetate was the first in a new class of antifungal agents, the echinocandins, to be launched for the parenteral treatment of invasive aspergillosis in patients refractive to other antifungal therapies. It inhibits the synthesis of 1,3-beta-D-glucans, which is present only in fungal cell walls, and thus is fungicidic. It is active both in vitro and in vivo against a range of *Candida* species and also *Aspergillus*. Its half-life in humans is around 10 h.

1.03.2.3.2 Flutrimazole

Trade name	Micetal
Manufacturer	Urlach
Country of origin	Spain
Year of introduction	1995

Country in which first launched	Spain
CAS registry number	119006-77-8
Structure	

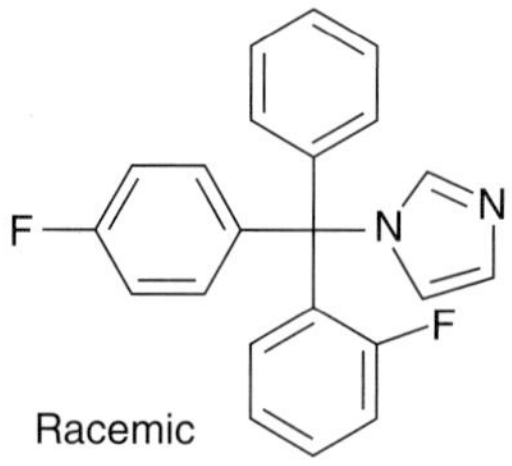

Racemic

Flutrimazole displays broad-spectrum activity against dematophytes, filamentous fungi, and yeasts, which are saprophytic and pathogenic to animals and humans. Mode of action is inhibition of fungal lanosterol 14α-demethylase. Its main indication is mycosis of the skin.

1.03.2.3.3 Lanoconazole

Trade name	Astat
Manufacturer	Nihon Nohyaku/Tsumura
Country of origin	Japan
Year of introduction	1994
Country in which first launched	Japan
CAS registry number	101530-10-3
Structure	

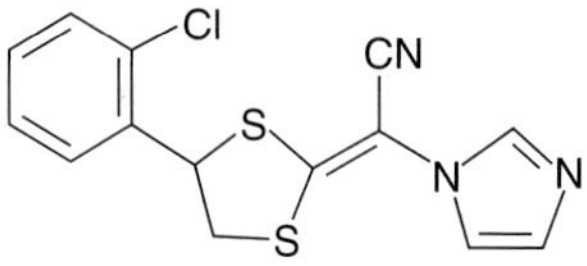

Racemic

Lanoconazole is an inhibitor of egosterol biosynthesis and thus affects the ultrastructure of the cell wall. It acts against a wide range of fungi and is used in the treatment of dermal mycoses.

1.03.2.3.4 Liranaftate

Trade name	Zefnart
Manufacturer	Tosoh
Country of origin	Japan
Year of introduction	2000
Country in which first launched	Japan
CAS registry number	088678-31-3
Structure	

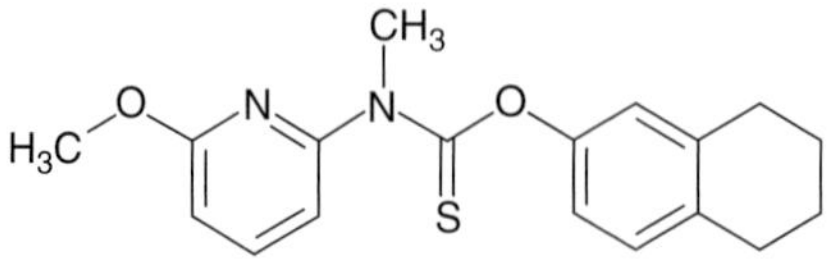

Liranaftate was launched as a topical antifungal for the treatment of dermatophycoses. It is a potent and specific inhibitor of squalene epoxidase, but has no effect on mammalian cholesterol biosynthesis.

1.03.2.3.5 Micafungin

Trade name	Funguard
Manufacturer	Fujisawa
Country of origin	Japan
Year of introduction	2002
Country in which first launched	Japan
CAS registry number	235114-32-6
Structure	

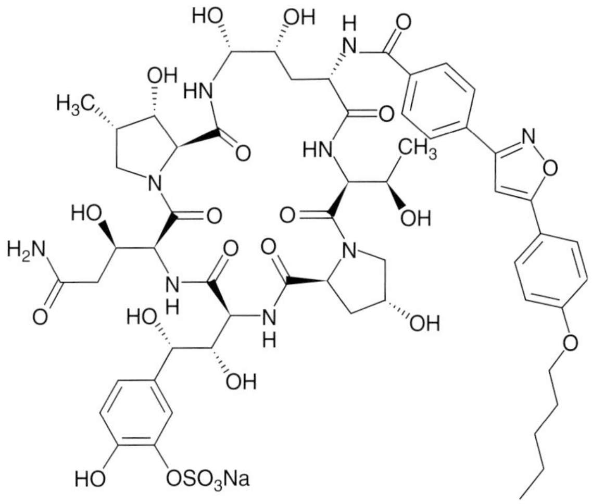

Micafungin was the second member of the echinocandin class of antifungal agents to be introduced for the parenteral treatment of various fungal infections caused by *Aspergillus* and *Candida*. Its half-life in humans is 11.7–15.2 h after injection.

1.03.2.3.6 Voriconazole

Trade name	Vfend
Manufacturer	Pfizer
Country of origin	USA
Year of introduction	2002
Country in which first launched	USA
CAS registry number	137234-62-9
Structure	

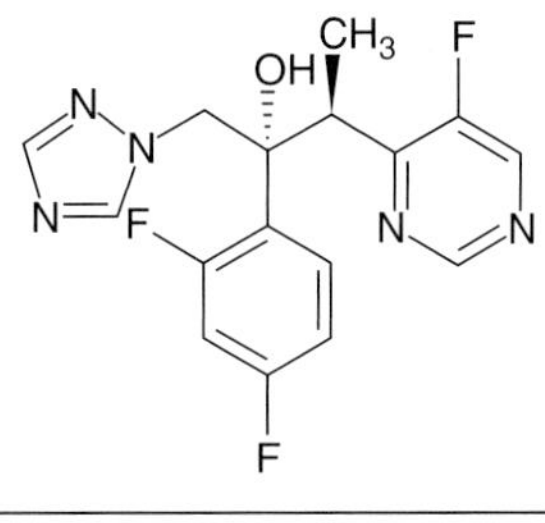

Voriconazole was introduced for the treatment of acute invasive aspergillosis, candidosis, and other emerging fungal infections seen in immunocompromised patients. It acts by inhibiting the cytochrome P450-dependent enzyme 14α-sterol demethylase of ergosterol biosynthesis. Voriconazole is active against a wide range of fungi including

Aspergillus, *Cryptococcus neoformans*, and various *Candida* species. It was effective in the treatment of neutropenic patients with acute invasive aspergillosis, non-neutropenic patients with chronic invasive aspergillosis, and HIV-infected patients with oropharyngeal candidiasis.

1.03.2.4 Antimalarials

1.03.2.4.1 Arteether

Trade name	Artemotil
Manufacturer	Central Drug Research Institute/Artecef BV
Country of origin	India
Year of introduction	2000
Country in which first launched	Netherlands
CAS registry number	075887-54-6
Structure	

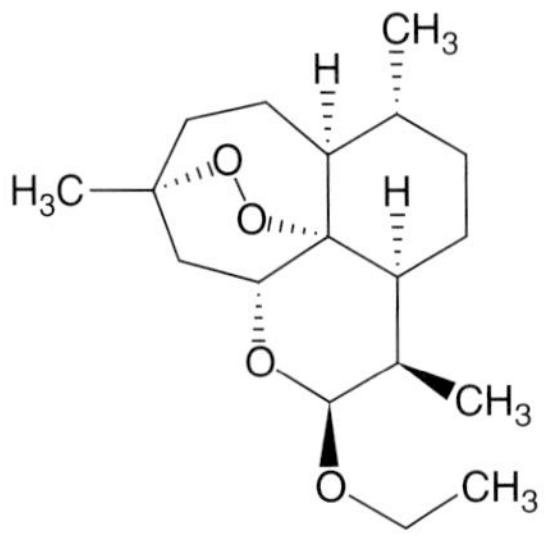

Arteether is marketed as a solution in sesame oil. It is administered by i.m. injection for the treatment of severe malaria infections in children and adolescents. It acts rapidly against *Plasmodium* during the early blood stage of its development. It also shows gametocytocidal activity against *Plasmodium falciparium*. The initial approval was only for treatment of young people, but clinical trials indicate that the drug acts rapidly and efficiently against *Plasmodium* in adults.

1.03.2.4.2 Bulaquine

Trade name	Ablaquin
Manufacturer	CDRI
Country of origin	India
Year of introduction	2000
Country in which first launched	India
CAS registry number	079781-00-3
Structure	

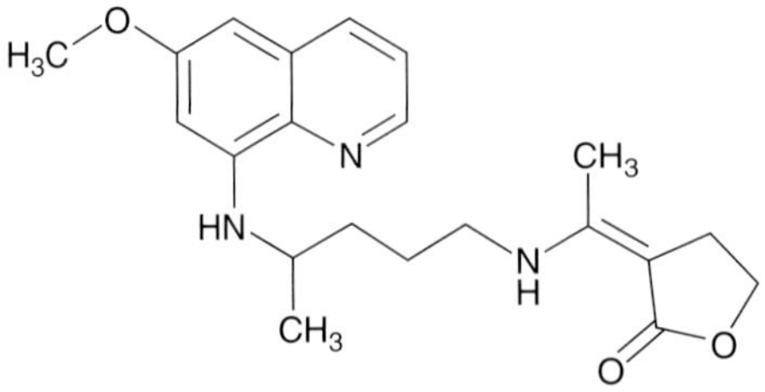

Bulaquine is used in combination with chloroquine in the treatment of malaria. It kills the latent tissue stage of the parasite *Plasmodium vivax*, which accumulates in the liver and is responsible for relapses.

1.03.2.5 Antisepsis

1.03.2.5.1 Drotrecogin alfa

Trade name	Xigris
Manufacturer	Lilly
Country of origin	USA
Year of introduction	2001
Country in which first launched	USA
CAS registry number	42617-41-4
Structure – recombinant protein	

Drotrecogin alfa was introduced as an intravenous treatment for the reduction of mortality in adult patients with severe sepsis associated with acute organ dysfunction. It is a recombinant activated protein C expressed in human kidney 293 cells. The majority of patients with sepsis have reduced levels of activated protein C, which acts as an antithrombic via inhibition of factor Va and VIIIa and as a promoter of fibrinolysis via inactivation of plasminogen-activator inhibitor-1, and exerts an anti-inflammatory effect by inhibiting the production of inflammatory cytokines.

1.03.3 Anticancer Agents

In this category, 21 anticancer drugs, 15 antineoplastic drugs, five antiemetic agents, one cytoprotective, one 5α-reductase inhibitor, one agent for treating chemotherapy-induced leukopenia, one agent β-treating hypercalcemia, and one anticancer adjuvant are included.

1.03.3.1 Anticancer Drugs

1.03.3.1.1 Alemtuzumab

Trade name	Campath
Manufacturer	Cambridge University/Millenium/LEX Oncology/Schering AG
Country of origin	UK
Year of introduction	2001
Country in which first launched	USA
CAS registry number	146705-13-7
Structure – humanized monoclonal antibody of molecular weight 21–28 kDa	

Alemtuzumab is a humanized monoclonal antibody of the IgG1 isotype specific for the glycoprotein CD52 expressed on the cell surface of over 95% of normal and malignant B and T lymphocytes and monocytes. It was introduced for the treatment of B-cell chronic lymphocytic leukemia in patients who have been treated with alkylating agents and have failed fludarabine therapy.

1.03.3.1.2 Alitretinoin

Trade name	Penretin
Manufacturer	Ligand
Country of origin	USA
Year of introduction	1999
Country in which first launched	USA
CAS registry number	5300-03-8

Structure

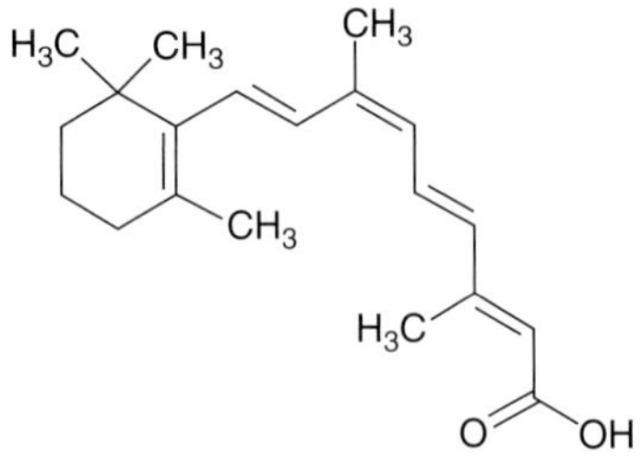

Alitretinoin was introduced for the treatment of cutaneous lesions in patients with Kaposi's sarcoma. It binds to all isoforms of the intracellular retinoid X and A receptors thus inducing cell differentiation, increasing cell apoptosis, and inhibiting cellular proliferation in experimental models of human cancer.

1.03.3.1.3 Arglabin

Trade name	Unknown
Manufacturer	NuOncology Labs
Country of origin	Kazakhstan
Year of introduction	1999
Country in which first launched	Russia
CAS registry number	84692-91-1
Structure	

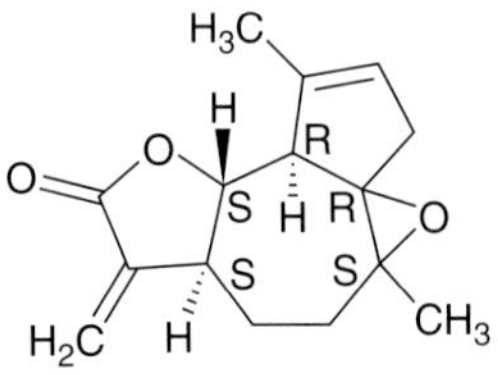

Arglabin is a potent and selective inhibitor of farnesyl transferase, which is critical to the function of the Ras oncogene in cancer cell reproduction. It was launched in Russia for the treatment of a range of cancers.

1.03.3.1.4 Bexarotene

Trade name	Targretin
Manufacturer	Ligand
Country of origin	USA
Year of introduction	2000
Country in which first launched	USA
CAS registry number	153559-49-0
Structure	

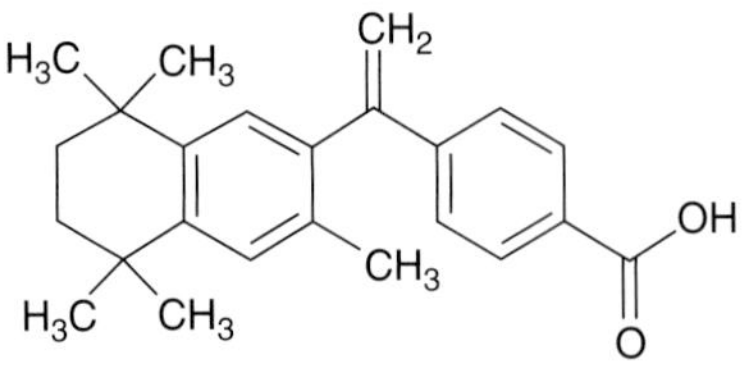

Bexarotene was launched for the treatment of manifestations of cutaneous T-cell lymphoma in patients who are refractory to at least one prior systemic therapy. It is selective for retinoid X rather than retinoid A receptors.

1.03.3.1.5 Bortezomib

Trade name	Velcade
Manufacturer	Millenium (LeukoSite, Proscript)
Country of origin	USA
Year of introduction	2003
Country in which first launched	USA
CAS registry number	179324-69-7
Structure	

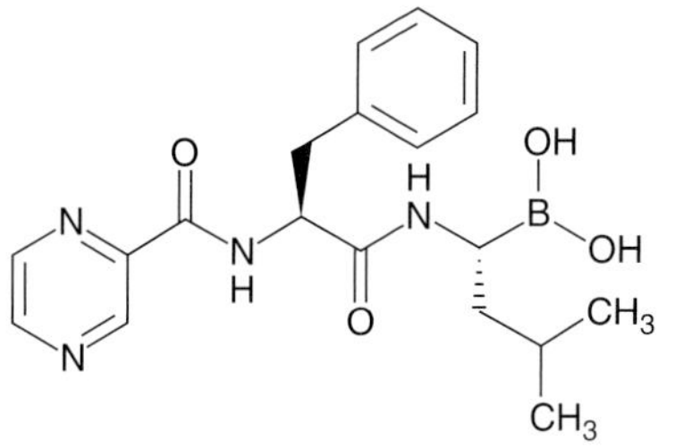

Bortezomib is a potent ubiquitin proteasome (26S) inhibitor that is marketed for the treatment of multiple myeloma.

1.03.3.1.6 Cetuximab

Trade name	Erbitux
Manufacturer	ImClone
Country of origin	USA
Year of introduction	2003
Country in which first launched	Switzerland
CAS registry number	205923-56-4
Structure – Monoclonal antibody of molecular weight 152 kDa	

Cetuximab is a human/mouse chimeric monoclonal antibody that blocks the epidermal growth factor receptor (EGFR). It was launched as a combination with irinotecan for the treatment of patients with colorectal cancer who no longer respond to standard chemotherapy.

1.03.3.1.7 Denileukin diftitox

Trade name	Ontak
Manufacturer	Ligand
Country of origin	USA
Year of introduction	1999
Country in which first launched	USA
CAS registry number	173146-27-5
Structure – recombination protein of molecular weight 58 kDa	

Denileukin diftitox is a fusion protein comprising diphtheria toxin fragment A–fragment B genetically fused to a human interleukin-2 (IL2) fragment. It was launched for the treatment of patients with persistent or recurrent cutaneous T-cell lymphoma.

1.03.3.1.8 Exemestane

Trade name	Aromasin
Manufacturer	Farmitalia Carlo Erba/Pharmacia
Country of origin	Italy
Year of introduction	2000
Country in which first launched	USA, Canada, Western Europe
CAS registry number	107868-30-4
Structure	

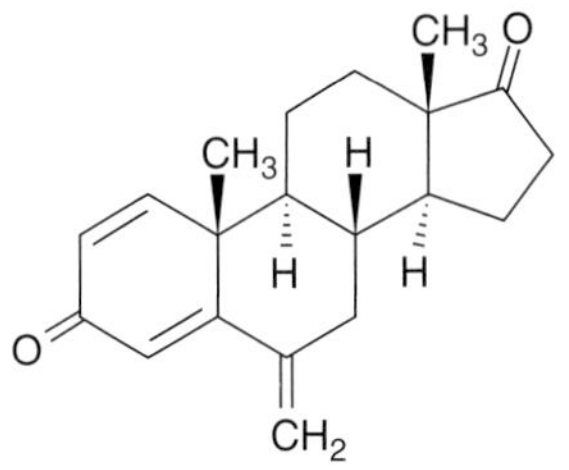

Exemestane is an irreversible inactivator of the aromatase enzyme system that was launched for the treatment of estrogen-dependent tumors and postmenopausal breast cancer. It has high activity in women failing antiestrogen therapy with tamoxifen.

1.03.3.1.9 Fulvestrant

Trade name	Faslodex
Manufacturer	AstraZeneca
Country of origin	UK
Year of introduction	2002
Country in which first launched	USA
CAS registry number	129453-61-8
Structure	

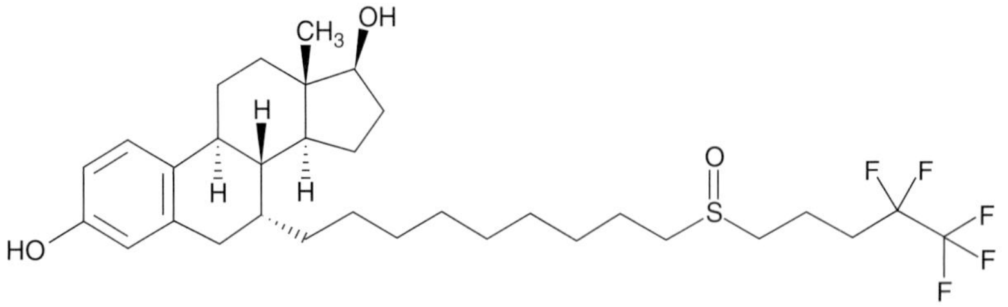

Fulvestrant binds to the estrogen receptor. It was launched as a once monthly injectable treatment of hormone receptor-positive metastatic breast cancer in postmenopausal women with disease progression following estrogen therapy. It completely blocks the cell growth in tamoxifen-resistant breast cancer cell lines and prevents growth of tamoxifen-resistant tumors in mice.

1.03.3.1.10 Gemtuzumab ozogamicin

Trade name	Myotarg
Manufacturer	Celltech Group/Wyeth-Ayerst Research
Country of origin	USA/UK

Year of introduction	2000
Country in which first launched	USA
CAS registry number	220578-59-6
Structure – recombinant protein of molecular weight 152 kDa	

Gemtuzumab ozogamicin was launched as the first antibody-targeted antineoplastic agent for the treatment of patients with acute myeloid leukemia. It is an immunoconjugate of anti-CD33 humanized mouse monoclonal antibody linked via a bifunctional link to the cytotoxic antibiotic calicheamicin.

1.03.3.1.11 **Ibritumomab tiuxetan**

Trade name	Zevalin
Manufacturer	IDEC/Syncor
Country of origin	USA
Year of introduction	2002
Country in which first launched	USA
CAS registry number	206181-63-7
Structure – monoclonal antibody of molecular weight 148 kDa	

^{90}Y-Ibritumomab tiuxetan is the first commercially available radiolabeled antibody for cancer therapy and more specifically for the treatment of relapsed or refractory low-grade, follicular, or transformed B-cell non-Hodgkin's lymphoma. Ibritumomab is a murine immunoglobulin G1 kappa isotype monoclonal antibody that targets CD20, a B-lymphocyte antigen. Conjugating with the isothiocyanatobenzyl derivative of diethylene triamine pentaacetate (DTPA) gives ibritumomab tiuxetan, which can then chelate radionucleotides.

1.03.3.1.12 **Letrozole**

Trade name	Femara
Manufacturer	Novartis
Country of origin	Switzerland
Year of introduction	1996
Country in which first launched	France
CAS registry number	112809-51-5
Structure	

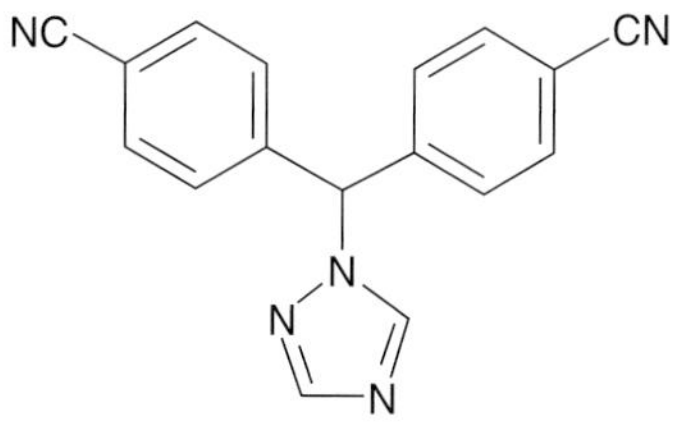

Letrozole is a third-generation aromatase inhibitor that specifically inhibits P450arom that catalyzes the conversion of androstenedione to estrone. It was launched for the second-line treatment of advanced breast cancer.

1.03.3.1.13 **OCT-43**

Trade name	Octin
Manufacturer	Otsuka
Country of origin	Japan
Year of introduction	1999

Country in which first launched	Otsuka
CAS registry number	159074-77-8
Structure – recombinant protein of molecular weight 17 kDa	

This protein is a recombinant variant of IL1beta that was launched for the treatment of mycosis fungoides. It is clinically useful in the treatment of aplastic anemia and myelodysplastic syndrome.

1.03.3.1.14 **Oxaliplatin**

Trade name	Eloxatin
Manufacturer	Bebiopharm/Sanofi
Country of origin	Switzerland
Year of introduction	1996
Country in which first launched	France
CAS registry number	61825-94-3
Structure	

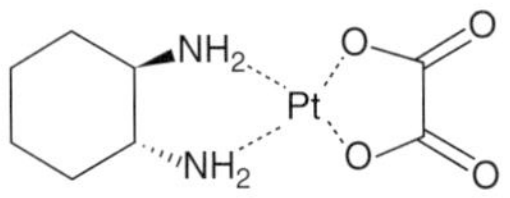

Oxaliplatin has an antitumor spectrum similar to cisplatin but is more effective against L1210 leukemia and cisplatin-resistant L1210. It was launched for second-line treatment of metastatic colorectal cancer. The mode of action involves binding to guanine-N7 leading to bidentate chelation, which results in the bending of DNA.

1.03.3.1.15 **Raltitrexed**

Trade name	Tomudex
Manufacturer	Zeneca
Country of origin	UK
Year of introduction	1996
Country in which first launched	UK
CAS registry number	112887-68-0
Structure	

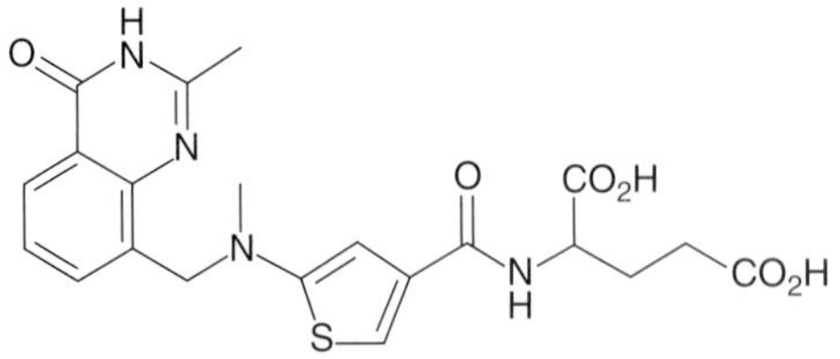

Raltitrexed is a highly selective inhibitor of thymidylate synthase, the key enzyme in the biochemical conversion of deoxyuridine monophosphate (dUMP) to deoxythymidine monophosphate (dTMP). It was launched for the treatment of advanced colorectal cancer and needs to be given only once every 3 weeks.

1.03.3.1.16 **SKI-2053R**

Trade name	Sunpla
Manufacturer	SK Pharm
Country of origin	Korea
Year of introduction	1999

Country in which first launched	Korea
CAS registry number	146665-77-2
Structure	

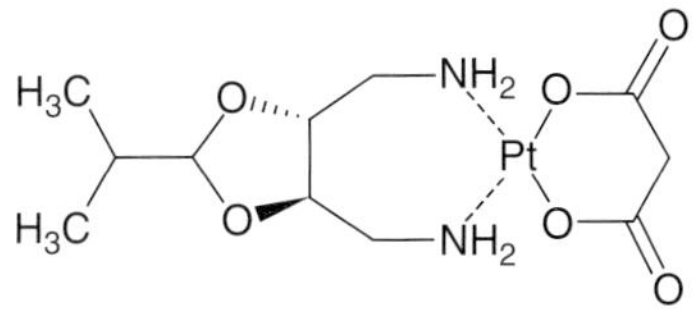

SKI-2053R is a third-generation platinum complex alkylating agent that is highly active against various cell lines including cisplatin-resistant tumor cell lines. It was launched for the treatment of unresectable or metastatic gastric adenocarcinoma.

1.03.3.1.17 Tasonermin

Trade name	Beromun
Manufacturer	Genentech
Country of origin	USA
Year of introduction	1999
Country in which first launched	Germany
CAS registry number	94948-59-1

Structure – Recombinant tumor necrosis factor of molecular weight 17 kDa

Tasonermin is the first tumor necrosis factor launched for the treatment of soft tissue sarcoma of the limbs.

1.03.3.1.18 Temozolomide

Trade name	Temodal
Manufacturer	CRC Technology/Schering-Plough
Country of origin	UK
Year of introduction	1999
Country in which first launched	UK
CAS registry number	85622-93-1
Structure	

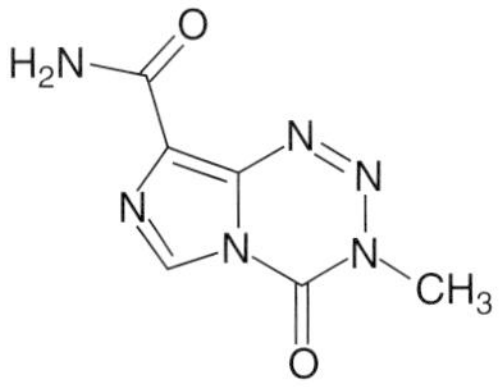

Temozolomide was introduced for the treatment of patients with glioblastoma multiforme showing recurrence or progression after standard therapy. The mode of action involves inhibition of *O*-6-alkylguanine-DNA alkyltransferase.

1.03.3.1.19 Topotecan hydrochloric acid (HCl)

Trade name	Hycamtin
Manufacturer	SmithKlineBeecham
Country of origin	UK
Year of introduction	1996

Country in which first launched	USA
CAS registry number	123948-87-8
Structure	

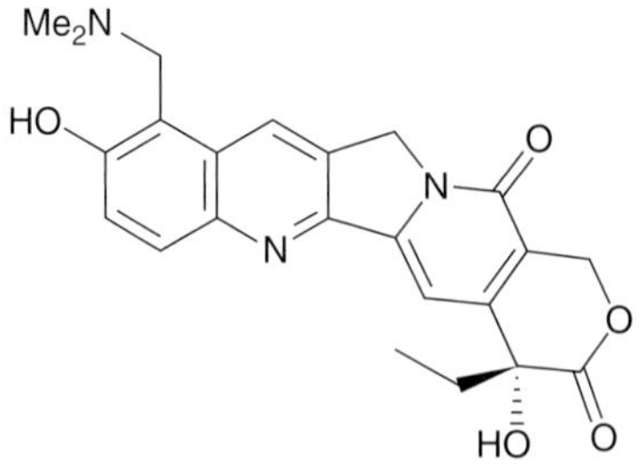

Topotecan is an inhibitor of topoisomerase 1 and inhibits its release from DNA where it relaxes super-coiled DNA giving rise to single-strand breaks. When the replication fork eventually reaches this complex double-strand breaks can occur thus signaling apoptosis and eventually giving rise to cell death. Topotecan was launched for the second-line treatment of ovarian cancer.

1.03.3.1.20 **Tositumomab**

Trade name	Bexxar
Manufacturer	Corixar/GlaxoSmithKline
Country of origin	USA
Year of introduction	2003
Country in which first launched	USA
CAS registry number	192391-48-3
Structure – monoclonal antibody of molecular weight 150 kDa	

Tositumomab is another radioimmunotherapeutic antibody for the treatment of B-cell non-Hodgkin's lymphoma. Treatment involves joint dosing of tositumomab and ^{131}I-tositumomab and thus combines the tumor-targeting ability of the cytotoxic antibody with the therapeutic potential of radiation with patient-specific dosing.

1.03.3.1.21 **Valrubicin**

Trade name	Valstar
Manufacturer	Anthra Pharm/Medeva
Country of origin	USA
Year of introduction	1999
Country in which first launched	USA
CAS registry number	56124-62-0
Structure	

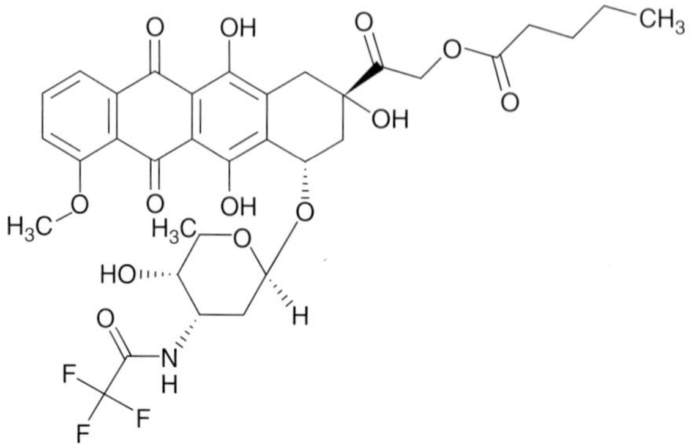

Valrubicin possibly acts by blockade of SV40 large T-antigen helicase. It was launched for the treatment of bladder cancer by intravesical installation.

1.03.3.2 Antineoplastic Drugs

1.03.3.2.1 Amribucin hydrochloride

Trade name	Calsed
Manufacturer	Sumitomo
Country of origin	Japan
Year of introduction	2002
Country in which first launched	Japan
CAS registry number	92395-36-5
Structure	

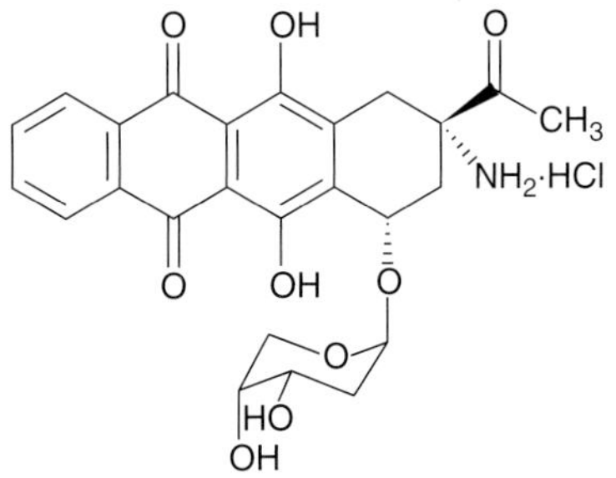

Amribucin inhibits topoisomerase II. It was introduced for the treatment of non-small-cell and small-cell lung cancers.

1.03.3.2.2 Anastrozole

Trade name	Arimidex
Manufacturer	Zeneca
Country of origin	UK
Year of introduction	1995
Country in which first launched	UK
CAS registry number	120511-73-1
Structure	

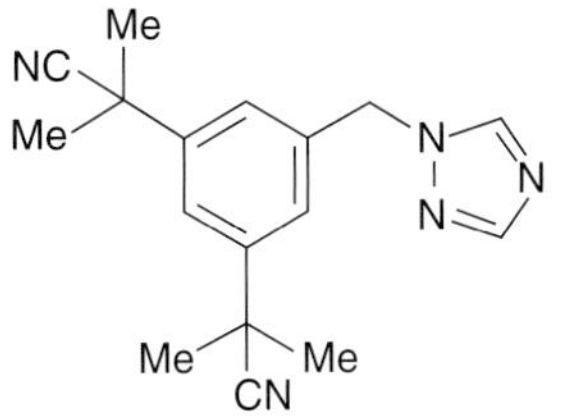

Anastrozole is a potent and selective aromatase inhibitor that was introduced for the treatment of advanced breast cancer in postmenopausal women.

1.03.3.2.3 Bicalutamide

Trade name	Casodex
Manufacturer	Zeneca

Country of origin	UK
Year of introduction	1995
Country in which first launched	UK
CAS registry number	90357-06-5
Structure	

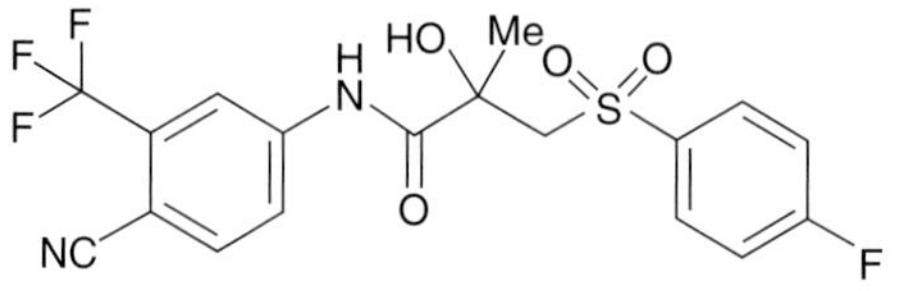

Racemic

Bicalutamide is a peripherally selective antiandrogen that was introduced for the treatment of advanced prostate cancer in combination with a luteinizing hormone-releasing hormone (LHRH) analog or surgical castration.

1.03.3.2.4 Capecitabine

Trade name	Xeloda
Manufacturer	Roche
Country of origin	Switzerland
Year of introduction	1998
Country in which first launched	Switzerland
CAS registry number	154661-50-9
Structure	

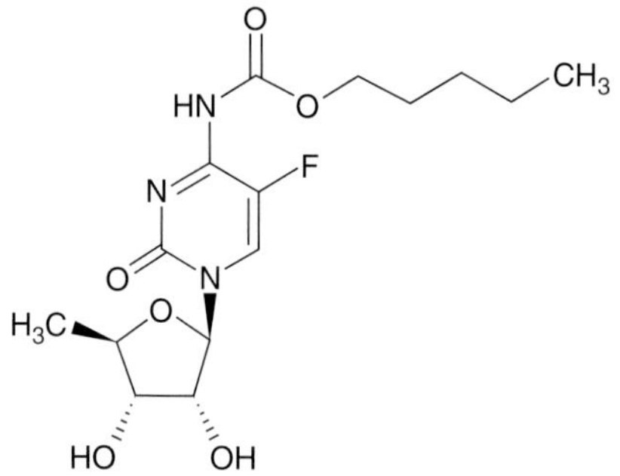

Capecitabine is a prodrug of doxifluridine from which it is released at the target tumor site. It is marketed for the treatment of advanced neoplastic disease including refractory metastatic breast cancer.

1.03.3.2.5 Docetaxel

Trade name	Taxotere
Manufacturer	Rhone-Poulenc-Rorer
Country of origin	France
Year of introduction	1995
Country in which first launched	South Africa
CAS registry number	114977-28-5

Structure

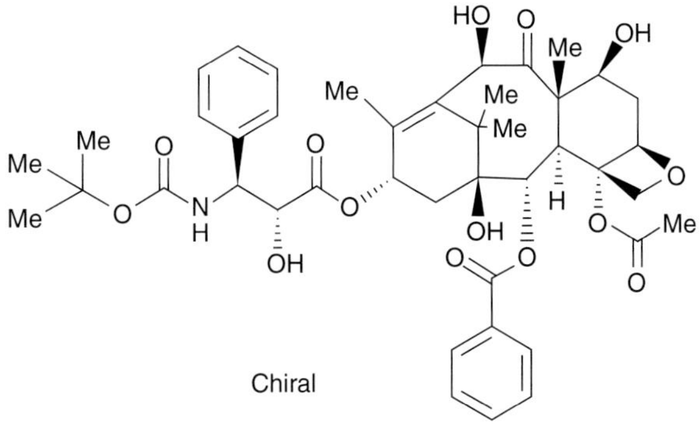

Docetaxel is a semisynthetic product from the taxoid family used for the treatment of ovarian, breast, and non-small-cell lung cancers.

1.03.3.2.6 Fadrazole hydrochloride

Trade name	Afema
Manufacturer	Ciba-Geigy
Country of origin	Switzerland
Year of introduction	1995
Country in which first launched	Japan
CAS registry number	102676-31-3
Structure	

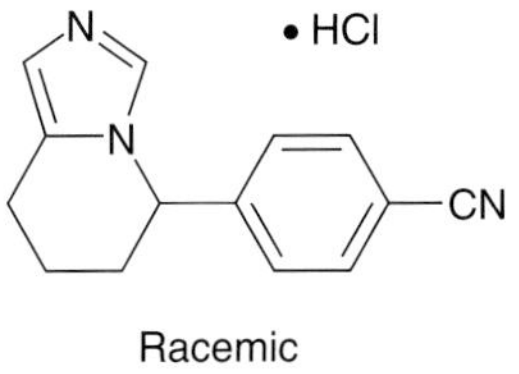

The potent and specific aromatase activity of Fadrazole led to its introduction for the treatment of postmenopausal breast cancer.

1.03.3.2.7 Gemcitabine hydrochloride

Trade name	Gemzar
Manufacturer	Lilly
Country of origin	USA
Year of introduction	1995
Country in which first launched	Netherlands, Sweden
CAS registry number	122111-03-9
Structure	

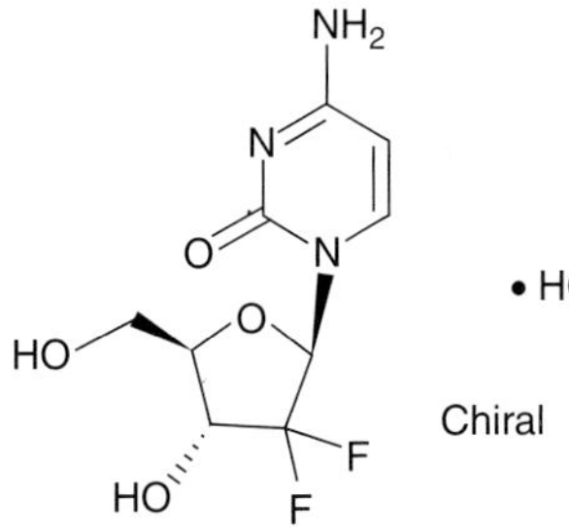

Following phosphorylation in vivo, gemcitabine acts on non-small-cell-lung cancer and pancreatic cancer. It inhibits processes required for DNA synthesis and metabolism and shows an extraordinary array of self-potentiating mechanisms that increase the concentration and prolong the retention of its active nucleotides in tumor cells.

1.03.3.2.8 Gefitinib

Trade name	Iressa
Manufacturer	AstraZeneca
Country of origin	UK
Year of introduction	2002
Country in which first launched	Japan
CAS registry number	184475-35-2
Structure	

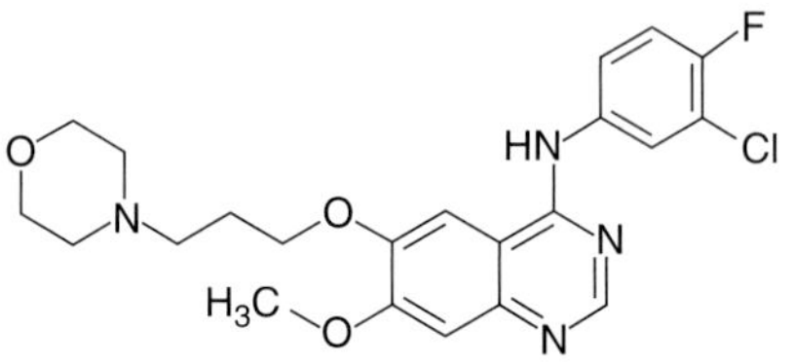

Gefitinib was introduced as a daily oral monotherapy for the treatment of inoperable or recurrent non-small-cell lung cancers. It reversibly inhibits the activity of the epidermal growth factor receptor tyrosine kinase.

1.03.3.2.9 Imatinib mesilate

Trade name	Gleevec, Glivec
Manufacturer	Novartis
Country of origin	Switzerland
Year of introduction	2001
Country in which first launched	USA
CAS registry number	220127-57-1
Structure	

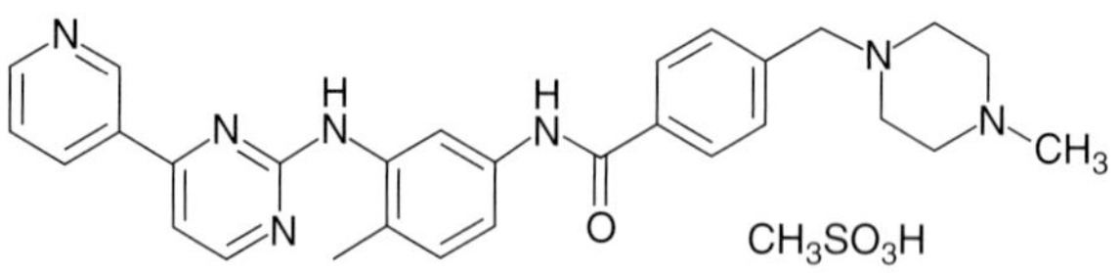

Imatinib mesilate was launched for the treatment of chronic myelogenous leukemia (CML) in blast crisis, accelerated phase or chronic phase after interferon-alpha failure. It is the first agent to act specifically on a constitutively active tyrosine kinase that causes CML.

1.03.3.2.10 Irinotecan hydrochloride

Trade name	Campto, Topotecin
Manufacturer	Yakult Honsha

Country of origin	Japan
Year of introduction	1994
Country in which first launched	Japan
CAS registry number	100286-90-6
Structure	

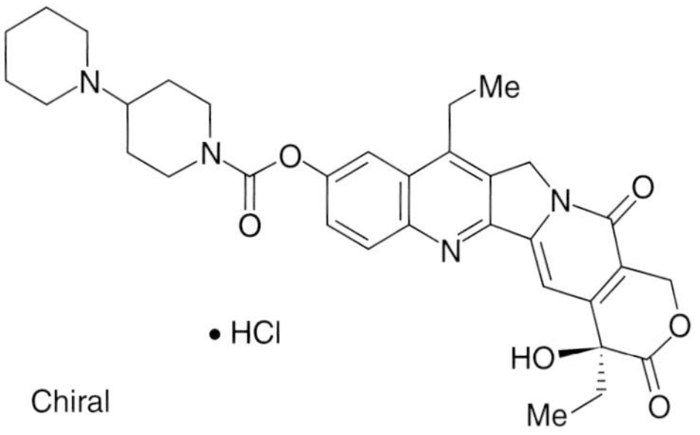

Irinotecan hydrochloride is a water-soluble derivative of the anticancer agent campothecin that was marketed for the treatment of lung, ovarian, and cervical cancers. It inhibits topoisomerase I, the enzyme involved in maintaining the topographic structure of DNA during the process of translation, transcription, and mitosis.

1.03.3.2.11 **Nedaplatin**

Trade name	Aqulpa
Manufacturer	Shionogi
Country of origin	Japan
Year of introduction	1995
Country in which first launched	Japan
CAS registry number	95734-82-0
Structure	

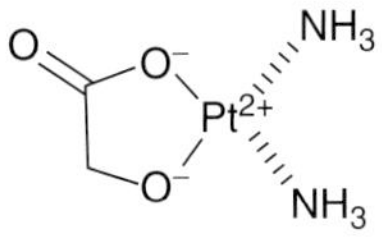

Nedaplatin is a novel, second-generation platinum complex that was marketed for the treatment of a variety of cancers including head and neck, small-cell and non-small-cell lung, osteophageal, prostatic, testicular, ovarian, cervical, bladder, and uterine cancers.

1.03.3.2.12 **Pegaspargase**

Trade name	Oncaspar
Manufacturer	Enzon/Rhone-Poulenc Rorer
Country of origin	USA
Year of introduction	1994
Country in which first launched	USA
CAS registry number	9015-68-3

Pegaspargase is a polyethylene glycol conjugate of L-asparaginase that was introduced for combination chemotherapy in acute lymphoblastic leukemia. The conjugation increases activity, prolongs half-life, and reduces immunogenicity.

1.03.3.2.13 Sobuzoxane

Trade name	Perazolin
Manufacturer	Zenyaku Kogyo
Country of origin	Japan
Year of introduction	1994
Country in which first launched	Japan
CAS registry number	98631-95-9
Structure	

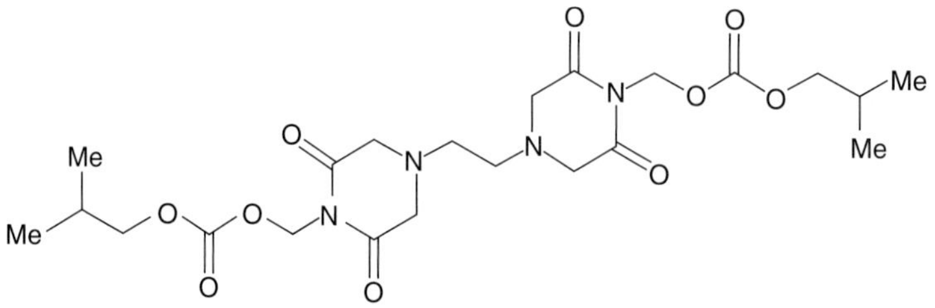

Sobuzoxane is a noncleavable complex-stabilizing topoisomerase II inhibitor that was introduced for the treatment of malignant lymphoma and adult T-cell leukemia/lymphoma that is resistant to chemotherapy and has a poor prognosis. It appears to significantly inhibit growth of human colon, lung, gastric, and breast cancers.

1.03.3.2.14 Temoporphin

Trade name	Foscan
Manufacturer	Quanta Nova
Country of origin	UK
Year of introduction	2002
Country in which first launched	UK
CAS registry number	122341-38-2
Structure	

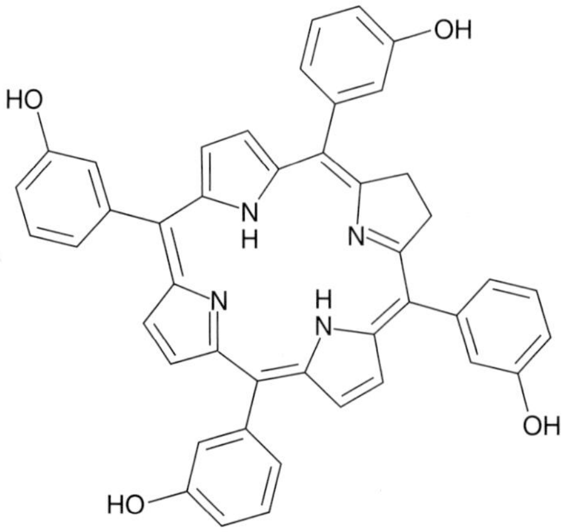

Temoporphin is a second-generation photosensitizer for the photodynamic therapy of advanced head and neck cancers. It is extremely sensitive to wavelengths of light that penetrate tissues thus resulting in lower light/dose and irradiation time.

1.03.3.2.15 Zinostatin stimalamer

Trade name	SMANCS
Manufacturer	Yamanouchi
Country of origin	Japan
Year of introduction	1994
Country in which first launched	Japan
CAS registry number	123760-07-6
Structure – conjugate with molecular weight of *c.* 16 kDa	

Zinostatin stimalamer is a conjugate of poly(styrene-maleic acid) and neocarzinostatin that was launched for the treatment of hepatoma. Conjugation improves the in vivo stability, the effectiveness in reducing tumor size, and in prevention of the development of new tumors, decreases toxic side effects, and extends the half-life.

1.03.3.3 Antiemetic Agents

1.03.3.3.1 Aprepitant

Trade name	Emend
Manufacturer	Merck
Country of origin	USA
Year of introduction	2003
Country in which first launched	USA
CAS registry number	170729-80-3
Structure	

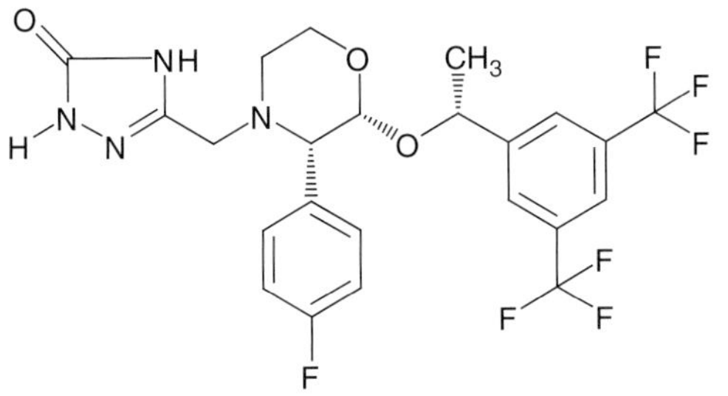

Aprepitant is a substance P receptor antagonist used for the treatment of chemotherapy-induced nausea and vomiting. It is effective against both the acute and delayed phase of cisplatin-induced emesis.

1.03.3.3.2 Dolasetron mesylate

Trade name	Anzemet
Manufacturer	Hoechst Marion Roussel
Country of origin	Germany
Year of introduction	1998
Country in which first launched	Australia
CAS registry number	115956-13-3
Structure	

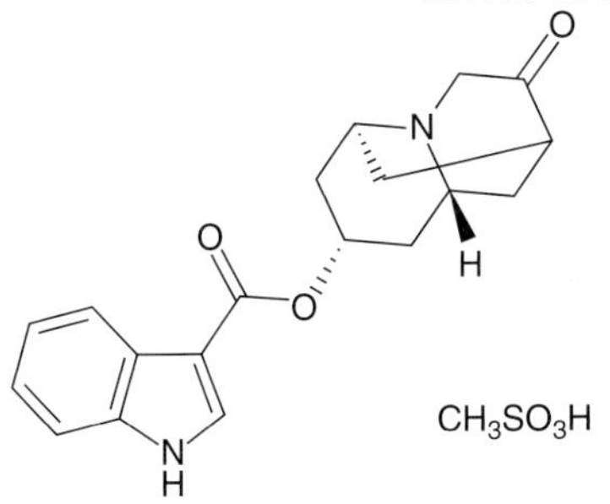

Dolasetron mesylate is a potent and very selective antagonist for $5HT_3$ receptors with potent antiemetic effects induced by cancer chemotherapy. Single doses of 10–50 mg prior to treatment with cisplatin effectively control emesis and nausea.

1.03.3.3.3 Nazasetron hydrochloride

Trade name	Serotone
Manufacturer	Yoshitomo/Japan Tobacco/Green Cross
Country of origin	Japan
Year of introduction	1994
Country in which first launched	Japan
CAS registry number	123040-16-4
Structure	

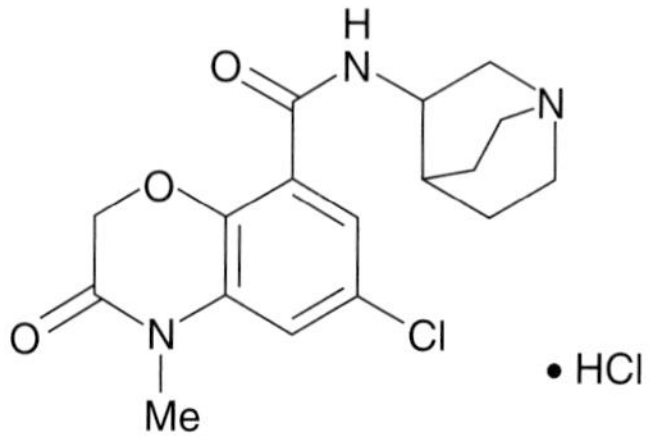

Nazasetron is a highly potent $5HT_3$ receptor antagonist that markedly reduces nausea and emesis following cancer chemotherapy and total body x-radiation.

1.03.3.3.4 Palonosetron

Trade name	Aloxi
Manufacturer	Syntex (Roche Bioscience)/MGI Pharma/Helsinn
Country of origin	USA
Year of introduction	2003
Country in which first launched	USA
CAS registry number	135729-62-3
Structure	

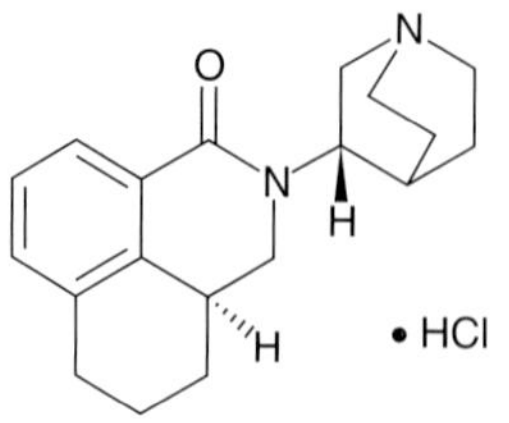

Palonosetron is a $5HT_3$ receptor antagonist used as an injectable agent for the prevention of acute and delayed nausea and vomiting associated with cancer chemotherapy. Because of its long half-life, the therapeutic effect lasts for several days. The antiemetic effect acts against a number of agents including cisplatin, cyclophosphamide, and dacarbazine.

1.03.3.3.5 Ramosetron

Trade name	Nasea
Manufacturer	Yamanouchi
Country of origin	Japan
Year of introduction	1996
Country in which first launched	Japan
CAS registry number	132907-72-3
Structure	

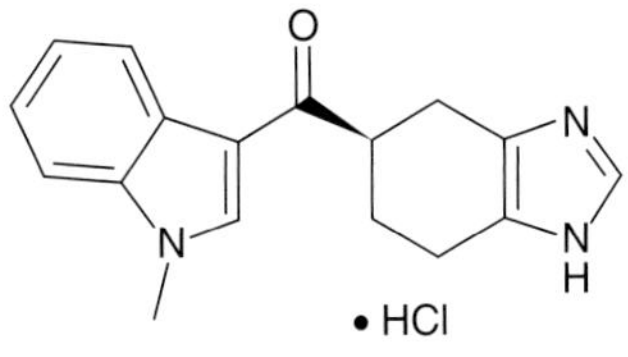

Ramosetron is a $5HT_3$ receptor antagonist that is active against cisplatin-induced emesis.

1.03.3.4 Cytoprotective Agents

1.03.3.4.1 Amifostine

Trade name	Ethyol
Manufacturer	US Bioscience/Schering Plough
Country of origin	USA
Year of introduction	1995
Country in which first launched	Germany
CAS registry number	63717-27-1
Structure	

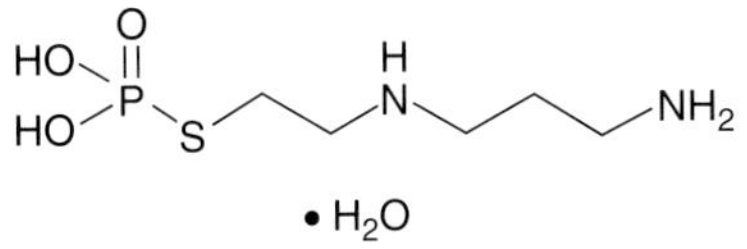

Amifostine is used to reduce cisplatin-induced renal toxicity in patients with advanced ovarian cancer. It is also a radioprotective agent.

1.03.3.5 5α-Reductase Inhibitor

1.03.3.5.1 Dutasteride

Trade name	Avodart
Manufacturer	GlaxoSmithKline
Country of origin	UK
Year of introduction	2002
Country in which first launched	USA
CAS registry number	164656-23-9

Structure

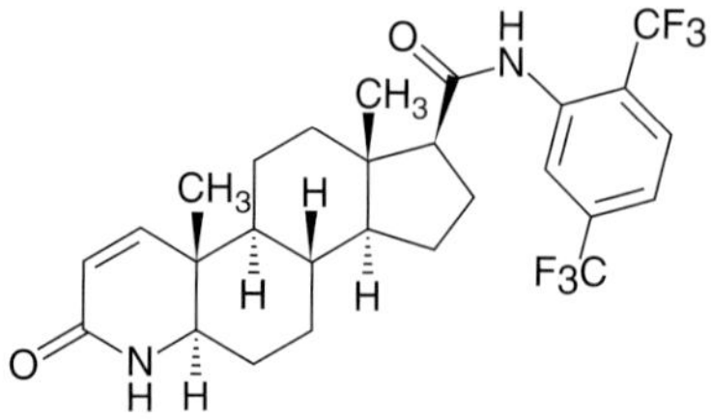

Dutasteride is marketed for the symptomatic treatment of benign prostatic hyperplasia. It is a dual inhibitor of type 1 and 2 isoforms of 5α-reductase and has a markedly prolonged half-life.

1.03.3.6 Chemotherapy-Induced Leukopenia

1.03.3.6.1 Nartograstim

Trade name	Neu-Up
Manufacturer	Kyowa Hakka
Country of origin	Japan
Year of introduction	1994
Country in which first launched	Japan
CAS registry number	134088-74-7

Nartograstim is a highly active mutein of the granulocyte colony-stimulating factor and is used in the treatment of chemotherapy-induced granulocytopenia, rhabdomyosarcoma, and anemia.

1.03.3.7 Anticancer Adjuvant

1.03.3.7.1 Angiotensin II

Trade name	Delibert
Manufacturer	Toa Elyo/Yamanouchi
Country of origin	Japan
Year of introduction	1994
Country in which first launched	Japan
CAS registry number	4474-91-3
Structure	

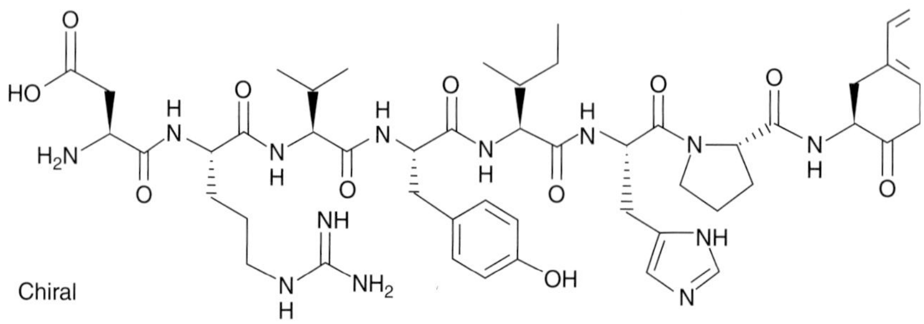

Angiotensin II was introduced to improve the efficacy of systemic chemotherapy by controlling blood flow to tumor tissues.

1.03.3.8 Hypercalcemia

1.03.3.8.1 Zoledronate

Trade name	Zometa
Manufacturer	Novartis
Country of origin	Switzerland
Year of introduction	2000
Country in which first launched	Canada
CAS registry number	165800-07-7
Structure	

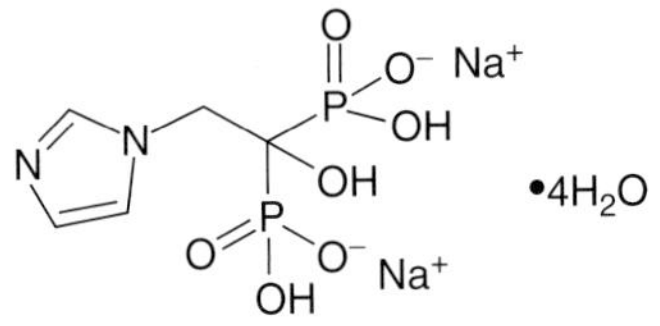

This bisphosphonate was introduced for the treatment of tumor-induced hypercalcemia.

1.03.4 Cardiovascular Agents

This category includes 22 antihypertensives, five antiarrhythmics, five antithrombotics, four agents for heart failure, two cardiostimulants, two anticoagulants, eight hypocholesterolemics, one iron chelator, one hematological agent, one subarachnoid hemorrhage agent, one hemophilia agent, one agent with cardiotonic activity, and a fibrinolytic agent.

1.03.4.1 Antihypertensives

1.03.4.1.1 Aranidipine

Trade name	Bec/Sapresta
Manufacturer	Maruko Seiyaku/Taiho
Country of origin	Japan
Year of introduction	1996
Country in which first launched	Japan
CAS registry number	86780-90-7
Structure	

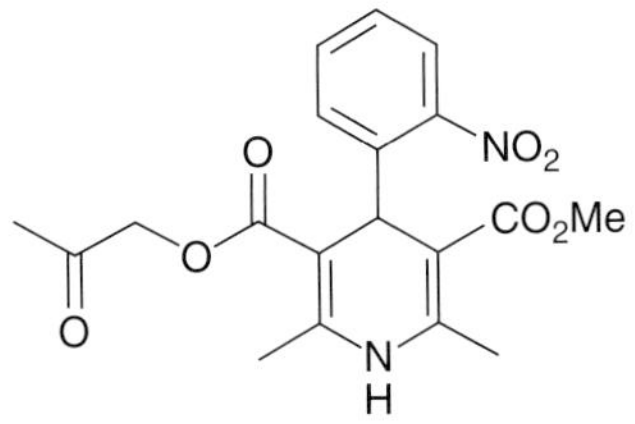

Aranidipine is a long-acting, potent antihypertensive agent that works by blocking Ca^{2+} entry into cells.

1.03.4.1.2 Azelnipidine

Trade name	Calblock
Manufacturer	Sankyo
Country of origin	Japan

Year of introduction	2003
Country in which first launched	Japan
CAS registry number	116574-11-9
Structure	

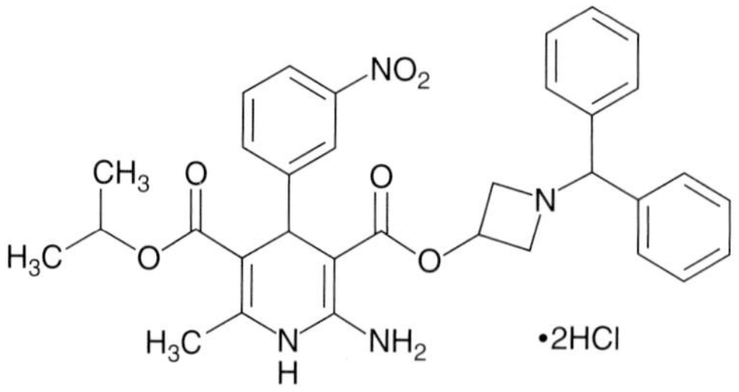

Azelnipidine is a blocker of L-type calcium channels and gives a sustained reduction in blood pressure in patients with mild to moderate hypertension.

1.03.4.1.3 Bosentan

Trade name	Tracleer
Manufacturer	Roche/Actelion/Genentech
Country of origin	Switzerland
Year of introduction	2001
Country in which first launched	USA
CAS registry number	147536-97-8
Structure	

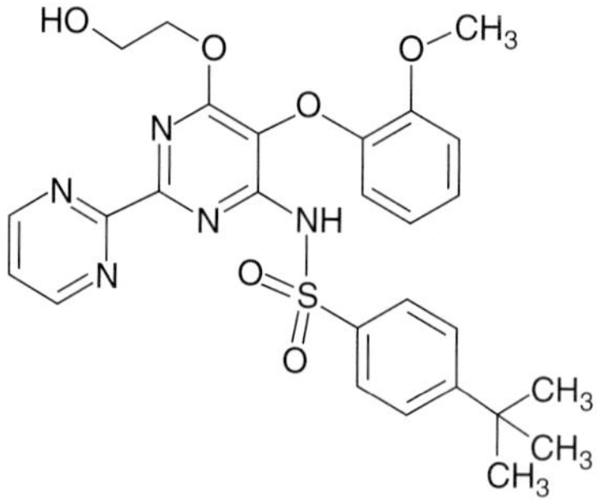

Bosentan was the first endothelin receptor antagonist to be launched and was introduced as a twice daily oral treatment for pulmonary arterial hypertension. It has demonstrated a beneficial selectivity for the pulmonary vasculature since it has no significant effect on mean aortic blood pressure and systolic vascular resistance.

1.03.4.1.4 Candesartan cilexetil

Trade name	Atacand
Manufacturer	Takeda/Astra
Country of origin	Japan
Year of introduction	1997
Country in which first launched	Sweden
CAS registry number	145040-37-5

Structure

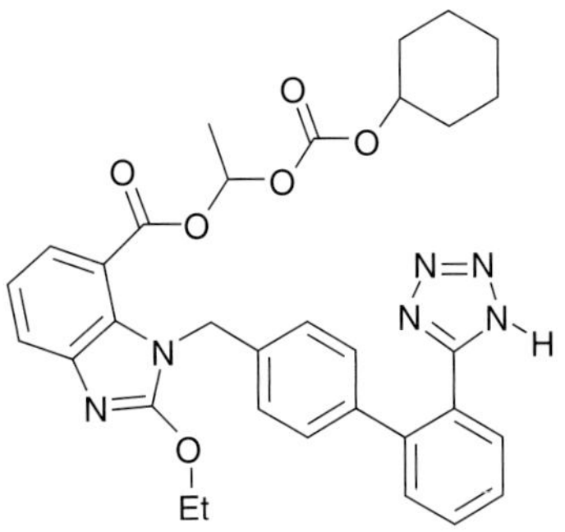

Candesartan cilexetil is a potent antagonist of angiotensin II type 1 receptors and thus acts as an antihypertensive. It is long acting and is more potent than angiotensin-converting enzyme (ACE) inhibitors.

1.03.4.1.5 **Clinidipine**

Trade name	Cinalong/Siscard/Atelec
Manufacturer	Fujirebio/Boehringer Ingelheim/Roussel-Morishita
Country of origin	Japan
Year of introduction	1995
Country in which first launched	Japan
CAS registry number	132203-70-4
Structure	

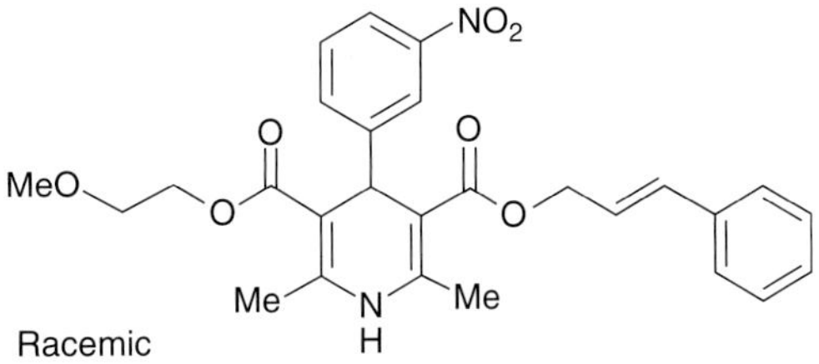

Clinidipine is dihydropyridine calcium antagonist intended for the treatment of essential and severe hypertension and hypertension associated with renopathy. Clinidipine has slow onset and long duration with less cardiodepressant activity.

1.03.4.1.6 **Efonidipine hydrochloride ethanol**

Trade name	Landel
Manufacturer	Nissan Chemical/Zeria
Country of origin	Japan
Year of introduction	1994
Country in which first launched	Japan
CAS registry number	111011-76-8
Structure	

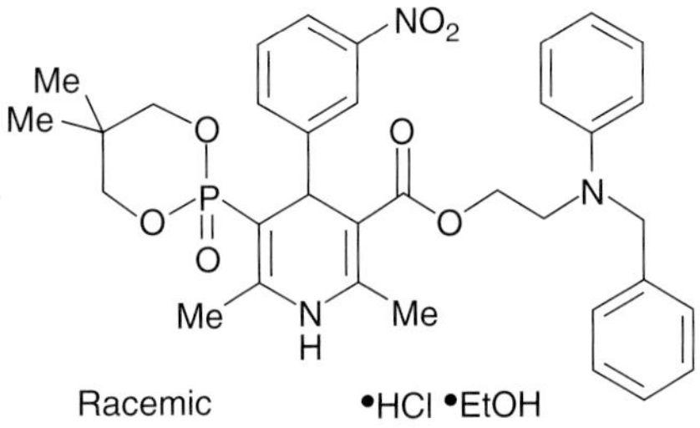

Efonidipine hydrochloride ethanol is the 14th dihydropyridine calcium channel blocker to be launched. It was marketed for the treatment of essential severe and renal hypertension.

1.03.4.1.7 Eplerenone

Trade name	Inspra
Manufacturer	Ciba-Geigy/Pfizer
Country of origin	USA
Year of introduction	2003
Country in which first launched	USA
CAS registry number	107724-20-9
Structure	

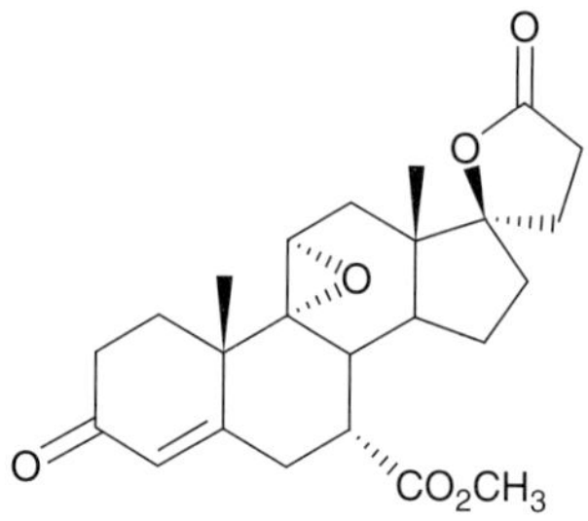

The antihypertensive effect of eplerenone is due to blockage of the binding of aldosterone at the mineralocorticoid receptor. It was approved for the oral treatment of hypertension but has been shown to also improve the survival of stable patients with left ventricular systolic dysfunction.

1.03.4.1.8 Eprosartan

Trade name	Teveten
Manufacturer	SmithKlineBeecham
Country of origin	UK
Year of introduction	1997
Country in which first launched	Germany
CAS registry number	133040-01-4
Structure	

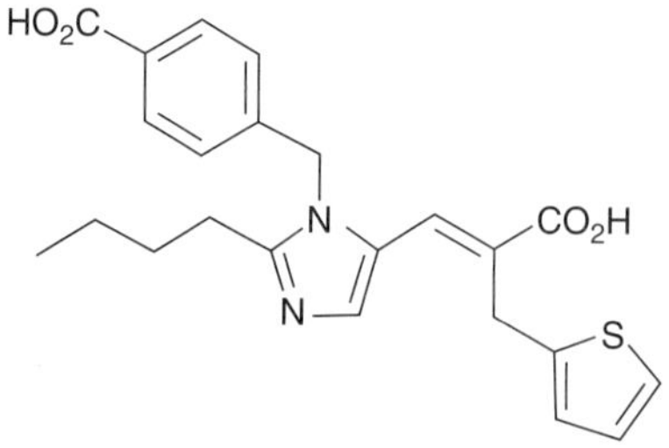

Teveten is a potent, highly selective competitive antagonist of the AT_1 receptor that was marketed for the treatment of hypertension.

1.03.4.1.9 Fenoldopam mesylate

Trade name	Neurex
Manufacturer	SmithKlineBeecham
Country of origin	USA
Year of introduction	1998
Country in which first launched	USA
CAS registry number	67227-57-0

Structure

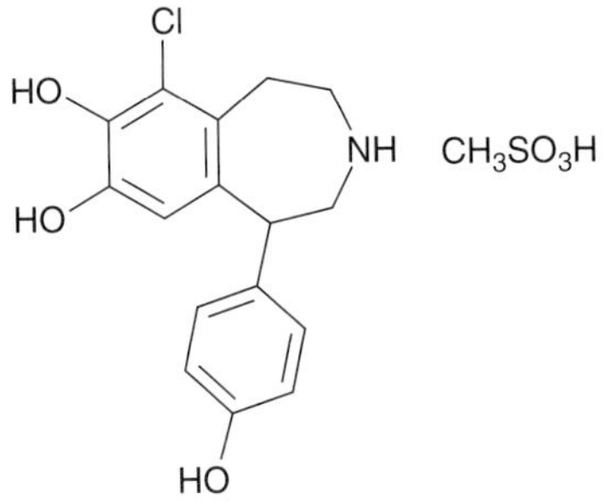

Fenoldopam is a potent dopamine D1 receptor agonist acting peripherally to produce systemic vasodilation. It is unable to cross the blood–brain barrier and thus has no central effects. In addition, it interacts with $5HT_{1c}$ and $5HT_2$ receptors. Fenoldopam is fast acting and maintains a long-lasting antihypertensive effect.

1.03.4.1.10 **Irbesartan**

Trade name	Aprovel/Avapro
Manufacturer	Sanofi
Country of origin	France
Year of introduction	1997
Country in which first launched	UK
CAS registry number	138402-11-6
Structure	

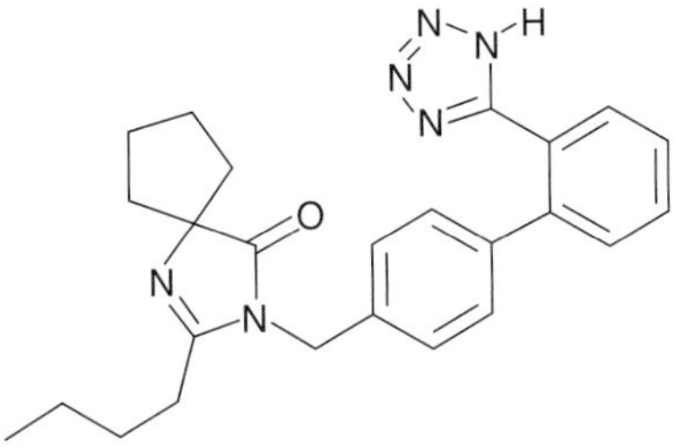

Irbesartan is an angiotensin II receptor antagonist that is noncompetitive and selective for AT_1 subtypes and has no AT_2 activity at postsynaptic receptors compared to presynaptic receptors. It was marketed as an antihypertensive.

1.03.4.1.11 **Lercanidipine**

Trade name	Lerdip
Manufacturer	Recordati/Byk Gulden
Country of origin	Italy
Year of introduction	1997
Country in which first launched	Netherlands
CAS registry number	100427-26-7
Structure	

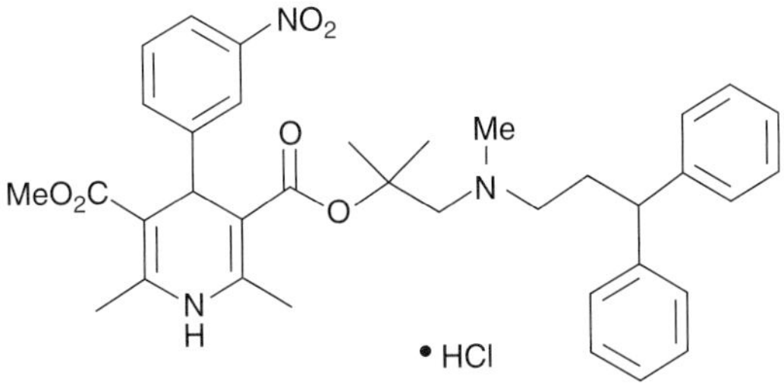

Lercanidipine is an antagonist of L-type calcium channels with no activity in smooth muscle cells, and has gradual onset with a long duration of activity as an antihypertensive agent.

1.03.4.1.12 Losartan potassium

Trade name	Cozaar
Manufacturer	DuPont Merck
Country of origin	USA
Year of introduction	1994
Country in which first launched	Sweden
CAS registry number	124750-99-8
Structure	

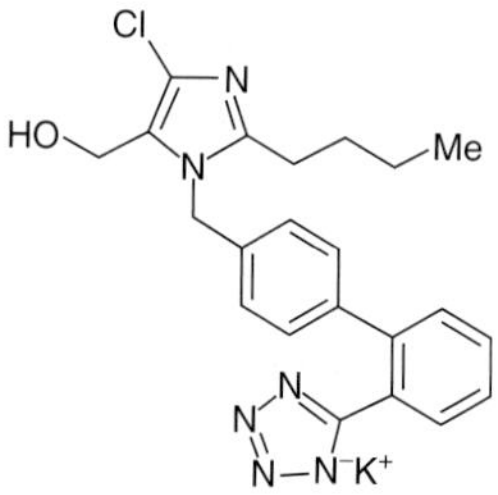

Losartan was the first potent and selective nonpeptide angiotensin II AT_1 receptor antagonist to be marketed. It acts as a potent and long-lasting once daily orally dosed antihypertensive.

1.03.4.1.13 Mibefradil hydrochloride

Trade name	Posicor
Manufacturer	Roche
Country of origin	Switzerland
Year of introduction	1997
Country in which first launched	USA
CAS registry number	116666-63-8
Structure	

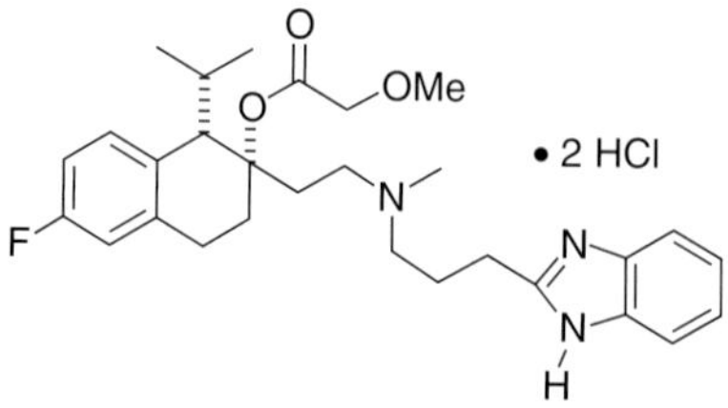

Mebefradil is a calcium channel blocker with selectivity for the T-type and is marketed for the treatment of mild to moderate hypertension.

1.03.4.1.14 Moexipril hydrochloride

Trade name	Univasc
Manufacturer	Warner-Lambert
Country of origin	USA
Year of introduction	1995
Country in which first launched	USA
CAS registry number	82586-52-5
Structure	

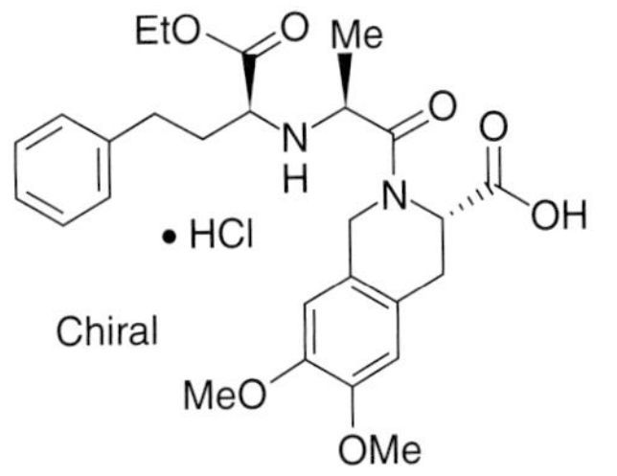

Moexipril is an ACE inhibitor that is marketed as a treatment for hypertension as a monotherapy and as a second-line therapy in combination with diuretics or calcium antagonists.

1.03.4.1.15 Nebivolol

Trade name	Nebilet
Manufacturer	Johnson & Johnson
Country of origin	USA
Year of introduction	1997
Country in which first launched	Germany
CAS registry number	99200-09-6
Structure	

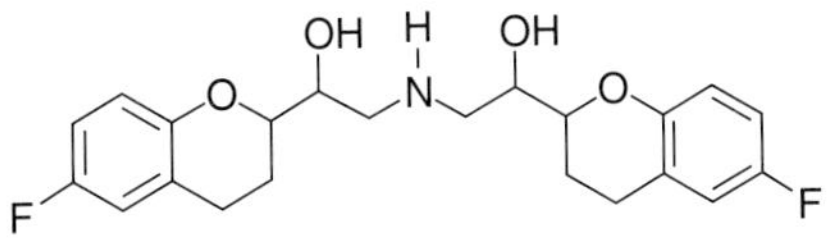

Nebivolol is a selective β_1-adrenergic receptor antagonist and is 50 times less potent at β_2-receptors. It is marketed as an antihypertensive agent and has vasodilating activity via the nitric oxide pathway. It causes an immediate fall in blood pressure and improves both left ventricular systolic and diastolic function and lowers peripheral blood resistance.

1.03.4.1.16 Olmesartan medoxomil

Trade name	Benicar
Manufacturer	Sankyo/Forest
Country of origin	Japan
Year of introduction	2002
Country in which first launched	USA
CAS registry number	144689-63-4

Structure

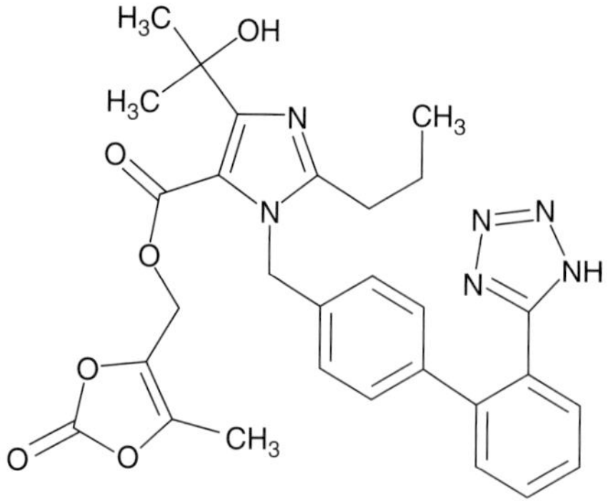

Benicar is a prodrug of olmesartan, which is a selective and competitive nonpeptide angiotensin II type 1 receptor antagonist. It was marketed as an orally active antihypertensive with a long half-life and few side effects.

1.03.4.1.17 **Spirapril hydrochloride**

Trade name	Renpress, Renormax
Manufacturer	Schering-Plough/ Sandoz
Country of origin	USA
Year of introduction	1995
Country in which first launched	Finland
CAS registry number	94841-17-5
Structure	

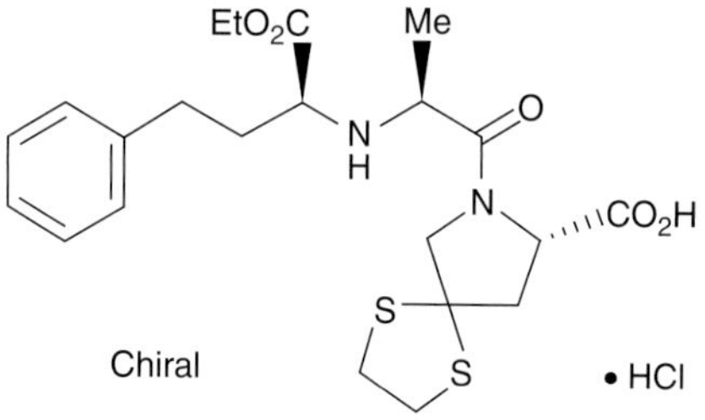

Spirapril is an orally active ACE inhibitor marketed for the treatment of hypertension.

1.03.4.1.18 **Telmisartan**

Trade name	Micardis
Manufacturer	Boehringer Ingelheim
Country of origin	Germany
Year of introduction	1999
Country in which first launched	USA
CAS registry number	144701-48-4
Structure	

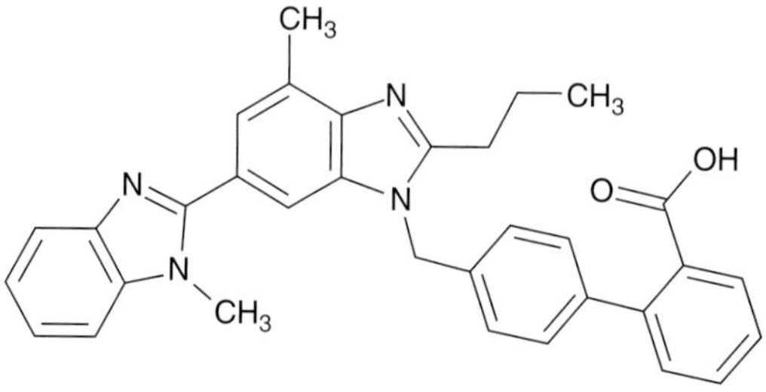

Telmisartan blocks the activity of angiotensin II and thus is a member of the 'sartan' class. It shows effective and sustained blood pressure lowering effects with a single daily oral dosage.

1.03.4.1.19 Temocapril hydrochloride

Trade name	Acecol
Manufacturer	Sankyo/Nippon Boehringer Ingelheim
Country of origin	Japan
Year of introduction	1994
Country in which first launched	Japan
CAS registry number	110221-44-8
Structure	

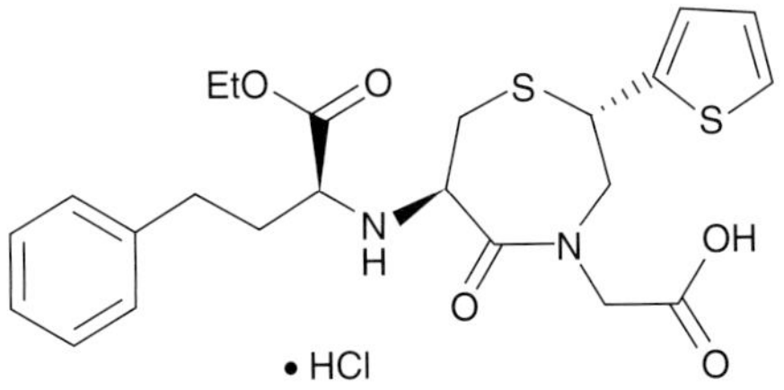

Chiral

Temocapril is an ACE inhibitor that is marketed for the treatment of hypertension. It is orally active and has a long duration of action. Because it is largely eliminated through the biliary route it is useful for the treatment of hypertensive patients with renal dysfunction.

1.03.4.1.20 Treprostinil sodium

Trade name	Remodulin
Manufacturer	Pharmacia/GSK
Country of origin	USA
Year of introduction	2002
Country in which first launched	USA
CAS registry number	289840-64-4
Structure	

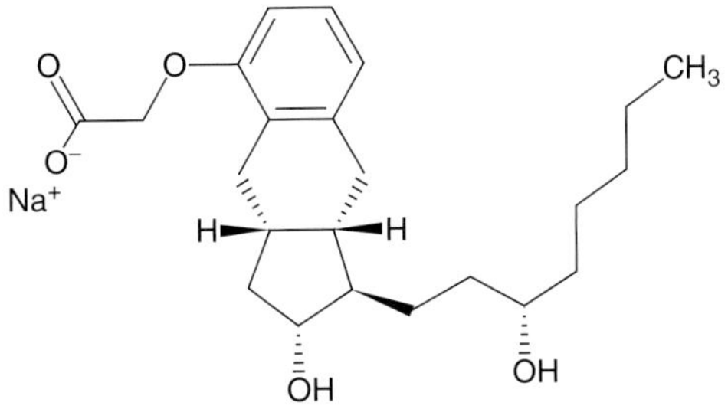

Treprostinil sodium, which appears to act at the prostacyclin receptor, was launched for the treatment of pulmonary hypertension. In patients with primary pulmonary hypertension, treprostinil causes a 22% improvement in cardiac output, a 24% significant decrease in peripheral vascular resistance, a decrease in mean pulmonary arterial pressure, and an improvement in New York Heart Association (NYHA) functional class.

1.03.4.1.21 Valsartan

Trade name	Diovan
Manufacturer	Norvartis
Country of origin	Switzerland
Year of introduction	1996
Country in which first launched	Germany
CAS registry number	137862-53-4
Structure	

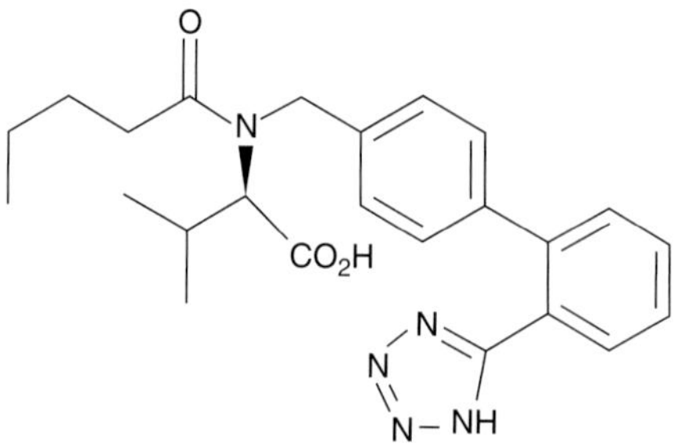

Valsartan is a highly specific antagonist of the AT_1 receptor that is potent and orally active.

1.03.4.1.22 Zofenopril calcium

Trade name	Zantipres
Manufacturer	Bristol-Meyers
Country of origin	USA
Year of introduction	2000
Country in which first launched	Italy
CAS registry number	81872-10-8; 81938-43-4
Structure	

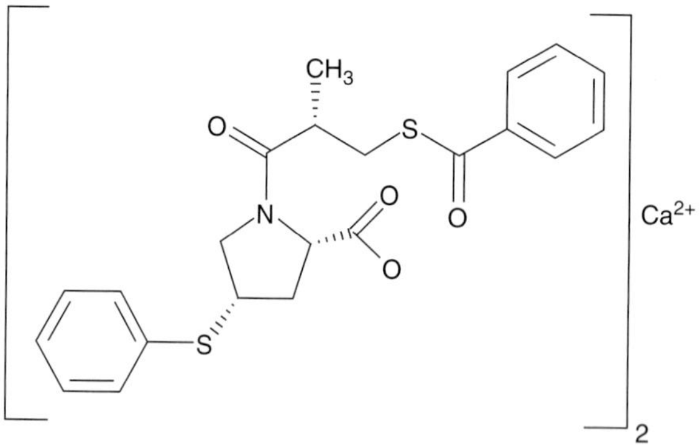

Zofenopril calcium was introduced as a second-generation ACE inhibitor for the treatment of acute myocardial infarction. It is a prodrug with the *S*-benzoyl group being rapidly hydrolyzed by cardiac esterase. It may also possess antioxidant effects, which could account for some of its strong anti-ischemic effects.

1.03.4.2 Antiarrhythmic

1.03.4.2.1 Dofetilide

Trade name	Tikosyn
Manufacturer	Pfizer
Country of origin	USA
Year of introduction	2000
Country in which first launched	USA
CAS registry number	115256-11-6

Structure

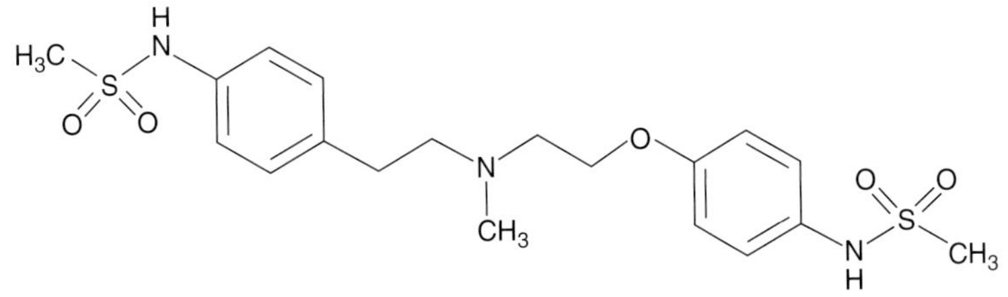

Dofetilide was launched for the treatment of cardiac patients with highly symptomatic atrial fibrillation. It potently and selectively inhibits a single potassium channel, Ikr, the rapidly acting component of the delayed rectifier potassium current. By blocking the open state of Ikr, dofetilide is able to prolong the effective refractory period in both atrial and ventricular myocardium and the monophasic action potential duration. Because it targets only one cardiac ion channel, it does not produce any effects on the sinus node, cardiac conduction system, and other extracardiac organs.

1.03.4.2.2 Ibutilide fumarate

Trade name	Corvert
Manufacturer	Pharmacia & Upjohn
Country of origin	UK
Year of introduction	1996
Country in which first launched	USA
CAS registry number	122647-32-9
Structure	

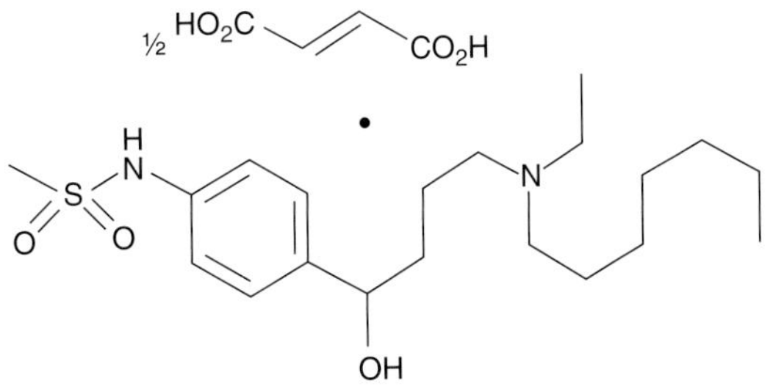

Ibutilide fumarate activates slow inward sodium channels and thus is able to prolong the action potential duration and lengthen the refractory period of myocardial tissue. It was launched for the treatment of atrial fibrillation and flutter.

1.03.4.2.3 Landiolol

Trade name	Onoact
Manufacturer	Ono Pharmaceutical
Country of origin	Japan
Year of introduction	2002
Country in which first launched	Japan
CAS registry number	133242-30-3
Structure	

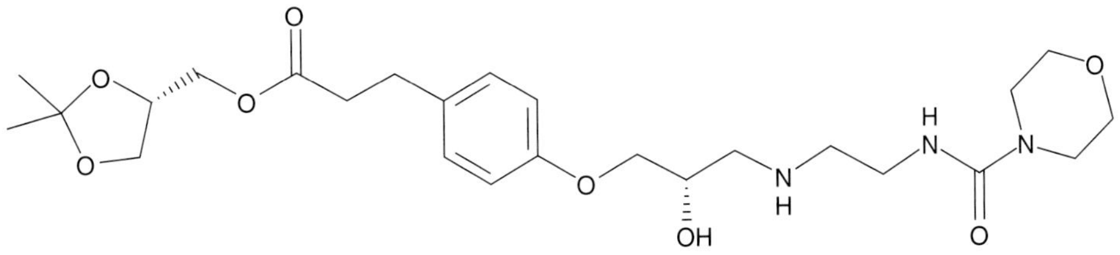

Landiolol is an ultra short-acting β1-adrenergic blocker that was launched for the treatment of tachyarrhythmia during surgery.

1.03.4.2.4 Nifekalant hydrochloride

Trade name	Shinbit
Manufacturer	Mitsui
Country of origin	Japan
Year of introduction	1999
Country in which first launched	Japan
CAS registry number	130656-51-8
Structure	

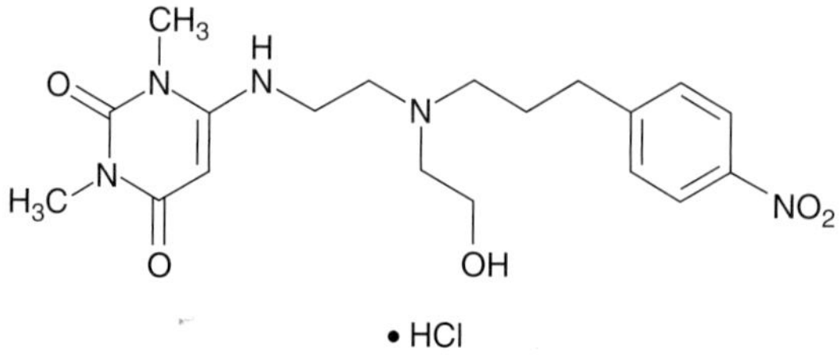

Nifekalant is a nonselective blocker of myocardial repolarizing potassium currents and is completely devoid of any β-adrenergic effects. It was launched for the treatment of serious ventricular arrhythmias.

1.03.4.2.5 Pirmenol hydrochloride

Trade name	Pimenol
Manufacturer	Warner-Lambert
Country of origin	USA
Year of introduction	1994
Country in which first launched	Japan
CAS registry number	61477-94-9
Structure	

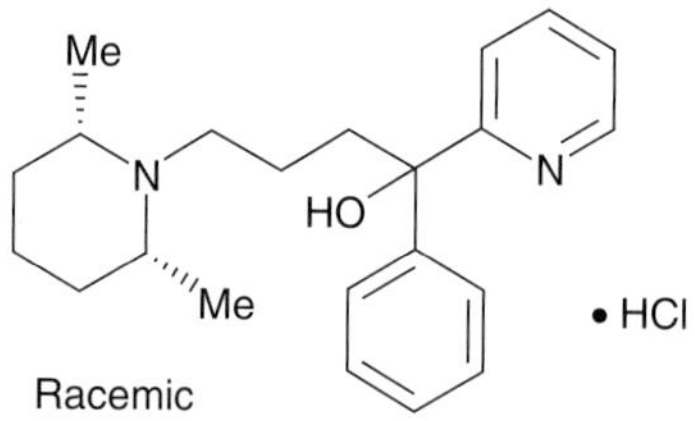

Pirmenol hydrochloride is a long-acting class Ia antiarrhythmic agent acting against atrial and ventricular arrhythmias of diverse etiology. It is highly active when given by both the oral and intravenous routes.

1.03.4.3 Antithrombotic Agents

1.03.4.3.1 Bivalirudin

Trade name	Angiomax
Manufacturer	Biogen/The Medicines Company
Country of origin	USA
Year of introduction	2000

Country in which first launched	New Zealand
CAS registry number	128270-60-0
Structure	

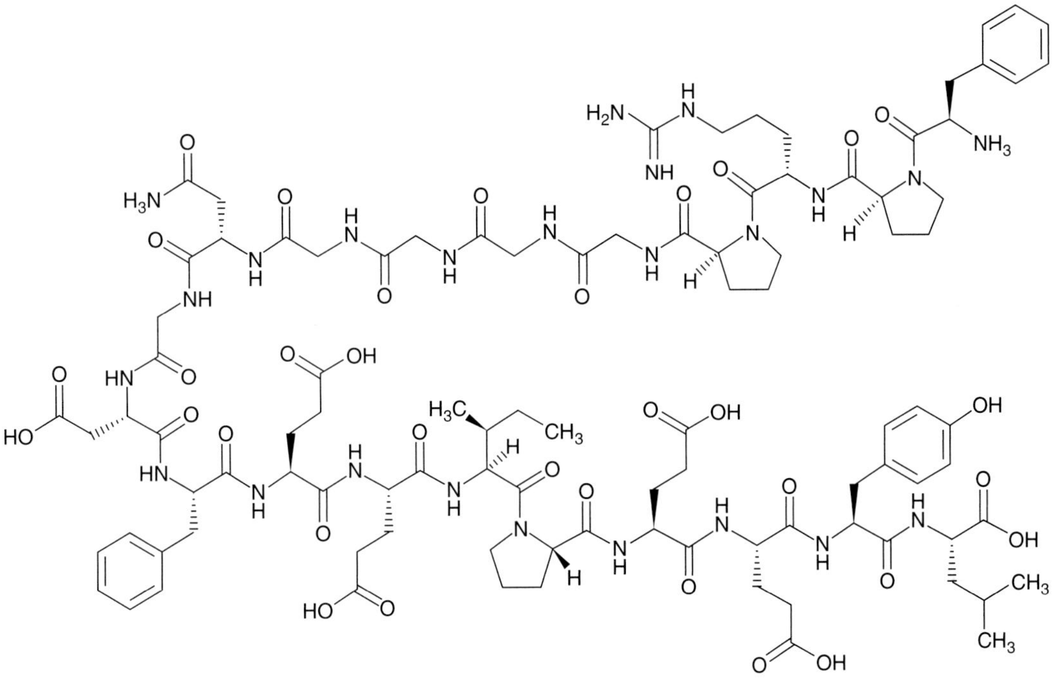

Bivalirudin was introduced as an intravenous treatment of patients with unstable angina undergoing percutaneous transluminal coronary angioplasty. It is a possible alternative to heparin treatment.

1.03.4.3.2 Clopidogrel hydrogensulfate

Trade name	Plavix, Iscover
Manufacturer	Sanofi/Bristol-Meyers Squibb
Country of origin	France
Year of introduction	1998
Country in which first launched	USA
CAS registry number	113665-84-2
Structure	

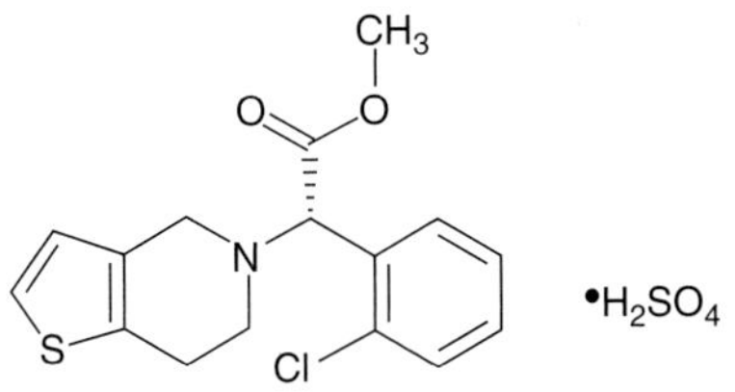

Clopidogrel is an adenosine diphosphate antagonist acting on the purinergic P2y receptor and was launched as a potent inhibitor of platelet aggregation for the preventative management of secondary ischemic events including myocardial infarction (MI), stroke, and vascular deaths.

1.03.4.3.3 Eptifibatide

Trade name	Integrilin
Manufacturer	Cor Therapeutics/Schering-Plough
Country of origin	USA
Year of introduction	1999
Country in which first launched	USA
CAS registry number	188627-80-7
Structure	

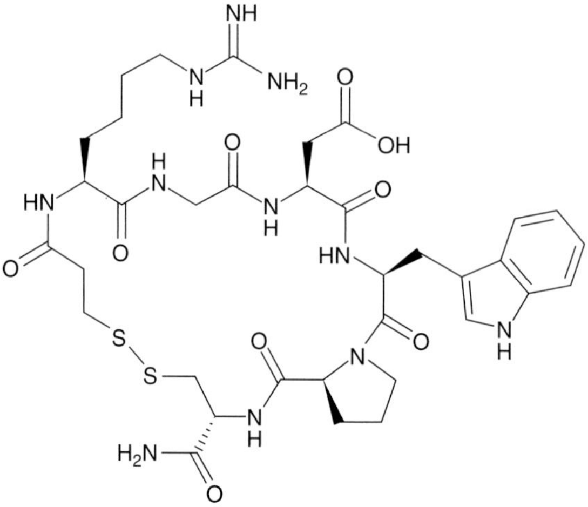

Eptifibatide was introduced for the treatment of acute coronary syndrome, in particular for patients at risk of abrupt vessel closure during or after coronary angioplasty.

1.03.4.3.4 Fondaparinux sodium

Trade name	Arixtra
Manufacturer	Sanofi-Synthelabo/Akzo Nobel
Country of origin	France
Year of introduction	2002
Country in which first launched	USA
CAS registry number	114870-03-0
Structure	

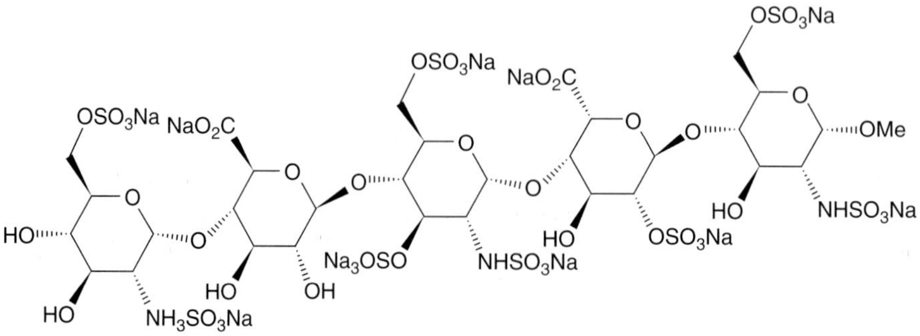

Fondaparinux sodium was introduced for prophylaxis of deep vein thrombosis, which sometimes leads to pulmonary embolism, following major orthopedic surgery.

1.03.4.3.5 Tirofiban hydrochloride

Trade name	Aggrastat
Manufacturer	Merck
Country of origin	USA
Year of introduction	1998
Country in which first launched	Switzerland/USA
CAS registry number	142373-60-2
Structure	

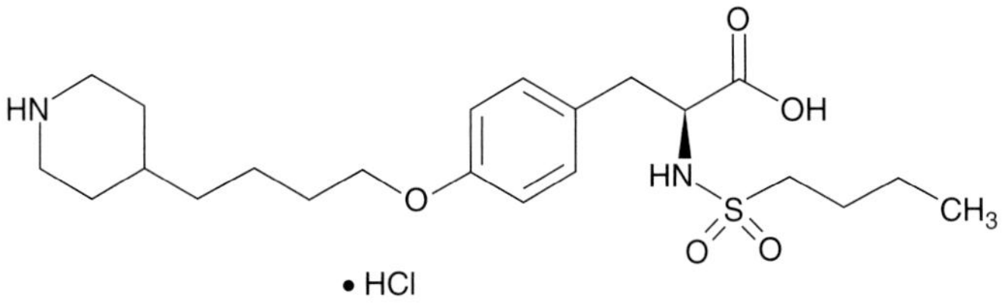

Tirofiban hydrochloride is a highly potent antiplatelet agent that inhibits the interaction of fibrinogen with GPIIb/IIIa. It was introduced for the treatment of patients with unstable angina or non-Q wave myocardial infarction to prevent cardiac ischemic events.

1.03.4.4 Heart Failure

1.03.4.4.1 Caperitide

Trade name	Hanp
Manufacturer	Suntory/Zeria
Country of origin	Japan
Year of introduction	1995
Country in which first launched	Japan
CAS registry number	89213-87-6
Structure – recombinant peptide of molecular weight 3080 Da	

Caperitide is the α-human atrial natriuretic peptide and is used for the treatment of congestive heart failure.

1.03.4.4.2 Levosimendan

Trade name	Simdax
Manufacturer	Orion
Country of origin	Finland
Year of introduction	2000
Country in which first launched	Sweden
CAS registry number	141505-33-1
Structure	

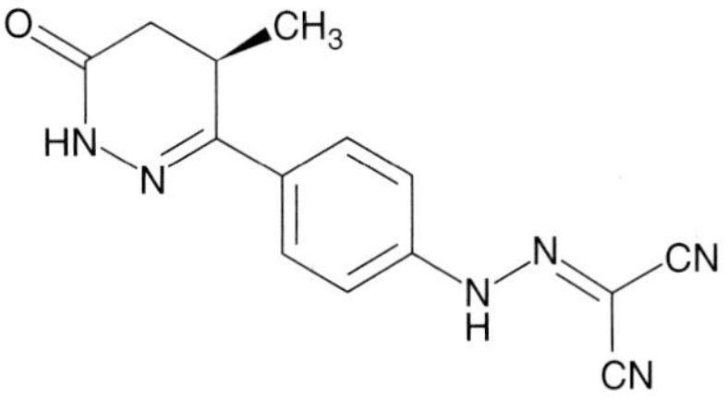

Levosimendan is a myofilament calcium sensitizer that increases myocardial contractility by selectively binding to the N-terminus of troponin C and by stabilizing the Ca^{2+}-bound conformation of this contractile protein. It was introduced as an intravenous infusion for the treatment of acute heart failure or refractory symptoms of chronic heart failure in cases where conventional treatment is not sufficient.

1.03.4.4.3 Neseritide

Trade name	Natrecor
Manufacturer	Scios/GSK
Country of origin	USA
Year of introduction	2001
Country in which first launched	USA
CAS registry number	124584-08-3
Structure – recombinant protein of molecular weight 3464 Da	

Neseritide is a recombinant form of the human vasodilatory B-type natriuretic peptide that was introduced for the treatment of patients with acutely decompensated congestive heart failure who have dyspnea at rest or with minimal activity.

1.03.4.4.4 Pimobendan

Trade name	Acardi
Manufacturer	Boehringer Ingelheim
Country of origin	Germany
Year of introduction	1994
Country in which first launched	Japan
CAS registry number	74150-27-9
Structure	

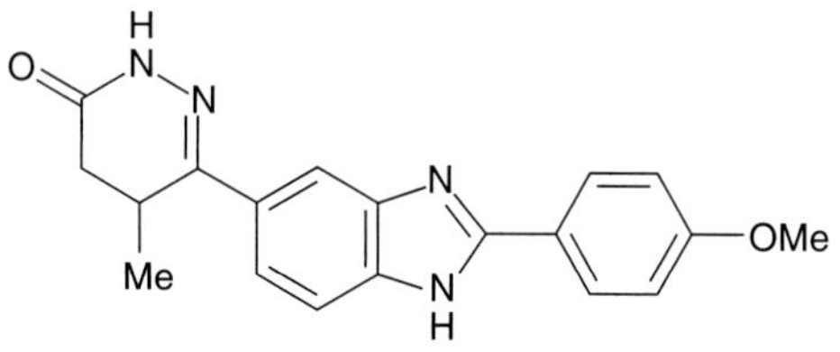

Racemic

Pimobendan is a phosphodiesterase III (PDE III) inhibitor that is able to enhance sensitization of myocardial contractile regulatory protein to calcium ions. It was introduced for the treatment of mild to moderate chronic heart failure.

1.03.4.5 Cardiostimulants

1.03.4.5.1 Docarpamine

Trade name	Tanadopa
Manufacturer	Tanabe Seiyaku
Country of origin	Japan
Year of introduction	1994
Country in which first launched	Japan
CAS registry number	74639-40-0

Structure

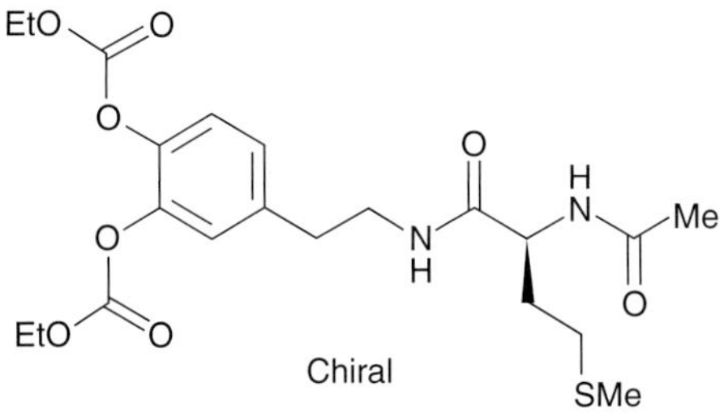

Docarpamine is an orally effective peripheral dopamine prodrug that was launched as a cardiostimulant with diuretic activity for the treatment of circulatory insufficiency.

1.03.4.5.2 Loprinone hydrochloride

Trade name	Coatec
Manufacturer	Eisai
Country of origin	Japan
Year of introduction	1996
Country in which first launched	Japan
CAS registry number	119615-63-3
Structure	

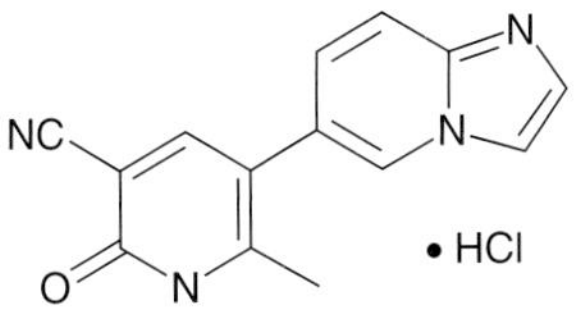

Loprinone is a potent and selective inhibitor of PDE III and a long-lasting, orally active positive inotropic agent. It is used for acute cardiac insufficiency in cases resistant to other treatments.

1.03.4.6 Anticoagulants

1.03.4.6.1 Duteplase

Trade name	Solclot
Manufacturer	Sumitomo
Country of origin	Japan
Year of introduction	1995
Country in which first launched	Japan
CAS registry number	120608-46-0
Structure – recombinant protein	

Duteplase is a recombinant tissue-type plasminogen activator that acts by converting the proenzyme plasminogen to the active enzyme plasmin. It was launched for the treatment of acute myocardial infarction.

1.03.4.6.2 Lepirudin

Trade name	Refludan
Manufacturer	Hoechst Marion Roussel
Country of origin	Germany

Year of introduction	1997
Country in which first launched	Germany
CAS registry number	8001-27-2
Structure – recombinant protein of molecular weight 6.9 kDa	

Refludan is an almost irreversible inhibitor of thrombin and was launched for the treatment of heparin-associated thrombocytopenia. It is used for myocardial infarcts, unstable angina, and cardiovascular events.

1.03.4.7 Hypocholesterolemic Agents

1.03.4.7.1 Atorvastatin calcium

Trade name	Lipitor
Manufacturer	Parke-Davis
Country of origin	USA
Year of introduction	1997
Country in which first launched	UK
CAS registry number	134523-03-8
Structure	

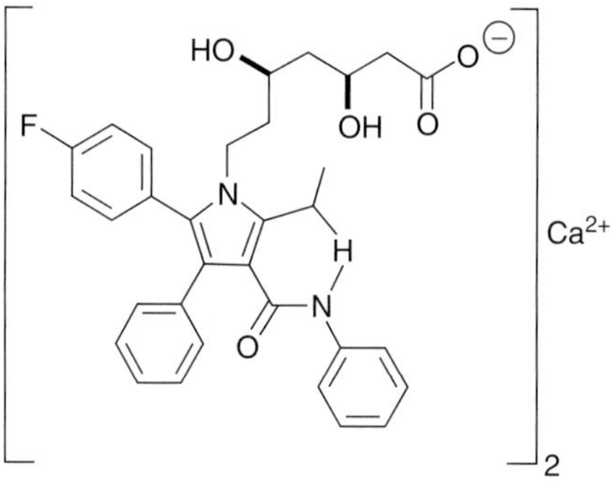

Atovastatin calcium is a liver-selective reversible competitive inhibitor of HMG-CoA reductase that was introduced as an orally active hypocholesterolemic agent. It became the first pharmaceutical product ever to make over US $1billion in sales in its first year.

1.03.4.7.2 Cerivastatin

Trade name	Lipobay
Manufacturer	Bayer
Country of origin	Germany
Year of introduction	1997
Country in which first launched	UK
CAS registry number	143201-11-0
Structure	

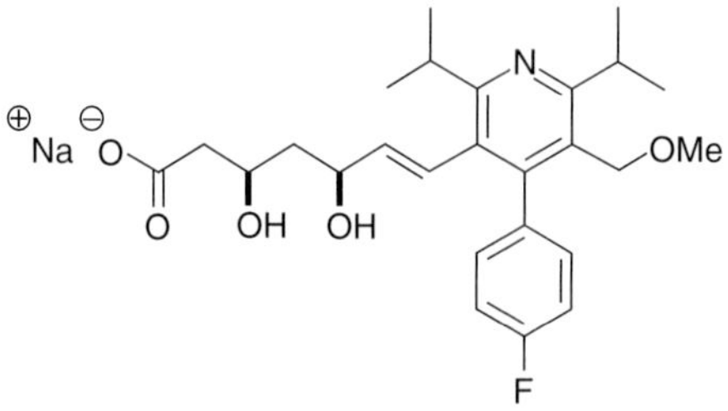

Lipobay is a 3-hydroxy-3-methylglutaryl-coenzyme A (HMG-CoA) reductase inhibitor with a high liver selectivity that was launched for the treatment of primary hypercholesterolemia.

1.03.4.7.3 Colesevelam hydrochloride

Trade name	Welchol
Manufacturer	GelTex, Sankyo, Pfizer
Country of origin	USA
Year of introduction	2000
Country in which first launched	USA
CAS registry number	182815-44-7
Structure	

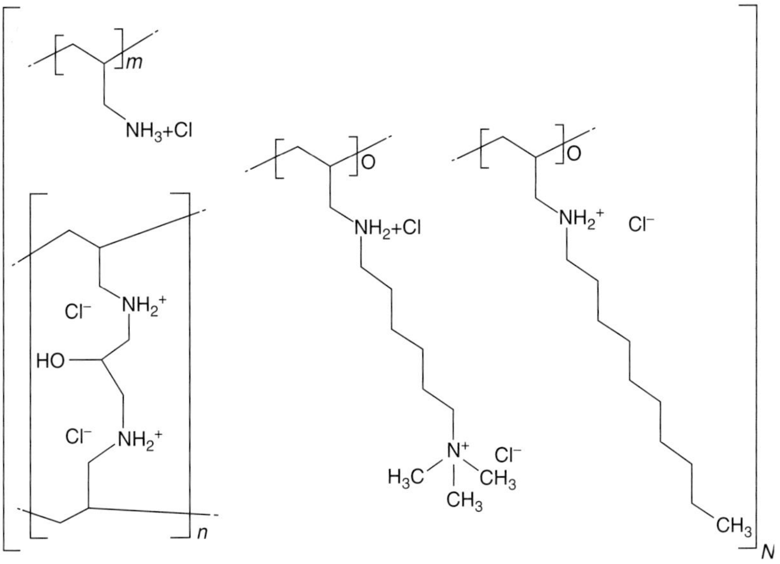

Colesevelam is a nonabsorbable, water-insoluble polymer of a hexanamium chloride with *N*-(2-propenyl) decanamine, 2-propene-1-amine hydrochloride, and chloromethyloxirane. It is used for the reduction of elevated levels of serum low-density lipoprotein (LDL) cholesterol. It can either be used as a monotherapy or as a dual therapy with statins.

1.03.4.7.4 Colestimide

Trade name	Cholebine
Manufacturer	Mitsubishi, Tokyo Tanabe, Yamanouchi
Country of origin	Japan
Year of introduction	1999
Country in which first launched	Japan
CAS registry number	95522-45-5
Structure	

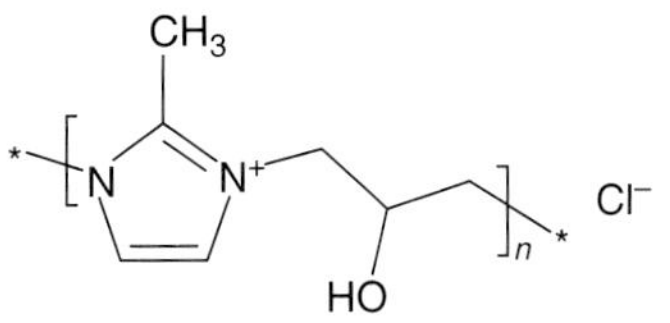

Colestimide is a methyl imidazole-epichlorin copolymer used as a bile acid sequestrant that leads to the enhancement of the rate of LDL removal. It can be used in monotherapy or coadministered with HMG-CoA reductase inhibitors.

1.03.4.7.5 **Ezetimibe**

Trade name	Ezetrol
Manufacturer	Schering-Plough, Merck
Country of origin	USA
Year of introduction	2002
Country in which first launched	Germany
CAS registry number	163222-33-1
Structure	

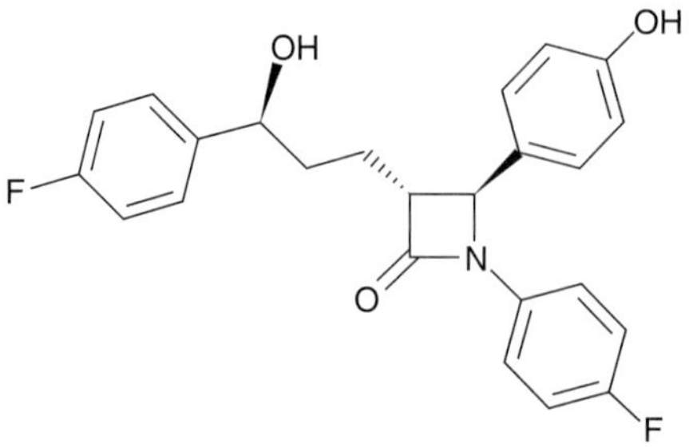

Ezetimibe is a once a day orally active cholesterol absorption inhibitor that is marketed as a hypolipidemic agent. When coadministred with statins it is more effective at reducing blood cholesterol levels.

1.03.4.7.6 **Fluvastatin sodium**

Trade name	Lescol
Manufacturer	Sandoz
Country of origin	Switzerland
Year of introduction	1994
Country in which first launched	UK
CAS registry number	93957-55-2
Structure	

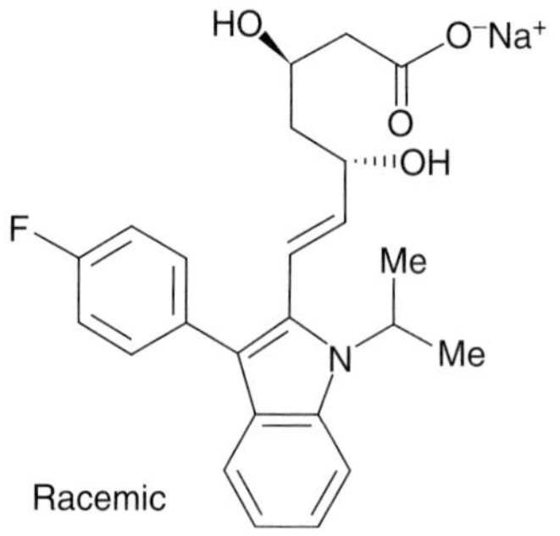

Fluvastatin sodium was the fourth HMG-CoA reductase inhibitor to be marketed for lowering total and LDL cholesterol.

1.03.4.7.7 **Pitavastatin**

Trade name	Livalo
Manufacturer	Nissan/Sankyo/Kowa
Country of origin	Japan
Year of introduction	2003

Country in which first launched	Japan
CAS registry number	147526-32-7
Structure	

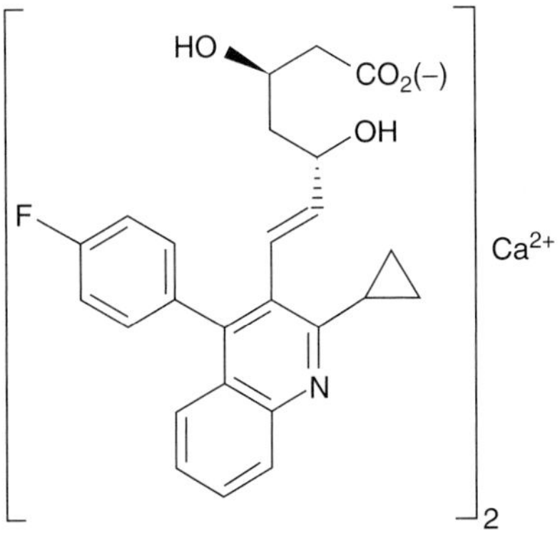

This is a second-generation statin that was introduced for the treatment of hypercholesterolemia. It is an inhibitor of HMG-CoA reductase. It is well absorbed, undergoes little metabolism, and is excreted via the bile.

1.03.4.7.8 Rosuvastatin

Trade name	Crestor
Manufacturer	Shionogi/AstraZeneca
Country of origin	Japan
Year of introduction	2003
Country in which first launched	Netherlands
CAS registry number	147098-20-2
Structure	

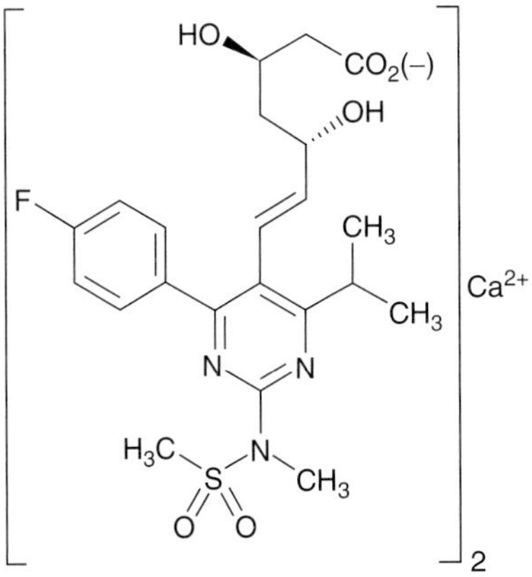

Rosuvastatin is a second-generation statin with high hepatoselectivity and more potent inhibitory effects on HMG-CoA reductase than the previously marketed statins. It is rapidly absorbed following oral dosing, minimally metabolized, and excreted via the bile. The length of action permits once daily dosing.

1.03.4.8 Iron Chelation

1.03.4.8.1 Deferiprone

Trade name	Kelfer
Manufacturer	Cipla
Country of origin	India

Year of introduction	1995
Country in which first launched	India
CAS registry number	30652-11-0
Structure	

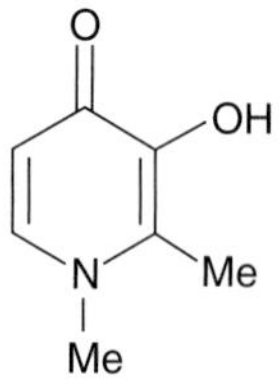

Deferiprone is the first oral iron chelator; it was marketed for the management of thalassemia. It appears also to act as a chelator for aluminum.

1.03.4.9 Hematological

1.03.4.9.1 Anagrelide hydrochloride

Trade name	Agrylin
Manufacturer	Roberts
Country of origin	USA
Year of introduction	1997
Country in which first launched	USA
CAS registry number	58579-51-4
Structure	

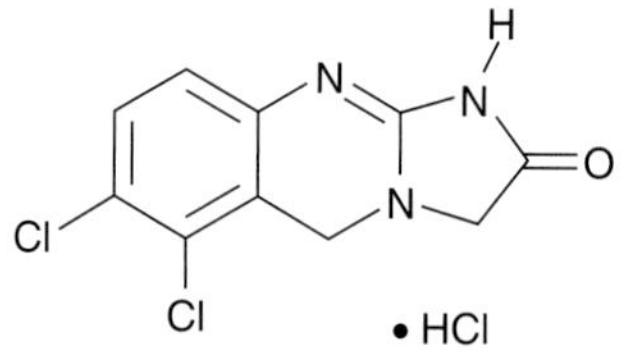

Anagrelide hydrochloride was launched for the treatment of thrombocytosis. It has anti-cAMP phosphodiesterase activity and was initially tested as a platelet aggregation inhibitor. Its actual mechanism of action is not fully understood. It does not shorten platelet survival but it does it appear to interfere with the maturation of megakaryocytes.

1.03.4.10 Subarachnoid Hemorrhage

1.03.4.10.1 Tirilazad mesylate

Trade name	Freedox
Manufacturer	Pharmacia & Upjohn
Country of origin	Sweden/USA
Year of introduction	1995
Country in which first launched	Austria
CAS registry number	110101-67-2
Structure	

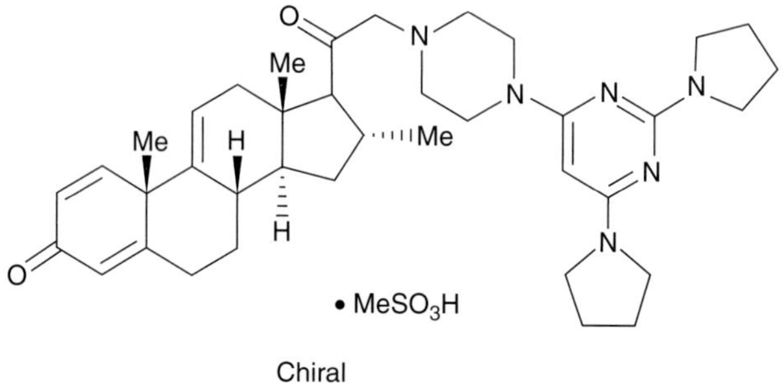

Tirilazad was marketed for the intravenous treatment of subarachnoid hemorrhage in male patients. It localizes into the cell membrane and is a potent inhibitor of lipid peroxidation induced by oxygen free radicals.

1.03.4.11 Hemophilia

1.03.4.11.1 Factor VIIa

Trade name	NovoSeven
Manufacturer	Novo Nordisk
Country of origin	Denmark
Year of introduction	1996
Country in which first launched	Denmark
CAS registry number	151821-07-7
Structure – recombinant human factor VIIa	

Factor VIIa was launched for the treatment of hemophila A and B. In vivo, it has inhibitor bypassing activity and thus can circulate throughout the body with no effect until it comes into contact with tissue factor at a site of injury where activation occurs.

1.03.4.12 Cardiotonic

1.03.4.12.1 Colforsin daropate hydrochloride

Trade name	Adele, Adehi
Manufacturer	Nippon Kayaku
Country of origin	Japan
Year of introduction	1999
Country in which first launched	Japan
CAS registry number	138605-00-2
Structure	

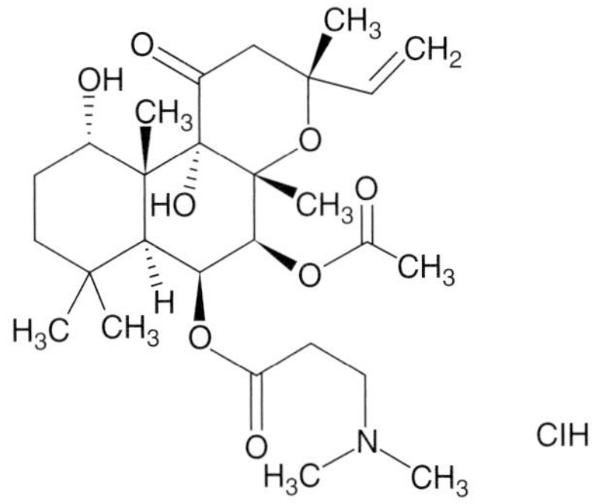

Colforsin daropate is a water-soluble and orally active prodrug of forskolin that was launched as a treatment for acute heart failure. It directly stimulates adenylate cyclase and increases the intracellular concentration of cAMP and inhibits calcium mobilization. Thus, it induces significant vasodilation and provides a moderate positive inotropic and chronotropic effect.

1.03.4.13 Fibrinolytic

1.03.4.13.1 Reteplase

Trade name	Retevase
Manufacturer	Boehringer Mannheim

Country of origin	Germany
Year of introduction	1996
Country in which first launched	USA
CAS registry number	133652-38-7
Structure – recombinant protein of molecular weight 39.5 kDa	

Reteplase is a single-chain recombinant form of tissue plasminogen activator that was launched for the treatment of acute myocardial infarction.

1.03.5 Nervous System (Both Central and Peripheral)

Fifty-five new drugs affecting some part of the central and peripheral nervous systems have been launched between 1993 and 2002. Heading the table in new introductions are seven compounds for depression, seven for Parkinson's disease, seven for migraine, six neuroleptics, four antiepileptics, three for neuroprotection, three antihistamines, three for muscle relaxation, two for Alzheimer's disease, two for multiple sclerosis, two for local anesthesia, and one each for dependence, narcolepsy, attention deficit hyperactivity, psychostimulation, inducing sleep, diagnosis of CNS disorders, sedation, anxiety, and stimulation.

1.03.5.1 Antidepressants

1.03.5.1.1 Escitalopram oxalate

Trade name	Cipralex
Manufacturer	Lundbeck
Country of origin	Denmark
Year of introduction	2002
Country in which first launched	Switzerland, UK, Sweden
CAS registry number	219861-08-2
Structure	

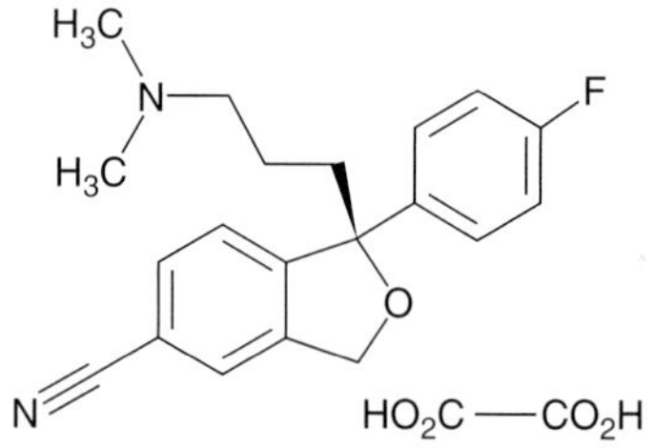

Escitalopram oxalate is the *S*-enantiomer of the selective serotonin reuptake inhibitor citalopram and was launched for the treatment of depression and panic disorder. It has a fast onset of action and long half-life (27–32 h).

1.03.5.1.2 Milnacipran

Trade Name	Ixel
Manufacturer	Pierre Fabre
Country of Origin	France
Year of Introduction	1997
Country in which first launched	France
CAS Registry number	92623-85-3

Structure

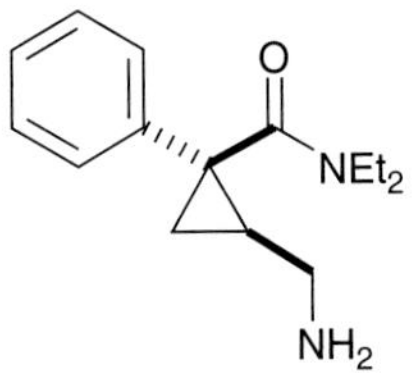

Milnacipran is a specific serotonin and noradrenaline reuptake inhibitor. It has as short (7 h) half-life, no active metabolites, and is not metabolized by cytochrome P450. It was launched for the treatment of severe depression.

1.03.5.1.3 Mirtazapine

Trade name	Remeron
Manufacturer	Organon
Country of origin	Netherlands
Year of introduction	1994
Country in which first launched	Netherlands
CAS registry number	61337-67-5
Structure	

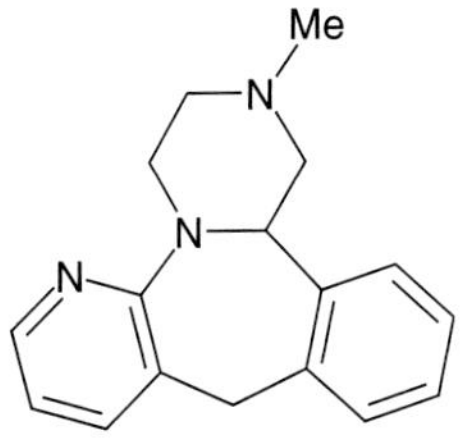

Racemic

Mirtazapine is a potent antagonist of presynaptic α_2-receptors and a moderately potent 5HT antagonist. In the clinic, it demonstrated antidepressant activity with low anticholinergic action and gastrointestinal side effects and low cardiovascular toxicity.

1.03.5.1.4 Nefazodone hydrochloride

Trade name	Serzone
Manufacturer	Bristol-Meyers Squibb
Country of origin	USA
Year of introduction	1994
Country in which first launched	Canada
CAS registry number	82752-99-6
Structure	

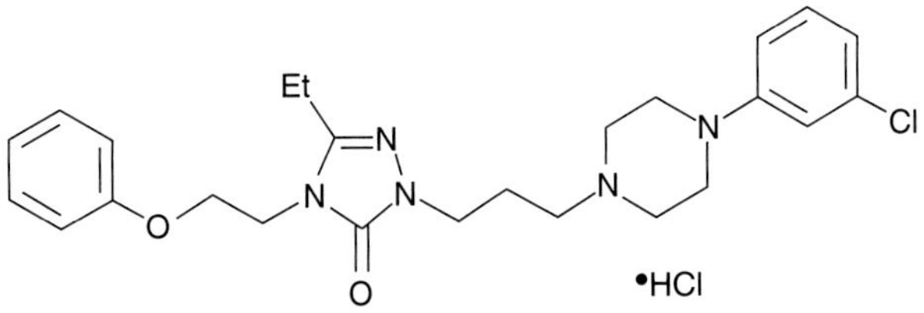

Nefazodone hydrochloride acts as both a potent $5HT_2$ receptor antagonist and as a serotonin reuptake blocker. The very selective serotonergic effects are accompanied by few side effects from cholinergic, histamine, and adrenergic systems.

1.03.5.1.5 Pivagabine

Trade name	Tonerg
Manufacturer	Angelini
Country of origin	Italy
Year of introduction	1997
Country in which first launched	Italy
CAS registry number	69542-93-4
Structure	

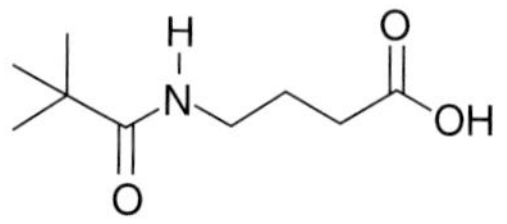

Pivagabine was launched for the treatment of acute stress, posttraumatic stress syndrome, and 'burnout' syndrome. It inhibits the release of corticotropin-releasing factor from the hypothalamus.

1.03.5.1.6 Reboxetine

Trade name	Edronax
Manufacturer	Pharmacia & Upjohn
Country of origin	Sweden
Year of introduction	1997
Country in which first launched	UK
CAS registry number	71620-98-8
Structure	

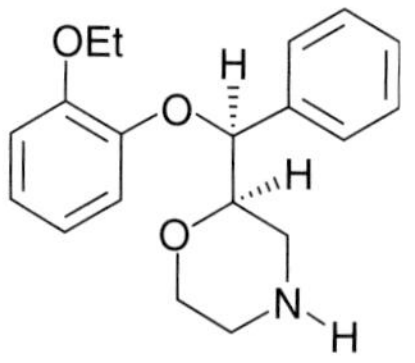

Reboxitine is a selective noradrenaline reuptake blocker that was shown to be effective in both the short and long term in the treatment of severe depression.

1.03.5.1.7 Venlafaxine hydrochloride

Trade name	Effexor
Manufacturer	Wyeth-Ayerst
Country of origin	USA
Year of introduction	1994
Country in which first launched	USA
CAS registry number	93413-69-5

Structure

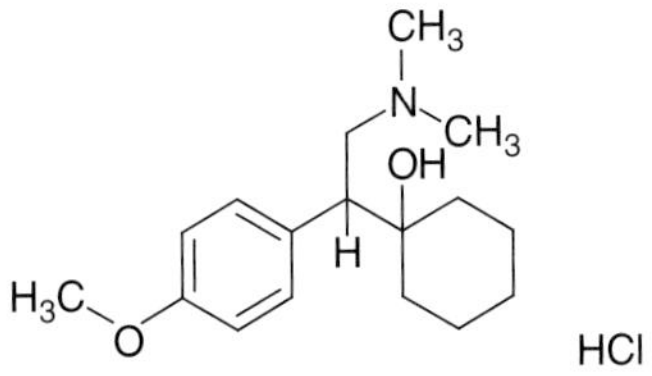

Venlafaxine is the first in the class of second-generation antidepressants with dual serotonin/norepinephrine reuptake inhibitory activity. It appears to have a rapid onset of action.

1.03.5.2 Antiparkinsonian Agents

1.03.5.2.1 Budipine

Trade name	Parkinsan
Manufacturer	Byk Gulden/Promonta Lundbeck
Country of origin	Germany
Year of introduction	1997
Country in which first launched	Germany
CAS registry number	57982-78-2
Structure	

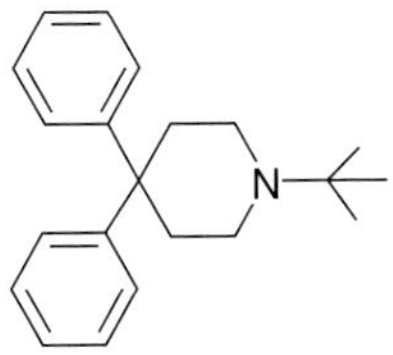

Budipine is effective for the treatment of parkinsonian tremors. It exhibits use-dependent, open channel, uncompetitive *N*-methyl-D-aspartate (NMDA) receptor antagonist activity and interacts with sigma-binding sites in the frontal cortex.

1.03.5.2.2 CHF-1301

Trade name	Levomet
Manufacturer	Chiesi Farmaceutici SPA
Country of origin	Italy
Year of introduction	1999
Country in which first launched	Italy
CAS registry number	7101-51-1
Structure	

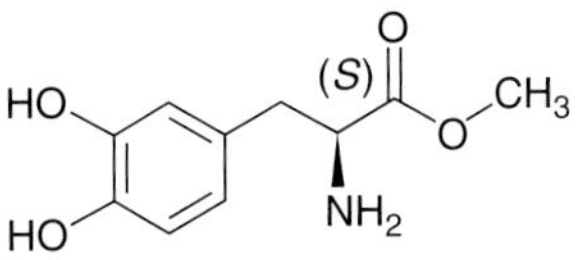

Levomet was introduced as an injection for patients with Parkinson's disease who experienced complications in their chronic treatment with levodopa.

1.03.5.2.3 Entacapone

Trade name	Comtess
Manufacturer	Orion Pharma
Country of origin	Finland
Year of introduction	1998
Country in which first launched	Finland, Germany, Sweden
CAS registry number	130929-57-6
Structure	

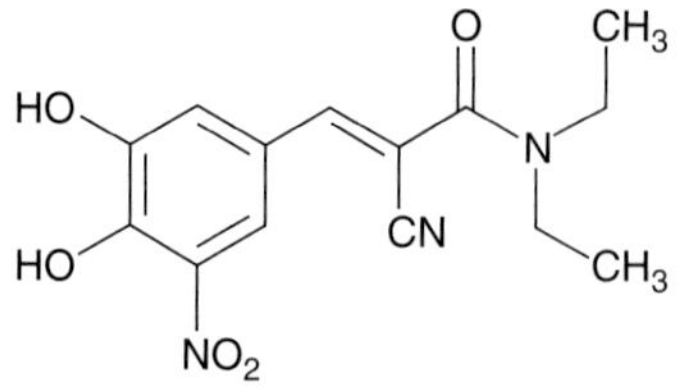

Entacapone is a peripherally selective, reversible, orally active catechol *O*-methyltransferase inhibitor intended to be used as an adjuvant to levodopa therapy for Parkinson's disease.

1.03.5.2.4 Pramipexole hydrochloride

Trade name	Mirapex
Manufacturer	Boehringer Ingelheim/Pharmacia & Upjohn
Country of origin	Germany
Year of introduction	1997
Country in which first launched	USA
CAS registry number	104632-25-9
Structure	

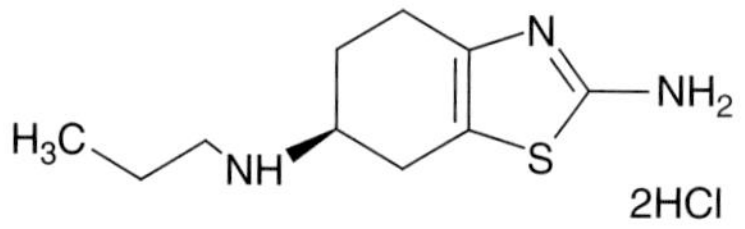

Pramipexole is a presynaptic dopamine D_2 autoreceptor agonist that also activates D_3 receptors.

1.03.5.2.5 Ropinirole hydrochloride

Trade name	ReQuip
Manufacturer	SmithKlineBeecham
Country of origin	UK
Year of introduction	1996
Country in which first launched	UK
CAS registry number	91374-20-8
Structure	

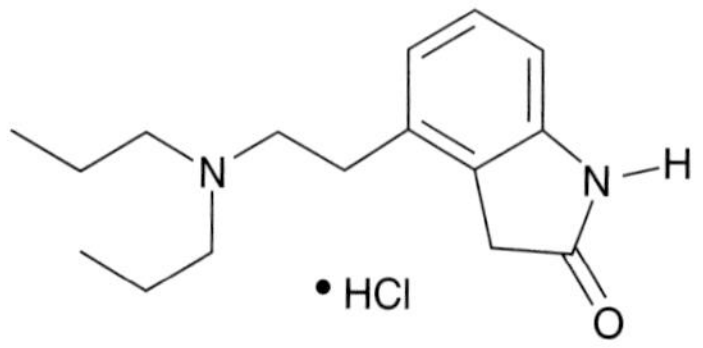

Ropinirole is a nonergot postsynaptic dopamine D_2 agonist with activity in the extrapyramidal system. It was introduced as a monotherapy or for use in combination with low-dose levodopa for the treatment of early stage idiopathic Parkinson's disease.

1.03.5.2.6 Talipexole

Trade name	Domin
Manufacturer	Boehringer Ingelheim
Country of origin	Germany
Year of introduction	1996
Country in which first launched	Japan
CAS registry number	101626-70-4
Structure	

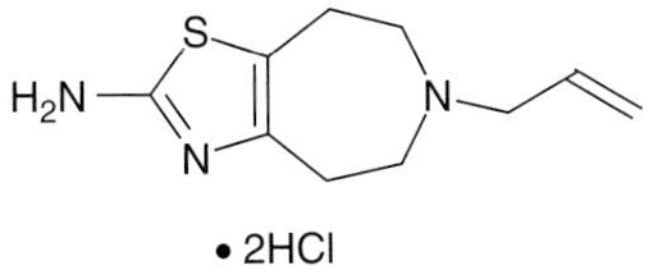

Talipexole is a selective agonist for presynaptic dopamine D_2 receptors with no D_1 receptor activity.

1.03.5.2.7 Tolcapone

Trade name	Tasmar
Manufacturer	Roche
Country of origin	Switzerland
Year of introduction	1997
Country in which first launched	Germany
CAS registry number	134308-13-7
Structure	

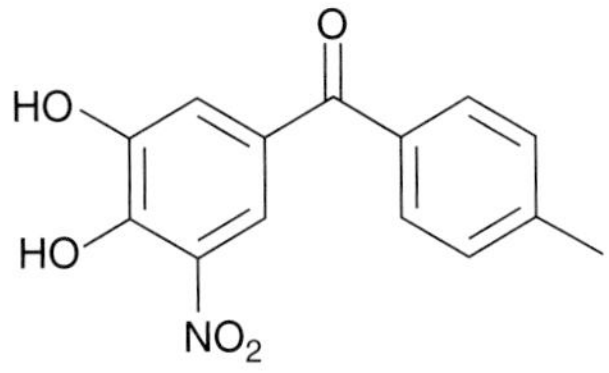

Tolcapone is a peripherally selective, reversible, orally active catechol *O*-methyltransferase inhibitor that was introduced as an adjuvant to levodopa therapy for Parkinson's disease.

1.03.5.3 Antimigraine Agents

1.03.5.3.1 Almotriptan

Trade name	Almogran
Manufacturer	Almirall Prodesfarma
Country of origin	Spain
Year of introduction	2000
Country in which first launched	Spain
CAS registry number	154323-57-6

Structure

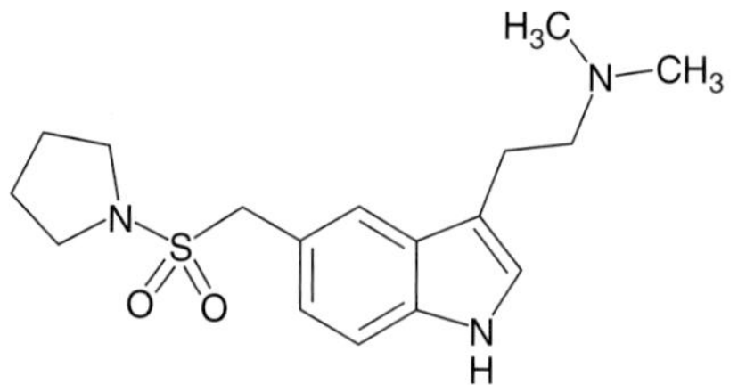

Almotriptan acts as a dual $5HT_{1D/1B}$ agonist with a 35- to 51-fold selectivity versus $5HT_{1A}$ and $5HT_7$ receptors, respectively, as well as having insignificant affinity for the most relevant nonserotonic receptors.

1.03.5.3.2 Electriptan

Trade name	Relpax
Manufacturer	Pfizer
Country of origin	USA
Year of introduction	2001
Country in which first launched	Switzerland
CAS registry number	143322-58-1
Structure	

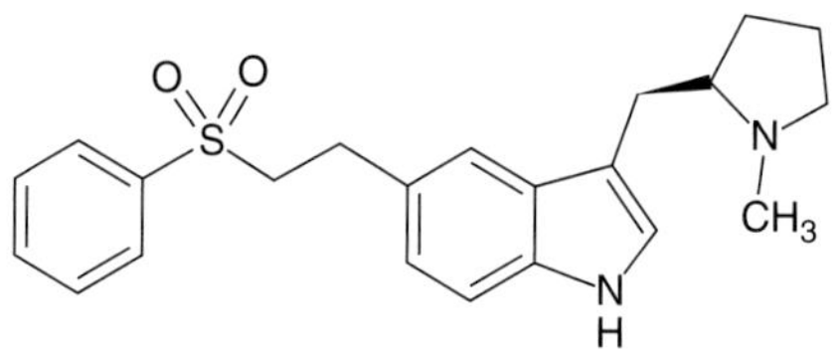

Electriptan is a 5HT receptor agonist that binds to $5HT_{1B}$, $5HT_{1D}$, and $5HT_{1F}$ receptors with high potency. It was found to be more potent than other triptans as an agonist at the $5HT_{1D}$ receptor. It acts more rapidly and effectively against migraine attacks than other triptans.

1.03.5.3.3 Frovatriptan

Trade name	Frova
Manufacturer	GlaxoSmithKline/Vernalis
Country of origin	UK
Year of introduction	2002
Country in which first launched	USA
CAS registry number	158930-09-7
Structure	

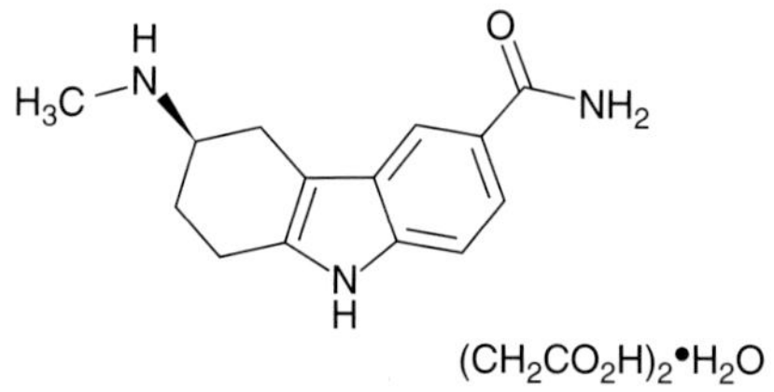

Frovatriptan was launched as an oral treatment for acute migraine attacks with or without aura in adults. It is the eighth member of the triptan class to be launched. It shows a very long half-life and a remarkably low level of headache recurrence rate.

1.03.5.3.4 Lomerizine hydrochloride

Trade name	Teranas, Migsis
Manufacturer	Akzo Nobel/Kanebo
Country of origin	Netherlands
Year of introduction	1999
Country in which first launched	Japan
CAS registry number	101477-54-7
Structure	

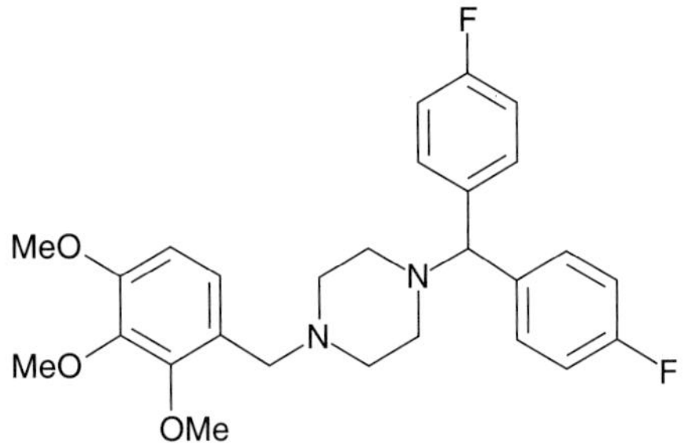

Lomerizine hydrochloride was the first in its class of dual sodium and calcium channel blockers to be marketed for migraine. It has demonstrated effectiveness in migraine treatment but it may also have utility in other neurological diseases such as cerebrovascular ischemia or cerebral infarction.

1.03.5.3.5 Naratriptan hydrochloride

Trade name	Naramig
Manufacturer	Glaxo Wellcome
Country of origin	UK
Year of introduction	1997
Country in which first launched	UK
CAS registry number	1433388-64-1
Structure	

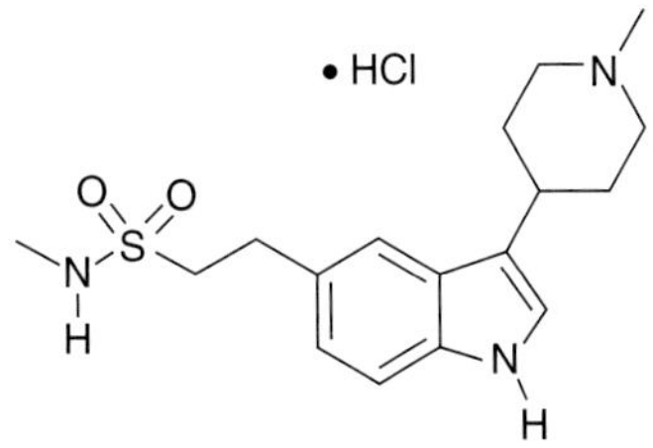

Naratriptan is a serotonin $5HT_{1B/1D}$ receptor antagonist with modest affinity for $5HT_{1a}$ and very weak affinity for $5HT_3$ receptors. It mediates vasoconstriction in cerebral vasculature, reduces neurogenic inflammation, and inhibits responses mediated by the trigeminal nerves. Naratriptan has no clinical effects on blood pressure or heart rate, and has a long duration of action with very good tolerability and a high oral bioavailability.

1.03.5.3.6 Rizatriptan benzoate

Trade name	Maxalt
Manufacturer	Merck
Country of origin	USA
Year of introduction	1998
Country in which first launched	Mexico
CAS registry number	145202-66-0

Structure

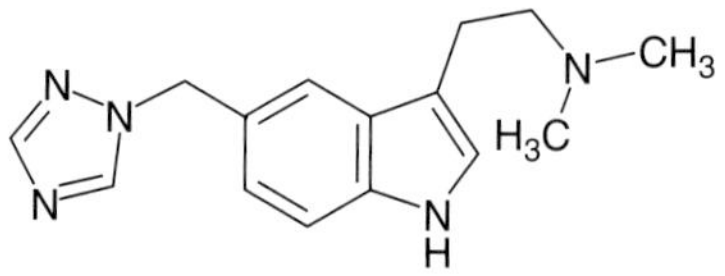

Rizatriptan is a full $5HT_{1B/1D}$ receptor agonist retaining a significant affinity at $5HT_{1A}$ sites with low affinity for non-5HT sites. In humans, it shows craniovascular selectivity for isolated middle meningeal arteries over coronary arteries.

1.03.5.3.7 Zomitriptan

Trade name	Zomig
Manufacturer	Zeneca
Country of origin	UK
Year of introduction	1997
Country in which first launched	UK
CAS registry number	139264-17-8
Structure	

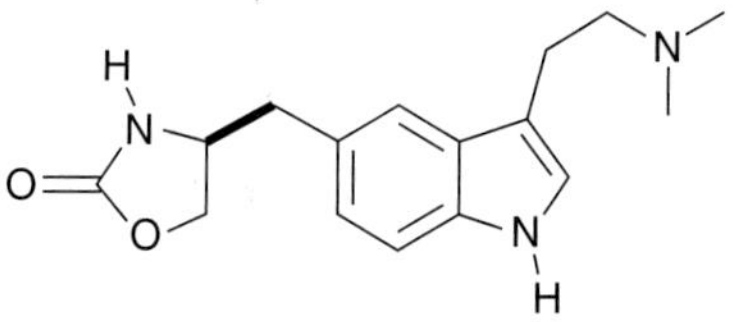

Zomitriptan is a serotonin $5HT_{1B/1D}$ receptor agonist with modest affinity for $5HT_{1A}$ and $5HT_{1F}$ receptors. It acts centrally on the trigeminal nucleus caudalis and peripherally on the trigeminovascular system.

1.03.5.4 Neuroleptics

1.03.5.4.1 Aripiprazole

Trade name	Abilify
Manufacturer	Otsuka/Bristol-Meyers Squibb
Country of origin	Japan
Year of introduction	2002
Country in which first launched	USA
CAS registry number	129722-12-9
Structure	

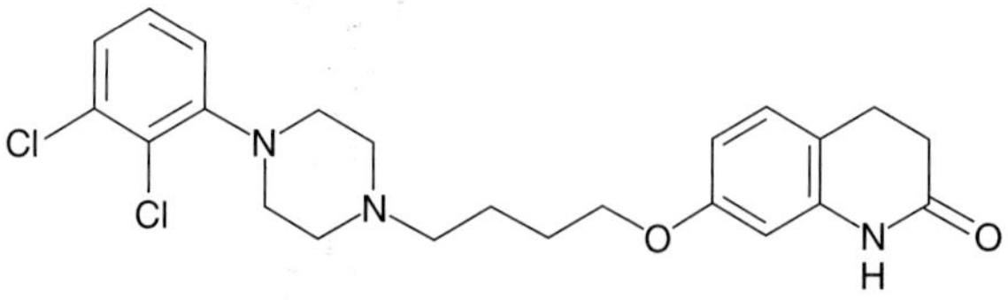

Aripiprazole is a partial D_2 receptor agonist that was launched for the treatment of psychoses including schizophrenia.

1.03.5.4.2 Olanzapine

Trade name	Zyprexa
Manufacturer	Lilly
Country of origin	USA
Year of introduction	1996
Country in which first launched	UK
CAS registry number	132539-06-1
Structure	

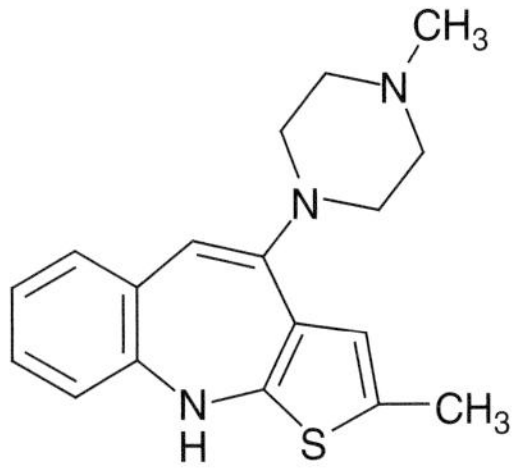

Olanzapine is a potent $5HT_2/D_2$ antagonist with anticholinergic activity that was launched as an antipsychotic agent.

1.03.5.4.3 Perospirone

Trade name	Lullan
Manufacturer	Sumitomo Pharm
Country of origin	Japan
Year of introduction	2001
Country in which first launched	Japan
CAS registry number	150915-41-6
Structure	

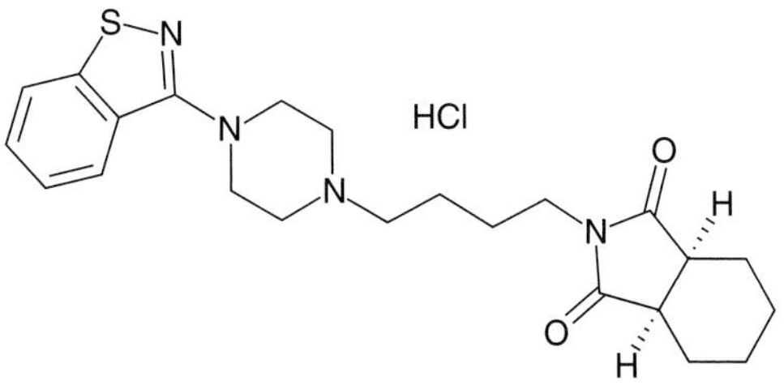

Perospirone was launched as a new treatment of schizophrenia and other psychoses. It seems to have a significantly lower propensity for the development of extrapyramidal symptoms and tardive schizophrenia.

1.03.5.4.4 Quetiapine fumarate

Trade name	Seroquel
Manufacturer	Zeneca
Country of origin	UK
Year of introduction	1997
Country in which first launched	UK
CAS registry number	111974-72-2

Structure

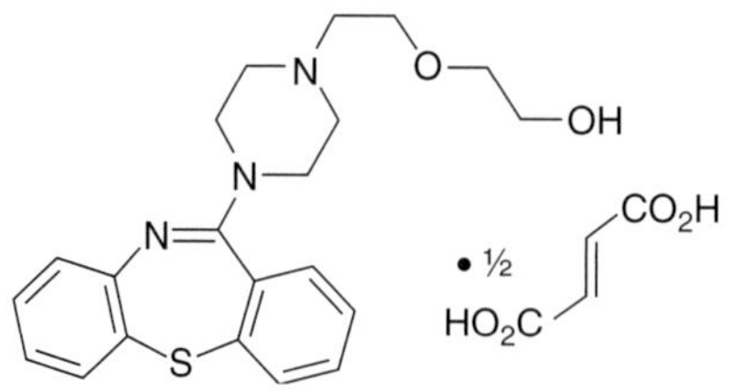

Quetiapine fumarate is selective for the limbic region with activity as a moderate dopamine receptor antagonist with greater affinity for D_2 over D_1 and D_4 receptors and a greater affinity for $5HT_{2A}$ than for D_2 receptors. It was launched for the treatment of schizophrenia.

1.03.5.4.5 **Sertindole**

Trade name	Serdolect
Manufacturer	Lundbeck
Country of origin	Denmark
Year of introduction	1996
Country in which first launched	UK
CAS registry number	106516-24-9
Structure	

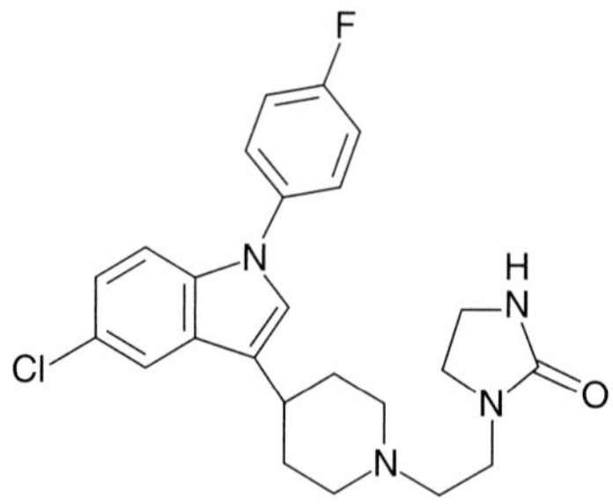

Sertindole is selective for the limbic areas of the brain and has a high affinity for the serotonin $5HT_2$ receptor where it acts as an antagonist. It was launched for the treatment of acute and chronic schizophrenia and schizoaffective psychoses.

1.03.5.4.6 **Ziprasidone hydrochloride**

Trade name	Zeldox
Manufacturer	Pfizer
Country of origin	USA
Year of introduction	2000
Country in which first launched	Sweden
CAS registry number	122883-93-6
Structure	

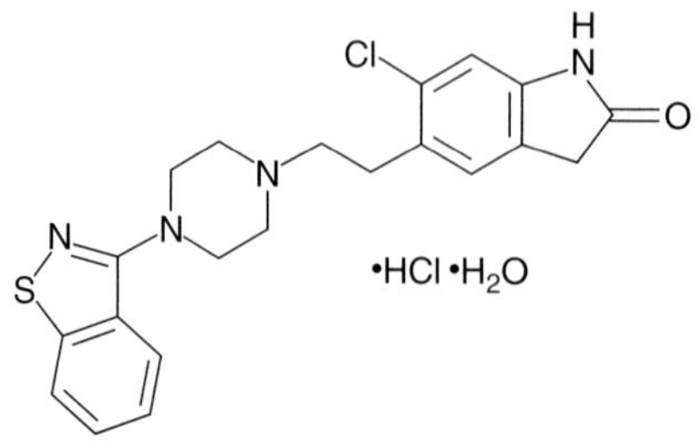

Ziprasidone was launched as the sixth atypical antipsychotic for the treatment of schizophrenia and agitated psychoses.

1.03.5.5 Antiepileptics

1.03.5.5.1 Fosphenytoin sodium

Trade name	Cerebyx
Manufacturer	Warner-Lambert
Country of origin	USA
Year of introduction	1996
Country in which first launched	USA
CAS registry number	93390-81-9
Structure	

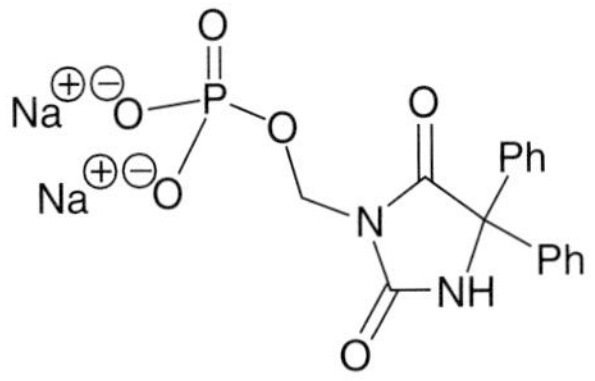

Fosphenytoin sodium is a prodrug that is rapidly converted by phosphatases to phenytoin. It was launched for the treatment of status epilepticus and for neurosurgery-derived seizures.

1.03.5.5.2 Levetiracetam

Trade name	Keppra
Manufacturer	UCB
Country of origin	USA
Year of introduction	2000
Country in which first launched	USA
CAS registry number	102767-28-2
Structure	

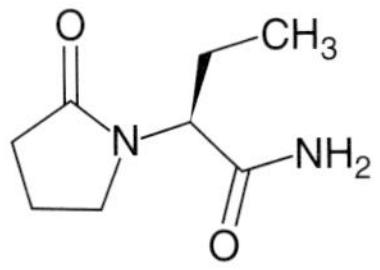

Levetiracetam is a second-generation analog of piracetam whose precise mode of action is not well established. However, it has been shown that it reversibly binds to a specific site predominantly present in the membranes of the brain. It was introduced as an adjunctive therapy in the treatment of partial onset seizures in adults with epilepsy.

1.03.5.5.3 Tiagabine

Trade name	Gabitril
Manufacturer	Novo Nordisk
Country of origin	Denmark
Year of introduction	1996
Country in which first launched	Denmark
CAS registry number	115103-54-3
Structure	

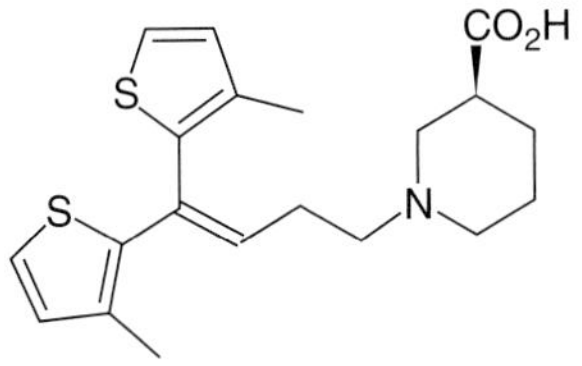

Tiagabine was launched as an add-on therapy in patients refractory to other epilepsy therapies and appears to work by potent and selective inhibition of GABA synaptosomal uptake. It is selective for the GAT-1 GABA transporter in neurons and glia, thus enhancing inhibitory GABAergic transmission.

1.03.5.5.4 **Topiramate**

Trade name	Topamax
Manufacturer	Johnson & Johnson
Country of origin	USA
Year of introduction	1995
Country in which first launched	UK
CAS registry number	97240-79-4
Structure	

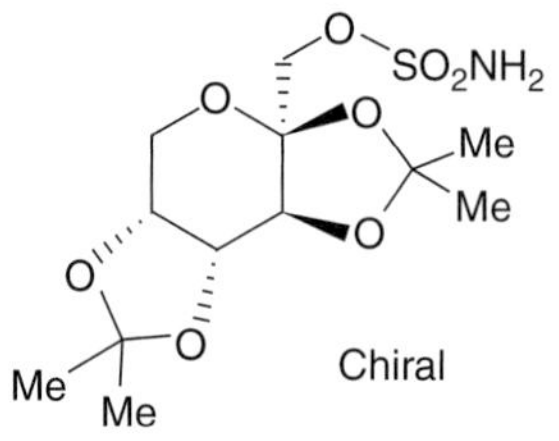

Topiramate was introduced as an adjunct therapy for use in partial seizures with or without secondary generalized seizures in adult patients inadequately controlled on conventional antiepileptics. It appears to act by blocking voltagesensitive sodium channels.

1.03.5.6 **Neuroprotective Agents**

1.03.5.6.1 **Edaravone**

Trade name	Radicut
Manufacturer	Mitsubishi Pharma
Country of origin	Japan
Year of introduction	2001
Country in which first launched	Japan
CAS registry number	89-25-8
Structure	

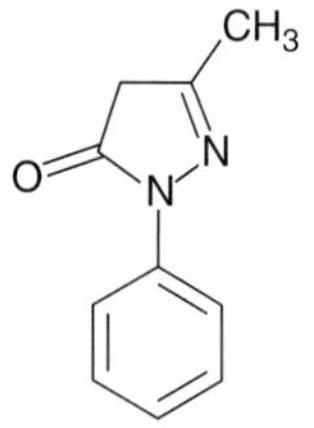

Edaravone was marketed for improving the neurologic recovery following acute brain infarction. It is an antioxidant with free radical scavenging activity.

1.03.5.6.2 **Fasudil hydrochloride**

Trade name	Eril
Manufacturer	Asahi Chemical
Country of origin	Japan

Year of introduction	1995
Country in which first launched	Japan
CAS registry number	105628-07-7
Structure	

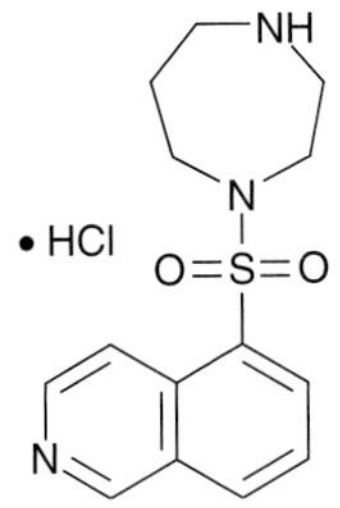

Fasudil is a calcium antagonist vasodilator that was marketed for the treatment of cerebral vasospasm following subarachnoid hemorrhage. It is also a potent inhibitor of myosin light chain kinase and protein kinase C.

1.03.5.6.3 Riluzole

Trade name	Rilutek
Manufacturer	Rhone-Poulenc Rorer
Country of origin	France
Year of introduction	1996
Country in which first launched	USA
CAS registry number	1744-22-5
Structure	

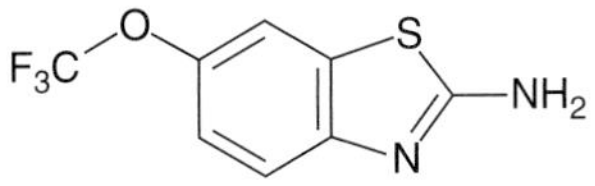

Riluzole was launched for the treatment of amyotrophic lateral sclerosis. It antagonizes excitatory amino acids and blocks presynaptic release of glutamate and quisqualate-induced increases in cGMP.

1.03.5.7 Antihistamines

1.03.5.7.1 Desloratadine

Trade name	Clarinex
Manufacturer	Sepracor/Schering-Plough
Country of origin	USA
Year of introduction	2001
Country in which first launched	UK
CAS registry number	100643-71-8
Structure	

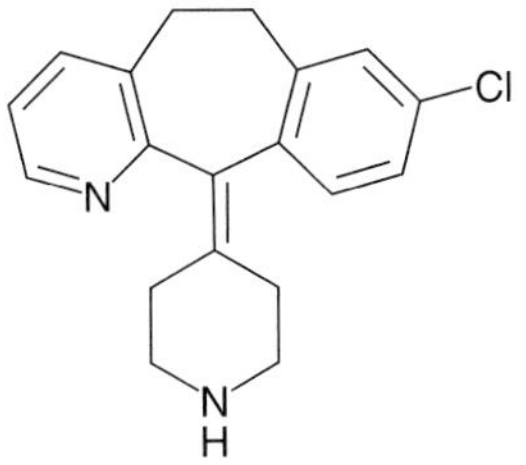

Desloratadine was launched for the treatment of nasal and non-nasal symptoms of seasonal allergic rhinitis. It is a nonsedating competitive histamine H_1-receptor antagonist with increased potency and improved safety.

1.03.5.7.2 Mizolastine

Trade name	Mizollen
Manufacturer	Synthelabo/Galderma
Country of origin	France
Year of introduction	1998
Country in which first launched	Germany, Switzerland
CAS registry number	108612-45-9
Structure	

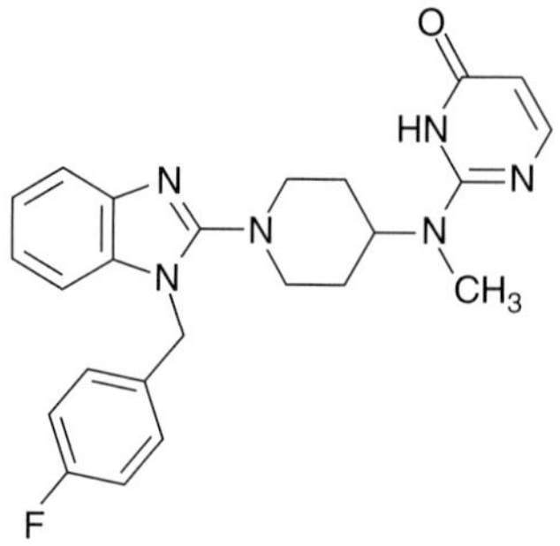

Mizolastine is a long-acting, orally active antihistamine that was introduced for the symptomatic relief of seasonal and perennial allergic rhinoconjuctivitis and urticaria. It selectively blocks peripheral H_1 receptors with little effect on brain receptors.

1.03.5.7.3 Levocetirizine

Trade name	Xusal
Manufacturer	Sepracor/UCB
Country of origin	USA
Year of introduction	2001
Country in which first launched	Germany
CAS registry number	130018-77-8
Structure	

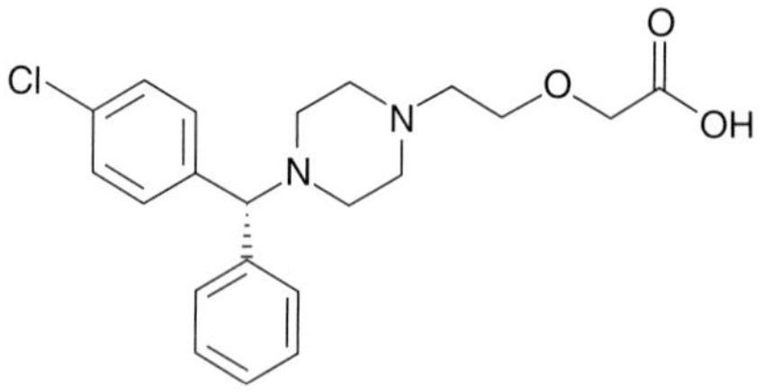

Levocetrizine is the *R*-isomer of the second-generation antihistamine cetirizine. It was launched for the treatment of seasonal allergic rhinitis, perennial allergic rhinitis, and chronic idiopathic urticaria.

1.03.5.8 Muscle Relaxation

1.03.5.8.1 Cisatracurium besylate

Trade name	Nimbex
Manufacturer	Glaxo Wellcome

Country of origin	UK
Year of introduction	1995
Country in which first launched	USA
CAS registry number	96946-42-8
Structure	

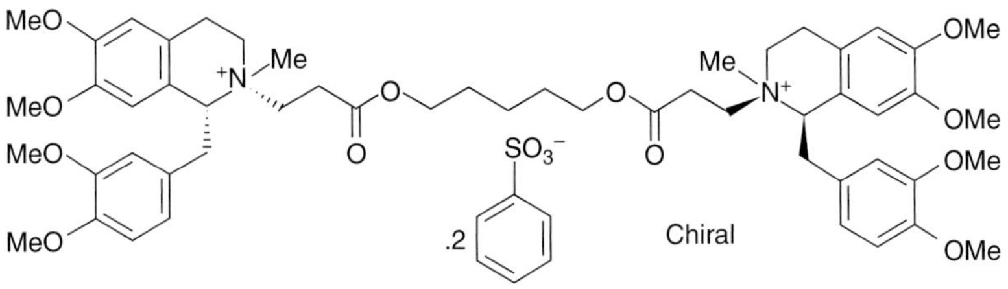

Cisatracurium besylate is a nondepolarizing muscle relaxant of intermediate duration that was launched for intubation and maintenance of muscle relaxation during surgery and intensive care.

1.03.5.8.2 Rapacuronium bromide

Trade name	Raplon
Manufacturer	Akzo Nobel
Country of origin	Netherlands
Year of introduction	1999
Country in which first launched	USA
CAS registry number	156137-99-4
Structure	

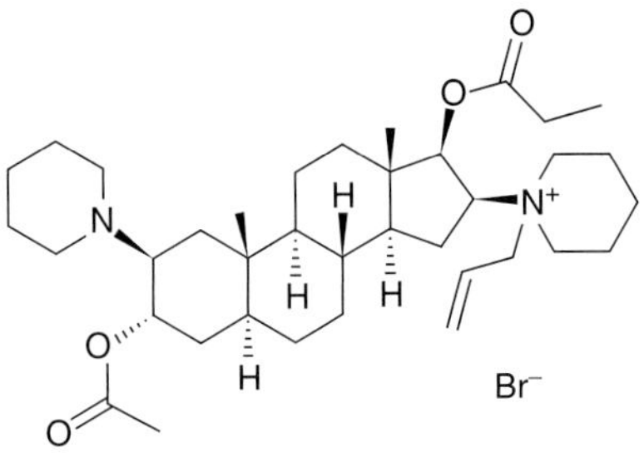

Rapacuronium bromide was introduced as a parenterally administered adjunct to general anesthesia during surgical procedures. It is a nondepolarizing neuromuscular blocking drug with a rapid onset and short duration of action.

1.03.5.8.3 Rocuronium bromide

Trade name	Zemuron, Esmeron
Manufacturer	Organon
Country of origin	Netherlands
Year of introduction	1994
Country in which first launched	USA, UK
CAS registry number	119302-91-9

Structure

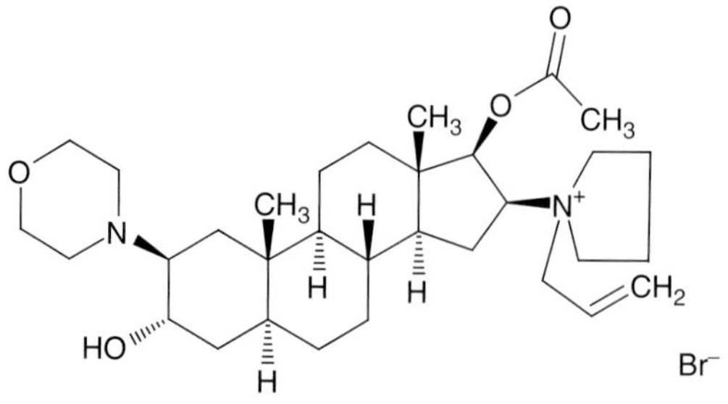

Rocuronium bromide is a short-acting, nondepolarizing steroidal neuromuscular blocker that was introduced for use as an adjunct to general anesthesia to facilitate endotracheal intubation and to provide skeletal muscle relaxation during surgery or mechanical ventilation.

1.03.5.9 Anti-Alzheimer's Disease

1.03.5.9.1 Donepezil hydrochloride

Trade name	Aricept
Manufacturer	Eisai/Pfizer
Country of origin	Japan
Year of introduction	1997
Country in which first launched	USA
CAS registry number	120011-70-3
Structure	

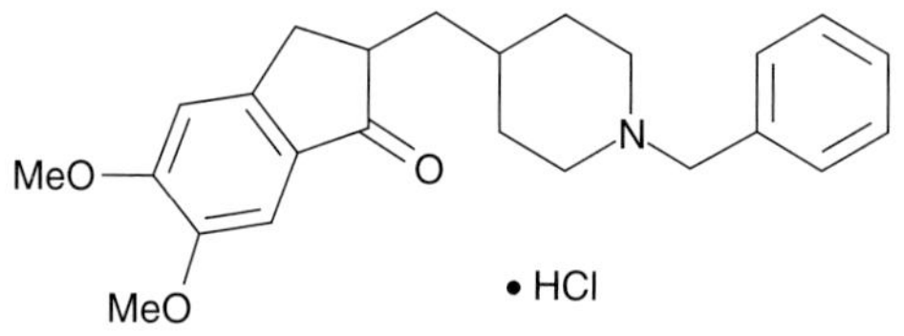

Aricept is a reversible, noncompetitive inhibitor of acetylcholinesterase that was introduced for the treatment of mild to moderate Alzheimer's disease and dementia.

1.03.5.9.2 Rivastigmin tartrate

Trade name	Exelon
Manufacturer	Norvartis
Country of origin	Switzerland
Year of introduction	1997
Country in which first launched	Switzerland
CAS registry number	129101-54-8
Structure	

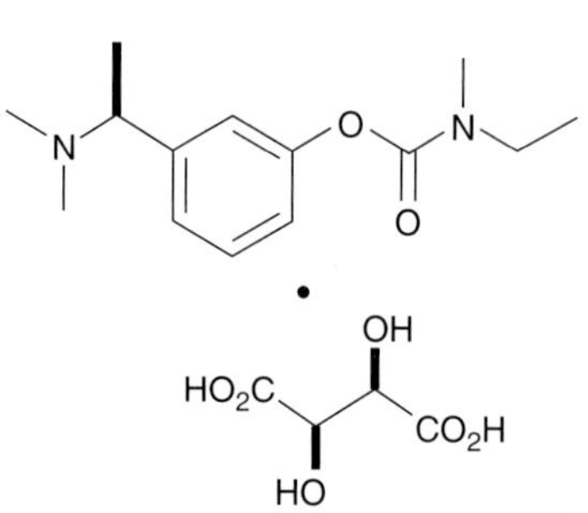

Rivastigmin is a centrally selective and long-lasting drug with anticholinesterase activity that facilitates cholinergic transmission in the cortex and hippocampus. It was launched for the treatment of mild to moderate Alzheimer's disease.

1.03.5.10 Multiple Sclerosis

1.03.5.10.1 Interferon β-1a

Trade name	Avonex
Manufacturer	Biogen
Country of origin	USA
Year of introduction	1996
Country in which first launched	USA
CAS registry number	145258-61-3
Structure – recombinant protein of molecular weight 20 025 Da	

Interferon β-1a was launched for the treatment of relapsing forms of multiple sclerosis.

1.03.5.10.2 Glatiramer acetate

Trade name	Copaxone
Manufacturer	Yeda
Country of origin	Israel
Year of introduction	1997
Country in which first launched	USA
CAS registry number	28704-27-0
Structure – recombinant protein of molecular weight 14–23 kDa	

Glatiramer acetate was launched for the treatment of relapsing-remitting multiple sclerosis.

1.03.5.11 Local Anesthesia

1.03.5.11.1 Levobupivacaine hydrochloride

Trade name	Chirocaine
Manufacturer	Chiroscience/Purdue Pharma
Country of origin	UK
Year of introduction	2000
Country in which first launched	USA
CAS registry number	27262-48-2
Structure	

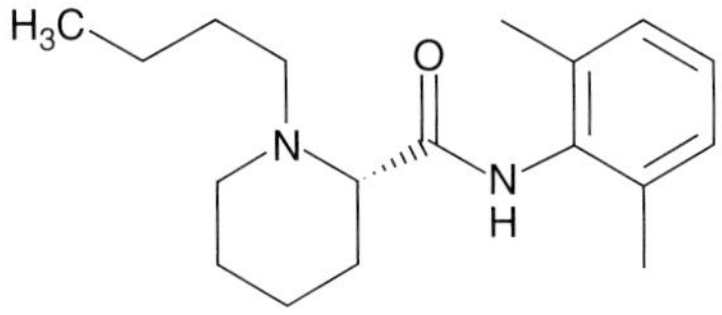

Levobupivacaine is the *S*-enantiomer of bupivacaine. It was launched for the production of local anesthesia in surgery and obstetrics and for postoperative pain management.

1.03.5.11.2 Ropivacaine

Trade name	Naropin
Manufacturer	Astra
Country of origin	Sweden
Year of introduction	1996

Country in which first launched	Australia
CAS registry number	84057-95-4
Structure	

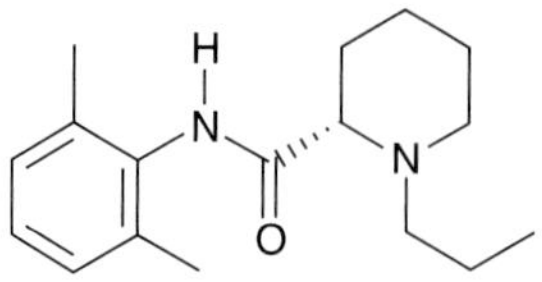

Ropivacaine is a sodium channel blocker that has a specific effect on the nerves involved in transmission of pain but has no effect on the nerve fibers responsible for motor function. It was launched as a local anesthetic.

1.03.5.12 Dependence Treatment

1.03.5.12.1 Nalmefene hydrochloride

Trade name	Revex
Manufacturer	IVAX/Ohmeda
Country of origin	USA
Year of introduction	1995
Country in which first launched	USA
CAS registry number	58895-64-0
Structure	

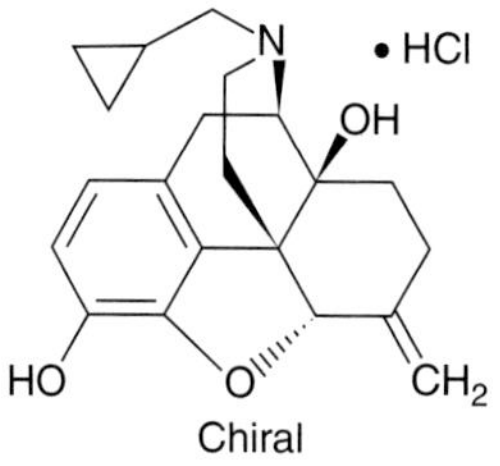

Nalmefene is an opioid antagonist that was introduced for opioid reversal following surgery and in the treatment of opioid overdoses and epidurally administered narcotics.

1.03.5.13 Narcolepsy

1.03.5.13.1 Modafinil

Trade name	Modiodal
Manufacturer	Lafon
Country of origin	France
Year of introduction	1994
Country in which first launched	France
CAS registry number	68693-11-8
Structure	

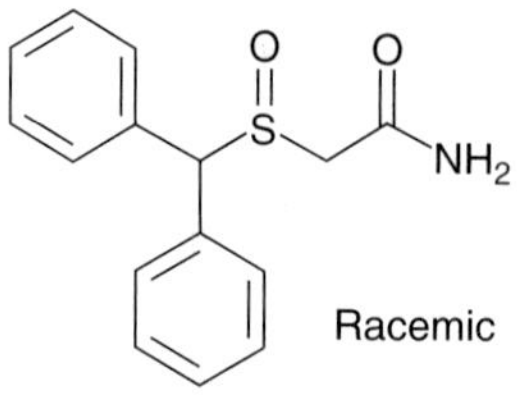

Modafinil is a centrally active α_1-adrenergic agonist that was marketed as a psychostimulant for the treatment of narcolepsy and idiopathic hypersomnia including Gelineau's syndrome.

1.03.5.14 Attention Deficit Hyperactivity

1.03.5.14.1 Atomoxetine

Trade name	Strattera
Manufacturer	Lilly
Country of origin	USA
Year of introduction	2003
Country in which first launched	USA
CAS registry number	082248-59-7
Structure	

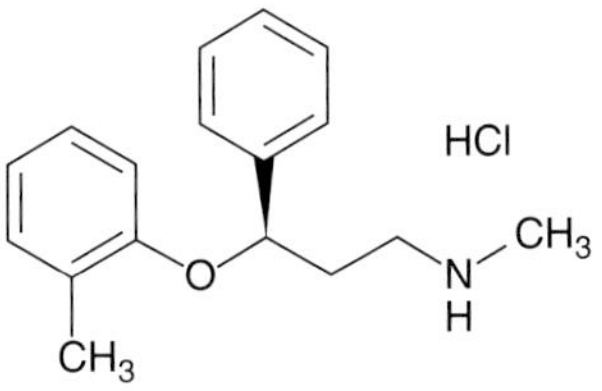

Atomexetine is the first nonstimulant to be markcted for this condition. It is a potent and selective norepinephrine uptake inhibitor and does not bind to monoamine receptors. It is about 63% orally bioavailable, is highly protein bound, and has a half-life of around 5.2 h.

1.03.5.15 Psychostimulants

1.03.5.15.1 Dexmethylphenidate hydrochloride

Trade name	Focalin
Manufacturer	Celgene/Norvartins
Country of origin	USA
Year of introduction	2002
Country in which first launched	USA
CAS registry number	19262-68-1
Structure	

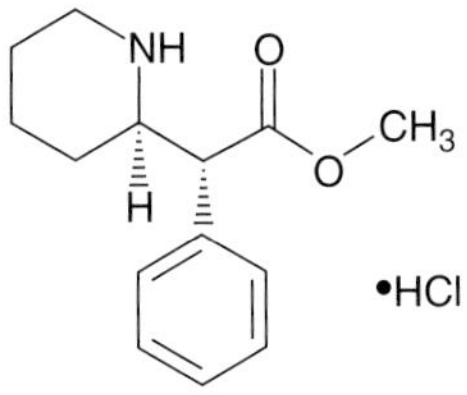

Dexmethylphenidate is the pharmacologically effective enantiomer of D,L-methyl phenidate. It was introduced as an improved treatment for children with attention deficit hyperactivity disorder.

1.03.5.16 Hypnotic

1.03.5.16.1 Zaleplon

Trade name	Sonata
Manufacturer	American Home Products
Country of origin	USA
Year of introduction	1999
Country in which first launched	Sweden, Denmark

CAS registry number	15319-34-5
Structure	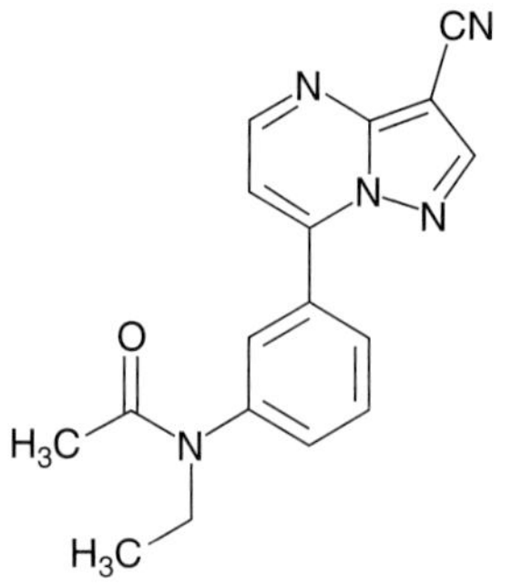

Zaleplon is a nonbenzodiazepine that reduces sleep latency and improves the quality of sleep in patients with chronic insomnia.

1.03.5.17 Central Nervous System Diagnosis

1.03.5.17.1 Ioflupane

Trade name	DatSCAN
Manufacturer	Research Biochemicals Int/Nycomed Amersham
Country of origin	UK
Year of introduction	2000
Country in which first launched	UK
CAS registry number	155798-07-5
Structure	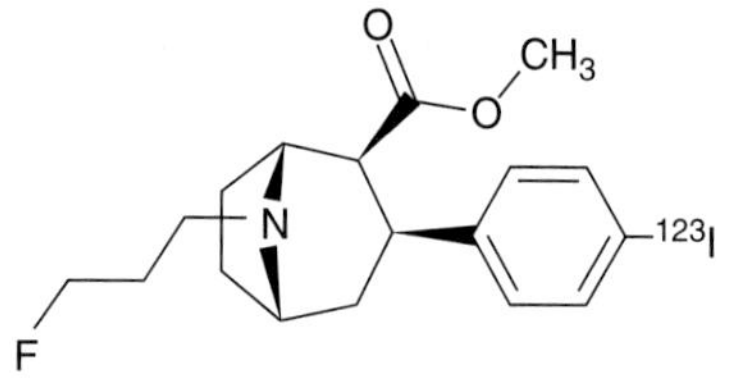

Ioflupane was introduced as an imaging agent for the investigation of the dopaminergic neurons and the early diagnosis of Parkinson's disease and related syndromes. The radioactive label is detected using single-photon emission computed tomography.

1.03.5.18 Sedative

1.03.5.18.1 Dexmedetomidine hydrochloride

Trade name	Precedex
Manufacturer	Orion/Abbott
Country of origin	Finland
Year of introduction	2000
Country in which first launched	USA
CAS registry number	145108-58-3
Structure	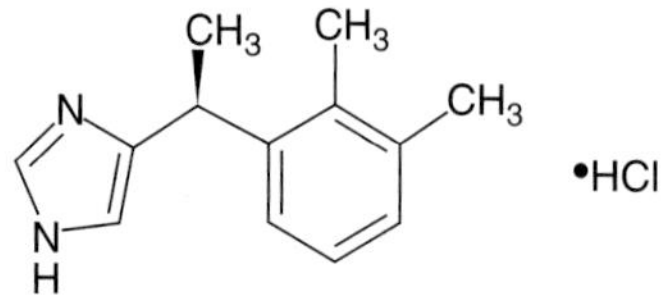

Dexmedetomidine was launched as an intravenous infusion for the sedation of initially intubated and mechanically ventilated patients during treatment in an intensive care unit. It is a full agonist of α_2-adrenoreceptors and has the unique property of being able to provide sedation, analgesia, and anxiolysis, yet allows patients to be easily awakened.

1.03.5.19 Anxiolytic

1.03.5.19.1 Tandospirone

Trade name	Sediel
Manufacturer	Sumitomo
Country of origin	Japan
Year of introduction	1996
Country in which first launched	Japan
CAS registry number	87760-53-0
Structure	

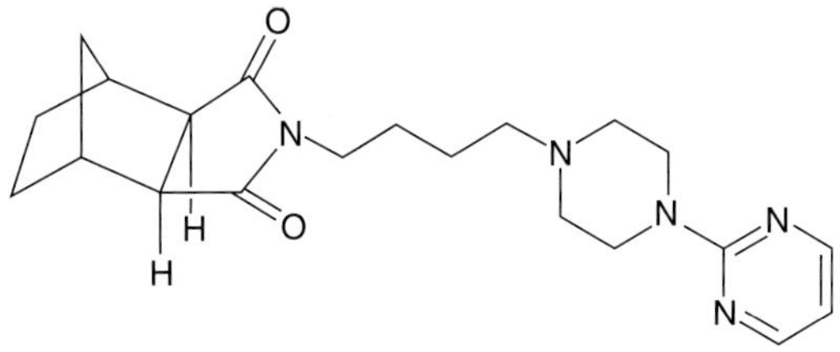

Tandospirone is a partial agonist of the postsynaptic $5HT_{1A}$ receptor. It was launched as an anxiolytic, which is as effective as the benzodiazepines but without the side effects.

1.03.5.20 Central Nervous System Stimulants

1.03.5.20.1 Taltirelin

Trade name	Ceredist
Manufacturer	Tanabe Seiyaku
Country of origin	Japan
Year of introduction	2000
Country in which first launched	Japan
CAS registry number	103300-74-9
Structure	

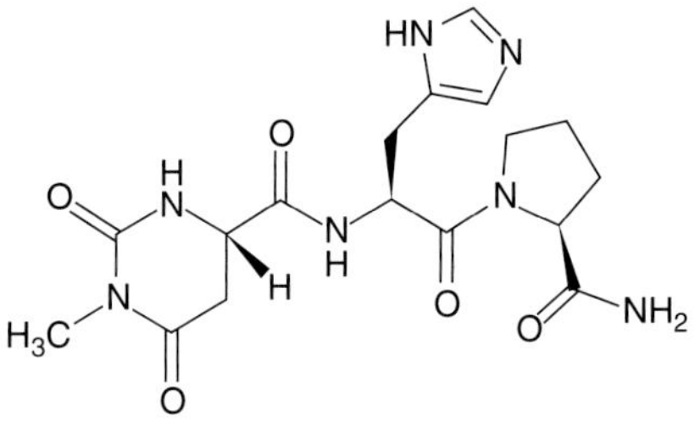

Taltirelin was launched as the first orally active drug for the treatment of neurodegenerative diseases, in particular the improvement of ataxia due to spinocerebellar degeneration.

1.03.6 Gastrointestinal Tract Related

Fourteen new therapeutics that have various effects on the gastrointestinal system were launched between 1993 and 2002. Nine of these have antiulcer effects, two have gastroprokinetic effects, two were launched for irritable bowel syndrome, and one was launched for ulcerative colitis.

1.03.6.1 Antiulcerative Agents

1.03.6.1.1 Dosmalfate

Trade name	Diotul
Manufacturer	Faes
Country of origin	Spain
Year of introduction	2000
Country in which first launched	Spain
CAS registry number	122312-55-4
Structure	

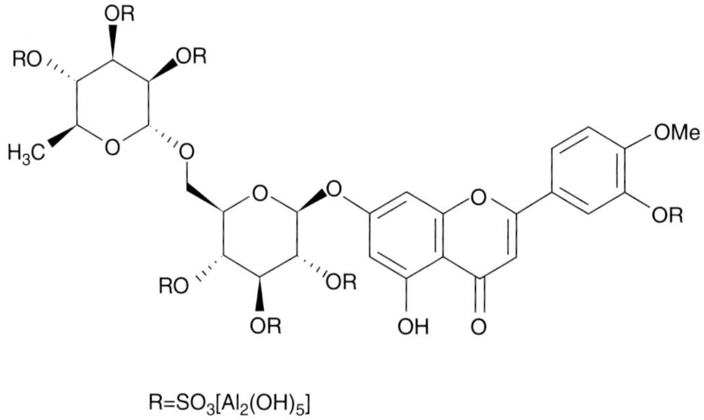

Dosmalfate was launched for the prevention and treatment of gastroduodenal lesions caused by prolonged treatment with nonsteroidal anti-inflammatory agents. Its mode of action does not appear to be defined.

1.03.6.1.2 Ebrotidine

Trade name	Ebrocit
Manufacturer	Ferrer
Country of origin	Spain
Year of introduction	1997
Country in which first launched	Spain
CAS registry number	100981-43-9
Structure	

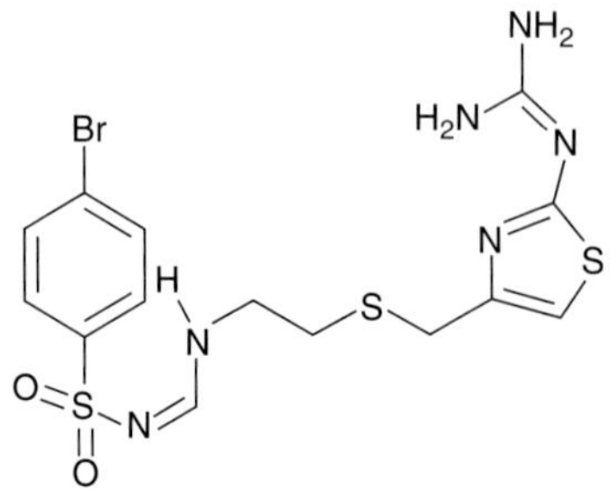

Ebrotidine was launched as a gastroprotective agent that is able to antagonize histamine H_2 receptors. It causes enhanced mucosal blood flow and has activity against *Helicobacter pylori*.

1.03.6.1.3 Egualen sodium

Trade name	Azuloxa
Manufacturer	Kotobuki Seiyaku
Country of origin	Japan
Year of introduction	2000
Country in which first launched	Japan
CAS registry number	97683-31-3
Structure	

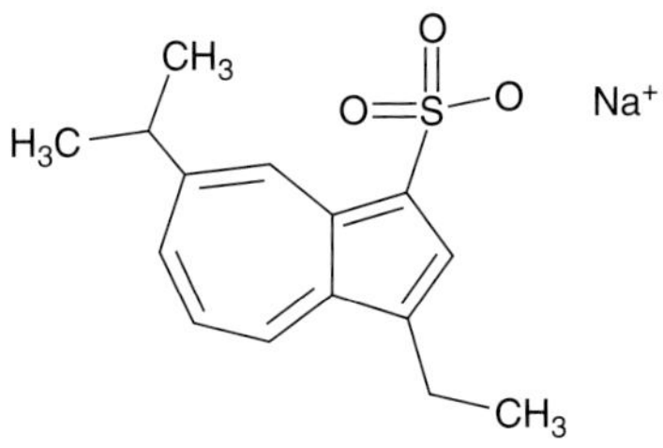

Egualen is a water-soluble sulfonate analog of guaiazulene, a natural azulene with antiulcer activity. Its mechanism of action does not appear to be well defined.

1.03.6.1.4 Esomeprazole magnesium

Trade name	Nexium
Manufacturer	AstraZeneca
Country of origin	UK
Year of introduction	2000
Country in which first launched	Sweden
CAS registry number	119141-88-7
Structure	

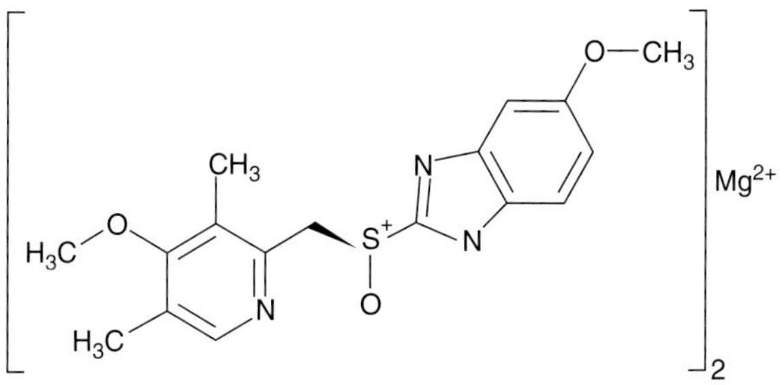

Esomeprazole magnesium was launched as a treatment for acid-related diseases such as gastroesophageal reflux disease including peptic ulcer disease and reflux esophagitis. It is the active *S*-enantiomer of omeprazole and irreversibly inhibits the gastric H^+/K^+ adenosine triphosphatase.

1.03.6.1.5 Lafutidine

Trade name	Stogar, Protecadin
Manufacturer	Fujirebio, UCB, Taiho
Country of origin	Japan
Year of introduction	2000
Country in which first launched	Japan
CAS registry number	118288-08-7

Structure

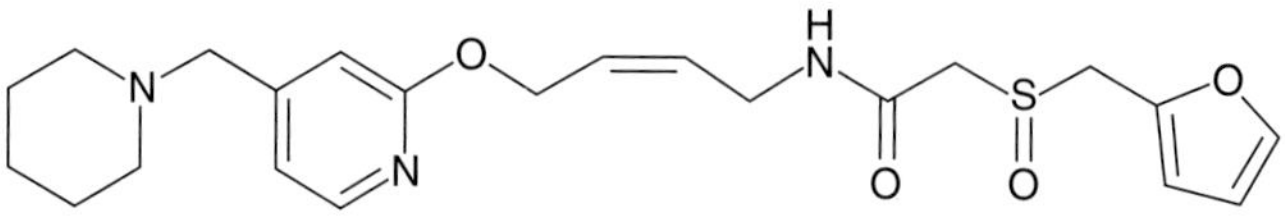

Lafutidine is a potent and long-acting H_2-receptor antagonist marketed for the treatment of gastritis, reflux esophagitis, and peptic ulcers.

1.03.6.1.6 Pantoprazole sodium

Trade name	Pantozol, Rufin
Manufacturer	Byk Gulden, Schwarz Pharm
Country of origin	Germany
Year of introduction	1994
Country in which first launched	Germany
CAS registry number	102625-70-7
Structure	

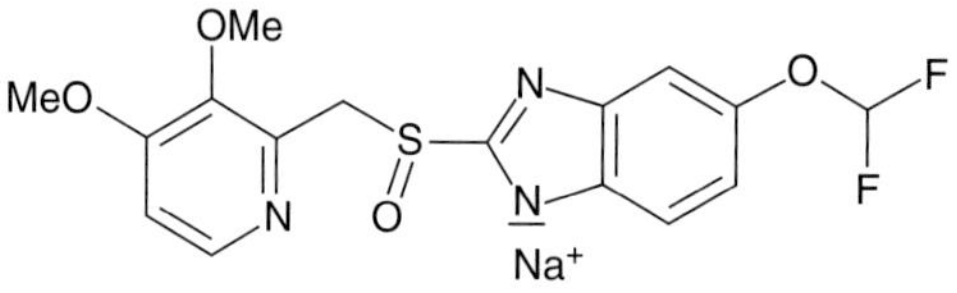

Racemic

Pantoprazole is an irreversible proton pump inhibitor that was marketed for acute treatment of gastric and duodenal ulcers and gastroesophageal reflux disease.

1.03.6.1.7 Polaprezinc

Trade name	Promac
Manufacturer	Hamari Chemicals
Country of origin	Japan
Year of introduction	1994
Country in which first launched	Japan
CAS registry number	107667-60-7
Structure	

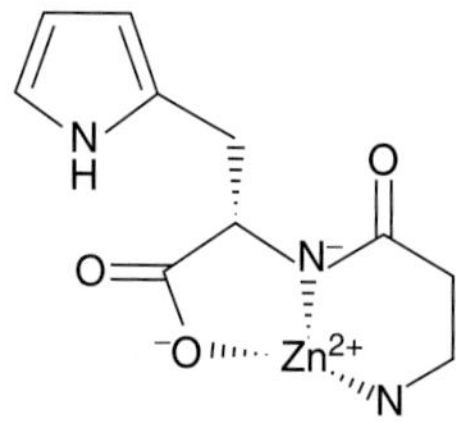

Chiral

It appears that this compound combines the cytoprotective properties of zinc with the ulcer healing properties of L-carnosine. It is effective in promoting the healing of gastric ulcers in addition to preventing ulcer relapse.

1.03.6.1.8 Rabeprazole sodium

Trade name	Pariet
Manufacturer	Eisai
Country of origin	Japan
Year of introduction	1998
Country in which first launched	Japan
CAS registry number	117976-90-6
Structure	

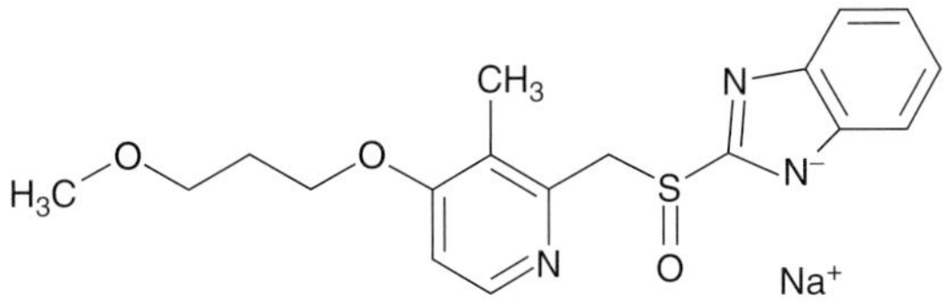

Rabeprazole is an inhibitor of gastric H^+/K^+ adenosine triphosphatase and also has antibacterial activity against *H. pylori*. It has a fast onset of action but a relatively short duration of activity.

1.03.6.1.9 Ranitidine bismuth citrate

Trade name	Pylorid
Manufacturer	Glaxo Wellcome
Country of origin	UK
Year of introduction	1995
Country in which first launched	UK
CAS registry number	128345-62-0
Structure	

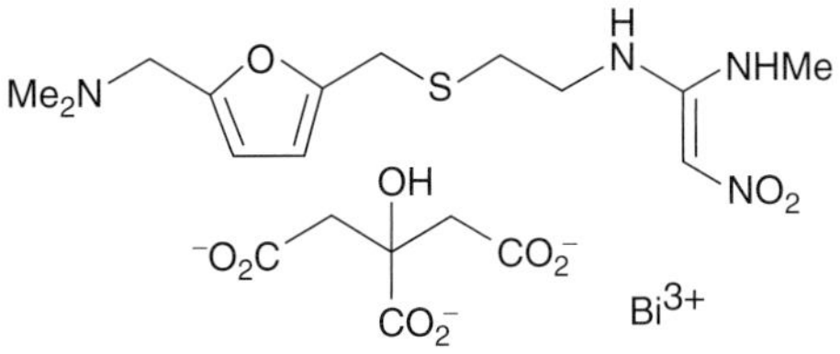

Ranitidine bismuth citrate is a novel salt formed between ranitidine and bismuth citrate complex. It was launched for the treatment of duodenal and benign gastric ulcers, for the eradication of *H. pylori*, and for the prevention of relapse of duodenal ulcer when administered in combination with an antibiotic such as clarithromycin or amoxicillin.

1.03.6.2 Irritable Bowel Syndrome

1.03.6.2.1 Alosetron hydrochloride

Trade name	Lotronex
Manufacturer	Glaxo Wellcome
Country of origin	UK
Year of introduction	2000
Country in which first launched	USA
CAS registry number	132414-02-9

Structure

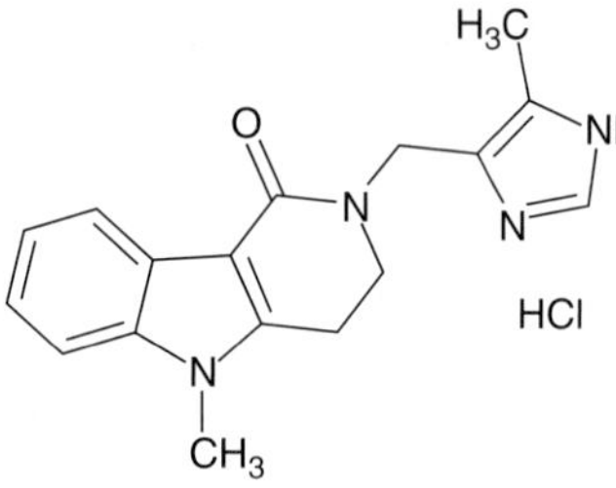

Alosetron is a potent and selective $5HT_3$ antagonist although its precise mechanism of action in the treatment of diarrhea-predominant irritable bowel syndrome in women for which it was launched is unclear.

1.03.6.2.2 Tegaserod maleate

Trade name	Zelmag
Manufacturer	Norvartis
Country of origin	Switzerland
Year of introduction	2001
Country in which first launched	Mexico
CAS registry number	189188-57-6

Structure

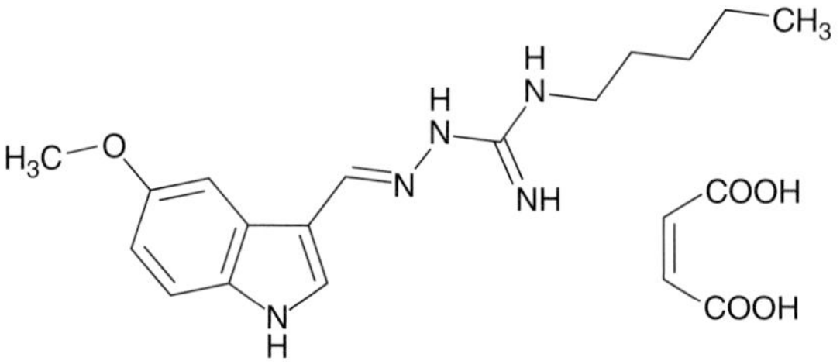

Tegaserod is a selective partial agonist of the excitatory $5HT_4$ receptor. It was launched for the treatment of constipation- or diarrhea-predominant irritable bowel syndrome.

1.03.6.3 Gastroprokinetic Agents

1.03.6.3.1 Itopride hydrochloride

Trade name	Ganaton
Manufacturer	Hokuirku
Country of origin	Japan
Year of introduction	1995
Country in which first launched	Japan
CAS registry number	122892-31-3

Structure

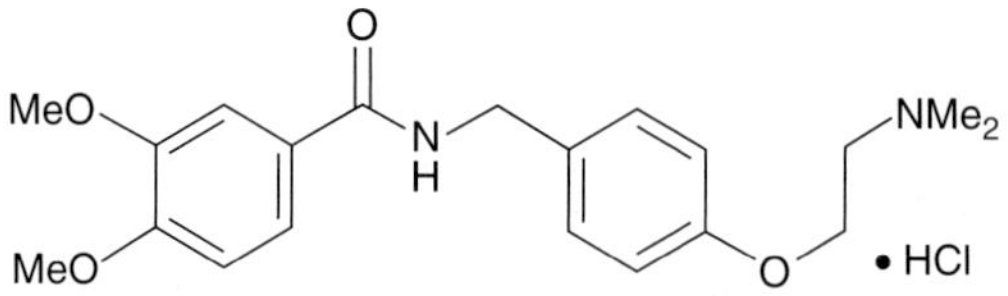

Itopride is a dopamine D_2-receptor antagonist that stimulates the release of acetylcholine on the postganglionic cholinergic neurons. It was launched for the relief of gastrointestinal symptoms in patients with chronic gastritis.

1.03.6.3.2 Mosapride citrate

Trade name	Gasmotin
Manufacturer	Dainippon
Country of origin	Japan
Year of introduction	1998
Country in which first launched	Japan
CAS registry number	112885-41-3
Structure	

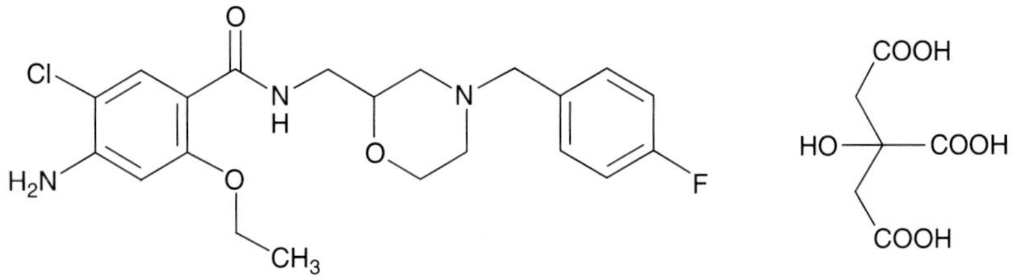

Mosapride was launched for the treatment of gastrointestinal symptoms in patients with chronic gastritis, gastrooesophageal reflux, dyspepsia, and postsurgery complications.

1.03.6.4 Ulcerative Colitis

1.03.6.4.1 Balsazide disodium

Trade name	Colazide
Manufacturer	Biorex, Astra
Country of origin	UK
Year of introduction	1997
Country in which first launched	UK
CAS registry number	80573-04-2
Structure	

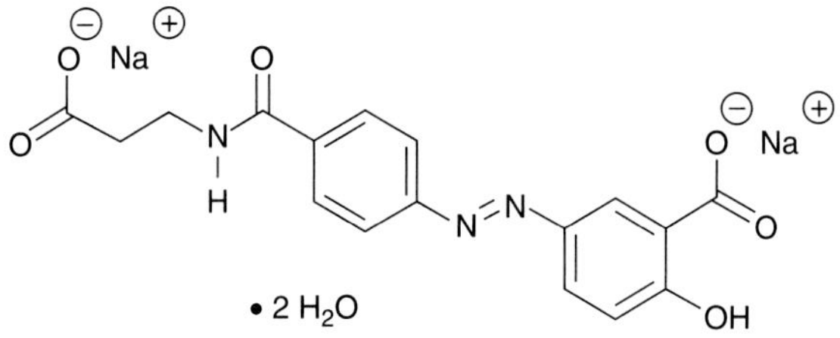

Balsazide has cytoprotective and anti-inflammatory properties. It was launched for mild to moderate acute attacks of ulcerative colitis.

1.03.7 Lung Related (Including Antiallergics)

Twenty-one compounds are described in this category, including nine antiallergics, six antiasthmatics, two expectorants, one bronchodilator, one mucopolysaccharide, one lung surfactant, and one agent for treating cystic fibrosis.

1.03.7.1 Antiallergics

1.03.7.1.1 Betotastine besilate

Trade name	Talion
Manufacturer	UBE, Tanabe Seiyaku
Country of origin	Japan
Year of introduction	2000
Country in which first launched	Japan
CAS registry number	190786-44-8
Structure	

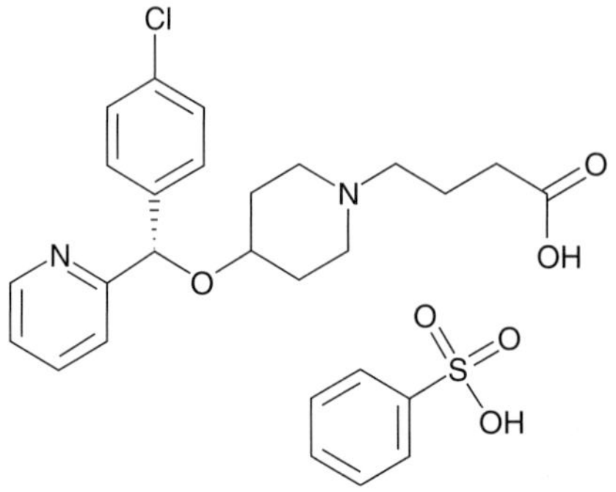

Betotastine besilate is a nonsedating histamine H_1-receptor antagonist that was marketed for the treatment of allergic rhinitis.

1.03.7.1.2 Epinastine hydrochloride

Trade name	Alesion
Manufacturer	Boehringer Ingelheim, Sankyo
Country of origin	Germany
Year of introduction	1994
Country in which first launched	Japan
CAS registry number	80012-43-7
Structure	

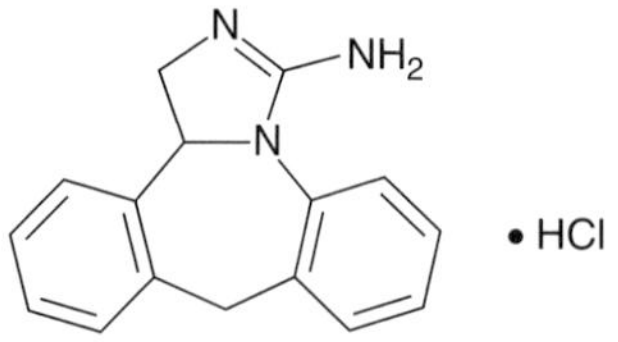

Racemic

Epinastine is a peripherally acting histamine H_1-receptor antagonist that was launched for the oral treatment of bronchial asthma, allergic rhinitis, urticaria, eczema, dermatitis, and psoriasis vulgaris. It is without sedative activity.

1.03.7.1.3 Fexofenadine

Trade name	Allegra
Manufacturer	Sepracor, Hoechst Marion Roussel
Country of origin	USA
Year of introduction	1996

Country in which first launched	USA
CAS registry number	138452-21-8
Structure	

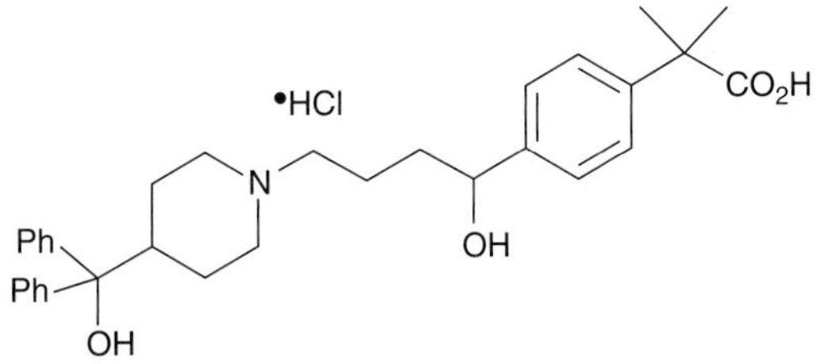

Fexofenadine is a metabolite of terfenadine and is a histamine H_1 receptor antagonist with less side effects. It is used for the treatment of seasonal allergies.

1.03.7.1.4 Loteprednol etabonate

Trade name	Lotemax, Ali
Manufacturer	Pharmos, Bausch & Lomb
Country of origin	USA
Year of introduction	1998
Country in which first launched	USA
CAS registry number	082034-46-6
Structure	

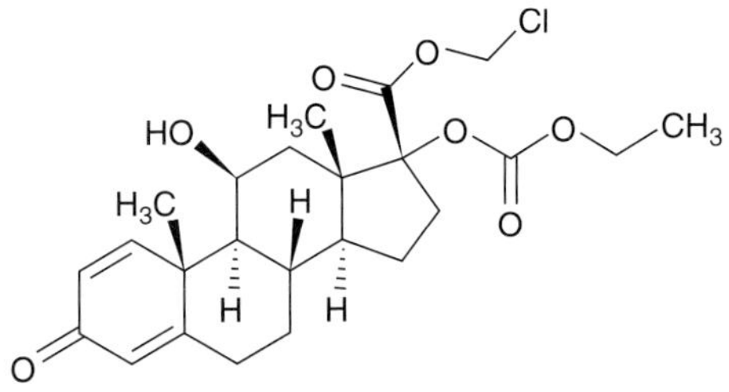

Loteprednol etabonate was introduced for the treatment of steroid responsive inflammatory conditions of the palpebral and bulbar cojunctiva, cornea, and anterior segment of the ocular globe.

1.03.7.1.5 Olopatadine hydrochloride

Trade name	Patanol
Manufacturer	Kyowa Hakko
Country of origin	Japan
Year of introduction	1997
Country in which first launched	USA
CAS registry number	140462-76-6
Structure	

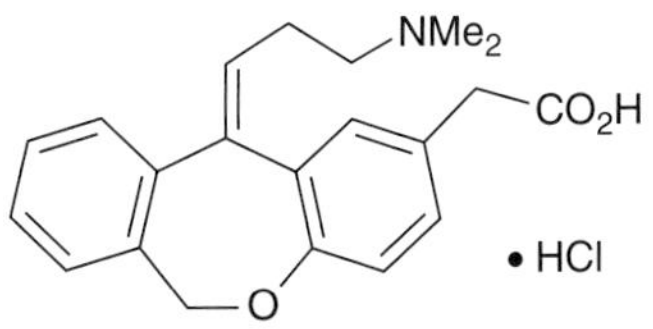

Olopatadine combines the ability to prevent human conjunctival mast cell mediator release with selective H_1-receptor antagonist activity to give potent activity in allergic conjunctivitis.

1.03.7.1.6 Omalizumab

Trade name	Xolair
Manufacturer	Genentech
Country of origin	USA
Year of introduction	2003
Country in which first launched	USA
CAS registry number	242138-07-04

Structure – humanized monoclonal antibody of molecular weight 149 kDa

Omalizumab is a recombinant humanized construct of murine IgG1k monoclonal antibody that was introduced for the treatment of allergic asthma. It forms complexes with circulating IgE, thus inhibiting the binding of IgE to the high-affinity IgE receptor on the surface of mast cells and basophils. Omalizumab is administered subcutaneously every 2–4 weeks. It is well tolerated and can replace steroidal treatment for asthma.

1.03.7.1.7 Ramatroban

Trade name	Baynas
Manufacturer	Bayer
Country of origin	Germany
Year of introduction	2000
Country in which first launched	Japan
CAS registry number	116649-85-5
Structure	

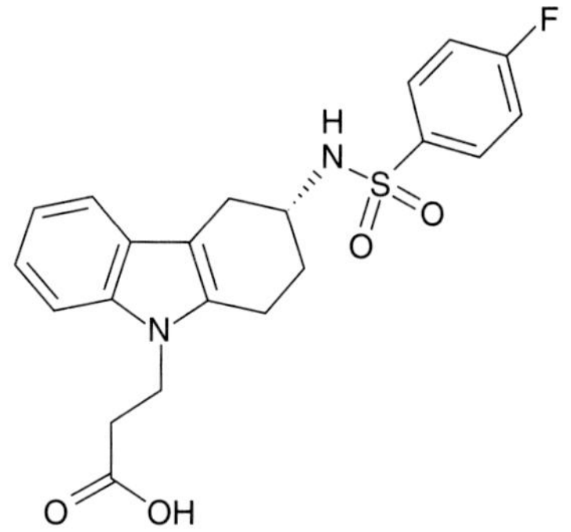

Ramatroban is a potent antagonist of prostaglandin receptors and thromboxane receptors that was marketed for the treatment of allergic rhinitis.

1.03.7.1.8 Rupatadine fumarate

Trade name	Rupafin
Manufacturer	Uriach
Country of origin	Spain
Year of introduction	2003
Country in which first launched	Spain
CAS registry number	182349-12-8

Structure

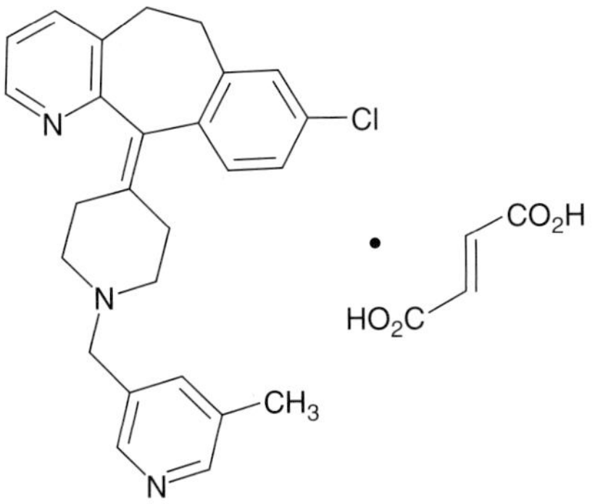

Rupatadine fumarate is an oral treatment for perennial and seasonal rhinitis. It acts as a nonsedating histamine H_1-receptor antagonist and platelet-activating factor antagonist.

1.03.7.1.9 Suplatast tosylate

Trade name	IPD
Manufacturer	Taiho
Country of origin	Japan
Year of introduction	1995
Country in which first launched	Japan
CAS registry number	94055-76-2
Structure	

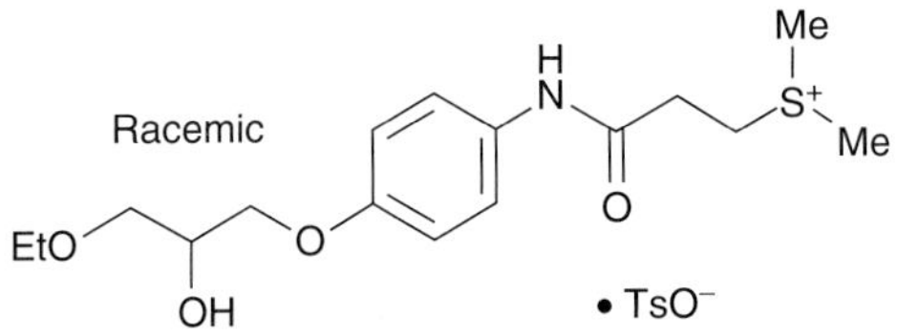

Suplatast is a potent inhibitor of IgE synthesis without affecting IgM and IgG. It was launched for the treatment of bronchial asthma, atopic dermatitis, and allergic rhinitis.

1.03.7.2 Antiasthmatic Agents

1.03.7.2.1 Levalbuterol hydrochloride

Trade name	Xopenex
Manufacturer	Sepracor
Country of origin	USA
Year of introduction	1999
Country in which first launched	USA
CAS registry number	34391-04-3
Structure	

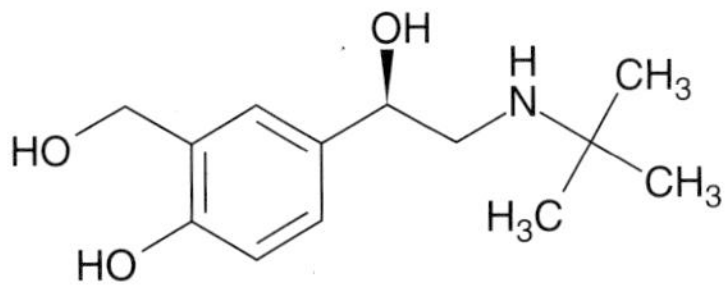

Levalbuterol is the single *R*-enantiomer of salbutamol. It was marketed for the treatment or prevention of bronchospasm in patients with reversible obstructive airway disease.

1.03.7.2.2 Montelukast sodium

Trade name	Singulair
Manufacturer	Merck
Country of origin	USA
Year of introduction	1998
Country in which first launched	Finland, Mexico
CAS registry number	151767-02-1
Structure	

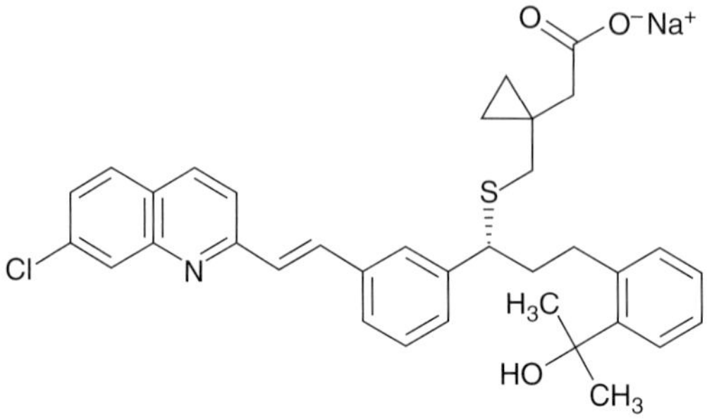

Montelukast is a potent, selective, and orally active antagonist of the leukotriene D_4 (LTD_4) receptor. It is marketed for the management of mild to moderate asthma that has been inadequately controlled by inhaled corticosteroids and short-acting $beta_2$ agonists.

1.03.7.2.3 Pranlukast

Trade name	Onon
Manufacturer	Ono
Country of origin	Japan
Year of introduction	1995
Country in which first launched	Japan
CAS registry number	103177-37-3
Structure	

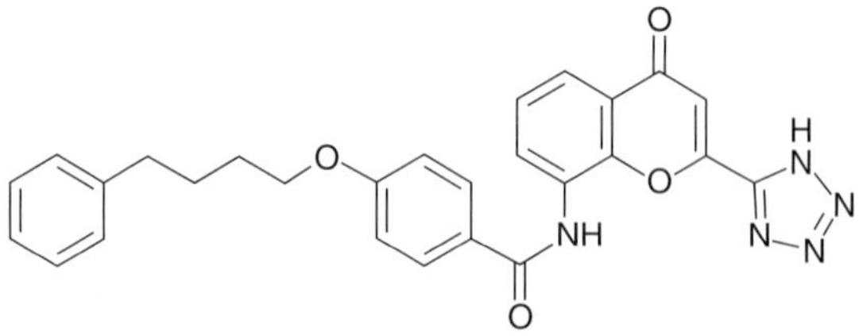

Pranlukast is a highly potent, selective, and competitive antagonist of peptidoleukotrienes with high affinity for the LTD_4 receptor. It was introduced for the treatment of bronchial asthma and allergic diseases.

1.03.7.2.4 Seratrodast

Trade name	Bronica
Manufacturer	Takeda
Country of origin	Japan

Year of introduction	1995
Country in which first launched	Japan
CAS registry number	112665-43-7
Structure	

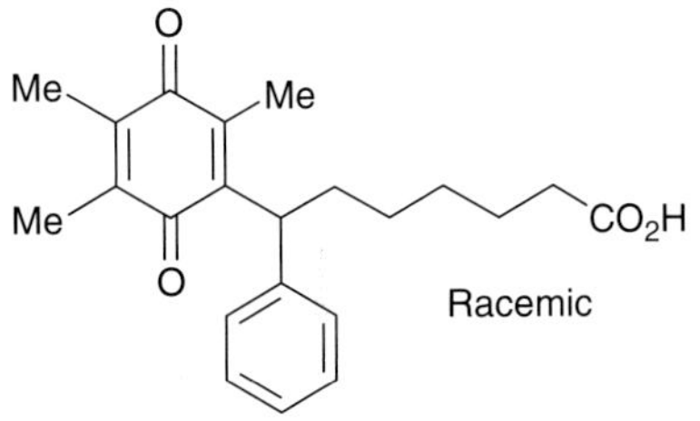

Seratrodast is a thromboxane A_2 receptor antagonist that was marketed for the treatment of bronchospastic disorders such as asthma.

1.03.7.2.5 **Zafirlukast**

Trade name	Accolate
Manufacturer	Zeneca
Country of origin	UK
Year of introduction	1996
Country in which first launched	USA
CAS registry number	107753-87-6
Structure	

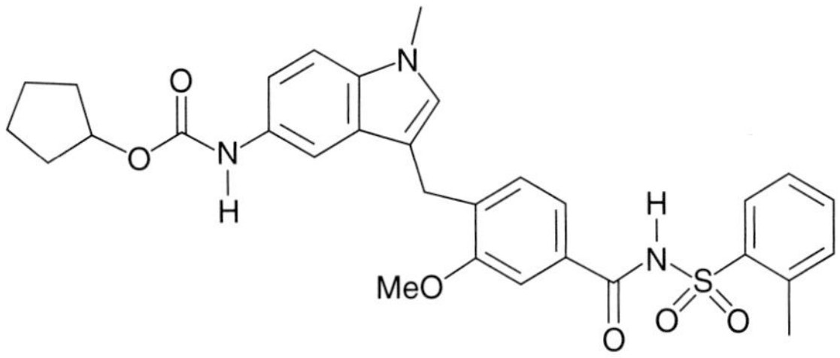

Zafirlukast is an LTD_4 antagonist that was launched for the treatment of asthma.

1.03.7.2.6 **Zileuton**

Trade name	Zyflo
Manufacturer	Abbott
Country of origin	USA
Year of introduction	1997
Country in which first launched	USA
CAS registry number	111406-87-2
Structure	

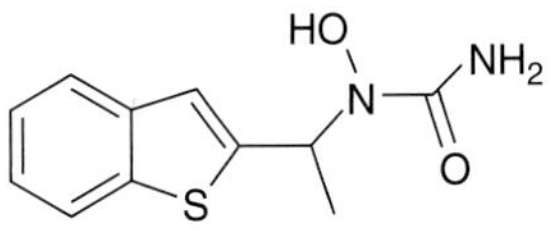

Zileuton is an orally active, reversible direct inhibitor of 5-lipoxygenase that was marketed for chronic asthma. The bronchodilatory effect occurs within 2 h of administration.

1.03.7.3 Expectorants

1.03.7.3.1 Erdosteine

Trade name	Edirel, Vectrine
Manufacturer	Refarmed, Pierre Fabre, Negma
Country of origin	Switzerland
Year of introduction	1995
Country in which first launched	France
CAS registry number	84611-23-4
Structure	

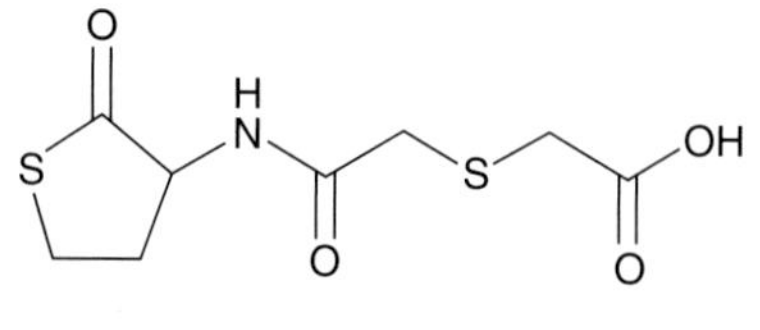

Racemic

Erdosteine has mucomodulator, mucolytic, mucokinetic, and free radical scavenging properties and was launched for the treatment of chronic bronchitis. It is also reported to possess local anti-inflammatory and antielastase activities, and enhances the penetration of antibiotics into bronchial mucus.

1.03.7.3.2 Fudosteine

Trade name	Cleanal
Manufacturer	SS Pharmaceutical/Mitsubishi Pharma
Country of origin	Japan
Year of introduction	2001
Country in which first launched	Japan
CAS registry number	13189-98-5
Structure	

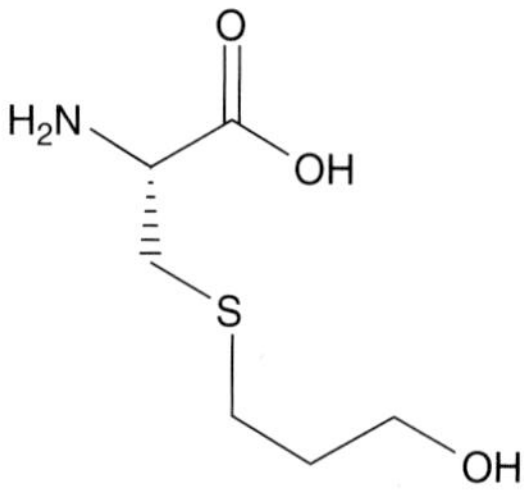

Fudosteine was introduced for the treatment of bronchitis and respiratory congestion. It is able to significantly reduce mucus glycoprotein hypersecretion and inhibit infiltration of airway mucosa by lymphocytes and inflammatory cells in bronchitic rats.

1.03.7.4 Bronchodilator

1.03.7.4.1 Tiotropium bromide

Trade name	Spiriva
Manufacturer	Boehringer Ingelheim/Pfizer
Country of origin	Germany
Year of introduction	2002
Country in which first launched	Netherlands, Philippines
CAS registry number	136310-93-5

Structure

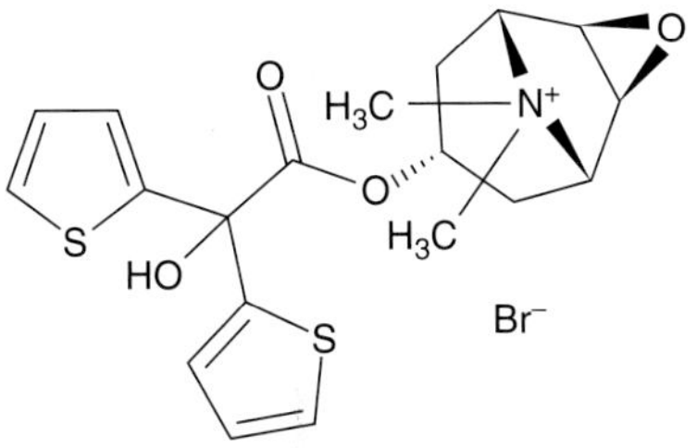

Tiotropium bromide is a long-acting inhaled muscarinic antagonist that was developed for the treatment of chronic obstructive pulmonary disease.

1.03.7.5 Mucopolysaccharides

1.03.7.5.1 Laronidase

Trade name	Aldurazyme
Manufacturer	BioMarin
Country of origin	USA
Year of introduction	2003
Country in which first launched	USA
CAS registry number	210589-09-6

Structure – recombinant form of human α-L-iduronidase

Mucopolysaccharidosis I is a genetic disorder caused by a deficiency of α-L-iduronidase, an enzyme required for the catabolism of dermatan sulfate and heparin sulfate. Sulfate laronidase is a recombinant form of this enzyme produced in Chinese hamster ovary cell lines. It is administered weekly as an intravenous infusion.

1.03.7.6 Lung Surfactant

1.03.7.6.1 Pumactant

Trade name	ALEC
Manufacturer	Britannia
Country of origin	UK
Year of introduction	1994
Country in which first launched	UK

Structure – mixture of dipalmitoylphosphatidylcholine and phosphatidylglycerol

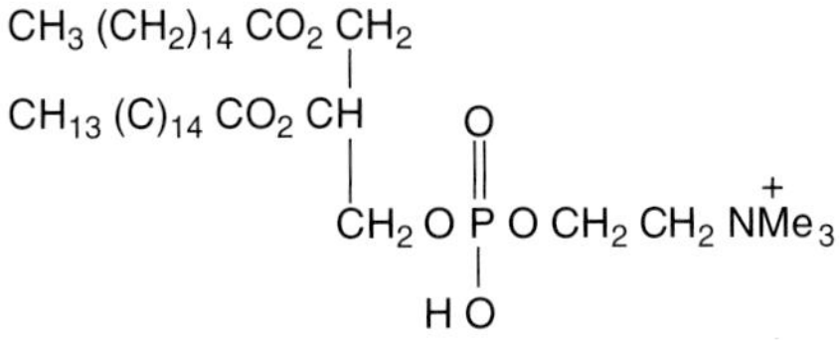

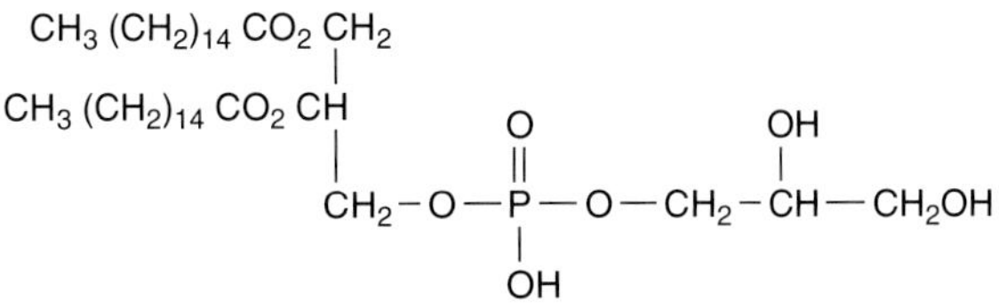

This is a mixture of dipalmitoylphosphatidylcholine and phosphatidylglycerol in a 7:3 ratio, used for the treatment of respiratory distress syndrome in premature infants.

1.03.7.7 Cystic Fibrosis

1.03.7.7.1 Dornase alfa

Trade name	Pulmozyme
Manufacturer	Genentech
Country of origin	USA
Year of introduction	1994
Country in which first launched	USA
CAS registry number	143831-71-4
Structure – recombinant human deoxyribonuclease	

Dornase alfa, the first new drug treatment for cystic fibrosis for 30 years, is effective in liquefying secretions from the lungs of cystic fibrosis patients. It achieves this by cleaving the extracellular DNA in purulent sputum, thereby reducing the viscoelasticity of the secretion.

1.03.8 Joints and Bones

Thirty-one new medicaments have been included in this category. As might be expected with the high incidence of arthritic conditions in an aging community, the largest group comprises 10 antiarthritic/rheumatic drugs including the cyclooxygenase-2 (COX-2) inhibitors, which were launched with such high hopes but, at the time of writing (summer 2005), are being viewed with greater caution, and seven anti-inflammatory agents. Six drugs have been launched for the treatment of osteoporosis and four vitamin D analogs, two analgesics, one osteoinductor, and one hypocalcemic agent have also appeared.

1.03.8.1 Antirheumatic Agents

1.03.8.1.1 Actarit

Trade name	Orcl, Mover
Manufacturer	Nippon Shinyaku, Mitsubishi, Kesei
Country of origin	Japan
Year of introduction	1994
Country in which first launched	Japan
CAS registry number	18699-02-0
Structure	

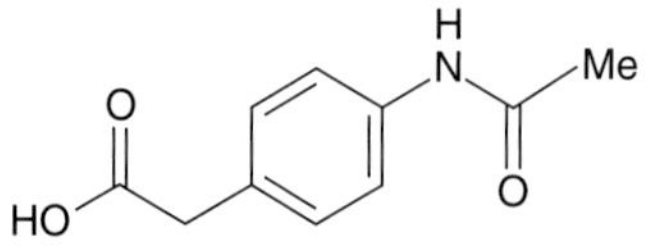

While the structure of actarit might suggest nonsteroidal anti-inflammatory activity, it actually has no effect on models of acute inflammation. It appears to work by the modulation of production and serum levels of interleukin-2, which enhances the production of suppressor T cells by the immune system.

1.03.8.1.2 Adalimumab

Trade name	Humira
Manufacturer	Cambridge Antibody Technology/ Abbott

Country of origin	USA
Year of introduction	2003
Country in which first launched	USA
CAS registry number	331731-18-1
Structure – recombinant humanized antibody of molecular weight 148 kDa	

Adalimumab was the first fully human neutralizing IgG1 monoclonal antibody specific for tumor necrosis factor alpha (TNF-α). Blockade of TNF binding to p55 and p75 cell surface TNF receptors decreases leukocyte migration and acute phase reactants. Adalimumab is administered subcutaneously once every 2 weeks; in combination with methotrexate it results in improved American College of Rheumatology (ACR) scores and reduced joint space narrowing.

1.03.8.1.3 **Anakinra**

Trade name	Kineret
Manufacturer	Amgen/University of Colorado
Country of origin	USA
Year of introduction	2001
Country in which first launched	USA
CAS registry number	143090-092-0
Structure – recombinant protein of molecular weight 17.3 kDa	

Anakinra is a recombinant nonglycosylated human interleukin-1 receptor antagonist that was introduced as a daily subcutaneous injection therapy for the reduction of signs and symptoms of moderate to severe rheumatoid arthritis in adults who have failed to respond to one or more disease-modifying antirheumatic drugs.

1.03.8.1.4 **Celecoxib**

Trade name	Celebrex
Manufacturer	Searle, Pfizer
Country of origin	USA
Year of introduction	1999
Country in which first launched	USA
CAS registry number	169590-42-5
Structure	

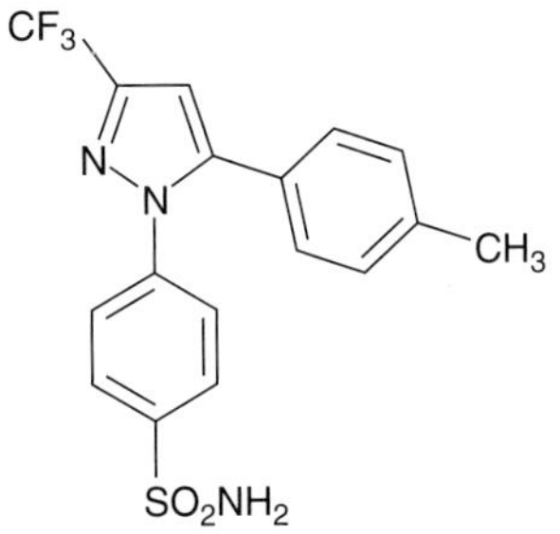

Celecoxib is a potent and highly selective inhibitor of COX-2, the inducible form of cyclooxygenase expressed during the inflammatory process. It does not block COX-1 thus suppressing the gastric and intestinal toxicity of most nonselective nonsteroidal anti-inflammatory drugs (NSAIDs). It was launched for the treatment of rheumatoid arthritis and osteoarthritis but has also been approved for use in patients with familial adenomatous polyposis.

1.03.8.1.5 **Etoricoxib**

Trade name	Arcoxia
Manufacturer	Merck

Country of origin	USA
Year of introduction	2002
Country in which first launched	Mexico
CAS registry number	202409-33-4
Structure	

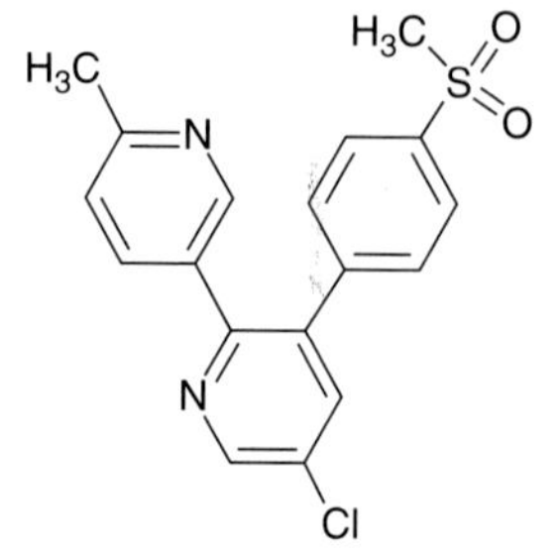

Etoricoxib was introduced as a follow-up to rofecoxib for the treatment of osteoarthritis, rheumatoid arthritis, dysmenorrhea, gout, ankylosing spondylitis, and pain. It is the most-selective COX-2 so far marketed.

1.03.8.1.6 Leflunomide

Trade name	Arava
Manufacturer	Hoechst Marion Roussel
Country of origin	Germany
Year of introduction	1998
Country in which first launched	USA
CAS registry number	75706-12-6
Structure	

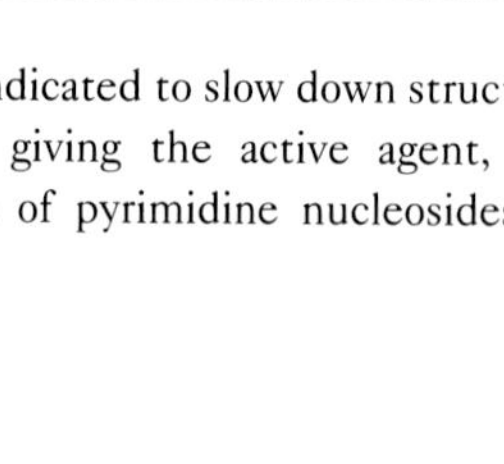

Leflunomide is the first and only drug to be indicated to slow down structural joint damage in rheumatoid arthritis. It is a prodrug with a metabolic ring opening giving the active agent, which inhibits the enzyme dihydroorotate dehydrogenase involved in the biosynthesis of pyrimidine nucleosides. It is an orally available disease-modifying antirheumatic drug.

1.03.8.1.7 Meloxicam

Trade name	Mobec
Manufacturer	Boehringer Ingelheim
Country of origin	Germany
Year of introduction	1996
Country in which first launched	UK
CAS registry number	71125-38-7

Structure

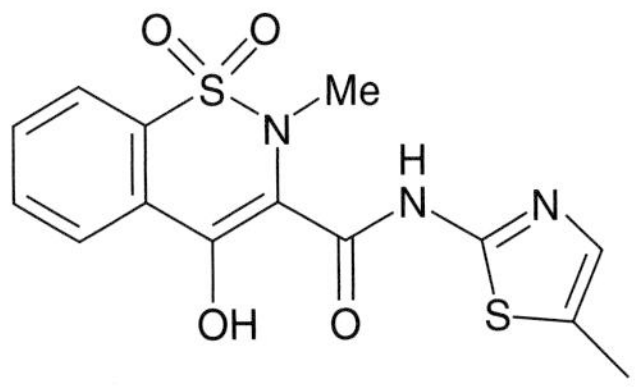

Meloxicam was launched as an NSAID for the treatment of osteo- and rheumatoid arthritis.

1.03.8.1.8 **Parecoxib sodium**

Trade name	Dynastat
Manufacturer	Pharmacia (Searle)
Country of origin	USA
Year of introduction	2002
Country in which first launched	UK
CAS registry number	197502-82-1
Structure	

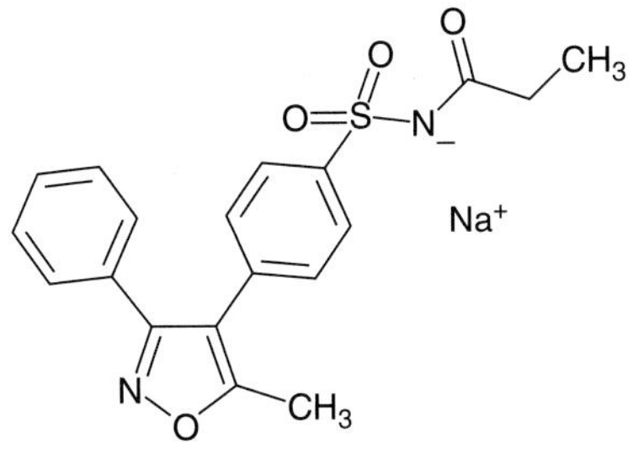

Parecoxib sodium is a prodrug of parecoxib that was launched as an injectable COX-2 inhibitor for the treatment of inflammation and acute pain, particularly postoperative pain.

1.03.8.1.9 **Rofecoxib**

Trade name	Vioxx
Manufacturer	Merck
Country of origin	USA
Year of introduction	1999
Country in which first launched	Mexico
CAS registry number	162011-90-7
Structure	

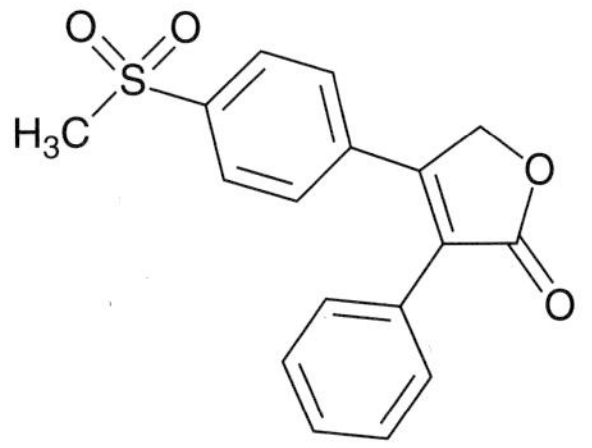

Rofecoxib was launched for the management of acute pain and the treatment of osteoarthritis and primary dysmenorrhea. It is a selective inhibitor of COX-2. Unanticipated circulatory problems have led to the withdrawal of Vioxx from the market.

1.03.8.1.10 Valdecoxib

Trade name	Bextra
Manufacturer	Pharmacia (Searle)
Country of origin	USA
Year of introduction	2002
Country in which first launched	USA
CAS registry number	181695-72-7
Structure	

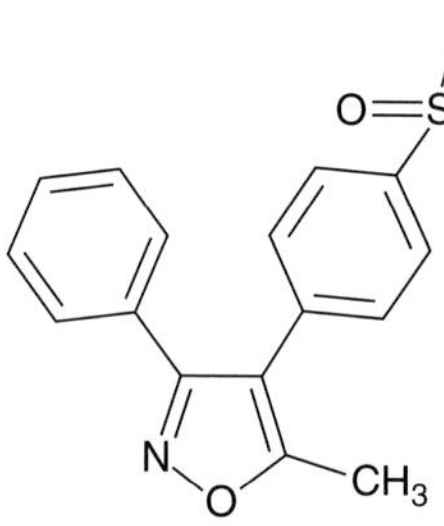

Valdecoxib was launched as a second-generation COX-2 inhibitor for the treatment of osteoarthritis, adult rheumatoid arthritis, and menstrual pain. It combines good activity against osteoarthritis, rheumatoid arthritis, and dysmenorrhea with lower levels of abdominal pain, dyspepsia, and constipation than seen with the traditional NSAIDs.

1.03.8.2 Anti-Inflammatory Agents

1.03.8.2.1 Ampiroxicam

Trade name	Nacyl, Flucam
Manufacturer	Pfizer
Country of origin	Japan
Year of introduction	1994
Country in which first launched	Japan
CAS registry number	99464-64-9
Structure	

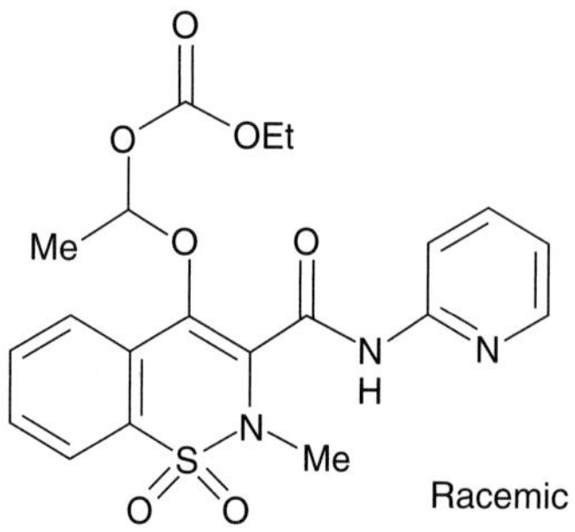

Ampiroxicam is a prodrug of piroxicam that was introduced as a once daily NSAID.

1.03.8.2.2 Betamethasone butyrate propionate

Trade name	Antebate, Antevate
Manufacturer	Mitsubishi Kasai, Torii

Country of origin	Japan
Year of introduction	1994
Country in which first launched	Japan
CAS registry number	5534-02-1
Structure	

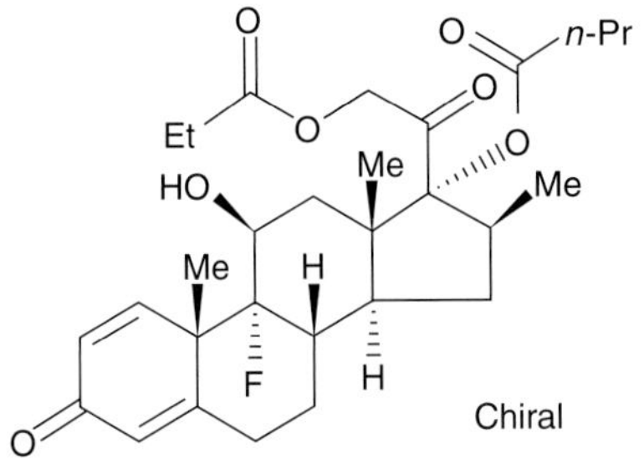

Betamethasone butyrate propionate was introduced as a topical anti-inflammatory agent. It has potent effects at the site of application with little or no topical or systemic side effects.

1.03.8.2.3 **Bromfenac sodium**

Trade name	Senju
Manufacturer	American Home Products, Senju
Country of origin	USA
Year of introduction	1997
Country in which first launched	USA
CAS registry number	120638-55-3
Structure	

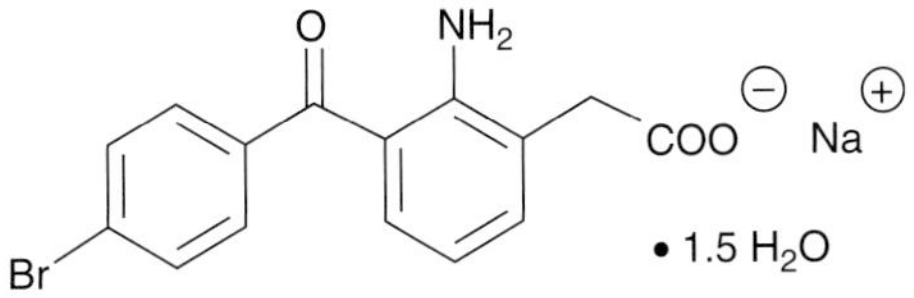

Bromfenac is a cyclooxygenase inhibitor that was launched as a potent, orally active, long-lasting peripheral analgesic with anti-inflammatory properties.

1.03.8.2.4 **Dexibuprofen**

Trade name	Seractil
Manufacturer	Gebro Broschek
Country of origin	Austria
Year of introduction	1994
Country in which first launched	Austria
CAS registry number	51146-56-6
Structure	

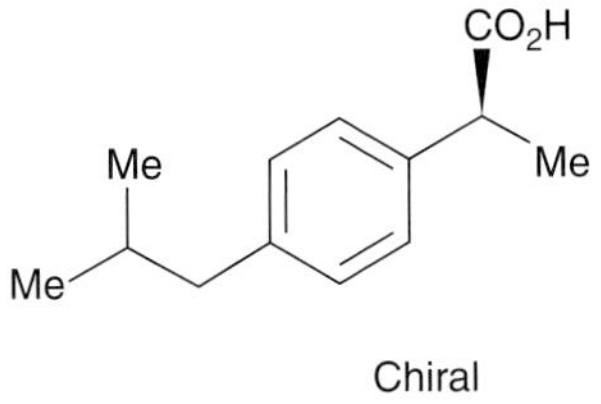

Dexibuprofen is the *S*-(+)-isomer of the widely used NSAID ibuprofen. It was launched for the treatment of rheumatoid arthritis.

1.03.8.2.5 Lornoxicam

Trade name	Xefo
Manufacturer	Nycomed Amersham
Country of origin	Norway
Year of introduction	1997
Country in which first launched	Denmark
CAS registry number	70374-39-9
Structure	

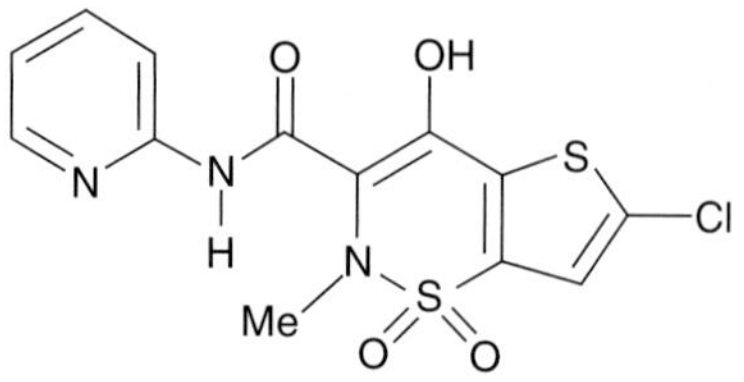

Lornoxicam was launched as an NSAID for mild to moderate pain and inflammation.

1.03.8.2.6 Rimexolone

Trade name	Vexol
Manufacturer	Akzo, Alcon
Country of origin	Netherlands
Year of introduction	1995
Country in which first launched	USA
CAS registry number	49697-38-3
Structure	

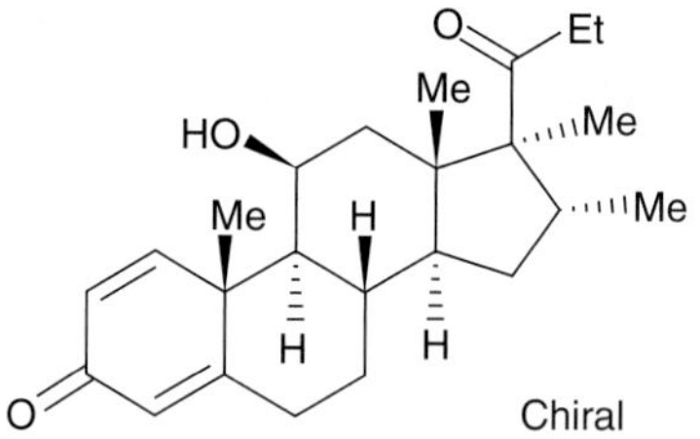

Rimexolone has a high corticoid receptor affinity and is a potent local anti-inflammatory agent with minimal systemic effects. It was launched for the treatment of postoperative inflammation following ocular surgery and anterior uveitis. It has also been approved for the treatment of rheumatoid arthritis.

1.03.8.2.7 Sivelestat

Trade name	Elaspol
Manufacturer	Ono Pharmaceutical
Country of origin	Japan
Year of introduction	2002
Country in which first launched	Japan
CAS registry number	201677-61-4

Structure

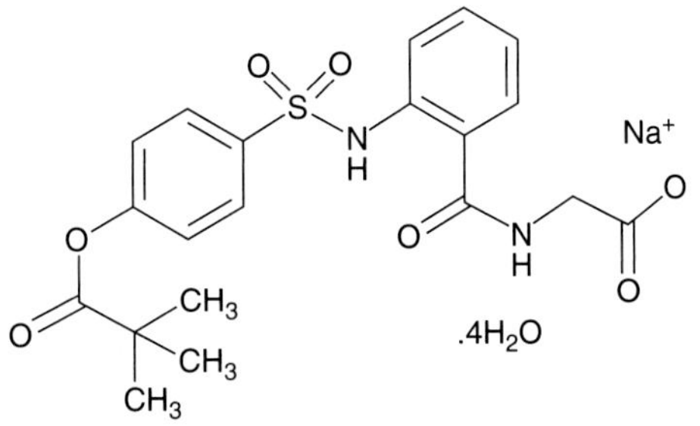

Sivelestat is an inhibitor of neutrophil elastase that was developed as an injectable formulation for the treatment of acute lung injury associated with systemic inflammatory response syndrome.

1.03.8.3 Osteoporosis

1.03.8.3.1 Ibandronic acid

Trade name	Bondronat
Manufacturer	Boehringer Mannheim
Country of origin	Germany
Year of introduction	1996
Country in which first launched	Germany
CAS registry number	114084-78-5
Structure	

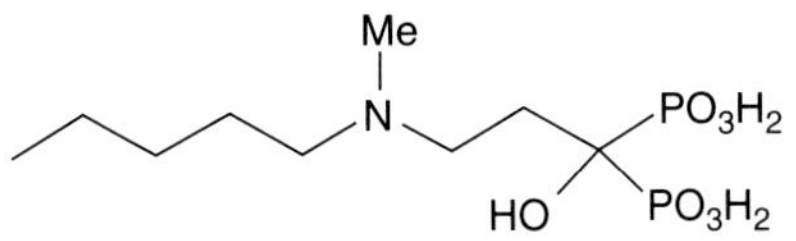

Ibandronic acid was introduced for the treatment of bone disorders such hypercalcemia in malignancy and osteolysis, Paget's disease, and osteoporosis. While the precise mode of action is unclear, it is known to be an inhibitor of osteoclast-mediated bone resorption and binds to hydroxyapatite crystals with a half-life in the skeleton of several years.

1.03.8.3.2 Incandronic acid

Trade name	Bisphonal
Manufacturer	Yamanouchi
Country of origin	Japan
Year of introduction	1997
Country in which first launched	Japan
CAS registry number	124351-85-5
Structure	

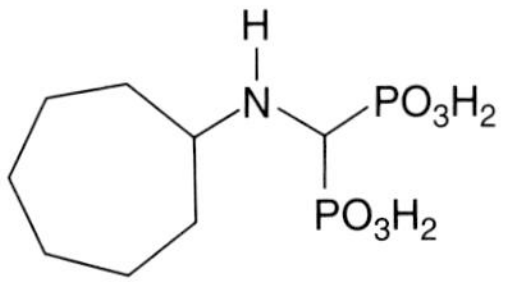

The mechanism of action of incandronic acid is not well understood but it is known to bind tightly to the calcified bone matrix and inhibit bone resorption. It was launched for the treatment of osteoporosis.

1.03.8.3.3 Raloxifene hydrochloride

Trade name	Evista
Manufacturer	Lilly
Country of origin	USA
Year of introduction	1998
Country in which first launched	USA
CAS registry number	82640-04-8
Structure	

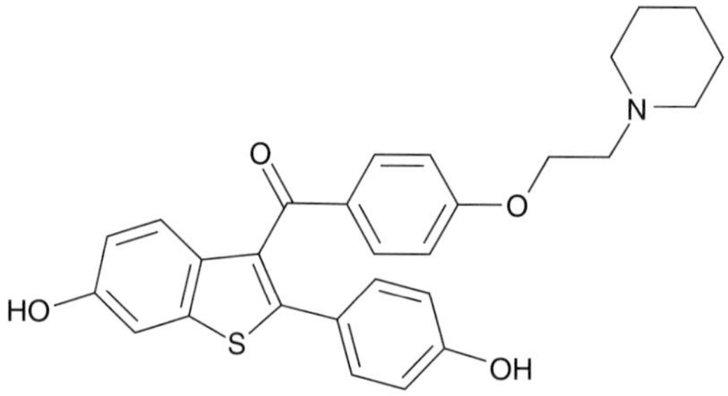

Raloxifene is a selective estrogen receptor modulator with estrogenic activity on bone metabolism. It was launched for the treatment of postmenopausal osteoporosis.

1.03.8.3.4 Risedronate sodium

Trade name	Actonel
Manufacturer	Procter Gamble, HMR
Country of origin	USA
Year of introduction	1998
Country in which first launched	USA
CAS registry number	115436-72-1
Structure	

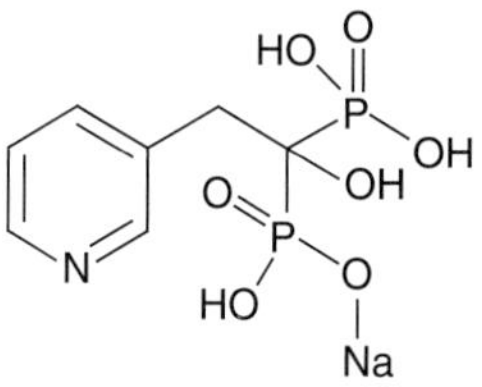

Risedronate is an orally active bisphosphonate with potent bone resorption properties. It was launched for the treatment of Paget's disease, a chronic disease of the elderly characterized by alteration of bone tissue, especially in the spine, shoulder, and pelvis.

1.03.8.3.5 Tiludronate disodium

Trade name	Skelid
Manufacturer	Sanofi
Country of origin	France
Year of introduction	1995
Country in which first launched	Switzerland
CAS registry number	149845-07-8
Structure	

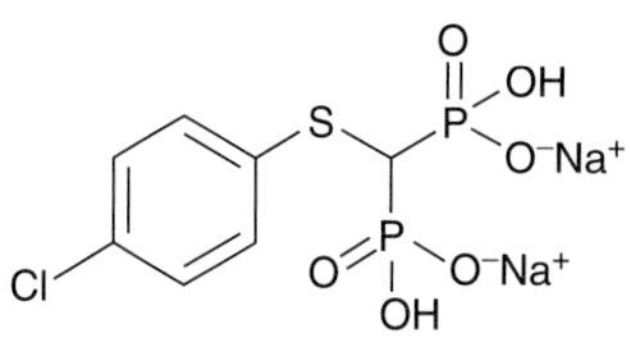

Tiludronate was introduced to treat Paget's disease. It is suggested to be a specific inhibitor of functioning osteoclasts through selective incorporation into the polarized osteoclast-like multinucleated cells and direct interference with the maintenance of the cytoskeletal structure.

1.03.8.3.6 Trimegestone

Trade name	Totelle Sekvens, Ondeva
Manufacturer	Aventis, Wyeth Pharmaceuticals
Country of origin	France
Year of introduction	2001
Country in which first launched	Sweden
CAS registry number	074513-62-5
Structure	

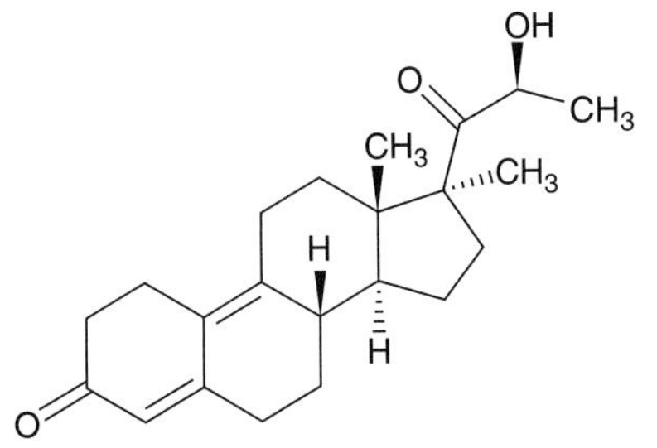

Trimegestone exhibits high specificity and affinity for the progesterone receptor with no affinity for the estrogen receptor and only weak binding to the androgen, glucocorticoid, and mineralocorticoid receptors. It was introduced in combination with 17-β-estradiol as hormone replacement therapy for, inter alia, the treatment of osteoporosis.

1.03.8.4 Vitamin D

1.03.8.4.1 Doxercalciferol

Trade name	Hectorol
Manufacturer	Bone Care Int.
Country of origin	USA
Year of introduction	1999
Country in which first launched	USA
CAS registry number	54573-75-0
Structure	

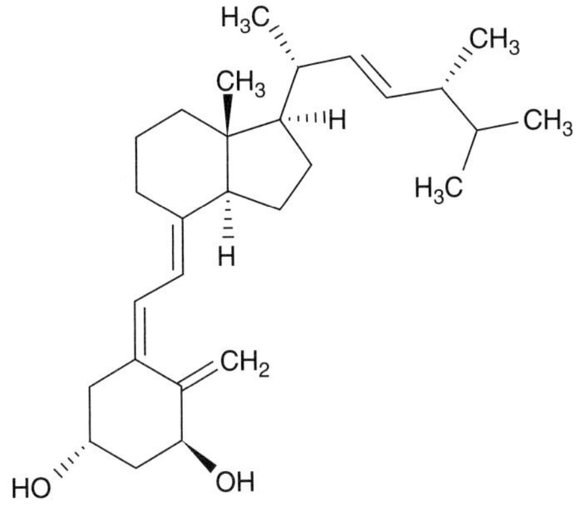

Doxercalciferol is an orally active, synthetic vitamin D_2 analog that was introduced for the treatment of secondary hyperparathyroidism in patients with end-stage renal failure.

1.03.8.4.2 Falecalcitriol

Trade name	Hornel, Fulstan
Manufacturer	University of Wisconsin, Kissei/Sumitomo/Taisho
Country of origin	USA
Year of introduction	2001
Country in which first launched	Japan
CAS registry number	83805-11-2
Structure	

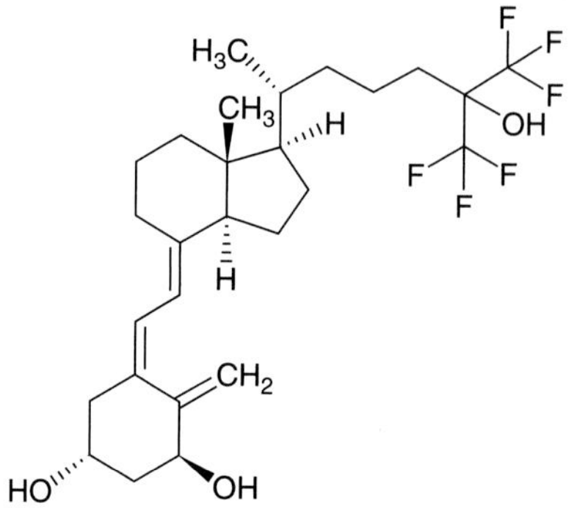

Falecalcitriol regulates the proliferation of parathyroid cells and parathyroid hormone synthesis possibly via binding to a nuclear receptor for vitamin D (VDR). It was launched for the treatment of secondary hyperthyroidism.

1.03.8.4.3 Maxacalcitol

Trade name	Prezios, Oxarol
Manufacturer	Chugai
Country of origin	Japan
Year of introduction	2000
Country in which first launched	Japan
CAS registry number	103909-75-7
Structure	

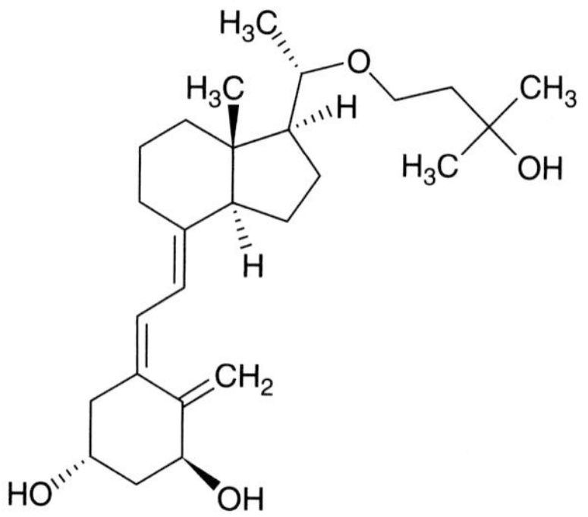

Maxacalcitol is a synthetic vitamin D analog that was introduced for the treatment of secondary hyperparathyroidism.

1.03.8.4.4 Paricalcitol

Trade name	Zemplar
Manufacturer	Abbott
Country of origin	USA
Year of introduction	1998

Country in which first launched	USA
CAS registry number	131918-61-1
Structure	

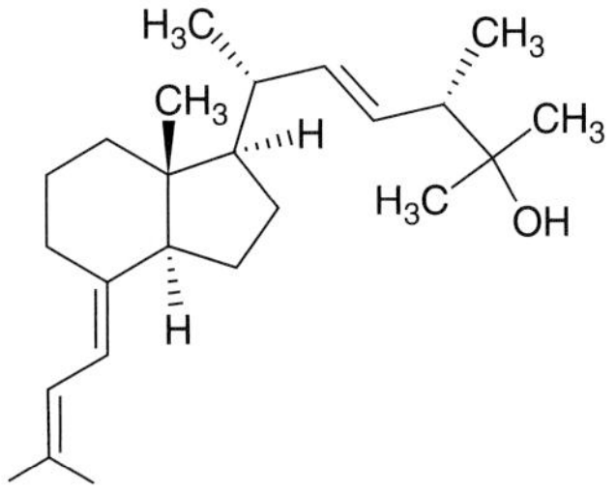

Paricalcitol is a synthetic vitamin D_2 analog that was launched for the treatment of secondary hyperthyroidism associated with chronic renal failure.

1.03.8.5 Analgesia

1.03.8.5.1 Mofezolac

Trade name	Disopain
Manufacturer	Pasteur Merieux, Taiho
Country of origin	France
Year of introduction	1994
Country in which first launched	Japan
CAS registry number	78967-07-4
Structure	

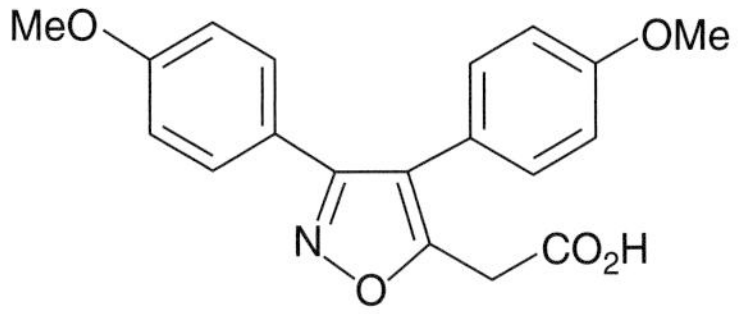

Mofezolac is an NSAID, which probably acts via the cyclooxygenase enzyme. It also has potent inhibitory activity on the algesic responses induced by the mechanical stimulus of the inflamed tissue. It was introduced for the treatment of postoperative and posttraumatic pain, acute upper respiratory tract pain, osteoarthritis, and lumbago.

1.03.8.5.2 Remifentanil hydrochloride

Trade name	Ultiva
Manufacturer	Glaxo Wellcome
Country of origin	UK
Year of introduction	1996
Country in which first launched	Germany
CAS registry number	132539-07-2
Structure	

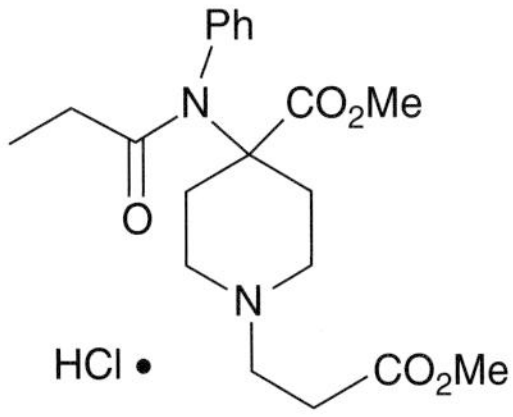

Remifentanil is a specific μ-opioid agonist with rapid onset of action and rapid offset independent of duration of administration. It was launched for use in general anesthesia and immediate postoperative pain management.

1.03.8.6 Osteoinductor

1.03.8.6.1 OP-1

Trade name	Novos
Manufacturer	Curis, Stryker Biotech
Country of origin	USA
Year of introduction	2001
Country in which first launched	Australia
CAS registry number	

Structure – recombinant bone morphogenetic protein of molecular weight 36 kDa

The OP-1 implant was introduced as an alternative to autograft in recalcitrant long bone nonunion fractures where the use of autograft is unfeasible and alternative treatments have failed. It is a mix of human recombinant osteogenic protein and a bovine-derived collagen carrier.

1.03.8.7 Hypocalcemic

1.03.8.7.1 Neridronic acid

Trade name	Nerixia
Manufacturer	Instituto Gentili
Country of origin	Italy
Year of introduction	2002
Country in which first launched	Italy
CAS registry number	79778-41-9
Structure	

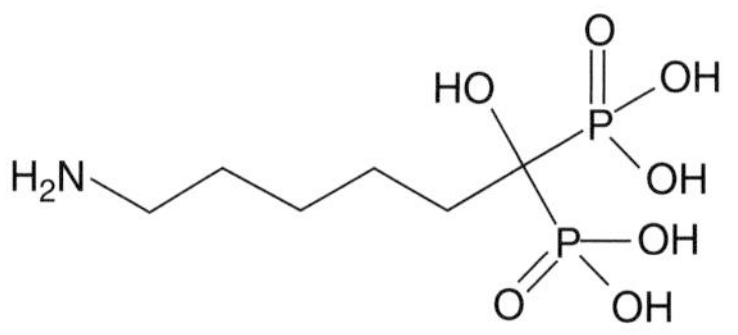

Neridronic acid was launched for the treatment of osteogenesis imperfecta, a disease in which the bones are characterized by extreme fragility. Clinical trials remarkably increased bone mass in patients, especially young growing individuals. It is a bisphosphonate, a class known to be potent inhibitors of bone resorption and to increase bone mineral density.

1.03.9 Immunology (Including Vaccines)

1.03.9.1 Immunosuppressants

1.03.9.1.1 Gusperimus trihydrochloride

Trade name	Spanidin
Manufacturer	Nippon Kayaku
Country of origin	Japan
Year of introduction	1994
Country in which first launched	Japan
CAS registry number	104317-84-2

Structure

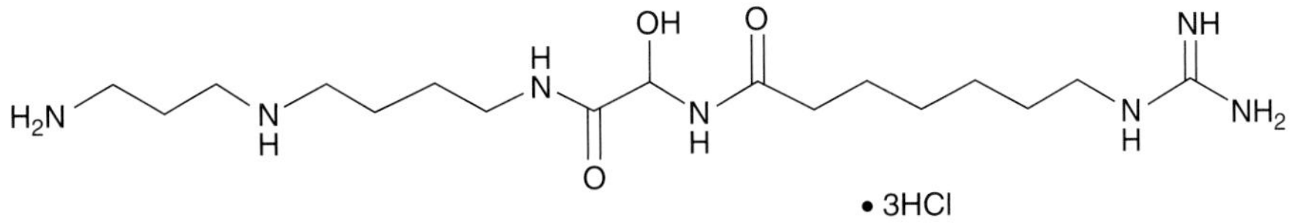

Gusperimus is a synthetic derivative of the antitumor antibiotic spergualin. It was introduced as a treatment for accelerated and acute rejection reactions after kidney transplants. In combination with cyclosporin it has been reported that gusperimus exhibits exceptional activity in the prevention of transplant rejection and in inducing long-term tolerance. It appears to work by inhibition of the differentiation and proliferation of effector cells, including cytotoxic T cells and antibody-producing B cells.

1.03.9.1.2 Mycophenolate mofetil

Trade name	Cellcept
Manufacturer	Roche
Country of origin	Switzerland
Year of introduction	1995
Country in which first launched	USA
CAS registry number	128794-94-5
Structure	

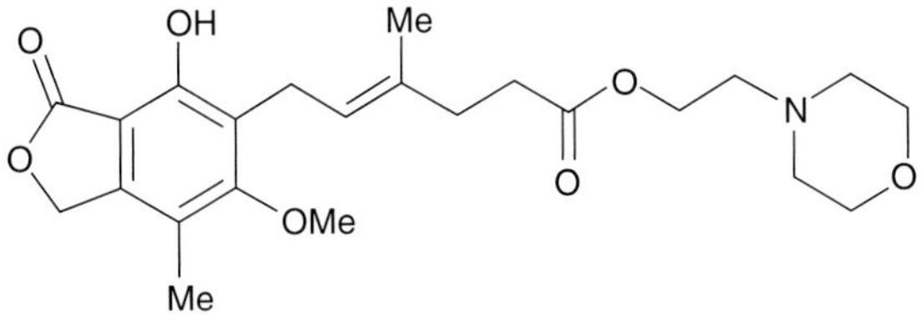

Mycophenolate mofetil is a prodrug of mycophenolic acid, a selective, reversible, noncompetitive inhibitor of inosinate dehydrogenase and guanylate synthetase. Mycophenolate mofetil has improved oral absorption and bioavailability and was introduced to prevent acute kidney transplant rejection in combination with other immunosuppressive therapy and to treat refractory acute kidney graft rejection.

1.03.9.1.3 Mycophenolate sodium

Trade name	Myfortic
Manufacturer	Novartis
Country of origin	Switzerland
Year of introduction	2003
Country in which first launched	Switzerland
CAS registry number	37415-62-6
Structure	

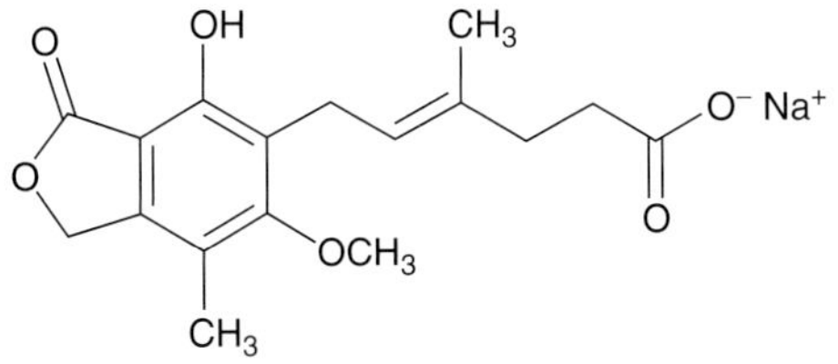

Mycophenolate sodium is formulated as an enteric coated tablet, which is absorbed in the upper intestine thus protecting the upper gastrointestinal (GI) tract from the side effects of mycophenolic acid. It was introduced as an oral treatment in combination with other immunosuppressants to prevent acute rejection in adult patients receiving allogeneic renal transplantation.

1.03.9.1.4 Pimecrolimus

Trade name	Elidel
Manufacturer	Novartis
Country of origin	USA
Year of introduction	2002
Country in which first launched	USA
CAS registry number	137071-32-0
Structure	

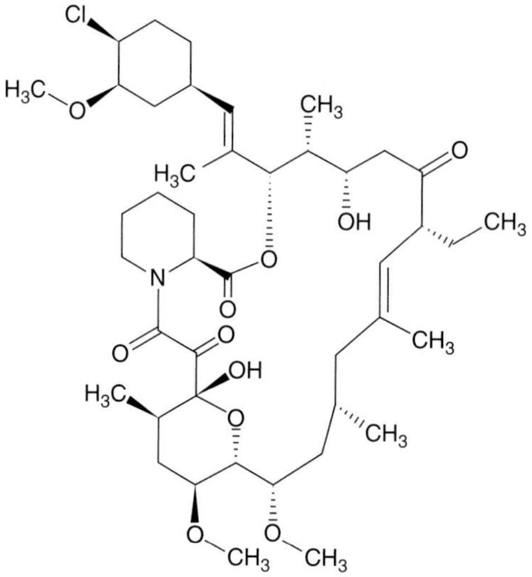

Pimecrolimus is a derivative of the immunosuppressant FK520. It was introduced as a topical formulation for the treatment of mild to moderate atopic dermatitis for patients of age 2 and over in whom the use of conventional therapies is inadvisable. It is an inflammatory cytokine inhibitor that works by selectively targeting T cells in the skin.

1.03.9.2 Immunostimulant

1.03.9.2.1 GMDP

Trade name	Likopid
Manufacturer	Peptech
Country of origin	Australia
Year of introduction	1996
Country in which first launched	Russia
CAS registry number	18194-24-6
Structure	

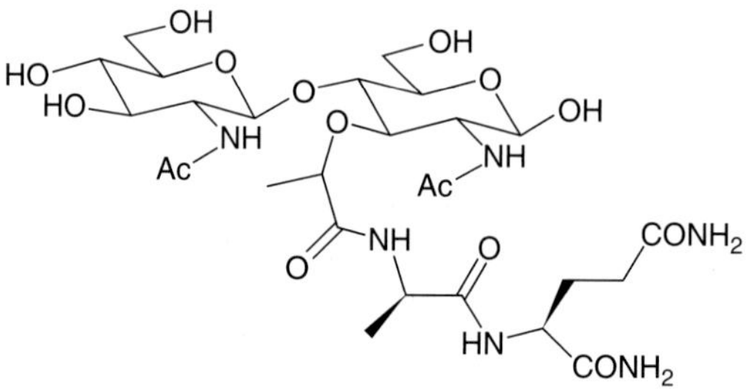

GMDP is an enzymatic degradation product of a bacterial peptidoglycan that was introduced for hospital-related infections, psoriasis, cervical precancerous lesions, and ophthalmic keratitis caused by herpes.

1.03.9.3 Vaccines

1.03.9.3.1 Influenza virus (live)

Trade name	FluMist
Manufacturer	Medimmune/Wyeth
Country of origin	USA
Year of introduction	2003
Country in which first launched	USA

This is a live attenuated influenza virus vaccine. It is a cold-adapted trivalent formulation that is administered nasally to provide active immunity against select influenza A and B strains in healthy people aged 5–49 years.

1.03.9.3.2 Lyme disease vaccine

Trade name	LYMErix
Manufacturer	Yale University, SmithKlineBeecham
Country of origin	USA
Year of introduction	1999
Country in which first launched	USA
CAS registry number	147519-65-1
Structure – recombinant protein of molecular weight 30 kDa	

This vaccine was introduced for protection against Lyme disease, a tick-borne disease caused by infection with the spirochete *Borrelia burgdorferi.*

1.03.10 Hormone Related

1.03.10.1 Antidiabetic

1.03.10.1.1 Glimipiride

Trade name	Amaryl
Manufacturer	Hoechst Marion Roussel
Country of origin	Germany
Year of introduction	1995
Country in which first launched	Sweden
CAS registry number	93479-97-1
Structure	

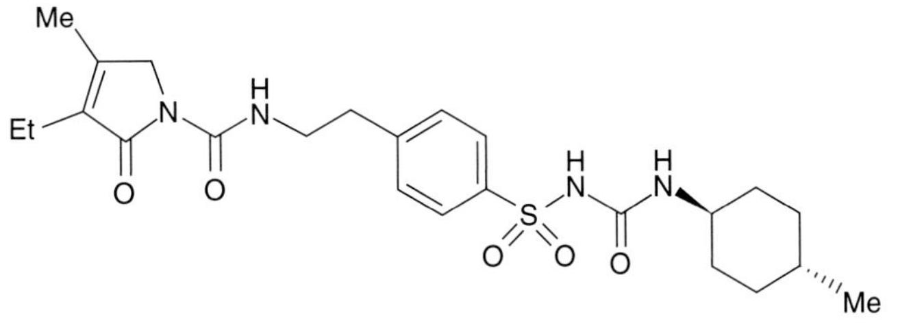

Glimipiride was claimed as a new generation sulfonylurea drug and was launched to reduce blood sugar levels in type 2 diabetic patients. Such agents function by direct stimulation of insulin release from glucose-insensitive pancreatic β-cells and glucose transporter (GLUT) translocation in insulin-resistant fat and muscle cells.

1.03.10.1.2 Insulin lispro

Trade name	Humalog
Manufacturer	Lilly
Country of origin	USA
Year of introduction	1996
Country in which first launched	USA
CAS registry number	133107-64-9
Structure – recombinant protein of molecular weight 5.8 kDa	

Humalog, a recombinant protein, is a form of human insulin in which amino acids 28 and 29 have been inverted. This reduces the tendency of the protein to dimerize and further aggregate to hexamers. Thus, it has a faster rate of absorption and shorter duration of action, and this has advantages in controlling blood glucose levels after meals.

1.03.10.1.3 Miglitol

Trade name	Diastabol
Manufacturer	Bayer, Sanofi
Country of origin	Germany
Year of introduction	1998
Country in which first launched	Germany
CAS registry number	72432-03-2
Structure	

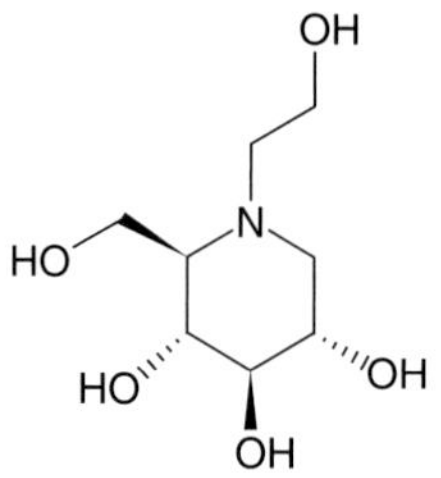

Miglitol is an α-D-glucosidase inhibitor that was marketed as an auxiliary treatment for type 2 diabetes mellitus.

1.03.10.1.4 Nateglinide

Trade name	Fastic, Starsis
Manufacturer	Ajinomoto, Yamanouchi, Roussel
Country of origin	Japan
Year of introduction	1999
Country in which first launched	Japan
CAS registry number	105816-04-4
Structure	

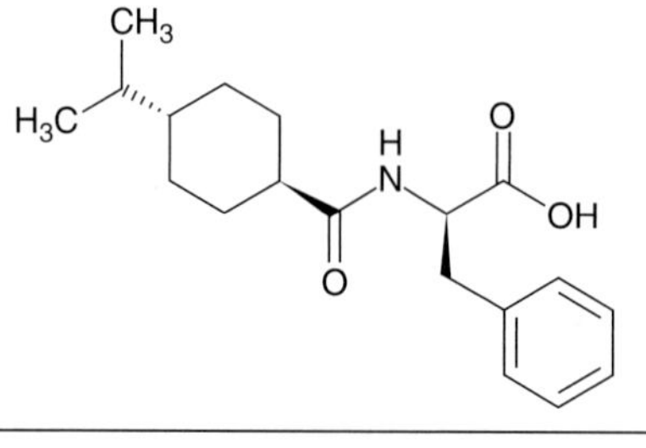

Nateglinide specifically blocks the ATP-sensitive K^+ channel in pancreatic β-cells, thus resulting in an increase in intracellular calcium concentrations, which causes an increase in insulin secretion. Hence, nateglinide was marketed as an insulinotropic agent for the treatment of type 2 diabetes mellitus.

1.03.10.1.5 **Pioglitazone hydrochloride**

Trade name	Actos
Manufacturer	Takeda, Lilly
Country of origin	Japan
Year of introduction	1999
Country in which first launched	USA
CAS registry number	112529-15-4
Structure	

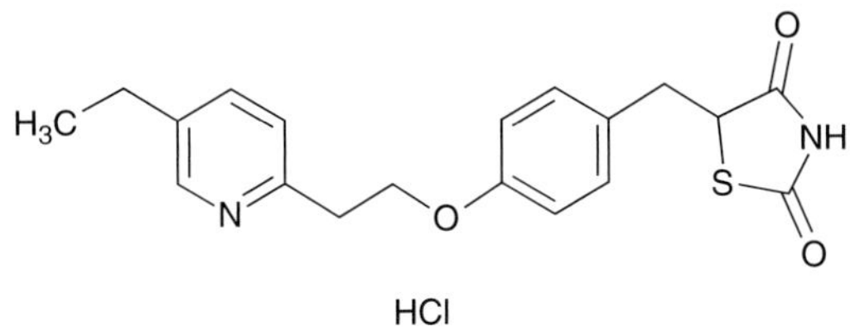

Pioglitzone is an orally active treatment for type 2 diabetes mellitus. It binds to peroxisome proliferator-activated receptor gamma, thus activating this nuclear receptor, which then influences carbohydrate metabolism.

1.03.10.1.6 **Repaglinide**

Trade name	Prandin
Manufacturer	Boehringer Ingelheim
Country of origin	Germany
Year of introduction	1998
Country in which first launched	USA
CAS registry number	135062-02-1
Structure	

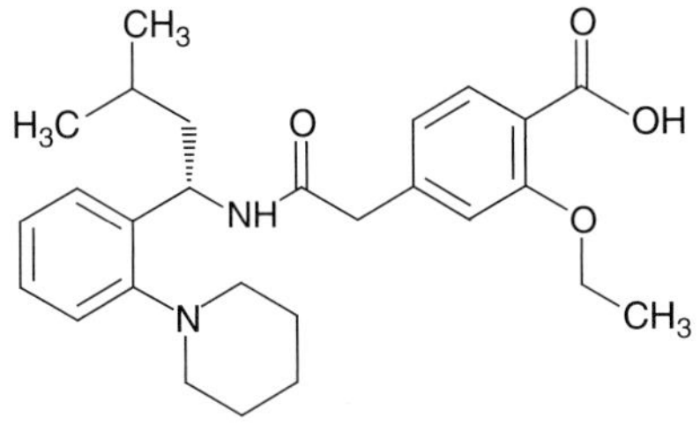

Repaglinide has an insulin-releasing effect mediated by pancreatic β-cells. It was marketed as an orally active hypoglycemic agent in patients with type 2 diabetes mellitus. The mechanism of action appears to be similar to that of nateglinide.

1.03.10.1.7 **Rosiglitazone**

Trade name	Avandia
Manufacturer	SmithKlineBeecham
Country of origin	USA
Year of introduction	1999
Country in which first launched	USA, Mexico
CAS registry number	155141-29-0

Structure

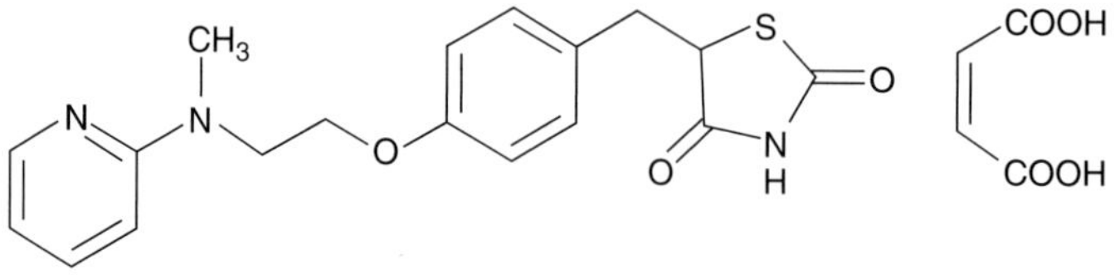

Rosiglitazone is a potent agonist of peroxisome proliferator-activated receptor gamma, a nuclear receptor involved in the differentiation of adipose tissue, without activating liver PPAR-alpha receptors. This could in turn mediate the downregulation of leptin gene expression. Rosiglitazone has been shown to normalize glucose metabolism and reduce the exogenous dose of insulin needed to achieve glycemic control.

1.03.10.1.8 **Troglitazone**

Trade name	Rezulin
Manufacturer	Sankyo
Country of origin	Japan
Year of introduction	1997
Country in which first launched	Japan
CAS registry number	97322-87-7
Structure	

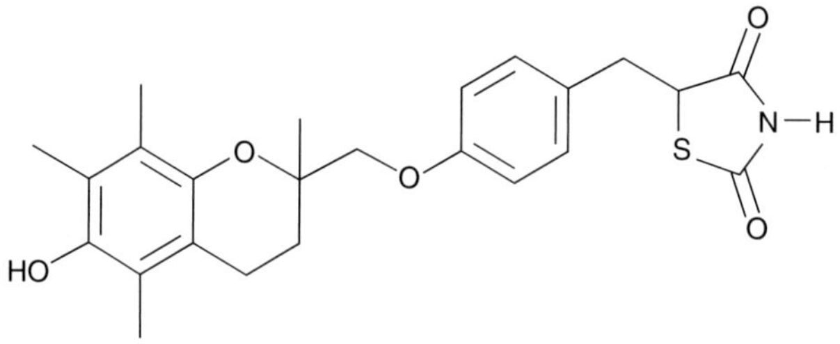

Troglitazone was introduced for the treatment of type 2 diabetes and apparently reduces glucose concentrations without affecting insulin secretion. Troglitazone binds to peroxisome proliferator-activated receptor gamma thus activating this nuclear receptor, which then influences carbohydrate metabolism. Unfortunately, liver toxicity problems have led to this drug being withdrawn from the market.

1.03.10.1.9 **Voglibose**

Trade name	Basen
Manufacturer	Takeda
Country of origin	Japan
Year of introduction	1994
Country in which first launched	Japan
CAS registry number	83480-29-9
Structure	

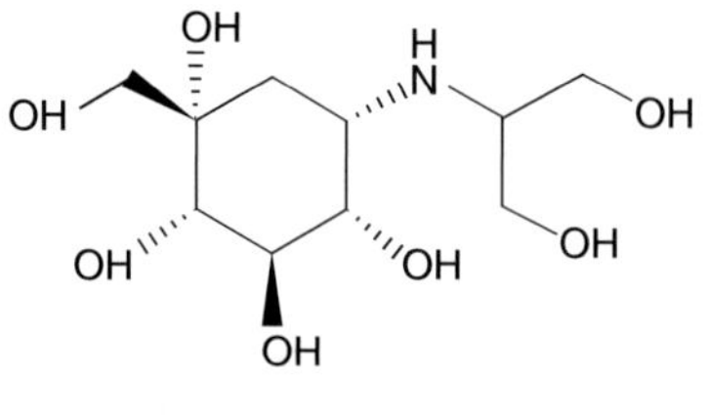

Voglibose is an orally active α-D-glucosidase inhibitor that was launched for the treatment of postprandial hyperglycemia in diabetic patients. It acts by decreasing the release of glucose from carbohydrates ingested in food.

1.03.10.2 Acromegaly

1.03.10.2.1 Lanreotide acetate

Trade name	Somatuline LP
Manufacturer	Beaufour-Ipsen
Country of origin	France
Year of introduction	1995
Country in which first launched	France
CAS registry number	127984-74-1
Structure	

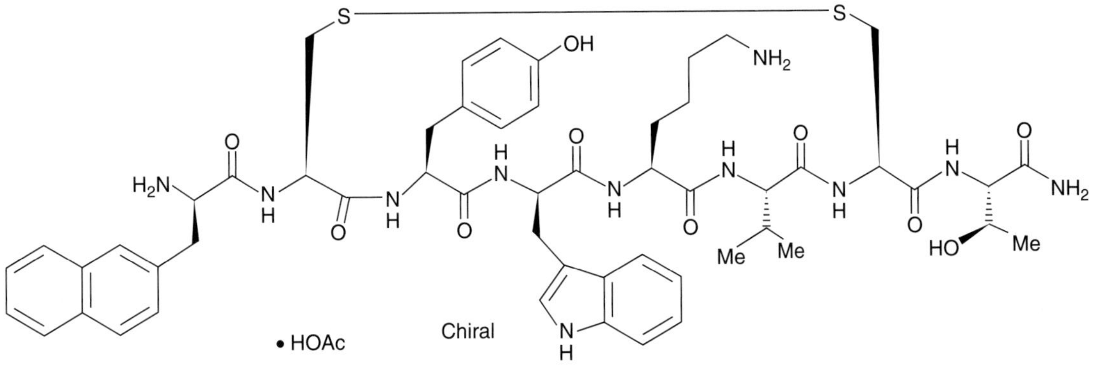

Lanreotide is an octapeptide somatostatin analog that was marketed for the treatment of acromegaly in cases where surgery or radiotherapy have failed to restore normal growth hormone secretion. It is a selective inhibitor of growth hormone and also reduces the secretion of growth hormone, thyrotropin, motilin, and pancreatic polypeptide. It is also being evaluated for the treatment of neuroendocrine tumors and hormone-responsive prostate cancer.

1.03.10.2.2 Pegvisomant

Trade name	Somavert
Manufacturer	Sensus, Pfizer
Country of origin	USA
Year of introduction	2003
Country in which first launched	USA
CAS registry number	218620-50-9
Structure – PEGylated recombinant protein of molecular weight	40–50 kDa

Acromegaly is a debilitating endocrine disease caused by the excessive secretion of growth hormone and increased production of insulin-like growth factor-I in middle-aged adults. Pegvisomant is a modified form of human growth hormone that acts as a highly selective growth hormone antagonist acting on cell surfaces where it blocks the binding of growth hormone.

1.03.10.3 Growth Hormone

1.03.10.3.1 Somatomedin-1

Trade name	Igef
Manufacturer	Biogen, Pharmacia
Country of origin	USA
Year of introduction	1994

Country in which first launched	Sweden
CAS registry number	67763-96-6
Structure – protein of molecular weight 7649 Da	

Somatomedin-1 is an insulin-like growth factor-I that was launched for the treatment of children with growth disorders caused by growth hormone insensitivity. It is also a potent hypoglycemic agent that resembles insulin in some respects and its use in diabetes is also being investigated.

1.03.10.3.2 Somatotropin

Trade name	Nutropin
Manufacturer	Genentech
Country of origin	USA
Year of introduction	1994
Country in which first launched	USA
CAS registry number	12629-01-5
Structure – recombinant protein of molecular weight 22 kDa	

Somatotropin is a second-generation and methionine-free human growth hormone that was launched for the treatment of growth failure in children due to chronic renal insufficiency before transplantation and for the long-term treatment of short stature in children with a lack of endogenous growth hormone secretion.

1.03.11 Reproduction and Fertility

1.03.11.1 Fertility Enhancer

1.03.11.1.1 Cetrorelix

Trade name	Cetrotide
Manufacturer	Asta Medica
Country of origin	Germany
Year of introduction	1999
Country in which first launched	Germany
CAS registry number	120287-85-6; 130289-71-3
Structure	

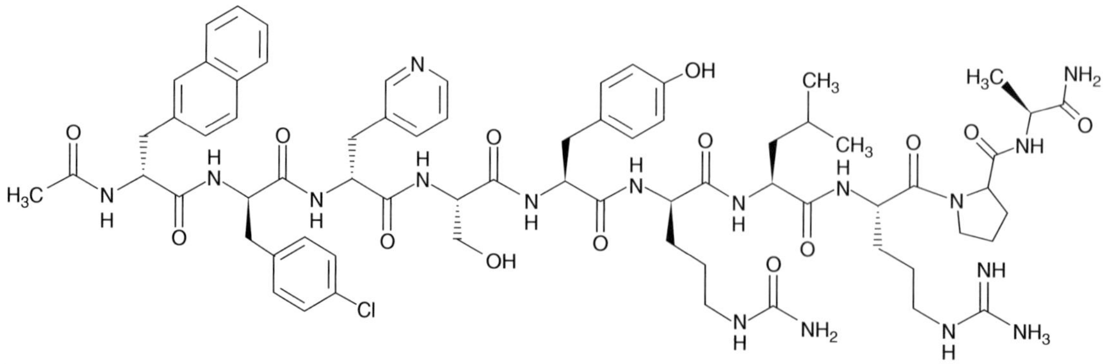

Cetrorelix is a structurally modified decapeptidic analog of luteinizing hormone-releasing hormone. It is an extremely potent and long-lasting gonadotropin releasing hormone antagonist, which blocks gonadotropins and sex steroid secretion immediately after administration.

1.03.11.1.2 **Follitropin alpha**

Trade name	Gonal-F
Manufacturer	Genzyme, Serono
Country of origin	Switzerland
Year of introduction	1996
Country in which first launched	Austria
CAS registry number	9002-68-0
Structure – recombinant protein	

Follitropin alpha is a recombinant form of human follical-stimulating hormone that was introduced for the treatment of infertility.

1.03.11.1.3 **Follitropin beta**

Trade name	Puregon
Manufacturer	Organon
Country of origin	Netherlands
Year of introduction	1996
Country in which first launched	Denmark
CAS registry number	9002-68-0
Structure – recombinant protein	

Follitropin beta is another recombinant form of human follicle-stimulating hormone that was introduced for the induction of ovulation in clomiphene-resistant anovulation and for controlled ovarian hyperstimulation. It appears to be more active than follitropin alpha in that a greater pregnancy rate is achieved.

1.03.11.1.4 **Ganirelix acetate**

Trade name	Orgalutran
Manufacturer	Roche Bioscience
Country of origin	USA
Year of introduction	2000
Country in which first launched	Germany
CAS registry number	129311-55-3
Structure	

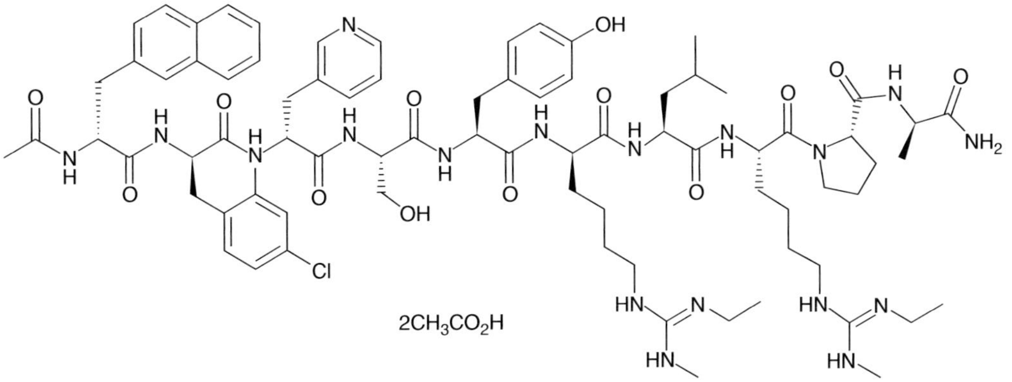

Ganirelix acetate is another decapeptide analog of luteinizing hormone-releasing hormone but is more water soluble than cetrorelix. It is highly bioavailable and immediately blocks the endogenous release of luteinizing hormone and follicle-stimulating hormone. It was launched as prefilled syringes for subcutaneous injections that inhibit premature luteinizing hormone surges in women undergoing controlled ovarian hyperstimulation.

1.03.11.2 Contraceptive

1.03.11.2.1 Drospirenone

Trade name	Yasmin
Manufacturer	Schering AG
Country of origin	Germany
Year of introduction	2000
Country in which first launched	Germany
CAS registry number	067392-87-4
Structure	

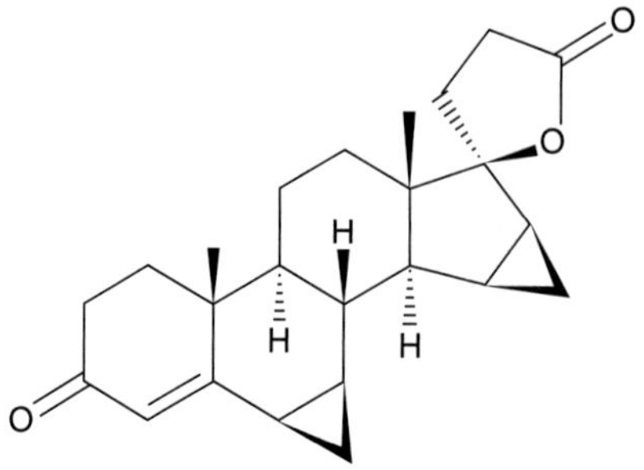

Drospirenone was launched in combination with ethinylestradiol as a novel oral contraceptive. Its receptor binding profile for steroid receptors is very similar to that of progesterone and thus it mimics the progestogen's agonistic activity as well as the antiandrogenic and antimineralocorticoid activity of the endogenous hormone.

1.03.11.2.2 Norelgestromin

Trade name	Ortho Evra
Manufacturer	Johnson & Johnson
Country of origin	USA
Year of introduction	2002
Country in which first launched	USA
CAS registry number	53016-31-2
Structure	

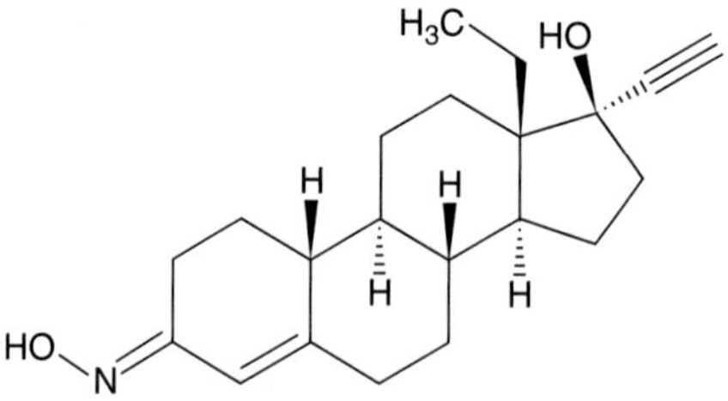

Norelgestromin is one of the components, along with ethinylestradiol, in the first birth control transdermal patch. This patch is changed weekly for 3 weeks, followed by a treatment-free week. Following application, norelgestromin rapidly appears in the serum and reaches a plateau after 48 h; this level is maintained during the patch-wearing period.

1.03.11.3 Hyperprolactinemia

1.03.11.3.1 Quinagolide hydrochloride

Trade name	Norprolac
Manufacturer	Sandoz

Country of origin	Switzerland
Year of introduction	1994
Country in which first launched	Netherlands
CAS registry number	94424-50-7
Structure	

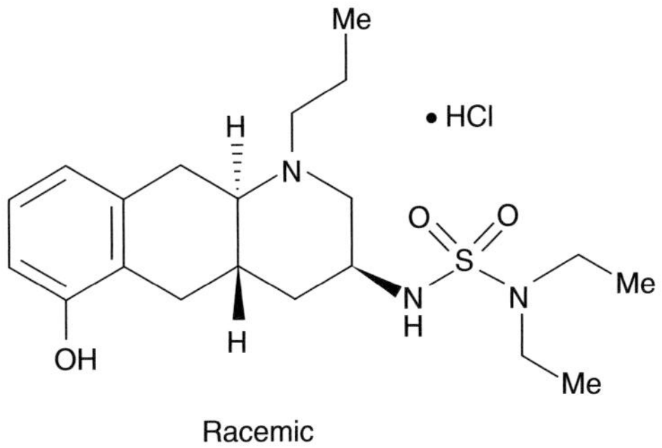

Quinagolide is a potent and specific nonergot dopamine D_2 agonist that was launched for the treatment of hyperprolactinemia. It is effective in inhibiting prolactin secretion by human pituitary tumors and can also provide relief from associated effects such as, inter alia, infertility.

1.03.11.4 Preterm Labor

1.03.11.4.1 Atosiban

Trade name	Tractocile, Antocin
Manufacturer	Ferring AG
Country of origin	Sweden
Year of introduction	2000
Country in which first launched	UK
CAS registry number	090779-69-4
Structure	

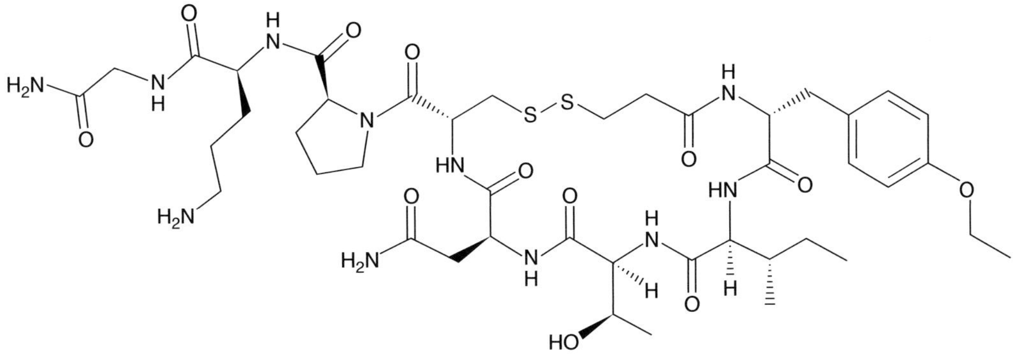

Atosiban is a peptidic oxytocin analog that acts as an antagonist of the vasopressin V_{1a} receptor and of the oxytocin receptor. It was introduced as an injectable inhibitor of preterm labor.

1.03.12 Skin Related

1.03.12.1 Plaque Psoriasis

1.03.12.1.1 Alefacept

Trade name	Amevive
Manufacturer	Biogen

Country of origin	USA
Year of introduction	2003
Country in which first launched	USA
CAS registry number	222535-22-0
Structure – recombinant protein of molecular weight 91.4 kDa	

Alefacept is a dimeric fusion protein consisting of the leukocyte function antigen-3 protein and Fc portion of human IgG1 that blocks the T-cell CD2 receptor thus preventing T-cell proliferation, a key mechanism in psoriasis.

1.03.12.1.2 Efalizumab

Trade name	Raptiva
Manufacturer	Xoma, Genentech
Country of origin	USA
Year of introduction	2003
Country in which first launched	USA
CAS registry number	214745-43-4
Structure – recombinant monoclonal antibody of molecular weight 150 kDa	

Efalizumab is a humanized monoclonal antibody that was marketed for the treatment of psoriasis. It is full-length IgG1 antibody developed through a murine antihuman CD11a mAb, where CD11a is the alpha-chain leukocyte function associate antigen that is expressed on the surface of T lymphocytes. CD11a also binds to the intercellular cell adhesion molecules ICAM-1, ICAM-2, and ICAM3, on endothelial cells, monocytes, keratinocytes, fibroblasts, and activated lymphocytes. Psoriasis is a disease that is mediated through inflammatory cells, primarily T cells expressing CD4 or CD8 markers and keratinocytes. Thus, by blocking binding, the ability of T cells to adhere, migrate, and be activated is blunted.

1.03.12.2 Wound Healing

1.03.12.2.1 Acemannan

Trade name	Acemannan hydrogel
Manufacturer	Carrington Laboratories
Country of origin	USA
Year of introduction	2001
Country in which first launched	USA
CAS registry number	110042-95-0
Structure – carbohydrate-derived polymer of molecular weight 1–2 million Da	

Acemannan is a complex water-soluble polymanno-galacto acetate derived from *Aloe vera* that was marketed as a wound-healing agent for the care of ulcers, burns, and postsurgical incisions.

1.03.12.2.2 Prezatide copper acetate

Trade name	Iamin gel
Manufacturer	ProCyte
Country of origin	USA
Year of introduction	1996
Country in which first launched	USA
CAS registry number	130120-57-9

Structure

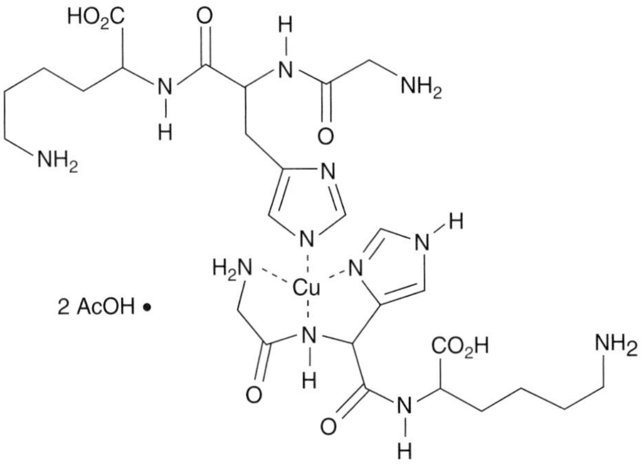

Prezatide copper acetate is a tripeptide–copper acetate complex where the sequence GHK is an endogenous growth factor that stimulates collagen synthesis and angiogenesis. The gel was launched for the treatment of chronic and acute wounds.

1.03.12.3 Photosenstization

1.03.12.3.1 Verteporfin

Trade name	Visudyne
Manufacturer	QLT
Country of origin	Canada
Year of introduction	2000
Country in which first launched	Switzerland
CAS registry number	129497-78-5
Structure	

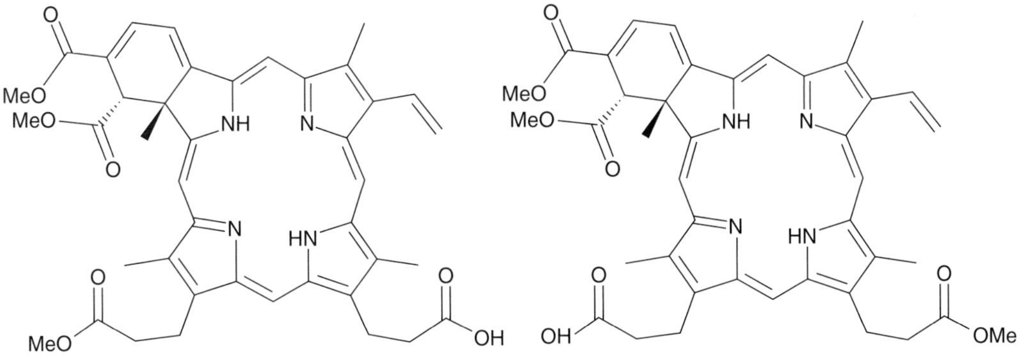

Verteporfin was launched as a photosensitizer for photodynamic therapy of wet age-related macular degeneration in patients with subfoveal choroidal neovascularization. It is injected intravenously as a liposomal formulation.

1.03.12.4 Dermatological

1.03.12.4.1 Kinetin

Trade name	Kinerase
Manufacturer	Senetek

Country of origin	UK
Year of introduction	1999
Country in which first launched	USA
CAS registry number	525-79-1
Structure	

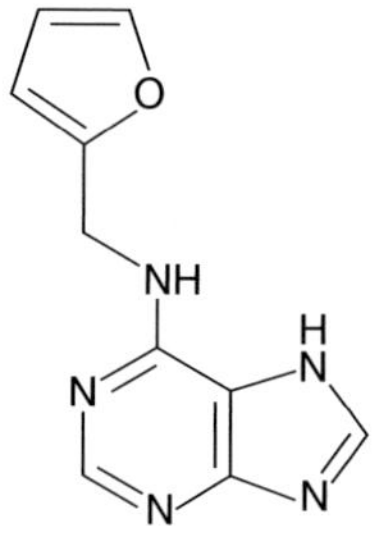

Kinetin is a synthetic cytokinin (a plant growth factor) that was introduced for the treatment of age-related photodamage of skin.

1.03.12.5 Antipsoriasis

1.03.12.5.1 Tazarotene

Trade name	Zorac
Manufacturer	Allergan
Country of origin	USA
Year of introduction	1997
Country in which first launched	Germany
CAS registry number	118292-40-3
Structure	

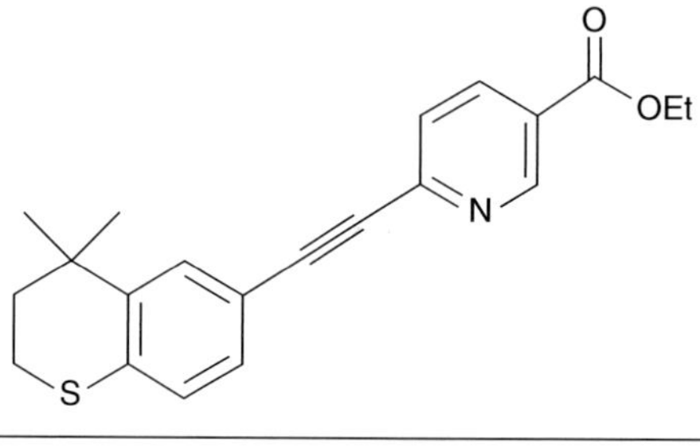

Tazarotene normalizes abnormal keratinocyte differentiation and proliferation and reduces the expression of inflammatory markers. It is a prodrug that is hydrolyzed in vivo to the acid, which has selectivity for retinoid acid receptors. Tazarotene was introduced for the treatment of psoriasis and acne.

1.03.13 Eye Related

1.03.13.1 Antiglaucoma

1.03.13.1.1 Bimatoprost

Trade name	Lumigan
Manufacturer	Allergan
Country of origin	USA
Year of introduction	2001
Country in which first launched	USA, Brazil
CAS registry number	155206-00-1

Structure

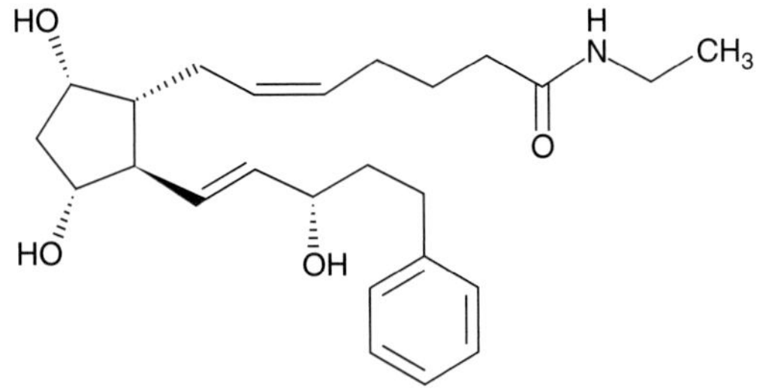

Bimatoprost is a PGF_{2a}-analog that was introduced for the reduction of elevated intraocular pressure in patients with open-angle glaucoma or ocular hypertension.

1.03.13.1.2 **Brimonidine**

Trade name	Alphagan
Manufacturer	Allergan
Country of origin	USA
Year of introduction	1996
Country in which first launched	USA
CAS registry number	59803-98-4
Structure	

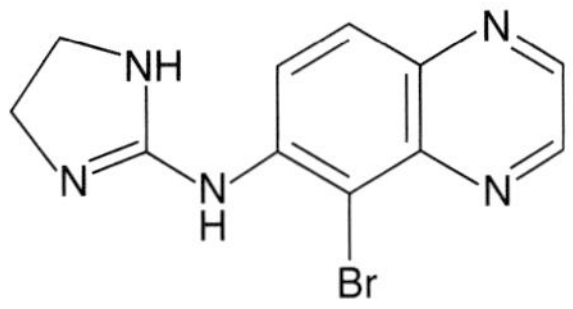

Brimonidine is a potent and relatively selective α_{2a}-adrenergic agonist with low affinity for the imidazoline I_1 receptor. It was introduced as a topical treatment for open-angle glaucoma and ocular hypertension.

1.03.13.1.3 **Brinzolamide**

Trade name	Azopt
Manufacturer	Alcon
Country of origin	USA
Year of introduction	1998
Country in which first launched	USA
CAS registry number	138890-62-7
Structure	

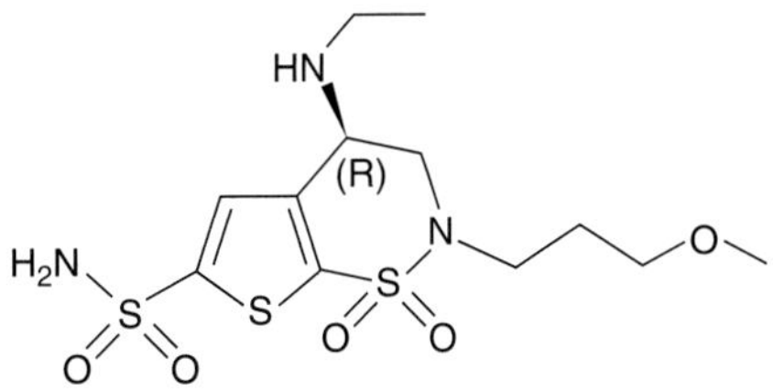

Brinzolamide is a potent inhibitor of the human carbonic anhydrase introduced for the treatment of elevated intraocular pressure in patients with ocular hypertension or open-angle glaucoma.

1.03.13.1.4 Dorzolamide hydrochloride

Trade name	Trusopt
Manufacturer	Merck
Country of origin	USA
Year of introduction	1995
Country in which first launched	USA
CAS registry number	130693-82-2
Structure	

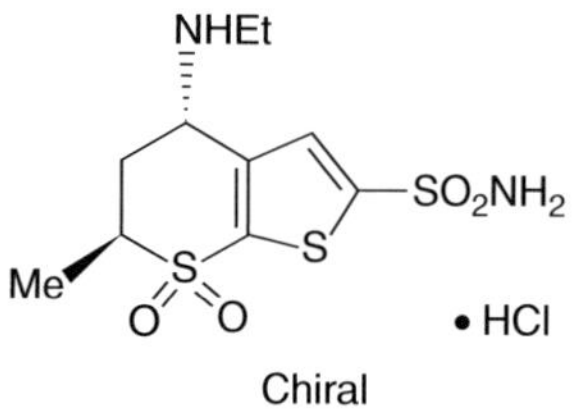

Dorzolamide is a carbonic anhydrase inhibitor that lowers intraocular pressure on topical administration. It was launched for the treatment of open-angle glaucoma and ocular hypertension.

1.03.13.1.5 Latanoprost

Trade name	Xalatan
Manufacturer	Pharmacia & Upjohn
Country of origin	UK
Year of introduction	1996
Country in which first launched	USA
CAS registry number	130209-82-4
Structure	

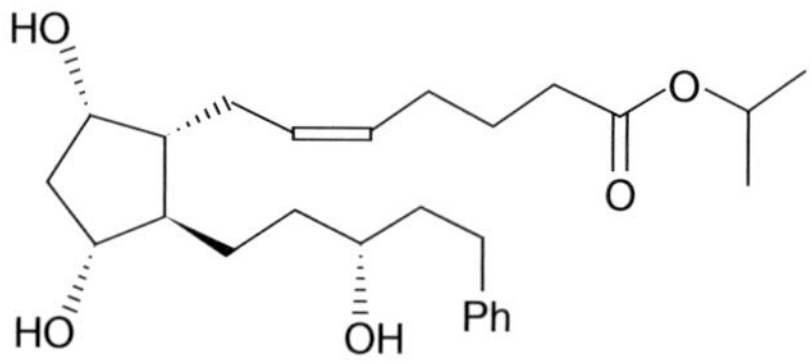

Latanoprost is a PGF_{2a}-analog with greater lipophilicity and therefore greater corneal penetration, which reduces intraocular pressure. It was launched for the treatment of glaucoma.

1.03.13.1.6 Travoprost

Trade name	Travatan
Manufacturer	Alcon
Country of origin	USA
Year of introduction	2001
Country in which first launched	USA
CAS registry number	157283-68-6

Structure

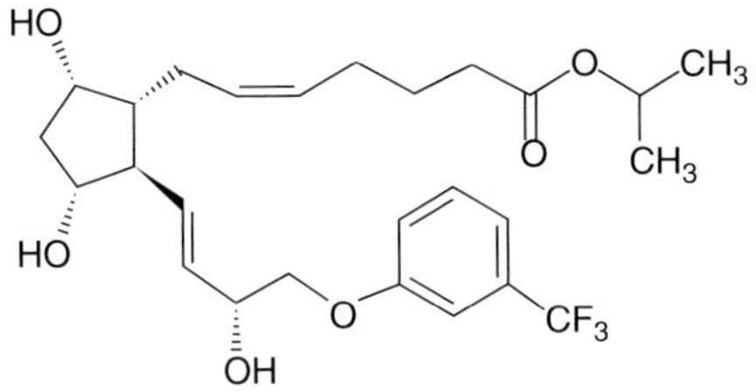

Travoprost is a PGF_{2a} analog that was launched as an ophthalmic solution administered topically for the treatment of elevated intraocular hypertension as a result of open-angle glaucoma, a common optic neuropathy, and a leading cause of blindness.

1.03.13.1.7 **Unoprostone ispropyl ester**

Trade name	Rescula
Manufacturer	Ueno, Fujisawa
Country of origin	Japan
Year of introduction	1994
Country in which first launched	Japan
CAS registry number	120373-24-2
Structure	

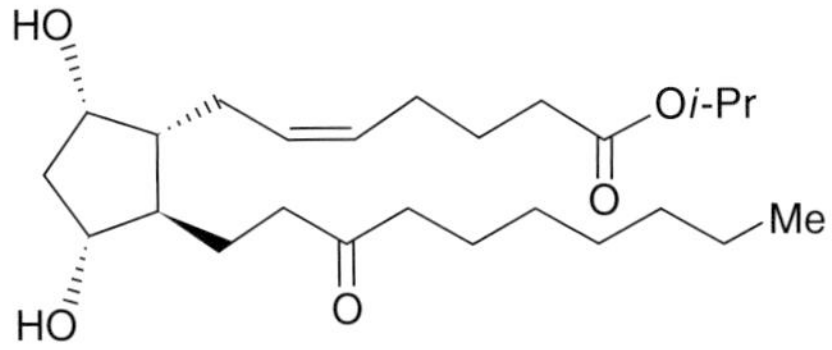

Unoprostone is a prostaglandin derivative launched for the treatment of glaucoma and ocular hypertension. It is suggested that it acts by increasing uveoscleral outflow or by decreasing episcleral venous pressure.

1.03.14 **Miscellaneous**

1.03.14.1 **Type 1 Gaucher's Disease**

1.03.14.1.1 **Imiglucerase**

Trade name	Cerezyme
Manufacturer	Genzyme
Country of origin	USA
Year of introduction	1994
Country in which first launched	USA
CAS registry number	154248-97-2
Structure – recombinant protein	

This is a mannose-terminated form of human placental glucocerebrosidase that catalyzes the hydrolysis of glucocerebroside and thus prevents the accumulation of this lipid in organs and tissues. It was launched for the treatment of type I Gaucher's disease.

1.03.14.1.2 **Miglustat**

Trade name	Zavesca
Manufacturer	G.D. Searle, Actelion
Country of origin	USA
Year of introduction	2003
Country in which first launched	UK
CAS registry number	72599-27-0
Structure	

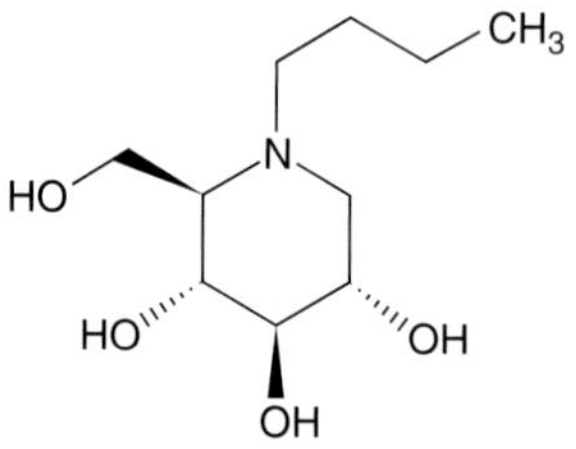

Miglustat is an inhibitor of glucosylceramide synthase that was launched as an oral treatment for mild to moderate type I Gaucher's disease in adult patients for whom enzyme replacement therapy is not a therapeutic option.

1.03.14.2 Antiobesity

1.03.14.2.1 **Dexfenfluramine**

Trade name	Redux
Manufacturer	Servier
Country of origin	France
Year of introduction	1997
Country in which first launched	France
CAS registry number	3239-45-0
Structure	

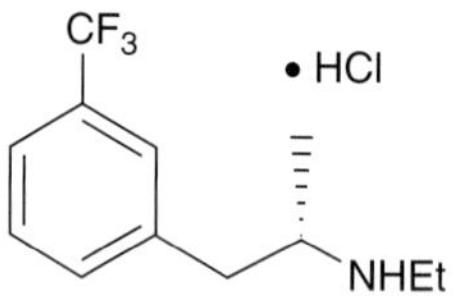

Dexfenfluramine is the (+)-isomer of fenfluramine; it has with a greater anorectic effect because of a greater selectivity for the serotonin system as a 5HT agonist with no dopaminergic or noradrenergic activity. It was launched for the treatment of obesity but was withdrawn due to side effects of primary pulmonary hypertension, brain serotonin neurotoxicity, and valvular heart disease.

1.03.14.2.2 **Orlistat**

Trade name	Xenical
Manufacturer	Roche
Country of origin	Switzerland
Year of introduction	1998
Country in which first launched	New Zealand
CAS registry number	096829-58-2

Structure

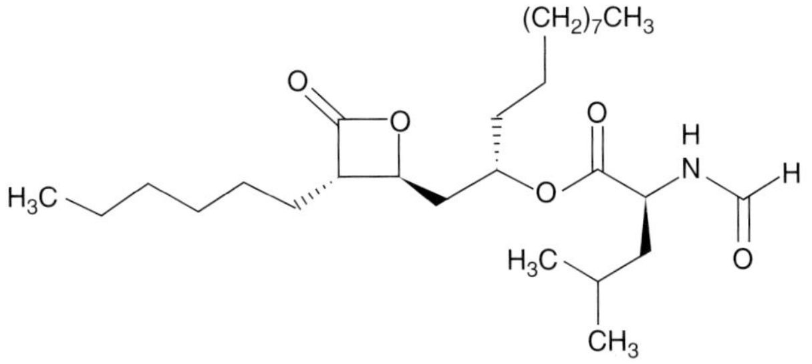

Orlistat is a potent inhibitor of gastrointestinal lipases required for the lypolysis and digestion of dietary fat; thus, it inhibits the absorption of about a third of the fat in food. It was introduced for the long-term treatment of obesity in conjunction with a moderately reduced calorie diet.

1.03.14.2.3 Sibutramine

Trade name	Meridia, Reductil
Manufacturer	Knoll
Country of origin	Germany
Year of introduction	1998
Country in which first launched	USA
CAS registry number	125494-59-9
Structure	

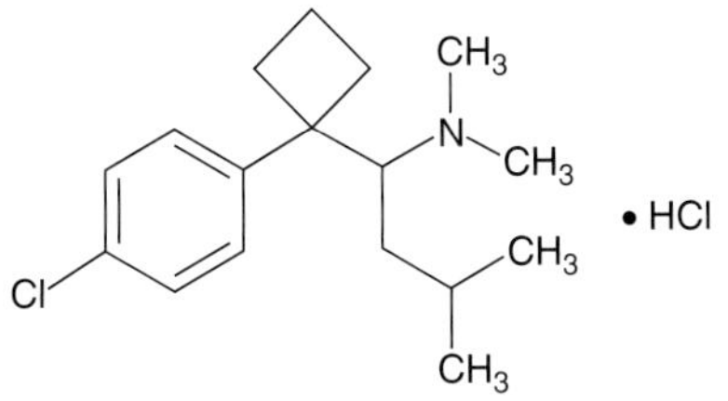

Sibutramine is a serotonin and noradrenaline reuptake inhibitor that reduces energy intake by creating a satiated feeling and increases energy expenditure by enhancing thermogenesis.

1.03.14.3 Antidote

1.03.14.3.1 Antidigoxin polyclonal antibody

Trade name	DigiFab
Manufacturer	Protherics, Savage Laboratories
Country of origin	UK
Year of introduction	2002
Country in which first launched	USA
CAS registry number	339086-83-8
Structure – polyclonal antibody of molecular weight 46 kDa	

Digoxin is widely used for the treatment of cardiac conditions such as atrial arrhythmias and congestive heart failure, but has a narrow therapeutic range. Intravenous infusion of antidigoxin polyclonal antibody was introduced to overcome digoxin poisoning.

1.03.14.3.2 Crotalidae polyvalent immune fab

Trade name	CroFab
Manufacturer	Protherics, Savage Laboratories
Country of origin	UK
Year of introduction	2001
Country in which first launched	USA

Crotalidae was launched for the treatment of North American crotalid snake envenomation.

1.03.14.3.3 Fomepizole

Trade name	Antizol
Manufacturer	Orphan Medical/Cambridge
Country of origin	USA
Year of introduction	1998
Country in which first launched	USA
CAS registry number	7554-65-6
Structure	

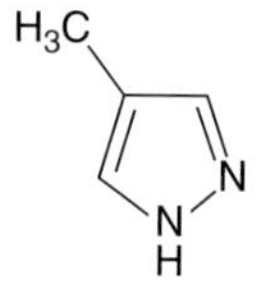

Fomepizole was introduced as an antidote for the treatment of ethylene glycol poisoning.

1.03.14.4 Male Sexual Dysfunction

1.03.14.4.1 Sildenafil citrate

Trade name	Viagra
Manufacturer	Pfizer
Country of origin	UK
Year of introduction	1998
Country in which first launched	USA
CAS registry number	139755-83-2
Structure	

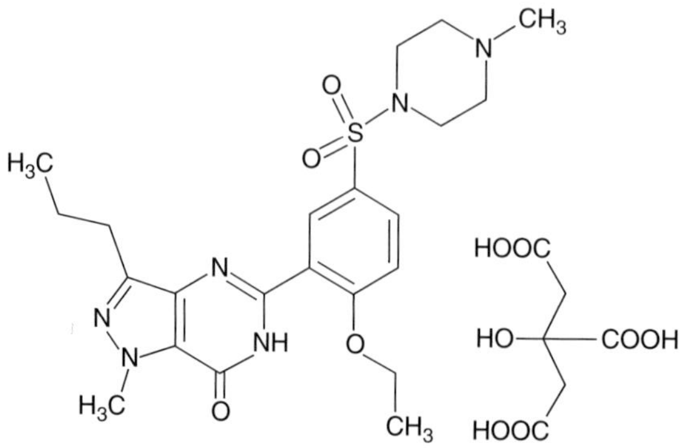

Sildenafil is a potent and selective inhibitor of type V cGMP phosphatase that was launched for the treatment of organic and/or psychological male erectile dysfunction.

1.03.14.4.2 **Tadalafil**

Trade name	Cialis
Manufacturer	Lilly/ICOS
Country of origin	USA
Year of introduction	2003
Country in which first launched	UK, Germany
CAS registry number	171596-29-5
Structure	

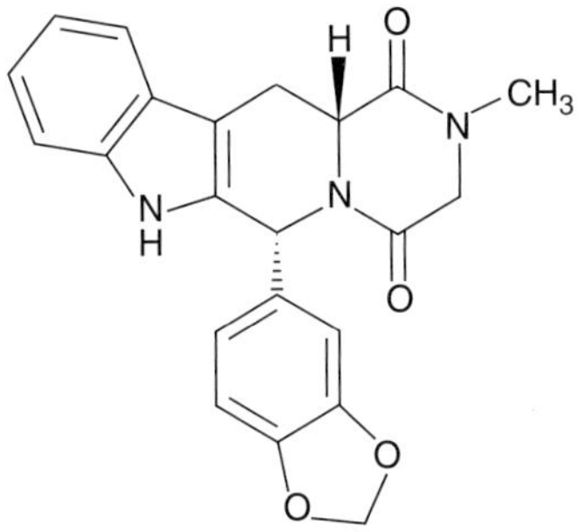

Tadalafil is a phosphodiesterase-5 (PDE5) inhibitor that was launched for the oral treatment of male erectile dysfunction.

1.03.14.4.3 **Vardenafil**

Trade name	Levitra
Manufacturer	Bayer AG, GSK
Country of origin	Germany
Year of introduction	2003
Country in which first launched	Germany
CAS registry number	224785-90-4
Structure	

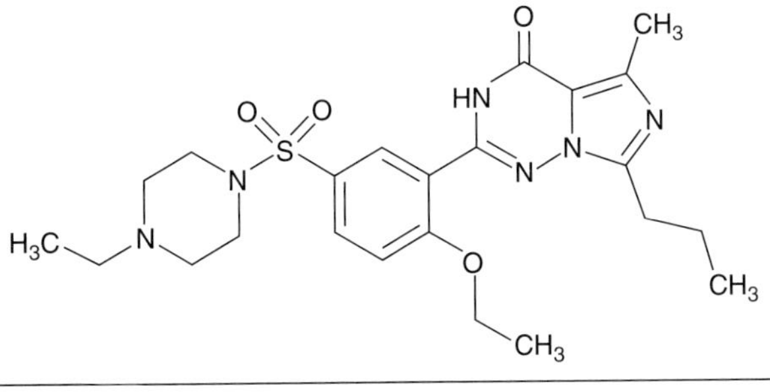

Vardenafil is another PDE5 inhibitor that was marketed for the treatment of male erectile dysfunction.

1.03.14.5 Hepatoprotectant

1.03.14.5.1 **Mivotilate**

Manufacturer	Yuhan Corp
Country of origin	South Korea
Year of introduction	1999
Country in which first launched	South Korea
CAS registry number	130112-42-4

Structure

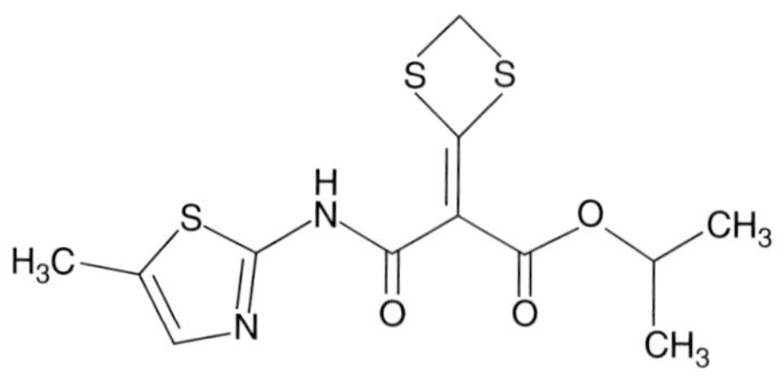

Mivotilate was launched as an orally active hepatoprotective agent for the treatment of liver cirrhosis and hepatitis B infection. The mode of action appears to involve inactivation of Kupffer cells.

1.03.14.6 Dysuria

1.03.14.6.1 Naftopidil

Trade name	Avishot, Flivas
Manufacturer	Boehringer Mannheim, Kanebo, Asahi
Country of origin	Germany
Year of introduction	1999
Country in which first launched	Japan
CAS registry number	57149-07-2
Structure	

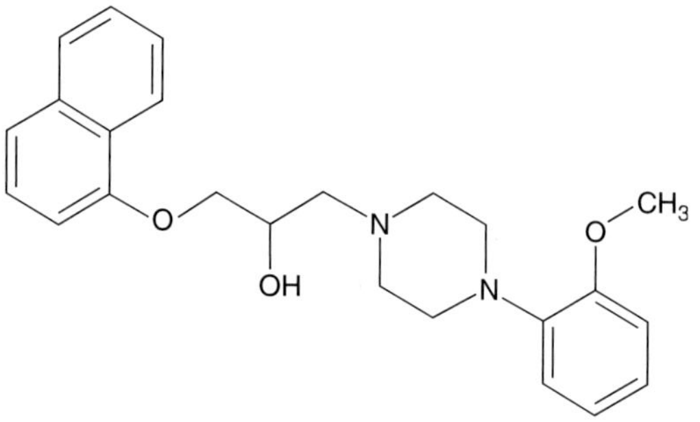

Naftopidil is a potent postsynaptic-selective alpha-1-antagonist that was launched for the treatment of dysuria associated with benign prostatic hypertrophy.

1.03.14.7 Antixerostomia

1.03.14.7.1 Cevimeline hydrochloride

Trade name	Evoxac
Manufacturer	Israel Institute for Biological Research/Snow Brand, Daiichi Pharmaceutical
Country of origin	Israel
Year of introduction	2000
Country in which first launched	USA
CAS registry number	153504-70-2
Structure	

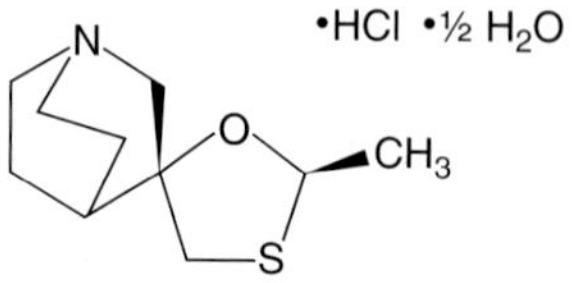

Cevimeline was originally developed as a cognition enhancer but was launched for the treatment of dry mouth symptoms in patients with Sjogren's syndrome. This action is due to stimulation of M_3 receptors in salivary and lacrimal glands.

1.03.14.8 Magnetic Resonance Imaging (MRI) Contrast Agent

1.03.14.8.1 Gadoversetamide

Trade name	OptiMARK
Manufacturer	Mallinckrodt
Country of origin	USA
Year of introduction	2000
Country in which first launched	USA
CAS registry number	131069-91-5
Structure	

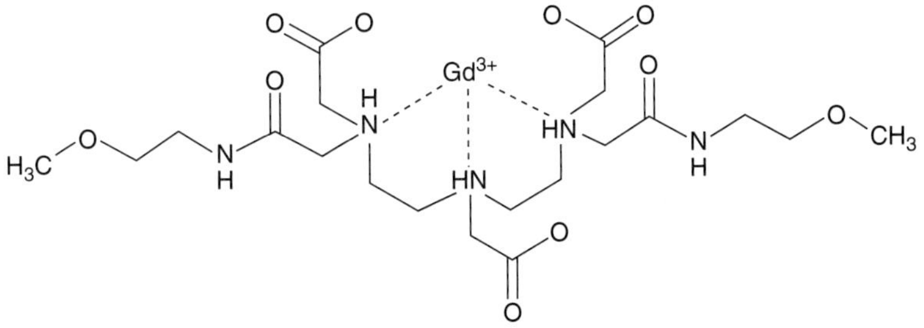

Gadoversetamide is a gadolinium(III) complex that was launched for intravenous injection prior to MRI in patients with anomalous blood–brain barrier or anomalous vascularity in the CNS or liver.

1.03.14.9 Fabry's Disease

1.03.14.9.1 Agalsidase alfa

Trade name	Replagal
Manufacturer	Transkaryotic Therapies
Country of origin	USA
Year of introduction	2001
Country in which first launched	Sweden
CAS registry number	104138-64-9
Structure – recombinant protein of molecular weight 100 kDa	

Fabry's disease is a genetic disorder of fat metabolism caused by a deficiency of the enzyme α-galactosidase A, which is involved in the biodegradation of lipids. Agalsidase alfa is a recombinant form of this enzyme that was launched as a twice-weekly intravenous infusion for the long-term treatment of Fabry's disease. Treatment for 6–12 months greatly reduces the accumulation of microvascular endothelial deposits of globotriaosylceramide in the kidneys, heart, and skin.

1.03.14.10 Antityrosinemia

1.03.14.10.1 Nitisinone

Trade name	Orfadin
Manufacturer	AstraZeneca, Rare Disease Therapeutics
Country of origin	UK

Year of introduction	2002
Country in which first launched	USA
CAS registry number	104206-65-7
Structure	

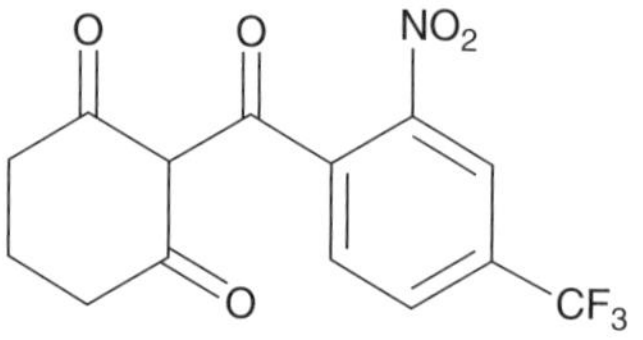

Nitisinone was introduced as an adjunct to dietary restriction of tyrosine and phenylalanine for the treatment of hereditary tyrosinemia type 1, an inborn error of metabolism. It acts as an inhibitor of the 4-hydroxyphenylpyruvate dioxygenase and prevents the formation of toxic metabolites such as succinylacetoacetate in the liver.

References

1. Craig, P. N. Drug Compendium. In *Comprehensive Medicinal Chemistry*; Hansch, C., Sammes, P. G., Taylor, J. B., Drayton, C. J., Eds.; Pergamon Press: Oxford, 1990; Vol. 6, pp 237–965.
2. Cheng, X. M. *Annu. Rep. Med. Chem.* **1995**, *30*, 295–317.
3. Cheng, X. M. *Annu. Rep. Med. Chem.* **1996**, *31*, 337–355.
4. Galatsis, P. *Annu. Rep. Med. Chem.* **1997**, *32*, 305–326.
5. Galatsis, P. *Annu. Rep. Med. Chem.* **1998**, *33*, 327–353.
6. Gaudilliere, B. *Annu. Rep. Med. Chem.* **1999**, *34*, 317–338.
7. Gaudilliere, B.; Berna, P. *Annu. Rep. Med. Chem.* **2000**, *35*, 331–356.
8. Gaudilliere, B.; Bernardelli, P.; Berna, P. *Annu. Rep. Med. Chem.* **2001**, *36*, 293–318.
9. Bernadelli, P.; Gaudilliere, B.; Vergne, F. *Annu. Rep. Med. Chem.* **2002**, *37*, 257–277.
10. Boyer-Joubert, C.; Lorthiois, E.; Moreau, F. *Annu. Rep. Med. Chem.* **2003**, *38*, 347–374.
11. Hegde, S.; Carter, J. *Annu. Rep. Med. Chem.* **2004**, *39*, 337–368.

Biography

Peter D Kennewell read Natural Sciences at Churchill College, Cambridge and then moved to the University of East Anglia to complete a PhD with Professor Alan Katritzky. This was followed by a postdoctoral fellowship with Professor Ned Heindel at Lehigh University, Bethlehem, Pennsylvania working on the synthesis of novel antimalarial agents. On his return to the UK, he joined Roussel Laboratories where he remained in a variety of managerial positions until the Swindon research unit was closed following the merger, which created Hoechst Marion Roussel. He then joined the Biotechnology and Biological Sciences Research Council working for the Biomolecular Sciences Committee and the Biochemistry and Cell Biology Committees where his particular interests were in the encouragement of the application of chemistry to biological problems. He finally retired from this organisation in June 2005.

Dr Kennewell acts as a course monitor and occasional lecturer on a number of on-line distance learning courses (Medicinal and Pharmaceutical Chemistry, Heterocyclic Chemistry) run by Lehigh University. He is author or co-author of 65 publications, 13 patents and, with John Taylor, co-authored 'Introductory Medicinal Chemistry and Modern Medicinal Chemistry.' He was a member of the Editorial Board and Editor of Volume I of 'Comprehensive Medicinal Chemistry.'

Comprehensive Medicinal Chemistry II
ISBN (set): 0-08-044513-6
ISBN (Volume 1) 0-08-044514-4; pp. 97–249

1.04 Epigenetics

W W Weber, University of Michigan, Ann Arbor, MI, USA

1.04.1 Introduction

Epigenetics is the study of heritable changes in gene expression that do not require, or do not generally involve, changes in genomic deoxyribonucleic acid (DNA) sequence. The term has been used in modern biology since at least the early 1940s, although it dates back as far as the 1800s in some reports. Initially, epigenetics referred mainly to developmental phenomena, but more recently the term has been applied more broadly to signify a relation to gene action while epigenetic inheritance signifies modulation of gene expression without modifying the DNA sequence.

Until recently, genetic inheritance has been regarded as the sole mode of transmission of information from one generation to the next. Despite the successes surrounding the unveiling of the human genome, a number of challenges remain in understanding the transmission of genetic information and gene expression.[1,2] Recently, epigenetic inheritance has been proposed as a mechanism complementing genetic inheritance to explain these phenomena. Epigenetic information is transmitted by way of direct modifications of DNA or of chromatin. In mammals, DNA methylation of cytosines is the only known physiological modification of DNA. On the other hand, numerous modifications of chromatin have been identified that affect its conformation, but they have proven much more difficult to sort out and their significance is as yet only partially understood.

Insight into the nature of chromatin in the modern era of biology stems from studies by the cytologist Heitz who proposed in 1928 that chromatin had certain genetic attributes. By following chromosomes through cell division cycles, Heitz recognized two classes of chromosomal material: euchromatin, which underwent a typical cycle of condensation and unraveling, and heterochromatin, which maintained its compactness in the nucleus. Nearly 40 years later, Spencer Brown referred to the investigation of chromatin as one of the most challenging and diffuse in modern biology in his first-rate review of the subject.[3] At that time, the repressive action of chromatin on gene action had already been recognized and the two states of chromatin were viewed as a visible guide to gene action during evolution and development. Brown believed that resolution of the properties of euchromatin and heterochromatin would eventually improve our understanding of the systems controlling gene action in higher organisms.

More recent research has greatly enhanced our knowledge of the chemical nature of epigenetic modifications and gene expression.[4–6] It is now known in affirmation of Brown's prediction that the genomes of many animals including humans are compartmentalized into either transcriptionally competent euchromatin or transcriptionally silent heterochromatin. Chromatin has been shown to be a polymeric complex that consists of histone and nonhistone proteins, and genomic DNA in all eukaryotic cells is known to be packaged in a folded, constrained, and compacted state in association with this polymer. The basic building block of chromatin is the nucleosome, which consists of approximately 146 base pairs of DNA wrapped around a histone octamer that contains two molecules each of core histones H2A, H2b, H3, and H4. These units are organized in arrays that are connected by histones of the H1 linker class. Repeating nucleosome cores are assembled into higher order structures, which are stabilized by linker DNA and histone H.[5] One of the functional consequences of chromatin packaging is to prevent access of DNA-binding transcription factors to the gene promoter. The N-termini of histones of chromatin are subject to a variety of posttranslational modifications such as acetylation, phosphorylation, and methylation, and as chromatin structure is plastic, various combinations of these modifications could lead to activation or repression of gene expression. Although it is clear that chromatin remodeling is closely linked to gene expression, the mechanism or mechanisms by which this occurs are not well understood.

Virtually every aspect of epigenetics has been examined in abundant detail, particularly within the last 5 years. A PubMed search under 'epigenetics' turned up only 346 citations of all kinds including 158 reviews; 134 of those reviews were published after 2000. It is evident that many more papers on epigenetics than responded to this search are in the literature, but it is also clear that much has been learned only recently. In assembling the timeline of epigenetic research shown in **Table 1**, an attempt was made to highlight conceptual advances and focus on key steps along the pathway to epigenetics. It is obvious from that table how the nature and rapidity of investigation has changed over time, particularly since the advent of molecular biology in the mid-1970s and the continuing development of new techniques for monitoring the biochemical aspects of epigenetic change. This chapter was written with the purpose of summarizing the origin, the foundations, and the current state of epigenetics. It brings together conceptual approaches, analytical technologies, and experimental evidence that led to the emergence of epigenetics as a field of basic biological inquiry primarily concerned with understanding the handling of genetic information by eukaryotic cells. It concludes with some perspectives on prospects of the field in health and disease.

1.04.1.1 The Origin of Epigenetics

As far back as the eighteenth century and well into the twentieth century, biologists had debated whether acquired (or adaptive) characters were heritable or not. They were often divided into two factions, the 'naturalists,' who believed they were heritable, and the 'geneticists,' who believed only in the inheritance of genetic variants that arose through natural selection. Beginning around 1910 and spanning several decades, the pioneering studies of heredity in fruit flies (*Drosophila*) initiated by Thomas Hunt Morgan and his students resulted in major conceptual and technical

Table 1 Epigenetics timeline

Year	*Event*	*Reference*
The origin and foundations of epigenetics		
1928	Euchromatin and heterochromatin are recognized as visible guides to gene action	3
1942	'Adaptively inducible characters' are transmitted from generation to generation in fruit flies	10,11
1948	Methylcytosine is identified genomic DNA	18
1949	'Sex chromatin body' is microscopically visible marker that is gender specific	12
1959	'Sex chromatin body' is tied to X chromosome inactivation	140
1960	Gametic imprinting marks functional differences in parental chromosomes	16,36
1961	A reduced frequency of CpG dinucleotides is observed in genomic DNA	17
	Studies of coat color as a genetic marker in mice show one of the two X chromosomes in inactivated in females	33
1962	Studies of G6PD as an X chromosome-linked marker show normal human females are a mosaic of X chromosome activity	15
1964	Histone acetylation is tied to gene transcription	64
1968	CpG doublet is by far the most highly methylated molecular species in mammalian DNA	19
1971	Methylcytosine in DNA is an unstable molecule that tends to mutate to thymine resulting in a 'directed mutation'	20
The effect of molecular biology on epigenetics		
Mid-1970s	Application of the principles of molecular biology to biological research begins with understanding the rules whereby cells read information encoded in DNA, and the invention of recombinant DNA techniques (such as RFLP analysis and the Southern blot)	38
1977	Salser affirms the instability of mCpG and its conversion of TpG to cause CpG deficiency	21
	n-Butyrate, as an inhibitor of DNA methyltransferase (DNMT), modifies histone acetylation in Friend erythroleukemia cells	141
1978	Restriction enzymes are used to identify tissue-specific variations in cytosine methylation of genomic DNA	39,40
	Restriction enzymes are used to describe the patterns and heritability of DNA methylation	142,143
1980	5-Azacytidine is found to perturb DNA methylation	24
1981	Treatment of cells with 5-azacytidine reactivates some X-linked genes in hamster–human hybrid cells	144
1982	Treatment of β thalassemia with 5-azacytidine yields a positive therapeutic response	133
1983	Altered DNA methylation patterns in individual genes are linked to human cancer	107
1984	Methylcytosine residues are recognized as hot spots for mutation of DNA	72
	Studies in mice show that both maternal and paternal contributions to the embryonic genome are needed for complete embryogenesis	31
1986	Neutralization of potentially damaging 'parasitic elements' (transposons, proviral sequences, etc.) is recognized as an ancestral function of DNA methylation	145
	Hypermethylation of CpG islands is associated with predisposition to human cancers	109
1987	Enzymatic DNA methylation is catalyzed by a Bi-Bi ordered mechanism	42
1988	CpG mutations within gene coding regions causing human disease are 42 times higher than random mutation predicts	74
	The first in the family of DNA methyltransferases, *Dnmt1*, is cloned, and sequenced	41
	Global hypomethylation is linked with human cancer	108
	Uniparental disomy is associated with human disease	32

continued

Table 1 Continued

Year	*Event*	*Reference*
1991	First molecular analysis of a human disease (fragile X syndrome) indicates the cause lies in faulty imprinting	85
1992	MeCP1 and MeCP2 are identified as DNA-binding proteins with affinity for methylated DNA. MeCP2 was found to be associated with chromatin	57
	DNA methyltransferase is associated with replication foci	47
	Targeted mutation of the DNA methyltransferase gene results in embryonic lethality	46
1993	Studies of many models systems indicate that regions of low and high density of DNA methylation exist that cause long-term repression of gene transcription	146
	Faulty imprinting is associated with cancer	112,113
1994	The *Igf2/MPR* gene is the first example of a primary gametic imprint shown to be methylated	34,36
	The toxic effects of 5-aza-2′-deoxycytidine, an analog of 5-azacytidine, are mediated by covalent trapping of DNA methyltransferase rather than by demethylation of DNA	25
1996	First evidence is presented in mammalian cells showing that a DNA methyltransferase is capable of de novo methylation of cellular and viral DNA in vivo	45
	Hypermethylation-associated inactivation indicates a tumor suppressor role for $p15^{INK4B1}$	147
1997	The crystal structure of the nucleosome, the basic unit of chromatin, is described	5
	Methyl-CpG-binding domain (MBD1), a component of MeCP1, was identified that binds specifically to methylated DNA sequences and represses transcription	58
	Beckwith–Wiedemann syndrome is most commonly associated with faulty imprinting of *IGF2* and *H19* genes	148
	Hypermethylation of $p15^{INK4B1}$ is associated with myelodysplasia syndromes	135
	A rapid method is devised for quantitation of methylation differences at specific sites using methylation-sensitive single nucleotide primer extension (Ms-SnuPE)	149
1998	A histone deacetylase inhibitor administered in combination with all-*trans* retinoic acid (ATRA) restores sensitivity to the antileukemic effects of ATRA in patients with refractory acute promyelocytic leukemia	129
	A family of methyl-CpG-binding proteins containing MBD2, MBD3, and MBD4 was identified and characterized. MBD3, like MBD2 and MBD4, contains a methyl-CpG-binding domain but does not bind methylated DNA in vitro or in vivo. In contrast, MBD2 and MBD4 both bind specifically to methyl-CpG in vitro and in vivo and are likely to mediate the biological consequences of the methylation signal	56
	The phenomenon of RNA interference (RNAi) was discovered	68
	DNA methyltransferases of mice (Dnmt3A and Dnmt3B) and humans (DNMT3A and DNMT3B) responsible for de novo methylation of DNA were identified, cloned, and characterized	50,53
	MeCP2 was found to exist in a complex with histone deacetylase. Studies indicated that genomic methylation patterns guide histone deacetylation to specific chromatin domains and that methylation-dependent transcriptional silencing relied on histone deacetylation. These studies were the first to link histone acetylation and cytosine methylation to silencing of gene transcription	59,60,62
1999	A human recombinant DNA methyltransferase was isolated and characterized that is capable of both de novo and maintenance methylation	43
	The synergy between DNA methylation and histone deacetylation was illustrated in the aberrant silencing of tumor suppressor genes. The study indicated that CpG island methylation in gene promoter regions was dominant over histone deacetylation in maintaining gene repression	137
	The fact that HeLa cells, which lack the known methylation-dependent repressor MeCP2, were capable of repressing transcription as determined by reporter constructs suggested that HeLa cells use an alternative pathway to silence methylated genes	64
	The autosomal recessive disorder ICF (immunodeficiency, centromere instability, and facial anomalies) is caused by mutations in DNMT3B	94
	Two reports indicate that Rett syndrome is caused by mutations in MeCP2	95,150
2000	Inactivation of the DNA repair gene O^6-methylguanine-DNA methyltransferase by promoter hypermethylation is a common event in primary human neoplasia, and a useful index of responsiveness of tumors to alkylating agents	124,125

Table 1 Continued

Year	*Event*	*Reference*
	Treatment of sickle cell anemia with 5-aza-2′-deoxycytidine resulted in a positive response	134
	A regional increase in the methylation of Lys9 in H3 by site-specific histone methyltransferases, in conjunction with a reduced acetylation of the H4 terminus, appears to be a key determinant in defining a 'histone code' for regulating higher order chromatin structure. The concept of a 'histone code' is introduced	66
2001	Distinct covalent modifications of histones, also called a 'histone code,' provide a mark on histone tails to recruit chromatin-modifing proteins, which then dictate transitions between transcriptionally active and inactive chromatin states. The combinatorial nature of the histone modifications suggests a 'histone code' that extends the information potential of the genetic code	151
	A pathway is defined in fission yeast wherein sequential histone modifications establish a 'histone code' essential for the epigenetic inheritance of heterochromatin assembly	67
2002	Deletion of part of the genetic machinery responsible for RNAi resulted in loss of histone H3 lysine-9 methylation, centromeric repeats, and impaired centromeric function. The results provide a possible link between RNAi and DNA methylation	69
	Myelodysplastic syndrome outcomes are improved by treatment with 5-azacitadine	136
2003	Evidence is presented for an epigenetic mechanism by which the chaperone protein, hsp90, acts as a capacitor for morphologic evolution. Evidence for and arguments against hsp90 to function in this capacity are discussed	11,152
	Development of microarrays of two different types for monitoring DNA methylation is described	80,81
	Epimutations in Prader–Willi and Angelman syndromes are attributed to faulty imprinting	105
2004	Hypermethylation is found to be a marker for prostate cancer	126
2005	Genomic methylation and gene silencing are associated with leukemia in mice	153,154
	The complete sequence of the X chromosome is reported, and the origin, evolution, and some other interesting features of the X chromosome are also discussed	155,156
	Patients with glioblastoma containing a methylated MGMT promoter benefited from temozolomide [an alkylating agent] whereas those without a methylated promoter had no such benefit	157,158
	An improved semiautomatic assay for monitoring DNA methylation was developed	79
	The salient features and unique identity of X chromosome inactivation are discussed	159,160
	Evidence is presented that global loss of monoacetylation and trimethylation of histones may be hallmark of human cancer	120
	A method of immunodetection is described for determining the spatial resolution of DNA methylation in metaphase chromosomes	83
	FDA approval of 5-azacytidine for treatment of myelodysplastic syndromes is discussed. This is the first such action by the FDA for an epigenetic therapeutic agent	161

breakthroughs in understanding of animal genomes,[7,8] and by the 1940s, the views of naturalists and their theories were in decline while those of geneticists were ascendant. Oswald, Avery, and McCarty had by then demonstrated that DNA and not protein was the genetic material while the double helix of DNA was established as the molecular basis of heredity during the 1950s. About the same time, protein polymorphism was shown to be a phenomenon of broad biological significance, the normal chromosome number of human cells was found to be 46, and several human pathological states were identified with aberrations in normal chromosomal number (e.g., Klinefelter's and Down's syndromes) or structure (e.g., the chimeric Philadelphia chromosome associated with chronic myelogenous leukemia). Biological research was experiencing rapid growth in new directions and undergoing redefinition as investigators asked old questions in new ways and generated new approaches that were relevant to the study of all living things including man. During this revolutionary period, coincident with and benefiting from advances in experimental biology, epigenetics originated as a series of isolated observations in three disparate areas — developmental biology, chromosomal biology, and molecular biology — and proceeded along various separate paths before their convergence in the 1980s.

With respect to developmental biology, we begin with the studies of C. H. Waddington in the 1940s.[9] Waddington had attempted to understand how the genotypes of evolving organisms responded to the environment.[10] He discovered that a *crossveinless* phenotype of *Drosophila* was induced in progeny when flies were heat shocked, and this prompted him to suggest a genetic mechanism for the apparent inheritance of acquired characters. After several generations he showed that nearly 100% of offspring of this phenotype had assimilated and expressed the *crossveinless* phenotype, even in the absence of heat shock. His studies led him to hypothesize that an adaptive response to the environment in earlier members of the evolutionary chain could eventually be assimilated and transmitted to progeny by an internal (genetic) factor during the course of evolution. By the 1960s, similar experiments had been performed in Rendel's laboratory as well as in several other laboratories in support of Waddington's hypothesis (reviewed by Ruden),[11] but the nature of the factor responsible for this phenomenon was not established.

As to the connection between epigenetics and chromosomal biology, we learn how a series of isolated discoveries made during the 1940s and 1950s began to shed light on X chromosome inactivation. In 1949, Murray Barr[12] demonstrated in somatic cells of animals that an anatomical distinction between male and females, the so-called 'sex chromatin body,' was easily visible under an ordinary microscope and could be used diagnostically to sort tissues and individuals into two groups according to gender without prior knowledge of the sex. In 1959, Ohno[13] explained this observation by showing that the X chromosomes of female cells were not alike in that one of the pair remained in an extended state during mitosis, while the other assumed a condensed state forming the Barr sex chromatin body. Shortly thereafter Lyon[14] and Beutler and colleagues[15] independently documented this observation at cellular and genetic levels using X chromosome-linked markers of coat color of mice and G6PD of red blood cells of humans, respectively, to show that X expression of these markers was a mosaic in normal females. Lyon and Beutler both suggested that through random inactivation, female cells became a mosaic consisting of cells with either an active paternal X chromosome or an active maternal X chromosome. They concluded that only one of the X chromosomes was active in each cell of females. However, the molecular basis of inactivation remained to be clarified.

While the studies on sex chromatin were in progress, another phenomenon bearing a relationship to X chromosome inactivation called parental imprinting or genomic imprinting was discovered. In her studies of the mealy bug *Sciara*, Crouse[16] first used the term 'imprinting' to describe "the chromosome which passes through the male germ line acquires an imprint that results in behaviour exactly opposite to the imprint conferred on the same chromosome by the female germ line." This was the earliest definition of a gametic imprint. After she introduced the term, occasional references to this concept appeared in the literature but nearly two decades elapsed before the nature of imprinting was better understood.

As a development of molecular studies of enzymic replication of DNA sequences ongoing in Arthur Kornberg's laboratory in the 1960s, Josse and colleagues[17] noticed that the frequency of cytosine in vertebrate genomes was much lower, about a quarter of that expected from the overall base composition. Actually, Roland Hotchkiss had reported the likely presence of methylcytosine (he called it 'epicytosine') in DNA many years earlier,[18] but several years elapsed before Grippo[19] in Scarano's laboratory pointed out that 5-methylcytosine (5mC) was the only methylated base in DNA, and that 90% of 5mC was present in CpG dinucleotide doublets in developing sea urchin embryos. In 1971, Scarano[20] suggested that 5mC was comparatively unstable, and would deaminate spontaneously to form thymine (**Figure 1**). Following on from Scarano's suggestion, Salser[21] showed in 1977 that the dinucleotide mCpG was indeed

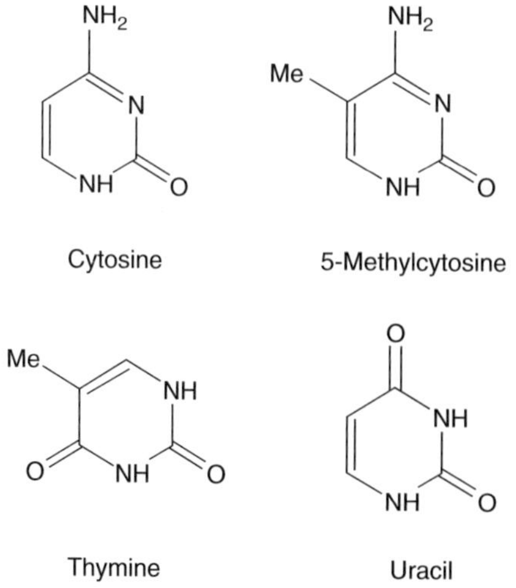

Figure 1 The chemical structures of cytosine, 5-methylcytosine, thymine, and uracil.

unstable, tending to deaminate to the dinucleotide TpG. In 1980, Adrian Bird drew attention to the fact that in the animal kingdom, DNA methylation ranged from very low levels in arthropods, through intermediate levels in many nonarthropod invertebrates, to high levels in vertebrates. He noticed that vertebrates, organisms with the most extreme CpG deficiency, also displayed the highest levels of DNA methylation. His analysis of nearest neighbor frequencies and the level of DNA methylation in animals provided further evidence for the suggestion that 5-methylcytosine tended to mutate abnormally frequently to thymine. Recent studies have established (see below) that methylation plays a major role in long-term gene silencing, but in the 1970s, the biological implications of direct modification of DNA by methylation were still unclear.

Observations regarding 5-methylcytosine raised many more questions than scientists could answer with the experimental tools available at the time. But we should recall that subsequent to the discovery of the DNA double helix and the genetic code, biologists were more inclined to think of hereditary transactions in terms of the flow of information from DNA to RNA to protein, and that the availability of molecular techniques for manipulating DNA and other nucleic acid molecules had kindled new interest in epigenetics.

1.04.2 Strengthening the Foundations of Epigenetics

1.04.2.1 Deoxyribonucleic Acid Methylation

In 1975, Riggs[22] and Holliday and Pugh[23] were attracted to the notion that DNA methylation was part of the system for controlling gene expression in mammalian cells. Even then, this idea was not entirely new because numerous authors had already suggested such a role.[22] Riggs [22] pointed out, however, that DNA methylation in eukaryotes had not been considered in light of recent accumulating evidence of changes in regulation in *Escherichia coli* involving bacterial DNA methylases, and of its advantages for bringing about permanent changes in regulation. Coincidentally, Holliday and Pugh[23] noted that the methyl groups in DNA were not randomly distributed, and because the CpG doublet occurred much less frequently than expected from the overall base composition, they believed that methylation of CpG doublets was exceptional. They concluded that a more thorough examination of this dinucleotide in a higher organism by both biochemical and genetic studies was necessary.

By the beginning of the 1980s, a general consensus had been reached on three points of epigenetic interest: that methylation of the cytosines of CpG dinucleotides was an established characteristic of genomic DNA; that deficiency of CpG doublets in genomic DNA was probably due to instability of 5-methylcytosine through its mutation to thymine; and that the distribution of CpGs in genomic DNA was not random. By the mid-1980s, additional studies had revealed a connection between DNA methylation and gene expression. In one early study, treatment of a variety of cell lines with the methylation inhibitor, 5-azacytidine, revealed that a large number of genes were reactivated,[24] although the chemical mechanism by which cytidine analogs altered at the 5 position perturbed established methylation patterns was not clear. Subsequently, cytidine analogs that had been altered at the 5 position, such as 5-azacytidine and 5-aza-2′-deoxycytidine, became important tools for studying the role of demethylation in gene expression. However, it was not realized until later that incorporation of the cytidine analog (5-aza-2′-deoxycytidine) into DNA led to covalent trapping of the DNA methyltransferase enzyme, thereby depleting the cells of enzyme activity resulting in DNA demethylation that led in turn to reactivation of the associated gene.[25]

Further comparisons across nonvertebrate and vertebrate genomic DNAs were also under way to explain the nonrandom distribution of CpGs. These studies showed that DNA methylation in nonvertebrate genomes was confined to a small fraction of nuclear DNA, and that this compartmentalization, termed low-density methylation, was also found in eukaryotes. Most DNA, some 98%, was methylated at low density in both nonvertebrate and vertebrate genomic DNA. But in the remaining 2% of genomic DNA, regions of high-density DNA methylation existed and their existence was particularly evident in vertebrate DNA. These regions rich in CpG nucleotides were identified as 'CpG islands,'[26] and more recent studies have shown they are commonly found at the promoters of genes, in exons and 3′-regions of genes.[27] But why the distribution pattern of methylated DNA cytosines in vertebrates was so strikingly different from the pattern in nonvertebrates remained an enigma.

Throughout much of the 1980s, this intriguing issue engaged the attention of many investigators who employed various model systems in attempting to assess the significance of DNA methylation. They had demonstrated that suppression of transposed elements and other 'parastic' elements – the ancestral function of DNA methylation – was thought to be the most likely function of DNA methylation. The strongest evidence in favor of this hypothesis came from studies outside the animal kingdom, namely slime molds and filamentous fungi. For example, Rothnie[28] had shown that DNA methylation of a single transposable element prevented damaging transposition events. The inference was that DNA methylation of the foreign element served to repress its function thereby preventing damaging transposition events. Others had demonstrated in different model systems that fully infectious proviruses were rendered harmless by methylation,[29] or that demethylation of DNA by 5-azacytidine reactivated quiescent proviral genomes.[30] On the basis of this information, Bird believed that invertebrate DNA methylation was more concerned with suppression of 'selfish' elements that could disrupt gene structure and function. He also thought it likely that vertebrates had not only retained the ancestral function of DNA methylation, but in addition had adapted methylation to serve as a repressor of endogenous promoters of genes.

1.04.2.2 X Inactivation and Genomic Imprinting

In 1984, McGrath and Solter[31] presented the first experimental evidence that maternal and paternal contributions to the mammalian embryonic genome were not equivalent – mouse embryos derived from purely maternal or paternal genomes failed to develop beyond implantation. By 1987, there was a growing realization that memories of the mother's and father's genes persisted throughout the development and life of the individual, residing in some form of imprinting imposed on the genetic material during gametogenesis. Interestingly, the report by Spence in 1988[32] of clinical disorders associated with uniparental disomy in humans was consistent with McGrath's observations in mice prompting investigators to explore potential molecular mechanisms of genomic imprinting. Though little was actually known about the molecular basis of parental imprinting, DNA methylation was doubtless the best candidate to explain this epigenetic modification.[33,34]

Evidence in support of the idea that stable chromatin states (modifications) occurred at a small number of chromosomal locations (reviewed in reference 35), and these chromatin states were believed to be controlled by small segments of methylated DNA.[34] Previous studies in Stöger's laboratory suggested that *Igf2r* was the only gene that had been described by classical mouse genetics as being imprinted so Stöger searched the mouse *Igf2r* locus for the presence of methylation modifications in a parental-specific manner, i.e., for modifications that could act as the imprinting signal permitting the cell's transcription machinery to distinguish genetically identical loci. *Igf2r* encoded the insulin-like growth factor receptor. Stöger and colleagues found that different levels of DNA methylation occurred between the maternal and paternal alleles. Two regions of the gene were identified. Region 1 contained the start of transcription and was methylated only on the silent paternal chromosome. Methylation of region 1 was acquired after fertilization, in contrast to methylation of region 2, which was inherited from the female gamete. The investigators proposed that methylation was implicated in expression of the *Igf2r*, and that methylation of region 2 might mark the maternal *Igf2r* locus as an imprinting signal. This was the first clear example of a primary gametic imprint shown to be methylated.[36]

1.04.3 The Effect of Molecular Biology on Epigenetics

The central dogma of molecular biology as formulated *c.* 1950 asserted that the cardinal function of gene action was synthesis of proteins. Protein synthesis proceeded according to a well-defined program of instructions encoded in the DNA, which was transcribed into RNA and subsequently translated into the primary protein sequence. Hence, the gene was deterministic of unidirectional gene expression. For more than 50 years most genomic research, including the Human Genome Project, has been guided by this model of genetic inheritance. With the maturation of this work, insights into human development, physiology, medicine, and evolution constitute a signal achievement in modern biology, but recent discoveries have revealed certain inadequacies in such a simplistic model of the gene–protein relationship. To begin with, reverse transcriptase exploded the idea of unidirectional gene expression while posttranslational protein modifications added another twist. More recently, it became apparent that some genes encoded just one protein, while other genes encoded more than one protein, and still others did not encode any protein confounding the predictive value of the genotype. The identification of previously unknown pathway components illustrated the complexity of cellular events, and recognition of the fact that gene expression could be altered at the translational, transcriptional, and posttranslational levels by a host of factors has necessitated a wider view of phenotypic expression and expansion of the basic principles of gene expression as originally formulated (**Figure 2**).

During the 1980s and 1990s, advances in molecular biology were brought into sharper focus by cloning and sequencing genes predictive of disease, expression of the proteins they encode, and fixing their chromosomal location in the human genome. Allelic variants that could only be inferred from familial inheritance patterns prior to the advent of molecular genetics could now be demonstrated by direct evidence. The polymerase chain reaction (PCR) combined with gene expression systems afforded well-defined recombinant proteins in quantities sufficient for biochemical and pharmacological characterization. Strategies to target and modify genes in a predictable manner could be used to create animal and cell models possessing knockout, overexpressed, and 'humanized' alleles in specific tissues and at specific developmental stages. In short, molecular biological approaches in all forms permeated and dominated biological research setting the stage

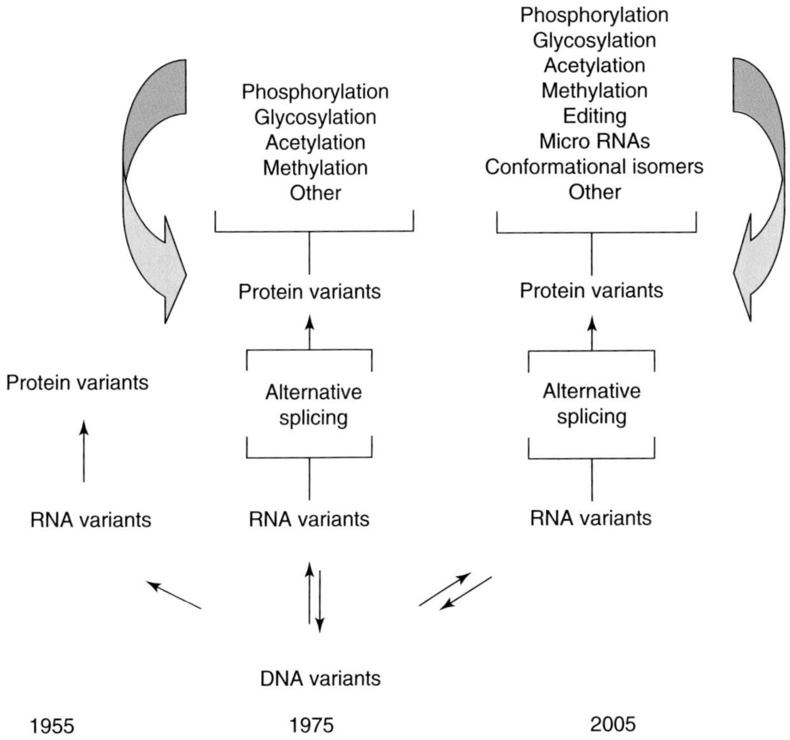

Figure 2 The expanding dogma of molecular biology. The central dogma of molecular biology that 'genes beget RNA, which in turn begets protein' was proposed in the 1950s, but advances have necessitated revisions of the original proposal as shown in this cartoon.

for the convergence of basic research and clinical medicine. These events not only solidified the foundations of epigenetics and provided novel insights into the multiplicity of factors affecting gene expression, but they also redoubled interest in human epigenetics enabling it to emerge as a bona fide discipline complementary to human genetics.

1.04.3.1 Clues to the Function of Methylated Deoxyribonucleic Acid: Restriction Enzymes

During the 1950s and 1960s, development of sensitive and specific instrumental, biochemical, and immunological technologies for the identification and quantification of methylated cytosines in DNA yielded important information about patterns of methylated DNA. However, none of these alone could define the distribution of methylated cytosines in DNA of eukaryotes or advance our understanding of its function.[37]

With the advent of molecular biology in the 1970s, assays capable of sequencing genomic DNA and of localizing genotypic differences in genomic targets evolved rapidly in response to demands in research and medicine, and restriction endonucleases were an integral component of many of these assays. Without these enzymes, many would say that recombinant technology would not have been possible. In a landmark paper published in 1975, Edwin Southern combined the specificity of restriction enzymes with gel electrophoresis to identify sequence variation in fragments of DNA.[38] He used restriction enzymes to generate predictable fragmentation of genomic DNA, and gel electrophoresis to separate and array the DNA fragments by size. Such arrays, termed 'Southern blots,' were quickly adopted by many laboratories as a reliably efficient and economical means of analyzing the variation within DNA fragments – 'restriction fragment length polymorphisms (RFLPs)' – with a variety of probes.

Restriction endonucleases are bacterial enzymes that make sequence-specific cuts in the phosphate-pentose backbone of DNA to yield 'restriction fragments' of the molecule. Typically, they recognize palindromic sequences 4–8 bases long. Several restriction enzymes include CpG in their recognition sequence, such as Hpa II (CCGG), Msp I (CCGG), Ava I (CPyCGPuG), Sal I (GTCGAC), and Sma I (CCCGGG). Interestingly, some of these enzymes (e.g., Hpa II) do not cut the DNA if the CpG sequence is methylated, while others (e.g., Msp I) cut the DNA regardless of the methylation state.[39] Investigators took advantage of this property to determine the pattern of methylation in specific regions of DNA.

In one of the first experiments of its kind, Waalwjik and Flavell[40] used Southern's technique to cut total rabbit genomic DNA with either Hpa II or Msp I followed by agarose electrophoresis, Southern blotting, and hybridization to a ^{32}P-labeled globin probe. Other investigators applied Southern's technique in experiments similar to that of Waalwjik and Flavell virtually simultaneously. The results of these experiments taken together showed that somatic DNA exhibited a definitive pattern of methylated cytosines, that methylated cytosines occurred in both strands of DNA, and that the pattern of DNA methylation was maintained through DNA replication (reviewed in Razin and Riggs).[37] This series of elegant experiments was the first to demonstrate the pattern and location of methylated cytosines at CpG residues in DNA. Subsequently, additional studies indicated that unmethylated DNA sequences generally remained unmethylated, and that methylated sequences retained their methyl moities even after 50 generations of growth and culture. They also indicated that DNA was clonally inherited, and that methylation patterns showed tissue specificity in further support of clonal inheritance.

1.04.3.2 Eukaryotic Deoxyribonucleic Acid Methylation Involves Two Dynamically Regulated Metabolic Pathways

Several DNA methylation processes are observed in cells: de novo cytosine methylation, maintenance methylation during replication of double-stranded DNA (dsDNA), active demethylation during the absence of replication, and spontaneous demethylation when maintenance methylation is suppressed. CpG sites are the primary sites of cytosine methylation in eukaryotic DNA, but methylation of other than CpG sites occurs.

In vertebrate genomes, approximately 70% of the CpG residues are methylated, the bulk of which occurs in eukaryotes during replication in the S-phase of the cell cycle. However, the regions of the genome termed 'CpG islands' were preferentially methylated while other areas were protected from methylation. Reasoning from the properties of DNA methyltransferases in bacteria, a combination of two distinct processes, de novo DNA methylation and maintenance DNA methylation, best explained the pattern of genomic methylated sites found in adult eukaryotic tissues. De novo methylation referred to the enzymatic transfer of a methyl group to CpG dinucleotides that were devoid of methyl moities and occurred mainly in the early embryo. The embryonic pattern of methylation was maintained by maintenance methylation, which referred to the enzymatic transfer of a methyl group to an unmethylated cytosine paired with a methylated cytosine, i.e., a CpG in which only one strand of DNA was methylated, sometimes referred to as 'hemimethylated CpG.' Thus, maintenance methylation converted the hemimethylated duplex into a symmetrically methylated form. At the next round of replication when a symmetrically methylated CpG duplex underwent semiconservative replication, hemimethylated sites were formed (**Figure 3**). Hence, the pattern of methylation in the parent nucleus was transmitted to the daughter nucleus by only one strand of the DNA double helix. These hemimethylated sites were rapidly converted to symmetrically methylated forms by maintenance methylation, ensuring faithful transmission of the methylation pattern from generation to generation.

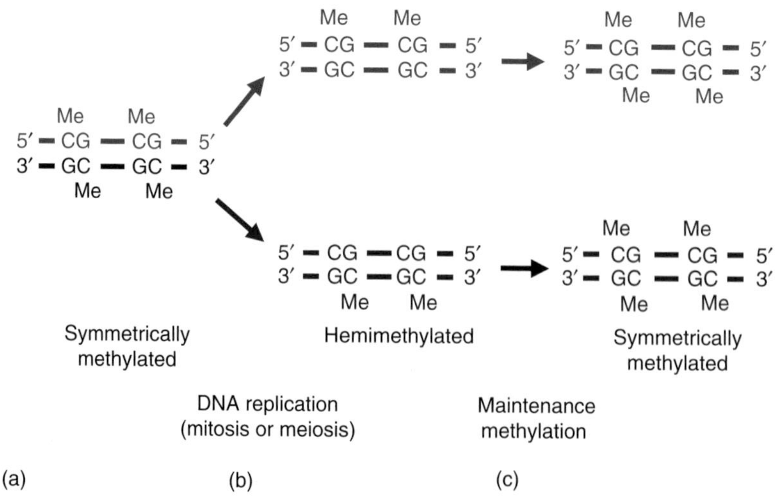

Figure 3 DNA maintenance methylation in eukaryotes. The chief function of DNA maintenance methylation in eukaryotes is the faithful transmission of the pattern of methylation from generation to generation. (a) Replication (at mitosis or meiosis) of symmetrically methylated DNA results in (b) hemimethylated DNA followed by (c) maintenance methylation catalyzed by DNA methyltransferases, methylate unmethylated cytosines in CG or CXG motifs to restore the original pattern of symmetrically methylated DNA.

1.04.3.3 Deoxyribonucleic Acid Methylases

Enzymatic methylation of the C^5-carbon position of cytosine residues in a DNA strand yields 5-methyl-2′-deoxycytidine monophosphate. Enzymes catalyzing this reaction belong to the family methyltransferases (EC 2.1.1, MTs).

Eukaryotic DNA methyltransferase (DNMT) was first cloned and sequenced by Bestor and Ingram in 1988.[41] They isolated well-resolved peptides from homogeneous DNA methyltransferase purified from mouse erythroleukemia cells, determined their amino acid sequences by Edman degradation, and used these sequences to design and synthesize a 19mer oligonucleotide hybridization probe. Screening of λgt11 complementary DNA (cDNA) libraries prepared from mouse cells with this probe revealed a predicted nucleotide sequence encoding a polypeptide of 1573 amino acid residues, which they named *Dnmt1*. The murine erythroleukemia cells used as a source of Dnmt1 actually contained three very similar species of the enzyme but their precise relationship was unclear.

1.04.3.3.1 The Dnmt1 family

DNA methyltransferase is a comparatively large molecule of approximately ~190 kDa containing 1620 amino acids. Enzymatically catalyzed DNA methylation is a covalent modification of DNA in which a methyl group is transferred from *S*-adenosylmethionine to the C-5 position of cytosine. Early studies[42] with a prokaryotic methyltransferase (HhaI) first showed that the methyl transfer reaction proceeds by an ordered Bi-Bi kinetic mechanism involving the transient formation of a covalent adduct between the enzyme and the methyl donor, *S*-adenosylmethionine. After transfer of the methyl group to cytosine of DNA the demethylated donor molecule dissociates followed by release of methylated DNA. All DNA methyltransferases that have been studied appear to follow a similar reaction mechanism.

More recently, expression, purification, and characterization of full-length recombinant human DNMT1[43] showed that the enzyme prefers hemimethylated DNA compared to unmethylated DNA as a substrate. Under optimal conditions, the preference for methylated DNA averaged about 15-fold greater than that for unmethylated DNA. DNMT1 was capable of both de novo and maintenance methylation at CG sites, and could also maintain methylation of some non-CG sites.

The C-terminal represents the catalytic domain, while the N-terminal has several functions including a targeting sequence that directs it to replication foci[44] (**Figure 4**). Targeting of the DNA methyltransferase to replication foci is believed to allow for copying of methylation patterns from parent to newly synthesized DNA of offspring. Functionally,[43] Dnmt1 was found to exhibit a 5–30-fold preference for hemimethylated DNA over completely unmethylated DNA. To

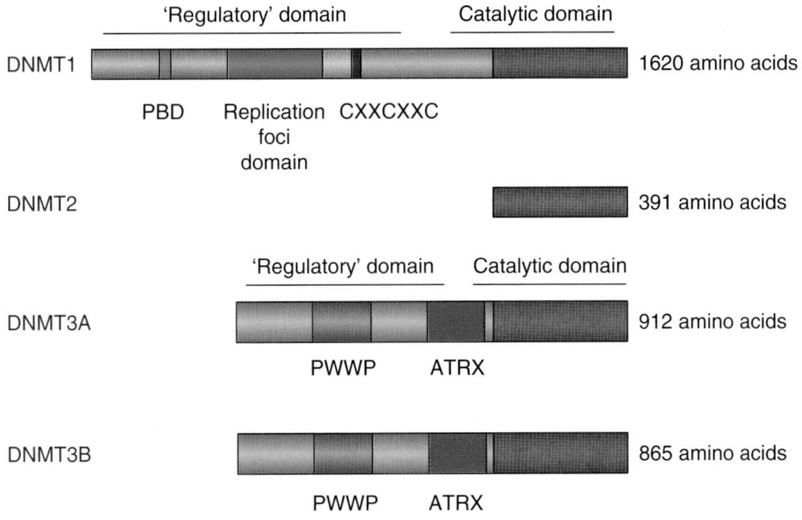

Figure 4 Human DNA methyltransferase (DNMTs) proteins. All known DNMTs share a highly conserved C-terminal catalytic domain. The regulatory domain located at the N-terminus of DNMT1 differs from those of DNMT3A and DNMT3B as shown in the cartoon: a proliferating cell nuclear antigen (PCNA) binding domain (PBD), a replication foci targeting domain (RTFD), a CXXCXXC domain implicated in DNA binding, a PWWP domain linked to protein targeting, and an α-thalassemia X-linked mental retardation syndrome (ATRX) domain implicated in histone deacetylase interactions. Human DNMT3A and DNMT3B are highly homologous and are probably products of gene duplication. Functionally, DNMT1 shows a strong preference for hemimethylated DNA (maintenance methylation) whereas DNMT3A and DNMT3B show equal activity for unmethylated and hemimethylated DNA (de novo methylation). Adapted from [44,53,63,75].

determine whether de novo methylation and maintenance methylation were performed by the same or different enzymes, a null mutation of this enzyme was generated in mice.[45,46] The null mutant was viable but retained some capacity ($\sim$1/3 of the wild-type) for de novo methyltransferase activity, suggesting the presence of one or more other DNA methyltransferase(s) in *Dnmt1* knockout cells.[46] The *Dnmt1* knockout mutation was also found to cause early embryonic lethality indicating that DNA methylation was crucial for normal mammalian development.[47,45] In addition, disruption of *Dnmt1* resulted in abnormal imprinting, and derepression of endogenous retroviruses (summarized in reference 48). Other targeted mutations in *Dnmt1* produced a number of additional unique phenotypes. The human homolog, *DNMT1*, mapped to chromosome 19p13.2–13.3.[49]

1.04.3.3.2 The Dnmt2 family

For a decade following its cloning and sequencing, *Dnmt1* was the only DNA methyltransferase identified in mammals. Several other groups sought new candidate DNA methyltransferases in mammals by searching expressed sequence tag (EST) databases. In 1998, Yoder and Bestor[50] reported another potential DNA methyltransferase (*pmt1*$^+$) in fission yeast, an organism not known to methylate its DNA. Disruption of *Dnmt2*, the mouse homolog of the yeast enzyme, had no discernible effect on methylation patterns of embryonic stem cells, nor did it affect the ability of such cells to methylate newly integrated retroviral DNA.[44]

About the same time, Van den Wygaert and colleagues reported the identification and characterization of the human *DNMT2* gene.[51] Sequence analysis indicated that *DNMT2* encoded 391 amino acid residues, but lacked a large part of the N-terminus, which is usually involved in the targeting and regulation of the MTAses. The protein overexpressed in bacteria did not show any DNA methyltransferase activity. Fluorescence in situ hybridization (FISH) mapping showed the *DNMT2* gene was located on human chromosome 10p13–22. Tissue-specific expression revealed the human enzyme was relatively high in placenta, thymus, and testis. However, *DNMT2* was overexpressed in several cancer cell lines consistent with the role of DNA methylation in cancer.

A more recent report has identified and analyzed the human homolog, *DNMT2*.[52] The purified enzyme had weak DNA methyltransferase activity at CG sites. Limited data indicated DNMT2 recognized CG sites in a palindromic TTCCGGAA sequence context (**Figure 4**).

1.04.3.3.3 The Dnmt3 family

A search by another group[53] for expressed sequence tags using full-length bacterial methyltransferase sequences as queries identified two homologous methyltransferase motifs in both human and mouse EST databases. The mouse genes were named Dnmt3a and Dnmt3b because they showed little sequence similarity to either Dnmt1 or Dnmt2. Mouse Dnmt3A and Dnmt3B cDNAs encode proteins of 908 and 859 amino acids, respectively. Mouse Dnmt3B also encodes two shorter polypeptides (840 and 777 amino acid residues) through alternative splicing.

Dnmt3a and Dnmt3b were both expressed abundantly in undifferentiated cells, but at low levels in differentiated embryonic stem cells and adult mouse tissues, and both showed equal activity toward hemimethylated and unmethylated DNA. The expression pattern plus the substrate selectivity suggested that Dnmt3a and Dnmt3b might encode de novo methyltransferases.[53]

The human homologs 3DNMT3A and DNMT3B were highly homologous to the mouse genes (**Figure 4**). They mapped to human chromosomes 2p and 20q, and encode proteins of 912 and 865 amino acids, respectively.[44] A human disorder has been attributed to mutations in DNMT3B (*see* Section 1.04.4.2.1).

1.04.3.4 Deoxyribonucleic Acid Methylation Silences Gene Expression, but How?

By 1980, analyses of tissue-specific genes and transfection studies strongly suggested that methylated DNA somehow affected the formation of chromatin and led to long-term silencing of gene activity during mammalian development. Structural studies had shown that the conversion of cytosine to 5mC places a methyl group in an exposed position in the major groove of the DNA helix. In this position, the methyl groups do not affect base pairing and hence do not impede DNA replication, but several studies had demonstrated that changes in the major groove affected binding of DNA to proteins. Hence, Riggs suggested that modification of protein–DNA interactions was an essential function of methylated cytosine.[22,37]

1.04.3.4.1 Methyl-cytosine guanine dinucleotide (CpG)-binding proteins

In 1991, Bird proposed two models[54] to explain how CpG methylation might cause transcriptional repression. Put simply, the 'direct' model postulated that essential transcription factors saw 5-methylcytosine as a mutation in their binding site and so were unable to bind, while the 'indirect' model postulated that methylated DNA sites were capable

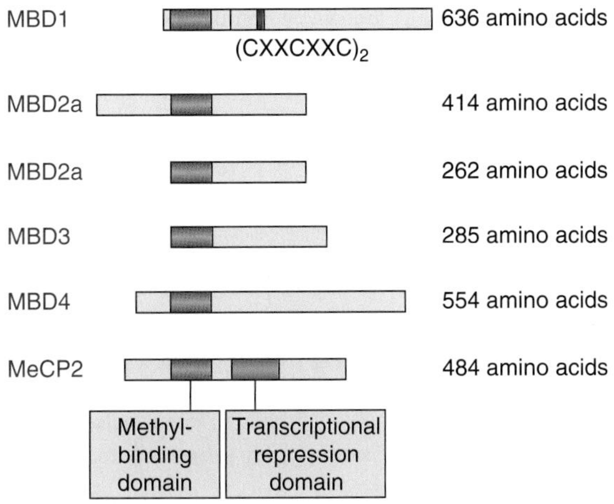

Figure 5 Murine methyl-CpG-binding (MBD) proteins. Comparison of the structures of methyl-CpG-binding domain (MBD) proteins. Two forms of MBD2 are shown corresponding to initiation of translation at either the first (MBD2a) or second (MBD2b) methionine codons. MeCP2 contains both an MBD and a transcriptional repression binding domain (TRD). Adapted from [63].

of binding nuclear protein species that blocked the binding of transcriptional factors. Subsequently, two proteins were identified, methyl-CpG-binding protein 1 (MeCP1) and MeCP2, that bound specifically to DNA-containing methyl-CpG (MBD1) sites.[54,55] Both proteins were widely expressed in mammalian cells, but MeCP2 was more abundant and more tightly bound in the nucleus than MeCP1, and most importantly was associated with chromatin. Subsequently, MeCP2 was shown to contain both a methyl-CpG-binding domain (MBD) and a transcriptional repression domain (TRD).[56] Bird and colleagues proposed that MeCP2 could bind methylated DNA in the context of chromatin and they suspected this protein somehow contributed to the long-term silencing of gene activity.[57] While this evidence supported inhibition of transcription by the indirect mechanism, it did not clarify the mechanism by which cytosine methylation affected the structure of chromatin.

Searches of the EST database with the MBD sequence as the query enabled Cross *et al.* to identify the MBD sequence, named MBD1, as a component of MeCP1.[58] Subsequently, Hendrich and Bird identified three new human and mouse proteins (MBD2, MBD3, and MBD4) that contain the methyl-CpG-binding domain[56] (**Figure 5**). MBD1, MBD2, and MBD4 were all shown to bind methylated DNA via its MBD region and to repress transcription. The precise significance of MBD3 in mammals was unclear.

1.04.3.4.2 Methyl-cytosine guanine dinucleotide-binding proteins, histone deacetylation, and chromatin remodeling

In 1998, Nan[59] and Jones[60] reported independently that MeCP2 resided in a complex with several histone deacetylases (HDACs). The complex also contained Sin3A, a corepressor in other deacetylation-dependent silencing processes, plus several additional unidentified proteins. Both laboratories demonstrated that transcriptional silencing could be reversed by trichostatin A, a specific inhibitor of histone deacetylases. Additionally, both obtained evidence that histone deacetylation was guided to specific chromatin domains by genomic methylation patterns, and that transcriptional silencing relied on histone deacetylation. Earlier contributions of Vincent Allfrey and of Vernon Ingram, whose laboratories were both engaged in studying biochemical events relevant to gene regulation, should also be acknowledged. These investigators had proposed several years earlier that acetylation and methylation of histones were linked to gene regulation[61] and that inhibition of histone deacetylation could alter cellular differentiation.[62]

In commenting on the studies of Nan[59] and Jones,[60] Bestor[62] suggested that deacetylation might exert two structural effects on chromatin that favor gene silencing. First, deacetylation of lysine ε-amino groups might permit greater ionic interactions between the positively charged N-terminal histone tails and the negatively charged phosphate backbone of DNA, which would interfere with the binding of transcription factors to their specific DNA sequences. Second, deacetylation might favor interactions between adjacent nucleosomes and lead to compaction of the chromatin. The higher affinity of deacetylated histone tails for DNA favored the first explanation while the crystal structure of the nucleosome favored the second. Whether the effects of histone acetylation acted at intra- or internucleosome levels was still unclear.[62] Nevertheless, the results of Nan *et al.*[59] and Jones *et al.*[60] clearly established a direct causal relationship between DNA methylation-transcriptional silencing and modification of chromatin.

The events of the 1990s can be summed up as follows. At the beginning of this decade, evidence from studies on many model systems had demonstrated that most regions of invertebrate genomes were free of methylation. In contrast, all regions of the vertebrate genome were subject to methylation primarily at sites rich in CpG dinucleotides. However, the methylated sites were distributed unevenly throughout the genome in patches of low and high density. At the beginning of this period, Dnmt1 was the only identified mammalian DNA methyltransferase and no methyl-CpG-binding protein was known. By the close of this decade, the mechanism of the methylation reaction had been established and four mammalian Dmnts and five methyl-CpG-binding proteins had been identified[63] (**Figures 4** and **5**).

With these new findings in hand, questions regarding the function of DNA methylation and the relation of histone acetylation and DNA methylation to chromatin remodeling and to repression of gene activity were beginning to yield to experimental scrutiny. Acetylation of conserved lysines on the N-terminals of the core histones was shown to be an important mechanism by which chromatin structure is altered. Histone acetylation was associated with an open chromatin conformation allowing for gene transcription while histone deacetylation maintained the chromatin in the closed, nontranscribed state. Aided by the tools of molecular biology, investigators had learned how CpG dinucleotides were targeted for methylation, how the patterns of methylation were read, maintained, and in most cases faithfully transmitted from one generation to the next.

1.04.3.4.3 Histone modifications in chromatin

Shortly after Nan *et al.*[59] and Jones *et al.*[60] reported the role of DNA methylation and modification of chromatin in transcriptional silencing, Ng[64] found that cells deficient in MeCP2 (e.g., HeLa cells) were capable of repressing transcription as determined by reporter constructs. Thus, MeCP2 was probably not the sole connection between the methylation of DNA and transcriptional silencing.

In 2001, Eric Selker and Hisashi Tamaru[65] reported that *dim-5*, a gene that encodes a histone methyltransferase for methylation of lysines of histone tails of chromatin in the fungus *Neurosparra crassa* was required for DNA methylation. They had accidentally generated a mutation in a previously unknown gene required for DNA methylation. In a series of elegant experiments, they mapped the fungal mutant gene to a region homologous to histone methyltransferases and demonstrated through biochemical tests on a recombinant form of the gene that the protein methylated histone H3. In characterizing the mutant gene they found a single nucleotide (C to G) change that generated a stop codon in the middle of a distinctive ~130 amino acid sequence motif called the SET domain,[66] which identified the gene (in *Drosophila*) required for heterochromatin formation, and possibly was one of the chromatin-associated SET methyltransferases. The implication that histone methylation controlled DNA methylation was inferred by demonstrating that substitution of lysine at position 9 with either leucine or arginine in histone H3 caused the loss of DNA methylation in vivo.

Nakayama and colleagues[67] provided additional evidence that lysine 9 of histone 3 (H3 Lys^9) was preferentially methylated at heterochromatin regions of fission yeast (*Schizosaccharomyces pombe*) and that modifications of histone tails were linked to heterochromatin assembly. They proposed that histone deacetylases and histone methyltransferases cooperate to establish a 'histone code' that would result in self-propagating heterochromatin assembly. On the basis of conservation of certain transacting proteins that affect silencing (Clr4/SUV39H1 and Swi6/HP1) and the presence of H3 Lys^9-methyl modification in higher eukaryotes, they also predicted that a similar mechanism might be responsible for higher order chromatin assembly in humans as well as in yeast (**Figure 6**). However, since the evidence obtained for this 'double methylation' system was limited to fungus (*N. crassa*) and fission yeast (*S. pombe*), the extent to which it applies to other eukaryotes, including humans, awaits confirmation.

1.04.3.4.4 Ribonucleic acid (RNA) interference regulates histone methylation in eukaryotic heterochromatin

While the experiments in Selker's laboratory (*see* Section 1.04.3.4.3) were in progress, Fire and colleagues[68] discovered that dsRNA was much more potent than sense or antisense single strands of RNA at silencing gene expression in *Caenorhabditis elegans*. They found that only a few molecules of dsRNA were required to cause silencing. They suggested that a catalytic or amplification step might be involved, and named this phenomenon RNA interference (RNAi).

In a study using the fission yeast (*S. pombe*) as a model, Shiv Grewal and Robert Martiennssen and colleagues[69] obtained evidence suggesting that histone modifications were guided by RNAi by deleting several genes (argonaut, dicer) and for AAN RNA-dependent RNA polymerase that encode parts of the molecular machinery of RNAi. Deletion resulted in aberrant accumulation of complementary transcripts from centromeric heterochromatic repeats, as well as transcriptional derepression of transgenes located at the centromere, loss of histone H3 Lys^9 methylation, and loss of centromere function. For intact cells, they explained their findings by the following mechanism: dsRNA derived from repeat

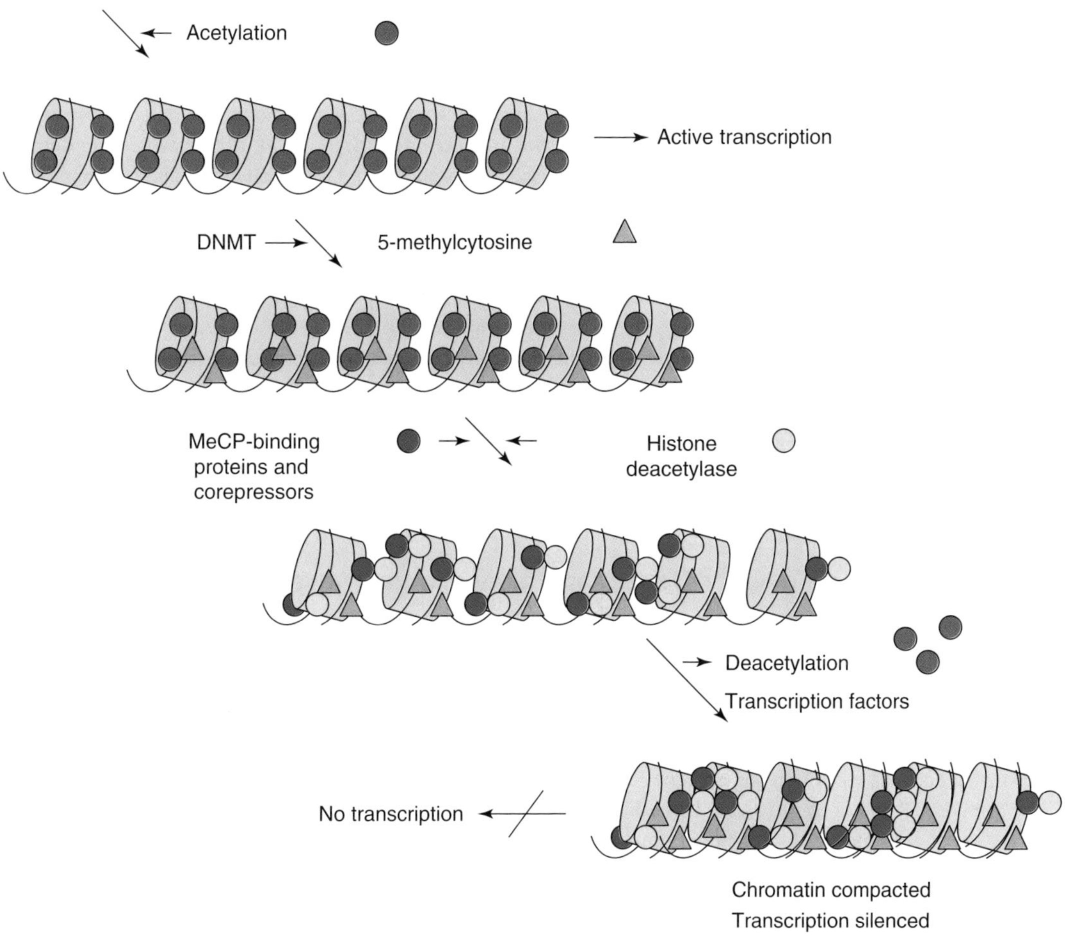

Figure 6 DNA methylation and histone deacetylation cooperate to repress transcription. Consider a region of DNA that is actively transcribing (i.e., the histones are acetylated) chosen for silencing. First, methylation of DNA catalyzed by DNMTs is acquired; then the methylated DNA recruits methyl-CpG-binding proteins, their associated corepressors, and histone deacetylases. Histone deacetylation permits the nucleosomes to become more compact. Compaction of the nucleosomes excludes access to transcription factors and gene transcription is silenced.

sequences in heterochromatin would trigger RNAi, which would initiate H3 Lys^9 methylation. The covalently modified histone would then signal DNA methylation. This mechanism could guide eukaryotic methyltransferases to specific regions of the genome, such as retroposons and other parasitic elements. They believed this arrangement could be reinforced by maintenance methyltransferase activity as well as by histone deacetylation guided by CpG-methyl-binding proteins.

RNAi is still a comparatively new model in regulatory biology,[70] and the mechanistic complexity of the process and its biological ramifications are only just beginning to be appreciated. The technique has been harnessed for the analysis of gene function in several diverse organisms and systems including plants, fungi, and metazoans but its use in mammalian systems has lagged behind somewhat.[70] The first indication that RNAi could induce gene silencing in mammals came from observations in early mouse embryos and numerous mammalian cell lines, but silencing in these systems was transient. By utilizing long, hairpin dsRNAs, Paddison and colleagues[71] have recently succeeded in creating stable gene silencing in mouse cell lines substantially increasing the power of RNAi as a genetic tool. The ability to create permanent cell lines with stable 'knock-down' phenotypes extends the utility of RNAi in several ways, one of which will be its application to research into epigenetics.

1.04.4 Epigenetic Perspectives on Health and Disease

Nearly three decades have elapsed since Salser observed[21] that 5-methylcytosine was responsible for mutation of CpG dinucleotides. This event was closely linked in time with the advent of molecular biology and the invention of recombinant DNA techniques. The simplicity and elegance of those technologies rapidly facilitated their adoption in many laboratories. New techniques were devised for characterizing the natural patterns of genomic methylation in DNA sequences, and for unraveling the mechanisms that initiate and maintain epigenetic silencing. In the course of screening for RFLPs in human DNA sequences, Barker and his associates were among the first to recognize the importance of polymorphisms at CpG islands. These investigators found that nine of 31 arbitrary loci were detectably polymorphic at CpG sites and estimated the likelihood that polymorphisms in CpG sequences might be some 10-fold higher than for other bases.[72] Global hypomethylation of various human cancers had just been discovered by Feinberg and Vogelstein as a feature common to many cancers of different types, and hypermethylation of CpG islands in various lymphomas and lung cancers reported subsequently by Baylin and associates was another feature common to various lymphomas and lung cancers.[73] Cooper and Yousoufian then found that CpG mutations within gene coding regions causing human disease were some 42-fold higher than random mutation predicted and they proposed methylation of DNA contributed to human disease.[74] As a consequence, the study of epigenetics and its health consequences held an attraction for investigators of human disease, particularly those with specialized interests in developmental abnormalities and cancer. They wanted to learn whether the natural patterns of DNA methylation were disturbed in human neoplasia and other human disorders, and why. During the last 5–10 years innovative approaches and technical refinements have greatly enhanced our understanding of epigenetic silencing as well as the diagnosis, characterization, and treatment of numerous human epigenetic disorders.[75–77]

1.04.4.1 Analytical Tools for Dissecting Epigenetic Pathways

Detailed studies have identified many of the components and basic principles of gene expression. Clearly, stable but reversible alteration of gene expression is mediated by patterns of cytosine methylation and histone modification, the binding of nuclear proteins to chromatin, and interactions between these networks. The positioning and occupancy of nucleosomes are likely to be important factors in gene expression because these structures may modulate the binding of transcription factors as well as the movements of transcribing RNA polymerases, and there may be other, as yet unknown, factors that contribute to epigenetic mechanisms of gene regulation. Over the years, the availability of incisive analytical tools has played and will continue to play a crucial role in dissecting these patterns and networks and advancing our understanding of their architecture and function. Some of the methods that have provided these insights are described in this section.

Analytical techniques for monitoring the methylation state of DNA may be divided into two groups: those designed to determine the level of global methylation of studied genomes, and those designed to determine regional patterns of methylation, mostly CpG islands, of studied genomes. Established methods for measuring the methylation status include routine chromatographic methods, electromigration methods, and immunoassays, and modifications of these older techniques continue to evolve.[78] Semiautomatic detection of methylation at CpG islands,[79] oligonucleotide-based microarrays, and tissue microarrays[80–82] are some of the newer methods that have been reported. In addition, immunodetection offers the possibility of obtaining spatially resolved information on the distribution of 5-methylcytosine on metaphase chromosomes.[83]

The choice of a proper method for a particular investigation rests with the aims and the specific requirements of the investigation. The important features of several widely used methods that employ 5-methylcytosine as a marker for DNA methylation are summarized below and in **Table 2**. For details about protocols, the reader is referred to the recent review of Havlis and Trbusek[78] as well as additional references cited in **Table 2**.

1.04.4.1.1 Chromatographic methods

This group includes thin-layer chromatography (TLC), high-performance liquid chromatography (HPLC), and affinity chromatography. Large-scale screening is the chief advantage of TLC; other advantages of TLC are its simple instrumentation, low cost, and speed. HPLC is the most commonly used chromatographic technique for global methylation analysis. It is sensitive, quantitative, highly reproducible, and can use fluorescent detection, but separations are slow. Affinity chromatography is an important component of a new technique for the identification of CpG islands exhibiting altered methylation patterns (ICEAMP),[84] a method that allows identification of methylation changes without the necessity of knowing the target sequence region and has no need for modification-sensitive restriction enzymes.

1.04.4.1.2 Slab gel electrophoresis

Slab gel analysis is a favorite separation technique of long standing used with modification-sensitive restriction enzymes (MSREs) sensitive to 5-methylcytosine for RFLP/Southern blot analyses or with specific sequencing protocols like

Table 2 Tools for monitoring DNA methylation in epigenetic disease

Method	*Detection*	*DNA amount needed*	*Sensitivity*	*Comments*	*Selected references*[a]
High-performance thin layer chromatography	Scintillation	5 μg	20 fmol	Simple, low cost, rapid, good for large scale screening	162
High-performance liquid chromatography	Optical–UV Scintillation Fluorescence MS	<1 μg	400 fmol	Quantitative, reproducible, sensitive	163
Capillary electrophoresis	MS	<1 μg	100 fmol	Automation possible, high sample throughput	164
Immunoassay	Fluorescence	NA	1.5 fmol	Spatial resolution on metaphase chromosomes previously stained by the Giemsa method	83
Modification-sensitive restriction enzymes (MSRE)	Gel electrophoresis, Southern blot	>5 μg	NA	Methylation site specific	85,165
Bisulfite sequencing	Gel electrophoresis	10 ng	2.5 fmol	Sensitive, easy, best for analysis of different sequences in a small number of samples	166,167
Bisulfite sequencing plus chloroaldehyde	Fluorescence	10 ng	175 fmol	Slow and chloroaldehyde is toxic, does not require extensive purification of DNA	168
Combined bisulfite restriction analysis [COBRA]	Gel electrophoresis	1 μg	125 fmol	Rapid, sensitive, quantitative and can be used with paraffin sections	86
Methylation-sensitive single nucleotide primer extension (MS-SnuPE)	Gel electrophoresis	5ng	500 fmol	Avoids MSRE and is automatable. Target sequence should contain only A, C, and T, while primer should contain only A, G, and T	149

[a] For a comprehensive list of references, see [78].

bisulfite and hydrazine/permanganate sequencing. It has often been replaced by tests that are more sensitive, less labor intensive, and automatable. In recent years, bisulfite sequencing has been used as the standard method to detect DNA methylation at CpG islands. Genomic DNA first reacted with bisulfite converts unmethylated cytosine to uracil while leaving 5-methylcytosine unchanged (**Figure 1**). The conversion to uracil is detected with specifically designed PCR primers. Methylation-specific restriction enzymes combined with PCR is a very sensitive, commonly used technique that requires much less DNA than traditional Southern blot analysis. A widely used combination of MSREs for detection of methylated cytosines is MspI/HpaII isoschizomers, although one or both of these have been replaced by other restriction enzymes. For instance, detection of the methylation state of the *FMR1* gene promoter used analysis of *EcoR1* and *Eag1* digests of DNA from fragile X patients to distinguish the normal genotype, the premutation, and the full mutation.[85] Bisulfite sequencing is straightforward and efficient and in the event that a site is only partially methylated, it has the added advantage of enabling determination of the proportion of cells that are methylated. Bisulfite sequencing has been combined with other methodologies such as 'combined bisulfite restriction analysis and amplification' (COBRA).[86] COBRA is easy to use, quantitatively accurate, and compatible with paraffin sections. Another method using bisulfite sequencing is 'methylation-sensitive single nucleotide primer extension' (MS-SNuPE).

This approach uses very small amounts of DNA, can be used with microdissected material, and avoids the use of restriction enzymes.

The MS-SNuPE technique has also been adapted to semiautomatic detection of DNA methylation at CpG islands.[79]

1.04.4.1.3 Immunological methods

Pfarr *et al.* have recently described a technique that renders possible immunodetection of 5-methylcytosine on human chromosome metaphase spreads.[83] A monoclonal antibody tagged with fluorescent dye is used to image the chromosomes after they have been previously stained by the Giemsa method. The technique can be used to obtain a fast and global overview of changes in genomic methylation patterns during the development of various types of cancer. Potential applications of the technique include characterization of disease-related methylation patterns on a genomic scale for diagnostic purposes and monitoring of patient responsiveness to therapy.

1.04.4.1.4 Microarrays

Many genes and signaling pathways controlling cell proliferation, death, and differentiation, as well as genomic integrity are implicated in the development of cancer and other diseases. New techniques that can measure the expression of thousands of genes simultaneously are needed to establish the diagnostic, prognostic, and therapeutic importance of each emergent cancer gene candidate. The development of microarrays for DNA methylation analysis may afford a potential solution to this problem.

Several microarray methods have been developed to map methylcytosine patterns in genomes of interest.[87] One set of methods uses methylation-sensitive restriction enzymes. For example, Shi and colleagues have devised an oligonucleotide-based microarray technique that measures hypermethylation in defined regions of the genome.[80] DNA samples from various tumor types were first treated with bisulfite and specific genomic regions are then PCR-amplified converting the modified UG to TG and conserving the originally methylated dinucleotide as CG. Primers were designed so that they contained no CpG dinucleotides and were complementary to the flanking sequences of a 200–300 bp DNA target. This permitted unbiased amplification of both methylated and unmethylated alleles by PCR. PCR amplified target DNAs were subsequently purified and labeled with fluorescent dyes (Cy5 or Cy3) for hybridization to the microarray. The fluorescently labeled PCR products were hybridized to arrayed oligonucleotides (affixed to solid supports, e.g., microscope slides) that could discriminate between methylated and unmethylated alleles in regions of interest. Shi *et al.* employed their technique to distinguish two clinical subtypes of non-Hodgkin's lymphomas, mantle cell lymphoma, and grades I/II follicular lymphoma based on the differential methylation profiles of several gene promoters.

This oligonucleotide technique for profiling methylation patterns[80] could also afford an alternative approach to predict and discover new classes of cancer in a manner similar to that pioneered by Golub[88,89] and other investigators.[90–93] Additionally, this technology might be used to monitor the effects of new pharmacological agents in patients under treatment and provide valuable information regarding the outcome of epigenetic therapies.

Chen and associates have developed another, somewhat different, microarray technique for analysis of hypermethylation.[81] This approach, which they called methylation target array (MTA), is similar to that developed by Kononen *et al.* for tissue microarrays.[82] Chen affixed methylation targets to a solid support and hybridized different CpG island probes to the array one at a time. The technique can interrogate hypermethylation of CpG islands in hundreds of clinical samples simultaneously. Its applicability to cancer-related problems was demonstrated by determining hypermethylation profiles of 10 promoter CpG islands in 93 breast tumors, 4 breast cancer cell lines, and 20 normal breast tissues. A panel of 468 MTA amplicons, which represented the whole repertoire of methylated CpG islands in these tissues and cell lines, was arrayed on nylon membrane for probe hybridization. Positive hybridization signals, indicative of DNA methylation, were detected in tumor amplicons, but not in normal amplicons indicating aberrant hypermethylation in tumor samples. The frequencies of hypermethylation were found to correlate significantly with the patient's hormone receptor status, clinical stage, and age at diagnosis. The fact that a single nylon membrane can be used repeatedly for probing with many CpG island loci may be advantageous clinically for rapid assessment of potential methylation markers that may be useful in predicting treatment outcome.

Transcription factors, nucleosomes, chromatin-modifying proteins and epigenetic markers together form extremely complex regulatory networks. Recently, two microarray approaches have been developed for genome-wide mapping of the binding sites of regulatory proteins and the distributions of methylation patterns and histone modifications.[87] In one of these, chromatin immunoprecipitation (ChIP) is combined with microarray detection. Cells are treated with a cross-linking reagent such as formaldehyde, the chromatin is isolated and fragmented, and immunoprecipitation is used to identify the protein of interest along with the attached DNA fragments. To identify the DNA fragments, the cross-links are reversed and the DNA fragments are labeled with a fluorescent dye and hybridized to microarrays with probes corresponding to regions of interest. The other method, called the DamID utilizes a different principle. In this case, a transcription factor or chromatin-binding protein of interest is fused to DNA adenine methyl transferase (Dam). When this fusion protein is expressed, Dam will be targeted to binding sites of its fusion partner resulting in methylation of adenines in DNA nearby the binding sites. To identify these sites, the methylated regions are either purified or selectively amplified from genomic DNA, fluorescently labeled, and hybridized to a microarray. Because adenine methylation does not occur endogenously in mammals, the binding sites of targeted methylation can be derived from the microarray signals. So far, only limited comparisons are available, but studies of one regulatory protein, the GAGA factor, indicate the two methods can yield similar results. However, ChIP-chip requires a highly specific antibody against the protein of interest, while DamID does not. On the other hand, DamID is suited for detection of histone or other posttranslational modifications. Additional merits and drawbacks of the two methods are considered elsewhere.[87]

1.04.4.2 Germline Mutations and Faulty Genomic Imprinting

1.04.4.2.1 Germline mutations

Germline mutations in genes that encode parts of the methylation machinery have been associated with two human diseases, ICF syndrome, the salient features of which are immunodeficiency, centromere instability, and facial anomalies, and Rett syndrome (**Table 3**). ICF is a recessive disorder of childhood that is associated with mutation in the DNA methyltransferase gene, *DNMT3B*. The mutations are largely confined to satellite DNA at the centromeric regions of chromosomes 1, 9, and 16. Normally, these satellites are heavily methylated, but in ICF syndrome they are almost completely unmethylated in all tissues.[94] Most affected patients succumb to infectious disease before adulthood.

Rett syndrome is a neurodevelopmental disorder that occurs almost exclusively in girls.[94–96] It is one of the commonest forms of intellectual disability in young girls and it is characterized by a period of early normal growth and development followed by regression, loss of speech and acquired motor skills, stereotypical hand movements, and seizures. Slowing of brain growth and growth failure frequently accompany the sporadic disorder. This disorder is associated with both nonsense and missense mutations of the X-linked gene encoding the methyl-CpG-binding protein MeCP2.

Recently, Shigematsu and colleagues have shown loss of silent chromatin looping and impaired imprinting of *DLX5* in individuals with Rett syndrome.[97] They identified *DLX5*, a maternally expressed gene, as a direct target of MeCP2. Its importance lies in production of γ-amino-butyric acid (GABA). In a Mecp2 null mouse model, they showed that repressive histone modification at Lys-9 and formation of a higher order chromatin loop structure was mediated by Mecp2 and specifically associated with silent chromatin at *Dlx5-Dlx6*. Because loss of imprinting of *DLX5* may alter GABAergic neuron activity, Shigematsu's finding suggests that dysregulation of *DLX5* by mutation of MeCP2 might contribute to some of the phenotypes of this syndrome.[97]

1.04.4.2.2 Faulty genomic imprinting

More than 25 imprinted genes have been identified in humans, and estimates based on mouse models indicate that as many as 100–200 may exist.[98] They are involved in many aspects of development including fetal and placental growth, cell proliferation, and adult behavior.[99] Several inherited disorders have been shown to be due to faulty genomic imprinting. Certain aberrations of human pregnancy show that faulty imprinting (i.e., loss of genomic imprinting, LOI) plays an important role in embryogenesis (**Table 3**). For example, ovarian dermoid cysts arise from LOI, which results in benign cystic tumors that contain two maternal chromosomes and no paternal chromosome. In contrast, the hydatidiform mole is characterized by a completely androgenic genome that arises from LOI so that these tumors contain two paternal chromosomes and no maternal chromosome.

More recently, numerous additional human genetic diseases and cancers (listed in **Table 3**) have been ascribed to faulty genomic imprinting. In 1991, the molecular basis of fragile X syndrome, a common form of heritable mental retardation, was shown to be associated with a massive expansion of CGG triplet repeats located in the 5′-untranslated region upstream of the *FMR1* (fragile X mental retardation) gene.[76,100] The fragile X site is located at Xq27.3. This was one of the first of about a dozen identified human disorders that are caused by unstable trinucleotide repeat expansions. In normal individuals, *FMR1* is a highly conserved gene that contains about 30 (range 7 to ~60) CGG repeats while over 230 repeats occur in most affected persons. In the normal transcript the repeats are unmethylated but in affected persons they are hypermethylated, which silences the *FMR1* gene and causes the absence of the FMR1 protein. Mental retardation is attributed to lack of proper protein expression in neurons during development. Owing to

Table 3 Epigenetics in human disease

Human disorder	*Salient features*	*Associated molecular pathology*	*Reference*
ICF syndrome	Recessive disorder of children with immunodeficiency, centromere instability, and mild facial anomalies causing most ICF patients to succumb to infectious disease before adulthood	Almost completely unmethylated satellite DNA in centromeric regions of chromosomes 1, 9, and 16 attributed to mutations in DMNT3B	94,169
Rett syndrome	Neurodevelopmental disorder in postnatal development of infant girls with variable clinical phenotype including loss of motor and communication skills, microcephaly, and stereotypic hand movements	Various germline MeCP2 mutations may affect the role of MeCP2 in higher order chromatin organization and imprinting	95–97,150
Fragile X syndrome	Mental retardation affecting males primarily. Other diagnostic criteria: long face, large everted ears, autism, hand biting, hyperactivity, and macroorchidism [large testicles]	Silencing of *FMRI* gene at Xq27.3 containing highly polymorphic, abnormally lengthy CCG repeats in the 5′ region; plus aberrant de novo methylation and histone deacetylation of the CpG island upstream of *FMRI*	76
Benign dermoid ovarian teratomas	Tumors contain many tissue types but no placental trophoblast	LOI results in tumors with two maternal chromosomes and no paternal contribution	99
Hydatidiform moles	Placental-derived extraembryonic tumors	LOI causes tumors with two paternal chromosomes with no maternal contribution	170
Wilms' tumors	Nephroblastoma of childhood	LOI causes preferential loss of maternal alleles on chromosome 11p15 in both Wilms' tumors and embryonal rhadomyosarcomas	101,102
Embryonal rhabdomyosarcoma	Tumors of striated muscle		103
Prader–Willi syndrome (PWS)	Deficiencies in sexual development and growth, behavioral and mental retardation. Major diagnostic criteria: hypotonia, hyperphagia and obesity, hypogonadism, and developmental delay	PWS and AS are neurogenic disorders caused by loss of function of imprinted genes at chromosome 15q11–q13. Approximately 70% of PWS and AS individuals have a 3–4 megabase deletion in their maternal or paternal chromosome 15q11–q13	104,105
Angelman syndrome (AS)	Deficiencies in sexual development and growth, behavioral and mental retardation. Major diagnostic criteria: ataxia, tremulousness, sleep disorders, seizures, and hyperactivity		104,105
Beckwith–Wiedemann syndrome	Pre- and postnatal overgrowth, macroglossia and other organomegaly, childhood tumors such as Wilms' tumor of the kidney, hypoglycemia, hemihypertrophy, and other minor complications	LOI results in biallelic expression of *IGF2* (80%), silencing or mutation of *H19* (35%), and silencing of *CDKNIC* (12%)	106,148

X-linkage affected males have more severe phenotypes than affected females, whose phenotype is modulated by the presence of the normal X chromosome.

Several other diseases that are due to faulty imprinting are also listed in **Table 3**. For example, Pal *et al.* observed preferential loss of the maternal alleles on chromosome 11 in 9 of 11 cases of Wilms' tumor where the parental origin of alleles could be followed.[101] Similar observations on five additional cases of Wilms' tumor were made by Schroeder *et al.*[102] Scrable and colleagues have found that embryonal rhabdomyosarcomas (malignant pediatric tumors of striated muscle origin) could arise from cells that were clonally isodisomic for paternal loci on chromosome 11.[103]

Glenn and associates demonstrated that the *SNRPN* gene, which encodes a small nuclear ribonucleoprotein subunit *SmN* thought to be involved in splicing of pre-mRNA, is expressed only from the paternally derived chromosome 15q11–q13 in humans with Prader–Willi syndrome.[104] More recently, Horsthemke and colleagues performed a molecular analysis at the *SNURF-SNRPN* locus in 51 patients with Prader–Willi syndrome and 85 patients with Angelman syndrome, which revealed the vast majority of these defects were epimutations. Seven patients with Prader–Willi syndrome (14%) and eight patients with Angelman syndrome (9%) had an imprinting center deletion. Sequence analysis of 32 Prader–Willi syndrome patients with no imprinting center deletion and 66 Angelman syndrome patients with no imprinting center deletion did not reveal any point mutation in imprinting center elements. In patients with Angelman syndrome, they found the imprinting defect occurred on the chromosome that was of maternal grandparental origin whereas in the patients with Prader–Willi syndrome and no imprinting center deletion, the imprinting defect occurred on the chromosome inherited from the paternal grandmother.[105] The fact that epimutations on the maternal chromosome were often present in a mosaic form suggested that in patients with Angelman syndrome and no imprinting center deletion, aberrant DNA methylation responsible for the imprinting defect occurred after fertilization.

Mannens *et al.* carried out cytogenetic and DNA analyses on patients with Beckwith–Wiedemann syndrome.[106] They localized the syndrome at chromosome 11p15.3-pter to two regions, BWSCR1 and BWSCR2. They found that LOI was involved in the etiology of the Beckwith–Wiedemann syndrome with BWSCR2 since all balanced chromosomal abnormalities observed at this region were maternally transmitted. Loss of imprinting can cause either biallelic expression (such as *IGF2*) or silencing (such as *CDKN1C*), which is found in most sporadic cases of Beckwith–Wiedemann syndrome (**Table 3**).

Another interesting aspect of imprinting is that imprinted genes tend to be clustered in the genome. In humans, the two major clusters are associated with the two major imprinting disorders (**Table 3**). The cluster on chromosome 15q11–13 is linked to the Prader–Willi and Angelman syndromes, and the one on 11p15.5 is linked to the Beckwith–Wiedemann syndrome.

1.04.4.3 Cancer

Cancer develops through a combination of genetic instability and selection, which results in clonal expansion of cells that have accumulated an advantageous set of genetic aberrations. Instability of genetic origin may occur as a result of point mutations, chromosomal rearrangements, DNA dosage abnormalities, and perturbed microsatellite sequences, while epigenetic instability may result from faulty imprinting as previously discussed, or from aberrant patterns of DNA or histone methylation. The abnormalities may act alone or in concert to alter the functions or expression of cellular components. Occasionally, cancers retain a history of their development but this may be difficult to decipher because some aberrations may be lost or obscured by subsequent events.

1.04.4.3.1 Epigenetic hallmarks of cancer

Salser's observation in 1977 that 5-methyl-cytosine was responsible for mutation of CpG dinucleotides prompted investigators to determine whether the natural patterns of methylation were disturbed in human neoplasia.[21] The first change to be reported for a number of cancers was loss of methylation at both the individual gene and globally.[107,108] In four of five patients, representing two different types of cancer, Southern blots revealed substantial hypomethylation in genes of cancer cells compared to their normal counterparts; in one of these patients hypomethylation was progressive in a metastasis.[107] Such a loss appeared to be ubiquitous for human neoplasms, and in benign neoplasms hypomethylation appeared to be the only type of change found. It was also noted that a generalized decrease in genomic methylation occurs as cells age, and this gradual loss could result in aberrant gene activation.

The second notable change observed in many human cancers was hypermethylation, particularly of CpG islands at gene promoters.[109] A very recent study has shown that promoter hypermethylation of genes involving important cellular pathways in tumorigenesis is a prominent feature of many major human tumor types.[110] Hypermethylation, which is often accompanied by global hypomethylation, could act as an alternative to mutations that inactivate tumor suppressor genes and it also could predispose to genetic alterations through inactivation of DNA repair genes.

The third important change is LOI. In cancer, LOI can lead to activation of growth-promoting genes, such as *IGF2*, and silencing of tumor suppressor genes[111–113] as already noted above.

The full range of epigenetic change that occurs in human cancers is not known, but hypomethylation, hypermethylation at CpG islands, and LOI occur most frequently.[114] An early study of *p53* illustrates what might be learned from a study of 5-methylcytosine mutations in human tissues.[115] *p53* is a well-known tumor suppressor gene that has been studied intensively. Normally in a cell, the expression of *p53* is kept at very low levels but DNA damage results in a rapid increase in levels and its activation as a transcription factor. As a transcription factor, *p53* either arrests the cells in the G1 phase of the cell cycle or triggers apoptosis.[116] More than 4500 mutations have been identified in the *p53* gene, and *p53* mutations are found in 50–55% of all human cancers. Three codons in *p53* (175, 273, and 248) are of particular interest because they are hot spots for point mutations that impair *p53* function in cancer. Sequencing indicates that all three of these codons contain 5-methylcytosine, and that they are mutated in various tumors. Rideout *et al.* found that as many as 43% (9 of 21) of the *p53* somatic mutations at these sites are due to 5-methylcytosine. There are 82 CpGs in the 2362 nucleotides of the double-stranded coding sequence of *p53*. The relevance of methylation to mutations in *p53* is brought out more clearly by the fact that no more than ~3.5% (82/2362) of the sequence contributed 33–43% of the point mutations, each of which was a transition from 5-methylcytosine to thymine (or a corresponding G to A)[115] (**Figure 1**).

1.04.4.3.2 Significance of faulty imprinting in cancer

Feinberg and colleagues believed that there was special significance attached to LOI in cancer and they initiated a search for the mechanism by which this epigenetic change might enhance the risk to these disorders. In the first of two papers, Cui *et al.* found that LOI of the insulin-like growth factor II (*IGF2*) gene, a feature of many human cancers, occurred in about 10% of the normal human population.[117] LOI in this segment of the population increased the risk of colorectal cancer 3.5–5-fold suggesting that faulty imprinting was related to the risk of cancer.

In the second paper, Sakatani *et al.* created a mouse model to investigate the mechanism by which LOI of *Igf2* contributed to intestinal cancer.[118] They knew from work of others that imprinting of *Igf2* was regulated by a differentially methylated region (DMR) upstream of the nearby untranslated *H19* gene, and that deletion of the DMR would lead to biallelic expression (LOI) of *Igf2* in the offspring. To model intestinal neoplasia, they used *Min* mice with an *Apc* mutation with or without a maternally inherited deletion, i.e., with or without LOI, and they designed the model to mimic closely the human situation where LOI caused only a modest increase in *IGF2* expression. They created their model of *Igf2* LOI by crossing female heterozygous carriers of deletion ($H19^{+/-}$) with male heterozygous carriers of the $Apc^{+/\mathrm{Min}}$. Their results showed that LOI mice developed twice as many intestinal tumors as control littermates, and they also showed a shift toward a less differentiated normal intestinal epithelium. In a comparative study of human tissues, a similar shift in differentiation was seen in the normal colonic mucosa of humans with LOI. These observations suggested that impairment of normal parental imprinting might interfere with cellular differentiation and thereby increase the risk of cancer. In more general terms, the results suggested that mutation of a cancer gene (*APC*) and an epigenetically imposed delay in cell maturation may act synergistically to initiate tumor development.[114]

1.04.4.3.3 Cytosine guanine dinucleotide island methylator phenotype

Despite their frequent occurrence in human cancers, the causes and global patterns of aberrant methylation remain poorly defined. To understand these patterns better, Toyota and colleagues examined the methylation status of CpG islands in a panel of 50 primary colorectal cancers and 15 adenomas.[119] A previous study had indicated that methylation of CpG islands of normal colonic mucosa was gradually lost as age advanced, and that aberrant methylation was associated with microsatellite instability. They found a majority of CpG islands methylated in colon cancer was also methylated in a subset of normal colonic cells as an age-related consequence of incremental hypermethylation. In contrast, methylation of the cancer-specific clones was found exclusively in a subset of colorectal cancers, which appeared to exhibit a CpG island methylator phenotype (CIMP). The CIMP^{+} tumors included the majority of sporadic colorectal cancers with microsatellite instability related to methylation of the mismatch repair gene hMLH1. The data suggested the existence of a pathway in colorectal cancer that was responsible for the risk of mismatch repair-positive sporadic tumors.[119]

1.04.4.3.4 Histone modifications of cancer cells

Considerable effort has been devoted to understanding the relevance of aberrant DNA methylation patterns to human cancer, but much less attention has been focused on histone modifications of cancer cells among many other layers of epigenetic control. Recently, Fraga and colleagues[120] characterized the profile of posttranslational modifications of one of the nucleosome core histones of chromatin, histone H4, in a comprehensive panel of normal tissues, cancer cell lines, and primary tumors. Using immunodetection, high-performance capillary chromatography, and mass spectrometry, Fraga found that cancer cells overall lost monoacetylation at H4-Lys16 and trimethylation of histone H4-Lys20. These are widely regarded as epigenetic markers of malignant transformation, like global hypomethylation and CpG island hypermethylation. In a mouse model of multistage carcinogenesis, these changes appeared early and accumulated during the tumorigenic process. They were also associated with hypomethylation of DNA repetitive sequences, a well-known feature of cancer cells. The data of Fraga *et al.* suggest that the global loss of monoacetylation and trimethylation of histone 4 might be another common hallmark of human cancer cells.

1.04.4.4 Therapeutic Potential for Epigenetic Disease

Silencing of key nonmutated genes such as tumor suppressor genes and mismatch repair genes is a common event in cancer progression[119–128] including hematological disorders.[129–131] Methylation of CpG islands located in promoter regions of cancer cell genes and conformational changes in chromatin involving histone acetylation are two processes that are associated with transcriptional silencing. Reversal of these processes and upregulation of genes important in preventing or reversing the malignant phenotype has thus become a therapeutic target in cancer treatment.[132]

1.04.4.4.1 Methyltransferase inhibitors and demethylating agents

One possible approach to promote expression of genes abnormally silenced by methylation is through inhibition of DNA methyltransferases or, alternatively, by agents capable of demethylating DNA.[132] This approach has been studied in hematological and myeloid disorders, although the data are limited. For example, in 1982 Ley *et al.* reported that the treatment of a patient with severe β-thalassemia with 5-azacytidine as a demethylating agent resulted in selective increases in γ-globin synthesis and in hemoglobin F. Measurement of pretreatment methylation levels compared to posttreatment levels revealed hypomethylation of bone marrow DNA in regions near the γ-globin and the ε-globin genes.[133] Subsequently, several studies examined the use of demethylating agents such as 5-aza-2′-deoxycytidine (decitabine) in the treatment of another heritable hemoglobinopathy, sickle cell anemia. Treatment of this disorder with 2-deoxy 5-azacytidine led to significant increases in hemoglobin F and γ-globin, which attained a maximum after 4 weeks of treatment and persisted for 2 weeks before falling below 90% of the maximum.[134] The mechanism of the therapeutic effect was not entirely clear but may have been caused by low pretreatment levels of methylation of the γ-globin gene and altered differentiation of stem cells induced by 2-deoxy 5-azacytidine.

Evidence also points to hypermethylation in the pathogenesis of the myelodysplastic syndromes. Patients with these disorders usually die from bone marrow failure or transformation to acute leukemia. Standard care for this disorder is supportive care. In one reported study, the cyclin-dependent kinase inhibitor $p15^{INK4b}$ was progressively hypermethylated and silenced in high-grade myelodysplasias, and treatment with 2-deoxy-5-azacytidine resulted in a decrease in *p15* promoter methylation and a positive clinical response in 9 of 12 myelodysplastic patients.[135] In another reported study involving 191 patients with high-risk myelodysplastic syndromes treated with 5-azacytidine (dose 75 mg $m^{-2} day^{-1}$) for 7 days every 4 weeks, statistically significant differences seen in the azacytidine group favored improved response rates, quality of life, reduced risk of leukemic transformation, and improved survival compared to supportive care.[136]

The potential reversal of epigenetic silencing by altering methylation levels with methyltransferase inhibitors or DNA demethylating agents has shown promise as a mode of therapy. In 2004, azacytidine was the first agent to receive FDA approval for treatment of several myelodysplastic syndrome subtypes. Cytidine analogs, such as 5-azacytidine and 5-aza-2′-deoxycytidine, achieve their therapeutic effects after a series of biochemical transformations. First, these agents are phosphorylated by a series of kinases to azacytidine triphosphate, which is incorporated into RNA, disrupting RNA metabolism and protein synthesis. Azacytidine diphosphate is reduced by ribonucleotide reductase to 5-aza-2′-deoxycytidine diphosphate, which is phosphorylated to triphosphate and incorporated into DNA. There it binds stoichiometrically DNA methyltransferases and causes hypomethylation of replicating DNA.[136] Most methyltransferase inhibitors are, however, not specific for a particular methyltransferase, and several of them have unfavorable toxicity profiles including severe nausea and vomiting. There are newer agents under development that may improve the targeting of methylation. Among these, for example, is MG98, an antisense oligonucleotide methyltransferase inhibitor that is specific for DNMT1. MG98 produces dose-dependent reduction of DNMT1 and demethylation of the *p16* gene promoter and re-expression of p16 protein in tumor cell lines. MG98 is currently in trial in patients with solid tumors. The drug is said to be well tolerated, but a decision on its efficacy awaits additional information.[132]

1.04.4.4.2 Histone deacetylase inhibitors

Acetylation of DNA-associated histones is linked to activation of gene transcription, whereas histone deacetylation is associated with transcriptional repression. Acute promyelocytic leukemia (APL) provides an excellent model to illustrate the modulation of gene transcription by acetylation and the therapeutic potential of histone deacetylase inhibitors. APL is a hematopoietic cancer that involves the retinoic acid receptor alpha (*RARα*) gene, which maps to the long arm of chromosome 17q21. Ninety-five per cent of cases arise from a translocation between chromosomes 15 and 17 (t15:17.q21), which leads to the formation of the fusion protein PML-RARα. PML-RARα results in a transcriptional block of the normal granulocytic differentiation pathway. RARα is a member of the nuclear hormone receptor family that acts as a ligand-inducible transcriptional activation factor by binding to retinoic acid response elements (RAREs) in a heterodimer with RXR, a related family of nuclear receptors. In the presence of a ligand (all-*trans* retinoic acid), the complex promotes transcription of retinoic acid-responsive genes. In the absence of ligand, transcription is silenced by a multistep process involving recruitment of transcriptional regulators, corepressors, and nuclear receptor core repressors such as Sin3 into a complex. Sin3 in turn recruits a histone deacetylase that causes condensation of chromatin and prevents accessibility of transcriptional machinery to target genes. The presence of ligand (all-*trans* retinoic acid) induces a conformational change in RAR enabling the dissociation of repressor complex and recruitment of coactivators (such as the *p160* family members). The coactivator molecules possess intrinsic histone acetylase activity that causes unwinding of DNA thereby facilitating transcription and promoting granulocyte differentiation. In an APL patient with a transcriptional block and refractoriness to all-*trans* retinoic acid resulting in a highly resistant form of APL, Warrell and colleagues showed that treatment with sodium butyrate, a histone deacetylase inhibitor, restored sensitivity to the antileukemic effects of all-*trans* retinoic acid.[129]

Evaluation of sodium phenyl butyrate (buphenyl) has demonstrated its beneficial effect in the treatment of other disorders including the hemoglobinopathy β-thalassemia, acute myelogenous leukemia, and prostate cancer. Phenyl butyrate is one of the older generation of histone deacetylase inhibitors and presently additional inhibitors are being tested in clinical trials.[132] Among these, the inhibitory agent suberoylanilide hydroxamic acid has shown differentiating effects in a bladder cancer cell line. A depsipeptide isolated from *Chrombacterium violaceum* has been demonstrated to have potent cytotoxic activity through several different mechanisms including histone deacetylase inhibition. This agent demonstrated activity against chronic myelogenous leukemia cells resulting in acetylation of histones H3 and N4 as well as expression of apoptotic proteins involving caspase pathways.[132]

1.04.4.4.3 **Hypermethylation and histone deacetylation**

The combined manipulation of histone acetylation and cytosine methylation in chromatin presents another strategy for gene-targeted therapy through epigenetic modification. These two epigenetic processes are linked as was shown by Nan *et al.*[59] and Jones *et al.*[60] (*see* Section 1.04.3.4.2) who showed that the repressive chromatin structure associated with dense methylation was also associated with histone deacetylation. Methylated DNA binds the transcriptional repressor, MeCP2, at the MBD, which recruits the Sin 3A/histone deacetylase complex to form the transcriptionally repressive chromatin. This process was reversed by trichostatin A, a specific inhibitor of histone deacetylase.

Since little was known about the importance of methylation relative to histone deacetylation in the inhibition of gene transcription, Cameron *et al.* examined this question.[137] They found that trichostatin alone did not reactivate several hypermethylated genes (*MLH1*, *TIMP3*, *CDKN2B* (*INK4B*, *p15*), and *CDKN2A* (*INK4*, *p16*)) under conditions that allowed reactivation of nonmethylated genes. These findings suggested that dense CpG island methylation in gene promoter regions was dominant over histone deacetylation in maintaining gene repression. They then induced partial CpG island demethylation by treatment with the demethylating agent 5-aza-2′-deoxycytidine, in the presence or absence of histone deacetylase inhibition. They observed robust expression (fourfold increase) of the genes tested by combined drug treatment (trichostatin plus 5-aza-2′-deoxycytidine) in an experiment in which low-level reactivation was seen with 5-aza-deoxycytidine treatment alone. These results indicated that histone deacetylation may not be needed to maintain a silenced transcriptional state, but histone deacetylase has a role in silencing when levels of DNA methylation are reduced. Bisulfite sequencing showed that the increase in gene expression brought about by the combination of the two drugs occurred with retention of extensive methylation in the genes tested. They also found that inhibition of deacetylase activity could induce gene expression without a large-scale change from repressive to accessible chromatin in agreement with the work of others. Taken together the data suggested that decreased methylation is a prerequisite for transcription following histone deacetylase inhibition.

In experiments similar to those of Cameron *et al.*, Chiurazzi and colleagues examined the relative roles of methylation and histone deacetylation in silencing the *FMR1* gene in fragile X syndrome.[138,139] Hypermethylation of CGG repeats in this disorder silences the *FMR1* gene to cause the absence of the FMR1 protein that subsequently leads to mental retardation (*see* Section 1.04.4.2). In their first paper, Chiurazzi *et al.* found that the demethylating agent 5-aza-2′-deoxycytidine partially restored FMR1 protein expression in B-lymphblastoid cell lines obtained from fragile X patients confirming the role of *FMR1* promoter hypermethylation in the pathogenesis of fragile

X syndrome.[138] In their second paper, they found that combining the 5-aza-2′-deoxycytidine with histone deacetylase inhibitors such as 4-phenylbutyrate, sodium butyrate, or trichostatin resulted in a two- to fivefold increase in FMR1 mRNA levels over that obtained with 5-aza-2′-deoxycytidine alone. The marked synergistic effect observed revealed that both histone hyperacetylation and DNA demethylation participate in regulating *FMR1* activity. These results may help pave the way for future attempts at pharmacologically restoring mutant *FMR1* activity in vivo.[139]

Methylation and histone deacetylation thus appear to act as layers for epigenetic silencing. Cameron *et al.* believe that one function of DNA methylation may be to firmly 'lock' genes into a silenced chromatin state.[137] They suggested that this effect may be involved in transcriptional repression of methylated inactive X chromosomal genes and imprinted alleles. They proposed that to achieve maximal gene reactivation, it might be necessary to block simultaneously both DNA methylation and histone deacetylation, both of which are essential to the formation and maintenance of repressive chromatin.

The application of drugs to the treatment of cancer and other diseases of epigenetic interest is a new area of clinical investigation. Studies to date involve older members of the first generation of methyltransferase inhibitors such as 5-azacytidine and sodium phenylbutyrate. In vitro studies and small clinical studies of these drugs have demonstrated intriguing results while several second generation drugs are in development or in the early stages of clinical trials.

References

1. Silverman, P. H. *The Scientist* **2004**, *18*, 32–33.
2. Baltimore, D. *Nature* **2001**, *409*, 814–816.
3. Brown, S. W. *Science* **1966**, *151*, 417–425.
4. Kimmins, S.; Sassone-Corsi, P. *Nature* **2005**, *434*, 583–589.
5. Luger, K.; Mader, A. W.; Richmond, R. K.; Sargent, D. F.; Richmond, T. J. *Nature* **1997**, *389*, 251–260.
6. Luo, R. X.; Dean, D. C. *J. Natl. Cancer Inst.* **1999**, *91*, 1288–1294.
7. Sturtevant A. H. *Am. Sci.* **1965**, *53*, 303–307.
8. Rubin, G. M.; Lewis, E. B. *Science* **2001**, *287*, 2216–2218.
9. Van Speybroeck, L. *Ann. NY Acad. Sci.* **2002**, *981*, 61–81.
10. Waddington, C. H. *Nature* **1942**, *150*, 563–565.
11. Ruden, D. M.; Garfinkel, M. D.; Sollars, V. E.; Lu, X. *Semin. Cell Dev. Biol.* **2003**, *14*, 301–310.
12. Barr, M. L.; Bertram, E. G. *Nature* **1949**, *163*, 676–677.
13. Ohno, S.; Kaplan, W. D.; Kinosita, R. *Exp. Cell Res.* **1959**, *18*, 415–418.
14. Lyon, M. F. *Cytogenet. Cell Genet.* **1998**, *80*, 133–137.
15. Beutler, E.; Yeh, M.; Fairbanks, V. F. *Proc. Natl. Acad. Sci. USA* **1962**, *48*, 9–16.
16. Crouse, H. *Genetics* **1960**, *45*, 1429–1443.
17. Josse, J.; Kaiser, A. D.; Kornberg, A. *J. Biol. Chem.* **1961**, *236*, 864–875.
18. Hotchkiss, R. D. *J. Biol. Chem.* **1948**, *175*, 315–332.
19. Grippo, P.; Iaccarino, M.; Parisi, E.; Scarano, E. *J. Mol. Biol.* **1968**, *36*, 195–208.
20. Scarano, E. *Adv. Cytopharmacol.* **1971**, *1*, 13–24.
21. Salser, W. *CSHL* **1977**, *XLVII*, 985–1003.
22. Riggs, A. D. *Cytogenet. Cell Genet.* **1975**, *14*, 9–25.
23. Holliday, R.; Pugh, J. E. *Science* **1975**, *187*, 226–232.
24. Jones, P. A.; Taylor, S. M. *Cell* **1980**, *20*, 85–93.
25. Juttermann, R.; Li, E.; Jaenisch, R. *Proc Natl. Acad. Sci. USA* **1994**, *91*, 11797–11801.
26. Gardiner-Garden, M.; Frommer, M. *J. Mol. Biol.* **1987**, *196*, 261–282.
27. Takai, D.; Jones, P. A. *Proc. Natl. Acad. Sci. USA* **2002**, *99*, 3740–3745.
28. Rothnie, H. M.; McCurrach, K. J.; Glover, L. A.; Hardman, N. *Nucleic Acids Res.* **1990**, *19*, 279–287.
29. Vardimon, L.; Kressmann, A.; Cedar, H.; Maechler, M.; Doerfler, W. *Proc. Natl. Acad. Sci. USA* **1982**, *79*, 1073–1077.
30. Groudine, M.; Eisenman, R.; Weintraub, H. *Nature* **1981**, *292*, 311–317.
31. McGrath, J.; Solter, D. *Cell* **1984**, *37*, 179–183.
32. Spence, J. E.; Perciaccante, R. G.; Greig, G. M.; Willard, H. F.; Ledbetter, D. H.; Hejtmancik, J. F.; Pollack, M. S.; O'Brien, W. E.; Beaudet, A. L. *Am. J. Hum. Genet.* **1988**, *42*, 217–226.
33. Monk, M. *Nature.* **1987**, *328*, 203–204.
34. Stoger, R.; Kubicka, P.; Liu, C. G.; Kafri, T.; Razin, A.; Cedar, H.; Barlow, D. P. *Cell* **1993**, *73*, 61–71.
35. Whitelaw, E.; Garrick, D. The Epigenome: Epigenetic Regulation of Gene Expression in Mammalian Species. In *Mammalian Genomics*, 2nd ed.; Ruvinsky, A, Graves, J. M., Eds.; CABI: Cambridge, MA, 2005; Chapter 7, pp 179–200.
36. Barlow, D. P. *Trends Genet.* **1994**, *10*, 194–199.
37. Razin, A.; Riggs, A. D. *Science* **1980**, *210*, 604–610.
38. Southern, E. M. *J. Mol. Biol.* **1975**, *98*, 503–517.
39. Waalwijk, C.; Flavell, R. A. *Nucleic Acids Res.* **1978**, *5*, 3231–3236.
40. Waalwijk, C.; Flavell, R. A. *Nucleic Acids Res.* **1978**, *5*, 4631–4634.
41. Bestor, T.; Laudano, A.; Mattaliano, R.; Ingram, V. *J. Mol. Biol.* **1988**, *203*, 971–983.
42. Wu, J. C.; Santi, D. V. *J. Biol. Chem.* **1987**, *262*, 4778–4786.
43. Pradhan, S.; Bacolla, A.; Wells, R. D.; Roberts, R. J. *J. Biol. Chem.* **1999**, *274*, 33002–33010.
44. Bestor, T. H. *Hum. Mol. Genet.* **2000**, *9*, 2395–2402.
45. Lei, H.; Oh, S. P.; Okano, M.; Juttermann, R.; Goss, K. A.; Jaenisch, R.; Li, E. *Development* **1996**, *122*, 3195–3205.

46. Li, E.; Bestor, T. H.; Jaenisch, R. *Cell* **1992**, *69*, 915–926.
47. Leonhardt, H.; Page, A. W.; Weier, H. U.; Bestor, T. H. *Cell* **1992**, *71*, 865–873.
48. Robertson, K. D.; Jones, P. A. *Carcinogenesis* **2000**, *21*, 461–467.
49. Yen, R. W.; Vertino, P. M.; Nelkin, B. D.; Yu, J. J.; el Deiry, W.; Cumaraswamy, A.; Lennon, G. G.; Trask, B. J.; Celano, P.; Baylin, S. B. *Nucleic Acids Res.* **1992**, *20*, 2287–2291.
50. Yoder, J. A.; Bestor, T. H. *Hum. Mol. Genet.* **1998**, 7, 279–284.
51. Van, d. W., I; Sprengel, J.; Kass, S. U.; Luyten, W. H. *FEBS Lett.* **1998**, *426*, 283–289.
52. Hermann, A.; Schmitt, S.; Jeltsch, A. *J. Biol. Chem.* **2003**, *278*, 31717–31721.
53. Okano, M.; Xie, S.; Li, E. *Nat. Genet.* **1998**, *19*, 219–220.
54. Boyes, J.; Bird, A. *Cell* **1991**, *64*, 1123–1134.
55. Boyes, J.; Bird, A. *EMBO J.* **1992**, *11*, 327–333.
56. Hendrich, B.; Bird, A. *Mol. Cell. Biol.* **1998**, *18*, 6538–6547.
57. Meehan, R. R.; Lewis, J. D.; Bird, A. P. *Nucleic Acids Res.* **1992**, *20*, 5085–5092.
58. Cross, S. H.; Meehan, R. R.; Nan, X.; Bird, A. *Nat. Genet.* **1997**, *16*, 256–259.
59. Nan, X.; Ng, H. H.; Johnson, C. A.; Laherty, C. D.; Turner, B. M.; Eisenman, R. N.; Bird, A. *Nature* **1998**, *393*, 386–389.
60. Jones, P. L.; Veenstra, G. J.; Wade, P. A.; Vermaak, D.; Kass, S. U.; Landsberger, N.; Strouboulis, J.; Wolffe, A. P. *Nat. Genet.* **1998**, *19*, 187–191.
61. Allfrey, V. G.; Faulkner, R.; Mirsky, A. E. *Proc. Natl. Acad. Sci. USA* **1964**, *51*, 786–794.
62. Bestor, T. H. *Nature* **1998**, *393*, 311–312.
63. Bird, A. P.; Wolffe, A. P. *Cell* **1999**, *99*, 451–454.
64. Ng, H. H.; Zhang, Y.; Hendrich, B.; Johnson, C. A.; Turner, B. M.; Erdjument-Bromage, H.; Tempst, P.; Reinberg, D.; Bird, A. *Nat. Genet.* **1999**, *23*, 58–61.
65. Tamaru, H.; Selker, E. U. *Nature* **2001**, *414*, 277–283.
66. Rea, S.; Eisenhaber, F.; O'Carroll, D.; Strahl, B. D.; Sun, Z. W.; Schmid, M.; Opravil, S.; Mechtler, K.; Ponting, C. P.; Allis, C. D. et al. *Nature* **2000**, *406*, 593–599.
67. Nakayama, J.; Rice, J. C.; Strahl, B. D.; Allis, C. D.; Grewal, S. I. *Science* **2001**, *292*, 110–113.
68. Fire, A.; Xu, S.; Montgomery, M. K.; Kostas, S. A.; Driver, S. E.; Mello, C. C. *Nature* **1998**, *391*, 806–811.
69. Volpe, T. A.; Kidner, C.; Hall, I. M.; Teng, G.; Grewal, S. I.; Martienssen, R. A. *Science* **2002**, *297*, 1833–1837.
70. Hannon, G. J. *Nature* **2002**, *418*, 244–251.
71. Paddison, P. J.; Caudy, A. A.; Hannon, G. J. *Proc. Natl. Acad. Sci. USA* **2002**, *99*, 1443–1448.
72. Barker, D.; Schafer, M.; White, R. *Cell* **1984**, *36*, 131–138.
73. Feinberg, A. P.; Tycko, B. *Nat. Rev. Cancer* **2004**, *4*, 143–153.
74. Cooper, D. N.; Youssoufian, H. *Hum. Genet.* **1988**, *78*, 151–155.
75. Brueckner, B.; Lyko, F. *Trends Pharmacol. Sci.* **2004**, *25*, 551–554.
76. Robertson, K. D.; Wolffe, A. P. *Nat. Rev. Genet.* **2000**, *1*, 11–19.
77. Egger, G.; Liang, G.; Aparicio, A.; Jones, P. A. *Nature* **2004**, *429*, 457–463.
78. Havlis, J.; Trbusek, M. J. *Chromatogr. B Analyt. Technol. Biomed. Life Sci.* **2002**, *781*, 373–392.
79. Hong, K. M.; Yang, S. H.; Guo, M.; Herman, J. G.; Jen, J. *BioTechniques* **2005**, *38*, 354, 356, 358.
80. Shi, H.; Maier, S.; Nimmrich, I.; Yan, P. S.; Caldwell, C. W.; Olek, A.; Huang, T. H. *J. Cell. Biochem.* **2003**, *88*, 138–143.
81. Chen, C. M.; Chen, H. L.; Hsiau, T. H.; Hsiau, A. H.; Shi, H.; Brock, G. J.; Wei, S. H.; Caldwell, C. W.; Yan, P. S.; Huang, T. H. *Am. J. Pathol.* **2003**, *163*, 37–45.
82. Kononen, J.; Bubendorf, L.; Kallioniemi, A.; Barlund, M.; Schraml, P.; Leighton, S.; Torhorst, J.; Mihatsch, M. J.; Sauter, G.; Kallioniemi, O. P. *Nat. Med.* **1998**, *4*, 844–847.
83. Pfarr, W.; Webersinke, G.; Paar, C.; Wechselberger, C.; *BioTechniques* **2005**, *38*, 527–528, 530.
84. Brock, G. J.; Huang, T. H.; Chen, C. M.; Johnson, K. J. *Nucleic Acids Res.* **2001**, *29*, E123.
85. Rousseau, F.; Heitz, D.; Biancalana, V.; Blumenfeld, S.; Kretz, C.; Boue, J.; Tommerup, N.; Van Der, H. C.; DeLozier-Blanchet, C.; Croquette, M. F. *N. Engl. J. Med.* **1991**, *325*, 1673–1681.
86. Xiong, Z.; Laird, P. W. *Nucleic Acids Res.* **1997**, *25*, 2532–2534.
87. van Steensel, B. *Nat. Genet.* **2005**, *37*, S18–S24.
88. Golub, T. R. *N. Engl. J. Med.* **2001**, *344*, 601–602.
89. Golub, T. R.; Slonim, D. K.; Tamayo, P.; Huard, C.; Gaasenbeek, M.; Mesirov, J. P.; Coller, H.; Loh, M. L.; Downing, J. R.; Caligiuri, M. A. et al. *Science* **1999**, *286*, 531–537.
90. Alizadeh, A. A.; Eisen, M. B.; Davis, R. E.; Ma, C.; Lossos, I. S.; Rosenwald, A.; Boldrick, J. C.; Sabet, H.; Tran, T.; Yu, X. et al. *Nature* **2000**, *403*, 503–511.
91. van de Vijver, M. J.; He, Y. D.; van't Veer, L. J.; Dai, H.; Hart, A. A.; Voskuil, D. W.; Schreiber, G. J.; Peterse, J. L.; Roberts, C.; Marton, M. J. et al. *N. Engl. J. Med.* **2002**, *347*, 1999–2009.
92. van't Veer, L. J.; Dai, H.; van de Vijver, M. J.; He, Y. D.; Hart, A. A.; Mao, M.; Peterse, H. L.; van der, K. K.; Marton, M. J.; Witteveen, A. T. et al. *Nature* **2002**, *415*, 530–536.
93. Camp, R. L.; Dolled-Filhart, M.; King, B. L.; Rimm, D. L. *Cancer Res.* **2003**, *63*, 1445–1448.
94. Xu, G. L.; Bestor, T. H.; Bourc'his, D.; Hsieh, C. L.; Tommerup, N.; Bugge, M.; Hulten, M.; Qu, X.; Russo, J. J.; Viegas-Pequignot, E. *Nature* **1999**, *402*, 187–191.
95. Amir, R. E.; Van, d. V., I; Wan, M.; Tran, C. Q.; Francke, U.; Zoghbi, H. Y. *Nat. Genet.* **1999**, *23*, 185–188.
96. Wan, M.; Lee, S. S.; Zhang, X.; Houwink-Manville, I.; Song, H. R.; Amir, R. E.; Budden, S.; Naidu, S.; Pereira, J. L.; Lo, I. F. et al. *Am. J. Hum. Genet.* **1999**, *65*, 1520–1529.
97. Horike, S.; Cai, S.; Miyano, M.; Cheng, J. F.; Kohwi-Shigematsu, T. *Nat. Genet.* **2005**, *37*, 31–40.
98. Barlow, D. P. *Science.* **1995**, *270*, 1610–1613.
99. Falls, J. G.; Pulford, D. J.; Wylie, A. A.; Jirtle, R. L. *Am. J. Pathol.* **1999**, *154*, 635–647.
100. Jin, P.; Warren, S. T. *Hum. Mol. Genet.* **2000**, *9*, 901–908.
101. Pal, N.; Wadey, R. B.; Buckle, B.; Yeomans, E.; Pritchard, J.; Cowell, J. K. *Oncogene* **1990**, *5*, 1665–1668.
102. Schroeder, W. T.; Chao, L. Y.; Dao, D. D.; Strong, L. C.; Pathak, S.; Riccardi, V.; Lewis, W. H.; Saunders, G. F. *Am. J. Hum. Genet.* **1987**, *40*, 413–420.

103. Scrable, H.; Cavenee, W.; Ghavimi, F.; Lovell, M.; Morgan, K.; Sapienza, C. *Proc. Natl. Acad. Sci. USA* **1989**, *86*, 7480–7484.
104. Glenn, C. C.; Porter, K. A.; Jong, M. T.; Nicholls, R. D.; Driscoll, D. J. *Hum. Mol. Genet.* **1993**, *2*, 2001–2005.
105. Buiting, K.; Gross, S.; Lich, C.; Gillessen-Kaesbach, G.; el Maarri, O.; Horsthemke, B. *Am. J. Hum. Genet.* **2003**, *72*, 571–577.
106. Mannens, M.; Hoovers, J. M.; Redeker, E.; Verjaal, M.; Feinberg, A. P.; Little, P.; Boavida, M.; Coad, N.; Steenman, M.; Bliek, J. *Eur. J. Hum. Genet.* **1994**, *2*, 3–23.
107. Feinberg, A. P.; Vogelstein, B. *Nature* **1983**, *301*, 89–92.
108. Feinberg, A. P.; Gehrke, C. W.; Kuo, K. C.; Ehrlich, M. *Cancer Res.* **1988**, *48*, 1159–1161.
109. Baylin, S. B.; Hoppener, J. W.; de Bustros, A.; Steenbergh, P. H.; Lips, C. J.; Nelkin, B. D. *Cancer Res.* **1986**, *46*, 2917–2922.
110. Esteller, M.; Corn, P. G.; Baylin, S. B.; Herman, J. G. *Cancer Res.* **2001**, *61*, 3225–3229.
111. Feinberg, A. P. *Proc. Natl. Acad. Sci. USA* **2001**, *98*, 392–394.
112. Rainier, S.; Johnson, L. A.; Dobry, C. J.; Ping, A. J.; Grundy, P. E.; Feinberg, A. P. *Nature* **1993**, *362*, 747–749.
113. Ogawa, O.; Eccles, M. R.; Szeto, J.; McNoe, L. A.; Yun, K.; Maw, M. A.; Smith, P. J.; Reeve, A. E. *Nature* **1993**, *362*, 749–751.
114. Klein, G. *Nature.* **2005**, *434*, 150.
115. Rideout, W. M., III; Coetzee, G. A.; Olumi, A. F.; Jones, P. A. *Science* **1990**, *249*, 1288–1290.
116. Levine, A. J. *Cell* **1997**, *88*, 323–331.
117. Cui, H.; Cruz-Correa, M.; Giardiello, F. M.; Hutcheon, D. F.; Kafonek, D. R.; Brandenburg, S.; Wu, Y.; He, X.; Powe, N. R.; Feinberg, A. P. *Science* **2003**, *299*, 1753–1755.
118. Sakatani, T.; Kaneda, A.; Iacobuzio-Donahue, C. A.; Carter, M. G.; Witzel, S. D.; Okano, H.; Ko, M. S.; Ohlsson, R.; Longo, D. L.; Feinberg, A. P. *Science* **2005**, *307*, 1976–1978.
119. Toyota, M.; Ahuja, N.; Ohe-Toyota, M.; Herman, J. G.; Baylin, S. B.; Issa, J. P. *Proc. Natl. Acad. Sci. USA* **1999**, *96*, 8681–8686.
120. Fraga, M. F.; Ballestar, E.; Villar-Garea, A.; Boix-Chornet, M.; Espada, J.; Schotta, G.; Bonaldi, T.; Haydon, C.; Ropero, S.; Petrie, K. et al. *Nat. Genet.* **2005**, *37*, 391–400.
121. Plumb, J. A.; Strathdee, G.; Sludden, J.; Kaye, S. B.; Brown, R. *Cancer Res.* **2000**, *60*, 6039–6044.
122. Ricciardiello, L.; Goel, A.; Mantovani, V.; Fiorini, T.; Fossi, S.; Chang, D. K.; Lunedei, V.; Pozzato, P.; Zagari, R. M.; De Luca, L. et al. *Cancer Res.* **2003**, *63*, 787–792.
123. Lee, W.-H.; Morton, R. A.; Epstein, J. I.; Brooks, J. D.; Campbell, P. A.; Bova, G. S.; Hsieh, W. S.; Isaacs, W. B.; Nelson, W. G. *Proc. Natl. Acad. Sci. USA* **1994**, *91*, 11733–11737.
124. Esteller, M.; Hamilton, S. R.; Burger, P. C.; Baylin, S. B.; Herman, J. G. *Cancer Res.* **1999**, *59*, 793–797.
125. Esteller, M.; Toyota, M.; Sanchez-Cespedes, M.; Capella, G.; Peinado, M. A.; Watkins, D. N.; Issa, J. P.; Sidransky, D.; Baylin, S. B.; Herman, J. G. *Cancer Res.* **2000**, *60*, 2368–2371.
126. Bastian, P. J.; Yegnasubramanian, S.; Palapattu, G. S.; Rogers, C. G.; Lin, X.; De Marzo, A. M.; Nelson, W. G. *Eur. Urol.* **2004**, *46*, 698–708.
127. Lynch, H. T.; de la Chapelle, A. *NEJM* **2003**, *348*, 919–932.
128. Baylin, S. B.; Herman, J. G. *Trends Genet.* **2000**, *16*, 168–174.
129. Warrell, R. P., Jr.; He, L. Z.; Richon, V.; Calleja, E.; Pandolfi, P. P. *J. Natl. Cancer Inst.* **1998**, *90*, 1621–1625.
130. Stirewalt, D. L.; Radich, J. P. *Hematology* **2000**, *5*, 15–25.
131. Chim, C. S.; Tam, C. Y.; Liang, R.; Kwong, Y. L. *Cancer* **2001**, *91*, 2222–2229.
132. Gilbert, J.; Gore, S. D.; Herman, J. G.; Carducci, M. A. *Clin. Cancer Res.* **2004**, *10*, 4589–4596.
133. Ley, T. J.; DeSimone, J.; Anagnou, N. P.; Keller, G. H.; Humphries, R. K.; Turner, P. H.; Young, N. S.; Keller, P.; Nienhuis, A. W. *N. Engl. J. Med.* **1982**, *307*, 1469–1475.
134. Koshy, M.; Dorn, L.; Bressler, L.; Molokie, R.; Lavelle, D.; Talischy, N.; Hoffman, R.; van Overveld, W.; DeSimone, J. *Blood* **2000**, *96*, 2379–2384.
135. Uchida, T.; Kinoshita, T.; Nagai, H.; Nakahara, Y.; Saito, H.; Hotta, T.; Murate, T. *Blood* **1997**, *90*, 1403–1409.
136. Silverman, L. R.; Demakos, E. P.; Peterson, B. L.; Kornblith, A. B.; Holland, J. C.; Odchimar-Reissig, R.; Stone, R. M.; Nelson, D.; Powell, B. L.; DeCastro, C. M. et al. *J. Clin. Oncol.* **2002**, *20*, 2429–2440.
137. Cameron, E. E.; Bachman, K. E.; Myohanen, S.; Herman, J. G.; Baylin, S. B. *Nat. Genet.* **1999**, *21*, 103–107.
138. Chiurazzi, P.; Pomponi, M. G.; Willemsen, R.; Oostra, B. A.; Neri, G. *Hum. Mol. Genet.* **1998**, *7*, 109–113.
139. Chiurazzi, P.; Pomponi, M. G.; Pietrobono, R.; Bakker, C. E.; Neri, G.; Oostra, B. A. *Hum. Mol. Genet.* **1999**, *8*, 2317–2323.
140. Wolffe, A. P. *Curr. Biol.* **1997**, *7*, R796–R798.
141. Riggs, M. G.; Whittaker, R. G.; Neumann, J. R.; Ingram, V. M. *Nature* **1977**, *268*, 462–464.
142. Bird, A. P.; Southern, E. M. *J. Mol. Biol.* **1978**, *118*, 27–47.
143. Bird, A. P. *J. Mol. Biol.* **1978**, *118*, 49–60.
144. Mohandas, T.; Sparkes, R. S.; Shapiro, L. J. *Science* **1981**, *211*, 393–396.
145. Bird, A. P. *Nature* **1986**, *321*, 209–213.
146. Bird, A. P. *Cold Spring Harb. Symp. Quant. Biol.* **1993**, *58*, 281–285.
147. Herman, J. G.; Jen, J.; Merlo, A.; Baylin, S. B. *Cancer Res.* **1996**, *56*, 722–727.
148. Reik, W.; Maher, E. R. *Trends Genet.* **1997**, *13*, 330–334.
149. Gonzalgo, M. L.; Jones, P. A. *Nucleic Acids Res.* **1997**, *25*, 2529–2531.
150. Hoffbuhr, K. C.; Moses, L. M.; Jerdonek, M. A.; Naidu, S.; Hoffman, E. P. *Ment. Retard. Dev. Disabil. Res. Rev.* **2002**, *8*, 99–105.
151. Jenuwein, T.; Allis, C. D. *Science* **2001**, *293*, 1074–1080.
152. Sollars, V.; Lu, X.; Xiao, L.; Wang, X.; Garfinkel, M. D.; Ruden, D. M. *Nat. Genet.* **2003**, *33*, 70–74.
153. Yu, L.; Liu, C.; Vandeusen, J.; Becknell, B.; Dai, Z.; Wu, Y. Z.; Raval, A.; Liu, T. H.; Ding, W.; Mao, C. et al. *Nat. Genet.* **2005**, *37*, 265–274.
154. Costello, J. F. *Nat. Genet.* **2005**, *37*, 211–212.
155. Ross, M. T.; Grafham, D. V.; Coffey, A. J.; Scherer, S.; McLay, K.; Muzny, D.; Platzer, M.; Howell, G. R.; Burrows, C.; Bird, C. P. et al. *Nature* **2005**, *434*, 325–337.
156. Gunter, C. *Nature* **2005**, *434*, 279–280.
157. Hegi, M. E.; Diserens, A. C.; Gorlia, T.; Hamou, M. F.; de Tribolet, N.; Weller, M.; Kros, J. M.; Hainfellner, J. A.; Mason, W.; Mariani, L. et al. *N. Engl. J. Med.* **2005**, *352*, 997–1003.
158. DeAngelis, L. M. *N. Engl. J. Med.* **2005**, *352*, 1036–1038.
159. Harsha, H. C.; Suresh, S.; Amanchy, R.; Deshpande, N.; Shanker, K.; Yatish, A. J.; Muthusamy, B.; Vrushabendra, B. M.; Rashmi, B. P.; Chandrika, K. N. et al. *Nat. Genet.* **2005**, *37*, 331–332.

160. Vallender, E. J.; Pearson, N. M.; Lahn, B. T. *Nat. Genet.* **2005**, *37*, 343–345.
161. Kaminskas, E.; Farrell, A. T.; Wang, Y. C.; Sridhara, R.; Pazdur, R. *Oncologist* **2005**, *10*, 176–182.
162. Gowher, H.; Leismann, O.; Jeltsch, A. *EMBO J.* **2000**, *19*, 6918–6923.
163. Havlis, J.; Madden, J. E.; Revilla, A. L.; Havel, J. *J. Chromatogr. B Biomed. Sci. Appl.* **2001**, *755*, 185–194.
164. Larsen, L. A.; Christiansen, M.; Vuust, J.; Andersen, P. S. *Comb. Chem. High Throughput Screen.* **2000**, *3*, 393–409.
165. Knox, M. R.; Ellis, T. H. *Mol. Genet. Genomics* **2001**, *265*, 497–507.
166. Frommer, M.; McDonald, L. E.; Millar, D. S.; Collis, C. M.; Watt, F.; Grigg, G. W.; Molloy, P. L.; Paul, C. L. *Proc. Natl. Acad. Sci. USA* **1992**, *89*, 1827–1831.
167. Thomassin, H.; Oakeley, E. J.; Grange, T. *Methods* **1999**, *19*, 465–475.
168. Oakeley, E. J.; Schmitt, F.; Jost, J. P. *BioTechniques* **1999**, *27*, 744–750, 752.
169. Hansen, R. S.; Wijmenga, C.; Luo, P.; Stanek, A. M.; Canfield, T. K.; Weemaes, C. M.; Gartler, S. M. *Proc. Natl. Acad. Sci. USA* **1999**, *96*, 14412–14417.
170. Kajii, T.; Ohama, K. *Nature* **1977**, *268*, 633–634.

Biography

Wendell W Weber is professor Emeritus (active) in Pharmacology at the University of Michigan. His teaching and research interests have centered on pharmocogenetics for more than 40 years. He has concentrated mainly on hereditary traits affecting human drug response and cancer susceptibility, particularly the metabolic polymorphisms in humans and experimental animal models. He is the author or coauthor of more than 175 research papers and book chapters, and has written two books on pharmacogenetics: *The Acetylator Genes and Drug Response* (Oxford University Press, 1987) and *Pharmacogenetics* (Oxford University Press, 1997).

Comprehensive Medicinal Chemistry II
ISBN (set): 0-08-044513-6

ISBN (Volume 1) 0-08-044514-4; pp. 251–278

1.05 Personalized Medicine

D Gurwitz, Tel-Aviv University, Tel-Aviv, Israel
V G Manolopoulos, Democritus University of Thrace, Alexandroupolis, Greece

1.05.1 Introduction: What is Personalized Medicine?

The term 'personalized medicine,' sometimes also called 'individualized medicine,' seems to include an inherent contradiction. Medicine, after all, is intended by definition to be individualized for each patient so that the cure can be fitted to her or his specific disease conditions. In the words attributed to Hippocrates, "The art [medicine] has three factors, the disease, the patient, the physician." In this famous quotation, Hippocrates was also alluding to the fact, well known to old scholars, that each patient is unique, the disease can be differently expressed for each patient, and as such, has to be treated in a special way individually matched for the particular patient. However, while no one doubts that each individual patient is unique, and his or her treatment must therefore take into account that individual's personal particulars, medicine has traditionally tended to view patients as groups rather than individuals, stratifying treatment protocols according to disease subtypes. Some claim that this fault lies with the pharmaceutical industry, which prefers selling "one size fits all" drugs, so that each drug enjoys the widest market possible, and profits are maximized. Pharmacotherapy is typically prescribed at certain doses for adults, at smaller doses for children (when applicable), and without additional dosage considerations (with the exception of patient's weight for some drugs that have a small therapeutic window). Other considerations, such as lifestyle and patient diet, are sometimes taken into account, but physicians are not always aware enough of the role played by such factors in patients' responses to drugs. In this context, it is notable to reflect on the words of E Hafter, written about 30 years ago: "Medical art, a notion difficult to define, is in danger of disappearing. It means a harmony of knowledge, skill experience, intuition and the predominant desire to help the patient. This means individualized medicine, which is only possible by sympathetic dialogue with the patient – not only by specialists in psychiatry or psychosomatics, but by every doctor. Will it be able to preserve medicine from inhumanity in spite of technology, rationalization and the computer?"[1] These insightful words seem as fitting today as they were a generation ago and maybe even more.

The term 'personalized medicine' in its newer sense covers the science and technologies of individualizing pharmacotherapy choices (both drug and dosage) according to each patient's genomics and proteomics information. The term was first featured in the scientific literature in its modern sense in a 1999 review article by Langreth and Waldholz,[2] who envisaged the use of patient genetic profiling for individualizing pharmacotherapy. In the same year, J C Stephens[3] delineated 'personalized medicine' as follows: "Recognition that there is a vast quantity of human genetic variation has had a pervasive impact on modern medicine, facilitating the identification of scores of genes that underlie monogenic clinical disorders, as well as genes involved in complex disease processes. The next logical step for human genetics is the exploration and elucidation of genes involved in differential pharmacological response: responders, nonresponders, and those with adverse side effects." With these words, indeed, individualized medicine, or personalized medicine, can be best defined: it represents the quest to tailor the best pharmacotherapy for each patient, so that adverse drug reactions (ADRs) are minimized, and drug efficacy is maximized. Thus, in other words, personalized medicine is about maximizing both drug safety and drug efficacy, by taking into consideration the uniqueness of the individual patient. This uniqueness is not necessarily encrypted in the patient's genome. However, as we shall see in this chapter, most interest in recent years has focused on the prospects of using genetic information and tools for individualizing patient care, as it currently seems to be able to offer the best prospects for individualizing healthcare.

Indeed, the website of the Personalized Medicine Coalition (PMC)[80] defines 'personalized medicine' as follows: "Personalized medicine uses new methods of molecular analysis to better manage a patient's disease or predisposition toward a disease. It aims to achieve optimal medical outcomes by helping physicians and patients choose the disease management approaches likely to work best in the context of a patient's genetic and environmental profile. Such approaches may include genetic screening programs that more precisely diagnose diseases and their sub-types, or help physicians select the type and dose of medication best suited to a certain group of patients." Remarkably, the 2005 British Royal Society Report entitled "Personalised Medicines: Hopes and Realities"[4] does not define the term itself, and instead defines pharmacogenetics and pharmacogenomics, which are widely perceived as the cornerstones for personalized medicine.

Hence, genetic information is often mentioned as the key for personalized medicine. Yet, we should bear in mind that many nongenomic factors also affect patients' response to drugs. These factors include environmental and lifestyle factors, such as exposure to environmental toxic agents; diet; physical activity; smoking; alcohol and drug abuse; stress; and family support. In addition, there are factors related to the patient life history, such as gender, age, presence, and history of concurrent diseases or injuries. Therefore, even though most of this chapter is focused on genetic factors, nongenetic factors are nonetheless important. Yet, genetic factors affecting pharmacotherapy have several advantages that make them more attractive for basic and applied research. Primarily, while environmental and lifestyle factors affecting the safety and efficacy of pharmacotherapy are difficult to study and interpret, and can dramatically change over a person's lifetime or even during the course of disease treatment in the scope of just few weeks or months, genetic information is stable (with the exception of malignant disease) and remains precisely the same throughout the patient's life. Thus, a single analysis can be useful for an entire lifetime. Moreover, due to the availability of molecular genetics tools, such information is highly accurate, unambiguous, and reliable, unlike the equivocal nature of some epigenetic (nongenetic) information related to drug safety and efficacy.

The US Food and Drug Administration (FDA) has long recognized the potential of genetic information for improving drug safety and efficacy, and has accordingly geared up toward the use of genetic information in the drug development process. As the FDA website states: "The use of genomic information, accelerated by the sequencing of the human genome and the advent of new tools and technologies, has opened new possibilities in drug discovery and development ... The FDA also engages in several applied research projects to support and promote the translation of pharmacogenomics from basic research, drug discovery and development into clinical practice, focused on ensuring its proper employment to protect public health."[81]

1.05.2 Why do we Need Personalized Medicine and What Should our Priorities be?

A short answer to this question is: because we must improve medical care, and in particular the safety of pharmacotherapy. In other words, we need to develop personalized medicine, so that the safety and efficacy of drugs, which are far from being satisfactory, may be improved; and the best way to improve drug safety and efficacy presently seems to be via the use of genomics and proteomics knowledge about the individual patient. That is, personalized medicine should not oppose the use of other types of data for improving healthcare. However, the availability of knowledge about human genome variation, and the relation of such variation to drug pharmacokinetics and pharmacodynamics, has created a unique opportunity for utilizing such data for improving healthcare. New pharmacogenomics knowledge would allow the incorporation of personalized medicine to the clinic, most likely starting with reductions in the alarmingly high rates of ADRs, currently estimated to account for about 6.5% of new hospital admissions to internal medicine wards, and about 4% of total bed occupancy.[5–7] A US study by Ernst and Grizzle[8] has estimated that overall, the cost of drug-related morbidity and mortality exceeded $177 billion in 2000. Hospital admissions accounted for nearly 70% ($121 billion) of total costs, followed by long-term care admissions, which accounted for 18% ($33 billion). Very likely, these alarming figures must have further increased since 2000; however at the time of writing this chapter, no updated figures were available (D Bates, personal communication). Such staggering figures clarify the urgent need to reduce ADRs rates. The money saved via reduced hospitalizations following ADRs could potentially be used to improve healthcare.

1.05.3 Brief History of Pharmacogenetics and Pharmacogenomics

The term 'pharmacogenetics' was first coined by Friedrich Vogel in 1959 in a review article[9] in German. It first appeared in the English scientific literature in 1961, in a well-cited review by Evans and Clarke.[10] A year later, a book entitled *Pharmacogenetics – Heredity and the Response to Drugs* was published by Werner Kalow.[11] However, the field of pharmacogenetics was delineated a few years earlier, when Arno Motulsky published his landmark review on "Drug reactions enzymes, and biochemical genetics."[12] In this review, Motulsky stated that "idiosyncratic drugs reactions might be caused by otherwise innocuous genetic traits and enzyme deficiencies." Indeed, it would be fit to view the year 1957 as the birthmark of modern pharmacogenetics, meaning that it is almost 50 years old at the time of writing this chapter.

Pharmacogenetics is often confused with the newer term, pharmacogenomics, coined in 1997.[13] The American Medical Association (AMA) website[82] defines pharmacogenomics as the discipline which 'examines how genetic makeup affects response to drugs.' This simple definition covers a very wide scope, as it is not only related to hereditary effects on drug pharmacokinetics, but also includes genomic aspects of drug development, such as identifying new drug targets using genomic tools. In recent years the term 'pharmacogenomics' has entered the scientific discussion more frequently, sometimes replacing the older term of pharmacogenetics, which has created some confusion about the extent of this discipline, and its relation with the older discipline of pharmacogenetics. The change has coincided with the completion of the Human Genome Project in 2003, and the coining of additional terms carrying the 'omics' ending, such as proteomics, toxicogenomics, neurogenomics, metabolomics, and immunogenomics. Indeed, some have claimed that the term 'pharmacogenomics' carries 'too much hype' and is 'being oversold' by commercial entities.[14]

The notion of individual variation in drug response was well known to old-time scholars. For example, in 510 BCE Pythagoras noted that some, but not all, individuals develop hemolytic anemia in response to fava bean ingestion. Today we know that the phenotype of hereditary toxicity from fava beans, known as favism, reflects a genetic deficiency in glucose-6-phosphate dehydrogenase.

Human variation was known already long ago to affect response to medical treatment. In 1892, the British physician Sir William Osler wrote that "if it were not for the great variability among individuals, medicine might as well be a science and not an art."[15] A few years later, Gorrod and Oxon[16] suggested that hereditary differences in biochemical processes were the cause of ADRs and interindividual differences in toxicity were due to enzyme deficiencies. Thirty years later, a noteworthy observation was made when for the first time Snyder[17] described an ethnic variation in a pharmacogenetic trait, i.e., the inability to taste phenylthiocarbamate. Such variability across different ethnic groups is now widely recognized as a common property of most pharmacogenetic traits.[18] However, it was the realization of genetic variation among individuals, rather than among ethnic groups, that would play the major part in accomplishing individualized pharmacotherapy. The reason being, that the scope of genetic variations among individuals is much larger than differences among ethnic groups.

During the 1960s and 1970s numerous further examples of ADRs related to inherited defects in metabolic enzymes were discovered. For example, it was realized that ADRs caused by debrisoquine, an agent used to treat hypertension, and sparteine, used to treat cardiac arrhythmia, are both related to an inherited deficiency in the liver enzyme CYP2D6, belonging to the cytochrome P450 (CYP) monooxygenase family (**Table 1**). Notably, this enzyme is involved in the metabolism of many drugs, including antidepressants, antipsychotics, beta-blockers, and opioids such as morphine, hydromorphine, and codeine.[19] Pharmacogenetic studies performed until the mid-1980s were done at the protein level, as they depended on analysis of the enzyme activities responsible for drug metabolism. However, toward the 1990s methods were becoming available for cloning and sequencing of human genes, and pharmacogenetics moved from the protein to the DNA era. Indeed, today most studies on inherited human variations are done at the DNA rather than the protein level.

When discussing the historic perspectives of pharmacogenetics, it is noteworthy to recall that blood transfusions according to blood groups have already been introduced in the clinical setting since the early years of the twentieth century. The first documented clinical matching of blood transfusion according to patient blood typing was reported in 1907 by Reuben Ottenberg.[20] Recipient-matched blood transfusions constitute a fine example of the successful

Table 1 Selected examples of human drug-metabolizing enzymes (DME) whose genes exhibit genetic polymorphisms with pharmacokinetic consequences

Enzyme	*Polymorphism*	*Examples of consequences*
CYP2C9	Low activity	Low-dose requirement for warfarin
CYP2C19	Low activity	Increased ADR risks from antiepileptic drugs
CYP2D6	Low activity	Increased side effects from antihypertensive and antidepressant drugs
CYP2D6	Very high activity	Lack of efficacy at normal dosage range with antihypertensive and antidepressant drugs; enhanced risk of toxicity from codeine due to faster metabolism to morphine
CYP2C8	Low activity	ADRs from antimalaria drugs
ALDH2	Null variant	More robust effects of alcohol; reduced risk of alcoholism
NAT2	Low activity	Increased side effects from antitubercular drug isoniazid (peripheral neuropathy); increased susceptibility to human bladder carcinogen 4-aminobiphenyl
TPMT	Low activity	Increased risk for 6-MP-induced myelosuppression
UGT1A	Low activity	Increased ADR risk of irinotecan

CYP, cytochrome P450; ALDH, aldehyde dehydrogenase; NAT, *N*-acetyltransferase; TPMT, thiopurine methyltransferase; UGT, UDP-glucuronosyltransferase.

practice of personalized medicine. The transfused blood can be viewed as a medicine, which is tailored for individual patients according to their own blood group. Obviously, the patient's blood group is identified with proteomics tools (antibodies) but is determined exclusively by her or his genes. Without such 'personalization,' as we all know, blood transfusion would be extremely dangerous rather than life-saving. It is important to keep this example in mind, because society tends to view new technologies with great suspicion and even fear (several examples come to mind here: x-rays; ultrasound imaging; nuclear magnetic resonance imaging; in vitro fertilization; and also nonmedical disciplines, including bioengineered plants; cellular phones). It is thus imperative, when discussing the prospects of personalized medicine with policy-makers and representatives of the public, to recall the life-saving capacity of personalized medicine as exemplified by the century-old technique of individualized blood transfusions.

It would be fitting to conclude this section on the history of pharmacogenetics and pharmacogenomics with an estimate on the timeframe needed for its implementation in the clinical setting. Various estimates have been given, but presently it seems premature to give a timescale, as there are too many barriers (*see* Section 1.05.9). Most novel clinical technologies had to wait 17 years on average before being incorporated into the clinical arena.[21] Thus, if we count this average time from the completion of the Human Genome Project, we can assume that personalized medicine would be widely implemented in the clinic by the year 2020. However, it is likely that some applications would arrive much sooner, following the availability of commercial diagnostic tests, such as CYP2D6 and CYP2C19, offered by the Roche AmiliChip P450, introduced to the market during 2004–2005,[83] and the Invader test for UGT1A1 polymorphisms (see **Table 3**).

1.05.4 The Scope of Human Genome Variation

The current estimate is that the human genome, composed of about 3.2 billion basepairs of DNA arranged in 23 chromosome pairs, contains between 20 000 and 25 000 genes (International Human Genome Sequencing Consortium 2004).[87] Current thinking is that virtually all the functional relevance of an organism is encoded within genes and within the various regulatory regions between genes, although we still know very little about the roles of DNA regions outside genes. Of note, humans have many more proteins than genes – as many as 10-fold more proteins by some estimates. This is due to complex alternative splicing mechanisms, as well as posttranslational modifications. Thus, one gene can encode and regulate the synthesis of several proteins, sometimes having different functions.

Soon after the completion of the Human Genome Project in April 2003, efforts were made for realization of the scope of human genome variation. In the first years of the twenty-first century it has become apparent that this variation is much larger than previously envisaged. We now know that our 3.2 billion base genetic code harbors about 11 million sites where a nucleotide present in most people is replaced by another one in at least 1% of the human population. Thus, on average, every 300th nucleotide in our genetic code can harbor such a polymorphic site where a single nucleotide is different in at least 1% of individuals in the population. Such polymorphic sites are known as single nucleotide polymorphisms (SNPs), and are the most common type of polymorphic alleles. If the definition of SNP was changed so that the cutoff is set, for example, at 0.5% of the population rather than the current definition of 1%, the number of SNPs in the genome would become much higher. Some experts have estimated that, in effect, any place in the genome where a SNP is not lethal when present in a single copy of the genome is likely present in few individuals on earth (as the number of individuals presently alive is about twice the number of nucleotides in our genome; and moreover, each individual has two complete sets of the human genome). In other words, although human DNA sequences are about 99.7% identical to each other, the remaining 0.3% of variation, mostly represented by SNPs, is the major biological reason that each one of us is a unique individual at the DNA level.

Unlike mutations, SNPs are very seldom linked with specific phenotypes. Yet, some SNPs can affect the individual phenotypic response to drugs. This could happen, for example, when the SNP occurs in an exonic region of a gene, and leads to a change in the amino acid sequence of the gene product. Such SNPs are known as 'coding SNPs,' unlike the 'silent SNPs' which can be outside genes, or in nonexonic regions of a gene, or even in exonic regions, but not affecting the protein's amino acid sequence (which often happens for many three-letter amino acid codes, when the third letter of the code is changed).

1.05.5 Human Genome Variation can Affect Drug Pharmacokinetics and Pharmacodynamics

1.05.5.1 Drug Pharmacokinetics

How is this large extent of human genome polymorphism linked to personalized medicine? If the coding SNP results in a change in the active site of a drug-metabolizing enzyme (DME) it could lead to less active enzyme, and hence to a phenotype of 'intermediate metabolizer.' When both gene copies contain SNPs (or more extensive polymorphic alleles leading to major alterations, such as deletions) affecting enzyme activity or expression levels, the consequence might be a 'poor metabolizer' phenotype for that gene product. Thus, certain polymorphic alleles in DMEs can dramatically affect drug pharmacokinetics. However, not only SNPs, but also larger polymorphic alterations, can affect the activity of DME. Other forms of genetic polymorphism include insertions and deletions of nucleotides, and repetitive sequences (microsatellites). For example, changes such as deletions or insertions in the intron/exon boundary region of a gene can lead to alternative splicing (exon skipping) so that a shorter enzyme protein is formed. The smaller protein might show reduced or zero activity, or lower affinity toward its substrates. Polymorphisms in other regions of the gene, most notably the promoter region, can also have a dramatic effect on drug metabolism, via lower levels of expression. Changes might also occur in transcriptional control regions located upstream of a gene; in such cases, the gene product could be normally produced under basal conditions, but inappropriately regulated by endogenous hormonal control.

Table 1 lists a few examples of genes whose polymorphism is related to drug pharmacokinetics. These genes code for human DMEs. It is beyond the scope of this chapter to supply extensive descriptions of such enzyme reactions, so here we will present only a few examples in detail and the reader is referred to several comprehensive reviews of the subject.[22–25]

The first example is the liver enzyme CYP2D6, belonging to the CYP gene family. Enzymes of this family are implicated in the phase I reactions of most drugs, in which drugs are oxidized as a first step toward their excretion by the kidneys. Excretion requires that drugs, which are typically lipophilic compounds, become more hydrophilic; this also requires phase II reactions, which follow drug oxidation, such as glucuronidation and sulfation (addition of glucose or sulfate groups, respectively), to render them more water-soluble for excretion. These enzymes have primarily evolved for detoxifying xenobiotics present in foods from plant origin, and some are also involved in the catabolism of steroid hormones. Indeed, it is believed that most natural drugs in plants have evolved as a means of plants to protect themselves from being eaten; in parallel, animals have developed P450 enzymes as a detoxification means, so that they can consume plants. Drugs, being xenobiotics, and often being closely similar to plant-derived natural compounds, are also oxidized by these enzymes. As most drugs are taken orally, a large part of their metabolism occurs during the first pass through the liver, following ingestion and absorption through the intestine walls. Humans have 57 different P450 genes, each of which codes for a protein that modifies a different subset of drugs. Different polymorphic alleles have been implicated in increased, decreased, or completely absent levels of metabolism of certain drug classes. Thus, both the use of the drug and its optimal dose may be affected by the individual's genotype for P450 enzymes.[22] Among the diverse range of genes that make up the CYP family, several have been identified as being particularly important in oxidative metabolism, including CYP2D6, CYP2C9, CYP2C19, and CYP3A4/A5.

The CYP2D6 gene is part of a cluster of three genes arranged in tandem on chromosome 22q13.1. The CYP2D gene cluster is composed of two pseudogenes, CYP2D8P and CYP2D7P, as well as the CYP2D6 gene. The only functional gene present in the human CYP2D gene locus is inactive in nearly 5% of Caucasian individuals because of detrimental mutations. More than 45 major polymorphic CYP2D6 alleles have been described. The frequencies for these alleles vary depending on the ethnicity of the individual.[26] The CYP2D6 enzyme is apparently the most crucial DME, being involved in the metabolism of about 25% of all prescribed medicines, including some beta-blockers used in the treatment of heart disease and high blood pressure, tricyclic antidepressants, some second-generation antidepressants, such as fluoxetine, paroxetine, venlafaxine, as well as several antipsychotic drugs. Therefore, identifying individuals who are 'poor metabolizers' for CYP2D6 would allow physicians to prescribe for them alternative drugs, metabolized by other routes, or to prescribe the same drugs but at much lower dosages, so that the occurrence of ADRs related to drugs metabolized by CYP2D6 would be substantially reduced.[27]

In the case of CYP2D6, it is crucial also to identify those individuals who are 'ultrarapid metabolizers.' Such individuals have extra intact copies of the CYP2D6 gene (an inherited phenomenon called 'gene duplication,' which is common in particular for this P450 enzyme) and hence, metabolize some drugs much faster than normally. Therefore, many drugs from the above-mentioned classes would not be effective in such individuals, because the therapeutic blood concentrations are unlikely to be reached at regular dosage. However, ultra-rapid metabolizers for CYP2D6 are not only at risk of having poor drug efficacy for many drugs: they are also at risk of a life-threatening ADR, namely, impaired breathing from the pain-relief drug codeine. Codeine is a prodrug that must first be converted from an inactive form to the active form, morphine, by CYP2D6. Thus, ultrarapid metabolizers of CYP2D6 would have too high blood levels of morphine soon after ingestion of the regular prescribed dose of codeine, along with the severe risks associated with morphine overdose. In contrast, poor metabolizers of CYP2D6 would not be at increased risk from codeine; rather, for them standard doses of codeine are unlikely to offer pain relief, due to the insignificant blood concentrations of morphine that are endogenously formed.

Another notable phase I metabolic enzyme whose polymorphism affects drug safety and efficacy is the cytochrome P450 2C9 (CYP2C9). CYP2C9 is a genetically polymorphic enzyme that is involved in the metabolism of phenytoin, *S*-warfarin, tolbutamide, losartan, torasemide, and many nonsteroidal antiinflammatory drugs, including diclofenac, ibuprofen, and flurbiprofen.[28] The CYP2C9 gene is located at chromosomal region 10q24, spanning approximately 55 kb with nine exons and encodes a protein of 490 amino acid residues. CYP2C9 is 92% homologous to CYP2C19, the expressed product of its neighboring gene (CYP2C19), differing by only 43 of 490 amino acids. However, the two enzymes have completely different substrate specificity. Within a 2.2 kb 5′ flanking region of the CYP2C9 gene, there are several consensus sequences for glucocorticoid response elements (GREs), and putative binding sites for transcription factors.[29] More than 50 SNPs have been described in the regulatory and coding regions of the CYP2C9 gene.[30] Two amino acid variants, $Arg_{144}Cys$, which result from a C_{430} to T nucleotide substitution in exon 3 (CYP2C9*2), and $Ile_{359}Leu$, produced by an A_{1075} to C substitution in exon 7 (CYP2C9*3), have reduced catalytic activity compared with the wild-type CYP2C9*1 and are rather common, with allelic frequencies of around 11% (*2) and 7% (*3) in Caucasians, and significantly lower frequencies in Africans and Asians.[31] The catalytic activity of the CYP2C9*3-encoded enzyme is much lower than those of CYP2C9*1 and CYP2C9*2. These two variants are known to reduce the metabolism of warfarin, for CYP2C9*2 by 30–50% and for CYP2C9*3 by around 90%.[29] The highest potential clinical impact of these polymorphisms has so far been related to the use of warfarin, a drug commonly used for the prevention of arterial and venous thromboembolism. Warfarin pharmacotherapy is extremely difficult to manage due to its very narrow therapeutic index and the seriousness of the bleeding complications it can cause. Several studies have provided evidence for a correlation of CYP2C9 gene variants and clinical outcomes. A recent meta-analysis of nine such primary clinical studies with data from 2775 patients solidified the idea that patients with CYP2C9*2 and CYP2C9*3 alleles have lower mean daily warfarin doses and a greater risk for bleeding.[32] The authors suggested that testing for gene variants of CYP2C9 could potentially alter clinical management in patients commencing warfarin treatment. However, many nongenetic factors, in particular diet, play a crucial role in the safety and efficacy of warfarin pharmacotherapy (*see* Section 1.05.6).

The third example concerns thiopurine *S*-methyltransferase (TPMT, EC2.1.1.67), which is a cytoplasmic enzyme that preferentially catalyzes the *S*-methylation of aromatic and heterocyclic sulfydryl compounds. This enzyme inactivates the drugs belonging to the thiopurine family such as 6-mercaptopurine, 6-thioguanine, and azathioprine. These drugs have been widely used in the treatment of leukemia, autoimmune disorders, and for immune suppression in organ transplant recipients.

Oral 6-mercaptopurine is routinely used in the maintenance treatment of acute lymphoblastic leukemia in children, which contributes to the remarkably high cure rates achieved for this type of malignancy. Azathioprine is widely used for the treatment of inflammatory bowel disease, autoimmune hepatitis, systemic lupus erythematosus, rheumatoid arthritis, dermatologic conditions, and organ transplantation. Individual differences in thiopurine drug metabolism, response, and toxicity in humans have been correlated with common polymorphism of the TPMT gene. TPMT is encoded by a 34-kb gene consisting of 10 exons that encodes for a 245-amino acid peptide with a molecular mass of 35 kDa. TPMT activity is inherited as an autosomal codominant trait with genetic polymorphism in all large populations studied to date.

Patients with intermediate activity are heterozygous at the TPMT gene locus and the TPMT-deficient subjects are homozygous for low-activity alleles. Altered TPMT activity predominantly results from SNPs. Patients with inherited very low levels of TPMT activity are at greatly increased risk for thiopurine-induced toxicity, which is often expressed by myelosuppression when treated with standard doses of these drugs. On the other extreme of the spectrum, thiopurine drugs might fail to show efficacy in subjects with very high TPMT activity. The wild-type allele is designated as TPMT*1 and, to date, at least 18 variant alleles of the TPMT gene have been reported. Based on the population phenotype–genotype studies performed to date, assays for the molecular diagnosis of TPMT deficiency have focused on the alleles named TPMT*2 (G238C), TPMT*3A (G460A/A719G), TPMT*3B (G460A), and TPMT*3C (A719G). These four mutant alleles together account for 80–95% of low-activity alleles in all studied human populations.[33,34] As seen for numerous other genes, the pattern and frequency of mutant TPMT alleles are different among various ethnic populations. The most prevalent low-activity TPMT allele in Caucasians is TPMT*3A, whereas TPMT*3C is the predominant low-activity allele in Chinese people, Egyptians, and African-Americans. Of note, at time of writing of this chapter, no commercial tests for TPMT genotyping are available, and testing for TMPT is not widespread, or is performed with biochemical tools, measuring erythrocytes' TPMT activity. However, leukemia patients often receive blood transfusions, which could mask the true TPMT phenotype in cases of poor TPMT metabolizers.

The final example of pharmacogenetic variation in molecules involved in drug pharmacokinetics also comes from oncology. It concerns the phase II metabolic enzymes UDP-glucuronosyltransferases which catalyze the glucuronidation of various lipophilic substances including drugs and environmental toxicants. UDP-glucuronosyltransferase 1A1 (UGT1A1) is a member of the UGT1A family and it inactivates 7-ethyl-10-hydroxycamptothecin (SN-38), the active metabolite of the anticancer drug irinotecan (Camptosar), to form SN-38 glucuronide (SN-38G). Irinotecan has potent antitumor activity against a wide range of tumors, and is one of the most commonly prescribed chemotherapy agents. SN-38 has been associated with severe diarrhea and myelosuppression, which are the main dose-limiting toxicities of irinotecan.[35] Variations in UGT1A1 activity most commonly arise from polymorphisms in the UGT1A1 promoter region that contains several repeating TA elements. The presence of 7 TA repeats (referred to as UGT1A1*28), instead of the wild-type number of 6, results in reduced UGT1A1 expression and activity.[36] Accordingly, UGT1A1*28 has been shown to be associated with reduced glucuronidation of SN-38, and increased clinical toxicity for patients treated with irinotecan.[37] The frequencies of UGT1A1*28 alleles vary significantly among different ethnic groups and they can be as high as 35% in Caucasians and African-Americans but much lower in Asians. It has been suggested that prospective screening of patients prior to chemotherapy selection may reduce the frequency of severe toxicities by allowing alternate-therapy selections for patients carrying the UGT1A1*28 polymorphism. Recently, the Invader UGT1A1 test has been approved by the FDA for use to identify patients who may be at increased risk of adverse reactions to irinotecan (Camptosar). In addition, Camptosar's package insert was recently relabeled to include dosing recommendations based on a patient's genetic profile.

1.05.5.2 Drug Pharmacodynamics

In addition to pharmacogenetic variations of DMEs, there are increasing numbers of examples of striking pharmacogenomic variations that influence drug pharmacodynamics as a result of inherited polymorphic alleles in drug target genes. In the long run, the importance of uncovering such variations is expected to increase because, even though it might be possible to minimize the impact of genetic variation on drug pharmacokinetics (by avoiding the development of drugs that are primarily metabolized by polymorphic enzymes), it will be much more difficult to avoid inherited variation in drug targets.[38]

The best-known example, which is already widely implemented in the clinic, is the selective application of the drug trastuzumab (Herceptin) for the targeted therapy of metastatic breast cancer. Trastuzumab is a humanized monoclonal antibody specific for HER-1/neu oncogene. This molecule is located on chromosome 17q, it encodes a transmembrane glycoprotein with intracellular tyrosine kinase activity, and it is overexpressed by 25–30% of breast cancers. These

HER-1/neu-overexpressing breast cancers define a subset of breast tumors that are characteristically more aggressive, and women who develop them have a shorter survival. Trastuzumab given alone to HER-1/neu-overexpressing breast tumors produces response rates similar to those of many single-agent chemotherapeutic agents and has limited toxicity, while in combination with standard chemotherapy, it can produce greater response rates and prolong the survival of women with advanced breast cancer.[39] Therefore, screening breast cancer patients for HER-2 is a prerequisite for treatment with trastuzumab. This drug is documented to benefit only patients whose tumors express high levels of the HER-2 gene product, which is the drug target for trastuzumab.

Asthma represents another area with hopes for meaningful exploration of polymorphisms of drug targets for improvement of drug efficacy. Pharmacogenetic studies of drugs used in the treatment of asthma have resulted in the identification of several situations of reduced response in patients carrying specific genotypes in genes involved in the action of all major classes of antiasthmatic drugs, including β_2-agonists, leukotriene modulators, and corticosteroids.[40,41] The β_2-agonists represent the most important bronchodilator drugs used in asthma treatment, and are at the same time the most commonly prescribed asthma medications. The β_2-adrenergic receptors (β_2-ARs) belong to the group of G protein-coupled receptors, and are present on many airway cells, including smooth-muscle cells, which are hyperreactive in asthma. At least 13 polymorphisms have been described in the gene, which is located on chromosome 5q31-32. From those, three coding polymorphisms have been most studied, located at amino acid positions 16, 27, 34, and 164. These functional polymorphisms appear to influence both disease susceptibility and treatment response in asthma. Several studies have shown that the β_2-AR SNPs Arg16Gly and Gln27Glu have significant effects in modulating responses to β_2-agonist therapy in asthma. Examples of the reported effects include a decrease in morning peak expiratory flow in patients with mild asthma who were Arg/Arg homozygotes at position 16 and who regularly use albuterol[42] and reduced responsiveness to salbutamol administration in position 16 Gly/Gly homozygotes in a study performed in 269 asthmatic children.[43] In addition, the conclusion of the large Beta-Agonist Response by Genotype study[44] indicated that Arg/Arg asthmatic patients might experience adverse effects from regular use of regular β-agonists rather than benefiting from the treatment. Unfortunately, despite all these reported associations with drug effects, there are two major drawbacks delaying translation of these findings into general clinical practice: (1) for every association that emerges from several studies, there also exists one or more published reports failing to replicate the observations; and (2) findings of studies have sometimes been inconclusive and contradictory. Even though clinical applications based on the pharmacogenetics of asthma are clearly not ripe for general clinical use, it is believed that the pharmacogenetic approach will eventually help clinicians to optimize and personalize antiasthmatic treatment, and will also provide useful information with regard to pre- and postmarketing evaluation of both effectiveness and side effects of newly introduced drugs.[45]

5-Hydroxytriptamine (5HT; serotonin) has been linked to the control of feeding behavior. Weight gain, a common consequence of treatment with antipsychotic drugs, appears to be related to the action of these drugs on 5HT receptors. The 5HT2C-receptor is of particular interest in this respect due to the obesity and increased feeding observed in the 5HT2C-knockout mouse. Interestingly, the 5HT2C-receptor gene has several SNPs and a C→T substitution at position -759 of the promoter region is found near the regulatory and transcription binding sites, which may alter gene expression. It has been shown in several studies that carriers of the T allele are relatively protected from weight gain following treatment with antipsychotics such as clozapine, risperidone, and chlorpromazine.[46] Additional examples for drug targets whose genes exhibit genetic polymorphisms with direct drug pharmacodynamics consequences are shown in **Table 2**. Some examples for FDA-approved diagnostic tests with potential use for personalized medicine are shown in **Table 3**.

1.05.6 Nongenomic Effects on Drug Metabolism

When discussing personalized medicine and its relation to pharmacogenetics and pharmacogenomics, it is important to bear in mind that many nongenetic variables contribute to drug metabolism and disposition and can often be related to ADRs or to lack of efficacy of drugs. These factors include environmental influences such as diet, alcohol consumption, cigarette smoking, use of illegal drugs, and exposure to environmental toxicants in air or water. In addition, lifestyle factors such as physical activity and stress can affect drug metabolism. For example, athletes tend to have higher liver enzyme activities, and hence are more likely to metabolize drugs faster; thus, the likelihood of nonefficacy of drugs due to insufficient blood levels could be more common in such individuals.

Drug metabolism and disposition are highly affected by diseases, in particular liver and kidney disorders. For this reason, drugs should often be prescribed at lower doses for aged individuals, for whom liver and kidney disorders are much more common. In addition, drug metabolism can be drastically affected by interactions with other drugs, as many drugs can inhibit the activity of certain liver enzymes, most notably, members of the CYP450 family. A well-known

Table 2 Selected examples of human drug targets whose genes exhibit genetic polymorphisms with pharmacodynamic consequences

Drug target	*Polymorphism*	*Examples of consequences*
B_2-AR	Arg16Gly and Gln27Glu	Efficacy of beta-agonists for asthma
EGFR	Gene copy number	Efficacy of some anticancer drugs
ER	Gene copy number	Efficacy of tamoxifen for breast and ovarian cancers
HER2	Gene copy number	Efficacy of trastuzumab (Herceptin) for recurrent breast cancer
5-HTT	Promoter length	Efficacy of SSRI antidepressant drugs
TS	Gene copy number	Efficacy of 5-fluorouracil for colorectal cancers
VDR	BsmI splice site	Efficacy of alendronate for treating osteoporosis

B_2-AR, β_2-adrenergic receptor; EGFR, epidermal growth factor receptor; ER, estrogen receptor; 5HTT, 5HT (serotonin) transporter; TS, thymidylate synthase; VDR, vitamin D receptor.

Table 3 FDA-approved diagnostic tests with potential use for personalized medicine (October 2005)

Test (maker; date approved by the FDA)	*Gene polymorphism tested*	*Examples of use*
AmpliChip P450 (Roche Diagnostics; Dec. 2004)	CYP2D6 and CYP2C19	Identification of individuals at increased ADR risk from antidepressant, antipsychotic, and antihypertensive drugs
Invader (Third Wave Technologies; Aug. 2005)	UGT1A1	Identification of individuals at increased risk from ADRs by irinotecan

example is the over-the-counter herbal drug, St John's wort (*Hypericum perforatum* L.), used extensively for the treatment of mild to moderate clinical depression (in particular in Europe). The active ingredients of this extract, hypericin, pseudohypericin, and hyperforin, were shown to inhibit CYP2C6 and, upon prolonged use, also induce higher expression levels of CYP3A4, CYP2D2, and CYP3A2.[47] Thus, it can increase the risk of ADRs from drugs metabolized by CYP2C6, such as tolbutamide, and reduce the efficacy of drugs metabolized by CYP3A4, CYP2D2, or CYP3A2, such as dextromethorphan and midazolam, respectively. St John's wort use was also reported to lower the efficacy of oral contraceptives.[48] The reason is that it causes an induction of ethinylestradiol-norethindrone metabolism due to the induced increase of CYP3A activity. Women taking oral contraceptives are therefore advised to consider adding a barrier method of contraception when taking St John's wort for depression.

In conclusion, it is crucial to remember that genetic variations among individuals cannot explain the entire range of ADRs or lack of efficacy of drugs. As is the case in most other areas of medicine, the phenotype is determined by a combination of heritable and environmental factors. Thus, pharmacogenetics and pharmacogenomics will never be able to eliminate ADRs entirely, but only reduce them. Protection from ADRs should include both genomics and nongenomic tools and considerations. Physicians and patients alike must be alert to these facts, not expect too much from genetic information, and yet be aware that in many cases genetic information can provide safer and more effective healthcare.

1.05.7 Unique Value of Personalized Medicine for Psychiatry

As became obvious from some of the previously mentioned examples, one field where personalized medicine is rapidly becoming a true success story is oncology, where measures are already being initiated to achieve the best chemotherapy for each patient. This is primarily based on genotyping, or rather, proteomics profiling of tumor biopsies rather than the patient's germline genetic make-up. Tumor genes mutate rapidly, allowing cancer cells to evade the immune system as well as chemotherapy, and identifying the new mutations helps in devising the most effective chemotherapy for many patients.

However, genetic profiling of the patients themselves will be required for potential therapeutic applications of pharmacogenomics in virtually all other medical disciplines. Among these, the most obvious discipline that is expected to benefit from the implementation of personalized medicine is psychiatry. This view reflects the many difficulties of current psychiatric practice, as compared with other medical disciplines.

On top of its well-known problems as a separate medical discipline, including stigmatization of practicing healthcare professionals, psychiatry lingers behind other medical disciplines with its limited understanding of disease biology. This is especially frustrating in view of the huge investment in neuroscience research during the 1990s, declared 'the decade of the brain.'[49] This paucity of knowledge about the biology of psychiatric illness is likely to remain a typical feature of psychiatry, given that the brain is definitely the most complex human organ, possibly too complex for humans ever to comprehend. This is illustrated by the fact that, although there are numerous clues for complex interrelations between mood and immune disorders, not a single immune system marker – or any other blood marker – is presently applied in the practice of psychiatry.

Limited new insight into genetic factors of psychiatric disorders, most notably schizophrenia, is coming from recent studies. Most notably, certain genes, including neuregulin, D-amino acid oxidase, p72, dysbindin, and catechol-*O*-methyltransferase, have been identified as contributing to disease risk, albeit with minute statistical effects for each risk allele.[50,51] Yet, the identification of such genes has not yet contributed to understanding of schizophrenia etiology and the mode of action of the antipsychotic medications. For example, it is by now evident that deficient glutamate signaling might be more crucial for schizophrenia etiology than previously recognized. The poor comprehension of schizophrenia biology is reflected by observations that, while most antipsychotic agents are primarily dopamine D_2-blockers, with some newer antipsychotic drugs also being potent 5HT2A-blockers, genes coding for various dopamine and serotonin receptors are unlikely to be among the major schizophrenia risk genes. Notably, the biological basis for major depression, the most frequent psychiatric disease, with estimated lifetime prevalence of between 10% and 20%, remains even more enigmatic than schizophrenia. Almost no genetic clues for the biological background of affective disorders are known although, like schizophrenia, they are also affected by genetic susceptibility.[52]

Another noteworthy difficulty for modern psychiatry, compared with other disciplines, is the lack of reliable and objective diagnostic tools. Indeed, diagnosis in psychiatry has barely changed in the last four decades, a striking fact when considering the huge advances in diagnostic tools, most notably computerized imaging technologies, which have entered the clinic in other medical disciplines. To date, diagnosis and follow-up in psychiatry continue to be based on clinician impressions from patient interviews and the physician subjective assessment, although some advance has been made following the introduction of the *Diagnostic and Statistical Manual* (DSM-IV-TR) criteria in 2000. This situation necessitates 'trial and error' therapeutic decisions, far more than is typical for other disciplines where, even when the arsenal of available drugs is at times insufficient, at least the diagnosis is more likely to be accurate, being based on objective biological markers. Unequivocal biological markers are also available, in most other disciplines, for following the efficacy of pharmacotherapy, and again this is with the distinct exception of psychiatry, where treatment efficacy must be determined by frequent clinician–patient interactions rather than by objective laboratory tests.

Collectively, these factors underlie the large ongoing difficulties in the practice of psychiatry. On the one hand, an impressive repertoire of new antipsychotic and antidepressant drugs has been introduced in recent years. Nonetheless, no tools for individualizing psychiatric treatment have emerged. Individualization of patient care in psychiatry largely remains a matter of trial-and-error decision process, with the first line of therapy choices often determined by local expertise and preferences in each hospital.

A notable example for the potential of genotyping-based personalized psychiatry is the treatment of major depression. Poor efficacy, estimated to affect up to 60% of depressed patients, typifies the treatment with current antidepressant drugs. Efficacy of a major class of these drugs, the selective serotonin reuptake inhibitors (SSRIs), has been shown in numerous independent trials to be strongly associated with a promoter-length polymorphism of the SSRI drug target gene, the serotonin transporter (5HTT). Presence of at least one copy of the longer promoter allele, containing 44 extra nucleotides, dictates higher expression levels for the transporter protein[53] and this somehow allows SSRIs to be more effective. In contrast, depressed individuals carrying two copies of the shorter promoter allele are unlikely to benefit from SSRI treatment. It remains to be seen if these patients would be more likely to benefit from other classes of antidepressant drugs, such as the classical tricyclic antidepressants, or the newer selective norepinephrine, or mixed norepinephrine/serotonin reuptake inhibitors, such as venlafaxine. Apparently, 5HTT promoter genotyping, possibly combined with genotyping of other yet-to-be determined genes, could be an effective tool for choosing the best therapy, e.g., choosing between SSRI or non-SSRI antidepressants, for depressed patients.

Personalized psychiatry may minimize adverse effects associated with some antipsychotics and antidepressants, most notably, those related to poor metabolizer alleles for CYP2D6, implicated in the metabolism of most current antipsychotic drugs, as well as most current antidepressants. Definitely, this aspect of personalized psychiatry is much closer to genuine implementation in the clinic compared with pharmacodynamic aspects of drug individualization.

Some researchers have taken the step of calculating specific dose recommendations for several antidepressant drugs, based on the patient CYP2D6 and CYP2C19 genotype.[54]

The need for genotyping-based treatment decision-making in schizophrenia is expected to increase alongside continued discovery of risk genes, and the projected discovery of newer drug targets following better comprehension of disease biology, including amino acid neurotransmitter and peptide hormone receptors. Choosing among a large projected repertoire of new drugs will likely require genotyping for identification of the specific disease risk or 'adverse event' risk gene alleles for each individual patient, allowing pharmacotherapy tailored for the individual psychiatric patient.

All told, it is evident that psychiatry is in great need of molecular tools to aid in the pharmacotherapy decision-making process for the individual patient. Together, the notions on the apparent weaknesses of psychiatry, with respect to both patient diagnosis and treatment, could be the foremost incentives for trying to improve its effectiveness by using pharmacogenomics techniques. In other words, the present tools for patient diagnosis, treatment, and follow-up decisions are so lacking that the potential for improvement in patient care based on genetic profiling is truly huge.

1.05.8 Regulatory Aspects Related to Drug Development of Individualized Medicines

Soon after the emergence of the concept of pharmacogenomics-based personalized medicine in the early years of the twenty-first century, strong expectations began to be voiced in the general media about the potential for using pharmacogenomics in the drug development process. New notions were mostly along the lines that the drug development process was soon about to undergo a transformation, so that drugs would be tested in clinical trials along with the collection and analysis of patients' genotypes. The notion was that, during new phase II clinical trials, the early data would be analyzed, to identify which patients are most likely to benefit from the new drug. Later on, during phase III trials, it would be possible to select only those patients whose genotypes predict them to be 'favorable responders' to the tested drug. Thus, it would be possible to show good drug safety and efficacy profiles using much smaller numbers of patients, thereby dramatically reducing the high costs of phase III trials, which play a major part in the continually rising drug costs.[55] Experts estimate that it might be sufficient to enlist a few hundreds of patients, instead of the typical 2000–3000 patients recruited for phase III trials.[56,57] Of course, this trend would also mean that later on, once the new drug is approved and on the market, it would only be prescribed for those patients whose genotypes 'fit' for the newly registered drug, along with an appropriate test. This would be expected to minimize adverse effects, improve efficacy, and thereby also improve patient compliance and reduce costs for healthcare providers. Thus, analysts have estimated that pharmaceutical companies would be able to charge higher costs for such 'genotype-tailored' drugs, owing to their favorable safety and efficacy profiles, and thereby averting some of the worries about smaller profits due to smaller market size expected when treating only a subpopulation of the patients who would otherwise be treated with similar drugs.[57–59]

However, such scenarios require that the drug regulatory authorities are geared up toward pharmacogenomics data submission, so that a new drug can be approved along with a genetic or proteomics-based test. Without such modifications in the drug review process, the vision of personalized medicine cannot succeed. Indeed, in the US the FDA has realized this fact, and has accordingly issued recommendations for pharmacogenomics data submissions; first as a draft, in November 2003, and later on as a final document, in March 2005. Issuing these new regulations was widely welcomed by the scientific community and drug industry alike, and heralded by many as "giving industry a strong political signal in favor of personalized medicines."[60] The main challenge for the FDA remains, however, to encourage industry to submit voluntary data to help refine the pharmacogenomics approval process and speed up the arrival of the next wave of such products to the market. At the time of writing of this chapter, this has only happened for a small number of drugs. The FDA guidelines outline the circumstances under which companies are required to submit pharmacogenomics data and the procedures for submitting them. Pharmacogenomics data must be submitted when they are based on valid biomarkers that have been rigorously tested by the scientific community and explicitly affect how trials for a product are designed. Using a number of examples, the document also defines a second category of data related to exploratory research. Data in this latter category do not have to be submitted; instead, the agency encourages companies to submit data voluntarily and in turn promises not to use it to make regulatory decisions. In addition to the document on pharmacogenomics data submission, in March 2005 the FDA issued a draft concept paper on co-developing gene-based diagnostics and therapeutics. A finalized document on this issue is expected to be published during 2006.[60–63]

Of note, while in the US the FDA has been increasingly proactive toward issuing regulations favoring pharmacogenomics data submission, this has not been the case in Europe. The European Medicines Agency (EMEA), the European equivalent of the FDA, has been rather cautious on these issues, seen by many experts as favoring the attitude of 'sit and wait' and, at the time of writing, the EMEA has not issued specific guidelines concerning the use of pharmacogenomics data in drug development,[64] although a draft of such guidelines has been released for external consultation from all interested parties in March 2005.[65] Indeed, the 2005 Royal Society report on personalized medicine[4] stated that an appropriate regulatory framework at a national and European level should be established by the UK Medicines and Healthcare Products Regulatory Agency and the EMEA. The report also suggested that these regulatory changes must include mandating some form of enforced postmarketing monitoring beyond phase III clinical trials. Such monitoring should explore links between genetic variability and clinical outcomes (with respect to both drug safety and drug efficacy).

Meanwhile the Council for International Organizations of Medical Sciences (CIOMS), affiliated with the World Health Organization (WHO), has set a working group on this issue, which has published its recommendations in 2005.[66] In addition, the European Commission, together with the European Society of Human Genetics, has issued a background paper on 'Polymorphic sequence variants in medicine: technical, social, legal and ethical issues. Pharmacogenetics as an example.'[67]

1.05.9 Current Barriers, Ethical Concerns, and Future Prospects for Individualized Medicine

Although pharmacogenetics is not a new discipline, as explained above, being nearly 50 years old, and although the entire human genome sequence has been available since 2003 (with a draft available since 2001), the high expectations of personalized medicine have not materialized so far, and are unlikely to be widely implemented in the clinic before 2020. What are the key barriers to the implementation of personalized medicine in the clinical setting? It seems that there are five key concerns that hinder its progress, as summarized in **Table 4**.

The prime barrier is definitely lack of sufficient knowledge about the clinical significance of human genetic variation and its effects on drug safety and efficacy. The solution to this key barrier must come from allocating more public funds for clinical studies on the utility of pharmacogenetics tests for reducing ADRs and improving drug efficacy.

A second concern is the lack of appropriate and affordable genotyping tools: as long as such tools are not widely available and approved by the appropriate authorities, pharmacogenetics testing is unlikely to be widely implemented. However, the pioneering tests are entering the market at the time of writing of this chapter, the first one being the Roche AmpliChip P450 (**Table 3**). Hopefully, additional tests will be approved in coming years. Their implementation would require National Health Service financing, and this in turn might stimulate public awareness and pressure, requiring decision-makers to make such tests widely available, as has been the case with certain drugs, most notably Herceptin for breast cancer.

Education in pharmacogenetics for health professionals is another key barrier to its successful clinical implementation. Most physicians who graduated before 1995 had very little formal education in molecular genetics, and are not familiar with the scope of human genome variation or with its clinical consequences. Thus, efforts to educate physicians, pharmacists, lab personnel, and nurses would be crucial so that there would be qualified healthcare personnel to apply pharmacogenetics diagnostic tests once they arrive in hospitals and community clinics. In the UK, the Royal Society report on "Personalised medicines: hopes and realities"[4] recommended to expend professional education in pharmacogenetics. Such efforts to educate the coming generation of physicians and health professionals

Table 4 Current barriers for clinical implementation of personalized medicine, and potential solutions

Barrier	*Potential solutions*
Knowledge	Allocating more public funds for clinical studies on the utility of pharmacogenetics tests for reducing ADRs and improving drug efficacy
Tools	Incentives for industry to develop genotyping tools; National Health Service financing for available tools
Education	Education in pharmacogenetics for health professionals; also for the general public via media efforts (public TV, etc.)
Regulatory	Regulations allowing co-development of drugs and genotyping tests; measures to allow similar developments for already marketed drugs
Ethical concerns	Legislation prohibiting genetic discrimination; developing tools to protect genetic information

are already being made at some medical schools, for example, at Tel-Aviv University, where pharmacogenomics has been taught since 2001 as part of the basic pharmacology class.[68] A dedicated graduate program was initiated at University of California San Francisco.[84] However, according to a recent global survey, most medical schools have not incorporated pharmacogenomics in their teaching curricula. Thus, the International Society for Pharmacogenomics[85] has recently issued a consensus article calling upon deans of education in medical, pharmaceutical, and health sciences schools to incorporate the teaching of pharmacogenomics to their school's curricula.[69] Hopefully, this call will soon be heard. If not, personalized medicine may join the long list of novel clinical technologies that had to wait 17 years on average before being incorporated into the clinical arena.[21] However, it can be expected that more widespread professional education will hasten this process, thereby reducing the alarmingly high costs to society of ADRs.

Educational efforts will also be needed for society, so that the general public becomes aware of the benefits of pharmacogenetics testing, and not fear the new technologies of genotyping as part of the drug prescription process. Public education would also help to ensure participation in clinical trials involving pharmacogenetics information. Such educational programs might involve media efforts, including public TV series, science museum exhibits, and health fairs.[27]

Regulatory aspects would also have a key role in the rate that personalized medicine is implemented. As discussed above, the FDA has already geared up for the co-development of drugs along with an associated genotyping test. Similar measures are likely to follow in other countries. However, to allow improvement in the individualized treatments with existing drugs, new regulations and international agreements might be required, so that patenting would become possible for co-development of drugs along with genotyping tests for drugs already on the market, even if their original patents have already expired. In the most constructive scenario, companies would be allowed to prolong patent lives for existing drugs, by issuing a new patent for the same drug when prescribed along with a genotyping test, and used according to the test results. This would create an incentive to drug-makers to study the genotyping information related to drug safety and efficacy, and thereby also improve the quality of healthcare. It is expected that such trends would further increase drug costs, as renewed patents would mean a license to charge more money for such drugs. However, very likely the extra costs would be justified in terms of money saved from reduced rates of ADRs; these are currently estimated to cost up to US$180 billion annually for the US.[8]

Ethical concerns related to the development of personalized medicine have been the focus of several group efforts. Discussions are well under way regarding medical, ethical, societal, and regulatory aspects of pharmacogenomics and its expected implementation into the clinic.[14,70–75] Entire public conferences have been devoted to bioethics and societal aspects of pharmacogenomics, including special working parties such as those set by the Nuffield Council on Bioethics[76] and the European Commission.[77] Global legislation prohibiting genetic discrimination, according to the model set forth by the Israeli Genetic Information Law,[78] which prohibits transfer of genetic data for any purpose other than treating the patient, and strictly prohibits use of such data by employers, financial or insurance companies, seems to be a key prerequisite for the successful implementation of pharmacogenetics.

In a July 2005 editorial of the *Boston Globe*,[79] Francis Collins, the leader of the concluded Human Genome Project and current head of the National Human Genome Research Institute, wrote as follows: "To realize the full potential of personalized medicine, we must venture beyond the fields of science and medicine and into the ethical, legal, and social arenas. For example, without legislative protections against genetic discrimination in health insurance and the workplace, many people will be reluctant to undergo potentially life-saving genetic tests or to participate in the clinical trials needed to develop genetically targeted therapies. Given that more than 800 genetic tests are now available and hundreds more are on the horizon, we need this legislation."

In summary, pharmacogenetics will not replace, but enhance, existing good medical practice. A deliberate approach starts with investing more in studies aimed at clarifying relations between genotypes and drug response phenotypes (both safety and efficacy); educating healthcare professionals by illustrating the benefits of pharmacogenomics; and educating society about the potential benefits for healthcare from the new genomics and proteomics technologies. As more knowledge is gained on genetic effects on drug safety, and more genotyping-based diagnostic tools for avoiding ADRs become available, better-educated societies would push for the inclusion of such tools in the National Health Service. In the long run, it may be the concern of health providers about potential lawsuits by patients harmed by ADRs that will push forward the accomplishment of personalized medicine tools in the clinical setting.

Many other open questions remain about how personalized medicine can be reconciled with equity in public healthcare. In a 2005 workshop on personalized medicine held at Tel-Aviv university, entitled "Personalized Medicine Europe: Health, Genes and Society,"[86] the following key questions were discussed:

- Should we oblige industry to do 'something' for people who do not have the right genotype for their drug? What should this 'something' include?
- Equality in access to new health technologies: who pays for genetic diagnostics?

- How far will individualized medicine change the interrelation of individual and public health, especially with regard to concepts of health responsibility?
- How can society ensure better equality in healthcare, along with individualization of pharmacotherapy?
- Will personalized genetic medicine be driven by the pharmaceutical industry in the future?
- What should our priorities be for incorporating personalized medicine into the clinical setting?

No consensus was reached in these discussions, and it seems that there will not be a clear-cut solution, but rather, solutions that keep evolving hand in hand with emerging pharmacogenetics knowledge, development of new diagnostics technologies, and the availability of new diagnostics tests. Clearly, further discussion on these key questions must continue, and being able to formulate these tough questions is an essential key step toward formulating the best answers.

In the words of Francis Collins,[79] "Other tough questions that we as a society need to ask ourselves are: Will access to genomic technologies be equitable? Will knowledge of human genetic variation reduce prejudice or increase it? What boundaries will need to be placed on this technology, particularly when applied to enhancement of traits rather than prevention or treatment of disease? Will we succumb to genetic determinism, neglecting the role of the environment and undervaluing the power of the human spirit?"

It is clear that the answers to these tough questions are still lacking, and many will remain so for quite some time. Fears that personalized medicine would obstruct equity in healthcare will continue, and must be addressed by proper regulation and legislation, to ensure that basic human rights are maintained.

It is nearly impossible to forecast at this time what direction will be taken by medicine by the middle of the twenty-first century. However, it is very likely that, by this time, many of the causes of ADRs would be largely eliminated thanks to genetic tests able to preidentify susceptible individuals. It remains to be seen if genetic information would also be able to allow better, more efficacious drugs that would be 'tailored' for patients, and to what extent this would change the practice of medicine globally. As these words are written (October 2005) we can only express hopes that personalized medicine will live up to its promises, and wonder whether it will lead to an improvement or a decline in equity of healthcare. Hopefully, personalized medicine will allow the establishment of better healthcare equity not only between individuals, but also between nations.

References

1. Hafter, E. *Schweiz. Med. Wochenschr.* **1976**, *106*, 262–266.
2. Langreth, R.; Waldholz, M. *Oncologist* **1999**, *4*, 426–427.
3. Stephens, J. C. *Mol. Diagn.* **1999**, *4*, 309–317.
4. Royal Society Working Group on Pharmacogenetics. Personalised Medicines: Hopes and Realities. The Royal Society, September **2005**. Available at: http://www.royalsoc.ac.uk/displaypagedoc.asp?id = 15874 (accessed June 2006).
5. Dormann, H.; Criegee-Rieck, M.; Neubert, A.; Egger, T.; Geise, A.; Krebs, S.; Schneider, T.; Levy, M.; Hahn, E.; Brune, K. *Drug Safe* **2003**, *26*, 353–362.
6. Dormann, H.; Neubert, A.; Criegee-Rieck, M.; Egger, T.; Radespiel-Troger, M.; Azaz-Livshits, T.; Levy, M.; Brune, K.; Hahn, E. G. *J. Intern. Med.* **2004**, *255*, 653–663.
7. Pirmohamed, M.; James, S.; Meakin, S.; Green, C.; Scott, A. K.; Walley, T. J.; Farrar, K.; Park, B. K.; Breckenridge, A. M. *Br. Med. J.* **2004**, *329*, 15–19.
8. Ernst, F. R.; Grizzle, A. J. *J. Am. Pharm. Assoc.* **2001**, *41*, 192–199.
9. Vogel, F. *Ergeb. Inn. Med. Kinderheilkd* **1959**, *12*, 52–125.
10. Evans, D. A.; Clarke, C. A. *Br. Med. Bull.* **1961**, *17*, 234–240.
11. Kalow, W. *Pharmacogenetics – Heredity and the Response to Drugs*; W B Saunders: Philadelphia, PA, USA, 1962.
12. Motulsky, A. G. *JAMA* **1957**, *165*, 835–837.
13. Marshall, A. *Nat. Biotechnol.* **1997**, *15*, 829–830.
14. Webster, A.; Martin, P.; Lewis, G.; Smart, A. *Nat. Rev. Genet.* **2004**, *5*, 663–669.
15. Osler, W. *The Principles and Practice of Medicine*; D. Appleton: New York, 1892.
16. Gorrod, A. E., Oxon, M. D. *Lancet* **1902**, 1616–1620.
17. Snyder, L. H. *Ohio J. Sci.* **1932**, *32*, 436–438.
18. Lash, L. H.; Hines, R. N.; Gonzalez, F. J.; Zacharewski, T. R.; Rothstein, M. A. *J. Pharmacol. Exp. Ther.* **2003**, *305*, 403–409.
19. Meyer, U. A. *Nat. Rev. Genet.* **2004**, *5*, 669–676.
20. Bass, M. H. *J. Mt Sinai Hosp. N. Y.* **1959**, *26*, 421–423.
21. Lenfant, C. *N. Engl. J. Med.* **2003**, *349*, 868–874.
22. Ingelman-Sundberg, M. *Trends Pharmacol. Sci.* **2004**, *25*, 193–200.
23. Evans, W. E.; Relling, M. V. *Science* **1999**, *286*, 487–491.
24. Evans, W. E.; McLeod, H. L. *N. Engl. J. Med.* **2003**, *348*, 538–549.
25. Weinshilboum, R.; Wang, L. *N. Engl. J. Med.* **2003**, *348*, 529–537.
26. Ingelman-Sundberg, M. *Pharmacogenomics J.* **2005**, *5*, 6–13.
27. Frueh, F. W.; Gurwitz, D. *Pharmacogenomics* **2004**, *5*, 571–579.
28. Kirchheimer, J.; Brockmoller, J. *Clin. Pharmacol. Ther.* **2005**, 77, 1–16.

29. Lee, C. R.; Goldstein, J. A.; Pieper, J. A. *Pharmacogenetics* **2002**, *12*, 251–263.
30. Ingelman-Sundberg, M.; Daly, A. K.; Nebert, D. W. Human cytochrome P450 allele nomenclature home page. 2004, http://www.imm.ki.se/CYPalleles (accessed June 2006).
31. Xie, H. G.; Prasad, H. C.; Kim, R. B.; Stein, C. M. *Adv. Drug Devel. Rev.* **2002**, *54*, 1257–1270.
32. Sanderson, S., Emery, J., Higgins, *J. Genet. Med.* **2005**, *7*, 97–104.
33. McLeod, H. L.; Siva, C. *Pharmacogenomics* **2002**, *3*, 89–98.
34. Sanderson, J.; Ansari, A.; Marinaki, T.; Duley, J. *Ann. Clin. Biochem.* **2004**, *41*, 294–302.
35. Gupta, E.; Lestingi, T. M.; Mick, R.; Ramirez, J.; Vokes, E. E.; Ratain, M. J. *Cancer Res.* **1994**, *54*, 3723–3725.
36. Bosma, P. J.; Chowdhury, J. R.; Bakker, C.; Gantla, S.; de Boer, A.; Oostra, B. A.; Lindhout, D.; Tytgat, G. N.; Jansen, P. L.; Oude Elferink, R. P. et al. *N. Engl. J. Med.* **1995**, *333*, 1171–1175.
37. Marsh, S.; McLeod, H. L. *Pharmacogenomics* **2004**, *5*, 835–843.
38. Weinshilboum, R.; Wang, L. *Nat. Rev. Drug Disc.* **2004**, *3*, 739–748.
39. Emens, L. A. *Am. J. Ther.* **2005**, *12*, 243–253.
40. Pignatti, P. F. *Pharmacol. Res.* **2004**, *49*, 343–349.
41. Israel, E. *J. Allergy Clin. Immunol.* **2005**, *115*, S532–S538.
42. Israel, E.; Drazen, J. M.; Liggett, S. B.; Boushey, H. A.; Cherniack, R. M.; Chinchilli, V. M.; Cooper, D. M.; Fahy, J. V.; Fish, J. E.; Ford, J. G. et al. *Am. J. Respir. Crit. Care Med.* **2000**, *162*, 75–80.
43. Martinez, F. D.; Graves, P. E.; Baldini, M.; Solomon, S.; Erickson, R. *J. Clin. Invest.* **1997**, *100*, 3184–3188.
44. Israel, E.; Chinchilli, V. M.; Ford, J. G.; Boushey, H. A.; Cherniack, R. M.; Craig, T. J.; Deykin, A.; Fagan, J. K.; Fahy, J. V.; Fish, J. et al. *Lancet* **2004**, *364*, 1505–1512.
45. Pelaia, G.; Vatrella, A.; Gallelli, L.; Cazzola, M.; Maselli, R.; Marsico, S. A. *Pulm. Pharmacol. Ther.* **2004**, *17*, 253–261.
46. Reynolds, G. P.; Templeman, L. A.; Zhang, Z. J. *Prog. Neuro-Psychopharmacol. Biol. Psychiatry* **2005**, *29*, 1021–1028.
47. Xie, H. G.; Kim, R. B. *Clin. Pharmacol. Ther.* **2005**, *78*, 19–24.
48. Hall, S. D.; Wang, Z.; Huang, S. M.; Hamman, M. A.; Vasavada, N.; Adigun, A. Q.; Hilligoss, J. K.; Miller, M.; Gorski, J. C. *Clin. Pharmacol. Ther.* **2003**, *74*, 525–535.
49. Morris, K. *Lancet* **2000**, *355*, 45.
50. Williams, M. *Curr. Opin. Invest. Drugs* **2003**, *4*, 31–36.
51. Gurwitz, D. *Drug Dev. Res.* **2003**, *60*, 71–74.
52. Maier, W.; Zobel, A.; Rietschel, M. *Pharmacopsychiatry* **2003**, *36*, 195–202.
53. Lesch, K. P.; Bengel, D.; Heils, A.; Sabol, S. Z.; Greenberg, B. D.; Petri, S.; Benjamin, J.; Muller, C. R.; Hamer, D. H.; Murphy, D. L. *Science* **1996**, *274*, 1527–1531.
54. Kirchheimer, J.; Brosen, K.; Dahl, M. K.; Gram, L. F.; Kasper, S.; Roots, I.; Sjoqvist, F.; Spina, E.; Brockmoller, J. *Acta Psychiatr. Scand.* **2001**, *104*, 173–192.
55. Binzak, B. A. *Food Drug Law J.* **2003**, *58*, 103–127.
56. Singer, E. *Nat. Med.* **2005**, *11*, 462.
57. Abbott, A. *Nature* **2003**, *425*, 760–762.
58. Jain, K. K. *Curr. Opin. Mol. Ther.* **2002**, *4*, 548–558.
59. Wallner, C. *Dis. Manage.* **2004**, 7, S17–S19.
60. Katsnelson, A. *Nat. Biotechnol.* **2005**, *23*, 510.
61. Little, S. *IDrugs* **2005**, *8*, 648–650.
62. Lewis, R. *Nature* **2005**, *436*, 746–747.
63. Harper, C. C.; Philip, R.; Robinowitz, M.; Gutman, S. I. *Exp. Rev. Mol. Diagn.* **2005**, *5*, 643–648.
64. Knudsen, L. E. *Toxicol. Appl. Pharmacol.* **2005**, *207*, 679–683.
65. EMEA. *Committee for Medicinal Products for Human Use (CHMP) Guidelines on Pharmacogenetics*; EMEA: London, 2005.
66. Council for International Organizations of Medical Sciences (CIOMS) Report. *Pharmacogenetics: Towards Improving Treatment with Medicines*; CIOMS: Geneva, 2005.
67. European Commission. Polymorphic Sequence Variants in Medicine: Technical, Social, Legal and Ethical Issues. Pharmacogenetics as an example, 2004. http://www.eshg.org/ESHG-IPTSPGX.pdf (accessed June 2006).
68. Gurwitz, D.; Weizman, A.; Rehavi, M. *Trends Pharmacol. Sci.* **2003**, *24*, 122–125.
69. Gurwitz, D.; Lunshof, J. E.; Dedoussis, G.; Flordellis, C. S.; Fuhr, U.; Kirchheiner, J.; Licinio, J.; Llerena, A.; Manolopoulos, V. G.; Sheffield, L. J. et al. *Pharmacogenomics J.* **2005**, *5*, 21–25.
70. Lipton, P. *Pharmacogenomics J.* **2003**, *3*, 14–16.
71. Wertz, D. C. *Pharmacogenomics J.* **2003**, *3*, 194–196.
72. Winkelmann, B. R. *Pharmacogenomics* **2003**, *4*, 531–535.
73. Weijer, C.; Miller, P. B. *Pharmacogenomics J.* **2004**, *4*, 9–16.
74. Breckenridge, A.; Lindpaintner, K.; Lipton, P.; McLeod, H.; Rothstein, M.; Wallace, H. *Nat. Rev. Genet.* **2004**, *5*, 676–680.
75. van Delden, J.; Bolt, I.; Kalis, A.; Derijks, J.; Leufkens, H. *Bioethics* **2004**, *18*, 303–321.
76. Nuffield Council on Bioethics. Pharmacogenetics: ethical issues. **2003**. http://www.nuffieldbioethics.org/fileLibrary/pdf/pharmacog_consultation (accessed June 2006).
77. European Commission. Ethical, legal and social aspects of genetic testing: research, development and clinical applications. European Commission Expert Group 2004. http://europa.eu.int/comm/research/-conferences/2004/genetic/pdf/report_en.pdf/
78. Israel Genetic Information Law 2000. www.justice.gov.il/NR/rdonlyres/46993742-CA48-41D6-A785-21CA8A0E2B52/0/GeneticInformation-LawEdited_050901.doc
79. Collins, F. *The Boston Globe*, July 17, 2005.
80. PMC: The Personalized Medicine Coalition. www.personalizedmedicinecoalition.org/index.php (accessed June 2006).
81. Clinical Trials Genomics at FDA. www.fda.gov/cder/genomics/default.htm (accessed June 2006).
82. AMA (Genetics) Pharmacogenomics. www.ama-assn.org/ama/pub/category/2306.html (accessed June 2006).
83. Roche Diagnostics AmpliChip P450 Test. www.roche-diagnostics.com/products_services/amplichip_cyp450.html (accessed June 2006).
84. UCSF graduate program in Pharmaceutical Sciences and Pharmacogenomics. http://www.ucsf.edu/dbps/pspg.html (accessed June 2006).
85. International Society for Pharmacogenomics. http://149.142.238.229/isp/default.asp (accessed June 2006).

86. Personalized Medicine Europe: Health, Genes and Society. (ESF-TAU workshop co-organized by the author; June 19–21, 2005). www.functionalgenomics.org.uk/sections/activitites/2005/Livshits/info.htm.www.functionalgenomics.org.uk/sections/activitites/Reports/report_TelAviv_2005.htm (accessed June 2006).
87. National Human Genome Research Institute. http://www.genome.gov/12513430 (accessed June 2006).

Biographies

David Gurwitz obtained his PhD from the Department of Biochemistry at Tel-Aviv University in 1986. His thesis delineated the interactions of brain muscarinic acetylcholine receptors with G proteins. From 1986 to 1989 he was a postgraduate fellow at the University of California Irvine. These studies have led to the discovery of the hormone-like properties of thrombin, the major blood coagulation enzyme, toward neuronal and astroglial cells, thereby opening up a new research field on the role of endogenous thrombin inhibitors in brain physiology and pathology.

In 1989, David joined the Israel Institute for Biological Research, where he led the biochemistry team in a project developing M1-selective muscarinic agonists as therapeutics. The lead compound, AF102B (Cevimeline), has received FDA approval for the treatment of Sjögren's syndrome, thereby becoming the second-ever Israeli drug receiving FDA approval.

In January 1995, David was appointed to direct the National Laboratory for the Genetics of Israeli Populations at the Sackler Faculty of Medicine at Tel-Aviv University. The laboratory is a national repository for human DNA samples and immortalized cell lines, representing the unique ethnic diversity of Jewish and Arab communities in Israel. Thousands of DNA samples distributed by the repository over the years have helped in numerous studies on human clinical genetics, population genetics, and pharmacogenetics.

Since 2001 David has been teaching pharmacogenetics to MD students at the Tel-Aviv University, as part of their pharmacology training. Since 2003 he also teaches a 30-hour graduate course entitled Pharmacogenetics: Toward Personalized Medicine. David has authored or co-authored over 100 scholarly articles in diverse disciplines including biochemistry, cell biology, and pharmacology. He is an associate editor or member of the editorial board of several scientific journals, including *Pharmacogenomics, Personalized Medicine, Drug Development Research,* and *Trends in Molecular Medicine.* Since 2004, David has been representing Israel on the steering committee on pharmacogenetics of the Organization for Economic Co-operation and Development (OECD), and since 2005 he is also a consultant for the European Commission's Institute for Prospective Technological Studies (IPTS) on pharmacogenetics.

Vangelis G Manolopoulos obtained his PhD from the Department of Pharmacology, at the Medical School of Patras University in Greece in 1991. Part of the experimental work for his thesis was done at the Department of Clinical Pharmacology, University of Groningen Medical School, Groningen, The Netherlands. His thesis was on the signal transduction cascades following alpha-adrenergic receptor stimulation in vascular smooth muscle. From January 1992 to August 1995 he was a postdoctoral fellow at the University of Wisconsin-Milwaukee Clinical Campus at Mt Sinai Hospital. His work was on the heterogeneity of the adenylate cyclase signaling pathway as well as the hormonal roles of thrombin in different types of endothelial cells.

From September 1995 until November 1998 Vangelis was at the Laboratory of Physiology, Medical School, Catholic University of Leuven, in Belgium, investigating the physiological roles of Volume-Regulated Anion Channels (VRAC) in endothelial cells. His finding that VRAC blockers inhibit angiogenesis led to the development of novel specific inhibitors of these channels currently in clinical development for antitumoral activity.

In December 1998, Vangelis joined the Laboratory of Pharmacology, at the Medical School of Democritus University of Thrace in Alexandroupolis, Greece as a faculty member. Since 2003 he is the Head of this Laboratory. Pharmacogenetics consist the major interest of his group, both from a research perspective and also in developing a pharmacogenetics clinical service at the Academic Hospital of Alexandroupolis with the intention of serving all North Greece.

For 3 years (2000–2002) Vangelis has also served as a scientific consultant to a major Greek Pharmaceutical Company, with the duty of establishing laboratories for in vitro preclinical research and designing and implementing a preclinical development program for novel proprietary compounds with antiatherosclerotic, hypolipidemic, and antioxidant properties. The development program was successfully brought to completion and two of the compounds are now in clinical development.

In addition to his regular teaching duties in Pharmacology, since 2001 Vangelis has been teaching an elective course in Pharmacogenetics to MD students at University of Thrace. He has authored or co-authored over 30 articles in diverse disciplines including pharmacology, physiology, cell and molecular biology. Since 2003 he is an elected member of the Board and Treasurer of the Greek Society of Pharmacology and Clinical Pharmacology, and since 2005 he is the deputy representative of Greece at the European Association of Clinical Pharmacology.

Comprehensive Medicinal Chemistry II
ISBN (set): 0-08-044513-6

ISBN (Volume 1) 0-08-044514-4; pp. 279–295

1.06 Gene Therapy

M Thanou, Imperial College London, London, UK
S Waddington, University College London, London, UK
A D Miller, Imperial College London, London, UK

1.06.1 Introduction

While many readers may wonder whether gene therapy deserves a place in the medicinal chemistry pantheon, it is hoped that this chapter will show that gene therapy is a potentially enormously valuable technology that still needs the input of creative medicinal chemists in order to fulfill its potential. Negative impressions about gene therapy are everywhere, even within the learned societies that exist to promote gene therapy! So, why should this chapter be written at all? The simple answer is that, surprising though it may seem, gene therapy may have troubles but it is probably here to stay. In other words, gene therapy is slowly emerging as a fully mature therapeutic modality in spite of setbacks. Only time will tell if this particular therapeutic modality will be sufficiently strong in the future to challenge other major branches of medicinal chemistry for supremacy, but the reader should be in no doubt that gene therapy is starting to make a serious contribution to therapy today, and seems not only able to provide treatments of disease complementary to existing treatments but may even be able to provide treatments for orphan diseases that are otherwise untreatable.

1.06.1.1 Gene Therapy Definitions

Gene therapy may be described as the use of genes as medicines to treat disease, or, more precisely, as the delivery of nucleic acids by means of a vector to patients for some therapeutic purpose. The vector is crucial for gene therapy. The vector represents the means of carriage of the therapeutic nucleic acids from their site of administration to their desired site of action in cells of interest within the target organ(s) of choice. Identifying the appropriate means of carriage for any gene therapy is the key rate-limiting step in the development of most promising gene therapy strategies for disease treatment. Viral vectors (adenovirus and retrovirus derived) have proved to be the most popular means of carriage in clinical trials (**Figure 1**), but physical nonviral approaches (such as direct injection, biolistics ('gene gun'), or electroporation) (see below) used in conjunction with naked plasmid deoxyribonucleic acid (pDNA) have become substantially more popular, followed by synthetic nonviral approaches using cationic liposome-based vectors (see below) to deliver pDNA to cells (lipofection) (**Figure 1**). Synthetic nonviral approaches using polymer-based vector systems to deliver pDNA have also begun to emerge in a few clinical trials. There is now even one adenoviral-based p53 gene therapy product on the market (Gendicine) licensed for use in China for the treatment of certain head and neck squamous cell cancers and applied to other later indications as well.[1a,2]

Viral vectors deliver either ribonucleic acid (RNA) or DNA to cells depending upon whether the virus is an RNA or DNA virus, respectively. In either event, delivered nucleic acids enter transcription that results in mRNA for translation into a therapeutically active protein, or nonoding RNAs (ncRNAs). These ncRNAs can be either micro-RNAs

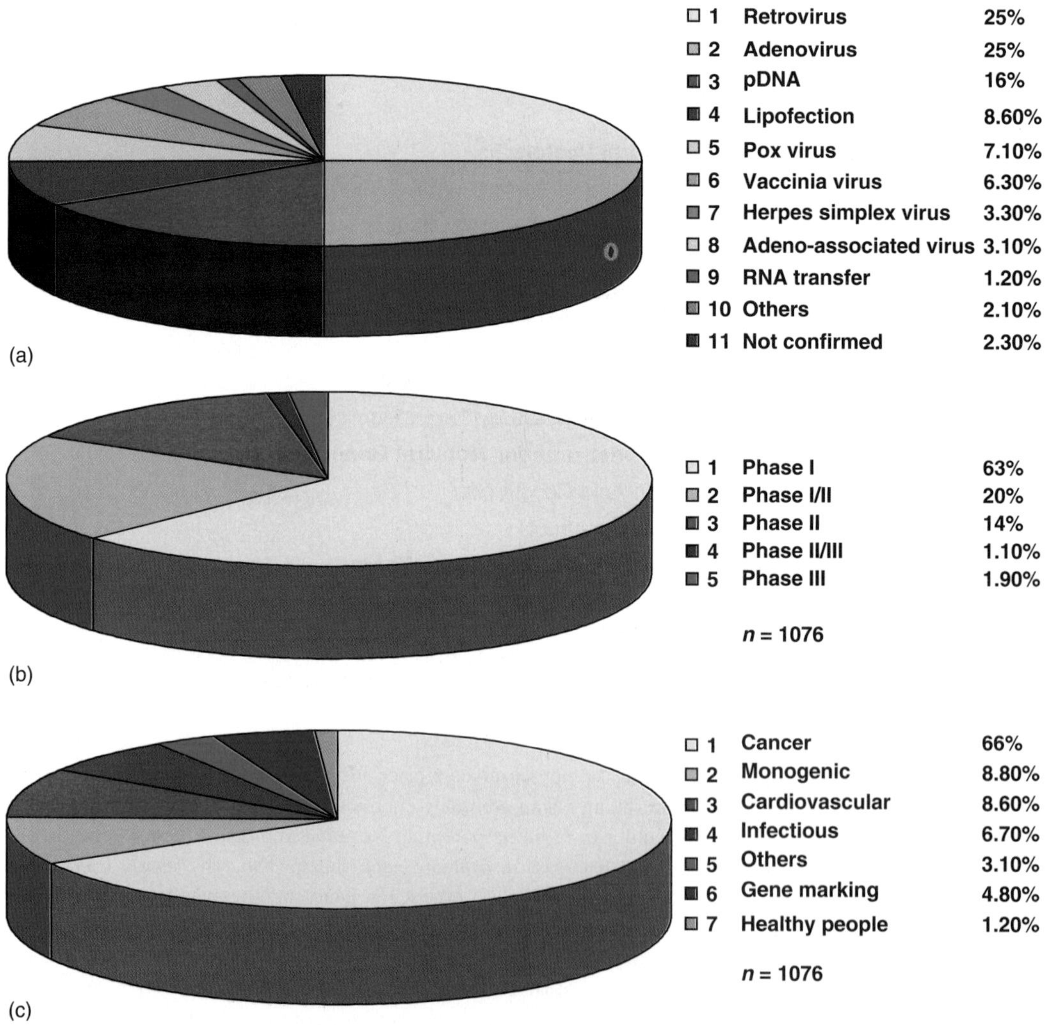

Figure 1 Summary of salient clinical trial data to September 2005. Clinical trial data has been analyzed according to (a) vector use, (b) clinical phase, and (c) disease indication. Data were derived from the Journal of Gene Medicine website.[1]

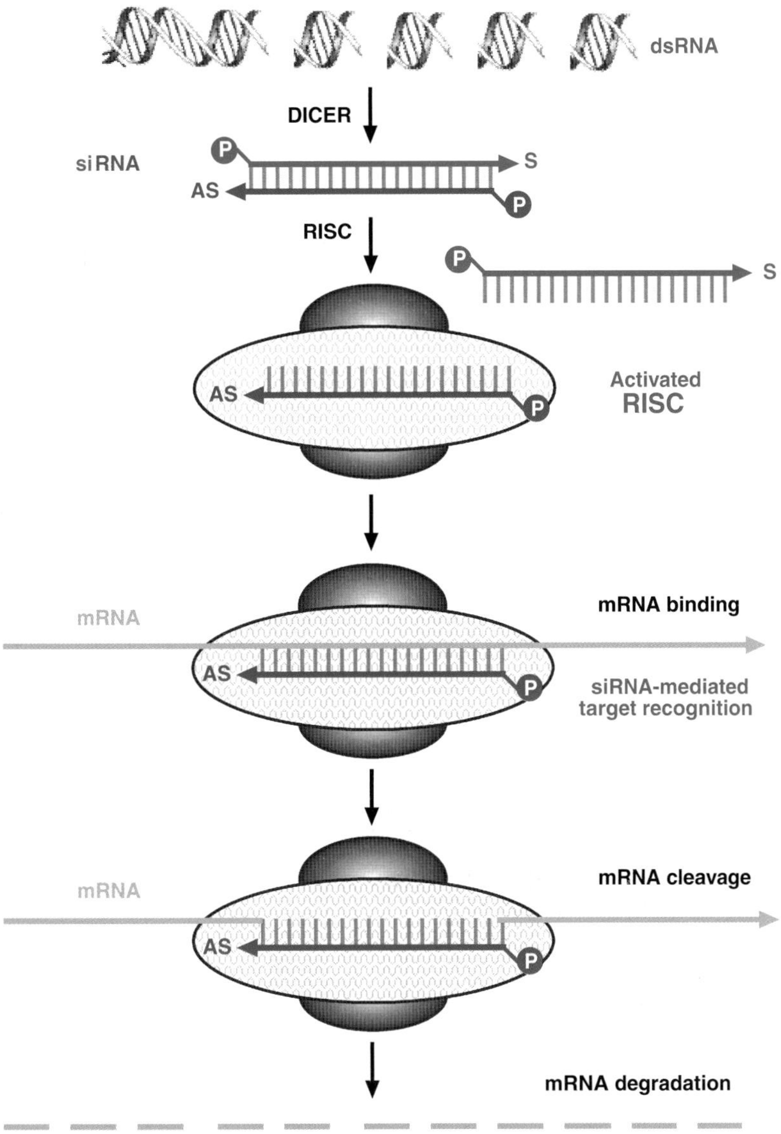

Figure 2 Summary of the siRNA mechanism of action. siRNA is derived from long double-stranded RNA (dsRNA) through the action of the DICER enzyme system. Interaction of siRNA molecules with the RISC enzyme system results in sense (S)/antisense (AS) strand separation, and the probable capture of individual AS strands by RISC. So activated, RISC recognizes mRNA molecules with Watson–Crick base pair complementary to bound AS strands. Once recognized and bound, mRNA is cleaved and then degraded.[3] (Reproduced with permission of Royal Society of Chemistry from Kostarelos, K.; Miller, A. D. *Chem. Soc. Rev.* **2005**, *34*, 970–994.)

(miRNAs) that are able to control gene transcription–translational processes, or small interfering/small interference RNAs (siRNAs) that target specific mRNA molecules for destruction (**Figure 2**). Both miRNAs and siRNAs are capable of therapeutic gene knockdown, leading to selective protein downregulation (see later).[3–8] By contrast with viral vectors, physical and synthetic vectors typically deliver pDNA-expressing therapeutic genes to cells. However, in principle these vectors can also mediate the delivery of many other types of nucleic acids such as artificial chromosomes, mRNA, or even actual ncRNAs themselves.

1.06.1.2 Gene Therapy Mechanisms

The putative mechanism of any given gene therapy depends upon the disease concerned and the nature of the nucleic acid delivered. Where monogenic disorders are concerned, gene therapy strategies for treatment that are now in various stages of clinical trials are either gene supplementation or replacement strategies. In other words, the gene therapy aims to supplement or replace existing, non- or partially functional mutant genes with normal wild-type genes able to express sufficient wild-type protein in cells for therapeutic correction. In the case of cancers, current clinical gene therapy strategies are designed to promote cellular destruction either by introducing genes for immunostimulation (cytokine and/or antigen genes) or genes for programmed cell death or necrosis (tumor suppressor, replication inhibitor, or suicide genes). Gene knockdown represents an alternative form of gene therapy, particularly for degenerative diseases and viral infections. In this context, the use of ncRNAs, particularly siRNA, could be particularly interesting and significant (see below).

1.06.1.3 Historical Background, Successes, and Failures

Gene therapy is a therapeutic modality with enormous promise, but one that is considered to have failed to deliver much of therapeutic significance in spite of all the apparent clinical interest (see **Figure 1**; the total number of gene therapy clinical trials approved worldwide is currently 1076, comprising only 12 Phase II/III trials and 20 Phase III trials). Clinical trial activity in gene therapy began in 1989, peaked in 1999, and is now currently declining. This decline has been marked by a number of clinical trial problems, including a death (Jesse Gelsinger), from toxic liver shock during an adenovirus-based clinical trial in 1999,[9] the anomalous appearance of a transgene in the gonads during adeno-associated virus-based preclinical trials in 2001,[10] signs of hypertension in lipofection clinical trials in 2005 (Pro-1),[11] and the development of leukemia in retrovirus-based clinical trials for ex vivo treatment of X-linked severe combined immune-deficiency (X-linked SCID).[12,13] Furthermore, pioneer monogenic diseases such as cystic fibrosis (CF) have proved refractory to gene therapy strategies in spite of the fact that proof of concept data was obtained at least as far back as 1993[14] that demonstrated effective treatment of the disease by gene therapy in transgenic CF animal models. These failures may look dismal, but they can be balanced by some notable successes as well. For instance, Gendicine in China has been administered to tumors in over 1000 patients to date without apparent side effects. Also, although X-linked SCID gene therapy trials have proved controversial owing to the appearance of leukemia in some patients, other patients have actually shown spectacular recovery to full health for over a year from a condition that is otherwise untreatable.[15–18] Expect more positive news as current research and development activity feeds into clinical trials in the future!

1.06.2 Current Genetic Therapies

1.06.2.1 Overview of Current Clinical Trials

Unsurprisingly, cancer diseases are by far the largest indication (66%) addressed by gene therapy in clinical trials, followed by monogenic disorders (8.8%), vascular diseases (8.6%), and infectious diseases (6.7%) in almost equal measure (see **Figure 1**). Clinical trials in cancer have largely been restricted to the development of recombinant anticancer cellular vaccines, ex vivo delivery of nucleic acids with viral vectors, or intratumoral administration of therapeutic nucleic acids using simple viral, synthetic, or physical vectors. There have been only a handful of more adventurous systemic administration protocols, none of which have progressed far to date.[19]

In the case of monogenic disorders, clinical trials in CF dominate, but trials with adenosine deaminase (ADA) deficiency and X-linked SCID have been more successful. Of these, CF gene therapy trials have required topical administration of nucleic acids to lungs using both simple viral and synthetic vectors.[20–22] In contrast, gene therapy clinical trials for the treatment of both ADA deficiency and X-linked SCID have only required ex vivo protocols using retrovirus to deliver nucleic acids to hemopoietic stem cells.[13,16] Where infectious disease gene therapy is concerned, trials have been largely focused around HIV treatment involving vaccination, although siRNA strategies are certain to lead to increasing gene therapy clinical trial activities aiming at the treatment of other viral diseases by siRNA mediated gene knockdown (see later).

1.06.2.2 Main Obstacles for Gene Therapies

In spite of the relatively undemanding nature of many of the selected routes of administration in clinical trials, results from many early stage clinical trials have been frequently disappointing due to the inadequacy of the vectors (see above). Arguably, researchers were seduced by the apparent simplicity of gene therapy approaches to treatment, leading to a drive for clinical applications before vector technologies had been adequately developed or understood.[23–25] For instance, current clinical data now suggest that virus-based vectors may be highly effective for nucleic acid delivery but suffer from a variety of problems, including immunogenicity, toxicity, and oncogenicity. By contrast, preclinical and current clinical trial data also suggest that synthetic nonviral vector systems are much less affected by immunogenicity, toxicity, and oncogenicity, but lack delivery efficacy. Hence, there is a consensus that virus-based vectors need to be engineered for improved safety profiles while synthetic nonviral vector systems should be engineered for improved efficacy. In consequence, a heavy burden of innovation has now been placed on the design of new and better vectors to bring about successful gene therapies.[26]

However, there is another way to frame this 'vector problem.' There can and should be an essential pragmatism to gene therapy if we seek to combine the very best of medical and surgical practice with the vector delivery of therapeutic nucleic acids. If each intended gene therapy can be adequately defined in terms of not just its molecular target but also in terms of the anatomical location of the target cells of interest within the organ(s) of choice, then that provides the critical starting point. From there, a rationally designed combination of appropriate surgical techniques and the selection of vectors whose delivery capabilities most closely match the requirements made of them by the process of migration from the point of administration to the site(s) of action should create opportunities for many gene therapies, assuming that there are no further problems created by the choice of therapeutic nucleic acid as well. Arguably, the extent of the invasive methods of administration required should then vary inversely with improvements in vector technologies (e.g., biological stability), allowing for greater and greater targeted access to the site(s) of action from less and less surgically demanding methods of administration.

The 'nucleic acid problem' in gene therapy can be broken down into fears over transgene biodistribution to the gonads and oncogenesis. Gene therapy should be seen as a somatic medicine that seeks to treat disease at a more fundamental level than most other therapeutic modalities are capable of. However, the delivery of nucleic acids by whatever vector is perceived to carry a finite risk of gene transfer to the gonads and genetic modification of gametes, placing future generations at risk from deleterious mutations. Viral vectors probably present a much greater risk of oncogenicity, particularly retroviral vectors that mediate insertion into actively expressing gene loci, thereby creating a high risk of oncogenesis. Adeno-associated virus (AAV) was widely thought to mediate site-selective chromosomal integration so avoiding the oncogenesis risk altogether, but this now no longer appears to be the case.[27,28] Importantly, oncogenesis risk may be avoided all together using synthetic or physical vectors that deliver pDNAs to cells. The reason for this is that in synthetic vectors pDNA typically expresses in an epichromosomal manner. However, in general, where chromosomal integration does not take place irrespective of whether a viral, synthetic, or physical vector is used for delivery, attenuation of gene expression is readily observed with increasing time and cumulative dose of nucleic acid. Accordingly, an ideal situation to minimize the oncogenicity risk should be to effect site-selective chromosomal integration of pDNA that has been delivered by either a synthetic or physical vector.[29] Certainly, technologies are advancing rapidly to open up this possibility (see below). Alternatively, pDNA could be modified in a number of ways to maintain long-term gene expression, as expression longevity still remains a problem. The only potentially serious problem as far as mRNA or ncRNA delivery is concerned appears to be the control of ncRNA specificity. Having written this, at least as far as siRNA is concerned, siRNA sequences may be designed that are both specific and effective for the target mRNA, but cross-react free with respect to alternate mRNAs (i.e., no off-target effects) (see below).

1.06.3 Viral Vectors

From 1989 to 2005, just over two-thirds of the reported 1076 clinical trials have involved viral vectors.[30] Six major types of virus vector have been used in these clinical trials, as summarized alongside data concerning the frequency of use of each virus vector in clinical trials and data concerning their frequency of use as cited in PubMed (**Table 1**). Adenovirus and retrovirus vectors clearly predominate, followed by AAV) lentivirus, pox virus, and herpes simplex virus (HSV) (**Figures 3** and **4**). Although the principal application of all six main virus vectors has been in cancer gene therapy, if we normalize the frequency of use across application and vector type, then viral vectors appear to have different use profiles in particular applications, commensurate with their different properties.

Without exception, no single viral vector used in clinical trials is a wild-type viral serotype. However, the extent to which viral vectors differ from their wild-type progenitors depends upon the specific application. Most wild-type viruses possess all the proteins and genetic material to result in infection of the host cell, replication of the viral genome, synthesis of the viral capsid, packaging of the viral genome, and release of viral particles from the cell. Viral vectors are genetically engineered to carry an expression cassette containing the desired gene. Although they are structurally similar to the wild-type progenitor virus, they generally lack some or all of the viral genes. Hence, their ability to replicate is frequently impeded or obliterated. Lately, conditional replicating viruses have appeared that do replicate, but only specifically within malignant tissue.[31]

Table 1 Viral vector usage in different therapeutic categories. The number and categorization of clinical trials (1989–2005) for the six most common viral vectors is shown. By comparison with values normalized across application and vector type, biases for certain vectors to specific applications have been calculated and marked (*). The number of citations from PubMed for each viral vector in a gene therapy context are also included

Vector	*Monogenic disease*	*Vascular disease*	*Cancer*	*Infectious disease*	*Gene marking*	*Total*	*PubMed citations*
Adenovirus	17	39*	205	3	1	265	5726
Retrovirus	41	1	140	39	47*	268	4123
Adeno-associated virus	18*	0	6	2	0	26	1418
Herpes simplex virus	0	0	35*	0	0	35	1525
Lentivirus	0	0	0	3*	0	3	1230
Pox virus	0	0	163*	17*	0	180	284
Other viruses	0	0	5	4	0	9	133
Total	76	40	554	68	48	786	14439

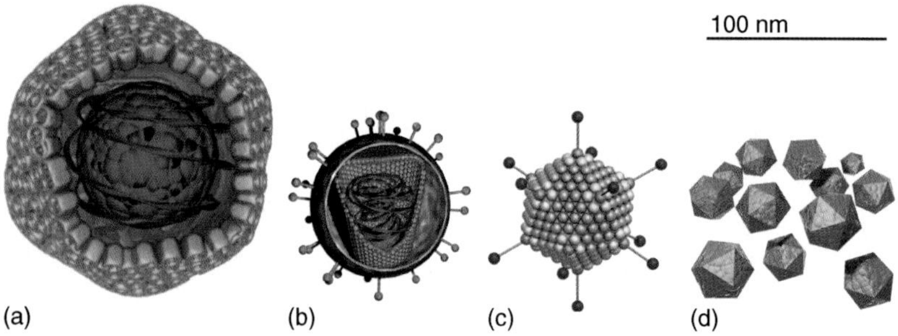

Figure 3 Relative sizes of viral vectors: (a) HSV, (b) retrovirus, (c) adenovirus, and (d) AAV. AAV vectors are among the smallest known (≈20 nm) whereas pox virus vectors are the largest (≈100 × 200 × 300 nm), and are visible by light microscopy.

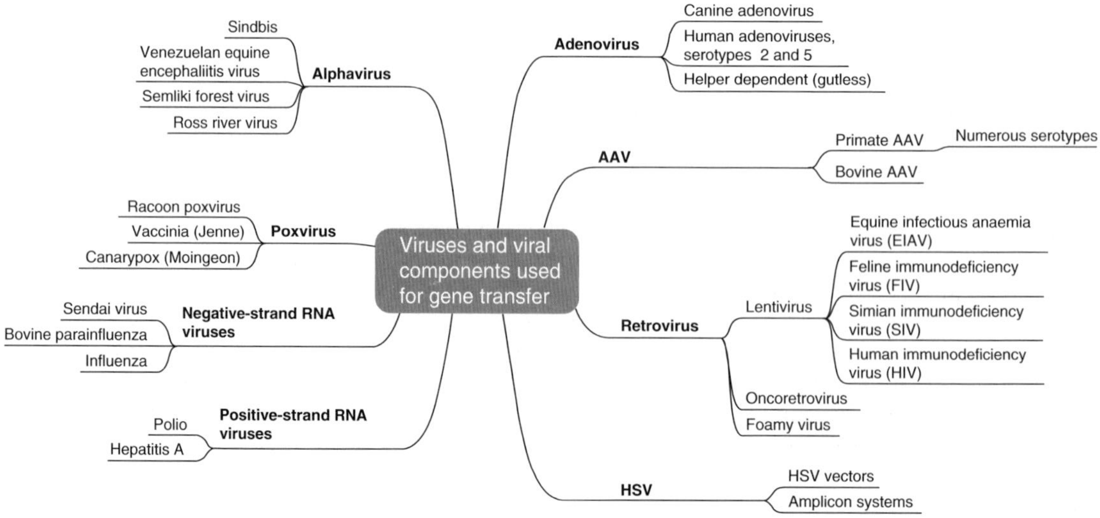

Figure 4 Geneology of viral vectors for gene therapy.

1.06.3.1 Application of Viral Vectors in Gene Therapy

Viral vectors have seen the most service in cancer applications, and real success has been reported with a serotype 5 adenovirus vector delivering the p53 tumor suppressor gene (Gendicine, the World's first commercial gene therapy product, licensed for use in China).[32] Aside from tumor suppressor genes, viral vectors have been used frequently to deliver genes to boost the immune response to tumor antigens. The viral delivery of suicide genes, particularly HSV thymidine kinase, which accounts for the next largest treatment group, is followed by the viral delivery of chemoprotective genes, where bone marrow cells are treated for protection from the toxic effects of chemotherapy.[33]

Viral vectors have also seen service in the treatment of monogenic diseases. In 2000, the World's first curative gene therapy trial was reported after bone marrow cells were isolated from patients with X-linked SCID. The wild-type gene for the γc cytokine receptor was introduced into the cells of the bone marrow of individual patients ex vivo by means of retroviral transduction before re-implantation.[34] As noted previously, some patients have since developed leukemia, but the majority appear to have been successfully cured. In contrast, viral vector clinical trials for CF have yielded little success to date in spite of the prodigious effort expended.[30]

Another notable arena of viral vector use has been in the development of viral vector gene therapy approaches to vascular disorders such as familial hypercholesterolemia, hypertension, restenosis, and the prevention of late vein graft failure.[35] The majority of clinical trials in this field have employed adenovirus vectors to deliver single genes appropriate to encourage the formation of new blood vessels to increase blood flow to ischemic regions.[36] This has particular relevance in coronary artery disease and peripheral artery disease that involve either myocardial or lower limb ischemia, respectively. Much effort has been expended, but poor vector targeting characteristics and limited surgical access have impeded progress.[37] In particular, although adenovirus may deliver genes to the vasculature with relative ease, systemically administered adenovirus has a strong tropism for liver hepatocyte cells, a tropism that needs to be restricted and/or redirected in order to enable any prospect of vascular gene therapy applications of adenoviral vectors in future (see below).[38]

The treatment of infectious diseases by viral vector gene delivery has been reported in numerous preclinical and clinical studies. The majority of clinical trials have been aimed at treating HIV infection, with regrettably scant success to date.[39] Alternatively, there are a number of preclinical studies underway for the treatment of hepatitis B and C, polio, dengue, and influenza A, where the use of RNA interference (RNAi) technologies may harmonize substantially with viral vector delivery.[40] Viral vector delivery of antigenic genes also represents a possible route to fast and effective vaccination.[41]

1.06.3.2 The Next Generation(s) of Viral Vectors

As the previous section makes clear, there is a substantial 'vector problem' (*see* Section 1.06.2.2) with the viral vectors used in clinical trials to date. Improvements are essential for viral vector gene therapy, but are being made.

1.06.3.2.1 Adenovirus vectors

The popularity of adenovirus vectors stems from the ease with which high-titer stocks may be generated that can achieve high levels of transgene expression postgene delivery. However, the adenovirus is nonenveloped, and carries 36 kbp double-stranded-DNA (dsDNA) that is used to replicate virions postviral transduction as an epichromosomal element. Consequently, the wild-type virus particle is highly immunogenic, and transgene expression can be very short lived (1–2 weeks posttransduction). First-generation vectors were engineered to be deficient in the early gene region E1, and sometimes also in E3. State-of-the-art vectors are now devoid of 11 adenoviral genes, and are therefore dependent upon helper viruses for production. Hence, they are termed, 'helper-dependent' or, occasionally, 'gutless' or 'high-capacity' viral vectors. These are substantially less immunogenic than earlier iterations, as viral 'late gene' products are no longer synthesized, and can thus provide more prolonged transgene expression.[42] For targeting tumors or the vasculature, several strategies have now been developed to ablate the unwanted natural tropism of adenovirus for hepatocytes.[35] These include artificial 'gene shuffling' between genetic elements in the genomes of the 50 or so known adenovirus serotypes. Studies have demonstrated that swapping the fiber or knob domains of various viral surface ligands results in substantially altered tissue tropism.[43] Alternatively, interactions between these surface ligands and their cognate receptors, such as the coxsackie adenovirus receptor (CAR), can be impaired by the introduction of mutations into these surface ligands by genetic engineering. Even bi-specific antibodies or cognate receptor fusion constructs have been used to bind simultaneously to surface ligands and also 'retarget' the vector to another receptor altogether.[44] For example, a bi-specific antibody for CAR and angiotensin-converting enzyme has been used to maximize adenovirus binding to and uptake by the pulmonary endothelium. The coating of adenovirus particles with cationic liposomes or synthetic polymers could also be of substantial future benefit to adenovirus gene delivery.[45,46]

1.06.3.2.2 Retrovirus vectors

Retrovirus types are numerous, but all are enveloped and carry single stranded RNA (ssRNA) that is used to replicate virions following integration in the host genome postmitosis. Consequently, transgene expression can show long duration (months to years), but the oncogenicity risk is high. Retrovirus delivery is also ineffective in vivo. Therefore, although retrovirus delivery was popular in the early days of gene therapy research,[47] this popularity is now declining rapidly, and next-generation innovations are largely incidental. Instead, there has been a marked shift toward lentivirus use. Lentiviridae are a genus of the retrovirus family, of which HIV is a subgenus.[48] In the case of lentiviral vectors, they do not require mitosis to replicate and hence have more utility.

1.06.3.2.3 Adeno-associated virus vectors

AAV types are nonenveloped and carry 4.7 kbp of ssDNA. Wild-type virus depends upon a helper virus (usually adenovirus) to proliferate and integrate into human chromosome 19. Consequently, transgene expression can be long (months to years), and the oncogenicity risk has been considered to be low. Indeed, AAV particle immunogenicity also appears to be low. AAV vectors are typically devoid of all viral coding regions and contain transgene DNA flanked by 145 bp-long inverted terminal repeats. These regions contain all *cis*-acting elements involved in genome rescue, replication, and packaging.[49]

AAV vectors have been used predominantly for the treatment of monogenic disorders, since they are relatively nonimmunogenic, and vector genomes persist for long periods. A number of AAV serotypes are now emerging. Intriguingly, the sequence diversity, but relative compatibility, of their capsid proteins is now allowing for a variety of cross-packaging capsid protein exchanges to be carried out between human AAV serotypes such as AAV2, and other human or nonhuman AAV isolates, in order to generate newer serotypes. Alternative serotypes have demonstrated tremendous efficiency in the transduction of liver, heart, and muscle in animal models of disease, raising the hope of successful clinical treatments of diseases such as hemophilia and Duchenne muscular dystrophy.[50,51] Indeed, AAV vector clinical trials for the treatment of hemophilia have demonstrated up to 6 weeks of expression of human factor IX, lending hope that AAV vectors could be powerful alternatives to the adenovirus and other virus vectors.[52]

Next-generation AAV vector technologies may involve the use of self-complementary ssDNA that may fold back on itself to form a dsDNA configuration. The formation of a dsDNA configuration is thought to be a rate- and efficiency-limiting step en route to gene expression. Therefore, such self-complementary vector ssDNA should allow for more efficient transgene expression. Another development exploits the tendency of linear AAV vector genomes to concatamerize through recombination of the free 5′ and 3′ ends after cell entry. Concatemerization may provide a means of doubling the size of the transgene available for expression in cells, albeit with reduced expression efficiency, thereby partially overcoming the size problem of AAV particles that only have restricted ssDNA packaging capacity within.

1.06.3.2.4 Other viral vectors

HSV is an enveloped virus that carries 152 kbp of linear, dsDNA that is used to replicate virions postviral transduction as an epichromosomal element. Wild-type HSV has given rise to two different viral vector systems known as recombinant HSV and HSV amplicons. First-generation recombinant HSVs were created by deletion of early genes such as the *vhs* genes from wild-type virus, creating the opportunity to accommodate a 30 kb transgene payload. First-generation HSV amplicon was created without the presence of viral genes except for the *cis*-acting sequences corresponding to the origin of viral DNA replication (*ori*) and the packaging and cleavage signal (*pac*).[53] Insufficient clinical trials have been performed to date, therefore the directions of future innovations are yet to be determined with HSV vector systems. The same is true of pox virus vectors. Wild-type pox viruses carry linear dsDNA, and are among the largest known viridae. The most common pox virus vector is based on Vaccinia virus that is enveloped and has been engineered to carry a dsDNA payload of at least 25 kbp. In distinction from other DNA viruses, pox virus remains in the cytoplasm for the entire infectious cycle.[54]

1.06.4 Synthetic Nonviral Vectors

1.06.4.1 Applications of Synthetic Vectors in Gene Therapy

Clinical trials with synthetic vectors represent less than 10% of the 1076 total trials. This percentage is attributed mainly to 93 Phase I/II lipofection clinical trials performed using cationic liposome-based vectors. At the level of Phase II, simple cationic liposome formulations of DOTMA/Chol (**Figure 5**) have been used to mediate interleukin-2 (IL2)

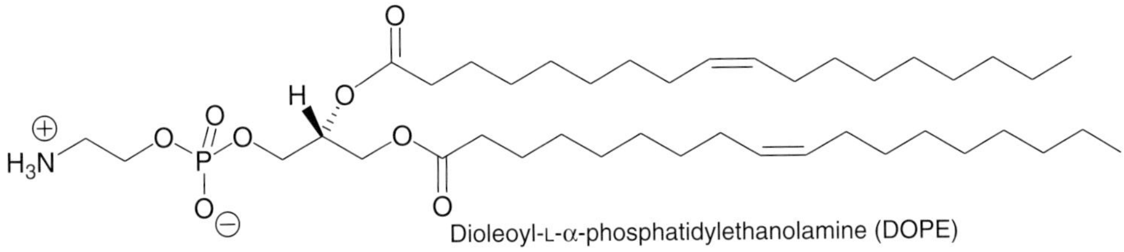

Dioleoyl-L-α-phosphatidylethanolamine (DOPE)

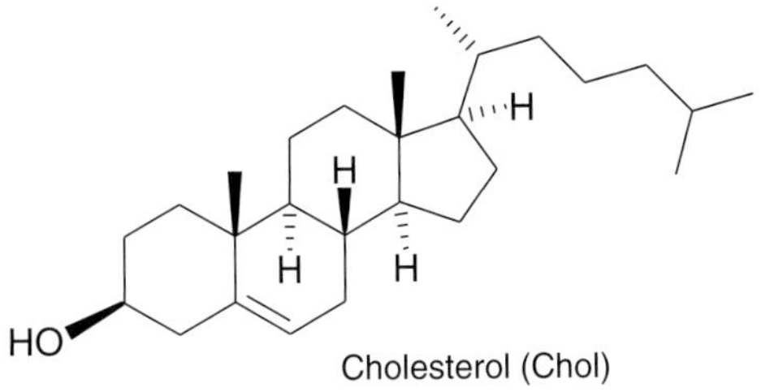

Cholesterol (Chol)

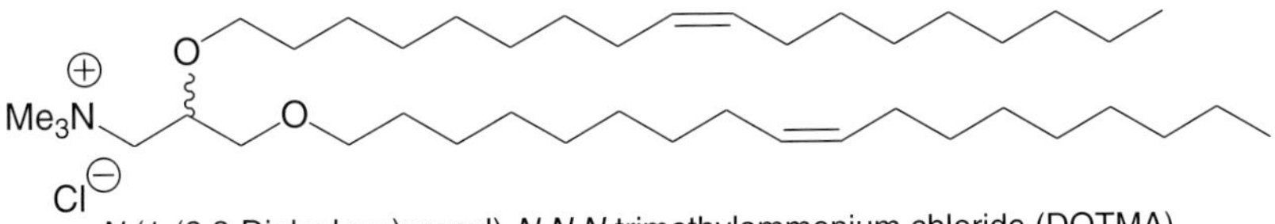

N-(1-(2,3-Dioleyloxy)propyl)-*N*,*N*,*N*-trimethylammonium chloride (DOTMA)

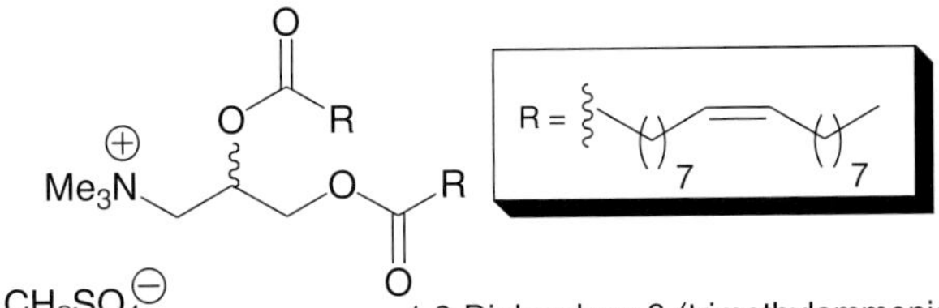

1,2-Dioleoyloxy-3-(trimethylammonio)propane (DOTAP)

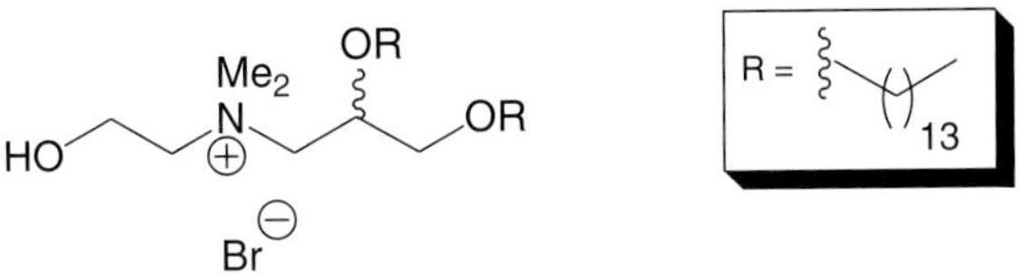

1,2-Dimyristyloxypropyl-3-dimethylhydroxyethylammonium bromide (DMRIE)

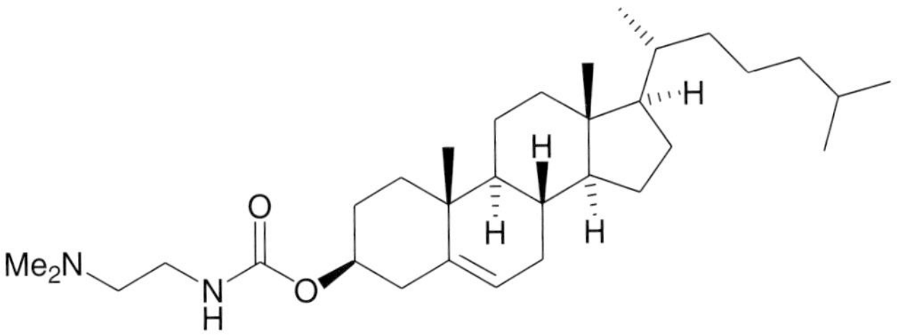

3β-(*N*-(*N*′,*N*′-Dimethylaminoethane)carbamoyl)cholesterol (DC-Chol)

Figure 5 Cytofectins, neutral lipids, and PEG–lipids in gene therapy clinical trials.

gene immunotherapy of cancer (**Table 2**),[55] and simple cationic liposome formulations of DC-Chol/DOPE (**Figure 5**) have been used to mediate adenoviral E1A gene therapy of ovarian cancer.[56] Furthermore, simple cationic liposome formulations of DMRIE/DOPE (**Figure 5**; allovectin-7) have experienced use in cancer gene therapy trials at the level of Phase III.[57] A very few more complex cationic liposome formulations are now also under evaluation at the level of Phase I clinical trials (**Table 3**).

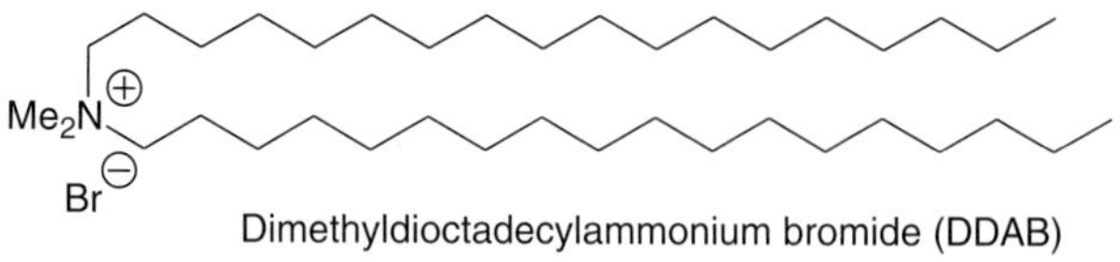

Dimethyldioctadecylammonium bromide (DDAB)

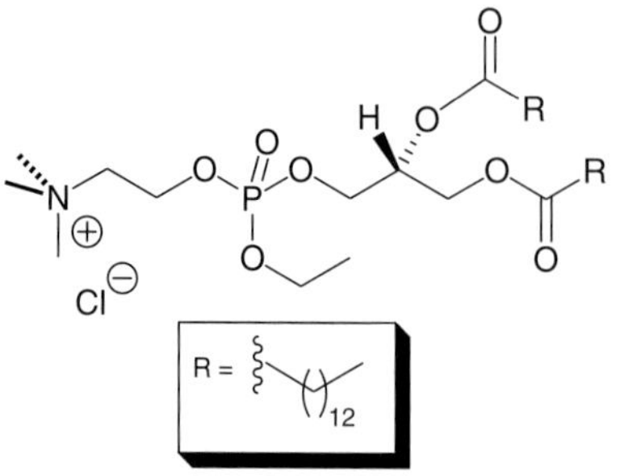

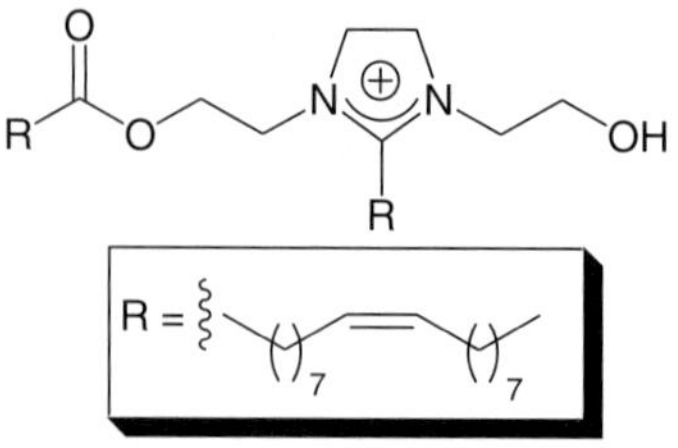

1-(2-(oleoyloxy)ethyl)-2-oleyl-3-(2-hydroxyethyl) imidazolinium chloride (DOTIM)

3-Ethyl-1,2-dimyristoyl-L-α-phosphatidylcholine (EDMPC)

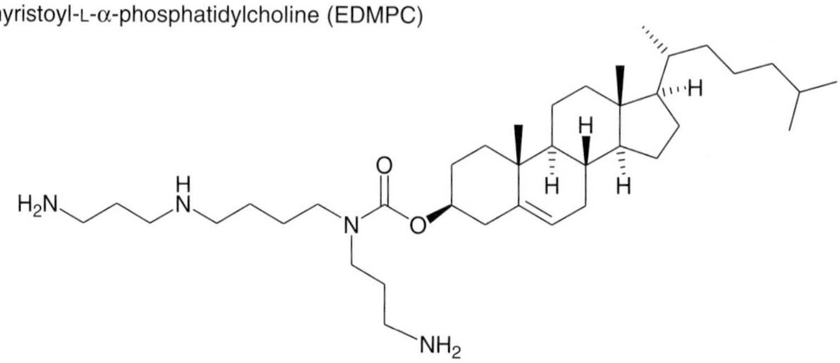

GL-67

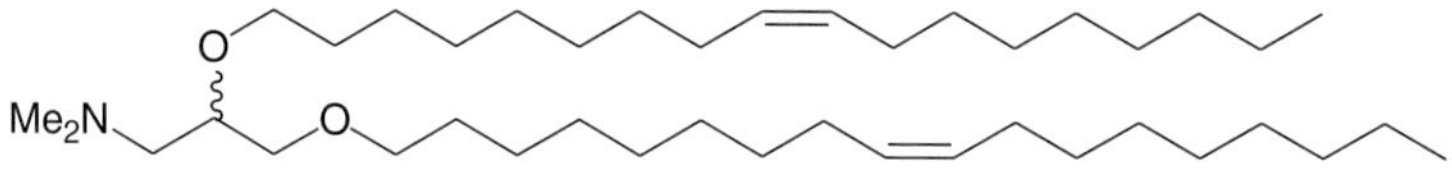

N-(1-(2,3-Dioleyloxy)propyl)-*N*, *N*-dimethylammonium chloride (DODMA)

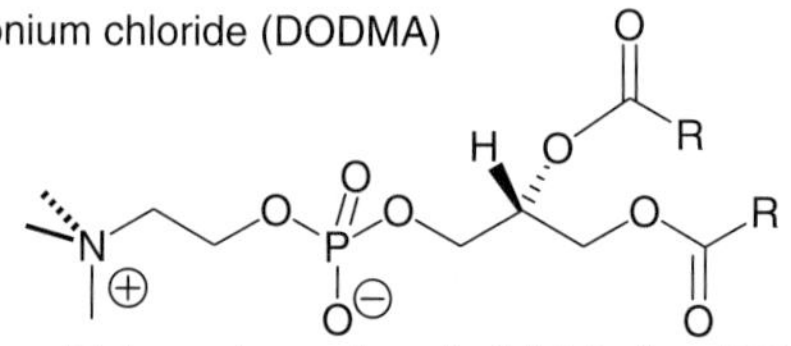

Distearoyl-L-α-phosphatidylcholine (DSPC)

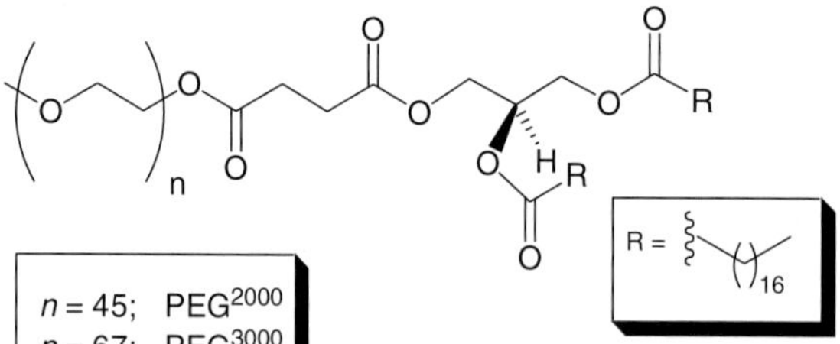

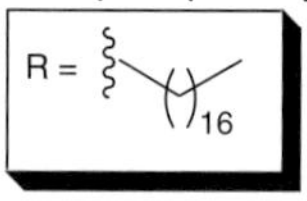

$n = 45$; PEG^{2000}
$n = 67$; PEG^{3000}
$n = 113$; PEG^{5000}

Polyethyleneglycol conjugate of distearoyl glycerol (PEG-DSG)

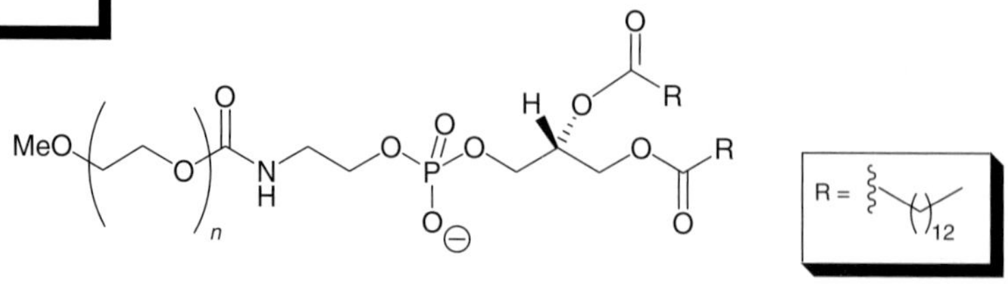

Polyethyleneglycol conjugate of dimyristoyl-L-α-phosphatidylethanolamine (DMPE-PEG^n)

Figure 5 Continued

Table 2 List of clinical trials indications per cationic liposome/micelle system[a]

Cationic liposome tested	*Disease*	*pDNA*	*Trials*	*Latest trial phase*
DMRIE/DOPE (allovectin-7)	Metastatic melanoma, Head and neck cancer	Bi-cistronic HLA-B7 and β_2-microglobulin	31	III
	CF, renal cell carcinoma, prostate cancer	PGT-1		
DC-Chol/DOPE (tgDCCE1A)	Ovarian, oropharyngeal cancer	E1A	13	II
	Coronary artery disease			
DOTMA/DOPE	Glioma, small cell lung cancer, malignant melanoma stage IV, ovarian cancer, breast cancer	IL2, neomycin resistance gene, HSV-1 thymidine kinase cDNA	6	I
VLTS-587 (Ad100 cell line transfected using DOTMA/DOPE)	Nonsmall cell lung cancer, lung adenocarcinoma (ex-vivo)		3	II
DDAB/DOPE	Breast cancer, ovarian epithelial cancer, prostate cancer	IL2	3	I
EDMPC/Chol	CF	CF transmembrane conductance regulator cDNA	1	I
DOTIM/Chol (pVS534)	Malignant melanoma	cIL2, manganese superoxide dismutase		
	Nonsmall cell lung cancer	cDNA	2	II
DOTMA/Chol	Oropharyngeal cancer stage unspecified	CIL2	2	II
	Squamous cell carcinoma			
DOTAP/Chol	Nonsmall cell lung cancer stage IIIB/IV	E1A	1	I
DC-Chol/DOPE	CF	CFTR	2	I
GL-67/DOPE	CF	CFTR	4	I

CFTR, CF transmembrane conductance regulator.
[a] Data from GeMCRIS[171] and the Gene Therapy Trials Worldwide database.[1]

Table 3 ABC or ABD cationic liposome systems entered into clinical trials[a]

Multicomponent system	*Disease*	*pDNA*	*Trials*	*Latest trial phase*
DSPC/DODMA/[PEG-DSG]/Chol	Malignant melanoma stage IV	HSV-1 thymidine kinase cDNA	1	I
DOTAP/DOPE complex with anti-transferrin receptor single-chain antibody Fv fragment	Carcinoma neoplasm	E1A	1	I
GL-67/DOPE/DMPE-PEG^{5000}	CF	CFTR plasmid	1	I

[a] Data from GeMCRIS[171] and the Gene Therapy Trials Worldwide database.[1]

Polymer and biomaterial scientists have been involved in the development of cationic or neutral polymer vectors as alternatives to cationic liposome-based vectors. In principle, cationic or neutral polymer vectors could be particularly beneficial to bind tightly therapeutic nucleic acids, thereby providing protection from nuclease degradation, but also allowing the polymer medium to serve as a controlled-release reservoir of therapeutic nucleic acids.[58] In spite of these potential advantages, polymer vectors have been tested in only a few clinical trials (**Table 4**). In the cases of both cationic liposome-based and polymer vectors, clinical results have been far from impressive, and synthetic vector gene therapies appear to be someway in the future, even more so than virus vector gene therapies.

1.06.4.2 The Next Generation(s) of Synthetic Vectors

There has been a great deal of soul searching about directions to take in the development of next-generation synthetic vectors in order to improve the efficacy of delivery. For this reason, we recently introduced the concept of the self-assembly ABCD nanoparticle as an appropriate structural paradigm for the design of successful synthetic vectors for the delivery of therapeutic nucleic acids ex vivo and in vivo (**Figure 6**).[23] The paradigm was originated for the understanding of cationic liposome-based vectors, but is equally well suited for the understanding and appreciation of cationic or neutral polymer-based vectors as well. Within self-assembly ABCD nanoparticles, nucleic acids (A) are condensed within functional concentric layers of chemical components purpose-designed for biological targeting (D), biological stability (C), and cellular entry/intracellular trafficking (B). The B layer components may either be cationic liposome or polymer in character. This paradigm was devised in order to help us improve our understanding of structure–activity correlations between different synthetic nucleic acid delivery systems, and to provide a simple way to correlate different synthetic nucleic acid delivery systems with therapeutic delivery opportunities. We have found this paradigm to be a useful way to frame our research activities, and hope that others will do so as well.

The origins of this paradigm are based in the growing experience acquired with time by researchers interested in the use of synthetic vectors for gene therapy. Core AB systems were the first to appear and enter clinical trials (as described above), wherein the B layer was cationic liposome based. The use of cationic liposomes containing synthetic cationic lipids (cytofectins) to mediate nucleic acid delivery to cells was first described Felgner and co-workers in 1987.[59] Many more cationic liposome/micelle-based systems have been developed since to facilitate nucleic acid delivery.[60] However, when core AB systems were introduced in vivo, they proved to have poor pharmacokinetic properties, were unstable with respect to aggregation, and quickly taken by the reticuloendothelial system (RES) (opsonization).[61,62] Therefore, in order to overcome these problems, there has been a realization that a stealth/biocompatibility polymer layer

Table 4 AB or ABC cationic and neutral polymeric systems reaching clinical trials[a]

Polymer tested	*Disease*	*pDNA*	*Trials*	*Latest trial phase*
Polyvinylpyrrolidone	Malignant melanoma	IFN-α/IL12	1	I
pLL-Cys-PEG	CF	CFTR	1	I
Poloxamers (188/407)	Intermittent claudication	Engineered zinc finger transcription factor	2	II
Poly(ethyleneimine)	Bladder carcinoma	Diptheria toxin A H19	1	I[102]

[a] Data from GeMCRIS[171] and the Gene Therapy Trials Worldwide database.[1]

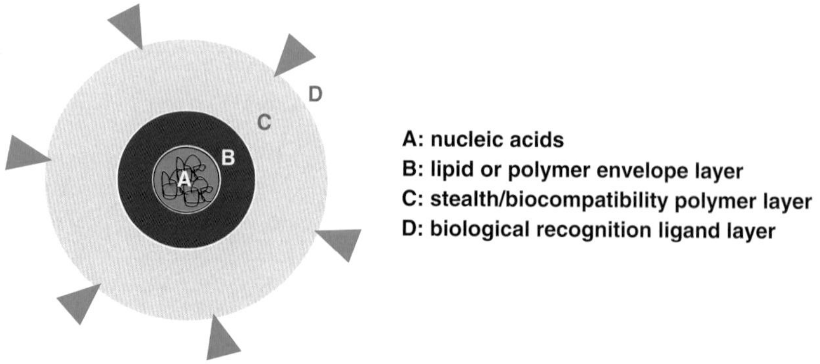

Figure 6 The self-assembly ABCD nanoparticle concept. Graphic illustration of an ABCD nanoparticle to show how therapeutic nucleic acids are condensed within functional concentric layers of synthetic chemical components purpose-designed for biological targeting (D), biological stability (C), and cellular entry/intracellular trafficking (B).

(C layer) may need to be attached to the outer surface of AB system core particles in order to provide biocompatibility and colloidal and structural integrity to AB core particles in biological fluids. At present, the stealth/biocompatibility polymer of choice is polyethyleneglycol (PEG), which is an FDA-approved inert polymer that allows lipid particles to go unnoticed by the RES by mimicking water structures.[23] Finally, ligands may also be required as part of an optional exterior coating (D layer) designed for the 'active targeting' of nanoparticles.[63] The D layer may be introduced to target cells expressing complementary receptors, and increase cell entry via receptor-mediated endocytosis.[64] D layer receptor targeting restores cell entry processes that have otherwise been mediated by the C layer.[23,65,66] Hopefully, the reader will appreciate that self-assembly ABCD nanoparticles are virus-like artificial vector particles designed for therapeutic nucleic acid delivery that are assembled from toolkits of purpose-designed synthetic chemical components. The paradigm illustrates one serious future possibility that chemistry itself could provide for clinical gene therapy through the provision of tailor-made delivery solutions for a huge variety of different nucleic acid payloads.[67]

1.06.4.2.1 Improved AB cationic liposome systems

AB core particles (lipoplexes) have been created mainly through the combination of simple cationic liposome/micelle systems with pDNA. Simple cationic liposome/micelle systems are themselves formed from either a single cytofectin or more commonly from the combination of a cytofectin and a neutral fusogenic lipid such as DOPE or Chol (see **Figure 5**). There are impressive numbers of cytofectins already described in the literature and available commercially,[24,68–71] but all have in common a hydrophobic moiety covalently attached to a hydrophilic moiety through a polar linker. While hydrophobic regions are structurally similar, polar linkers and cationic head groups vary quite substantially. These have formed key components of AB systems mediating gene delivery in clinical trials, mostly against various types of cancers or else against CF.[72,73] Routes of administration have been intravenous or intratumoral for cancer patients, and intranasal or pulmonary (aerosol) delivery in the case of CF patients. Results from clinical trials (see **Table 2**) have not been very impressive on the whole,[55,74–78] but core AB particles could possibly be improved for successful local or regional delivery of nucleic acids in vivo through rational engineering of cationic liposome structure by introducing structural changes to the fusogenic co-lipid and adjustments to the co-lipid/cytofectin mole ratios.[79,80]

1.06.4.2.2 ABC and ABCD cationic liposome systems

In general, most researchers have now concluded that ABC or ABCD multicomponent nanoparticle systems are the best proposition for future clinical gene therapy applications. Given the well-known stealth/biocompatibility properties of PEG.,[68,81] Safinya and co-workers have recently demonstrated that only PEG with a molecular weight of 2000 Da and above gives adequate stealth/biocompatibility characteristics.[82] Hong *et al.* reported one of the first attempts to generate a self-assembly ABC system using PEG for the C layer.[83] A number of key systems have been developed since. First among these was the ABC system prepared from GL-67/DOPE/DMPE-PEG^{5000} cationic liposomes and pDNA expressing wild-type CFTR protein for CF gene therapy. Pulmonary administration resulted in a significant degree of correction of the chloride abnormality in the patients, indicating that this ABC system could at least overcome some of the physiological and biological barriers in CF patients.[20] Then best among these are currently the stabilized plasmid–lipid particle (SPLP) systems, known more widely now as stabilized nucleic acid–lipid particle (SNALP) systems. In many ways, almost perfect ABC systems in terms of formulation, the latest SPLP system (Pro-1) has been built from DSPC/DODMA/[PEG-DSG]/Chol cationic liposomes (see **Table 3** and **Figure 5**). These SPLP systems are the epitome of a cationic liposome-based synthetic vector stabilized with PEG in aqueous media.[84–87] However, at the time of writing, there have been disturbing reports from a Phase I cancer clinical trial using Pro-1 that hypertension was a serious side effect of the vector used and its repeated administration (see above).

Really impressive preclinical results have been reported using ABCD systems known as Pegylated immunoliposome (PIL) systems, wherein the D layer is comprised of anti-transferrin receptor (TfR) monoclonal antibody. The TfR is over-expressed by cells at the blood–brain barrier, and also in peripheral organs such as the liver and spleen,[88,89] therefore, PIL particles have been shown convincingly to enter murine brain. However, studies, though impressive, have not advanced past the preclinical level either with PILs or any other ABCD system to date.[90–94]

1.06.4.2.3 Improved AB polymer systems

In this case, AB core particles (polyplexes) have been created mainly through the combination of simple cationic polymers with pDNA. Cationic polymers have been used to facilitate gene delivery as early as simple cationic liposome/micelle systems.[95] Cationic polymers are the least efficient synthetic vectors, hence the low number of these types of synthetic vectors in clinical trials. However, their versatility and chemical potential keeps gene therapy researchers and material scientists interested. Cationic polymers include poly-L-lysine (pLL), poly(ethyleneimine) (PEI),

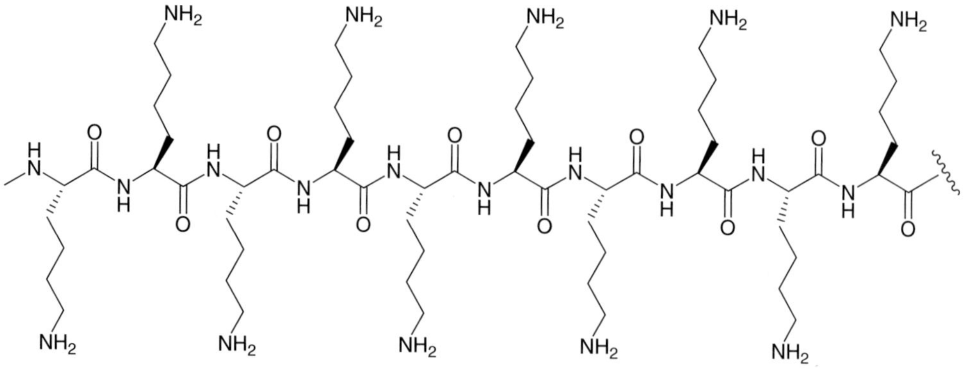

Poly-L-lysine (pLL)

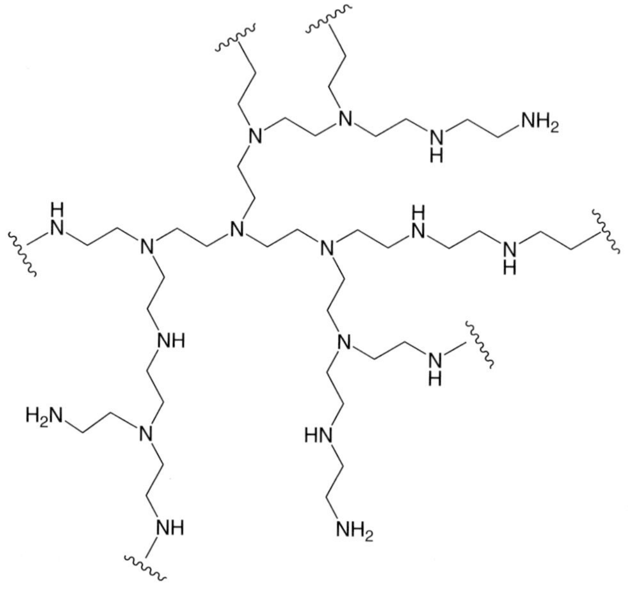

Poly(ethyleneimine) (PEI)

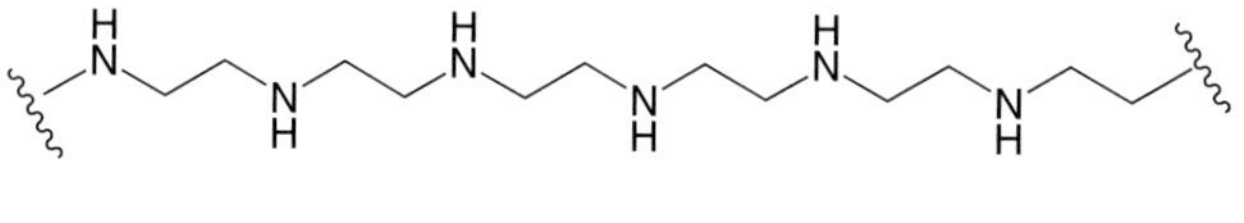

Linear PEI

Figure 7 The main cationic polymers used in gene therapy research.

polyamidoamine (PAMAM) dendrimers, and chitosan polymers (**Figure 7**). Of all these, PEI appears to be the most effective polymer vector for gene delivery polymer studied to date, and has been used extensively since 1995.[96] Linear PEI (22 kDa) seems to outperform the branched polymer in epithelial cells and after intranasal instillation.[97] PEI has also been used successfully in vivo to deliver genes to the central nervous system,[98] lung,[99,100] and tumours,[101] and is even now in Phase I clinical trial (see **Table 4**).[102] However, PEI shows significant cytotoxicity, leading researchers to search for less toxic alternatives.[103–105] PEI has been subjected to various chemical treatments and modifications to increase gene transfer and lower toxicity.[106,107]

PAMAM dendrimers show substantial cytotoxicity that currently limits applications all round.[108] In contrast, chitosan polymer looks much more promising. Chitosan is poly(2-deoxy-2-amino-β-D-glucopyranose), and consists of (β1→4)-linked 2-deoxy-2-amino-glucopyranose residues, so it is therefore a polymer of natural origin with reported

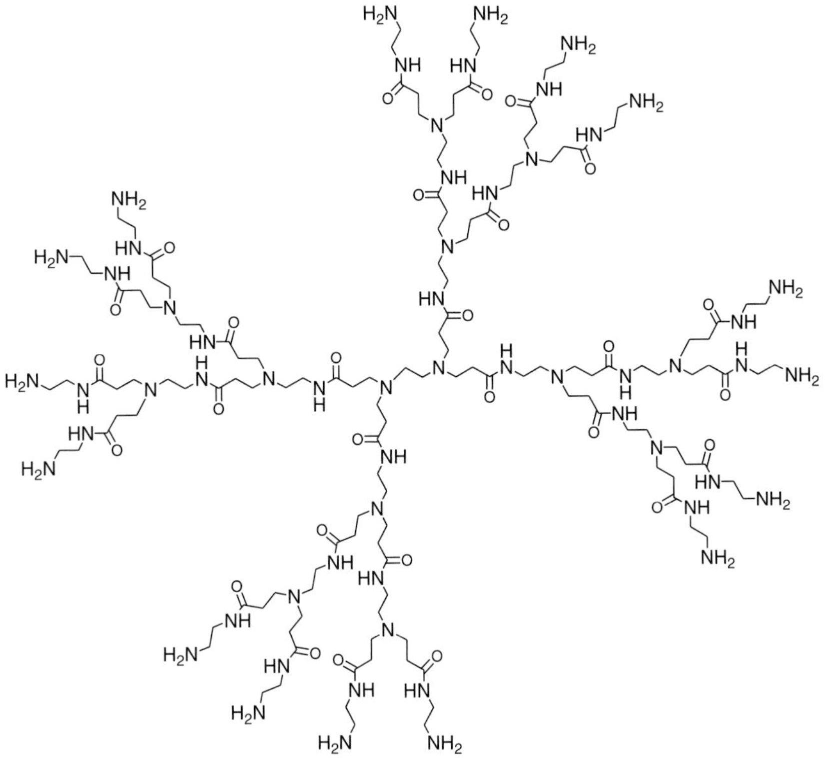

Polyamidoamine dendrimer (PAMAM)

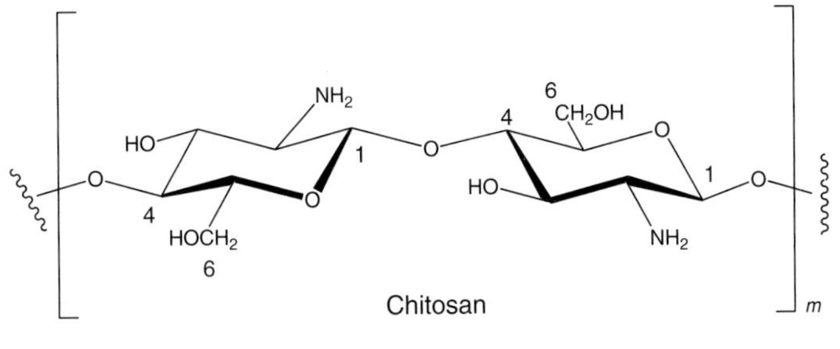

Chitosan

Figure 7 Continued

biocompatibility and lack of cytotoxicity.[109] Chitosan and its derivatives have been reported for gene transfer; however, it is known to aid DNA vaccination.[110] Unfractionated or oligomeric chitosan will mediate gene delivery to mouse lung comparable in efficacy to PEI.[111,112] Quaternized chitosan oligomers also appear to be potent agents for gene delivery comparable with PEI.[113,114] In view of much written in this review, it is perhaps ironic that polar neutral polymers may also be valuable for nucleic delivery. Polymer systems such as collagen, poly-anhydride, or polyethylene acetate are able to bind therapeutic nucleic acids, and may act as controlled-release reservoirs of therapeutic nucleic acids when implanted into tissue.[115] In addition, pluronic block co-polymers (poloxamer) have been used successfully to deliver pDNA under slow-release conditions to skeletal muscle, and such polymers can be conjugated with other cationic polymers to improve their performance.[116] Poloxamer 188 is now in Phase I clinical trials (**Table 4**).

1.06.4.2.4 ABC cationic polymer systems

Preclinical studies have been performed with a variety of derivatized cationic polymers. First among these is the mixed polymer (pLL-Cys-PEG) formed from pLL attached to PEG by means of two flanking cysteine residues. The ABC system formed in combination with pDNA was evaluated by intranasal administration in CF clinical trials, giving some evidence of gene delivery and partial correction on the nasal potential difference (**Table 4**).[117] PEG–PEI polymers are now also in existence.[105,118,119] In general, polymer structures may be diverse and plentiful, but progress in their use toward therapy has been sporadic, perhaps owing to the difficulties in establishing firm structure–activity rules for gene delivery, in contrast to cationic liposome/micelle-based systems. In an attempt to elucidate this issue, the Langer's group has prepared libraries of various polymers, and tested them for toxicity and gene delivery. Selected polymers were further tested in vivo in mice, showing improved efficiency compared with PEI.[120–122] More extended studies of this type should provide necessary information to predict the best polymers for each route of administration and clinical gene therapy application.

1.06.5 Physical Nonviral Methods of Delivery

Physical nonviral methods of delivery (i.e., physical vectors) are as simple as direct injection into tissue, or involve some form of medical device. The primary device that has appeared in clinical trials to date is the gene gun (biolistics).[123] Exposed surfaces are bombarded with carrier particles to which nucleic acids may be bound, allowing for the kinetic carriage of bound nucleic acids into cells. Both this device and direct injection enable the delivery of naked pDNA to skin, epithelium, or muscle. All five clinical trials (Phase I) involving the gene gun have been focused on the treatment of melanoma.[124] However, the gene gun also enables the delivery of pDNA for DNA vaccination, although the clinical results of naked DNA vaccination have not been so promising to date.[125] Electroporation is a counterpart technology to the gene gun.[123] Electroporator devices use pulsed electrical discharges from needle clusters inserted into tissue such as muscle to enable transient access of co-administered naked pDNA into tissue cells. This technique may be of real clinical value provided that pain generated by the pulsed electrical discharges in tissue does not become too severe a side effect problem. Finally, an emerging physical vector technique could be the hydrodynamic delivery of pDNA.[126–128] If naked pDNA is administered by means of a high-speed, high-volume injection into murine circulation, then some form of hydrodynamic shock process results in substantial uptake of pDNA by liver hepatocytes. Direct duplication of this process for clinical use is out of the question, but modified protocols may well become available soon that can be clinically acceptable. If so, then the potency of this method of delivery could lead to real clinical applications for a variety of liver conditions.[128]

1.06.5.1 Opportunities with Naked Deoxyribonucleic Acid

There have been 174 clinical trials involving naked pDNA in which pDNA is usually administered by direct injection into tissue (for instance, myocardium or limb muscle) or tumor. Numbers of these trials (usually Phase I or II) have involved either the cellular or direct introduction of pDNA into tissue (intramuscular or subcutaneous) for therapeutic vaccination against cancers or against infectious agents such as HIV.[128] Clinical trials for peripheral artery disease or coronary artery diseases involving the appropriate intra-arterial or intramuscular (limb or myocardium) administration of pDNA expressing growth factor genes (e.g.,, vascular endothelial growth factors or hepatocyte growth factor) have achieved some significance too, and are advancing through Phase II trials in some cases.[128–130]

1.06.6 Nucleic Acid Constructs for Nonviral Gene Therapy

As noted previously, viral vectors by their very nature harbor nucleic acid constructs that enter transcription, resulting in mRNA for translation into therapeutic proteins, or in ncRNAs for therapeutic transcription/translational control. These nucleic acid constructs are easy to prepare, and necessary viral control sequences are already present for the control and manipulation of expression. The primary restraint on the preparation of these nucleic acid constructs is the allowed size limit of the included transgene, as determined by the physical dimensions of the viral capsid.

The preparation of useful nucleic acid constructs for nonviral gene therapy is altogether a more complex and varied proposition. Broadly speaking, pDNA has been the most common form of nucleic acid construct delivered by synthetic or physical vectors. Typically, pDNA expression takes place postnuclear delivery, and declines to background levels between 7 and 14 days posttransfection. This is known as plasmid silencing. Plasmid silencing is unhelpful for most projected in vivo applications of gene therapy. Curiously, the reasons for plasmid silencing do not appear to be plasmid shedding (loss of pDNA from cells) or plasmid CpG methylation, as might be expected,[131] suggesting that other mechanisms are involved such as chromatin remodeling (nuclear protein condensation of pDNA leading to inactivation of gene expression). While research into plasmid silencing would be of undoubted use in the design of long-term expression plasmids such as plasmid minicircles[132] molecular biology has not stood still, and the flexibility of synthetic or physical vectors to deliver a wide range of nucleic acid constructs for future therapies can be exploited to find alternatives to simple epichromosomal pDNA.

1.06.6.1 Deoxyribonucleic Acid Constructs

The pDNA may be modified itself for long-term expression by the introduction of sequence elements involved in the control of gene expression. Obviously, just as open-reading frames (ORFs) (genes or sections of genes) only comprise a fraction of chromosomal DNA in any one cell, so ORFs do not comprise the entirety of any one plasmid. There are now known to be elements such as locus control regions (LCRs) and ubiquitous chromatin opening elements (UCOEs) that act to sustain associated ORFs in states appropriate for transcription (into mRNA) and promote long-term expression.[133] Novel pDNA constructs with scaffold/matrix-attached regions are also emerging that are clearly enabled for long-term gene expression as well.[134–136] These could be very potent tools for gene therapy.[137]

Otherwise, the arrival of mammalian transposons (transposable elements) now looks remarkably promising. Transposons are stretches of DNA (either linear or circular) that are capable of insertion into chromosomal DNA at defined sites with the assistance of a transposase enzyme.[138,139] A transposon active known as Sleeping Beauty is able to insert a transgene embedded in the transposon sequence into mammalian chromosomal DNA, leading to long-term transgene expression (months).[140–142] Proof-of-principle studies with nonviral delivery of Sleeping Beauty transposon to cells have been accomplished,[143] although there are concerns that this transposon integrates into too many sites in chromosomal DNA and therefore may be cancer-inducing (oncogenic) in the same way that retroviral vectors can be. The physicochemical properties of Sleeping Beauty transposon integration sites in chromosomal DNA are known (palindromic AT repeat), and are theoretically numerous.[144] However, transposase enzymes do not integrate transposons into genes under active transcription, unlike retroviral vectors. Therefore, the oncogenicity risk should be much reduced. Accordingly, there have been proposals for the construction of chimeric transposase enzymes engineered with binding domains (such as zinc finger domain proteins) with high affinity for select DNA sequences that could guide the transposon to site-specific integration.[29] This idea is seductive but requires complete validation. An alternative approach has been suggested and validated using the integrase enzyme ΦC31 from bacteriophage. In this instance, the enzyme integrates an alternative transposable element that interacts with chromosomal DNA at binding sites less prevalent than the Sleeping Beauty integration sites, and consequently is perceived to minimize the risks of oncogenicity in comparison.[145–148] This approach is still in the early stages of technical development but appears potentially useful.

Artificial chromosomes represent the main alternative to pDNA, and may be delivered with synthetic vectors to cells owing to the fact that these vectors are able to deliver any nucleic acid construct irrespective of size. Clearly, artificial chromosomes (50 kbp to 1 Mbp) are much larger than pDNA (typically 4–7 kbp), but have been constructed to express genes in an epi-chromosomal manner supported by all the main features of a chromosome (such as the centromere) so that they operate as pseudo-chromosomes. Depending upon the source of sequences and genes, there are bacterial artificial chromosomes,[149–151] P1-derived artificial chromosomes,[151–153] yeast artificial chromosomes,[154,155] mammalian artificial chromosomes,[15,156,157] and human artificial chromosomes.[158,159] Proof of concept has been demonstrated using synthetic vectors and artificial chromosomes, resulting in long-term expression in cells in vitro,[151,152,156] and even in vivo.[149,150] At the other end of the spectrum, DNA aptamers and ribozyme DNAs have also been delivered successfully to cells in vitro. HIV-1 gene expression was successfully inhibited by the intervention of anti-HIV Rev-binding aptamer (RBE(apt)), and a ribozyme directed against the HIV-1 *env* gene.[160]

1.06.6.2 Ribonucleic Acid Constructs

In the case of RNA, we might consider the delivery of mRNA, but the complex heterogeneous secondary structure of such molecules and perceived vulnerability to hydrolysis have ensured that mRNA has seen little application. The concept of siRNA has risen with incredible speed within the past 2 years.[3] The phenomenon of RNAi has a provenance stretching back to at least 1995, when large double-stranded RNAs (dsRNAs) were found to silence genes in nematodes by a mechanism that is only now being properly appreciated (see **Figure 2**). According to this mechanism, dsRNA is broken down into siRNA duplexes (typically comprising 2 base overhangs at each 3′-end and a central antiparallel 19 bp double-helical region) by an enzyme system known as DICER. The siRNAs then associate with a protein complex (RNA-induced silencing complex, RISC) that interacts in an asymmetric manner with each siRNA, separating sense (S) and antisense (AS) strands from each other and preferentially adopting the AS over the S strand as a 'template' to bind target mRNA. Target mRNA is singled out for destruction by this activated RISC through siRNA-mediated target recognition apparently provided by the AS strand of siRNA bound to RISC that presumably makes complementary Watson–Crick base pair interactions with a corresponding region in target mRNA.[161,162]

In principle, synthetic or physical vectors can deliver either siRNA or DNA expressing short hairpin RNA interference (shRNAi). Both ncRNAs are expected to operate in the same way as described above. Viral vectors are restricted to the delivery of nucleic acid constructs for the ultimate expression of shRNAi. What makes siRNA or shRNAi constructs so potent is that numbers of these can be found with the capacity to complement corresponding regions in a target mRNA of interest, leading to activated RISC-catalyzed mRNA destruction. The S and AS strand base sequences of these siRNA or shRNAi constructs may then be sifted with defined rule sets so as to identify those siRNA or shRNAi constructs (approximately 1–3% per gene) with base sequences that are optimal for RISC-mediated destruction of the target mRNA of interest.[163] Furthermore, these sequences may then be screened at one higher level by means of high-end bioinformatics analyses (such as siDIRECT analysis), ensuring that they have a minimal likelihood of cross-reactivity with other off-target mRNA sequences and hence little likelihood of eliciting undesirable cellular toxicities.[164] The role of siRNA therapeutics in gene therapy now seems increasingly credible, especially in the light of the ongoing Phase I clinical trial for acute macular degeneration treatment involving the physical administration of chemically optimized siRNA (Sirna-027 and anti-VEGFR-1). Other trials involving siRNA therapeutics are also promised for other diseases such as respiratory syncytial virus[165,166] and hepatitis B virus,[167] where siRNA is delivered to mediate specific viral genome knockdown. Conceivably, ncRNAs such as miRNAs may offer alternative anticancer and antiviral gene therapy strategies as well.[4] However, such strategies remain the stuff of research and development, and are yet to enter clinical trials. Moreover, siRNA applications are increasingly beginning to be troubled by off-target and immune stimulation effects[167,168] that will have to be properly characterized, understood, and accounted for going forward if the full potential of siRNA therapeutics is to be properly harnessed.

1.06.7 Current and Future Needs in Gene Therapy

Gene therapy research has evolved considerably since the first clinical trials of this technology, moving from ex vivo and local administration to systemic administration. The range of therapeutic targets has also expanded from the treatment of monogenetic disorders to the prevention and treatment of acquired diseases (cancer), and so has the number and range of possible therapeutic nucleic acids. Gene therapy is now not just about replacing a missing or defective gene within patient's target cells but is also increasingly about supplementing the body with the production of proteins that can prevent or treat diseases as well, or the selective knockdown of genes whose expression is otherwise deleterious to good health.[169] Therefore, like any other therapeutic modality, gene therapy will require multiple and repeat dosing regimes to maintain effectiveness according to the disease indication concerned.

However, in spite of the widening range of opportunities under consideration, the main challenge for gene therapy remains the vector problem. Therapeutic nucleic acids will always have to be formulated in such a way that somatic and local pharmacokinetics guarantee the delivery of the therapeutic nucleic acid to the target cells of interest within the organ(s) of choice, at the right time for the right duration. Innovations in vector design are sure to be needed for a while to come, but as each innovation surfaces, there will be the earnest hope that more and more gene therapy clinical gene therapies will become practicable. Better vector design will require a better understanding of delivery problems, both at the level of intracellular trafficking of pDNA to the nucleus and at the level of biological fluid stability and tissue penetration. Ideally, each delivery problem should be assessed in the round from the site of administration to target cells of interest. A thorough knowledge of biological barriers assisted by studies with humanized models and tissue samples can then be reduced to selection criteria for vectors and the design of tailor-made delivery systems for the selected therapeutic nucleic acid of choice. Proper knowledge of nucleic acid mechanisms then becomes critical to maximize on therapeutic delivery. An appropriate knowledge of nucleic acid state,[170] tissue-specific promoters, enhancers, LCRs, UCOEs, and the like is then essential to sustain therapeutic effects. Gene therapy is complicated, and its promise is not yet even remotely fulfilled. However, diligence and hard work will yield results, and we believe that gene therapy has the power to become a dominant therapeutic modality in the future – but all in good time.

References

1. *Journal of Gene Medicine*. Gene Therapy Trials Worldwide: http://www.wiley.co.uk/genmed/clinical/ (accessed Aug 2006).
1a. Guan, Y. S.; Liu, Y.; Zhou, X. P.; Li, X.; He, Q.; Sun, L. *Gut* **2005**, *54*, 1318–1319.
2. Luo, J.; Sun, M. H.; Kang, Q.; Peng, Y.; Jiang, W.; Luu, H. H.; Luo, Q.; Park, J. Y.; Li, Y.; Haydon, R. C. et al. *Curr. Gene Ther.* **2005**, *5*, 167–179.
3. Dykxhoorn, D. M.; Novina, C. D.; Sharp, P. A. *Nat. Rev. Mol. Cell Biol.* **2003**, *4*, 457–467.
4. Jopling, C. L.; Yi, M.; Lancaster, A. M.; Lemon, S. M.; Sarnow, P. *Science* **2005**, *309*, 1577–1581.
5. Lu, C.; Tej, S. S.; Luo, S.; Haudenschild, C. D.; Meyers, B. C.; Green, P. J. *Science* **2005**, *309*, 1567–1569.
6. Mattick, J. S. *Science* **2005**, *309*, 1527–1528.
7. Vaughn, M. W.; Martienssen, R. *Science* **2005**, *309*, 1525–1526.
8. Zamore, P. D.; Haley, B. *Science* **2005**, *309*, 1519–1524.
9. Marshall, E. *Science* **2000**, *288*, 953.
10. Arruda, V. R.; Fields, P. A.; Milner, R.; Wainwright, L.; De Miguel, M. P.; Donovan, P. J.; Herzog, R. W.; Nichols, T. C.; Biegel, J. A.; Razavi, M. et al. *Mol. Ther.* **2001**, *4*, 586–592.

11. MacLachlan, I.; Cullis, P.; Graham, R. W. *Curr. Opin. Mol. Ther.* **1999**, *1*, 252–259.
12. Cavazzana-Calvo, M.; Thrasher, A.; Mavilio, F. *Nature* **2004**, *427*, 779–781.
13. Gaspar, H. B.; Thrasher, A. J. *Expert Opin. Biol. Ther.* **2005**, *5*, 1175–1182.
14. Alton, E. W.; Middleton, P. G.; Caplen, N. J.; Smith, S. N.; Steel, D. M.; Munkonge, F. M.; Jeffery, P. K.; Geddes, D. M.; Hart, S. L.; Williamson, R. et al. *Nat. Genet.* **1993**, *5*, 135–142.
15. Antoine, C.; Muller, S.; Cant, A.; Cavazzana-Calvo, M.; Veys, P.; Vossen, J.; Fasth, A.; Heilmann, C.; Wulffraat, N.; Seger, R. et al. *Lancet* **2003**, *361*, 553–560.
16. Cavazzana-Calvo, M.; Lagresle, C.; Hacein-Bey-Abina, S.; Fischer, A. *Annu. Rev. Med.* **2005**, *56*, 585–602.
17. Fischer, A.; Hacein-Bey-Abina, S.; Cavazzana-Calvo, M. *Semin. Hematol.* **2004**, *41*, 272–278.
18. Hacein-Bey-Abina, S.; Fischer, A.; Cavazzana-Calvo, M. *Int. J. Hematol.* **2002**, *76*, 295–298.
19. MacLachlan, I.; Cullis, P. *Adv. Genet.* **2005**, *53*, 157–188.
20. Alton, E. W.; Stern, M.; Farley, R.; Jaffe, A.; Chadwick, S. L.; Phillips, J.; Davies, J.; Smith, S. N.; Browning, J.; Davies, M. G. et al. *Lancet* **1999**, *353*, 947–954.
21. Harvey, B. G.; Leopold, P. L.; Hackett, N. R.; Grasso, T. M.; Williams, P. M.; Tucker, A. L.; Kaner, R. J.; Ferris, B.; Gonda, I.; Sweeney, T. D. et al. *J. Clin. Invest.* **1999**, *104*, 1245–1255.
22. Ostedgaard, L. S.; Rokhlina, T.; Karp, P. H.; Lashmit, P.; Afione, S.; Schmidt, M.; Zabner, J.; Stinski, M. F.; Chiorini, J. A.; Welsh, M. J. *Proc. Natl. Acad. Sci. USA* **2005**, *102*, 2952–2957.
23. Kostarelos, K.; Miller, A. D. *Chem. Soc. Rev.* **2005**, *34*, 970–994.
24. Miller, A. D. *Curr. Med. Chem.* **2003**, *10*, 1195–1211.
25. Tagawa, T.; Manvell, M.; Brown, N.; Keller, M.; Perouzel, E.; Murray, K. D.; Harbottle, R. P.; Tecle, M.; Booy, F.; Brahimi-Horn, M. C. et al. *Gene Ther.* **2002**, *9*, 564–576.
26. Ferber, D. *Science* **2001**, *294*, 1638–1642.
27. Kay, M. A.; Nakai, H. *Nature* **2003**, *424*, 251.
28. Nakai, H.; Montini, E.; Fuess, S.; Storm, T. A.; Grompe, M.; Kay, M. A. *Nat. Genet.* **2003**, *34*, 297–302.
29. Kaminski, J. M.; Huber, M. R.; Summers, J. B.; Ward, M. B. *FASEB J.* **2002**, *16*, 1242–1247.
30. Edelstein, M. L.; Abedi, M. R.; Wixon, J.; Edelstein, R. M. *J. Gene Med.* **2004**, *6*, 597–602.
31. Nemunaitis, J.; Edelman, J. *Cancer Gene Ther.* **2002**, *9*, 987–1000.
32. Peng, Z. *Hum. Gene Ther.* **2005**, *16*, 1016–1027.
33. McCormick, F. *Nat. Rev. Cancer* **2001**, *1*, 130–141.
34. Cavazzana-Calvo, M.; Hacein-Bey, S.; de Saint Basile, G.; Gross, F.; Yvon, E.; Nusbaum, P.; Selz, F.; Hue, C.; Certain, S.; Casanova, J. L. et al. *Science* **2000**, *288*, 669–672.
35. Baker, A. H. *Prog. Biophys. Mol. Biol.* **2004**, *84*, 279–299.
36. Rissanen, T. T.; Rutanen, J.; Yla-Herttuala, S. *Adv. Genet.* **2004**, *52*, 117–164.
37. Beck, C.; Uramoto, H.; Boren, J.; Akyurek, L. M. *Curr. Gene Ther.* **2004**, *4*, 457–467.
38. Baker, A. H.; Kritz, A.; Work, L. M.; Nicklin, S. A. *Exp. Physiol.* **2005**, *90*, 27–31.
39. Strayer, D. S.; Akkina, R.; Bunnell, B. A.; Dropulic, B.; Planelles, V.; Pomerantz, R. J.; Rossi, J. J.; Zaia, J. A. *Mol. Ther.* **2005**, *11*, 823–842.
40. Shlomai, A.; Shaul., Y. *Liver Int.* **2004**, *24*, 526–531.
41. Lee, J. S.; Hadjipanayis, A. G.; Parker, M. D. *Adv. Drug Deliv. Rev.* **2005**, *57*, 1293–1314.
42. Palmer, D. J. *P. Ng. Hum. Gene Ther.* **2005**, *16*, 1–16.
43. Noureddini, S. C.; Curiel, D. T. *Mol. Pharm.* **2005**, *2*, 341–347.
44. Everts, M.; Curiel, D. T. *Curr. Gene Ther.* **2004**, *4*, 337–346.
45. Ma, Z.; Mi, Z.; Wilson, A.; Alber, S.; Robbins, P. D.; Watkins, S.; Pitt, B.; Li, S. *Gene Ther.* **2002**, *9*, 176–182.
46. Mok, H.; Palmer, D. J.; Ng, P.; Barry, M. A. *Mol. Ther.* **2005**, *11*, 66–79.
47. Barquinero, J.; Eixarch, H.; Perez-Melgosa, M. *Gene Ther.* **2004**, *11*, S3–S9.
48. Sinn, P. L.; Sauter, S. L.; McCray., P. B., Jr. *Gene Ther.* **2005**, *12*, 1089–1098.
49. Goncalves, M. A. *Virol. J.* **2005**, *2*, 43.
50. Wang, L.; Herzog, R. W. *Curr. Gene Ther.* **2005**, *5*, 349–360.
51. Athanasopoulos, T.; Graham, I. R.; Foster, H.; Dickson, G. *Gene Ther.* **2004**, *11*, S109–S121.
52. High, K. A. *Semin. Thromb. Hemost.* **2004**, *30*, 257–267.
53. Epstein, A. L.; Marconi, P.; Argnani, R.; Manservigi, R. *Curr. Gene Ther.* **2005**, *5*, 445–458.
54. Shen, Y.; Nemunaitis, J. *Mol. Ther.* **2005**, *11*, 180–195.
55. O'Malley, B. W., Jr.; Li, D.; McQuone, S. J.; Ralston, R. *Laryngoscope* **2005**, *115*, 391–404.
56. Madhusudan, S.; Tamir, A.; Bates, N.; Flanagan, E.; Gore, M. E.; Barton, D. P.; Harper, P.; Seckl, M.; Thomas, H.; Lemoine, N. R. et al. *Clin. Cancer Res.* **2004**, *10*, 2986–2996.
57. Kaushik, A. *Curr. Opin. Investig. Drugs* **2001**, *2*, 976–981.
58. Pack, D. W.; Hoffman, A. S.; Pun, S.; Stayton, P. S. *Nat. Rev. Drug Disc.* **2005**, *4*, 581–593.
59. Felgner, P. L.; Gadek, T. R.; Holm, M.; Roman, R.; Chan, H. W.; Wenz, M.; Northrop, J. P.; Ringold, G. M.; Danielsen, M. *Proc. Natl. Acad. Sci. USA* **1987**, *84*, 7413–7417.
60. Chesnoy, S.; Huang, L. *Annu. Rev. Biophys. Biomol. Struct.* **2000**, *29*, 27–47.
61. Zhang, J. S.; Liu, F.; Huang, L. *Adv. Drug Deliv. Rev.* **2005**, *57*, 689–698.
62. Yan, X.; Scherphof, G. L.; Kamps, J. A. *J. Liposome Res.* **2005**, *15*, 109–139.
63. Schaffer, D. V.; Lauffenburger, D. A. *Curr. Opin. Mol. Ther.* **2000**, *2*, 155–161.
64. Bartsch, M.; Weeke-Klimp, A. H.; Meijer, D. K.; Scherphof, G. L.; Kamps, J. A. *J. Liposome Res.* **2005**, *15*, 59–92.
65. Gunther, M.; Wagner, E.; Ogris, M. *Curr. Med. Chem. Anti-Cancer Agents* **2005**, *5*, 157–171.
66. Wickham, T. J. *Nat. Med.* **2003**, *9*, 135–139.
67. Mahato, R. I.; Smith, L. C.; Rolland, A. *Adv. Genet.* **1999**, *41*, 95–156.
68. Miller, A. D. *Angew. Chem. Int. Ed.* **1998**, *37*, 1768–1785.
69. Miller, A. D. *Curr. Res. Mol. Ther.* **1998**, *1*, 494–503.
70. Ilies, M. A.; Balaban, A. T. *Expert Opin. Ther. Patents* **2001**, *11*, 1729–1752.
71. Nicolazzi, C.; Garinot, M.; Mignet, N.; Scherman, D.; Bessodes, M. *Curr. Med. Chem.* **2003**, *10*, 1263–1277.

72. Schatzlein, A. G. *Anticancer Drugs* **2001**, *12*, 275–304.
73. Cockett, M. I. *Curr. Opin. Mol. Ther.* **1999**, *1*, 279–283.
74. Hortobagyi, G. N.; Ueno, N. T.; Xia, W.; Zhang, S.; Wolf, J. K.; Putnam, J. B.; Weiden, P. L.; Willey, J. S.; Carey, M.; Branham, D. L. et al. *J. Clin. Oncol.* **2001**, *19*, 3422–3433.
75. Bergen, M.; Chen, R.; Gonzalez, R. *Expert Opin. Biol. Ther.* **2003**, *3*, 377–384.
76. Reszka, R. C.; Jacobs, A.; Voges, J. *Methods Enzymol.* **2005**, *391*, 200–208.
77. Gill, D. R.; Southern, K. W.; Mofford, K. A.; Seddon, T.; Huang, L.; Sorgi, F.; Thomson, A.; MacVinish, L. J.; Ratcliff, R.; Bilton, D. et al. *Gene Ther.* **1997**, *4*, 199–209.
78. Hyde, S. C.; Southern, K. W.; Gileadi, U.; Fitzjohn, E. M.; Mofford, K. A.; Waddell, B. E.; Gooi, H. C.; Goddard, C. A.; Hannavy, K.; Smyth, S. E. et al. *Gene Ther.* **2000**, *7*, 1156–1165.
79. Fletcher, S.; Ahmad, A.; Perouzel, E.; Heron, A.; Miller, A. D.; Jorgensen, M. R. *J. Med. Chem.* **2006**, *49*, 349–357.
80. Fletcher, S.; Ahmad, A.; Perouzel, E.; Jorgensen, M. R.; Miller, A. D. *Org. Biomol. Chem.* **2006**, *4*, 196–199.
81. Lasic, D. D.; Papahadjopoulos, D. *Science* **1995**, *267*, 1275–1276.
82. Martin-Herranz, A.; Ahmad, A.; Evans, H. M.; Ewert, K.; Schulze, U.; Safinya, C. R. *Biophys. J.* **2004**, *86*, 1160–1168.
83. Hong, K.; Zheng, W.; Baker, A.; Papahadjopoulos, D. *FEBS Lett.* **1997**, *400*, 233–237.
84. Wheeler, J. J.; Palmer, L.; Ossanlou, M.; MacLachlan, I.; Graham, R. W.; Zhang, Y. P.; Hope, M. J.; Scherrer, P.; Cullis, P. R. *Gene Ther.* **1999**, *6*, 271–281.
85. Zhang, Y. P.; Sekirov, L.; Saravolac, E. G.; Wheeler, J. J.; Tardi, P.; Clow, K.; Leng, E.; Sun, R.; Cullis, P. R.; Scherrer, P. *Gene Ther.* **1999**, *6*, 1438–1447.
86. Tam, P.; Monck, M.; Lee, D.; Ludkovski, O.; Leng, E. C.; Clow, K.; Stark, H.; Scherrer, P.; Graham, R. W.; Cullis, P. R. *Gene Ther.* **2000**, *7*, 1867–1874.
87. Ambegia, E.; Ansell, S.; Cullis, P.; Heyes, J.; Palmer, L.; MacLachlan, I. *Biochim. Biophys. Acta* **2005**, *1669*, 155–163.
88. Shi, N.; Boado, R. J.; Pardridge, W. M. *Pharm. Res.* **2001**, *18*, 1091–1095.
89. Shi, N. Y.; Pardridge, W. M. *Proc. Natl. Acad. Sci. USA* **2000**, *97*, 7567–7572.
90. Zhang, Y. F.; Boado, R. J.; Pardridge, W. M. *Pharm. Res.* **2003**, *20*, 1779–1785.
91. Zhang, Y.; Boado, R. J.; Pardridge, W. M. *J. Gene Med.* **2003**, *5*, 1039–1045.
92. Hofland, H. E.; Masson, C.; Iginla, S.; Osetinsky, I.; Reddy, J. A.; Leamon, C. P.; Scherman, D.; Bessodes, M.; Wils, P. *Mol. Ther.* **2002**, *5*, 739–744.
93. Reddy, J. A.; Low, P. S. *J. Control. Release* **2000**, *64*, 27–37.
94. Shi, G.; Guo, W.; Stephenson, S. M.; Lee, R. J. *J. Control. Release* **2002**, *80*, 309–319.
95. Zhang, S.; Xu, Y.; Wang, B.; Qiao, W.; Liu, D.; Li, Z. *J. Control. Release* **2004**, *100*, 165–180.
96. Boussif, O.; Lezoualc'h, F.; Zanta, M. A.; Mergny, M. D.; Scherman, D.; Demeneix, B.; Behr, J. P. *Proc. Natl. Acad. Sci. USA* **1995**, *92*, 7297–7301.
97. Wiseman, J. W.; Goddard, C. A.; McLelland, D.; Colledge, W. H. *Gene Ther.* **2003**, *10*, 1654–1662.
98. Abdallah, B.; Hassan, A.; Benoist, C.; Goula, D.; Behr, J. P.; Demeneix, B. A. *Hum. Gene Ther.* **1996**, *7*, 1947–1954.
99. Goula, D.; Benoist, C.; Mantero, S.; Merlo, G.; Levi, G.; Demeneix, B. A. *Gene Ther.* **1998**, *5*, 1291–1295.
100. Ferrari, S.; Pettenazzo, A.; Garbati, N.; Zacchello, F.; Behr, J. P.; Scarpa, M. *Biochim. Biophys. Acta* **1999**, *1447*, 219–225.
101. Coll, J. L.; Chollet, P.; Brambilla, E.; Desplanques, D.; Behr, J. P.; Favrot, M. *Hum. Gene Ther.* **1999**, *10*, 1659–1666.
102. Ohana, P.; Gofrit, O.; Ayesh, S.; Al-Sharef, W.; Mizrahi, A.; Birman, T.; Schneider, T.; Matouk, I.; de Groot, N.; Tavdy, E. et al. *Gene Ther. Mol. Biol.* **2004**, *8*, 181–192.
103. Moghimi, S. M.; Symonds, P.; Murray, J. C.; Hunter, A. C.; Debska, G.; Szewczyk, A. *Mol. Ther.* **2005**, *11*, 990–995.
104. Neu, M.; Fischer, D.; Kissel, T. *J. Gene Med.* **2005**, *7*, 992–1009.
105. Lungwitz, U.; Breunig, M.; Blunk, T.; Gopferich, A. *Eur. J. Pharm. Biopharm.* **2005**, *60*, 247–266.
106. Forrest, M. L.; Meister, G. E.; Koerber, J. T.; Pack, D. W. *Pharm. Res.* **2004**, *21*, 365–371.
107. Thomas, M.; Lu, J. J.; Ge, Q.; Zhang, C.; Chen, J.; Klibanov, A. M. *Proc. Natl. Acad. Sci. USA* **2005**, *102*, 5679–5684.
108. Jevprasesphant, R.; Penny, J.; Jalal, R.; Attwood, D.; McKeown, N. B.; D'Emanuele, A. *Int. J. Pharm.* **2003**, *252*, 263–266.
109. Guang Liu, W.; De Yao, K. *J. Control. Release* **2002**, *83*, 1–11.
110. Alpar, H. O.; Papanicolaou, I.; Bramwell, V. W. *Expert Opin. Drug Deliv.* **2005**, *2*, 829–842.
111. Koping-Hoggard, M.; Tubulekas, I.; Guan, H.; Edwards, K.; Nilsson, M.; Varum, K. M.; Artursson, P. *Ther. Gene* **2001**, *8*, 1108–1121.
112. Koping-Hoggard, M.; Varum, K. M.; Issa, M.; Danielsen, S.; Christensen, B. E.; Stokke, B. T.; Artursson, P. *Gene Ther.* **2004**, *11*, 1441–1452.
113. Thanou, M.; Florea, B. I.; Geldof, M.; Junginger, H. E.; Borchard, G. *Biomaterials* **2002**, *23*, 153–159.
114. Kean, T.; Roth, S.; Thanou, M. *J. Control. Release* **2005**, *103*, 643–653.
115. Luo, D.; Saltzman, W. M. *Nat. Biotechnol.* **2000**, *18*, 33–37.
116. Kabanov, A.; Zhu, J.; Alakhov, V. *Adv. Genet.* **2005**, *53*, 231–261.
117. Konstan, M. W.; Davis, P. B.; Wagener, J. S.; Hilliard, K. A.; Stern, R. C.; Milgram, L. J.; Kowalczyk, T. H.; Hyatt, S. L.; Fink, T. L.; Gedeon, C. R. et al. *Hum. Gene Ther.* **2004**, *15*, 1255–1269.
118. Kichler, A. *J. Gene Med.* **2004**, *6*, S3–S10.
119. Lee, M.; Kim, S. W. *Pharm. Res.* **2005**, *22*, 1–10.
120. Anderson, D. G.; Akinc, A.; Hossain, N.; Langer, R. *Mol. Ther.* **2005**, *11*, 426–434.
121. Anderson, D. G.; Lynn, D. M.; Langer, R. *Angew. Chem. Int. Ed. Engl.* **2003**, *42*, 3153–3158.
122. Anderson, D. G.; Peng, W.; Akinc, A.; Hossain, N.; Kohn, A.; Padera, R.; Langer, R.; Sawicki, J. A. *Proc. Natl. Acad. Sci. USA* **2004**, *101*, 16028–16033.
123. Rochlitz, C. F. *Swiss Med. Wkly.* **2001**, *131*, 4–9.
124. Steitz, J.; Buchs, S.; Tormo, D.; Ferrer, A.; Wenzel, J.; Huber, C.; Wolfel, T.; Barbacid, M.; Malumbres, M.; Tuting, T. *Int. J. Cancer* **2006**, *118*, 373–380.
125. Koide, Y.; Nagata, T.; Yoshida, A.; Uchijima, M. *Jpn. J. Pharmacol.* **2000**, *83*, 167–174.
126. Fritz, J.; Katopodis, A. G.; Kolbinger, F.; Anselmetti, D. *Proc. Natl. Acad. Sci. USA* **1998**, *95*, 12283–12288.
127. Liu, F.; Tyagi, P. *Adv. Genet.* **2005**, *54*, 43–64.
128. Wolff, J. A.; Budker, V. *Adv. Genet.* **2005**, *54*, 3–20.
129. Nakagami, H.; Kaneda, Y.; Ogihara, T.; Morishita, R. *Expert Rev. Cardiovasc. Ther.* **2005**, *3*, 513–519.

130. Isner, J. M. *Nature* **2002**, *415*, 234–239.
131. Walker, W. E.; Miller, A. D. *Mol. Ther.* **2004**, *9*, 184.
132. Bigger, B. W.; Tolmachov, O.; Collombet, J. M.; Fragkos, M.; Palaszewski, I. *C. J. Coutelle. Biol. Chem.* **2001**, *276*, 23018–23027.
133. Chow, C. M.; Athanassiadou, A.; Raguz, S.; Psiouri, L.; Harland, L.; Malik, M.; Aitken, M. A.; Grosveld, F.; Antoniou, M. *Gene Ther.* **2002**, *9*, 327–336.
134. Jenke, A. C.; Eisenberger, T.; Baiker, A.; Stehle, I. M.; Wirth, S.; Lipps, H. J. *Hum. Gene Ther.* **2005**, *16*, 533–539.
135. Jenke, A. C.; Stehle, I. M.; Herrmann, F.; Eisenberger, T.; Baiker, A.; Bode, J.; Fackelmayer, F. O.; Lipps, H. J. *Proc. Natl. Acad. Sci. USA* **2004**, *101*, 11322–11327.
136. Schaarschmidt, D.; Baltin, J.; Stehle, I. M.; Lipps, H. J.; Knippers, R. *EMBO J.* **2004**, *23*, 191–201.
137. Glover, D. J.; Lipps, H. J.; Jans, D. A. *Nat. Rev. Genet.* **2005**, *6*, 299–310.
138. Coates, C. J.; Kaminski, J. M.; Summers, J. B.; Segal, D. J.; Miller, A. D.; Kolb, A. F. *Trends Biotechnol.* **2005**, *23*, 407–419.
139. Fritz, G. N. *DNA Seq.* **1998**, *8*, 215–221.
140. Izsvak, Z.; Ivics, Z.; Plasterk, R. H. *J. Mol. Biol.* **2000**, *302*, 93–102.
141. Schouten, G. J.; van Luenen, H. G.; Verra, N. C.; Valerio, D.; Plasterk, R. H. *Nucleic Acids Res.* **1998**, *26*, 3013–3017.
142. Yant, S. R.; Meuse, L.; Chiu, W.; Ivics, Z.; Izsvak, Z.; Kay, M. A. *Nat. Genet.* **2000**, *25*, 35–41.
143. Ortiz-Urda, S.; Lin, Q.; Yant, S. R.; Keene, D.; Kay, M. A.; Khavari, P. A. *Gene Ther.* **2003**, *10*, 1099–1104.
144. Vigdal, T. J.; Kaufman, C. D.; Izsvak, Z.; Voytas, D. F.; Ivics, Z. *J. Mol. Biol.* **2002**, *323*, 441–452.
145. Groth, A. C.; Fish, M.; Nusse, R.; Calos, M. P. *Genetics* **2004**, *166*, 1775–1782.
146. Groth, A. C.; Calos, M. P. *J. Mol. Biol.* **2004**, *335*, 667–678.
147. Olivares, E. C.; Hollis, R. P.; Chalberg, T. W.; Meuse, L.; Kay, M. A.; Calos, M. P. *Nat. Biotechnol.* **2002**, *20*, 1124–1128.
148. Ortiz-Urda, S.; Thyagarajan, B.; Keene, D. R.; Lin, Q.; Calos, M. P.; Khavari, P. A. *Hum. Gene Ther.* **2003**, *14*, 923–928.
149. Magin-Lachmann, C.; Kotzamanis, G.; D'Aiuto, L.; Cooke, H.; Huxley, C.; Wagner, E. *J. Gene Med.* **2004**, *6*, 195–209.
150. Magin-Lachmann, C.; Kotzamanis, G.; D'Aiuto, L.; Wagner, E.; Huxley, C. *BMC Biotechnol.* **2003**, *3*, 2.
151. Westphal, E. M.; Sierakowska, H.; Livanos, E.; Kole, R.; Vos, J. M. *Hum. Gene Ther.* **1998**, *9*, 1863–1873.
152. Compton, S. H.; Mecklenbeck, S.; Mejia, J. E.; Hart, S. L.; Rice, M.; Cervini, R.; Barrandon, Y.; Larin, Z.; Levy, E. R.; Bruckner-Tuderman, L. et al. *Gene Ther.* **2000**, 7, 1600–1605.
153. Mecklenbeck, S.; Compton, S. H.; Mejia, J. E.; Cervini, R.; Hovnanian, A.; Bruckner-Tuderman, L.; Barrandon, Y. *Hum. Gene Ther.* **2002**, *13*, 1655–1662.
154. Huertas, D.; Howe, S.; McGuigan, A.; Huxley, C. *Hum. Mol. Genet.* **2000**, *9*, 617–629.
155. Vassaux, G.; Manson, A. L.; Huxley, C. *Gene Ther.* **1997**, *4*, 618–623.
156. de Jong, G.; Telenius, A.; Vanderbyl, S.; Meitz, A.; Drayer, J. *Chromosome Res.* **2001**, *9*, 475–485.
157. Stewart, S.; MacDonald, N.; Perkins, E.; DeJong, G.; Perez, C.; Lindenbaum, M. *Gene Ther.* **2002**, *9*, 719–723.
158. Csonka, E.; Cserpan, I.; Fodor, K.; Hollo, G.; Katona, R.; Kereso, J.; Praznovszky, T.; Szakal, B.; Telenius, A.; deJong, G. et al. *J. Cell Sci.* **2000**, *113* (Part 18), 3207–3216.
159. Mejia, J. E.; Willmott, A.; Levy, E.; Earnshaw, W. C.; Larin, Z. *Am. J. Hum. Genet.* **2001**, *69*, 315–326.
160. Konopka, K.; Rossi, J. J.; Swiderski, P.; Slepushkin, V. A.; Duzgunes, N. *Biochim. Biophys. Acta* **1998**, *1372*, 55–68.
161. Khvorova, A.; Reynolds, A.; Jayasena, S. D. *Cell* **2003**, *115*, 209–216.
162. Schwarz, D. S.; Hutvagner, G.; Du, T.; Xu, Z.; Aronin, N.; Zamore, P. D. *Cell* **2003**, *115*, 199–208.
163. Ui-Tei, K.; Naito, Y.; Takahashi, F.; Haraguchi, T.; Ohki-Hamazaki, H.; Juni, A.; Ueda, R.; Saigo, K. *Nucleic Acids Res.* **2004**, *32*, 936–948.
164. Naito, Y.; Yamada, T.; Ui-Tei, K.; Morishita, S.; Saigo, K. *Nucleic Acids Res.* **2004**, *32*, W124–W129.
165. Bitko, V.; Musiyenko, A.; Shulyayeva, O.; Barik, S. *Nat. Med.* **2005**, *11*, 50–55.
166. Zhang, W.; Yang, H.; Kong, X.; Mohapatra, S.; San Juan-Vergara, H.; Hellermann, G.; Behera, S.; Singam, R.; Lockey, R. F.; Mohapatra, S. S. *Nat. Med.* **2005**, *11*, 56–62.
167. Morrissey, D. V.; Lockridge, J. A.; Shaw, L.; Blanchard, K.; Jensen, K.; Breen, W.; Hartsough, K.; Machemer, L.; Radka, S.; Jadhav, V. et al. *Nat. Biotechnol.* **2005**, *23*, 1002–1007.
168. Hornung, V.; Guenthner-Biller, M.; Bourquin, C.; Ablasser, A.; Schlee, M.; Uematsu, S.; Noronha, A.; Manoharan, M.; Akira, S.; de Fougerolles, A. et al. *Nat. Med.* **2005**, *11*, 263–270.
169. Rolland, A. *Adv. Drug Deliv. Rev.* **2005**, *57*, 669–673.
170. Schleef, M.; Schmidt, T. *J. Gene Med.* **2004**, *6*, S45–S53.
171. NIH Genetic Modification Clinical Research Information System (GeMCRIS): http://www.gemcris.od.nih.gov (accessed Aug 2006).

Biographies

Maya Thanou is a Royal Society research fellow/lecturer in the Genetic Therapies Centre, Department of Chemistry at Imperial College London, UK, where she leads research on receptor targeting of gene-delivery systems for cancer therapies. Before this post, she was a lecturer in Drug Delivery at the Centre of Polymer Therapeutics in Cardiff University, UK, where she focused on polymeric synthetic vectors. She received her PhD in Drug Delivery (2000) from Leiden University and the Leiden/Amsterdam Centre for Drug Research in the Netherlands. She has been a consultant for biotechnology companies. She has been awarded by the Research Committee of Cardiff University and the American Association of Pharmaceutical Sciences (AAPS). She is a member of the Controlled Release Society, AAPS, and the British, European, and American Societies of Gene Therapy.

Simon Waddington is a lecturer at University College London, UK, in the Katharine Dormandy Haemophilia Centre & Haemostasis Unit, investigating novel therapies for hereditary diseases of hemostasis. Prior to this, he pioneered preclinical models of in utero gene transfer for study and the potential treatment of a broad range of diseases, including hemophilia, cystic fibrosis, and Duchenne muscular dystrophy under the mentorship of Prof Charles Coutelle at Imperial College London, UK. He gained his PhD at St Mary's Hospital Medical School, London, UK, studying immune diseases of the kidney. He is a member of the British, European, and American Societies of Gene Therapy.

Andrew David Miller is the Professor of Organic Chemistry & Chemical Biology at Imperial College London, UK, and the founding director of the Imperial College Genetic Therapies Centre (GTC). He began his chemistry education at the University of Bristol, UK, from where he graduated in 1984 with a BSc degree. His PhD thesis research was carried out at the University of Cambridge, UK, in the research group of Professor Alan Battersby, after which he joined the research group of Professor Jeremy Knowles at Harvard University, USA. Since 1990, Miller has been a member of the academic staff in the Chemistry Department of Imperial College London, UK, where he has been researching into synthetic nonviral vector systems to enable gene therapy and siRNA therapeutics, the chemistry of stress, and the proteomic code. In his career, he has received several awards and fellowships, including the Thomas Malkin and Robert Garner prizes for chemistry (Bristol University), an Emmanuel College research scholarship (Emmanuel College, Cambridge), a Lindemann Trust Fellowship, and, more recently, a Leverhulme Trust Fellowship. In 2000, he was awarded the Novartis Young Investigator Award in Chemistry, and was elected President of the International Society of Cancer Gene Therapy. He was appointed in 2005 to serve on the Nonviral Gene Transfer Vectors committee of the American Society of Gene Therapy. Miller has cofounded two GTC spin-out companies, Proteom Ltd in September 1999 and IC-Vec Ltd in December 2001. He maintains strong scientific links with IC-Vec, owing to his personal interest in bringing the fruits of academic research to market in the form of laboratory reagents and therapeutic products.

Comprehensive Medicinal Chemistry II
ISBN (set): 0-08-044513-6

ISBN (Volume 1) 0-08-044514-4; pp. 297–319

1.07 Overview of Sources of New Drugs

D Brown, Alchemy Biomedical Consulting, Cambridge, UK

1.07.1 Sources of Drugs: Prehistory to the Early Twentieth Century – The Age of Natural Products from Traditional Medicines

Before the beginning of synthetic medicinal chemistry in the nineteenth century, drugs were derived by extraction of plant, animal, or marine products, with varying degrees of purification. Societies have used these sources since prehistoric times, and the medicinal properties were discovered by trial and error through human ingestion (*see* 1.09 Traditional Medicines). Although 'prescriptions' are recorded back to Sumerian times, 2100 BC, the earliest major compendium is ascribed to the Egyptian Ebers Papyrus which listed some 800 'remedies.' Use of these was linked to religious practices and the placebo effect was undoubtedly of importance in their use. Sneader,[1] in the first edition of Comprehensive Medicinal Chemistry, provided a detailed account of drug usage in ancient societies.

In the Western world at least it was not until the eighteenth century that, with the ascendancy of scientific method, controlled experiments were designed to try to assess whether 'folklore' products did actually confer therapeutic benefit.[2,3] The dawn and growth of science in the seventeenth and eighteenth centuries plus the development of synthetic chemistry in the eighteenth and (particularly) the nineteenth and twenteenth centuries allowed thorough assessment of ancient remedies and their hidden ingredients, which were at last extracted and synthesized in quantity and exposed to critical evaluation.

The result of controlled experiments was to show that some of the products were efficacious, though all too often the trials demonstrated limited or no efficacy and maybe important adverse effects. Early trials were undoubtedly confused by the relatively poor understanding of the placebo effect and also by poor use of the statistical methods that must be applied to interpret data correctly. Where some efficacy did appear to exist, attempts to purify active ingredients to boost activity of the medicine very often resulted instead in demonstration of the lack of usefulness of the preparation. This led to the abandonment of many previously valued 'remedies' and prepared the way for synthetic chemistry to become the major source of new medicines in the late nineteenth century and the twenteenth century and, in some cases, into the twenty-first century.

However, some ancient remedies have survived both the test of time and modern clinical trials. A few remain among the best treatments we have for certain conditions, or alternatively they have acted as good leads for discovery of drugs during the twenteenth century. Among these are morphine, papaverine, caffeine, quinine, pilocarpine, physostigmine, curare derivatives, ephedrine, digitalis, cannabis, and salicylic acid. The following paragraphs illustrate the original sources of these drugs and their impact on rational drug discovery later in the nineteenth and twenteenth centuries.

Morphine (**1**), derived from the opium poppy, remains a mainstay of treatment for severe pain; and its methylated derivative codeine (**2**) is used widely in some countries for moderate pain. Morphine was isolated by Serturner in 1805, probably in an impure form.[4,5] Codeine was isolated in 1832.[6] However, a century was to pass before Gulland and Robinson[7] in 1923 determined the correct structure of morphine, enabling an expansion of rational chemistry around the biological activity of this molecule. Chemists today are still trying to 'improve' morphine via synthesis and clinical trial of derivatives,[8] or trying to find alternative mechanisms and drugs for the treatment of severe pain.[9,10] However, this ancient remedy dating back to antiquity remains unbeaten as a treatment for severe pain despite its well-described adverse effects.

Papaverine (**3**) was first isolated in 1848, from the mother liquors remaining after extraction of morphine from opium, but its spasmolytic activity on smooth muscle was not recognized until 1917.[11] Then it was rapidly introduced into clinical use and remained a mainstay until introduction of synthetic atropine analogs in the 1930s.

Caffeine (**4**) was isolated by Runge in 1819 from coffee beans[12] and in the same year he isolated quinine (**5**) from cinchona bark. Caffeine, apart from its own stimulant properties, has given a lead into a rich source of pharmacologically active substances; and quinine soon found use as an antimalarial.

The leaf of the cocoa plant *Erythroxylon coca*, long used by native societies, also proved a source of bioactive drugs. Cocaine (**6**), a tropane alkaloid, was obtained pure and crystalline as early as 1860 by Niemann.[13]

Pilocarpine (**7**) was isolated in 1875 by Hardy[14] and Gerrard.[15] It remains in use today for reduction of intraocular pressure in glaucoma. Also used for this indication is physostigmine (**8**) which was isolated from calabar bean (*Physostigma venenosum*) in 1864 by Jobst and Hesse.[16]

Curare derivatives, prepared from extracts used by native South American indians to paralyze their prey, led eventually to isolation of the essential ingredient D-tubocurarine (**9**) in 1935, which led to the development of neuromuscular blocking agents such as gallamine (**10**).[17]

Ephedrine (**11**) was isolated in 1897 by Nagai from the Chinese plant *Ephedra vulgaris*, and its sympathomimetic activity was recognized in the 1920s. Its synthetic derivative amphetamine (**12**), originally prepared in 1887, found clinical use as an inhaled decongestant, in part because it was easier and cheaper to make.[18]

Extracts of digitalis were long used as a stimulant and by the Romans as a diuretic, heart tonic, and emetic.[19] Isolation of an active principle, digitoxin (**13**), occurred in 1841 but a half century passed before its structure was solved. Widespread clinical use of a digitalis-derived drug awaited the discovery in 1930 of digoxin (**14**), an analog with better pharmacodynamic properties.[19]

The medicinal properties of *Cannabis sativa* were known in China 4000 years ago[20] yet active ingredients such as cannabinol (**15**) were isolated as late as the1960s. In recent years debate has emerged about the role of cannabis in treating patients with multiple sclerosis, a disease in which patients themselves have self-medicated despite legal barriers. Cannabis remains a controlled substance in many countries and an unresolved debate continues about its safety, particularly about its liability to precipitate schizophrenia. The recent discovery of two receptor subtypes and of natural ligands for cannabinoid receptors has led to increased interest such that in 2005 some 20 synthetic agonist and antagonist molecules were listed as under development for a broad range of diseases including pain, migraine, obesity, multiple sclerosis, nausea, arthritis, and addiction.[20]

Perhaps the most successful drug of modern times is aspirin (**16**). It has its origins in the ancient remedy of willow bark extract (*Spirea ulmaria*). In 1829 Leroux isolated Salicin (**17**) which was introduced by Maclagan[21] as a treatment for rheumatic conditions some 50 years later. Maclagan later found that a metabolite of salicin, salicylic acid (**18**), was more efficacious in treating rheumatic conditions and also about this time its antipyretic properties were discovered. As salicylic acid became more widely used, its adverse effects on the stomach became apparent, which led to a search for derivatives lacking this effect. Hoffmann[22] prepared aspirin, the acetyl derivative of salicylic acid, in 1898, and it proved more acceptable to patients, though some gastric problems remained.

Aspirin remains one of the most widely used drugs in the world, with its range of utility expanding in recent years through discovery of its beneficial effects in conditions additional to rheumatic conditions, such as cancer, stroke, angina, and heart failure. It is ironic that the marketing information from manufacturer Bayer in the 1920s included the assurance 'Does not affect the heart' because of concerns of the day that some drugs could damage the myocardium. Today, aspirin's beneficial effects on the heart are well recognized and it is estimated that of the 80 million aspirin tablets Americans take each day, most are taken not for inflammation or pain but to reduce the risk of heart disease. This new use followed studies showing aspirin's usefulness in treating cardiovascular disease, including heart attack and stroke, and the granting of approval by the Food and Drug Administration (FDA) for its use in these conditions. The labeling for aspirin now includes use in the following conditions: stroke in those who have had a previous stroke or who have had a warning sign called a transient ischemic attack (ministroke); heart attack in those who have had a previous heart attack or experience angina; complications from a heart attack if the drug is taken at the first signs of a heart attack; recurrent blockage for those who have had heart bypass surgery or other procedures to clear blocked arteries, such as balloon angioplasty or carotid endarterectomy.

The mechanism of action of aspirin was resolved in the early 1970s by Sir John Vane, for which he received the Nobel Prize in Medicine. He showed that aspirin's ability to reduce the body's production of certain prostaglandins is the reason for both its effectiveness in reducing pain and inflammation and its protective effects in the vascular system.

The effectiveness of aspirin has inspired during the twentieth and early twenty-first centuries the discovery of a range of anti-inflammatories such as the NSAIDs (nonsteroidal anti-inflammatory drugs) and COX-2 inhibitors (cyclooxygenase subtype 2).

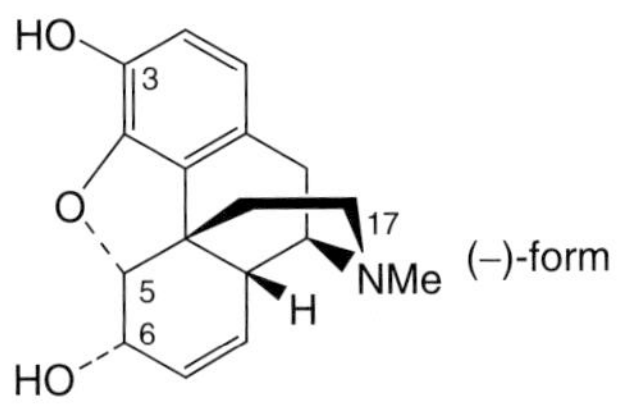

1 Morphine

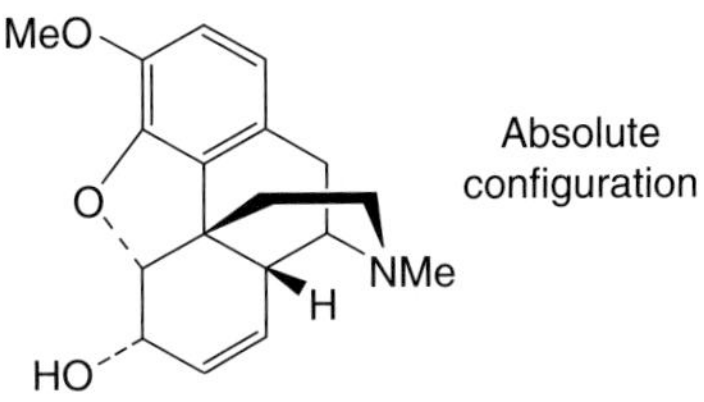

2 Codeine

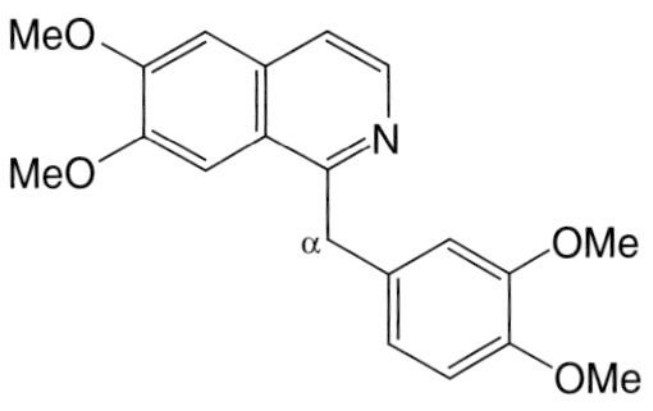

3 Pavaverine

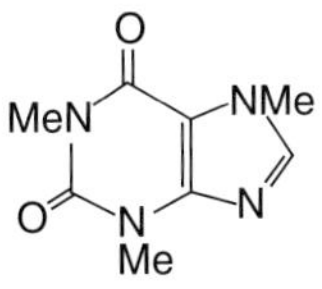

4 Caffeine

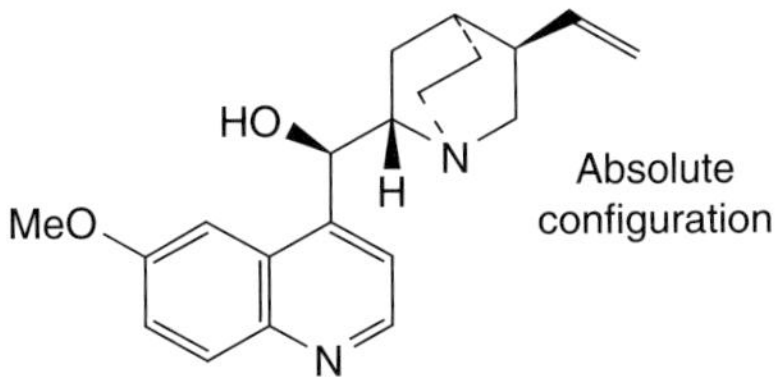

5 Quinine

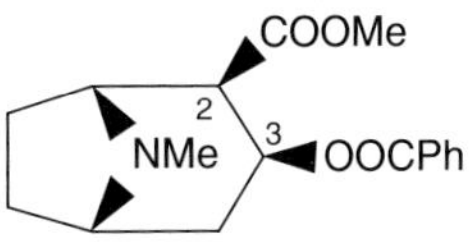

6 Cocaine

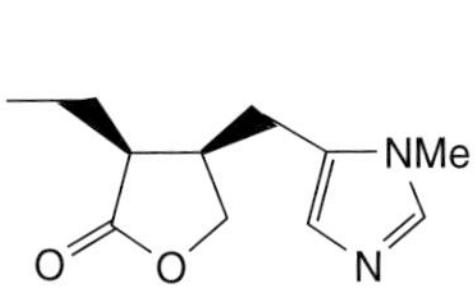

7 Pilocarpine

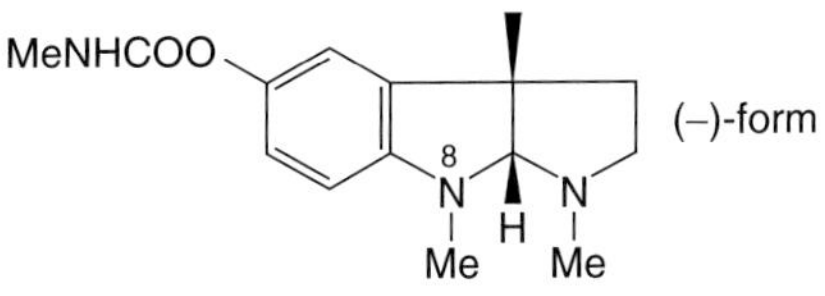

8 Physostigmine

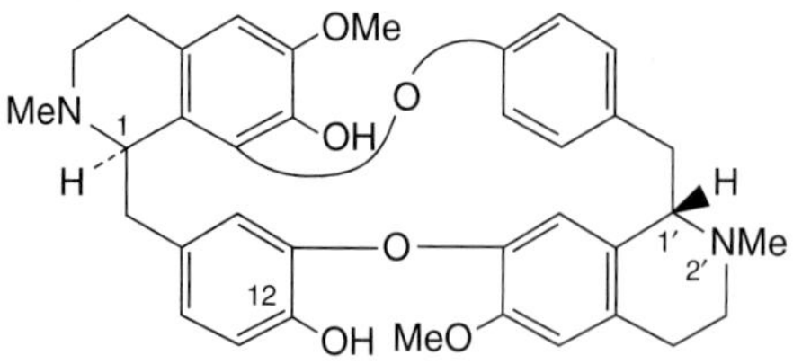

(+)-form

9 Tubocuranine

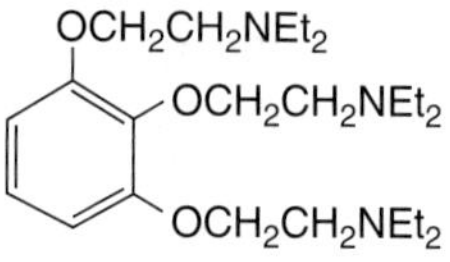

10 Gallamine

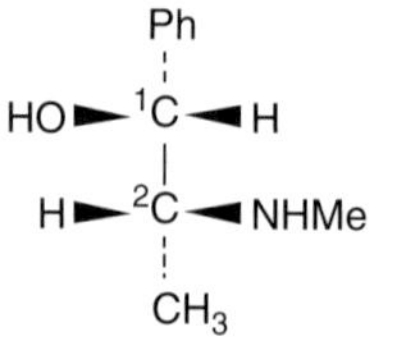

(1*R*, 2*R*)-form

11 Ephedrine

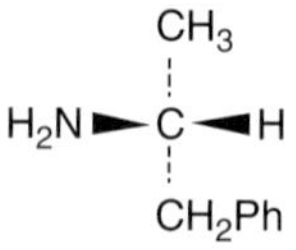

(*R*)-form

12 Amphetamine

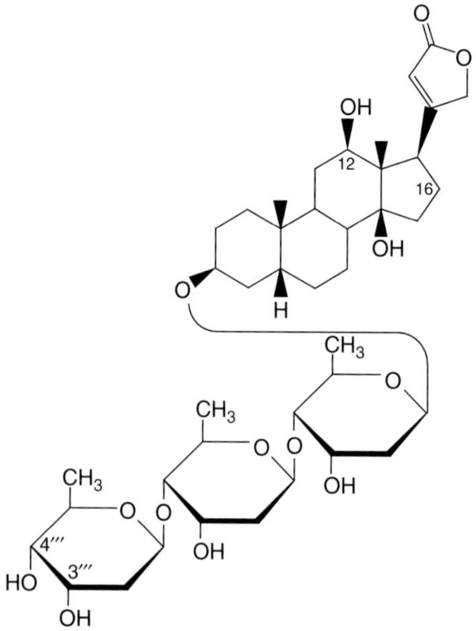

13 Digitoxin is 12-deoxydigoxin and **14** Digoxin

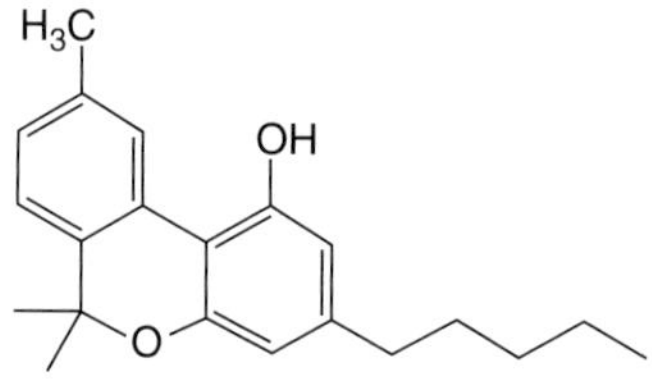

15 Cannabinol

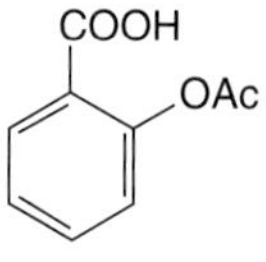

16 Aspirin

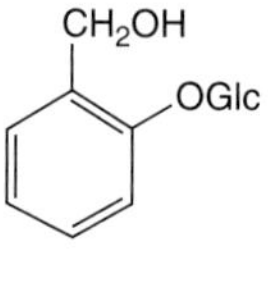

17 Salicin

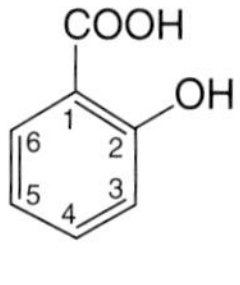

18 Salicylic acid

During the second half of the nineteenth century the foundations of modern chemistry were established, and their practical application to biological problems such as drug discovery became feasible. Avogadro's atomic theory had been accepted and Mendeleev's periodic table of elements was developed. These advances allowed some predictive ability. The puzzle of aromatic structure was resolved by Kekule in 1865 and in part as a result of this and the widespread availability of coal tar (which was a by-product of the industrialization of Western countries) as a source of chemicals for experimentation, a large dye industry came into existence which was to have an important role in the biology and medicine of the time. Paul Erlich developed the crucial idea that differential effects on 'chemoreceptors' of pathogens versus host might be exploited by drugs. Steady advances were made in analytical sciences and in particular the advent of x-ray crystallography proved to be a key element in confirming the chemical structure of bioactive molecules. In parallel to these advances in the chemical sciences, pharmacology was developing as a discipline during the final quarter of the nineteenth century. Chemical and dye companies founded pharmaceutical divisions and a new industry was in the process of being born.

With this historical background we will now consider the role of synthetic medicinal chemistry in the discovery of drugs over the past century.

1.07.2 Sources of Drugs: The Twentieth and Early Twenty-First Centuries - The Age of Modified Natural Products, Synthetic Small Molecules, and Recombinant Biologicals

1.07.2.1 Small-Molecule Products

1.07.2.1.1 Drugs from modified natural products

Following the gradual exploration of the value of traditional medicines and isolation of their active components during the late nineteenth and early twentieth centuries, additional useful medicines were derived from natural product sources based on observations made in research laboratories. Perhaps the most famous is the discovery of the antibiotic penicillin (**19**). The great success of penicillin, followed quite rapidly by the emergence of resistance to the drug, led to application of synthetic chemistry to modify the molecule to produce new drugs with improved resistance profiles. Learning from this history, the pharmaceutical industry invested heavily during the 1960s to 1980s in the screening of natural products, both in pure form and as semipurified extracts. From this effort emerged cephalosporins such as cefuroxime (**20**) (Zinacef), aminoglycosides such as streptomycin (**21**), polyenes such as amphotericin B (**22**), and other useful antibiotics. In each class synthetic chemistry was applied to improve the properties of the original lead molecule. Screening of natural products was extended to other disease areas, and notable successes have been attained in oncology with the discovery of paclitaxel (**23**) (Taxol); in lipid disorders with the discovery of 3-hydroxy 3-methyl glutaryl (HMG) CoA inhibitors such as atorvastatin (**24**) (Lipitor) based on the original discovery in Japan of compactin (**25**) (Mevastatin); and also in immunology with the discovery of FK506/tacrolimus (**26**) (Protopic). Overall, less success was attained in other disease areas and the scale of investment in screening of natural products has been reduced in recent decades based on cost-effectiveness calculations, the move toward high-throughput screening (HTS) of small organic molecules against defined targets, and the preference for 'druglike' leads. So whereas close to 100% of drugs were derived from natural products up till late in the nineteenth century, now natural products account for 5% of the new chemical entities (NCEs) approved by the FDA between 1981 and 2002 with another 23% having their origins in natural product sources.[23]

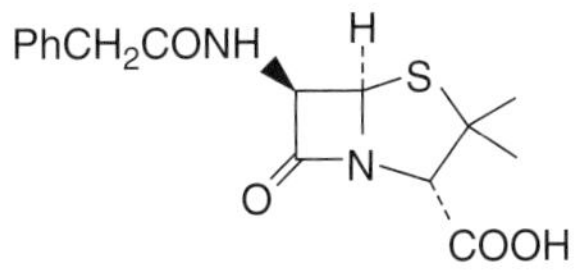

19 Penicillin G

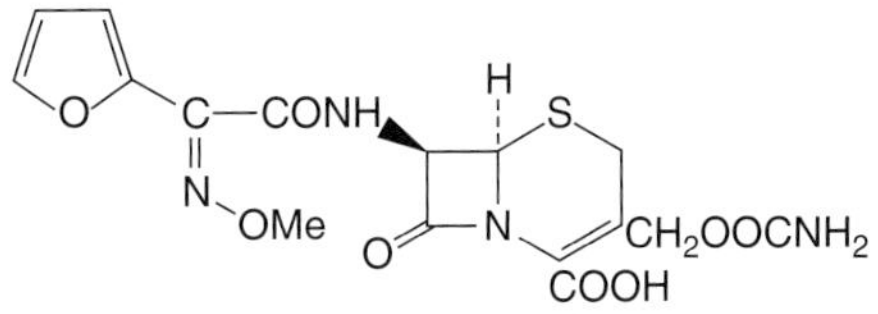

20 Cefuroxime

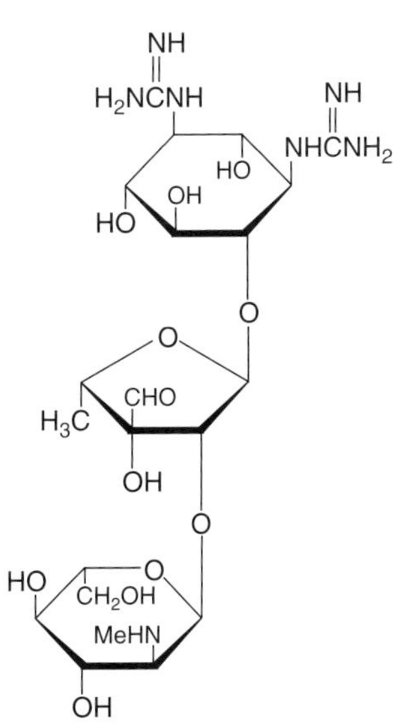

21 Streptomycin

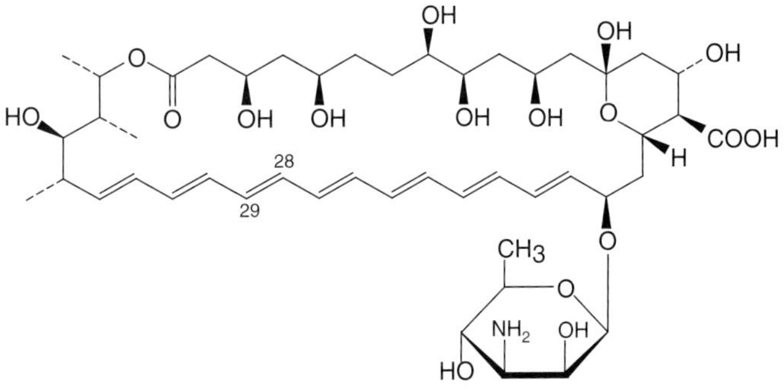

22 Amphotericin B

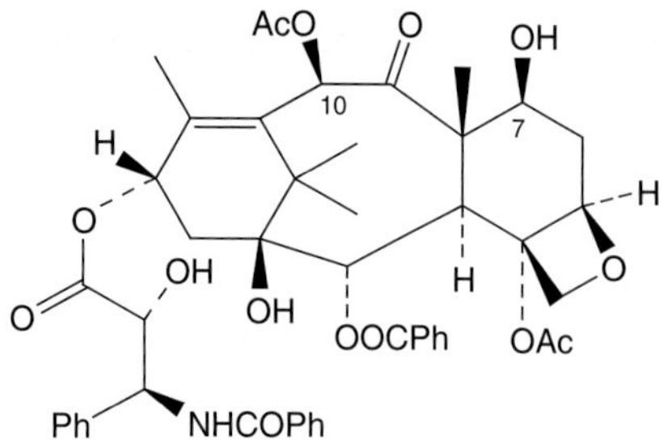

23 Paclitaxel

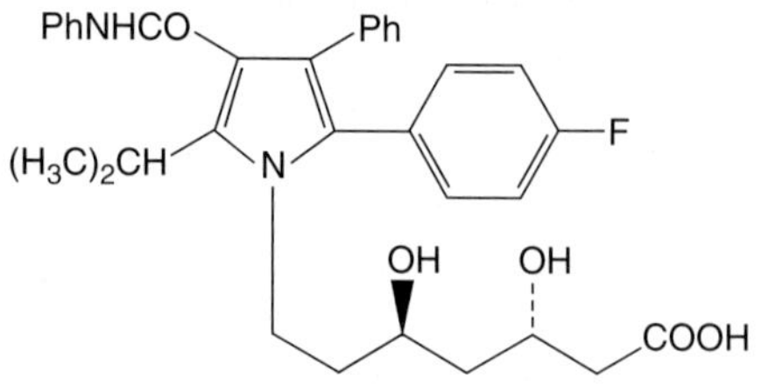

24 Atorvastatin

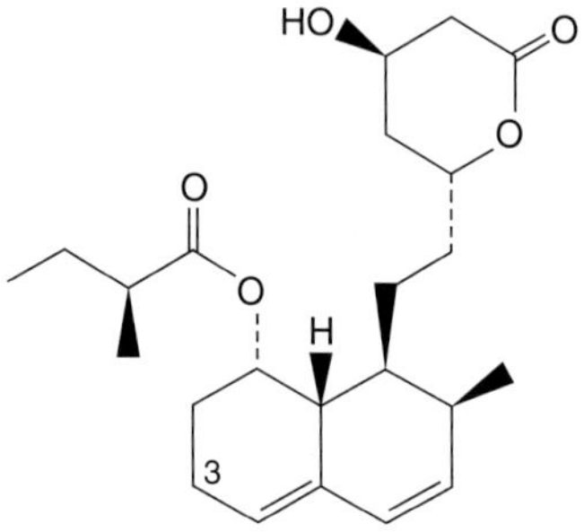

25 Compactin

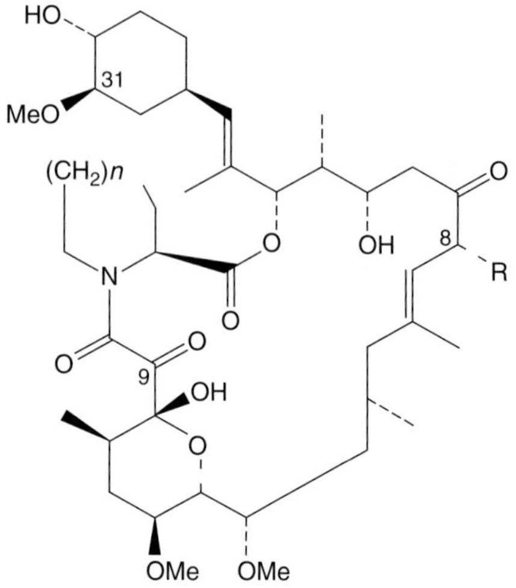

$R = —CH_2CH{=}CH_2$
$n = 2$

26 Tacrolimus

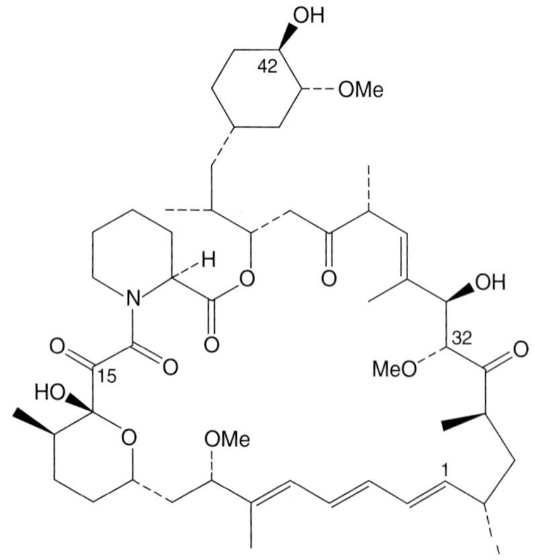

27 Sirolimus

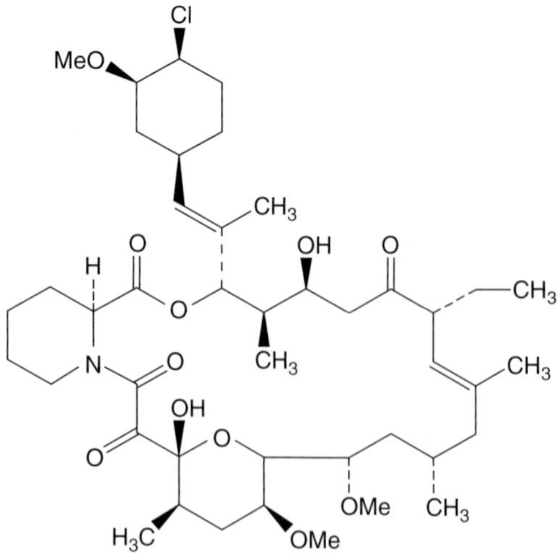

28 Pimecrolimus

However, we may see a resurgence of interest in natural product sources of leads as the search for chemical diversity intensifies, driven by the urgent need to find good leads against the many targets that genetic technologies are now producing (*see* 1.08 Natural Product Sources of Drugs: Plants, Microbes, Marine Organisms, and Animals). Libraries of synthetic molecules have been the main source of chemical matter in screens in recent years but there appears to be quite widespread concern that these have given fewer good leads than expected. Certainly, many targets are still seeking good leads. Natural product drugs do differ significantly compared with their synthetic drug counterparts, having a broader distribution of molecular properties such as molecular mass, lipophilicity, and diversity of ring structures; and greater steric complexity and a higher number of hydrogen bond donor and acceptor functions.[24] Debate continues whether this is an advantage (structural diversity) or a liability (druggability).

One apparent advantage for natural products is in disrupting protein–protein interactions, which is notoriously difficult with small molecules. HTS against this type of target has given very poor results using conventional small-molecule or combinatorial libraries. Success against protein–protein targets has been obtained with natural products perhaps by virtue of their increased size against a target with binding points spread over a large surface area. Examples include FK506/tacrolimus (**26**), rapamycin/sirolimus (**27**), (Rapamune), and ascomycin/pimecrolimus (**28**) (Elidel). These are in clinical use for liver and kidney transplantation (FK506 and rapamycin) and atopic dermatitis (ascomycin).[25]

A second advantage for natural products is the ability to exploit biological chemistry in variation of the structural core of families of proven valuable lead series. This biosynthetic approach has been well reviewed by Khosla and Keasling.[26] The rapidly increasing understanding of gene clusters that encode the building-blocks of natural products is being exploited to engineer diverse analogs of natural products at low cost.[27] While still in its infancy, this approach is likely to lead to new drugs and also to lowered cost of production of known natural product drugs. For instance, the antimalarial compound artemesinin is difficult to synthesize or isolate. This leads to supply problems and it also means it is quite costly for use in developing countries. Bioengineering methods are being applied to produce this drug at a fraction of current costs, ensuring both supply and affordability in poorer countries.[28]

An additional reason for renewed interest in natural products is likely to be a resource soon to be provided by the World Health Organization (WHO) Centre for Health Development in Kobe, Japan. This organization is cataloging all indigenous medicines, creating a Global Atlas of Traditional Medicine. At the time of writing, this is intended to be a two-volume reference book, which will be used to improve healthcare systems, clinical research, training, and the quality of traditional medicine around the world. This reference work will catalog all preparations from the folklore of the world's societies that now have evidence of efficacy. It is highly likely that this compendium will open up new ideas for drug discovery. As examples of the potential, reports of the efficacy of traditional medicines are now appearing regularly.

For instance, a randomized, double-blind, reference-controlled, multicenter phase III clinical trial investigated the antidepressant efficacy of an extract of the herb St John's Wort versus the antidepressant paroxetine (**29**) (Plaxil). The trial, in 251 patients with acute major depression, which was published in the *British Medical Journal*,[29] showed the traditional medicine to be at least as effective as paroxetine in the treatment of moderate to severe major depression. When compared to paroxetine, treatment with St John's Wort showed advantage in efficacy scores. This clinically relevant superiority was additionally supported by a responder rate of 70% versus 60% and a remission rate of 50% versus 35% for St John's Wort and paroxetine, respectively. All secondary efficacy measures showed moderate to large advantages in favor of St John's Wort.

Another example comes from studies with green tea. Numerous epidemiologic and animal studies have suggested that green tea extract is efficacious in several human cancers, including bladder cancer. It has been shown to induce death in cancer cells, as well as inhibiting angiogenesis, the development of the independent blood supply that cancers develop so they can grow and spread. One mechanism by which green tea may produce this effect was demonstrated in a study[30] with bladder cancer cells lines. The investigators showed that green tea extract interrupts actin remodeling, an event associated with cell movement. Actin remodeling is regulated by complex signaling pathways, including the Rho pathway. By inducing Rho signaling, the green tea extract increased the maturation rate of the cancer cells leading to increased cell adhesion which inhibited their mobility.

A final example: an herb used in traditional Indian medicine to treat diabetes lowers blood sugar and insulin levels by mechanism similar to prescription drugs.[31] Extracts of the herb *Salacia oblonga* decreased insulin and blood glucose levels by 29% and 23%, respectively. The extracts of *S. oblonga*, which is native to regions of India and Sri Lanka, bind to intestinal enzymes that break down carbohydrates in the body. These enzymes, called alpha-glucosidases, metabolize glucose from complex carbohydrates, and are the target of the prescription drug acarbose (**30**) (Glucobay).

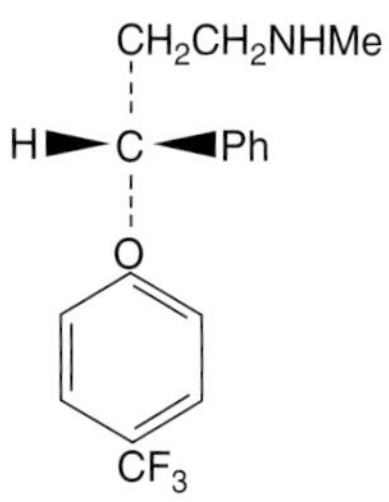

(R)-form

29 Fluoxetine

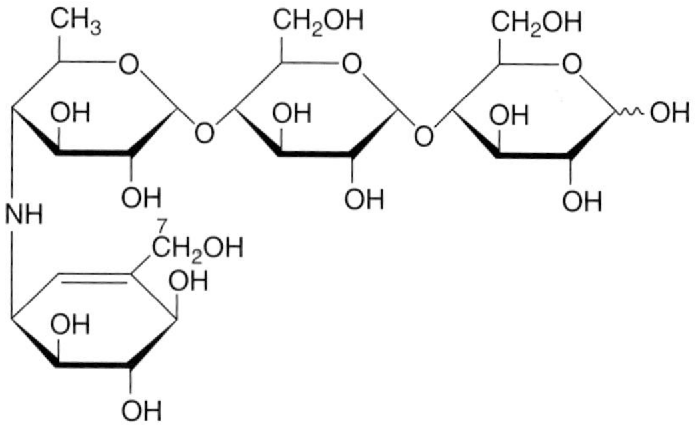

30 Acarbose

While all three of these studies must be validated by further trials, they illustrate the seriousness with which traditional medicines are now being reexamined. It is likely that valuable medicines will be introduced as a result in coming years

1.07.2.1.2 Drugs from rational design

Historically, rational design of drugs also has its origin in natural products. Rational design followed from the development of methods to isolate, then purify, analyze, and synthesize natural components of mammalian systems. This opened up a new source of potential drugs because many of the natural components identified were small molecules with properties ideal as leads in drug discovery. Progress awaited the development of sensitive chemical technologies in the early twentieth century as these natural chemicals exist in quite small amounts, consistent with their role as chemical messengers. Bloodborne messengers acting at membrane-bound receptors were the easiest to isolate and these led to a great wave of drug introductions from the middle of the twentieth century. The range of drug opportunities proved to be significantly greater than a mere count of the isolated natural compounds, because many of them interact at multiple receptors, now recognized as 'subtypes,' offering multiple drug opportunities for both agonists and antagonists at each receptor subtype. As a result of the existence of receptor subtypes, the isolated natural components proved to exhibit very complex pharmacology when administered exogenously, so characterization of their potential as drugs themselves, or as leads for drugs after chemical modification, was a slow and difficult process. Many earlier drugs derived by this route lacked needed selectivity because of the poor state of knowledge of receptor subtypes. Nevertheless, in the late nineteenth century and particularly during the twentieth century many such natural compounds were discovered and their receptors/subtypes were elucidated. Many of the drugs in current use have come from this source.

The story[32] of discovery of drugs acting at adrenergic receptors illustrates how this approach has provided a very rich source of new medicines. Epinephrine (adrenaline) (**31**) was first obtained in pure form in 1901 and a viable synthetic route was devised soon afterward. Epinephrine was rapidly marketed as a hemostatic and vasoconstrictor drug. Over the next 30 years a series of related drug introductions occurred, including phenylephrine (**32**) and isoprenaline (**33**) though without much understanding at the time of the complex pharmacology displayed by these molecules. However, the role of alpha and beta adrenergic receptor subtypes was gradually determined, particularly by Ahlquist in classical studies in the late 1940s, and differential tissue distribution of these receptors also played a role in opening up new therapeutic opportunities for a range of agonists and antagonists at both alpha and beta receptor subtypes. As better tools emerged from chemists' laboratories it became apparent that alpha and beta receptors can be further subtyped, into alpha 1 and 2 and beta 1, 2, and 3, which opened up further therapeutic opportunities. First to emerge were antagonists at the alpha adrenergic receptor such as phentolamine (**34**), followed later by tolazoline (**35**) (Priscoline) and prazosin (**36**) (Minipress). Black and colleagues[33] introduced beta-selective antagonists, particularly propranolol (**37**) (Inderal) which remains a mainstay of therapy today, though this drug has no selectivity for subtypes of the beta receptor. Then Salbutamol (**38**) (Ventolin) emerged in 1967 as the first agent selective for the beta 2 adrenergic receptors of the lung, having negligible effect on heart beta 1 receptors. To complete the range of agents, some selectivity for beta receptors of the heart over the lung was obtained with sotalol (**39**) (Betapace) and atenolol (**40**) (Tenormin) among other agents. Beta 3 selective agents have also been pursued though none has achieved clinical success to date. An excellent overview of the development of this field is provided by Weiner.[32]

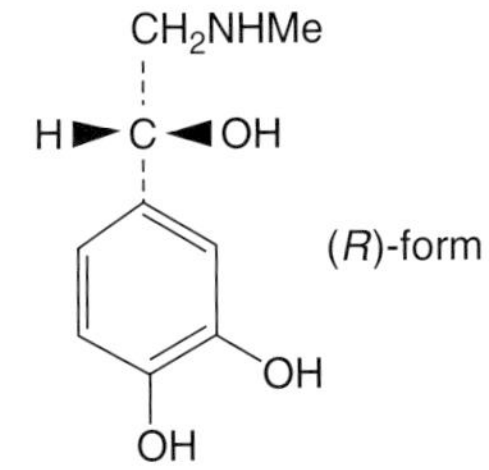

31 Epinephrine

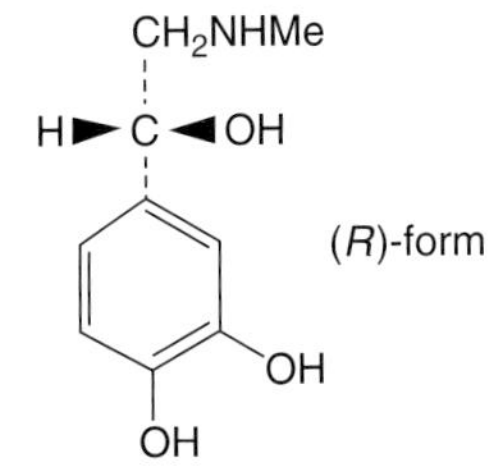

32 Phenylephrine

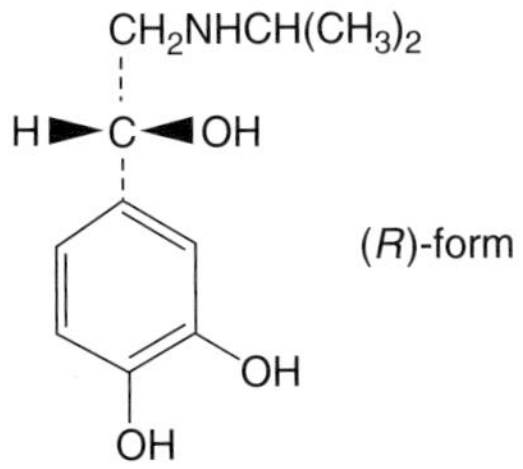

33 Isoprenaline

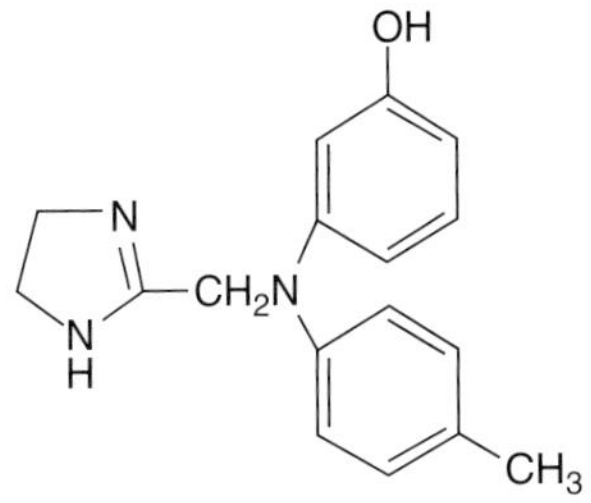

34 Phentolamine

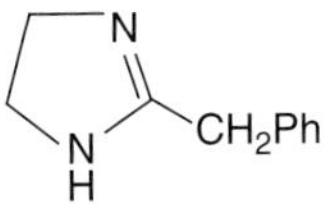

35 Tolazoline

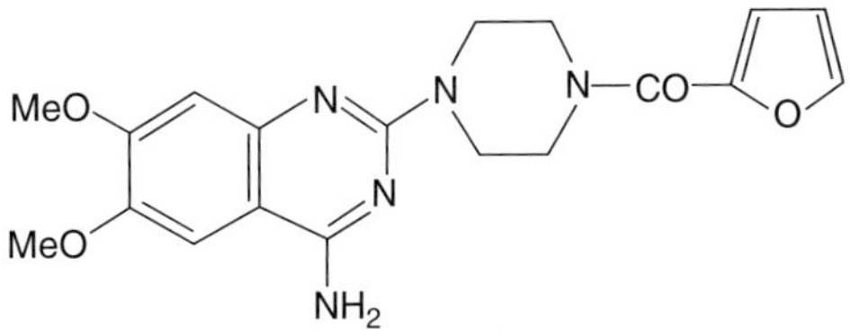

36 Prazosin

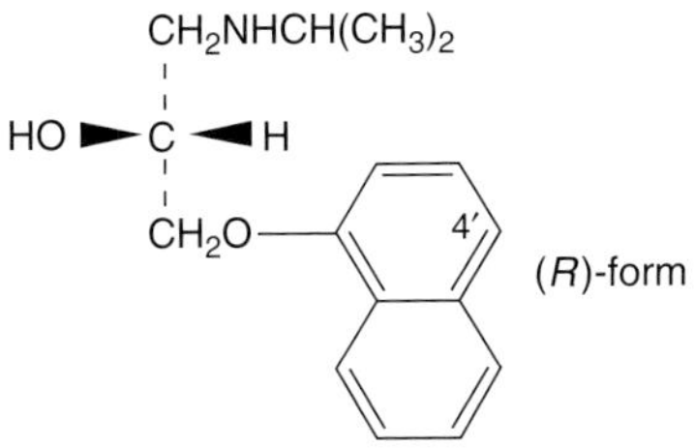

37 Propranolol

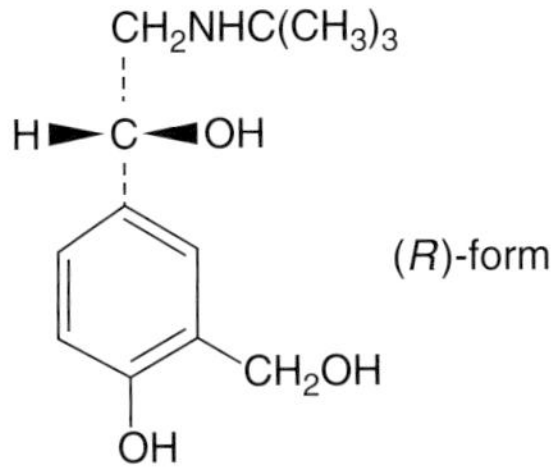

38 Salbutamol

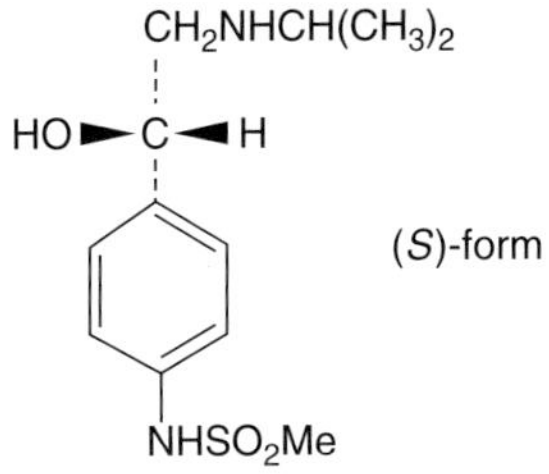

39 Sotalol

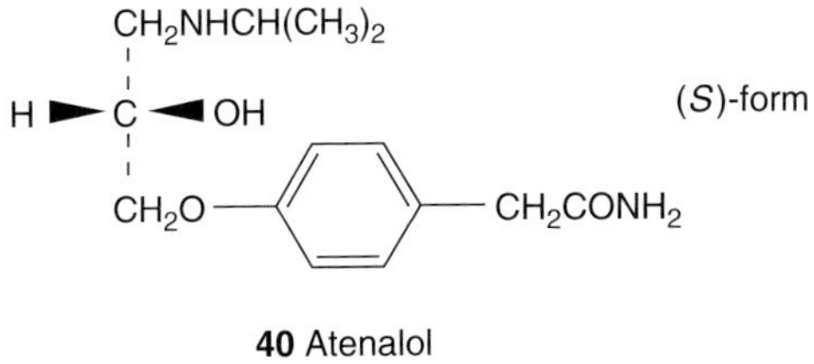

40 Atenalol

Over the past half century the story of discovery of adrenergic agonists and antagonists and their receptor subtypes has been repeated many times for other natural small-molecule neurotransmitter and hormone receptors, including serotonin, dopamine, histamine, etc. This campaign has been a very notable success for academic research and for the pharmaceutical industry such that many of the most effective and safe drugs in use in the early twenty-first century have come from this approach.

In the early 1980s it became possible to identify and characterize peptide neurotransmitters. Several hundred of these are now known, and many of these also have associated receptor subtypes that might be exploited. Drugs are emerging, for instance, that interact at receptors for bradykinin, substance P, neuropeptide Y, etc. A distinction must be made between the methods of drug discovery for small-molecule transmitters and peptide neurotransmitters. Small-molecule transmitters have physicochemical properties close to those required in a molecule intended for oral dosing as a drug, and so the neurotransmitters themselves may be used as good leads and drugs may be found by careful modification of their structure to build in desired potency, selectivity, and pharmacokinetic properties. However, this is not true of peptide neurotransmitters, which with few exceptions have properties that make them unsuitable for chemical modification to give drugs with usable properties. Moreover, solution of x-ray structures of membrane-bound receptors, to guide lead design, has been a difficult problem though encouragingly in the past decade we have recently seen publication of high-resolution structures for part of the bacterial ribozyme,[34–37] a G protein-coupled receptor (GPCR),[38] and an ion channel.[39] In this case, screening of compound collections is the best route to lead discovery, though experience has shown that leads may be quite difficult to obtain. Progress in finding drugs acting at peptide neurotransmitter and hormone receptors has been slower than the outstanding success achieved with small-molecule transmitters. Some targets which have relatively small peptides as their ligands are amenable to rational design of drugs, however, and these have been among the first to yield drugs. Examples include goserelin (**41**) (Zoladex), a luteinizing hormone-releasing hormone (LHRH) superagonist, designed by chemical modification of the nonapeptide ligand; and tirofiban (**42**) (Aggrastat) which binds the glycoprotein IIb/IIIa receptor on platelets by mimicking and the tripeptide RGD motif of fibrinogen alpha chain and thereby blocks platelet aggregation (*see* 2.14 Peptide and Protein Drugs: Issues and Solutions; 2.15 Peptidomimetic and Nonpeptide Drug Discovery: Receptor, Protease, and Signal Transduction Therapeutic Targets).

Rational design is sometimes also possible for enzyme targets, particularly if the substrate is known, if the mechanism of the enzyme is well characterized, and if the targeted active site has functionality suitable for binding an inhibitor. These conditions, particularly the last one, are not always met and knowledge of binding pockets may depend on a protein x-ray structure being available. The introduction of recombinant technology and improved methods of protein purification in the 1980s brought within reach the opportunity to crystallize many enzymes such that chemists had access to detailed descriptions of the active site they were targeting. However, it should be noted that, in vivo, enzymes exist in a heavily solvated state and in dynamic form, neither of which is true of the conditions under which protein x-ray structures are obtained, so the x-ray derived representations must be used with some caution (*see* 3.21 Protein Crystallography; 3.24 Problems of Protein Three-Dimensional Structures). This limitation has led to exploration of protein nuclear magnetic resonance (NMR) derived structures though these generally provide much less detailed information than do x-ray derived structures and are not widely used (*see* 3.22 Bio-Nuclear Magnetic Resonance).

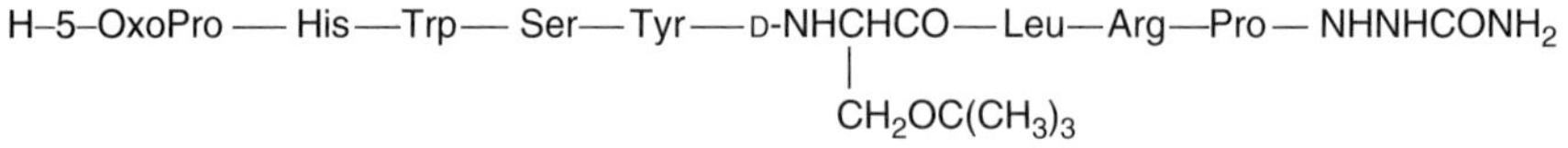

41 Goserelin

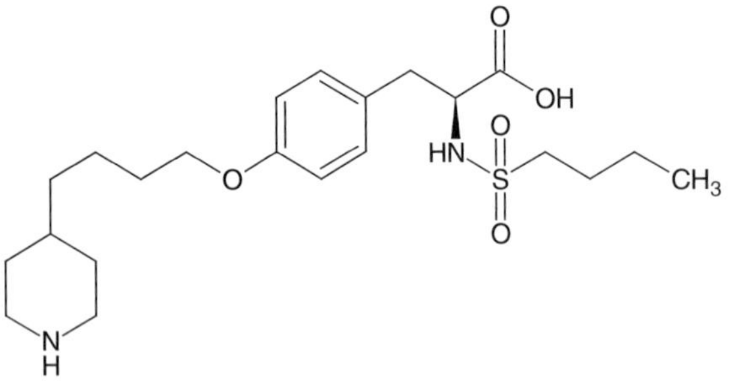

42 Tirofiban

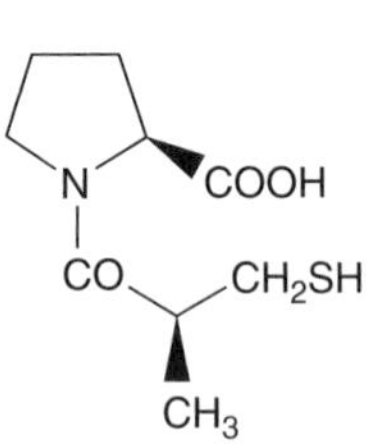

43 Captopril

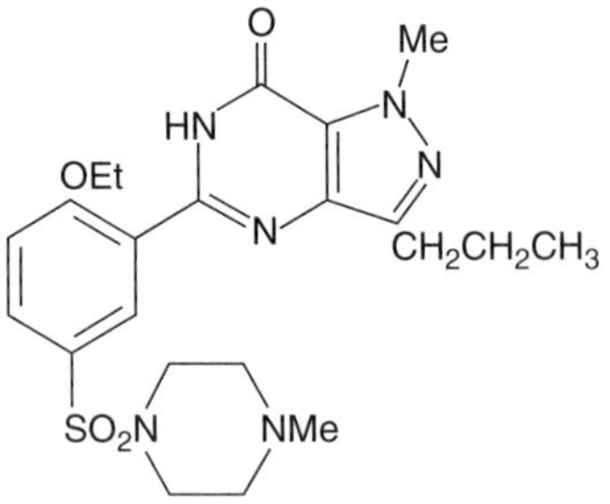

44 Sildenafil

Design of enzyme inhibitors based on knowledge of the substrate alone has given a good range of breakthrough drugs over the past 30 years including inhibitors of angiotensin-converting enzyme (ACE) inhibitors such as captopril (**43**) (Capoten), and cGMP phosphodiesterase type 5 inhibitors such as sildenafil (**44**) (Viagra).

In addition to GPCRs and enzymes, ion channels (*see* 2.21 Ion Channels – Voltage Gated; 2.22 Ion Channels – Ligand Gated) and nuclear receptors comprise two additional classes of target that have been successfully exploited in drug discovery in recent years. However, for these two classes, rational design has played less part in discovery of the original lead series, which have come mostly from random screening, the subject of the next section.

1.07.2.1.3 **Screening compound libraries**

From the early 1990s the pharma industry invested heavily in HTS. This was in recognition that what most often determined success in a drug discovery project was the availability of a good chemical lead, both to validate the target and to optimize into a drug. Yet an increasing number of targets lacked good leads, in part due to the nature of the targets that were coming to the fore. These included peptides and proteins acting at membrane-bound GPCRs, ion channels, and nuclear receptors; for most of these, rational design was not feasible and no suitable leads existed for related targets that might be exploited.

HTS now typically may involve 500 000 to several million compounds being screened rapidly against a biochemical target, which is often an isolated enzyme or a receptor or ion channel cloned into a membrane fragment (*see* 3.32 High-Throughput and High-Content Screening). Whole cell screening is sometimes used though throughputs may be lower (*see* 3.29 Cell-Based Screening Assays). HTS is often complemented by lower-throughput screening of focused libraries specifically designed to contain structural motifs known to interact with the target class. For some targets this can be more cost-effective and quicker at giving leads. Whether high-throughput or focused screening is used, data collection and initial data analysis are highly automated. In addition to the screening facility, several other operations must run at similar capacity and speed. Protein supply is a crucial issue (*see* 3.19 Protein Production for Three-Dimensional Structural Analysis) and requires dedicated laboratories. Very large numbers of chemical compounds are required to feed the screens, and a whole subindustry has grown to meet this need. These compounds need to be available for screening accurately weighed, accurately diluted and stable in solution, and accurately pipetted to assays. Each of these processes is also highly automated (*see* 3.26 Compound Storage and Management). Moreover, experience rapidly showed that very careful selection of compounds for 'chemical diversity' was required and a parallel science has grown up around this need.

Methods for screening for leads are undergoing rapid development. A recent innovation is the development of methods for high-resolution screening (HRS).[40] These methods allow selection of actives from complex mixtures of compounds and thus allows access to expanded chemical diversity. A prime use may be in assessment of natural product extracts where several main components may be present.[41] Also, screening of focused combinatorial libraries can benefit from this approach because it reduces the problems arising from false positives due to impurities. Solution-phase libraries are particularly prone to impurity problems, necessitating expensive and time-consuming purification strategies. This is a particular problem if such libraries are used in lead optimization, a phase during which interpretation of small differences in activity with change in structure is critical yet can be undermined by impure test compounds. HRS provides efficient screening of mixtures by coupling in a single instrument the three activities of compound separation by high-pressure liquid chromatography, structural analysis by mass spectrometry (MS) and biochemical screening. It is likely that this or similar methods will become central to renewed interest in assessing bioactivity of natural products, particularly assessment of extracts long-used by indigenous peoples (*see* Section 1.07.2.1.1 above).

As an adjunct to HTS, preselection of libraries can play a vital role in determining success rates. Virtual screening (in silico)[42,43] has, in recent years, become a viable partner to the physical screening of molecules, and is often used to preselect among those compounds available for screening or to predict which libraries of compounds or individual compounds should be purchased for screening. This approach takes a protein with an experimentally determined structure and docks potential 'hit' molecules. Large numbers can be prescreened in silico before a smaller number is selected for physical screening. This enables molecules not actually owned by the screening organization to be assessed before purchase. Virtual screening based on protein structure has been successful in particular for enzyme targets. However, the major target class of currently used drugs is membrane-spanning GPCRs, and there are few useful available accurate, experimentally determined structures for targets of this type. Ion channels also constitute a large target class, again with little available structural information sufficiently accurate for virtual docking experiments. Membrane transporters are also emerging as a target class and it is unlikely that structures will be available for the foreseeable future.

In these cases where sufficiently accurate experimentally determined structures are not available, virtual screening must rely on three-dimensional ligand-based pharmacophore models, developed from medicinal chemistry input of structure–activity relationships (SARs) for lead series versus the target. Since structural information is not available for the majority of drug discovery targets, ligand-based virtual screening is a widely used complement to the efforts of project medicinal chemists. This approach however suffers from the weakness that docking accuracy cannot be experimentally determined to refine methodology. However, recently, methods for three-dimensional ligand-based virtual screening have been reported to yield enrichments in hit rate similar to those obtained using molecular docking into structurally detailed protein targets.

Improvement in methods combined with availability of modestly priced hardware has ensured that virtual screening is used widely in most industrial companies and in academic laboratories interested in drug discovery. It finds use in lead optimization also and in particular three-dimensional ligand-based screening is heavily used in this phase of drug discovery because the availability of detailed structure activity data from the ongoing medicinal chemistry program supports development of detailed pharmacophore models for three-dimensional ligand-based screening. A comprehensive and critical independent evaluation of methods is provided by Jain[42] and by Blake and Laird.[43]

Lipinski and colleagues have made key contributions to improving the quality of molecules going into HTS, and that should be accepted as leads.[44] By analyzing the properties of successful drugs recorded in the World Drug Index, they identified several key properties common to small molecules suitable for oral administration. These properties, known as the Lipinski Rules or the 'rule of five,' are molecular mass less than 500 daltons; number of hydrogen-bond donors less than 5; number of hydrogen bond acceptors less than 10, calculated octanol/water partition coefficient less than 5. Poor absorption or permeation is highly likely if these guidelines are not met. By selecting for screening molecules that meet the guidelines, probability is increased that any 'hits' in the screens will be optimizable. However, these guidelines do not cover natural products, for which other absorption mechanisms may often be involved (*see* 4.18 Lead Discovery and the Concepts of Complexity and Lead-Likeness in the Evolution of Drug Candidates; 4.20 Screening Library Selection and High-Throughput Screening Analysis/Triage).

Another key consideration determining the success of HTS is whether the targets chosen for screening are essentially druggable. If they are nondruggable it means that it just may not be possible to find lead molecules with druglike properties. This is the subject of much debate between chemists and biologists. If a target is inherently nondruggable, then merely increasing the numbers of compounds screened is unlikely to produce breakthroughs. For some and maybe for many targets, their structure may not offer opportunities for small-molecule binding. It is understood from Lipinski's work that 'hits' must not only bind but also have the potential for suitable pharmacokinetic properties to be developed during lead optimization, so merely increasing the complexity of the molecule to get 'hits' is not finally productive. This debate is highlighting that target discovery cannot be dissociated from other phases of drug discovery, particularly from lead discovery and optimization.

1.07.2.1.4 **Target discovery**

For most of the history of drug discovery selection of targets was not considered a separate activity from selection of a lead. For example, the activity of a natural product extract was determined by its chemical contents, proven first to be useful, then the target was identified at a later time. Even until the 1970s a lot of drug discovery was driven by a similar 'phenotypic screening' approach in which compounds were tested for a desired readout in cell or animal systems, then the mechanism might be determined if useful effects were found. Where a mechanistic approach was possible (for instance the examples discussed in Section 1.07.2.1.2 above starting from endogenous ligands such as epinephrine), there would be extensive dialogue between chemists and biologists. Projects were chosen slowly and carefully and in small numbers. This began to change in the late 1980s and early 1990s as biochemistry advanced; and in particular it was thought that the sequencing of the human genome would provide a rich source of drug targets. It was expected that thousands of genes would be shown to be linked to diseases of interest. One influential publication in the mid-1990s stated "The number of potential drug targets may lie between 5,000 and 10,000,"[45] though others have questioned this conclusion (as discussed in Section 1.07.4.1.1 below). With this expectation many pharmaceutical companies adjusted their approach to drug discovery, instituting 'industrialization' of scale and a 'process' approach which begins with specialists selecting targets and continues within other specialist groups with HTS then lead optimization.

Target discovery and validation methods have been developed at a rapid pace over the past decade. It is a daunting task to assign molecular and cellular function to many thousands of newly predicted gene products and select those of most value as drug targets. Ian Humphrey-Smith, cofounder of the Human Proteome Organization (HUPO), has estimated that the 30 000–40 000 human genes can each be expected to produce between 1 and 70 products, resulting in a total of about 400 000–500 000 different protein isoforms. This is accounted for by alternative splicing of about 40% of all human genes and the generation of different protein isoforms due to posttranslational modifications. Proteomic pathway mapping is beginning to show that yet-greater complexity may exist through formation of protein complexes[46] or even diversity of function for a protein depending on its localization or cell type. For example, phosphoglucose isomerase, in addition to its role in glycolysis, also serves as the nerve growth factor neuroleukin. Therefore, genetic approaches to target validation need complementing with cell-based or even organism-based methods that probe the predicted gene product under a number of in situ circumstances. This is important not only in determining the potential efficacy of a drug versus that target but also its potential side effects. Inactivation of a single gene may have widespread effects on different protein functions, complicating assessment. Protein discovery methods in use at the time of writing include: gene expression studies, differential display, expressed sequence tag (EST) databases; proteomics approaches using two-dimensional gel electrophoresis of diseased versus normal tissue, and pathway mapping technologies. These technologies are supported by powerful bioinformatics systems.[47] Among the broad range of methods to determine protein function we now have peptidomimetics, RNA interference, aptamers, ribozymes, monoclonal antibodies, antisense RNA, gene knockout and knockin methods, and dominant negative mutants.[48,49]

It is essential that reliable methods are developed to predict the success of a target before drug discovery begins. Encouragingly, a retrospective study[50] using mouse knockouts (ko's) to assess the validity of the targets of the 100 best-selling drugs demonstrated that in the vast majority of cases there is a direct correlation between the ko phenotype and the proven clinical efficacy of drugs which modulate that specific target. A further analysis[51] of clinical development pipeline drugs addressing novel targets from the top 10 pharmaceutical companies also suggested that over 85% demonstrated a sound biological rationale for the selected disease indication on the basis of ko phenotypes. Most of these drugs would have been discovered without the benefit of mouse ko's at the time of initiation of the project.

The need for improved confidence in developing drugs against novel mechanistic targets is illustrated by the observation that over the past decade we have had fewer innovative targets per year turning into successful drugs. Zambrowicz and Sands[50] have presented data showing that only 24 innovative drugs with new targets were launched between 1994 and 2001 (**Table 1**). The industry is marketing only three new mechanisms a year on average, which constitutes less than 10% of total drug launches.

Table 1 FDA accepted breakthrough or innovator targets

Year	*Drug*	*Innovator target*
1994	Glucophage	Perhaps acetyl CoA carboxylase2
1995	Precose	alpha-Glucosidase
1995	Cozaar	Angiotensin receptor AT1
1995	Cellcept	Inosine monophosphate dehydrogenase
1995	Fosamax	Perhaps farnesyl diphosphate synthase
1996	Accolate	Leukotriene receptor
1997	Plavix	Platelet $P2Y_{12}$ receptor
1997	Rezulin	Peroxisome proliferator activated receptor
1998	Celebrex	Cyclooxygenase 2
1998	Aggrastat	Platelet glycoprotein IIb/IIIa receptor
	Integrelin	
1998	Viagra	Phosphodiesterase type 5
1998	Enbrel, Remicade	Tumor necrosis factor-alpha recombinant receptor or antibody
1998	Herceptin	Erb-B2 (also known as HER2/neu)
1999	Rapamune	FK-binding protein 12 and target of
		rapamycin (TOR kinase)
1999	Xenical	Gastrointestinal lipase
1999	Targretin	Retinoid X receptors
2000	Mylotarg	Antibody to CD33
2001	Gleevec	BCR-Abl
2001	Tracleer	Endothelin receptor
2001	Natrecor	Recombinant B-type natriuretic peptide
2001	Xigris	Recombinant activated protein C
2001	Kineret	Recombinant interleukin 1 receptor antagonist

Reproduced from Zambrowicz, B. P.; Sands, A. T. *Nat. Rev. Drug Disc.* **2003**, *2*, 38–51, with permission from Nature Reviews Drug Discovery, copyright (2003) Macmillan Magazines Ltd.

1.07.2.2 Biological Products

1.07.2.2.1 Protein products

In contrast to the small number of small-molecule drugs with innovative targets reaching market in recent years, there has been considerable innovation in protein products. In part this is because this is a new field with low-hanging fruit to pick. Over the past quarter century, advances in molecular biology and genetic engineering led to rapid growth in the availability of recombinant-engineered protein products. Paul Berg, the 1980 Nobel Prize winner in chemistry, first produced recombinant DNA in the early 1970s. A second key event was the invention in 1980 of the polymerase chain reaction (PCR), a technique for multiplying DNA sequences. Boyer, who in 1972 transformed *Escherichia coli* cells with recombinant plasmid, went on to found Genentech, and was instrumental in production of the first recombinant human protein drug, human insulin (Humulin, Lilly, 1986). Today the rDNA protein industry comprises over 100 companies, over 70 marketed products and over 100 potential new products in clinical development.[52,53] One estimate, based on an analysis of historic success rates of clinical development of rDNA-based protein products, suggests that another 33 of the drugs currently in clinical trial may achieve FDA approval.[54] The top-selling recombinant protein products are listed in **Table 2**.

Why does one choose to pursue a protein drug rather than a small molecule? The simplest case is when a missing or malfunctioning protein causes a disease, such as in hemophilia; then a replacement protein is the obvious choice. Also, protein therapeutics can mimic a natural agonist and so they may be the preferred choice if an agonist action is required and obtaining this with a small molecule appears difficult – which is often the case. With current technology, a key requirement is that the target must be extracellular, either a cell surface receptor or soluble factor, because proteins are too large to enter a cell directly. If therapy requires an antagonist then an antibody is often the preferred approach.

Discovery of therapeutic proteins conceptually involves the same stages as small-molecule discovery. An optimal lead may have efficacy that closely matches the native human protein. It should have high affinity, high specificity, good stability, and also a high level of expression in cell culture. This latter point is a major difference from small-molecule discovery. The eventual drug will be produced from a cellular system and this must be factored into the discovery stage. Protein sequences must be selected that have certain properties that make them more likely to express well, fold and secrete correctly, be subject to the right posttranslational modifications, and remain stable for longer times and soluble at higher concentrations. SARs are developed in much the same way as for small molecules but with these added factors. Because binding surfaces involved are so large, it may be possible to obtain much more potent and specific interactions between a therapeutic protein and its target than with a small molecule. Also, it may be possible to get long pharmacokinetic half-lives: in pursuit of this goal 'pegylation' strategies have been developed that offer once-weekly or even once-monthly dosing. Anchoring a protein to a large polymer protects it from enzymic degradation and also slows clearance by the kidney. It also provides controlled release properties. Since protein products are dosed parenterally

Table 2 Top 10 selling recombinant proteins 2003

Generic name	*US trade name*	*Company*	*Indication*	*Sales (2003) $ million*
Epoetin alpha	Procrit	Johnson & Johnson	Anemia	3986
Epoetin alpha	Epogen	Amgen	Anemia, neutropenia	2435
Insulin systemic	Novolin	Novo Nor disk	Diabetes	2235
Peg-interferon alpha	Peg-Intron A	Schering Plough	Hepatitis C	1851
Darb-epoetin alpha	Aranesp	Amgen	Anemia	1544
Epoetin beta	NeoRecormon	Roche	Anemia	1318
Etanercept	Enbrel	Amgen	Rheumatoid arthritis	1300
Pegfilgrastim	Neulasta	Amgen	Neutropenia	1255
Filgrastim	Neupogen	Amgen	Neutropenia	1268
Interferon beta-1a	Avonex	Biogen IDEC	Multiple sclerosis	1170

Adapted from Pavlou, A. K.; Reichert, J. M. *Nat. Biotechnol.* **2004**, *22*, 1513–1519 © Nature Publishing Group.

this is an important development and we can expect to see other polymers, in addition to polyethyleneglycol (PEG), explored in the future.[55]

Although the protein discovery stage may have similarities with small-molecule discovery, the development stage is distinctive. Process development involves seeking expression vectors that produce high levels of protein. Cultivated mammalian cells have become the preferred means to produce recombinant proteins for human clinical use because of their capacity for proper protein folding, assembly, and posttranslational modification. Today about two-thirds of all recombinant proteins are produced in mammalian cells.[56]

A major problem in the development of protein products, responsible for many failures in clinical trials, is the production of an immune reaction in the patient. So far, the only way to know for sure if a protein, even a humanized one, is immunogenic is to test it clinically. Currently we lack in vitro or in vivo assays able to predict accurately the immunological potential of a protein. One implication of this issue is that the drug manufacturing process must be identical to that used for clinical trials to ensure that exactly the same protein product is being marketed as that proven safe versus immunological responses in controlled trials.

Since delivery is perhaps the major hurdle for protein drugs, new delivery systems are being sought. Proteins break down quickly in the stomach, and so must usually be injected. The delivery problem, coupled with the fact that protein drugs cannot yet act on targets inside cells, often leads companies to pursue in parallel both protein drugs and small-molecule drugs, to increase the probability of success. In the 1980s and 1990s there was a view that parenterally administered protein products would be vulnerable to erosion of sales as small-molecule competitors versus the same targets entered the marketplace. This has not proved to be a problem for manufacturers of protein products because it has proved extraordinarily difficult to mimic the agonist action of large proteins with small molecules. And in the arena of strength of small molecules, antagonism, antibody products are now regarded as complementary therapy to small molecules where these are available. Particularly in diseases such as cancer, with significant mortality rates, usage of both types of drug in parallel, even if acting within the same biochemical pathway, is being explored.

1.07.2.2.2 Monoclonal antibody products

Monoclonal antibody products have appeared more recently than protein-based drugs. The first to be approved by the FDA, in 1986, was muronomab-CD3, a murine antibody (**Table 3**). All following approvals were of antibodies incorporating some degree of 'humanization,' for reasons discussed below. It took until the mid-1990s for the first of these to be marketed, and by March 2004 the FDA had approved 18 therapeutic monoclonal antibodies (mAbs). Of these 3 were murine, 5 chimeric, 9 humanized, and 1 human[57] and 9 of these 18 were approved in the European Union. mAbs have proven to be the second wave of innovation for the biotechnology industry, following the success of recombinant proteins. It is notable that even though a decade ago there were only two mAbs approved by the FDA, now, not only are there 18 approved products, but there are also at any one time between 100 and 150 mAb products in clinical development: this suggests a rapid increase in the number of available marketed products is imminent.

Antibody therapy has its origin in 1895 when two French physicians attempted a radical approach to cancer treatment. Charles Richet and Jules Hericourt administered an antiserum derived from dogs to patients with advanced cancer.[58] Some patients improved, although significant immunogenicity problems occurred. Richet and Hericourt's antiserum was actually an antibody soup, containing antibodies that targeted antigens on both diseased and healthy cells. What was required was a single antibody, preferably human, targeting a specific antigen, but 80 years passed before the scientific breakthrough occurred that enabled such molecules to be produced.

In 1975, César Milstein and Georges J. F. Köhler reported a way to produce monoclonal antibodies.[59] The first products were murine mAbs but these were problematic as therapeutics for chronic indications because they were highly antigenic. Even a single dose may trigger a vigorous immune system response that neutralizes the activity of the administered antibodies. Only one murine mAb (muromonab-CD3) entered into the pharmacopeia.

Discovery of these early monoclonal antibodies relied on injecting the target antigen into mice, collecting their antibodies, and assaying them for target binding. Spleen or lymph cells from chosen mice are fused with human myeloma cells into immortal antibody-producing cells, called hybridomas. The problem of immune reaction in humans to the foreign mouse antibody protein requires the antibodies to be further engineered to increase the percentage of human sequence present. Design of these 'chimeric' antibodies depends on computational technology that can align the sequence of a mouse antibody with human antibodies to find the closest human match. Critical regions of the human antibody are then transferred to the mouse antibody. Recombinant DNA technology is used to fuse the mouse variable regions (VH and VL) to the human immunoglobulin (Ig)-constant domain. With mAbs that are two-thirds human and one-third mouse, these chimeric molecules substantially reduce human-antimouse antibody (HAMA)

Table 3 Monoclonal antibody products approved by the FDA (March 2005)

Generic name	*US trade name*	*Monoclonal antibody type*	*Therapeutic application*	*Company (US approval date)*
Muromomab-CD3	Orthoclone OKT3	Murine	Immunological	Johnson & Johnson (1986)
Abciximab	ReoPro	Chimeric	Hemostasis	Centocor (1994)
Rituximab	Rituxan	Chimeric	Antineoplastic	Biogen IDEC (1997)
Daclizumab	Zenapax	Humanized	Immunological	Protein Design Labs (1997)
Basiliximab	Simulect	Chimeric	Immunological	Novartis (1998)
Palivizumab	Synagis	Humanized	Anti-infective	Medimmune (1998)
Infliximab	Remicade	Chimeric	Immunological	Centocor (1998)
Trastuzumab	Herceptin	Humanized	Antineoplastic	Genentech (1998)
Gemtuzumab ozogamicin	Mylotarg	Humanized	Antineoplastic	Wyeth (2000)
Alemtuzumab	Campath	Humanized	Antineoplastic	Millennium/ILEX (2001)
Ibritumomab tiuxetan	Zevalin	Murine	Antineoplastic	Biogen IDEC (2002)
Adalimumab	Humira	Human	Immunological	Abbott (2002)
Omalizumab	Xolair	Humanized	Immunological	Genentech (2003)
Tositumomab-I131	BEXXAR	Murine	Antineoplastic	Corixa (2003)
Efalizumab	Raptiva	Humanized	Immunological	Genentech (2003)
Cetuximab	Erbitux	Chimeric	Antineoplastic	Imclone (2004)
Bevacizumab	Avastin	Humanized	Antineoplastic	Genentech (2004)
Natalizumab	Tysabri[a]	Humanized	Multiple sclerosis	Biogen IDEC (2004)

Reproduced from Lesko, L.; Woodcock, J. *Nat. Rev. Drug Disc.* **2004**, *3*, 763–769, with permission from Nature Reviews Drug Discovery, copyright (2004) Macmillan Magazines Ltd.
[a]Withdrawn February 2005.

responses. However, though improved, the HAMA response is not completely eliminated and some of these drugs have been associated with serious allergic reactions in particular patients.

This observation led to a search for 'humanized' mAbs, prepared by inserting only the mouse complementary-determining regions (CDRs) – those regions of the antibody that bind antigen – into a human antibody framework. These antibodies are 90% to 95% human and they produce far fewer HAMA responses – though this strategy does not eliminate the HAMA response totally. In addition, 'humanization' can lead to reduced affinity of the antibody for its target.

The ease of discovery of totally human mAb drugs was greatly advanced in 1990 when a group led by Greg Winter, working in the same Medical Research Council (MRC) laboratory in the UK as had Milstein and Köhler earlier, developed a system to produce highly specific human antibody fragments.[60] This method completely avoids the 'humanization' problem. The MRC team incorporated genes encoding human antibody variable regions (V genes) into a bacteriophage. Each phage carries a different human V gene, which directs infected bacteria to make the human antibody fragments, expressed on the viruses' surface. The resulting phage display library contains billions of bacteriophages, each carrying a unique human antibody fragment. These libraries can show equivalent antibody diversity to the human immune system and allow researchers to rapidly isolate antibodies that recognize specific antigens. Phage display methods are excellent for producing pure antigen. However, because the phage are sticky, and because a small number of specificities quickly come to dominate in the phage population, classical mAbs, produced by cell fusion, are still the method of choice for analysis of complex antigens, such as those of a cell surface.

In the next wave of products humanized mAbs look likely to predominate during the middle years of this decade, then fully human mAbs will come to the fore in the later years of the decade. One, adalimumat (Humira, Abbott) is already in use (**Table 3**). Many of these products are targeted at two therapeutic categories, oncology, and arthritis/immune/inflammatory disorders, though other diseases such as multiple sclerosis are receiving attention.

Monoclonal antibodies have limitations as drugs. A major one is that current mAbs cannot enter cells and this limits targets that can be tackled. (It also may provide some specificity of course, reducing the opportunity for side effects.) In addition, many biologically important epitopes have proven to be poor targets against which to raise mAbs. These include sugars, lipids, and GPCRs. The cost of mAbs as therapeutic agents is also a concern and limits their use, certainly in poorer countries but also in developed Western countries. And all mAbs currently marketed must be injected: technology for oral or inhaled formulations is being researched.

1.07.3 Issues with Current Sources of Drugs

1.07.3.1 Drugs for Diseases of Less Developed Countries

Infectious diseases remain a major cause of morbidity and mortality in poorer countries. While infectious diseases cause only 1 out of 10 deaths in the world's richest countries, among the poorest people 6 in 10 still die of infectious diseases, according to the WHO. Therefore these figures suggest that up to 5 in 10 deaths, fully one half, could be prevented in poorer countries by application of the hygiene, vaccines, and medicines available to richer countries (*see* 1.20 Health Demands in Developing Countries).

Every year more than $100 billion is spent worldwide on health research and development by the public and private sectors. Research into developing country diseases, which comprise 90% of the world's health problems, attracts approximately £3.5 billion, about 3% of global health research expenditure. Despite calls and stated intentions to improve the situation, this low figure actually represents a worsening from the position 10 years ago when the figure was approximately 10%. Much of the current investment in research toward treating Third World diseases comes from private foundations, governments, and charities. As a result of this level of spending, barely 1% of new drugs marketed in the past quarter century have been targeted at diseases of poorer countries. Most of these have been discovered by major pharmaceutical companies (**Table 4**) using methods of drug discovery discussed in previous sections.

Preventable diseases still kill millions of people. The main killers identified by the WHO are summarized here.

- Lower respiratory infections such as pneumonia are the largest killer, causing the deaths of 3–4 million, many of them children under 5 years old. The Disability Adjusted Life Year burden (DALY; the number of healthy years of life lost due to premature death and disability) is estimated at 91 million. Most of these lives could be saved with drugs available in richer countries.
- Malaria, now eliminated from most of the developed world, remains a serious problem in many poor countries, especially sub-Saharan Africa, where 90% of the world's cases occur. Malaria kills 1–2 million people each year. Globally, an estimated 2.4 billion people are at risk and malaria infects at least 500 million people worldwide every

Table 4 Drugs marketed in past quarter century for tropical diseases

Drug	*Year of registration*	*Disease*	*Source*
Praziquantel	1980	Schistosomiasis	Bayer
Mefloquine	1984	Malaria	Hoffman-La Roche, WRAIR
Ivermectin	1987	Onchocerciasis	Merck
Halofantrine	1988	Malaria	Smith Kline Beecham, WRAIR
Eflornithine	1991	African Trypnosomiasis	Marion Merrel Dow
Liposomal amphotericin B	1994	Leishmaniasis (kala-azar)	NeXstar
Artemether	1997	Malaria	Rhone Poulenc Rorer, Kunming
Artemether-umefantrine	1999	Malaria	Novartis
Artemotil (beta-arteether)	2000	Malaria	Artecef, WRAIR, Dutch Ministry of Development
Miltefosine	2002	Leishmaniasis (kala-azar)	Zentaris, Indian CMR
Chlorproguanildapsone	2003	Malaria	GlaxoSmithKline, DFID

year. The DALY is estimated at 42 million. There is still no effective vaccine for malaria despite many attempts using modern vaccine-discovery methods. This represents a notable and tragic failure for modern methods.
- Tuberculosis (TB) kills 2–3 million annually, 90% of them in developing countries. About one-third of the world's population is infected. The DALY is estimated at 35 million. Human immunodeficiency virus (HIV) epidemics in many countries have led to a new wave of TB infection because coinfection of HIV and *Mycobacterium tuberculosis* is more likely to reactivate latent TB, often lethally.
- HIV/acquired immune deficiency syndrome (AIDS) affects almost 40 million people worldwide. In poorer countries, most patients do not have access to life-saving drugs. HIV now kills nearly 3 million annually, mostly in poorer countries, and the DALY is estimated at 85 million.
- Diarrhea, through causing severe dehydration, kills 2 million annually, many of these being children. There are over 1 billion cases worldwide annually, mostly in developing countries, according to the WHO. Although global diarrheal disease mortality declined from 1980 to 1990, due probably to the wide-spread introduction of oral replacement therapy (ORT), diarrhea mortality rates in published reports from 1990 through 2000 show no decrease. Diarrhea is more serious than widely realized because children who survive often suffer retardation in physical and mental growth due to repeated chronic episodes of diarrhea. Of all childhood infectious diseases, diarrheal diseases within the first 2 years of life are thought to have the greatest quantifiable long-term impact on growth, fitness, and cognitive function.[61] The DALY is estimated at well over 100 million, making this one of the most damaging afflictions.[62]
- Measles kills an estimated 700 000 annually, even though it can be prevented with a cheap, effective vaccine.
- Leishmaniasis, in its most deadly form, called visceral leishmaniasis, afflicts some 1.5 million people, primarily in the Indian subcontinent, Ethiopia, Kenya, Sudan, and Brazil. Untreated, the disease is more than 90% fatal. In India alone, as many as 200 000 die every year.
- Trypanosomiasis, or sleeping sickness, infects half a million people worldwide and causes 50 000 deaths annually. Drug resistance has rendered many current treatments ineffective.

Apart from immediate human suffering, these illnesses undermine economic progress of underdeveloped countries and contribute to a vicious circle of disease and poverty. Foreign investment is hindered. It has been estimated by the WHO that Africa's annual gross domestic product (GDP) could be $100 billion higher if malaria alone is eliminated. From the perspective of the topic of this chapter, no new methods are required to discover effective treatments for most of these diseases. The knowledge and capability exist already in the laboratories of the developed world. For some developing country diseases the drugs and vaccines are already available in developed nations, and where this is not the case the required treatments can be sourced if we can find ways to encourage investment. The issue is that although the numbers of afflicted peoples is high, their countries have limited ability to pay for the healthcare their citizens. Those drugs that are available are often not affordable by those in need. Pharmaceutical R&D is a long, expensive, and risky activity with the industry's cost of capital requiring a return that is dependent on peak sales higher than can be attained for these diseases.

Some strategies for stimulating R&D on neglected diseases (and orphan drugs) have been summarized by Grabowski *et al.*[63] They concluded that 'push programs' of R&D cost-sharing or subsidy coupled with regulatory 'fast track' treatment have not proved effective in stimulating provision of drugs and vaccines for neglected diseases, particularly Third World diseases. They recommend that governments and international agencies should focus on 'pull programs' to compensate for low expected sales. This would include a purchase fund that guarantees a price to companies that produce for example a new vaccine for malaria, TB, or AIDS. Firms would obtain a transferable patent right in the US market as an incentive for developing a new drug for a neglected disease. They argue that this would provide an incentive to firms with already established blockbuster products in US that are suitable for Third World use. In addition, the firm could receive a right for priority review by the FDA with shortened review time and reduced costs for a new drug application in the USA as an incentive for developing a new product for a neglected disease. This topic was reviewed again recently in some depth with similar conclusions.[64]

So how will drugs likely to be sourced in future for poorer countries? A very encouraging recent development is the initiation of public–private partnerships (PPP) focused on neglected diseases. These include Aeras Global TB Vaccine Foundation, the Malaria Vaccine Initiative (MVI), the International AIDS Vaccine Initiative (IAVA), the Global Alliance for Vaccines and Immunization (GAVI), the Global Alliance for TB (GATB), and the Medicines for Malaria Venture (MMV). Many of these are funded by the Bill and Melinda Gates Foundation which has emerged as a major donor of initiatives to confront Third World illness,[65] and this and other donors are providing significant levels of finance to organizations focused on clinical development of treatments for the major diseases of the Third World. In the Western world, nonprofit pharmaceutical companies, for example the Institute for OneWorld Health, have been established specifically to develop drugs and vaccines, including those for neglected diseases such as TB, malaria, and diarrhea.

These initiatives complement the drug donation programs of the large pharmaceutical companies. Merck has provided the drug ivermectin (**45**) (Mectizan) for onchocerciasis (river blindness) and treated more than 25 million individuals in 33 countries since 1987. GlaxoSmithKline, through its membership of the Global Alliance to Eliminate Lymphatic Filariasis (elephantiasis) is committed to eradicating a disease which currently disfigures and disables more than 120 million people in Asia, Africa, and Latin America. GlaxoSmithKline has committed to donate as much of the antiparasitic drug albendazole (**46**) (Helmintal) as required to treat the 1 billion people at risk of the disease in an effort to rid the world of the disease by 2020. To date, over 80 million people, including 30 million children, have received free preventive treatment. GlaxoSmithKline has also initiated a major TB research collaboration with GATB. Pfizer donates its antibiotic azithromycin (**47**) (Zithromax) as part of a large, integrated five-country effort to control trachoma, another disease that causes blindness. And the Novartis Foundation for Sustainable Development has since 1986 been involved in leprosy programs in Asia, Africa, and Latin America in partnership with local health authorities, the WHO, and nongovernmental agencies, in an effort to eliminate leprosy, with the company providing a multidrug treatment regimen.

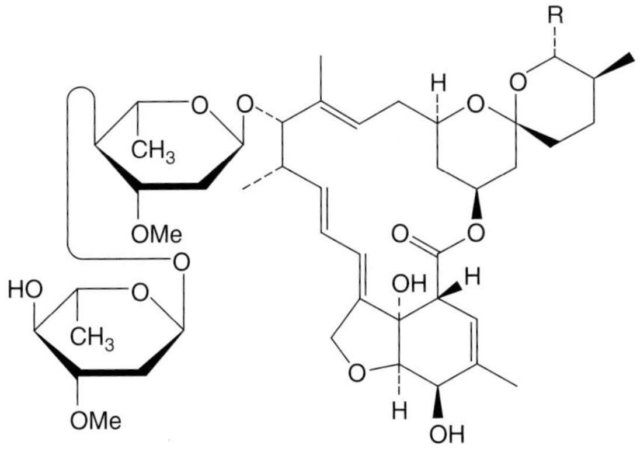

B_{1a} R = $CH(CH_3)CH_2CH_3$ (H, CH_3 / CCH_2CH_3)

B_{2a} R = $CH(CH_3)_2$

45 Ivermectin

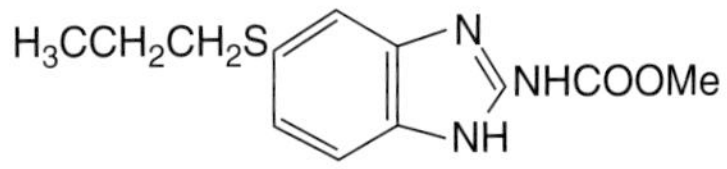

46. Albendazole

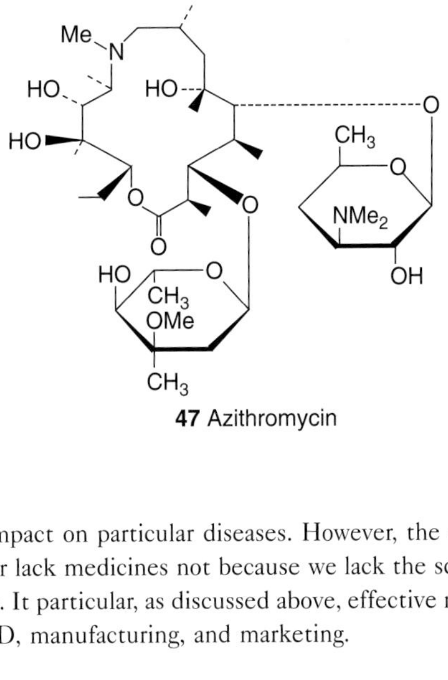

47 Azithromycin

Undoubtedly these programs have had or will have significant impact on particular diseases. However, the overall expenditure on Third World diseases remains pitifully low. The poor lack medicines not because we lack the science, but because we lack the will to apply the resources that already exist. It particular, as discussed above, effective market mechanisms must be established that encourage the necessary R&D, manufacturing, and marketing.

1.07.3.2 The Efficiency of Current Drug Discovery

In contrast with the situation in poorer countries, the range of medicines for treatment or prevention of disease available to patients in developed countries has expanded rapidly over the past half century. These products are a factor in the steady increase in longevity of peoples in the developed countries. By any standard, pharmaceutical scientists have been remarkably successful in their endeavors. Moreover, in recent years scientific knowledge has advanced at the fastest rate in history, and this has been complemented by tremendous innovation in technologies supporting drug discovery. We now have available a range of possible sources of drug discovery perhaps unimaginable only 20 years ago and one would expect that future drugs will come from a greater range of sources than is currently the case. Yet doubts have arisen recently about whether the sources of drugs that we are currently exploiting are indeed effective and efficient. There is currently a strong opinion that 'something has gone wrong' in drug discovery in recent years[66,67] and particularly over the past decade, although the output of new molecule entities (NMEs) appears to have been in long-term decline since the early 1970s.[68]

In part this concern is a reaction to slowing growth rates of the large pharmaceutical companies. During the decades of the1960s–1980s the industry enjoyed remarkable commercial success and companies grew rapidly. The industry was stable compared to most others because the rate of new medicine discovery supported strong growth of most individual companies. There were relatively few mergers and scientists in the large companies enjoyed a high level of job security. It was an industry characterized by research productivity, commercial success, corporate growth, and consistent increase in shareholder value. However, from the early to mid 1990s, there was a slowing of sales and profit growth, coupled with increasing external pressures of regulatory and political nature. Many companies suffered significant disruption in growth as their blockbuster drugs lost patent protection.

Underlying these developments is the stark fact that overall the pharmaceutical and biotechnology industry submitted almost 50% fewer new drug approval applications to the FDA in 2002–03 than it did in 1996–97, despite a 2.5-fold increase in annual expenditure over that period. **Figure 1** illustrates the rate of investigational new drug submissions and new drug approvals granted by the FDA over the past one to two decades.

Over the years 1996–2003 roughly the same number of molecules per annum entered clinical development (investigational new drugs), yet failure rates climbed well above historic averages such that the number of molecules reaching new drug approval fell by 50%. Critically, the failure rate in phase III trials, the time of highest expenditure, has risen dramatically such that approximately half of drugs have failed in this phase in recent years.[69] Some 4% of drugs have been withdrawn after new drug approval and marketing (*see* 1.02 Drug Discovery: Historical Perspective, Current Status, and Outlook).

Although the figures quoted above cannot reflect the eventual commercial value of the products, an increasing rate of company mergers and takeovers indicates that commercial returns have not been adequate. Inevitably mergers have been accompanied by cost-cutting, including significant numbers of jobs lost. While distressing to individuals, this has helped drive explosive growth in the start-up of small biotechnology companies, staffed in part by executives and scientists leaving the larger companies either voluntarily or involuntarily. Significantly, these companies have often been the force behind introduction of new methods for small-molecule drug discovery. Also, they have been the source of many of the new generation of biological products and a few of the small-molecule drugs that have appeared over the past two decades. Some striking figures are that in the early 1990s there were over 30 large pharmaceutical companies and fewer than 200 biotechnology companies worldwide. In not much more than a decade there are maybe now only 15 large pharmaceutical companies and over 3500 biotechnology companies. And this trend continues to develop. This must have major implications for the source of future drugs.

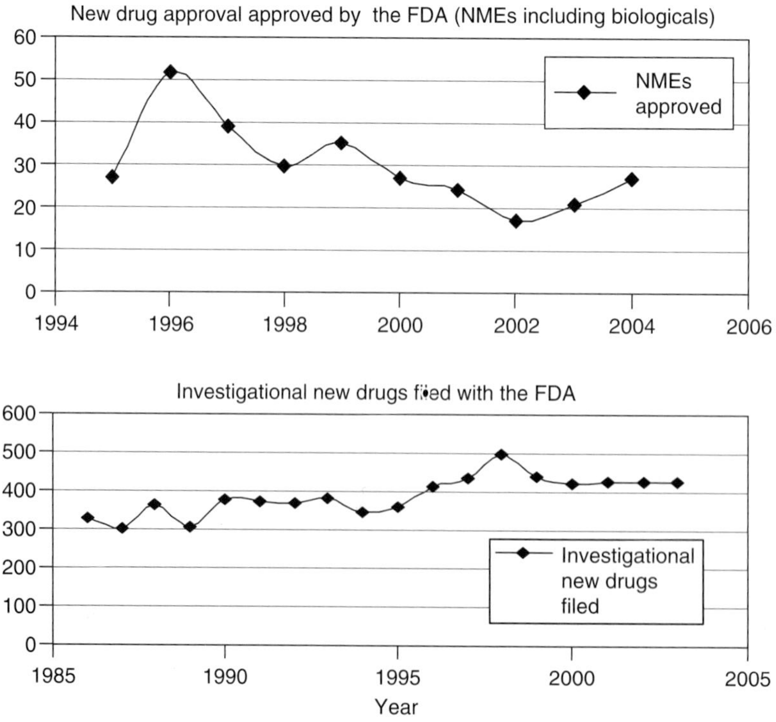

Figure 1 New drug approval and investigational new drug productivity.

1.07.3.3 Problems with Both Observation- and Hypothesis-Led Drug Discovery

To some extent the developments considered above reflect the success of previous efforts in drug discovery; effective medicines are now available, often off-patent as generics at low cost, to treat many diseases. Pharmaceutical scientists face increased barriers in that now the drugs required are often either 'third-generation' drugs for diseases now moderately well treated; or 'first generation' drugs for 'difficult' diseases. Either way, scientists face higher hurdles.

Some observers argue however that another factor is important. For most of the history of drug discovery, from prehistory until the 1980s, the major method to discover new drugs was by an observation-led approach. However, over the past 20 years, and accelerating over the past decade, there has been a rapid switch from observation-led to a hypothesis-led (mechanism-based) approach.

The observation-led approach was the only feasible means before the mid-1980s because the then current state of knowledge of human biochemistry and disease was inadequate in most cases to pinpoint a single biochemical mechanism whose modulation would favorably impact the disease. So, at least during most of the twentieth century, a biologist would screen compounds versus tissues, cells or directly in whole animals with a readout chosen to be indicative of the response desired in humans. In modern terminology, the biologist then selected 'actives' eliciting the 'phenotype' of interest and chemists modified the structure to optimize activity. Therefore drug discovery projects always started with a lead compound and an effect of interest in a physiological system. If possible, attempts were made to identify the mechanism of action of the leads but this was rarely successful, at least during the time-lines of the discovery project. If the mechanism could be found this was of great benefit in further refining the drug either for the initial market entry or for a 'second-generation' entry. With knowledge of the mechanism the new project would then be hypothesis-led but often with the advantage that this second-generation project was supported by clinical data indicative of effectiveness of the first-generation molecule, and also highlighting which properties required improvement.

The observation-led approach has major weaknesses. Among these, the lack of knowledge of the mechanism of action of leads and eventual clinical development molecules means there is inadequate knowledge of the relevance of the mechanism to man; clinical efficacy trials can be something of a gamble. Another consequence is that mechanism-based toxicity cannot be predicted. Moreover, leads may interact with several targets potentially leading to non-mechanism-based toxicity. Overall with this approach it was often difficult to tell whether toxicity was or was not mechanism-based, and therefore whether to continue the project in the face of adverse data. However, this approach also potentially had the consequence of a positive polypharmacy with resulting increased effectiveness. Lead seeking was very low throughput compared with modern methods, e.g., a few dozen compounds might be screened each week; and with no mechanistic assay, SAR was often complex for chemists to optimize, though it must be said that they often did succeed. Good disease models were essential to this approach and some observers of modern methods of sourcing drugs have suggested that this should be reemphasized today. Modern methods for 'finding the target' have led to something of a renaissance in the observation-led approach in recent years.[70] Phenotypic screens may be followed by target identification using two fundamentally different approaches; first, the lead molecule may be used to bind directly the target of interest, as detected by affinity purification, expression or display cloning, three-hybrid systems, or the probing of protein microarrays. Second, indirect approaches may be used in which compound-induced changes in mRNA, protein, or metabolite expression profiles are compared with profiles obtained with well-known target compounds or following genetic manipulation. Both approaches to determination of mechanism have strengths and weaknesses which are well reviewed by Hart.[71]

The hypothesis-led approach became the main paradigm from the early 1990s in response to the difficulties mentioned above, as biochemical knowledge and methods advanced. Critical developments were the availability of fast protein liquid chromatography (FPLC) for protein purification and the rapid spread of cloning methods. In this approach (described in detail in Section 1.07.2.1.3 above), a biochemist screens compounds on purified protein, often using highly automated equipment and data analysis, at the rate of many thousands per week, and 'hits' are assessed in functional assays for in vitro efficacy. The number of compounds required for screening has escalated dramatically with most being sourced by providers external to the discovery company. Natural product screening became de-emphasized perhaps because of the relative lack of pure compounds – many were available as impure 'fractions' – and because of the perceived difficulty and cost of optimizing leads from this source. Following some lead optimization, compounds are assessed in animal models of disease, if these are available. In recent years, with availability of human genome data and exploration of genetic links to disease (discussed in Section 1.07.4.1 below) an earlier step may be generation of a hypothesis by this means, and it seems likely that an increasing number of drugs may eventually derive from this source over the next decade or two.

The 'hypothesis-led' approach to sourcing drugs does itself have important weaknesses.[72] The drug discovery process now has several extra steps before identifying a lead with an effect of interest in a physiological system. Targets selected may have poor disease linkage, unsupported by any animal or clinical data; they are 'unvalidated' and this can lead to a high failure rate downstream in the discovery phase or in clinical trials. Hits and leads are often difficult to find for the targets selected. The ability to predict off-target effects has been poorer than expected, despite this being one of the perceived advantages over the observation-led approach. And 'physiology' is eliminated until late in the process; most of the early phase of a project is based on drug–target interaction not on drug–organism interaction.

Also, because of the emphasis on targets it appears that many companies may more often now be working on the same targets than maybe they were in the days of the observation-led approach, with consequences for industry-wide failure if these targets prove fruitless. Concerns about the hypothesis-led (or target-based) approach are summarized in a published comment from one industry manager: 'For the past decade the pharmaceutical industry has experienced a steady decline in productivity and a striking observation is that the decline coincided with the introduction of target based drug discovery.'[73] Whether one agrees with this or not, it certainly seems true that far-reaching changes have been made quite rapidly over the past decade to how we source new drugs without evidence of the likely success of the new approaches.

The reduction in productivity of new drug approvals was discussed in Section 1.07.3.2. Analyses have also been made of productivity in earlier stages of R&D, based on study of attrition rates at each phase of the R&D process. Reasons for attrition have also been studied and will now be considered here. Note that these analyses come from large pharmaceutical companies; there is no current evidence to indicate whether small biotechnology companies are more or less efficient in their R&D efforts.

The cumulative attrition in the discovery phase is approximately 80%; only about 1 in 5 projects gets as far as selecting a compound for clinical trials (**Table 5**). In clinical development, attrition between selecting a compound and marketing is >90%. Fewer than 1 in 10 clinical projects gets a product to market. Overall throughout the R&D process fewer than 1 in 50 projects succeeds in getting a drug to market.[72]

Analysis of the reasons for project failure indicates four or five main reasons.[73,74] These include the choice of target – which supports the comment quoted above regarding the hypothesis led, target based approach. The key issue is that the chosen target mechanism fails in animal or clinical studies. A second reason is the choice of leads, either total failure to find a lead that can be optimized, or the selection of leads that eventually prove nonoptimizable. A third reason is drug safety, when the final drug candidate selected from the lead series fails to pass regulatory toxicology requirements. In addition to these three issues, failures in clinical development may also occur because of adverse events or poor pharmacokinetics not predicted by animal studies, though overall these types of failure appear to have reduced as preclinical studies have improved in quality (**Figure 2**). Unfortunately late failure of compounds, in phase IIb and phase III clinical trials, appears to have increased in recent years, the major reason being failure to demonstrate the efficacy expected from earlier smaller trials. Increasingly companies are being asked to demonstrate pharmacoeconomic benefit for their drug to justify reimbursement of drug costs by central payors, and this hurdle is likely to increase the pressure on failure rate late in the R&D process.

Have the reasons for R&D project failure changed over the past decade with the introduction of new methods? A comparison of the reasons for drug failure in 1991 and a decade later in 2001 is made in **Figure 2**. (Note that equivalent data analyzing reasons for failure of biological products is not available, possibly due to the recent development of these types of product and the small sample size; however, early data indicate that overall survival rates are higher.[75,76]) It appears that failures for pharmacokinetic reasons have reduced, and many in the industry will be well aware of significant improvements in knowledge and practice that has led to this positive result.

What's not improving? Most of the causes of R&D project failure have not improved over the past decade (**Table 2**). Losses due to drug efficacy have not improved after a decade of the widespread use of target-based drug discovery,

Table 5 Success rates in research and development[72]

	Target assessment	*Lead identification*	*Lead optimization*	*Preclinical*	*Phase I*	*Phase II*	*Phase III*	*Registration*	*On market*
Rate (% median)	63	60	63	57	58	45	58	85	
Cumulative success rate	1.7	2.7	4.6	7.3	13	22	49	85	
Projects (for one drug approval)	57	36	22	14	8	4.5	2	1.2	1

Reprinted from Brown, D.; Supert-Furga, G. F. *Drug Disc. Today* **2003**, *8*, 1067–1077, copyright (2003) with permission from Elsevier.

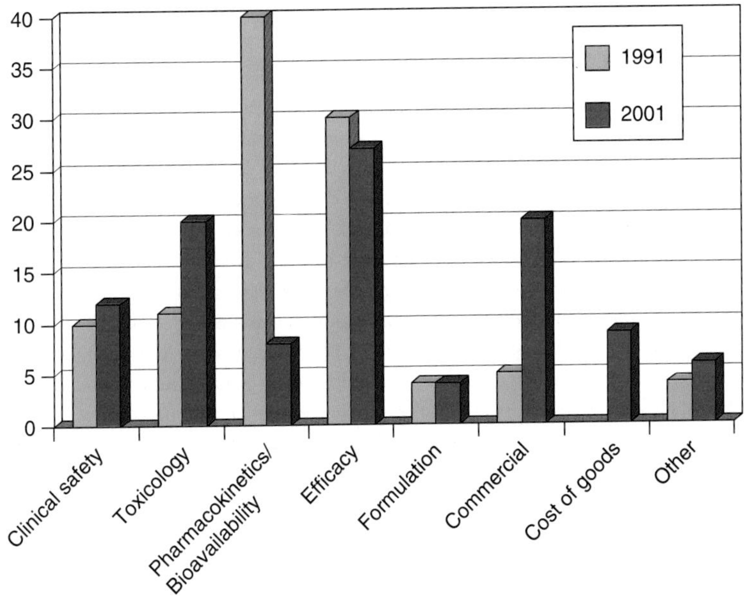

Figure 2 Comparison of the reasons for R&D project losses in 1991 and 2001. (Reproduced from Kola, I.; Landis, J. *Nat. Rev. Drug Disc.* **2004**, *3*, 711–715, with permission from Nature Reviews Drug Discovery, copyright (2004) Macmillan Magazines Ltd.)

which was particularly intended to reduce the incidence of this cause of loss. Similarly, the approach was intended to reduce losses due to problems with toxicology and clinical safety, yet both these causes of loss appear to have increased over the decade. And project terminations for commercial reasons have increased dramatically; possibly this is in part because of elimination of projects as companied have merged and reduced overall R&D personnel numbers, but other factors may be important too, such as focus on larger market sizes leading to termination of some projects. And cost of goods appears to be of rising concern as targets and drugs become more complex and commercial pricing pressures increase.

In summary, it appears that there are unsolved problems with both the observation-led and the hypothesis-led approaches and these are holding back the discovery of new medicines. It appears very important that these are resolved if small-molecule drug discovery is to regain the productivity levels of the mid to late twentieth century.

1.07.4 Future Perspectives

1.07.4.1 The Impact of Pharmacogenomics and Pharmacogenetics

Pharmacogenomics includes the study of individual variations in whole-genome or candidate gene single nucleotide polymorphism (SNP) maps, haplotype markers, and alteration in gene expression or inactivation. Pharmacogenetics by contrast is narrower in definition and refers to the study of interindividual variations in DNA sequence related to drug absorption and disposition or drug action. This includes polymorphic variation in genes that encode transporters, drug metabolizing enzymes, and drug targets such as receptors. There is some overlap in these definitions and it should be noted that this occasionally can lead to some flexibility of terms used in the literature.

There can be little doubt that these new disciplines will have a significant impact on both discovery and use of new medicines. The scope and timing is less certain, though. In this section we will consider the likely impact on the discovery of new targets for drug discovery, on finding alternative uses for existing drugs, on drug metabolism and toxicology, and on diagnostics and personalized medicine.

1.07.4.1.1 Targets

The new approaches to drug discovery provided in the early twenty-first century by genomics and genetics are being widely deployed and will be an important source of new drugs. There is widespread hope that this will increase the number of truly innovative new medicines that will be discovered and reverse the apparent declining productivity of pharmaceutical scientists. However, the task is not easy, despite press hype surrounding completion of the Human

Genome Project. It must be remembered that disease is a very complicated process and is rarely due to a simple genetic 'cause.'

Diseases caused by a single mutation to a gene are called monogenic diseases, and there are some 6000 of these. They were well described before the Human Genome Project and yet it is a striking fact that knowledge of the genetic link has not yet been instrumental in the development of a single drug for any of these diseases. Where treatments are available they have come from other means.

It can be argued that these diseases are relatively rare, and that the more widespread polygenic diseases will attract much more research and eventual success. Challenges ahead are many however. One includes the observation that "Gene Association Studies are Typically Wrong," to quote the heading of a leading article in one journal in 2004,[77] an assertion based on the observation that only one in three gene associations first published have been subsequently confirmed by later studies. And when the link is confirmed, typically it is weaker than first reported.[78,79] This indicates that improvement in methodology and/or standards required for publication is required, particularly since new high-throughput analysis techniques now enable researchers to study many gene–disease associations quickly and cheaply which increases the likelihood of the reporting of spurious links through chance occurrences. Another challenge is that several genes may be implicated in the causation of a disease, as well as environmental factors. For true positives, there are several subsequent steps to be navigated, each with high attrition, before a drug discovery project can begin. Which protein is coded by the gene? Do we need to modulate function up or down? Does the gene product contribute to disease at the time of treatment? Do we have the required biological models or transgenics to prove that modulating function produces the desired effects? Will modulation lead to side effects? Is the target chemically tractable? Is efficiency likely to be competitive with established products, if any are available? It may be that modulation of another protein in the pathway pinpointed by the gene product offers a better therapy. Or it may be that modulation of a compensatory pathway offers a better therapy. Note that several issues that have held back target-based drug discovery over the previous decade are equally important here. One advantage of gene-association-based approaches may be improved confidence in the mechanistic hypothesis, though much additional evidence of appropriateness for drug discovery will be needed, as outlined above. Even with a genetic linkage, target validation is still necessary in a disease model, and preferably also in a mouse knockout model whenever possible (*see* Section 1.07.2.1.4 above). In addition, finding good leads will remain a challenge; and the potential for mechanism-based side effects will need careful consideration. Where a genetic link is strongly established but the identified target proves nondruggable for whatever reason, the information may still prove useful if the biochemical pathway surrounding the target can be mapped to identify alternative druggable targets. Technologies to enable full mapping of all components of biochemical pathways are now available.[46]

It is estimated that 99.9% of DNA sequence is conserved between individual humans, with the remaining 0.1% determining different susceptibility to genetically determined diseases and different response to drug therapy. Understanding the genetic basis of disease has proved quite difficult and currently there are relatively few well-validated cases of drug discovery or even earlier stage new target discovery by this route. However, based on the observation that selection of targets using animal models, simple organism genetics, and genetically manipulated animals has been less successful than hoped for, use of patients to select targets for their diseases has merit.[80] Genomic approaches used for target identification include positional cloning (a hypothesis-free approach which follows the transmission of polymorphic markers through families to identify those cosegregating with the disease), and the candidate gene approach, described below.[81]

Positional cloning has been very successful for mutations that underlie single-gene Mendelian disorders because the gene identified explains a high proportion of the phenotypic variation giving a high correlation between genotype and phenotype. However, for the most prevalent diseases, multiple genes are involved and in addition there is complex interplay with environmental factors. Application of whole-genome linkage analysis to these complex diseases has so far identified a small number of disease genes, mostly coding for targets that have historically been regarded as of low drug tractability.[82] However, the link between the target 5-lipoxygenase activating protein (FLAP) and heart disease (discussed in Section 1.07.4.1.3 below) did come from this approach.[83]

In an alternative to linkage studies, the hypothesis-led candidate gene approach has been widely applied to identify complex disease susceptibility genes.[84] Genetic diversity in the form of single nucleotide DNA polymorphisms (SNPs) contributes to variable disease susceptibility and drug response. The candidate gene approach uses a priori knowledge of pharmacological action, drug disposition, and disease pathogenesis to identify genes relevant to drug response. The process of candidate gene study involves selecting pharmacogenetic candidate genes, identifying all potential alleles in ethnically diverse populations for each candidate gene, and testing in a clinically relevant population for statistical association of the SNPs with a selected pharmacogenetic trait. TS and SNP genotyping technologies allow the study of thousands of candidate genes and the identification of those involved in drug efficacy and toxicity. The SNP Consortium

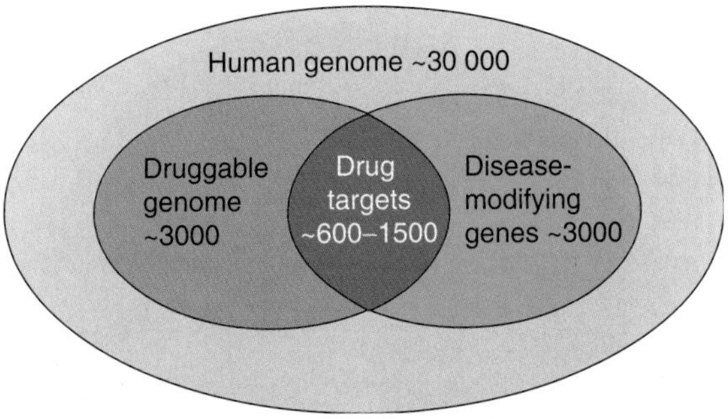

Figure 3 Number of drug targets. The effective number of exploitable drug targets can be determined by the intersection of the number of genes linked to disease and the 'druggable' subset of the human genome. (Reproduced from Hopkins, A. L.; Groom, C. R. *Nat. Rev. Drug Disc.* **2002**, *1*, 727–730, with permission from Nature Reviews Drug Discovery, copyright (2002) Macmillan Magazines Ltd.)

coupled with sequencing of the human genome has been beneficial in identifying some 1800 potential targets in the GPCR and kinase classes.[80,85] Expression-based genomic technologies such as DNA microarrays and proteomics also add to understanding of important biological and pharmacological pathways, thus identifying more candidate genes for SNP studies. Candidate-gene-based pharmacogenetics is becoming established as an important approach to improved drug development, improved clinical trial design, and therapeutics tailored to individual genotypes.[86]

Both positional cloning and candidate gene methods may frequently identify nondruggable targets, so a third approach being explored is 'target-class genetics' in which the candidate gene approach is applied only to drug target gene families (e.g., GPCRs, proteases, ion channels).[87]

Another approach to target discovery is the use of ESTs. This approach identifies genes undergoing expression in a particular tissue by identification of mRNA fragments. This does not directly link a gene to a disease, but it can lead to discovery of expressed proteins that might themselves be useful as therapeutic proteins in treating disease. This approach has been particularly successful in identifying many of the therapeutic proteins that are currently used in clinical practice or that are currently in clinical trials.

This search for new targets within the genome raises the question: 'How many druggable targets are there?' This is a difficult question to answer and all attempts must be approached with caution. In considering the possible number of new drug classes that might emerge we must remember the inherent redundancy of biological networks. Blocking an enzyme, receptor, or channel will not necessarily close down a biological process; biological pathways often have compensatory mechanisms that maintain function, as a way to preserve life. So, experimental drugs that look excellent in an in vitro environment may fail to demonstrate the expected efficacy when tested in more complex systems in vivo.

Despite this caution there have been several attempts to give an estimate of the likely number of druggable targets represented in the genome. Through a comprehensive review of the total portfolio of the pharmaceutical industry[45,88,89] Drews identified 483 targets that have been or are currently being utilized, and concluded that there could be 5000–10 000 potential targets on the basis of an estimate of the number of disease-related genes. However, large-scale mouse ko studies have indicated that only about 10% of all gene knockouts might have the potential to be disease-modifying,[90] which suggests a much lower number of useful targets. Another approach suggested[91] is to assess the number of ligand-binding domains as a measure of the number of potential points at which small-molecule therapeutic agents could act. The authors suggest that this figure could be even greater than the 5000–10 000 suggested by Drews though again this figure does not take into account the potential for these targets to be disease-relevant. In another analysis Hopkins and Groom[92] suggest that there will be 600–1500 targets in total. This figure is derived from the calculation that 3000 (10%) of the 30 000 genes in the human genome will be disease-relevant, that a similar number of genes will offer gene products that are druggable, and that the intersection of these subsets represents the actual set of druggable targets (**Figure 3**).

The paper by Hopkins and Groom also provides useful data on gene family distribution and target classes represented by current drugs and experimental drugs, and the percentage representation of each target class predicted in their full druggable genome analysis (**Figure 4**).

What is the evidence that genome data is helping drug discovery? Target validation is a complex process[93] which is followed by another lengthy process of lead discovery and optimization, so it is perhaps too early to assess the impact of genome data on the discovery of small-molecule drugs. However, these are quicker for biological products. Protein and

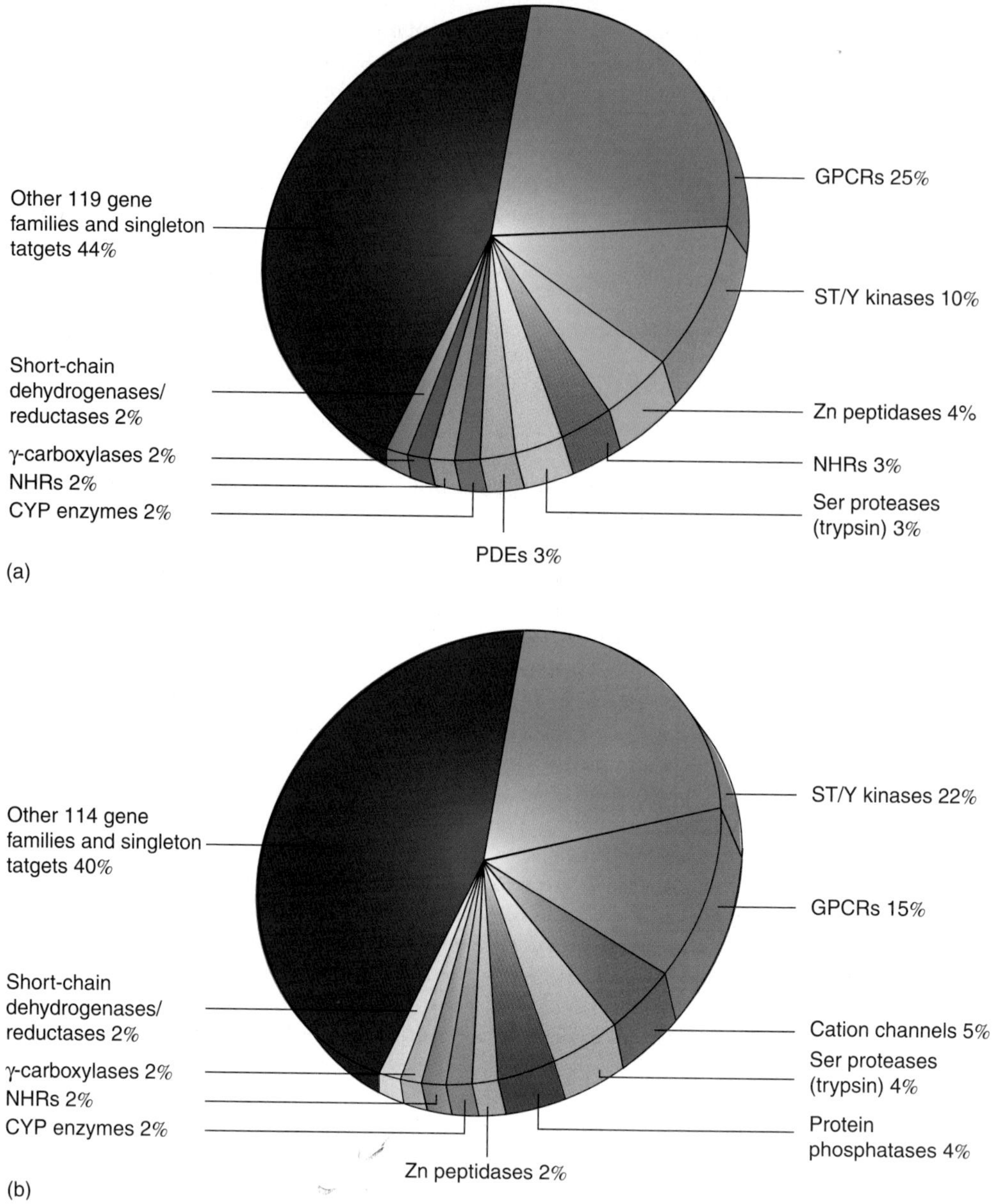

Figure 4 Drug-target families. Gene-family distribution of (a) the molecular targets of current rule-of-five-compliant experimental and marketed drugs, and (b) the druggable genome. Serine (Ser)/threonine and tyrosine protein kinases are grouped as one gene family (ST/Y kinases), as are class 1 and class 2 G protein-coupled receptors (GPCRs). CYP, cytochrome P450; Cys, cysteine; NHR, nuclear hormone receptor; PDE, phosphodiesterase; Zn, zinc. (Reproduced from Hopkins, A. L., Groom, C. R. *Nat. Rev. Drug Disc.* **2002**, *1*, 727–730, with permission from Nature Reviews Drug Discovery, copyright (2002) Macmillan Magazines Ltd.)

antibody products in particular are proving to be the early wins in the age of the genome, and these are increasingly being complemented by a range of diagnostics as discussed later.

The potential of antibodies as drugs was outlined in (*see* Section 1.07.2.2.2). Antibodies as antagonists of chemically intractable targets, and even for some tractable targets, are attracting increased interest, for a number of reasons. Target identification and target assessment are easier than for small-molecule targets. Antibodies usually have greater selectivity of pharmacological effect, and more predictable safety. Pharmacokinetic properties are more predictable too. There are limitations on target type, as currently antibodies do not penetrate cells at all well, and limitations too on the route of delivery of the product and this has focused attention on serious, life-threatening diseases with high unmet medical need.

Another source of biological products may come from RNA interference approaches. RNA interference is being explored as a target validation tool, and also as a potential source of drugs products in their own right. The very rapid growth in interest in RNA interference is because an ideal approach to utilizing the genome for target validation could be to specifically silence an individual gene by blocking its expression, and RNA interference provides this capability.[94,95] Short double-stranded RNA fragments (siRNA) around 22 nucleotides in length are prepared either synthetically or from a vector. The sequence is complementary to the target mRNA and binds cellular proteins to produce a RNA-induced silencing complex. Within the complex the two RNA strands separate and the whole complex is guided to the target mRNA by the complementary sequence on the siRNA. On binding to the target the complex cleaves and completely degrades the target mRNA, abolishing expression of the encoded protein. siRNA appears to provide better specificity and higher efficacy than antisense approaches, though recent studies have demonstrated that gene silencing is not absolute. This 'knockdown' effect can be used to reversibly block expression of any specific protein rapidly. Expression blocking is normally around 70% which means that lethality can be avoided where a gene is essential to survival. An RNA interference knockdown model can be produced in a fraction of the time and cost previously involved with antisense technology, which means that the latter has been largely superseded. This technology is transforming target validation and future developments will undoubtedly enhance its utility further.

The use of RNA interference directly as a drug approach to treat disease is the subject of intense research. Many diseases are being targeted and a number of products are in late preclinical research.[96] At the time of writing it is unclear when or if the first RNA interference-based drug will succeed. There are a number of obstacles. Direct targeting of the desired cell or tissue may not be easy for most diseases. In addition, stability is a key issue as the inhibitory RNAs must remain intact long enough to work. Questions on specificity must be further investigated, as must concerns about whether RNA interference induces an interferon response. Delivery is a problem as successful administration of siRNAs often requires use of plasmids or viral vectors, which can lead to random integration into chromosomal DNA with potential for toxicity. Some of these issues can be avoided if researchers employ local administration strategies instead of systemic ones, though this will limit the range of diseases that might be treated.

1.07.4.1.2 Drug metabolism and toxicology

Poor drug metabolism leading to higher than expected drug exposure is a significant cause of drug side effects. The potential for pharmacogenomics to reduce this problem is illustrated by one estimate suggesting that more than half of the 27 drugs frequently cited as causing side effects are linked to at least one enzyme with a variant allele (one of two or more forms of a gene that may occur at a given site on a chromosome) known to cause poor metabolism.[97] The ability to diagnose the presence of these alleles predicting poor metabolism in patients would be of great benefit.

Over 50% of the population is believed to have detectable genetic variations of the 8 to 10 enzymes in the liver responsible for metabolism of the majority of drugs. Genetic variations of CYP3A4 and CYP2D6, for example, may be particularly relevant for clearance of many major classes of drugs, and diagnostics for these are of particular importance.

In addition to variation in drug exposure, adverse drug reactions caused by differential reactive metabolite formation have resulted in a number of drug withdrawals.[98] Methods are being assessed to identify potential problems before committing to development of a molecule, including CYP-expressing cell lines and nuclear receptor reporter gene assays.

A related topic is toxicogenomics, which is the use of gene expression techniques for understanding and predicting toxicological effects of a compound. Usually toxicogenomic methods are applied once toxicity of a compound has been demonstrated in animal models either during pharmacology experiments or during drug safety assessments. Gene expression profiles are based on RNA extracted from the target organs of the toxicity. However, it is also possible to employ gene expression profiling based on compound dosing to cells, for in vitro studies. Toxicogenomics is used both for in vitro profiling to predict the toxicity potential of compounds, or to understand the mechanism of proven toxicity of a compound. Increases or decreases in gene expression in organs from dosed animals caused by the compound provide evidence of genes and pathways that have been affected in addition to those targeted for efficacy by the compound. These clues can be followed up by molecular studies to evaluate the hypothesis. Once understanding is gained it may be possible to use gene expression profiling of analogs of lead compounds to eliminate early those likely to be associated with toxicity. However, predictive toxicogenomics is a new science and methods still require much validation before they can be used with confidence to reject potential drug molecules. Many companies are investing in building reference databases that will eventually increase the value of employing predictive toxicogenomics.

1.07.4.1.3 Second medical use: reprofiling old drugs and rescuing failed clinical drugs

Disease association data can offer opportunities for rapid discovery of effective new medicines if currently available drugs prove to have utility in diseases in addition to those for which they are currently marketed. Also, drugs that have proved safe but ineffective in clinical trials may find alternative uses.[99]

The Icelandic company Decode identified a variant gene that can double the risk of heart attack among Icelanders[83] and on this basis they licensed from Bayer a molecule that was originally in clinical development as a treatment for asthma. The molecule, DG-031, inhibits the same target, 5-lipoxygenase activating protein (FLAP), coded by the gene identified by Decode. Bayer had hoped that inhibition of FLAP might provide a treatment for asthma but, while proving safe in clinical tests, the drug was ultimately ineffective against the lung disorder. By licensing a product which had already completed safety tests in another disease area, Decode could reduce its risk of failure of a molecule for safety reasons, and also be much closer to market than if they had initiated a discovery project on the basis of their observations.

It is likely that such an approach will be the source of a number of new drugs over the next decade. Drugs discovered by reprofiling[99] and developed for second medical use have substantially reduced risk of failure compared with novel molecules because they have often been through several stages of clinical development and have substantial record of safety in humans. Moreover, the pharmacokinetic properties in humans may be well characterized, and chemical manufacturing and drug formulation experience helps reduce cost and time in redeveloping the drug for an alternative use. Indeed, repositioning of drugs has become a major strategy for venture capital firms seeking to reduce the risk of their investment in biotechnology start-up companies: a pipeline can be built quickly and at lower risk than with research-based R&D.

It is also likely that genomics will be used also 'after the event' to understand unexpected beneficial effects detected empirically during usage of current drugs. It is a classic approach to reposition drugs by capitalizing on unexpected observations in clinical trials or after marketing of the drug. There are numerous examples, and some interesting examples are given here to show the range of opportunity. A detailed review has been provided by Ashburn and Thor[100] in which they list some 35 drugs either in clinical development or now marketed for a second medical use. These include the well-known examples of sildenafil (**44**), from angina to erectile dysfunction; minoxidil (**48**) (Rogaine) from hypertension to hair loss; finasteride (**49**) (Propecia) from benign prostatic hypertrophy to hair loss; thalidomide (**50**) (Distaval) from nausea and insomnia to multiple myeloma; Zidovudine/AZT (**51**) (Retrovir) from oncology to HIV/AIDS.

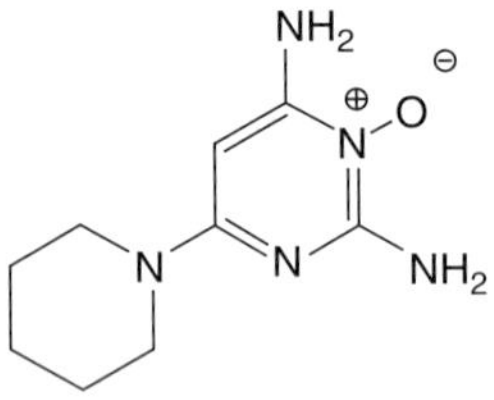

48 Minoxidil

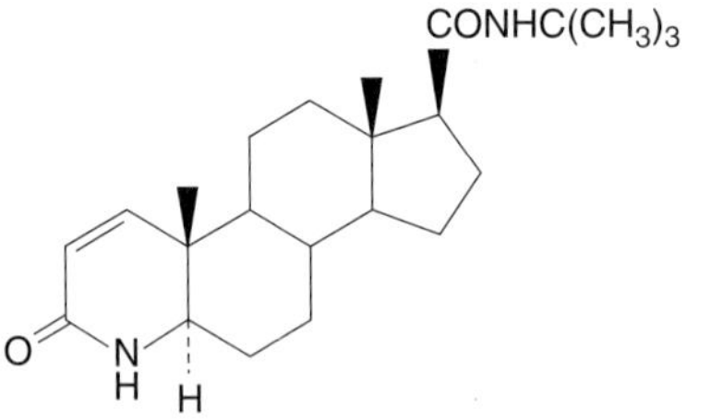

49 Finasteride

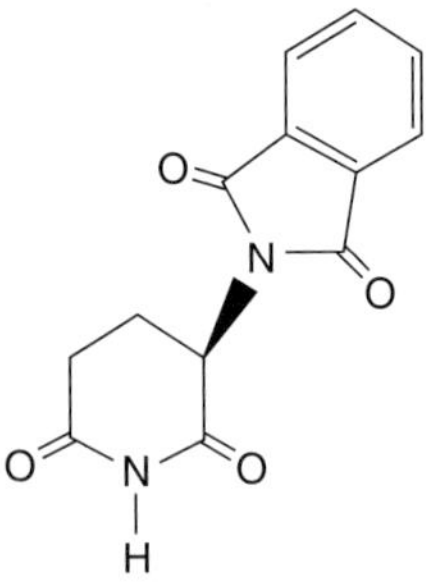

(*R*)-form

50 Thalidomide

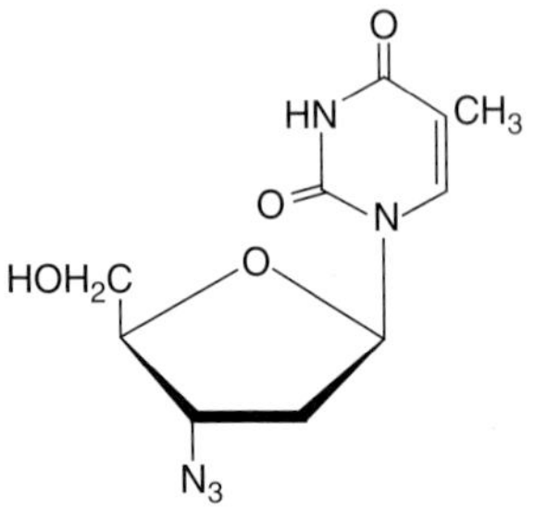

51 Zidovudine

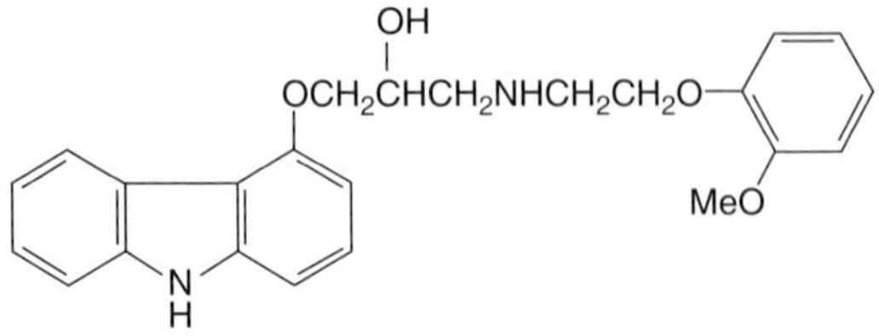

52 Carvedilol

One striking example in recent years is provided by carvedilol (**52**) (Coreg), a beta adrenergic blocker originally marketed for treatment of hypertension. This drug, like its whole class, was originally contraindicated for patients with congestive heart failure on the basis that beta adrenergic blockade of the myocardium in these patients could be harmful and possibly lethal. However, the drug was given to one such patient (to treat a comorbid condition) and the physician involved in the treatment noted that the expected deterioration in the patient's condition did in fact occur for a short time but that subsequently the patient's cardiac function improved markedly. Eventually a large trial was funded, the Carvedilol Post Infarction Survival Control in Left Ventricular Dysfunction (CAPRICORN) study, with the positive outcome that subsequently the FDA expanded the indications for carvedilol to include reduction in the risk of death among patients with left ventricular dysfunction following myocardial infarction. Treatment with carvedilol reduced the risk of death for any reason by 23% and was associated with a 41% reduction in the risk of recurrent nonfatal myocardial infarction.

Reprofiling of current drugs in animal models is also gaining in interest.[101] The approach resembles the phenotypic screening used earlier in the industry (discussed in Sections 1.07.2.1.4 and 1.07.3.3 above) but differs in that species with well-defined genomes are subjected to genome-wide mutagenesis by a chemical agent such as *N*-ethyl *N*-nitrosourea (ENU), followed by selection of a subset of mutant animals displaying a disease-relevant phenotype. This provides model organisms with homogenous genetic backgrounds, a distinction from earlier phenotypic models. Moreover, subsequent identification of the genetic alteration provides guidance on the likely mechanism of action of active compounds, which was a major barrier to previous-generation use of phenotypic screens. This approach might be used if the drug being reprofiled is expected to produce its effect by its known mode of action, but it also allows serendipitous discovery of a new mechanism. Model organisms in common use are mouse, zebrafish, fruitfly, and worm (*Caenorhabditis elegans*).

1.07.4.1.4 Diagnostics and personalized medicine

It is widely reported that drugs may often work in only 40–60% of the target population. Moreover, a minority of patients within both the 'effective' and 'ineffective' groups may experience significant side effects. With traditional methods of drug discovery and clinical development, it has been very difficult to predict which patients will benefit from the drug and which patients may experience side effects. Phase III trials may be very large, involving many thousands of patients, but even this size is in adequate to detect rare but very serious side effects. Adverse drug reactions (ADRs) are a serious matter. One US meta-analysis of the year 1994 estimated that over 2 million hospital admissions in the USA were due either to ADRs or inappropriate prescribing.[102] Over 100 000 died as a result. ADRs may be between the fourth and sixth leading cause of mortality in the USA. Similarly, in the UK more than 800 000 patients using the National Health Service experienced ADRs, with 68 000 patients dying and 50 000 permanently disabled.[103]

Use of pharmacogenomics, pharmacogenetics, and pharmacoproteomics will likely improve this situation by enabling better definition of disease and better distinction of individual variability between patients so that the right drug can be targeted much more often to the right patient. This will likely mean that each drug may have a smaller target patient population, at least for its primary indication, though each drug may find use in a broader range of diseases (*see* Section 1.07.4.1.3 above; and 1.05 Personalized Medicine).

Another development in personalized medicine is the advent of biomarkers. These are proteomics approaches used to identify differentially expressed proteins that correlate with a disease state or drug response, and are currently being applied early in drug development to improve dose selection, definition of treatment regime, and for early assessment of efficacy.[104] Currently, the greatest impact of biomarkers is via use of diagnostics at the point of care; however, their use in preclinical and clinical studies is likely to grow rapidly in importance. Biomarkers developed for use during these early phases may eventually form the basis of companion diagnostics that guide the use of the drug once the latter is marketed. The FDA has given guidance on the ideal biomarker, stating that it must have a predictable expression level and be specifically associated with a particular disease or disease state and be able to differentiate between similar physiological conditions. Simple, accurate, low-invasive detection of the biomarker is required.[105]

An important factor in personalized medicine is racial differences. Careful studies are essential to determine the origins of racial health disparities and drug responses, which can involve both genes and environment and the interactions between them. A particular genetic input or particular environmental input may have quite different consequences in different individuals. One well-studied case is that of African-Americans whose propensity to high blood pressure is an example of how the relative inputs of genetics and the environment to a disease state can vary across cultures. High blood pressure affects about 65% of African-American elders between the ages of 65 and 74. Analysis of data from a twin study, the Carolina African American Twin Study of Aging, found that a large proportion of the individual variability in blood pressure for African-American adults arose from environmental sources. Previous studies on other populations had shown that, although environment does impact blood pressure, genetic factors played a larger role in determining the individual differences in blood pressure.

So a number of factors are coming together to enable personalized medicine to become a reality: the redefinition of diseases into discrete entities based on genetics rather than phenotype; the genetic profiling of patients; development of diagnostics predicting individual patients' absorption, distribution, metabolism, and excretion (ADME) response to a drug; and discovery of a broader range of drugs to treat each disease together with methods to select the appropriate drug for each patient. Supporting this is a growing bioinformatics resource that is critical to the knowledge-based future of healthcare. In addition, we can expect over the next decade to see gradual introduction of changes in healthcare driven by these technical advances: introduction of screening using diagnostics to evaluate an individuals susceptibility to particular diseases; prophylactic use of drugs, vitamins, exercise, and diet for those at risk; molecular markers to define precisely disease that is established and to measure its severity and rate of progression.

One success for personalized medicine is already on the market. The antibody trastuzumab (Herceptin, Roche) would, in all likelihood, never have received regulatory approval except for a test which helps to select the one in three breast cancer patients for whom it is likely to provide benefit.[106]

Among examples for drugs currently in clinical trials, Allen Roses of GlaxoSmithKline reports[107] that a molecule under trial for treating obesity, with weight loss as the endpoint, showed significant efficacy in 25% of treated patients. Some patients gained weight. Using SNPs from candidate genes related to the mechanism of the molecule, several homozygous SNPs were found in the group of efficacy patients, whereas the weight-gain group were homozygous for the other allele. Heterozygotes segregated in between. So both cohorts of patients were identified by distribution of SNP alleles. Thus it may be possible to avoid treatment of patients very likely to be unresponsive to the drug.

Pharmocogenomics also has a key role in drug safety during clinical trials and the following example illustrates how such data may be used in coming years. During a phase III trial a number of treated patients exhibited hyperbilirubinemia. This adverse event correlated with a variant of the UTG1 A1 gene. Fifty percent of patients carrying the 7-7 polymorphism became hyperbilirubinemic on drug treatment. Controls on placebo with the 7-7 polymorphism did not exhibit hyperbilirubinemia.[108] This illustrates how pharmacogenomics data is likely to be utilized in future to reduce adverse reactions to drugs.

It is likely that regulatory authorities will increasingly demand such data over the coming decade. The perspective of the FDA is recorded in a review by Lesko and Woodcock[109] which succinctly summarizes the opportunities and challenges. In addition, the FDA issued in 2002 a forward-looking paper[110] that provides a regulatory perspective on the opportunities and challenges of integrating pharmacogenomics into drug development and regulatory decision-making and has coordinated a number of workshops on this topic with client companies. The proceedings of these workshops have been published and these provide a valuable resource on the status of pharmacogenomics in drug development, and what is needed to advance these tools.[111–113]

1.07.4.2 The Future for the Human Genome Project

Longer term, into the second decade of the twenty-first century, further opportunities for discovery of new medicines will arise as more detailed information becomes available from genome studies. There is much still to do.[114–117] A key task is to develop a definitive catalog of all protein-coding genes. Currently, the number is estimated to be between 20 000 and 25 000, a quite broad range which reflects the limitations of current gene-prediction software. Use of comparative genomics, aligning the human genome with the genomes of other animals, may help resolve this issue, because natural selection ensures that functional regions are more highly conserved than nonfunctional ones. So this approach can highlight regions coding candidate proteins.

Also, some other functional elements remain poorly defined. These include gene promoters, which control the timing and level of expression of genes; and micro-RNAs, which have been implicated as regulatory agents of many developmental processes. Elucidation of these is very likely to trigger opportunities for drug discovery.

Finally, also ahead is the task of sequencing the remaining 20% of the genome that lies within heterochromatin, the gene-poor, highly repetitive sequence that is implicated in the processes of chromosome replication and maintenance. The repetitiveness of heterochromatin means that it cannot be tackled using current sequencing methods; new technologies will be required.

References

1. Sneader, W. In *Comprehensive Medicinal Chemistry*, 1st ed.; Elsevier: Oxford, UK, 1990; Vol. 1, pp 8–76.
2. Lind, J. In *Treatise on Scurvy*; Stewart, C. P., Guthrie, D., Eds.; Edinburgh University Press: Edinburgh, UK, 1953, p 63.

3. Aronson, J. K. *An Account of the Foxglove and its Medical Uses 1785-1985*; Oxford University Press: Oxford, UK, 1985.
4. Coenen, H. *Arch. Pharmacol* **1954**, *287*, 165.
5. Lockemann, G. J. *Chem. Ed.* **1951**, *28*, 277.
6. Robiquet, P. J. *Ann. Chim.* **1832**, *51*, 259.
7. Gulland, J. M.; Robinson, J. *J. Chem. Soc.* **1923**, *123*, 980.
8. *Anethesiology* **2005**, *102*, 815–821.
9. Butera, J. A.; Brandt, M. R. *Annu. Rep. Med. Chem.* **2003**, *38*, 1–10.
10. Kowalul, E. A.; Lynch, K. J.; Jarvis, M. F. *Annu. Rep. Med. Chem.* **2002**, *35*, 21–30.
11. Macht, D. I. *Ter. Arkh.* **1916**, *17*, 786.
12. Anft, B. *J. Chem. Ed.* **1952**, *32*, 566.
13. Niemann, A. *Ueber eine neue organisches Base in den Cocablattern*; Huth: Gottingen, 1860.
14. Hardy, E. *Bull. Soc. Chim. Fr.* **1875**, *24*, 497.
15. Gerrard, A. W. *Pharm. J.* **1875**, *5*, 865.
16. Rodin, F. H. *Am. J. Opthalmol.* **1947**, *30*, 19.
17. Taylor, P. In *The Pharmacological Basis of Therapeutics*, 7th ed.; Goodman, L. S., Gilman, A. G., Eds.; Macmillan: London, 1985, pp 222–235.
18. Pines, G.; Miller, H.; Alles, G. *J. Am. Med. Assoc.* **1930**, *94*, 790.
19. Hoffman, B.; Bigger, J. T. In *The Pharmacological Basis of Therapeutics*, 7th ed.; Goodman, L. S., Gilman, A. G., Eds.; Macmillan: London, 1985, pp 716–747.
20. Hensen, B. *Drug Disc. Today* **2005**, 7, 459–462.
21. Maclagan, T. *Br. Med. J.* **1876**, *1*, 627.
22. Rodnan, G. P.; Benedek, T. G. *Arthritis Rheum.* **1978**, *13*, 145.
23. Newman, D. J.; Cragg, G. M.; Snader, K. M. *J. Nat. Prod.* **2003**, *66*, 1022–1037.
24. Koehn, F. E.; Carter, G. T. *Nat. Rev. Drug Disc.* **2005**, *4*, 206–220.
25. Takahashi, N. In *Macrolide Antibiotics*, 2nd ed.; Omura, S., Ed.; Academic Press: London, 2002, pp 577–621.
26. Khosla, C.; Keasling, J. D. *Nat. Rev. Drug Disc.* **2003**, *2*, 1019–1025.
27. Clardy, J.; Walsh, C. *Nature* **2004**, *432*, 829–837.
28. Martin, V. J. J.; Pitera, D. J.; Withers, S. T.; Newman, J. D.; Keasling, J. D. *Nat. Biotechnol.* **2003**, *21*, 796–802.
29. Szegedi, A; Kohnen, R; Dienel, A; Kieser, M. *Br. Med. J.* **2005**, *330*, 503.
30. Lu, Q. Y.; Jin, Y. S.; Pantuck, A.; Zhang, Z. F.; Heber, D. *J. Clin. Cancer Res.* **2005**, *11*, 1675–1683.
31. Heacock, P. M.; Hertzler, S. R.; Williams, J. A.; Wolf, B. W. *J. Am. Diet. Assoc.* **2005**, *105*, 65–71.
32. Weiner, N. In *The Pharmacological Basis of Therapeutics*, 7th ed.; Goodman, L. S., Gilman, A. G., Eds.; Macmillan: London, 1985, pp 145–214.
33. Black, J. W.; Stephenson, J. S. *Lancet* **1962**, *2*, 311.
34. Ban, N.; Nissen, P.; Hansen, J.; Moore, P. B.; Steitz, T. A. *Science* **2000**, *289*, 905–920.
35. Schluenzen, F. *Cell* **2000**, *102*, 615.
36. Wimberley, B. T. *Nature* **2000**, *407*, 327–339.
37. Carter, A. P. *Nature* **2000**, *407*, 340–348.
38. Palczewski, K. *Science* **2000**, *289*, 739–745.
39. Doyle, D. A. *Science* **1998**, *280*, 69–77.
40. Irth, H.; Long, S.; Schenk, T. *Curr. Drug Disc.* **2004**, *8*, 19–23.
41. van Elswijk, D. A.; Irth, H. *Phytochem. Rev.* **2003**, *1*, 427–439.
42. Jain, A. N. *Curr. Opin. Drug Disc. Dev.* **2004**, 7, 396–403.
43. Blake, J. F.; Laird, E. R. *Annu. Rep. Med. Chem.* **2003**, *38*, 305–314.
44. Lipinsky, C. A.; Lombardo, F.; Dominy, B. W.; Feeney, P. J. *Adv. Drug Deliv. Rev.* **1997**, *23*, 3–25.
45. Drews, J. *Nat. Biotechnol.* **1996**, *14*, 1516–1518.
46. Gavin, A.-C.; Bösche, M.; Krause, R.; Grandi, P.; Marzioch, M.; Bauer, A.; Schultz, J.; Rick, J. M.; Michon, A. M.; Cruciat, C. M. et al. *Nature* **2002**, *415*, 141–147.
47. Golden, J. *Curr. Drug Disc.* **2003**, 7, 17–20.
48. Henning, S. W.; Beste, G. *Curr. Drug Disc.* **2002**, May, 17–21.
49. Kramer, R.; Cohen, D. *Nat. Rev. Drug Disc.* **2004**, *3*, 965–972.
50. Zambrowicz, B. P.; Sands, A. T. *Nat. Rev. Drug Disc.* **2003**, *2*, 38–51.
51. Zambrowicz, B. P.; Turner, C. A.; Sands, A. T. *Curr. Opin. Pharmacol.* **2003**, *3*, 563–570.
52. Reichert, J. M.; Paquette, C. *Curr. Opin. Mol. Ther.* **2003**, *5*, 139–147.
53. Reichert, J. M. *Reg. Affairs J. Pharma.* **2004**, *15*, 491–497.
54. Pavlou, A. K.; Reichert, J. M. *Nat. Biotechnol.* **2004**, *22*, 1513–1519.
55. Whelan, J. *Drug Disc. Today* **2005**, *10*, 301.
56. Wurm, F. M. *Nat. Biotechnol.* **2004**, *22*, 1393–1398.
57. Reichert, J.; Pavlou, A. K. *Nat. Rev. Drug Disc.* **2004**, *3*, 383–384.
58. Hericourt, J.; Richet, C. *C. R. Hebd. Seanc. Acad. Sci. Paris* **1895**, *121*, 567.
59. Köhler, G.; Milstein, C. *Nature* **1975**, *256*, 495–497.
60. McCafferty, J. *Nature* **1990**, *348*, 552–554.
61. Guerrant, R. L.; Kosek, M.; Moore, S.; Lorntz, B.; Brantley, R.; Lima, A. A. *Arch. Med. Res.* **2002**, *33*, 351–355.
62. Guerrent, R. L.; Kosek, M.; Lima, A. A. M.; Lotntz, B.; Guyatt, H. *Trends Parasitol.* **2002**, *18*, 191–193.
63. Grabowski, H.; Vernon, J.; DiMasi, J. A. *PharmacoEconomics* **2002**, *20*, 11–29.
64. *Biocentury* **2005**, March 28, lead article.
65. Klausner, R. D. *Nat. Rev. Drug Disc.* **2004**, *3*, 470.
66. Roses, A. D.; Burns, D. K.; Chissoe, S.; Middleton, L.; Jean, P. S. *Drug Disc. Today* **2005**, *10*, 177–189.
67. Fishman, M. C. *Nat. Rev. Drug Disc.* **2004**, *3*, 292.
68. Booth, B.; Zemmel, R. *Nat. Rev. Drug Disc.* **2004**, *3*, 451–456.
69. Gilbert, J. *In Vivo* **2003**, Nov. 73–80.
70. Austen, M.; Dohrmann, C. *Drug Disc. Today* **2005**, *10*, 275–282.

71. Hart, P. *Drug Disc. Today* **2005**, *10*, 513–519.
72. Brown, D.; Supert-Furga, G. F. *Drug Disc. Today* **2003**, *8*, 1067–1077.
73. Sams-Dodd, F. *Drug Disc. Today* **2005**, *10*, 139–147.
74. Kola, I.; Landis, J. *Nat. Rev. Drug Disc.* **2004**, *3*, 711–715.
75. Booth, B.; Zemmel, R. *Nat. Rev. Drug Disc.* **2004**, *3*, 451–456.
76. Reichert, J. M.; Paquette, C. *Curr. Opin. Mol. Ther.* **2003**, *5*, 139–147.
77. *The Scientist* **2004**, 24.
78. Lohmueller, K. E. *Nat. Genet.* **2003**, *33*, 177–182.
79. Ioannidid, J. P. *Nat. Genet.* **2001**, *29*, 306–309.
80. Roses, A. D.; Burns, D. K.; Chissoe, S.; Middleton, L.; Jean, P. S. *Drug Disc. Today* **2005**, *10*, 177–189.
81. Hirschhorn, J. *Genet. Med.* **2002**, *4*, 45–61.
82. Allen, M. J.; Carey, A. H. *Drug Disc. Today* **2004**, *3*, 183–190.
83. Helgadottir, A.; Manolescu, A.; Thorleiffson, S. *Nat. Genet.* **2004**, *36*, 233–239.
84. Tabor, H. K. *Nat. Rev. Genet.* **2002**, *3*, 391–397.
85. Sachidanandam, R. *Nature* **2001**, *409*, 928–933.
86. Ring, H. Z.; Kroetz, D. L. *Pharmacogenomics* **2002**, *3*, 47–56.
87. Allen, M. J.; Carey, A. H. *Drug Disc. Today* **2004**, *3*, 183–190.
88. Drews, J.; Ryser, S. *Nat. Biotechnol.* **1997**, *15*, 1318–1319.
89. Drews, J. *Science* **2000**, *287*, 1960–1964.
90. Walke, D. W. *Curr. Opin. Biotechnol.* **2001**, *12*, 626–631.
91. Bailey, D.; Zanders, E.; Dean, P. *Nat. Biotechnol.* **2001**, *19*, 207–209.
92. Hopkins, A. L.; Groom, C. R. *Nat. Rev. Drug Disc.* **2002**, *1*, 727–730.
93. Jackson, P. D.; Harrington, J. J. *Drug Disc. Today* **2005**, *10*, 53–60.
94. Vanhecke, D.; Janitz, M. *Drug Disc. Today* **2005**, *10*, 205–212.
95. Chen, X.; Dudgeon, N.; Shen, L.; Wang, J. H. *Drug Disc. Today* **2005**, *10*, 587–593.
96. Beal, J. *Drug Disc. Today* **2005**, *10*, 169–172.
97. Phillips, K. A.; Veenstra, D. L. *J. Am. Med. Assoc.* **2001**, *286*, 2270–2279.
98. Plant, N. *Drug Disc. Today* **2005**, *10*, 462–464.
99. *Biocentury* **2005**, *13*, A1–A9.
100. Ashburn, T. T.; Thor, K. B. *Nat. Rev. Drug Disc.* **2004**, *3*, 673–683.
101. Austen, M.; Dohrmann, C. *Drug Disc. Today* **2005**, *10*, 275–282.
102. Lazarou, J.; Pomeranz, B. H. *J. Am. Med. Assoc.* **1998**, *279*, 1200–1205.
103. Hall, C. *Daily Telegraph,* 2 March 2001.
104. Lewin, D.; Weiner, M. P. *Drug Disc. Today* **2004**, *9*, 976–983.
105. Bodovitz, S.; Patterson, S. *Drug Disc. World* **2003**, 67–78.
106. Knowles, J. K. C. *Nat. Rev. Drug Disc.* **2004**, *3*, 822.
107. Roses, A. D. *Nat. Rev. Drug Disc.* **2004**, *3*, 382.
108. Roses, A. D. *Nat. Rev. Drug Disc.* **2002**, *1*, 541–549.
109. Lesko, L.; Woodcock, J. *Nat. Rev. Drug Disc.* **2004**, *3*, 763–769.
110. Lesko, L. J.; Woodcock, J. *Pharmacogenomics* **2002**, *2*, 20–24.
111. Lesko, L. J. *J. Clin. Pharmacol.* **2002**, *43*, 342–358.
112. Salemo, R. A.; Lesko, L. J. *Pharmacogenomics* **2004**, *5*, 25–30.
113. Salemo, R. A.; Lesko, L. J. *Pharmacogenomics* **2004**, *5*, 503–505.
114. International Human Genome Sequencing Consortium. *Nature* **2001**, *409*, 860–921.
115. International Human Genome Sequencing Consortium. *Nature* **2004**, *431*, 931–945.
116. Venter, J. C. *Science* **2001**, *291*, 1304–1351.
117. She, X. *Nature* **2004**, *431*, 927–930.

Biography

David Brown, PhD, FRSC, serves as Managing Director of Alchemy Biomedical Consulting Ltd, providing services to the pharmaceutical and biotechnology industries from his base in Cambridge, UK. He has over 30 years experience in the pharmaceutical/biotechnology industry both in research and in senior executive roles. He served with four of the top ten pharma companies: Zeneca, Pfizer, Glaxo, and Hoffman La Roche; and also as President, CEO, and board member of Cellzome AG, a biotechnology company headquartered in Heidelberg, Germany. At Roche, Dr Brown was Global Head of Drug Discovery, based in Basel, Switzerland, and in this role he had responsibility for the output of clinical candidate drugs from approximately 2000 scientists across Roche's five research sites.

Dr Brown has extensive experience of pharmaceutical R&D in all phases from early discovery through clinical trials, and he has experience of leading research in all major disease areas. While at Pfizer he was named coinventor on the patent for Viagra, a treatment for male impotence, and he led the team that developed Viagra through to proof of concept (clinical efficacy) in humans. He was also instrumental in the discovery of Relpax, a treatment for migraine. At Roche, in addition to his role in leading Global Drug Discovery, Dr Brown also served on the committee responsible for clinical drug development and he was a core member of the Business Development Committee responsible for in-licensing of products and for technology agreements and acquisitions. Since 2005 Dr Brown has pursued a career as an independent consultant.

Dr Brown received his PhD in chemistry from the University of Bristol, and his BSc in chemistry with first-class honors from the University of East Anglia, where he was named top student in his year. He is an Editor-in-Chief of *Current Opinion in Drug Discovery and Development*, and is author of over 100 primary and review publications, patents, and invited presentations.

Comprehensive Medicinal Chemistry II
ISBN (set): 0-08-044513-6
ISBN (Volume 1) 0-08-044514-4; pp. 321–353

1.08 Natural Product Sources of Drugs: Plants, Microbes, Marine Organisms, and Animals

G M Cragg and D J Newman, NCI, Frederick, MD, USA

Published by Elsevier Ltd.

1.08.1 Introduction

Throughout the ages humans have relied on nature for such basic needs as the production of foodstuffs, shelter, clothing, means of transportation, fertilizers, flavors and fragrances, and, not least, medicines. Plants have formed the basis of sophisticated traditional medicine systems that have been in existence for thousands of years. Some of the first records, written on hundreds of clay tablets in cuneiform, are from Mesopotamia and date from about 2100 BC. Among the approximately 1000 plant-derived substances used at that time were oils of *Cedrus* species (cedar) and *Cupressus sempevirens* (cypress), *Glycyrrhiza glabra* (licorice), *Commiphora* species (myrrh), and *Papaver somniferum* (poppy juice), all of which are still in use today for the treatment of ailments ranging from coughs and colds to parasitic infections and inflammation.[1,2] Egyptian medicine dates from about 2900 BC, but the best-known Egyptian pharmaceutical record is the '*Ebers Papyrus*' dating from 1500 BC; this documents over 700 drugs (mostly plants, though animal organs were included together with some minerals) and includes formulas such as gargles, snuffs, poultices, infusions, pills, and ointments, with beer, milk, wine, and honey being commonly used as vehicles.

The Chinese materia medica has been extensively documented over the centuries,[3] with the first record dating from about 1100 BC (*Wu Shi Er Bing Fang*, containing 52 prescriptions), followed by works such as the *Shennong Herbal* (~100 BC; 365 drugs) and the *Tang Herbal* (659 AD; 850 drugs). Likewise, documentation of the Indian Ayurvedic system dates from before 1000 BC (*Charaka* and *Sushruta Samhitas* with 341 and 516 drugs, respectively).[4,5] In the ancient Western world, the Greeks contributed substantially to the rational development of the use of herbal drugs. The philosopher and natural scientist, Theophrastus (~300 BC), in his *History of Plants*, dealt with the medicinal qualities of herbs, and noted the ability to change their characteristics through cultivation. Dioscorides, a Greek physician (100 AD), during his travels with Roman armies throughout the then 'known world,' accurately recorded the collection, storage, and use of medicinal herbs, and is considered by many to be the most important representative of the science of herbal drugs in 'ancient times.' Galen (130–200 AD), who practiced and taught pharmacy and medicine in Rome, and published no less than 30 books on these subjects, is well known for his complex prescriptions and formulas used in compounding drugs. These were based on the Hippocratic theory that all illnesses arise from an imbalance of four primary 'humors'; such formulas were sometimes extremely complex and contained dozens of ingredients (the so-called 'galenicals').

During the Dark and Middle Ages, from roughly the fifth to the twelfth centuries, the monasteries in countries such as England, Ireland, France, and Germany preserved the remnants of this Western knowledge. However, it was the Arabs who were responsible for the preservation of much of the Greco-Roman expertise, and for expanding it to include the use of their own resources, together with Chinese and Indian herbs unknown to the Greco-Roman world. The Arabs were the first to establish privately owned drug stores in the eighth century, and the Persian pharmacist, physician, philosopher, and poet, Avicenna, contributed much to the sciences of pharmacy and medicine through works such as *Canon Medicinae*, which is regarded as 'the final codification of all Greco-Roman medicine.' This was subsequently succeeded by the comprehensive compilation known as the *Corpus of Simples* by Ibn al-Baitar who practiced in Malaga during the Moorish occupation of Spain. This document combined the data of Dioscorides with works from the Middle and Far East.[349]

These, and many other works, were formally codified at least in the UK by the publication in 1618 of the *London Pharmacopoeia* and the idea of 'pure' compounds as drugs may be traced to the isolation of the active principles of commonly used plants and herbs, such as quinine, strychnine, morphine, atropine, and colchicine in the early 1800s. These isolations were then followed by the isolation of morphine (generally considered the first commercial pure natural product) by E. Merck in 1826 and aspirin, the first semisynthetic pure drug based on a natural product, by Bayer in 1899.[6,349]

1.08.2 The Role of Traditional Medicine in Drug Discovery

The use of plants in the traditional medicine systems of many cultures has been extensively documented (*see* 1.09 Traditional Medicines).[7,8] These plant-based systems continue to play an essential role in healthcare, and it has been estimated by the World Health Organization that approximately 80% of the world's inhabitants rely mainly on traditional medicines for their primary healthcare.[9] Plant products also play an important role in the healthcare systems of the remaining 20% of the population, mainly residing in developed countries. An analysis of data on prescriptions dispensed from community pharmacies in the US from 1959 to 1980 indicated that about 25% contained plant extracts or active principles derived from higher plants, and at that time at least 119 chemical substances, derived from 90 plant species, could be considered as important drugs in use in one or more countries.[9] Of those 119 drugs, 74% were discovered as a result of chemical studies directed at the isolation of the active substances from plants used in traditional medicine. A more recent study using US-based prescription data from 1993, demonstrated that natural products were still playing a major role in drug treatment, as over 50% of the most-prescribed drugs in the US had a natural product either as the drug, or as a 'forebear' in the synthesis or design of the agent.[10]

1.08.3 Natural Product Drug Discovery and Development: A Multidisciplinary Process

The continuing valuable contributions of nature as a source of potential chemotherapeutic agents has been reviewed by Newman *et al.*[11] An analysis of natural products as sources of new drugs over the period 1981–2002 indicates that 67% of the 877 small molecule, new chemical entities (NCEs) are formally synthetic, but 16.4% correspond to synthetic molecules containing pharmacophores derived directly from natural products.[12] Furthermore, 12% are actually modeled on a natural product inhibitor of the molecular target of interest, or mimic (i.e., competitively inhibit) the endogenous substrate of the active site, such as ATP. Thus, only 39% of the 877 NCEs can be classified as truly synthetic in origin (**Figure 1**). In the area of anti-infectives (antibacterial, antifungal, antiparasitic and antiviral), close to 70% are naturally derived or inspired, while in the area of cancer treatment 67% fall into this category.

In recent years, there has been a steady decline in the output of the research and development programs of the pharmaceutical industry, and the number of new active substances, or NCEs, hit a 20-year low of 37 in 2001.[13] Further evidence of this fall in productivity is evident from the report that only 16 new drug applications were received by the US Food and Drug Administration (FDA) in 2001, down from 24 applications the previous year. This downturn was attributed to various factors, but it is significant that the past 10–15 years have seen a decline in interest in natural products on the part of major pharmaceutical companies in favor of reliance on new chemical techniques such as combinatorial chemistry for generating molecular libraries. The realization that the number of NCEs in drug development pipelines is declining may have led to the recent interest in 'rediscovering natural products.'[14]

As stated by one authority commenting on natural products:

> ...we would not have the top-selling drug class today, the statins; the whole field of angiotensin antagonists and angiotensin-converting-enzyme inhibitors; the whole area of immuno-suppressives; nor most of the anticancer and antibacterial drugs. Imagine all of those drugs not being available to physicians or patients today.

It is clear that nature has played, and will continue to play, a vital role in the drug discovery process.

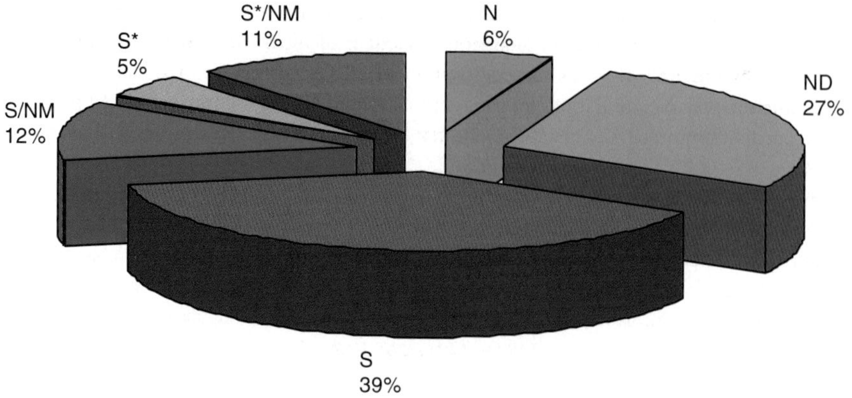

Figure 1 Sources of new chemical entities 1981–2002 ($n = 877$). N, natural product; ND, derived from a natural product and is usually a semisynthetic modification; S, synthetic drug, often found by random screening/modification of existing agent; S*, made by synthesis, but the pharmacophore is/was from a natural product; NM, natural product mimic (where the compound, though nominally synthetic, is a direct competitor of the normal substrate).

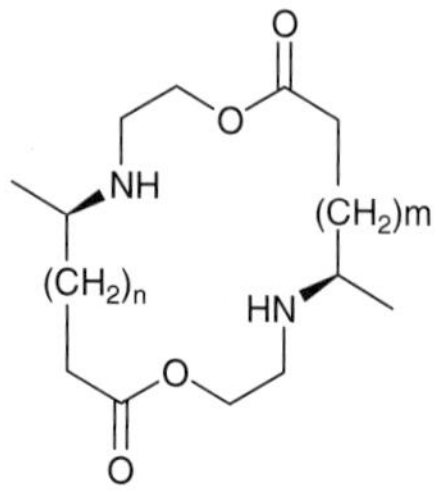

1

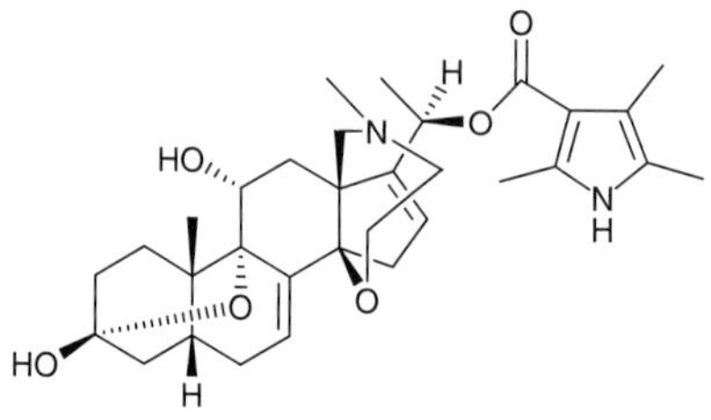

2

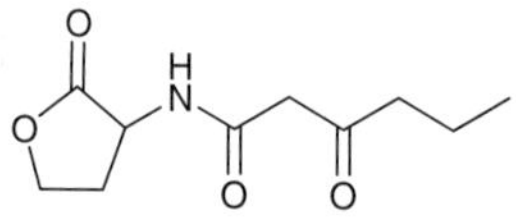

3

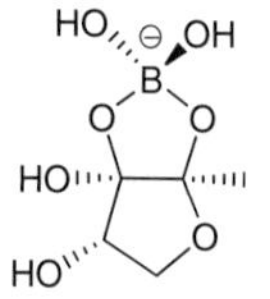

4

Figure 2 Some of nature's defensive molecules.

Before discussing the application of natural products as leads for the development of drugs for the treatment of human diseases, it is instructive briefly to consider natural product sources and some of the advances that have occurred in the approaches to natural product-derived drug discovery. While space does not permit a detailed discussion of these issues, interested readers are referred to the excellent articles and reviews that are cited for more comprehensive discussions. In addition, examples of the applications of these advances will be discussed in sections dealing with specific diseases.

1.08.3.1 Why Natural Products?

While the contributions of natural secondary metabolites to modern medicine are abundantly clear, the reasons for the production of these inherently biologically active compounds by organisms are still debated. Initially they were regarded as waste products, but it seems reasonable to assume that, in many instances, the production of these complex and often toxic chemicals has evolved over eons as a means of chemical defense by essentially stationary organisms such as plants and many marine invertebrates, against predation and consumption (e.g., herbivory).

For instance, pupae of the coccinellid beetle, *Epilachna borealis*, secrete droplets from their glandular hairs containing a library of hundreds of large-ring (up to 98 members) macrocyclic polyamines (see **Figure 2** for the simplest example with the generic formula **1**), which appear to act as a chemical defensive mechanism against predators.[15] These libraries are built up from three simple (2-hydroxyethylamino)-alkanoic acid precursors, and are clear evidence that combinatorial chemistry has been pioneered and widely used in nature for the synthesis of biologically active compound libraries.

It is interesting to note that some forms of chemical defense have been shown to be acquired through the diet of the host macroorganisms. Thus, the neurotoxic steroidal alkaloids, the batrachotoxins (**2**; **Figure 2**), first isolated from poison-dart frogs (genus *Phyllobates*) in Colombia, and later from certain passerine birds of Papua New Guinea, have been found in beetles of the genus *Choresine* belonging to the cosmopolitan Melyridae family.[16] Analysis of the stomach contents of *Pitohuii* birds in Papua New Guinea revealed these beetles in their diet, and related species could be the source of these alkaloids in the Colombian frogs. Likewise, the anticancer agent dolastatin 10, first isolated from the mollusk *Dolabella auricularia* in minute yields, has been isolated from a marine cyanobacterium of the genus *Simploca* known to be grazed on by the mollusk.[17] The microbial influence is further highlighted by the observation that analogs of the potent guanidinium alkaloids of the tetrodotoxin class are found in frogs of the genus *Atelopus* found in Costa Rica and Panama, but are also isolated from marine bacteria (*Moraxella*), anaerobic bacteria (*Enterobacter* and *Klebsiella* species), and freshwater cyanobacteria (e.g., *Anabaena* and *Lyngbya* species).[18] Thus, symbiotic/commensal bacteria are possibly responsible for the production of these alkaloids in the frogs, especially in view of the fact that frogs raised in captivity appear to have no detectable presence of the alkaloids.

Microorganisms are also reported to produce and excrete antimicrobial toxins as a means of killing sensitive strains of the same or related species (cf. Czaran *et al.*[19] and references cited therein). This is similar to allelopathy in which plants release toxic compounds in order to suppress the growth of neighboring plants.[20,21] Bacteria also control their density of population growth and so-called biofilm formation through a cell-to-cell signaling mechanism known as quorum-sensing involving the excretion of quorum-sensing compounds. The best studied of these are the acyl homoserine lactones (AHLs), with the compounds from *Vibrio fisheri* being examples, *N*-3-oxohexanoyl-l-homoserine lactone (**3**; **Figure 2**), and a previously unidentified furanone boronate diester that appears to be a universal signal (**4**; **Figure 2**); they signal the activation of genes promoting virulence, spore formation, biofilm formation, and other phenomena.[22,23]

Natural products may be used for purposes of both predation and defense. Thus, species of the cone snail genus *Conus* stun their prey prior to capture by injection of venom composed of combinatorial libraries of several hundred peptides[24]; the venom may also be used for defense against predators. The same may apply to snake and at least some arachnid venoms, which rather than using relatively small peptides contain complex mixtures of proteins, which may well be evolutionary variants of proteins that originally had important functions in other animals, but when modified and expressed as venom components, exert lethal effects when injected into their prey.[25] Thus, variants of the ancestral liver enzymes, factors X and V from chickens, which aid in blood clotting in animals, when injected into prey (humans and other animals) by snake bite, lead to millions of microscopic clots that consume key blood factors, leading to profuse bleeding within the prey.

1.08.3.2 Classical Natural Sources: Untapped Potential

Despite the intensive investigation of terrestrial flora, it is estimated that only 5–15% of the approximately 300 000 species of higher plants have been systematically investigated, chemically and pharmacologically,[26,27] while the potential of the marine environment as a source of novel drugs remains virtually unexplored.[28]

1.08.3.3 The Unexplored Potential of Microbial Diversity

Until recently, microbiologists were greatly limited in their study of natural microbial ecosystems due to an inability to cultivate most naturally occurring microorganisms. It has been estimated that less than 1% of microorganisms seen microscopically have been cultivated, and 'a handful of soil contains billions of microbial organisms.'[29] Adding to this observation, the assertion that 'the workings of the biosphere depend absolutely on the activities of the microbial world,'[30] provides a clear indication that the microbial universe is a vast untapped resource for drug discovery. In addition, advances in the understanding of the gene clusters that encode the multimodular enzymes, such as polyketide synthases (PKSs) and/or nonribosomal peptide synthetases (NRPSs), both involved in the biosynthesis of a multitude of microbial secondary metabolites, has enabled the detailed analysis (genome mining) of the genomes of long-studied microbes such as *Streptomyces avermitilis*. These studies have revealed the presence of additional PKS and NRPS clusters for as yet unidentified secondary metabolites not detected in standard fermentation isolation processes.[31] Such analyses have led to the prediction and isolation of a novel antifungal agent from *Streptomyces aizunensis*.[32]

1.08.3.3.1 Improved culturing procedures

Recent developments of procedures for cultivating and identifying microorganisms are aiding microbiologists in their assessment of the earth's full range of microbial diversity. Application of a technique using 'nutrient-sparse' media under conditions simulating the original natural environment, for the massive parallel cultivation of gel-encapsulated single cells (gel microdroplets; GMDs) derived from microbes separated from environmental samples (sea water and soil)[33] 'permits the simultaneous and relatively noncompetitive growth of both slow- and fast-growing microorganisms.' This process, a modification of the late 1980s technique pioneered by a small, now defunct biotechnology company known as 'One-Cell Systems,' prevents the overgrowth by fast-growing 'microbial weeds,' and has resulted in the identification of previously undetected species (using 16S ribosomal RNA (rRNA) gene sequencing), and the culturing and scale-up cultivation of previously uncultivated microbes.

1.08.3.3.2 Extraction of environmental samples

Procedures based on the extraction of nucleic acids (the metagenome) from environmental samples permits the identification of uncultured microorganisms through the isolation and sequencing of rRNA or rDNA (genes encoding for rRNA). Samples from soils and seawater are currently being investigated,[34,35] and in a study by Craig Venter *et al.*, whole-genome shotgun sequencing of environmental-pooled DNA obtained from water samples collected in the Sargasso Sea off the coast of Bermuda, indicated the presence of at least 1800 genomic species, which included 148 previously unknown bacterial phylotypes.[35] These methods may be applied to other habitats, such as the microflora of insects and marine animals, and there is a recent report of an 'Air Genome Project' being launched in Manhattan, New York where samples of air are being analyzed for content of DNA from bacteria, fungi, and other microbes.[36]

Valuable products and information are certain to result from the cloning and understanding of the novel genes discovered through these processes. Heterologous expression of gene clusters encoding the enzymes involved in biosynthetic pathways in viable host organisms, such as *Escherichia coli*, should permit the production of novel metabolites produced from as yet uncultured microbes. A recent example of heterologous expression of genomic DNA is the production of the antibiotic pantocin A (**5**; **Figure 3**) from the bacterium *Pantoea agglomerans*.[37] Low titers and the complexity of the mixture of metabolites produced made production of pantocin A by the microbe grown in liquid culture impractical; however, expression of a genomic DNA library from *P. agglomerans* in *E. coli* provided access to reasonable quantities of the small molecule antibiotics of interest.

1.08.3.3.3 Extremophiles

Extreme habitats harbor a host of extremophilic microbes (extremophiles), such as acidophiles (acidic sulfurous hot springs), alkalophiles (alkaline lakes), halophiles (salt lakes), piezophiles (barophiles) and thermophiles (deep-sea vents),[38–40] and psychrophiles (arctic and antarctic waters, alpine lakes).[41] While investigations thus far have focused on the isolation of thermophilic and hyperthermophilic enzymes (extremozymes),[42–46] these extreme environments will also undoubtedly yield novel bioactive chemotypes.

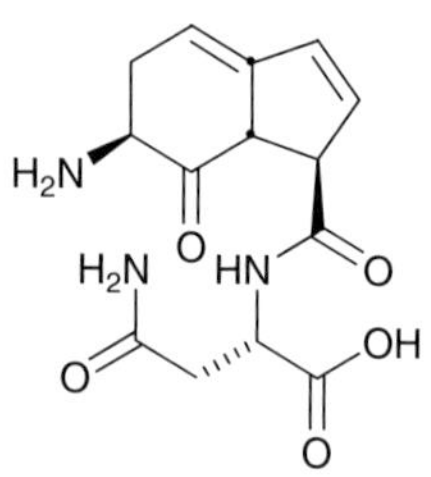

5

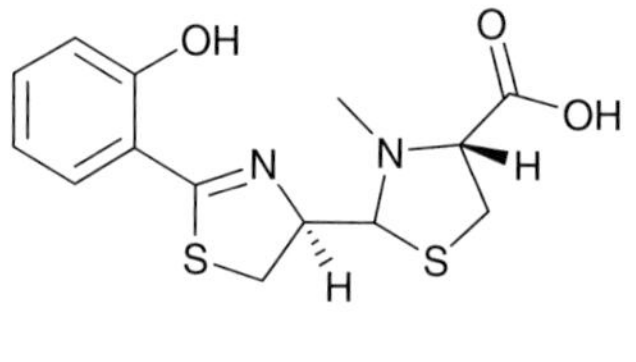

6

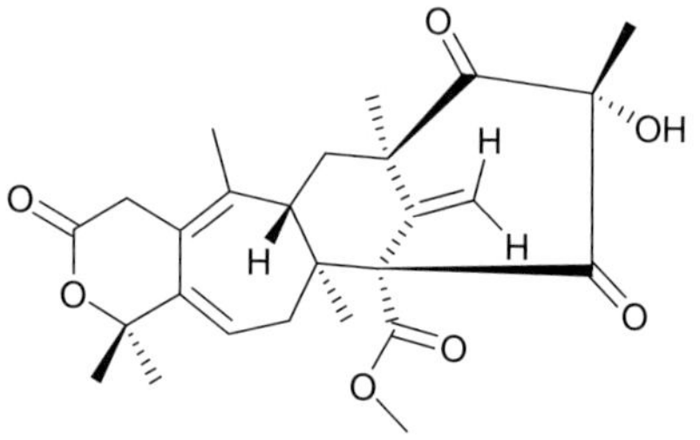

7

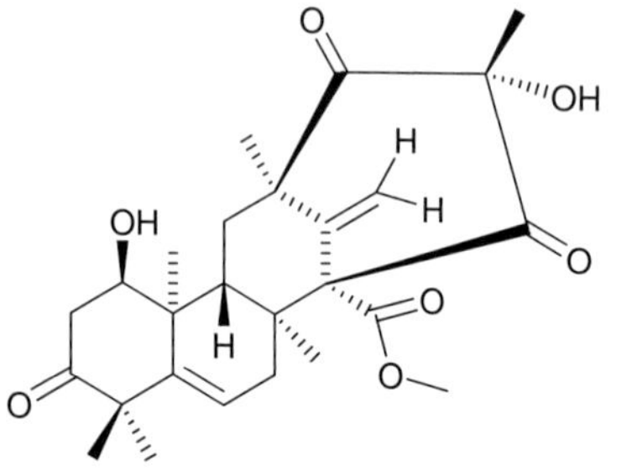

8

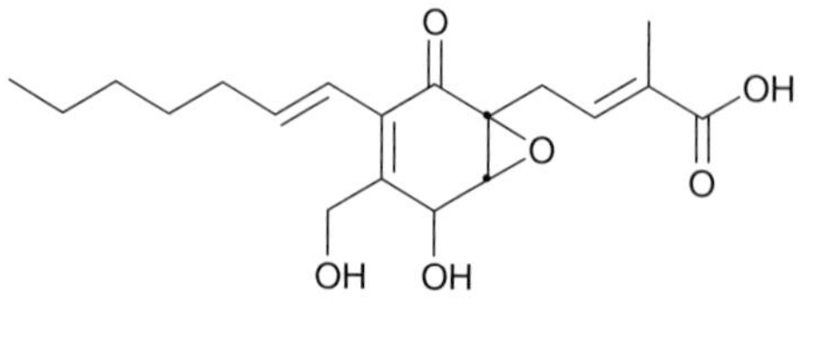

9

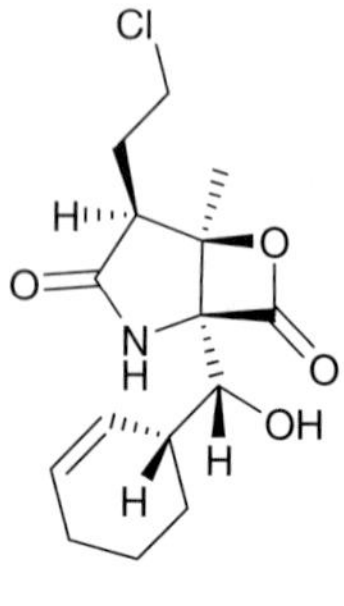

10

Figure 3 Products of extremophiles, free and epiphytic microbes.

In this respect, antifungal screening of extracts of 217 extremophiles isolated from diverse locales has yielded pyochelin (**6**; **Figure 3**), a known compound, as the antifungal active agent isolated from a novel thermophilic *Pseudomonas* species.[47] An unusual group of acidophiles that thrive in acidic, metal-rich waters has been found in abandoned mine-waste disposal sites, polluted environments that are generally toxic to most prokaryotic and eukaryotic organisms.[48] *Penicillium* species have been isolated from the surface waters of Berkeley Pit Lake in Montana, and have yielded the novel sesquiterpenoid and polyketide-terpenoid metabolites berkeleydione (**7**; **Figure 3**) and berkeleytrione (**8**; **Figure 3**) showing activity against metalloproteinase-3 and caspase-1, activities relevant to cancer, Huntington's disease, and other diseases.[49,50]

1.08.3.3.4 Endophytes

While plants have received extensive study as sources of bioactive metabolites, the endophytic microbes that reside in the tissues between living plant cells have received scant attention. The relationship established between the endophytes and their host plants may vary from symbiotic to pathogenic, and limited studies have revealed an interesting realm of novel chemistry.[51,52] Among the new bioactive molecules discovered are: novel wide-spectrum antibiotics, kakadumycins, isolated from an endophytic *Streptomycete* associated with the fern leafed grevillea (*Grevillea pteridifolia*) from the Northern Territory of Australia[53]; ambuic acid (**9**; **Figure 3**), an antifungal agent, which has been recently described from several isolates of *Pestalotiopsis microspora*, found in many of the world's rainforests[54]; and peptide antibiotics, the coronamycins, from a *Streptomyces* species associated with an epiphytic vine (*Monastera* species) found in the Peruvian Amazon.[55]

1.08.3.3.5 Marine sediments

Recent research has revealed that deep ocean sediments are a valuable source of new actinomycete bacteria that are unique to the marine environment. Based on combined culture and phylogenetic approaches, the first truly marine actinomycete genus named *Salinospora* has been described.[56] Members of the genus are ubiquitous, and are found in sediments on tropical ocean bottoms and in more shallow waters, often reaching concentrations of up to 10^4 per mL of sediment. They also appear on the surfaces of numerous marine plants and animals. They can be cultured using the appropriate selective isolation techniques, and significant antibiotic and cytotoxic activity has been observed, leading to the isolation of the very potent cytotoxin salinosporamide A (**10**; **Figure 3**), which is a very strong proteasome inhibitor ($IC_{50} = 1.3$ nM).[57]

1.08.3.3.6 Microbial symbionts

There is mounting evidence that many bioactive compounds isolated from various macroorganisms are actually metabolites synthesized by symbiotic bacteria.[58] These include the anticancer maytansanoids, originally isolated from several plant genera of the Celastraceae family, the pederin class of antitumor compounds isolated from beetles of genera *Paederus* and *Paederidus*, which have also been isolated from several marine sponges,[59–61] and a range of antitumor agents isolated from marine organisms that closely resemble bacterial metabolites. Some of these will be discussed in more detail in subsequent sections.

1.08.3.4 Combinatorial Biosynthesis

Advances in the understanding of bacterial aromatic polyketide biosynthesis have led to the identification of multifunctional polyketide synthase enzymes (PKSs) responsible for the construction of polyketide backbones of defined chain lengths, the degree and regiospecificity of ketoreduction, and the regiospecificity of cyclizations and aromatizations, together with the genes encoding for the enzymes.[31,62,63] Likewise, nonribosomal peptide synthases (NRPSs) are responsible for the biosynthesis of nonribosomal peptides (NRPs).[31] The rapidly increasing analysis of microbial genomes has led to the identification of a multitude of gene clusters encoding for polyketides, NRPs, and hybrid polyketide-NRP metabolites, providing the tools for engineering the biosynthesis of novel 'nonnatural' natural products through gene shuffling, domain deletions, and mutations.[31,64] The recent review on the combinatorial biosynthesis of anticancer natural products by Shen *et al.* gives examples of novel analogs of anthracyclines, ansamitocins, epothilones, enediynes, and aminocoumarins produced by combinatorial biosynthesis of the relevant biosynthetic pathways.[65]

A recent example of the power of this technique when applied to natural products is the development of an efficient method for scale-up production of epothilone D (**11**; **Figure 4**), currently undergoing clinical trials as a potential anticancer agent. Epothilone D is the most active of the epothilone series isolated from the myxobacterium

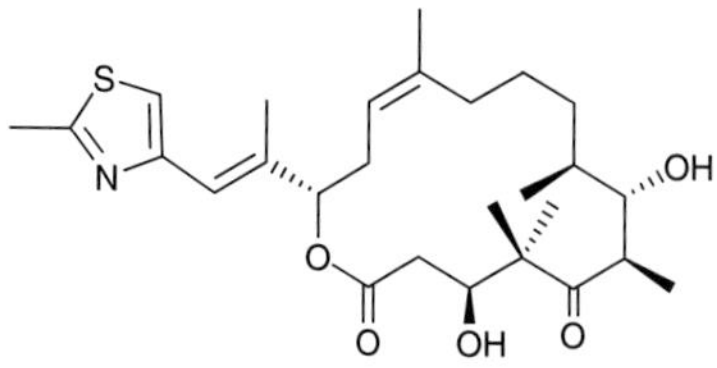

11

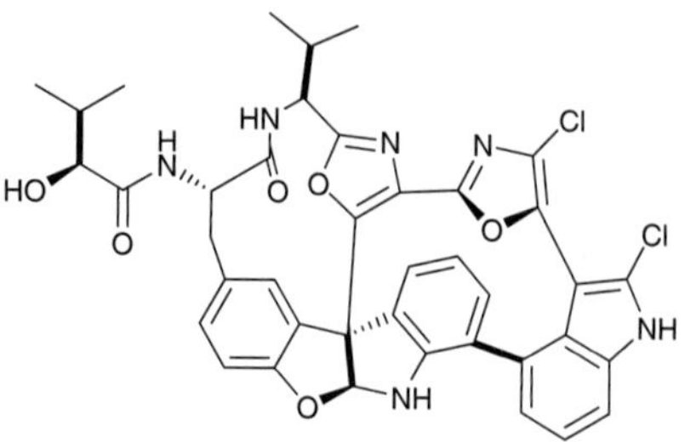

12

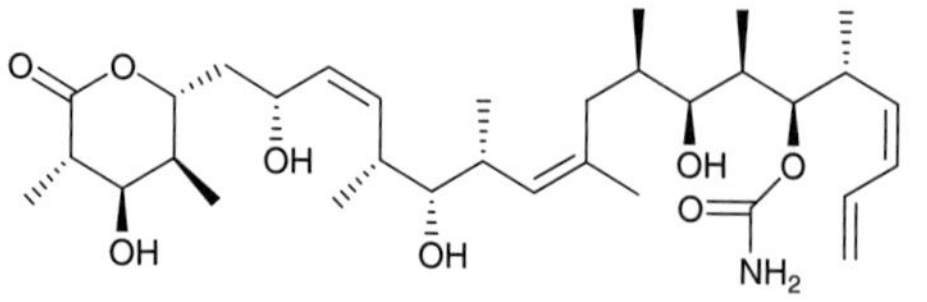

13

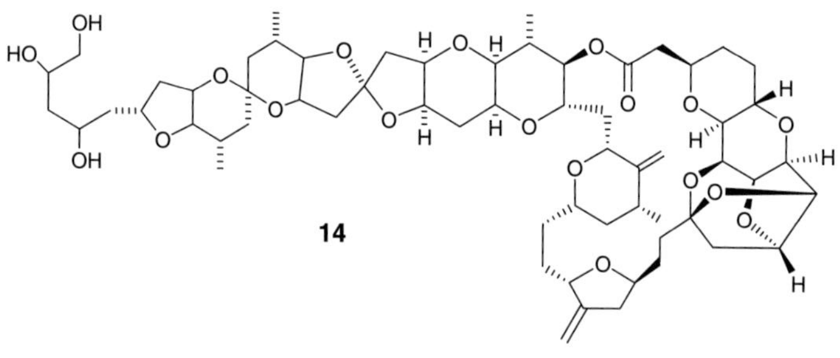

14

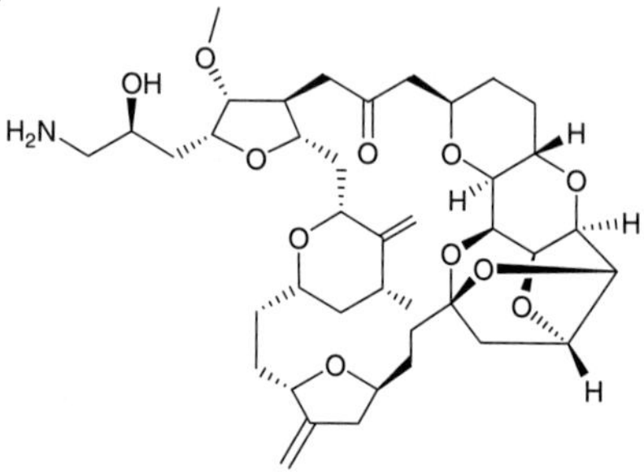

15

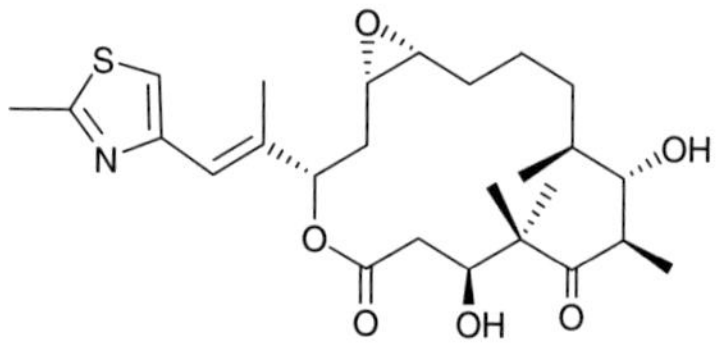

16

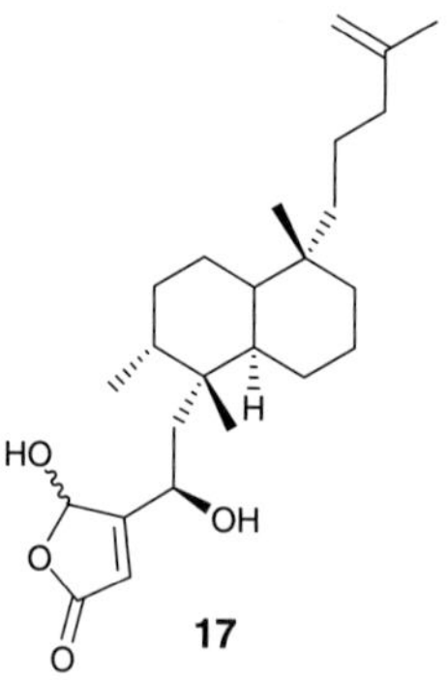

17

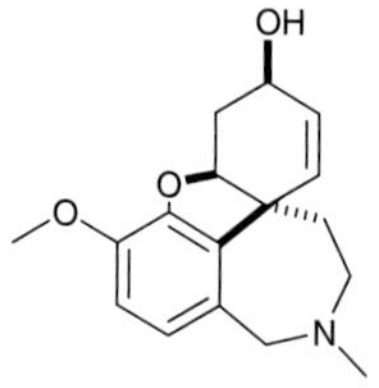

18

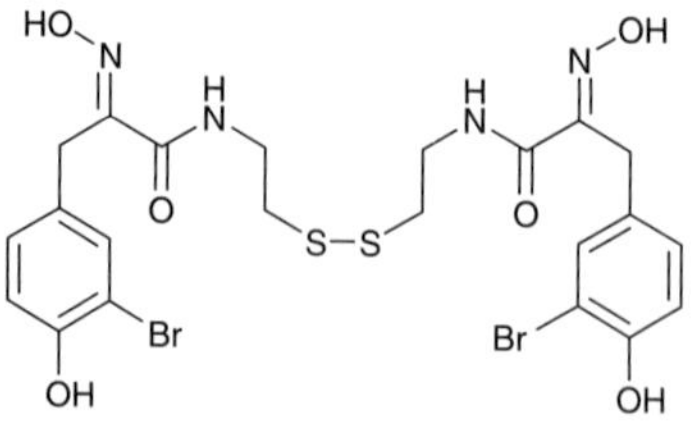

19

Figure 4 Products of total synthesis, combinatorial biosynthesis, and diversity-oriented synthesis.

Sorangium cellulosum, and is the des-epoxy precursor of epothilone B (*see* Section 1.08.8). The isolation and sequencing of the polyketide gene cluster producing epothilone B from two *S. cellulosum* strains has been reported,[66,67] and the role of the last gene in the cluster, epoK, encoding a cytochrome P450, in the epoxidation of epothilone D to epothilone B has been demonstrated. Heterologous expression of the gene cluster minus the epoK into *Myxococcus xanthus* resulted in large-scale production of crystalline epothilone D.[68]

1.08.3.5 Total Synthesis

The total synthesis of complex natural products has long posed challenges to the top synthetic chemistry groups worldwide, and has led to the discovery of many novel reactions and to developments in chiral catalytic reactions.[69] In a number of instances, the synthesis of natural products whose structures had been previously published has revealed errors in their structural elucidation and has led to the correct structural assignment[70]; a notable example discussed by Nicolaou and Snyder is that of the marine-derived antitumor compound, diazonamide A (**12**; **Figure 4**).

The efforts of some groups have been focused on the synthesis and modification of drugs that are difficult to isolate in sufficient quantities for development, such as the marine-derived anticancer agent discodermolide (**13**; **Figure 4**) (*see* Section 1.08.8.1).[71–76] In the process of total synthesis, it is often possible to determine the essential features of the molecule necessary for activity (the pharmacophore), and, in some instances, this has led to the synthesis of simpler analogs having similar or better activity. A notable example is that of the marine-derived antitumor agent halichondrin B (**14**; **Figure 4**) (*see* Section 1.08.8), where, following the report of the synthesis of both halichondrin B and norhalichondrin B in 1992,[77] the synthetic schemes were utilized to synthesize a large number of variants of halichondrin B, particularly smaller molecules that maintained the biological activity but were intrinsically more chemically stable. One of the compounds, E7389 (**15**; **Figure 4**), is now in phase II clinical trials.[78]

1.08.3.6 Diversity-Oriented Synthesis, Privileged Structures, and Combinatorial Chemistry

The human genome, as well as advances in the description of the genomes of pathogenic microbes and parasites,[79] is permitting the determination of the structures of many of the proteins associated with disease processes. With the development of these new molecular targets there is an increasing demand for novel molecular diversity for screening. Combinatorial chemistry is a technique originally developed for the synthesis of large chemical libraries for high-throughput screening against such targets,[80] which coupled to the development of robotic systems and tools, such as solid-phase synthesis and new immobilization strategies involving novel resins, reagents, and linkers, has permitted high-throughput parallel approaches to the synthesis of very large libraries (millions) of compounds. While there are claims that new leads are being found,[80] the declining numbers of new NCEs[13] indicate that the use of de novo combinatorial chemistry approaches to drug discovery over the past decade have been disappointing, with some of the earlier libraries being described as "poorly designed, impractically large, and structurally simplistic."[80]

As stated in this chapter,[80] "an initial emphasis on creating mixtures of very large numbers of compounds has largely given way in industry to a more measured approach based on arrays of fewer, well-characterized compounds" with "a particularly strong move toward the synthesis of complex natural-productlike compounds – molecules that bear a close structural resemblance to approved natural-product-based drugs."

The importance of natural productlike scaffolds in the generation of meaningful combinatorial libraries has been further emphasized in a recent article by Borman entitled 'Rescuing Combichem. Diversity-oriented synthesis (DOS) aims to pick up where traditional combinatorial chemistry left off.'[81] In this chapter it is stated that 'the natural productlike compounds produced in DOS have a much better shot at interacting with the desired molecular targets and exhibiting interesting biological activity.' The synthesis of natural productlike libraries is exemplified by the work of the Schreiber group who have combined the simultaneous reaction of maximal combinations of sets of natural productlike core structures (latent intermediates) with peripheral groups (skeletal information elements) in the synthesis of libraries of over 1000 compounds bearing significant structural and chiral diversity.[82,83]

Over the last few years, detailed analyses of active natural product skeletons have led to the identification of relatively simple key precursor molecules that form the building blocks for use in combinatorial synthetic schemes that have produced a number of potent molecules, thereby enabling structure–activity relationships (SARs) to be probed. Several reviews[84–90] discuss small libraries built through the solid-phase synthesis of molecules such as epothilone A (**16**; **Figure 4**), dysidiolide (**17**; **Figure 4**), galanthamine (**18**; **Figure 4**), and psammaplin (**19**; **Figure 4**). Thus, in the study of the SARs of the epothilones, solid-phase synthesis of combinatorial libraries was used to probe regions of the molecule important to retention or improvement of activity.[91]

The combinatorial approach, using an active natural product as the central scaffold, can also be applied to the generation of large numbers of analogs for structure–activity studies, the so-called parallel synthetic approach as with the syntheses around the sarcodictyin (**20**; **Figure 5**) scaffold by Nicolaou *et al.*[92] The importance of natural products as leads for combinatorial synthetic approaches is embodied in the concept of 'privileged structures,' originally

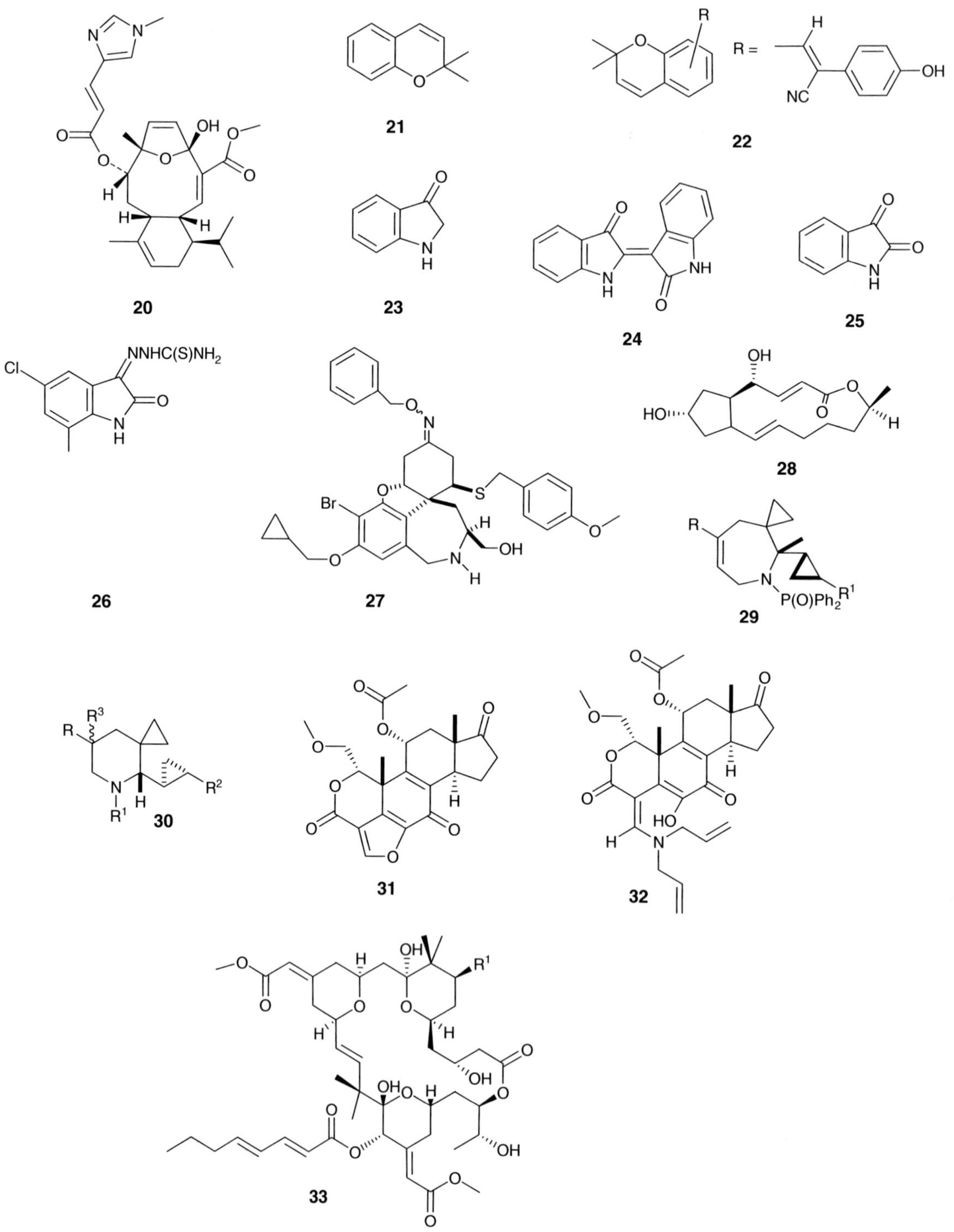

Figure 5 Privileged structures and chemical genetics.

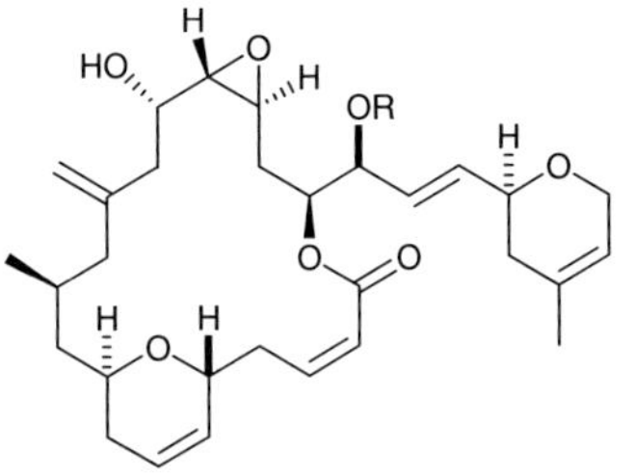

34: R = H
45: R = CH_3

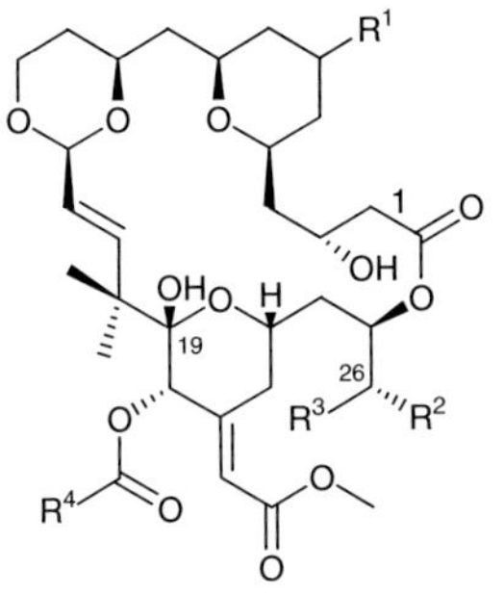

Compound	R^1	R^2	R^3	R^4
35	H	OH	CH_3	C_7H_{15}
39	H	OH	H	C_7H_{15}

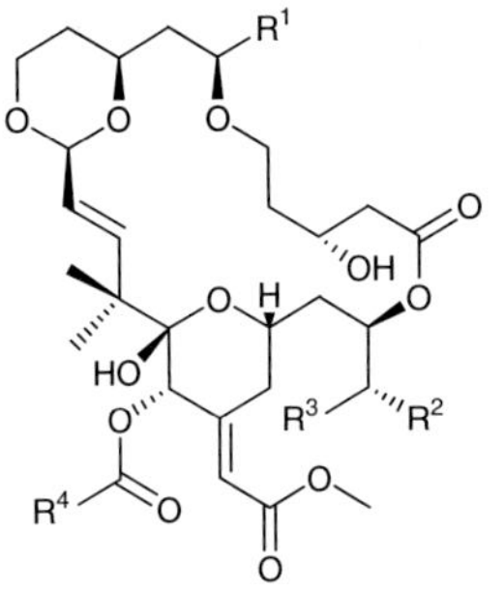

Compound	R^1	R^2	R^3	R^4
36	*t*-Bu	OH	CH_3	C_7H_{15}
37	H	OH	CH_3	C_7H_{15}

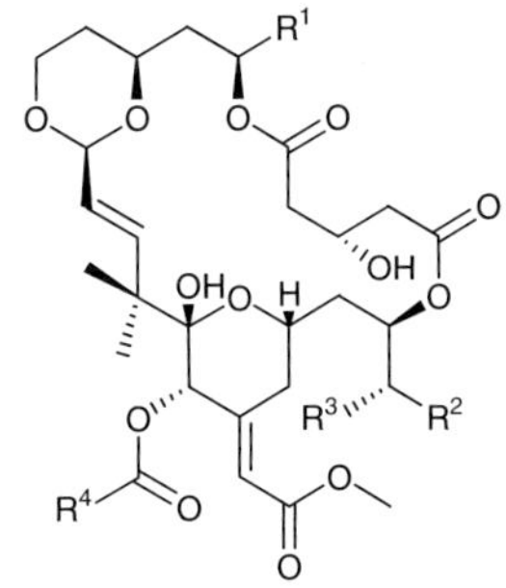

Compound	R^1	R^2	R^3	R^4
38	$C(CH_3)_2C_2H_5$	OH	CH_3	C_7H_{13}

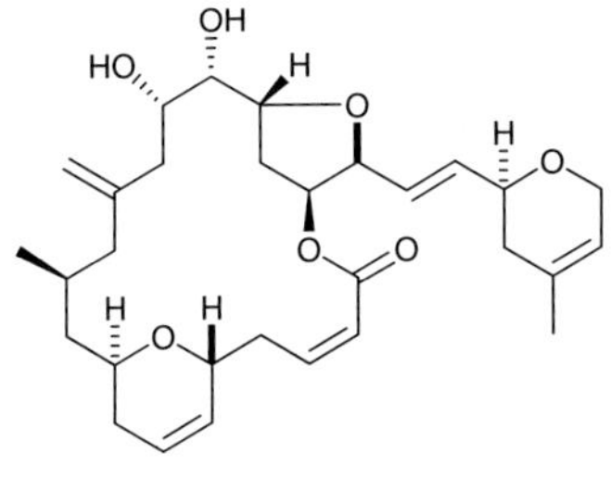

40

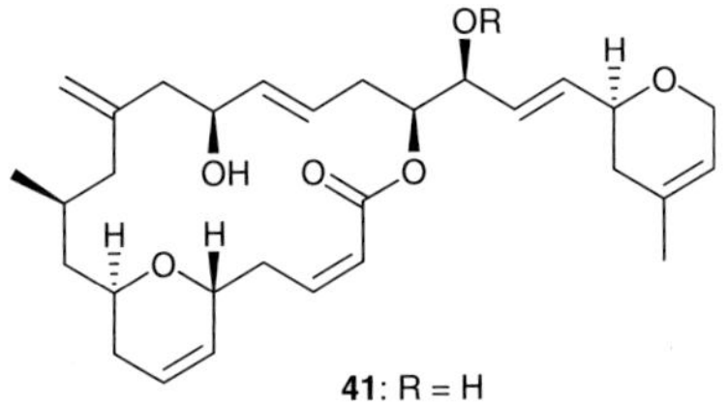

41: R = H
44: R = CH_3

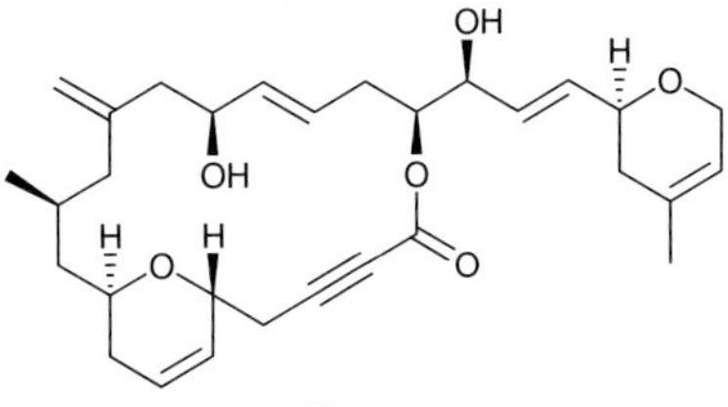

42

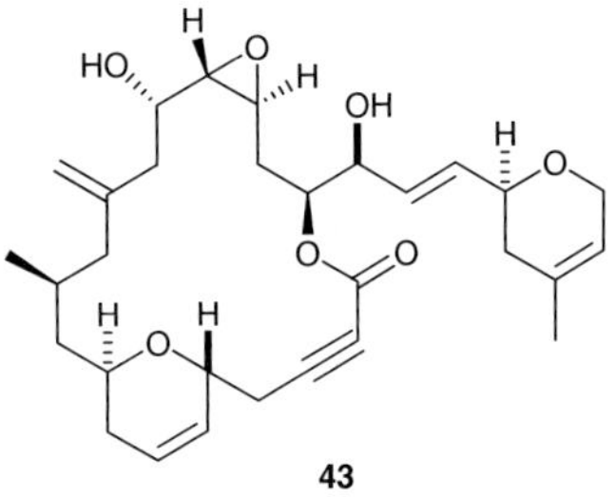

43

Figure 5 Continued

proposed by Evans *et al.*[93] and then advanced recently by Nicolaou *et al.*[94–96] and the Waldmann group.[87,90] Nicolaou[94] stated the underlying thesis as follows:

> "We were particularly intrigued by the possibility that using scaffolds of natural origin, which presumably have undergone evolutionary selection over time, might confer favorable bioactivities and bioavailabilities to library members."

A search of the literature yielded nearly 4000 2,2-dimethyl-2*H*-benzopyran moieties (**21**; **Figure 5**), with another 8000 structures identified through the inclusion of a slight modification of the search. Nicolaou's group then proceeded to develop the necessary solid-phase synthetic methods by modifying a reagent that they had reported in the literature a couple of years earlier, a polystyrene-based selenyl bromide resin.[97] Application of this methodology has led to the identification and subsequent optimization of benzopyrans with a cyanostilbene substitution (**22**; **Figure 5**) that are effective against vancomycin-resistant bacteria, together with modifications of other natural product structures that have generated leads to both antimicrobial agents and inhibitors of mammalian cell processes.[98–103]

Indole derivatives such as indoxyl (**23**; **Figure 5**), indirubin (**24**; **Figure 5**), and isatin (**25**; **Figure 5**) are also 'privileged' scaffolds with examples being the synthesis of thiosemicarbazone derivatives (**26**; **Figure 5**) of isatin[104] that have activity against parasitic cysteine proteases identified in trypanosomes and the malarial parasites, and derivatives of indirubins as cyclin and glycogen synthase kinase-3 (GSK-3) inhibitors extensively studied by Meijer and collaborators (discussed at length by Kingston and Newman[105] and Newman and Cragg[106] in recent reviews).

1.08.3.7 Chemical Genetics: Natural Products as Molecular Probes

The approach of probing complex biological processes by altering the function of proteins through binding with small molecules has been called chemical genetics.[107] This technique has been applied to the modification[108] (by Shair's group) of the natural product galanthamine (**18**; **Figure 4**) using combinatorial techniques, and then assaying the products using a novel screen that looked for inhibition of protein trafficking in the Golgi apparatus. They identified a modified nucleus derived from galanthamine, which they named secramine (**27**; **Figure 5**), that inhibited this process. The only other known agents to do this were based on the microbial product brefeldin (**28**; **Figure 5**), an entirely different structure in a formal sense.

The original concept of using small synthetic molecules has probably been overtaken by a combination of what is known today as diversity-oriented synthesis (DOS, see above) and use of natural product skeletons (NPSs). The latter, the coupling of the synthetic processes to a known active structure, currently seems to be the most-effective method, though synthetic chemists are still producing novel structures that have natural product characteristics with multiple stereocenters and heterocyclic atoms, in contrast to those initially used by Schreiber, with the recent publication from the Wipf group on azaspirocycles (**29**, **30**; **Figure 5**) being an excellent example.[109] Some examples of compounds based on natural product skeletons have been given above but another example would be the modification of wortmannin (**31**; **Figure 5**) by Wipf's group at the University of Pittsburgh in collaboration with Powis' group at the Univerity of Arizona, leading to PX-866 (**32**; **Figure 5**), a novel inhibitor of phosphoinositide-3-kinase (PI3K) signaling, that is currently under preclinical evaluation by ProlX Pharmaceuticals.[110,111] Further examples are the multiple compounds reviewed by Lyon *et al.* derived from microbial and marine sources (some of which might well have involved microbial commensals in their production) that have both general and specific activities against dual specificity phosphatases and have potential as leads to antitumor therapies.[112]

In addition to these formalized compounds based on NPSs there is another interesting source of compounds that is based upon the natural product but is a spin-off from the more classical methods of synthesis of a compound. We have already alluded to the production of E7389 (the halichondrin B-based Eisai compound), but there are at least two series of compounds available based upon active marine-sourced natural products that are simpler renditions of the base molecules and whose biological activities are under investigation in a variety of assay systems.

These are the so-called 'bryologs' based on the well-known *Bugula neritina* metabolite bryostatin 1 (**33**; **Figure 5**) and the 'laulilogs' based on laulimalide (an antimitotic from Pacific marine sponges; **34** in **Figure 5**), synthesized by Wender and colleagues.

1.08.3.7.1 Bryostatin

In 1986, Wender *et al.* analyzed the potential binding site of the phorbol esters on protein kinase C (PKC) with the aim of designing simpler analogs of these agents.[113] In 1988, the work was expanded by modeling bryostatin 1 onto the same site as initial results showed that bryostatin 1 interacted with PKC.[114] From these results, further refinements led to the production of three simpler analogs (**35**, **36**, and **37**; **Figure 5**) that maintained the putative binding sites at the oxygen atoms at C_1 (ketone), C_{19} (hydroxyl), and C_{26} (hydroxyl) in bryostatin. These molecules demonstrated nanomolar binding constants when measured in displacement assays of tritiated phorbol esters, with the figures being in the same general range as bryostatin 1, and analogs **35** and **36** had activities in in vitro cell line assays close to those demonstrated by bryostatin 1 itself.[115–118] With this data in hand, a second lactone was introduced giving compound **38** (**Figure 5**), with 8 nM binding affinity and an ED_{50} of 113 nM against P388.[119] When different fatty acid esters were made of the base analog (**35**), they also exhibited binding affinities for PKC isozymes in the 7–232 nM range depending upon the fatty acid used.[120] Recently, Wender *et al.* published a simple modification where by removal of a methyl group in the C_{26} side chain from compound **35** to produce **39** (**Figure 5**), the binding affinity to PKC was increased to the picomolar level[121]; the compound demonstrated greater potency than bryostatin 1 in in vitro cell line assays. Finally, improved methods of synthesis of the molecules have been published, which might well permit further exploration of these analogs.[122,123]

1.08.3.7.2 Laulimalide

The cytotoxic agents laulimalide (**34**) and isolaulimalide (**40**; **Figure 5**) were reported by a variety of investigators in the late 1980s but it was not until 1999 that Mooberry *et al.*[124] reported that these compounds demonstrated paclitaxel-like activity and that this was probably the reason for the cytotoxicities reported. Although a microtubule-stabilizing agent, work from Hamel's group at NCI[125], indicated that laulimalide, the more potent compound, might well bind at a site different from the taxanes, and in late 2003, Mooberry *et al.*[126] reported that laulimalide, like other microtubule stabilizers, has an additional mechanism independent of mitotic arrest whereby G_1 aneuploid cells are formed due to aberrant mitotic events at 5–7 nM; these concentrations are approximately 30% of those required for mitotic accumulation. Very recent work has suggested yet another effect of this agent: a combination of 100 fM each of laulimalide and taxotere inhibited HUVEC migration by 70%, which is at least two orders of magnitude below the IC_{50} values of 10 pM seen for each agent when used alone.[127]

Over 10 synthetic routes to laulimalide have been published, plus many more that give methods of synthesis of 'subassemblies' of the overall molecule. For a thorough discussion of the results of these endeavors, readers are referred to the excellent synthetic paper from Multzer's group[128] and the 2003 review by Multzer and Ohler.[129] Subsequent papers have reported both the synthesis of the base molecule by Uenishi and Ohmi,[130] and then work on modified molecules (mainly 11-desmethyl variants) based on modeling studies by Paterson *et al.* and subsequent syntheses by the same group,[131,132] with similar molecules being reported from investigators at the Eisai Research institute in the US in a similar time frame.[133,134]

Finally, as mentioned earlier, in 2003 Wender's group reported[135] the syntheses of five laulimalide analogs (laulilogs), with modifications in the epoxide, the C_{20} alcohol, and the C_1–C_3 enoate of the base natural product structure (**41–45**; **Figure 5**). These were poor substrates for the phosphoglycoprotein (P-gp) that is a major mechanism of drug resistance in mammalian cells, and thus these compounds were active against paclitaxel-resistant cell lines. Further biological data was given in an extended report in 2004.[136]

1.08.4 Role of Natural Products in Treatment of Diseases

The following general areas of disease are covered in subsequent sections: microbial and some parasitic areas; neoplastic; cardiovascular and hypertension; and pain/CNS. They will be discussed from the perspective of the disease rather than the more customary division by source of organism. This approach was chosen because of the crossover that is now being seen where agents originally isolated and purified as a result of their ability to inhibit or modulate one type of biological activity are now being utilized in other entirely different areas.

Perhaps the best example of this crossover is the work undertaken over the last 30 or so years on the microbial product rapamycin (**46**; **Figure 6**). Originally isolated as an antifungal agent, it was approved as the immunosuppressive drug Sirolimus in 1999 and a derivative is now in trials as an antineoplastic agent. Thus, the use of a disease area permits agents from all sources to be discussed in a group. For earlier discussions of the role of natural products in drug discovery readers should consult the comprehensive reviews by Newman *et al.*[11] and Buss and Waigh.[137]

1.08.5 Anti-Infective Agents (Including Antimalarials)

We have not attempted to make the following discussion comprehensive in its coverage of the various anti-infective classes. The guiding principle has been to show areas where the natural product has led to "improved materials by semisynthetic modifications of a base molecule that a medicinal chemist would not even dream of making." Rather than a discussion of the 'conventional antibiotics,' which have been covered in detail in many recent and not so recent reviews, we will cover only synthetic work related to the vancomycins and erythromycins, methods generally used to produce modified agents that have activity against resistant organisms. It is clear, however, that natural products have played a dominant role in the development of anti-infective agents, with only some 30% being truly synthetic in origin.[12] Interestingly, of the two major synthetic classes, the azole-based antifungals and the quinolone-based antibacterials, the origin of the latter class can be traced back to the large-scale synthesis of the antimalarial agent chloroquin (a synthetic mimic of quinine), where by-products of the synthesis based on oxoquinolines serendipitously were shown to possess antibacterial properties.[138]

A recent review (March, 2005) by Butler[139] has thoroughly covered the antibacterial and antifungal agents based upon natural product skeletons that are either in clinical trials or have been approved since the late 1990s. In addition to Butler's excellent review, readers interested in some of the history surrounding antibiotics should consult the article by Buss and Waigh,[137] the historical coverage given by Newman *et al.*[11] through to 1999, and the 1998 review by Shu,[140] the precursor to Butler's review. Therefore, we will not go into any significant details on antibacterial and antifungal drugs except to mention some specific molecules as exemplars of what may be feasible using modern synthetic and biosynthetic techniques.

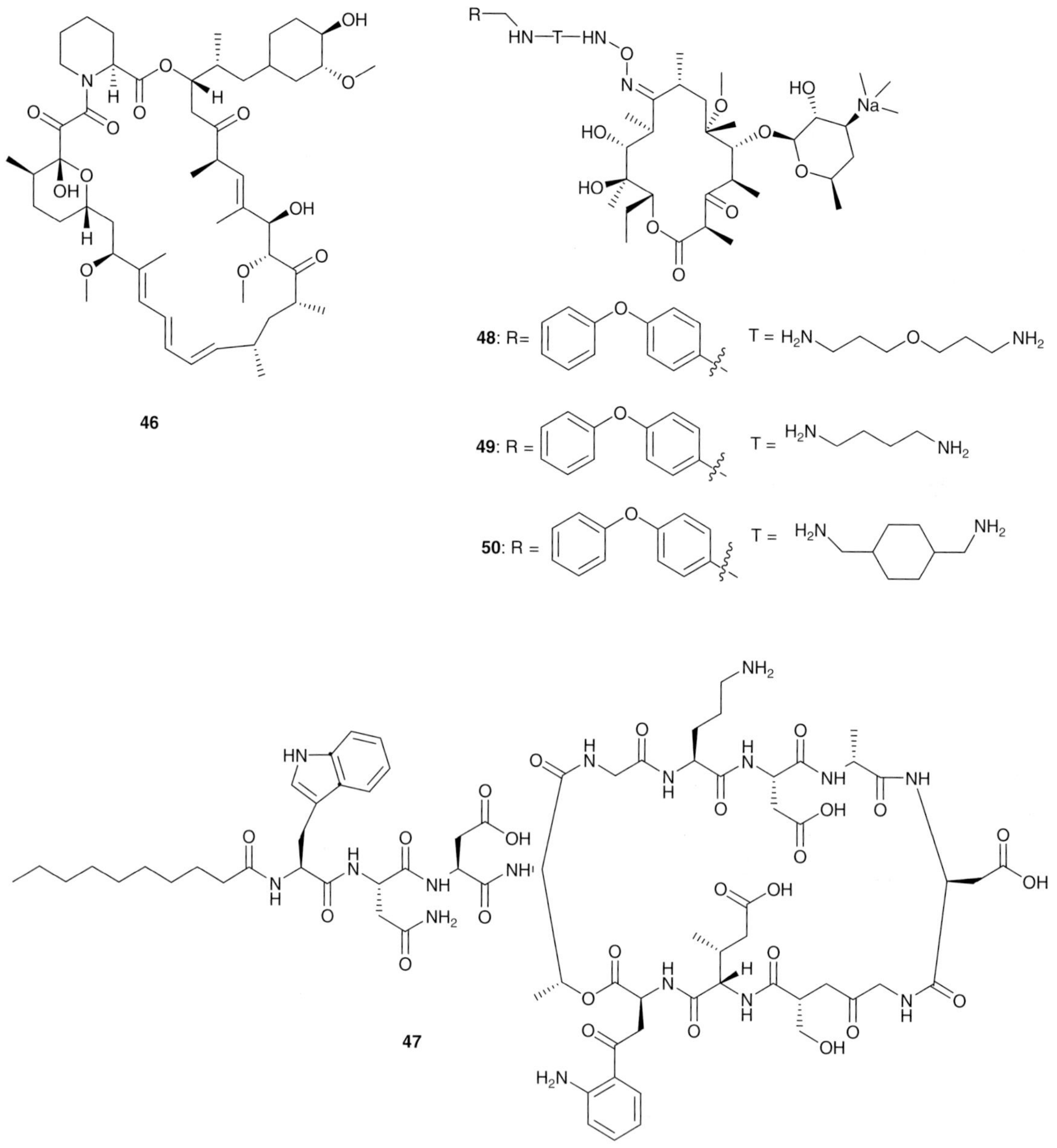

Figure 6 Daptomycin, erythromycin analogs.

1.08.5.1 Antibacterials (*see* Volume 7)

1.08.5.1.1 Lipopeptides: daptomycin

Daptomycin was the first lipopeptide (**47**; **Figure 6**) approved (as Cubicin in 2003) for use in humans against *Staphylococcus aureus* (both methicillin-susceptible and methicillin-resistant strains), a variety of streptococci and vancomycin-sensitive *Enterococcus faecalis*. Its mechanism of action is unique among antibacterials in that insertion of the lipophilic tail in the presence of calcium ions into the membrane structure causes depolarization of the cell membrane and potassium ion efflux, and is bactericidal with treated cells demonstrating >99.9% cell kill within 60 min.[141–144]

The original workers with this compound at Eli Lilly reported in 1998 that they had successfully cloned and mapped the daptomycin gene cluster, then also known as a component of the complex A21978C and as LY146032, prior to its licensing to Cubist Pharmaceuticals.[145] Subsequently, the gene cluster has been worked on by scientists at Cubist and their collaborators and in 2004, Marahiel's group in Germany reported a chemoenzymatic route to daptomycin-like antibiotics by utilizing the recombinant cyclization domain of the *Streptomyces coelicolor* calcium-dependent antibiotic nonribosomal peptide synthetase, to cyclize linear peptide/nonpeptide chains produced by nonenzymatic methods.[146]

1.08.5.1.2 Glycopeptides: vancomycins

Antibiotics of the vancomycin class are enormously important as being the 'antibiotics of last resort' in the treatment of drug-resistant infections; therefore, the development of improved vancomycin analogs is a goal of many workers in the field. Since 1996, at least three groups have reported combinatorial techniques whereby libraries are built up in the presence of the target, with the aim of optimizing the chemistries to initially produce the tightest binding compound(s), a form of chemical genetics. In 1997, Huc and Lehn[147] reported the use of what they called 'dynamic target-driven combinatorial synthesis,' followed 2 years later with the publication of a more complete study.[148] Concomitantly, Fesik's group utilized nuclear magnetic resonance (NMR) to screen for binding to a target, followed by covalent attachment in a separate experiment.[149]

In 2000, Ellman's group reported using an ELISA assay rather than NMR to demonstrate binding, which they followed up with a separate tethering experiment.[150] With vancomycin, they used a split and pool technique to look for synthetic peptidic-based molecules that might bind more avidly to vancomycin-resistant organisms than vancomycin itself. Following the screening of a library containing 39 304 members,[151] they discovered two molecules that bound more avidly to L-Lys-D-Ala-D-Lac than to the normal bacterial tripeptide where there is a terminal D-Ala, but the molecules were not useful as drug leads.

Nicolaou, in an extension of his selenium-based chemistry for construction of polymeric linkers,[97] applied the Lehn concept to the syntheses of tethered vancomycin analogs in the presence of the acetylated tripeptide Ac-Lys-(D)-Ala-(D)-Ala; this tripeptide is the preferred substrate for glycopeptides of the vancomycin type. In a series of three papers, he gave the preliminary results and then the full details of both solution-phase and solid-phase combinatorial syntheses of linked vancomycin dimers[102,152] finishing up with a demonstration of the biological power of the technique by showing activity against Synercid (name of a recent Gram-positive active antibiotic approved for MRSA treatment) and vancomycin-resistant *Enterococcus* species with minimum inhibitory concentrations (MICs) at the $<1\,\mu g\,mL^{-1}$ level.[152] These papers should be consulted for the structural details as they are rather complex.

Recently, workers from China demonstrated that the vancomycin analog *N*-demethyl-vancomycin, which had been used in China since 1967, and whose N-terminal amino acid is leucine rather than *N*-methyl-leucine, can be modified by initially binding the N-terminal to a solid support. This permitted modification of the amino group on the vancosamine using solid-state peptide techniques followed by substitution at the N-terminal leucine if desired. From this work, they reported that substituting the vancosamine with a bulky hydrophobic substituent increased activity against vancomycin-resistant *E. faecalis*.[153]

It will be interesting to follow these chemical modifications and to compare the activities seen with those demonstrated by the newer glycopeptide antibiotics of this class that are now entering the antibacterial arena, both as approved and as clinical candidates. For further information the reader should consult the recent reviews by Butler,[139] Anstead and Owens,[154] and Bradley.[155]

1.08.5.1.3 Macrolides (*see* Volume 7)

Continuous use of macrolide antibiotics has resulted in the development of macrolide-resistant bacteria. In order to address this resistance, combinatorial libraries have been built around the erythromycin and structurally related cores giving a series of tethered molecules based on either the C-3 or the C-9 region of the macrolide ring.[156] An extension of the structures to include oxime analogs based on the C-9 position gave a series of compounds of which at least three (**48**, **49**, and **50**; **Figure 6**) have demonstrated promising activity against resistant strains of *S. aureus* and *Streptomyces pneumoniae*.

1.08.5.2 Antifungals: Nonlipopeptides (*see* Volume 7)

Using a high-throughput screen directed against protein synthesis in *Candida*, Glaxo-Wellcome discovered that an analog of the previously known molecule sordarin (**51**; **Figure 7**) was an effective in vitro inhibitor. Following a long program of mutation and medicinal chemistry, semisynthetic derivatives of sordarin were produced with the analog GM237354 (**52**; **Figure 7**), which is a selective inhibitior of the yeast elongation factor 2 entering preclinical development.[157–160] Subsequently, following GlaxoSmithKline's decision to stop natural product-based programs, the compounds were licensed in the middle of 2003 to Diversa for further development.

Although farnesyl and geranyl transferases are candidate targets for antitumor therapies, groups in various parts of the world have reported potential leads from both microbial and marine sources that have effects on the fungal

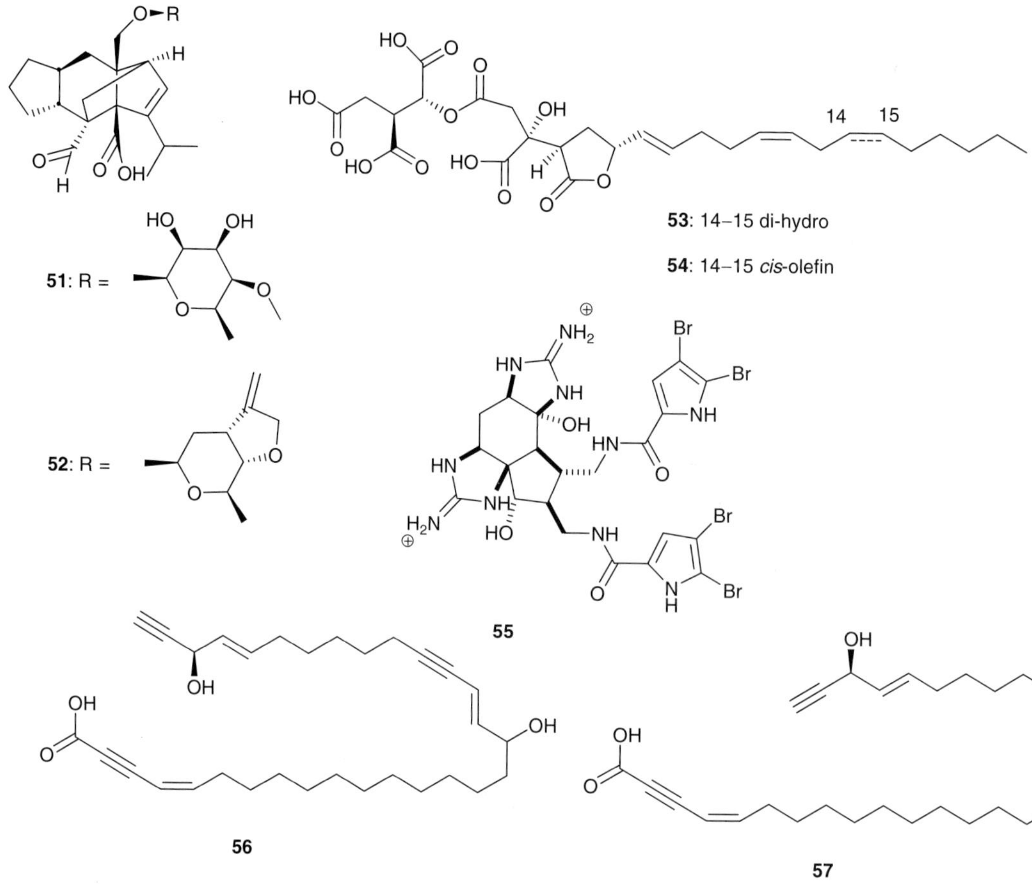

Figure 7 Some novel nonlipopeptide antifungal agents.

geranylgeranyltransferase I (GGTase I) enzyme. Work at Merck led to the isolation of the antifungal metabolites citrafungins A and B (**53**, **54**; **Figure 7**) from a fungus isolated from a cow dung sample collected in Alaska. These compounds showed activity against fungal GGTase I, and moderate to potent activity against various pathogenic fungal species, including *Candida albicans*, *C. neoformans*, and *Aspergillus fumigatus*.[161] Other natural products reported to be GGTase I inhibitors are two marine-derived compounds, the alkaloid massadine (**55**; **Figure 7**)[162] and the polyacetylenic acid derivatives corticatic acids D and E (**56**, **57**; **Figure 7**).[163]

1.08.5.3 Antimalarials (*see* Volume 7)

The quintessential antimalarial lead was quinine. Originally isolated from *Cinchona* bark, it acted as the template for the synthetic agents of the chloroquine/mefloquine type.[137] With the increase in the number of parasites resistant to these agents came the search for other synthetic and natural product-based agents. Inspection of data from Chinese herbal remedies led to the investigation of extracts of *Artemisia annua* (wormwood) as this particular plant had been used for centuries in China as an antimalarial[164] (known as 'Quinghaosu'); in 1972, the active agent was isolated and identified as a sesquiterpene endoperoxide named artemisinin (**58**; **Figure 8**). Tan *et al.*[165] reported that this particular agent is not limited to the species *A. annua*, but is also found in at least two other species. Using the base structure of artemisinin, many semisynthetic compounds were made with the aim of optimizing the pharmacology of the base molecule. This led to the identification of artemether (**59**; **Figure 8**), a relatively simple modification of the actual active metabolite dihydroartemisinin (**60**; **Figure 8**), as a potent antimalarial agent that is now used throughout the world. Other simple modifications have been made to improve solubility and distribution but in all cases, the active constituent ends up being the same molecule, so these modifications can all be considered to be 'prodrugs' of dihydroartemisinin (cf. table IV in De Smet).[166]

Recently, a thorough review of the discovery and development of the artemisinins, as well as their proposed mechanisms of action, has been published by O'Neill and Posner,[167] and it is clear that their activity resides in the

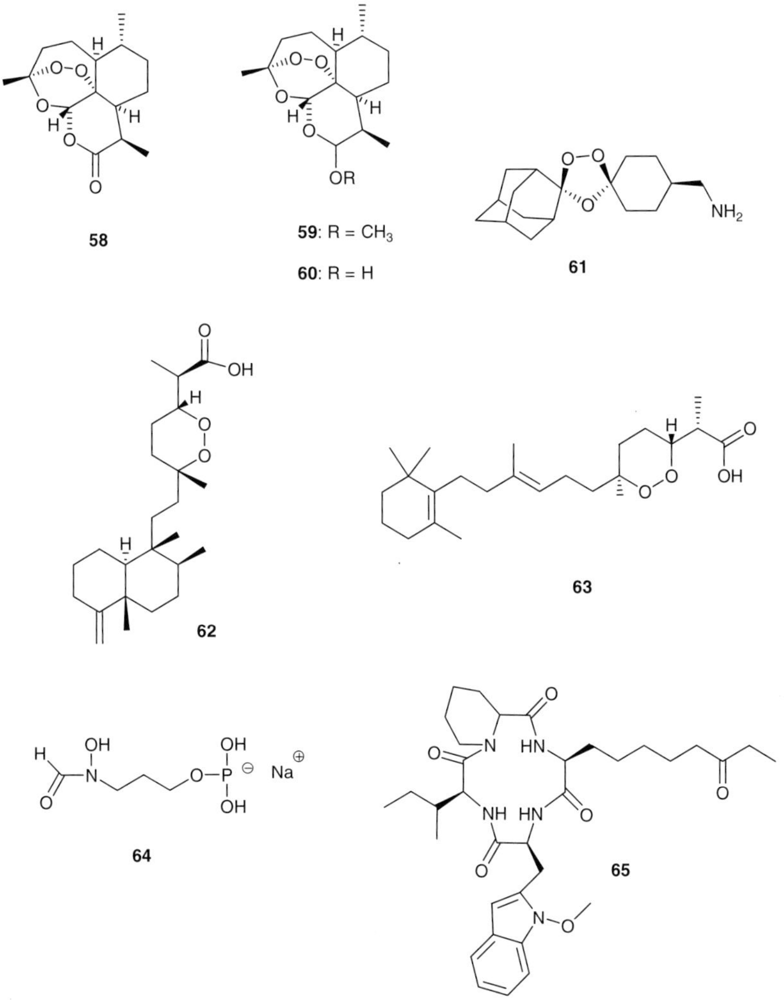

Figure 8 Antimalarials.

peroxy bridge. It is proposed that this is converted to a cytotoxic radical species by ferrous iron formed by the reduction of hematin to heme, and that this radical species then alkylates heme and other proteins that are specific to the malarial parasites (e.g., *Plasmodium falciparum*). With the recognition that the peroxy bridge is essential for activity came the synthesis of a host of analogs incorporating this feature (see review by O'Neill and Posner).[167] A significant development has been the synthesis of ozonides (1,2,4-trioxolanes) incorporating an adamantine ring (**61**; **Figure 8**); these are chemically stable, orally available in rats, and are more potent and possess longer half-lives than the artemisinin derivatives.[168]

Natural products from marine sources also contain the peroxy bridge found in the artemisins; examples are the norsesterterpenes sigmosceptrellin[169] (**62**; **Figure 8**) and muqubilin[170] (**63**; **Figure 8**) from the Red Sea sponges *Sigmosceptrella* and *Prianos*, respectively. Other similar molecules isolated from a South African species of *Plakortis* have been reported by Kashman's group.[171] At that time, the potential of marine organisms as leads to antimalarials was not appreciated, though later work by Hamann's group at the University of Mississippi[172,173] and Konig's group at Braunschwieg[174] has shown that a variety of marine-derived structures may have potential as leads to other antimalarial structures. It will be interesting to see how these molecules act as leads compared to those from plant sources.

Jomaa *et al.*[175] have demonstrated that a very simple series of microbial metabolites, the fosmidomycins (**64**; **Figure 8**), originally reported by Fujisawa[176,177] as *Streptomyces* metabolites, have antimalarial activity both in vitro and in vivo in rodent models. These simple molecules appear to inhibit isoprenoid biosynthesis that is dependent upon the DOXP pathway (from 1-deoxy-D-xylulose 5-phosphate) rather than the more usual HMG-coenzyme A (CoA) reductase route. This discovery of a nonmammalian isoprenoid biosynthetic route in the parasite and the identification of an inhibitor may well open up an entirely new route to antimalarials, particularly as synthetic routes to this type of molecule have been published in detail[178] and they appear to lend themselves to parallel syntheses to produce a wide variety of congeners. The apicidins, a novel class of cyclic tetrapeptides isolated from the fungus *Fusarium pallidoroseum* and related species, have been shown to exhibit potent activity against the protozoal apicomplexan family, including *Plasmodium* spp., *Toxoplasma gondii*, *Cryptosporidium* spp., and *Eimeria* spp.[179] Their activity has been associated with reversible inhibition of histone deacetylase (HDAC), with the parent compound, apicidin (**65**; **Figure 8**), demonstrating in vivo activity against *Plasmodium berghii* in mice.

1.08.5.4 Antivirals

It might be said that there is a paucity of effective antifungal agents, but the antiviral area is positively barren by comparison, in spite of the vast number of chemical compounds, derived from natural sources, semi- or total synthesis, that have been tested for their efficacy as antiviral agents. One major reason for this is the very nature of a viral disease, in that the virus, irrespective of type, effectively takes over an infected cell and hence there are very few specific viral targets for small molecules to interact with. With the advent of molecular cloning techniques, however, the situation is changing as it is now possible to identify specific viral-related proteins, clone them, express them, and then use them in rapid screening systems, looking for specific interactions in the absence of the host proteins.

Rather than deal with each type of viral infection, we have elected to show how a serendipitous discovery of a series of natural products has led to the plethora of similar compounds that are now in preclinical or clinical evaluation and, in some cases, in clinical use.

1.08.5.4.1 Nucleoside analogs

From 1950 to 1956, Bergmann *et al.*[180–182] reported on two compounds they had isolated from marine sponges, spongouridine (**66**; **Figure 9**) and spongothymidine (**67**; **Figure 9**). What was significant about these materials was that they demonstrated for the first time that naturally occurring nucleosides could be found using sugars other than ribose or deoxyribose, thus challenging the then current dogma that only nucleosides with ribose derivatives were bioactive.

These two compounds can be thought of as the prototypes of all of the modified nucleoside analogs made by chemists that have crossed the antiviral and antitumor stages since then. Once it was realized that biological systems would recognize the base and not pay too much attention to the sugar moiety, chemists began to substitute the 'regular pentoses' with acyclic entities, and with cyclic sugars with unusual substituents.

These experiments led to a vast number of derivatives that were tested extensively as antiviral and antitumor agents over the next 30 years or more. Suckling in a 1991 review[183] showed how such structures evolved in the (then) Wellcome laboratories, leading to azidothymidine (AZT) and, incidentally, to Nobel Prizes for Hitchens and Elion, though no direct mention was made of the original arabinose-containing leads from natural sources.

Showing that 'Mother Nature' may follow chemists rather than the reverse, or conversely that it was always there but the natural compound had not, as yet, been isolated, arabinosyladenine (Ara-A or Vidarabine: **68** in **Figure 9**) was synthesized in 1960 as a potential antitumor agent,[184] but was later produced by fermentation[185] of *S. griseus* and isolated, together with spongouridine,[186] from the Mediterranean gorgonian *Eunicella cavolini* in 1984. Of the many compounds derived from these early discoveries, some such as Ara-A, Ara-C, acyclovir (**69**; **Figure 9**), and later azidothymidine and DDI have gone into clinical use, but most have simply become entries in chemical catalogs.

However, two excellent reviews have been published (in 2004 and 2005) that give thorough coverage of naturally occurring purine[187] and pyrimidine[188] nucleosides and their derivatives. These should be consulted for the multiplicity of structures, including in some cases, carbocycles in place of the sugar moieties from marine, microbial, and plant sources, which may be used as leads to new agents in a variety of diseases (including antivirals).

1.08.5.4.2 Human immunodeficiency virus (HIV) protease inhibitors

As mentioned above, there are few targets that are 'virus only' in nature when it comes to screening, but in the case of human immunodeficiency virus-1 (HIV-1), a specific target is the aspartic protease, which is an essential part of the processing of the viral proteins *pol* and *gag* that permit the virus to replicate in the host cell. The initial work on this protease by the Merck group demonstrated that it was an aspartic proteinase and could be inhibited by the microbial pepsin inhibitor, pepstatin. In fact, inhibition by this peptide could be seen in both isolated enzymic and in whole cell assays.[189]

Pepstatin (**70**; **Figure 9**) contains an unusual hydroxy-amino acid, statine, which can be thought of as a mimic of a putative transition state intermediate, where the hydroxy group takes the place of a water molecule that is the second

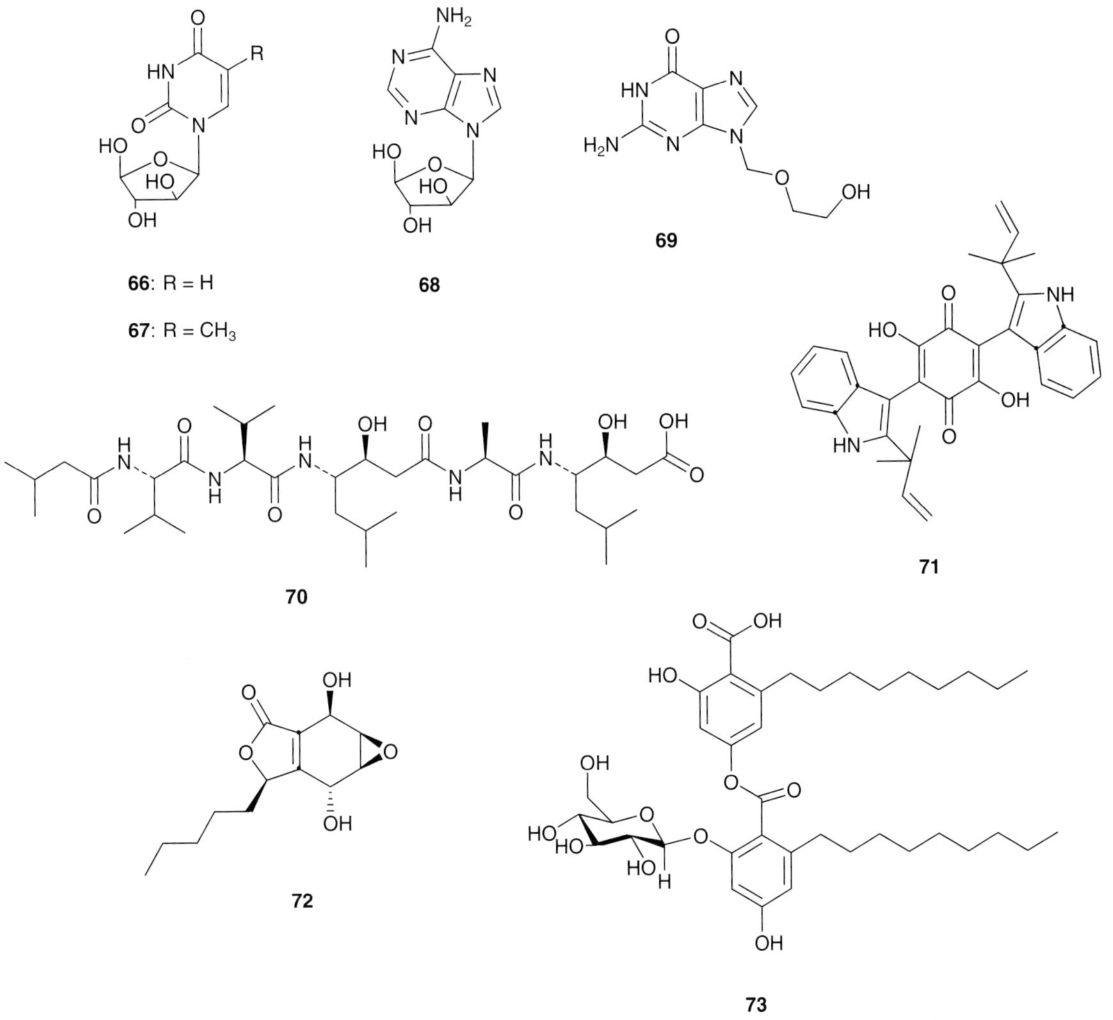

Figure 9 Some anti-HIV protease inhibitors.

substrate for the hydrolytic reaction.[190] Using this hypothesis, plus the idea that statine is also acting as a dipeptide replacement, two groups, one at University of Wisconsin and the other at Merck, had earlier collaborated to produce renin inhibitors containing a hydroxyethylene isostere that gave activity comparable to that of pepstatin as a pepsin inhibitor.[190] Concomitantly, studies with replacement of amino acids in aspartic proteinase substrates (in general, 6–8 residues in length) with statine or isosteres led to the production of potent inhibitors of the aspartic proteinases renin and elastase.[190] Up until the early part of 2005, over 150 citations have been made to this review article, covering the use of peptidomimetics from natural sources as leads to inhibitors of many enzyme systems.

Once the investigators realized from the pepstatin results that the activity of HIV-1 protease was due to the presence of the same (or similar) catalytic site to that of renin, the collection of renin inhibitors was tested, leading to identification of potential HIV-1 inhibitors. Using similar techniques to those that proved fruitful with the renin and elastase systems, variations around those inhibitors and/or others based on a short peptide that was the consensus substrate of HIV-1 protease were synthesized using isosteric replacements for a variety of the amino acids, but, in most cases, keeping the 'statine mimic' aligned with the geometry of the active site.

These synthetic exercises based on a natural product model have led to successful drug entities[191,192] that are now available for the treatment of HIV infections, both clinically and experimentally, though the molecules used (e.g., Crixivan) show no formal structural relationship to the original natural product inhibitor, pepstatin.[193–196]

In a very recent paper, the scaffolds that mimic the natural substrate, but act as inhibitors, were further extended to include aminohydroxysulfones and pyrrolidinemethaneamines. This paper by Specker *et al.*[197] should be consulted to see how x-ray binding studies and, most importantly, experimental verification can extend the 'privileged skeletons' that will act as the basis for novel inhibitors of HIV protease specifically, but, in theory, can be applied to any aspartic protease.

Though not a peptidomimetic, recently a plant-associated fungus, *Nodulisporium hinnuleum*, obtained from the leaves of *Quercus coccifera* yielded the bis-indolyl quinone hinnuliquinone (**71**; **Figure 9**), which demonstrated roughly equal potency ($\sim 2\,\mu M$) against wild-type and mutant HIV-1 protease, thus providing another template for the synthesis of agents preferentially targeting resistant forms of HIV-1 protease.[198]

1.08.5.4.3 Human immunodeficiency virus integrase inhibitors

HIV-1 integrase plays a critical role in the replication of HIV-1, catalyzing assembly, endonucleolytic cleavage of the viral DNA, and strand transfer of the viral DNA into the host cell DNA. It is only in recent years that this enzyme has been used as a target for the discovery of novel anti-HIV agents, and a number of moderately active lead compounds have been isolated from natural sources, particularly by the Merck group. Integrasone (**72**; **Figure 9**) is a novel polyketide isolated from an unidentified fungus and having modest activity (IC_{50} of $41\,\mu M$),[199] while exophillic acid (**73**; **Figure 9**) (IC_{50} of $68\,\mu M$) is a novel dimeric 2,4-dihydroxy-alkylbenzoic acid isolated from the fungus *Exophiala pisciphila*.[200] Other polyketide- and terpenoid-derived natural product integrase inhibitors of terrestrial fungal origin are reported in a review by the Merck group,[201] and include known napthaquinones, bi- and triphenyls, a variety of simple aromatics, and linear, di- and sesterterpenoids.

1.08.6 Cardiovascular (*see* Volume 6)

In our usage, cardiovascular diseases will cover the following general areas: control of the β-adrenergic nervous system, which will include some aspects of respiratory function; control of cholesterol/lipid metabolism; and control of angiotensin levels. The underlying physiological principle that is addressed in these areas is the homeostatic control of blood pressure.

1.08.6.1 The β-Adrenergic Amines

The role played by epinephrine (adrenaline; **74** in **Figure 10**) and ephedrine (**75**; **Figure 10**) isolated from the Chinese medicinal plant *Ephedra sinaica* in the development of bronchodilators, such as isoprenaline (**76**; **Figure 10**), and subsequently the beta-blockers propanolol (**77**; **Figure 10**), atenolol (**78**; **Figure 10**), metaprolol (**79**; **Figure 10**), and their lineal chemical descendents has been discussed in the review by Newman *et al.*[11] Thus, starting with an agonist structure (ephedrine) that significantly affects blood pressure due to its effect on cardiac output and on release of other sympathomimetic amines, related structures are now available that are specific blockers of such activities on cardiac tissue and are excellent 'reducers of hypertension.'

1.08.6.2 Cholesterol-Lowering Agents

Another major cause of elevated blood pressure is the physical blockage of arteries by plaques of cholesterol/lipoproteins (atherosclerotic plaque). Since humans synthesize about 50% of their requirement for cholesterol, inhibition of that synthesis causing a reduction in overall cholesterol may reduce the deleterious effects of this steroid.

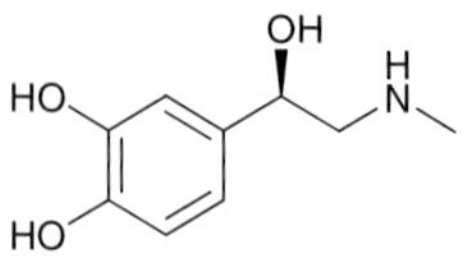

74

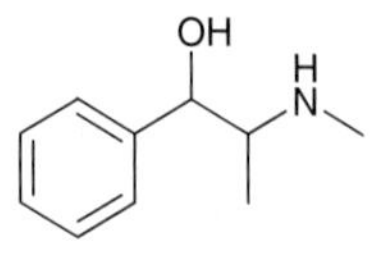

75

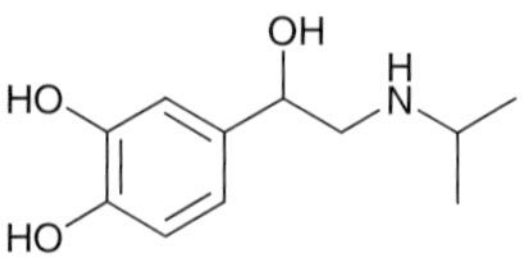

76

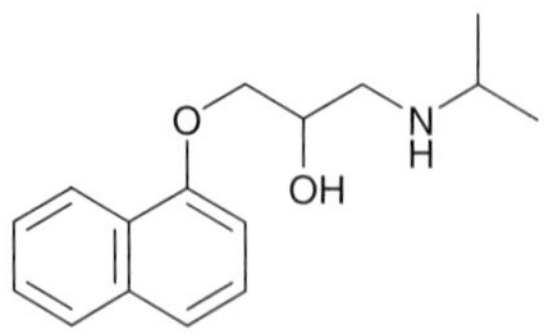

77

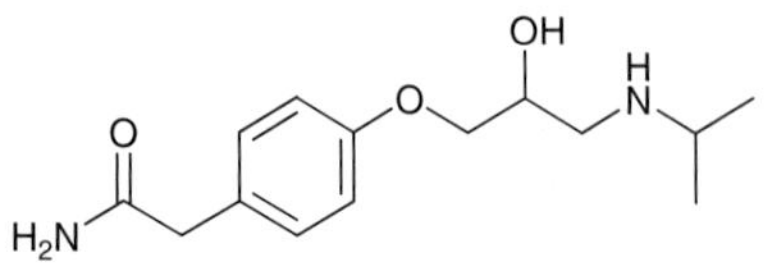

78

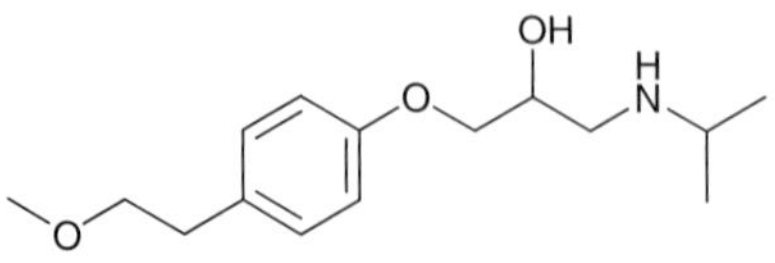

79

Figure 10 β-Adrenergic amines.

A potential site for inhibition of cholesterol biosynthesis in eukaryotes is at the rate-limiting step in the system, the reduction of hydroxymethylglutaryl CoA by HMG-CoA reductase to produce mevalonic acid (**80**; **Figure 11**). The discovery of the inhibition of this reduction step by the fungal metabolites compactin (**81**; **Figure 11**) and mevinolin (**82**; **Figure 11**) led initially to the (bio)chemically modified simvastatin (**83**; **Figure 11**) and pravastatin (**84**; **Figure 11**).

Comparison of the ring-opened or -closed lactone structures common to all of the 'statins' shows the resemblance to mevalonic acid in either its linear or closed forms and this recognition led to the synthesis of the original three 'synthetic' clinical products fluvastatin (**85**; **Figure 11**), atorvastatin (Lipitor; **86** in **Figure 11**), and cerivastatin (**87**; **Figure 11**). The obvious role played by nature in the development of these blockbuster cholesterol-lowering drugs (worldwide sales for Lipitor exceeded US$10 billion in 2004) is reviewed by Newman *et al.*[11] and as shown in **Figure 11**, 15 of the 16 agents (**81–95**) that have entered preclinical or clinical trials, with some now reaching commercialization, have the same basic natural product-derived 'warhead.' The one nonmevalonate, crilvastatin (**96**; **Figure 11**) does not appear to have progressed since 1997.

1.08.6.3 Angiotensin Metabolism and Receptor Inhibitors

The angiotensin I to angiotensin II cascade is an essential mechanism in the maintenance of blood pressure in humans and it was realized approximately 40 years ago that if one could inhibit the conversion of the decapeptide (angiotensin I) to the biologically potent octapeptide (angiotensin II), then such compounds might control blood pressure. The discovery that fractions from the venom of the pit viper, *Bothrops jararaca*, inhibited the degradation of the mammalian nonapeptide bradykinin, and the isolation of a nonapeptide, teprotide (**97**; **Figure 12**), as the active principle that also showed activity as an angiotensin-converting enzyme (ACE) inhibitor, led to a definition of the active site of ACE and the synthesis of more specific and potent ACE inhibitors. The first commercialized product was captopril (Capoten; **98** in **Figure 12**), which was modeled on teprotide and the use of carboxypeptidase B as a surrogate for the ACE.

These initial successes, which aided in a greater understanding of one of the controlling systems related to hypertension, led to the development of more potent ACE inhibitors, such as the prodrugs enalapril (Vasotec; **99** in **Figure 12**), quinapril (Accuprit; **100** in **Figure 12**), and fosinopril (Monopril; **101** in **Figure 12**).[11] Subsequent work by a large number of investigators has further extended the knowledge of ACE with the identification of at least two different binding modes for potential inhibitors, leading to subtly different in vitro and in vivo responses. Readers should consult the excellent review by Achayra *et al.*[202] for specific details on structures, interactions, and active site models of the current compounds being investigated as ACE inhibitors.

Another example that demonstrates the evolution of a synthetic product from natural product structural information is that of the prototypical angiotensin II receptor (AT1R) blocker, losartan (**102**; **Figure 12**). In the discussion of ACE-related inhibitors there is a potential for confusion, in that the conventional shorthand biochemical designation for the pharmacologically active octapeptide that results from the action of ACE upon the decapeptide angiotensin I (or AT I) is AT II, whereas in biochemical pharmacology nomenclature, the receptor for this octapeptide ligand is designated as the angiotensin 1 receptor (AT1R). Thus, AT1R is the receptor for the octapeptide AT II, the active ligand produced by ACE action upon angiotensin I (AT I), not, as some may expect, the receptor for the ACE substrate, AT I.

From SAR studies of multiple peptide analogs of the octapeptide AT II, whose formal sequence is H_2N-Asp^{1}-Arg^{2}-Val^{3}-Tyr^{4}-Ile^{5}-His^{6}-Pro^{7}-Phe^{8}-CO_2H, there were suggestions that the His^{6} residue was required for receptor recognition and that the agonist activity required the phenyl ring of the Phe^{8}, the hydroxyl group of the Tyr^{4}, and the C-terminal carboxylate. Thus, a working hypothesis for the binding pocket in AT1R for the ligand AT II would be a positively charged site, a lipophilic pocket or pockets, and a hydrogen bond acceptor.[203]

The first leads to a nonpeptidic structure that demonstrated AT1R inhibition were three microbial metabolites (**103–105**; **Figure 12**), reported in 1982, that had low potency as antihypertensive agents.[204] Using simple modeling methods, both Dreiding models and simple computerized techniques, workers at DuPont postulated that these compounds, which at high concentrations demonstrated a small reduction in blood pressure via blockade of AT1R, bound to the receptor in a manner such that the carboxylic acid was equivalent to the C-terminal carboxylate of AT II, the imidazole nitrogens were comparable with the histidine residue, and the benzyl group pointed toward the N-terminus of AT II, with the *para* position of that residue holding the most promise for a systematic extension toward the N-terminus of AT II.

By making the (correct) assumption that a second carboxylate in the *para* position of the phenyl ring would give a negative charge in the vicinity of the Tyr^{4} hydroxyl and the Asp^{1} β-carboxylic acid, a compound (**106**; **Figure 12**) was prepared that demonstrated a 10-fold increase in binding affinity. The subsequent derivation of what finally became

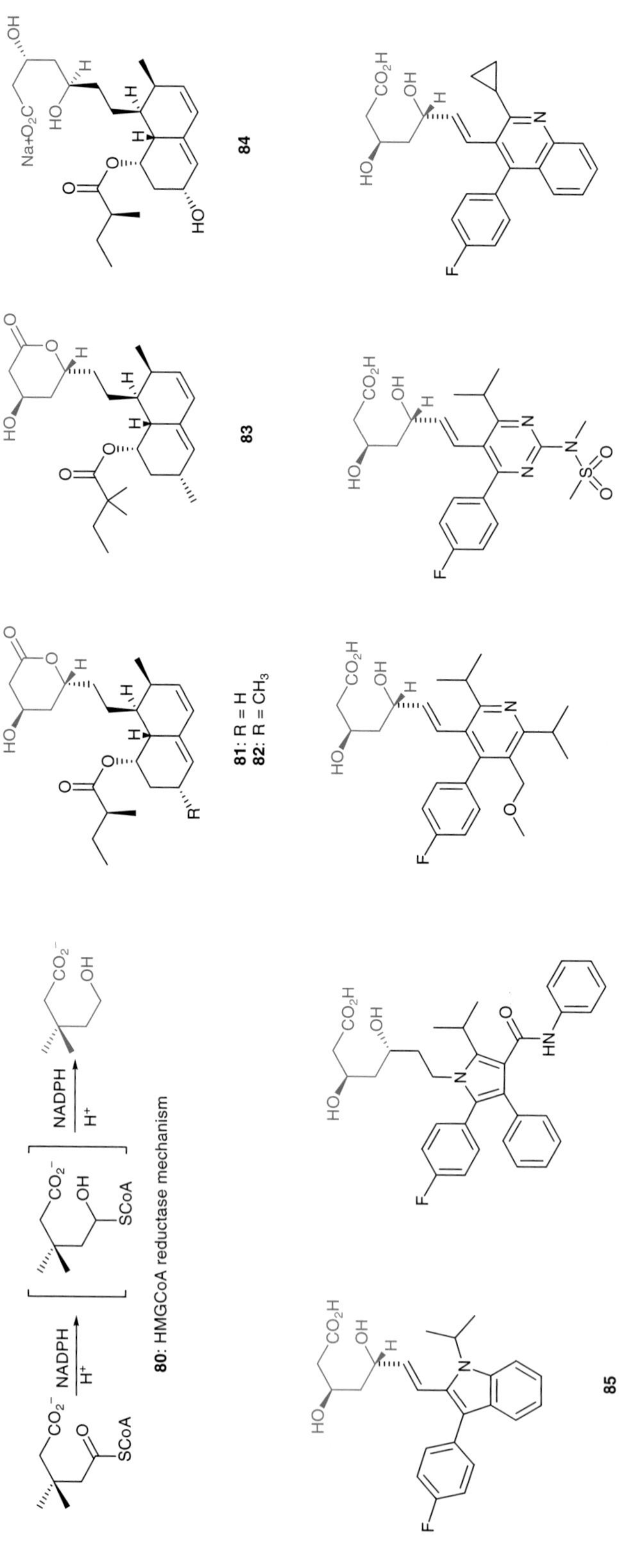

Figure 11 Statins.

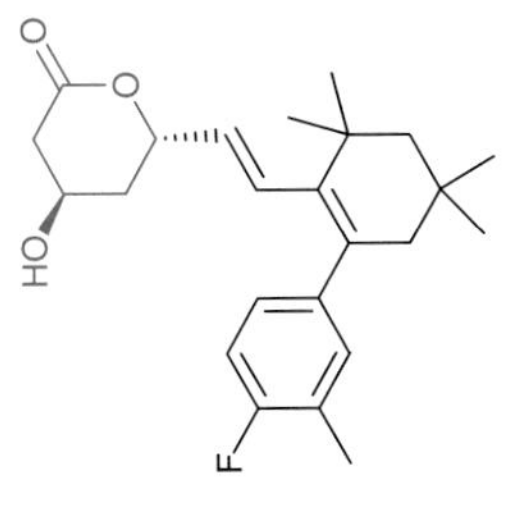

90

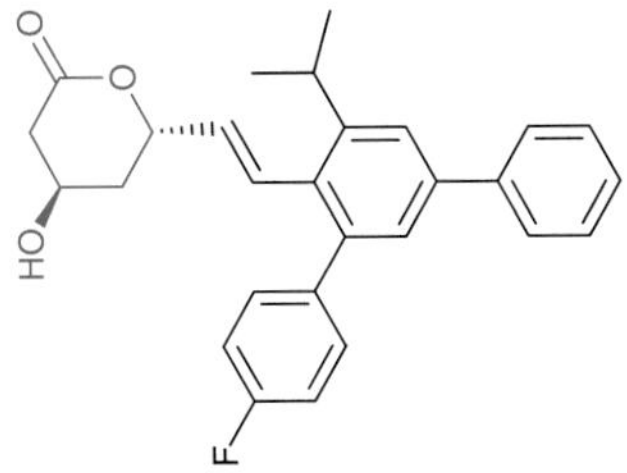

91

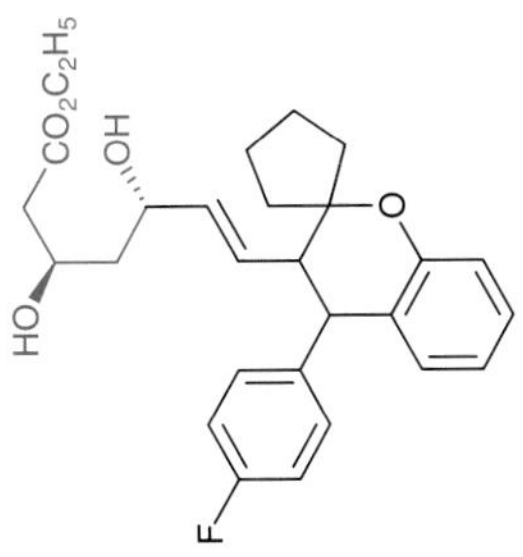

92

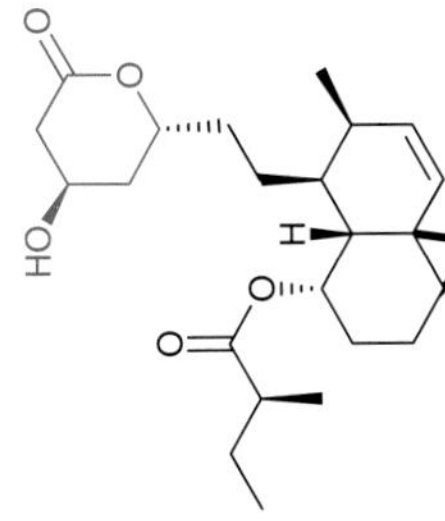

93

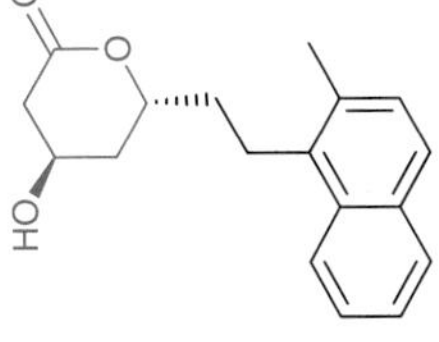

94

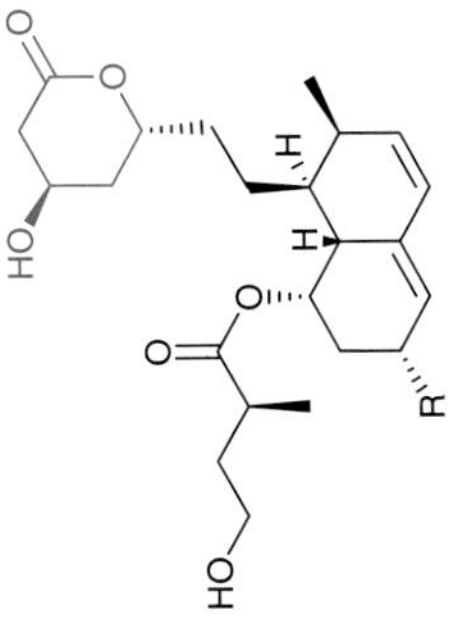

95

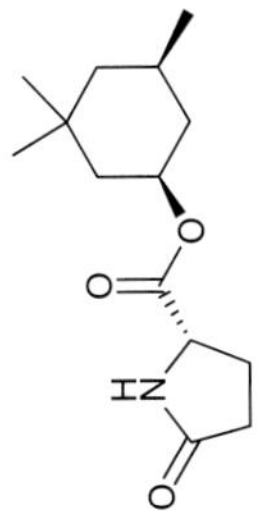

96

Figure 11 Continued

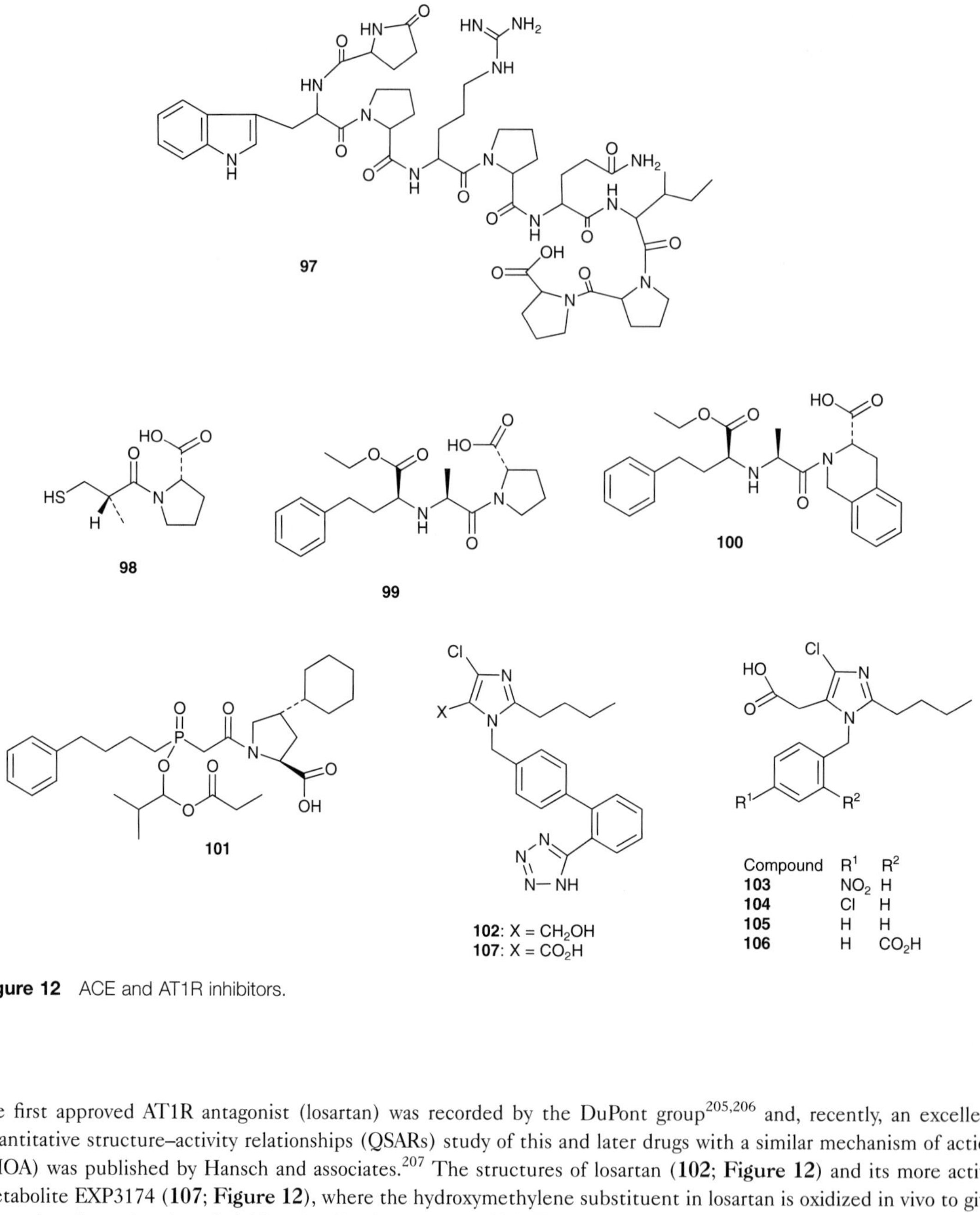

Figure 12 ACE and AT1R inhibitors.

the first approved AT1R antagonist (losartan) was recorded by the DuPont group[205,206] and, recently, an excellent quantitative structure–activity relationships (QSARs) study of this and later drugs with a similar mechanism of action (MOA) was published by Hansch and associates.[207] The structures of losartan (**102**; **Figure 12**) and its more active metabolite EXP3174 (**107**; **Figure 12**), where the hydroxymethylene substituent in losartan is oxidized in vivo to give the carboxylate, thereby mimicking the 'first' derivative (**106**; **Figure 12**) of the microbial metabolites referred to earlier, are shown.

1.08.7 Pain/Central Nervous System (*see* Volume 6)

The conquest of pain was one of the major uses of medicinal plants in the ancient world, with the possible use of the crude extract of the opium poppy (*Papaver somniferum*) dating from around 6000 years ago in Sumeria. However, the earliest undisputed reference to the use of the 'juice of the poppy' is from Theophrastus approximately 2300 years ago. The use of opium, or laudanum, as Paracelsus named it, in the sixteenth century was popular in both Europe and particularly the Orient, leading to the so-called 'Opium Wars' between the UK and China in the 1840s.

1.08.7.1 The Opiates

The history of the discovery of morphine (**108**; **Figure 13**) and codeine (**109**; **Figure 13**), and the attempts to develop nonaddictive analogs have been reviewed[11,137] and will not be discussed here. With the exception of the semisynthetic compound buprenorphine (**110**; **Figure 13**), which is approximately 25–50 times more potent than morphine and has a lower addiction potential, none of the compounds made to date from modifications around the phenanthrene structure of morphine have exceeded the pain control properties without a concomitant addiction potential.

1.08.7.2 The Conotoxins

The extremely complex libraries of short (usually 10–35 residue) peptide toxins elaborated by snails of the genus *Conus* have turned out to be a treasure trove of pharmacologically active materials, initially in the area of analgesia but now in inflammation and neurochemistry as well. In addition, separation of the individual peptides has led to fundamental biochemical studies in voltage-gated channels of all types. A recent paper by Olivera and collaborators (the person who has probably done more to realize the potential of these peptides than any other) should be consulted as a guide to the potential for these peptides in a variety of pharmacologic areas.[208]

Ziconotide (**111**; **Figure 13**) is a 25-residue peptide with three interlocking cystinyl bridges that was originally isolated by Olivera's group from *Conus magus* and was known as MVIIA toxin. It demonstrated a potent activity against voltage-gated Ca^{2+} channels, and because of its novel binding characteristics, Olivera coined the phrase 'Janus ligand' for this and other similar peptidic agents, as they appeared to have both a 'docking face' and a 'locking face' at the receptor level. The peptide demonstrated significant effects as an analgesic and was licensed to Neurex Inc., who then proceeded to synthesize over 200 variations on the structure, before eventually deciding that the original structure was optimal. Ziconotide was approved by the Food and Drug Administration (FDA) at the very end of 2004 for the treatment of intractable (phantom limb) pain under the trade name of Prialt.

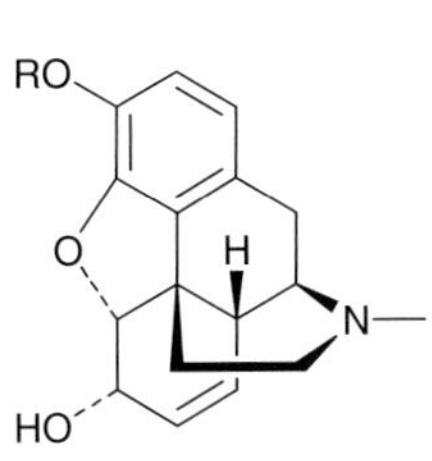

108: R = H
109: R = CH_3

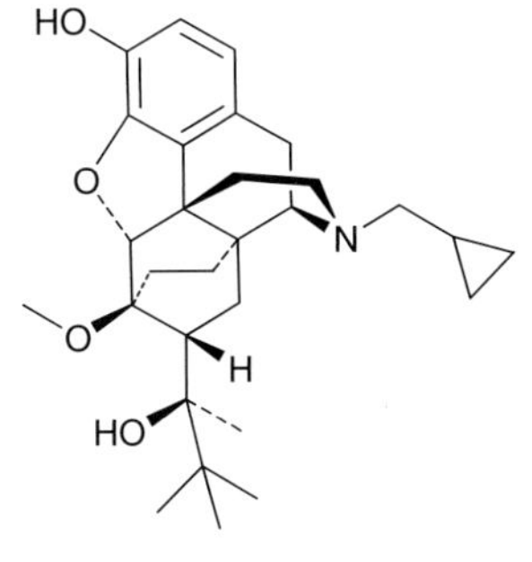

110

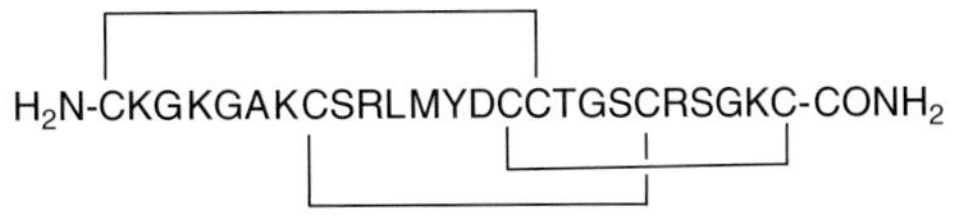

111

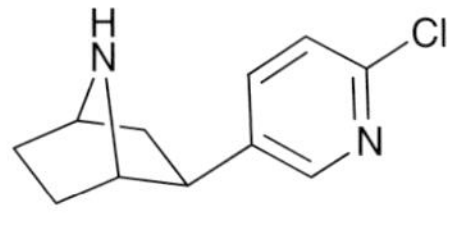

112

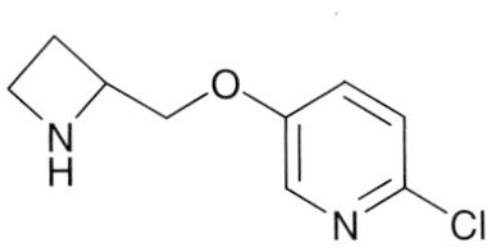

113

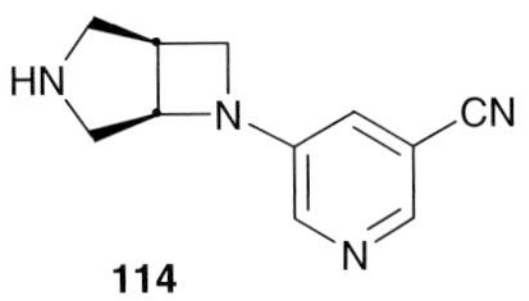

114

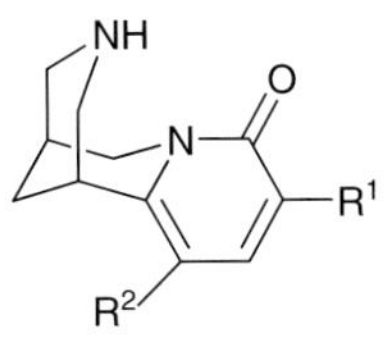

Compound	R^1	R^2
115	H	H
116a	Cl	H
116b	Br	H
116c	I	H
117a	H	Cl
117b	H	Br

Figure 13 Analgesic agents.

Other conatoxins are in various stages of clinical or preclinical development. These include CGX-1007 (conantokin G) for neuropathic pain and intractable epilepsy, CGX-1160 (contulakin G from *Conus geographus*, a neurotensin agonist), R-conotoxin Vc1.1 from *Conus victoriae* for neuropathic pain, ω-conotoxin CVID from *Conus catus* for severe morphine-resistant pain, and χ-conotoxin MRIA/B for neuropathic pain.

1.08.7.3 The Epibatidines and Cytisine

As described in the review by Newman *et al.*[11] the elucidation of the structure of epibatidine (**112**; **Figure 13**) by Daly's group at the National Institutes of Health (NIH) led Abbott Laboratories to synthesize a series of related compounds with one molecule, ABT-594 (**113**; **Figure 13**), that mediated the nicotinic acetylcholine receptor (nAChR) locus entering clinical trials as an analgesic. In 2001, using techniques similar to those used in parallel syntheses, Carroll *et al.*[209] reported the biological results from testing a series of epibatidine derivatives where the 2-substituent of the pyridine ring was successively replaced with halogens and amino, dimethylamino, hydroxyl and trifluoromethane-sulfonate groups. Surprisingly, no significant differences were seen for any of the halogeno or *nor*-halogeno derivatives as far as binding affinities and inhibition of epibatidine binding to the receptor was concerned, though the hydroxysubstituted derivative was 10 000-fold less active than epibatidine as an analgesic in the assay used. That this particular pharmacophore is still a molecule of interest in both academic and industrial circles is shown by two recent (2005) papers reporting further studies on the synthesis of epibatidine analogs.[210,211] Clinical trials of ABT-594 were discontinued at the phase II level due to unacceptable side effects, but another analog developed by Abbott, A 366833 (**114**; **Figure 13**), is currently in phase I trials as a nonaddictive pain killer.

Cytisine (**115**; **Figure 13**), a toxic alkaloid isolated from the seeds of *Laburnum anagyroides medicus*, is similar to nicotine in its physiological responses, but binds strongly (with subnanomolar affinity) to the α4β2 nAChR subtype and crosses the blood–brain barrier. There have been some attempts to make other derivatives for SAR studies, but Seitz *et al.*[212] chose to incorporate halogeno substituents and modify the pyridone to a pyridothione (functionally replacing a lactam with a thiolactam). They found that conversion of the lactam to the thiolactam, although it reduced overall potency in a binding assay by a factor of about 8, actually increased selectivity for the receptor subtype by 20-fold.

Studies in cell lines expressing neuromuscular α1β1γδ, ganglionic α3β4, and central neuronal α3β4 subtypes indicated that halogenation in the 3-position (**116a–116c**; **Figure 13**) conferred greater potency with a preference for the bulkier, less electronegative iodide, while 5-halogenation (**117a–117b**; **Figure 13**) resulted in decreased potency, and 3,5-dihalogenation gave inactive compounds except for the di-iodo derivative, which had much reduced activity; the degree of activity is therefore associated with the electron-withdrawing ability of the halogen, possibly due to the reduced hydrogen bonding capability of the carbonyl functionality.[213]

1.08.8 Antineoplastics (*see* Volume 7)

It is in the treatment of cancers and in anti-infective areas that natural products have made their major impact as templates or direct treatments. In the cancer area, of the 131 small molecule drugs (updated figures from the 2003 review by Newman *et al.*[12]) commercially available or approved worldwide through 2004, approximately 62% can be related to a natural product origin.

A common feature of many of the natural product-derived anticancer drugs now in clinical use is that the original natural product was too toxic. Therefore, chemists had to make semisynthetic compounds based on the natural product structures that had the more deleterious properties of the natural product diminished by selective modification. In some cases, notably that of mitoxantrone (**118**; **Figure 14**), a totally synthetic product evolved but with the attributes of two or more parent natural product structures in one molecule.[214]

In contrast to our earlier review,[11] in the following discussion, the anticancer agents are categorized according to their proposed mechanisms of action as opposed to their natural source (plant, marine organism, etc.). One reason for this is the increasing evidence that active metabolites are synthesized by microbes associated with the host macroorganism (*see* Section 1.08.3.3.6). Mechanistic details are not discussed in detail, but interested readers are referred to relevant articles. For authoritative discussions on the discovery and development of the majority of the anticancer agents in current clinical and preclinical development readers are referred to the recent book edited by Cragg *et al.*[215]

1.08.8.1 Tubulin Interactive Agents (TIAs)

The majority of the TIAs in development up to 2003, from preclinical studies to clinical launch, have been discussed in detail in an early 2004 review[216] and a 2005 book.[215] The TIAs covered in the latter volume include taxanes,[217] epothilones,[218] combretastatins,[219] vinca alkaloids,[220] dolastatins,[221] discodermolide,[222] halichondrins,[78]

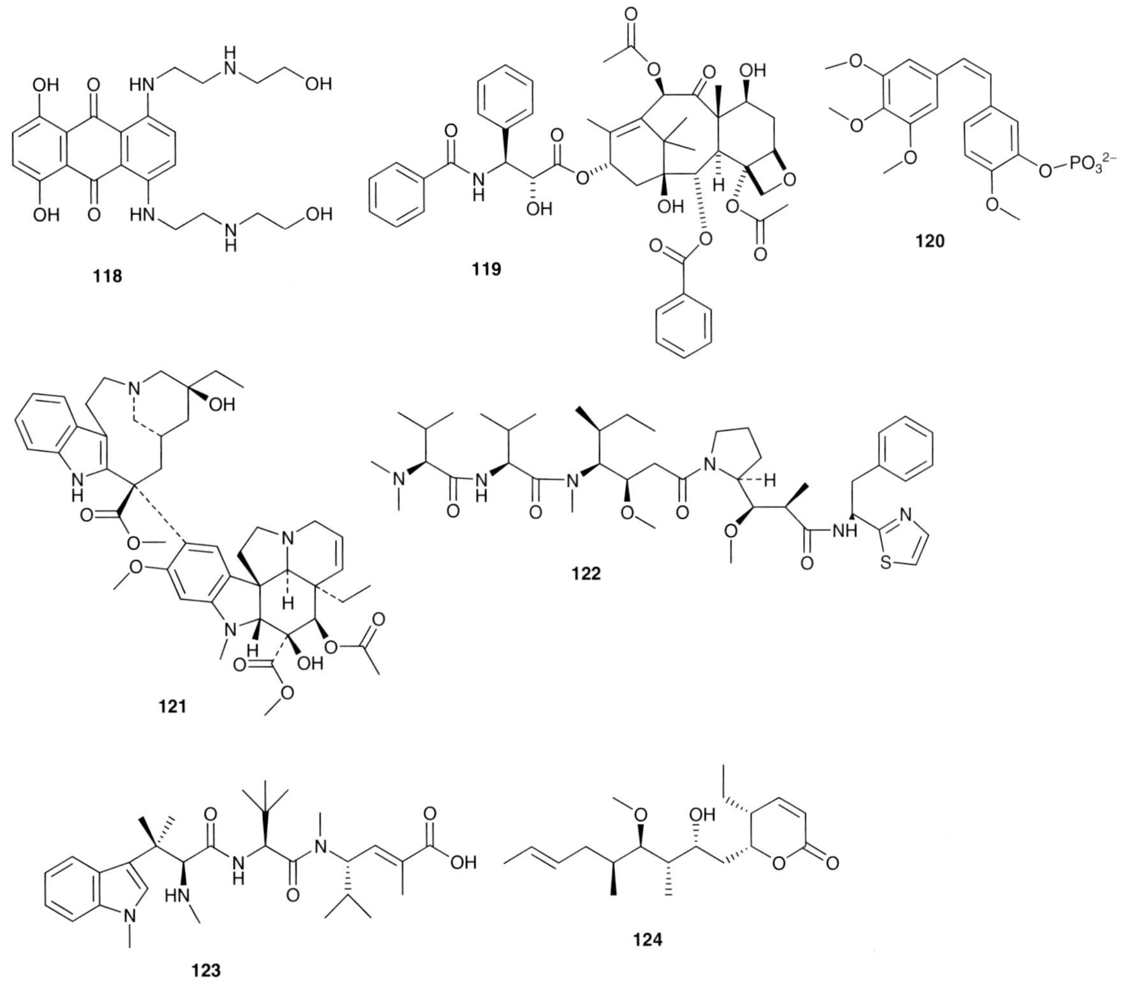

Figure 14 Examples of tubulin interactive agents (**119–124**).

and hemiasterlins[223] (see structures **119**, **11** and **16**, **120**, **121**, **122**, **13**, **14** and **15**, and **123**, respectively) and agents derived from or synthetically modeled on those initial structures to develop drug candidates with improved solubilities, pharmacodynamics, or metabolic patterns, compared with the original natural products. The interested reader should consult this review and the book chapters cited, together with the references given therein for a discussion of the multiplicity of structures that have been developed from natural product lead compounds.

Another report in this field covers the discovery, by Kondon *et al.*, that pironetin (**124**; **Figure 14**), a TIA derived from a *Streptomyces* species whose initial biological activity was reported in 1998,[224] is the only TIA identified so far that covalently binds to the α-tubulin chain at Lys^{352}. Lys^{352} is an amino acid located at the entrance of a small pocket in α-tubulin that faces the β-tubulin of the next dimer.[225] No pironetin derivatives have as yet been reported as candidate leads.

1.08.8.2 Inhibitors of Topoisomerases I and II

A review by Cragg and Newman[216] reporting new developments in the field of topoisomerase inhibitors appeared in a special issue of the *Journal of Natural Products* in early 2004 honoring Monroe Wall and Mansukh Wani, the codiscoverers of both Taxol and camptothecin, and should be consulted for the history of camptothecin. Although the majority of new topoisomerase I inhibitors are based on the camptothecin pharmacophore (**125**; **Figure 15**),[226] the protein kinase inhibitor staurosporine (**126**; **Figure 15**) is also a weak topoisomerase inhibitor and various derivatives of the basic staurosporine scaffold demonstrate inhibitory activity for both topoisomerase I and II.[227] Another class of compounds that are important drugs that act via inhibition of topoisomerase II are the many anthracyclines, with doxorubicin (**127**; **Figure 15**) being a suitable example, though it should be pointed out that almost all of the clinically useful compounds

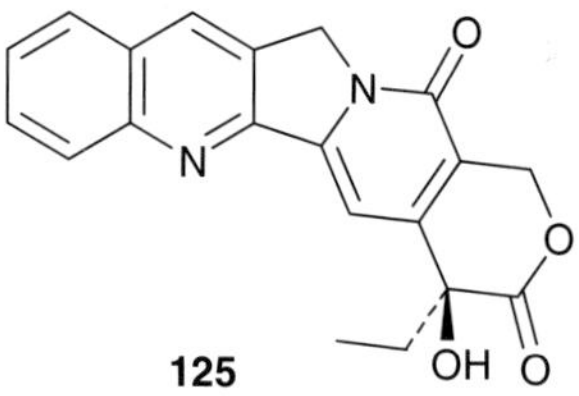

125

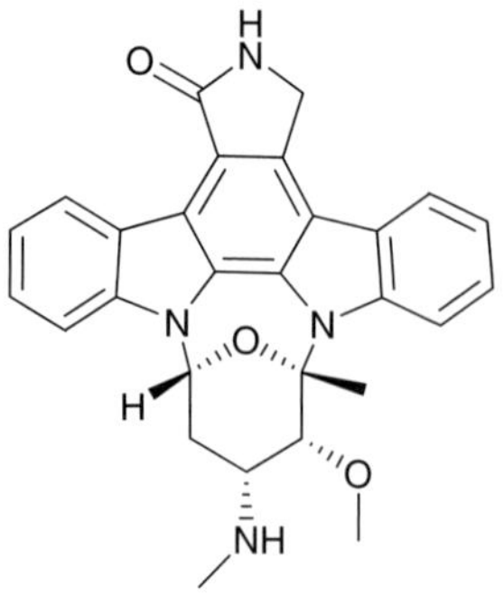

126

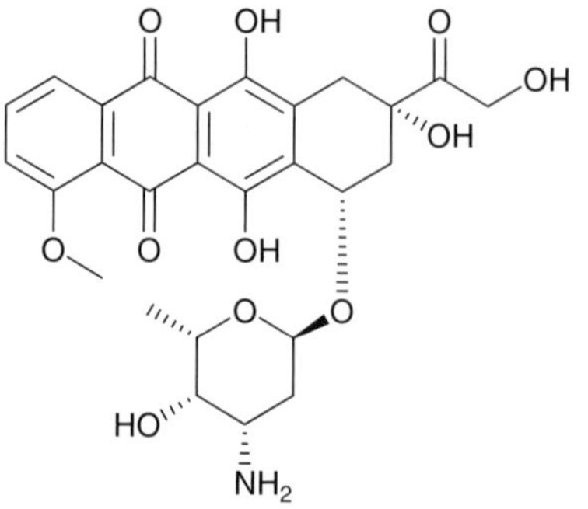

127

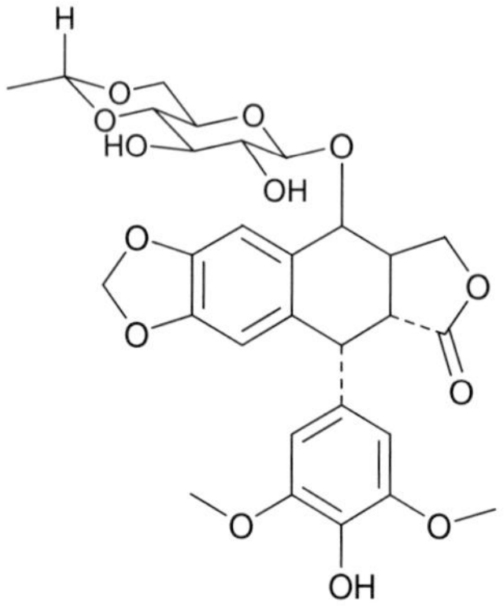

128

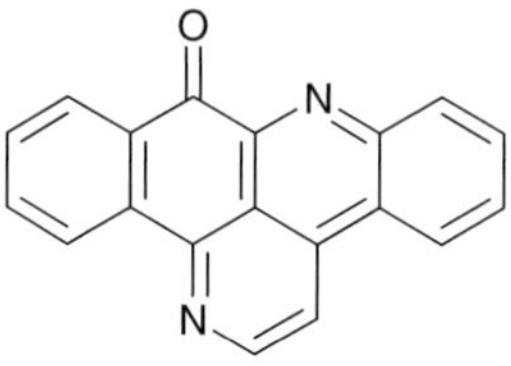

129

Figure 15 Naturally derived topoisomerase inhibitors.

of this chemical class were developed as a result of their cytotoxic activities and not with prior knowledge of this mechanism of action.[228] Similarly, the podophyllotoxin derivative, etoposide (**128**; **Figure 15**) was developed by the then Sandoz company by modification of *epi*-podophyllotoxin without prior knowledge of the mechanism.[229]

Since the publication of the 2004 review,[216] two important papers related to chemical derivatives of (mainly) natural products with this MOA have been published: Denny reviewed[230] the anticancer activity of some new topoisomerase inhibitors including six topoisomerase I, 12 topoisomerase II, and six dual topoisomerase inhibitors; and Marshall *et al.*[231] reported on AK-37 (**129**; **Figure 15**). This compound (AK-37) was modified from a marine-derived pyridoacridine, and stabilizes the topoisomerase I cleavable complex in a manner comparable to that of 9-nitrocamptothecin (the precursor to 9-aminocamptothecin), which is currently in phase III clinical trials. The interested reader should also consult the recent review by Marshall and Barrows on the pyridoacridines in general for a more 'in-depth' discussion of the wide variety of structures and activities in this class of natural products.[232]

1.08.8.3 Caspase Activation and Apoptosis Induction

The relatively simple naphthoquinone β-lapachol (**130**; **Figure 16**) is a well-known compound obtained from the bark of the lapacho tree *Tabebuia avellanedae* and other species of the same genus. These trees are native to South America and β-lapachol and other plant components are extensively used as ethnobotanical treatments in the Amazonian region. β-Lapachol was advanced to clinical status by the National Cancer Institute (NCI) in the 1970s, but was later withdrawn due to unacceptable levels of toxicity. Its close relative β-lapachone (**131**; **Figure 16**), however, has demonstrated interesting molecular target activity, with one mechanism of action being the induction of apoptosis in transformed cells.[233] Evidence of the further involvement of β-lapachone in transcription processes was reported by Choi *et al.*,[234] demonstrating that the agent induced activation of caspase-3, inhibition of nuclear factor kappa B (NFκB), and subsequent downregulation of *bcl-2*. Currently, β-lapachone is in phase II clinical trials in the US, and further information on the background of these agents is given in a 2004 review by Ravelo *et al.*, which should be consulted for further information.[235]

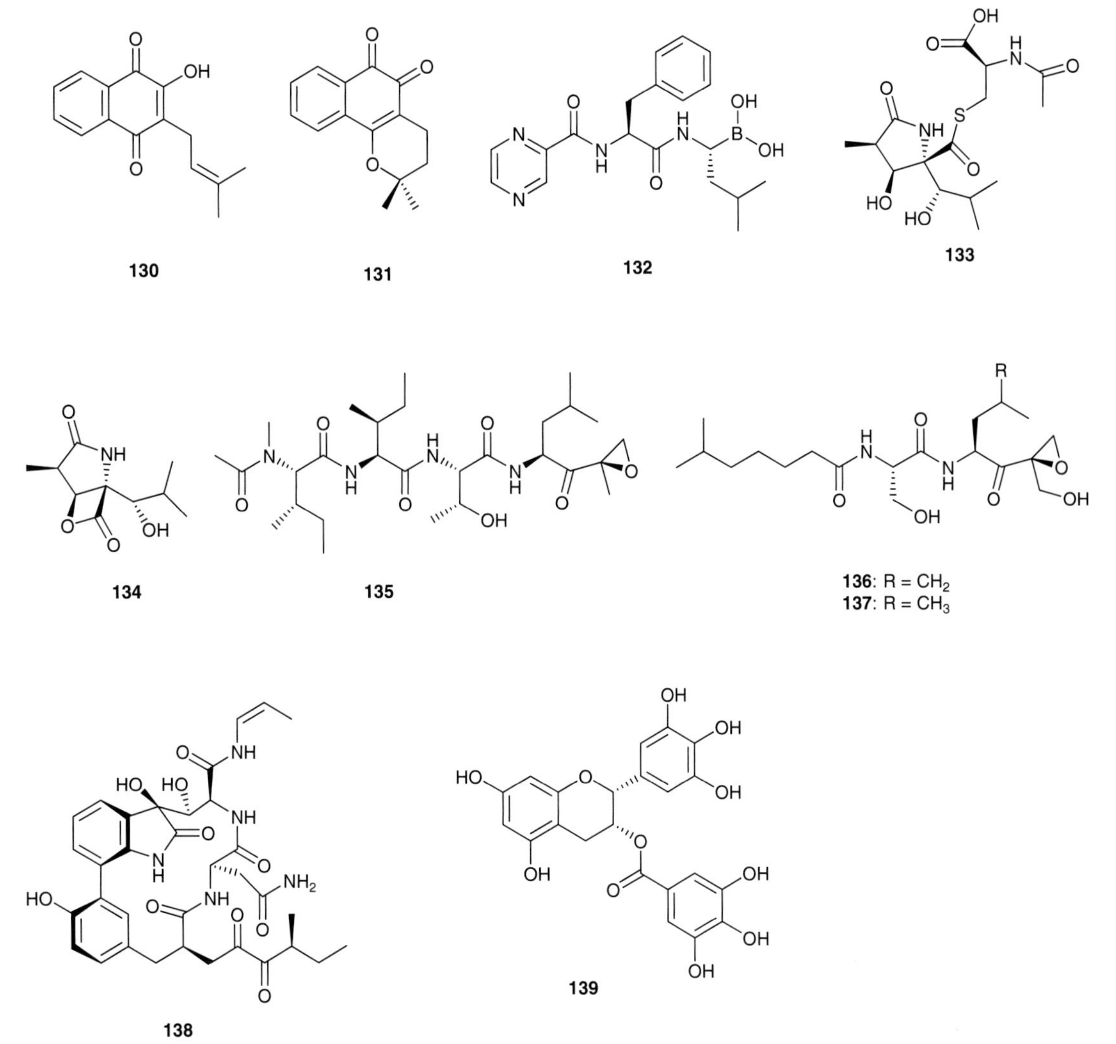

Figure 16 Caspase and proteasome inhibitors.

1.08.8.4 Proteasome Inhibitors

The proteasome is a multienzyme complex, which is involved in the ubiquitin-proteasome pathway control of cell cycle progression, in the termination of signal transduction cascades, and in the removal of mutant, damaged, and misfolded proteins, making it a promising therapeutic target. The background to the proteasome as a potential target is described in a review by Kisselev and Goldberg,[236] and the first clinical drug that employs this MOA bortezomib (**132**; **Figure 16**) is now in use. The compound is a synthetic dipeptidyl peptide boronate[237] based upon a natural product structure that inhibited chymotrypsin. The development of this compound has been described by the original inventor and will not be further commented on here.[238] There are, however, a significant number of other compounds from nature, and their derivatives, that have led to a greater understanding of the intricacies of this multienzyme complex.

The 20S proteasome in mammals has three closely linked proteolytic activities, which are termed trypsin-, chymotrypsin-, and caspase-like from their substrate profiles, though it should be emphasized that the characteristic 'active triad' in the namesake enzymes of these activities have had the catalytic serine residue replaced by threonine in the proteasome. Another major difference between the proteasome and the single enzymes is that the complex only acts as a concerted whole; individual activities are not demonstrable. In fact, if a suitable compound inhibits the chymotrypsin-like activity then a large reduction in the rate of protein degradation is observed. In contrast, if the sites corresponding to the other nominal activities are modified, the overall rate of hydrolysis of proteins is not significantly changed. Owing to the substrate specificity of chymotryptic sites, most inhibitors are hydrophobic, whereas in the case of the other two active sites, their 'peptide-based' substrates/inhibitors tend to be charged. As a result, almost all of the proteasome inhibitors tend to have chymotrypsin-like activities with some overlapping, but weaker, effects on the other sites.

In 1991, Omura *et al.* reported that the microbial metabolite lactacystin (**133**; **Figure 16**) induced neuritogenesis in neuroblastoma cells.[239] This paper was then followed by further reports from Fenteany *et al.*[240] and Craiu *et al.*[241] which demonstrated that radiolabeled lactacystin selectively modified the β5(X) subunit of the mammalian proteasome, and irreversibly blocked activity. In subsequent studies, it was demonstrated by Dick *et al.*[242,243] that the actual inhibitor in vitro was the β-lactone, *clasto*-lactacystin-β-lactone (**134**; **Figure 16**) and that this substance was formed spontaneously when lactacystin was exposed to neutral aqueous media. In 1999, Corey and Li reported on the synthesis of the parent compound and other analogs, suggesting that *clasto*-lactacystin-β-lactone should be named omuralide (**134**; **Figure 16**).[244,245]

The marine bacterial metabolite salinosporamide A (*see* Section 1.08.3.3.5; **10**) demonstrates activity as a cytotoxic proteasome inhibitor[57] and has been synthesized.[246] Interestingly, the salinosporamide structure, compared with omuralide, is uniquely functionalized, with a cyclohexene ring replacing the isopropyl group found at the C(5)-position in omuralide. The significance of this substitution is that the isopropyl group in omuralide is essential for activity, and it was found by Corey *et al.*[245] that replacement by a phenyl ring abolished activity. Thus, salinosporamide A might interact with the 20S proteasome in a modified manner, compared to that of omuralide and, unlike omuralide, salinosporamide A appears to be the true metabolite and not a derivative from a precursor. Currently, this molecule is undergoing preclinical evaluation at Nereus Pharmaceuticals as a proteasome inhibitor with the express aim of entering clinical trials in late 2006.

In 1999, Crews *et al.* at Yale University reported that the epoxyketone microbial metabolites epoxomicin (**135**; **Figure 16**) and eponemycin (**136**; **Figure 16**) exhibited cytotoxic activities as a result of proteasome inhibition.[247,248] Similarly to the omuralide-based inhibitors, epoxomicin reacted predominately with the chymotrypsin-like site, while the less potent eponemycin and its semisynthetic analog dihydroeponemycin (**137**; **Figure 16**) had approximately equal activity against both the chymotrypsin-like and the caspase-like sites, with these epoxyketones being the most selective proteasome inhibitors reported to date.

There are reports of other natural products active as proteasome inhibitors but with different mechanisms to those above. Thus, the cyclic peptide TMC-95-A (**138**; **Figure 16**), isolated from *Apiospora montagnei* is a potent chymotrypsin-like inhibitor, with activity against the other sites as well[249] that apparently binds noncovalently to active sites through an array of hydrogen bonds, and (−)-*epi*-gallocatechin 3-gallate (EGCG; **139**; **Figure 16**) is a potent covalent inhibitor of the 20S proteasome, due apparently to acylation of the active site threonines through threonine cleavage of the ester linkage in EGCG.[250]

1.08.8.5 Histone Deacetylase Inhibitors

The role of HDACs in the regulation of gene expression, oncogenic transformation, cellular differentiation, and the promotion of angiogenesis is discussed by Kingston and Newman[105] and references cited therein. Suffice it to say, the inhibition of HDAC activity can exert a significant role in suppression of the neoplastic process.

HDAC inhibitors have been described as tripartite materials, containing an enzyme binding group that is frequently aromatic, a hydrophobic spacer group, and an inhibitor group.[251–253] Such a model is well illustrated by the natural product trichostatin A (TSA; **140**; **Figure 17**), in which the structure mimics the Lys side chain of the substrate (the 'linker'). The inhibitory end is the zinc-chelating hydroxamic acid and the aromatic enzyme-binding group is the 4-dimethylamino-benzoyl group. This molecule together with its congeners (trichostatin B, C, and D) was first isolated as an antifungal agent,[254] and approximately a decade after the structural characterization of the trichostatins, they were found to have potent differentiation-inducing and antiproliferative activities in Friend erythroleukemia cells. TSA was subsequently demonstrated in vitro and in vivo to be a highly potent inhibitor (nanomolar range) of the class I and class II HDACs with a slight specificity for HDAC1 and HDAC6 compared with HDAC4. The *S* enantiomer of TSA was inactive, and neither enantiomer had any activity against the class III enzymes. The full mechanism of action has not yet been elucidated, but a large series of effects were observed in signal transduction systems, including induction of apoptosis, when healthy and tumor cells from many different sources were treated with this agent. More detail and references to the specific events observed with this compound at the molecular level can be found in tables 2 and 3 of the recent review by Vanhaecke *et al.*[255]

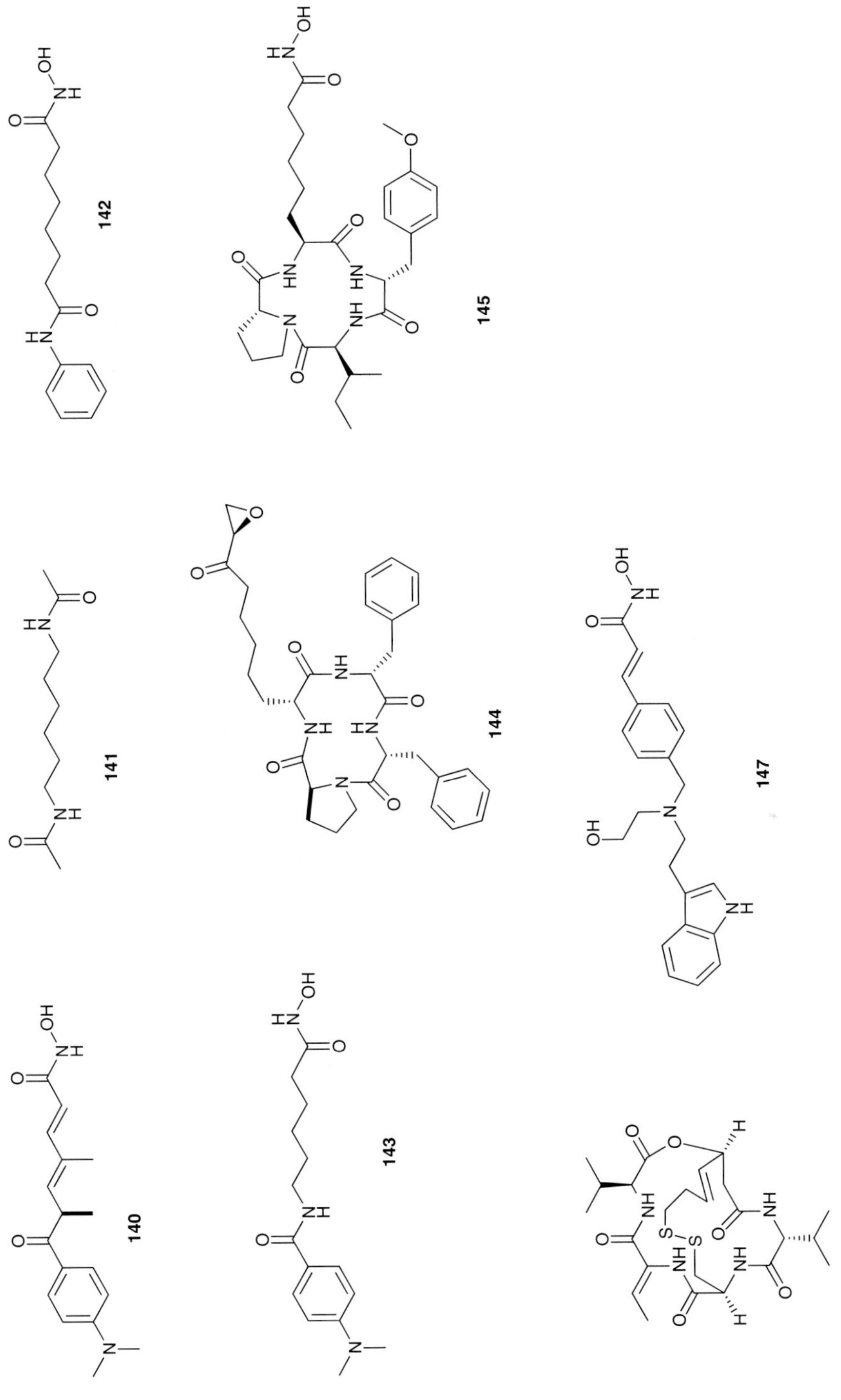

Figure 17 HDAC inhibitors.

Once the basic structural features of TSA and its initial activities were identified, research began on the synthesis of compounds that were more stable and had improved water solubility. Prior to this TSA research, hexamethylene bisacetamide (HMBA; **141** in **Figure 17**), from the family of molecules known as hybrid polar compounds (HPCs), was demonstrated to induce hyperacetylation of histone H4 in healthy keratinocytes. HMBA also induced histone hyperacetylation in squamous cell carcinoma derived from these cells, but did not inhibit their growth in vitro,[256] and induced a wide variety of other pathway modulations. Since the levels required for activity in vivo were in the millimolar range, toxic side effects, such as thrombocytopenia occurred and further development ceased.[257] These results, combined with a knowledge of the basic structure of TSA, were used by a variety of investigators to develop a series of second-generation HPCs, which were tested as HDAC inhibitors. The lead compound from these studies, suberoylanilide hydroxamic acid (SAHA; **142** in **Figure 17**), is currently in phase II/III clinical trials and has demonstrated antitumor activity in phase I clinical trials. An oral formulation is currently undergoing evaluation in a phase II study of patients with a variety of refractory tumors, both solid and leukemic in nature.[258]

Since TSA is difficult to synthesize and the yield is low, (*R*)-TSA is generally obtained from natural sources. In an attempt to resolve this issue, Jung *et al.* devised a simple four-step strategy for the synthesis of achiral amide analogs of the natural product,[259] where the analogs consisted of a hydroxamate function, a benzamide, and an aliphatic spacer, and the maximal inhibitory activity was observed with a five-carbon linker chain. The resulting lead compound was 6-(4-dimethylaminobenzoyl)-aminocaproic acid hydroxamide (**143**; **Figure 17**), and though the antitumor and cell transduction activities of these compounds have been reported relatively recently, no in vivo data has yet been published.[260]

The natural product trapoxin (**144**; **Figure 17**), a cyclic tetrapeptide with an epoxy side chain, was reported to be an irreversible inhibitor of HDACs in 1993 by Kijima *et al.*[261] In contrast to TSA, these investigators found that trapoxin demonstrated some selectivity against class I and class II HDACs, inhibiting HDAC1 and HDAC4 but not HDAC6. This particular compound and its 'warhead' acted as the structural motifs for the synthesis of a series of cyclic hydroxamic acid-containing compounds (CHAPs), in which the tetrapeptide motif was maintained, but with significant changes in the amino acids and with the epoxide replaced by a hydroxamate.[253] The lead compound CHAP-31 (**145**; **Figure 17**) demonstrated reversible inhibition of HDACs, again with a preference for HDAC1 and HDAC4 over HDAC6, and demonstrated growth arrest in various cell lines and growth inhibition in mouse xenograft models. Unfortunately, toxicity was observed on repeated dosing at the level required for long-term growth regression; therefore, CHAP-31 was not advanced to clinical trials, although other derivatives may yet be good candidates.[255]

The microbially derived depsipeptide FR-901228 (**146**; **Figure 17**), which is a four amino acid-containing cyclic macrolide with a second ring system containing a dithio linkage, was originally identified as a result of its potent antitumor activity, and is now known to be active in signal transduction as a result of its HDAC activity[262]; it is currently in phase II and III clinical trials.

The final compound in this section, NVP-LAQ-824 (dacinostat; **147** in **Figure 17**) is totally synthetic but its structure results from a combination of chemical and biological information from three natural products and (semi-) synthetic variations. The structure of this compound comprises the structural features from the marine natural product psammaplin A (**19** in **Figure 4**), a potent HDAC inhibitor, in combination with trapoxin and TSA. The full history of the evolution of NVPLAQ-824 has been reviewed by Remiszewski.[263]

1.08.8.6 Heat Shock Protein 90 (hsp90) Inhibitors

1.08.8.6.1 Ansamycins: geldanamycin (GA) derivatives

The development of the ansamycins leading to the 17-substituted analogs has recently been reviewed by Snader.[264] 17-Allylaminogeldanamycin (17-AAG; **148** in **Figure 18**) is currently in phase II clinical trials while the more soluble 17-dimethylamino-ethylaminogeldanamycin (17-DMAG; **149** in **Figure 18**) is in phase I clinical trials; a number of other 17-substituted derivatives have been prepared as potential alternative candidates.[265,266]

Over the past 3–4 years, the interactions of GA derivatives and radicicol (monorden; **150** in **Figure 18**) with hsp90 have been under study due to at least two apparent anomalies in their responses. Thus,

1. the difference in response to these drugs between healthy and tumor cells, despite the fact that both require hsp90 for cellular function; and
2. the affinity of these drugs for recombinant hsp90 (rhsp90) being much lower than the levels required for responses in tumor cell lysates.

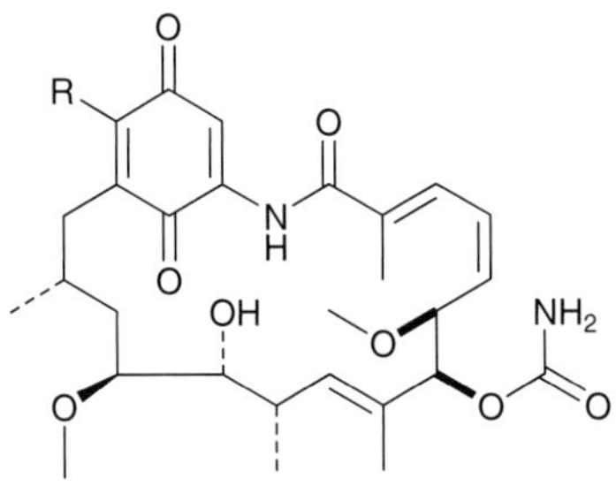

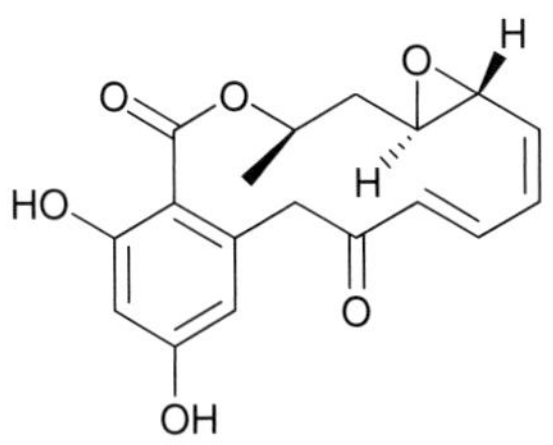

150

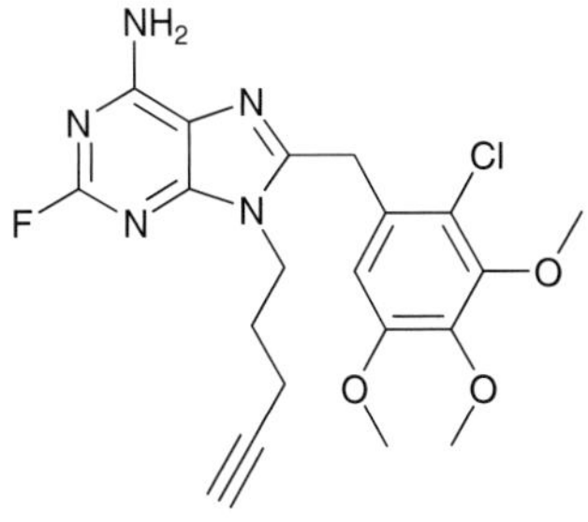

151

148: R = $NHCH_2{=}CH_2$
149: R = $NHCH_2CH_2N(CH_3)_2$

Figure 18 HSP90 inhibitors.

In late 2003, some potential reasons for these anomalies were reported by Kamal *et al.*[267] The higher binding affinity was attributed to the existence of other co-chaperones in tumor cells that are not expressed in healthy cells, and this effect was demonstrated by the addition of such proteins to rhsp90. In addition, x-ray crystallography studies have demonstrated that the structure of GA in the unbound form has a *trans*-configuration at the amide bond between the benzoquinone and the rest of the ansa ring, whereas when bound to hsp90, GA displays the *cis*-configuration at this center.[268] Similarly, Jez *et al.* reported that the closely related GA derivative, 17-DMAG, requires both a macrocyclic ring conformational change and a *trans–cis* isomerization of the amide bond in order to bind to hsp90.[269]

There is one important chemical point that is relevant in any discussion of GA derivatives; if the structures provided for the base GA structure in many of the references are compared, significant differences in the macrocyclic ring stereochemistries are observed for what is nominally the same molecule. These differences are not simply due to a complete stereochemical inversion around the ring, where the relative stereochemistries are maintained, but are quite different renditions from different research groups. The problems involved in representing these compounds, and closely related ansamycins, have been alluded to in the review by Snader.[264]

1.08.8.6.2 Nonansamycin inhibitors

In an attempt to overcome the supply and toxicity problems associated with GA derivatives, and the supply problems of radicicol, Chiosis *et al.* suggested a simple substituted adenine derivative as a potential base molecule. What was most impressive about this series of compounds, aside from their activities, was that they were not derived by computerized modeling but were the product of contemplating which particular substructures might provide ATP mimics with improved binding characteristics.

Knowledge of the requirements for the ATP-binding pocket of hsp90, and demonstration that a small molecule could function as a cytostatic agent,[270] provided the intellectual impetus to design the purine-based PU class of compounds.[271] Rational changes in the substituents in both rings and alteration of the length and rigidity of the linker gave rise to PU24FCl (**151**; **Figure 18**),[272] which, although not the most active in the series, was utilized to further investigate hsp90 inhibition in both healthy and tumor cells. In a recent paper, Chiosis *et al.* reported the extensive effects exhibited by both healthy and tumor tissues when exposed to the compound.[273] As with 17-AAG and GA, PU24FCl exhibited at least 10-fold (brain, pancreas, lung) to 50-fold (heart, kidney, liver) lower affinity for hsp90s from healthy tissues compared with those from transformed cells.

Thus, this particular class of molecules, designed by medicinal chemists from knowledge of the natural products GA and ATP and their binding sites, without the use of computerized modeling, proved to be highly successful.

1.08.8.7 Protein Kinase Inhibitors

Over the past few years, a number of agents derived directly from nature or incorporating key structural features from natural products have moved into clinical trials or into the market as drugs. Thus, both imatinib mesylate (Gleevec; **152** in **Figure 19**) and gefitinib (Iressa; **153** in **Figure 19**) can be traced back to ATP mimicry; the history of the former is given in a review by Newman *et al.*,[12] and the history of the latter is similar.

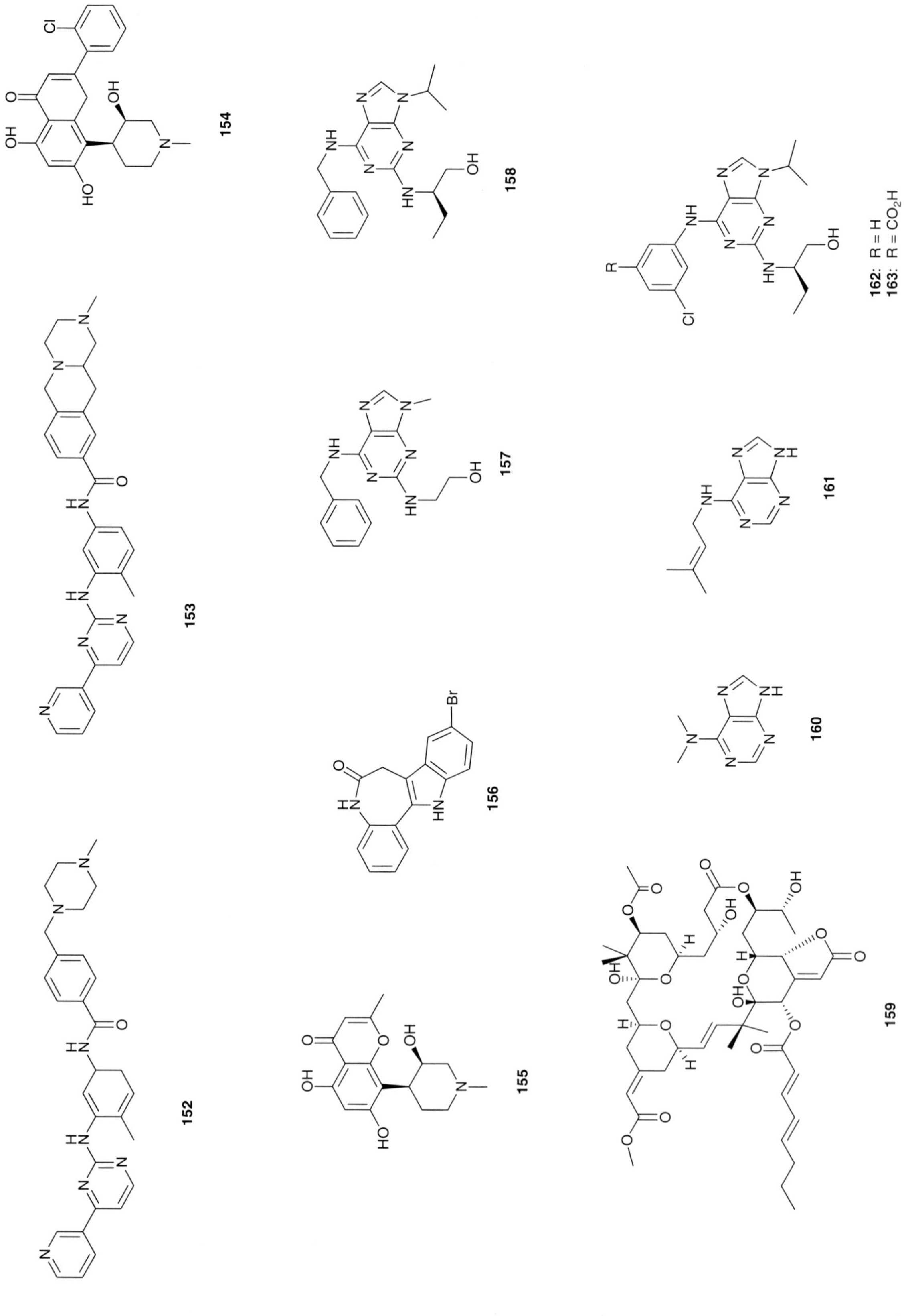

Figure 19 Protein kinase inhibitors.

1.08.8.7.1 Flavopiridol (Alvocidib)

The flavone, flavopiridol (**154**; **Figure 19**), has a very interesting mechanism of action in that it was originally considered to be an inhibitor of cyclin-dependent kinases (the regulators of the G_2 to M transition in the cell cycle), and was entered into phase I and then phase II clinical trials against a broad range of tumors,[274] though the single agent clinical trials appeared to have ceased in 2004 (Prous Integrity database). Very recently, late 2005, trials against some leukemias are being restarted at Ohio State University with the help of Aventis (pers. commun. Dr M Grever). While flavopiridol is totally synthetic, the basis for its novel structure is the natural product rohitukine (**155**; **Figure 19**) isolated from *Dysoxylum binectariferum*. This discovery has led to a series of compounds (the paullones as exemplified by structure **156** in **Figure 19**) from the NCI open compound database that, though not natural products, would not have been discovered but for the use of the natural product-derived agent as a 'seed compound' in the NCI's COMPARE algorithm. This story has been described in detail by Sausville *et al.*[275]

However, as mentioned above, flavopiridol as a monotherapy appears to have been discontinued, but it has now been reported that flavopiridol is, like the olomucine (**157**; **Figure 19**) derivative roscovitine (selicicib) (**158**; **Figure 19**), a very potent inhibitor of cyclin-dependent kinase 7 (CDK-7) and CDK-9, the kinases primarily responsible for promoting RNA polymerase II (RNAP II) activity, thus, involving these agents in the transcription process. Their potential for combination therapies with cytotoxic agents is being investigated. The very recent review by Fischer and Gianella-Borradori[276] and the report by Mayer *et al.*[277] should be consulted by the interested reader for fuller details regarding the molecular targets/interactions involved in the transcription processes and flavopiridol interactions.

1.08.8.7.2 Bryostatins

The bryostatins are a class of highly oxygenated macrolides originally isolated by the Pettit group under the early NCI program (1955–82) designed to discover novel antitumor agents from natural sources. The initial discovery of bryostatin 3 (**159**; **Figure 19**) was indirectly reported in 1970.[278] Subsequent developments leading to the report of the isolation and x-ray structure of bryostatin 1 (**33**) in 1982,[279] and the multi-year program that culminated in the isolation and purification of (currently) 20 bryostatin structures, have been well documented by a variety of authors over the years.[280–286] These reviews show that the bryostatins, and in particular bryostatin 1 (from which most of the experimental and all clinical data have been derived), have signal transduction activities; this feature is currently employed clinically as a component of sequential experimental treatments. A recent review by Newman[287] should be consulted for further details on these clinical trials.

At present, the total synthesis of bryostatin 1 is not a feasible process for the production of this agent even though three of the naturally occurring bryostatins have been synthesized. Enantioselective total synthesis of bryostatin 7 was achieved in 1990 by Masamune's group,[288] followed by bryostatin 2 in 1999 by Evans *et al.*[289] This was then followed by the synthesis of bryostatin 3 in 2000 by Nishiyama and Yamamura.[290] In addition to these papers, three excellent review articles (covering results up to 2002) on the synthesis of these three and other partial bryostatin structures, including bryostatin 1, have been published, and should be consulted for specific details of reaction schemes and comparison of routes.[284,286,291] None of these methods, however, are viable for the synthesis of any of the bryostatins for further work.

However, synthesis of a simpler analog with comparable activity might well be a viable option. Fortuitously, in 1986 Wender *et al.* had analyzed the potential binding site of the phorbol esters on PKC as a guide to the design of simpler analogs of these agents.[113] The work that led to these simpler molecules, known colloquially as 'bryologs' (**35–39**), has been covered in Section 1.08.3.7.

1.08.8.7.3 Adenine derivatives

Meijer and Raymond demonstrated that substituted purines, particularly 6-dimethylamino-purine (6-DMAP; **160** in **Figure 19**) and isopentenyladenine (**161**; **Figure 19**), from *Castanea* species, showed weak inhibition of the mitotic histone H1 kinase, better known as CDK1/cyclin B.[292] On searching for other purine-derived compounds, this research group tested a plant secondary metabolite, subsequently named olomucine (**157**), that was originally isolated from the cotyledons of the radish, but had been synthesized in Australia in 1986 by Parker *et al.*[293] Olomucine disproved the existing dogma that that no specific kinase inhibitors could be found for ATP-binding sites as they would be swamped by the excess of ATP, since it demonstrated an improved efficacy ($IC_{50} = 7\,\mu M$) and selectivity for CDKs and, to some extent, MAP kinases, by direct competition with ATP. Further development of this series led to roscovitine (**158**), and finally to purvalanol A (**162**; **Figure 19**) and purvalanol B (**163**; **Figure 19**) from a focused library synthesized using combinatorial chemistry techniques. The purvalanols demonstrated improved potency, with IC_{50} values in the 4–40 nM range, compared to 450 nM for roscovitine.[294] The *R*-isomer of roscovitine is currently in phase II clinical trials in Europe. As with other signal transduction inhibitors (STIs) in clinical trials, sequential treatment with cytotoxins is also being used and/or considered. Although some beneficial effects are observed with the STIs alone, complete or partial responses tend only to be demonstrated when sequential treatments of STI/cytotoxin are used.

1.08.8.7.4 Indigo and the indirubins

Indigo and the indirubins are based upon the simple bicyclic heterocycle indole. If indole is hydroxylated in the 3-position, presumably by a suitable cytochrome P450, then it is tautomeric with the 3-keto analog indoxyl (**23**). Various levels of oxidation then lead to a mixture of indigo, indirubin (**24**), and their isomers, which is commonly used as the source of indigo dyestuffs for either clothing or was used as a body paint (the ancient Celtic 'woad'). This mixture is obtained from the plant *Isatis tinctoria,* which has been found to contain an indigo precursor.[295]

Although usually thought of as plant products, indigo and the indirubins have been reported from four nominally independent sources: a variety of plants,[295] a number of marine mollusks, usually belonging to the *Muricidae* family of gastropods,[296] natural or recombinant bacteria,[297] and human urine.[298] The indirubins have been identified as the major active ingredient of the traditional Chinese medicine formulation known as Danggui Longhui Wan, which has been used for many years to treat chronic myelogenous leukemia (CML) in China.[299]

Importantly, from both a natural product and a pharmacological perspective, Meijer *et al.* recognized that the indirubins were both inhibitors of several CDKs and potent inhibitors of GSK-3.[300] This study included 6-bromoindirubin (**164**; **Figure 20**) and its chemically modified oxime derivative BIO (bromoindirubin oxime) (**165**; **Figure 20**). 6-Bromoindirubin had not been isolated from natural sources prior to the research by Meijer *et al.* with the mollusk *Hexaplex trunculus.*[292] 6-Bromoindirubin and BIO demonstrated at least a fivefold specificity versus CDK1/cyclin B and/or CDK/p25, and significantly greater specificity against a wide range of other kinases. A later paper from the same research group provided full details of the chemistry involved, and established SARs using x-ray crystallography and molecular modeling techniques.[301]

Using the same basic suite of compounds it was demonstrated that indirubins serve as ligands for the 'orphan receptor' known as the arylhydrocarbon receptor (AHR).[302] No other natural ligands have been identified for AHR to date, even though, contrary to earlier beliefs, it has existed for over 450 million years. Significantly, GSK-3 is also an important target in both Alzheimer's disease and type 2 diabetes, and although there have been no reports in the literature of indole derivatives associated with pharmacological intervention in these specific disease areas, their

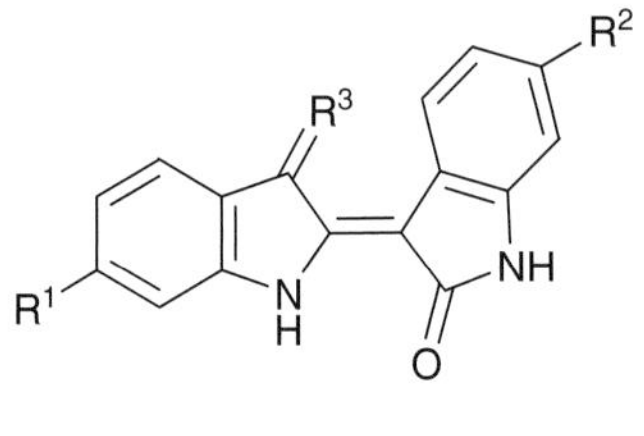

164: R^1 = H, R^2 = Br, R^3 = O
165: R^1 = H, R^2 = Br, R^3 = N-OH

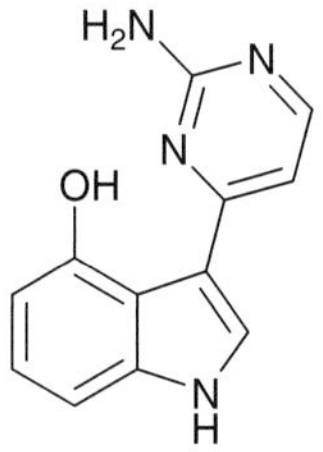

166

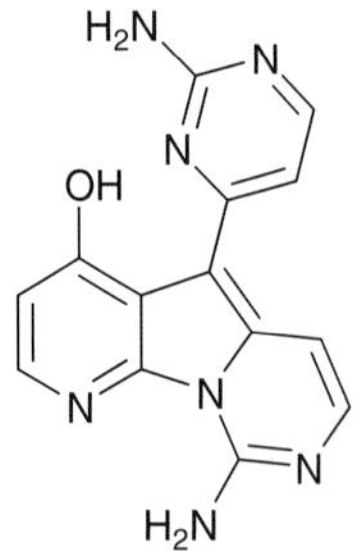

167

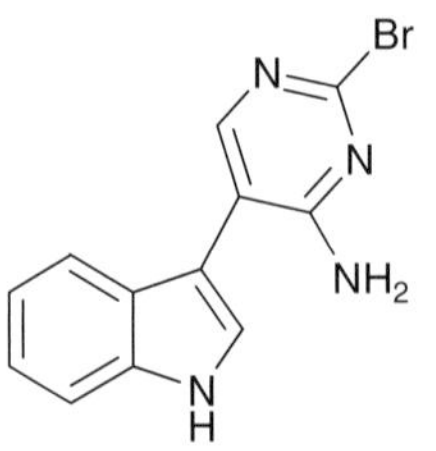

168

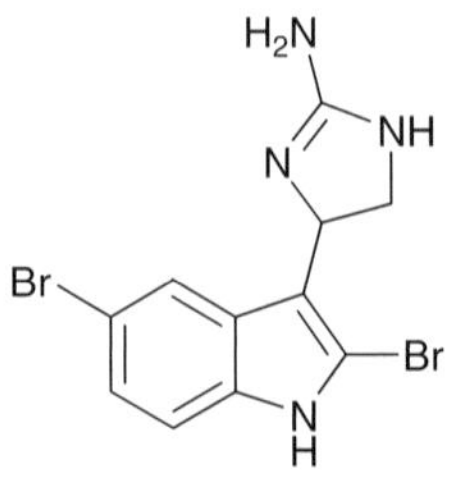

169

Figure 20 Indole-related inhibitors.

potential must be considered quite high. An excellent overview of the treatment potential for inhibitors of GSK-3 has recently been published, and should be consulted by the interested reader, particularly as other natural product-related structures are listed in the paper as possible inhibitors in these disease states.[303]

Other natural products with indirubin-like kinase inhibitory activities include the meridianins (e.g., meridianin A; **166** in **Figure 20**), a group of halogenated indole derivatives that are closely related to the base structures of variolin B (**167**; **Figure 20**), the psammopemmins (e.g., psammopemmin A; **168** in **Figure 20**), and discodermindol (**169**; **Figure 20**). Unlike variolin B, psammopemmins, and discodermindol, which were isolated from sponges, the meridianins were isolated from the ascidian *Aplidium meridianum*.[304]

1.08.8.7.5 Protein folds and kinase inhibitors

A significant amount of effort has been, and continues to be, directed at the 'fitting of structures to the ATP-binding sites' to develop novel kinase inhibitors, and this approach has been relatively successful in developing structures for clinical trials.[305] However, a variation on this theme has been successfully developed over the last few years by Waldmann *et al.* in which, rather than initially concentrating on the specifics of the ATP-binding site, two other fundamental premises were used to search for kinase and other enzyme inhibitors.

The first is that biologically active natural products are viable, biologically validated starting points for library design, permitting the discovery of lead compounds with an enhanced probability of success if included in high-throughput screening.[86,87] The second is that although there are estimates of between 100 000 and 450 000 proteins in humans, the number of topologically distinct 'shapes' defined as 'protein folds' is actually much lower, with estimates of 600–8000.[306]

Waldmann *et al.* considered, therefore, that if an inhibitor of a specific 'protein fold' from nature could be found, then it could be used as a prototype from which closely related structures could be developed that might inhibit proteins with similar 'folds,' and even allow the discovery of specificity. These concepts are fundamentally similar to the 'privileged structure' concept (Section 1.08.3.6), but the Waldmann approach has the added dimension of using protein folding patterns as the basis for subsequent screens.

The success of this approach was demonstrated by the derivation of inhibitors of Tie-2, insulin-like growth factor 1 receptor (IGF-1R), and vascular endothelial growth factor receptor 2 (VEGFR-2) and VEGFR-3 from the original discovery of the c-ErbB2 inhibitor nakijiquinone C (**170**; **Figure 21**).

Nakijiquinone C, derived from a marine sponge and first reported by Kobayashi *et al.* in 1995, was demonstrated to be an inhibitor of epidermal growth factor receptor (EGFR), c-ErbB2, and PKC, in addition to having cytotoxic activity against L1210 and KB cell lines.[307] Using the basic structure of the sesquiterpene, Waldmann *et al.* built a library of 74 compounds and, on testing against a battery of kinases with similar protein domain folds, were able to identify seven new inhibitors with low micromolar activity in vitro.[85] These included one VEGFR-2 inhibitor (**171**; **Figure 21**) and four inhibitors of Tie-2 kinase (**172–175**; **Figure 21**), a protein intimately involved in angiogenesis, and for which, at the beginning of the study, no inhibitors were known. During the study, the first natural product inhibitor of Tie-2 kinase was reported[308] (**176**; **Figure 21**) from the plant *Acacia aulacocarpa*, and a set of four papers from another research group demonstrated the activity of synthetic pyrrolo[2,3-*d*]pyrimidines as inhibitors of the same class of kinases.[309–312] The details of the models used, the chemistry leading to the nakijiquinone-based compounds, and the ribbon structures of the kinase domain of the insulin receptor, with the corresponding homology domains of the as yet uncrystallized VEGFR-2 and Tie-2, are reported in a series of papers from the Waldmann research group, with a full review being published in 2003[313] and an update in 2005.[90]

1.08.8.8 Hypoxia Inducible Factor (HIF)

Hypoxia inducible factor-1 (HIF-1) is composed of two subunits: an oxygen-sensitive inducible factor (HIF-1α) and the constitutive HIF-1β (also known as AHR nuclear translocator; ARNT). This complex may prove to be an important target in diseases that have a hypoxic component such as cancer (where the interior of a tumor is anoxic compared with the outer surfaces), heart disease, and/or stroke. The involvement of HIF proteins with a variety of inhibitors (not necessarily direct inhibition, but alteration of transduction pathways upstream and downstream) have been reviewed by Giaccia *et al.*[314] Included in the review are well-known materials with natural product 'backgrounds,' such as Taxol, vincristine, 2-methoxyestradiol, rapamycin, GA, quinocarmycin, and the IP3K inhibitors wortmannin (**31**) and LY-294002.

Of significance from a natural product perspective was the initial realization that inhibition of thioredoxin reductase 1 (TRX-1) may act indirectly on HIF-1α. By using the NCI COMPARE algorithm, Powis *et al.* discovered that the fungal natural product pleurotin (**177**; **Figure 22**) exhibited a similar killing pattern to the simple thio compound PX-12 (**178**; **Figure 22**), which is currently undergoing phase I clinical trials.[315] Lazo *et al.* conducted research into a small but focused combinatorial library based upon the palmarumycin naphthoquinone acetals[316] (e.g., palmarumycin

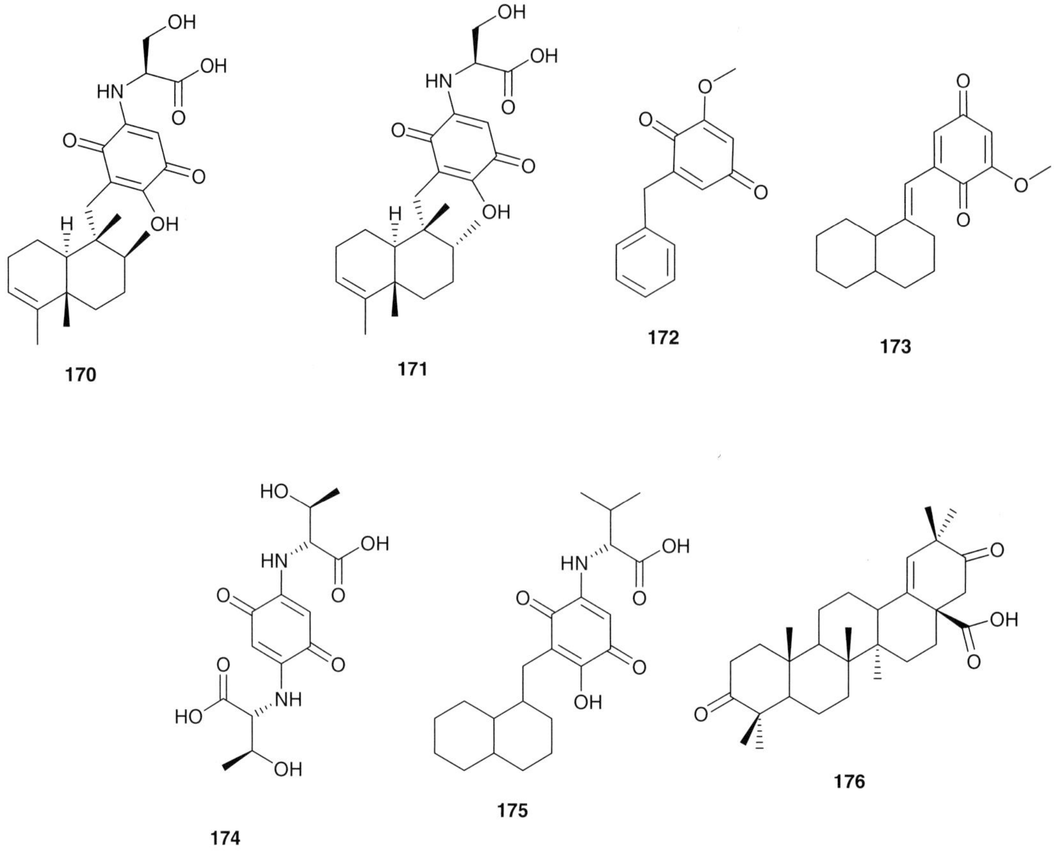

Figure 21 Nakjikaquinone derivatives.

CP1; **179** in **Figure 22**), which include diepoxins (e.g., diepoxin; **180** in **Figure 22**) and deoxypreussomerins (e.g., deoxypreussomerin A; **181** in **Figure 22**).

The palmarumycins and derivatives initially reported by Lazo *et al.* were exposed to many pharmacological screens, but although they showed potent cytotoxicity, their potential targets were unidentified at that time. Approximately 1 year after the submission of that paper, Wipf *et al.* reported that palmarumycin CP1 had inhibitory activity comparable to that of pleurotin, with IC_{50} values in the 170–350 nM range when assayed against thioredoxin/thioredoxin reductase, and demonstrated that certain aspects of the base structure, in particular the enone system, were required for activity in this assay.[317]

In 2003, Welsh *et al.* reported evidence for direct inhibition of HIF-1α by both pleurotin and PK12, helping to demonstrate that the cytotoxicity of these compounds was likely due to these interactions.[318] This paper was followed in 2004 by another paper from the Wipf research group, in collaboration with scientists from the Instituto Nacional de Biodiversidad (INBio), Costa Rica, who had discovered further palmarumycins in extracts from the fermentation broth of an unidentified ascomycete from the Guanacaste region of Costa Rica.[319] These compounds, although effectively inactive in the assays used, provided important SAR information, leading in turn to further modifications of the base structure, yielding the simple analogs S-11 (**182**; **Figure 22**) and S-12 (**183**; **Figure 22**), which exhibited biological activities comparable to pleurotin in both the thioredoxin enzyme system and (most importantly) in the cytotoxicity assays. Further biological studies, including in vivo experiments, are planned and will be reported in due course.

1.08.8.9 Miscellaneous Target Inhibitors

There are a number of agents, particularly from marine sources, whose 'initial' molecular targets have been identified, though it is highly probable that over the next few years these initial targets will be refined as new methods and other information becomes available. Two of these agents that are currently in clinical trials are discussed below.

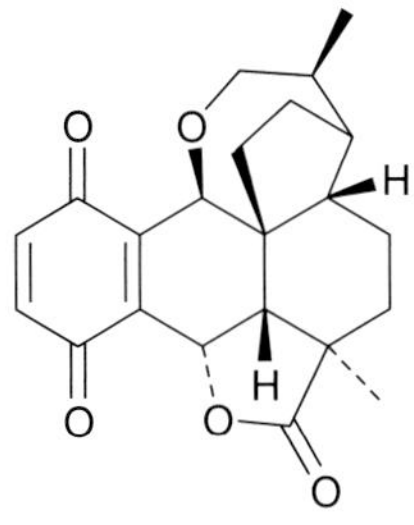

177

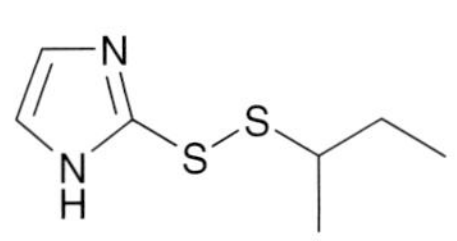

178

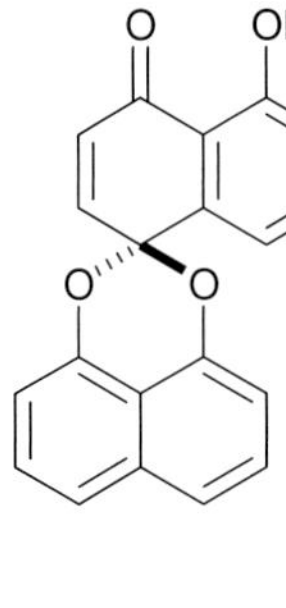

179

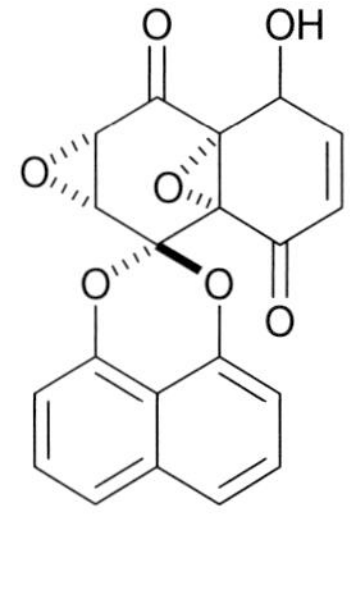

180

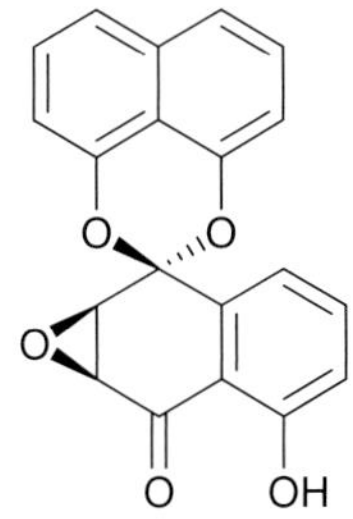

181

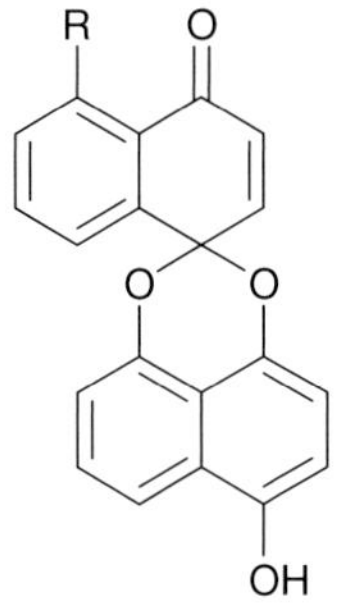

182: R = H
183: R = OH

Figure 22 HIF and thioredoxin inhibitors.

1.08.8.9.1 Ecteinascidin 743

The antitumor activity of extracts from the ascidian *Ecteinascidia turbinata* had been reported as early as 1969 by Sigel *et al.*[320] but it was not until 1990 that the structures of the active components, a series of related alkaloids, were published simultaneously by Rinehart *et al.*[321] and Wright *et al.*,[322] with the most stable member of the series being ecteinascidin 743 (Et743; **184** in **Figure 23**). The base structure, without the exocyclic isoquinoline group, is a well known chemotype[323] originally reported from microbes, where the compound classes are saframycins, naphthyridinomycins, safracins, and quinocarcins. Similar molecules were reported from marine mollusks, i.e., jorumycin (**185**; **Figure 23**) from the nudibranch *Jorunna funebris*[324] and from sponges; the renieramycins, with the variation renieramycin S (**186**; Fig. XXIII), were recently reported by Amnuoypol *et al.*[325] However, with Et743, the exocyclic substituent was novel as was the bridging sulfur. The work by many research groups leading up to the production of Et743 in quantities and quality good enough for human clinical trials has been reported in a number of reviews and these should be consulted for those aspects of the story.[28,326]

Over the last few years, a considerable number of reports have been published in the literature giving possibilities as to the MOA(s) of Et743 when tumor cells are treated in vitro. A significant problem with some of the reports is that the concentration(s) used in the experiments are often orders of magnitude greater than those that demonstrate activity in vivo. These levels are in the low nanomolar to high picomolar range and thus care should be taken when evaluating published work on the MOA of this compound.

At physiologically relevant concentrations the MOAs of Et743 have been shown to include the following: effects on the transcription-coupled nucleotide excision repair process (TC-NER)[327,328] and interaction between the Et743 DNA adduct and DNA transcription factors, in particular the NF-Y factor.[329] In a review published in 2003 by van Kesteren *et al.*[330] other possible mechanisms are given in their table 1; the references cited therein should be consulted for in-depth information and discussion on other potential MOAs ascribed to Et743.

Since the publication of that review, a number of reports on other potential mechanisms, including genes related to apoptosis, cell cycle, transcription factors, growth factors/receptors and cyclin D1/D3, GRO1, and NFκB pathways have also been published with the majority being in abstract form (cf. references in the review by Newman and Cragg[28]),

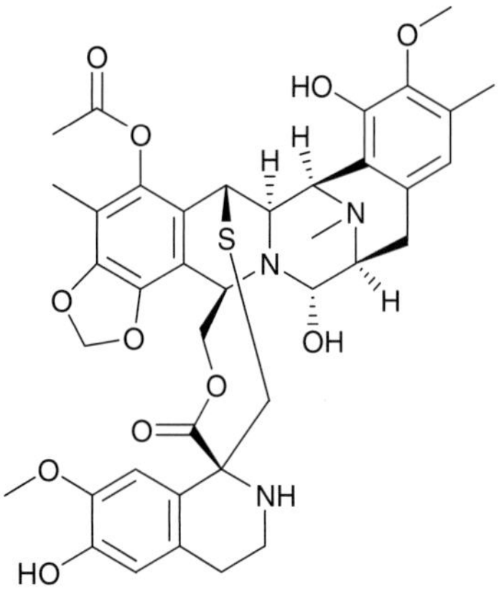

184

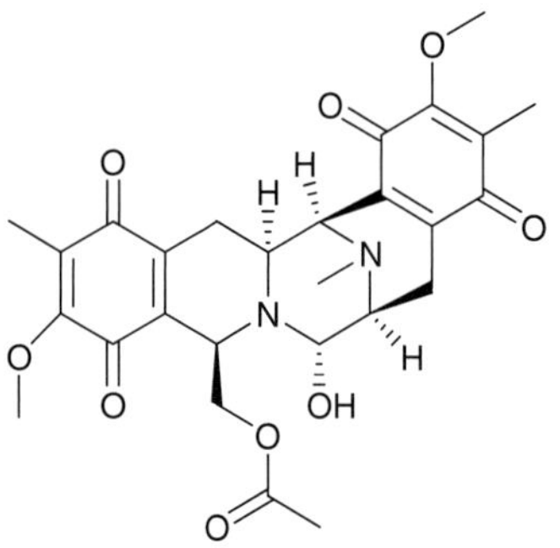

185

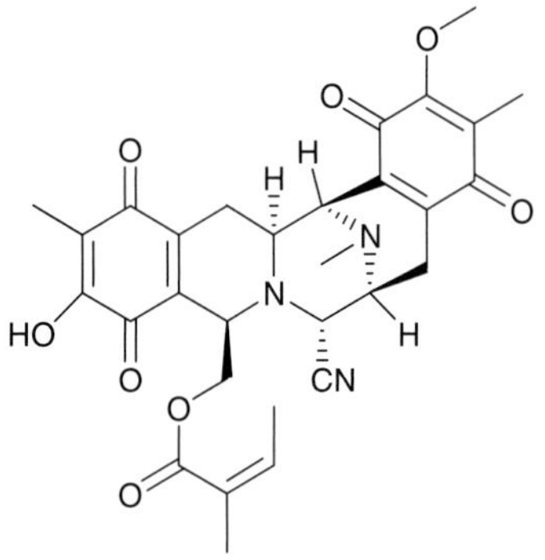

186

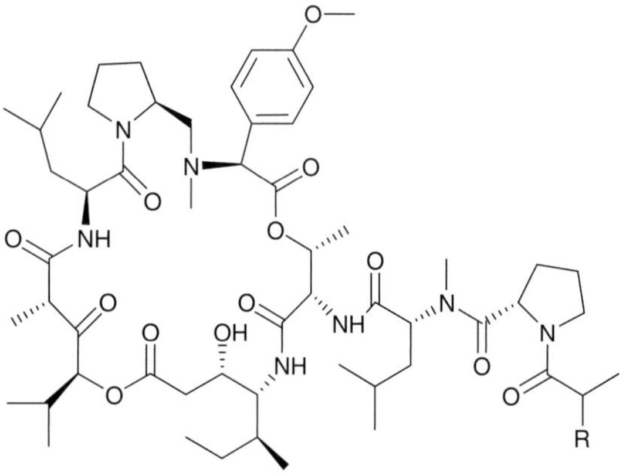

187: R = O
188: R = OH

Figure 23 Miscellaneous targets; inhibitors.

though a recent paper by Dziegielewska *et al.* sheds further light on the DNA alkylation properties of Et743 in the inhibition of simian virus 40 (SV40) DNA replication where it is at least 10 times more active in cells than saframycin (a chemical cousin) and its DNA adducts (replication intermediates) may be blocked in fork progression.[331] In the abstracts for the 2005 American Association for Cancer Research (AACR) meeting, there is one abstract by Mandola *et al.* that suggests that poly (ADP-ribose) polymerase (PARP) interactions with Et743 in vivo may play an important role in the cytotoxicity mechanism(s) of this agent.[332]

As alluded to earlier, although there are a number of other mechanisms postulated, on careful inspection, these are usually shown to occur at concentrations of drug well above (i.e., > ~250 nM) those that are physiologically relevant.[323]

Et743 was entered for human clinical trials while these mechanisms were being worked out, and many reports have been published in both abstract and full paper formats in the last 4–5 years. One of the latest full papers is a report by Le Cesne *et al.* on a phase II European trial in advanced sarcomas, where the authors considered that Et743 was a very promising drug candidate for specific soft tissue sarcomas.[333] In another paper by Sessa *et al.* a 43% objective response rate in resistant ovarian carcinomas was reported.[334] A potential problem with Et743 was the reports of hepatotoxicity in toxicology studies in animals, particularly the rat. However, clinical results suggest that human hepatotoxicity is controllable. Beumer *et al.* have recently provided a thorough analysis of this problem as reported in the literature to date, and conclude that human hepatoxicity is controllable.[335]

1.08.8.9.2 **Aplidine**

This compound (formally dehydrodidemnin B), an agent with multiple targets (**187**; **Figure 23**), was first reported in a patent application in 1989, with a UK patent issued in 1990 and was then referred to in the 1996 paper from Rinehart's group on SARs among the didemnins.[336] In 1996, its antitumor potential was reported by PharmaMar scientists[337,338] the total synthesis was reported in a patent application[339] in 2000 and the patent was issued in 2002.

The compound (generic name aplidine or dehydrodidemnin B; tradename Aplidin) was entered into phase I clinical trials in 1999 under the auspices of PharmaMar in Canada, Spain, France, and the UK for treatment of both solid tumors and non-Hodgkin's lymphoma. The results of these trials through to early 2004 are given in a report by Newman and Cragg together with a discussion of the possible mechanisms of action involved.[28]

Since then, further evidence has been published confirming the inhibition of endothelial cell functions related to angiogenesis including blockade of formation of matrix metalloproteinases (MMP-2 and MMP-9) at concentrations achievable during patient treatment.[340] Furthermore, by using dominant-negative kinase mutants of mouse embryo fibroblasts, Cuadrado *et al.* recently demonstrated that aplidine targets the essential kinase *JNK*.[341] Further confirmation of this pathway being the probable primary target for aplidine, at least in its induction of the apoptosis mechanism, came from a recent paper from PharmaMar where investigation of the reasons for an aplidine-resistant HeLa cell line being more than 1000-fold resistant implicated the bypassing of the MAPK activation pathway.[342]

In an abstract presented at the 2005 AACR meeting, Menon *et al.* reported that in studies with human leukemic cells (K562, CCRF-CEM, and SKI-DLCL) where aplidine exhibited synergy with cytarabine, when aplidine-treated cells were analyzed using the U133 GeneChip from Affymetrix, multiple cellular targets were implied, including downregulation of ribosomal 18S and 28S mRNA expression, and upregulation of various TNF-related ligands.[343]

It should be noted that clinical trials of the very close aplidine analog didemnin B (**188**; **Figure 23**) were discontinued because of the toxicities observed, including significant immunosuppression. However, in contrast, evidence for a lack of myelosuppression by aplidine, Et743, and kahalide F, compounds currently in phase II, II/III, and II trials, respectively, was reported by the PharmaMar group using a murine competitive repopulating model as the test system. However, confirmation of this effect in human bone marrow cells has not been published to date.[344]

What is very interesting, both chemically and pharmacologically, is that the removal of two hydrogen atoms, i.e., conversion of the lactyl side chain to a pyruvyl side chain, appears to significantly alter the toxicity profile, as this is the only formal change in the molecule when compared to didemnin B, though the comments on dosage regimens in the excellent review of didemnin B and analogs by Vera and Joullie[345] should be taken into account when such comparisons are made in the future.

Similarly, the resemblance to didemnin B is emphasized by the recent work of Cardenas *et al.* who reported[346] that in DMSO solution, aplidine, like didemnin B, does not exhibit a formal β-turn in its side chain in approximately 20% of its solution conformers, suggesting that the presence of such a turn is not required for biological activity. However, as the authors point out, there may well be other as yet unrecognized minor conformers that are responsible for some or all of the biological activities demonstrated.

The validity of these latter comments are supported by data given in the recent paper by Gutierrez-Rodriguez *et al.* on the modeling of aplidine and tamandarin A analogs with spirolactam β-turn mimetics, which implied the possibility of a peptidylprolyl *cis/trans* isomerase in the MOA of aplidine. This is another potential mechanism in addition to the results described above in the discussion on *JNK* activation, and the known inhibition of protein synthesis and ornithine decarboxylase activities.[347]

1.08.9 Conclusion

It can be seen from the discussion above that natural products from all sources still have the potential to lead chemists of all types into areas of drug discovery and development that would never have been considered if the 'privileged structures from Nature' had not been isolated, purified, and used as probes of cellular and molecular mechanisms. In spite of the discussions in the early to late 1990s about the vast potential of combinatorial chemistry as a discovery tool, it is now quite evident that this technique, except in the very special cases of peptides and nucleosides (which are actually 'privileged structures' in their own right), is not the panacea that it was once thought to be.

However, the use of combinatorial syntheses around a skeleton from a privileged structure as beautifully demonstrated by Nicolaou, Waldmann, and Wipf, to name but a few, demonstrates that the use of both techniques will lead to novel agents with potential as drug entities in many disease states, with the very recent review from Waldmann's group being a prime example of the processes involved.[348]

References

1. Anon. *Herbalgram* **1998**, 33.
2. Borchardt, J. K. *Drug News Perspect.* **2002**, *15*, 187.
3. Huang, K. C. *The Pharmacology of Chinese Herbs*, 2nd ed.; CRC Press: Boca Raton, FL, 1999.

4. Kapoor, L. D. *CRC Handbook of Ayurvedic Medicinal Plants*; CRC Press: Boca Raton, FL, 1990.
5. Dev, S. *Environ. Health Persp.* **1999**, *107*, 783–789.
6. Grabley, S.; Thiericke, R. *Adv. Biochem. Eng. Biotechnol.* **1999**, *64*, 101–154.
7. Johnson, T. *CRC Ethnobotany Desk Reference*; CRC Press: Boca Raton, FL, 1999.
8. Moerman, D. E. *Medicinal Plants of Native America*; University of Michigan Museum of Anthropology: Ann Arbor, MI, 1986; Vol. 1.
9. Farnsworth, N. R.; Akerele, R. O.; Bingel, A. S.; Soejarto, D. D.; Guo, Z. *Bull. World Health Org.* **1985**, *63*, 965–981.
10. Grifo, F.; Newman, D.; Fairfield, A. S.; Bhattacharya, B.; Grupenhoff, J. T. In *Biodiversity and Human Health*; Grifo, F., Rosenthal, J., Eds.; Island Press: Washington, DC, 1997, pp 131–163.
11. Newman, D. J.; Cragg, G. M.; Snader, K. M. *Nat. Prod. Rep.* **2000**, *17*, 215–234.
12. Newman, D. J.; Cragg, G. M.; Snader, K. M. *J. Nat. Prod.* **2003**, *66*, 1022–1037.
13. Class, S. *Chem. Eng. News* **2002**, *80*, 39–49.
14. Rouhi, M. *Chem. Eng. News* **2003**, *81*, 1377–1391.
15. Schröder, F. C.; Farmer, J. J.; Attygalle, A. B.; Smedley, S. R.; Eisner, T.; Meinwald, J. *Science* **1998**, *281*, 428–431.
16. Dumbacher, J. P.; Wako, A.; Derrickson, S. R.; Samuelson, A.; Spande, T. F.; Daly, J. W. *Proc. Natl. Acad. Sci. USA* **2004**, *101*, 15857–15860.
17. Luesch, H.; Moore, R. E.; Paul, V. J.; Mooberry, S. L.; Corbett, T. H. *J. Nat. Prod.* **2001**, *64*, 907–910.
18. Yotsu-Yamashita, M.; Kim, Y. H.; Dudley, S. C., Jr.; Choudhary, G.; Pfahnl, A.; Oshima, Y.; Daly, J. W. *Proc. Natl. Acad. Sci. USA* **2004**, *101*, 4346–4351.
19. Czaran, T. L.; Hoekstra, R. F.; Pagie, L. *Proc. Natl. Acad. Sci. USA* **2002**, *99*, 786–790.
20. Fitter, A. *Science* **2003**, *301*, 1337–1338.
21. Bais, H. P.; Vepachedu, R.; Gilroy, S.; Callaway, R. M.; Vivanco, J. M. *Science* **2003**, *301*, 1377–1380.
22. Rice, S. A.; McDougald, D.; Kumar, N.; Kjelleberg, S. *Curr. Opin. Investig. Drugs* **2005**, *6*, 178–184.
23. Borman, S. *Chem. Eng. News* **2005**, *83*, 38.
24. Bulaj, G.; Buczek, O.; Goodsell, I.; Jiminez, E. C.; Kranski, J.; Nielsen, J. S.; Garrett, J. E.; Olivera, B. M. *Proc. Natl. Acad. Sci. USA* **2003**, *100*, 14562–14568.
25. Fry, B. G. *Genome Res.* **2005**, *15*, 403–420.
26. Balandrin, M. F.; Kinghorn, A. D.; Farnsworth, N. R. In *Human Medicinal Agents from Plants;* American Chemical Society Symposium Series 534; Kinghorn, A. D., Balandrin, M. F., Eds.; American Chemical Society: Washington, DC, 1993, pp 2–12.
27. Raskin, I.; Ribnicky, D. M.; Komarnytsky, S.; Ilic, N.; Poulev, A.; Borisjuk, N.; Brinker, A.; Moreno, D. A.; Ripoll, C.; Yakoby, N. et al. *Trends Biotech.* **2002**, *20*, 522–531.
28. Newman, D. J.; Cragg, G. M. *J. Nat. Prod.* **2004**, *67*, 1216–1238.
29. Pace, N. R. *Science* **1997**, *276*, 734–740.
30. Madigan, M. T.; Martinko, J. M.; Parker, J. *Brock Biology of Microorganisms,* 8th ed.; Prentice-Hall: Upper Saddle River, NJ, 1996.
31. Walsh, C. T. *Science* **2004**, *303*, 1805–1810.
32. McAlpine, J. B.; Bachmann, B. O.; Piraee, M.; Tremblay, S.; Alarco, A.-M.; Zazopoulos, E.; Farnet, C. M. *J. Nat. Prod.* **2005**, *68*, 493–496.
33. Zengler, K.; Toledo, G.; Rappe, M.; Elkins, J.; Mathur, E. J.; Short, J. M.; Keller, M. *Proc. Natl. Acad. Sci. USA* **2002**, *99*, 15681–15686.
34. Rondon, M. R.; August, P. R.; Bettermann, A. D.; Brady, S. F.; Grossman, T. H.; Liles, M. R.; Loiacono, K. A.; Lynch, B. A.; MacNeil, I. A.; Minor, C. et al. *Appl. Environ. Microbiol.* **2000**, *66*, 2541–2547.
35. Venter, J. C.; Remington, K.; Heidelberg, J. F.; Halpern, A. L.; Rusch, D.; Eisen, J. A.; Wu, D.; Paulsen, I.; Nelson, K. E.; Nelson, W. et al. *Science* **2004**, *304*, 66–74.
36. Holden, C. *Science* **2005**, *307*, 1558.
37. Jin, M.; Liu, L.; Wright, S. A. I.; Beer, S. V.; Clardy, J. *Angew. Chem. Int. Ed.* **2003**, *42*, 2898–2901.
38. Persidis, A. *Nature Biotechnol.* **1998**, *16*, 593–594.
39. Rossi, M.; Ciaramella, M.; Cannio, R.; Pisani, F. M.; Moracci, M.; Bartolucci, S. *J. Bacteriol.* **2003**, *185*, 3683–3689.
40. Abe, F.; Horikoshi, K. *Trends Biotechnol.* **2001**, *19*, 102–108.
41. Cavicchioli, R.; Siddiqui, K. S.; Andrews, D.; Sowers, K. R. *Curr. Opin. Biotechnol.* **2002**, *13*, 253–261.
42. Gomes, J.; Steiner, W. *Food Technol. Biotechnol.* **2004**, *42*, 223–235.
43. Hoyoux, A.; Blaise, V.; Collins, T.; D'Amico, S.; Gratia, E.; Huston, A. L.; Marx, J. C.; Sonan, G.; Zeng, Y. X.; Feller, G.; Gerday, C. *J. Biosci. Bioeng.* **2004**, *98*, 317–330.
44. Wiegel, J.; Kevbrin, V. V. *Biochem. Soc. Trans.* **2004**, *32*, 193–198.
45. van den Burg, B. *Curr. Opin. Microbiol.* **2003**, *6*, 213–218.
46. Schiraldi, C.; De Rosa, M. *Trends Biotechnol.* **2002**, *20*, 515–521.
47. Phoebe, C. H.; Combie, J.; Albert, F. G.; Van Tran, K.; Cabrera, J.; Correira, H. J.; Guo, Y. H.; Lindermuth, J.; Rauert, N.; Galbraith, W.; Selitrennikoff, C. P. *J. Antibiot.* **2001**, *54*, 56–65.
48. Johnson, D. B.; Hallberg, K. B. *Res. Microbiol.* **2003**, *154*, 466–473.
49. Stierle, A. A.; Stierle, D. B.; Kemp, K. *J. Nat. Prod.* **2004**, *67*, 1392–1395.
50. Stierle, D. B.; Stierle, A. A.; Hobbs, D.; Stokken, J.; Clardy, J. *Org. Lett.* **2004**, *6*, 1049–1052.
51. Strobel, G. A.; Daisy, B.; Castillo, U.; Harper, J. *J. Nat. Prod.* **2004**, *67*, 257–268.
52. Tan, R. X.; Zou, W. X. *Nat. Prod. Rep.* **2001**, *18*, 448–459.
53. Castillo, U.; Harper, J. K.; Strobel, G. A.; Sears, J.; Alesi, K.; Ford, E.; Lin, J.; Hunter, M.; Maranta, M.; Ge, H. et al. *FEMS Microbiol. Lett.* **2003**, *224*, 183–190.
54. Li, J. Y.; Harper, J. K.; Grant, D. M.; Tombe, B. O.; Bashyal, B.; Hess, W. M.; Strobel, G. A. *Phytochemistry* **2001**, *56*, 463–468.
55. Ezra, D.; Castillo, U. F.; Strobel, G. A.; Hess, W. M.; Porter, H.; Jensen, J. B.; Condron, M. A.; Teplow, D. B.; Sears, J.; Maranta, M. et al. *Microbiology* **2004**, *150*, 785–793.
56. Mincer, T. J.; Jensen, P. R.; Kauffman, C. A.; Fenical, W. *Appl. Environ. Microbiol.* **2002**, *68*, 5005–5011.
57. Feling, R. H.; Buchanan, G. O.; Mincer, T. J.; Kauffman, C. A.; Jensen, P. R.; Fenical, W. *Angew. Chem. Int. Ed.* **2003**, *42*, 355–357.
58. Piel, J. *Nat. Prod. Rep.* **2004**, *21*, 519–538.
59. Piel, J.; Hui, D.; Wen, G.; Butzke, D.; Platzer, M.; Fusetani, N.; Matsunaga, S. *Proc. Natl. Acad. Sci. USA* **2004**, *101*, 16222–16227.
60. Piel, J.; Hofer, I.; Hui, D. *J. Bacteriol.* **2004**, *186*, 1280–1286.
61. Piel, J.; Butzke, D.; Fusetani, N.; Hui, D.; Platzer, M.; Wen, G.; Matsunaga, S. *J. Nat. Prod.* **2005**, *68*, 472–479.
62. Khosla, C. *J. Org. Chem.* **2000**, *65*, 8127–8133.

63. Staunton, J.; Weissman, K. J. *Nat. Prod. Rep.* **2001**, *18*, 380–416.
64. Clardy, J.; Walsh, C. T. *Nature* **2004**, *432*, 829–837.
65. Thomas, M. G.; Bixby, K. A.; Shen, B. In *Anticancer Agents from Natural Products*; Cragg, G. M., Kingston, D. G. I., Newman, D. J., Eds.; Taylor & Francis: Boca Raton, FL, 2005, pp 519–552.
66. Molnar, I.; Schupp, T.; Ono, M.; Zirkle, R. E.; Milnamow, M.; Nowak-Thompson, B.; Engel, N.; Toupet, C.; Stratmann, A.; Cyr, D. D. et al. *Chem. Biol.* **2000**, 7, 97–109.
67. Julien, B.; Shah, S.; Ziemann, R.; Goldman, R.; Katz, L.; Khosla, C. *Gene* **2000**, *249*, 153–160.
68. Lau, J.; Frykman, S.; Regentin, R.; Ou, S.; Tsuruta, H.; Licari, P. *Biotechnol. Bioeng.* **2002**, *78*, 280–288.
69. Service, R. F. *Science* **1999**, *285*, 186.
70. Nicolaou, K. C.; Snyder, S. A. *Angew. Chem. Int. Ed.* **2005**, *44*, 1012–1044.
71. Freemantle, M. *Chem. Eng. News* **2004**, *82*, 33–35.
72. Mickel, S. J.; Niederer, D.; Daeffler, R.; Osmani, A.; Kuesters, E.; Schmid, E.; Schaer, K.; Gamboni, R.; Chen, W. C.; Loeser, E. et al. *Org. Process Res. Dev.* **2004**, *8*, 122–130.
73. Mickel, S. J.; Sedelmeier, G. H.; Niederer, D.; Daeffler, R.; Osmani, A.; Schreiner, K.; Seeger-Weibel, M.; Berod, B.; Schaer, K.; Gamboni, R. et al. *Org. Process Res. Dev.* **2004**, *8*, 92–100.
74. Mickel, S. J.; Sedelmeier, G. H.; Niederer, D.; Schuerch, F.; Grimler, D.; Koch, G.; Daeffler, R.; Osmani, A.; Hirni, A.; Schaer, K. et al. *Org. Process Res. Dev.* **2004**, *8*, 101–106.
75. Mickel, S. J.; Sedelmeier, G. H.; Niederer, D.; Schuerch, F.; Koch, G.; Kuesters, E.; Daeffler, R.; Osmani, A.; Seeger-Weibel, M.; Schmid, E. et al. *Org. Process Res. Dev.* **2004**, *8*, 107–112.
76. Mickel, S. J.; Sedelmeier, G. H.; Niederer, D.; Schuerch, F.; Seger, M.; Schreiner, K.; Daeffler, R.; Osmani, A.; Bixel, D.; Loiseleur, O. et al. *Org. Process Res. Dev.* **2004**, *8*, 113–121.
77. Aicher, T. D.; Buszek, K. R.; Fang, F. G.; Forsyth, C. J.; Jung, S. H.; Kishi, Y.; Matelich, M. C.; Scola, P. M.; Spero, D. M.; Yoon, S. K. *J. Am. Chem. Soc.* **1992**, *114*, 3162–3164.
78. Yu, M. J.; Kishi, Y.; Littlefield, B. In *Anticancer Agents from Natural Products*; Cragg, G. M., Kingston, D. G. I., Newman, D. J., Eds.; Taylor & Francis: Boca Raton, FL, 2005, pp 241–266.
79. Morel, C. M.; Toure, Y. T.; Dobrokhotov, B.; Oduola, A. M. J. *Science* **2002**, *298*, 79.
80. Borman, S. *Chem. Eng. News* **2003**, *81*, 45-52.
81. Borman, S. *Chem. Eng. News* **2004**, *82*, 32–40.
82. Burke, M. D.; Berger, E. M.; Schreiber, S. L. *Science* **2003**, *302*, 613–618.
83. Burke, M. D.; Schreiber, S. L. *Angew. Chem. Int. Ed.* **2004**, *43*, 46–58.
84. Thutewohl, M.; Kissau, L.; Popkirova, B.; Karaguni, I.-M.; Nowak, T.; Bate, M.; Kuhlmann, J.; Muller, O.; Waldmann, H. *Angew. Chem. Int. Ed.* **2002**, *41*, 3616–3620.
85. Stahl, P.; Kissau, L.; Mazitschek, R.; Giannis, A.; Waldmann, H. *Angew. Chem. Int. Ed.* **2002**, *41*, 1174–1178.
86. Brohm, D.; Metzger, S.; Bhargava, A.; Muller, O.; Lieb, F.; Waldmann, H. *Angew. Chem. Int. Ed.* **2002**, *41*, 307–311.
87. Breinbauer, R.; Vetter, I. R.; Waldmann, H. *Angew. Chem. Int. Ed.* **2002**, *41*, 2878–2890.
88. Breinbauer, R.; Manger, M.; Scheck, M.; Waldmann, H. *Curr. Med. Chem.* **2002**, *9*, 2129–2145.
89. Ganesan, A. *Curr. Opin. Biotech.* **2004**, *15*, 584–590.
90. Koch, M. A.; Waldmann, H. *Drug Disc. Today* **2005**, *10*, 471–483.
91. Nicolaou, K. C.; Roschangar, F.; Vourloumis, D. *Angew. Chem. Int. Ed. Engl.* **1998**, *37*, 2014–2045.
92. Nicolaou, K. C.; Kim, S.; Pfefferkorn, J. A.; Xu, J.; Oshima, T.; Hosokawa, S.; Li, T. *Angew. Chem. Int. Ed.* **1998**, *37*, 1418–1421.
93. Evans, B. E.; Rittle, K. E.; Bock, M. G.; DiPardo, R. M.; Fredinger, R. M.; Whitter, W. L.; Lundell, G. F.; Veber, D. F.; Anderson, P. S.; Chang, R. S. L. et al. *J. Med. Chem.* **1988**, *31*, 2235–2246.
94. Nicolaou, K. C.; Pfefferkorn, J. A.; Roecker, A. J.; Cao, G.-Q.; Barluenga, S.; Mitchell, H. J. *J. Am. Chem. Soc.* **2000**, *122*, 9939–9953.
95. Nicolaou, K. C.; Pfefferkorn, J. A.; Barluenga, S.; Mitchell, H. J.; Roecker, A. J.; Cao, G.-Q. *J. Am. Chem. Soc.* **2000**, *122*, 9968–9976.
96. Nicolaou, K. C.; Pfefferkorn, J. A.; Mitchell, H. J.; Roecker, A. J.; Barluenga, S.; Cao, G.-Q.; Affleck, R. L.; Lillig, J. E. *J. Am. Chem. Soc.* **2000**, *122*, 9954–9967.
97. Nicolaou, K. C.; Pastor, J.; Barluenga, S.; Winssinger, N. *Chem. Commun.* **1998**, 1947–1948.
98. Nicolaou, K. C.; Pfefferkorn, J. A.; Schuler, F.; Roecker, A. J.; Cao, G.-Q.; Casida, J. E. *Chem. Biol.* **2000**, 7, 979–992.
99. Nicolaou, K. C.; Roecker, A. J.; Barluenga, S.; Pfefferkorn, J. A.; Cao, G.-Q. *ChemBioChem* **2001**, 460–465.
100. Nicolaou, K. C.; Hughes, R.; Pfefferkorn, J. A.; Barluenga, S.; Roecker, A. J. *Chem. Eur. J.* **2001**, 7, 4280–4295.
101. Nicolaou, K. C.; Hughes, R.; Pfefferkorn, J. A.; Barluenga, S. *Chem. Eur. J.* **2001**, 7, 4296–4310.
102. Nicolaou, K. C.; Cho, S. Y.; Hughes, R.; Winssinger, N.; Smethurst, C.; Labischinski, H.; Endermann, R. *Chem. Eur. J.* **2001**, 7, 3798–3823.
103. Nicolaou, K. C.; Evans, R. M.; Roecker, A. J.; Hughes, R.; Downes, M.; Pfefferkorn, J. A. *Org. Biomol. Chem.* **2003**, *1*, 908–920.
104. Chiyanzu, I.; Hansell, E.; Gut, J.; Rosenthal, P. J.; McKerrow, J. H.; Chibale, K. *Bioorg. Med. Chem. Lett.* **2003**, *13*, 3527–3530.
105. Kingston, D. G. I.; Newman, D. J. *Curr. Opin. Drug Disc. Dev.* **2005**, *8*, 207–227.
106. Newman, D. J.; Cragg, G. M. *Curr. Drug Targ.* **2006**, in press.
107. Schreiber, S. L. *Bioorg. Med. Chem.* **1998**, *6*, 1127–1152.
108. Pelish, H. E.; Westwood, N. J.; Feng, Y.; Kirchausen, T.; Shair, M. D. *J. Am. Chem. Soc.* **2001**, *123*, 6740–6741.
109. Wipf, P.; Stephenson, C. R. J.; Walczak, M. A. A. *Org. Lett.* **2004**, *6*, 3009–3012.
110. Wipf, P.; Minion, D. J.; Halter, R. J.; Berggren, M. I.; Ho, C. B.; Chiang, G. G.; Kirkpatrick, L.; Abraham, R.; Powis, G. *Org. Biomol. Chem.* **2004**, *2*, 1911–1920.
111. Ihle, N. T.; Williams, R.; Chow, S.; Chew, W.; Berggren, M. I.; Paine-Murrieta, G.; Minion, D. J.; Halter, R. J.; Wipf, P.; Abraham, R. et al. *Mol. Cancer Ther.* **2004**, *3*, 763–772.
112. Lyon, M. A.; Ducruet, A. P.; Wipf, P.; Lazo, J. S. *Nature Rev. Drug Disc.* **2002**, *1*, 961–976.
113. Wender, P. A.; Koehler, K. F.; Sharkey, N. A.; Dell'Aquila, M. L.; Blumberg, P. M. *Proc. Natl. Acad. Sci. USA* **1986**, *83*, 4214–4218.
114. Wender, P. A.; Cribbs, C. M.; Koehler, K. F.; Sharkey, N. A.; Herald, C. L.; Kamano, Y.; Pettit, G. R.; Blumberg, P. M. *Proc. Natl. Acad. Sci. USA* **1988**, *85*, 7197–7201.
115. Wender, P. A.; De Brabander, J.; Harran, P. G.; Hinkle, K. W.; Lippa, B.; Pettit, G. R. *Tetrahedron Lett.* **1998**, *39*, 8625–8628.
116. Wender, P. A.; De Brabander, J.; Harran, P. G.; Jimenez, J.-M.; Koehler, M. T. F.; Lippa, B.; Park, C.-M.; Shiozaki, M. *J. Am. Chem. Soc.* **1998**, *120*, 4534–4535.

117. Wender, P. A.; De Brabander, J.; Harran, P. G.; Jimenez, J.-M.; Koehler, M. T. F.; Lippa, B.; Park, C.-M.; Siedenbiedel, C.; Pettit, G. R. *Proc. Natl. Acad. Sci. USA* **1998**, *95*, 6624–6629.
118. Wender, P. A.; Hinkle, K. W.; Koehler, M. T. F.; Lippa, B. *Med. Res. Rev.* **1999**, *19*, 388–407.
119. Wender, P. A.; Lippa, B. *Tetrahedron Lett.* **2000**, *41*, 1007–1011.
120. Wender, P. A.; Hinkle, K. W. *Tetrahedron Lett.* **2001**, *41*, 6725–6729.
121. Wender, P. A.; Baryza, J. L.; Bennett, C. E.; Bi, F. C.; Brenner, S. E.; Clarke, M. O.; Horan, J. C.; Kan, C.; Lacote, E.; Lippa, B.; Nell, P. G.; Turner, T. M. *J. Am. Chem. Soc.* **2002**, *124*, 13648–13649.
122. Wender, P. A.; Mayweg, A. V. W.; VanDeusen, C. L. *Org. Lett.* **2003**, *5*, 277–279.
123. Wender, P. A.; Koehler, M. T. F.; Sendzik, M. *Org. Lett.* **2003**, *5*, 4549–4552.
124. Mooberry, S. L.; Tien, G.; Hernandez, A. H.; Plubrukarn, A.; Davidson, B. S. *Cancer Res.* **1999**, *59*, 653–660.
125. Pryor, D. E.; O'Brate, A.; Bilcer, G.; Diaz, J. F.; Wang, Y.; Wang, Y.; Kabaki, M.; Jung, M. K.; Andreu, J. M.; Ghosh, A. R. et al. *Biochemistry* **2002**, *41*, 9109–9115.
126. Mooberry, S. L.; Leal, R. M.; Randall-Hlubeck, D. A.; Davidson, B. S. *Clin. Canc. Res.* **2003**, *9*, A274 (Abs.).
127. Lu, H.; Murtagh, J.; Schwartz, E. L. *Proc. Am. Assoc. Cancer Res.* **2005**, *46*, 2989 (Abs.).
128. Ahmed, A.; Hoegenauer, E. K.; Enev, V. S.; Hanbauer, M.; Kaehlig, H.; Ohler, E.; Mulzer, J. *J. Org. Chem.* **2003**, *68*, 3026–3042.
129. Mulzer, J.; Ohler, E. *Chem. Rev.* **2003**, *103*, 3753–3786.
130. Uenishi, J.; Ohmi, M. *Angew. Chem. Int. Ed.* **2005**, *44*, 2756–2760.
131. Paterson, I.; Bergmann, H.; Menche, D.; Berkessel, A. *Org. Lett.* **2004**, *6*, 1293–1295.
132. Paterson, I.; Menche, D.; Hakansson, A. E.; Longstaff, A.; Wong, D.; Barasoain, I.; Buey, R. M.; Diaz, J. F. *Bioorg. Med. Chem. Lett.* **2005**, *15*, 2243–2247.
133. Gallagher, B. M., Jr.; Fang, F. G.; Johannes, C. W.; Pesant, M.; Tremblay, M. R.; Zhao, H.; Akasaka, K.; Li, X.-y.; Liu, J.; Littlefield, B. A. *Bioorg. Med. Chem. Lett.* **2004**, *14*, 575–579.
134. Gallagher, B. M., Jr.; Zhao, H.; Pesant, M.; Fang, F. G. *Tetrahedron Lett.* **2005**, *46*, 923–926.
135. Wender, P. A.; Hegde, S. G.; Hubbard, R. D.; Zhang, L.; Mooberry, S. L. *Org. Lett.* **2003**, *5*, 3507–3509.
136. Mooberry, S. L.; Randall-Hlubeck, D. A.; Leal, R. M.; Hegde, S. G.; Hubbard, R. D.; Zhang, L.; Wender, P. A. *Proc. Natl. Acad. Sci. USA* **2004**, *101*, 8803–8808.
137. Buss, A. D.; Waigh, R. D. In *Burger's Medicinal Chemistry, Drug Discovery*, 5th ed.; Wolff, M. E., Ed.; Principles and Practice, Vol. 1; John Wiley & Sons, Inc.: New York, 1995, pp 983–1033.
138. Goss, W. A.; Cook, T. M. In *Antibiotics III, Mechanism of Action of Antimicrobial and Antitumor Agents*; Corcoran, J. W., Hahn, F. E., Eds.; Springer-Verlag: Berlin, 1975; Vol. 3, pp 174–196.
139. Butler, M. S. *Nat. Prod. Rep.* **2005**, *22*, 162–195.
140. Shu, Y.-Z. *J. Nat. Prod.* **1998**, *61*, 1053–1071.
141. Jung, D.; Rozek, A.; Okon, M.; Hancock, R. E. W. *Chem. Biol.* **2004**, *11*, 949–957.
142. Jeu, L.; Fung, H. B. *Clin. Ther.* **2004**, *26*, 1728–1757.
143. Steenbergen, J. N.; Alder, J.; Thorne, G. M.; Tally, F. P. J. *Antimicrob. Chemother.* **2005**, *55*, 283–288.
144. Hancock, R. E. W. *Lancet Infect. Dis.* **2005**, *5*, 209–218.
145. McHenney, M. A.; Hosted, T. J.; Dehoff, B. S.; Rosteck, P. R., Jr.,; Baltz, R. H. *J. Bacteriol.* **1998**, *180*, 143–151.
146. Grunewald, J.; Sieber, S. A.; Mahlert, C.; Linne, U.; Marahiel, M. A. *J. Am. Chem. Soc.* **2004**, *126*, 17025–17031.
147. Huc, I.; Lehn, J.-M. *Proc. Natl. Acad. Sci. USA* **1997**, *94*, 2106–2110.
148. Lehn, J.-M. *Chem. Eur. J.* **1999**, *5*, 2455–2463.
149. Shuker, S. B.; Hajduk, P. J.; Meadows, R. P.; Fesik, S. W. *Science* **1996**, *274*, 1531–1534.
150. Maly, D. J.; Choong, I. C.; Ellman, J. A. *Proc. Natl. Acad. Sci. USA* **2000**, *97*, 2419–2424.
151. Xu, R.; Greiveldinger, G.; Marenus, L. E.; Cooper, A.; Ellman, J. A. *J. Am. Chem. Soc.* **1999**, *121*, 4898–4899.
152. Nicolaou, K. C.; Hughes, R.; Cho, S. Y.; Winssinger, N.; Labischinski, H.; Endermann, R. *Chem. Eur. J.* **2001**, 7, 3824–3843.
153. Yao, N.-H.; Liu, G.; He, W.-Y.; Niu, C.; Carlson, J. R.; Lam, K. S. *Bioorg. Med. Chem. Lett.* **2005**, *15*, 2325–2329.
154. Anstead, G. M.; Owens, A. D. *Curr. Opin. Infect. Dis.* **2004**, *17*, 549–555.
155. Bradley, J. S. *Curr. Opin. Pediatr.* **2005**, *17*, 71–77.
156. Akritopoulou-Zanze, I.; Phelan, K. M.; Marron, T. G.; Yong, H.; Ma, Z.; Stone, G. G.; Daly, M. M.; Hensey, D. M.; Nilius, A. M.; Djuric, S. W. *Bioorg. Med. Chem. Lett.* **2004**, *14*, 3809–3813.
157. Dominguez, J. M.; Kelly, V. A.; Kinsman, O. S.; Marriott, M. S.; Gomez De Las Heras, F.; Martin, J. J. *Antimicrob. Agents Chemother.* **1998**, *42*, 2274–2278.
158. Dominguez, J. M.; Martin, J. J. *Antimicrob. Agents Chemother.* **1998**, *42*, 2279–2283.
159. Herreros, E.; Martinez, C. M.; Almela, M. J.; Marriott, M. S.; Gomez De Las Heras, F.; Gargallo-Viola, D. *Antimicrob. Agents Chemother.* **1998**, *42*, 2863–2869.
160. Martinez, A.; Regadera, J.; Jimenez, E.; Santos, I.; Gargallo-Viola, D. *Antimicrob. Agents Chemother.* **2001**, *45*, 1008–1013.
161. Singh, S. B.; Zink, D. L.; Doss, G. A.; Polishook, J. D.; Ruby, C.; Register, E.; Kelly, T. M.; Bonfiglio, C.; Williamson, J. M.; Kelly, R. *Org. Lett.* **2004**, *6*, 337–340.
162. Nishimura, S.; Matsunaga, S.; Shibazaki, M.; Suzuki, K.; Furihata, K.; van Soest, R. W. M.; Fusetani, N. *Org. Lett.* **2003**, *5*, 2255–2257.
163. Nishimura, S.; Matsunaga, S.; Shibazaki, M.; Suzuki, K.; Harada, N.; Naoki, H.; Fusetani, N. J. *J. Nat. Prod.* **2002**, *65*, 1353–1356.
164. Clark, A. M. *Pharm. Res.* **1996**, *13*, 1133–1144.
165. Tan, R. X.; Zheng, W. F.; Tang, H. Q. *Planta Med.* **1998**, *64*, 295–302.
166. De Smet, P. A. G. M. *Drugs* **1997**, *54*, 801–840.
167. O'Neill, P. M.; Posner, G. H. *J. Med. Chem.* **2004**, *47*, 1–20.
168. Vennerstrom, J. L.; Arbe-Barnes, S.; Brun, R.; Charman, S. A.; Chiu, F. C. K.; Chollet, J.; Dong, Y.; Dorn, A.; Hunziker, D.; Matile, H. et al. *Nature* **2004**, *430*, 900–904.
169. Albericci, M.; Braekman, J. C.; Daloze, D.; Tursch, B. *Tetrahedron* **1982**, *38*, 1881.
170. Kashman, Y.; Rotem, M. *Tetrahedron Lett.* **1979**, *20*, 1707–1708.
171. Rudi, A.; Talpir, R.; Kashman, Y.; Benayahu, Y.; Schleyer, M. *J. Nat. Prod.* **1993**, *56*, 2178–2182.
172. El Sayed, K. A.; Dunbar, D. C.; Goins, D. K.; Cordova, C. R.; Perry, T. L.; Wesson, K. J.; Sanders, S. C.; Janus, S. A.; Hamann, M. T. *J. Nat. Toxins* **1996**, *5*, 261–285.

173. Donia, M.; Hamann, M. T. *Lancet Infect. Dis.* **2003**, *3*, 338–348.
174. Konig, G. M.; Wright, A. D. *Planta Med.* **1996**, *62*, 193–211.
175. Jomaa, H.; Wiesner, J.; Sanderbrand, S.; Altincicek, B.; Weidemeyer, C.; Hintz, M.; Turbachova, I.; Eberl, M.; Zeidler, J.; Lichtenthaler, H. K.; et al. *Science* **1999**, *285*, 1573–1576.
176. Iguchi, E.; Okuhara, M.; Kohsaka, M.; Aoki, H.; Imanaka, H. *J. Antibiot.* **1980**, *33*, 19–23.
177. Okuhara, M.; Kuroda, Y.; Goto, K.; Okamoto, M.; Terano, H.; Kohsaka, M.; Aoki, H.; Imanaka, H. *J. Antibiot.* **1980**, *33*, 13–17.
178. Ohler, E.; Kanzler, S. *Synthesis* **1995**, 539–543.
179. Singh, S. B.; Zink, D. L.; Liesch, J. M.; Mosley, R. T.; Dombrowski, A. W.; Bills, G. F.; Darkin-Rattray, S. J.; Schmatz, D. M.; Goetz, M. A. *J. Org. Chem.* **2002**, *67*, 815–825.
180. Bergmann, W.; Feeney, R. J. *J. Am. Chem. Soc.* **1950**, *72*, 2809–2810.
181. Bergmann, W.; Feeney, R. J. *J. Org. Chem.* **1951**, *16*, 981–987.
182. Bergmann, W.; Burke, D. C. *J. Org. Chem.* **1956**, *21*, 226–228.
183. Suckling, C. *J. Sci. Prog. Edin.* **1991**, *75*, 323–360.
184. Lee, W. W.; Benitez, A.; Goodman, L.; Baker, B. R. *J. Am. Chem. Soc.* **1960**, *82*, 2648–2649.
185. Davis, P. *Chem. Abstr.* **1969**, *71*, 79757z: UK 1159290, July 23, 1969.
186. Cimino, G.; De Rosa, S.; De Sttefano, S. *Experientia* **1984**, *40*, 339–400.
187. Rosemeyer, H. *Chem. Biodiv.* **2004**, *1*, 361–401.
188. Lagoja, I. M. *Chem. Biodiv.* **2005**, *2*, 1–50.
189. Darke, P. L.; Leu, C. T.; Davis, L. J.; Heimbach, J. C.; Diehl, R. E.; Hill, W. S.; Dixon, R. A.; Sigal, I. S. *J. Biol. Chem.* **1989**, *264*, 2307–2312.
190. Wiley, R. A.; Rich, D. H. *Med. Res. Rev.* **1993**, *13*, 327–384.
191. Rana, K. Z.; Dudley, M. N. *Pharmacotherapy* **1999**, *19*, 35–59.
192. Gago, F. *IDrugs* **1999**, *2*, 309.
193. Vacca, J. P.; Guare, J. P.; Desolms, S. J.; Sanders, W. M.; Giuliani, E. A.; Young, S. D.; Darke, P. L.; Zugay, J.; Sigal, I. S.; Schleif, W. A. et al. *J. Med. Chem.* **1991**, *34*, 1225–1228.
194. Abdel-Rahman, H. M.; Al-karamany, G. S.; El-Koussi, N. A.; Youssef, A. F.; Kiso, Y. *Curr. Med. Chem.* **2002**, *9*, 1905–1922.
195. Mimoto, T.; Hattori, N.; Takaku, H.; Kisanuki, S.; Fukazawa, T.; Terashima, K.; Kato, R.; Nojima, S.; Misawa, S.; Ueno, T. et al. *Chem. Pharm. Bull.* **2000**, *48*, 1310–1326.
196. Sohma, Y.; Hayashi, Y.; Skwarczynski, M.; Hamada, Y.; Sasaki, M.; Kimura, T.; Kiso, Y. *Biopolymers (Peptide Sci.)* **2004**, *76*, 344–356.
197. Specker, E.; Bottcher, J.; Lilie, H.; Heine, A.; Schoop, A.; Muller, G.; Griebenow, N.; Klebe, G. *Angew. Chem. Int. Ed.* **2005**, *44*, 3140–3144.
198. Singh, S. B.; Ondeyka, J. G.; Tsipouras, N.; Ruby, C.; Sardana, V.; Schulman, M.; Sanchez, M.; Pelaez, F.; Stahlhut, M. W.; Munshi, S. et al. *Biochem. Biophys. Res. Comm.* **2004**, *324*, 108–113.
199. Herath, K. B.; Jayasuriya, H.; Bills, G. F.; Polishook, J. D.; Dombrowski, A. W.; Guan, Z.; Felock, P. J.; Hazuda, D. J.; Singh, S. B. *J. Nat. Prod.* **2004**, *67*, 872–874.
200. Ondeyka, J. G.; Zink, D. L.; Dombrowski, A. W.; Polishook, J. D.; Felock, P. J.; Hazuda, D. J.; Singh, S. B. *J. Antibiot.* **2003**, *56*, 1018–1023.
201. Singh, S. B.; Jayasuriya, H.; Dewey, R.; Polishook, J. D.; Dombrowski, A. W.; Zink, D. L.; Guan, Z.; Collado, J.; Platas, G.; Pelaez, F. et al. *J. Ind. Microbiol. Biotechnol.* **2003**, *30*, 721–731.
202. Acharya, K. R.; Sturrock, E. D.; Riordan, J. F.; Ehlers, M. R. W. *Nat. Rev. Drug Disc.* **2003**, *2*, 891–902.
203. Duncia, J. V.; Chiu, A. T.; Carini, D. J.; Gregory, G. B.; Johnson, A. L.; Price, W. A.; Wells, G. J.; Wong, P. C.; Calabrese, J. C.; Timmermans, P. B. M. W. M. *J. Med. Chem.* **1990**, *33*, 1312–1329.
204. Furukawa, Y.; Kishimoto, S.; Nishikawa, K. USP4355040, Oct 19, 1982.
205. Carini, D. J.; Duncia, J. V.; Johnson, A. L.; Chiu, A. T.; Price, W. A.; Wong, P. C.; Timmermans, P. B. M. W. M. *J. Med. Chem.* **1990**, *33*, 1330–1336.
206. Carini, D. J.; Duncia, J. V.; Aldrich, P. E.; Chiu, A. T.; Johnson, A. L.; Pierce, M. E.; Price, W. A.; Santella, J. B., III; Wells, G. J.; Wexler, R. R. et al. *J. Med. Chem.* **1991**, *34*, 2525–2547.
207. Kurup, A.; Garg, R.; Carini, D. J.; Hansch, C. *Chem. Rev.* **2001**, *101*, 2727–2750.
208. Ferber, M.; Sporning, A.; Jeserich, G.; DeLaCruz, R.; Watkins, M.; Olivera, B. M.; Terlau, H. *J. Biol. Chem.* **2003**, *278*, 2177–2183.
209. Carroll, F. I.; Liang, F.; Navarro, H. A.; Brieaddy, L. E.; Abraham, P.; Damaj, M. I.; Martin, B. R. *J. Med. Chem.* **2001**, *44*, 2229–2237.
210. Carroll, F. I.; Ma, W.; Yokota, Y.; Lee, J. R.; Brieaddy, L. E.; Navarro, H. A.; Damaj, M. I.; Martin, B. R. *J. Med. Chem.* **2005**, *48*, 1221–1228.
211. Wei, Z.-L.; Xiao, Y.; Yuan, H.; Baydyuk, M.; Petukhov, P. A.; Musachio, J. L.; Kellar, K. J.; Kozikowski, A. P. *J. Med. Chem.* **2005**, *48*, 1721–1724.
212. Imming, P.; Klaperski, P.; Stubbs, M. T.; Seitz, G.; Gundisch, D. *Eur. J. Med. Chem.* **2001**, *36*, 375–388.
213. Fitch, R. W.; Kaneko, Y.; Klaperski, P.; Daly, J. W.; Seitz, G.; Gündisch, D. *Bioorg. Med. Chem. Lett.* **2005**, *15*, 1221–1224.
214. Cragg, G. M.; Newman, D. J.; Snader, K. M. *J. Nat. Prod.* **1997**, *60*, 52–60.
215. Cragg, G. M.; Kingston, D. G. I.; Newman, D. J.; *Anticancer Agents from Natural Products*; Taylor & Francis Group: Boca Raton, FL, 2005.
216. Cragg, G. M.; Newman, D. J. *J. Nat. Prod.* **2004**, *67*, 232–244.
217. Kingston, D. G. I. In *Anticancer Agents from Natural Products*; Cragg, G. M., Kingston, D. G. I., Newman, D. J., Eds.; Taylor & Francis: Boca Raton, FL, 2005, pp 89–122.
218. Hofle, G.; Reichenbach, H. In *Anticancer Agents from Natural Products*; Cragg, G. M., Kingston, D. G. I., Newman, D. J., Eds.; Taylor & Francis: Boca Raton, FL, 2005, pp 413–450.
219. Pinney, K. G.; Jelinek, C.; Edvardsen, K.; Chaplin, D. J.; Pettit, G. R. In *Anticancer Agents from Natural Products*; Cragg, G. M., Kingston, D. G. I., Newman, D. J., Eds.; Taylor & Francis: Boca Raton, FL, 2005, pp 23–46.
220. Gueritte, F.; Fahy, J. In *Anticancer Agents from Natural Products*; Cragg, G. M., Kingston, D. G. I., Newman, D. J., Eds.; Taylor & Francis: Boca Raton, FL, 2005, pp 123–136.
221. Flahive, E.; Srirangam, J. In *Anticancer Agents from Natural Products*; Cragg, G. M., Kingston, D. G. I., Newman, D. J., Eds.; Taylor & Francis: Boca Raton, FL, 2005, pp 191–214.
222. Gunasekera, S. P.; Wright, A. E. In *Anticancer Agents from Natural Products*; Cragg, G. M., Kingston, D. G. I., Newman, D. J., Eds.; Taylor & Francis: Boca Raton, FL, 2005, pp 171–190.
223. Andersen, R. J.; Roberge, M. In *Anticancer Agents from Natural Products*; Cragg, G. M., Kingston, D. G. I., Newman, D. J., Eds.; Taylor & Francis: Boca Raton, FL, 2005, pp 267–280.
224. Kondoh, M.; Usui, T.; Kobayashi, S.; Tsuchiya, K.; Nishikawa, K.; Nishikiori, T.; Mayumi, T.; Osada, H. *Cancer Lett.* **1998**, *126*, 29–32.

225. Usui, T.; Watanabe, H.; Tada, Y.; Kanoh, N.; Kondoh, M.; Asao, T.; Takio, K.; Watanabe, H.; Nishikawa, K.; Kitahara, T. et al. *Chem. Biol.* **2004**, *11*, 799–806.
226. Rahier, N. J.; Thomas, C. J.; Hecht, S. M. In *Anticancer Agents from Natural Products*; Cragg, G. M., Kingston, D. G. I., Newman, D. J., Eds.; Taylor & Francis: Boca Raton, FL, 2005, pp 5–22.
227. Prudhomme, M. In *Anticancer Agents from Natural Products*; Cragg, G. M., Kingston, D. G. I., Newman, D. J., Eds.; Taylor & Francis: Boca Raton, FL, 2005, pp 499–518.
228. Arcamone, F. M. In *Anticancer Agents from Natural Products*; Cragg, G. M., Kingston, D. G. I., Newman, D. J., Eds.; Taylor & Francis: Boca Raton, FL, 2005, pp 299–320.
229. Lee, K.-H.; Xiao, Z. In *Anticancer Agents from Natural Products*; Cragg, G. M., Kingston, D. G. I., Newman, D. J., Eds.; Taylor & Francis: Boca Raton, FL, 2005, pp 71–88.
230. Denny, W. A. *Expert Opin. Emerg. Drugs* **2004**, *9*, 105–133.
231. Marshall, K. M.; Holden, J. A.; Koller, A.; Kashman, Y.; Copp, B. R.; Barrows, L. R. *Anti-Cancer Drugs* **2004**, *15*, 907–913.
232. Marshall, K. M.; Barrows, L. R. *Nat. Prod. Rep.* **2004**, *21*, 731–751.
233. Li, Y.; Sun, X.; LaMont, J. T.; Pardee, A. B.; Li, C. J. *Proc. Natl. Acad. Sci. USA* **2003**, *100*, 2674–2678.
234. Choi, B. T.; Cheong, J.; Choi, Y. H. *Anti-Cancer Drugs* **2003**, *14*, 845–850.
235. Ravelo, A. G.; Estevez-Braun, A.; Chavez-Orellana, H.; Perez-Sacau, E.; Mesa-Siverio, D. *Curr. Topics Med. Chem.* **2004**, *4*, 241–265.
236. Kisselev, A. F.; Goldberg, A. L. *Chem. Biol.* **2001**, *8*, 739–758.
237. Kyle, R. A.; Rajkumar, S. V. *N. Engl. J. Med.* **2004**, *351*, 1860–1873.
238. Adams, J. *Cancer Cell* **2003**, *5*, 417–421.
239. Omura, S.; Fujimoto, T.; Otoguro, K.; Matsuzaki, K.; Moriguchi, R.; Tanaka, H.; Sasaki, Y. *J. Antibiot.* **1991**, *44*, 113–116.
240. Fenteany, G.; Standaert, R. F.; Lane, W. S.; Choi, S.; Corey, E. J.; Schreiber, S. L. *Science* **1995**, *268*, 726–731.
241. Craiu, A.; Gaczynska, M.; Akopian, T.; Gramm, C. F.; Fenteany, G.; Goldberg, A. L.; Rock, K. L. *J. Biol. Chem.* **1997**, *272*, 13437–13445.
242. Dick, L. R.; Cruikshank, A. A.; Grenier, L.; Melandri, F. D.; Nunes, S. L.; Stein, R. L. *J. Biol. Chem.* **1996**, *271*, 7273–7276.
243. Dick, L. R.; Cruikshank, A. A.; Destree, A. T.; Grenier, L.; McCormack, T. A.; Melandri, F. D.; Nunes, S. L.; Palombella, V. J.; Parent, L. A.; Plamodon, L. et al. *J. Biol. Chem.* **1997**, *272*, 182–188.
244. Corey, E. J.; Li, W.-D. Z. *Chem. Pharm. Bull. (Tokyo)* **1999**, *47*, 1–10.
245. Corey, E. J.; Li, W.-D. Z.; Nagamitsu, T.; Fenteany, G. *Tetrahedron* **1999**, *55*, 3305–3316.
246. Reddy, L. R.; Saravan, P.; Corey, E. J. *J. Am. Chem. Soc.* **2004**, *126*, 6230–6231.
247. Meng, L. H.; Kwok, B. H. B.; Sin, N.; Crews, C. M. *Cancer Res.* **1999**, *59*, 2798–2801.
248. Meng, L. H.; Mohan, R.; Kwok, B. H. B.; Elofsson, M.; Sin, N.; Crews, C. M. *Proc. Natl. Acad. Sci. USA* **1999**, *96*, 10403–10408.
249. Koguchi, Y.; Kohno, J.; Nishio, M.; Takahashi, K.; Okuda, T.; Ohnuki, T.; Komatsubara, S. *J. Antibiot.* **2000**, *53*, 105–109.
250. Nam, S.; Smith, D. M.; Dou, Q. P. *J. Biol. Chem.* **2001**, *276*, 13322–13330.
251. Yoshida, M.; Furumai, R.; Nishiyama, M.; Komatsu, Y.; Nishino, N.; Horinouchi, S. *Cancer Chemother. Pharmacol.* **2001**, *48*, S20–S26.
252. Remiszewski, S. W. *Curr. Opin. Drug. Disc. Dev.* **2002**, *5*, 487–499.
253. Yoshida, M.; Matsuyama, A.; Komatsu, Y.; Nishino, N. *Curr. Med. Chem.* **2003**, *10*, 2351–2358.
254. Tsuji, N.; Kobayashi, N.; Nagashima, K.; Wakisaka, Y.; Koizumi, K. *J. Antibiot.* **1976**, *29*, 1–6.
255. Vanhaecke, T.; Papeleu, P.; Elaut, G.; Rogiers, V. *Curr. Med. Chem.* **2004**, *11*, 1629–1643.
256. Brinkmann, H.; Dahler, A. L.; Popa, C.; Sereweko, M. M.; Parsons, P. G.; Gabrielli, B. G.; Burgess, A. J.; Saunders, N. A. *J. Biol. Chem.* **2001**, *276*, 22491–22499.
257. Andreeff, M.; Stone, R.; Michaeli, J.; Young, C. W.; Tong, W.; Sogoloff, H.; Ervin, T.; Kufe, D.; Rifkind, R. A.; Marks, P. A. *Blood* **1992**, *80*, 2604–2609.
258. Secrist, J. P.; Zhou, X.; Richon, V. M. *Curr. Opin. Investig. Drugs* **2003**, *4*, 1422–1427.
259. Jung, M.; Brosch, G.; Kolle, D.; Scherf, H.; Gerhauser, C.; Loidl, P. *J. Med. Chem.* **1999**, *42*, 4669–4679.
260. Remiszewski, S. W.; Sambucetti, L. C.; Atadja, P.; Bair, K. W.; Cornell, W. D.; Green, M. A.; Cornell, W. D.; Green, M. A.; Howell, K. L.; Jung, M. et al. *J. Med. Chem.* **2002**, *45*, 753–757.
261. Kijima, M.; Yoshida, M.; Sugita, K.; Horinouchi, S.; Beppu, T. *J. Biol. Chem.* **1993**, *268*, 22429–22435.
262. Sandor, V.; Senderowicz, A.; Mertins, S.; Sackett, D.; Sausville, E.; Blagosklonny, M. V.; Bates, S. E. *Br. J. Cancer* **2000**, *83*, 817–825.
263. Remiszewski, S. W. *Curr. Med. Chem.* **2003**, *10*, 2393–2402.
264. Snader, K. M. In *Anticancer Agents from Natural Products*; Cragg, G. M., Kingston, D. G. I., Newman, D. J., Eds.; Taylor & Francis: Boca Raton, FL, 2005, pp 339–356.
265. Le Brazidec, J.-Y.; Kamal, A.; Busch, D.; Thao, L.; Zhang, L.; Timony, G.; Grecko, R.; Trent, K.; Lough, R.; Salazar, T. et al. *J. Med. Chem.* **2004**, *47*, 3865–3873.
266. Tian, Z.-Q.; Liu, Y.; Zhang, D.; Wang, Z.; Dong, S. D.; Carreras, C. W.; Zhou, Y.; Rastelli, G.; Santi, D. V.; Myles, D. C. *Bioorg. Med. Chem.* **2004**, *12*, 5317–5329.
267. Kamal, A.; Thao, L.; Sensintaffar, J.; Zhang, L.; Boehm, M. F.; Fritz, L. C.; Burrows, F. J. *Nature* **2003**, *425*, 407–410.
268. Lee, Y.-S.; Marcu, M. G.; Neckers, L. *Chem. Biol.* **2004**, *11*, 991–996.
269. Jez, J. M.; Chen, J. C.; Rastelli, G.; Stroud, R. M.; Santi, D. V. *Chem. Biol.* **2003**, *10*, 361–368.
270. Chiosis, G.; Timaul, M. N.; Lucas, B.; Munster, P. N.; Zheng, F. F.; Sepp-Lorenzino, L.; Rosen, N. *Chem. Biol.* **2001**, *8*, 289–299.
271. Chiosis, G.; Lucas, B.; Huezo, H.; Solit, D.; Basso, A.; Rosen, N. *Curr. Cancer Drug Targ.* **2003**, *3*, 371–376.
272. Chiosis, G.; Lucas, B.; Shtil, A.; Huezo, H.; Rosen, N. *Bioorg. Med. Chem.* **2002**, *10*, 3555–3564.
273. Vilenchik, M.; Solit, D.; Basso, A.; Huezo, H.; Lucas, B.; He, H.; Rosen, N.; Spampinato, C.; Modrich, P.; Chiosis, G. *Chem. Biol.* **2004**, *11*, 787–797.
274. Christian, M. C.; Pluda, J. M.; Ho, T. C.; Arbuck, S. G.; Murgo, A. J.; Sausville, E. A. *Semin. Oncol.* **1997**, *24*, 219–240.
275. Sausville, E. A.; Zaharevitz, D.; Gussio, R.; Meijer, L.; Louarn-Leost, M.; Kunick, C.; Schultz, R.; Lahusen, T.; Headlee, D.; Stinson, S. et al. *Pharmacol. Ther.* **1999**, *82*, 285–292.
276. Fischer, P. M.; Gianella-Borradori, A. *Expert Opin. Investig. Drugs* **2005**, *14*, 457–477.
277. Mayer, F.; Mueller, S.; Malenke, E.; Kuczyk, M.; Hartmann, J. T.; Bokemeyer, C. *Investig. New Drugs* **2005**, *23*, 205–211.
278. Pettit, G. R.; Day, J. F.; Hartwell, J. L.; Wood, H. B. *Nature* **1970**, *227*, 962–963.
279. Pettit, G. R.; Herald, C. L.; Doubeck, D. L.; Herald, D. L.; Arnold, E.; Clardy, J. *J. Am. Chem. Soc.* **1982**, *104*, 6846–6848.

280. Suffness, M.; Newman, D. J.; Snader, K. M. In *Bioorganic Marine Chemistry*; Scheuer, P., Ed.; Springer-Verlag: Berlin-Heidelberg, 1989, Vol. 3, pp 131–167.
281. Pettit, G. R. In *Progress in the Chemistry of Organic Natural Products*; Hertz, W., Kirby, G. W., Steglich, W., Tamm, C., Eds.; Springer-Verlag: New York, 1991, Vol. 57, pp 153–195.
282. Pettit, G. R. *J. Nat. Prod.* **1996**, *59*, 812–821.
283. Newman, D. J. In *Bryozoans in Space and Time*; Gordon, D. P., Smith, A. M., Grant-Mackie, J. A., Eds.; NIWA: Wellington, NZ, 1996, pp 9–17.
284. Mutter, R.; Wills, M. *Bioorg. Med. Chem.* **2000**, *8*, 1841–1860.
285. Pettit, G. R.; Herald, C. L.; Hogan, F. In *Anticancer drug development*; Baguley, B. C., Kerr, D. J., Eds.; Academic Press: San Diego, 2002, pp 203–235.
286. Hale, K. J.; Hummersone, M. C.; Manaviazar, S.; Frigerio, M. *Nat. Prod. Rep.* **2002**, *19*, 413–453.
287. Newman, D. J. In *Anticancer Agents from Natural Products*; Cragg, G. M., Kingston, D. G. I., Newman, D. J., Eds.; Taylor & Francis: Boca Raton, FL, 2005, pp 137–150.
288. Kageyama, M.; Tamura, T.; Nantz, M. H.; Roberts, J. C.; Somfai, P.; Whritenour, D. C.; Masamune, S. *J. Am. Chem. Soc.* **1990**, *112*, 7407–7408.
289. Evans, D. A.; Carter, P. H.; Carreira, E. M.; Charette, A. B.; Prunet, J. A.; Lautens, M. *J. Am. Chem. Soc.* **1999**, *121*, 7540–7552.
290. Ohmori, K.; Ogawa, Y.; Obitsu, T.; Ishikawa, Y.; Nishiyama, S.; Yamamura, S. *Angew. Chem. Int. Ed.* **2000**, *39*, 2290–2294.
291. Norcross, R. D.; Paterson, I. *Chem. Rev.* **1995**, *95*, 2041–2114.
292. Meijer, L.; Raymond, E. *Acc. Chem. Res.* **2003**, *36*, 417–425.
293. Parker, C. W.; Entsch, B.; Letham, D. *Phytochemistry* **1986**, *25*, 303–310.
294. Chang, Y. T.; Gray, N. G.; Rosania, G. R.; Sutherlin, D. P.; Kwon, S.; Norman, T. C.; Sarohia, R.; Leost, M.; Meijer, L.; Schultz, P. G. *Chem. Biol.* **1999**, *6*, 361–375.
295. Maugard, T.; Enaud, E.; Choisy, P.; Legoy, M. D. *Phytochemistry* **2001**, *58*, 897–904.
296. Cooksey, C. *J. Molecules* **2001**, *6*, 736–769.
297. MacNeil, I. A.; Tiong, C. L.; Minor, C.; August, P. R.; Grossman, T. H.; Loiacono, K. A.; Lynch, B. A.; Phillips, T.; Narula, S.; Sundaramoorthi, R. et al. *J. Mol. Microbiol. Biotechnol.* **2001**, *3*, 301–308.
298. Adachi, J.; Mori, Y.; Matsui, S.; Takigam, H.; Fujino, J.; Kitagawa, H. C. A.; Miller, C. H., III; Kato, T.; Saeki, K.; Matsuda, T. *J. Biol. Chem.* **2001**, *276*, 31475–31478.
299. Xiao, Z.; Hao, Y.; Liu, B.; Qian, L. *Leuk. Lymphoma* **2002**, *43*, 1763–1768.
300. Meijer, L.; Skaltsounis, A.-L.; Magiatis, P.; Polychronopoulos, P.; Knockaert, M.; Leost, M.; Ryan, X. P.; Vonica, C. A.; Brivanlou, A.; Dajani, R. et al. *Chem. Biol.* **2003**, *10*, 1255–1266.
301. Polychronopoulos, P.; Magiatis, P.; Skaltsounis, A.-L.; Myrianthopoulos, V.; Mikros, E.; Tarricone, A.; Musacchio, A.; Roe, S. M.; Pearl, L.; Leost, M. et al. *J. Med. Chem.* **2004**, *47*, 935–946.
302. Knockaert, M.; Blondel, M.; Bach, S.; Leost, M.; Elbi, C.; Hager, G. L.; Nagy, S. R.; Han, D.; Denison, M.; Ffrench, M. et al. *Oncogene* **2004**, *23*, 4400–4412.
303. Cohen, P.; Goedert, M. *Nature Rev. Drug Disc.* **2004**, *3*, 479–487.
304. Gompel, M.; Leost, M.; De Kier Joffe, E. B.; Puricelli, L.; Franco, L. H.; Palermo, J.; Meijer, L. *Bioorg. Med. Chem. Lett.* **2004**, *14*, 1703–1707.
305. Fischer, P. M. *Curr. Med. Chem.* **2004**, *11*, 1563–1583.
306. Koonin, E. V.; Wolf, Y. I.; Karev, G. P. *Nature* **2003**, *420*, 218–223.
307. Kobayashi, J.; Madono, T.; Shigemori, H. *Tetrahedron* **1995**, *51*, 10867–10974.
308. Zhou, B.-N.; Johnson, R. K.; Mattern, M. R.; Fisher, P. W.; Kingston, D. G. I. *Org. Lett.* **2001**, *3*, 4047–4049.
309. Arnold, L. D.; Calderwood, D. J.; Dixon, R. W.; Johnston, D. N.; Kamens, J. S.; Munschauer, R.; Rafferty, P.; Ratnofsky, S. E. *Bioorg. Med. Chem. Lett.* **2000**, *10*, 2167–2170.
310. Burchat, A. F.; Calderwood, D. J.; Hirst, G. C.; Holman, N. J.; Johnston, D. N.; Munschauer, R.; Rafferty, P.; Tometzki, G. B. *Bioorg. Med. Chem. Lett.* **2000**, *10*, 2171–2174.
311. Burchat, A. F.; Calderwood, D. J.; Friedman, M. M.; Hirst, G. C.; Li, B.; Rafferty, P.; Ritter, K.; Skinner, B. S. *Bioorg. Med. Chem. Lett.* **2002**, *12*, 1687–1690.
312. Calderwood, D. J.; Johnston, D. N.; Munschauer, R.; Rafferty, P. *Bioorg. Med. Chem. Lett.* **2002**, *12*, 1683–1686.
313. Kissau, L.; Stahl, P.; Mazitschek, R.; Giannis, A.; Waldmann, H. *J. Med. Chem.* **2003**, *46*, 2917–2931.
314. Giaccia, A.; Siim, B. G.; Johnson, R. S. *Nature Rev. Drug Disc.* **2003**, *2*, 1–9.
315. Kunkel, M. W.; Kirkpatrick, D. L.; Johnson, J. I.; Powis, G. *Anticancer Drug. Des.* **1997**, *12*, 659–670.
316. Lazo, J. S.; Tamura, K.; Vogt, A.; Jung, J.-K.; Rodriguez, S.; Balachandran, R.; Day, B. W.; Wipf, P. *J. Pharmacol. Exp. Therap.* **2001**, *296*, 364–371.
317. Wipf, P.; Hopkins, T. D.; Jung, J.-K.; Rodriguez, S.; Birmingham, A.; Southwick, E. C.; Lazo, J. S.; Powis, G. *Bioorg. Med. Chem. Lett.* **2001**, *11*, 2637–2641.
318. Welsh, S. J.; Williams, R. R.; Birmingham, A.; Newman, D. J.; Kirkpatrick, D. L.; Powis, G. *Mol. Can. Therap.* **2003**, *2*, 235–243.
319. Wipf, P.; Lynch, S. M.; Birmingham, A.; Tamayo, G.; Jimenez, A.; Campos, N.; Powis, G. *Org. Biomol. Chem.* **2004**, *2*, 1651–1658.
320. Sigel, M. M.; Wellham, L. L.; Lichter, W.; Dudeck, L. E.; Gargus, J.; Lucas, A. H. In *Food-Drugs from the Sea Proceedings*; Younghen, H. W., Jr., Ed.; Marine Technology Society: Washington, DC, 1969, pp 281–294.
321. Rinehart, K.; Holt, T. G.; Fregeau, N. L.; Stroh, J. G.; Kiefer, P. A.; Sun, F.; Li, L. H.; Martin, D. G. *J. Org. Chem.* **1990**, *55*, 4512–4515.
322. Wright, A. E.; Forleo, D. A.; Gunawardana, G. P.; Gunasekera, S. P.; Koehn, F. E.; McConnell, O. J. *J. Org. Chem.* **1990**, *55*, 4508–4512.
323. Scott, J. D.; Williams, R. M. *Chem. Rev.* **2002**, *102*, 1669–1730.
324. Fontana, A.; Cavaliere, P.; Wahidulla, S.; Naik, C. G.; Cimino, G. *Tetrahedron* **2000**, *56*, 7305–7308.
325. Amnuoypol, S.; Suwanborirux, K.; Pummangura, S.; Kubo, A.; Tanaka, C.; Saito, N. *J. Nat. Prod.* **2004**, *67*, 1023–1028.
326. Simmons, T. L.; Andrianasolo, E.; McPhail, K.; Flatt, P.; Gerwick, W. H. *Mol. Cancer Ther.* **2005**, *4*, 333–342.
327. Zewail-Foote, M.; Li, V.-S.; Kohn, H.; Bearss, D.; Guzman, M.; Hurley, K. H. *Chem. Biol.* **2001**, *8*, 1033–1049.
328. Takebayashi, Y.; Pourquier, P.; Zimonjic, D. B.; Nakayama, K.; Emmert, S.; Ueda, T.; Urasaki, Y.; Akiyama, S.-I.; Popescu, N.; Kraemer, K. H. et al. *Nature Med.* **2001**, *7*, 961–966.
329. Bonfanti, M.; La Valle, E.; Fernandez Sousa-Faro, J.-M.; Faircloth, G.; Caretti, G.; Mantovani, R.; D'Incalci, M. *Anticancer Drug Des.* **1999**, *14*, 179–186.
330. van Kesteren, C.; de Vooght, M. M. M.; Lopez-Lazaro, L.; Mathot, R. A. A.; Schellens, J. H. M.; Jimeno, J. M.; Beijnen, J. H. *Anticancer Drugs* **2003**, *14*, 487–502.

331. Dziegielewska, B.; Kowalski, D.; Beerman, T. A. *Biochemistry* **2004**, *43*, 14228–14237.
332. Mandola, M. V.; Kolb, E. A.; Scotto, K. W. *Proc. Am. Assoc. Cancer Res.* **2005**, *46*, 4122 (Abs).
333. Le Cesne, A.; Blay, J. Y.; Judson, I.; Van Oosterom, A.; Verweij, J.; Radford, J.; Lorigan, P.; Rodenhuis, S.; Ray-Coquard, I.; Bonvalot, S. et al. *J. Clin. Oncol.* **2005**, *23*, 576–584.
334. Sessa, C.; De Braud, F.; Perotti, A.; Bauer, J.; Curigliano, G.; Noberasco, C.; Zanaboni, F.; Gianni, L.; Marsoni, S.; Jimeno, J. et al. *J. Clin. Oncol.* **2005**, *23*, 1867–1874.
335. Beumer, J. H.; Schellens, J. H. M.; Beijnen, J. H. *Pharmacol. Res.* **2005**, *51*, 391–398.
336. Sakai, R.; Rinehart, K. L.; Kishore, V.; Kundu, B.; Faircloth, G.; Gloer, J. B.; Carney, J. R.; Manikoshi, M.; Sun, F.; Hughes, R. G., Jr. et al. *J. Med. Chem.* **1996**, *39*, 2819–2834.
337. Faircloth, G.; Rinehart, K.; Nunez de Castro, I.; Jimeno *J. Ann. Oncol.* **1996**, 7, 34.
338. Urdiales, J. L.; Morata, P.; Nunez de Castro, I.; Sanchez-Jimenez, F. *Cancer Lett.* **1996**, *102*, 31–37.
339. Cuevas, C.; Cuevas, F.; Gallego, P.; Mandez, P.; Manzanares, I.; Munt, S.; Polanco, C.; Rodriguez, I. WO 02022596, Oct 1, 2002.
340. Taraboletti, G.; Poli, M.; Dossi, R.; Manenti, L.; Borsotti, P.; Faircloth, G. T.; Broggini, M.; D'Incalci, M.; Ribatti, D.; Giavazzi, R. *Br. J. Cancer* **2004**, *90*, 2418–2424.
341. Cuadrado, A.; Gonzalez, L.; Suarez, Y.; Martinez, T.; Munoz, A. *Oncogene* **2004**, *23*, 4673–4680.
342. Losada, A.; Lopez-Oliva, J. M.; Sanchez-Puelles, J. M.; Garcia-Ferenandez, L. F. *Br. J. Cancer* **2004**, *91*, 1405–1413.
343. Menon, L. G.; Saydam, G.; Longo, G. S. A.; Jimeno, J. M.; Bertino, J. R.; Banerjee, D. *Proc. Am. Assoc. Cancer Res.* **2005**, *46*, 5104 (Abs.).
344. Gomez, S. G.; Bueren, J. A.; Faircloth, G. T.; Jimeno, J.; Albella, B. *Exp. Hematol.* **2003**, *31*, 1104–1111.
345. Vera, M.; Joullie, M. M. *Med. Res. Rev.* **2002**, *22*, 102–145.
346. Cardenas, F.; Caba, J. M.; Feliz, M.; Lloyd-Williams, P.; Giralt, E. *J. Org. Chem.* **2003**, *68*, 9554–9562.
347. Gutierrez-Rodriguez, M.; Martin-Martinez, M.; Garcia-Lopez, M. T.; Herranz, R.; Cuevas, F.; Polanco, C.; Rodriguez-Campos, I.; Manzanares, I.; Cardenas, F.; Feliz, M. et al. *J. Med. Chem.* **2004**, *47*, 5700–5712.
348. Balamurugan, R.; Dekker, F. J.; Waldmann, H. *Mol. Biosyst.* **2005**, *1*, 36–45.
349. National Library of Medicine (NLM) Home Page. www.nlm.nih.gov (accessed March 2006).

Biographies

Gordon M Cragg obtained his undergraduate training in chemistry at Rhodes University, South Africa, and his DPhil (organic chemistry) from Oxford University in 1963. After two years of postdoctoral research at the University of California, Los Angeles, he returned to South Africa to join the Council for Scientific and Industrial Research. In 1966, he joined Chemistry Department at the University of South Africa, and transferred to the University of Cape Town in 1972. In 1979, he returned to the US to join the Cancer Research Institute at Arizona State University. In 1985, he moved to the National Cancer Institute (NCI) in Bethesda, Maryland, and was appointed Chief of the Natural Products Branch in 1989. He retired in December, 2004, and is currently serving as an NIH Special Volunteer. His major interests lie in the discovery of novel natural product agents for the treatment of cancer and AIDS, and he has given over 100 invited talks at conferences in many countries worldwide. He has been awarded the NIH Merit Awards for his contributions to the development of taxol (1991), inspirational leadership in establishing international collaborative research in biodiversity and natural products drug discovery (2004), and contributions to developing and teaching NIH technology transfer courses (2004). In 1998–99, he was President of the American Society of Pharmacognosy, and was elected to Honorary Membership in 2003. He has established collaborations between the NCI and organizations in many countries promoting drug discovery from their natural resources. He has published over 140 papers related to these interests.

David J Newman was born in Grays, Essex, UK. Initially trained as a chemical analyst, he received an MSc in Organic Chemistry (Liverpool) and then, after a time in the UK chemical industry, a D Phil in Microbial Chemistry from Sussex in 1968. Following postdoctoral studies at the University of Georgia, USA, he worked for SK&F (now GSK) in Philadelphia, PA, as a biological chemist, mainly in the area of antibiotic discovery. Following the discontinuance of antibiotic discovery programs at SK&F, he worked for a number of US companies in natural products-based discovery programs in anti-infective and cancer treatments, joining the Natural Products Branch of the NCI in 1991, working with Gordon Cragg prior to his retirement at the end of 2004. His scientific interests are in the discovery and history of novel marine and microbial natural products as drug leads in the anti-infective and cancer areas and in the application of information technologies to drug discovery. In conjunction with Gordon Cragg, he established collaborations between the NCI and organizations in many countries promoting drug discovery from their natural resources. In 2003, he received the NIH Merit Award for development of marine and microbial antitumor agents and, in addition to over 50 invited lectures, has published over 100 papers, presented over 60 abstracts, holds 16 patents that are related to these interests, is both a UK Chartered Chemist and a UK Chartered Biologist, and is also an adjunct full professor at the Center of Marine Biotechnology, University of Maryland.

Comprehensive Medicinal Chemistry II
ISBN (set): 0-08-044513-6

ISBN (Volume 1) 0-08-044514-4; pp. 355–403

1.09 Traditional Medicines

L A Mitscher, University of Kansas, Lawrence, KS, USA

1.09.1 Overview

Traditional medical systems depend heavily upon herbal remedies for treatment. What is an herbal? The definition put forth by the late Verro Tyler is apt.[1] Herbals are crude drugs of vegetable origin utilized for the treatment of disease states, often of a chronic nature, or to attain or to maintain a condition of improved health. Clearly, although usually regulated differently, herbal agents are drugs. Despite the application of various euphemisms for a variety of purposes, that simple fact remains. Partly as a result of their different categorization the status accorded to herbal remedies is significantly less in western nations than is that of prescription medications. Until fairly recently health professionals in many wealthy nations were slower to embrace the usage of herbal remedies than were the general population.

Herbals are more complex mixtures than the usual synthetic drug. Their properties are no more or less magical than those of synthetic drugs. Some herbals are safe and some unsafe. They are nonetheless potential sources of new remedies.

The use of herbs for medicinal purposes is very ancient and is persistent. Many of the best-known drugs in use today have had their origin in natural sources. Indeed, well into the twentieth century a significant percentage of the medications in the official compendia of scientifically sophisticated lands were derived from the herbal lore of native peoples. Furthermore, the World Health Organization today notes that, of 119 plant-derived medications employed today, no less than 74% are used in medicine in ways that correlate directly with their uses in native medical systems. They further estimate that 80% of the world's population today uses herbal medicines for some aspects of their primary medical care needs.[2]

Throughout the vast majority of human experience herbal medicine and prescription medicine were identical. The picture began to change slowly at about 1805 with the discovery that much of the medicinal value of opium resided in its content of morphine. In 1828 Woehler synthesized urea (an animal-derived substance) from nonorganic ammonium cyanate, destroying the belief that substances of natural origin were somehow magically different from synthetic substances. A segment of lay society has not yet adjusted to this reality and it is reflected in their attitude toward herbals today. Scientists do not share this faith in the magic and intrinsic safety of natural substances. A practical example can clarify the situation further. Caffeine is a naturally occurring alkaloid with stimulatory properties. As a pure chemical ingredient in pharmaceutical preparations it is commonly so used. It is also very frequently used for basically the same purposes in the form of soft drinks and beverages prepared from tea leaves and coffee beans. When taken for stimulatory purposes caffeine of synthetic or of plant origin is undeniably a drug and is identical in its properties. Despite the total lack of any evidence to the contrary, there are still people today who believe that synthetic caffeine is somehow different from the caffeine present in vegetable material. This belief in vitalism can most charitably be described as quaint.

Another fantasy held by a segment of the consuming public is that herbs, being 'organic,' are somehow safe, as opposed to 'chemical,' which has become a pejorative term in their minds. True believers in the virtues of organic preparations forget or are unaware that strychnine and botulism toxin are among the many natural substances of biological origin that are dangerous when used in even modest doses. Scholars are well aware that a number of otherwise useful herbal remedies do contain toxic substances. Comfrey serves as an example. This venerable and popular herbal often contains toxic pyrrolizidine alkaloids that make its chronic use potentially dangerous due to liver damage.[3–6] Kava is another popular herbal product and its use is associated with liver damage.[7]

It has been estimated that about 13 000 plant species have been used at one time and place or another for medicinal purposes. Undoubtedly others have yet to be discovered. The obvious question is: why did herbals fall from official favor? Foremost was the comparative absence of strongly supportive scientific data attesting to their value. Among the other reasons is that they cannot be standardized for potency and identity as conveniently as single chemical remedies and preparations containing them are thus more difficult to prepare reproducibly. Advances in scientific methodology have largely overcome this worry in many cases but synthetics are still easier to define. Also intrinsic variation can be controlled much more efficiently and effectively when only a single active ingredient is present. This also simplifies the task of agencies charged with regulating their commerce and protecting the consumer. Medicaments containing a single active component are still present in variously complex formulations containing many other ingredients such as flavoring, coloring, stabilizing, hydration controlling and the like, but the level of complexity is generally less than that seen with herbal remedies. The distinction between herbals and prescription remedies on the basis of complexity is not sharp, however. In fact, such frequently prescribed medications as Premarin (a mixture of estrogenic hormone metabolites isolated from pregnant mare's urine) and Kaletra (a mixture of lopinavir and ritonavir) contain more than one active ingredient. Their description and standardization are somewhat more complex than single agents but are well within the capacities of modern science. Furthermore, therapy of viral diseases and cancer, for example, now frequently employ mixtures of powerful active ingredients (and formulation ingredients) to extend the power and decrease the effect of mutations to resistance. Thus the distinction between herbals and officially sanctioned medications is becoming more and more blurred. The increasing trend to change the legal status of prescription pharmaceuticals to remedies available over the counter further blurs the distinction between them and self-selected herbals.

Failure of herbals to receive more scientific study can also be attributed to the costs. One notes that estimates of the costs of clinical evaluation of single pharmaceutical agents prior to introduction as prescription medications range anywhere from 250 million to 800 million US dollars. No one will undertake such a study unless there is a reasonable expectation of patent protection and therefore a period of market exclusivity in which to return a profit.

Beginning about 200 years ago and continuing to this date a transition from employment of crude drugs (today referred to as herbals) in medicinal practice to the use of single pure chemical entities began to take place. This transition, though pronounced in wealthy nations, is not complete and, indeed, contemporarily there is a partial reversion to the use of herbals. In many less wealthy lands herbal use today makes up nearly 80% of the medications employed.[2] Worldwide the commerce in herbal or botanical medicines is variously estimated but appears to amount to about $100 billion.[2] Herbal agents are most commonly self-selected in the US and a very significant commerce involves these materials. In Europe and in certain Asiatic lands the use of herbals is even more pronounced. Population growth in poorer regions of the globe outpaces that in wealthier lands; consequently it is likely that the use of herbal remedies in those regions will increase disproportionately.

In the last couple of generations teaching about herbals in western academic circles has largely fallen out of the curriculum but the needs of society and the desire of students for authoritative information about herbals has now risen to the level that a number of schools have reintroduced material about them either as stand-alone courses or have integrated this material into existing courses. Scientific study of herbals has become more common as newer techniques develop and their utilization increases, so it seems likely that a number of herbals will morph into single pure medicinal agents and others will become sanctioned by medical authorities in the foreseeable future. Thus a study of their properties is appropriate.

Herbal remedies, despite some euphemism, are primarily drugs so also should possess closely similar characteristics. These include the following:

- sustainable access to the material
- agreed-upon (ideally official and therefore enforced) standards of quality and identity
- employment of an end-product use-related chemical assay or bioassay
- adequate human and animal experimentation establishing safety and, ideally, efficacy
- registration and policing of vendors and suppliers
- freely available and reliable public and professional information.

1.09.2 Some Definitions

1.09.2.1 Herb

An herb in the botanical sense is a plant lacking a permanent woody stem that produces seeds and flowers and that dies down after its growing season. In a culinary or gardening sense these materials possess in addition strong flavors and fragrances and therefore make foods more attractive to consume. These definitions are broader than that applied in traditional medical systems. In a traditional medical sense an herb is a small, nonwoody, plant valued for its medicinal, savory, or aromatic qualities. An herbal preparation is, then, a natural remedy derived from herbs.

1.09.2.2 Functional Food

There is no standard definition for 'functional food.' Even so, this term is more readily interpreted in the correct way than are the alternative terms 'designer food' and 'nutraceutical.' These are considered to be foods that possess potential medicinal benefits going beyond those provided by their content of basic nutrients. Green and black tea, berries and cruciferous vegetables (with their content of antioxidants), soy (with its content of phytoestrogens and/or its content of cholesterol-lowering protein), certain fish and fish oils (with their content of *n*-3 fatty acids), orange juice (with its content of ascorbic acid), whole oats (with its content of β-glucan), grape juice and red wine (which contain resveratrol), and the like are examples of nutritive materials usually included in lists of functional foods. The putative health benefit is often intrinsic to the food as grown and is not necessarily a function of materials subsequently added to them, although supplemented foods also qualify. Functional foods, for example, can be prepared from normal foods by adding or subtracting or combining certain ingredients normally found in food products such as vitamins, essential amino acids, flavanoids, etc. As with herbals, in some cases the health benefits are amply documented by scientific experimentation and in others it is quite vague. A good discussion of functional foods can be found in the website cited in reference.[211] The Bureau of Nutritional Sciences of the Food Directorate of Health Canada has proposed in addition the definition of a 'nutraceutical' as a product isolated or purified from foods that is generally sold in medicinal forms not usually associated with food. This definition makes it difficult to distinguish clearly between nutraceuticals and some drugs and some herbals.

1.09.2.3 Dietary Supplement

A dietary supplement is legally defined in the USA under the Dietary Supplement Health and Education Act of 1994 (DSHEA) as a product that supplements the diet when taken orally and whose label clearly states that it is a dietary supplement. It is clearly not a drug but is considered to be a food. Examples of dietary supplements include vitamins, minerals, certain herbs, botanicals, other plant-derived substances, amino acids and concentrates, organ tissues, metabolites, constituents, and extracts of these substances. They are intended for ingestion in pill, capsule, tablet, or liquid form and are not recommended for use as conventional foods or as the sole item of a meal or diet. Tobacco and tobacco products are specifically excluded. For more information, see the websites given in [212,213].

1.09.2.4 Complementary Medicine

Complement medicine is a group of medical and healthcare systems, practices, and products that are not considered at this time to be part of conventional medicine, primarily because there is insufficiently convincing evidence of their efficacy, but are used together with conventional medicine. Some of these practices employ herbs.

1.09.2.5 Alternative Medicine

Alternative medicine is a group of medical and healthcare systems, practices, and products that are not considered at this time to be part of conventional medicine primarily because there is insufficiently convincing evidence of their efficacy. These are used in place of conventional medical practices. Dietary supplements and herbals are employed in some of these practices. An example is the use of shark cartilage to treat cancer.

1.09.2.6 Integrative Medicine

Integrative medicine involves the use of complementary and/or alternative medical materials, and practices for which there is sufficiently convincing evidence of utility that they may be integrated with mainstream medical practices. Herbals are employed in some of these practices.

1.09.2.7 Homeopathy

Homeopathy is an alternative or complementary medical practice based on the belief that 'like cures like.' Typically small, highly diluted quantities of medicinal substances, often herbals, are given to cure symptoms in cases where higher concentrations of the same substance would produce the very symptoms of the condition being treated.

1.09.2.8 Naturopathy

Naturopathy is an alternative or complementary medical practice based on the belief that there is an intrinsic healing power in the body that establishes, maintains, and restores health. Dietary supplements, herbals, homeopathy, and the like are often employed along with other modalities to supplement this hypothetical intrinsic healing power of the body.

1.09.3 Historical Aspects

It is not easy to provide a historical context for the use of herbals that does not include the history of biologically active natural products because until comparatively recently they were essentially the same. From the earliest of times, long before recorded history, humanity has been dependent upon natural sources for its needs. The discovery of the healing power of certain herbs cannot have been long in coming. The ability to remember and to communicate these things was a vital activity for survival and continuity of human existence. Therefore natural medical systems were identical with official medicine and this situation prevailed until well into the 1800s. Lest it be considered that there are no more drugs to be discovered in herbal remedies, one should take note that the anticancer drug taxol is present in various *Taxus* species known to be cytotoxic for centuries in various cultures, including, for example, Indian ayurvedic medicine. Less than two decades ago taxol was introduced into prescription medicine as an effective antitumor agent.[8,9] One also notes that artemesinin, a potent antimalarial agent, was recently uncovered in the traditional Chinese medicament including *Artemesia* sp.,[10] as was the male contraceptive gossypol from Chinese *Gossypium* species.[11] One can only speculate about the number and kind of additional remedies awaiting discovery following additional scientific study.

The first preserved medical records in our possession consist of cuneiform writing on clay tablets dating from about 4600 years ago. These were found in Mesopotamia (now known as Iraq) and about 1000 different medicinal plants are listed. Reading the tablets requires some assumptions in that the descriptions are not those we would employ today; nevertheless we are confident that we recognize descriptions of several medicinal plants, including *Glycyrrhiza glabra* (licorice) and *Papaver somniferum* (opium). Remarkably, these materials are still utilized in today's medical practice.

Other ancient descriptions of medicinal plants come from many scattered regions of the earth. The Ebers papyrus describes medicinal practice employing about 700 plants in Egypt from about 3500 years ago. Chinese prescriptions (about 52 such) were recorded in the *Wu Shi Er Bing Fang* about 3100 years ago. In India ayurvedic medical practice records the use of nearly 1000 natural remedies about 3000 years ago.

There is little doubt that such practices took place in all corners of the earth, although surviving contemporaneous records of this are scarce. These ancient writings record practices that were already venerable when recorded. To this day interaction with isolated societies lacking previous contact with outside influences reveals flourishing indigenous medical systems that have been preserved by oral tradition.

Eons undoubtedly passed before modifications were made to enhance the utility of natural products for human purposes. Subsequent but still early Greek (e.g., Theophrastus, Dioscorides, Galen) and Arabic (Avicenna and Ibn al-Baitar) writings describe mixtures of plants, teas, and alcoholic extracts, often flavored with honey to enhance their palatability, prepared from medicinal plants.

Until about 1800 this form of medication remained the case even in advanced societies. In lesser-developed regions this activity persists to this day. Approximately four-fifths of today's world population relies heavily or exclusively on traditional medical practices with a heavy emphasis on the use of plants and plant extracts.[2]

Unfortunately in the US the latter decades of the nineteenth century were characterized by unfettered and unregulated commerce in herbals. Opportunists rushed in and too many products were promoted based upon unscientific assertions. Extravagant and unsupportable claims were made for many rather questionable patent medicines, leading to skepticism and cynicism about herbals that lingers to this day. The negative consequences for serious scientific study have been devastating.

After decades of comparative neglect the use of herbal medicines in Europe and the US has become popular again and exists side by side with the products of the pharmaceutical industry.

A parallel but less commonly recognized set of activities took place in the animal kingdom. Zoopharmacognosists recognize that many animal species partake, apparently knowingly, of herbal materials for a number of medicinal purposes such as control of intestinal parasites, gastrointestinal upsets, and fertility problems. This lore is passed on to their offspring at least by example.[12–14]

The evolution of herbal medicine into modern prescription medicine took place during the last 200 years. The story of how this took place is fascinating but lies beyond the scope of this treatment of traditional medicines (*see* 1.08 Natural Product Sources of Drugs: Plants, Microbes, Marine Organisms, and Animals).

Where can one find novel herbals today? The density of different species is highest in tropical rainforests. In part this is due to the presence of many microenvironments and the intense competition for living room. Fortunately it is in these various regions that herbal medicine continues to flourish. In a country like Panama, for example, the seashore is flanked by a relatively narrow plain which is itself soon replaced by mountain slopes. Further it is an isthmus separating North American and South American landmasses, so transient populations traverse it. Like so many others, it is under human population pressures threatening habitats and thus significant extinctions are taking place. Local populations quite understandably prefer to use the land for agricultural and other purposes, even though they derive a significant proportion of their medicaments from the same areas. Who knows, then, what we are to miss because of extinction of species and losses of habitat. The situation in nearby Costa Rica is rather similar. Whereas these nations stand out compared to some of their neighbors in their efforts to strike a reasonable balance between preservation and development, recently it has become more and more complex to acquire bioprospecting rights in developing countries.

It is understandable that desperately poor peoples would like to share in the profits that can amount to billions for a blockbuster drug. Such individuals often do not realize, however, that the majority of newly introduced drugs fail to recover their research and development costs. Previously indigenous peoples were generally either unaware of the amounts of money that can in principle be made or were altruistic to the extent that they believed that God provided such products for the benefit of all peoples. No more. The desire to share in profits provides a disincentive for collecting as the desire to share does not readily take into account the mammoth expense in time and treasure and the risks that lie between the average hit and the very unusual big score. Unfortunately, all too often the lands with the most promising biota are not the lands with the necessary expertise and capital to develop them for medicinal purposes. With good will on both sides these issues can be resolved but the negotiations can be frustrating to both sides.

Fermentations, on the other hand, are attractive because they sidestep territoriality, ownership questions, and seasonal fluctuations as well as often being more readily scalable and occasionally have intersected in interesting ways with the herbal world. Microbes produce an astonishing array of complex natural products and were recognized in the mid decades of the twentieth century to be prolific producers of antimicrobials and antitumor agents. The first penicillin preparations examined for medical potential were crude enough to be classified as herbals today. These, and like materials, matured into the prescription medications. As demonstrated particularly dramatically in Japan by Umezawa and Omura and their coworkers, many other therapeutic areas subsequently profited from screening of fermentation broths. The statins, billion-dollar drugs valuable for control of cholesterol metabolism, resulted from such screens.[15] Some crude natural fungal products (red yeast–rice) used in traditional Chinese medicine contain a variety of monacolins, including lovastatin, and were shown to be effective in lowering cholesterol in humans and were briefly marketed in the US as a herbal remedy.[16–18] This is another example wherein the differentiation between a prescription medication and an herbal product is a question of the degree of purification employed before marketing.

Considering this brief recapitulation, one can perceive that herbals do in favorable cases lead to very pronounced medicinal benefits or to individual compounds having such properties if examined by intelligent chemical and biological methods.

1.09.4 The Contemporary Scene

After decades of comparative neglect, herbals are enjoying a significant resurgence of popularity in wealthy countries. Part of the impetus for this probably stems from the desire of a significant percentage of the population to play a more active role in their own health, utilizing means under their own control.

In the US today the herbal medicine and dietary supplement industry is only loosely regulated, even though it amounts to about $17 billion in sales annually and is growing rapidly with each passing year. This is particularly impressive since insurance firms affording third-party payments for medical treatment are particularly scornful of alternative medicine so patients must usually pay for these products from their own pockets. The worldwide utilization of herbals is much more pronounced.[1,2,19–23]

The medical systems in developing countries today simultaneously involve both traditional medical systems and official medicine. The choice between them by health professionals as well as by the laity is based on personal preference and economics. Where faith exists in traditional herbal medicine it is based upon experience and belief. The latter includes elements of the tangible and the putative spiritual properties of vegetable remedies. Only one of these elements can be measured objectively by modern scientific methods. Intrinsic to many such systems of health is the belief that health requires a harmony between mind, body, and environment. Thus cure of disease requires restoration of balance between them. Clearly scientific medicine shares these goals in diagnosis but emphasizes different considerations and utilizes a different approach to treatment. It is not clear how one might bridge the gap between the molecular and the spiritual beliefs of the two systems. At the very least the systems agree on the need for safety, efficacy, quality, and rationality.

The barrier between traditional medicine and official medicine is not insurmountable and there are many instances where traditional medicines have developed into prescription medicines when studied and evaluated appropriately. The reverse happens less frequently. Crossover has decreased in the last half-century and will likely never occur completely. Taxol, artemesinin, and gossypol are examples of success. Cholestin, however, is an interesting and instructive example of a failure. In Chinese traditional medicine for centuries the key ingredients in cholistin were prepared by fermenting premium rice with the fungus *Monascus purpureus*. The resulting red yeast–rice mixture contains unsaturated fatty acids, amino acids, naturally occurring red pigments, and a collection of fermentation products that lower cholesterol levels and thus rationally could be used as a dietary supplement for cardiovascular health. Unfortunately, scientific investigation of these components identified agents related to compactin (initially investigated as an antifungal agent but subsequently valued more for cholesterol biosynthesis inhibition).[17,18] Analogs of compactin and lovastatin have become important prescription drugs used to control cholesterol biosynthesis by inhibition of hydroxymethylglutaryl-coenzyme A (HMG-CoA) reductase. This provided a convincing rationale for traditional use of red yeast fungus but triggered a lawsuit claiming that cholestin was nothing more than impure or adulterated compactin and thus infringed patent rights. The court upheld this view and the product had to be removed from interstate commerce. One wonders in this case whether the firm marketing cholestin would not have been better off not doing any scientific analysis of its product but simply marketing it as a vaguely characterized nostrum possessing unknown active constituents! The resulting herbal would have been cheaper for the consumer but less powerful than pure prescription HMG-CoA reductase inhibitors. The situation is being muddied even further today by moves on the part of proprietary drug manufacturers to have these prescription products reclassified so that they could be self-selected by the customer and sold over the counter. So far Mevacor has twice been rejected by the US Food and Drug Administration (FDA) for this indication. Who knows what the future will bring.

St John's wort (*Hypericum perforatum* L.) has been shown clinically to be as effective as sertraline or imipramine for the treatment of mild depression (although it is no more effective than a placebo for major depression).[24] Fortunately for the manufacturers it does not contain either of these constituents but rather owes its action, apparently, to its content of monoamine oxidase inhibitors belonging to the xanthone and flavanoid classes.[25,26]

1.09.5 Secondary Metabolism

Secondary metabolism encompasses the majority of natural products and natural drugs, including herbals.

Primary metabolism of living creatures involves the production of the metabolic products essential for life. These include fats, proteins, carbohydrates, nucleic acids, vitamins, bone, and the like. Study of the processes involved is the primary occupation of biochemists and physiological chemists. These metabolites are nearly universally distributed in living organisms.

Secondary metabolism involves the production of metabolic products of idiosyncratic distribution whose value to the producing organisms is often conjectural. Study of the processes involved in secondary metabolism is primarily carried out by organic and natural product chemists. The processes and enzymes of secondary metabolism are nearly universal and involve much of the same biochemical machinery as is seen in primary metabolism but the substrates and the resulting products usually have narrow distributions in biota. For example, taxol is found primarily in a few species of the genus *Taxus*. The majority of naturally derived drugs are secondary metabolites and the use of herbals depends primarily upon their content of these substances.

The purpose that secondary metabolism serves for the producing organism is highly speculative. Many believe that the products are defensive secretions assisting plants or animals in securing a survival niche in the face of competition with other species or for help in resisting predation. Some appear to assist in capturing and killing food and a smaller group has healing properties. While logical, it seems more likely that such uses are consequences rather than causes. Evolutionary pressures would, in this view, lead to the preservation of accidental products that provided a competitive advantage to the producing organism. In any case, those comparatively few metabolites that serve us as medicaments are almost certainly also the result of a fortunate coincidence. As an example, it is not credible to believe that the opium poppy produces morphine and its analogs so that humans would achieve rest and comparative freedom from pain. This use is surely a fortunate byproduct of its formation for it is not obvious that the opium poppy suffers from the same maladies and thus derives the same benefit that we do from its presence. More likely, the presence of opioids in the poppy helps to prevent the plants from being consumed by foraging animals.

Some secondary metabolites appear to be processed further by a consuming species before achieving their final form. These provide a fascinating glimpse into the process of coevolution. The cardioactive principles of Panamanian tree frogs serve as examples as they apparently originate as precursors and/or metabolites in arthropods that the tree frogs eat.[27] Monarch butterflies, for another example, consume plant materials containing cardioactive glycosides. When retained, these make the butterflies toxic to birds that might otherwise eat them.[28] Their attractive colors, in this view, are a form of advertisement that lessens the likelihood that they will be consumed accidentally by predators.

There is also gathering evidence that a number of interesting metabolites of higher plants are products cooperatively following infection by endophytic fungi.[29–36] At the present, deliberate infection of herbals with fungi would seem to be perhaps the only realistic method for producing novel secondary metabolites without necessarily isolating individual constituents first.

1.09.6 Use of Pure Single Agents versus Combination Therapy

The majority of herbal remedies are complex mixtures of unknown metabolites in which are included one or more active ingredients. By comparison many prescription medications contain a single active component embedded in a more or less complex mixture of added ingredients making up a formulation. The primary difference is that the components of prescription medications are largely known, having been deliberately added.

The use of single, pure medicinal ingredients formulated for therapeutic purposes is the norm in prescription medicine. The advantages of this practice are sufficiently compelling that this is readily understood. Ease of chemical assay, reproducibility, and content uniformity of formulations, simplification of the process of preparation, and reproducibility of action are prominent among the reasons. Herbal remedies on the other hand are almost always complex mixtures. Further, in many cases the identity of the active constituent(s) is not yet known and the nonactive components are even less known. Those herbals that are standardized are often assayed for the content of indicator substances characteristic of the biological source but these may well not have the activity for which the herbal is prized.

The biota from which herbals are derived are often variable, depending on difficult-to-control factors such as climate, processing, storage, season, and genetic variability of the plant taken, and so on. Arriving at reproducible composition under these circumstances is complex and may require blending batches or spiking the product with various amounts of pure components. This problem is potentially more severe with plant material collected from wild stands rather than those cultivated for the purpose. These factors add to uncertainties about the nature of the active component(s) or even how to perform a biological assay, so complicating the work of the herbalist. As if this were not potential trouble enough, in traditional Chinese medicine and in ayerveda, for example, a common practice is to mix together a group of herbals for treatment of a given disease. This practice increases the chance of accidental substitution and also compounds the problem of identifying and assaying the most important constituent(s). These factors contribute to a reluctance to give official sanction to the prescription of herbals and bedevil attempts to develop useful methods of standardization.

On the other hand, a significant number of regulated prescription medicines are also used as mixtures. Until very recently Premarin (a mixture of the estrogenic metabolites excreted in pregnant mare's urine that is used to control postmenopausal symptoms) was the most frequently prescribed medication in the US. (The current list of the top 200 most frequently prescribed medications in the US can be obtained from the website listed in [214].) Gentamicin (an important antibiotic used for the treatment of severe infections by Gram-negative bacteria) is a mixture of several related aminoglycosides. Bactrim/Septra is a mixture of sulfamethoxazole and trimethoprim (used for the treatment of human bacterial infections). Kaletra (a mixture of lopinavir and ritonavir, two inhibitors of the protease of the human immunodeficiency virus, the causative agent of acquired immunodeficiency disease syndrome) is used to delay or prevent the emergence of resistance by a devastating virus. Many mixtures of anticancer agents are employed to delay the emergence of resistance and to kill transformed cells in nonsynchronously dividing tumors. Many other examples could be listed. All of these are rendered even more complex by formulation. Thus the distinction between herbal mixtures and officially recognized drugs based upon molecular singularity is not sharp and is often a matter of the degree to which other substances are present and to which the precise composition of all of the contents is known. The growing popularity of using mixtures of prescription drugs to delay or prevent drug resistance will likely continue and make the distinction based on complexity less and less compelling. One can readily envision a future in which herbals may have become progressively less complex and human medicines more complex.

1.09.7 Economic Issues

Compared to prescription medications, herbals are relatively cheap and for the desperately poor may represent the only realistic therapeutic option. For the comparatively well-off, when properly validated they represent an economically attractive possible alternative.

The general lack of patent protection for herbals leads to intensive marketing competition, tending to drive prices down. These agents are also not as tightly regulated, so that the price of entering the marketplace is significantly less. Herbals are usually not processed as extensively as are prescription medications, also reducing the costs of goods. Finally, distribution to mass outlets places them before a mass consumer base. Advertising costs in part negate some of these advantages but economic factors clearly contribute to the attractiveness of herbals to customers.

On the other hand, commercial firms have been slow to provide reliable clinical information. Why? The reasons are often economic. First and foremost, the majority of herbal remedies have been in use for long periods of time – often for centuries. This makes it very difficult to patent them or their use. Without the protection from competition that this would provide, the very large expenditure required to prove safety and efficacy to the satisfaction of the FDA and register them as actual drugs will not be forthcoming. If one supposes that a private, smallish firm would take such an action altruistically despite the opposition of its investors, then it would be punished in the marketplace for having done so. Other, perhaps more realistic firms would sit on the sidelines and wait until definitive proof of efficacy was in hand and then simply take advantage of the information without having incurred any of the costs of developing it. Clearly this is an overwhelming disincentive. Such research, under present circumstances, would have to be done in academia with funds provided by federal agencies. This would require a significant attitudinal change on the part of study sections evaluating grant proposals in comparison with proposals addressing officially sanctioned medications. Recognizing this, the National Institute of Complementary and Alternative Medicine has funded several national centers for such studies and has also begun to fund individual investigator-originated grants. Progress should now take place at a faster rate.

A practical illustration of the impact of economic considerations lies in the chemotherapy of malaria. This devastating disease afflicting tropical lands is estimated to kill a child every 20 s! Whereas quinine and synthetic compounds derived from it have been employed in antimalarial chemotherapy for a long time, resistance development is all too common and the expense is high for impoverished populations. The herbal qing hao (in the form of a tea prepared from *Artemisia annua*) has been employed successfully in China for centuries in the treatment of malaria and its active ingredient and analogs derived from it are known to be effective against resistant malaria. The tea contains a variety of related compounds that are believed to be synergistic, suggesting that it might be more effective than a single pure constituent derived from it, and that resistance to all of the components might be more difficult for the plasmodia. The tea can be prepared for pennies from locally available plants whereas the pure ingredient is scarce and costs dollars for a course of therapy. It appears logical that a carefully controlled clinical trial comparing qing hao tea and established chemotherapies is called for. The motivations for this are humanitarian and also economic.

1.09.8 Ethical Issues

Safety of herbals is foremost, sometimes even outranking efficacy as a consideration.

The ancient rule in medicine is 'first do no harm.' Obviously the gains to be derived from treatment should outweigh the risks therefrom. This applies with full force to the use of herbals. All too often the benefits are harder to quantify than is their toxicity, particularly when the intension is to prevent a condition that might not develop in the first place. Under those circumstances the risks associated with the use of herbals must be unusually slight. Acute risks are commonly easier to measure than are the often more subtle risks that can accompany long-term use. This applies with greater force when one is trying to prevent or delay the deterioration accompanying age, for example, where chronic use of drugs is common. This contrasts with the different level of risk associated with short-term acute use as when, for example, treating a bacterial infection. The infection will either resolve in a short time or other treatments must be instituted. This provides a shorter timeframe for toxicity to occur and to be detected.

Toxicologists are very conversant with the dictum that toxicity is a function of dose. When agents, like herbals, are self-selected and self-administered, dosage is very hard for healthcare-givers to control. Facing this reality, one usually takes comfort from the venerable history of safe use associated with many herbals such as green tea. The generally innocuous nature of such products greatly lessens the danger of toxicity due to overdose.

This experience leads to a common assumption among the laity that herbals are safer than prescription drugs. Whereas this is most likely justified in a number of cases, one should bear in mind that diagnosis is dramatically refined over earlier times and also that modern humanity lives strikingly longer lives than did their ancestors. One consequence of this is that chronic toxicities that might not have been recognized in earlier generations might well be seen now. Further, the contemporary practice is for patients frequently to be taking a number of different medicaments to control a variety of conditions. Drug–drug interactions not possible or even imagined by earlier generations could be experienced now, particularly among the elderly. Thus the question of safety in a modern context remains paramount and must be intelligently revisited from time to time. This places a heavy burden upon the developers and manufacturers of herbals. A heavy burden is also placed upon the healthcare-giver to be conversant with the properties of herbals as well as prescription medications and their possible interactions. Much careful research is needed to fill in the present gaps in our knowledge in this area. Universities are only now beginning to address the information gap about herbals opened up by dropping courses in natural medications from the curriculum a generation ago.

Purity and identity are also an ethical responsibility of the purveyors of herbals. The label must adequately identify the contents and the state of purity of the material being presented for use. Further, the patient should be able to be confident that heavy-metal concentrations, insecticides, herbicides, insect parts, microorganisms, and other extraneous materials are either absent or are present only in trivial amounts. Without regulatory oversight there are few guarantees that this will be the case. Fortunately the DSHEA addresses some of these concerns.

Unfortunately, even with prescription medications, regulation usually follows scandal and unanticipated toxicity. The recent charges that improper use of ephedrine-containing dietary supplements to enhance athletic performance are an example where laisser-faire oversight may have proven to be inadequate.[37–39] Ephedrine had been used in the form of ma huang in China for centuries and in western countries for over a century, before its use in herbal form led to public awareness that cardiovascular damage could occur with excessive dosage. Of course, healthcare-providers had long been aware of this. The proper boundary between the FDA and the herbal industry was sooner or later going to have to be defined more clearly for herbals and the ephedrine preparation case provided a significant field of battle for this. The FDA ruled that ephedrine preparations were not safe (without defining what doses constituted this risk) and required them to be removed from the market. The herbal industry declared this to be an overreaction and the removal has apparently been reversed by court order. The central issue is the belief that the burden of proof of lack of safety is placed on the FDA by the provisions of the DSHEA. The act is based upon the belief that herbals that have been long used are presumed to be safe unless proven otherwise and that the manufacturer does not have the burden of proving this. The FDA is apparently not empowered to remove a preparation by assertion of lack of safety but must define at what dosages lack of safety exists. The final resolution of this case will have profound implications for the future of the herbal industry and for the public safety.

1.09.9 Bioprospecting

Concerns about respecting intellectual property rights associated with exploitation of folk medicinal practices have become pronounced in the last couple of decades. Previously it was generally assumed that folk remedies were in the public domain, to be enjoyed by all.[40] The practice of patenting even genes for proprietary advantage, coupled with persistent poverty in regions where herbal medicine flourishes, has altered this picture dramatically. Considerations of ownership and a proper compensation for divulging carefully preserved lore were previously largely neglected but now are the subject of heated debate and acrimonious dispute. Certain nations and learned societies have promulgated conventions to attempt to resolve the worst of these problems but the conventions often lack the force of law, are weakly enforced, and are not universally subscribed to.[40] Commerce that has grown up over a long time and for which, therefore, there is an established history largely escapes the worst aspects of exploitation. Commerce in coffee comes to mind in this context. More recently discovered agents, such as the vinca antitumor alkaloids, have, on the other hand, become divisive issues. For example, the country where vinca grew naturally has demanded very large (and unrealistic) payments from the firm that commercialized alkaloids from it for the treatment of cancer. The demands fail to take into account the enormous research and development costs incurred before these materials could reach medical use. The plant had only horticultural value without this investment. Some compensation might be justified, but the demands have often been unrealistic. The National Cancer Institute, US,[41] and the American Society for Pharmacognosy have taken helpful and leading stances in trying to make these negotiations more reasonable. One of the unfortunate consequences of these disputes is that it is increasingly difficult to obtain permission to bioprospect for research purposes. Unfortunately the regions where promising biota grow often do not have the scientific expertise or the expertise to take advantage of the contents of these herbs. As a consequence, research and development are hindered, so no one profits.

1.09.10 Relationship between Traditional Medicines and Official Medicines

Despite strongly defended differences by partisans of each class, the two systems are complementary.

It is interesting to note the reciprocal relationship between animal and human testing in comparing traditional medical systems with official medicine. In traditional medical systems humans most often have consumed the agent for a long time so there is already a presumption of safety and a possibility of efficacy. When scientists become interested in them and wish to validate the usage, then animal or even cell-free testing is undertaken. Experimentation then proceeds back to human testing. In official medical systems, on the other hand, the validation process usually starts with cell-free experimentation, progresses to animal studies, and only then does human testing begin. The two classes of drugs then differ dramatically with respect to when human data become available.

Many promising substances originating from opportunistic synthesis fail to reach the market based on the outcome of clinical trials in humans due to lack of safety or efficacy. This happens late in the process when substantial sums have been expended. Failure in the clinic is very costly. With herbals human experimentation is already in hand, even if it is not in the form usually employed for presentation to regulatory agencies. There is a smaller possibility that surprising negative findings will be encountered in the clinic with such materials.

News writers enjoy reporting about instances of toxic reactions to herbals following marketing. A balanced account would point out that today the package insert for virtually every prescription medication contains a lengthy recital of side effects – some potentially quite serious. This has not been prevented by the extensive safety evaluations undertaken before clearance for marketing is given. It is estimated that adverse drug reactions in US hospitals is the fifth leading cause of death. As a partial explanation for this level of adverse effects, one notes that relatively uncommon adverse effects are not likely to be detected in clinical trials unless very large numbers of patients are involved. Cost considerations often prevent this from happening. It is food for thought to wonder if resort to widely used natural medicines may be no worse, if not even less likely to produce such deleterious results, even though their premarketing scrutiny is less. Reports of adverse reactions to either herbals or prescription medications have value to the public if not presented in the spirit of scandal and shame.

1.09.11 Processing and the Border between Herbal and Official Medicines

Most herbal remedies receive significantly less processing than prescription medications before reaching the consumer. Typically, prescription medications are refined to molecular singularity before formulation. Herbals, on the other hand, are typically much less refined. In many instances they may be purchased as ground plant powders. In other cases, they may be purchased as whole plant parts. In some cases the manufacturer partially refines the material for the convenience of the consumer. Green tea, for example, can be purchased in tea bags or in capsules containing primarily the isolated crude flavanoids free from caffeine. Resveratrol is popularly consumed in the form of red wines. Obviously, therefore, herbals are typically much more complex mixtures than are prescription pharmaceuticals. The reasons underlying these practices are largely economic and also are responsive to patient preferences.

Clearly, then, there is a border between these classes of medicaments that may not be crossed. If an herbal was to be refined to molecular singularity, it would then be classified as a prescription medication and be subject to a very much more stringent degree of regulation. It would also be much more costly to produce and to purchase.

1.09.12 Natural Medicinal Systems

There are traditional medical systems in every part of the world having regular habitation by humans.

Virtually every human society has a history, if not a present reality, of the use of herbal medicines that is based upon empiricism and spiritualism. Before the advent of scientific medicine the vast majority of medicines were herbals. These natural medical systems are the result of many centuries of trial-and-error investigations, often using herbals, and the results are preserved and transmitted to future generations by shamans – traditional herbal practitioners.

Among the most developed of these systems one notes traditional Chinese medicine, kempo, and ayurveda. In many of these systems, careful observation and critical evaluation of the results are central and a significant degree of refinement has taken place over time. Thus there is a significant degree of reliability involved, although the correlation with modern scientific beliefs is not always easy. Almost all natural medicinal systems involve a belief that health is normal and disease is a departure from harmony between body, mind, and external influences. The use of herbal remedies to restore harmony is not fundamentally antithetical to taking prescription drugs to cure disease. The rationales and approaches are, however, very different. Lacking a clear understanding of normal physiology and the departures leading to disease, ancient societies codified their approaches to therapy in terms much different from those employed in modern medicine. Such descriptors as hot, cold, sweet, and sour are not easy to apply to arthritis, for example, although the term 'inflammatory disease' betrays lingering elements of the earlier empiric system. The reader will readily perceive that inflammation is a description of either the result of the disease or at least a correlated event. It does not reveal much at all about the underlying pathology of the disease and therefore does not readily lead to a logical curing remedy, even if an herbal that cools might give relief. Sweet urine appears closely related to sugar diabetes but is also a symptom, not a cause.

The spiritual aspects of traditional medical systems relate most likely to the well-known concept of placebo in scientific medicine.

1.09.12.1 Ayurveda

Ayurveda is the traditional healing system of India and is more than 5000 years old. The Sanskrit word 'ayur' means life in English and 'ved' means knowledge. This system regards health as a normal state resulting from a suitable balance in mind, body, and emotions brought about by proper attention to diet, herbal supplements, and physical and meditative actions. This brings ether, air, earth, fire, and water into a harmonious relationship, blending three humors (doshas). Individuals are born with different intrinsic blends of the humors. Disease, in this view, is an excess or deficiency of any of these elements from the inherited balance and must be adjusted. The origins of imbalances producing disease can be from either external or internal sources and are usually treated differently. Individuals displaying the same symptoms may have different dosha imbalances so may need to be treated differently. In this sense ayurveda has an emphasis on health rather than disease, cause rather than symptom, the individual rather than the symptom, and herbal remedies are one of the adjusting elements, but not the only one. A relationship to Chinese traditional medicine and to kempo is clear. The relationship between ayurveda and prescription medications is not always easy to discern. When herbals are employed in ayurveda, they are almost never given as a single herb but rather are given in mixtures that can be quite complex.

1.09.12.2 Traditional Chinese Medicine

Traditional Chinese medicine is an ancient Chinese healthcare system based on the concept of balanced qi (a vital energy believed by practitioners to flow throughout the body that regulates a person's spiritual, emotional, mental, and physical balance). Qi is influenced by the opposing forces of yin (negative energy) and yang (positive energy). In this system disease is believed to result when the flow of qi is disrupted and yin and yang become unbalanced. Herbal and nutritional therapeutic interventions supplement physical exercises, meditation, acupuncture, and massage in restoring health. These concepts are not readily reconciled with the beliefs of modern medicine. The herbals that are employed are often complex mixtures of different plants added together to supplement each other's actions.

1.09.12.3 Kampo Medicine

Kampo medicine is a variant of traditional Chinese medicine that also includes an intensive reliance on herbs. The Japanese word 'kan' literally means China and 'po' means medicine. It is clear, therefore, that kampo is derived from traditional Chinese medicine, although it includes, in addition, traditional Japanese materials. Partly as a consequence of its popularity, herbal consumption in Japan is probably the highest in the world. The concepts of kampo are not readily reconciled with those of western medicine as they consider that health involves a proper balance between positive and negative thoughts (yin–yang), strong–weak, and hot–cold influences, and disease is an imbalance that must be adjusted. The emphasis is on the body as a whole rather than on specific organ system failures. Diagnosis is noninvasive and based upon appearance, touch, and smell, patient statements, palpation, body sounds, pulse and the like, and so appears to have a large intuitive component. Treatment often involves administration of complex mixtures of herbals, some of which contain reinforcing constituents and others counteract otherwise deleterious effects of the others. Often there is a principal ingredient responsible for the main therapeutic response (called jun). A second ingredient enhances the effect of the main ingredient (called chen). A third ingredient (called zuo) is used to counteract side effects or to modulate the potency of the main ingredient. Finally a fourth ingredient (called shi) directs the action of the prescription to the site of the problem.

The source of the disease influences the remedies chosen. It is particularly important for practitioners to determine whether they believe the malady to be due to an external influence or to an internal weakness in the body. Drugs are classified according to the expected response: removing external pathogens, resolving phlegm and other agents associated with cough, reduction of fever, antirheumatics, providing warmth to combat internal coldness, resolving dampness, enhancing removal of wastes by use of diuretics and purgatives, regulating the flow of vital energy, regulating blood-clotting and circulation, stimulating consciousness, inducing repose with sedatives and tranquilizers, reducing hepatic hyperactivity, use of tonics and aphrodisiacs, controlling blood flow with astringents and hemostatics, digestives and evacuants, laxatives, expelling invasive organisms with anthelmintics, and dealing with skin problems by employing drugs for external use. All of these are examples. Potential applications in western medicinal systems are readily visualized for many of these. Consequently these treatments are under intensive investigation with an aim to identify single pure ingredients that can be co-opted.

The relationship between these three oriental medical systems is apparent but extremely difficult to reconcile with the intellectual and evidentiary basis of prescription medicine. The herbal treatments clearly elicit body responses of actual or potential value but the rationales for the responses are very different. Translating between the systems is thus very complex.

African and Hispano-American medical systems are also venerable and complex but have been much less extensively codified so are more difficult to discuss in general terms.

1.09.13 Regulatory Status of Traditional Medicines

Herbals are less well regulated than are prescription medications. The recent passage of the DSHEA in the US was an attempt to bring order into an otherwise somewhat chaotic market-place and to accord a higher status to these remedies.

Legally enforcible regulation is vital to ensure the wholesomeness, reproducibility, identity, and safety of herbal products. The DSHEA was promulgated in order to regulate commerce in herbal medicines in the US. The term 'dietary supplement' includes vitamins, amino acids, and minerals, as well as herbal remedies in the eyes of this law. For our purposes the discussion will center about herbals. Historically the FDA regulated dietary supplements (including herbals) mainly in the same manner as they did foods. The objective was to ensure safety and wholesomeness and that the labeling/advertising was truthful. Claims should be no more misleading than other forms of advertisement. All new ingredients (post 1958) were required to be safe. This changed with the DSHEA in that herbals and herbal constituents were now examined in a somewhat different light. Before DSHEA dietary supplements were mainly considered to consist of essential nutrients (vitamins, minerals, and proteins). After DSHEA the definition of 'dietary supplement' was broadened to include herbals and mixtures thereof. Basically now a dietary supplement is a product (exempting tobacco) that is intended to supplement the diet and that bears or contains one or more of such ingredients as a vitamin, a mineral, an herb or other botanical, an amino acid, a dietary supplement for use by humans to supplement the diet by increasing the total daily intake, or a concentrate, metabolite, constituent, extract, or a combination of these ingredients. It is intended for ingestion in a pharmaceutical form (pill, capsule, tablet, or liquid). It is not represented for use as a conventional food or as the sole item of a meal or a diet. It is explicitly labeled as a dietary supplement. It includes products such as approved new drugs, certified antibiotics, or licensed biologics that were marketed as a dietary supplement or food before approval, certification, or license, unless explicitly exempted by the Secretary of Health and Human Services.

Under DSHEA, a dietary supplement will be declared adulterated if it or one of its ingredients presents a significant or unreasonable risk of illness or injury when used as the label directs or when used in the customary manner for this material if no directions are given. A product containing a new ingredient (one marketed subsequently to 15 October 1994) may be declared adulterated if there is inadequate information about its safety. The responsibility for providing such assurance lies with the manufacturer but the responsibility for the safety judgment lies with the Secretary of Health and Human Services, US.

Retail outlets may distribute or sell 'third-party' informational materials to assist customers in becoming knowledgeable about the putative health-related benefits of dietary supplements provided that the information contained is not false or misleading, does not promote a specific product, is displayed with similar materials presenting a balanced view, is displayed separate from the supplements, and does not have other information attached to it (such as promotional literature).

DSHEA allows three types of claims to be made. Health claims may imply a relationship between a dietary supplement and a disease or health-related situation. Such a claim might legally state something like: "diets high in calcium may reduce the risk of osteoporosis." Legal structure/function claims may relate consumption of a product containing an ingredient to maintenance of a healthy structure or function. Such a claim might legally state something analogous to: "diets rich in calcium help build and maintain strong bones." Nutrient content claims may associate consumption of an ingredient with establishment or maintenance of general health. Such a claim might legally assert something like: "daily consumption of calcium helps maintain bone health." What is not permitted is a claim that the dietary supplement can be used to diagnose, prevent, mitigate, treat, or cure a specific disease unless the product has successfully gone through the new drug approval process with the FDA. Therefore it cannot make statements like: "this dietary supplement cures cancer or treats arthritis or prevents strokes." The permissible boundary is therefore somewhat subtle.

The FDA does not evaluate and certify these claims. Instead it is the responsibility of the manufacturer to ensure the accuracy and validity of such claims. The manufacturer is also required to inform the consumer by appropriate labeling that: "This statement has not been evaluated by the Food and Drug Administration. This preparation is not intended to diagnose, treat, cure, or prevent any disease." This means that substances performing these excluded functions are drugs and are subject to a separate category of regulations but the consumer is not 'bothered' with such an explanation. Generally the FDA does not intervene in herbal commerce unless mislabeling or toxic effects are encountered.

Unfortunately, whereas labeling must state the name and quantity of ingredients, it is not required to give botanical binomials. This can lead to an undesirable degree of ambiguity. Ginseng, for example, may not in fact be *Panax ginseng* (the Asiatic variety that is mostly used) but could also be American ginseng (*P. quinquefolius* L.) or san qui ginseng (*P. pseudo-ginseng* Wallach). These materials are not pharmacologically identical. The situation with a number of other herbals is equally unclear.

The DSHEA gives the FDA authority to establish good manufacturing practices governing the preparation, packing, and holding of dietary supplements under conditions that ensure their safety. This may include unannounced inspection visits.

Clearly the DSHEA is a step in the right direction in attempting to regulate the marketplace for herbals in the public interest, but it is clearly not ideal. No standardization is mentioned and the language is loose enough that sharp practices and cynicism are not entirely obviated.

More information and a deeper treatment of the DSHEA can be found at the websites listed in references.[215,216] Should one decide to improve the regulatory system for herbals, one needs to recognize that there are differences between them that must be taken into account. What classification system should be used? The efficacy classification advanced by Tyler[1] seems the most clear-headed and sensible but should be expanded. If so, there would be five main choices and three subchoices. The main choices would address the question of activity. Class 1 would be those for which absolute proof has been established by reliable clinical trials. Class 2 would be those for which reasonable proof is available based upon traditional reputation but which has not been sufficiently verified by clinical study. Class 3 would be those for which a 'reasonable certainty' based on all available information is available. Class 3 would be the most practical since it would not be remotely as expensive as class 1. Class 4 would include those herbals for which little or no evidence can be found validating utility. Class 5 would include those remedies for which evidence indicates lack of activity.

These classes, however, do not address proof of safety. Here there would be three subclasses. Class A would contain those where careful study has revealed no toxicity. Class B would contain those presumed to be safe based on the bulk of available evidence. Class C would contain those shown to be unsafe. The last group would be prohibited. Herbals belonging to class 1A would be the very best.

Unfortunately, much of the clinical information available about many herbal remedies is flawed, particularly evidence based upon older studies. Many such studies are not adequately controlled, not blinded, involve too few patients for the questions raised, involve patients who have been inadequately diagnosed, have endpoints that are vague or patient-reported, use herbal materials of unreliable origin or that are inadequately described, or are based upon population studies in which the various activities of large populations are inadequately controlled or characterized. At the time of writing, approximately 40 clinical trials involving herbals are under way exploring various questions of efficacy and safety. Imperfections exist in the design of some of these studies also.

1.09.14 Sources of Material

Herbals can be found wherever plants grow.

A significant number of users of herbal medicines believe that herbals should be collected from wild-growing specimens in order to be of top value. Most scientists do not share this belief. Chemists abandoned the belief in vitalism following Woehler's demonstration nearly two centuries ago that urea could be produced both physiologically and in the laboratory from inanimate materials and that the properties of urea are identical regardless of its origin. No subsequent evidence has successfully contradicted this finding. Among a significant segment of the laity, however, such beliefs persist. An example of this belief is that cultivated plants are somehow 'unnatural.' It would be hard to find a chemist who believes that morphine obtained from wild or cultivated poppies is somehow different from morphine obtained from cultivated plants or even from morphine prepared by partial or total laboratory synthesis.

Among the more persuasive considerations in favor of cultivated plants is the fear that wild stocks will not be sufficient to supply the quantities needed for sustained commerce. The recent search for adequate supplies of taxol-bearing plants provides an example.[8] Camptothecin provides another.[42] An interesting further example results from excessive browsing by deer. Without control of deer populations an extinction of American ginseng-bearing plants might take place at some future date.[43] Appropriate farming practices could address this worry.

When cultivated for the purpose, the best strains of many plants can be utilized, contamination can be minimized, and optimum growth conditions can be utilized. Further, sustainable cultivation methodologies can ensure that the species will not be lost.

1.09.15 Dosage Forms

Raw plants are inconvenient to consume for medical purposes. Historically they are prepared in various ways to enhance their desirability and efficacy. Those dosage forms that mimic the manner in which indigenous populations use herbals are most likely to be useful.

In their simplest forms, herbal remedies come on to the market as whole plant parts or as ground powders prepared from them. The powders are often also placed in gelatin capsules for convenience and standardization of dosing. Less often they are compressed along with a variety of formulation ingredients into tablets. The bulk of material that often has to be consumed makes this inconvenient. Many herbals are consumed in the form of aqueous teas. These are often prepared extemporaneously by the consumer to preserve freshness and to prevent microbial contamination. Such teas are prepared by stirring (steeping or soaking) in hot water for a period (infusions) or by boiling in water or having boiling water poured over them (decoctions). The final concentration of ingredients is variable when made in these manners but the preparations are usually consumed completely, so the dose is dictated by the amount of plant material taken in the first place, the time of contact with hot water, and the quantity drunk. Beverage tea is a well-known example of a tea and coffee is likewise a well-known example of a decoction. Both are prized, among other things, for their caffeine content.

Fluid extracts are aqueous or hydroalcoholic solutions adjusted to a fixed concentration. For example, each milliliter of a fluid extract typically contains the material extractable from 1 g of plant material.

Tinctures are hydroalcoholic solutions that are usually more dilute than fluid extracts. A typical concentration would be the amount of material that can be extracted from 1 g of plant material dissolved in 10 mL of final solution.

Extracts are dried preparations containing the residue left following evaporation of the extracting solution. These are sometimes mixed with talc or such solids to render these otherwise often gummy and sticky masses more readily handled.

Homeopathic solutions are often so dilute that evaporation leaves a barely measurable quantity of extractive/active ingredient.

Officially defined fluid extracts, extracts, and tinctures of herbs largely disappeared from the *US Pharmacopoeia* and the *US National Formulary* following World War II.

1.09.16 Standardization

Compared to prescription medications, herbals are much less thoroughly standardized.

In their heyday herbals were identified by careful botanical descriptions of the plants from which they were derived and standardized for potency by bioassays. These monographs had the power of law behind them, being published in the *US Pharmacopoeia* and the *US National Formulary.* This fell away after about 1950 when herbals were largely supplanted by synthetic prescription medications. There is no obvious reason why this system of standardization and regulation could not be reintroduced other than the lingering antipathy toward herbals shown by some medical and governmental authorities. Methods of analysis have advanced very considerably in the last half-century so chemical standardization in many cases would now be possible. Aside from the obvious benefits in regulating the commerce in these materials in the public interest, this provides the impetus for the continued scientific investigation of these substances in cases where suitable methods of analysis are not yet available. Objective scientific investigation into their merits or lack of merit would no longer be regarded as at least slightly disreputable. In countries that have taken this step, notably Germany, it has been shown that it works and prevents the more extreme frauds, misbrandings, and cynicism that lack of such action can encourage. Surely the strong clarifying light of modern scientific exploration under critical peer review is the best way to dispel dangerous and untrue or partially true assertions! In the absence of objective scientific study and consensual standards for comparison, herbals are in danger of being loosely monitored by ill-prepared persons. Purchasers are even less scientifically sophisticated and are mostly dependent upon a chain of personages to stand in for their best interests by performing with skill and integrity. One might suppose that physicians and pharmacists would be reliable sources of objective opinion on these matters. Unfortunately, this is not so. Formal training in herbal medicine has all but completely disappeared until very recently from medical and pharmacy curricula.

In some cases at present it is not known what active ingredient, if any, is responsible for the putative properties of even such widely used herbals as echinacea.[44–46] It is well within the capacity of modern science to deal with this kind of problem. With identification of active principles and determination of their mode of action, regulation and standardization would readily follow.

The obtaining of wholesome, pure products for internal consumption is a goal that few customers would fail to support. With prescription medications stringent standards and regulatory oversight stand behind drugs and violations are punished. The public is scarcely aware of the details of this but derives comfort from knowing that it must be going on somewhere. With herbal products the situation is quite different. The regulations are much looser. A product may not be sold if the Secretary of the Department of Health and Human Services declares it to contain one or more ingredients presenting a significant or unreasonable risk of illness or injury when used as the label directs or in the manner customary for such a product (if there are no specific instructions given as to use). This regulation is silent about the quantity of any specific ingredient that must be present or the complexity of accompanying substances provided that they are not overtly harmful. The efficacy of the herbal product is also not certified.

The labeling does have to indicate the nature of the contents and its quantity but does not have to certify its quality. Many manufacturers do specify the quantity of specific contents, finding that the consuming public derives comfort from such representations. Unfortunately, however, there is still significant scientific doubt as to the identity of active ingredients in a number of herbals, let alone about how much of such an 'active' ingredient should be present. Manufacturers often measure the quantity of an ingredient that is associated with the genus and species being purveyed, but this does not guarantee that this represents anything reliable about the identity of the active constituent or the possible potency and efficacy of the product. Such representations may well be misleading. As discussed in the previous section, a degree of vagueness even about the specific species contained is also possible. Echinacea could be one of several species. Ginseng also could represent one or more of several species. Not all of the possible species involved have equal potential herbal value. Thus, even though there is a degree of regulation in the public interest in the US, it falls short of being satisfactory. More strict regulation will surely come in time. Unfortunately, regulation mostly follows scandal and may overshoot the mark. A far better situation would result if trade organizations took it upon themselves to regulate and police their membership intelligently. A seal of approval could be awarded to complying firms and their products and the public could take comfort in knowing that a sophisticated group is looking out for their interests. This is perhaps too idealistic but one fears that without somewhat more definitive regulation there could be a return to some of the unpleasant aspects of 'snake oil' hucksterism that characterized the late years of the nineteenth century trade in herbal medicines. The result would be another period of disdain.

The situation in certain other countries is much more satisfactory. In Germany, for example, the Commission E has set forth standards for many herbal products that have the force of law.[47] In 1967 the government of the Federal Republic of Germany defined herbal remedies in the same manner as other drugs. Specifically these were identified as plants, plant parts, or preparations from plants in either the processed or crude state that are intended to cure, alleviate, or prevent disease, suffering, physical injury, or symptoms of illness, or to influence the nature, state, or function of the body or the mind. In 1978 the Federal Health Agency (Bundesgesundheitsamt) established an expert committee on herbal remedies to evaluate the safety and efficacy of herbals. This group, composed of a variety of specialists and lay members, evaluates bibliographic data and makes conclusions about the degree of trustworthiness of this information with regard to safety and efficacy. When this is done, a draft monograph is prepared stating positively or negatively whether the material is suitable for medicinal use. Following a period in which comments and suggestions are received and considered, the final monographs are published in the Federal Register (Bundesanzeiger). The monographs were prepared with the intension that they be included as package insert information for physicians, pharmacists, other healthcare professionals, and for customers. These monographs have been translated into English and published under the sponsorship of the American Botanical Council.[47]

A smaller number of herbals are official in the German *Pharmacopoeia* (hawthorn leaf with flower, hawthorn fluid extract, horse chestnut seed, standardized horse chestnut seed extract, lemon balm, and milk thistle fruit).

The European *Pharmacopoeia* contains monographs for witch hazel leaf, German chamomile flower, senna leaf, Alexandrian senna pods, Tinnevelly senna pods, and valerian root.

In Canada herbal commerce is under the overview of Health Canada, whose stated purpose is to ensure that consumers have trustworthy information and that the herbal products available are safe, of high quality, and effective. Since 1 July 1997 each establishment that manufactures, distributes, or imports must have a license certifying that it meets standards of cleanliness, quality, and record-keeping.

The United Kingdom follows the European practice of accepting substances that have a long history of use under the doctrine of 'reasonable certainty.' That is, if there are years of use with apparent positive effects and no evidence of detrimental side effects or scientific evidence to the contrary, then a product can be sold. This agrees with the World

Health Organization's guidelines that state that a substance's historical use is a valid way to document safety and efficacy unless there is scientific evidence to the contrary.

In France traditional medicines can be sold with labeling consonant with their traditional use. They are licensed by the French Licensing Committee following approval by the French Pharmacopoeia Committee.

In the orient (China, Japan, and India in particular), as in many other lands, patent herbal remedies are the custom and are readily available to the consuming public.

1.09.17 Safety Issues

Generally, herbals are no safer or more dangerous than prescription medications.

Since herbals are mostly self-selected and self-administered, safety issues are of paramount importance. Clearly, overtly dangerous products are not and should not be tolerated. This is easier to control in cases where toxicity is acute and cause and effect are closely linked temporally. It is also a truism that every substance, even water, is toxic if used in excessive quantities. By way of example, the use of ephedrine- and pseudoephedrine-containing products for respiratory problems and other purposes is venerable.[48] Use in prescription medications for asthma symptoms, colds, and other problems grew out of this. Further uses in weight control also became popular. As long as reasonable doses are utilized, no particular harm comes from this. Unfortunately, overdose can lead to cardiovascular damage and such instances have become too common in the US. Athletes have also taken to consuming such products to enhance their performance, sometimes in dangerous quantities. Consequently use of such materials is now forbidden or tightly controlled.[49,50]

Adulterated products contaminated with potentially harmful ingredients such as heavy metals, for example, are occasionally encountered and are forced off the market.[51–70] These examples fall within the umbrella of current regulations and need not concern us further, other than to point out the prevalence of the problem.

Rather more insidious problems can arise that are more difficult to deal with. For example, there is a segment of the public that has faith in shark cartilage[71] or other nostrums as curative for cancer. While resort to such materials may provide a degree of comfort to desperate individuals and they are not harmful by themselves, they can be seriously harmful to the extent that they delay or prevent resort to established medicinal treatments. It is difficult to see how one could prevent this except through education.

It is now well known that ingestion of a number of herbal remedies elicits body responses in order to process the materials. Oxidation by the P450 microsomal system of the liver is often required in order to detoxify or to prepare lipoidal constituents for more facile excretion. The capacity of this system is not infinite. When certain herbals are coingested with certain prescription medications that also utilize this system, interference with the intended properties of both preparations can result. This arises primarily in terms of unintended high blood levels of one or both such products and this can reach toxic levels. An example of this is found with St John's wort and other medications, such as oral contraceptives and immunosuppressive drugs.[72,73] It is important that labels indicate when this might take place and that patients should inform their healthcare-providers of all medications being used and that purveyors of such products should also caution patients. This requires a significant level of knowledge of such matters on the part of healthcare-givers.

Chronic toxicity is more difficult to identify and to control due to the separation of the injury temporally from the development of the symptoms. In the early stages the development of symptoms and pathology may be too subtle to detect. The long-term effects of chronic use of high doses of cardiostimulatory substances (including caffeine) can be heart and kidney failure and chronic overconsumption of sugars can lead to diabetes. The threshold for developing these problems is lowered when these preparations are consumed along with other materials having the same general biochemical and pharmacological effects.

An example of a once popular herbal with a venerable history of use that is now banned due to chronic toxicity is comfrey (*Symphytum officinale*). Comfrey has now been shown to be hepatotoxic in both humans and rats due to its content of pyrrolizidine alkaloids.[3–6,74,75] Chronic use, even in comparatively low doses. leads to veno-occlusive disease, in which small hepatic veins are destroyed, leading to cirrhosis and eventual liver failure. Use of comfrey is now banned in Germany and Canada and the FDA has recommended a voluntary removal of comfrey from the market in the US.

The popular press has recently characterized a number of herbal remedies as the 'dirty dozen' based upon reports of toxicity.[76] Herbals containing aristolochic acid (all *Aristolochia* species as well as wild ginger, the rhizome of *Asarum canadense*) and germander (the aerial parts of *Teucrium chamaedrys* that contain furanoditerpenes) are definitely hazardous. The furanoditerpenes are converted to liver toxins by the action of P450 enzymes.[77–79] At least 30 cases of kidney failure and a number of significant cases of kidney damage were encountered in Belgium when persons used adulterated herbals taken for weight control (purported to be *Magnolia officinalis* and *Stephania tetrandra*, but actually consisting of *Aristolochia fangchi*).[80–84] Some of these patients later developed urothelial carcinomas due to conversion of aristolochic acid to a carcinogen by cytochrome P450 enzymes.[85]

Comfrey (see immediately above), kava (the roots/rhizomes of *Piper methysticum* associated with severe liver toxicity),[7] chapparal (*Larrea tridentate* leaves that are associated with toxic hepatitis),[86] pennyroyal oil (the volatile oil from the tops and leaves of *Hedeoma pulegioides* is hepatotoxic and nephrotoxic due to its content of pulegone),[87] yohimbe (the bark of *Corynanthe yohimbe* used for erectile dysfunction and known to cause hypertension),[88] lobelia (the aerial parts of *Lobelia inflata* that are used for bronchitis or asthma and are known also to cause nausea, vomiting, dizziness, stupor, arrhythmias, and breathing problems),[89] and bitter orange peel (*Citrus aurantium* is used as a bitter tonic and sedative but contains synephrine, so may cause hypertension and arrhythmia)[89] are regarded as very likely hazardous. More research is needed on these agents.

Specific information about drug–drug interactions and acute and chronic toxicity properties of specific herbals can be found below under the individual agents.

1.09.18 Drug–Herbal Interactions

There are well-known drug–herbal interactions, herbal–herbal interactions, and drug–drug interactions. These interactions are both chemical and pharmacological in nature.

Dislike of herbal remedies on the part of many health professionals in the US leads many patients to conceal their use of such substances from their healthcare-givers. Since there are several well-known drug–drug interactions between certain herbals and certain prescription medicines as well as significant side effects, this is unfortunate.

Given the molecular complexity of herbal preparations it would be astonishing if there were not well-known instances of deleterious interactions with prescription drugs because of induction of P450 oxidative-metabolizing enzymes by St John's wort.[72,90–100] The induced enzymes not only affect the action of the wort constituents but also affect the blood levels of a number of prescription medications that the patient might also be employing. Some of these interactions can have serious consequences for the patient. Such drugs include coumadin, certain antihistamines, and macrolide antibiotics.

In addition to these metabolic incompatibilities, herbal agents can have the opposite pharmacological activity to that of a prescription medication and thus be pharmacologically incompatible. The importance of a primary caregiver being made aware of the use of herbal medications in this context is clear.

1.09.19 Bioavailability Issues

Every drug has potential bioavailability concerns that must be dealt with satisfactorily in order to ensure that blood levels of the active ingredient are adequate for therapeutic purposes. Herbals are no exceptions.

With prescription medications, careful clinical studies are required to have been carried out under scientific conditions and safe and effective doses have been established from them. Formulations must be made that produce such levels. This is also true of subsequently introduced generic versions. All of this is carefully regulated. A second manufacturer who wishes to market a competing product must establish that the new product will also produce the same safe and effective level of the medication following oral administration.

With herbals the situation is quite different. Here the amount of material to be administered is most commonly based upon experience over many years. This is generally a safe practice. What is less clear is whether this will lead to a useful therapeutic result. Clinical trials of echinacea, for example, give very variable results.[46] It is not entirely clear what clinical endpoint should be measured and the very nature of the product employed is often very vague. Under these circumstances it is exceedingly difficult to imagine how one would determine what ingredient or ingredients should be delivered to the blood and for what period of time. Since this is a function of dosage it is understandable that problems can arise in achieving a satisfactory medical endpoint. Another example of the potential difficulties faced lies with the consumption of green tea. How many cups should be consumed daily and what should their spacing be? This is dependent, for example, on just how thoroughly the tea powder or leaves are extracted and what quantity is taken (to say little about the reproducible quality of the tea taken). Also, an oriental tea cup is rather smaller on average than a western tea cup.

These vexing considerations relate to what goes in the body but says little to nothing about what happens to the material in the body. How much is absorbed? How much is metabolized? How much is excreted? It is not possible to set a reasonable regulation for use of herbals without answers to some of these complex questions. Current regulations are silent or vague on many of these points and so do not elicit enquiry into them.

1.09.20 Chemo-Prevention and Chemo-Cure

Prevention of disease is clearly more desirable than its cure. Disease is an indication of some kind of failure.

Restoration to vigorous health achieved by appropriate treatment is clearly the ideal goal of all forms of curative therapy. Whereas this is a reasonable objective and is often achieved with prescription medications, this is not always the case even here. Disease modification or at least tolerance of the symptoms is the best we can do in conditions such as rheumatoid arthritis, for example, or essential hypertension, for another. In these cases the state of medicine is such that long-term maintenance therapy is the best one can hope for. With herbal preparations it is not common to effect a cure. Indeed, the DSHEA explicitly requires the pretense that a cure is not being sought. Mostly these materials are used for chemo-prevention or maintenance rather than for chemo-cure. Thus the use of these materials diverges most commonly with respect to cure but overlaps considerably with respect to prevention or maintenance.

1.09.21 Lifestyle Issues

These have become prominent in recent years and herbals are among the devices used to deal with them.

There are many individuals who believe that natural medications are intrinsically safer than prescription medications. To the extent that this belief is rooted in the generally weaker potency of herbals, there is something to this. However, many carry this belief to extremes. Strychnine, for example, is present in 'dog buttons' and is highly toxic.[101] Aflatoxin is a deadly carcinogen present in certain moldy peanuts.[102,103] These are examples of natural products that are potentially very harmful. All generalizations (including this one!) are suspect and fail to be universally applicable. The use of conjugated estrogens derived from pregnant mare's urine or phytoestrogens from soya to control the physiological symptoms accompanying menopause is effective but not in some fundamental way superior to the use of natural human estrogens themselves. The use of St John's wort instead of Prozac produces clinically reliable results in mild cases of mental disease and with a different set of side effects.[104] In such cases the form of therapy chosen can be based on lifestyle issues related to one's convictions. One need not take an opposing view except as a reflection of one's own belief system. The use of shark skin instead of prescription medications to treat cancer is quite another situation. Here resort to silly practices does cause harm. The degree to which society should be tolerant in such cases can be debated.

Expansion of interest in lifestyle drugs also plays a role in choice of herbals. Television viewers have become all too familiar with advertisements for prescription and natural substances intended to enhance sexual pleasure as well as to enhance breast or penis size. Compared to disease treatment, these uses may be considered to be trivial but the underlying conditions, real or perceived, leading to resort to such substances can be sources of significant distress to individuals if not attended to. If the agents are safe and their use is a comfort to individuals who can afford them, then it appears to be essentially harmless.

1.09.22 Problems in Evaluating the Earlier Literature on Herbal Drugs

Even today the herbal literature covers the full range, from unsupported testimonials to carefully validated scientific studies. The earlier literature is less trustworthy.

Much herbal literature was developed before modern science reached its present stage of development. The earlier literature often contains imprecise terms and is overlaid with cultural and belief systems that defy modern interpretation. The names of the material in use are confusing. Snake root, for example, is usually derived from the gnarled and serpentine shape of roots when dug. A great many herbals have this morphology and it has no obvious relationship to the medicinal constituents contained. A great many plants came to be known as snake root by indigenous populations, so being told that snake root is used for a clinical situation does convey precise information. Obtaining an herbarium specimen and identification using proper modern botanical binomial nomenclature by a taxonomist form part of an essential record-keeping exercise.

Diagnosis was often imprecise as well. Just what is meant in modern terms by statements such as 'to purify the blood,' 'for feminine weakness,' etc.? The therapeutic response was often not carefully recorded. 'The patient felt improved' can all too often be ascribed to the placebo effect. In many of the rather vague complaints for which treatment was sought, the improvement may well be assigned to mental relief associated with receiving treatment from one upon whom one relies with confidence and in the form in which one has belief. The most common complaints for which herbals are taken include gastrointestinal upset, bile problems, psychological problems, kidney and bladder disorders, and colds. These are mostly notoriously subject to placebo effects. The placebo effect (Latin: to please) confounds clinical trials of pure ingredients as well. In cases where mental processes play a significant role the placebo effect can be as high as one in three patients. The precise nature of the placebo effect is still unknown but there is no doubt that mind and body processes interact in complex ways. Thus nonobjective testimonials even from prominent individuals must be regarded with a degree of skepticism that makes much herbal literature hard to credit.

Clinical trials have often also been faulty. In a number of cases the nature and the history of the herbal being given are missing. The patient population may not have been carefully diagnosed, doses not standardized, the number of patients may not be sufficient to answer the questions posed with statistical significance, the clinical endpoint may be vague, and so on. In many cases clinical information is based on population studies alone. Given the many differences in lifestyle between individuals and the many possibly confounding events that would not have been noted, these studies lack the precision one would like.

For these reasons, even some venerably used and highly regarded herbals can require reexamination with more precise modern methods before unqualified acceptance can be accorded to them.

1.09.23 Examples of Frequently Utilized Herbal Medicinal Products

Despite the reservations expressed in the previous section, many herbals have gained acceptance and are extensively employed. Some of the more popular are discussed briefly in this section.

1.09.23.1 Bilberry

Bilberry is the fruit of *Vaccinium myrtillus*. It originates in northern and central Europe. It is administered as a tea, historically for the amelioration of diarrhea and for mouth and throat inflammations. During World War II, British Royal Air Force pilots ate bilberry preserves to improve their vision on night sorties. Following the war, investigation confirmed that the extracts could improve visual acuity and accommodation between light and dark. The antidiarrheal properties are attributed to tannins that develop upon drying, the antiinflammatory action to a decrease in vascular permeability attributed to anthocyanoside cross-linking of collagen, and the visual effects are associated with the anthocyanidins.[105–107]

1.09.23.2 Black Cohosh

Black cohosh is made from the roots and rhizomes of *Cimicifuga racemosa*/*Actaea racemosa* (members of the buttercup family) native to North America. It is used as tablets containing the extract or from a hydroalcoholic solution for the treatment of hot flashes and other postmenopausal symptoms. The preparations are standardized for their content of a triterpene saponin, 26-deoxyactein. Other constituents (fukinolic acid) have been shown to have estrogenic properties in animals and these may be responsible for its medical properties. It was adopted into formal medicine from its historic use for a variety of complaints by Amerindians. Clinical studies have produced ambiguous results, with some indicating value and others not. Side effects are generally mild and include stomach discomfort, weight problems, and headaches.[108–119]

1.09.23.3 Cat's Claw

Cat's claw (uña de gato) is a tropical vine scientifically named *Uncaria tomentosa* (Wild.) DC and *Uncaria guianensis* (Aubl.) (Gmel.) (fam. Ribiaceae), growing primarily in South America, although members of the genus grow in the tropics everywhere. It should not be confused with Asiatic *U. gambir* (Hunter) Roxb., that is widely used in tanning. Cat's claw has a long folkloric history of use in wound-healing, against cancer, as a contraceptive, and for the treatment of intestinal distress. Many of the putative therapeutic indications appear to be due to the presence of a multitude of pentacyclic oxindole alkaloids. The actions appear to include antioxidant, antiinflammatory, and immunostimulatory actions. A few small clinical trials support this action but fail to be totally convincing. It appears to be without significant toxicity when used in the usual doses.[120]

1.09.23.4 Chaste Berry

Chaste berry is the dried, ripe fruit of *Vitex agnus-castus* L. fam. Verbenaceae. It is a small tree or shrub growing along the river banks in southern Europe and the Mediterranean. The plant has been recognized since ancient times and is presently used for treatment of breast pain and menstrual irregularities. Some vendors promote it for use in bust enlargement. It also has a reputation for use in menopause, depression, low libido, vaginal dryness, premenstrual syndrome, and other reproductive problems. Many constituents are present, some of which (linoleic acid, for example) bind to progesterone receptors, suggesting a rationale for some of its uses.[121,122]

1.09.23.5 Echinacea

Echinacea is the whole plant of *Echinacea angustifolia* DC, *E. purpurea* (L.) Moench, and *E. pallida* (Nutt.) Britton fam. Compositae. It is native to the High Plains region of Central North America. It is often, unfortunately, adulterated with inactive *Parthenium integrefolium* L. The plant was in use by Amerindian tribes at the time of European contact and was quickly assimilated by settlers. It is primarily used for wound-healing and in teas as an immunostimulant. Its popularity decreased with the introduction of antibiotics but it has had a recent resurgence of use. The plant contains numerous ingredients, several of which are believed to contribute to its action. Recent work supports a stimulant effect of the water-soluble glycans on the nonspecific immune system. Its action appears to be more preventive than curative against colds.[44–46]

1.09.23.6 Ginkgo

Ginkgo is derived from the leaves and/or roots of *Ginkgo biloba* L. (fam. Ginkgoaceae), an ancient tree found in Japan and grown elsewhere. The tree is apparently essentially unchanged from the time of the dinosaurs, so can be characterized as a living fossil. Extracts contain flavonone glycosides, determined as kaempferol and quercetin and terpene lactones (ginkgolides, bilobalide, and ginkgolic acids), and are used for treating fatigue, memory deficits, and lapses in concentration, depression, dizziness, tinnitus, and headache. Its value in Alzheimer's disease is conjectural. These effects are associated with enhanced blood flow in central nervous system capillaries. Side effects include hypersensitivity along with occasional stomach or intestinal upsets and headaches.[123–152]

1.09.23.7 Green Tea

Green tea is derived from the leaves of *Camellia sinensis* and *C. assamica* (L.) O. Kuntze (fam. Theaceae), a shrub or tree indigenous to China and now grown in at least 30 different countries. Present worldwide commerce utilizes more than 2 million tons annually. Teas (green, black, and oolong) prepared from this plant have been used since ancient times as a stimulating beverage. Extracts contain caffeine and catechins belonging to the flavanoid family. The catechins are strongly antioxidant and also inhibit quinol oxidase (NOX), properties associated with potential cancer chemoprevention. Epidemiological studies support a chemopreventive action of green tea as a beverage against breast and prostate cancer. Convincing clinical trials are not yet available but a mass of animal and population studies are supportive. Side effects are few and rare.[153,154]

1.09.23.8 Gymnema

Gymnema is prepared from the leaves of *Gymnema sylvestre* R. Br. (fam. Asclepiadaceae), indigenous to Africa and India but now distributed worldwide. It is an ayurvedic herb used for at least 2000 years for the treatment of sugar diabetes. The active constituents appear to be the gymnemic acids, a group of at least nine triterpene glycosides, and a 35-mer peptide named gurmarin. Some limited clinical trials support the use of gymnema for the treatment of both type 1 and type 2 diabetes, but the trials are small and lack rigor.[155,156]

1.09.23.9 Horse Chestnut

Horse chestnut is derived from a variety of *Aesculus* plants, including *Aesculus hippocastanum* L., *A. californica* Nutt., and *A. glabra* Willd., commonly found in North America. An extract prominently contains a mixture of triterpene glycosides, notably aescin, which decrease capillary permeability. The bark contains aesculin which improves vascular resistance, thus aiding in toning vein walls. This action is useful in the treatment of hemorrhoids, varicose veins, leg ulcers, and frostbite. Preparations of the plant are also believed useful in edema.[157–160] The herb is of uncertain safety, so caution should be exercised.

1.09.23.10 Milk Thistle

Milk thistle is generally the seeds of *Silybum marianum* (L.) Gaertn. fam. Compositae, originating in India but subsequently introduced into Europe and North America. Its medicinal use, particularly as a liver protectant, against cirrhosis, and against hepatitis, extends for over 2000 years. The antioxidant and hepatoprotective action is presently attributed primarily to the flavonolignan silybin, whose lack of water solubility results in a pronounced enterohepatic recirculation providing significant liver exposure over time. The liver-protectant action appears consistent with its stimulation of protein biosynthesis at the DNA/RNA level, thus promoting liver regeneration. Additional actions consistent with its purported value are its antioxidant action, thus sparing glutathione, and also an alteration of outer liver cell membranes blocking the entrance of toxins up to and including amanita toxin. This last is useful in cases of ingestion of the death's cap mushroom. Few adverse effects have been noted.[161–169]

1.09.23.11 St John's Wort

St John's wort is an aggressive weed, *Hypericum perforatum* L. fam. Hypericaceae, native to Europe but subsequently introduced into North America. It has been used since the Middle Ages for its antiinflammatory, analgesic, diuretic, and wound-healing properties. In more recent times it has gained favor as a treatment for anxiety and depression and is now quite popular for this. Hyperforin, a phloroglucinoid, is now believed to be responsible for the plant's action due to its ability to inhibit serotonin, noradrenaline, and dopamine reuptake, thus increasing their concentration in the synaptic cleft. At the same time it enhances free intracellular sodium ion concentration, an effect not seen with the usual selective serotonin reuptake inhibitor. Clinical trials support the efficacy of St John's wort in mild depression and anxiety. It is, however, not useful in severe cases. When used carefully, side effects are reported to be rare.[104,170,171]

At one time the colorful hypericins, naphthodianthrones, were thought to be responsible for the action of this herb but this position has now been abandoned.

1.09.23.12 Saw Palmetto

Saw palmetto berries are derived from *Serenoa repens* (Bartram) Small fam. Palmaceae, a low, scrubby palm growing in the coastal plains of the southern US. Originally used as a food, its use for urinary tract and prostate problems dates from about the 1870s. Today its use is primarily for benign prostatic hyperplasia and to promote more nearly normal micturition. Its molecular mode of action is not well understood, although its rich content of plant sterols may play a role. Immunostimulant and antihormonal actions have been noted, leading to cautions about using saw palmetto preparations when hormonal therapies are being used at the same time.[172–180]

1.09.23.13 Soya Isoflavones

Soya isoflavones are derived from *Glycine max* L. fam. Fabaceae. Soya beans have been cultivated in China for at least 30 centuries. It was introduced into Japan and subsequently into Europe and North America, where it is extensively cultivated today. It is mainly used as a food but it is known to contain phytoestrogens as well that are believed to be of value in cancer chemoprevention, treatment of menopausal symptoms, including osteoporosis, treatment of cardiovascular disease, and gastrointestinal problems. The hormone active constituents are primarily isoflavones that bind to estrogen receptors normally binding estradiol. The active isoflavones undergo extensive metabolism in the body, complicating their study.[181–210]

1.09.24 Future Prospects

Herbals will continue to evolve. Traditional medicinal systems have persisted for eons and are unlikely to disappear in the foreseeable future.

Market predictions are notoriously inaccurate but projections for the sale of herbal products have them predicted to rise about 6% annually through 2008 from a level of \$46.7 billion in 1999 to about \$46.7 billions in 2002. The fastest growth took place between the years 1999 and 2000. Sales then slumped and are predicted to slump somewhat further as the market reaches equilibrium unless a blockbuster introduction (presently unanticipated) arises. The worldwide economic downturn and intensive interfirm competition must also have played a significant role.

References

1. Tyler, V. E. *Herbs of Choice*; Haworth Press: New York, 1994, pp xv–xvi, 1–187.
2. Wijesekera, R. O. B. *The Medicinal Plant Industry*; CRC Press: Boston, 1991.
3. Abbott, P. J. *Med. J. Aust.* **1988**, *149*, 678–682.
4. Winship, K. A. *Adv. Drug React. Toxicol. Rev.* **1991**, *10*, 47–59.
5. Roitman, J. N. *Lancet* **1981**, *1*, 944.
6. Mattocks, A. R. *Lancet* **1980**, *2*, 1136–1137.
7. Humberston, C. L.; Akhtar, J.; Krenzelok, E. P. *J. Toxicol. Clin. Toxicol.* **2003**, *41*, 109–1113.
8. Suffness, M. *Taxol: Science and Applications*; CRC Press: Boston, 1995, pp i–417.
9. Kingston, D. G. I.; Molinero, A. A.; Rimoldi, J. M. The Taxane Diterpenoids. In *Progress in the Chemistry of Organic Natural Products*; Herz, W., Kirby, G. W., Moore, R. E., Steglich, W., Tamm, C., Eds.; Springer-Verlag: New York, 1993; Vol. 61, pp 1–201.
10. Trevett, A.; Lalloo, D. *P. N. G. Med. J.* **1992**, *35*, 264–269.
11. Anonymous. *Chin. Med. J. (Engl.)* **1978**, *4*, 417–428.
12. Koshimizu, K.; Ohigashi, H.; Huffman, M. A. *Physiol. Behav.* **1994**, *56*, 1209–1216.
13. Rodriguez, E.; Aregullin, M.; Nishida, T.; Uehara, S.; Wrangham, R.; Abramowski, Z.; Finlayson, A.; Towers, G. H. *Experientia* **1985**, *41*, 419–420.
14. Rodriguez, E.; Wrangham, R. *Zoopharmacognosy: The Use of Medicinal Plants by Animals*; Plenum Press: New York, 1993, pp 89–105.

15. Nakamura, C. E.; Abeles, R. H. *Biochemistry* **1985**, *24*, 1364–1376.
16. Heber, D. *Am. J. Clin. Nutr.* **1999**, *70*, 106–108.
17. Heber, D.; Yip, I.; Ashley, J. M.; Elashoff, D. A.; Elashoff, R. M.; Go, V. L. *Am. J. Clin. Nutr.* **1999**, *69*, 231–236.
18. Havel, R. J. *Am. J. Clin. Nutr.* **1999**, *69*, 175–176.
19. Bisset, N. G.; Wichtl, W. *Herbal Drugs and Phytopharmaceuticals*, 2nd ed.; CRC Press: New York, 2001, p 553.
20. Tyler, V. E. *The Honest Herbal*, 3rd ed.; Haworth Press: New York, 1993, p 353.
21. Packer, L.; Ong, C. N.; Halliwell, B. H. *Herbal and Traditional Medicine*; Marcel Dekker: New York, 2004, p 903.
22. Yaniv, Z.; Bachrach, U. *Handbook of Medicinal Plants*; Haworth Press: New York, 2005, p 300.
23. DerMarderosian, A.; Beutler, J. A. *The Lawrence Review of Natural Products*; Facts and Comparisons: St Louis, MO, 2004.
24. Brenner, R.; Azbel, V.; Madhusoodanan, S.; Pawlowska, M. *Clin. Ther.* **2000**, *22*, 411–419.
25. Kaehler, S. T.; Sinner, C.; Chatterjee, S. S.; Philippu, A. *Neurosci. Lett.* **1999**, *262*, 199–202.
26. Chatterjee, S. S.; Bhattacharya, S. K.; Wonnemann, M.; Singer, A.; Muller, W. E. *Life Sci.* **1998**, *63*, 499–510.
27. Daly, J. W.; Kaneko, T.; Wilham, J.; Garraffo, H. M.; Spande, T. F.; Espinosa, A.; Donnelly, M. A. *Proc. Natl. Acad. Sci. USA* **2002**, *99*, 13996–14001.
28. Reichstein, T.; von Euw, J.; Parsons, J. A.; Rothschild, M. *Science* **1968**, *161*, 861–866.
29. Rodrigues, K. F.; Hesse, M.; Werner, C. J. *Basic Microbiol.* **2000**, *40*, 261–267.
30. Stierle, A.; Strobel, G.; Stierle, D. *Science* **1993**, *260*, 214–216.
31. Strobel, G.; Daisy, B.; Castillo, U.; Harper, J. *J. Nat. Prod.* **2004**, *67*, 257–268.
32. Strobel, G. A.; Hess, W. M. *Chem. Biol.* **1997**, *4*, 529–536.
33. Wang, J.; Li, G.; Lu, H.; Zheng, Z.; Huang, Y.; Su, W. *FEMS Microbiol. Lett.* **2000**, *193*, 249–253.
34. Wagenaar, M. M.; Clardy, J. *J. Nat. Prod.* **2001**, *64*, 1006–1009.
35. Krohn, K.; Florke, U.; Rao, M. S.; Steingrover, K.; Aust, H. J.; Draeger, S.; Schulz, B. *Nat. Prod. Lett.* **2001**, *15*, 353–361.
36. Brady, S. F.; Bondi, S. M.; Clardy, J. *J. Am. Chem. Soc.* **2001**, *123*, 9900–9901.
37. Loizou, L. A.; Hamilton, J. G.; Tsementzis, S. A. *J. Neurol. Neurosurg. Psychiatry* **1982**, *45*, 471–472.
38. Pentel, P. *JAMA* **1984**, *252*, 1898–1903.
39. LeBayon, A.; Castelnovo, G.; Briere, C.; Labauge, P. *Presse Med.* **2000**, *29*, 1702.
40. Baker, J. T.; Borris, R. P.; Carte, B.; Cordell, G. A.; Soejarto, D. D.; Cragg, G. M.; Gupta, M. P.; Iwu, M. M.; Madulid, D. R.; Tyler, V. E. *J. Nat. Prod.* **1995**, *58*, 1325–1357.
41. Cragg, G. M. *P. R. Health Sci. J.* **2002**, *21*, 97–111.
42. Cragg, G. M.; Newman, D. J. *J. Nat. Prod.* **2004**, *67*, 232–244.
43. McGraw, J. B.; Furedi, M. A. *Science* **2005**, *307*, 920–922.
44. Bauer, R.; Wagner, H. Echinacea-Species as Potential Immunostimulatory Drugs. In *Economic and Medicinal Plant Research*; Farnsworth, N., Wagner, H., Eds.; Academic Press: San Diego, 1991.
45. Bauer, R. Echinacea: Biological Effects and Active Principles. In *Phytomedicines of Europe: Chemistry and Biological Activity*; Lawson, L.; Bauer, R., Eds.; American Chemical Society: Washington, D.C., 1998; ACS Symposium Series 691, pp 140–157.
46. Mitscher, L. A.; Cooper, R. Echinacea and Immunostimulation. In *Herbal and Traditional Medicine*; Packer, L., Ong, C. N., Halliwell, B. H., Eds.; Marcel Dekker: New York, 2004, pp 721–756.
47. Busse, W. R.; Goldberg, A.; Gruenwald, J.; Hall, T.; Riggins, C. W.; Rister, R. S. *The Complete German Commission E Monographs*; American Botanical Council: Austin, TX, 1998, pp ix–xix, 3–685.
48. Haller, C. A. Ephedra. In *Herbal and Traditional Medicine: Molecular Aspects of Health*; Packer, L., Ong, C. N., Halliwell, B. H., Eds.; Marcel Dekker: New York, 2004, pp 703–720.
49. Lipman, A. G. *J. Pain Palliat. Care Pharmacother.* **2004**, *18*, 1–4.
50. Phillips, G. C. *Curr. Sports Med. Rep.* **2004**, *3*, 224–228.
51. Ang, H. H.; Lee, E. L.; Cheang, H. S. *Int. J. Toxicol.* **2004**, *23*, 65–71.
52. Ang, H. H.; Lee, E. L.; Matsumoto, K. *Hum. Exp. Toxicol.* **2003**, *22*, 445–451.
53. Ang, H. H.; Lee, K. L.; Kiyoshi, M. *Int. J. Environ. Health Res.* **2004**, *14*, 261–272.
54. Saper, R. B.; Kales, S. N.; Paquin, J.; Burns, M. J.; Eisenberg, D. M.; Davis, R. B.; Phillips, R. S. *JAMA* **2004**, *292*, 2868–2873.
55. Parab, S.; Kulkarni, R.; Thatte, U. *Ind. J. Gastroenterol.* **2003**, *22*, 111–112.
56. Gogtay, N. J.; Bhatt, H. A.; Dalvi, S. S.; Kshirsagar, N. A. *Drug Safe.* **2002**, *25*, 1005–1019.
57. Ernst, E. *Eur. J. Clin. Pharmacol.* **2002**, *57*, 891–896.
58. Ernst, E. *Trends Pharmacol. Sci.* **2002**, *23*, 136–139.
59. Ernst, E. *Am. J. Med.* **2004**, *117*, 533; author reply 533.
60. A Khan, I.; Allgood, J.; Walker, L. A.; Abourashed, E. A.; Schlenk, D.; Benson, W. H. *J. AOAC Int.* **2001**, *84*, 936–939.
61. Garvey, G. J.; Hahn, G.; Lee, R. V.; Harbison, R. D. *Int. J. Environ. Health Res.* **2001**, *11*, 63–71.
62. Ao, P.; Hu, S.; Zhao, A. *Zhongguo Zhong Yao Za Zhi* **1998**, *23*, 42–43, 63.
63. Ko, R. J. *N. Engl. J. Med.* **1998**, *339*, 847.
64. Koh, H. L.; Woo, S. O. *Drug Safe* **2000**, *23*, 351–362.
65. Chuang, I. C.; Chen, K. S.; Huang, Y. L.; Lee, P. N.; Lin, T. H. *Biol. Trace Elem. Res.* **2000**, *76*, 235–244.
66. Au, A. M.; Ko, R.; Boo, F. O.; Hsu, R.; Perez, G.; Yang, Z. *Bull. Environ. Contam. Toxicol.* **2000**, *65*, 112–119.
67. Bateman, J.; Chapman, R. D.; Simpson, D. *Scott. Med. J.* **1998**, *43*, 7–15.
68. Shaw, D.; Leon, C.; Kolev, S.; Murray, V. *Drug Safe* **1997**, *17*, 342–356.
69. Olujohungbe, A.; Fields, P. A.; Sandford, A. F.; Hoffbrand, A. V. *Postgrad. Med. J.* **1994**, *70*, 764.
70. Wong, M. K.; Tan, P.; Wee, Y. C. *Biol. Trace Elem. Res.* **1993**, *36*, 135–142.
71. Ostrander, G. K.; Cheng, K. C.; Wolf, J. C.; Wolfe, M. J. *Cancer Res.* **2004**, *64*, 8485–8491.
72. Hall, S. D.; Wang, Z.; Huang, S. M.; Hamman, M. A.; Vasavada, N.; Adigun, A. Q.; Hilligoss, J. K.; Miller, M.; Gorski, J. C. *Clin. Pharmacol. Ther.* **2003**, *74*, 525–535.
73. Hennessy, M.; Kelleher, D.; Spiers, J. P.; Barry, M.; Kavanagh, P.; Back, D.; Mulcahy, F.; Feely, J. *Br. J. Clin. Pharmacol.* **2002**, *53*, 75–82.
74. Stickel, F.; Seitz, H. K. *Public Health Nutr* **2000**, *3*, 501–508.
75. Williams, L.; Chou, M. W.; Yan, J.; Young, J. F.; Chan, P. C.; Doerge, D. R. *Toxicol. Appl. Pharmacol.* **2002**, *182*, 98–104.
76. Anonymous. *Consumer's Rep.* May, **2004**.

77. Fau, D.; Lekehal, M.; Farrell, G.; Moreau, A.; Moulis, C.; Feldmann, G.; Haouzi, D.; Pessayre, D. *Gastroenterology* **1997**, *113*, 1334–1346.
78. Lekehal, M.; Pessayre, D.; Lereau, J. M.; Moulis, C.; Fouraste, I.; Fau, D. *Hepatology* **1996**, *24*, 212–218.
79. Loeper, J.; Descatoire, V.; Letteron, P.; Moulis, C.; Degott, C.; Dansette, P.; Fau, D.; Pessayre, D. *Gastroenterology* **1994**, *106*, 464–472.
80. Arlt, V. M.; Pfohl-Leszkowicz, A.; Cosyns, J.; Schmeiser, H. H. *Mutat. Res.* **2001**, *494*, 143–150.
81. Stengel, B.; Jones, E. *Nephrologie* **1998**, *19*, 15–20.
82. Vanherweghem, J. L. *Bull. Mem. Acad. R. Med. Belg.* **1994**, *149*, 128–135, discussion 135–140.
83. Vanherweghem, L. J. *J. Altern. Complement Med.* **1998**, *4*, 9–13.
84. Violon, C. *J. Pharm. Belg.* **1997**, *52*, 7–27.
85. Nortier, J. L.; Martinez, M. C.; Schmeiser, H. H.; Arlt, V. M.; Bieler, C. A.; Petein, M.; Depierreux, M. F.; De Pauw, L.; Abramowicz, D.; Vereerstraeten, P. et al. *N. Engl. J. Med.* **2000**, *342*, 1686–1692.
86. Brinker, F. *Altern. Med. Alert* **2004**, 7, 89–91.
87. Anderson, I. B.; Mullen, W. H.; Meeker, J. E.; Khojasteh Bakht, S. C.; Oishi, S.; Nelson, S. D.; Blanc, P. D. *Ann. Intern. Med.* **1996**, *124*, 726–734.
88. De Smet, P. A.; Smeets, O. S. *Br. Med. J.* **1994**, *309*, 958.
89. Brinker, F. *The Toxicity of Botanical Medicines*, 3rd ed.; Eclectic Medical Publications: Sandy, OR, 2000.
90. Pal, D.; Mitra, A. K. *Life Sci.* **2006**, *78*, 2131–2145.
91. Durr, D.; Stieger, B.; Kullak-Ublick, G. A.; Rentsch, K. M.; Steinert, H. C.; Meier, P. J.; Fattinger, K. *Clin. Pharmacol. Ther.* **2000**, *68*, 598–604.
92. Henderson, L.; Yue, Q. Y.; Bergquist, C.; Gerden, B.; Arlett, P. *Br. J. Clin. Pharmacol.* **2002**, *54*, 349–356.
93. Izzo, A. A. *Int J Clin. Pharmacol. Ther.* **2004**, *42*, 139–148.
94. Jang, E. H.; Park, Y. C.; Chung, W. G. *Food Chem. Toxicol.* **2004**, *42*, 1749–1756.
95. Komoroski, B. J.; Zhang, S.; Cai, H.; Hutzler, J. M.; Frye, R.; Tracy, T. S.; Strom, S. C.; Lehmann, T.; Ang, C. Y.; Cui, Y. Y. et al. *Drug Metab. Dispos.* **2004**, *32*, 512–518.
96. Markowitz, J. S.; DeVane, C. L. *Psychopharmacol. Bull.* **2001**, *35*, 53–64.
97. Markowitz, J. S.; Donovan, J. L.; DeVane, C. L.; Taylor, R. M.; Ruan, Y.; Wang, J. S. *JAMA* **2003**, *290*, 1500–1504.
98. Moore, L. B.; Goodwin, B.; Jones, S. A.; Wisely, G. B.; Serabjit-Singh, C. J.; Willson, T. M.; Collins, J. L.; Kliewer, S. A. *Proc. Natl. Acad. Sci. USA* **2000**, *97*, 7500–7502.
99. Wang, Z.; Gorski, J. C.; Hamman, M. A.; Huang, S. M.; Lesko, L. J.; Hall, S. D. *Clin. Pharmacol. Ther.* **2001**, *70*, 317–326.
100. Zhou, S.; Chan, E.; Pan, S. Q.; Huang, M.; Lee, E. J. *J. Psychopharmacol.* **2004**, *18*, 262–276.
101. Chan, T. Y. *Hum. Exp. Toxicol.* **2002**, *21*, 467–468.
102. Omer, R. E.; Bakker, M. I.; van't Veer, P.; Hoogenboom, R. L.; Polman, T. H.; Alink, G. M.; Idris, M. O.; Kadaru, A. M.; Kok, F. J. *Nutr. Cancer* **1998**, *32*, 174–180.
103. Nguyen, H. N. *Dtsch Med. Wochenschr.* **1999**, *124*, 299.
104. Kasper, S.; Schulz, V. *Schweiz. Rundsch. Med. Prax.* **2000**, *89*, 2169–2177.
105. Canter, P. H.; Ernst, E. *Surv. Ophthalmol.* **2004**, *49*, 38–50.
106. Kramer, J. H. *Surv. Ophthalmol.* **2004**, *49*, 618, author reply 618.
107. Muth, E. R.; Laurent, J. M.; Jasper, P. *Altern. Med. Rev.* **2000**, *5*, 164–173.
108. McKee, J.; Warber, S. L. *South. Med. J.* **2005**, *98*, 319–326.
109. Pockaj, B. A.; Loprinzi, C. L.; Sloan, J. A.; Novotny, P. J.; Barton, D. L.; Hagenmaier, A.; Zhang, H.; Lambert, G. H.; Reeser, K. A.; Wisbey, J. A. *Cancer Invest* **2004**, *22*, 515–521.
110. Fugate, S. E.; Church, C. O. *Ann. Pharmacother.* **2004**, *38*, 1482–1499.
111. Anonymous. *Harv. Womens Health Watch* **2004**, 11, 2–3.
112. Anonymous. *Menopause* **2004**, *11*, 11–33.
113. Amato, P.; Marcus, D. M. *Climacteric* **2003**, *6*, 278–284.
114. Burdette, J. E.; Liu, J.; Chen, S. N.; Fabricant, D. S.; Piersen, C. E.; Barker, E. L.; Pezzuto, J. M.; Mesecar, A.; Van Breemen, R. B.; Farnsworth, N. R. et al. *J. Agric. Food Chem.* **2003**, *51*, 5661–5670.
115. Kligler, B. *Am. Fam. Phys.* **2003**, *68*, 114–116.
116. Philp, H. A. *Altern. Med. Rev.* **2003**, *8*, 284–302.
117. Huntley, A.; Ernst, E. *Menopause* **2003**, *10*, 58–64.
118. Kang, H. J.; Ansbacher, R.; Hammoud, M. M. *Int. J. Gynaecol. Obstet.* **2002**, *79*, 195–207.
119. Kronenberg, F.; Fugh-Berman, A. *Ann. Intern. Med.* **2002**, *137*, 805–813.
120. Kiefer, D. *Altern. Med. Alert* **2005**, *8*, 20–24.
121. Fugh-Berman, A. *Obstet. Gynecol.* **2003**, *101*, 1345–1349.
122. Liu, J.; Burdette, J. E.; Sun, Y.; Deng, S.; Schlecht, S. M.; Zheng, W.; Nikolic, D.; Mahady, G.; van Breemen, R. B.; Fong, H. H. et al. *Phytomedicine* **2004**, *11*, 18–23.
123. Elsabagh, S.; Hartley, D. E.; Ali, O.; Williamson, E. M.; File, S. E. *Psychopharmacology (Berl.)* **2005**, *179*, 437–446.
124. Lee, T. F.; Chen, C. F.; Wang, L. C. *Phytother. Res.* **2004**, *18*, 556–560.
125. Wheatley, D. *Hum. Psychopharmacol.* **2004**, *19*, 545–548.
126. Colciaghi, F.; Borroni, B.; Zimmermann, M.; Bellone, C.; Longhi, A.; Padovani, A.; Cattabeni, F.; Christen, Y.; Di Luca, M. *Neurobiol. Dis.* **2004**, *16*, 454–460.
127. Hoffman, J. R.; Donato, A.; Robbins, S. J. *Pharmacol. Biochem. Behav.* **2004**, 77, 533–539.
128. Nathan, P. J.; Tanner, S.; Lloyd, J.; Harrison, B.; Curran, L.; Oliver, C.; Stough, C. *Hum. Psychopharmacol.* **2004**, *19*, 91–96.
129. Petkov, V. D.; Belcheva, S.; Petkov, V. V. *Am. J. Chin. Med.* **2003**, *31*, 841–855.
130. Persson, J.; Bringlov, E.; Nilsson, L. G.; Nyberg, L. *Psychopharmacol. (Berl.)* **2004**, *172*, 430–434.
131. Stackman, R. W.; Eckenstein, F.; Frei, B.; Kulhanek, D.; Nowlin, J.; Quinn, J. F. *Exp. Neurol.* **2003**, *184*, 510–520.
132. Hartley, D. E.; Heinze, L.; Elsabagh, S.; File, S. E. *Pharmacol. Biochem. Behav.* **2003**, *75*, 711–720.
133. van Dongen, M.; van Rossum, E.; Kessels, A.; Sielhorst, H.; Knipschild, P. *J. Clin. Epidemiol.* **2003**, *56*, 367–376.
134. Andrieu, S.; Gillette, S.; Amouyal, K.; Nourhashemi, F.; Reynish, E.; Ousset, P. J.; Albarede, J. L.; Vellas, B.; Grandjean, H. *J. Gerontol. A Biol. Sci. Med. Sci.* **2003**, *58*, 372–377.
135. Lin, C. C.; Cheng, W. L.; Hsu, S. H.; Chang, C. M. *Neuropsychobiology* **2003**, *47*, 47–51.
136. Arnold, K. R. *JAMA* **2003**, *289*, 546, author reply 547–548.

137. Cheuvront, S. N.; Carter, R., III, *JAMA* **2003**, *289*, 547; author reply 547–548.
138. Doraiswamy, P. M.; Pomara, N. *JAMA* **2003**, *289*, 547, author reply 547–548.
139. Nathan, P. J.; Harrison, B. J.; Bartholomeusz, C. *JAMA* **2003**, *289*, 546, author reply 547–548.
140. Wheatley, D. *JAMA* **2003**, *289*, 546–547, author reply 547–548.
141. Mix, J. A.; Crews, W. D., Jr. *Hum. Psychopharmacol.* **2002**, *17*, 267–277.
142. Polich, J.; Gloria, R. *Hum. Psychopharmacol.* **2001**, *16*, 409–416.
143. Soulie, C.; Nicole, A.; Christen, Y.; Ceballos-Picot, I. *Cell Mol. Biol. (Noisy-le-grand)* **2002**, *48*, 641–646.
144. Abad-Santos, F.; Novalbos-Reina, J.; Gallego-Sandin, S.; Garcia, A. G. *Rev. Neurol.* **2002**, *35*, 675–682.
145. Hadjiivanova, Ch. I.; Petkov, V. V. *Phytother. Res.* **2002**, *16*, 488–490.
146. Luo, Y. *J Alzheimers Dis.* **2001**, *3*, 401–407.
147. Solomon, P. R.; Adams, F.; Silver, A.; Zimmer, J.; DeVeaux, R. *JAMA* **2002**, *288*, 835–840.
148. Kennedy, D. O.; Scholey, A. B.; Wesnes, K. A. *Physiol. Behav.* **2002**, *75*, 739–751.
149. McKenna, D. J.; Jones, K.; Hughes, K. *Altern. Ther. Health Med.* **2001**, *7*, 70–86, 88–90.
150. Stough, C.; Clarke, J.; Lloyd, J.; Nathan, P. J. *Int. J. Neuropsychopharmacol.* **2001**, *4*, 131–134.
151. van Dongen, M. C.; van Rossum, E.; Kessels, A. G.; Sielhorst, H. J.; Knipschild, P. G. *J. Am. Geriatr. Soc.* **2000**, *48*, 1183–1194.
152. Wesnes, K. A.; Ward, T.; McGinty, A.; Petrini, O. *Psychopharmacol. (Berl.)* **2000**, *152*, 353–361.
153. Krochmal, R.; Hardy, M. L. *Altern. Med. Alert* **2004**, *7*, 78–83.
154. Mitscher, L. A.; Jung, M.; Shankel, D.; Dou, J. H.; Steele, L.; Pillai, S. P. *Med. Res. Rev.* **1997**, *17*, 327–365.
155. Cicero, A. F.; Derosa, G.; Gaddi, A. *Acta Diabetol.* **2004**, *41*, 91–98.
156. Udani, J.; Hardy, M. L. *Altern. Med. Alert* **2005**, *8*, 28–32.
157. Siebert, U.; Brach, M.; Sroczynski, G.; Berla, K. *Int. Angiol.* **2002**, *21*, 305–315.
158. Pittler, M. H.; Ernst, E. *Arch. Dermatol.* **1998**, *134*, 1356–1360.
159. Simini, B. *Lancet* **1996**, *347*, 1182–1183.
160. Diehm, C.; Trampisch, H. J.; Lange, S.; Schmidt, C. *Lancet* **1996**, *347*, 292–294.
161. Tanamly, M. D.; Tadros, F.; Labeeb, S.; Makld, H.; Shehata, M.; Mikhail, N.; Abdel-Hamid, M.; Shehata, M.; Abu-Baki, L.; Medhat, A. et al. *Dig. Liver Dis.* **2004**, *36*, 752–759.
162. Kravchenko, L. V.; Morozov, S. V.; Tutel'yan, V. A. *Bull. Exp. Biol. Med.* **2003**, *136*, 572–575.
163. Boerth, J.; Strong, K. M. *J. Herb Pharmacother.* **2002**, *2*, 11–17.
164. Lieber, C. S.; Leo, M. A.; Cao, Q.; Ren, C.; DeCarli, L. M. *J. Clin. Gastroenterol.* **2003**, *37*, 336–339.
165. Jacobs, B. P.; Dennehy, C.; Ramirez, G.; Sapp, J.; Lawrence, V. A. *Am. J. Med.* **2002**, *113*, 506–515.
166. He, Q.; Osuchowski, M. F.; Johnson, V. J.; Sharma, R. P. *Planta Med.* **2002**, *68*, 676–679.
167. Giese, L. A. *Gastroenterol. Nurs.* **2001**, *24*, 95–97.
168. Ferenci, P.; Dragosics, B.; Dittrich, H.; Frank, H.; Benda, L.; Lochs, H.; Meryn, S.; Base, W.; Schneider, B. J. *Hepatol.* **1989**, *9*, 105–113.
169. De Martiis, M.; Fontana, M.; Assogna, G.; D'Ottavi, R.; D'Ottavi, O. *Clin. Ter.* **1980**, *94*, 283–315.
170. Vanoni, C. *Schweiz. Rundsch. Med. Prax.* **2000**, *89*, 2163–2177.
171. Muller, W. E.; Singer, A.; Wonnemann, M. *Schweiz. Rundsch. Med. Prax.* **2000**, *89*, 2111–2121.
172. Arruzazabala, M. L.; Carbajal, D.; Mas, R.; Molina, V.; Rodriguez, E.; Gonzalez, V. *Drugs Exp. Clin. Res.* **2004**, *30*, 227–233.
173. Gerber, G. S.; Fitzpatrick, J. M. *BJU Int.* **2004**, *94*, 338–344.
174. Gong, E. M.; Gerber, G. S. *Am. J. Chin. Med.* **2004**, *32*, 331–338.
175. Kaplan, S. A. *J. Urol.* **2003**, *170*, 340–341.
176. Onega, T. *JAAPA* **2002**, *15*, 59–60, 63–64.
177. Preuss, H. G.; Marcusen, C.; Regan, J.; Klimberg, I. W.; Welebir, T. A.; Jones, W. A. *Int. Urol. Nephrol.* **2001**, *33*, 217–225.
178. Goldmann, W. H.; Sharma, A. L.; Currier, S. J.; Johnston, P. D.; Rana, A.; Sharma, C. P. *Cell Biol. Int.* **2001**, *25*, 1117–1124.
179. Marks, L. S.; Partin, A. W.; Epstein, J. I.; Tyler, V. E.; Simon, I.; Macairan, M. L.; Chan, T. L.; Dorey, F. J.; Garris, J. B.; Veltri, R. W. et al. *J. Urol.* **2000**, *163*, 1451–1456.
180. McPartland, J. M.; Pruitt, P. L. *J. Am. Osteopath. Assoc.* **2000**, *100*, 89–96.
181. Stokes, M. J. *Altern. Med. Alert* **2004**, *7*, 49–55.
182. Fischer, L.; Mahoney, C.; Jeffcoat, A. R.; Koch, M. A.; Thomas, B. E.; Valentine, J. L.; Stinchcombe, T.; Boan, J.; Crowell, J. A.; Zeisel, S. H. *Nutr. Cancer* **2004**, *48*, 160–170.
183. Russo, R.; Corosu, R. *Acta Biomed. Ateneo Parmense* **2003**, *74*, 137–143.
184. Graf, M. C.; Geller, P. A. *Clin. J. Oncol. Nurs.* **2003**, *7*, 637–640.
185. Cassidy, A. *J. Br. Menopause Soc.* **2003**, *9*, 17–21.
186. Chiechi, L. M.; Putignano, G.; Guerra, V.; Schiavelli, M. P.; Cisternino, A. M.; Carriero, C. *Maturitas* **2003**, *45*, 241–246.
187. Messina, M.; Hughes, C. *J. Med. Food* **2003**, *6*, 1–11.
188. Adlercreutz, H. *J. Steroid Biochem. Mol. Biol.* **2002**, *83*, 113–118.
189. Brewer, D.; Nashelsky, J.; Hansen, L. B. *J. Fam. Pract.* **2003**, *52*, 324–325, 329.
190. Burke, G. L.; Legault, C.; Anthony, M.; Bland, D. R.; Morgan, T. M.; Naughton, M. J.; Leggett, K.; Washburn, S. A.; Vitolins, M. Z. *Menopause* **2003**, *10*, 147–153.
191. Cassidy, A. *Int. J. Vitam. Nutr. Res.* **2003**, *73*, 120–126.
192. Valtuena, S.; Cashman, K.; Robins, S. P.; Cassidy, A.; Kardinaal, A.; Branca, F. *Br. J. Nutr.* **2003**, *89*, S87–S99.
193. Bhathena, S. J.; Velasquez, M. T. *Am. J. Clin. Nutr.* **2002**, *76*, 1191–1201.
194. Adlercreutz, H. *Lancet Oncol* **2002**, *3*, 364–373.
195. Albert, A.; Altabre, C.; Baro, F.; Buendia, E.; Cabero, A.; Cancelo, M. J.; Castelo-Branco, C.; Chantre, P.; Duran, M.; Haya, J. et al. *Phytomedicine* **2002**, *9*, 85–92.
196. Clarkson, T. B. *J. Nutr.* **2002**, *132*, 566S–569S.
197. Van Patten, C. L.; Olivotto, I. A.; Chambers, G. K.; Gelmon, K. A.; Hislop, T. G.; Templeton, E.; Wattie, A.; Prior, J. C. *J. Clin. Oncol.* **2002**, *20*, 1449–1455.
198. Busby, M. G.; Jeffcoat, A. R.; Bloedon, L. T.; Koch, M. A.; Black, T.; Dix, K. J.; Heizer, W. D.; Thomas, B. F.; Hill, J. M.; Crowell, J. A. et al. *Am. J. Clin. Nutr.* **2002**, *75*, 126–136.
199. Tsourounis, C. *Clin. Obstet. Gynecol.* **2001**, *44*, 836–842.

200. Sirtori, C. R. *Drug Safe* **2001**, *24*, 665–682.
201. Setchell, K. D.; Brown, N. M.; Desai, P.; Zimmer-Nechemias, L.; Wolfe, B. E.; Brashear, W. T.; Kirschner, A. S.; Cassidy, A.; Heubi, J. E. *J. Nutr.* **2001**, *131*, 1362S–1375S.
202. Swift, M.; West, J. *Menopause* **2001**, *8*, 76–77.
203. Vincent, A.; Fitzpatrick, L. A. *Mayo Clin. Proc.* **2000**, *75*, 1174–1184.
204. Wiseman, H. *Exp. Opin. Invest. Drugs* **2000**, *9*, 1829–1840.
205. Scambia, G.; Mango, D.; Signorile, P. G.; Anselmi Angeli, R. A.; Palena, C.; Gallo, D.; Bombardelli, E.; Morazzoni, P.; Riva, A.; Mancuso, S. *Menopause* **2000**, *7*, 105–111.
206. Sirtori, C. R. *Lancet* **2000**, *355*, 849.
207. Quella, S. K.; Loprinzi, C. L.; Barton, D. L.; Knost, J. A.; Sloan, J. A.; LaVasseur, B. I.; Swan, D.; Krupp, K. R.; Miller, K. D.; Novotny, P. J. *J. Clin. Oncol.* **2000**, *18*, 1068–1074.
208. Clarkson, T. B.; Anthony, M. S.; Williams, J. K.; Honore, E. K.; Cline, J. M. *Proc. Soc. Exp. Biol. Med.* **1998**, *217*, 365–368.
209. Knight, D. C.; Eden, J. A. *Obstet. Gynecol.* **1996**, *87*, 897–904.
210. Cassidy, A.; Bingham, S.; Setchell, K. *Br. J. Nutr.* **1995**, *74*, 587–601.
211. American Dietetic Association. http://www.eatright.org (accessed May 2006).
212. Supplement Quality. http://www.supplementquality.com/glossary.html (accessed May 2006).
213. VitaMedica (the Science of Natural Health). http://www.vitamedica.com(accessed May 2006).
214. RxList (The Internet Drug Index). http://www.rxlist.com/top200.htm (accessed May 2006).
215. HealthWorld Online. http://www.healthy.net/public/legal-lg/fedregs/S784_ENR.HTM (accessed May 2006).
216. Tax Help – Government Information. http://vm.cfsan.fda.gov (accessed May 2006).

Biography

Lester A Mitscher received his PhD degree in Chemistry in 1958 from Wayne State University, Detroit, USA, where he worked on the structure of coffee oil diterpenes and on optical rotatory dispersion. He continued his work on natural product chemistry at Lederle Laboratories where he rose to group leader in antibiotic discovery until he accepted an assistant professorship in Natural Products Chemistry at Ohio State University (1967), rising soon to the position of Professor. In 1975 he accepted a University Distinguished Professorship and Chairmanship in the Department of Medicinal Chemistry at Kansas University. He returned to the faculty in 1992, and remains there. His academic studies have centered on spectroscopy, synthesis, screening, and structure determination, primarily of naturally occurring antimicrobial and antimutagenic agents. He has published actively in the field of herbal medicines. He consults extensively in the pharmaceutical industry and is a member of the National Institutes of Health Drug Discovery and Antimicrobial Resistance Study Section. His research awards include the Smissman Award in Medicinal Chemistry (American Chemical Society), Volweiler Award (American Association for Pharmaceutical Education), the Research Achievement Award in Natural Products Chemistry (American Pharmaceutical Association), the Award in Medicinal Chemistry, Medicinal Chemistry Division, ACS, and he is an Elected Fellow of the American Association for the Advancement of Science.

Comprehensive Medicinal Chemistry II
ISBN (set): 0-08-044513-6
ISBN (Volume 1) 0-08-044514-4; pp. 405–430

1.10 Biological Macromolecules

J K Osbourn, Cambridge Antibody Technology, Cambridge, UK

1.10.1 Biological Macromolecules

1.10.1.1 Introduction

While pharmaceutical agents have been used to treat and cure disease and illness for centuries, it is only in the past 100 years that routine drug research has resulted in the discovery and development of many of the medicines that we use today. Traditionally, the pharmaceutical industry had its basis in chemicals: research on coal-tar derivatives, in particular dyes, gave rise to chemotherapy; and analytical chemistry techniques in the nineteenth century were used to isolate and purify naturally occurring therapeutic molecules from plants.[1] These biochemical and microbiological disciplines have shaped the drug discovery process to the extent that, until recently, most pharmaceutical agents were the result of chemical synthesis of molecules with a small molecular weight. These small molecules have been found empirically to act on receptors, ion channels, and enzymes to elicit a cellular response. The consequent characterization of receptors throughout the human body has produced a vast number of diverse pharmacological agonists and antagonists, although it is estimated that current small-molecule therapies act on only a small fraction of the available molecular targets.[2]

In recent years, drug discovery has moved from its chemical background to a more biological focus. Advances in molecular biology have allowed the processing and utilization of extensive amounts of genetic data and the consequent

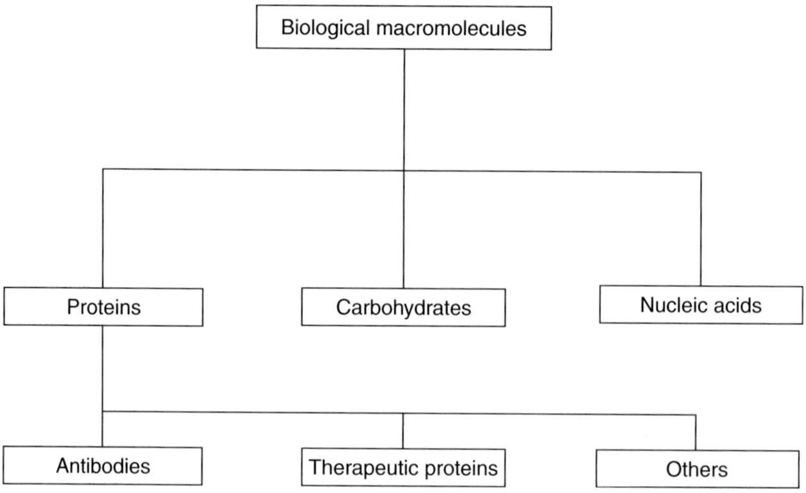

Figure 1 Biological macromolecule drug types.

development of genetically engineered therapeutic proteins, based on the substances that the human body already produces. The use of protein-based drugs, however, is not necessarily a new approach in the treatment of disease. Insulin, a well-known peptide, was originally isolated in the early 1920s, and was the first available treatment for patients with type 1 diabetes. Likewise, the production of heparin, a glycosaminoglycan, which is used for the treatment and prevention of thromboembolic disease, first came about in the 1930s. Nevertheless, it is only since the 1990s that these biological drugs have come to the fore in the marketplace. The chief difference between biological drugs and small molecules is their molecular weight: biological drugs are generally much larger than small-molecule drugs, and are therefore often termed biological macromolecules (**Figure 1**). In general terms, biological macromolecules can be broken down into three categories: proteins, carbohydrates and nucleic acids. Here, we will focus on the proteinaceous group, which are most applicable to therapeutic regimens.

1.10.1.2 A Brief History of Biological Macromolecules

As mentioned above, therapeutic biological macromolecules are not necessarily novel drugs. Naturally occurring proteins, such as insulin and growth hormone, have been purified from animal sources for many years. The advent of molecular biological techniques has allowed for the recombinant production of human insulin, growth hormone, and many other therapeutic proteins, such as epogen, in a comparatively safe and efficient manner. Other proteins, such as streptokinase (a thrombolytic), followed, and different techniques allowed the production of more specialized recombinant proteins.

The hybridoma technique, first reported in 1975,[3] brought forth monoclonal antibodies (mAbs) as potentially useful therapeutic tools, and they now contribute a significant proportion (approximately 15%) of biologics on the market. mAbs are so called because they are created by the immortalization of a single B lymphocyte (white blood cells that elicit a specific immune response when exposed to antigens). Immortalization of B lymphocytes (B cells) from an immunized mouse is achieved by fusion of a mouse myeloma cell. The fusion cells can secrete unlimited quantities of the antibody produced by the B cell. While this technology initiated interest in mAbs, it took several decades to refine the production techniques and bring the first mAbs to market. The first mAbs were completely murine, and as such were often recognized as foreign by the patient's immune system. This led to the development of human antimurine antibody responses. During the past two decades, B-cell-derived molecular biology and display technologies have been developed to generate humanized and fully human antibodies, and the problems associated with immunogenicity have diminished.[4]

1.10.1.3 Biological Therapeutic Areas

The therapeutic areas where biologicals can be of benefit are potentially numerous. Recombinant proteins are already used in diverse areas such as hemostasis, treated with the clotting factor, activated factor VIIa (NovoSeven), or anemia (caused by renal disease) treated with recombinant erythropoietin. In these cases, the drugs are used as replacements for absent or low concentrations of endogenous proteins. It is often the opposite situation with mAb therapy in that

Table 1 Therapeutic mAb products approved by the US Food and Drug Administration (FDA) or European Agency for the Evaluation of Medicinal Products (EMEA)

Brand name	*Generic name*	*Company*	*Indication*	*Year approved*		*Sales 2003 (M $)*
				FDA	*EMEA*	
Avastin	Bevacizumab	Genentech/Roche	Colorectal carcinoma	2004		499
Erbutix	Cetuximab	ImClone	Colorectal carinoma	2004		263
Raptiva	Efalizumab	Genentech	Psoriasis	2003		1
Xolair	Omalizumab	Genentech	Asthma	2003		25
Bexxar	Tositumomab	Corixa	NHL	2003		27
Humira	Adalimumab	Abbott	Rheumatoid arthritis	2002	2003	280
Zevalin	Ibritumomab tiuxetan	IDEC	NHL	2002	2004	22
Campath	Alemtuzumab	Millenium & ILEX	CLL	2001	2001	72
Mylotarg	Gemtuzumab	Wyeth	AML	2000		45
Synagis	Palivizumab	MedImmune	Respiratory infection	1998	1999	849
Herceptin	Trastuzumab	Genentech	HER2 + breast cancer	1998	2000	1002
Simulect	Basiliximab	Novartis	Transplant rejection	1998	1998	62
Remicade	Infliximab	Centocor	Crohn's disease	1998	1999	1729
Rituvan	Rituximab	Genentech	NHL	1997	1998	1753
Zenapax	Diclizumab	Roche	Transplant rejection	1997	1999	33
ReoPro	Abciximab	Centocor	PCI	1994		384
Orthoclone	Muromonab	Ortho Biotech	Transplant rejection	1986		52

NHL, non-Hodgkin's lymphoma; CLL, chronic lymphatic leukemia; AML, acute myeloid leukemia; PCI, percutaneous coronary intervention.

these drugs are used to target molecules produced by the body that have an unwanted effect. For mAb therapy the main therapeutic areas in which medicines have been marketed are disorders or conditions with an immunological basis, including inflammatory responses, cancer, infectious diseases, and acute coronary syndromes (**Table 1**).

One of the critical properties that mAbs possess is their specificity: they can be produced to recognize any number of therapeutic targets, and their potency and/or affinity can be engineered. There are nine isotypes of human antibodies found in the blood (immunoglobulin IgG1, IgG2, IgG3, IgG4, IgM, IgA1, IgA2, IgD, IgE), the most common of which is IgG1. Each isotype has distinctive characteristics and can therefore be selected for specific targets and to minimize the risk and type of adverse effects. IgG1 antibodies are currently the most suitable for therapeutic use as they have a long biological half-life (>20 days) and the ability to interact with the human immune system. Antibodies consist of two different domains, Fab and Fc; antigens bind to the Fab domain and the Fc domain determines the effector functions of the antibody. By engineering differences in these domains, the valency, biodistribution, half-life, and immunogenicity of the antibody can be altered, thus producing a tailor-made drug, suitable for treatment in a diverse range of therapeutic areas.

Via competitive antagonism of certain immunomodulatory molecules, mAbs can be used in immunosuppression regimens following renal graft transplants. The mAb basiliximab (Simulect) has been shown to be more effective than placebo in controlled trials examining the rates of rejection of grafts 6 months after transplant.[5,6]

Other mAbs also have roles in immunomodulation and inflammation, by targeting parts of the body's immune response, such as T or B cells, or cytokines, especially in autoimmune diseases such as Crohn's disease. mAbs are also used in the treatment of rheumatoid arthritis (RA) by targeting tumor necrosis factor alpha (TNF-α), a cytokine with wide-ranging activities in inflammation.[7] Phage display techniques led to the development of one of the first human mAbs on the market. Adalimumab (Humira) is used in the treatment of RA; it is a fully human recombinant IgG1 anti-TNF-α antibody. Further developments in antibody technology have also allowed for novel proteins to be constructed; for example, the chimeric protein, etanercept (Enbrel), is a fusion of a TNF-α receptor linked to a part of the IgG1 molecule, and is used in the treatment of RA and psoriasis. mAbs are also used in the treatment of RA by target TNF-α, a cytokine with wide-ranging activities in inflammation.[7]

The first mAb to be approved in the treatment of cancer was rituximab (Mabthera). This is a chimeric IgG antibody that targets a specific protein on the surface of B cells present in certain types of cancer. Rituximab has had clinical success in relapsed low-grade and follicular non-Hodgkin's lymphoma.[8] Another approach in cancer treatment with mAbs has been to target growth factor receptors present on tumor cells. Trastuzumab (Herceptin) is the first mAb available for the treatment of certain types of breast cancer.[9]

mAbs have also shown a clinical benefit in antiplatelet therapy. Platelet aggregation occurs in acute coronary syndromes, and this can lead to thromboses and myocardial infarction. However, platelet activation and aggregation can be reduced by glycoprotein IIb/IIIa receptor antagonists, which interfere with the pathway involved with platelet aggregation. Abciximab (Reo-Pro) is one such antagonist, and data have shown that the incidence of ischemic complications, such as myocardial infarction, is reduced when used by patients undergoing angioplasty.[10] Finally, the use of mAbs has been explored in infectious diseases. Palivizumab (Synagis) is a humanized mAb used to treat and prevent respiratory syncytial virus (RSV) in bronchopulmonary dysplasia.[11]

These initial clinical studies have shown the potential of mAb therapy in several therapeutic areas, and further studies are currently under way with other experimental mAbs in different disease conditions.

1.10.1.4 Current Status of Biologics

The discovery and development of biological pharmaceuticals seem to be emerging from their infancy and currently represent an exciting new field for therapeutics. This is reflected by the established commercial success of several marketed biologics and the continued growth of biotechnology companies.

Current global pharmaceutical sales have been estimated to be almost US$500 billion and, as part of this market, biologic medicines have been estimated to generate US$32 billion in revenue. Forecasts for 2010 estimate this figure will increase to US$53 billion. This projected development is supported by the dramatic increase in the annual growth of protein therapeutic sales in recent years. Annual growth has been estimated to be 20% from 1998 to 2002, with a 24% increase alone from 2001 to 2002.[12] In particular, the value of the global therapeutic mAb market grew by 37% to US$5.4 billion between 2001 and 2002 and is projected to increase to US$16.7 billion in 2008.[13]

This escalation in the growth of biologics is, in part, mirrored by the number of pipeline drugs and expected approvals: the number of biologic drug approvals by the US Food and Drug Administration (FDA) increased to six in 2003 from two in 1999. In contrast, small-molecule new medical entities have experienced a decrease in approvals: from 24 in 1999 to 19 in 2003. With a focus on mAb, there are estimated to be 132 products in development, with 16 new mAb therapies expected to be approved in the next 4 years.[13]

There seems to be a difference in the pipeline development of drugs produced by traditional pharmaceutical companies and the growing number of biotechnology companies. During the past 10 years there has been an increase in research and development costs for new drugs, but a decrease in the number of overall drugs put forward to the FDA for approval. Consequently, in order to boost pipeline productivity, many pharmaceutical companies have entered into partnerships with biotechnology companies, thus shifting their focus from small-molecule medicines to the novel biological therapies. This has allowed both pharmaceutical companies and biotechnology companies to share knowledge and expertise, expand product portfolios and pipelines, and at the same time share the risks involved with the drug discovery process.

1.10.2 Biologic Drug Discovery and Production

1.10.2.1 Introduction

Pharmaceutical companies spend billions annually on the identification and development of molecules that have the potential to target disease processes at a cellular level. There is increasing pressure to find better drugs with increased efficacy and fewer side-effects than those currently on the market, especially for those illnesses where few effective therapies exist, such as many cancers and neurodegenerative conditions like Alzheimer's and Parkinson's disease.

The processes involved in the discovery of a novel drug are becoming increasingly complex. Advances in molecular biology have had an enormous impact on the research and development of protein-based drugs and traditional small-molecule medicines. There are many similarities between the fundamental drug discovery process for small molecules and biological macromolecules; the key process being the generation of libraries of potential drug candidates, which are screened for particular characteristics required of the drug candidate. Such characteristics may include binding to the drug target and functional activity in a range of biochemical and biological assays, and may extend to ease of manufacture and formulation. This section will describe some of the current technologies involved in the development of biological macromolecules and highlight how these differ from small-molecule drug discovery.

1.10.2.2 Phage Display

Phage display was pioneered by George Smith in 1985.[14] It is a simple and effective method of producing very large libraries of proteins (up to 1×10^{11}) in a selectable format. These libraries are analogous to chemical combinatorial libraries, the key differences being the larger sizes of the libraries and the ability to select molecules from these phage libraries on the basis of desired drug phenotype, such as binding to a target. The phage system enables linkage of the protein phenotype with the genetic information encoding that phenotype. Hence it is a rapid and efficient way to identify lead-drug candidates. Phage libraries of variants of antibodies or other proteins can also be used to characterize antibody–antigen interaction sites, to investigate interactions between receptors and binding molecules, and in particular to optimize the affinity of interactions between protein variants and their cognate ligands.

A bacteriophage (phage) is a virus that infects bacterial cells. Its structure consists of a 'head' (icosahedral or filamentous) and in some phage a 'tail.' Filamentous phage are rod-shaped and contain circular strands of DNA; they are generally used in phage display techniques. DNA sequences that encode proteins are inserted into the genome of the phage and the protein is displayed as a fusion product to one of the phage coat proteins. Consequently, the key feature of this technique is that the genotype and phenotype are linked. A'library' of filamentous phage can be used to display several billion proteins simultaneously.

In the case of phage antibody libraries, this is achieved by introducing the phage library into *Escherichia coli* cells, where they can replicate and assemble phage particles. When the *Escherichia coli* cells lyse, phage particles are released that have antibody coat fusion proteins, whilst encapsulating the antibody DNA. The phage that express antibodies, which bind to a specific antigen of interest, are isolated from the background population of nonbinding phage particles by a binding capture process. Unbound phage are washed away and discarded.[15,16] Phage antibody libraries can be generated in a number of formats, the most widely used being Fab and single-chain Fv. Both of these formats involve the isolation of the variable domains of antibody fragments, which are critical in directing antigen binding.[17] These fragments can then be reformatted into full immunoglobulins when combined with the conserved antibody region (Fc) and expressed in mammalian cell lines.

The key advantage of this technique, compared with chemical library screening, is that very large, diverse libraries can be used as a starting point, from which panels of specific antibodies can be identified quickly. Selection of leads from the library is iterative and can be carried out on the basis of more than one criterion; for example, by firstly selecting for binding to the human antigen, followed by selecting the resultant panel for binding to the mouse homologue, or to another human isoform. Once leads have been selected, further phage libraries can be created from variants of this lead candidate. Selection from such libraries is made on the basis of affinity for the target.

1.10.2.3 Ribosome Display

Ribosome display is conceptually similar to phage display, again utilizing the ability to link genotype with phenotype. Instead of having to introduce the DNA into a host phage, the ribosome itself becomes the link between the gene and the displayed protein. The genetic information encoded within DNA is transcribed into the form of messenger RNA (mRNA) within the cell nucleus, and this is then used as the template to produce the protein, using the ribosome as the translational machinery to assemble the full-length protein molecule. Normally, once the full-length protein has been produced, the ribosome dissociates from the mRNA, but if the signal instructing the ribosome to dissociate is missing, the ribosome stops at the end of the mRNA, providing a link between the mRNA and the newly produced protein.[18]

In the case of antibody libraries, selection for lead-drug candidates in the ribosome display system is again very similar to the phage concept: ribosomal complexes that bind to the target of interest are captured and nonbinding complexes washed away. The binding complexes are then subjected to a process called reverse transcription. This turns the population of selected mRNA sequences back into DNA, which can be amplified by the polymerase chain reaction (PCR) and used as a starting point for further selection. Genetic variation can be introduced at the PCR step to allow sampling of variants of the selected lead-drug candidates. This repetitive process enables the refinement of selection of target antibodies, with the potential to produce highly potent candidates.[19]

A key difference between ribosome display and phage display is that ribosome display is a cell-free system. mRNA ribosome translational extracts are assembled in vitro and this reduces the constraints of the cloned library size, which are associated with the phage system. The theoretical library size using a ribosome display system is about 1×10^{14}, and hence provides further potential diversity in the population of entities for initial selection and screening. A bigger library means a greater probability of finding an appropriate antibody. Ribosome libraries can also be established quickly and be selective of candidates with improved potency compared with those identified via phage display.[18–21]

1.10.2.4 Hybridoma Technology

Hybridoma cell lines are widely used to create mAbs. This is achieved by immunizing a mouse with a target antigen, thereby eliciting an immune response. The B lymphocytes, taken from the immunized mouse spleen, produce antibodies to the antigen. Each B lymphocyte is then fused with an immortal myeloma cell line, allowing the production of unlimited quantities of antibodies. This hybridoma of murine B lymphocytes and myeloma cells is then screened to ascertain whether a potential therapeutic antibody is being expressed.

The problem with this method is that it takes months to cultivate and harvest the B lymphocytes from the immunized mice, and also relies on the mouse generating specific antibodies of interest. Another drawback to the classical application of the method is the murine origin of the antibodies. Any exogenous protein used therapeutically has the potential to elicit an immune response in humans,[22] which raises three potential problems: (1) allergic reactions – at worst, anaphylaxis; (2) reduction in efficacy; and (3) induction of autoimmunity against a patient's own endogenous proteins. In the case of mouse mAbs, human antimurine antibody responses led to the development of humanized or completely human mAbs.

To reduce the potential immune response to a murine antibody, the constant region of a human antibody can be combined with the variable region of the mouse antibody to produce a chimeric mouse–human antibody. Even so, this chimera may still produce an immune response such as headache, chills, and fever.[23,24] Completely human antibodies can be created from mice by introducing human antibody genes (the heavy and light chains) into the mouse genome. The mice are modified so that they do not produce their own antibodies but instead generate totally human versions after being immunized with specific antigens.[25,26] The B lymphocytes are again taken from the spleen and used to generate human hybridoma cell lines. This technique has advantages over classic hybridization as a transgenic mouse can produce a variety of specific antibodies, from which candidates can be isolated, and it is likely that these will be less immunogenic than the chimeric antibodies.[16,25,26]

1.10.2.5 Transgenic Animals

A transgenic animal has had an exogenous gene introduced into its genome, or conversely, a gene selectively removed (knockout) to resemble a disease state. Transgenic animals are very important in the pharmaceutical industry; they can function as models for translating basic science into therapeutic products and can be used to produce potential biological macromolecule drugs. The advances in microinjection and recombinant DNA technology paved the way for the rearing of genetically modified animals. Today, genetic modification is possible by several methods including[27]:

- DNA microinjection – recombinant DNA is injected into a one-celled embryo (a zygote), thus allowing the 'transgene' to be incorporated into virtually every cell in the developing organism.
- Retrovirus-mediated gene transfer – recombinant retroviruses are used to infect 4- to 16-cell stage embryos; the virus produces a DNA copy of its RNA genome via reverse transcription, and this integrates into the host genome.
- Embryonic stem cell technology – pluripotent embryonic stem cells are taken from early embryos and are subjected to various in vitro manipulations that allow insertion of DNA sequences. The cells are then transplanted back into new host embryos.
- Nuclear transfer – this is commonly referred to as cloning. A donor nuclei is transferred into an enucleated oocyte (the embryo before it is fertilized).

Using these methods, recombinant proteins can be expressed and secreted by transgenic animals. A recent advance in the production of proteins has been the secretion of therapeutic molecules in the milk of certain mammals. Transgenic goats have been engineered to produce a recombinant form of human antithrombin in this way.[27]

Another significant example of the use of transgenic animals to produce protein-based drugs has been the use of mice to generate human antibodies, as discussed above.

1.10.2.6 Sexual Polymerase Chain Reaction

PCR was invented in the 1980s by Kary Mullis and colleagues[28] and represented a major breakthrough in the field of molecular biology. PCR enables large amounts of DNA to be produced from short segments.

It is possible that novel proteins, with different properties and characteristics and potential therapeutic applications, may be constructed from the starting point of naturally occurring DNA. This can be achieved by amplifying segments of DNA using the PCR technique, but with the introduction of deliberate variation in the order of segments, resulting in changes to the sequence of amino acids. This technique was first called 'sexual PCR' by Willem Stemmer.[29] A biological target is identified and its genes and related genes are isolated and 'shuffled' into novel combinations via sexual PCR. The process is repeated until a whole library of novel genes is created with potentially improved characteristics from the parental clone. The novel genes are expressed as proteins, which are then screened for desirable therapeutic properties and functions. In essence, this technique speeds up genetic evolution by testing new combinations of DNA segments.[30]

1.10.2.7 Structure–Activity Relationships

New protein-based drugs can be produced by rational modification of the structure of original human proteins. There are often considerable structural homologies in families of proteins, and these sequences can be determined via cloning and protein expression and purification. The three-dimensional structure can then be clarified by x-ray crystallography. Highly conserved regions between proteins highlight the areas of structural and functional importance. These areas can then be modified to enhance potency or confer particular characteristics to the protein.

One of the most common examples of this is the refinement of the protein, insulin. Originally, insulin was extracted from animal pancreata for the treatment of patients with diabetes, but the preparations were impure and, being animal in origin, were immunogenic. The sequencing of the insulin gene and the advent of recombinant DNA technology enabled the production of purified preparations of human recombinant insulin in the 1980s. However, it became apparent that the pharmacological profile of subcutaneously injected human insulin was inadequate compared with the secretion profile of endogenous insulin in people without diabetes: during mealtimes there is a sharp rise in insulin secretion to control the rise in glucose levels.

After injection, human insulin does not get absorbed into the bloodstream quickly enough to combat post-prandial glucose effectively. Therefore, to have a pharmacological profile similar to endogenous insulin secretion, it has to be injected 30 min before eating. Consequently, modifications to the insulin structure were undertaken to engineer a more rapid-acting insulin for administration at mealtimes. The analog, insulin aspart, is based on the structure of human insulin, so that it retains insulin receptor-binding properties, but has modifications to the amino acid sequence that enable it to enter the circulation after injection more rapidly than human insulin.[31] Conversely, another insulin analog, insulin detemir, has been engineered to provide a protracted duration of action, more suited to the overnight and daytime insulin requirements of patients with diabetes.[32] In this instance, the amino acid sequence of insulin has been changed so the molecule self-associates after injection, thereby slowing its entry into the circulation. Another modification is the attachment of a fatty-acid side chain, which allows the molecule to bind to albumin, further slowing the time–action profile. However, these modifications do not affect the insulin–receptor interaction profile.

1.10.2.8 Traditional Drug Discovery

Originally, drug discovery was a hit-and-miss affair. Natural products known to have therapeutic properties were identified, and structurally similar chemical compounds were randomly screened or synthesized to see if they had any desirable therapeutic characteristics. Typically, this is a laborious and time-consuming (and therefore expensive) process, which involves thousands of test compounds. For example, the development of piroxicam, a nonsteroidal anti-inflammatory drug (NSAID), began in 1962, but the drug did not reach the market until 1980.[33] Its original development was based on the chemical class of carboxylic acids, several of which (e.g., aspirin, indomethacin, ibuprofen) were already on the market at that time for the treatment of inflammatory conditions. The quest started to find a novel chemical compound that had acidic properties and was highly anti-inflammatory, but was not a carboxylic acid, and was suitable for long-term treatment. Chemists synthesized potentially suitable compounds, following activity testing and serum half-life testing in dogs. In vivo testing in animal models of inflammation followed, with several classes of compounds failing over a 5-year period of investigation until the 'oxicam' family of molecules were revealed to be the most suitable candidates. Further refinement of several hundred analogous chemical structures narrowed down the search to three development drugs, of which piroxicam successfully completed clinical trial testing.

Recent developments in genomic sciences, rapid DNA sequencing, combinatorial chemistry, assays and screening have initiated some progress in traditional drug discovery of chemical compounds. Moving towards a more selective approach, like protein drug discovery, traditional drug discovery has recently focused on locating a biological target that is linked to a disease phenotype and then finding a suitable molecule that modifies the target.

Hypothetical biological targets (e.g., enzymes or receptors) are incorporated into in vitro assays and exposed to large numbers of structurally related chemical compounds. High-throughput screening (HTS) is an automated process that enables researchers to investigate very rapidly thousands of different chemical entities against particular targets. 'Hits' (molecules that were identified to modulate the biological target) can then be used to develop leads that may have effects in more complex disease models (e.g., animals). However, this exponential increase in the number of compounds that have been tested has, so far, not resulted in increased numbers of products reaching the market, although these concepts in chemical drug discovery are still novel.[1]

1.10.3 Opportunities and Limitations of Biological Macromolecules

1.10.3.1 Introduction

Analysis of the fundamental differences between traditional small-molecule drugs and biological drugs, highlighted previously in this chapter and explored further here, illustrates very different pharmacology profiles between the two types of medicine. Consequently, there are advantages and disadvantages for both, and this section will discuss some of the opportunities and limitations of biological drugs, especially in comparison with their chemical drug counterparts.

1.10.3.2 Why Are Biologics Different from Traditional Medicines?

In order to appreciate the advantages and disadvantages of biological macromolecules, we need to understand their pharmacology and the context in which they are developed.

Traditionally, pharmaceutical companies have tended to focus on the most common disease and therapy areas, while biotechnology companies have pursued more difficult-to-treat populations. However, this distinction is now blurring and with it the implications regarding cost of drugs. Small-molecule drug manufacture is relatively well defined, and enables large-scale uniform production. Consequently, traditional chemical drug costs are relatively cheap (notwithstanding research costs). On the other hand, biologics tend to have complex and small-scale production processes, and often new facilities have to be set up to produce these novel medicines. Couple this with the smaller patient populations who have the less common diseases, which the biologics have so far focused on, and consequently biologic drug costs are relatively expensive (again, notwithstanding research costs). There are also the issues of stability, chemical purity, storage, and distribution that need to be overcome.

However, manufacturing centers equipped to produce biologics are now well established, and it is becoming clear that patients, healthcare providers, and governmental bodies are prepared to use novel treatments in less common therapy areas or alternative treatment options where standard treatment has failed.

The pharmacology of biological macromolecules will also have an impact on the potential of biologics. These are large polypeptides and therefore cannot be taken orally. Thus, until other modes of administration are developed, biologics have to be injected or infused. Compared with oral treatments, this is obviously a barrier to patient compliance and convenience. Because of the protein nature of biologics they are degraded and not metabolized like small-molecule medicines. The large size of biologics also means that, once administered, they cannot enter certain body compartments; for example, they cannot normally cross the blood–brain barrier, which has both advantages (no central nervous system (CNS) effects occur) and disadvantages (increases the difficulties in the application of biologics for any CNS illnesses or disorders).

The specificity with which biological drugs are designed potentially makes them better suited than small-molecule medicines in terms of safety/tolerability; side-effects related to nonmechanistic pathways are also rare. Many biological candidates identified by drug discovery processes safely progress through preclinical and early-phase clinical testing because of the highly specific nature of their biological interactions and, therefore, lack of systemic side-effects. The specificity of biologics, especially mAbs, has made them particularly useful not only as therapeutic agents but also as delivery vehicles.[34] For example, mAbs can 'deliver' cytotoxic agents to cancer cells and promise to expand the oncologist's future armamentarium.[35]

Another advantageous aspect of the pharmacology of biologics is their pharmacokinetic/pharmacodynamic profiles. Generally, biologics have relatively long serum half-lives, and therefore dosing is potentially far less frequent

Table 2 Difference between biological macromolecules and traditional small-molecule medicines

	Traditional	*Biologic*
Molecular weight	Low	High
Structure	Single fixed chemical formula	Complex, can be heterogeneous
Dose	Maximum tolerated dose	Optimal biological dose
Mechanisms of action	Pleiotropic, not always understood	Specific, often understood
ADME	Metabolized	Degraded
Production	Chemical synthesis	Via cellular systems
Manufacturing	Contamination easily avoided, detected, and resolved	Contamination hard to detect and resolve

ADME, absorption, distribution, metabolism, and excretion.

than small-molecule medicines, which often have to be dosed daily. Despite the fact that biologics currently have to be injected, treatment is infrequent, and therefore injection may pose less of a problem for patient compliance and convenience than was first thought. Furthermore, with the debilitating and serious diseases that biologics often treat, the problem of injection is inconsequential.

Some of the differences between the two genres of therapies are summarized in **Table 2** and discussed further in two case studies of a successful small-molecule medicine and a burgeoning mAb.

1.10.3.2.1 Omeprazole case study

Heartburn, resulting from gastroesophageal reflux disease (GERD) and stomach ulcers, is a widespread health problem. Around 45% of people recently questioned in a survey reported the occurrence of one or more upper gastrointestinal symptom during the past 3 months; heartburn was the second most commonly reported symptom.[36] There are clear economic and quality-of-life implications to these findings: sufferers miss work and leisure/household days because of the symptoms associated with GERD.[36,37] Indeed, it is estimated that the annual direct and indirect medical costs attributed to GERD total $10 billion in the USA.[38] Consequently, the pharmaceutical market for gastric acid-related disorders is huge.

Antacids (magnesium and aluminium salts) and histamine (H_2) receptor antagonists (cimetidine and ranitidine) are the traditional treatments of heartburn. Both classes of drugs are oral agents, and are available as over-the-counter (OTC) medications. Antacids have a rapid effect in the treatment of symptoms of heartburn but, because of their short duration of action, may have to be dosed several times a day. H_2 receptor antagonists were originally hailed as a major breakthrough in the treatment of GERD, and opened the market for inhibitors of acid secretion. They are generally dosed twice a day, and their efficacy can be improved by increasing the dose and dosing frequency. However, this may reduce compliance, and will increase the cost of treatment.[39] The efficacy of H_2 receptor antagonists for the prophylaxis and treatment of GERD was not as great as expected,[39] and therefore further therapies were investigated for this potentially massive market.

Omeprazole (Losec/Prilosec) is a member of the proton pump inhibitor (PPI) class of drugs, which are used in the treatment of heartburn. These drugs act by irreversibly blocking the proton pump (H^+/K^+-ATPase) found in the stomach. Parietal cells secrete hydrochloric acid (HCl; pH < 1) via active transport of Cl^-, accompanied by K^+, into the stomach lumen. The proton pump exchanges K^+ ions, from the lumen of the stomach, for H^+ ions from within the parietal cell; the net result is secretion of HCl (gastric acid). The discovery of the proton pump in 1968[40] led to its investigation as a potential therapeutic target. Cimetidine, a substituted imidazole, was already in use as a regulator of acid secretion. This chemical structure served as the basis for the development of omeprazole: a benzimadazole moiety was added to pyridine-2-thioacetamide, a known acid secretion inhibitor.[41] The resulting congeners included omeprazole.

Omeprazole is a weak base and therefore accumulates in the very-low-pH areas surrounding the gastric parietal cells that secrete stomach acid. The main component of omeprazole is a substituted benzimidazole (**Figure 2**): protonation of the benzimadazole and pyridine nitrogens forms a sulfenamide, and this binds to exposed cysteine residues of a subunit of the proton pump.[42] The pump now becomes blocked and gastric acid secretion is inhibited. Omeprazole is generally administered orally, but it degrades rapidly at a low pH. Consequently, enteric-coated granules deliver the drug to the stomach and thereby avoid inactivation. It is metabolized into inactive metabolites and eliminated rapidly,

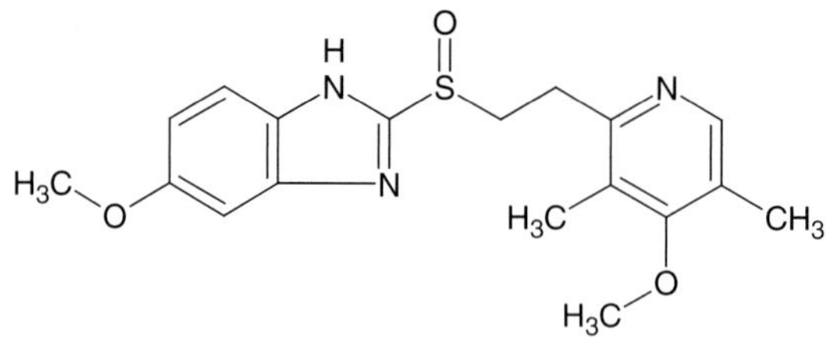

Figure 2 Omeprazole.

with a half-life of approximately 1 h. However, given as a single daily dose, it can affect acid secretion for up to 3 days because of its covalent binding to the proton pump.[43]

Omeprazole was the first approved PPI and has been marketed in more than 100 countries worldwide since 1988. It is one of the most frequently prescribed medications and is also available as an OTC medication. It can be called a 'blockbuster,' a term given to a drug that can generate worldwide sales of more than US$1 billion yearly. Losec/Prilosec came off patent in 2001, and omeprazole is now available as a generic drug.

Omeprazole's success obviously has something to do with the vast market for gastrointestinal disorders and the clinical needs unmet by previous treatments. Its mode of action, by irreversibly binding to the proton pump, means that it can be dosed once-daily, as opposed to antacids and H_2 receptor antagonists. This is especially important for the compliance of patients with chronic conditions. Also, because omeprazole is a weak base, it accumulates around the parietal cells, therefore preferentially acting on these cells as opposed to others. In comparison, H_2 receptors are also found in cardiac muscle, and while there is no evidence to suggest that H_2 receptor antagonists act in the heart, cimetidine can cause antiandrogenic effects, due to its binding of androgen receptors, and has an inhibiting effect on estradiol catabolism.[44] Cimetidine inhibits the cytochrome P450 pathway, which is responsible for the biotransformation of many drugs. Consequently, cimetidine is implicated in potential clinically relevant interactions with other drugs, such as coumarins, cilostazol, ciclosporin, ergotamine, lidocaine, procainamide, propafenone, quinidine, sertindole, carbamazepine, theophylline, phenytoin, and valproate. While omeprazole, with its chemical similarity to cimetidine, does have an effect on the cytochrome P450 pathway, it is to a lesser degree and has little effect on estradiol pathways.[45] Consequently, omeprazole is unlikely to cause estrogen-related side-effects such as gynecomastia in men (as has been reported with chronic cimetidine use)[46] and has potential clinically relevant interactions with coumarins and cilostazol only.

Accordingly, the success of omeprazole, compared with other previously available gastric acid inhibitors, is intricately related to its pharmacology. It can be taken orally, which is important for patient compliance, especially given its daily dosing schedule. Omeprazole is also cheap, particularly now it has come off patent. Fundamental to omeprazole's success is the significant pharmacoeconomic benefit the drug has produced. Omeprazole has reduced costs associated with gastritis, ulcers, and related surgery.[47] Another factor in its success is an increased understanding of the physiology of gastric acid secretion, which allowed the identificaion of a therapeutically relevant target (the proton pump).

1.10.3.2.2 Adalimumab case study

RA is the most common form of chronic joint inflammation in the world. Prevalence of RA has been changing, but a survey in the UK estimated prevalence to be 1.2% in women and 0.4% in men,[48] and worldwide prevalence has been reported to be approximately 1%.[49] The risk of developing RA increases with age, and women have a higher risk than men. It can be an incredibly debilitating disorder that severely affects quality of life for sufferers, and of course places an economic burden on healthcare systems: annual indirect and direct medical costs have been estimated to be almost €12 000 and €38 000 per patient, respectively.[50]

Traditional treatments for RA have included NSAIDs, and drugs that suppress the rheumatic disease process nonbiologic disease-modifying antirheumatoid drugs (DMARDs) such as gold, penicillamine, hydroxychloroquine, sulfasalazine, methotrexate, and corticosteroids. NSAIDs reduce the symptoms associated with RA, but have no effect on the underlying cause. Long-term use is restricted by adverse events, particularly gastrointestinal side-effects. Nonbiologic DMARDs are often used in patients whose disease has progressed. For example, methotrexate is generally used in patients with active moderate-to-severe RA. Methotrexate is a folate antagonist, and inhibits dihydrofolate reductase (an enzyme important in the reaction that is vital for the function of folates as coenzymes in DNA synthesis and cell division). This action also results in inhibition of de novo purine and pyrimidine synthesis, and it is this effect that is thought to be methotrexate's mechanism of action in RA.[51] In turn, this results in an accumulation of adenosine (which has immunomodulatory effects).[52,53] Unusually, for a small-molecule medicine, methotrexate's mechanism of action in RA is less well defined than in the other diseases in which it is indicated (cancer chemotherapy and psoriasis).

It does have one of the highest response rates of any nonbiologic DMARDs, with its symptom modification and erosion-slowing activity.[54] However, many patients fail to sustain their response to methotrexate or other DMARDs.

Methotrexate is usually given orally once weekly, and maximum blood plasma concentration occurs 1–5 h after administration.[43] About 50% of the drug is plasma protein-bound and is therefore distributed throughout body fluids; methotrexate is taken up into cells by the folate transport system. It is mainly excreted unchanged in the urine. Consequently, there is a risk of nephrotoxicity with high doses, as methotrexate can precipitate in the renal tubules. Indeed, methotrexate is a very toxic drug, and while it is the most successful of all nonbiologic DMARDs, its long-term administration and efficacy are still limited by its tolerability.[55–59] Gastrointestinal side-effects are common. There is a risk of pulmonary toxicity and patients have to be monitored regularly for rises in liver enzymes as an indication of hepatic toxicity.

Subsequently, long-term treatment with methotrexate, like other DMARDs and NSAIDs, is not ideal and has encouraged the search for new therapies in RA. In particular, interest has focused on agents directed against cytokines involved in immune responses.

TNF-α is an inflammatory cytokine that is present in synovial fluid. It stimulates cell proliferation, metalloproteinase expression, adhesion molecule expression, secretion of other cytokines, and prostaglandin production.[49] Elevated levels of TNF-α are found in the synovial fluid of patients with RA[60] and the release of the above-named proinflammatory molecules via TNF-α action causes joint destruction.[61] Regulation of TNF-α is provided through two soluble receptors (p55 and p75), which occur in the synovial fluid. These soluble receptors compete with p55 and p75 cell surface-bound receptors to inhibit TNF-α binding.

Several mAbs, or mAb-based, molecules have been produced that bind to TNF-α: infliximab, etanercept, and adalimumab. Infliximab (Remicade) is a mouse/human chimeric protein that joins the variable region of a mouse antibody to the constant region of human IgG1. Etanercept (Enbrel) is a dimeric fusion protein that joins the human p75 TNF-α receptor to the Fc domain of human IgG1. Adalimumab (Humira) is a fully human IgG1 antibody (**Figure 3**). All of these agents are novel biologic DMARDs. Data from clinical trials have shown that they are effective in treating RA.[62–64] Adalimumab is the newest of the three products to market and is currently experiencing a rapid period of growth in sales. While not yet a 'blockbuster,' as defined by small-molecule medicine standards, it is proving to have significant potential.

Adalimumab binds specifically to TNF-α and blocks its interaction with cell surface-bound p55 and p75 TNF-α receptors. Consequently, adalimumab modulates the effects of TNF-α, such as controlling the levels of adhesion molecules that are responsible for leukocyte influx into inflamed joints.[65]

As it is a fully human antibody identical in structure to IgG1, adalimumab has a very similar half-life to IgG1 of approximately 2 weeks.[66] Consequently, this enables adalimumab to be dosed by subcutaneous injection every 2 weeks. Adalimumab is also very specific for TNF-α; it does not bind to other cytokines, including TNF-α. This specificity is important as it means that adalimumab does not have any nonmechanistic activity. However, it does mean that TNF-α is modulated in other immune or disease states. In particular, TNF-α appears to be involved in the immunological defence against tuberculosis (TB). There have been cases of TB in patients taking TNF-α inhibitor therapy,[67] and while there have been no reports with adalimumab, it is reasonable to assume that the development of active TB may be a class effect with TNF-α blockers.[68]

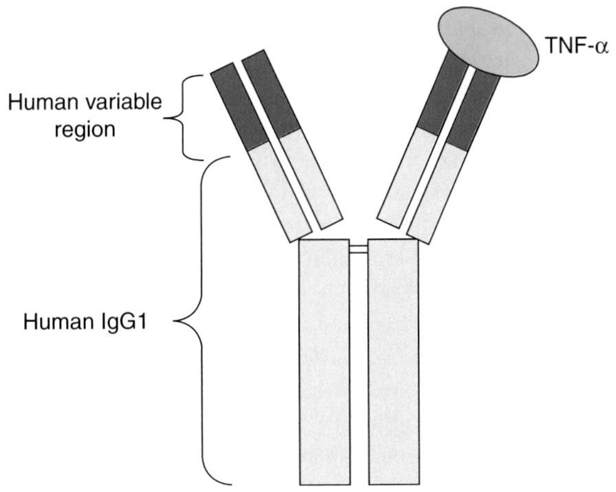

Figure 3 Adalimumab (Humira).

Other drawbacks with TNF-α inhibition include the risk of infection or developing cancer due to suppression of the immune system. However, patients with RA are at an increased risk of infection and cancer, especially with concomitant immunosuppression, such as with methotrexate. To date, this does not appear to be an issue with adalimumab: a 24-week safety trial has shown that adalimumab in conjunction with other standard RA therapies does not result in an increased risk of infection compared with placebo[69]; there are no postmarketing studies yet available to investigate any incidence of lymphoma with adalimumab.[68]

Finally, the issue of immunogenicity is one that should be explored with TNF-α blockers (or any mAb). Adalimumab is expected to elicit a lesser immunogenic response than other animal chimeric antibodies as it is a fully human antibody. However, all TNF-α blockers are recombinant foreign proteins, and the human body may develop human antihuman antibodies to adalimumab. Results from trials so far indicate that there is a low incidence of antibody formation in patients treated with adalimumab, especially in comparison with other TNF-α blocker therapy (although direct comparison between trials would be unfair).[70]

The potential success of adalimumab in RA could be due to a number of factors. While this agent is administered by subcutaneous injections, they are infrequent. Although adalimumab is more expensive than other nonbiologic DMARDs, it does offer effective treatment options for those patients who have not responded or who have failed to sustain their response to standard therapies.

1.10.3.3 Summary

The two drugs outlined above share some common sources to their success, for example, an increased understanding of the disease state enabled therapeutic targets to be identified, and previous therapies in those areas were inadequate. However, there are also differences between the two types of drug that work for and against their success in the marketplace. For example, biological drugs are protein in nature and so currently cannot be given orally, as they would be broken down before being absorbed; alternatively, small-molecule medicines can be delivered via several routes, including oral administration. Conversely, small-molecule medicines can have systemic side-effects via interaction with receptors in parts of the body not affected by a specific condition, while biologic drugs, such as mAbs, generally do not have nonmechanistic effects because of the highly specific nature of their binding interactions.

1.10.4 Biologics and Small Molecules: The Future

1.10.4.1 Introduction

The science of biologics is very much in its infancy compared with small-molecule development. With the advent of the genomic and proteomic age, both biologic and small-molecule drugs are entering an exciting new phase of discovery and rational design.

There are many challenges to be faced, and opportunities to be grasped, by both traditional large pharmaceutical companies and biotechnology companies in their search for new drugs.

1.10.4.2 The Future of Small-Molecule Drugs

One new method that is currently being used by pharmaceutical firms is fragment-based lead discovery. This approach to discovering new small-molecule medicines is being taken alongside more traditional screening methods. Fragment-based lead discovery involves the selection of compounds with a very small molecular mass. These ‘fragments’ have a molecular mass of between 150 and 250 and low binding affinities (mmol L^{-1} to 30 μmol L^{-1}). Consequently, the fragments are screened at very high concentrations, and the number of ‘hits’ is much higher than HTS.[71]

The fragments are identified using x-ray crystallography or nuclear magnetic resonance, which also allows structure–function relationship information to be obtained; this information is very useful in optimizing the small fragment into a viable therapeutic compound. Fragments that are identified against particular targets can be ‘added’ together in order to ‘tailor’ the evolving compound into a specific lead.

Even though the fragments identified initially have low absolute affinities, the fragments as ligands are very potent for their size and this potency increases with the concurrent increase in molecular size, as more fragments are put together. Interestingly, the molecular size of a small-molecule medicine is inversely proportional to its success in completing clinical trials and reaching market (**Figure 4**).[72]

Wenlock and colleagues commented that “the mean molecular weight of orally administered drugs in development decreases on passing through each of the phases and appears to gradually converge toward the mean molecular weight of marketed oral drugs data set.” Consequently, supporters of fragment-based lead discovery argue that it is more

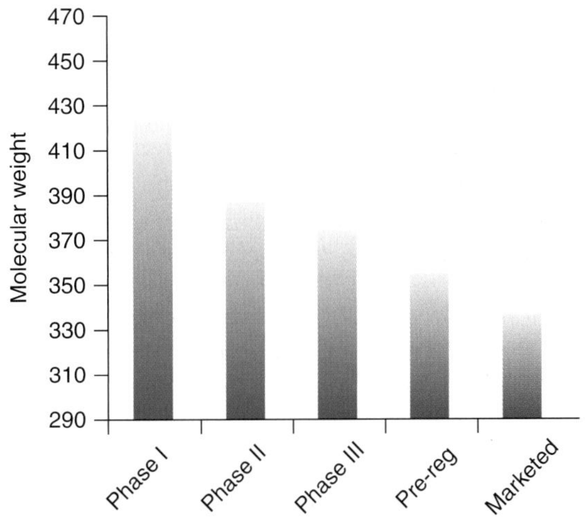

Figure 4 Mean molecular weight for drugs in different phases. (Adapted from Wenlock, M. C.; Austin, R. P.; Barton, P.; Davis, A. M.; Leeson, P. D. *J. Med. Chem.* **2003**, *46*, 1250–1256.)

efficient to start small and 'build up.'[72] HTS identifies molecules with a wider and larger range of molecular masses (Mr: 250–600) than fragment-based screening, yet these compounds often have to be reduced in size. However, they still need to retain their potency in order to produce viable leads, and this can be problematic.[71] Workers using fragment-based lead discovery have also been able to identify compounds for particular targets that were not identified by traditional HTS.[73–75] The potential benefits of fragment-based lead discovery have yet to be fulfilled, but structure-based drug design (such as this technique) is being used by several companies, and is anticipated to be the main driving force in the research and development of small-molecule medicines during the next decade.[76]

1.10.4.3 Looking Ahead with Biological Macromolecules

As biologic development continues to grow, the biotechnology industry is facing many of the challenges that the large pharmaceutical companies have faced previously: the need for refinement of medicines and to overcome problems with delivery; the expansion of treatments into different therapeutic areas; the need to enhance research and development pipelines, possibly through mergers or collaborations; and the threat of generic competition.

Biological macromolecules, particularly mAbs, are easier to refine than small-molecule medicines because of their specific nature and design. As mentioned previously, each human antibody has specific characteristics and these can be selected for particular targets to optimize efficacy and minimize side-effects.

Ongoing research continues to identify new proteins that are implicated in the pathogenesis of disease and illness, and therefore, hypothetically, can provide a multitude of new targets against which therapeutic mAbs can be evolved.

One current and future consideration is the method of delivery of mAbs. They are currently administered via injection; however, other methods of delivery have already been investigated. In particular, local delivery of mAbs and controlled-release formulations have been mentioned in the scientific literature. Examples include controlled-release implants of antibodies; topical applications such as dermal, buccal, nasal, vaginal, or rectal; or oral formulations of antibodies contained within polymeric vehicles that protect the mAbs from gastrointestinal degradation.[77] Topical immunotherapy with antibodies is not a totally new idea; dermal immunoglobulin treatment has been used against dermatitis,[78] and the transmission of herpes simplex virus has been prevented in animal models.[79] Controlled-release oral and implantable formulations of antibodies have been shown to be useful in the treatment of infections in animal models.[80–82] Local applications of antibodies co-administered with intravenous antibiotics have also demonstrated a synergistic effect in the same models.[82] Nasal and inhaled formulations of antibodies have also been explored in the treatment or prophylaxis of pulmonary infections or diseases[83–85]; this method of delivery also has the potential to deliver systemic doses of mAbs.[77]

With these new methods of delivery the pharmacokinetics and pharmacodynamics of mAbs can be altered; thus the dosing requirements can be refined and the treatment tailored for the benefit of the patient.[77] It is vital that different routes of administration other than subcutaneous injection are developed, so that costs can be reduced and convenience increased for the patient. Ongoing research in this area is likely to improve formulations and further the understanding of the structure and function of antibodies.

Findings from the above-mentioned animal studies have shown that antibodies have potential use in therapeutic areas other than those with a noninfectious immunological component. Obviously, the proliferation of infectious disease is intricately linked to our immune system, but this is an area in which the use of mAbs is still very much at an early stage.[86] Currently, only one mAb has been licensed to prevent an infectious disease: palivizumab for RSV. The specificity and versatility of mAbs could be used particularly in this area. They work independently of the patient's immune system, and therefore could be used effectively in immunocompromised patients.[87] The synergistic effects of antibodies and antibiotics previously reported in animal models may also provide extra defense against acutely dangerous pathogens that are becoming more resistant to current antimicrobial treatment.[87] There is also the interesting observation that mAbs may provide immediate immunity against biological weapons.[87]

Currently, mAbs are also being employed in the fight against cancer. Several mAbs are licensed for use in leukemias (alemtuzumab and gemtuzumab), non-Hodgkin's lymphoma (ibritumomab, rituximab, tositumomab) and colorectal cancer (bevacizumab, infliximab). It is expected that further refinement of mAb treatment will provide more effective therapies for cancer.[35] mAbs may also have potential in treating cardiovascular disease. Inflammatory processes are implicated in cardiovascular conditions such as atherosclerosis and chronic heart failure,[88] and thus make these conditions a target for therapeutic mAb treatment. Indeed, in biotechnology start-up companies, research into cardiovascular disease is taking top priority; established, disease-specialized companies have previously focused on cancer, infection, the CNS, and autoimmune research.[89]

Coupled with the need to diversify into different therapeutic areas is the need to improve research and development and enhance drug pipelines for both biotechnology and traditional pharmaceutical companies. Mergers between targeted biotechnology companies could result in the emergence of specialized, small pharmaceutical companies.[90] It is thought that large biotechnology companies will be at the forefront of drug discovery, and that small companies will specialize in specific therapeutic areas.[90] Furthermore, new biotechnology companies will emerge as new technologies are discovered and applied.

Biotechnology companies tend to be good at exploiting niches and the larger ones are able to develop and market their own products effectively. However, some smaller companies lack experience and capital when it comes to clinical development and marketing of products, and this is where collaborations and mergers between biotechnology companies and pharmaceutical companies occur.[91] Large pharmaceutical companies have recognized the skills and experience of the manufacturers of biological products and are keen to incorporate that knowledge base into their own research and development.[91]

The increasingly successful biotechnology companies may soon see acquisitions by smaller pharmaceutical companies (who have the experience of taking products to market); there has been a large increase in the number of profitable biotechnology companies during the past 20 years,[89] and it is expected that firms will attempt to build their own development pipelines and launch products by themselves or with the help of other biotechnology companies.[89] The biotechnology industry may become the fastest-growing area in the healthcare sector, and this could be achieved by continuing collaboration with the traditional, large pharmaceutical companies, and the consolidation and growth of existing biotechnology firms.[89]

Similar to the large pharmaceutical companies, one of the biggest issues facing the developers of biological macromolecules is the emergence of generic competition. However, there are fundamental differences between small-molecule generics and biological generics. The characteristics of biological products mean that it is very difficult for two different companies to produce two identical products. Generally 'generic' biologics are termed 'biosimilar' products. They are the same as small-molecule generics in that they have the same mechanism of action and therapeutic indication as the original product, and may even be identical in terms of molecular structure. However, the complex production processes and complex nature of biologics mean that biosimilar products may not be exactly the same as the initial biologic. Consequently, regulatory approval for biosimilar products is very different from regulatory approval for small-molecule generics. Shorter approval times for small-molecule generics (by cutting out some preclinical and clinical trials) is possible because it is easy to show that the generic is chemically and therapeutically identical to the original and will therefore have the same efficacy and safety profile.

The efficacy and safety of biosimilar products have to be evaluated by the necessary preclinical and clinical trials before approval, thus making development far more costly and lengthy than for a small-molecule generic. The importance of treating biosimilar products as separate drug applications was highlighted in the case of human erythropoietin, used in the treatment of anemia in patients with kidney disease. One brand of erythropoietin was associated with an unusually high incidence of pure red-cell aplasia (PRCA). The increase in cases of PRCA with this erythropoietin was attributed to subtle changes in the manufacture, distribution, and administration of the product.[92] Consequently, the development of 'generic' and second-generation erythropoietin formulations has to be closely monitored and subject to the same strict testing as the original product to ensure the same safety and efficacy profiles.

It is a contentious issue whether biosimilar products should be able to bypass the preclinical and clinical testing currently necessary for approval, especially with the current high cost of biologics. However, costs of biological macromolecules are expected to drop because of improvements in manufacturing and increasing research into new products, and it may be that biosimilar products will be less of a problem for manufacturers of original biological macromolecules than chemical generics are for current small-molecule medicines.

1.10.4.4 Summary

Despite the major differences between biological macromolecules and small-molecule medicines, there is a lot to be gained from the development of both types of product when we look at the history, discovery, research, and future of the industries that nurture them. It would be difficult to say that biological macromolecules are definitely the future of medicine when there is still so much to be offered by small-molecule medicines. Nevertheless, biological macromolecules certainly have a very healthy future and represent an exciting new category of drugs that can only complement, add to, and enhance current medicines in the fight against disease and illness.

Indeed, the concomitant use of biological macromolecules and small-molecule medicines is becoming increasingly common, and there are growing reports of synergistic effects from using both types of medicine. Biological macromolecules and traditional agents are currently being used in combination in the area of oncology. Erythropoietin treatment has consistently shown a benefit in outcomes in the treatment of cancer-related anemia; physical effects are an increase in hemoglobin levels without the need for blood transfusions, as well as related increases in well-being and quality-of-life outcomes.[93] Some preclinical evidence also points to the agent being synergistically effective in improving chemotherapy treatment against tumors,[94] but this has yet to be evaluated fully. Oncology is an area in which mAbs are being increasingly used as adjunctive therapies, especially in patients with advanced disease.[35,95] They are also being utilized as cytotoxic drugs or radiolabeled delivery agents in the treatment of cancer.[35,96] In fact, antibody-targeted chemotherapy research has resulted in a biologic/small-molecule combined treatment (or immunoconjugate) for acute myeloid leukemia.[97] Gemtuzumab ozogamicin represents a real clinical advancement in the management of malignant disease.

Other therapeutic areas can surely benefit from this 'holistic' approach to treatment. A mixture of biologic and systemic agents is increasingly being used in the treatment of psoriasis, with the expectation that physicians will be able to treat the symptoms of the disease more rapidly and effectively, while decreasing the possibility of side-effects.[98] The maximum potential of combining mAbs, such as alefacept or etanercept, with traditional drugs, such as methotrexate or ciclosporin, in the treatment of psoriasis has yet to be defined. There is a pressing need for more studies investigating the combined efficacy and safety profiles of biologics and small-molecule medicines in this and every other relevant therapeutic area.

The future of biological macromolecules is growing alongside that of small molecules. Both have benefits and disadvantages, and perhaps the optimum approach to drug discovery and development is a complementary and synergistic direction utilizing the best of both medicines.

References

1. Drews, J. *Science* **2000**, *287*, 1960–1964.
2. Drews, J.; Ryser, S. *Nat. Biotechnol.* **1997**, *15*, 1530.
3. Kohler, G.; Milstein, C. *Nature* **1975**, *256*, 495–497.
4. Chowdhury, P. S.; Wu, H. *Methods* **2005**, *36*, 11–24.
5. Nashan, B.; Moore, R.; Amlot, P. et al. *Lancet* **1997**, *350*, 1193–1198.
6. Kahan, B. D.; Rajagopalan, P. R.; Hall, M. *Transplantation* **1997**, *63*, 33–38.
7. Weinblatt, M. E.; Keystone, E. C.; Furst, D. E. et al. *Arthritis Rheum.* **2003**, *48*, 35–45.
8. McLaughlin, P.; Grillo-López, A. J.; Link, B. K.; Levy, R.; Czuczman, M. S.; Williams, M. E.; Heyman, M. R.; Bence-Bruckler, I.; White, C. A.; Cabanillas, F. et al. *J. Clin. Oncol.* **1998**, *16*, 2825–2833.
9. Slamon, D. J.; Leyland-Jones, B.; Shak, S. et al. *N. Engl. J. Med.* **2001**, *344*, 783–792.
10. Topol, E. J.; Califf, R. M.; Weisman, H. F. et al. *Lancet* **1994**, *343*, 881–886.
11. Impact-RSV Study Group. *Pediatrics* **1998**, *102*, 531–537.
12. Projan, S. J.; Gill, D.; Zhijian, L.; Herrmann, S. *Exp. Opin. Biol. Ther.* **2004**, *4*, 1345–1350.
13. Reichert, J.; Pavlou, A. *Nat. Rev. Drug. Disc.* **2004**, *3*, 383–384.
14. Smith, G. P. *Science* **1985**, *14*, 1315–1317.
15. Azzazy, H.; Highsmith, W. *Clin. Biochem.* **2002**, *35*, 425–445.
16. Osbourn, J.; Jermutus, L.; Duncan, A. *Drug Disc. Today* **2003**, *8*, 845–851.
17. McCafferty, J.; Griffiths, A.; Winter, G.; Chiswell, D. *Nature* **1990**, *348*, 552–554.
18. Groves, M.; Osbourn, J. *Exp. Opin. Biol. Ther.* **2005**, *5*, 125–135.
19. Zahnd, C.; Spinelli, S.; Luginbuhl, B.; Amstutz, P.; Cambillus, C.; Pluckthun, A. *J. Biol. Chem.* **2004**, *279*, 18870–18877.

20. Hanes, J.; Pluckthun, A. *Proc. Natl. Acad. Sci.* **1997**, *94*, 4937–4942.
21. Hanes, J.; Jermutus, L.; Weber-Bornhauser, S.; Bosshard, H.; Pluckthun, A. *Proc. Natl. Acad. Sci.* **1998**, *95*, 14130–14135.
22. Schellekens, H. *Clin. Ther.* **2002**, *24*, 1720–1740.
23. Weber, R. W. *Curr. Opin. Allergy Clin. Immunol.* **2004**, *4*, 277–283.
24. Gottlieb, A. B.; Evans, R.; Li, S.; Dooley, L. T.; Guzzo, C. A.; Baker, D.; Bala, M.; Marano, C. W.; Menter, A. *J. Am. Acad. Dermatol.* **2004**, *51*, 534–542.
25. Green, L. *J. Immunol. Methods* **1999**, *231*, 11–23.
26. Kellermann, S.-A.; Green, L. *Curr. Opin. Biotechnol.* **2002**, *13*, 593–597.
27. Dunn, A.; Kooyman, D.; Pinkert, K. *Drug Disc. Today* **2005**, *10*, 757–767.
28. Mullis, K.; Faloona, F.; Scharf, S.; Saiki, R.; Horn, G.; Erlich, H. *Cold Spring Harb. Symp. Quant. Biol.* **1986**, *51*, 263–273.
29. Stemmer, W. *Proc. Natl. Acad. Sci.* **1994**, *91*, 10747–10751.
30. Hall, B. *FEMS Microbiol. Lett.* **1999**, *178*, 1–6.
31. Chapman, T. M.; Noble, S.; Goa, K. L. *Drugs* **2002**, *62*, 1945–1981.
32. Kurtzhals, P. *Int. J. Obes. Rel. Metab. Disord.* **2004**, *28*, S23–S28.
33. Lombardino, J. G.; Lowe, J. A., III. *Nat. Rev. Drug Disc.* **2004**, *3*, 853–862.
34. Keler, T.; He, L.; Graziano, R. F. *Curr. Opin. Mol. Ther.* **2005**, 7, 157–163.
35. Stern, M.; Herrmann, R. *Crit. Rev. Oncol. Hematol.* **2005**, *54*, 11–29.
36. Camilleri, M.; Dubois, D.; Coulie, B. et al. *Clin. Gastroenterol. Hepatol.* **2005**, *3*, 543–552.
37. Henke, C. J.; Levin, T. R.; Henning, J. M.; Potter, L. P. *Am. J. Gastroenterol.* **2000**, *95*, 788–792.
38. American Gastroenterologists Association. http://www.gastro.org/clinicalRes/pdf/burden-report.pdf (accessed April 2006).
39. Gilmache, J. P.; Letessier, E.; Scarpignato, C. *Br. Med. J.* **1998**, *316*, 1720.
40. Sachs, G.; Collier, R. H.; Shoemaker, R. L.; Hirschowitz, B. I. *Biochim. Biophys. Acta* **1968**, *162*, 210–219.
41. Berkowitz, B. A.; Sachs, G. *Mol. Interv.* **2002**, *2*, 6–11.
42. Horn, J. *Clin. Ther.* **2000**, *22*, 266–280.
43. Rang, H.; Dale, M.; Ritter, J. *Pharmacology*, 3rd ed. Churchill Livingstone: Edinburgh, 1995.
44. Galbraith, R. A.; Michnovicz, J. J. *N. Engl. J. Med.* **1989**, *321*, 269–274.
45. Galbraith, R. A.; Michnovicz, J. J. *Pharmacology* **1993**, *47*, 8–12.
46. Garcia Rodriguez, L. A.; Jick, H. *Br. Med. J.* **1994**, *308*, 503–506.
47. Myrvold, H. E.; Lundell, L.; Miettinen, P. et al. *Gut.* **2001**, *49*, 488–494.
48. Symmons, D.; Turner, G.; Webb, R. et al. *Rheumatol. (Oxf.)* **2002**, *41*, 793–800.
49. Lee, D. M.; Weinblatt, M. E. *Lancet* **2001**, *358*, 903–911.
50. Rat, A. C.; Boissier, M. C. *Joint Bone Spine* **2004**, *71*, 518–524.
51. Johnston, A.; Gudjonsson, J. E.; Sigmundsdottir, H.; Ludviksson, B. R.; Valdimarsson, H. *Clin. Immunol.* **2005**, *114*, 154–163.
52. Morabito, L.; Montesinos, M. C.; Schreibman, D. M. et al. *J. Clin. Invest.* **1998**, *101*, 295–300.
53. Cronstein, B. N.; Naime, D.; Ostad, E. *J. Clin. Invest.* **1993**, *92*, 2675–2682.
54. Gardner, G.; Furst, D. E. *Drugs Aging* **1995**, 7, 420–437.
55. Alarcon, G. S.; Tracy, I. C.; Blackburn, W. D. *Arthritis Rheum.* **1989**, *32*, 671–676.
56. McKendry, R. J.; Dale, P. *J. Rheumatol.* **1993**, *20*, 1850–1856.
57. Felson, D. T.; Anderson, J. J.; Meenan, R. F. *Arthritis Rheum.* **1990**, *33*, 1449–1461.
58. Kremer, J. M.; Phelps, C. T. *Arthritis Rheum.* **1992**, *35*, 138–145.
59. Weinblatt, M. E.; Maier, A. L.; Fraser, P. A.; Coblyn, J. S. *J. Rheumatol.* **1998**, *25*, 238–242.
60. Steiner, G.; Tohidast-Akrad, M.; Witzmann, G. et al. *Rheumatol. (Oxf.)* **1999**, *38*, 202–213.
61. Brennan, F.; Maini, R.; Feldmann, M. *Br. J. Rheumatol.* **1992**, *31*, 293–298.
62. Moreland, L. W.; Schiff, M. H.; Baumgartner, S. W. et al. *Ann. Intern. Med.* **1999**, *16*, 478–486.
63. Maini, R. N.; Breedveld, F. C.; Kalden, J. R. et al. *Arthritis Rheum.* **1998**, *41*, 1552–1563.
64. van de Putte, L. B.; Atkins, C.; Malaise, M. et al. *Ann. Rheum. Dis.* **2004**, *63*, 508–516.
65. den Broeder, A. A.; Wanten, G. J.; Oyen, W. J.; Naber, T.; van Riel, P. L.; Barrera, P. *J. Rheumatol.* **2003**, *30*, 232–237.
66. Weisman, M. H.; Moreland, L. W.; Furst, D. E.; Weinblatt, M. E.; Keystone, E. C.; Paulus, H. E.; Teoh, L. S.; Velagapudi, R. B.; Noertersheuser, P. A.; Granneman, G. R. et al. *Clin. Ther.* **2003**, *25*, 1700–1721.
67. Gardam, M. A.; Keystone, E. C.; Menzies, R. et al. *Lancet Infect. Dis.* **2003**, *3*, 148–155.
68. Hochberg, M. C.; Lebwohl, M. G.; Plevy, S. E.; Hobbs, K. F.; Yocum, D. E. *Semin. Arthritis Rheum.* **2005**, *34*, 819–836.
69. Furst, D. E.; Schiff, M. H.; Fleischmann, R. M. *J. Rheumatol.* **2003**, *30*, 2563–2571.
70. Anderson, P. *Semin. Arthritis Rheum.* **2005**, *34*, 19–22.
71. Carr, R.; Congreve, M.; Murray, C.; Rees, D. *Drug Disc. Today* **2005**, *10*, 987–992.
72. Wenlock, M. C.; Austin, R. P.; Barton, P.; Davis, A. M.; Leeson, P. D. *J. Med. Chem.* **2003**, *46*, 1250–1256.
73. Erlanson, D. A.; McDowell, R. S.; O'Brien, T. *J. Med. Chem.* **2004**, *43*, 463–482.
74. Rees, D. C.; Congreve, M.; Murray, C. W.; Carr, R. *Nat. Rev. Drug Disc.* **2004**, *3*, 660–672.
75. Hajduk, P. J.; Huth, J. R.; Fesik, S. W. *J. Med. Chem.* **2005**, *48*, 2518–2525.
76. Mountain, V. *Chem. Biol.* **2003**, *10*, 95–98.
77. Grainger, D. *Exp. Opin. Biol. Ther.* **2004**, *4*, 1029–1044.
78. Burek-Kozlowska, A.; Morell, A.; Hunziker, T. *Int. Arch. Allergy Immunol.* **1994**, *104*, 104–106.
79. Zeitlin, L.; Whaley, K. J.; Sanna, P. P. et al. *Virology* **1996**, *225*, 213–215.
80. Poelstra, K. A.; Barekzi, N. A.; Rediske, A. M.; Felts, A. G.; Slunt, J. B.; Grainger, D. W. *J. Biomed. Mater. Res.* **2002**, *60*, 206–215.
81. Barekzi, N. A.; Poelstra, K. A.; Felts, A. G.; Rojas, I. A.; Slunt, J. B.; Grainger, D. W. *Antimicrob. Agents Chemother.* **1999**, *43*, 1609–1615.
82. Barekzi, N. A.; Felts, A. G.; Poelstra, K. A.; Slunt, J. B.; Grainger, D. W. *Pharm. Res.* **2002**, *19*, 1801–1807.
83. Graham, B. S.; Tang, Y. W.; Gruber, W. C. *J. Infect. Dis.* **1995**, *171*, 1468–1474.
84. Lang, A. B.; Cryz, S. J., Jr.; Schurch, U.; Ganss, M. T.; Bruderer, U. *J. Immunol.* **1993**, *151*, 466–472.
85. Ramisse, F.; Deramoudt, F. X.; Szatanik, M. et al. *Clin. Exp. Immunol.* **1998**, *111*, 583–587.
86. Birch, J. R.; Onakunle, Y. *Methods Mol. Biol.* **2005**, *308*, 1–16.
87. Casadevall, A.; Dadachova, E.; Pirofski, L. A. *Nat. Rev. Microbiol.* **2004**, *2*, 695–703.

88. Torre-Amione, G., *Am. J. Cardiol.* **2005**, 6, 3C–8C, discussion 38C–40C.
89. Drews, J. *Drug Disc. Today* **2001**, *6*, 21–26.
90. Drews, J. *Drug Disc. Today* **2003**, *8*, 411–420.
91. Salfeld, J. *Best Pract. Res. Clin. Rheum.* **2004**, *18*, 81–95.
92. Casadevall, N.; Rossert, J. *Best Pract. Res. Clin. Haematol.* **2004**, *18*, 381–387.
93. Desai, J.; Demetri, G. D. *Best Pract. Res. Clin. Haematol.* **2005**, *18*, 389–406.
94. Ludwig, H.; Drews, J. *Oncologist* **2004**, *9*, 48–54.
95. Chong, G.; Cunningham, D. *Eur. J. Surg. Oncol.* **2005**, *31*, 453–460.
96. Lambert, J. M. *Curr. Opin. Pharmacol.* **2005**, *5*, 1–7.
97. Damle, N. K.; Frost, P. *Curr. Opin. Pharmacol.* **2003**, *3*, 386–390.
98. Stebbins, W. G.; Lebwohl, M. G. *Dermatol. Ther.* **2004**, *17*, 432–440.

Biography

Jane K Osbourn obtained a BA in Natural Sciences (Biochemistry) from the University of Cambridge. Following this she went on to complete a PhD at the John Innes Centre for Plant Science Research in Norwich, UK, focusing on plant responses to viral infections. Jane then moved into medical research, taking a British Heart Foundation Post Doctoral Fellowship to study the regulation of gene expression in smooth muscle cells at the Department of Medicine at Addenbrooke's Hospital in Cambridge. In 1993 she moved to Cambridge Antibody Technology (CAT) where she has held a number of positions involved in product and technology development. Her current position is Vice President of Biologics Discovery within the Drug Discovery department.

Comprehensive Medicinal Chemistry II
ISBN (set): 0-08-044513-6

ISBN (Volume 1) 0-08-044514-4; pp. 431–447

1.11 Nutraceuticals

P A Lachance and Y T Das, Rutgers – The State University of New Jersey, New Brunswick, NJ, USA

1.11.1 Introduction

Modern drug discovery is based on unraveling the biochemical mechanism of abnormality (disease) and foresightedly designing chemicals that can counteract the abnormality. This is in contrast with the ancient practice of medicine, wherein crude 'medicine' or 'poison' (Greek *pharmakon*) was employed to cure a disease in an apparently unexplainable but ingenious manner. Diseases other than those due to infections and genetic disorders are usually attributable to either a deficiency or imbalance of bioactive chemical, usually a nutrient. The challenge is then to ascertain the 'disease–nutrient' interrelationship and develop a nutritional regimen that can reverse the course of the disease. Nutraceuticals (*nutr*itive + pharm*aceuticals*) hold a key position in this approach and offer great potential for new drugs.

Lachance[1] defined nutraceuticals as naturally occurring (usually botanical) bioactive compounds that have health benefits. This definition is clearly more acceptable than an earlier definition by DeFelice[2] that accommodates nutraceuticals with claims of prevention and treatment of disease. In Japan, the Nutrition Improvement Law Enforcement Regulations (Ministerial Ordinance no. 41, July 1991), a definition for functional foods was established as Foods for Specified Health Use (FOSHU).[3] In Europe, phytopharmaceuticals were defined as drugs whose active constituents are exclusively plant-based (containing plant parts, extracts, plant juices, or distillates) used in rational phytotherapy, excluding isolated plant constituents (e.g., morphine, digitoxin, quinone, etc.). Phytopharmaceuticals were distinguished from herbal drugs, which are plants or plant parts that have been conserved for storage by drying, including products without organ structure (such as gums, fatty oils, and volatile oils).[4] By now it will be obvious that the initial definition of nutraceuticals, as derived only from food plants, has been widened to include bioactive compounds from nonfood sources. Thus, nutraceuticals may be consumed in any one form, or combination of several forms, as categorized in **Figure 1**. In contrast, pharmaceuticals (drugs) are usually synthesized pure chemicals, or isolated as pure chemicals from natural sources. Some examples of the latter are Taxol (anticancer drug from *Taxus breviofolia*), aspirin (analgesic drug based on salicin and salicylic acid from *Salix alba* and *Filipendula* spp.), and quinine

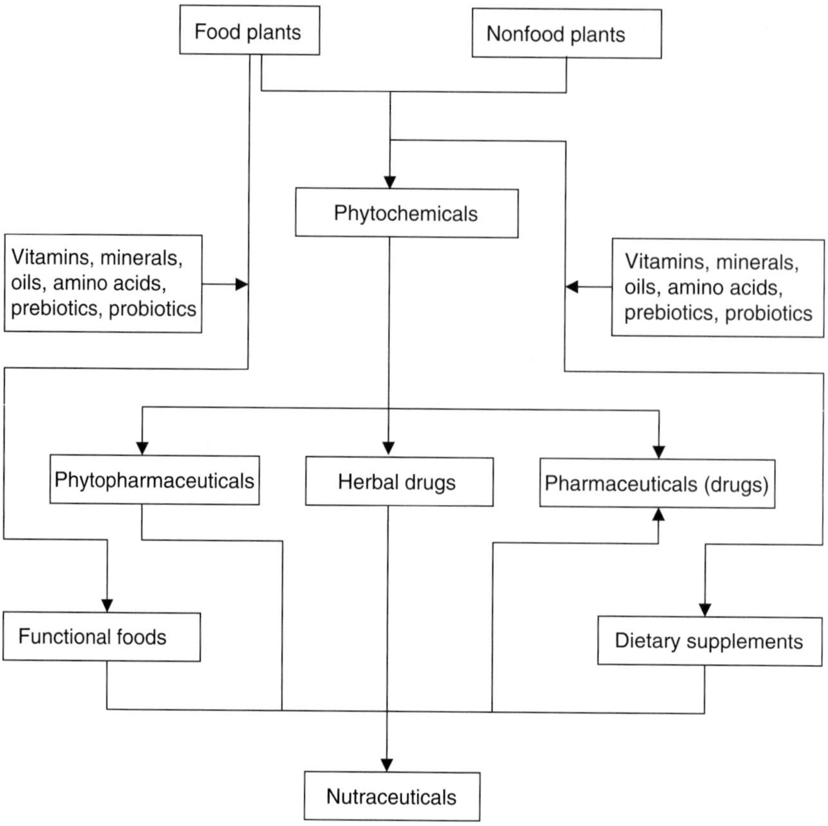

Figure 1 Schematic diagram representing the origin of nutraceuticals. See text for individual descriptions.

(antimalarial drug from *Cinchona officinalis*). These are usually the result of extensive research on their biological activity and clinically proven ability to prevent or cure disease. Synthetic drugs are usually associated with undesirable side effects, some of which are severe and serious, while adverse side effects from nutraceuticals are usually minimal.

1.11.2 Chemical Classes of Nutraceuticals

Guhr and Lachance[5] classified health-promoting phytochemicals into nine classes (**Table 1**). In turn, Della Penna[6] provided an estimate of the number of compounds in four of the classes as: 4000 mono- and polyphenolics, 700 carotenoids, 200 phytoestrogens, and 100 glucosinolates. It is impossible to cover all these compounds in one chapter. We have, therefore, elected to illustrate some information about the mode of action and clinical activity of two classes of nutraceuticals – polyphenolics and carotenoids. Not included in this chapter are certain other nutraceutical compounds classified by Wildman[7] as carbohydrates, fatty acids, amino acids, minerals, prebiotics, and probiotics.

1.11.2.1 Phenolic Compounds

There are an estimated 4000 phenolic and polyphenolic compounds in plants, including benzoic acid, vanillic acid, and gallic acid. The hierarchy of the polyphenolic compounds according to their Trolox equivalent antioxidant capacity (TEAC) values is presented in **Table 2**. The role of the number of free phenolic OH groups (discounting the sugar moiety) governing the antioxidant activity is clear. "Besides the 2,3 double bond, other conditions, such as 3–OH, 4–oxo functions and the o-dihydroxy structure in the B ring, are required for the higher antioxidant activity."[8] The ingestion of mono- and polyphenols in a typical day is estimated to be greater than that of vitamin C, vitamin E, and carotenoids combined.

1.11.2.1.1 Bioavailability of phenolic compounds

In a review of 97 studies on bioavailability of polyphenols, Manach *et al.*[9] reported that gallic acid and isoflavones are the best-absorbed polyphenols. These are followed by catechins, flavanones, and quercetin glucosides, but with different

Table 1 List of some phytochemical groups, their sources, and organoleptic/aesthetic or nutritive properties

Phytochemical group	*Some phytochemicals in the phytochemical group*	*Sources*	*Organoleptic/aesthetic or nutritive properties*
1. Phenols	Simple phenols, phenolic acids, hydroxy cinnamic acid derivatives, flavonoids	Almost all fresh fruits and vegetables, cereal grains, tea (black and green), nuts	Flavor, color, or aroma
2. Indoles	Indole-3-carbinol, indole-3-acetonitrile, L-tryptophan	Cruciferous vegetables (including brussel sprouts, kale, cabbage, broccoli, cauliflower, spinach, watercress, turnip, radish)	Pungent flavor
3. Isothiocyanates	Phenethyl isothiocyanate, benzyl isothiocyanate, sulforaphane	Cruciferous vegetables (including brussel sprouts, kale, cabbage, broccoli, cauliflower, spinach, watercress, turnip, radish)	Pungent flavor
4. Allylic sulfur compounds	Diallyl sulfide, diallyl disulfide, *S*-allyl propyl disulfide, ajoene	*Allium* vegetables (including garlic, onion, leek, shallot, chive, scallion)	Flavor
5. Monoterpenes	D-limoneone, D-carvone	Citrus oils, vegetable oils	Flavor, aroma
6. Monoterpene-like	Perryl alcohol	Cherries	Flavor
7. Carotenoids	α-carotene, β-carotene, α-cryptoxanthin, lutein, lycopene, zeaxanthin	Most red to yellow fruits and vegetables	Color
8. Antioxidant vitamins	Vitamin C, vitamin E	Fruits and vegetables, whole cereal grains	Nutrient
9. Antioxidant mineral	Selenium	Garlic, whole cereal grains	Nutrient

Reproduced with permission from Guhr, G.; Lachance, P. A. Role of Phytochemicals in Chronic Disease Prevention. In *Nutraceuticals, Designer Foods III, Garlic, Soy and Licorice*; Lachance, P. A., Ed.; Food and Nutrition Press: Trumbull, CT, **1997**, pp 311–364 © Elsevier.

kinetics. Relatively less absorbed are the proanthocyanins, the galloylated tea catechins, and the anthocyanins. **Table 3** is a compilation of pharmacokinetic data for 97 studies. Only 18 polyphenols could be compared. Some polyphenols, such as epigallocatechin-3-gallate and genistein, are excreted in bile and, therefore, absorption was underestimated.

Polyphenols have been classified into extractable and nonextractable.[10] Nonextractable polyphenols have high molecular weights and are insoluble in normal solvents. The oxidation and polymerization of soluble polyphenols lower their antioxidant capacity. In food systems, this occurs with enzymatic browning and probably accounts for the difference in the antioxidant capacity of green and black teas.

One of the most documented polyphenolics is the catechins of tea. The dried leaves from *Camellia sinensis* contribute to the most common beverage in the world, second only to water. It has been estimated that up to 30% of the dry leaf weight of green tea consists of polyphenols.[11] The major polyphenols in green tea are epicatechin, epicatechin-gallate, epigallocatechin and epigallocatechin-3-gallate. The levels of catechins vary with the age of the tea and method of processing.

1.11.2.1.2 Plausible mechanisms of action of polyphenolics

Polyphenols have been widely recognized for their free radical scavenging and metal-chelating abilities.[10,12] Oxidative stress is considered a fundamental phenomenon in a plethora of human diseases.[13] Conditions disturbing the 'balance' of higher oxidation and inflammation conditions are believed to promote cell proliferation, such as in cancer and malfunctions in heart disease, diabetes, autoimmune diseases, as well as in mental disorders. The excessive presence of reactive oxygen species is believed to modulate gene expression via signaling pathways and to lead to DNA damage via mutations. Selected phytochemicals are able to alter human cellular signaling and gene expression.[14] Thus dietary nutraceuticals function not only as individual attributes, such as antioxidants, but can also act synergistically with other nutraceuticals, such as omega-3 fatty acids in cell function and integrity. Most nutraceuticals are not recognized as nutrients, yet may function synergistically with nutrients. Lachance[15] believes that certain nutrients such as ascorbic acid, vitamin E, and beta-carotene, as well as selenium, should be recognized as nutraceuticals, since they function as antioxidants synergistically with non-nutrients such as lycopene, lutein, omega-3 fatty acids, and many flavonoids.

Table 2 Hierarchy of trolox equivalent antioxidant capacity of flavonoids and related phenoloc compounds

Compound	*The ring structure*	*OH and substituents*			*Glycosylated position*	*TEAC*	
		Position	*No. of OH*	*1*		*Obsed.*	*Calcd.*
Flavanol							
1. Epicatechin gallate		5,7,3′,4′,3″,4″,5″	7	2		4.90	3.69
2. Epigallocatechin gallate (EGCG)	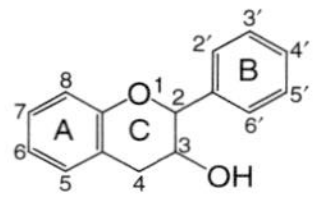	5,7,3′,4′,5′,3″,4″,5″	8	2		4.80	4.15
3. Epigallocatechin (EGC)		3,5,7,3′,4′,5′	6	2		3.80	3.23
4. Epicatechin (EC)		3,5,7,3′,4′	5	2		2.50	2.77
5. Taxifolin		3,5,7,3′,4′	5	2		1.90	2.77
6. Catechin		3,5,7,3′,4′	5	2		2.40	2.77
Aflavins							
7. Theaflavine		R = − OH, R′ = − OH	9	1		2.94	(4.29)
8. Theaflavin-3-monogallate	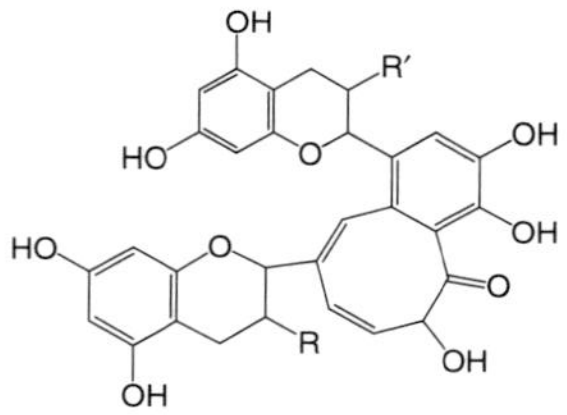	R = gallate, R′ = − OH	11	1		4.65	5.21
9. Theaflavin-3′-monogallate		R = − OH, R′ = gallate	11	1		4.78	5.21
10. Theaflavin-3,3′-digallate		R = gallate, R′ = gallate	13	1		6.18	6.12
Flavonol							
11. Quercetin		3,5,7,3′,4	5	3		4.70	(3.09)
12. Myricetin	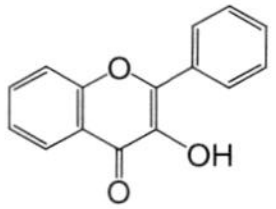	3,5,7,3′,4′,5′	6	3		3.10	3.55
13. Morin		3,5,7,2′,4′	5	2		2.55	2.77
14. Rutin		5,7,3′,4′	4	3	3-Rutinose	2.40	2.63
15. Kaempferol		3,5,7,4′	4	2		1.34	2.31
Flavone							
16. Luteolin		5,7,3′,4′	4	3		2.10	2.63
17. Luteolin-4′-glucoside	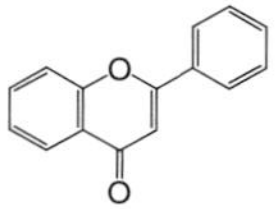	5,7,3′	3	2	4′-Glucose	1.74	1.85
18. Apigenin		5,7,4′	3	2		1.45	1.85
19. Chrysin		5,7	2	2		1.43	1.39
20. Luteolin-3′, 7-diglucoside		5,4′	2	1	3′,7-Diglucose	0.79	1.07

Flavanone							
21. Naringenin	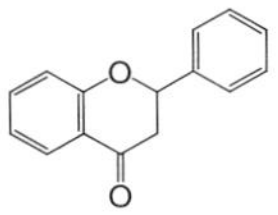	5,7,4′	3	1		1.53	1.53
22. Hesperetin		5,7,3′	3	1	4′-OCH_3	1.37	1.53
23. Hesperidin		5,3′	2	0	4′-OCH_3	1.08	1.21
24. Narirutin		5,4′	2	1	7-Rutinose	0.76	0.75
25. Dihydrokaempferol		3,5,7,3′,4′	4	1	7-Rutinose	1.39	1.99
26. Eriodictycol		3,5,7,4′	4	2		1.80	2.31
Anthocyanidin							
27. Delphinidin	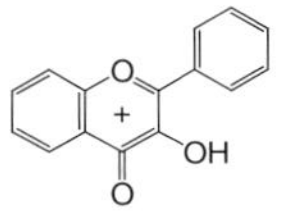	3,5,7,3′,4′,5′	6	2		4.44	3.23
28. Cyanidin		3,5,7,3′,4′	5	2		4.40	(2.77)
29. Apigenidin		5,7,4′	3	1		2.35	1.53
30. Peonidin		3,5,7,4′	4	1	3′-OCH_3	2.22	1.99
31. Malvidin		3,5,7,4′	4	1	3′,5′-di-OCH_3	2.06	1.99
32. Pelargonidin		3,5,7,4′	4	1		1.30	1.99
33. Keracyanin		5,7,3′,4′	4	2	3-Rutinose	3.25	2.31
34. Ideain		5,7,3′,4′	4	2	3-Galactose	2.90	2.31
35. Oenin		5,7,4′	3	1	3′,5′-di-OCH_3	1.78	1.53
Isoflavone					3-Glucose		
36. Genistein	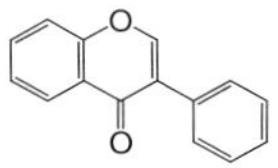	5,7,4′	3	2		2.90	1.85
37. Genistin		5,4′	2	1	7-Glucose	1.24	1.07
38. Daidzein		7,4′	2	1		1.25	1.07
39. Daidzin		4′	1	1	7-Glucose	1.15	0.61
40. Biochamin A		5,7	2	2	4′-OCH_3	1.16	1.39
41. Formononetin		7	1	1	4′-OCH_3	0.11	0.61
42. Ononin			0	1	7-Glucose; 4′-OCH_3	0.05	0.15

Reproduced from Lien *et al.*[8]

Table 3 Compilation of pharmacokinetic data from 97 bioavailability studies

	T_{max} (h)		C_{max} ($\mu mol\,L^{-1}$)		AUC ($\mu mol\,h^{-1}\,L^{-1}$)		Urinary excretion (% of intake)		Elimination half-life (h)	
	Mean	*Range*	*Mean*	*Range*	*Mean*	*Range*	*Mean*	*Range*	*Mean*	*Range*
Daidzin	6.3±0.6	4.0–9.0	1.92±0.25	0.36–3.14	21.4±6.5	2.7–38.6	42.3±3.0	21.4–62.0	5.3±0.8	3.4–8.0
Daidzein	4.9±1.0	3.0–6.6	1.57±0.52	0.76–3.00	12.2±2.9	7.5–17.4	27.5		8.5±0.8	7.7–9.3
Genistin	6.5±0.6	4.4–9.3	1.84±0.27	0.46–4.04	23.7±6.7	6.2–45.1	15.6±1.8	6.8–29.7	7.8±0.7	5.7–10.1
Genistein	4.1±0.6	3.0–5.2	2.56±1.00	1.26–4.50	19.8±6.5	10.4–32.2	8.6		7.1±0.3	6.8–7.5
Glycitin	5.0		1.88±0.38	1.50–2.26	7.9		42.9±12.0	19.0–55.3	8.9	
Hesperidin	5.5±0.1	5.4–5.8	0.46±0.21	0.21–0.87	2.7±0.7	1.9–4.1	8.6±4.0	3–24.4	2.2	
Naringin	5.0±0.2	4.6–5.5	0.50±0.33	0.13–1.50	3.7±1.5	0.9–7.0	8.8±3.17	1.1–30.2	2.1±0.4	1.3–2.7
Quercetin glucosides	1.1±0.3	0.5–2.9	1.46±0.45	0.51–3.80	9.8±1.9	5.7–16.0	2.5±1.2	0.31–6.4	17.9±2.2	10.9–28.0
Rutin	6.5±0.7	4.3–9.3	0.20±0.06	0.09–0.52	2.9±0.9	1.6–5.5	0.7±0.3	0.07–1.0	19.9±8.1	11.8–28.1
(Epi)catechin	1.8±0.1	0.5–2.5	0.40±0.09	0.09–1.10	1.1±0.3	0.5–2.0	18.5±5.7	2.1–55.0	2.5±0.4	1.1–4.1
EGC	1.4±0.1	0.5–2.0	1.10±0.40	0.30–2.70	2.0±0.8	1.0–3.6	11.1±3.5	4.2–15.6	2.3±0.2	1.7–2.8
EGCG	2.3±0.2	1.6–3.2	0.12±0.03	0.03–0.38	0.5±0.1	0.2–0.9	0.06±0.03	0.0–0.1	3.5±0.3	2.5–5.1
Gallic acid	1.6±0.2	1.3–1.5	4.00±0.57	2.57–4.70			37.7±1.0	36.4–39.6	1.3±0.1	1.1–1.5
Chlorogenic acid	1.0		0.26				0.3			
Caffeic acid	1.4±0.6	0.7–2.0	0.96±0.26	0.45–1.35			10.7			
Ferulic acid	2.0		0.03				27.6±17.6	3.1–61.7		
Anthocyanins	1.5±0.4	0.7–4.0	0.03±0.02	0.001–0.20			0.4±0.3	0.004–5.1		
Proanthocyanidin dimers	2.0		0.02±0.01	0.008–0.03						

All data were converted to correspond to a supply of 50 mg aglycone equivalent. T_{max} time to reach C_{max}; AUC, area under the plasma concentration–time curve; EGC, epigallocatechin. From Manach *et al.*[9]

Various clinical studies with whole fruits and vegetables have shown short-term benefits in the oxidation of DNA.[16] Esterbauer and Ramos[17] conducted oxidation of low-density lipoprotein experiments in vitro and demonstrated that the protective function of antioxidants occurs in a sequence from ascorbic acid to tocopherols to carotenoids. However, the interactions of nutritive and non-nutritive, hydrophilic and hydrophobic compounds with the various polyphenols have only been recently studied.[18]

A Mayo Clinic phase II trial found that green tea consumption was able to decrease prostate-specific antigen values, while the placebo group experienced a 48% rise in prostate-specific antigen values.[19] Black (fermented) tea also has beneficial effects, albeit at a lower intensity. The implications of polyphenols in several maladies have been reviewed by Lachance *et al.*[20] and Zern and Fernandez.[21]

1.11.2.2 Carotenoids

Without carotenoids, photosynthesis and all life in an oxygen atmosphere would be impossible.[22] Photosynthesis makes possible the synthesis of phytochemicals such as flavonoids and carotenoids.[13] The polyisoprenoid lycopene, modified in a variety of ways during photosynthesis, leads to alpha- and beta-carotenes which in turn lead to lutein and zeaxanthin and the other natural carotenoids. The synthesis occurs in plants and some microorganisms, algae, and fungi, totaling over 600 carotenoids. Carotenoids interact structurally with lipid membranes and proteins.[23] "All carotenoids react rapidly with oxidizing agents and free radicals."[22] The overall shape, size, and hydrophobicity of a carotenoid are features that affect the fit into subcellular structures. For example, carotenoids can bind to the protein tubulin in the retina and thus affect visual acuity.[23]

Although beta-carotene was initially incriminated in lung cancer in smokers, it was later questioned, based on more definitive analyses.[24,25] In the Alpha Tocopherol and Beta Carotene (ATBC) study, 29 133 male smokers supplemented for 5–8 years with 50 mg day^{-1} alpha-tocopherol or 20 mg day^{-1} beta-carotene or both had an 18% increase in lung cancer compared with placebo (confidence interval 3–36%) and an 8% increase in total mortality, but this is not observed in smokers of <20 cigarettes per day. The lungs were vulnerable to impending cancer (premalignant lesions) due to heavy smoking. No other sites of cancer in the body were seen. More importantly, lung cancer only occurs when beta-carotene subjects show evidence of heavy alcohol consumption (compromised liver enzymes and evidence of liver pathology). When the data were corrected for the presence of liver disease, no increase in lung cancer occurred that could be attributed to beta-carotene. This fact was not reported to the public and, therefore, the original finding remained unchallenged in the news media.

In the Carotene and Retinol Efficacy Trial (CARET), 18 314 smokers, ex-smokers, and asbestos workers were supplemented with either a combination of 50 mg day^{-1} beta-carotene and 25 000 IU day^{-1} of retinol in comparison to placebo and were thus exposed to no less than 4 years of the toxic level of retinol (no less than 37 500 000 IU), creating liver toxicity, further aggravated by the possible conversion of 30% of the beta-carotene to vitamin A. Cancer was not seen in any other organs, nor was it observed in former smokers, but lung cancer was aggravated by heavy alcohol consumption. It may be recalled here that liver pathology is identical in alcohol and vitamin A toxicity. In the light of the severe liver toxicity caused by vitamin A, beta-carotene per se could not be incriminated as the cause of the lung cancer incidence, and the study was brought to an abrupt close.

1.11.2.3 Applicable Metabolic Principles and Bioavailability

The principles of absorption, distribution, metabolism, and excretion (ADME) serve to guide the reporting of the applicable facts. The primary route of administration for nutraceuticals is oral, whether the nutraceutical is present by itself (e.g., polyphenols in tea) or present as an additive to a food ingredient (e.g., sitosterol added to margarine). Although the oral route of administration poses a potential problem of bioavailability and metabolism in the liver, it appears that most of the nutraceuticals that are popular in today's market are effective enough through oral administration.

Reasonable ADME data are available for only three categories of phenolic nutraceuticals: (1) catechins in tea; (2) the antioxidants in spices; and (3) the isoflavones (phytoestrogens) in soy products. Obviously, more research is needed in this area.

1.11.3 Search for New Drugs of Plant Origin

1.11.3.1 Inference from Historical Usage

Many more plant sources have been utilized for their medicinal properties in traditional Chinese medicine (TCM) and ayurvedic medicines than are cultivated for food use.[5] It is estimated that approximately 1% of the roughly

300 000 species of higher plants that now exist have a history of food use. However, the use of 10–15% is documented in 'traditional' medicine. Ayurvedic medicine uses 2000–3000 plant species, while TCM uses 5000–6000 plant species. In contrast, allopathic (western) medicine draws upon approximately 500 plant species. In other words, there is available to the consumer considerably more 'functional' products predominantly ingested by mouth containing bioactive nutraceutical attributes than are actually used in clinical practice in western medicine. Thus, nutraceuticals can be viewed as a viable source of bioactive compounds that have disease-preventive and/or medicinal benefits, and they can be explored from either a regulatory viewpoint as a 'drug' or, at lower cost, as a nutraceutical and/or functional food.

1.11.3.1.1 Ayurvedic remedies

Ayurveda (Sanskrit *ayur*, life + *veda*, knowledge) is based on the principle that individual life is a microcosm of the cosmos. Ayurvedic medicine dates back more than 5000 years, and is practiced in at least eight major clinical specialties that are still the same in modern medicine. It also has certain warnings about the incompatibility of various food items. Although the last ayurvedic medicine degree to be conferred in the US was in 1959, certain states still provide for licensed practice in ayurvedic medicine. Certain assessment procedures are used to understand the pathogenesis and etiologic factors of the individual. There are six progressive stages of disease treatment options. Ayurveda has a pharmacopoeia that uses only whole herbs. Active ingredients are not separated from the whole plant or plant part. Therefore, standardization based on any active ingredient or a marker is difficult.

1.11.3.1.2 Traditional Chinese medicine

TCM is believed to have been in existence for the past 3000 years. It is based on the doctrine that the human body is comprised of five elements (metal, wood, water, fire, and earth) and that the relative arrangement of the body organs/tissues are as yin and yang. The treatment is based on the principle of differentiation of symptoms. The number of herbals in use today in TCM is estimated to be about 5700. Because of this high number, standardization is limited. However, collections are being maintained and preserved. The fact that animal tissues are used in some Chinese medicinal products may limit the use of TCM in individuals with certain food restrictions.

1.11.3.1.3 Tibetan medicine

Tibetan medicine was believed to have been taught in India by Lord Buddha himself when he was 71 years old (*c*. 552 BC). It was documented in 156 chapters and divided into four subjects: (1) explanation of disease; (2) anatomy and physiology, birth and death; (3) cause and nature of disease, disease classification, and treatment; and (4) methods of diagnosis, pharmacology, and external treatments. The Tibetan practice of medicine is for the benefit of others, and forbids doctors from personal gain. The Tibetan Medical and Astrological Institute of His Holiness the Dalai Lama was reestablished in exile in India in 1961. The medicines prescribed are often blends of crude (raw/dried) plant materials, usually from Himalayan mountain regions. Several bioactive ingredients are believed to work together to remedy signs and symptoms of the presenting condition. Adverse effects are usually minimal. By comparison with ayurvedic and Chinese medicine, Tibetan medicine is relatively recent. It is, therefore, unclear how much of it was novel at the time.

1.11.3.2 Isolation and Assays

The most extensive data available on nutraceuticals are for chemical composition, followed by in vitro results for a possible mode of action. There are several reviews available that provide comprehensive coverage and a database on herbs and nutraceuticals.[29–46]

1.11.3.3 Screening Techniques

The practice of human cell line-directed fractionation[26] of food or herbal components in order to concentrate on elucidating the structure and composition of the most active bioactive components is rapidly emerging. The fractionation sequence is guided at each stage by the intensity of the response (tumor attenuation) in the presence of human breast or prostate or colon human cells, and avoids the issues of digestion, absorption, and first-pass metabolism. Subsequently, animal and human trials can be designed and conducted in a more intelligent and cost-effective manner. **Table 4** lists some representative poster presentations at a national scientific meeting in 2005 wherein the research

Table 4 Recent developments in cell line/gene interface research of nutraceuticals

Title	*Author(s)*	*Facility*
Ginger and orange peel extracts inhibit pro-inflammatory genes (COX-2 and iNOS) in mouse macrophage (RAW 264.7) cell line	M. M. Rafi; L. Rakhlin	Rutgers–The State University of New Jersey, New Brunswick, NJ 08901, US
Broccoli sprout extracts inhibit bladder cancer cell proliferation	R. A. Rosselot; S. K. Clinton; S. J. Schwartz; Q. Tian	Ohio State University, Columbus, OH 43210, US
Mechanism of inhibition of platelet aggregation by lenthionine, a flavor component from shiitake mushroom	T. Shibuya; S. Shimada; H. Sakurai; H. Kumagai	Nihon University, Fujisawa-shi, 252-8510, Japan
Phenolics in cherry and their protective effects on oxidative stress in PC 12 cells	D. O. Kim; C. Y. Lee; H. J. Heo; Y. J. Kim; H. S. Yang	Cornell University, Geneva, NY 14456, US
Potential of cranberry-based phytochemical synergies for diabetes and hypertension management	E. Apostolidis; Y. I. Kwon; Y. T. Lin; K. Shetty	University of Massachusetts, Amherst, MA 01003, US
Radical scavenging activity of phenolic compounds isolated from mulberry (*Morus* spp.) fruit	S. Choi; H. K. No; Y. Shin; S. Lee; S. Rhee	Catholic University of Daegu, Gyeongsan, 712-702, South Korea
Protective effect of *Rosa laevigata* against amyloid beta peptide-induced neuronal cell death	S. J. Choi; M. J. Kim; H. J. Heo; H. Y. Cho; B. S. Hong; H. K. Kim; D. H. Shin	Hanseo University, ChungNam, 356-706, South Korea
Inhibitory effects of flavonoids on 3T3-L1 preadipocytes for the mechanisms of proposed antiobesity	C. L. Hsu; G. C. Yen	National Chung Hsing University, Taichung 40227, Taiwan
Induction of apoptosis by anthocyanins in human gastric tumor AGS cells	P. H. Shih; G. C. Yen	National Chung Hsing University, Taichung 40227, Taiwan

Selected reports from Institute of Food Technologists Annual Meeting, New Orleans, LA (July 12–20, 2005).

findings of the response of certain nutraceutical fractions in a cell line model system are reported. These low costs and patentable discoveries are available to the food and pharmaceutical industries for further development as dietary supplements or drugs.

1.11.3.4 Nutragenomics

This is a newly designated area of research, focusing on the understanding of fundamental processes at the molecular level and the effect on genes responsible for disease susceptibility. More importantly, the research focuses on the role of nutraceuticals in counteracting disease susceptibility. It is well known that individual genetic variations are commonplace among people, as evident, for example, that the same food leads to obesity in some yet has no such effect on others. Once a responsible gene has been identified, its presence or absence in an individual can be investigated, and a nutraceutical regimen could be customized for each individual.

1.11.4 Practical Considerations of Nutraceuticals

Based on the extent to which the natural plant is subjected to processing, it is possible that a partial loss of biological activity may occur. Also, alterations of the plant tissue microenvironment and loss of certain micronutrients during processing may cause some loss of complementarity (synergism) between the various ingredients within the plant matrix. Because of this holistic situation, there is a potential for frustration in attempting to isolate and purify active ingredients from plants. The emerging trends in alternative/complementary medicine are likely to provide more understanding and tolerance of the holistic approach of ancestral medicines. It is important to recognize that nutraceuticals in a food delivery system may complement a concomitant drug and serve as a source of physiological and/or enhanced pharmacological actions and interactions. However, a drug is invariably a defined chemical entity (United States Pharmacopeia, USP) with a targeted intent and administered at doses that can evoke serious side effects. The cost of discovery and extensive safety testing and clinical trials for one new drug requires millions of dollars over several years. The testing required is relatively the same (in the US) when petitioning for a new food additive, unless it has already been deemed generally recognized as safe (GRAS).

1.11.4.1 Intelligent Product Delivery System

Security concerns demand the traceability of all herbs and foods, nutraceuticals, and ethical pharmaceuticals, not only for quality control but also to prevent the covert introduction of chemical and microbiological hazards as agents of terrorism or counterfeiting. Of first concern are errors in identity. Problems (some fatal) have arisen with confusion emanating from the use of a common name descriptor rather than the specific descriptor in Latin or a Chemical Abstract Service (CAS) nomenclature. For example, guang fang ji (root of *Aristolochia fangchi*) was mistaken for fang ji (root of *Stephania tetrandra*) in Belgium in 1993. Likewise, guang mu tong (stem of *Aristolochia manshuriensis*) was mistaken for chuan mu tong (stem of *Clematis armandii*) and mu tong (stem of *Akebia* sp.) in the UK in 1999. In both cases, the products were unintentionally made with a toxic contaminant, aristolochic acid (a carcinogen and nephrotoxin).

A further concern is gaps in traceability. Safety through traceability[27] (**Table 5**) can be assured by coupling three existing technologies: (1) global positioning of source and monitoring of lots during transport phases; (2) sophistication of bar-coding from radiofrequency identity to chip coding; and (3) hazard analysis and critical control point (HACCP) management. These may be coupled with the development of rapid nanotechnology-based marker assays.[15]

1.11.4.2 Regulatory Considerations

Natural products chemistry and the synthetic chemistry discovery of the new candidate drugs are not analytically very different from for the discovery of the bioactive fractions and compounds from nutraceutical sources. Since extensive safety testing and clinical trials are not mandatory in the US to meet the regulatory demands under the Dietary Supplement Health and Education Act of 1984, the dietary supplement industry has quickly grown into an estimated $50 billion business. The industry estimates that the number of dietary supplements has jumped from about 4000 in 1994 to about 29 000 by 2004, with an anticipated addition of about 1000 new products every year. The market is driven by consumer concern about how healthcare is managed, administered, and priced. There is frustration with expensive high-technology treatment approaches that are predominant in allopathic (western) medicine. While less willing to tolerate the side effects of synthetic pharmaceuticals, the consumer is generally unaware that the marketing of various prototypical nutraceuticals is vastly ahead of the scientific basis required to assure quality, safety, efficacy, and defined

Table 5 Standardization needs of a quality management plan for nutraceuticals

Quality management *To assure safety and quality, standardization must be established:*
(a) Plant materials identified by Latin name, and chemical entity by Chemical Abstract Service (CAS) number, source, or synthesis location
(b) Global Information System (GIS) descriptors of plant material
(c) Sample verification of plant material systematic botanist or chemical certification
(d) Certification of below tolerence levels of:
organics (e.g., pesticide residues)
heavy metals (e.g., lead, mercury)
microbiological pathogens
(e) Content of bioactive(s)/biomarker(s), if known
(f) Digital documentation of good manufacturing practices (GMPs), compliance with hazard analysis and critical control points (HACCPs), and monitoring of US FDA database for adverse effects

Reproduced from Lachance and Saba.[27]

mode of action. Some adverse health effects of usnic acid, ephedrine alkaloids, and androstenedione that were present in certain dietary supplements have only surfaced after the fact. This situation has created an urgency for increased regulatory oversight and for monitoring the products at various stages of manufacturing for quality and integrity of the products.

Development and implementation of safeguards – such as good manufacturing practice(s), TruLabel, quality certification, product reviews, and quality demonstration/verification – have since been intensified. Unlike the synthetic drugs that are preapproved by the US Food and Drug Administration for treating specific disease(s), the dietary supplements are essentially regulated as foods. There has also been a misunderstanding or misinterpretation that the dietary supplement industry is deregulated. Most of the dietary supplement products are formulated as complex mixtures of a variety of natural ingredients that are not amenable to tests routinely applied to the synthetic drug approval process. Regulatory procedures are, therefore, likely to be different for the dietary supplements but not necessarily less effective than they are for synthetic drugs in assuring public safety. The current regulatory framework has recently been reviewed.[28]

References

1. Lachance, P. A. *The Wall Street Transcript* **2003**, *159*, 132–137.
2. DeFelice, S. L. *Regul. Affairs* **1993**, *5*, 163–167.
3. FOSHU System of Japan: Nutrition Improvement Law No. 248 of 31 July 1952 (amended by Law No. 101 of 24 May 1995); Nutritional Improvement Law Enforcement Regulations, Ministerial Ordinance No. 41 of July 1991.
4. Wichtl, M., Ed. *Herbal Drugs and Phytopharmaceuticals*, 3rd ed.; Medpharm: Stuttgart, Germany, 2004.
5. Guhr, G.; Lachance, P. A. Role of Phytochemicals in Chronic Disease Prevention. In *Nutraceuticals, Designer Foods III, Garlic, Soy and Licorice*; Lachance, P. A., Ed.; Food and Nutrition Press: Trumbull, CT, **1997**, pp 311–364.
6. Della Penna, A. A. *Science* **1999**, *285*, 375–379.
7. Wildman, R. E. C., Ed. *Handbook of Nutraceuticals and Functional Foods*; CRC Press: Boca Raton, FL, 2001.
8. Lien, E. J.; Ren, S.; Bui, H.-H.; Wang, R. *Free Radical Bio. Med.* **1999**, *26*, 285–294.
9. Manach, C.; Williamson, G.; Morand, C.; Scalbert, A.; Rémésy, C. *Am. J. Clin. Nutr.* **2005**, *81*, 230S–242S.
10. Bravo, L. *Nutr. Rev.* **1998**, *56*, 317–332.
11. Graham, H. N. *Prev. Med.* **1992**, *21*, 334–350.
12. Robak, J.; Gryglewski, R. J. *Biochem. Pharmacol.* **1988**, *37*, 387–481.
13. Demming-Adams, B.; Adams, W. *Science* **2002**, *298*, 2149–2153.
14. Allen, R. G.; Tresini, M. *Free Radical Bio. Med.* **2000**, *23*, 463–499.
15. Lachance, P. *Toxicol. Lett.* **2004**, *150*, 25–27.
16. Young, J. F.; Dragstedt, L. O.; Haroldsdottir, J.; Daneshwar, B.; Kal, M. A.; Loft, S.; Nilsson, L.; Nielson, S. E.; Mayer, B.; Skibsted, L. H. et al. *Br. J. Nutr.* **2002**, *87*, 343–355.
17. Esterbauer, H.; Ramos, P. *Rev. Physiol. Biochem. Pharmacol.* **1996**, *127*, 31–64.
18. Wu, X.; Pittman, H. E.; Mckay, S.; Prior, R. L. *J. Nutr.* **2005**, *135*, 2417–2424.
19. Jatoi, A.; Ellison, N.; Burch, P. A.; Sloan, J. A.; Dakhil, S. R.; Novotny, P.; Tan, W.; Fitch, T. R.; Rowland, K. M.; Young, C. Y. et al. *Cancer* **2003**, *97*, 1442–1446.
20. Lachance, P. A.; Nakat, Z.; Jeong, W.-S. *Nutrition* **2001**, *17*, 835–838.
21. Zern, T. L.; Fernandez, M. L. *J. Nutr.* **2005**, *135*, 2291–2294.
22. Britton, G. *FASEB J.* **1995**, *9*, 1551–1558.
23. Krinsky, N. J. *J. Nutr.* **2002**, *132*, 549S.
24. Albanes, D.; Heinonen, O. P.; Taylor, P. R.; Virtamo, J.; Edwards, B. K.; Rautalahti, M.; Hartman, A. M.; Palmgren, J.; Freedman, L. S.; Haapakoski, J. et al. *J. Natl. Cancer Inst.* **1996**, *88*, 1560–1570.
25. Omenn, G. S.; Goodman, G. E.; Thornquist, M. D.; Balmes, J.; Cullen, M. R.; Glass, A.; Keogh, J. P.; Meyskens, F. L.; Valanis, B.; Williams, J. H. et al. *J. Natl. Cancer Inst.* **1996**, *88*, 1550–1559.
26. Ghai, G.; Boyd, C.; Csiszar, K.; Ho, C.-T.; Rosen, R. *Method of Screening Foods for Nutraceuticals.* U.S. Patent 5,955,269, September 21, 1999.
27. Lachance P. A.; Saba, R. Intelligent Product-Delivery Systems and Safety through Traceability. In *Quality Management of Nutraceuticals*; Ho, C.-T, Zheng, Q., Eds.; American Chemical Society: Washington, DC, 2002; series no. 803, pp 2–9.
28. Taylor, C. L. *Nutr. Rev.* **2004**, *2*, 55–59.
29. Barrett, M. L., Ed. *The Handbook of Clinically Tested Herbal Remedies*; Haworth Herbal Press: New York, 2004; Vols 1 and 2.
30. Brendler, T.; Gruenwald, J.; Jaenicke, C.; Eds.; *Herbal Remedies CD-ROM*, 5th ed.; Culinary and Hospitality Industry Publications Services (CHIPS): Weimer, TX, 2003.
31. DerMarderosian, A.; Beutler, J.A., Eds. *The Review of Natural Products*; Facts and Comparisons, St. Louis, MO, 2005.
32. Duke, J. A. *CRC Handbook of Medicinal Spices*; CRC Press: Boca Raton, FL, 2002.
33. Duke, J. A. *Handbook of Edible Weeds*; CRC Press: Boca Raton, FL, 1992.
34. Duke, J. A. *Handbook of Medicinal Herbs*; CRC Press: Boca Raton, FL, 2001.
35. Duke, J. A. *Handbook of Phytochemical Constituents of GRAS Herbs and Other Economic Plants*; CRC Press: Boca Raton, FL, 2001.
36. DerMarderosian, A.; Beutler, J. A., Eds. *A Guide to Popular Natural Products*; Facts and Comparisons: St. Louis, MO, 2003.
37. Jo, J. Y.; Gonzalez de Mejia, E.; Lia, M. A. *J. Agric. Food Chem.* **2005**, *53*, 2489–2498.
38. Kelly, J. P.; Kaufman, D. W.; Kelley, K.; Rosenberg, L.; Anderson, T. E.; Mitchell, A. A. *Arch. Int. Med.* **2005**, *65*, 281–286.
39. Lachance, P. A. *Science* **1996**, *272*, 1860–1861.
40. Liu, C.; Tseng, A.; Yang, S. *Chinese Herbal Medicine: Modern Applications of Traditional Formulas*; CRC Press: Boca Raton, FL, 2005.
41. Neeser, J.-R.; German, J. B., Eds. *Bioprocesses and Biotechnology for Functional Foods and Nutraceuticals*; Marcel Dekker: New York, 2004.

42. *PDR for Herbal Medicines*, 3rd ed.; Thompson PDR: Des Moines, IA, 2004.
43. Scalbert, A.; Williamson, G. *J. Nutr.* **2000**, *130*, 2073S–2085S.
44. *Supplement to The American Journal of Clinical Nutrition*. The American Society for Clinical Nutrition: Bethesda, MD, January 2005, pp 215S–335S.
45. Van Wyk, B.-E.; Wink, M. *Medicinal Plants of the World: An Illustrated Scientific Guide to Important Medicinal Plants and Their Uses*; Timber Press: Portland, 2004.
46. Yaniv, Z.; Bachrach, U. *Handbook of Medicinal Plants*; Food Products Press: New York, 2005.

Biographies

Paul A Lachance, PhD, is currently the Director of the Nutraceuticals and Functional Foods Institute of the Center for Advanced Food Technology. He received his PhD in Biology–Nutrition from the University of Ottawa, Canada. His areas of expertise include changes in micronutrients during food processing and preparation; food nutrification; nutraceuticals; and food/nutrition policy issues.

Dr Lachance was the first Flight Food and Nutrition coordinator for NASA (1963–67), where he established the Gemini/Apollo flight food systems, formulated the concept of hazard analysis critical control point (HACCP) and co-authored the first studies of bone mass loss and negative nitrogen balance in astronauts. His contributions earned him an honorary Doctor of Science (DSc) degree from his alma mater, St Michael's College in Vermont.

He is a fellow of the American Society for Clinical Nutrition, American College of Nutrition, and the Institute of Food Technologists.

Professor Lachance has mentored over 60 MS and/or PhD recipients. He has published more than 200 technical papers, scholarly chapters, and articles. He is cited in *Who's Who in Medicine and Health Care, Who's Who in America*, and *Who's Who in the World*.

Yesu T Das, PhD, is the President of ISSI Laboratories and an Adjunct Faculty Member in the Department of Food Science of Rutgers University. He received his PhD from Rutgers in 1975 and has since worked at Rutgers, US Department of Defense, and ISSI Laboratories. He was a Ford Foundation Scholar and a Senior Research Associate of the National Research Council. He is a United Nations Technical Cooperation Mission Expert with IAEA.

His areas of expertise include isolation and identification of phytochemicals of medicinal importance, and structural characterization of unknown metabolites by application of modern analytical instrumentation and techniques. He pioneered the development of nonmammalian models in biomedical research by establishing techniques for the rapid screening of poison antidotes using the housefly.

He is a Member of the American College of Nutrition (CAN) and National Nutritional Foods Association (NNFA). He has published over 30 journal articles and conducted over 300 research projects on proprietary nutraceutical, pharmaceutical, and agricultural compounds dealing with the metabolism and identification of unknown substances.

Comprehensive Medicinal Chemistry II
ISBN (set): 0-08-044513-6

ISBN (Volume 1) 0-08-044514-4; pp. 449–461

1.12 Alternatives to Animal Testing

R D Combes, FRAME, Nottingham, UK

1.12.1 Introduction

1.12.1.1 Replacement Alternatives

Replacement alternatives are methods that permit a given purpose to be achieved without conducting experiments or other scientific procedures on sentient animals.[8] Systems based on computer (in silico) modeling (such as (quantitative) structure–activity relationships ((Q)SARs) and expert systems) are omitted as they have been adequately covered in several recent reviews[1–6] and there is currently much dispute as to their validation status and regulatory acceptance.[7] The range of replacement alternative methods and approaches includes: (1) the improved storage, exchange, and use of information about animal experiments already carried out, so that unnecessary repetition of animal procedures can be avoided; (2) predictions based on the physical and chemical properties of molecules; (3) mathematical and computer models; (4) the use of in vitro tissue culture methods; (5) 'lower' organisms with limited sentience; (6) early developmental stages of vertebrates before they reach the point at which their use in experiments and other scientific procedures is regulated; and (7) human studies (both prospective and retrospective).

1.12.1.2 Advantages and Disadvantages

Replacement methods have several advantages over using animals. First, they can provide information in a more cost-effective and time-saving manner. Second, information generated from replacement methods can sometimes be adequate to produce more reliable and reproducible data than can be obtained in animals. In vitro systems are very useful for mechanistic investigations, as well as for target organ and target species toxicity studies. The ability to use human tissues in vitro obviates the need for cross-species extrapolation, thereby allowing the production of data that are more relevant to the human situation. Lastly, an absence of complex body systems, which might act as confounding factors, can permit research that cannot be investigated in animals.

However, the use of in vitro systems has several limitations: (1) the lack of integrated systemic mechanisms of absorption, distribution, metabolism, and excretion; (2) an absence of the complex interactive effects of the immune, blood, endocrine, and nervous systems; (3) lack of appropriate nonanimal models for all tissues and organs, or for all diseases and toxicity endpoints; and (4) the potential for artefactual conditions in tissue culture media.

1.12.1.3 Replacement Methods for Toxicity Testing

There has been considerable effort directed to finding replacement methods for toxicity testing, even though the proportion of animals used for this purpose is considerably lower than that used for other reasons. This effort is due to: (1) the existence of legally enforceable regulatory animal test guidelines; (2) test protocols often requiring administration of toxic doses of test materials, resulting in unavoidable suffering; (3) the necessity to use large group sizes in some tests; (4) the need to validate new methods according to strict criteria; and (5) the relevance of developing and validating new toxicity tests to achieving progress in finding new approaches to biomedical research.

The development and validation of new toxicity test methods have recently received much added impetus by the introduction of legislation in the 7th Amendment to the EU Cosmetics Directive[9] for a marketing ban on cosmetics that have been tested, or have had any of their ingredients newly tested, on animals, and also by impending EU legislation that seeks to harmonize the testing of new and existing chemicals.[10] In response to the latter, the European Centre for the Validation of Alternative Methods (ECVAM) produced a comprehensive document that presents the current status of alternatives with respect to chemicals testing, and discusses what is necessary to develop tests further so that they can be validated for screening chemicals, according to the timetable envisaged for testing.[11]

1.12.1.4 The Screening Concept

There is also a need for rapid screening systems, particularly in the pharmaceutical industry, for testing small quantities of numerous newly synthesized candidate molecules derived rapidly by combinatorial chemistry, for biological activity.

Screening involves testing a large range of candidate chemicals for a particular purpose in nonanimal systems (especially computer prediction models and in vitro tissue culture).[12] Those chemicals with desirable biological activity (efficacy) and devoid of undesirable activity (toxicity) are then tested in animals (partial replacement), having eliminated those chemicals that do not show the required activity profiles, on the basis of likely toxicity and lack of efficacy. In this way, it is possible to reduce the numbers of animal tests required to assess a given number of chemicals, and also the severity of testing on animals should be minimized by screening out substances that are likely to be toxic at an early stage.

1.12.1.5 Validation

Validation refers to the process whereby the reliability and relevance of an assay are assessed for a particular purpose.[13,14] Thus, validation is intended to assess whether a new method is going to provide useful data for predicting toxicity, and whether the same information can be obtained whenever and wherever the test is conducted by using appropriate equipment, materials, and by personnel with relevant training and expertise. It is also necessary for there to be a prediction model available to allow the information from the nonanimal test to be interpreted in relation to the data obtained from the animal test that is being replaced **Figures 1** and **2**.

The scale and complexity of validation are dictated by the purpose of the assay. For example, where a method is to be used for in-house purposes only, then validation can be conducted on a limited scale. If, however, an assay is designed to replace an animal method and used widely for regulatory purposes (e.g., safety testing), then validation trials are conducted at several different laboratories in international collaborative studies.

ECVAM was established by the European Commission in 1993 in northern Italy, with the express purpose of undertaking research into alternative methods and facilitating and organizing their validation. More recently, ECVAM, together with another organization in the USA, the Interagency Co-ordinating Committee on the Validation of Alternative Methods (ICCVAM), and in conjunction with the Organization for Economic Co-operation and Development (OECD), agreed upon some internationally harmonized criteria for validation, which are intended to facilitate the regulatory acceptance of alternative methods,[15] and these were discussed further in the light of more recent needs to expedite the validation of new animal test methods.[14,16] These criteria were based on the experiences gained from earlier validations and a significant change in approach was the introduction of a prevalidation stage prior to formal validation. This enabled a more rigorous assessment of the suitability of a proposed test method for validation, by ensuring that it was likely to provide relevant information and be transferable between laboratories.

1.12.1.6 Principal Sources of Information

The British Toxicology Society produced a report on the use of in vitro methods for toxicity testing in 1997[17] and the Third FRAME Toxicity Committee has also published a comprehensive discussion on the development of replacement methods for toxicity testing over the last decade.[18] Readers are also referred to the following references: Castell and Gómez-Lechón (1997)[19] and Knight and Breheny (2002),[20] the comprehensive series of ECVAM workshop reports, published in *Alternatives to Laboratory Animals* (*ATLA*) and reviewed by Combes,[21] and the recent EU survey of available nonanimal methods for cosmetics testing.[22]

Lastly, an overview of EU and other regulations influencing regulatory toxicity testing can be found in Schiffelers *et al.*[23]

1.12.2 Status of Alternatives for Regulatory Safety Testing

1.12.2.1 Pyrogenicity

A substance with pyrogenic activity, such as a drug in liquid formulation or a medical fluid like a saline drip, is able to induce fever in animals and humans. All batches of such materials are routinely tested. Pyrogenicity arises from the presence in these fluids of lipopolysaccharides (LPS) that act as endotoxins, arising from contaminating Gram-negative and Gram-positive bacteria. The fever reaction is due to the presence of even very small concentrations of pyrogens that initiate a cascade of events involving the release of several cytokines by monocytes, especially interleukin (IL)-1β, IL6, and tumor necrosis factor alpha (TNF-α). These substances lead to further reactions, in particular potentially fatal fever symptoms.

The rabbit pyrogenicity in vivo assay was the only method available for detecting pyrogens until the development of the limulus amebocyte lysate (LAL) test.[24] The LAL, which measures coagulation of amebocytes in blood samples extracted from the horseshoe crab, is widely used, and has led to a substantial reduction (*c*. 80%) in the number of animal tests required. However, it has several disadvantages, being: (1) based on a reaction in an arthropod; (2) insensitive to LPS of Gram-negative bacteria only; (3) unable to detect the LPS of Gram-negative bacteria with a molecular weight of <8000 Da; and (4) unable to distinguish between different types of endotoxins from Gram-negative bacteria, with potentially very different fever-inducing potential.[24]

These drawbacks to the LAL prompted the development of tests based on the use of whole blood.[25] The first to be developed was the human whole-blood test (WBT).[26] The WBT involves enzyme-linked immunosorbent assay (ELISA) detection of IL-1β by human monocytes. The test is less expensive and more sensitive than the rabbit test, and: (1) measures fever-inducing potency directly in human blood; and (2) is sensitive to endotoxins, lipoteichoic acids, fungi, and superantigens such as staphylococcus enterotoxin B antibody (SEB) (the enterotoxin of *Staphylococcus aureus*). The WBT has become commercially available as PyroCheck.[27]

However, a variant of the WBT, the WBT in rabbits, has been developed to overcome the problem that responses to pyrogens in the WBT can be subjected to donor variability of the blood sample. The WBT in rabbits can used to explain false-positive and false-negative results obtained in the WBT compared with the rabbit in vivo results.

The WBT and several other related tests have been subjected to an extensive validation study, which has now been endorsed by a peer review undertaken by the ECVAM Scientific Advisory Committee (ESAC).

1.12.2.2 Eye Irritation

Many in vitro and ex vivo alternatives exist for predicting eye irritation (**Table 1**), although no test, or battery of tests, has been validated and accepted for regulatory use as a complete replacement for the Draize test in rabbits. However, the available tests provide information that can be directly related to some index of ocular irritation at different levels.[20,28,29] The most complex are ex vivo tests involving eyes from recently killed animals, such as the rabbit enucleated eye or chicken enucleate eye tests (the REET and CEET respectively). These assays involve direct measurements of corneal opacity, swelling, and histopathology after topical exposure to the eye.

In the REET, eyes are taken from rabbits previously used in dermal irritation studies, as well as from control animals from any study not involving pathology of the eye. The corneal surface is exposed to test material for 10 s, and then washed, after which corneal opacity and thickness, and barrier function are determined. The latter is achieved by using the fluorescein leakage assay (FLL). The FLL was originally developed to assess the ability of substances to compromise barrier function by using cells in monolayer culture.[30,31] In its original form, the FLL involved Madin–Darby canine kidney (MDCK) confluent cell monolayers for the assessment of both ocular and dermal irritancy potential. The effects of chemicals on tight junctions (similar to those found between cells in the corneal epithelium), and hence barrier function, are predicted by the amount of leakage of fluorescein across the monolayer.

Slightly less complex than the REET is the bovine corneal opacity and permeability (BCOP) assay, in which just the isolated cornea is used. The REET and the BCOP have each performed variably in validation studies, and they need to be improved.[29]

The next lower level of complexity of tests for eye irritation are three-dimensional tissue constructs, usually comprised of the corneal epithelium only.[29,32] These models generally have cytotoxicity as an endpoint that can be correlated with the extent of injury to the epithelium, although barrier function can also be measured by the FLL. Cytotoxicity can be assessed in a variety of ways that indicate different types of toxicity, both lethal and nonlethal. The resazurin assay assesses biochemical activity (the conversion of resazurin to resorufin) colorimetrically, and is becoming increasingly popular. It is preferable to couple the cyotoxicity assay with the FLL test as leakage of the dye could be due to toxicity to the cells, rather than to a specific effect on gap junctions.[33]

There is increasing interest in developing complex models that include other cell types in co-culture systems, for example between sensory neurons of the ND7/23 cell line and corneal epithelial cells.[34] This model has been used to assess the effects on viability and barrier function of repeat exposures to surfactants. Since resazurin and fluorescein are nontoxic themselves, the same cells can be repeatedly assessed and then challenged with toxic chemicals for the determination of: (1) initial toxicity; (2) level of recovery from this toxicity; and (3) the time taken for recovery. The ability of such models to distinguish between mild and moderate irritants can be improved by considering the capacity of cells to recover from toxic damage.[35]

The use of monolayer cultures of human or animal cells is the simplest way of measuring eye irritation: by using similar methods to those involved in multilayer systems. Many other assays have been developed to look at various endpoints other than direct tissue damage, for example vascular endpoints, such as hemorrhage and coagulation, to the in vitro armamentarium can be assessed by using the hen's egg test-chorioallantoic membrane[36,37] (HET-CAM) and the chorioallantoic membrane vascular assay[38] (CAMVA). Membrane damage can also be measured in the red blood cell (RBC) assay.[39] Other less obvious examples of tests for eye irritation include the pollen tube and slug mucosal assays, but these are unsuitable for further development and validation.[40]

Table 1 In vitro methods for regulatory toxicity testing discussed

Toxicity endpoint	*Test*	*Comments*
Pyrogenicity	**Limulus amebocyte lysate (LAL) test**	Measures coagulation of amebocytes in blood samples extracted from the horseshoe crab, accepted by pharmacopoeias
	Human whole-blood test (WBT)	ELISA detection of IL-1β release by human monocytes; validated, awaiting peer review; commercial version: PyroCheck
	WBT in rabbits	To overcome some problems with WBT
Eye irritation	*Rabbit (REET) and chicken enucleated eye tests (CEET)*	Direct measurement of corneal opacity, swelling and histopathology, variable validation performance
	Fluorescein leakage assay (FLL)	Nontoxic measurement of barrier function of eye
	Bovine corneal opacity and permeability (BCOP) assay	Use of just isolated cornea; variable validation performance
	Three-dimensional tissue constructs	Corneal epithelium only; cyotoxicity and FLL measured, as well as cytokine release; can be derived from immortalized human corneal cells
	Co-culture systems	e.g., sensory neurons of the ND7/23 cell line and corneal epithelial cells
	Hen's egg test–chorioallantoic membrane (HET-CAM) and chorioallantoic membrane vascular assay (CAMVA)	Vascular endpoints, such as hemorrhage and coagulation
	Red blood cell (RBC) assay	Membrane damage
Genotoxicity	***Salmonella typhimurium* his reversion mutation assay (Ames test); *Escherichia coli* tryp reversion assay**	Well-established tests in core battery
	Numerical (polyploidy and aneuploidy) and structural chromosomal aberrations in cultured mammalian cells	Chinese hamsters (e.g., ovary (CHO) or lung (CHL)) and primary human lymphocytes usually used; well-established tests in core battery when structural changes mainly scored
	Sister chromatid exchanges (SCE) and micronucleus induction in cultured mammalian cells	Adjunct tests currently, although latter being retrospectively validated and peer-reviewed, and has an OECD draft TG
	Unscheduled DNA synthesis (UDS), mitotic recombination, and gene conversion/mutation in yeast (Saccharomyces cerevisiae): the COMET assay	Adjunct assays for detecting DNA damage (e.g., strand breaks; induction of DNA repair).
	Gene mutation in cultured mammalian cells, e.g., L5178Y mouse lymphoma tk assay	Forward mutation at the thymidine kinase (tk) locus in L5178Y cells which are heterozygous for this locus and mutate to trifluorothymidine resistance; sometimes included in the core battery of tests
Phototoxicity – photoirritation/ photosensitization	**3T3 Neutral red uptake phototoxicity test (3T3-NRU-PT)**	Uses mouse fibroblasts (human keratinocytes can be used instead); regarded as complete replacement test
	RBC phototoxicity test (RBC-PT) and human 3-D skin model phototoxicity test (H3D-PT); organotypic skin models (e.g., EpiDerm)	Suggested to overcome limitations of 3T3-NRU-PT
Phototoxicity – photogenotoxicity	**Photo-Ames (P-Ames) test (bacterial mutation)**, the photo-thymidine kinase test (P-TKT) (mammalian cell gene mutation), **the photo-chromosome aberration test (P-CAT)**, the photo-micronucleus test (PMNT)), and photo-comet assay (P-Comet)	Applied in similar fashion to genotoxicity tests, but with and without UV irradiation

continued

Table 1 Continued

Toxicity endpoint	*Test*	*Comments*
Skin penetration	**Excised pig or human skin in flow-through or static diffusion cells**	Skin strips are used, without need for being viable for routine screening; OECD TG 428 regarded as complete replacement for in vivo method
Skin (eye) corrosivity	**Transepidermal resistance (TER), Corrositex, the skin.[2] ZK1350 corrosivity test and Episkin)**	Validation studies resulted in OECD TG 404 for skin corrosion/irritation testing
	Organotypic 3D skin models; Advanced-Skin-Test 2000 (AST-2000); fully grown reconstructed epidermal model (Epidermal-Skin-Test: EST-1000)	Proposed as models with improved barrier function
Skin irritation	*Reconstituted skin models (EpiDerm, EPISKIN, PREDISKIN, and SkinEthic)*	Assessed in ECVAM prevalidation and validation studies; final outcome still to be reported; involves cytotoxicity and induction of release of cytokines of inflammatory cascade
	Excised skin models (the pig ear and mouse skin integrity function tests (SIFT)	Assessed in ECVAM prevalidation and validation studies; final outcome still to be reported; involves cytotoxicity and induction of release of cytokines of inflammatory cascade
	Cell culture models	Assessed in ECVAM prevalidation and validation studies; final outcome still to be reported; involves cytotoxicity and induction of release of cytokines of inflammatory cascade
Skin sensitization	Primary monocyte-derived dendritic cells	Changes in expression of HLA-DR, CD86, and CD83 cell surface markers
	Dendritic cells	IL-1β release; these assays are used in tier-testing strategy, involving expert systems, reactivity (protein-binding) tests, and in vitro skin permeability
Neurotoxicity	Human SH-SY5Y and mouse NB41A3 neuroblastoma cell lines	To study organophosphate (OP) neuropathy through the inhibition of neurotoxic esterase (NTE)
	Human SH-SY5Y and mouse N1E115 cells	To study axonopathic and calcium-mediated processes for several neurotoxicants
	Primary astrocyte cultures (rat/mouse) or human and rodent astrocytoma cells	Quantification of glial fibrillary acidic protein (GFAP) to assess toxicant-induced damage
	Primary cortical neuronal cultures or rodent PC12 cells	To assess drug interactions and aminoacidergic transmission, by measuring calcium perturbations, early gene expressions and other endpoints
	Whole-brain organotypic spheroid culture systems (derived from rat and hen)	To study developing, and more differentiated, nervous tissues
	Co-culture models of the blood–brain barrier (BBB) (e.g., glial/endothelial co-cultures and complex cultures of brain spheroids, endothelial cells and Madin–Darby canine kidney (MDCK) cells)	To study barrier penetration of potential neurotoxicants
	Chick embryonic neuronal retina and retinal pigment epithelium cell cultures	To investigate toxicant-induced retinopathies
Embryotoxicity	*Frog embryo teratogenesis assay (Xenopus (FETAX)); rat limb bud micromass test, MM; the postimplantation rat whole-embryo culture test, WEC; and the embryonic stem cell test, EST*	Validated as screens, more work being undertaken as part of new EU ReproTect project

Table 1 Continued

Toxicity endpoint	*Test*	*Comments*
Reproductive toxicity	Germ cells and associated somatic cells in monoculture systems	e.g., testis slices, Leydig, Sertoli, follicle, sperm, spermatocyte, seminiferous tubule cell culture systems; immortalized testicular somatic and germ cells with specific toxicity markers
	Germ cells and associated somatic cells in co-culture systems	Sertoli and various other cells (e.g., germ cells)
Endocrine toxicity	*Subcellular hormone receptor ligand binding*	Usually to purified receptors (competitive binding) to give binding affinity
	Induction of proliferation (mitogenesis)	In hormone-responsive mammalian cell lines expressing various hormone and other receptors
	Transactivation systems in yeast and mammalian cell lines	Induction of transcription by binding to hormone receptor in reported gene assays associated with hormone response element
Acute systemic toxicity	*Basal cell cytotoxicity in a cultured mammalian cell line or primary cell culture (e.g., 3T3 NRU)*	Coupled with biokinetics; biotransformation; penetration of internal membranes (e.g., gut wall, BBB)
	Basal cell cytotoxicity in target organ-differentiated cells	Coupled with biokinetics; biotransformation; penetration of internal membranes (e.g., gut wall, BBB). Further development and prevalidation being undertaken in new EU Acutetox project
Chronic toxicity	Metabolically competent hepatocytes maintained as reconstructed collagen sandwich monolayers	Extended viability and differentiation status
	Single cell or co-culture-based systems in systems that allow replacement of depleted, with fresh, media (e.g., hollow-fiber systems; inert perfusion membrane cultures EpiFlow and Minucell)	Coupled with repeat dosing and recovery measurements by using nontoxic markers (e.g., resazurin and FLL); the latter being considered for validation
Carcinogenicity	**Genotoxicity tests (see earlier)**	Will not detect nongenotoxic carcinogens (NGCs)
	Morphological cell transformation based on rodent cell lines, primary cells (e.g., Balb/c 3T3 cells; C3H10T1/2; SHE cell assay)	Can detect genotoxic carcinogens and NGCs and some can detect tumor promoters
	Bhas 42 cells	Specific for promoters
	Gap junctional intercellular communication inhibition assay (GJIC)	For NGCs (but not specific)
	Miscellaneous	Binding to nuclear and other receptors in a variety of cells (e.g., hepatocytes: measurement of peroxisome proliferation)

Italics indicate a test that has been subjected to an interlaboratory study and/or a formal validation study; bold indicates a test that has been accepted for regulatory testing.

The results of the EC/UK Home Office validation study of nine tests for eye irritation led to the conclusion that none was suitable as a replacement method. A subsequent European Cosmetic, Toiletry and Perfumery Association (COLIPA) validation study[41] of 10 methods with 55 test chemicals (23 ingredients and 32 formulations) was conducted. In this study, although the FLL, RBC, and a tissue-equivalent assay performed quite well, none of the methods was concluded to be a valid replacement for the Draize test. However, some reassessments of the validation studies mentioned earlier, and analyses of the data by using different statistical methods, suggest that some tests can distinguish between strong and weak irritants.[18]

The search for an assay, or test battery, to act as a complete replacement for in vivo eye irritation testing has therefore been unsuccessful so far. There are three main reasons for this: (1) a failure to model the fundamental molecular phenomena that cause eye irritation and corneal opacity; (2) a traditional reluctance to develop methods that take account of the reversible nature of irritation and the ability of cells to recover from the toxicity induced; and (3) the lack of neural cells in the cellular systems thus far developed, thereby compromising the ability to detect neurogenic inflammation. The latter is considered to be important in determining the susceptibility of the eye to irritants as the corneal epithelium in vivo is heavily innervated.[42,43] It is noteworthy that none of the test systems models the fundamental mechanism involved in the induction of corneal opacity, namely hydration levels of the endothelial stromal cells in the cornea, which in turn are controlled by the sodium/potassium ATPase system.[44,45] However, some progress is being made with using organotypic multilayered models derived from immortalized human corneal cells with improved barrier function.[46–48]

Irritation involves induction of the inflammatory cascade that includes the release of cytokines (as discussed in more detail elsewhere in this review). Śladowski *et al.*[32] found that the effects of LPS on IL-1β release by human corneal epithelium cell cultures were augmented by the complement system, normally found in tears, although the corneal cells were highly resistant to complement.

Despite the fact that there have been difficulties in developing replacement tests for eye irritancy, the successful validation of tests for eye corrosion[49–54] has resulted in the acceptance by the OECD of a revised guideline, TG 405 (acute eye irritation/corrosion), which incorporates a tiered-testing strategy for eye irritation and corrosion.[55] The strategy in OECD TG 405 includes the use of SARs for eye corrosion/irritation and for skin corrosion coupled with pH.[56] The use of the Draize test is obviated if a substance has been classified as systemically toxic, a severe dermal irritant, or a skin corrosive.

It is very likely that a battery of complementary tests will be required for modeling eye irritation, since it is caused by several different mechanisms. At the present time, such a screening battery for eye irritation might usefully include the BCOP, HET-CAM, and isolated chicken eye (ICE) assays. Maximum use should be made of dermal irritation data in any hierarchical approach.[17] For example, it is suggested that test materials that exhibit severe effects on the skin should be classified as eye irritants, without any further testing. However, regulatory authorities need to harmonize their attitudes toward the regulatory use of the BCOP, HET-CAM, isolated rabbit eye, and ICE assays, and other nonanimal tests for eye irritation.

1.12.2.3 Genotoxicity

Many in vitro methods are available for the testing of genotoxicity[57–59](the induction of DNA damage). These are divided into core and ancillary assays: the former are generally used in a tier-testing battery, and designed to indicate a potential for genotoxicity, while the latter are used to clarify unexpected or anomalous results in assays of the core battery. Unlike other toxicity endpoints, short-term in vivo studies are then routinely employed to assess if any potential for genotoxicity detected in vitro is actually expressed in the whole animal.

Core in vitro genotoxicity tests are based on the detection of mutagenicity in bacteria and chromosomal damage in mammalian cells. Bacterial gene mutation assays comprise treating tester/indicator strains which are hypermutable, due to their possession of defective cell envelopes (to enhance uptake of test chemicals), and to their deficiency in one or more aspects of DNA repair. Both reverse and forward mutation systems are available; the most widely used being the Ames test. This involves selecting histidine (*his*) revertants of *Salmonella typhimurium* on minimally supplemented agar. Revertant colonies that arise by growth of the mutated bacteria are then counted and mutagenic potency can be obtained from dose–response data.[60] A battery of different tester strains, carrying different *his* mutations, have to be used to detect chemicals that induce mutagenesis by specific mechanisms.[61] A less widely used bacterial mutation system is based on *trp* reversion in *Escherichia coli* strain WP2, and some repair-defective derivatives of this strain.

Chromosomal damage comprises both structural and numerical chromosomal abnormalities and these are usually detected in metaphase-spread preparations from exposed mammalian cells and scored after staining under light microscopy, although a variety of staining techniques and other visual aids, increasingly based on molecular biological techniques,[62] has become available to allow chromosomal painting. Permanent cell lines, usually derived from Chinese hamsters (e.g., ovary or lung) and primary cells usually of human origin (e.g., lymphocytes) are used.[58] Exposure can be either in culture suspension using tubes or flasks or on slides of coverslips in dishes. Cells must undergo DNA replication following exposure to the test substance for chemically induced DNA lesions to be detected as chromosomal damage resulting in structural aberrations (such as gaps and breaks, induced by clastogens), micronuclei, or sister chromatid exchanges (SCEs). Numerical aberrations (polyploidy and hypodiploidy/hyperdiploidy), indicative of aneuploidy, are also important to score.[62]

Ancillary in vitro genotoxicity assays[57,59] include gene mutation in a variety of cells, including cultured mammalian cells, differential DNA repair, unscheduled DNA synthesis, mitotic recombination, and gene conversion/mutation in yeast (*Saccharomyces cerevisiae*) and the COMET assay. The COMET assay detects DNA strand breaks which, when subjected to electrophoresis, result in migration of DNA fragments out of the nucleus to form the tail of a comet-like structure. The extent of migration of the DNA fragments (the size of the comet's tail) indicates the amount of DNA damage. Double strand breaks are detectable under conditions of DNA denaturation, such as high alkalinity. Benefits of the COMET assay include: (1) it can be conducted in any mammalian cell line or tissue; (2) proliferating cells are not required; and (3) it is very sensitive, able to detect strand breaks in individual cells.[63]

The International Conference on Harmonization (ICH)[64] suggested the inclusion of mammalian cell gene mutation assays, particularly the mouse lymphoma gene mutation assay, in the core battery of tests. This requirement was endorsed by the results of a collaborative trial demonstrating the equivalence of both cytogenetics and mouse lymphoma gene mutation assays for detecting the activity of a range of chemicals.[65,66]

The mouse lymphoma assay involves measuring forward mutation at the thymidine kinase locus in L5178Y cells which are heterozygous for this locus. Such cells are sensitive to the nucleoside base analog, trifluorothymidine, which is converted by thymidine kinase to a toxic form. Homozygous mutant cells for the tk locus, arising from forward mutation at the wild-type allele, are resistant to trifluorothymidine, and therefore grow in its presence. Such mutations can arise as a result of a variety of genotoxic molecular events, including both point mutations (basepair changes) and chromosomal damage.[58]

Micronuclei can be scored in cultured mammalian cells and this has stimulated the development and validation of an in vitro micronucleus assay.[67–69] Initial protocols for the test have been modified to avoid apoptosis interfering with scoring of micronuclei, by using the CTLL 2 cell line transfected with the *bcl2* gene inhibiting apoptosis.[70] If the in vitro micronucleus test became an acceptable substitute for the chromosomal aberration test, it would provide a way of detecting aneugens (by the induction of spindle disruption, resulting in micronuclei-containing centromeres) and clastogens (by the induction of micronuclei-lacking centromeres) in one test, while additionally being considerably less labor-intensive than metaphase analysis.[71]

The assay is currently undergoing retrospective validation by ECVAM prior to being peer-reviewed by the ESAC. Meanwhile, the OECD currently has a draft test guideline (OECD TG 487) under review, and data from the assay are being accepted by some regulatory agencies.

Several formal testing strategies have been proposed for genotoxicity. The UK Department of Health[72] scheme starts with the use of physicochemical properties combined with in silico modeling. Further testing as required is conducted by using the two core tests above. A third test, stipulated on a case-by-case basis in the strategy, particularly when human exposure is likely to be high, is the ICH-approved mouse lymphoma assay. When all three tests are negative no further testing is necessary. However, if one of them is positive or equivocal, further testing by using short-term in vivo assays is required to confirm the in vitro data. Any further in vivo testing is limited to those substances that are considered to pose a risk to the germ cells.

Worth and Balls[11] describe a genotoxicity testing strategy that involves four steps: (1) (Q)SAR modeling; (2) validated core in vitro tests, with the results being used for classification; (3) short-term in vivo testing; and (4) only rarely, germline mutagenicity tests and biokinetic data to confirm finally the mutagenic potential of the test substance.

1.12.2.4 Phototoxicity

1.12.2.4.1 Photoirritancy and photosensitization

A phototoxin is a substance that is rendered toxic or more toxic in the presence of light. Various methods for measuring phototoxicity in vitro have been developed, by investigating the effects of ultraviolet (UV) A (320–400 nm) and UVB (280–320 nm), separately and variously in combination with each other, and with test substances, such as a sunscreen.[20,73] The 3T3 Neutral Red Uptake Phototoxicity Test (3T3-NRU-PT) was subjected to an EU/COLIPA validation study, and was subsequently accepted for phototoxicity testing by the Scientific Committee on Cosmetic Products and Non-Food Products (SCCNFP) in 1998, and recommended by the European Agency for the Evaluation of Medicinal Products/Committee for Proprietary Medicinal Products as a test for acute phototoxicity. The test was also accepted as method no. 41 in Annex V to Directive 67/548/EEC in 2000, and as a new Test Guideline 432 In Vitro 3T3 NRU Phototoxicity Test by the OECD in 2002.

The assay involves comparing the cytotoxicity of a substance when tested in the presence and absence of a noncytotoxic dose of UV light, expressed as photoirritation factor (PIF) and mean photo effect (MPE) values. OECD TG 432 requires the application of a prediction model to convert the above data into a measure of phototoxicity as follows: a PIF of <2 or an MPE <0.1 indicates 'no phototoxicity'; a PIF of >2 and <5, or an MPE of >0.1 and <0.15 indicates 'probable phototoxicity,' and a PIF of >5 or an MPE of >0.15 indicates 'phototoxicity.'

However, this test has several limitations: (1) it does not allow an assessment of phototoxic potency; (2) it does not model bioavailability of a chemical to the skin; and (3) it does not involve human cells which are in vivo targets for phtotoxicity. Attempts have been made to address these concerns. The RBC Phototoxicity Test (RBC-PT) and the Human 3-D Skin Model Phototoxicity Test (H3D-PT) have been suggested as additional methods.[73] The former test is resistant to high levels of UVB, and also provides information on the mechanism of phototoxicity. The H3D-PT involves the use of a reconstructed organotypic culture comprising primary human skin, to assess compound bioavailability. This assay is being used by industry, although it has not been formally validated. Jírová *et al.*[74] used the three-dimensional human skin model (EpiDerm) in conjunction with the 3T3-NRU-PT to demonstrate the phototoxicity of ingredients of bitumous tars. The results obtained with the organotypic culture correlated with the activity of human patch test data. Clothier *et al.*[75] and Combes *et al.*[76] have shown that the use of human keratinocytes instead of 3T3 cells in the NRU test produces results broadly comparable to those generated by the validated method.

1.12.2.4.2 Photogenotoxicity

Photomutagenicity is a more recent toxicity endpoint required by the SCCNFP for the safety assessment of cosmetics.[59] However, only chemicals which are not completely stable to solar-simulated radiation (UVA and UVB) for at least 10 h have to be tested. This legislation followed efforts to adapt standard genotoxicity test methods for photogenotoxicity and these subsequently led to the SCCNFP proposing that a bacterial test for gene mutation and an in vitro test for chromosomal aberrations in mammalian cells should be performed in the presence of UV radiation. Several minimal criteria were established for the methods following a COLIPA ring trial involving seven laboratories.[77]

None of the protocols used in the above studies involved an exogenous source of mammalian biotransformation enzymes, despite the need to take this into account, as stated in the SCCNFP guidelines. Utesh and Splittgerber[78] showed, however, that different results can be obtained with and without S9 mix for photomutagenicity of 12-dimethylbenzanthracene.

Many other in vitro genotoxicity tests have now been, or are being, adapted to detect photogenotoxicity,[73] and all the available tests are: the photo-Ames test (bacterial mutation), the photo-thymidine kinase test (mammalian cell gene mutation), the photo-chromosome aberration test and the photo-micronucleus test (PMNT): for clastogenicity and, in the case of the PMNT, also aneuploidy), and the photo-Comet assay. Some of these methods are in the process of being validated.

1.12.2.5 Skin Permeability

In vitro methods for skin permeability were extensively discussed by ECVAM workshop 13 and a tier-testing strategy was recommended.[79,80] The strategy involved a preliminary prediction based on physicochemical information to derive a skin permeability coefficient, followed by further studies based on in vitro skin strip testing and, if necessary, an in vivo confirmatory assay. For the routine testing of cosmetics, it was suggested that it would be sufficient to use a simple approach (for example, skin strips) without the need to take skin biotransformation into account. This approach is the one subsequently recommended, following peer review, and appears in a draft protocol that was submitted to the OECD before the workshop. The strategy was incorporated into the EU Annex V guidelines, and the resulting OECD TG 428 (skin absorption: in vitro method) includes the use of excised pig or human skin in flow-through or static diffusion cells.

It is noteworthy that the National Coordinators for the OECD TGs formally accepted the TG for in vitro percutaneous absorption simultaneously with an in vivo assay in May 2002. However, the reason why the latter was developed is obscure since the OECD regards the in vitro TG as being a full replacement for TG 427 (the in vivo method for skin absorption).[81,82]

1.12.2.6 Skin (Eye) Corrosivity

Corrosion, unlike irritation, is a severe, irreversible process of cell necrosis and lethality. As such, the mechanisms involved are more straightforward to model than those of irritation, and unsurprisingly, the development and validation of replacement methods for corrosivity have been more successful. Thus, ECVAM prevalidation and validation studies assessed the abilities of four tests (TER, Corrositex, the skin^2 ZK1350 corrosivity test, and Episkin) to discriminate corrosives from noncorrosives for a range of 60 coded chemicals and to identify correctly chemicals that had been classified according to United Nations packing group categories I, II, and III in three different laboratories.[83]

Descriptions of these assays are found in Knight and Breheny[20] and Zuang *et al.*[84] The TER assay and Episkin met all the performance criteria and are considered valid for the above purposes. Subsequently, a further assay, the Epiderm skin corrosivity test, was successfully validated in a catch-up study. All three assays are now accepted as valid for regulatory purposes, particularly when conducted according to OECD TG 404 for skin corrosion/irritation testing.[85]

This TG now stipulates a tiered testing strategy, comprising the use of two screens: (1) existing information; and (2) pH and SAR,[86] followed by validated, OECD-approved in vitro models for skin corrosion, when a positive is not indicated by the screens.

Hoffmann *et al.*[87] found a new fully grown reconstructed epidermal model (Epidermal Skin Test: (EST-1000)), consisting of proliferating and differentiating keratinocytes, to be highly comparable to normal human skin histologically and immunohistochemically, and in its barrier function characteristics. The model also possessed KI-67, CK 1/10/5/14, transglutaminase, collagen IV, involucrin, and β_1 integrin markers associated with human skin. The authors concluded that EST-1000 shows a very high predictive potential and recommended it to complement the established full-thickness model Advanced Skin Test 2000 (AST-2000), originally developed by Noll *et al.*,[88] for skin corrosivity testing.

1.12.2.7 Skin Irritation

No alternative tests for skin irritation have been fully validated and accepted as replacements for the Draize test, mainly because this endpoint is very complex, involving several different cell types involved in the inflammatory cascade. This sequence of events involves specific interactions between cells of the immune system, initiated following stimulation by the release of chemokines from initial target cells in the skin epithelium on exposure to an irritant. These immune cells include mast cells, which release histamine and prostaglandins, and macrophages, which release cytokines like TNF-α and IL1, in particular, on activation. These cytokines then activate endothelial cells of the blood vessels to undergo vasodilation and influx of fluid, leading to erythema and edema respectively.

As irritation is a reversible process it is also important, as with eye irritation, to consider measuring recovery levels from repeat dose exposures rather than just initial toxicity to single acute doses.

Prevalidation studies have been conducted on five in vitro tests based on three types of systems[20,83,89,90]: (1) reconstituted skin models (EpiDerm, Episkin, Prediskin, and SkinEthic); (2) excised skin models (the pig ear and mouse skin integrity function tests (SIFT); and (3) cell culture models measuring induction of inflammatory mediator release (e.g., arachidonic acid). Not surprisingly, in view of the mechanism of irritation discussed above, assays based on the release of various cytokines of the inflammatory cascade (e.g., IL-1α, IL6, IL8, IL10, TNF-α, and arachidonic metabolites) show promise for predicting dermal irritation.[91]

However, none of the tests was considered ready for progression to a formal validation study, and, following their modification, a common EpiDerm/Episkin protocol and a modified SIFT protocol are in the process of being formally validated. This second phase of the ECVAM skin irritation study has now been completed, and was discussed at a meeting of the Validation Management Committee. EpiSkin performed better than EpiDerm, giving about 80% sensitivity, although specificity was disappointingly lower (P Botham, personal communication, 2005). This result was obtained when cytotoxicity of the three-dimensional culture was the endpoint used. However, surprisingly, the performance of the assay, particularly its specificity, dropped further when IL-1α was measured as the endpoint. This could have been due to measuring it at the wrong time during culture growth, and further experiments are planned to investigate this possibility. The report of phase II of the study should be ready for peer review during late 2006.

In the meantime, the use of animals for dermal irritancy testing can sometimes be avoided by implementing a sequential testing strategy, as stipulated in regulatory test guidelines, which involves: (1) physicochemical properties and chemical reactivity, e.g., substances of either pH of <2 or >11.5 may not require testing, taking into account other factors such as alkaline or acidic reserve; (2) if convincing evidence of severe effects in well-validated in vitro tests exists; and (3) if no skin irritation is detected in a dermal acute toxicity test when a substance has been dosed at the limit dose level (assume no irritancy) or when a high level of toxicity is detected (assume irritancy).

1.12.2.8 Skin Sensitization

Currently, no test or testing strategy exists to obviate the need for the Magnusson & Kligman guinea-pig maximization test of EU Method B6,[92] the nonadjuvant Buehler method,[93] for EU notification and classification, or the more recently accepted local lymph node assay, although this assay is a reduction and a refinement of the conventional animal methods.

There has, however, been progress in modeling some of the essential mechanisms and processes involved in sensitization.[94] These are: (1) dermal penetration and covalent binding to immunological proteins in the skin; (2) specific stimulation of Langerhans cells by epidermal cytokines; and (3) stimulation of T lymphocytes. After stage 2, haptenated proteins are processed by Langerhans cells that then migrate to draining lymph nodes and interact with

T cells; resulting in recognition of the hapten. Stage 1 can be modeled by using in vitro skin penetration methods (as discussed earlier) and by determining and/or predicting reactivity by using biochemical assays (e.g., covalent binding to nucelophiles) and in silico methods (based on known structural alerts), including expert systems such as Deductive Estimation of Risk from Existing Knowledge (DEREK).[2]

Modeling the later stages of sensitization, especially antigen presentation, is complicated by difficulties in obtaining sufficient numbers of Langerhans cells in vitro. Instead, the current focus is on using cells derived from monocyte dendritic cells (DCs), originating from bone marrow and umbilical cord blood samples.[94] Thus, Verstraelen *et al.*[95] used monocyte-derived DC from seven different donors to investigate the induction of changes in expression of human leukocyte antigen (HLA)-DR, CD86, and CD83 cell surface markers by LPS and diesel exhaust particles.

As sensitizers modulate cytokine release by target cells, tests are being developed, based on assessing the effects of exposures of immunologically active cells, such as DC, on cytokine release, especially of IL-1β, as this is considered to be a crucial stage,[96] and such a method is part of a new validation study called Sensitivit being organized by ECVAM.

More research is urgently required, however, on the development of a skin biotransformation system, to take account of the existence of prohaptens that need to be activated before they can penetrate the skin and/or react with protein.[97] The potential contribution of metabolism to sensitizing potential might be assessed by using a combination of expert systems such as DEREK and METEOR to predict the reactivity of both the parent molecule and its likely skin metabolites.[98]

The tier-testing strategy proposed by Worth and Balls[11] is based on the above known mechanisms involved in sensitization, and was derived from one originally developed by Unilever that additionally involved the empirical determination of skin penetration.[56]

1.12.2.9 Toxicokinetics

Biokinetic modeling can provide information that is extremely useful for interpreting in vitro and in vivo data, and for using such data for risk assessment by helping to: (1) relate external and internal dose effects; and (2) undertake species extrapolation. In its more elaborate form of physiologically-based biokinetic (PBPK) modeling, predicting biokinetics has become a technique used for the above purpose,[99] and involves representing organs or groups of organs as discrete compartments interconnected with physiological volumes and blood flows.

PBPK modeling was used in a recent investigation in which an integrated testing scheme was applied for predicting the toxicities of a number of chemicals.[100] The application of modeling and comparison of IC_{50} values with indirect liver concentrations (corrected for bound fraction) yielded a number of correct predictions for acute oral toxicity. However, while several of the chemicals already had a PBPK model available, the general applicability of this approach to chemicals is currently limited by the need to develop such models. Nevertheless, this problem is being addressed through the development of software programs and databases for the rapid generation of new models (George Loizou, personal communication).[175]

1.12.2.10 Neurotoxicity

No single in vitro test, or package of tests, has been adequately evaluated and validated for neurotoxicity testing, although there has been considerable progress in finding potential prescreens for these endpoints.[17,18,97,101–103] The main reasons for lack of progress in this area are: (1) the great diversity of cell types in the nervous system; (2) the complex cellular interplay necessary for maintaining normal functioning of the nervous system; (3) the need to take account of differences and similarities between the central and peripheral nervous systems; and (4) the need to model the blood–brain barrier.[104]

Over the last decade, increased understanding of cell systems, toxic mechanisms, and endpoints for assessment has resulted in the emergence of many in vitro models for neurotoxicity, and these models are summarized in **Table 1**.[17,18]

Some of these models have been incorporated into tier-testing schemes, such as the one proposed for validation by ECVAM workshop 3.[18] The schemes start with simple screens/endpoints (such as cytotoxicity in cell lines), with later tiers involving more complex systems like organotypic/spheroid/blood–brain barrier cultures from different species, and a range of cell-specific and neurotoxicological markers, in conjunction with biokinetic modeling.

The development of assays for the first tier of the suggested three-tiered hierarchical test screen have been completed, with encouraging results.[18] However, there is a need for much more research before a scheme will emerge that is suitable for validation. In the meantime, as a consequence of the potential high severity to animals of exposure to neurotoxicants, it is recommended that screening prior to undertaking neurotoxicity studies (TG 418, TG 419, and TG 424) should be mandatory.[81]

1.12.2.11 Embryotoxicity, Reproductive and Endocrine Toxicity

The reproductive cycle is also sufficiently complex for there to be a need for a battery of complementary in vitro assays to be developed and validated, to form part of an integrated testing strategy.[105] Many potential mechanisms exist for affecting the reproductive system and the embryo, including direct effects on: cellular development, signaling pathways, inhibition, and perturbation of developmental and differentiation processes, reactive binding of chemicals with nucleic acids and proteins leading to mutagenesis and carcinogenesis, disruption of tubulin, microfilaments, and the cytoskeleton, membrane disruption, effects on the endocrine and nervous systems, as well as indirect effects relating to maternal toxicity.

1.12.2.11.1 Embryotoxicity

The report of ECVAM Workshop 12 on reproductive toxicity screens recommended[105] that the frog embryo teratogenesis assay (*Xenopus* (FETAX)), a chick embryo assay, a micromass culture system, mammalian whole embryos, and methods involving mammalian embryonic stem cells should be further developed, optimized, and validated. FETAX has been subjected to several validation studies and been evaluated by ICCVAM. In addition an ECVAM-sponsored validation study was undertaken on three in vitro embryotoxicity tests.[106,107] These assays were the rat limb bud micromass test, MM; the postimplantation rat whole-embryo culture test, WEC; and embryonic stem cell, which was initiated in 1997. The assays proved reliable and relevant for predicting the embryotoxicity of a range of chemicals, and are suitable for using as screens. The more widespread use of these methods, however, will largely depend on how predictive the tests are for indirect-acting agents tested in the presence of a metabolic activation system[108] (e.g., cyclophosphamide in the MM). Further development of the embryonic stem cells along these lines is planned in a recently initiated EU ReProTect validation project. It should be noted that the embryonic stem cell is the only method that could be adapted for relatively high throughput, and that it does not involve the sacrifice of large numbers of pregnant animals, although the MM requires some donor animals for primary cells, and these can come from species other than rodents.[109]

1.12.2.11.2 Reproductive toxicity

Many different cell-based systems exist for investigating toxicity to the reproductive system and these comprise both germ cells and associated somatic cells either in mono- or co-culture systems. For example, a recent review lists some 14 cell types that are available for investigating effects on fertility at different stages of cell development.[110] These range from testis slices to Leydig, Sertoli, follicle, sperm, spermatocyte, and seminiferous tubule cell culture systems. In addition, immortalized testicular somatic and germ cells with specific toxicity markers, and co-cultures of Sertoli and various other cells (e.g., germ cells), are available, in particular for improving the longevity of the cells in culture and for modeling the blood–testis barrier.[17,111]

1.12.2.11.3 Endocrine toxicity

More than 50 different tests (including those based on subcellular, in vitro, and animal systems) have been developed[82,112–114] for the detection of endocrine disruptors, and evaluation of many of these tests is ongoing at the US Environmental Protection Agency (EPA), and as part of the work of the OECD in validating methods for endocrine disruptors in an effort to meet legislation in the US to implement the recommendations of the Endocrine Disruptor Screening and Testing Advisory Committee (EDSTAC).[21]

The main in vitro tests are: (1) subcellular hormone receptor ligand binding; (2) induction of proliferation (mitogenesis) in hormone-responsive mammalian cell lines; and (3) transactivation systems in yeast and mammalian cell lines (**Figure 1**).[113,115] These in vitro tests are generally very sensitive to a wide range of natural and synthetic endocrine disruptors (e.g., to nM and pM concentrations of estradiol), and they can be used to rank chemicals according to potency, and to detect agonistic and antagonistic effects. Strains of indicator cells have been isolated that express all the known relevant hormone receptors and also both subtypes of the estrogen receptor and they are usually easy to culture and genetically engineer, with no hormone receptor background. A good example of a recent sensitive in vitro assay is the T47D human breast cancer cell line developed by Wilson *et al.*,[116] that stably expresses a luciferase reporter gene system and both estrogen receptor subtypes, and is responsive to estrogen agonists and antagonists.

However, in vitro methods can give rise to intra- and interlaboratory differences in data depending on subclone, protocol, culture conditions (e.g., endocrine disruptors in serum), and also time-dependent specific effects. Moreover, some cell lines become unstable and estrogen-independent with repeated subculturing. Nonspecific binding of potential endocrine disruptors with tissue culture medium components such as serum, and/or lack of cell permeability (e.g., yeast) can be the cause of false-negative data. In some cases, the mechanistic basis of the assays and the relevance of specific gene activation is obscure.[112] Lastly, the omission of metabolism from in vitro test methods for endocrine disruptors is a serious problem and has been reviewed elsewhere.[117]

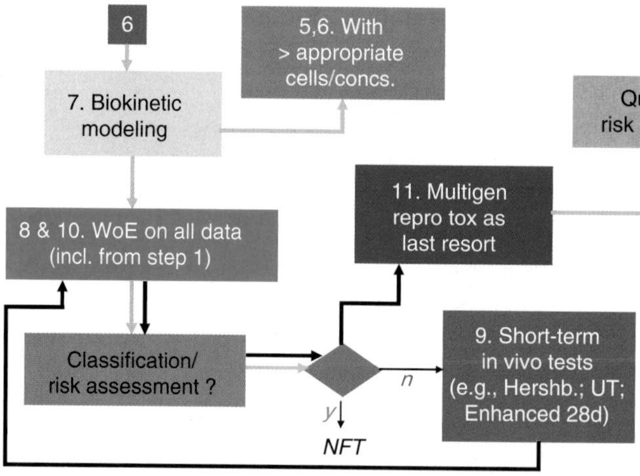

Figure 1 Decision tree integrated testing scheme for endocrine disruptors. Hershb, Hershberger assay; NFT, no further testing; UT, uterotrophic assay; WOE, weight of evidence evaluation.

1.12.2.12 Acute Systemic Toxicity

Alternative strategies for predicting acute systemic toxicity have made substantial progress since the outcome of the Multicenter Evaluation of In Vitro Cytotoxicity (MEIC) study showed that basal cytotoxicity is a good general predictor of systemic lethal blood concentrations in humans.[118] This overall correlation applied irrespective of: (1) cell culture systems or (2) the technique used for measuring cytotoxicity, and verified the hypothesis that such a correlation would be expected since acute systemic toxicity is primarily due to impairment of the function of a key target organ, such as the nervous system, arising from the death of a critical number of target cells.[119]

The MEIC study involved 59 laboratories, 50 reference test chemicals with human acute toxicity poisoning data, and a large number of in-house protocols.[118] The observed correlation between lethal blood concentrations and basal cytotoxicity showed several outliers, as would be expected. It has been suggested that these could be corrected by taking into account the following properties of the test substance: (1) in vivo biokinetics; (2) biotransformation; (3) ability to pass across internal membranes; and (4) target organ specificity due to specific effects on different cell types.[120]

These factors are being taken into account in a large multidisciplinary integrated EU project, A-Cute-Tox, involving some 40 laboratories and several different work packages, with the objective of developing and ultimately prevalidating an optimized in vitro testing strategy for predicting acute toxicity.

One of the basal cell cytotoxicity assays in the MEIC study, and widely used in in vitro toxicity assay, is the NRU assay with 3T3 cells (the basis of photoirritation; see earlier). This assay has been the subject of an ECVAM/ICCVAM validation study,[121] to replace the in vivo studies required to define starting doses for in vivo acute systemic toxicity testing.[122]

Further to the outcome of the MEIC study, several strategies proposed for acute systemic toxicity involving in vitro methods have been proposed. Seibert *et al.*[123] advocated the use of: (1) basal cytotoxicity (to give a rough indication of acute toxicity); (2) hepatocyte-specific toxicity, to account for any biotransformation required; and (3) target cells in culture (to include an indicator of selective cytotoxicity to target tissues, as indicated by likely target organs). A further similar test strategy was described by Worth and Balls,[11] comprising a series of in silico and in vitro steps, with some aspects of in vitro biotransformation tests, cell-specific toxicity tests, and culminating in an in vivo study if toxicity cannot be determined at an earlier stage in the scheme.

1.12.2.13 Chronic Toxicity

Prospects for replacing animal use for chronic and repeat-dose toxicity are, at present, very limited and there are no alternative repeat-dose/subchronic tests accepted for regulatory testing. ECVAM Workshop 45 identified the need for the development of new techniques, including methods for the long-term maintenance of in vitro systems and refined perfusion culturing, for undertaking long-term and repeat-dose toxicity studies.[124–126] There is a need for more focused and coordinated research in this area, but a number of initiatives have already been taken.

Single cell or co-culture-based systems, in which in vivo rates of perfusion are mimicked such that the bioaccumulation of reactive metabolites is avoided, and the metabolic kinetics more closely reflect the in vivo situation, are in development. For example, metabolically competent hepatocytes maintained as reconstructed collagen sandwich monolayers show extended viability such that repeat-dose, long-term, and reversible effects can be studied.[127] Also, tissue engineering techniques and co-culturing have been used to extend longevity in culture.[128] Similarly, cells can be grown in hollow-fiber bioreactors through which the culture medium flows to and from the cells and is replaced regularly, or on perfusable inert polymer support membranes or in culture plate inserts, such as two-compartment human skin models, where cells cultured on an inert filter can be washed and analyzed after an initial exposure using a nontoxic indicator, then rechallenged and reanalyzed.[129] Methods for long-term in vitro studies are being developed: the perfusion culture systems EpiFlow and Minucell are under evaluation.[126] An in vitro long-term cytotoxicity study,[130] in which 27 of the chemicals from the original MEIC study were tested on HEp-G2 cells, found that the results obtained correlated quite well with acute human toxicity data ($r^2 = 0.709$) when linear regression analysis was used.

1.12.2.14 Carcinogenicity

Owing to the severity of the rodent bioassay, the numbers of animals used, the costs and length of time involved, and concerns about the relevance of the data to human hazard, there has been much interest in finding alternatives for this endpoint. Moreover, there are many concerns about the relevance of bioassay data for identifying carcinogens of human relevance,[131,132] and also the protocols and methods involved in the relevant OECD TG are in urgent need of being updated.[80,133] Recently, when Knight *et al.*[131] compared the International Agency for Research on Cancer (IARC) and US EPA databases for human carcinogens, they found significant differences in human carcinogenicity classifications for the same chemicals: the IARC classification is more conservative and less reliant on animal data. This implies that the true predictivity for human carcinogenicity of animal bioassay data is even poorer than indicated by EPA figures alone. The same authors[132] also analyzed the same 160 carcinogens in the EPA database, classified as human carcinogens solely on the basis of animal data. Wide differences in species, routes of administration, and affected target organs were noted, together with discordant data and the use of very high and totally unrealistic dose levels, thereby confounding extrapolation of the data to human hazard and risk. The authors concluded that such animal bioassay data are unsuitable for characterizing human carcinogens.

Depending on the intended use of the test compound and the likely level of human exposure, most regulatory requirements stipulate the need for in vivo carcinogenicity testing, which still includes at least one rodent bioassay, either alone or coupled with a transgenic mouse assay, for risk assessment,[64] and particularly when there is a need to be sure of lack of carcinogenicity. This situation is mainly due to an inability to identify nongenotoxic carcinogenic activity by using nonanimal approaches, to a sufficient level of confidence. The existence of nongenotoxic carcinogens became obvious following analysis of data from the National Toxicology Program (NTP),[57] in which carcinogenicity was often observed in one species, one sex, or even one specific tissue, and usually only at high dosages for chemicals that were devoid of genotoxicity. Thus, whereas carcinogens acting via a genotoxic mechanism can be identified with high levels of certainty by a combination of in vitro and short-term in vivo assays (*see* Section 1.12.2.3), and labeled as such without the need for long-term animal testing, this is not the case for nongenotoxic carcinogens.[134]

Kirkland *et al.*[135] have evaluated the predictive performance of a test battery, comprising the Ames, mouse lymphoma, and the in vitro micronucleus assay, for 700 chemicals for which rodent carcinogenicity is available. While the sensitivity (ability to predict carcinogenicity correctly) of the battery was high (93% of chemicals with reliable genotoxicity data yielding positive data in at least one of the tests), the specificity was variable depending on the test used (73.9% for the Ames test, but <45% for the two mammalian cell-based assays). In fact, 75–95% of noncarcinogens were incorrectly positive in at least one test in the in vitro battery, and only 19 chemicals that were carcinogenic in vivo were consistently negative in vitro, either because they were very weak carcinogens or they acted via a nongenotoxic mechanism. This unsatisfactorily low level of specificity suggests that the data from in vitro tests for predicting carcinogenicity need to be interpreted with care.

The major problem with developing in vitro models for nongenotoxic carcinogens is the wide diversity of their potential modes of action, some of which involve binding with a range of different types of intracellular receptors, and others that involve epigenetic changes coupled with cell proliferation in tissues.[134] The inability to detect the carcinogenic activity of nongenotoxic carcinogens by using the standard battery of genotoxicity assays poses major problems for regulatory bodies.

The main alternative nonanimal methods for modeling carcinogenicity involve morphological cell transformation,[136,137] including the induction of certain phenotypic alterations in cells in tissue culture that are characteristic of tumorigenic cells (i.e., they behave like cells taken from tumours when cultivated in vitro and they can elicit tumours when implanted into host animals). The main cell transformation systems are based on rodent cell lines, e.g., Balb/c 3T3 cells, immortalized fibroblast cell lines of rodent or human origin, e.g., C3H10T1/2 cells or primary cells, such as those used in the Syrian hamster embryo (SHE) cell assay.[138]

Some transformation assays are designed to detect initiators of carcinogenesis, acting by a genotoxic mechanism via electrophilic interactions with cellular macromolecules such as DNA, while others can be used in a two-stage process to detect tumour promoters additionally via nongenotoxic mechanisms. Also, an in vitro assay specifically for tumor promoters has become available.[139] This is based on the creation of Bhas 42 cells by transfecting Balb/c T3 cells with the v-*Ha*-ras oncogene.

Cell transformation assays have been shown to detect variously some well-known animal and human carcinogens, as well as tumor promoters, particularly when new protocols have been developed for the Balb/c 3T3 and SHE cell assays. The results of a collaborative study on the use of a new protocol for the Balb/c 3T3 assay have now been published, and this provides an improvement in the reliability of these assays for detecting rodent carcinogens,[136] particularly for the SHE cell assay which exhibits high sensitivity for genotoxic and nongenotoxic rodent and human carcinogens.

However, in order to improve the identification of carcinogens it will be necessary to: (1) develop systems that take more account of the need for chemical biotransformation; and (2) develop a larger battery of in vitro assays. With regard to the former, exogenous biotransformation systems for the detection of procarcinogens are not routinely used.[97] When biotransformation systems have been added, there have been some problems with the assays, although it is known that SHE cells can express some metabolic competence.[140] An ECVAM workshop[136] concluded that "a thorough investigation is needed to ascertain the necessity of using exogenous metabolising systems in the various cell transformation assays." Unfortunately, such an investigation has not been conducted, and it is suggested that it should form part of the design of a prevalidation study of the cell transformation assay, involving the use of SHE and Balb/c 3T3 cells currently being planned by ECVAM. With regard to the requirement for a wider battery of tests, there is a need to take account of nongenotoxic carcinogens that act via highly tissue-specific effects (for example, thyroid carcinogenesis and peroxisome proliferation). In this regard, attempts are being made, for example, to use several cell culture methods for investigating the tissue-specific induction of proliferation, such as MCF-7 cells to detect hormonally active chemicals that are the linked with the etiology of breast cancer.[134] Moreover, several receptors for nongenotoxic hepatocarcinogenesis have been identified, and this should facilitate the development of ligand-binding assays. In addition, it should not be overlooked that some endpoints measured by methods used in mutagenicity testing batteries are relevant to the modes of action of nongenotoxic carcinogens, such as changes in chromosomal number (polyploidy or aneuploidy) and, as a consequence, some authorities have called for the development and deployment of specific tests for aneuploidy.

There have been suggestions that nongenotoxic carcinogens can affect intercellular communication between cells, although a correlation between the two endpoints has been difficult to demonstrate unequivocally.[141,142] However, a gap junction intercellular communication (GJIC) assay has been developed,[142,143] and there is more evidence from SAR of a relationship between substances able to inhibit GJIC and rodent carcinogencity, although no relationship exists between gap junction inhibition and genotoxicity. However, the GJIC assay requires further evaluation.

Various integrated testing schemes have been proposed for carcinogenicity. One strategy[11] commences with a determination of genotoxicity by methods discussed earlier, with a positive result possibly allowing a classification as a genotoxic carcinogen. A negative result should trigger further testing, which might firstly involve the sequential use of in silico, and then in vitro, models for nongenotoxic carcinogens, culminating in an in vivo test, possibly a rat bioassay and a mouse transgenic carcinogenicity assay. It might be feasible to use information from shorter, chronic testing, particularly that regarding effects on endocrine function and cell proliferation in target tissues, without resorting to the bioassay. Combes *et al.*[81] suggested the following strategy: (1) no further testing is necessary when a chemical proves to be unequivocally genotoxic (backed up if necessary by a short-term in vivo assay); (2) in the case of nongenotoxic chemicals, the need for, and type of, further in vivo testing should be based on a case-by-case, flexible approach, depending on likely human exposure and other information, including the use of nonvalidated tests (for example, cell transformation assays) for chemical prioritization; and (3) rodent bioassays should only be conducted as a last resort, when the rat should be the first species of choice.

Knight *et al.*[144] have proposed a four-stage strategy, starting with the use of all existing information to enable the selection of relevant test methods. Stage 2 comprises the use of initial tests (e.g., an expert system modeling, standard core and ancillary genotoxicity assays, cell transformation and cDNA microarrays). Stage 3 involves human studies with specific biomarkers of exposure and effect and biokinetic modeling. Lastly, stage 4 comprises volunteer and patient clinical trials.

1.12.2.15 Prospective Human Studies

The ethical use of humans, particularly for efficacy and safety testing of pharmaceuticals, household products, and cosmetics,[145] is an important way of reducing the use of animals. Such studies are intended to confirm observations that have been obtained in earlier safety testing investigations. The necessity for prior animal testing was banned several years ago in Europe for new finished cosmetics products containing reformulations of existing ingredients for which animal test data are available. This was based on the premise that alternative methods (as described earlier) can be used for predicting the overall toxicity of combinations of ingredients, and the toxicity of mixtures can be predicted by taking into account possible antagonistic and synergistic effects between different components. Skin absorption and skin patch testing studies[146] can be performed according to standard protocols (OECD TG 428 in the case of the former endpoint).

In the UK, in particular, first administration to humans (phase I testing) of a candidate drug, to assess absorption, distribution, metabolism, and excretion (ADME) properties, is possible at a very early stage of testing with further, longer-term clinical studies being dependent on the results of subsequent animal testing involving an increased number of endpoints. The possibility of using volunteers more extensively and earlier, with less prior animal testing, is being investigated by applying the technique of microdosing. The idea is to administer very low doses of a drug to humans, below threshold dose levels for toxicity, but still detectable by hypersensitive analytical techniques to enable ADME to be undertaken.[147] In this way, the likely pharmacological and safe first dose for phase I studies can be established with only minimal risk to the volunteer. There has been much interest in this approach, such that it has become known as phase 0 testing. This activity has led to one of the principal analytical techniques, accelerated mass spectrometry (AMS) being subjected to a trial (CREAM: Consortium for Resourcing and Evaluting AMS Microdosing), in which predictions for the pharmacological doses for three of five drugs were useful; the data for the other two drugs were considered less useful but for reasons that did not detract from the utility of the AMS technique.[148] The potential for microdosing is gaining interest from a number of regulatory bodies.

The Declaration of Helsinki is the principal ethical guideline relevant to the conduct of clinical research worldwide. There are national guidelines covering the conduct of such research in various countries. However, in contrast to the situation with pharmaceuticals, there is much variation between what is permitted in different countries with regard to cosmetics and household products testing. ECVAM Workshop 36 recommended that a standard ethical protocol should be defined for the testing of cosmetics on humans.[149] The SCCNFP has approved new ethical guidelines on human volunteers. This committee has also discussed a draft guideline on human testing for the irritancy of finished cosmetic products.

1.12.3 The Application of New Approaches

1.12.3.1 Increased Use of Human Cells

The use of human cells in tissue culture has the great advantage over animal cells that species extrapolation problems are obviated.[150–152] This is because human cells are better models than animal cells for phenomena occurring in humans (since they possess human enzymes and targets such as receptors). Moreover, there is no need to use donor animals. Human cells can be used in a wide variety of ways, although they suffer from a number of technical disadvantages, particularly their short longevity in culture, and their reluctance to undergo immortalization spontaneously. In addition, their acquisition poses logistical and ethical problems.

The principal clinical sources of human cells and tissues for research are: (1) patients undergoing routine surgical operations (living donors); (2) patients who die in intensive care; (3) patients who die in Accident and Emergency units (brainstem-dead, heart-beating donors); and (4) post-mortem examinations. Cells are derived from tissue that is deemed unsuitable for transplantation (surgical waste). The tissues most frequently obtained for research in this way are liver, kidney, lungs, tracheobronchial tract, skin, and heart, from human research tissue banks.[153]

1.12.3.2 Genetically Engineered Cells

1.12.3.2.1 Reporter gene constructs

Many new screening systems are being developed in a variety of cell types and organisms, including yeast, and mammalian and human cell lines, by using genetically engineered cells with specific reporter genes.[154] DNA sequences coding for a receptor, and introduced on a vector molecule into recipient cells, are linked to the corresponding DNA response element sequence. The latter stimulates specific gene expression in response to the binding of the receptor to a chemical ligand (e.g., the test substance). New transcription is visualized by the promoter region of the response element being ligated to the promoter region of a reporter gene, whose product can be detected colorimetrically.[134]

1.12.3.2.2 Cells with biotransforming potential

There are many different cell lines that have been genetically engineered to express various phase I and phase II enzymes, and also a combination of both, for use in toxicity testing.[97,98] Some of these cell lines have a number of benefits, including a stable diploid karyotype and no biotransforming enzyme background activity, and they can be transfected with rodent and human genes encoding enzymes. This allows an investigation of the contribution of specific isozymes to metabolism and also studies on species specificity.

1.12.3.3 Using Stem Cells

Stem cells retain totipotency and can in principle be stimulated (reprogrammed) to develop into any cell type in the body.[155,156] They can be obtained from either embryos (embryonic stem cells) or from differentiated somatic tissues (adult stem cells, or ASTs). There has been much research into finding ways to stimulate the reprogramming of embryonic stem cells and ASTs to generate cells of different types, with varied success. The requirements for ASTs are generally more stringent than for embryonic stem cells.

Apart from the use of embryonic stem cells in embryotoxicity models, stem cells are becoming increasingly used in neurotoxicity.[99] Thus, stem cells from human umbilical cord blood (HUCB)[154] can produce a neural stem cell-like subpopulation which proliferates in culture for several months. Such cells can be induced in the presence of trophic factors, mitogens, and neuromorphogens to produce neuronal or glial cell markers and become responsive to neurogenic signals (i.e. to develop into neuronal, astrocytic, and oligodendroglial lineages). Bużańska *et al.*[157] have used two-dimensional and three-dimesional cultures of a HUCB-derived neuronal cell line with a stable karyotype, which forms neurospheres in serum-free medium as a model for neurotoxicity testing. Growth rate and cell differentiation of the cultures were measured in 96-well plates by using the 3-(4,5-dimethylthiazol-2-yl)-2,5-diphenyltetrazolium bromide (MTT) test, and differentiation into neural-like cells was assessed by immunocytochemistry.

There has also been considerable interest in trying to derive hepatocytes from adult stem cells.[97,98] Thus, advances are being made in stimulating the maturation of stem cells derived from human adult bone marrow to differentiate into hepatocytes.[149] Several potential sources of stem cells have been considered for generating differentiated hepatocytes, including embryonic stem cells and intrahepatic progenitors. Under appropriate culture conditions and stimulation by cytokines, growth factors or chemical reagents, embryonic stem cells can differentiate into hepatocyte-like cells expressing markers such as albumin and other plasma proteins, CK8/18, glycogen, and xenobiotic-mediated induction of cytochrome P-450 genes.[158] Similarly, intrahepatic progenitor cells can generate hepatocyte-like cells expressing the same liver-specific markers. However, it appears that the routine production of differentiated hepatocytes from stem cells remains elusive for the time being.

1.12.3.4 Toxicogenomics and Proteomics

1.12.3.4.1 Principles

The use of this collection of new techniques[154,159] is based on the principle that toxicity affects the patterns of gene expression in cells in tissues. After exposure to a test substance, gene expression profiles at either the transcriptional (genomic) level or the translational (proteomic) level are compared with those detected after exposure to a known toxicant.

Differential transcription is measured using microarrays of oligonucleotides which are used to detect cDNA and cRNA copies of new transcripts extracted from cells. Specific hybridization to the oligonucleotides is visualized by scanning under laser light, resulting in several thousand measurements of gene expression. Highly specific panels of oligonucleotides are used to determine tissue-dependent temporal and spatial patterns of gene expression to be measured.

1.12.3.4.2 Constraints

Currently, there are constraints on the use of microarray data, since huge amounts of information are generated rapidly and their interpretation is proving extremely difficult. This is a particular problem with the analysis of global patterns of gene expression data, and this has prompted the development of new disciplines, such as bioinformatics, to alleviate the problem, involving the use of online access to complex databases. However, identification of the induction of specific patterns of gene expression could facilitate the classification of agents with different mechanisms of action.[160]

It is, nevertheless, more likely that in the short term, toxicogenomic approaches will provide more useful data when they become focused on specific areas of toxicity, for which mechanisms of action of toxic chemicals are well characterized. Also microarrays can be used that target the likely genes involved. For example, the induction of heat shock proteins as an indicator of toxicity is being investigated.[161,162]

Moreover, the validation of these techniques will require careful consideration,[163] in view of their complexity. This has to some extent been already achieved with studies on sensitization and in the area of endocrine disruption.[164]

1.12.3.4.3 Use for sensitization testing

An example of a focused study on toxicogenomics concerns the recent investigation of Schoeters *et al.*,[165] who used cDNA microarrays to assess patterns of transcriptional activity of 11 000 human genes in human DCs derived from $CD34^{+}$ progenitor cells exposed to the sensitizer dichloronitrobenzene (DNCB) or solvent for 3, 6, and 12 h. Compared to control gene expression, changes larger than $\pm$ twofold were observed for 241 genes after exposure to DNCB, of which 137 were upregulated and 104 downregulated. Interestingly, 20 of the genes were known to encode proteins that are related to the immune response.

1.12.3.4.4 Use for endocrine disruptor testing

Toxico-genomic and -proteomic analyses have been used to detect endocrine disruptors[166,167] and Terasaka *et al.* have developed a DNA microarray comprising 172 genes that were shown to be estrogen-responsive in previous studies with MCF-7 cells. The microarray was more sensitive than conventional tests to the effects of some endocrine disruptors, and was used to analyze the effects of a number of naturally occurring and synthetic chemicals, with highly predictive results.

1.12.3.4.5 Proteomics

Proteomics involves the analysis of total cellular protein, using standard methods for this purpose. The methods for detecting and quantifying new proteins, as a result of upregulation, have been greatly improved recently, and use of this technique is likely to increase. Moreover, more useful information could be obtained from a combined use of microarrays and global protein analysis.[164] A new EU project (Predictomics STREP (ECVAM): Short term in vitro assay for long term toxicity) has been initiated, and this involves integrating the use of genomics and proteomics with advanced cell culture technology to detect early biomarkers of cellular injury.[164]

1.12.4 Some Important Issues

1.12.4.1 Taking Account of Biotransformation

The potential biotransformation of chemicals in in vitro assays needs to be taken into account since it could be the cause of false-positive data (due to lack of detoxification) or false-negative data (due to lack of activation).[98] Apart from genotoxicity testing and hepatoxicity studies (usually by default when primary hepatocytes are used),this factor has largely been ignored in the designing of validated in vitro toxicity testing methods.[11] In such cases, the good predictivity of test methods must have been due to: (1) toxicity being induced by the parent substance; (2) the residual biotransformation by the toxicity indicator cells; or (3) biotransformation being nonessential. Embryoxicity, endocrine disruption, sensitization, and acute toxicity are key endpoints for which there is an urgent need for new tests to be developed that include a means of assessing the involvement of biotransformation in toxicity.[97,116]

1.12.4.2 Assessing Risk by Using In Vitro Data

The use of data from nonanimal toxicity methods in risk assessment has been limited primarily to hazard identification (HI). This is because: (1) most existing replacement test methods are not appropriate for hazard characterization (HC); and (2) quantitative data from nonanimal methods are considered impossible to use to define threshold doses and toxic potency.[133] However, there is no reason why hazard data from in vitro assays should not be used for hazard classification, for example, where a positive result is assumed, or accepted, to be definitive. Such data can also help to elucidate a toxic mechanism of action which can facilitate risk assessment (e.g., distinguishing a genotoxic carcinogen from a nongenotoxic carcinogen, leading to assumptions of no threshold and threshold dose levels, respectively).

For a test to be useful for HC it should: (1) be mechanistically based with a biologically plausible relationship between the endpoint measured and the phenomenon being modeled (i.e., preferably based on human cells, for reasons discussed earlier); (2) ideally have been validated against human data; (3) have a relevant endpoint (one occurring in the target species); and (4) have a prediction model that is readily related to toxicity occurring in the target species.[133] Interpreting in vitro dose–response data could be facilitated by using target cells in culture and toxicokinetic modeling to predict in situ concentrations of test compound.[168]

There is a need for more tests suitable for HC to be developed and validated, and for information on the differences between cells in culture and in tissues, and the effects of dosing in vitro and in vivo, to develop realistic and meaningful uncertainty factors to allow in vitro information to be used for risk assessment in its own right, and in conjunction with animal data. In the meantime, the most realistic use of nonanimal methods for risk assessment in large-scale risk assessment programs, such as Registration Evaluation and Authorisation of Chemicals (REACH), is for them to be integrated into hierarchical integrated tier-testing strategies in conjunction with animal test methods.

1.12.4.3 Integrated Tier-Testing Strategies

The application of nonanimal methods to toxicity testing has evolved from nonregulatory, in-house investigative and screening studies, through the use of other partial replacement approaches comprising test batteries, to hierarchical testing schemes, and eventually to the use of complete replacement methods for regulatory purposes.

Batteries of complementary in vitro tests have been developed and used to overcome the problem that a process in vivo might occur via several different mechanisms. Each test in the battery is based on detecting an endpoint which involves at least one of the mechanisms known or suspected to be relevant to the overall biological effect. These tests can all be nonanimal, as well as animal, assays, which are applied in a sequential or hierarchical approach, beginning with the nonanimal methods. Various strategies for individual toxicity endpoints have been discussed earlier.

The use of integrated testing strategies and schemes for prioritization are particularly suitable for large-scale screening and testing, for example in food toxicity testing[169] and for chemicals testing, and tier-testing schemes have been proposed for the latter purpose to satisfy the EU REACH legislation.[170,171]

Integrated testing strategies generally follow similar steps: (1) maximum use of pre-existing data (e.g., by data-sharing and effective information retrieval); (2) the use of in silico and physicochemical chemical approaches (e.g., (Q)SAR modeling, expert systems, read-across approaches[7] and metabolism screening); (3) the use of tissue culture methods to identify toxic hazard; and (4) the use of in vivo tests and human data. The different published schemes differ in the relative proportions of each of the stages, depending on the toxicity endpoint and the extent to which nonanimal methods exist (**Figure 2c**). Some, like a British Union for the Abolition of Vivisection (BUAV) overall scheme for REACH, omit the use of animals completely.[172] Recently, a study was concluded in which a series of integrated testing strategies for a range of toxicity endpoints required for REACH were developed in the form of individual decision trees (**Figure 2a** and **2b**).[173]

1.12.5 Discussion and Conclusions

Three principal barriers to the increased use of in vitro tests for regulatory testing exist: the need for: (1) new, improved methods, particularly for toxic endpoints for which mechanisms of toxicity are complex and/or obscure; (2) such new methods to be validated, according to internationally agreed criteria, and accepted by regulatory agencies; and (3) the more widespread and acceptable use of in vitro test data for risk assessment. While the first requires a dual approach of more fundamental and basic research, the second two could be greatly facilitated by: (1) the increased use of human cells of different types; (2) the better maintenance of differentiated cells in culture for long periods; (3) the judicious use of genetically engineered cells with useful characteristics; (4) the further development of complex organotypic cell systems involving co-cultures of relevant cell types; (5) the development of techniques for long-term culturing, repeat dosing, and assessment of recovery; and (6) finding ways to use the new technologies to maximum effect. Advances are being made on all these fronts, but a key rate-limiting factor is our lack of knowledge concerning the basis of the intrinsic differences between cells in culture and in situ in tissues that determine their susceptibilities to toxicity. These differences are determined by changes in many factors, including: cell surface receptors, cell membrane, cell signaling mechanisms, transporter proteins, transport systems to and from cells, and the cytoskeleton, for which new imaging techniques have been reported.[174] Such information will be crucial in enabling the development of realistic and meaningful uncertainty factors to allow in vitro information to be used for risk assessment in its own right, and in conjunction with animal data.[133]

Clearly, there has been much progress with the development, validation, and acceptance of in vitro test systems for regulatory safety assessment over the last decade. However, there is much still to be achieved and, although this will be facilitated by the availability of new technologies, it is crucial that a strictly scientific and multidisciplinary approach is adopted so that new mechanisms of toxicity can be discovered, and the resulting knowledge used to identify useful new models and test systems.

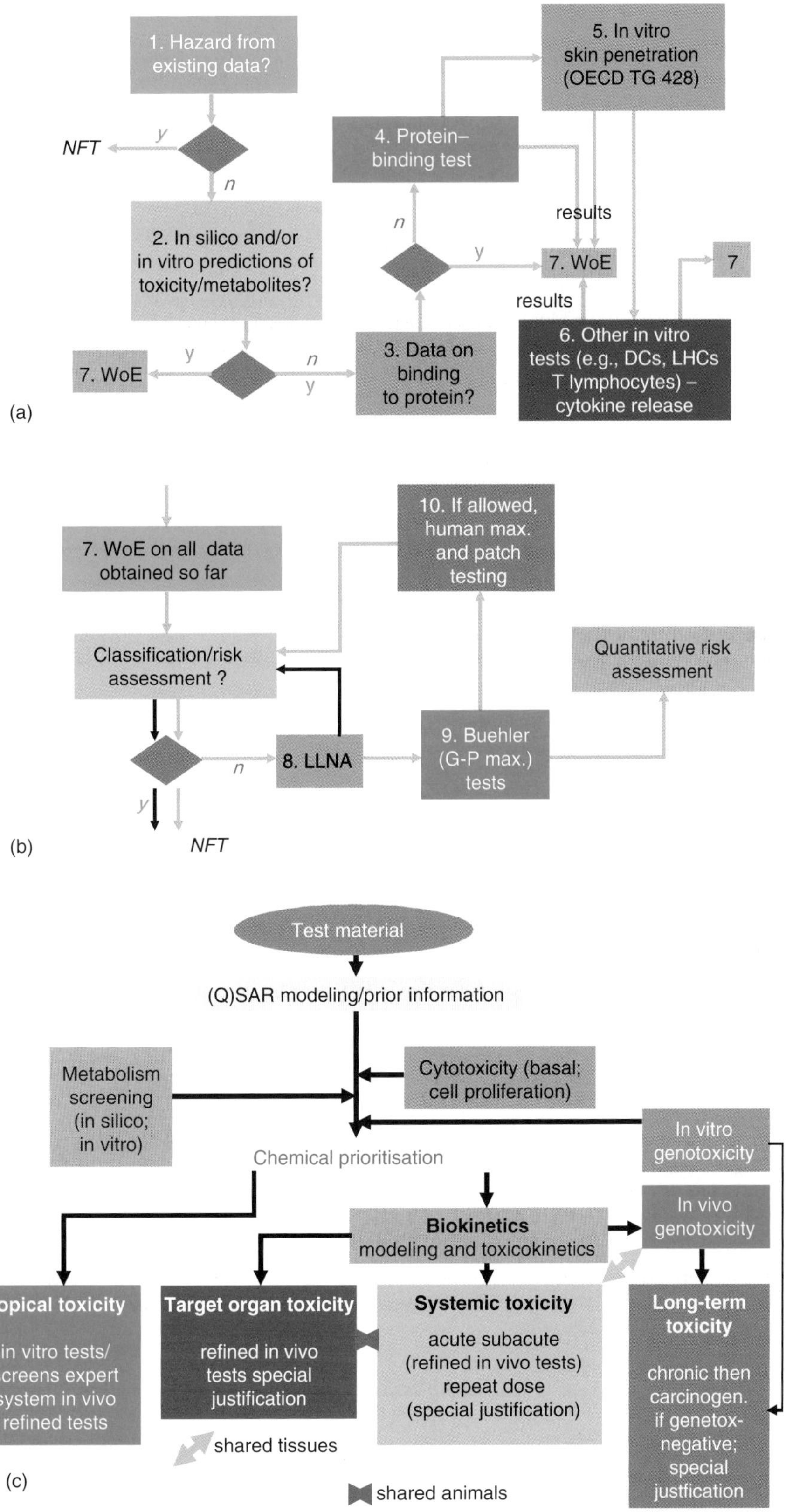

Figure 2 (a) and (b) Decision tree integrated testing scheme for sensitization. (c) Integration of in vitro toxicity tests with in vivo methods for chemicals testing. LHC, Langerhans cell; LLNA, local lymph node assay; G-P, guinea pig.

References

1. OECD. *Report from the Expert Group on (Quantitative) Structure–Activity Relationships [(Q)SARs] on the Principles for the Validation of (Q)SARs*. OECD Environment Health and Safety Publications, Series on Testing and Assessment: Paris, France, 2004; p 206.
2. Combes, R. D.; Rodford, R. A. The Use of Expert Systems for Toxicity Prediction: Illustrated with Reference to the DEREK Program. In *Predicting Chemical Toxicity and Fate*; Cronin, M. T. D., Livingstone, D. J., Eds.; CRC Press: Boca Raton, FL, 2004, pp 193–204.
3. Cronin, M. T. D.; Livingstone, D. J. *Predicting Chemical Toxicity and Fate*; CRC Press: Boca Raton, FL, 2004.
4. Cronin, M. T. D.; Jaworska, J. S.; Walker, J. D.; Coomber, M. H. I.; Watts, C. D.; Worth, A. P. *Environ. Hlth Persp.* **2003**, *111*, 1391–1401.
5. Dearden, J. C.; Barratt, M. D.; Benigni, R.; Bristol, D. W.; Combes, R. D.; Cronin, M. T. D.; Judson, P. M.; Payne, M. P.; Richard, A. M.; Tichy, M. et al. *ATLA* **1997**, *25*, 223–252.
6. Helguera, A. M.; Pérez, M. A. C.; Combes, R.; González, M. P. *Curr. Comput.-Aided Drug Design* **2005**, *1*, 65–72.
7. Combes, R. D.; Balls, M. *ATLA* **2005**, *33*, 289–297.
8. Balls, M. *AATEX* **2005**, *11*, 4–14.
9. Anonymous. *7th Amendment to the EU Cosmetics Directive*, Directive 2003/15/EC of the European Parliament and of the council of 27 February 2003, amending Council Directive 76/768/EEC on the approximation of the laws of the Member States relating to the cosmetic products was issued on March 11, 2003, in the official Journal of the European Union, Brussels, 2003.
10. Anonymous. *Proposal for a Regulation of the European Parliament and of the Council concerning the Registration, Evaluation, Authorisation and Restriction of Chemicals (REACH), establishing a European Chemicals Agency and amending Directive 1999/45/ec and Regulation (EC) {on Persistent Organic Pollutants}*. COM(2003) 644 final; EU: Brussels, 2003.
11. *ATLA* **2002**, *30*, 1–125.
12. van de Waterbeemd, H. *Curr. Opin. Drug Disc. Dev.* **2002**, *5*, 33–43.
13. Balls, M.; Blaauboer, B. J.; Bruner, L.; Combes, R. D.; Eckwall, B.; Fentem, J. H.; Fielder, R.; Guillouzo, A.; Lewis, R.; Lovell, D. et al. *ATLA* **1995**, *23*, 129–147.
14. Combes, R.; Balls, M. *ATLA* **2003**, *31*, 225–232.
15. OECD. *Report of the OECD Workshop on Harmonization of Validation and Acceptance Criteria for Alternative Toxicological Test Methods.* January 22–24, *Solna, Sweden*. OECD: Paris, France, 1996.
16. Hojke, R. A.; Ankley, G. T. *Environ. Toxicol. Chem.* **2005**, *24*, 2677–2690.
17. Fielder, R. J.; Atterwill, C. K.; Anderson, D.; Boobis, A. R.; Botham, P.; Chamberlain, M.; Combes, R.; Duffy, P. A.; Lewis, R. W.; Lumley, C. E. et al. *Hum. Exp. Toxicol.* **1997**, *16*, S1–S40.
18. Combes, R.; Balls, M.; Bansil, L.; Barratt, M.; Bell, D.; Botham, P.; Broadhead, C.; Clothier, R.; George, E.; Fentem, J. et al. *ATLA* **2002**, *30*, 365–406.
19. Castell, J. V.; Gómez-Lechón, M. J. *In vitro Methods in Pharmaceutical Research*; Academic Press: Basingstoke, 1997, pp 318–352.
20. Knight, D. J.; Breheny, D. *ATLA* **2002**, *30*, 7–22.
21. Combes, R. D. *ATLA* **2002**, *30*, 151–165.
22. Eskes, C.; Zuang, V., Eds. *ATLA* **2005**, *33*, 1–228.
23. Schifflers, M. -J.; Hagelstein, G.; Harreman, A.; van der Spek, M. Regulatory Animal Testing – A Survey of the Factors Influencing the Use of Animal Testing to Meet Regulatory Requirements; University of Utrecht: Netherlands, 2005.
24. Hartung, T. *ATLA* **2002**, *30*, 49–51.
25. Langezaal, I.; Hoffmann, S.; Hartung, T.; Coecke, S. *ATLA* **2002**, *30*, 581–595.
26. Hartung, T.; Wendel, A. *In Vitro Toxicol.* **1996**, *9*, 353–359.
27. Fennrich, S.; Wendel, A.; Hartung, T. *ALTEX* **1999**, *16*, 146–149.
28. Balls, M.; Berg, N.; Bruner, L. H.; Curren, R. D.; de Silva, O.; Earl, L. K.; Esdaile, D. K.; Fentem, J. H.; Liebsch, M.; Ohno, Y. et al. *ATLA* **1999**, *27*, 53–77.
29. Curren, R. D.; Harbell, J. W. *ATLA* **2002**, *30*, 60–74.
30. Tchao, R. Trans-Epithelial Permeability of Fluorescein in Vitro as an Assay to Determine Eye Irritants. In *Alternative Methods in Toxicology*, Goldberg, A. M., Ed.; *Progress in In Vitro Toxicology*; Mary Ann Liebert, Inc.: New York, 1988; Vol. 6; pp 271–283.
31. Shaw, A. J.; Clothier, R. H.; Balls, M. *ATLA* **1990**, *18*, 145–151.
32. Śladowski, D.; Liberek, I.; Lipski, K.; Ozga, T.; Olkowska-Truchanowicz, J.; Szaflik, J. *Toxicol. in Vitro* **2005**, *19*, 875–878.
33. Clothier, R.; Starzec, G.; Pradel, L.; Baxter, V.; Jones, M.; Cox, H.; Noble, L. *ATLA* **2002**, *30*, 493–504.
34. Moore, P.; Ogilvie, J.; Horridge, E.; Mellor, I. R.; Clothier, R. H. *Eur. J. Cell Biol.* **2005**, *84*, 581–592.
35. Combes, R. D. *Toxicol. in Vitro* **1999**, *13*, 853–857.
36. Spielmann, H.; Kalweit, S.; Liebsch, M.; Wirnsberger, T.; Gerner, I.; Bertram-Neis, E.; Krauser, K.; Kreiling, R.; Miltenburger, G.; Pape, W.; Steiling, W. *Toxicol. in Vitro* **1993**, 7, 505–510.
37. Spielmann, H.; Liebsch, M.; Kalweit, S.; Moldenhauer, F.; Wirnsberger, T.; Holzhutter, H. -G.; Schneider, B.; Glaser, S.; Garner, I.; Pape, W. J. W. et al. *ATLA* **1996**, *24*, 741–858.
38. Institute for In Vitro Sciences website. www.iivs.org (accessed Sept 2006).
39. Pape, W. J. W.; Pfannenbecker, U.; Hoppe, U. *Mol. Toxicol.* **1987**, *1*, 525–536.
40. Balls, M.; Combes, R. D. *ATLA* **2005**, *33*, 299–308.
41. Brantom, P. G.; Bruner, L. H.; Chamberlain, M.; DeSilva, O.; Dupuis, J.; Earl, L. K.; Lovell, D. P.; Pape, W. J. W.; Uttley, M.; Bagley, D. M. et al. *Toxicol. in Vitro* **1997**, *11*, 141–179.
42. Garle, M. J.; Fry, J. R. *ATLA* **2003**, *31*, 295–316.
43. Forsby, L. J. A. *ATLA* **2004**, *32*, 339–343.
44. Cejkova, J.; Lojda, Z.; Brunova, B.; Vacik, J.; Michalek, J. *Histochemistry* **1988**, *89*, 91–97.
45. Ubels, J. L.; Prius, R. M.; Sybesma, J. T.; Casterton, P. L. *(2002) Toxicol. in Vitro* **2000**, *14*, 379–386.
46. Papini, S.; Rosellini, A.; Nardi, M.; Giannarini, C.; Revoltella, R. P. *Differentiation* **2005**, *73*, 61–68.
47. Clothier, R.; Orme, A.; Walker, T. L.; Ward, S. L.; Kruszewski, F. H.; DiPasquale, L. C.; Broadhead, C. L. *ATLA* **2000**, *28*, 293–302.
48. Wilkinson, P. J.; Clothier, R. H. *ATLA* **2005**, *33*, 509–518.
49. ECVAM. *ATLA* **1998**, *26*, 277–280.
50. ECVAM. *ATLA* **1998**, *26*, 275–277.
51. ECVAM. *ATLA* **2000**, *28*, 365–366.

52. ECVAM. Statement on the application of the Corrositex® assay for skin corrosivity testing. http://ecvam.jrc.it/index.htm, 2000.
53. NIH. *Corrositex®: An In Vitro Test Method for Assessing Dermal Corrosivity Potential of Chemicals. The Corrositex® Assay Peer Review Meeting Final Report, ICCVAM and NICEATM*. NIH publication no. 99–4495. US National Institute of Environmental Health Sciences: Research Triangle Park, NC, USA, 1999.
54. ICCVAM. Expert Panel Report: Evaluation of the Current Validation Status of In vitro Test Methods for Identifying Ocular Corrosives and Severe Irritants. http://iccvam.niehs.nih.gov/methods/ocudocs/EPreport/ocureport.htm; 2005.
55. OECD. *OECD Guideline for the Testing of Chemicals. Acute Eye Irritation/Corrosion*; OECD: Paris, 2002.
56. Barratt, M. D. *ATLA* **1995**, *2*, 111–122.
57. Combes, R. D. *ATLA* **1995**, *23*, 352–379.
58. Mitchell, I.; de, G.; Combes, R. D. In Vitro Genotoxicity and Cell Transformation Assessment. In *In Vitro Methods in Pharmaceutical Research*; Castell, J. V., Gómez-Lechón, M. J., Eds.; Academic Press: Basingstoke, 1997, pp 318–352.
59. Combes, R. D. Mutagenicity. In *Cosmetics: Controlled Efficacy Studies and Regulation*; Elsner, P., Merk, H., Mailbach, H. I., Eds.; Springer-Verlag: Berlin, 1999, pp 291–308.
60. Combes, R. D. *Toxicol. in Vitro* **1997**, *11*, 683–687.
61. Dillon, D.; Combes, R.; Zeiger, E. *Mutagenesis* **1998**, *13*, 19–26.
62. Parry, E. M.; Parry, J. M. In Vitro Cytogenetics and Aneuploidy. In *Environmental Mutagenesis*; Phillips, D. H., Venitt, S., Eds.; Bios Scientific Publishers: Oxford, 1995, pp 121–139.
63. Anderson, D.; Plewa, M. J. *Mutagenesis* **1998**, *13*, 67–73.
64. ICH. www.ich.org/cache/compo/276-254-1.html. The ICH Secretariat (Geneva).
65. Honma, M.; Hayashi, M.; Shimada, H.; Tanaka, N.; Wakuri, S.; Awogi, T.; Yamamoto, K. I.; Kodani, N.; Nishi, Y.; Nakadate, M. et al. *Mutagenesis* **1999**, *14*, 5–52.
66. Cole, J.; Harrington-Brock, K.; Moore, M. M. *Mutagenesis* **1999**, *14*, 265–270.
67. Garriott, M. L.; Phelps, J. B.; Hoffman, W. P. *Mutat. Res.* **2002**, *517*, 123–134.
68. Kirsch-Volders, M.; Sofuni, T.; Aardema, M.; Albertini, S.; Eastmond, D.; Fenech, M.; Ishidate, M., Jr.; Kirchner, S.; Lorge, E.; Morita, T. et al. *Mutat. Res.* **2003**, *540*, 153–163.
69. Phelps, J. B.; Garriott, M. L.; Hoffman, W. P. *Mutat. Res.* **2002**, *521*, 103–112.
70. Meintieres, S.; Biola, A.; Pallardy, M.; Marzin, D. *Environ. Mol. Mutat.* **1995**, *41*, 14–27.
71. Combes, R. D.; Stopper, H.; Caspary, W. J. *Mutation Res.* **1995**, *5*, 403–408.
72. DOH. *Guidance on a Strategy for Testing of Chemicals for Mutagenicity. Committee on Mutagenicity of Chemicals in Food, Consumer Products and the Environment*. 2000. http://www.doh.gov.uk/com/htm.
73. Liebsch, M.; Spielmann, H.; Pape, W.; Krul, C.; Deguercy, A.; Eskes, C. UV-Induced Effects. In *Alternative (NonAnimal) Methods for Cosmetics Testing: Current Status and Future Prospects*; Eskes, C.; Zuang, V., Eds.; *ATLA* **2005**, *33*, 131–146.
74. Jírová, D.; Kejlová, K.; Bendová, H.; Ditrichová, D.; Mezuláníková, M. *Toxicol. in Vitro* **2005**, *19*, 931–934.
75. Clothier, R.; Starzec, G.; Stipho, S.; Kwong, Y. C. *Toxicol. in Vitro* **1999**, *13*, 713–717.
76. Combes, R.; Manners, J.; Boavida, P.; Clothier, R. The Effects of UVA on Human Corneal and Keratinocyte Cell Lines. In *Alternatives to Animal Testing II – Proceedings of the Second International Scientific Conference Organised by the European Cosmetic Industry*; Clarke, D. G., Lisansky, S. G., Macmillan, R., Eds.; CPL Press: Newbury, UK, 1999, pp 152–155.
77. Henderson, L.; Gocke, E.; Utesch, D.; Brendler, S.; Gahlmann, R.; Pungier, J.; Gorelick, N.; Dean, S.; Clare, C.; Jones, E. Development of Test Methods for Assessing the Photomutagenic Potential of Sunscreens. In *Alternatives to Animal Testing – Proceedings of an International Scientific Conference Organised by the European Cosmetic Industry*; Lisansky, S. G., Macmillan, R., Dupuis, J., Eds.; CPL Press: Newbury, UK, 1996, pp 303–305.
78. Utesh, D.; Splittgerber, J. *Mutat. Res.* **1996**, *361*, 41–48.
79. Howes, D.; Guy, R.; Hadgraft, J.; Heylings, J.; Hoeck, U.; Kemper, F.; Maibach, H.; Marty, J. -P.; Merk, H.; Parra, J. et al. *ATLA* **1996**, *24*, 81–106.
80. Diembeck, W.; Eskes, C.; Heylings, J. R.; Langley, G.; Rogiers, V.; van de Sandt, J. J. M.; Zuang, V. *ATLA* **2005**, *33*, 105–107.
81. Combes, R. D.; Gaunt, I.; Balls, M. *ATLA* **2004**, *32*, 163–208.
82. Combes, R. D. In Vitro Testing for Endocrine Disruptors. In *Issues in Environmental Science and Technology*; Hester, R., Ed.; The Royal Society of Chemistry: London, 2006, Vol. 23 (in press).
83. Fentem, J. H.; Botham, P. A. *ATLA* **2002**, *30*, 61–67.
84. Zuang, V.; Alonso, M. -A.; Botham, P. A.; Eskes, C.; Fentem, J.; Liebsch, M.; van de Sandt, J. J. M. *ATLA* **2005**, *33*, 35–46.
85. OECD (2002) OECD guideline for the testing of chemicals. Acute dermal irritation/corrosion. www.oecd.org/home/.
86. Worth, A. P.; Cronin, M. T. D. *Toxicology* **2001**, *169*, 119–131.
87. Hoffmann, J.; Heisler, E.; Karpinski, S.; Losse, J.; Thomas, D.; Siefken, W.; Ahr, H. -J.; Vohr, H. -W.; Fuchs, H. W. *Toxicol. in Vitro* **2005**, *19*, 925–929.
88. Noll, M.; Merkle, M. -L.; Kandsberger, M.; Matthes, T.; Fuchs, H.; Graeve, T. *ATLA* **1999**, *27*, 302.
89. Botham, P. A. *Toxicol. Lett.* **2004**, *149*, 387–390.
90. Zuang, V.; Balls, M.; Botham, P. A.; Coquette, A.; Corsini, E.; Curren, R. D.; Elliott, G. R.; Fentem, J. H.; Heylings, J. R.; Liebsch, M. et al. *ATLA* **2002**, *30*, 109–129.
91. Welss, T.; Basketter, D. A.; Schroder, K. R. *Toxicol. in Vitro* **2004**, *18*, 231–243.
92. De Silva, O.; Basketter, D. A.; Barratt, M. D.; Corsini, E.; Cronin, M. T. D.; Das, P. K.; Degwert, J.; Enk, A.; Garrigue, J. L.; Hauser, C. et al. *ATLA* **1996**, *24*, 683–705.
93. Buehler, E. V. *Food Chem. Toxicol.* **1994**, *32*, 97–101.
94. Kimber, I.; Cumberbatch, M.; Betts, C. J.; Dearman, R. J. *Toxicol. in Vitro* **2004**, *18*, 195–202.
95. Verstraelen, S.; Van Den Heuvel, R.; Nelissen, I.; Witters, H.; Verheyen, G.; Schoeters, G. *Toxicol. in Vitro* **2005**, *19*, 903–907.
96. Casati, S.; Aeby, P.; Basketter, D. A.; Cavani, A.; Gennari, A.; Gerberick, G. F.; Griem, P.; Hartung, T.; Kimber, I.; Lepoittevin, J. -P. et al. *ATLA* **2005**, *33*, 47–62.
97. Coecke, S.; Eskes, C.; Gartlon, J.; Kinsner, A.; Price, A.; van Vliet, E.; Prieto, P.; Boveri, M.; Bremer, S.; Adler, S. et al. *Environ. Toxicol. Pharmacol.* **2005**, *21*, 153–167.
98. Coecke, S.; Ahr, H.; Blaauboer, B. J.; Bremer, S.; Casati, S.; Castell, J.; Combes, R.; Corvi, R.; Crespi, C. L.; Cunningham, M. J. et al. *ATLA* **2006**, *34*, 49–84.

99. DeJongh, J.; Forsby, A.; Houston, J. B.; Beckman, M.; Combes, R.; Blaauboer, B. J. *Toxicol. in Vitro* **1999**, *13*, 549–554.
100. Gubbels-van Hal, W. M. L. G.; Blaauboer, B. J.; Barentsen, H. M.; Hoitink, M. A.; Meerts, I. A. T. M.; van der Hoeven, J. C. M. *Reg. Toxicol. Pharm.* **2005**, *42*, 284–295.
101. Tahti, H.; Nevala, H.; Toimela, T. *ATLA* **2003**, *31*, 273–276.
102. Tiffany-Castiglioni, E., Ed. *In Vitro Neurotoxicology*; Humana Press: Totowa, NJ, 2004, p 331.
103. Aschner, M.; Syversen, T. *ATLA* **2004**, *32*, 323–327.
104. Garberg, P. *ATLA* **1998**, *2*, 821–847.
105. Brown, N. A.; Spielmann, H.; Bechter, R.; Flint, O. P.; Freeman, S. J.; Jelinek, R. J.; Koch, E.; Nau, H.; Newall, D. R.; Palmer, A. K. et al. *ATLA* **1995**, *23*, 868–882.
106. Bremer, S.; Pellizzer, C.; Adler, A.; Paparella, M.; Lange, J. *ATLA* **2002**, *30*, 107–109.
107. Bremer, S.; Cortvrindt, R.; Daston, G.; Eletti, B.; Mantovani, A.; Maranghi, F.; Pelkonen, O.; Ruhdel, I.; Spielmann, H. *ATLA* **2005**, *33*, 183–209.
108. Wiger, R.; Trygg, B.; Holme, J. A. *Teratologia* **1989**, *40*, 603–613.
109. L'Huillier, N.; Pratten, M. K.; Clothier, R. H. *Toxicol. in Vitro* **2002**, *16*, 433–442.
110. Bremer, S.; Balduzzi, D.; Cortvrindt, R.; Daston, G.; Eletti, B.; Galli, A.; Huhtaniemi, I.; Laws, S.; Lazzari, G.; Liminga, U. et al. *ATLA* **2005**, *33*, 391–416.
111. Steinberger, A.; Klinefelter, G. *Reprod. Toxicol.* **1993**, *7S1*, 23–37.
112. Combes, R. D. *ATLA* **2000**, *28*, 81–118.
113. Baker, V. *Toxicol. in Vitro* **2001**, *14*, 413–419.
114. Harvey, P. W.; Everett, D. J. *Best Pract. Res. Clin. Endocrinol. Metab.* **2006**, *20*, 145–165.
115. ICCVAM (2002) Development of stably transfected cell lines to screen endocrine disrupters. http://iccvam.niehs.nih.gov/methods/endodocs/final/erta_brd/ertaappx/erta_b4.pdf.
116. Wilson, V. S.; Bobseine, K.; Lambright, C. R.; E Gray, L., Jr. *Toxicol. Sci.* **2002**, *66*, 69–81.
117. Combes, R. D. *ATLA* **2004**, *32*, 121–135.
118. Ekwall, B. *Toxicol. in Vitro* **1999**, *13*, 665–673.
119. Halle, W. *ATLA* **2003**, *31*, 89–198.
120. Clemedson, C.; Nordin-Andersson, M.; Bjerregaard, H. F.; Clausen, J.; Forsby, A.; Gustafsson, H.; Hansson, U.; Isomaa, B.; Jørgensen, C.; Kolman, A. et al. *ATLA* **2002**, *30*, 313–321.
121. NIH. *Report of the International Workshop on In Vitro Methods for Assessing Acute Systemic Toxicity*; NIH publication 01–4499; NIEHS: Research Triangle Park, NC, USA, 2001.
122. Spielmann, H.; Genschow, E.; Liebsch, M.; Halle, W. *ATLA* **1999**, *27*, 957–966.
123. Seibert, H.; Gulden, M.; Voss, J. -U. *Toxicol. in Vitro* **1994**, *8*, 847–850.
124. Hanley, A. B.; McBride, J.; Oehlschlager, S.; Opara, E. *Toxicol. in Vitro* **1999**, *13*, 847–851.
125. Pfaller, W.; Balls, M.; Clothier, R.; Coecke, S.; Dierickx, P.; Ekwall, B.; Hanley, B. A.; Hartung, T.; Prieto, P. et al. *ATLA* **2001**, *29*, 393–426.
126. Prieto, P. *ATLA* **2002**, *30*, 101–105.
127. Canová, N.; Kmoníckova, E.; Lincova, D.; Kitek, L.; Farghali, H. *ATLA* **2004**, *32*, 25–35.
128. Bhandari, R. N. B.; Riccalton, L. A.; Lewis, A. L.; Fry, J. R.; Hammond, A. H.; Tendler, S. J. B.; Shekesheff, K. M. *Tissue Eng.* **2001**, 7, 345–357.
129. Pazos, P.; Fontaner, S.; Prieto, P. *ATLA* **2002**, *30*, 515–523.
130. Scheers, E. M.; Ekwall, B.; Dierickx, P. J. *Toxicol. in Vitro* **2001**, *15*, 153–161.
131. Knight, A.; Bailey, J.; Balcombe, J. *ATLA* **2006**, *34*, 19–27.
132. Knight, A.; Bailey, J.; Balcombe, J. *ATLA* **2006**, *34*, 29–38.
133. Combes, R. D. *Toxicol. in Vitro* **2006**, *34*, 39–48.
134. Combes, R. D. *Toxicol. in Vitro* **2000**, *14*, 387–399.
135. Kirkland, D.; Aardema, M.; Henderson, L.; Muller, L. *Mutat. Res.* **2005**, *584*, 1–256.
136. Combes, R.; Balls, M.; Curren, R.; Fischbach, M.; Fusenig, N.; Kirkland, D.; Lasne, C.; Landolph, J.; LeBoeuf, R.; Marquardt, H. et al. *ATLA* **1999**, *27*, 745–767.
137. OECD. *Detailed Review Paper on NonGenotoxic Carcinogens Detection: The Performance of In-Vitro Cell Transformation Assays*. Environment Health and Safety Publications Series on Testing and Assessment no. 31; OECD: Paris, 2001.
138. LeBoeuf, R. A.; Kerckaert, K. A.; Aardema, M. J.; Isfort, R. J. Use of Syrian Hamster Embryo and Balb/c 3T3 Cell Transformation for Assessing the Carcinogenicity Potential of Chemicals. In *The Use of Short- and Medium-term tests for Carcinogenic Hazard Evaluation*; IRAC Scientific Publications no. 146, McGregor, D. B., Rice, J. M., Venitt, S, Eds., IRAC: Lyon, France, 1999; pp 409–425.
139. Sasaki, K.; Mizusawa, H.; Ishidate, M.; Tanaka, N. *Toxicol. in Vitro* **1990**, *4*, 657–659.
140. Stuard, S. B.; Kerckaert, G. A.; Lehman-McKeeman, L. D. *Toxicol. Sci.* **1999**, *48*, 366.
141. Toraason, M.; Bohrman, J. S.; Krieg, E.; Combes, R. D.; Willington, S. E.; Zajac, W.; Langenbach, R. *Toxicol. in Vitro* **1992**, *6*, 165–174.
142. Rivedal, E.; Mikalsen, S. O.; Sanner, T. *Toxicol. in Vitro* **2000**, *14*, 185–192.
143. Blaha, L.; Kapplova, P.; Vondracek, J.; Upham, B.; Machala, M. *Toxicol. Sci.* **2002**, *65*, 43–51.
144. Knight, A.; Bailey, J.; Balcombe, J. *ATLA* **2006**, *34*, 39–48.
145. Elsner, H. M.; Mailbach, H. I., Eds. *Cosmetics: Controlled Efficacy Studies and Regulation*; Springer-Verlag: Berlin.
146. Griffiths, H. A.; Wilhelm, K. P.; Robinson, M. K.; Wang, X. M.; McFadden, J.; York, M.; Basketter, D. A. *Food Chem. Toxicol.* **1997**, *35*, 255–260.
147. Combes, R. D.; Barratt, M.; Balls, M. *ATLA* **2003**, *31*, 7–19.
148. Willis, R. C. *Scientist* **2005**, *19*, 38.
149. Rogiers, V.; Balls, M.; Basketter, D.; Berardesca, E.; Edwards, C.; Elsner, P.; Ennen, J.; Lévêque, J. L.; Lóden, M.; Masson, P. et al. *ATLA* **1999**, *27*, 515–537.
150. Anderson, R.; O'Hare, M.; Balls, M.; Brady, M.; Brahams, D.; Burt, A.; Chesne, C.; Combes, R.; Dennison, A.; Garthoff, B. et al. *ATLA* **1998**, *26*, 763–777.
151. MacGregor, J. T.; Collins, J. M.; Sugiyama, Y.; Tyson, C. A.; Dean, J.; Smith, L.; Andersen, M.; Curren, R. D.; Houston, J. B.; Kadlubar, F. F. et al. *Toxicol. Sci.* **2001**, *59*, 17–36.
152. Combes, R. D. *ATLA* **2004**, *32*, 43–49.
153. Orr, S.; Alexandre, E.; Clark, B.; Combes, R.; Fels, L. M.; Gray, N.; Jnsson-Rylander, A. -C.; Helin, H.; Koistinen, J.; Oinonen, T. et al. *Cell Tissue Bank* **2002**, *3*, 133–137.

154. Bhogal, N.; Grindon, C.; Combes, R.; Balls, M. *Trends Biotechnol.* **2005**, *23*, 300–307.
155. Rolletschek, A.; Blyszczuk, P.; Wobus, A. M. *Toxicol. Lett.* **2004**, *149*, 361–369.
156. Cezar, G. G.; Thiede, M.; Strom, S.; Miki, T.; Trosko, J. *Toxicol. Sci.* **2004**, *79*, 214–223.
157. Bużańska, L.; Habich, A.; Jurga, M.; Sypecka, J.; Domańska-Janik, K. *Toxicol. in Vitro* **2005**, *19*, 991–999.
158. Maurel, P.; Duret, C.; Gerbal-Chaloin, S.; Daujat-Chavanieu, M. Differentiation of Hepatocyte-Like Cells from Human Embryonic Stem Cells and Adult Liver Progenitors. In *ALTEX Proceedings of the Fifth World Congress on Alternatives and Animal Use in the Life Sciences, Berlin, August, 2006* (in press).
159. Atterwill, C. K.; Goldfarb, P.; Purcell, W.; *Approaches to High Throughput Toxicity Screening*; Taylor & Francis: London, UK, 2000.
160. Aardema, M.; MacGregor, J. T. *Mutat. Res.* **2002**, *499*, 13–25.
161. Farzaneh, P.; Allameh, A.; Pratt, S.; Moore, N.; Travis, L.; Gottschalg, E.; Kind, C.; Fry, J. *ATLA* **2005**, *33*, 105–110.
162. Sõti, C.; Nagy, E.; Giricz, Z.; Vígh, L.; Csermely, P.; Ferdinandy, P. *Br. J. Pharmacol.* **2005**, *146*, 769–780.
163. Corvi, R.; Ahr, H. -J.; Albertini, S.; Blakey, D. H.; Clerici, L.; Coecke, S.; Douglas, G. R.; Gribaldo, L.; Groten, J. P.; Haase, B. et al. *Environ. Health Persp.* **2006**, *114*, 420–429.
164. Fentem, J. H.; Chamberlain, M.; Sangster, B. *ATLA* **2004**, 617–623.
165. Schoeters, E.; Verheyen, G. R.; Van Den Heuvel, R.; Nelissen, I.; Witters, H.; Van Tendeloo, V. F. I.; Schoeters, G. E. R.; Berneman, Z. N. *Toxicol. in Vitro* **2005**, *19*, 909–913.
166. Pennie, W. D.; Aldridge, T. C.; Brooks, A. N. *J. Endocrinol.* **1998**, *158*, R11–R14.
167. Terasaka, S.; Aita, Y.; Inoue, A.; Hayashi, S.; Nishigaki, M.; Aoyagi, K.; Sasaki, H.; Wada-Kiyama, Y.; Sakuma, Y.; Akaba, S. et al. *Environ. Health Persp.* **112**, 7, 773–781.
168. Gülden, M.; Seibert, H. *Toxicol. In Vitro* **1997**, *11*, 479–483.
169. Eisenbrand, G.; Pool-Zobel, B.; Baker, V.; Balls, M.; Blaauboer, B. J.; Boobis, A.; Carere, A.; Kevekordes, S.; Lhuguenot, J.-C.; Pieters, R. et al. *Food Chem. Toxicol.* **2002**, *40*, 193–236.
170. Hofer, T.; Gerner, I.; Gundert-Remy, U.; Liebsch, M.; Schulte, A.; Speilmann, H.; Vogel, R.; Wettig, K. *Arch. Toxicol.* **2004**, *78*, 549–564.
171. Combes, R. D.; Berridge, T.; Connelly, J.; Eve, M. D.; Garner, R. C.; Toon, S.; Wilcox, P. *Eur. J. Clin. Pharmacol.* **2003**, *19*, 1–11.
172. Langley, G. *The Way Forward – Action to End Animal Toxicity Testing*. British Union for the Abolition of Vivisection and European Coalition to End Animal Experiments. BUAV: London, UK, 2001.
173. Grindon, C.; Combes, R.; Cronin, M. T. D.; Roberts, D.; Garrod, J. *ATLA* **2006**, *34*, 149–158.
174. Fumarola, L.; Urani, C.; Crosta, G. F. *(2005) Toxicol. in Vitro* **2005**, *19*, 935–941.
175. http://www.hsl.gov.uk/capabilities/pbpk-jip.htm.

Biography

Robert D Combes is Director of FRAME (Fund for the Replacement of Animals) and an honorary Special Professor in the School of Biomedical Sciences, University of Nottingham Medical School. His main interests are mechanisms of toxicity, metabolic activation/detoxification, structure–activity relationships, and developing and validating short-term and alternative assays for toxicity testing.

Comprehensive Medicinal Chemistry II
ISBN (set): 0-08-044513-6

ISBN (Volume 1) 0-08-044514-4; pp. 463–487

1.13 The Role of Small- or Medium-Sized Enterprises in Drug Discovery

C G Newton, BioFocus DRI, Saffron Walden, UK

1.13.1 Definitions of Small- or Medium-Sized Enterprises (SMEs)

In the US, the Asia-Pacific Economic Cooperation organization defined small- or medium-sized enterprises (SMEs) as "manufacturing companies with less than 500 employees, or nonmanufacturing companies with sales less than $5 M."[1] Similarly, the UK Government Small Business Service (an agency of the Department of Trade and Industry) in February 2005 defined "SME as a business with less than 250 staff."[2] In Japan the 2005 definition is sector-defined (**Table 1**),[3] but is broadly comparable to the above.

For the purpose of this chapter, an SME is defined as a company with fewer than 500 people in staff, but turnover is ignored. Readers should recognize that the fluid nature of the pharmaceutical industry (frequent start-ups, mergers, takeovers, and closures) indicates that companies described in this chapter may well not be around in their current forms even in 1–5 years' time. Conversely, some companies that were SMEs 10 or 20 years ago whose activities have impacted drug discovery greatly have now expanded beyond that definition. The impact of these (relatively few) companies which originated as SMEs is consequently great and is discussed below.

Table 1 Japan's sector definition of what constitutes an SME in 2005

Industries	*Capital size (million yen)*	*Number of employees*
Manufacturing and others	300 or less	300 or less
Wholesale	100 or less	100 or less
Retail		50 or less
Services	50 or less	100 or less

http://www.chusho.meti.go.jp/sme_english/outline/02/01.html.

Traditional large, multinational pharmaceutical companies (big pharma) and small-sized regional or national pharma companies are not included in this chapter, except insofar as how their dealings with SMEs have impacted their own pipelines.

1.13.1.1 Definitions of Drug Discovery

For the purpose of this chapter, drugs are considered to include not only small molecules, but also macromolecules, proteins, antisense agents, and antibodies – vaccines are also mentioned where pertinent to this chapter.

1.13.1.2 Definitions of Drugs

The definition of what constitutes 'drug discovery' also needs delineating to put the chapter into context. For the purpose of this chapter, the author has taken activities up to phase IIa (proof of concept) clinical trial, when examining the roles SMEs have played over the past 10 years in drug discovery. One particular impact of SMEs on drug discovery (as far as the patient is concerned) is what proportion of the dugs that the patient takes actually originated within SMEs – consequently, a survey of the origin of such new chemical entities (NCEs) has been conducted.

1.13.2 Origins of Small- or Medium-Sized Enterprises Engaged in Drug Discovery

Each SME (which in this industry mostly tend to fall under the generic banner of the biotechnology or 'biotech' sector) originated with a group of scientists and entrepreneurs who have an idea, technology, intellectual property (IP), or vision that has been considered worthwhile converting to a business. Three types of origin can be discerned. The founders frequently originate from:

1. academic research institutes;
2. pharma companies that are divesting IP, research centers, and scientists; and
3. pharma companies looking to develop technologies outside the parent company.

1.13.2.1 Small- or Medium-Sized Enterprises Originating from Universities and Academic Institutions

Public funding of medical research over the past 20 years has been concentrated in the clinical, genetic, genomic, molecular biology, and biochemistry fields rather than in straightforward medicinal chemistry/drug discovery. Such research to commercial ends tends to be regarded as beyond the proper use of public funds. Hence, discoveries and inventions that might lead to the discovery of a new drug made in the institutions conducting such academic research, have tended to be licensed to traditional pharma, possibly via a university technology transfer office. However, in many cases, particularly entrepreneurial scientists have elected to commercialize the IP themselves by founding a company, and a new SME (biotechnology sector company) is born.

Two examples are used to illustrate how universities and university/academic research clusters have spawned SMEs engaged in drug research.

Biotechnology, published in May 2001 by the Office of Economic Research of the California Trade and Commerce Agency,[4] described California as being home to 75 publicly funded research institutions that are now surrounded by 2500 biotechnology companies. These companies then employed some 212 700 scientists and co-workers, and had a turnover of about $212 billion. At that time, California was also in receipt of $1.7 billion in National Institute of Health grants.

Imperial College of Science, Technology and Medicine (IC) in London is one of Europe's largest centres of cross-disciplinary research; the medical school alone is a conglomerate of the former Chelsea and Westminster Hospitals, Charing Cross Hospital, the Hammersmith Hospital, North West London Hospitals, the Royal Brompton Hospital, and St Mary's Hospital.[5] Allied to existing chemistry, biochemistry, and molecular biology research departments, strong cross-disciplinary ties have led to recognition that IP generation may be managed by the creation of spinout companies from Imperial College. To this end, Imperial College Innovations was set up to facilitate spinouts as well as licencing IP to third parties. **Table 2** lists the Imperial College spinout companies (SMEs) as listed on the Imperial College Innovation website in 2005 that are engaged in drug research.[6]

Table 2 Imperial College spinout companies (SMEs) engaged in drug discovery, as listed on the Imperial College innovation website

Imperial College spinout company	*Year founded*	*Activity*	*Location*	*Fate by 2005*
Amedis Pharmaceuticals	2001	In silico prediction of pharmacokinetic, organosilicon drugs	Cambridge, UK	Acquired by Paradigm Therapeutics 2005
Argenta Discovery	2000	Respiratory drug discovery, contract medicinal chemistry research	Harlow, UK	Independent
Atazoa	2004	Conducts research and development based upon technology emanating from Imperial College's Institute of Developmental and Reproductive Biology, and Cedars-Sinai Medical Center (California)	London, UK	Independent, but majority owned by Mitsubishi
Biogeny (Implyx)	2004	Preparing and purifying large proteins which will be capable of passing across the cell membrane to deliver drugs directly to individual cells		
Cerestem	N/K	Stem cell company: develops mitogenic growth factors that can send stem cell differentiation down certain paths, e.g., neuro- or muscular lineages		
D-Gen	2000	Develops screening tests and diagnostics for prion diseases in humans and animals		
IC VEC	2001	Research and development of siRNA therapeutics using proprietary synthetic nanoparticle technology (CONZENTRx) developed for customized delivery of nucleic acids		Majority owned by Mitsubishi
Lipoxen Technologies	1997	Drug and vaccine delivery technologies: molecule engineering, particle engineering, for large and small molecules	London	Independent
Lorantis Holdings	2000	Therapies to treat immune diseases and improve transplantation success	Cambridge	Independent
Metabometrix	2002	Characterizes the key biochemical changes caused by drug toxicity in animals and humans	London	Independent
Microscience	1996	Development of vaccines for infectious disease, in particular the identification and production of vaccines against group B streptococcus in pregnancy/newborns. The company is also working to determine genes critical to virulence (ability to cause disease) in bacteria	London	Private independent
NanoBioDesign	2001	High-throughput screening systems based on human and bacterial P450 enzymes	London	Independent
Polytherics	N/K	Develops novel polymers capable of enhancing drug delivery	London	Independent
Proteom	1999	Bioinformatics company designing in silico peptide ligands	Cambridge	
Protexeon	2002	Investigating use of xenon as anesthetic and to treat nerve damage	London	
Riotech	2003	Vaccines, biological antivirals, and small-molecule antivirals to hepatitis A, B, and C and other viral infections		

Table 2 Continued

Imperial College spinout company	*Year founded*	*Activity*	*Location*	*Fate by 2005*
Sterix	1998	Research and development of a new generation of steroid-based therapeutic products for use in oncology and metabolism or endocrine-related diseases	Bath	Acquired by Ipsen Beaufour 2004
Synovis	2004			
Thiakis	2004	Novel medicines for the treatment of obesity and metabolic disease: exclusive licensee of four patent families relating to use of PYY3-36 and oxyntomodulin for the treatment of obesity and associated conditions		

http://www.imperialinnovations.co.uk/

1.13.2.2 Small- or Medium-Sized Enterprises with Origins in Big Pharma Rationalizations

The continued amalgamation of large pharmaceutical companies and subsequent rationalization of research groups have been a fertile source of new start-up companies from scientists determined to forge an independent path in research aimed at generating new drugs. The 2000 merger of Rhône-Poulenc and Hoechst-Marion-Roussel to create Aventis Pharmaceuticals led to the creation of Scynexis Inc. (Research Triangle Park, NC) and Argenta (Harlow, UK) from groups of scientists formerly employed at Aventis' rationalized sites at Research Triangle Park (NC) and Dagenham (UK), respectively. Similarly, the formation of the then Glaxo Wellcome in 1997 created BioFocus (initially located in Sittingbourne, UK), Arrow Therapeutics (London, UK), and Triangle Pharmaceuticals (NC). Closure of a midwest Amgen research site (Boulder, CO) gave birth to Array Biosystems. Major rationalizations of drug research groups by big pharma in Italy have led to the creation of Nikem and Newron Pharmaceuticals (both in Milan, Italy).

1.13.2.3 Small- or Medium-Sized Enterprises with Other Origins

In 2001, RJ Reynolds, the tobacco giant, spun out a pharmaceutical business called Targacept to continue work on the therapeutic indications of ligands binding to nicotinic acid receptors. Aventis Pharmaceuticals spun off its bone research group and assets at the Romainville site in Paris, France, which became known as Proskelia – this was merged into a similar company in Scotland, known as the Strachan Group – the merged entity now being known as the ProStrachan group. More recently, Sanofi-Aventis has spun off its antiinfectives group into the Paris-based company known as Novexel.[7] Novexel inherited an advanced portfolio of antiinfective programs and IP, and received €40 million in financing from life-science investors (Atlas Venture, Sofinnova, 3i, Abingworth, and Novo). Galapagos was a joint venture between Crucell of the Netherlands and Tibotec of Belgium and was founded in 1999 with funding and IP from both parent organizations.

1.13.3 Funding of Small- or Medium-Sized Enterprises

Fledgling biotech SMEs from whatever origin have a need for cash that is significant compared to SMEs in other industry sectors. Financing may come from private investors (business angels, venture capital (VC) companies) or from public investment. Typically, one, two, or three rounds of private funding may have occurred before a company has the IP assets, drug prospects, or revenues to make it an attractive proposition to public investors. Such public investment based upon a proper legal and financial prospectus is usually described as a flotation or initial public offering (IPO). Finally, investment income for drug discovery activities can also be supplemented or replaced if the SME has an income stream from services or products of its own.

1.13.3.1 Funding from Business Angels and Venture Capitalists

Business angels typically invest a few hundreds of thousands of dollars into a new business idea – within a short timeframe, a successful small company will then require a more substantive investment of several millions, up to tens of millions, of dollars, in exchange for equity in the company or as a loan or as a combination of both. Several rounds of VC investments (known as series A, series B, etc.) may well have occurred before the several years of research and development (R&D) necessary to translate ideas into assets have passed. Such rounds may be funded entirely by existing investors, or new investors may join – each round will have dilutive effects on existing shareholders and timings are usually critical to getting best terms and conditions for both existing and new investors. VC companies recognize the longevity of investments (3–7 years is typical of the time required to realize a substantial profit on investment). There are many VC companies that specialize in investing in life-science companies: 3i, Abingworth, Advent, Advent International, Merlin, and MVM are typical UK healthcare company financiers. In some cases, VC companies are associated with parent organization – MVM had its origins in the UK Medical Research Council's desire to exploit its own inspired IP.

The value of VC-funded SMEs to drug discovery was examined in a 2004 report commissioned by Pacific Bridge Life Sciences and the Weinberg Group.[8] Stolis and Goodman[8] examined the impact of VC investment on the treatment of chronic diseases and leading causes of death in the US, including cancer, heart disease, stroke, diabetes, asthma, and arthritis. The authors assert that venture-backed medical innovations are developed and made available to patients as much as three times faster than a 'bootstrapping' approach to product development. The report claims that more than 100 million (one in three) Americans have benefited from venture-backed medical innovations developed during the past 20 years.

1.13.3.2 Funding from Initial Public Offerings and Stock Market Listings

Refunding at VC exit usually takes one of two directions. The first is an IPO, where the public is invited to subscribe to the shares of the company and the venture capitalists are then able to liquidate a part, or all, of their shareholding. The timing of IPOs is very market-sensitive, not only to the individual company prospects but also to the public and analysts' perception of the industry sector as a whole. The so-called IPO windows can open and close in very short timeframes. Despite good news on a company's prospects (for example, Edinburgh (Scotland)-based Ardana Biosciences released a press release claiming that its prostate cancer treatment, Teverelix, was successful in phase II clinical trials[9]), sometimes IPOs fail to raise the money expected or needed. Ardana Bioscience achieved an IPO on Alternative Investment Market (AIM) in March 2005 but the share price achieved (£1.28) was apparently disappointing. It did, however, net Ardana some $38 million to $32 million after expenses. In certain cases, IPOs can be withdrawn at short order when a healthcare bad-news story impacts public confidence in share subscriptions, as occurred with the UK-based company Phytopharm (a company specializing in natural product research) in March 2005. Boston-based US SME/biotech CombinatoRx was due for an IPO in March 2005, and Valera Pharmaceuticals registered on Monday 14 March 2005 for an IPO of up to $74.75 million in common stock, according to a filing with the US Securities and Exchange Commission.[10] The second typical VC exit from an SME is described in Section 1.13.3.3, below.

A few SMEs raise their initial cash requirements by a listing on a public stock exchange – usually an exchange that is created for riskier ventures, and that can impose less regulation on the reporting of such listed companies. Examples of such exchanges are Ofex and the AIM (both UK), TecDAX (Germany), and NASDAQ (US). Examples are UK-based contract research organization BioFocus (UK), which raised its initial funding requirements through a placing on Ofex in 1997, and Cambridge (UK)-based structure-based drug discovery company Sareum, which listed on AIM in 2004.

1.13.3.3 Funding by Revenue Generation

A third source of revenue for SMEs engaging in drug discovery is to sell or license assets or technologies and services to other companies within the drug discovery sector. Typically offered are late-stage research or development products that big pharma considers essential to supplement in-house pipelines and for which SMEs struggle to resource the required cost of development. Typical deals can run into hundreds of millions of dollars (including upfront payments, milestone payments, and royalties). A listing of the impact that such in-licensing deals have had on the pipeline of the world's largest pharmaceutical company (as of January 2005), Pfizer, will be discussed in Section 1.13.9.1.

Alternatively, SMEs can seek to earn revenue by offering services to big pharma or other SME companies in technologies or areas of resource where such companies are deficient. For big pharma, such offerings can be for new technologies or in areas of expertise that, for whatever reason, the larger organization have elected to consider better outsourced in whole or in part. The trade between SMEs themselves tends to be much greater, albeit SMEs have generally less cash to spend on deals and often seek partnerships described as 'shared-risk,' whereby each partner contributes resources to a project with the objective of sharing the rewards later. Typical SME technology and service companies in the drug discovery sector include those selling technologies to perform gene sequencing, target identification and validation, hit-finding, lead optimization compound profiling, good laboratory practice and good manufacturing practice preclinical and clinical studies, and pharmaceutical profiling, as well as chemical scale-up. Typical examples of such companies are discussed in Section 1.13.8.

The ultimate in revenue generation for a company is also the second alternative exit for VC investors. This is a trade sale of the company to another within the industry – such trade sales may be to big pharma or to other biotechnology companies. In each case, the incentive is a hunger for an asset, technology, workforce or, in some cases, cash, if the SME being acquired has a substantial reserve that the investors consider would be better placed elsewhere. As might be expected, the deep pockets of large pharma have financed some spectacular deals in the past 10 years – a survey of typical acquisitions by the main established pharmaceutical players is shown in **Table 3**. It is instructive to note that certain companies have been very active acquiring SMEs to bolster their internal drug pipelines or technology bases; however, other major pharma companies are conspicuous by their absence.

1.13.3.4 Funding by Grants

Public bodies (governments, charities) will make grants available to companies setting up with new technologies, research projects, or in locations deemed useful to the public/national interest. Recent examples are the Wellcome Trust grant to Microscience (London, UK) for the development of a drinkable typhoid vaccine (2005), and in April 2004, the US National Institute of Health awarded the vaccine company Antigen a $192 000 grant to support development of its prostate cancer cell vaccine based on modulation of major histocompatibility complex class II expression.

1.13.4 Locations of Small- or Medium-Sized Enterprises

The rise of the biotechnology industry in the world has caused a profound shift in the geographical localization of the industry away from the traditional pharma centers, often located in large industrial cities, to clusters around major academic and publicly funded research foundations.

1.13.4.1 North America

On the west coast of the US, two very large thriving clusters can be distinguished. The first of these clusters, which many regard as the birthplace of the biotechnology industry, is located around the San Francisco Bay area, California. This geographical concentration, termed 'Biotech Bay' by the US website Biospace, is clustered around the University of California (San Francisco, Davis, and Berkeley campuses). There are 262 companies and associated research institutions listed in this area as of March 2005.[11] Notable companies that began life as SMEs in this area are Genentec and Chiron, both of which are now major players and compete with big pharma in the worldwide pharmaceutical market. Other notable SMEs in this thriving region are Abgenix, Chemocentryx, CV Therapeutics, Discovery Partners International, Exelixis, Theravance (once known as Advanced Medicines), and Gilead.

The second US west-coast cluster is located in the south of California, around San Diego and Los Angeles and near the sites of the University of California (San Diego, Los Angeles, Santa Barbara, and Irvine campuses) together with the California Institute of Technology, the Salk Institute, and the Scripps Research Institute. Scientists from or who have trained at these institutions have launched or now work in a large number of local companies. This geographical concentration (266 companies and research institutions are listed) has been termed 'Biotech Beach' by the American website Biospace.[12] As one example, chemical research technologies associated with Professor KC Nicolaou at the Scripps Research Institute (radiofrequency tag-monitoring and barcode monitoring of vials used in combinatorial chemistry) were commercialized by the SME Irori (subsequently acquired by Discovery Partners International). The San Diego area of California has its own Biotech Discussion Group, facilitating intercompany exchanges of information.[13] Notable companies of the San Diego/Los Angeles area are: Amgen (now grown far greater than an SME), Agouron (now part of Pfizer), Idun (acquired by Pfizer in 2005), Cortex, IDEC (now part of the major Biotech company Biogen Idec), Invitrogen, Medigene, Mycogen, Sequenom, and Vical.

On the east coast of the US around Boston is a huge cluster of SME biotechnology companies, many of which owe their origins to scientists trained at Harvard University, Massachusetts Institute of Technology (MIT), Brandeis

Table 3 SMEs involved in drug discovery acquired by big pharma 1995–2005

Acquirer	*Target*	*Year*	*Deal size*	*Type of acquisition*
Abbott	–	–	–	–
AstraZeneca	KuDOS Pharmaceuticals	2005	$210 million	Cancer therapeutics company based in UK
Amgen	Tularik	2004	$1300 million	Tularik, founded in 1991, was a biopharmaceutical company engaged in the research and development of drugs that regulate gene expression and focuses on cancer, immunology, and metabolic disorders
Bayer	–	–	–	–
BMS	–	–	–	–
GSK	Affymax	1995	$450 million	Automated synthesis and screening technologies
Lilly	Applied Molecular Evolution	2004	$400 million	Protein and peptide technologies
	Sphinx Technologies	1995		Automated synthesis and screening technologies
J&J	Transform Pharmaceuticals	2005	$230 million	Optimization of crystal form and formulation of drugs
	Egea Biosciences	2004	N/K	Biological design and molecular engineering company creating therapeutic proteins for the treatment of cancer, metabolic, and immunological diseases
	3-D Pharmaceuticals	2003	$88 million	Automated synthesis and screening technologies and structural
	Scios	2003	$2400 million	Cardiovascular drug research company with two marketed products for the treatment of heart disease: Natrecor (nesiritide), for the treatment of acutely decompensated congestive heart failure, and Retavase
	Tibotec-Virco	2002	$320 million	Human immunodeficiency virus diagnostic services and products
	Discovery Laboratories	2000		(Increased ownership) Development of surfactant replacement therapies based on engineered lung surfactant technology for the potential treatment of respiratory diseases
	Centocor	1999	$4900 million	Monoclonal antibodies
Merck	Aton Pharma	2004		HDAC inhibitors
	Rosetta Inpharmatics	2001		Analysis of gene data to predict how medical compounds will interact with different kinds of cells in the body, therefore potentially allowing scientists to select drug targets more accurately and speed up the development process
Novartis	Idenix Pharmaceuticals	2003	> $255 million	Antiviral drugs
	Grandis Biotech	2000		Human growth hormone
Pfizer	Idun Pharmaceuticals	2005	Not disclosed	Idun's technology is focused on the control of caspase activity. Idun developed therapeutic applications focused on inhibiting caspase activity as potential treatments for liver disease and inflammation

Table 3 Continued

Acquirer	*Target*	*Year*	*Deal size*	*Type of acquisition*
	Angiosyn	2005	\$527 million	Biopharmaceutical company developing novel proprietary biologics for controlling angiogenesis
	Meridica	2004	\$125 million	Drug delivery company, based in Cambridge, UK, specializing in inhaled, nasal, and parenteral drug administration routes
	Esperion Therapeutics	2003	\$1300 million	Formed in July 1998. Esperion was a company specialized in developing high-density lipoprotein (HDL)-targeted therapies to treat cardiovascular disease, including atherosclerosis
Roche	–	–	–	Roche bought Genentech in 1990
Sanofi Aventis	–	–	–	–
Schering-Plough	Neogenesis	2005	Not yet disclosed (21/2/2005)	NeoGenesis Drug Discovery was dedicated to the development of applied genomics and combinatorial chemistry
	Canji	1996	\$54.5 million	Founded in 1990, Canji was a privately held gene therapy company before it was acquired
	DNAX	1982		DNAX is a biotechnology research institute that was founded in 1980
Takeda	Syrrx	2005	\$270 million	Syyrx was founded in 1999 – high-throughpout structural biology technology and drug discovery company
Wyeth	–	–	–	–

University, and Boston University, together with research ideas spawned from scientists at institutions such as the Harvard Medical School, Brigham and Women's Hospital, the Dana-Farber Cancer Institute, and Massachusetts General. The Biospace website has dubbed this concentration of SME/biotech companies 'Genetown' and in 2005 listed some 206 companies in the area. Notable companies include Biogen (now merged with San Diego-based IDEC to form Biogen Idec), Vertex, Millennium, Sepracor (all of these biotechnology companies have grown beyond the scope of this chapter's definition of an SME), Affymetrix, Ariad, Arqule, Cubist, Curis, En Vivo Pharmaceuticals, Leukocyte (acquired by Millennium), Momenta, Oscient, and Repligen. In the US, the geographical concentration in the Boston area has become so important to the world drug discovery science base that the desire of the traditional large pharma companies to overcome geographical inertia and move their scientists to this biotech/university cluster has occurred. Both Merck and Novartis have built large research centers in Boston, MA since 2000.

Other US clusters are in New Jersey, Pennsylvania, and Michigan, and there is also a significant cluster in the Montreal/Quebec area of Canada.

1.13.4.2 Europe

Geographical clusters are also apparent in Europe. In the UK, the Eastern Region Biotechnology Industry (ERBI) association[14] represents SMEs in the Cambridge area, close to Cambridge University, the John Innes Research Centre, University of East Anglia, and the Human Genome Campus at Abington. Notable companies include Astex, Argenta Discovery, BioFocus Discovery, De Novo Pharmaceuticals, Celltech (which grew to a size greater than this chapter's SME definition before being acquired by the Belgian UCB Group in 2004), Arachnova, Biotica, Biowisdom, Cambridge Antibody, Cambridge Combinatorial (which became part of Millennium in 2001), Cellzome, Cytomix, Domantis, DanioLabs, De Novo Pharmaceuticals, Huntingdon Life Sciences, Kudos (acquired by AstraZeneca in December 2005), and Pharmagene. Some notable figures obtained from the ERBI website state that the region is home to almost 200 biotech companies together with some 250 specialist service providers with biotech expertise. These SMEs are co-located around more than 30 research institutes and universities and four leading hospitals involved in research. The region is home to half of the UK's top 15 London Stock Exchange (LSE)-quoted biotech companies and venture capitalists have invested more than $1.8 billion of funds.[14]

In Europe, a remarkable cluster has grown up on either side of the Øresund, separating Sweden from Denmark. This SME/institution cluster, known as 'Medicon Valley,' has its own website.[16] Main academic centers in this area include the Royal Danish School of Pharmacy and the University of Copenhagen, both in Copenhagen (Denmark) and Lund University in Sweden. The Group of Povl Krogsgaard–Larsen at the Royal Danish School of Pharmacy has long conducted research into the treatment of neurodegenerative disorders, including research into muscarinic and nicotinic acid receptors. In 2002, the Boston Consulting Group published a 70-page report on Medicon Valley, which is available on the Medicon Valley website.[16] This report shows that this particular biotech cluster also owes much of its existence to a concentration of large and medium-sized pharma in the region: Pharmacia (much of which is now Biovitrum), AstraZeneca, Leo, Lundbeck and NovoNordisk, which have spawned a large number of SME companies. The Boston Consulting Group report[16] considers that the Medicon Valley diabetes research cluster is probably the strongest in the world, and there is also a strong emphasis on antiinflammatory and neuroscience drug discovery. Notable drug discovery and associated companies that are SMEs in Medicon Valley are 7-TM Pharma, Acadia Pharmaceuticals, BioImage, Zealand Pharmaceuticals, Maxygen, Pharmexa, Topotarget, and Active Biotech.

Other notable European clusters are located in the Oxford area of the UK, around Munich in Germany, Leiden in the Netherlands, and around Paris in France.

1.13.4.3 Rest of the World

In the 1990s, Japan was one of the most important sources of new drugs, with origins in major companies like Yamanouchi, Fujisawa, and Takeda being paralleled by drug origins in much smaller research groups in Kirin Brewery, Nippon Flour Mills, Teijin, etc. A combination of past success and the reluctance of the industry to subject itself to merger and acquisition activities has probably been one reason why the emergence of biotech clusters as seen in Europe and in the US has not yet happened. The mergers of Fujisawa and Yamanouchi, announced in 2004, and the ambition of companies like Takeda may allow the emergence of similar concentrations of SMEs, but hitherto there are few such companies. One homegrown Japanese SME engaged in drug discovery, Sosei, is located in Tokyo – the founder of this company was Shinichi Tamura, the former chief executive officer of Genentech Japan.[17] It is interesting that one Japanese pharmaceutical company, Tanabe, has deliberately created a Biotech spinout, MediciNova, but this is based in San Diego.

Government assistance in the location of drug discovery SMEs should also not be ignored. Many governments recognize the assets that a strong biotechnology business can bring to their country, and tax advantages are often given to companies locating in certain territories. Few countries have been as ambitious as Singapore, which has created a large science park for this end.[18] The academic research centers that are at the core of the science park are: the Institute of Molecular and Cell Biology, the Natural Product Research (CNPR: privatized in 2003 to become MerLion Pharmaceuticals), the Genome Institute of Singapore (formed in 2000) and the Johns Hopkins Institute – an offshoot of the US-based institute of the same name. Of the major companies located in Singapore, GlaxoSmithKline and Novartis are present, whilst Chiron has established S*BIO, formed through a partnership with the Singapore Economic Development Board Investments (EDBI). S*BIO is a chemistry-driven, product-oriented company, with a focus on cancer, with the objective of identifying and developing novel, small-molecule anticancer drug candidates from its own leading-edge research, as well as through joint collaborations with external partners. Another notable company with a location in Singapore is Albany Molecular, the contract research organization headquartered in Albany (NY, US).

Clusters of high-tech SME companies engaged in drug discovery are also notable in the Bangalore region of India, and around Shanghai in China.

1.13.5 Fate of Small- or Medium-Sized Enterprises Engaged in Drug Discovery

1.13.5.1 Small- or Medium-Sized Enterprises That Have Grown to be Major Players in Drug Discovery

Those companies that have grown to such a size as to challenge the major traditional pharma companies represent the 'most aspired to' ambition for companies that began as SME biotech companies. Of the many thousands of start-up companies, only a handful have reached this exalted position, but their success has been so phenomenal it has been one of the major causes why investors seek to fund SMEs that perform drug research.

Table 4 SMEs that have become major players in the marketplace

	Amgen	*Biogen Idec*	*Cephalon*	*Chiron*	*Genentech*	*Gilead*	*Ligand*	*Millennium*	*Sepracor*	*Vertex*
Year of founding	1980	1978	1987	1981	1976	1987	1987	1993	1984	1989
Drugs on market	12	5	7	13	13	8	6	3	1	2
Market capitalization ($ million)	77 994	12 518*	2954	6657	50 633	14 965	728	2882	6225	962
Turnover ($ million)	8356	679	714	1658	3300	867	141	434	344	69
Profit (loss) ($ million)	2280	−875	92	222	610	−72	−32.5	222	−135	−192

*Biogen Idec's market capitalization fell to this value in May 2005 (a drop of $10 000 million from the value in February 2005) after the disappointing news on its Tysabri product.[89]

Traditional big pharma prior to the 1980s was (with the exception of vaccine businesses) almost exclusively small-molecule-based. The rise of biologicals as pharmaceutical therapies based upon the molecular biology technologies of the 1980s and 1990s allowed several companies (Amgen, Genentech, Biogen Idec, and Chiron) that began life as SMEs to become world-class players in their own rights before or despite the best efforts of big pharma to compete with them or acquire them. Indeed, with the exception of Genentech (now majority-owned by Roche), the remainder have maintained their independence. It is, however, salient to realize that all these SMEs that have become big pharma players in their own rights were established to research and develop biological agents (antibodies, proteins) rather than (in the first instance) small molecules. Only subsequently, after success gained by getting biological products on the markets, have some turned to small-molecule research. The major success stories of SMEs turning into major competitors of traditional big pharma are shown in **Table 4**. Clearly, in the first tier (market capitalizations over $25 billion) are Amgen and Genentech (actually majority-owned by Roche but still separately quoted); in the second tier (market capitalizations greater than $5 billion) are Biogen Idec, Gilead, Chiron (at the end of 2005 Chiron had agreed to be acquired by Novartis), and Sepracor, and in tier 3 ($1–5 billion) are Millennium, Vertex, and Cephalon. It is instructive to note which products of these companies have reached the marketplace, and thus contributed to those companies achieving fully integrated pharmaceutical company (FIPCO) status.

Genentec. Genentec was founded in San Francisco in 1976 and was acquired by Roche in 1990 (with an initial investment of $1537.2 million in return for one-half of Genentech's then outstanding common stock). Genentech is a free-standing entity within Roche and now has an impressive range of (mostly) biological products.[19] These are

- Actimmune (IFN-γ1b), launched for the treatment of infections in patients with chronic granulomatous disease
- Alteplase (recombinant tissue plasminogen activator, marketed in the US in 1987 for the treatment of acute myocardial infarction, in 1990 for use in the treatment of acute pulmonary embolism, and in June 1996 for the treatment of acute ischemic stroke within 3 h of symptom onset)
- Bevacizumab (intravenously administered antivascular endothelial growth factor monoclonal antibody), an antiangiogenesis therapy for the treatment of colorectal cancer
- Dornase alfa, a DNase which has been developed and launched for the treatment of cystic fibrosis
- Efalizumab, a humanized anti-CD11a monoclonal antibody as a once-weekly subcutaneous formulation, for the treatment of moderate-to-severe plaque psoriasis
- Kogenate, a recombinant human factor VIII preparation used for treating hemophiliacs
- Omalizumab, an anti-IgE humanized monoclonal antibody for the treatment of asthma
- Recombinant somatotropin, first launched in 1987 for the treatment of growth hormone deficiency in children, children with chronic renal insufficiency, and children with short stature associated with Turner's syndrome
- Tenecteplase, an injectable, slower-clearing, fibrin-specific tissue plasminogen activator for the treatment of myocardial infarction, first launched in the US in June 2000 and in 2001 in the EU
- Trastuzumab, a humanized version of the anti-HER-2/neu monoclonal antibody, 4D5, and marketed for the treatment of metastatic breast cancer since 1998
- Roferon-A, a recombinant interferon-α_{2a} that has been developed and used for the treatment of various lymphomas, leukemias, and other neoplasms, and for the treatment of chronic hepatitis B and C virus (HBV and HCV) infections
- Ituximab, a mouse/human chimeric anti-CD20 monoclonal antibody, used for the treatment of B-cell nonHodgkin's lymphoma, including relapsed, refractory low-grade, or follicular lymphomas

- Erlotinib (**1**), a small-molecule, orally active epidermal growth factor receptor tyrosine kinase inhibitor (discovered by but relinquished by Pfizer after the merger with Warner-Lambert), for the treatment of nonsmall-cell lung cancer.

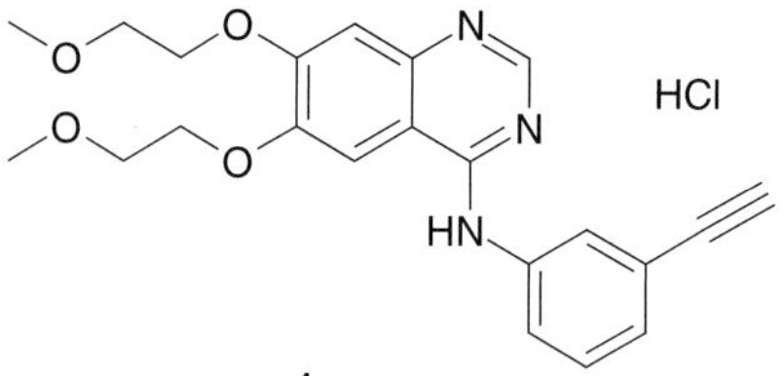

1

Amgen. Amgen (formerly known as Applied Molecular Genetics; AMGen) was established in 1980 and is headquartered in Thousand Oaks, CA. Amgen was originally a biotechnology company focused on the development and commercialization of proteins, antibodies, and small molecules in the areas of oncology, inflammation, hematology and nephrology, neurology, metabolic diseases, and osteoporosis. Amgen has done its own share of acquisitions, including the purchase of Kinetix in 2000 for $170 million, Immunex in 2002 for $13 billion and Tularik in 2004 for $1.3 billion. Amger has an extensive portfolio of products.

- Anakinra is an interleukin-1 receptor antagonist, developed and launched extensively by Amgen for the treatment of rheumatoid arthritis. The drug was originally developed in collaboration with researchers at the University of Colorado
- Ancestim, a recombinant human stem cell factor, has been approved for the mobilization of stem cells in the treatment of cancer
- Darbopoetin alfa is a synthetic recombinant novel erythropoiesis-stimulating protein, for the treatment of anemia associated with renal disease
- Epoetin alfa is a recombinant human erythropoietin marketed in the US since 1989 as an orphan drug for the treatment of anemia resulting from kidney failure and was the first erythropoietin preparation to be approved for the treatment of anemia in premature infants
- Filgrastim is a granulocyte colony-stimulating factor product – the initial indication was reduction in the incidence of infection, as manifested by febrile neutropenia, in patients undergoing myelosuppressive chemotherapy
- Interferon alfacon-1 (consensus interferon) is indicated for use in the treatment of viral infections, particularly hepatitis
- Pegfilgrastim is a sustained-duration, pegylated form of recombinant granulocyte colony-stimulating factor used for the treatment of chemotherapy-induced neutropenia
- Etanercept is a soluble TNF receptor (TNFR) fusion protein, for the treatment of adult and juvenile RA and psoriatic arthritis
- Cinacalcet (**2**) is a calcimimetic for the treatment of primary and secondary hyperparathyroidism.

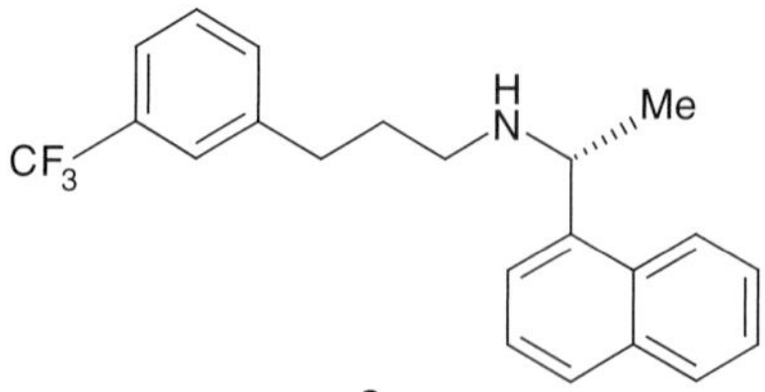

2

Biogen Idec. Biogen Idec was established in 1978 in Cambridge, MA and merged with IDEC (formed in 1985 in San Diego) in 2003, to form Biogen Idec. The number of marketed compounds was five in February 2005.[20] Marketed products from Biogen Idec are:

- Alefacept, a T-cell-inhibiting, CD2 antagonist, LFA3-IgG1 fusion protein, launched in 2003 for the treatment of psoriasis
- Avonex (interferon-β_1), used since 1996 as a treatment for relapsing-remitting multiple sclerosis

- Ibritumomab tiuxetan, an anti-CD20 murine monoclonal antibody conjugated to either yttrium-90 or indium-111 for the radioimmunotherapy of relapsed or refractory low-grade, follicular, or transformed B-cell non-Hodgkin's lymphoma
- Rituximab, a mouse/human chimeric anti-CD20 monoclonal antibody, used for the treatment of B-cell non-Hodgkin's lymphoma, including relapsed, refractory low-grade or follicular lymphomas.

Chiron. Chiron was founded in 1981. In 1984, the company cloned several HBV antigens and sequenced the human immunodeficiency virus (HIV) genome. In 1987, it sequenced the HCV genome. This IP has formed the basis of several Chiron products and also generates income through licensing. As of January 2004, Novartis owned 42% of Chiron's outstanding common stock, following an alliance established in January 1995. In February 2005, Chiron had 18 products on the market, of which the majority were vaccines. The nonvaccine products are:

- Aldesleukin, a recombinant interleukin-2 compound for the treatment of cancer
- Betaseron, a recombinant interferon β_{1b} for the treatment of relapsing-remitting multiple sclerosis
- Tobi, an inhalant formulation of the antibiotic tobramycin, for the treatment of chronic pseudomonal lung infections in patients with cystic fibrosis
- Pamidronate disodium (**3**), a calcium metabolism inhibitor discovered by Henkel (Düsseldorf, Germany) and licensed to both Novartis and Amgen for the treatment of moderate to severe Paget's disease and for hypercalcemia of malignancies such as multiple myeloma
- Dexrazoxane (**4**) (licensed from the British Technology Group), used against chemotherapy-induced cardiotoxicity in several major markets, including the US
- Pilocarpine hydrochloride (oral formulation) stimulates salivary and lacrimal secretion for the treatment of xerostomia and keratoconjunctivitis sicca arising from a number of conditions, including autoimmune diseases and radiotherapy to the head and neck.

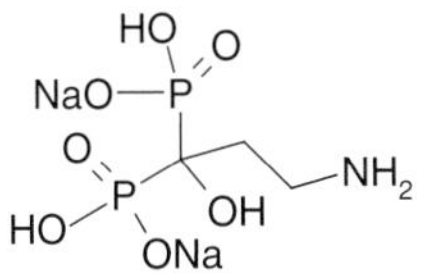

3

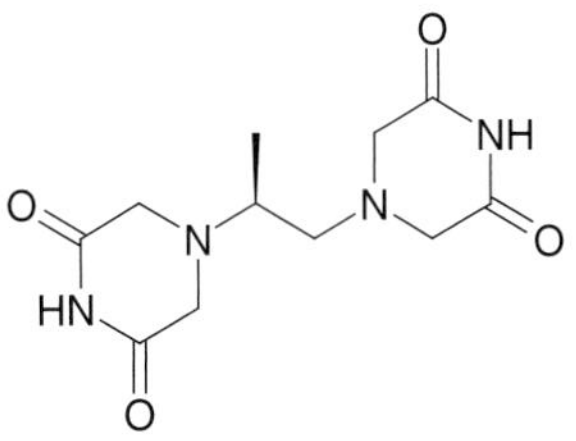

4

Chiron announced that it was to be acquired by Novartis for $5.1 billion in cash on October 31, 2005.

Gilead. In contrast to the biology-oriented SME successes, no SME that began life as a small-molecule drug discovery house has yet become a global competitor to traditional big pharma, although Gilead, Ligand, Sepracor, Vertex, and Millennium are probably the closest to achieving that ambition and have already grown well beyond the SME definitions given above. Gilead Sciences, headquartered in Foster City, CA, was founded in 1987 and has more than 1500 employees in 2005, making it too large to be fairly considered an SME. Gilead has several antiinfectious products on the market:

- Emtricitabine, a nucleoside analog licensed from Emory University that is structurally related to 3TC for the treatment of HIV infection
- Cidofovir, a specific nucleotide inhibitor of viral replication, which was launched in an injectable formulation in 1996
- Oseltamivir (Tamiflu: **5**), a neuraminidase inhibitor, for the treatment and prevention of influenza virus type A or B infections
- Tenofovir disoproxil fumarate (**6**), an oral prodrug of the intravenously administered antiviral agent tenofovir used for the treatment and prophylaxis of HIV infection
- AmBisome, a liposomal formulation of amphotericin B, jointly developed by NeXstar and Lyphomed, marketed for systemic fungal infections and in the US for cancer and acquired immunodeficiency syndrome (AIDS) patients with blood-borne fungal infections
- Adefovir dipivoxil, an analog of the reverse transcriptase inhibitor adefovir for the treatment of HBV infection.

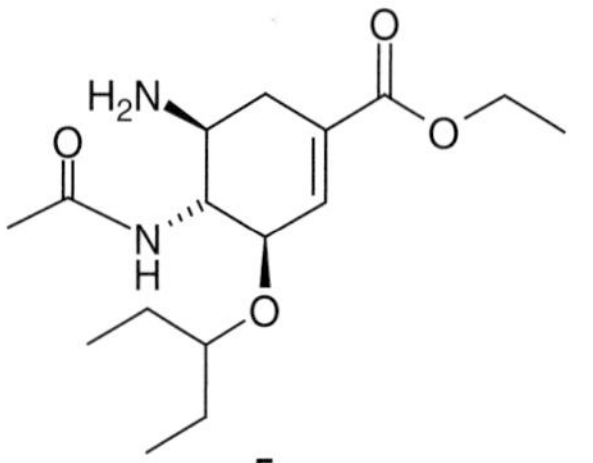
5

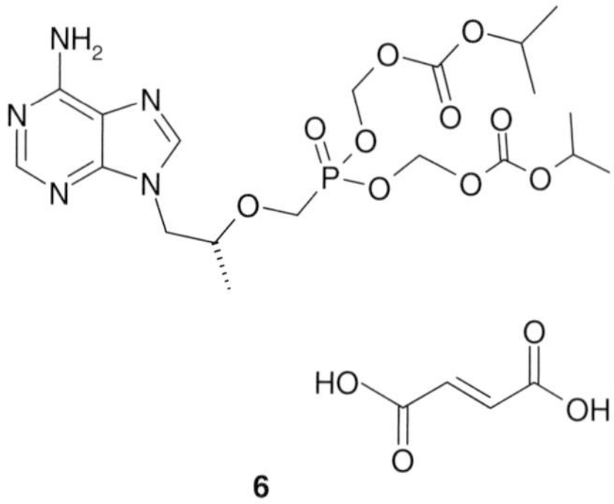
6

Gilead has made two major acquisitions of SMEs to bolster both its technology base and antiviral pipeline, NeXstar (1999) and Triangle Pharmaceuticals (Research Triangle Park (RTP), NC). In 2002 Gilead acquired Triangle Pharmaceuticals, formed by Dr David Barry, formerly of Burroughs Wellcome in 1995, at a cost of $464 million. Triangle had specialized in nucleoside analogs as antiviral agents, including IP licensed from the Universities of Emory and Georgia.

Ligand Pharmaceuticals. Ligand Pharmaceuticals (San Diego, CA) was founded in 1987 to develop gene transcription technologies, particularly an intracellular receptor (IR) technology and signal transducers and activators of transcription (STATs). The company has applied these technologies to discover and develop novel small-molecule drugs for the treatment of cancer and osteoporosis, as well as gynecologic, dermatologic, cardiovascular, and inflammatory diseases. Ligand is a company that has achieved the status of having launched products, and as of 2005, its portfolio included the following US Food and Drug Administration-approved drugs: morphine sulfate extended-release capsules, alitretinoin (**7**), bexarotene, and denileukin diftitox.

Vertex Pharmaceuticals. Vertex Pharmaceuticals, founded in 1989 and initially focused on a structure-based approach to drug discovery, is headquartered in Cambridge, MA, and became a publicly owned biotechnology company in 1991. By July 2001, Vertex completed an agreement signed in April 2001 to acquire Aurora Bioscience (a screening technology company) in a stock-for-stock transaction, where the fully diluted equity value of the transaction was $592 million. As of February 2005, Vertex had only launched one product by itself: Fosamprenavir (**8**), a prodrug of the existing HIV protease inhibitor amprenavir, for the treatment of HIV infection. In addition, Vertex's first original drug to be invented, amprenavir, is an oral, nonpeptidic HIV protease inhibitor and was licensed to and developed and launched extensively by Glaxo Wellcome (now GlaxoSmithKline), for the treatment of HIV infection and AIDS.

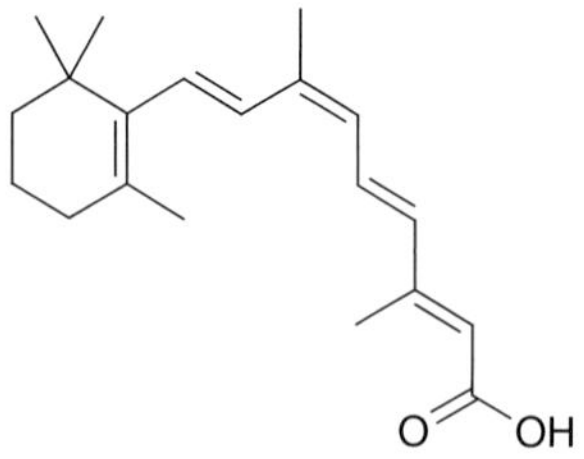
7

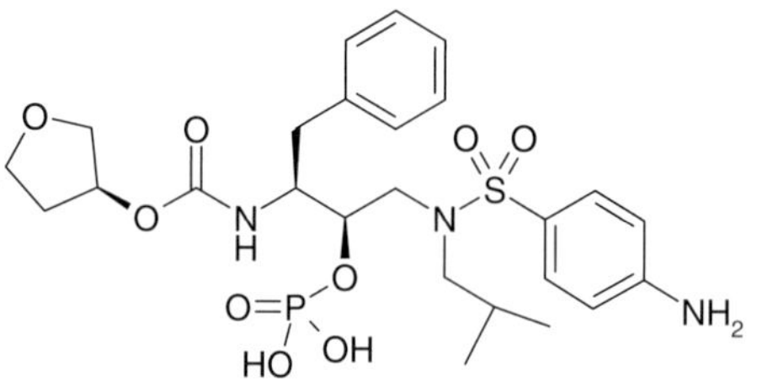
8

Millennium Pharmaceuticals. Millennium Pharmaceuticals was founded in 1993, and is a biopharmaceutical company focused on the discovery and development of small molecules, biotherapeutics (antibodies and proteins), and predictive medicine products. In February 2005, three products were reported to be marketed[21]:

- Alemtuzumab (Campath) is a lymphocyte-depleting humanized monoclonal antibody against CD52, for the treatment for chronic lymphocytic leukemia
- Eptifibatide, a synthetic peptide glycoprotein IIb/IIIa antagonist and platelet aggregation inhibitor derived from rattlesnake venom, is an intravenous treatment for acute coronary syndrome
- Bortezomib (**9**) is a ubiquitin proteosome inhibitor for the treatment of multiple myeloma.

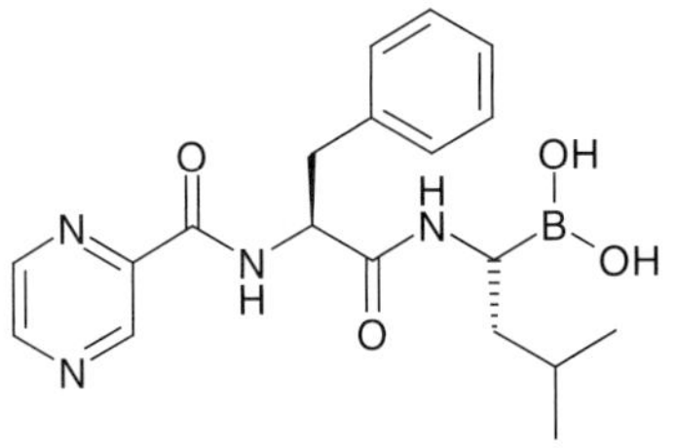

9

1.13.5.2 Small- or Medium-Sized Enterprises That Have Been Acquired by Major Pharma

It is salutary to note that, in contrast to the biological drug houses listed above, big pharma has been quicker to recognize the value of small-molecule drug research houses and thus has prevented many from achieving the roles of FIPCOs due to them becoming acquisition targets of big pharma companies, including some of the new biotech majors listed above. In many cases SMEs have welcomed the takeover from major pharma companies as the only way to realize the value inherent in their IP portfolio. Illustrative acquisitions by the major pharma players are shown in **Table 3.**

1.13.5.3 Small- or Medium-Sized Enterprises That Have Been Acquired by Other Small- or Medium-Sized Enterprises

Many SMEs would have failed to realize any potential due to cash exigencies and have been encouraged to merge, or be acquired by, other SMEs, when the combination of the two companies appears synergistic and is the only way to realize further investment. Biotech hunger to acquire capabilities beyond its own, often specialized and narrow capabilities, equally encourages vertical mergers. Acquiring chemistry capability has been the logic behind the fusions of SMEs, including those of Lexicon Genetics with Coelacanth in the US in 2001, Etiologics with Argenta Discovery of the UK in 2004, and Paradigm with Amedis in the UK in 2005.

The acquiring of IP assets by a service company determined to become a therapeutics player can also engender acquisitions: Evotec-OAI purchased Evotec Neurosciences in 2005. Merging two therapeutics companies together to cut costs whilst maintaining synergies can also be an important cause of mergers: Ribotargets, British Biotechnology, and Vernalis performed a three-way merger in 2004, creating a single entity, now known only as Vernalis. Similarly, in the US, VI Technologies (Vitex) is the surviving company from the November 1999 merger of Pentose Pharmaceuticals and VI Technologies. Vitex merged again with Panacos Pharmaceuticals in early 2005 with a private placement of $20 million, allowing financial compliance with NASDAQ requirements. The deal would combine Vitex's antiinfective capabilities into therapeutic products, including Panacos' pipeline of antiviral drugs, and its antiviral drug discovery platforms.

1.13.5.4 Small- or Medium-Sized Enterprises That Have Failed

Finally, some companies – a merciful few – go out of business. Axxima AG's (Munich, Germany) filing for insolvency in December 2004 was followed by the acquisition of the company by GPC Biotech AG in 2005; however, the bankruptcy and total closure of Cambridge (UK)-based Axis Genetics in 1999 was a salutary experience for those in the industry.[22]

1.13.6 Small- or Medium-Sized Enterprises That Are Drug Discovery Houses

1.13.6.1 Multitherapeutic Disease Area Companies

Under this heading is grouped a multitude of companies, which are neither constrained by therapeutic area nor have as prime reason for existence research in a particular technological axis.

Aderis Pharmaceuticals. Aderis Pharmaceuticals (formerly Discovery Therapeutics) of Hopkinton, MA is a privately owned pharmaceutical company, founded in 1994, created upon the IP/research programs from Whitby Research, a division of Ethyl. Aderis focuses on the R&D of small-molecule therapeutics for use in the treatment of central nervous system (CNS), cardiovascular, and renal diseases. Rotigotine (licensed to Schwarz Pharma, Germany) has been developed for early stage Parkinson's disease. Selodenoson (**10**) is a selective adenosine A1 agonist formulated for intravenous administration to control heart rate during acute attacks, and formulated for oral administration for the chronic management of atrial fibrillation. King Pharmaceuticals (Bristol, TN) is the partner on the binodenoson project (adenosine A2A receptor agonist for cardiac pharmacologic stress-imaging).

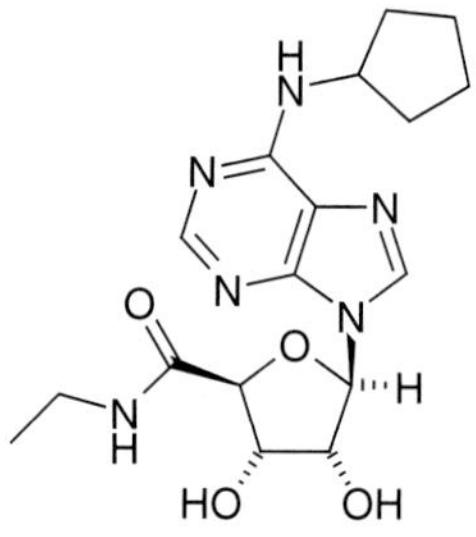

10

Arakis. Arakis (Saffron Walden, UK, founded January 2000) is a pharmaceutical company focusing on the treatment of inflammatory disease and oncology adjunctive therapy. The most advanced drugs in development are AD237 for chronic obstructive pulmonary disease (a muscarinic antagonist); AD 452, in development for rheumatoid arthritis, and apparently the single isomer of a racemic drug marketed for an unrelated indication; AD923, being developed for pain associated with cancer; and AD337, for emesis associated with cancer.[23] Arakis was acquired by the Japanese biotech Sosei in July 2005.

Curagen. Curagen (a public company located in New Haven, CT) was formed as a genomic company (single nucleotide polymorphism technology) that has more recently moved to a therapeutic axis. It is now a multitherapeutic axis company that has recently changed strategic direction to become a preclinical and clinical development company – a move to advance its pipeline of genomics-based therapeutics. Curagen is focusing on CG-53135 (human FGF-20 (fibroblast growth factor)) for the treatment of oral mucositis. Curagen is working with Abgenix on CR-002, a fully human monoclonal antibody that specifically recognizes and blocks the active form of PDGF-D (platelet-derived growth factor), for the potential treatment of kidney inflammation. CR-011 is a potential treatment of metastatic melanoma. Curagen and Bayer have a long-standing joint interest in new diabetic treatments. Curagen has also in-licensed the Topotarget HDAC inhibitor (itself acquired from the former UK SME Prolifix) PXD-101, for the potential treatment of cancer, and is also investigating it for the potential treatment of inflammation.

Curis. Curis, a public company headquartered in Cambridge (MA), was established in July 2000 through the merger of three biotechnology SME companies: Creative BioMolecules, Ontogeny, and Reprogenesis. The company specializes in the development of proteins and small molecules to modulate the regulatory pathways which control repair and regeneration of cells. This technology has been used to produce potential therapeutics for kidney disease, neurological disorders, cancer, hair growth regulation, and cardiovascular disease. Curis has specialized in therapeutic treatments that might ensue through interference with the so-called hedgehog pathways. Products include both biologicals – BMP-7 is a signaling protein that was discovered by scientists from Curis, and treatment with BMP-7 ameliorates two major complications of chronic kidney disease – and small-molecule hedgehog agonists and antagonists.

Exelixis. Exelixis (San Francisco, CA) is another genomics-based (model systems for pharmaceuticals include the fruitfly, nematode worm, zebrafish, and mouse) drug discovery company focusing mainly on developing therapies for cancer and other proliferative diseases. Scientists at Exelixis have done a large amount of work on receptor tyrosine kinases, and other enzymes important in angiogenesis and vascularization, e.g., the ADAM-10 metalloprotease. The three nuclear receptors – liver X receptor (LXR), farnesoid X receptor (FXR) and mineralocorticoid receptor (MR) – are the targets for small molecules for modulation. Such modulation is implicated in various metabolic and cardiovascular disorders. Oncology programs are focused on the inhibition of the RAF, Akt/S6k, and IGF1R kinases that are implicated in cancer and other therapeutic areas. In October 2004, Exelixis acquired another SME, X-Ceptor, once a private corporation that was formed by Ligand Pharmaceuticals in July 1999 to research and identify therapeutic products that target orphan nuclear receptors using technology under exclusive license from Ligand. X-Ceptor focused on the R&D of drugs for the treatment of cancer, cardiovascular disease, endocrine disorders, metabolic disease and inflammation. Exelixis and GlaxoSmithKline have an agreement on an R&D collaboration under which GlaxoSmithKline pays milestone payments to Exelixis upon filing of three investigational new drugs or successful completion of one phase I trial by the end of 2005 for products from the Exelixis stable.

Fulcrum Pharmaceuticals. IP from the University of Pennsylvania (Philadelphia, PA) and the Johns Hopkins University (Baltimore, MD) created Fulcrum Pharmaceuticals (New York) founded in 2002, a biotechnology company focused on the development of small molecules for the treatment of inflammatory disease, osteoporosis, infectious disease. and oncology. Lead programs include small-molecule inhibitors of the tumor necrosis factor receptors and small-molecule inhibitors of the receptor activator of NFκB (RANK)/RANK ligand (RANKL) system.

Idun Pharmaceuticals. Idun Pharmaceuticals (San Diego, CA), founded in 1993, was a private company researching the caspase enzymes, a group of cellular proteases involved in the pathway of apoptosis and inflammation. IDN-6556 (**11**) is a first-in-class pancaspase inhibitor which is in phase II clinical trials of patients having had liver transplantation and in patients infected with hepatitis C virus. In 2005 the company was purchased by Pfizer, probably following encouraging results of its lead product in clinical trials for liver fibrosis.[24]

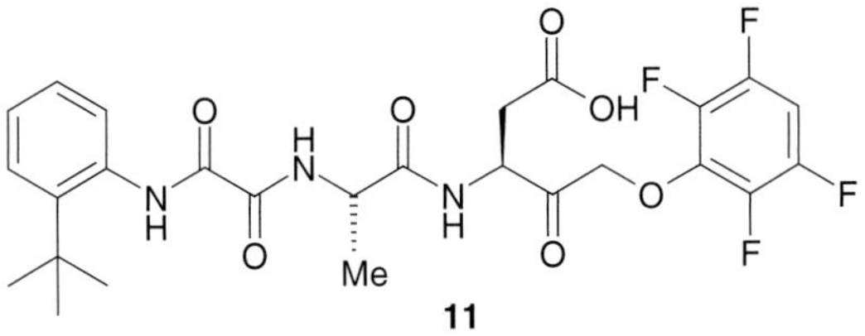

11

Jerini. The private company Jerini (Berlin, Germany) was spun off from the Medical Faculty at the Humboldt University, Berlin, in 1994. It uses a platform technology (SPOT) of high-density protein, peptide, and small-molecule arrays to map novel targets, leading to the identification and efficient optimization of lead compounds for drug discovery. Jerini has worked with Alcon, Baxter Healthcare, and Merck KGaA on eyecare and oncology projects. Jerini is developing the selective bradykinin β_2-antagonist icatibant (**12**) (under license from Sanofi-Aventis and previously known as Hoe 140) as a potential treatment for decompensated liver cirrhosis with resistant ascites, angioedema, and edema in severe burn patients.

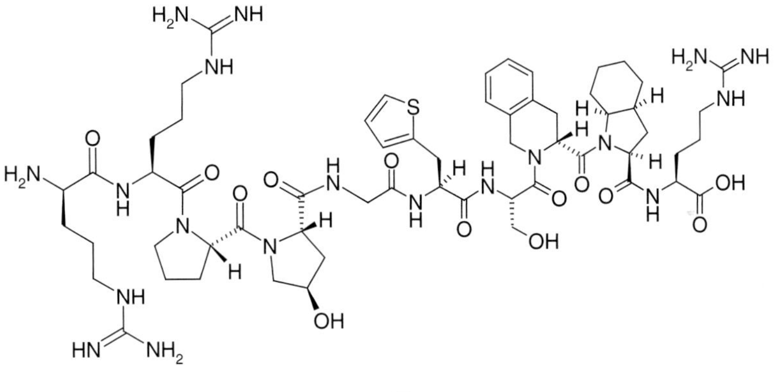

12

MediciNova. MediciNova was founded in September 2000 by Tanabe Seiyaku of Japan (which has a controlling stake) and is located not in Japan but in San Diego, CA. Products and projects include MN-001, which is a novel, orally bioavailable compound for the treatment of bronchial asthma. MN-001 (**13**) has actions including leukotriene (LT) receptor antagonism and inhibition of phosphodiesterase IV. The ANG-600 series of benzimidazole carbamate vascular-targeting agents are potential angiogenesis inhibitors for the treatment of cancer.

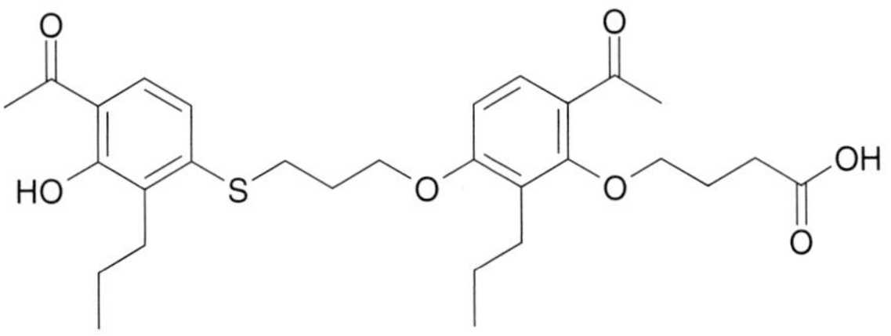

13

Myriad. Myriad Genetics and Myriad Pharmaceuticals are two parts of the Salt Lake City (UT)-located SME. The two Myriad companies are publicly held and focus on gene discovery and analysis, the development of prophylactics and therapeutics (through its subsidiary, Myriad Pharmaceuticals), predictive mutations of the *BRCA1* breast cancer gene, and the discovery of the *HPC2* prostate cancer gene on chromosome 17p. The *HCP2* gene was to form the basis of Polaris, Myriad's predictive test for assessing a man's risk of developing prostate cancer. Drug discovery efforts target cancer, rheumatoid arthritis, acute thrombosis, Alzheimer's disease, HIV/AIDS, and other viral diseases.

Oxagen. Oxagen (Abingdon, UK) was formed in April 1997 as a spinout genomics company from the Wellcome Trust Centre for Human Genetics, but then mutated into a biopharmaceutical company with a full drug pipeline for the treatment of metabolic diseases, such as type 2 diabetes, osteoporosis and endometriosis, and inflammatory diseases, such as rheumatoid arthritis, inflammatory bowel disease, psoriasis, and asthma. Oxagen is investigating antagonists of the chemoattractant G protein-coupled receptor (GPCR) expressed on Th2 cells (CRTH-2) for the potential treatment of asthma and allergy. A large second fundraising was completed in the second quarter of 2005.

ReOx. A 2004 spinout company, from Oxford University (UK), is ReOx, a drug discovery company whose technology is based on the body's biological response to a lack of oxygen (hypoxia) and in particular the mechanism by which the activity of the master regulator, hypoxia-inducible factor, is activated. The company will build on the research of the academics involved to develop therapies for a range of diseases in which it could be beneficial to regulate the body's response to oxygen.[25]

Rigel. Rigel (San Francisco, CA) has pioneered the inhibition of syk kinase as potential treatment of respiratory diseases. In 2004 Rigel entered into collaboration with Pfizer to pursue the clinical development of such molecules. Other compounds from the Rigel syk kinase portfolio are being developed for the possible treatment of rheumatoid arthritis. Rigel is also developing R-803, a small-molecule, non-nucleoside HCV RNA polymerase inhibitor, for the potential treatment of HCV infection. Merck is in collaboration with Rigel on inhibition of ubiquitin ligases as potential therapies in oncology, and is also developing inhibitors of aurora kinase as potential cancer drugs.

Sosei. Sosei was founded in 1990 and is based in Tokyo, Japan. It is a biopharmaceutical company that has projects for contraception, cancer, incontinence, asthma, and other allergies. The company's main strategy is to use its drug-reprofiling program to optimize or find new indications for discontinued drug candidates, mainly from Japanese companies. Sosei is unusual – it is one of the few Japanese SME biotech companies, and it has a policy of in-licensing compounds that are already marketed or at the late stages of development, from US/EU companies and of in-licensing compounds from Japanese companies via its drug-reprofiling platform.[26] Sosei acquired the UK biotech Arakis in July 2005.

Tularik. Tularik (San Francisco, CA) was founded in 1991, and was a biopharmaceutical company engaged in the R&D of drugs that regulate gene expression and focused on cancer, immunology, and metabolic disorders. Amgen acquired Tularik in 2004. From the Tularik stable, Amgen now has batabulin (**14**) in development. Batubulin is a tubulin polymerization inhibitor, for the potential treatment of various cancers.

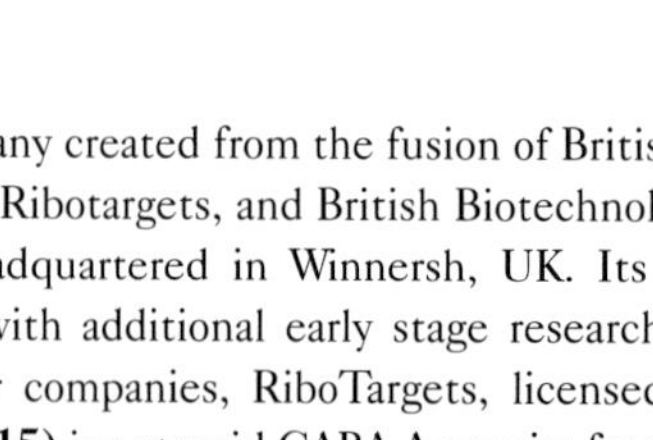

14

Vernalis. Vernalis is the name of a company created from the fusion of British SME biotechnology companies Vernalis (already incorporating Vanguard Medica), Ribotargets, and British Biotechnology (founded as British Biotech in 1986). Vernalis is a biotechnology company headquartered in Winnersh, UK. Its R&D focus is on migraine, thrombotic diseases, pain, and Parkinson's disease, with additional early stage research in oncology, obesity, inflammation, and depression. One of Vernalis' predecessor companies, RiboTargets, licensed alfadolone (**15**) for development from Monash University, Australia. Alfadolone (**15**) is a steroid GABA A agonist for the potential treatment of pain, especially that associated with cancer. Vernalis has done deals with Novartis on an hsp90 oncology target project. Under another agreement, Biogen IDEC received exclusive worldwide rights to Vernalis' adenosine A2A receptor antagonist program. Under this agreement, Biogen IDEC received exclusive worldwide rights to develop and commercialize Vernalis' program including the compound VER 6323 (**16**).

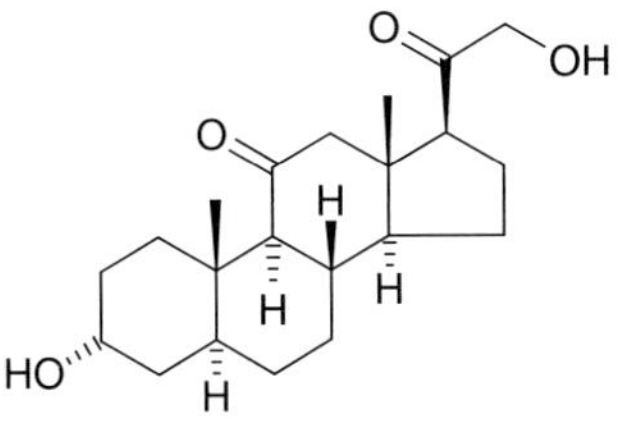

15

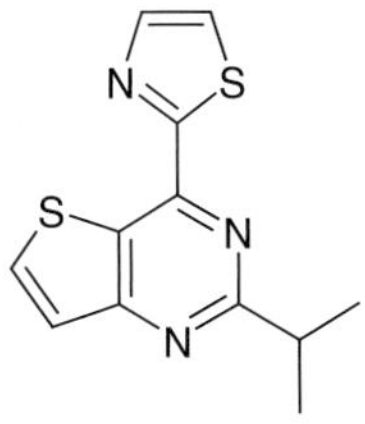

16

1.13.6.2 Antiinfectious Diseases Companies

It is a truism to say that, by 2005, traditional big pharma has largely abandoned the previously lucrative antibacterial field to the biotechnology sector – despite the rise in cases of patients with hospital-acquired infections of methicillin-resistant *Staphylococcus aureus* (MRSA) and vancomycin-resistant enterococci (VRE). The public perception of danger is not yet a commercial driver sufficient to warrant reinvestment of large pharma research groups and, in this field, discovery of new treatments will largely be due to the efforts of SMEs in the next few years. Antifungal and antiviral drug discovery, however, remain the preserve of both traditional large pharma and of biotech.

Rib-X. Rib-X (New Haven, CT) is a spinout of the University of Yale (founders Dr Thomas Steitz and Dr Peter Moore) that focuses on its licensed proprietary knowledge of the structure of the human ribosome to discover new antibacterial agents of the macrolide family. Structural information has been obtained for a number of antibiotics complexed with the 50S subunit and reveals that 50S-targeting antibiotics bind in a tunnel in the central part of the ribosomal subunit. This binding thus blocks access to the pocket where peptide bond formation in growing proteins takes place.

Athelas. Athelas (Geneva, Switzerland) is a private company founded in 2002 by Professor Pierre Cosson and Dr Jean-Pierre Paccaud (University of Geneva). The company focuses on the development of bactericidal antivirulence drugs, the identification of bacterial genes involved in the virulence mechanisms, and their use as new drug targets with a proprietary cell-based screening technology called DiVi.[27]

Novexel. Novexel (Paris, France) is an SME spinoff from Aventis after its merger with Sanofi-Synthelabo. Novexel inherited a portfolio of antiinfective drugs in phase I clinical trial, including NXL-103, an oral antibiotic against bacterial respiratory infections, an oral streptogramin consisting of a 70/30 combination of RPR-202868 (type 1 streptogramin), and RPR-132552 (type 2 streptogramin) for the potential treatment of community-acquired pneumonia and respiratory tract infections, and also NXL-201, which is being developed for the treatment of severe fungal infections.

Oscient Therapeutics. Oscient Therapeutics (Cambridge, MA) incorporated what was Genome Therapeutics, an antiinfectives company, in 2004. This company originally focused on the discovery of new antibacterial drugs and had a long-standing (9–year) research agreement with Schering-Plough to discover new antibacterial targets using its proprietary genetics technology base, including a database licensed in by several companies, PathoGenome. Oscient has changed business focus as from 2004, becoming a more commercial organization. It has one launched product – gemifloxacin (**17**) (licensed from LG Life Sciences in 2002).

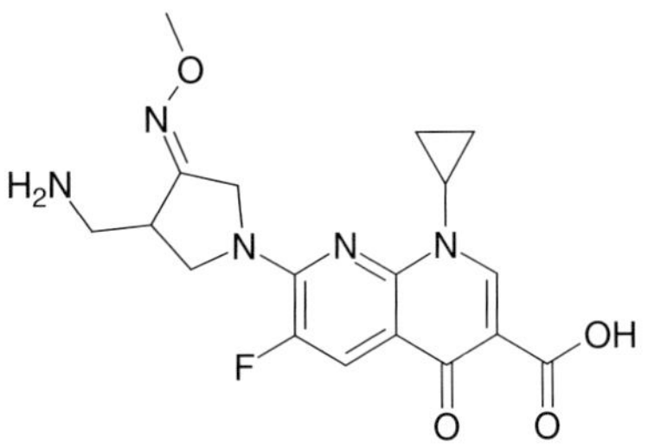

17

Arrow Therapeutics. Arrow Therapeutics (London, UK), founded in 1998, was created after the acquisition of Wellcome by Glaxo and the closure of the former Wellcome research facilities in south London. Arrow is a privately owned antiinfective drug discovery company, focused on targets for antiviral and antibacterial therapies utilizing a genome-based approach. In the antiviral sector, a lead compound against respiratory syncytial virus infection completed phase I trials in June 2004. Arrow also has a hepatitis C program, which entered preclinical studies in March 2004, with clinical trials planned for the first half of 2005.[28]

1.13.6.3 Cardiovascular Disease Companies

Cardiovascular disease is an area of focus for both major pharma and biotech companies – most SMEs have arisen from academic origins.

Cardiome. Cardiome (formerly Nortran Pharmaceuticals of Vancouver, Canada) is a drug discovery and development company focused on the treatment and prevention of cardiac diseases. Cardiome focuses its development within two areas of cardiac disease: arrhythmia and congestive heart failure. In clinical trial is RSD 1235 **(18)** for arrhythmia and also oxypurinol for congestive heart failure. Ion channel research is a specialty of this company.

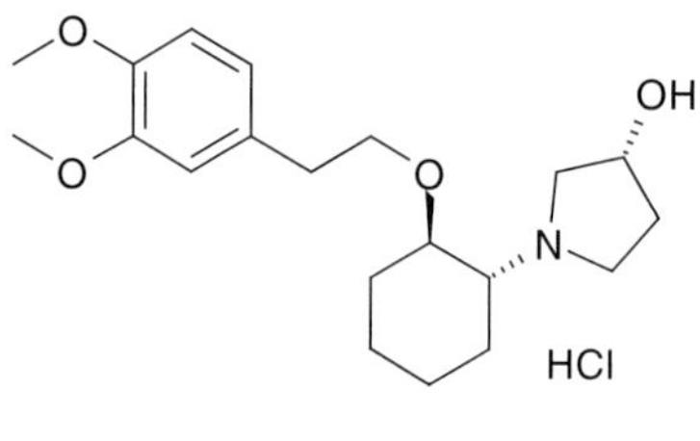

18

CV Therapeutics. CV Therapeutics (Palo Alto, CA), which was founded in 1990, performs R&D in cardiovascular disease. Regadenoson **(19)** is a selective A2A-adenosine receptor agonist for potential use as a pharmacologic stress agent in myocardial perfusion imaging studies, developed by CV Therapeutics and the Japanese company Fujisawa. Tecadenoson **(20)** is a selective A1-adenosine receptor agonist for the potential reduction of rapid heart rate during atrial arrhythmias. Adentri is a selective A1-adenosine receptor antagonist for the potential treatment of heart failure, and has been licensed to Biogen Idec. CV Therapeutics, under license from Roche, is also developing ranolazine, a metabolic modulator and a partial fatty acid oxidation inhibitor, for the potential treatment of angina and acute coronary syndromes.

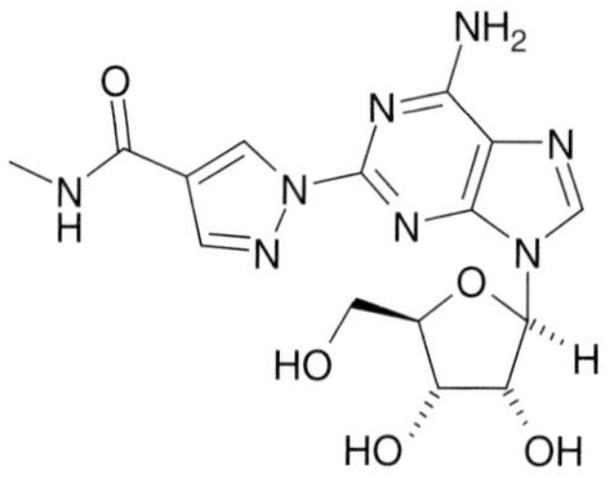

19

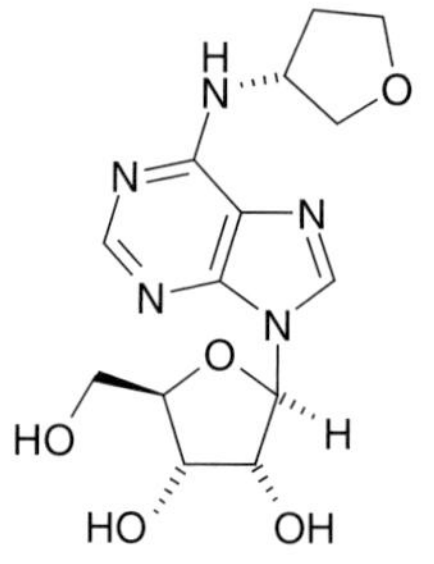

20

Artesian Therapeutics. Artesian Therapeutics (Gaithersburg, MD) is a privately held biopharmaceutical company focused on novel disease-modifying and disease-reversing therapeutics for the treatment of cardiovascular disease. Artesian is another company developing 'dual-pharmacophore' molecules (which contain the individual pharmacophoric moieties needed to inhibit two targets linked together by a flexible chemical tether into a single molecule).

Myogen. Myogen, founded in 1996 and based in Westminster (CO), is a biopharmaceutical company engaged in the discovery, development, and commercialization of small-molecule therapeutics for the treatment of cardiovascular disorders, based on research undertaken at the University of Colorado Health Sciences Center. Myogen, under license from Aventis (now Sanofi-Aventis), is developing an oral formulation of the vasodilating phosphodiesterase III inhibitor enoximone **(21)** for the potential treatment of advanced heart failure. Under license (from Abbott) is an oral endothelin A antagonist darusentan **(22)** for the potential treatment of uncontrolled hypertension. Under license from an Abbott research group (formerly BASF Pharma) is ambrisentan **(23)**, an endothelin ET-A antagonist, for the potential treatment of cardiovascular diseases. In terms of target discovery, scientists at Myogen have discovered that nonfailing human ventricular myocardium contains a larger amount of the fast-contracting alpha-myosin heavy chain (alpha-MyHC) isoform mRNA. In contrast, in myocardial failure, alpha-MyHC mRNA is markedly downregulated, and the slow-contracting isoform beta-MyHC is upregulated. Thus, it might be expected that agents that can cause reversal of the MyHC gene switch would provide therapeutic benefit.

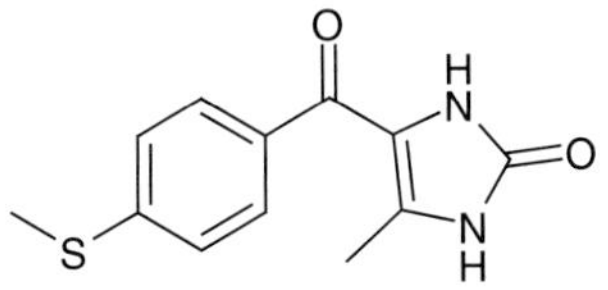

21

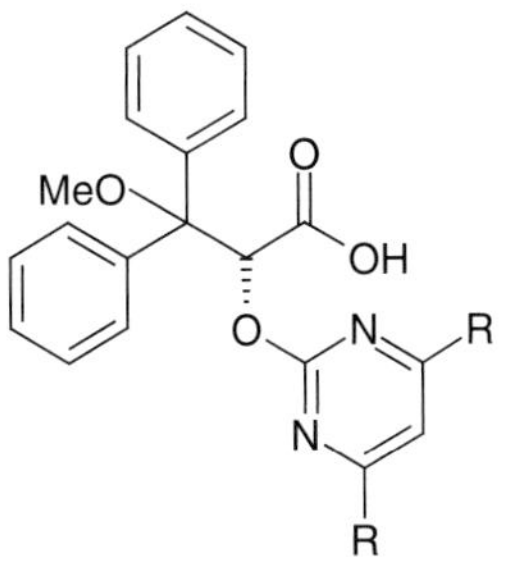

22 R = MeO

23 R = Me

Zealand Pharma. Zealand Pharma (Glostrup, Denmark) was founded in 1998 and is a private biopharmaceutical company that focuses its R&D efforts on developing peptides by investigating gap junction modulation and peptide modification chemistry using its structure-inducing probe technology. The company's therapeutic focuses are type 2 diabetes, acute heart failure, and arrhythmia.

1.13.6.4 Central Nervous System Disease and Antiaging Companies

The unmet need for new therapies to treat CNS disorders is one that is especially attractive to scientists, clinicians, and investor backers. Consequently, a large number of companies have been created to examine many new approaches for a huge variety of CNS indications. Some particularly informative examples of the impact these SMEs are having on CNS drug discovery are listed below.

The treatment of pain has barely changed from the two centuries' use of morphine and its long-known derivative diamorphine, with all the well-known addiction and respiratory depression side effects these drugs can cause. Companies like CeNeS (Cambridge, UK) have pioneered the use of morphine-7-glucuronide (**24**), whilst Exton (PA)-based Adolor (the name is derived from the Spanish meaning 'no pain') launched itself on the back of an IP estate licensed in from Eli Lilly and has researched a range of approaches in the field. In 2003, Adolor struck a deal with GlaxoSmithKline to codevelop and comarket Adolor's μ-opiate receptor antagonist product alvimopan (**25**) to treat the gastrointestinal effects associated with prolonged use of opiate analgesics.

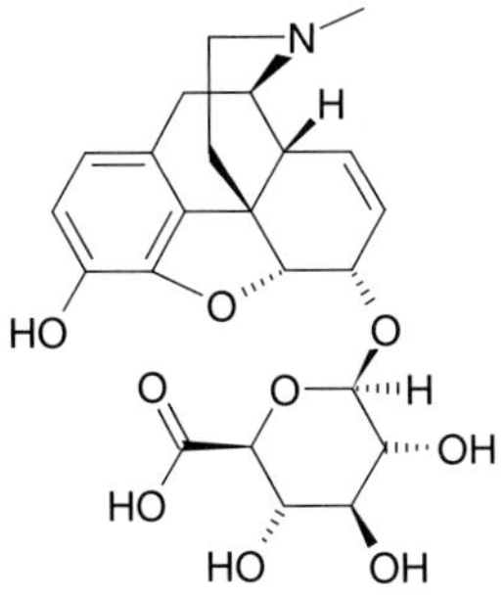

24

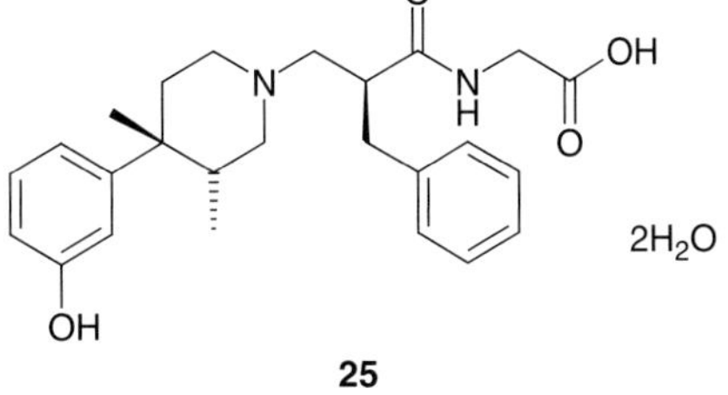

25

Another Pennsylvania-based company, Theraquest Biosciences, was established in 2000 and researches into nonsteroidal antiinflammatory drugs for the treatment of pain.

Ionix Pharmaceuticals. Cambridge (UK)-based Ionix Pharmaceuticals conducted research into nonopiate treatments of pain associated with using proprietary drug targets that are involved in the perception and signaling of pain in the peripheral nervous system. Such treatments might be useful for pain associated with chronic debilitating diseases such as osteoarthritis, rheumatoid arthritis, multiple sclerosis, and diabetic neuropathy. Ionix was a typical SME having to outsource much of its research to partners – it collaborated with Evotec and with Tripos on the design and synthesis of drug-like chemical compounds for evaluation as potential inhibitors of its own targets, and with Xenome on the design, synthesis, and screening of toxins as potential inhibitors of proprietary Ionix drug targets. The IX-2000 series are small-molecule blockers of the calcium ion channel that are able to enter the spinal cord. In preclinical pharmacokinetic investigations in a neuropathic pain model, administration of IX-2100 resulted in an almost identical concentration in the spinal cord and plasma and > 40% reversal of neuropathic hyperalgesia was observed for almost 6 h. Ionix was acquired by Vernalis of the UK in July 2005.

Chronogen. Chronogen (Montreal, Canada) was founded in 1998 to develop therapeutics to counter the effects of aging. Chronogen has focused on the identification and characterization of longevity genes, which were cloned from *Caenorhabditis elegans* mutants that exhibited a long lifespan phenotype. Those genes that were the most relevant to human diseases were selected as the basis of the company's drug discovery programs. In 2002, Chronogen secured an exclusive worldwide license to the proteins CLK-1, CLK-2, and ISP-1, in addition to a series of targets, from McGill University (Montreal, Canada). Targets under investigation include isp-1, which regulates levels of reactive oxygen species, and clk-1, which regulates ubiquinone synthesis.

Targacept. Targacept (RTP, NC) is a 2000 spinout from the tobacco company RJ Reynolds and has products under research for postoperative pain (TC-2696) and Alzheimer's disease. Targacept is researching novel approaches to treat pain, including $\alpha_4\beta_2$ neuronal nicotinic acetylcholine receptors that are known to have pain-relieving effects in animals.

Neurogen. Neurogen, incorporated in 1987, is a neuropharmaceutical company focused on the R&D of small-molecule therapeutics for the treatment of neurological, inflammatory, pain, and metabolic disorders. Neurogen is developing NGD-2000-1, the lead in a series of oral C5a antagonists, for the potential treatment of inflammatory conditions, including rheumatoid arthritis and asthma. As well as treatment of inflammatory conditions, Neurogen is developing NG-2-73, a selective GABA modulator, for the potential treatment of insomnia. In December 2004, the company initiated a phase I oral dose escalation trial with the drug. Neurogen and Pfizer are also developing the GABA-A agonist, NGD-96-3 (**26**) for the potential treatment of insomnia.

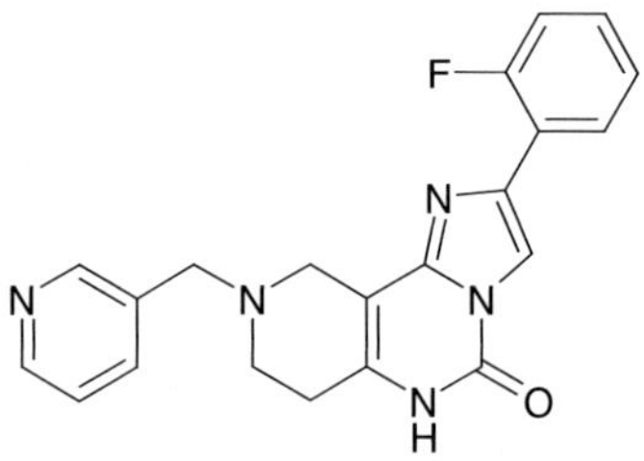

26

En Vivo Pharmaceuticals. The treatment of Alzheimer's disease is a severe unmet medical need and novel approaches are being examined by a variety of companies. En Vivo Pharmaceuticals (Cambridge, MA) founded in 2001, is a biopharmaceutical company dedicated to discovering and developing drugs for CNS disorders, with its initial focus on Alzheimer's, Parkinson's, and Huntington's diseases and spinocerebellar ataxias. Using technology licensed from the Baylor College of Medicine (Houston, TX), the company inserts known human brain disease genes into *Drosophila* (fruitfly) and examines the effects of candidate drugs on the whole species. In 2004, En Vivo pharmaceuticals struck a deal with Methylgene (Quebec, Canada) to examine the use of the latter's portfolio of histone deacetylase inhibitors (which hitherto have been almost exclusively an oncology therapy) in its Alzheimer's disease models.

Evotec Neurosciences. Evotec Neurosciences (Hamburg, Germany) was founded in May 1999 by Evotec BioSystems and Professor Nitsch. Evotec Neurosciences received a €25 million Series A funding in 2004 from Evotec OAI and five venture capitalists (TVM Techno Venture Management (Munich, Germany), 3i (London, UK), MVM (London, UK), Ventech (Paris, France), and Star Ventures (Munich, Germany)). Research focus was on Alzheimer's disease and other neurological disorders. Licensed in from Roche is a series of NR1/NR2B subtype-selective *N*-methyl-D-aspartate antagonists for the potential treatment of neurodegenerative diseases, including stroke.[29] Evotec Neurosciences was reacquired in its entirity from its VC investors by Evotec OAI in March 2005.

DOV Pharmaceutical. New drugs to treat depression and Parkinson's disease are a speciality of DOV Pharmaceutical, a biopharmaceutical company formed in 1996. The products of this company have attracted the attention of Merck, which is developing under license DOV-21947 (possible structure shown as (**27**)), a triple (serotonin, norepinephrine, and dopamine) reuptake inhibitor, for the potential treatment of depression. Merck licensed exclusive worldwide rights to DOV-21947 for all therapeutic indications and exclusive worldwide rights to DOV-216303 for the treatment of depression, anxiety, and addiction. The size of the deal struck with Merck is impressive: DOV would receive a $35 million upfront licensing payment, $300 million for achieving certain milestones, and up to $120 million upon achievement of certain sales thresholds. Merck would assume responsibility for the full development, manufacturing, and commercialization of DOV-21947. DOV is investigating DOV-51892, the lead in a series of GABA-A modulators, for the potential treatment of panic disorder. A further CNS product from DOV Pharmaceutical is ocinaplon (DOV-273547) for the potential treatment of generalized anxiety disorder.

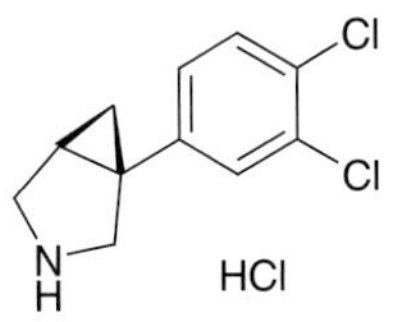

27

Corcept. Corcept (San Francisco, CA) is an SME studying the use of mifepristone (RU-486) (**28**) which (as of March 2005) is in phase III clinical trials for the treatment of the psychotic features of psychotic major depression, a disorder that affects approximately 3 million people in the US each year and for which there are no Food and Drug Administration-approved treatments. Corcept has sponsored further research in conjunction with Argenta Discovery (Harlow, UK) on this indication.

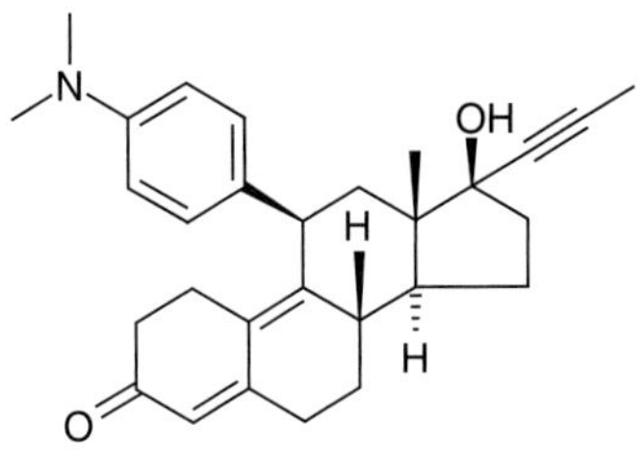

28

Neurocrine Biosciences. Neurocrine Biosciences (San Diego, CA) was founded in 1992 and focuses on drug discovery of small-molecule therapeutics to treat diseases of the central nervous, immune, and endocrine systems. Neurocrine is developing NBI-30702, a second-generation corticotrophin-releasing factor receptor antagonist, for the potential treatment of cerebrovascular ischemia leading to stroke. NBI-30702 is one of a series of compounds resulting from Neurocrine's work with excitatory amino acid transporter modulators.[30]

Memory Pharmaceuticals. Memory Pharmaceuticals (Montvale, NJ), founded in 1998, also researches and develops potential drugs to treat Alzheimer's disease, depression, schizophrenia, vascular dementia, mild cognitive impairment, and memory impairments associated with aging. The company (in alliance with big pharma partners, including Roche) is looking for compounds that will modulate the L-type calcium channel. Such compounds modulate neuronal calcium channels and regulate calcium ion flow into neurons, preventing the deleterious consequences of excessive levels of calcium entry. One of the benefits of this modulation is that neurons remain more responsive to incoming signals, counteracting the reduced activity that normally occurs during aging. In March 2005, at least one such compound (MEM-1003) had entered clinical trials. Research workers at Memory are also looking at an unusual use of phosphodiesterase type IV inhibitors for mild cognitive impairment.[31] One such compound, MEM 1414, has been reported as effective in hippocampus-dependent memory tasks over a very wide dose range. Memory also has a project examining the use of nicotinic α_7 receptor agonists, including MEM 3454, for schizophrenia and Alzheimer's disease.[31]

Newron Pharmaceuticals. Newron Pharmaceuticals is a 1999 spinout formed from the closure of the former Pharmacia & Upjohn's research center in Milan (Italy). Newron is a private biopharmaceutical company funded privately – more than €25 million was injected in two rounds of financing led by Atlas Venture and Apax Partners in 2002. This funding has been supplemented by another €23 million in 2005. Newron focuses on the discovery and development of diagnostics and therapeutics of the nervous system, particularly epilepsy, neurodegenerative disorders, and pain. The company runs preclinical research up to the early phases of clinical development, then pursues partners for continued development. The lead compound, safinamide (**29**), was licenced from Pharmacia & Upjohn and is a sodium-channel blocker, calcium-channel modulator, glutamate release inhibitor, and a dopamine metabolism modulator (monoamine oxidase B inhibition/dopamine uptake inhibitor), with potential for the treatment of epilepsy, Parkinson's disease, and restless-legs syndrome.

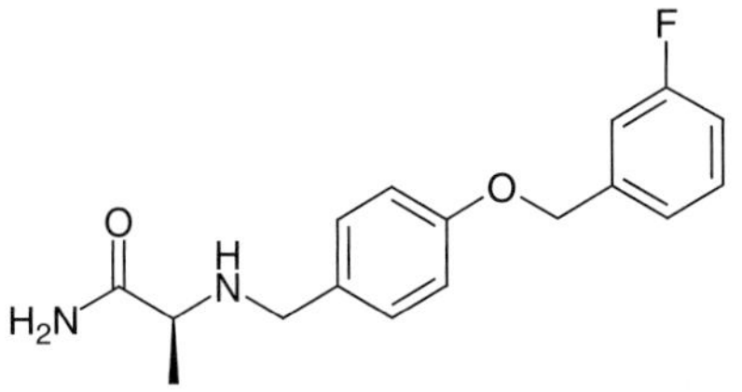

29

1.13.6.5 Genitourinary Tract Diseases and/or Reproductive Health Companies

Ardana Bioscience. Ardana (Edinburgh, Scotland) was formed in 2000 and has exclusive rights to commercialize research developed by the Medical Research Council Human Reproductive Sciences Unit (Edinburgh, Scotland). Ardana became a public company in 2005 as a result of a London IPO, raising $38 million. Products and projects include: a testosterone replacement therapy for male hypogonadism (confirmed by clinical features and biochemical tests), which was launched in June 2004; a compound for prostate cancer, benign prostatic hyperplasia, and infertility related to endometriosis and uterine fibroids; and a compound to prevent premature ovulation in ovulation induction.[32]

1.13.6.6 Metabolic Disease Companies

Metabolex. Metabolex (Hayward, CA), founded in 1991, is a private biopharmaceutical company focused on developing treatments for diabetes and related metabolic disorders. Metabolex works closely with Pfizer and Yamanouchi to develop treatments for type 2 diabetes, selecting targets from its proprietary database. With Pfizer, a deal was struck whereby Metabolex was to receive more than $50 million to develop its β-cell program designed to target insulin secretion defects. With Yamanouchi, a program began in March 2002 to develop drugs for type 2 diabetes, insulin resistance, impaired glucose tolerance, and obesity by evaluating 100 targets selected from Metabolex's database. Metabolex is developing MBX-2044, the lead in a series of peroxisome proliferator-activated receptor modulators and insulin sensitizers for the potential treatment of type 2 diabetes. In March 2005, the drug was shown to be well tolerated in a phase I trial.[33]

Arexis. Arexis AB (Molndal, Sweden) is a drug discovery and development company focusing on metabolic and inflammatory diseases, founded by Professors Holger Luthman, Rikard Holmdahl, Leif Andersson, and Dr Vidar Wendel-Hansen (who had common interests in genetics and genomics) in 1999. The company is biology-driven, with proprietary technology and a strong focus on product development, and uses forward genetics for a high-resolution molecular dissection of causative disease mechanisms. A typical project is to enhance the action of bile salt-stimulated lipase, a naturally occurring human enzyme (found in mature pancreas and in mother's milk) with a key function of degrading a large spectrum of lipids in food. For pharmaceutical use, recombinant human bile salt-stimulated lipase will be administered orally and manufactured in a cell culture system.[34]

1.13.6.7 Musculoskeletal Disease Companies

The Prostrakan Group. Prostrakan (Galashiels, Scotland) is the merged entity created between the Strakan Group and the former Aventis spinout SME, Proskelia (Paris, France), created in 2004. Proskelia was the former bone research group of Hoechst Marion Roussel, located at the former Romainville site in Paris (France). Proskelia now forms the research arm of this women's health and musculoskeletal company. In 2005 the Prostrakan Group employed some 280 people, making the combined entity still an SME by definition. The company is essentially European in focus, with projects around two anabolic pathways: bone morphogenetic proteins and the high-bone-mass gene (*LRP5*).[35]

1.13.6.8 Oncology Companies

There are many SME companies specializing in drug discovery to treat cancer, reflecting a large area of unmet medical need, and the availability of funding for the research into treatments with major economic potential. Big pharma retains major interest in this therapeutic area, and consequently few spinouts originate from former big pharma research groups.

The perceived ease of clinical development has persuaded several former drug discovery service SME companies to change business model and wholly or partly engage in proprietary therapeutic research in this field, notably Arqule and Array Biosciences. Thus, Arqule (Woburn, MA) is developing ARQ-501 (**30**), which is a topoisomerase I inhibitor isolated from the lapacho tree that causes cell cycle arrest at G1/S, for potential use in chemotherapy of breast, ovarian, colorectal, prostate, and other cancers. Array Biopharma (Boulder, CO) has investigated the potential of MEK inhibitors for the potential treatment of cancer.

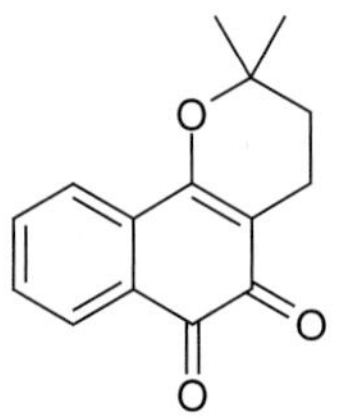

30

Sugen. One of the original SMEs engaged in discovering inhibitors of kinases as anticancer therapies was Sugen (San Francisco, CA), which was acquired by Pharmacia (now part of Pfizer) in 1999. The former research of Sugen has been responsible for a number of clinical drug candidates now in development by the parent organization. For example, SU-11248 (**31**) is an orally active inhibitor of vascular endothelial growth factor receptor kinase (VEGFR2), PDGFR-β, KIT, and Flt3 tyrosine kinase signaling pathways that is being developed as a potential anticancer agent.

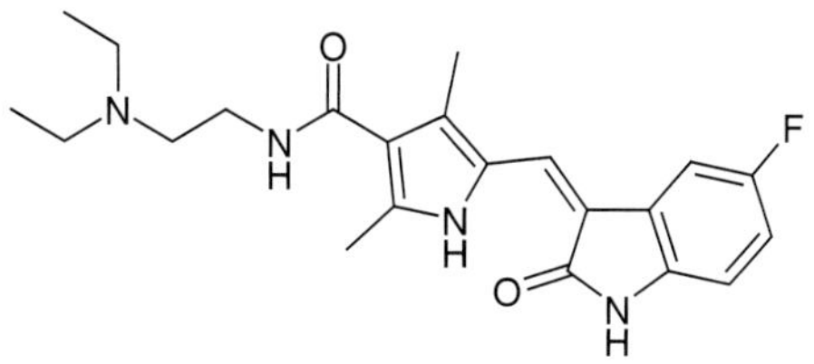

31

Aton Pharmaceuticals. Aton Pharmaceuticals (Tarrytown, NY) developed cancer drugs based on research by Professor R Breslow of Columbia University Department of Chemistry and Sloan Kettering. In 2004 Merck acquired Aton for its portfolio of drugs, including its clinical candidate SAHA (**32**), which inhibits histone deacetylase enzymes.

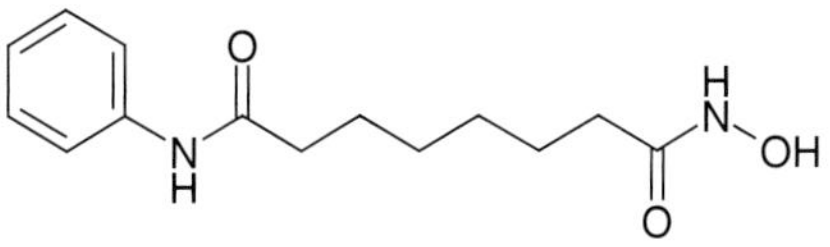

32

Biotica Technology. Biotica Technology (Saffron Walden, UK) is a spinout from the University of Cambridge (UK) based on intellectual property generated partly in Professor James Staunton's group. Biotica is focused on the production of novel biopharmaceuticals through targeted alteration of biosynthetic pathways leading to polyketides. An initial focus on polyketides as antiinfectious agents appears to have been displaced by potential use in oncology. Biotica has polyketide mammalian target of rapamycin (mTOR) inhibitors, 90 kDa heat shock protein inhibitors and novel angiogenesis inhibitors. Biotica's technology platform comprises methods for the genetic engineering of polyketide synthase and associated postpolyketide synthase genes.

Cytokinetics. Cytokinetics, established in 1998 and headquartered in San Francisco (CA), is a biopharmaceutical company focused on the R&D and commercialization of novel small-molecule drugs. In June 2001, Cytokinetics and GlaxoSmithKline entered into strategic collaboration for the R&D and commercialization of novel small-molecule therapeutics targeting mitotic kinesins for applications in the treatment of cancer and other diseases. Cytokinetics and GlaxoSmithKline are developing CK-0238273 (SB-71599), a mitotic kinesin spindle protein inhibitor, for the potential treatment of various cancers. Cytokinetics and GlaxoSmithKline are developing SB-743921, the lead from a series of mitotic kinesin spindle protein inhibitors, for the potential treatment of cancer.

EiRx Therapeutics. Another company with an interest in kinases as potential treatments for cancer is genomics-based EiRx Therapeutics (Cork, Ireland), a public (LSE-AIM) company founded in 1999 and researching apoptotic mechanisms of cell death. The company has identified over 200 apoptosis-associated gene targets using its proprietary genomics model system (ALIBI), and has silencing RNA technology.[36]

GPC Biotech. GPC Biotech (Munich, Germany), formed in 1997 and headquartered in Munich, Germany, is a publicly owned genomics and proteomics-based company focused on the discovery, development, and commercialization of anticancer drugs – antiinfectives research was abandoned in 2003. The company changed its name to GPC Biotech in March 2000, following its acquisition of US-based SME Mitotix. GPC Biotech acquired all the assets of the insolvent Axxima, a specialist kinase company, in March 2005.

KuDOS. KuDOS (Cambridge, UK) is a private oncology company that has access to IP from Professor Stephen Jackson's laboratory at the University of Cambridge (UK). KuDOS and Cancer Research UK are developing lomeguatrib, an alkyltransferase inhibitor (discovered by the Paterson Institute for Cancer Research and Trinity College, Dublin). KuDOS also has interest in an in-licensed (originator De Montfort University) candidate drug, AQ4N, a hypoxia-selective cytotoxic prodrug, as a potential oncology treatment.[37] KuDOS was acquired by AstraZeneca at the end of 2005.

MethylGene. MethylGene, formed in 1996 (Quebec, Canada), is an SME engaged in the R&D and commercialization of enzyme inhibitors for the treatment of cancer (and also infectious diseases). The company was formed in 1996 as a joint venture between Hybridon and three Canadian VC partners. MethylGene has IP in the fields of histone deacetylase inhibitors and DNA methyltransferase inhibitors. Methylgene and MGI Pharma are developing MG-98, a second-generation antisense oligonucleotide which inhibits expression of the DNA methyltransferase-1 gene, for the potential treatment of cancer.

*S*BIO.* S*BIO (*see* Section 1.13.3) is another cancer company, partly funded by Chiron, located in Singapore.[38]

Spectrum Pharmaceuticals. Spectrum Pharmaceuticals (formerly NeoTherapeutics) of Irvine (CA), a listed company is now a specialist oncology in-licensing company – the CNS research axis having been curtailed in 2002. Spectrum is developing the former EORTC (European Organisation for the Research and Treatment of Cancer) compound apaziquone (**33**) (first synthesized in 1987) for the potential treatment of bladder cancer and as a radiation sensitizer for solid tumors. Licensed from BMS is elsamitrucin (**34**), a topoisomerase I and II inhibitor with anti-DNA gyrase activity and an actinomycete fermentation product, for the potential treatment of non-Hodgkin's lymphoma. Spectrum has also licensed in a series of endothelin B agonists from Chicago Labs, which in turn obtained them from the University of Illinois.

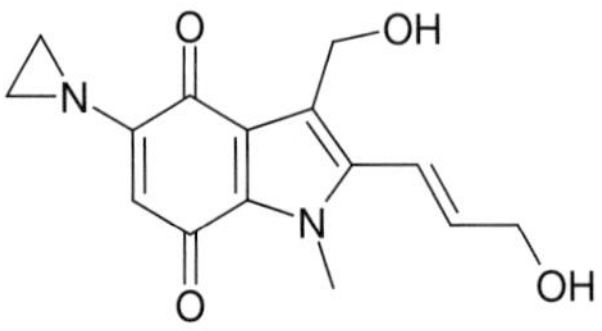

33

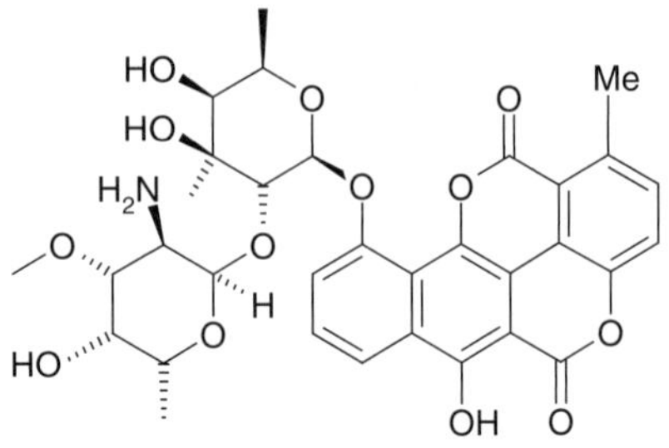

34

Sterix. Sterix is formerly a spinout of Imperial College of Science, Technology and Medicine, located in Bath (UK). Sterix was focused on steroid-based therapeutics for hormone-dependent cancers, women's health, and other hormone-associated diseases. Sterix was acquired by the French medium-sized pharma company, Ipsen Beaufour, in 2004.

Strida Pharma. Strida Pharma, a spinoff from McGill University (Quebec, Canada) and founded in 2002, develops small-molecule therapeutics based on the anticancer target, methylenetetrahydrofolate reductase (MTHFR). Strida used antisense technologies to validate MTHFR as an oncology target.

Topotarget. Topotarget (Copenhagen, Denmark), founded in 2000, is a specialist oncology company focused on the discovery and development of small-molecule drugs for cancer. Its lead compound is a topoisomerase II inhibitor for a niche indication of extravasation. Topotarget acquired the Abingdon (UK) oncology SME Prolifix in 2003, along with PXD101, an oncology product with action against histone deacetylase.

1.13.6.9 Respiratory Disease and Inflammation Companies

New small-molecule drugs to treat rheumatoid arthritis, inflammatory bowel disease, and respiratory diseases are rare, even from the big pharma stables, some of which (but not all) have abandoned interests in these domains. Bayer's withdrawal from the respiratory therapeutics area, for example, initiated the formation of two spinoff companies by former Bayer scientists (Etiologics, UK: see below) and Aerovance (2004, Berkeley, CA). Aerovance began with two ex-Bayer products, a recombinant human interleukin-4 that is both an antagonist at interleukin-4 and interleukin-13 receptors for severe asthma, and a recombinant human protease inhibitor which inhibits a sodium ion channel, that may be a therapy for cystic fibrosis patients.[39]

Theravance. Theravance (formerly Advanced Medicines) of San Francisco, US, although not entirely specialized in the respiratory therapeutic area, has performed major research into long-acting β-agonists and multimer pharmacophore approaches to new respiratory drugs and struck a major deal to codevelop drugs in this domain with GlaxoSmithKline in 2004. Under this agreement, in March 2005, Theravance licensed its novel dual-acting β_2 antagonist/muscarinic agonist molecules to GlaxoSmithKline.[40]

Argenta Discovery. Argenta Discovery (Harlow, UK) is originally a spinout from IC (London, UK) and the former Rhône-Poulenc group at Dagenham. After its merger with Etiologics in 2004, Argenta declared itself an internal discoverer of new drugs for chronic obstructive pulmonary disease by virtue of Etiologics' prior acquisition of the preclinical research group of Bayer, UK, in 2003.[41]

Avidex. Avidex (Oxford, UK) is a private drug discovery company which was formed in 1999 as a spinout from Oxford University to exploit technology arising in the Institute of Molecular Medicine. The company focuses on developing

T-cell receptor-based therapeutics for the treatment of cancer and autoimmune diseases. Avidex has a license from Active Biotech for the exclusive development and marketing rights to CD80 antagonists for the potential treatment of autoimmune diseases, including rheumatoid arthritis.[42]

Topigen. Topigen (Quebec, Canada) was founded in 2000, and is a biotechnology company engaged in the R&D of antisense drugs for asthma, allergic rhinitis, and other respiratory diseases such as chronic obstructive pulmonary disease.[43]

CoTherix. CoTherix (formerly Exhale Therapeutics), based in San Francisco (CA), is a privately held biopharmaceutical company and was formed in February 2000 to commercialize research performed at Columbia University (New York). CoTherix has exclusive US development and marketing rights to iloprost, a synthetic prostacyclin analog for the treatment of primary pulmonary hypertension, first developed by Schering. Before the company changed its name, Exhale was focused on developing therapies to treat inflammatory lung disease.[44]

1.13.7 Small- or Medium-Sized Enterprises That Specialize in Drug Discovery in Specific Target Classes or Disciplines

1.13.7.1 G Protein-Coupled Receptor Companies

Arena Pharmaceuticals. Arena Pharmaceuticals (San Diego, CA), founded in 1997, specializes in the discovery and development of therapeutics, particularly those that act at orphan GPCRs using its constitutively activated receptor technology (and melanophore technology. Arena is developing a series of orally active, small-molecule 5HT2c agonists for the potential treatment of obesity and diabetes, and a 5HT2a inverse agonist, for the potential treatment of insomnia. Arena has agonists of the GPCR 21AX for the potential treatment of inflammatory diseases, including asthma, pulmonary fibrosis, and inflammatory bowel disease. Arena Pharmaceuticals and Ortho-McNeil are investigating orally active agonists of the GPCR 19AJ for the potential treatment of type 2 diabetes.[45]

7TM Pharma. 7TM Pharma (Horshoom, Denmark) is investigating a series of antagonists of the CRTH2 receptor (chemoattractant receptor-homologous molecule expressed on T-helper type 2 cells) for the potential treatment of allergy, asthma, and inflammation.[46] 7TM has also identified a ureidobenzamide compound as a melanin-concentrating hormone receptor type 1 antagonist. The company also has an interest in ghrelin receptor agonists as potential treatments for obesity.[47]

Adenosine Therapeutics. Adenosine Therapeutics (Charlottesville, WV) is a biotechnology company focused on the R&D of agonists and antagonists of adenosine receptor subtypes for medical imaging and for the potential treatment of inflammation, chronic obstructive pulmonary disease, diabetes, asthma, Parkinson's disease, pain, angiogenesis, and epilepsy.[48]

Juvantia Pharma. Juvantia Pharma (Turku, Finland) was founded in 1997. It is a company that discovers and develops small-molecule pharmaceuticals active in transmembrane signaling via GPCRs. Its programs include developing treatments for neurodegenerative diseases, mental disorders, and vascular wall disorders. Projects include an α_2 adrenoceptor antagonist (fipamezole) (**35**), for the potential treatment of Parkinson's disease, and a G protein-coupled α_{2b} adrenoceptor antagonist for the potential treatment of cardiovascular ailments.[49]

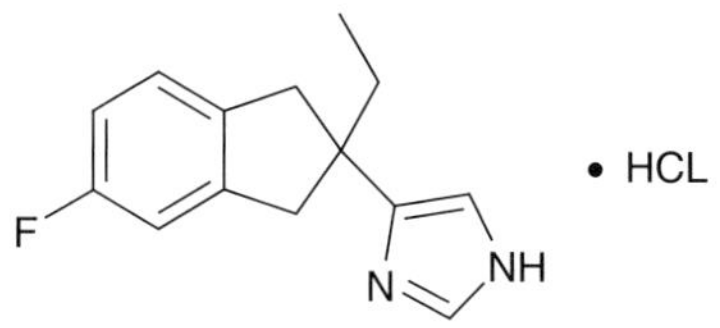

35

1.13.7.2 Chemokine Receptor Companies

ChemoCentryx (San Francisco, CA) is a private company engaged in seeking therapeutics for infectious and inflammatory diseases based on the biology of chemokines and chemokine receptors. ChemoCentryx has products against the CCR9 receptor for inflammatory bowel disease, the CCR2 receptor for cardiovascular indications, and the CXCR4 and CXCR7 receptors for cancer. In addition, ChemoCentryx has partnered programs: a CCR1 antagonist for autoimmune disorders with Forest Laboratories and a small molecule targeting CXCR3 with Tularik (now Amgen).[50]

1.13.7.3 Ion Channel Companies

Xention (Cambridge, UK) is a 2002-founded company that acquired the CeNeS ion channel drug discovery business, Channelwork. This included the AutoPatch high-throughput target-screening technology, an automated patch-clamp system. Xention has patented novel thienopyrimidine derivatives, their salts, and their use as potassium channel inhibitors for the prevention and treatment of arrhythmia. The compounds are particularly useful as Kv1.5 channel inhibitors.[51]

1.13.7.4 Kinase Enzyme Companies

Kinases are a very popular target class for drug discovery, particularly in the oncology field, and also in inflammation. Consequently, many companies already discussed as oncology specialty companies could have equally been placed in this section.

Sugen. Sugen (San Francisco, CA) was a major independent SME player in kinases until acquired by Pharmacia (now part of Pfizer) in 1999.

Activx. Activx (San Diego, CA) is a company specializing in kinase and protease drug discovery. Activx Biosciences originally licensed technology from Scripps Research Institute that formed the basis of its proteomics platform. Activx is now a subsidiary of the Japanese company Kyorin.[52]

Axxima. Axxima (Munich, Germany) specialized in research into kinases, but was subsumed into GPC Biotech AG, a specialist anticancer company, in March 2005.[53]

Cellular Genomics. Cellular Genomics (Branford, CT) is a company specializing in the fields of functional genomics, target validation, and drug discovery. Among its assets, CGI has exclusive licenses to proprietary gene and protein analysis technologies from Yale and Princeton Universities. Cellular Genomics has generated lead series of kinase inhibitors that demonstrate concurrent inhibition of a critical set of angiogenesis and cancer-related kinases, including VEGFR, Tie2, and c-Kit.[54]

PIramed. PIramed (2003, Slough, UK) was formed as a spinout of several academic cancer research organizations such as the Ludwig Institute for Cancer Research, Cancer Research UK, and the Institute of Cancer Research UK. PIramed's drug discovery research is on signal transduction inhibitors, principally the superfamily of lipid kinases, exemplified by phosphatidylinositol 3-kinase.[55]

Upstate Biotechnology. Upstate Biotechnology (a subsidiary of Serologicals of Atlanta (GA) offers a broad-spectrum kinase-screening service to the industry – showing kinase selectivity in kinase projects is often required due to the presence of approximately 600 kinase enzymes in the superfamily.[56]

1.13.7.5 Zinc Finger-Binding Domain Companies

Sangamo Biosciences. Sangamo (Richmond, CA), specializes in finding drugs that act on zinc finger-binding domains, for example drugs that upregulate VEGF and VEGF receptors to treat coronary artery disease and peripheral arterial disease. Sangamo has also entered into a joint collaboration with Edwards to develop ZFP-Therapeutics for the downregulation of phospholamban, a gene target with a role in calcium flux in heart muscle and that is believed to be directly involved in congestive heart failure.

1.13.7.6 Antibody, Protein, siRNA Therapy, and Gene Therapy Companies

Given the success that SME companies specializing in new biological entities have had in becoming powerful companies in the pharmaceutical industry, it is instructive to see what new companies in this field are doing.

Abgenix. Abgenix (Fremont, CA) was incorporated in June 1996 as a wholly owned subsidiary of Cell Genesys, but subsequently became independent. Its valuation in February 2005 was $774 million, ranking it well below Biogen Idec, but better than many small-molecule companies. Abgenix uses its XenoMouse technology to enable the rapid generation of high-affinity, fully human antibody product candidate medicines. Abgenix has alliances with AstraZeneca (36 cancer targets), with Sosei (CRTH2), with Amgen (panitumumab – a human monoclonal antibody against the epidermal growth factor receptor for use as a monotherapy and in combination with other agents in the treatment of solid tumors), and with CuraGen (CR-002 – a fully human monoclonal antibody that specifically recognizes and blocks the active form of platelet-derived growth factor D, for the potential treatment of immunoglobulin A nephropathy).

Alligator Bioscience. Alligator Bioscience (Lund, Sweden) was founded in 2001, and develops new and optimizes existing therapeutic and diagnostic products. In the Alligator pipeline are potential treatments for rheumatoid arthritis; for acute myocardial infarction; and for acute inflammation. Alligator has a proprietary Fragment-INduced Diversity

(FIND) technology to conduct protein optimization. This technology involves fragmenting DNA encoding a particular protein of interest, recombining it and testing expressed proteins for improvement over the original.

Domantis. Domantis (Cambridge, MA, US, and Cambridge, UK), a private company founded in 2000 using IP from Sir Greg Winter and Dr Ian Tomlinson from the Medical Research Council Laboratory of Molecular Biology, is a drug discovery company developing antibody molecules. Human domain antibodies are therapeutic molecules that are small and highly stable, but, like human antibodies, can be designed to have specificity and high affinity for the biological target of interest. Domantis has more than a dozen proprietary domain antibody therapeutic programs, primarily in the fields of inflammation and oncology, including human domain antibodies that neutralize cytokines.[57]

Medarex. Medarex (Princeton, NJ), a public US company, was founded in 1987 and is a biopharmaceutical company focused on the discovery and development of fully human antibody-based therapeutics to treat life-threatening and debilitating diseases, including cancer, inflammation, autoimmune, and infectious diseases. Using its UltiMab human antibody development system, the company can create human antibodies using transgenic mice. Medarex has collaborations with BMS on the investigational fully human anti-CTLA-4 antibody, MDX-010, which is in clinical trial for metastatic melanoma, and with Amgen.[58]

MorphoSys. MorphoSys, based in Martinsried/Planreg, Germany, was founded in 1992 and is a biotechnology company focused on the development of antibody-based products for treating infectious diseases, cancer, and inflammation. The company's proprietary technology includes the Human Combinatorial Antibody Library. Projects declared on the Morphosys website include MOR 101 and MOR 102, which are fully human HuCAL Gold antibodies directed against the inflammation target intercellular adhesion molecule-1, also known as CD54. MOR 101, a Fab fragment, will be developed to treat skin burn (deep dermal burn), and MOR 102, an immunoglobulin G antibody, is intended for psoriasis.[59]

XOMA. XOMA (Berkeley, CA) is a biopharmaceutical company that develops and manufactures genetically engineered protein, peptide, and monoclonal antibody pharmaceuticals. Targets include cancer, immunological disorders, inflammatory disorders, and infectious diseases. Its primary drug development platform is bactericidal/permeability-increasing protein, a human host defence protein with multiple antiinfective properties. With its partners Genentech and Serono, XOMA has now marketed a new biological entity: efalizumab, a therapy for continuous control of chronic moderate-to-severe plaque psoriasis.

Cambridge Antibody Technology. Cambridge Antibody Technology (CAT; Cambridge, UK) began life as a subsidiary of Peptide Technology (now Peptech) in January 1990. CAT was spun-off as a freestanding entity in 1996, and in 1997 completed its IPO. CAT develops novel human monoclonal antibody-based therapeutic products based on its proprietary antibody phage display technology platform, as well as licensing its technology and capabilities to others to develop products. CAT's first product, adalimumab, was isolated and optimized in collaboration with Abbott and has been approved for marketing as a treatment for rheumatoid arthritis in 51 countries. CAT also has a broad collaboration with Genzyme for the development and commercialization of antibodies directed against transforming growth factor-ß, a family of proteins associated with fibrosis and scarring.[60] In May 2006, CAT was under offer to be acquired by AstraZeneca.

Isis Pharmaceuticals. Isis Pharmaceuticals (Carlsbad, CA) was founded in 1989 in order to explore the potential of antisense technology. The company's antisense drugs are designed to treat bone, cardiovascular, infectious, metabolic, and inflammatory diseases, as well as cancer. A research area where Isis has a strong focus is ribonuclease H. Isis research workers have published on a small interfering RNA motif consisting entirely of 2′-O-methyl and 2′-fluoro nucleotides, which display enhanced plasma stability and increased in vitro potency.[61]

Oxford BioMedica. Oxford BioMedica (Oxford, UK) was established in 1995 as a spinout of Oxford University and is a biopharmaceutical company involved in developing gene therapy treatments with a focus on oncology and neurotherapy. Oxford BioMedica has developed a lentivirus, lacking many accessory genes, which is claimed to cause fewer unwanted side effects when used as a gene therapy. Oxford BioMedica is investigating lentivirus vectors for potential use in gene therapy. The vectors have been used in the company's ProSavin, ChanEx, and ProCaStat candidate gene therapies for Parkinson's disease, cystic fibrosis, and prostate cancer, respectively.

1.13.7.7 Vaccine Companies

Antigen Express. Antigen Express (Worcester, MA) has been a subsidiary of Generex Biotechnology Corporation (Toronto, Canada) since 2003. The company's technology focuses on modulating immune responses mediated by T-helper cells. Antigen is developing vaccine formulations based on the expression and use of the invariant chain protein – a regulator of antigen presentation by major histocompatibilty class molecules – for the treatment of breast and prostate cancer, HIV infection, and severe acute respiratory syndrome (SARS). Development of a smallpox vaccine is being conducted in 2005 with Emory University (Atlanta, GA) and Imperial College (London).

Microscience. Microscience is an IC (London) spinout specializing in vaccine research (**Table 2**). In 2005, the Wellcome Trust charity awarded Microscience an award of $3.6 million, the largest it has ever made, for the development of its drinkable typhoid vaccine.

Oxxon Therapeutics. Oxxon Therapeutics (formerly Oxxon Pharmaccines), a biotechnology company developing immunotherapeutics for cancer and chronic infectious diseases, is headquartered in Oxford (UK) but also operates in Boston (MA). Its IP is based on its PrimeBoost technology.[62] Oxxon has established five development programs in hepatitis, melanoma, and HIV, two of which are in phase II clinical trials.[63]

1.13.8 Small- or Medium-Sized Enterprises That Focus on Providing Services on Part of the Drug Discovery Value Chain

1.13.8.1 Target Validation Companies

DanioLabs. DanioLabs was established in May 2002 and is located in Cambridge, UK. A major part of the DanioLabs platform involves the use of zebrafish as an experimental species to enable the identification of in vivo activity of experimental compounds.

Deltagen. Deltagen is a privately funded biotechnology company focused on using knockout mouse molecular genetics to unlock the in vivo function of novel genes in order to develop new therapeutic targets. Deltagen has agreements with Roche, Merck, Tularik (now Amgen), and Pfizer to provide this service.

DeVgen. DeVgen, founded in 1997 and headquartered in Ghent, Belgium, achieved an IPO in June 2005 and is a privately owned, genomics-based, biotechnology company, focused on the use of the model organism *C. elegans* in target validation in drug discovery.

Galapagos. Galapagos (Cambridge, UK) was established in 1999 and is a functional genomics company. Galapagos has two technologies based on the PER.c6 adenoviral vector system, and a high-throughput miniaturization and automation platform. Galapagos achieved an IPO in May 2005 and acquired BioFocus (Saffron Walden, UK) in October 2005 establishing itself as a broad-based target and drug discovery company.

Jurilab. Jurilab (Kuopio, Finland) is a privately owned company founded in 1999 from the Research Institute of Public Health at the University of Kuopio, Finland. Jurilab has developed a proprietary method of identifying and verifying DNA variations known as hierarchical phenotype-targeted sequencing, a means of finding new functionally important mutations in humans and other species. The technology has been used to identify novel variants, associated with prostate cancer and type 2 diabetes, which have the potential to be used in both predictive tests and as therapeutic targets.

Lexicon Genetics. Lexicon Genetics (The Woodlands, TX) was founded in 1995 with gene targeting and embryonic stem cell technologies IP from Baylor College of Medicine. Lexicon uses this proprietary gene knockout technology to identify suitable targets from the human genome and then investigates ligands for these proteins to develop treatments for metabolic disorders, cardiovascular diseases, immunological disorders, neurological disorders, and cancer. In July 2001, Lexicon completed a merger with the chemistry service company Coelacanth, which has since become the drug discovery arm of the original target discovery company.

Paradigm Therapeutics. Paradigm Therapeutics is a discovery company focused on novel drug targets identified from the human genome. Paradigm was started as a spinout company from the University of Cambridge and it has since established a platform based on mouse transgenic technology for defining the biological functions of previously uncharacterized human druggable proteins. Paradigm has a collaboration with Medivir to identify novel protease drug targets and to discover small-molecule protease inhibitor drugs. Paradigm acquired a chemistry arm in 2005 (Amedis Pharmaceuticals).[64]

Pharmagene. Pharmagene (Royston, Cambridge) was founded in April 1996 as the first UK company to focus entirely on the use of human tissue for drug discovery and focusing on expression and function of genes and gene products. It has offered its target validation service TargetEvaluator to provide information on the expression of genes in a range of diseased and nondiseased tissues to a large number of companies.[65]

1.13.8.2 Computer-Aided Drug Design Companies

Computer-aided drug design plays a major part in hit-finding activities (in silico screening by docking of ligands into protein cavities), optimization, and prediction of potencies and properties of molecules. Computer-aided prediction is now often sufficient to be a guide to the value of making a compound or not.

4SC. 4SC is a company based in Martinsried (near Munich, Germany), founded in 1997, which uses its technologies to predict and rank biological properties. Martinsried has provided such services for major companies like Sanofi-Aventis and for smaller companies like Estève (Spain).

Accelrys. Accelrys (San Diego, CA) has been a long-standing supplier of both software and service to the worldwide pharmaceutical drug discovery. It has undergone a series of name changes and acquisitions, and its products have formerly been marketed under the Molecular Simulations, Synopsis, Pharmacopeia, and Oxford Molecular names.[66]

Cresset Biomolecular Discovery. Cresset Biomolecular Discovery (Letchworth, UK) is a recently created SME with a field-based virtual screening technology, which has been used, for example, by partners such as the James Black Foundation (London) and BioFocus.[67]

De Novo Pharmaceuticals. De Novo Pharmaceuticals is a spinout of the pharmacology department at the University of Cambridge, UK (2000), and has proprietary computational software for structure-based drug design to create novel, patentable lead molecules as candidates for drug development.[68] The scientists at De Novo have recently advanced the understanding of hydrophobic groups in drug–protein docking interactions.[69]

Inpharmatica. Inpharmatica (London, UK) was spun out of University College London in 1998, with the aim of identifying novel drug targets by integrating computer-modeling techniques with genomics data. As well as providing such services to other drug discovery players, to signal the success of its techniques, workers at Inpharmatica have reported the identification of the molecular basis for the biological role of nicastrin, a potential new protein drug target for Alzheimer's disease. Using its proprietary database, the company found that a portion of the protein has significant similarity to some known proteases, suggesting that this region confers peptide-binding function to nicastrin.[70]

Libraria. Libraria (San Jose, CA), has proprietary software to capture and evaluate structure–activity and chemical synthesis relationships to automate the recognition of relevant patterns that permit the design of new patentable small molecules. The company's initial focus is on kinases.

Tripos. Tripos (St Louis, MO) is a longstanding (founded 1979) SME supplier of computer-aided drug design services and also works in chemistry drug discovery using its UK subsidiary Tripos Discovery Research. Most recently, Tripos has pioneered the idea of 'lead hopping' to get patentable series.

1.13.8.3 Hit-Finding Companies

A large number of companies offer services to the industry in hit-finding techniques. In 2005, those companies offering hit-finding technologies and products (including companies specializing in providing large libraries of novel compounds made by automated methods to big pharma) include DPI (from June 2006 BioFocus DPI), Evotec-OAI, Array Biopharma, Pharmacopeia (Princeton, NJ), and Albany Molecular (Albany, NY). However, commodification of this area is seeing a profound shift of service provider companies to the Far East, either from indigenous organizations (Dr Reddy's subsidiary, Aurigene (Bangalore, India), GVK Bio (Hyderabad, India), Wuxi Pharmatech (Shanghai, China)), or by the previously established companies electing to set up subsidiaries taking advantage of high-technology – and cheaper – workforces (Albany Molecular, Singapore). The apparent improvement of the ability of companies to enforce IP rights in India and China is partly responsible for the increased shift in outsourcing of drug discovery activities to these countries. One consequence of this shift has been the willingness for some US service providers to set up satellite operations in India or in Singapore (Pharmacopeia) to take advantage of reduced cost bases in competitive markets.

High-technology biological screening engines have long been sourced from expert engineering companies: Technology Partnership (Melbourn, UK) formed a consortium with Rhône-Poulenc and AstraZeneca to develop its laser-screening engine for cellular screening. This instrument is in competition with another SME-inspired device, the optical high-throughput screening engine developed by Axon Instruments (which was acquired on July 1, 2004 by Molecular Devices (Union City, CA)). Developing this theme, SMEs have continued to develop difficult screening technologies, particularly ion channel screening, where the required personnel with biophysics disciplines are rare in typical drug discovery companies; these include CeNeS (Cambridge, UK) and Axon Instruments (CA). The latter has developed two products for ion channel screening: (1) a parallel electrophysiology system in which ion channel activity in eight oocytes at a time can be recorded automatically; and (2) an instrument for the parallel recording of ion channels in mammalian cell lines using the patch-clamp technique.

SME companies performing high-throughput screening services include BioFocus Discovery and also Evotec Technologies (an affiliate of Evotec-OAI) (Germany). Evotec pioneered ultrahigh-throughput screening in the mid-1990s; development was supported by a consortium including GlaxoSmithKline and Novartis. Evotec has performed contract high-throughput screening on targets from Novartis.[71]

1.13.8.4 Hit and Lead Optimization Companies

Companies offering services in medicinal chemistry and biological screening include Cerep (France), Discovery Partners International (DPI) (San Diego, CA), Array Biosciences (Boulder, CO), Albany Molecular (Albany, NY), Pharmacopeia

(US), BioFocus DPI (Saffron Walden, UK), Argenta Discovery (Harlow, UK), Evotec-OAI (Hamburg, Germany and Oxford, UK) and Nikem (Milan, Italy). Big pharma partners include GlaxoSmithKline (Nikem, Argenta), J&J (BioFocus DPI), and Novartis (Argenta Discovery). In general, big pharma outsources such activities because of internal capacity constraints, rather than a perceived lack of technology – the critical factor is generally recognized that medicinal chemistry optimization is a stage where experience is very valuable. To this end, companies like Biofocus Discovery have concentrated on projects where substantial in-house expertise has been built up – projects involving kinase targets, ion channel targets, and 7-TM targets. In some cases, niche technology areas can be acquired by SMEs performing contract drug discovery. Big pharma has largely abandoned natural products as a source of hits and drugs in the late 1990s; nonetheless, natural products SMEs exist and in one case (Biofrontera), the company has recently been acquired by DPI (March 2005).[72] It may be that natural product research will be kept alive in the future by specialist SME companies rather than in big pharma research centers.

1.13.8.5 Drug Candidate Validation Companies

Companies offering expert pharmacological validation that compete with established major players include Renasci (Nottingham, UK)[73] in metabolic, obesity, and CNS disease areas, Oncodesign (Dijon, France)[74] in oncology, and Argenta Discovery[41] in respiratory disease.

1.13.8.6 Other Technology Service Providers

A huge number of other companies offer services to those interested in drug discovery (in fact there are few, if any, disciplines that cannot be purchased). Some important adjunct service providers are listed. These include protein x-ray crystallography to aid drug discovery from crystalline drugable targets. Some examples of structural biology companies are: (1) Astex Technologies, Cambridge, UK, a structural biology company formed in July 1999 that has recently moved into proprietary oncology drug discovery[75]; (2) Syrrx (1999, San Diego, CA) with internal IP in oncology targets including histone deacetylase – this company was purchased outright by Takeda of Japan in 2005; (3) Crystal Genomics (Korea)[76]; (4) Structural GenomiX (San Diego, CA)[77]; and (5) Plexxicon (Berkeley, CA). Plexxicon, like many service companies, has also moved into proprietary drug discovery. Plexxicon has a c-Kit inhibitor developed project in collaboration with Phenomix for the treatment of cancer and inflammation.[78]

Custom radiosynthesis to assist in compound validation can be offered by SMEs like Selcia (Ongar, UK), which became independent of its former parent, Scynexis in 2006.

1.13.9 Value of Small- or Medium-Sized Enterprises in Drug Discovery

1.13.9.1 The Last 10 Years

An analysis of the effect that SMEs have had on drug discovery, of course, should be taken from the patient point of view. It is difficult to give an overall assessment of which recently launched drugs have had any form of input from an SME organization. It is, however, possible to perform an analysis of all drugs that have reached the market in the past 10 years for those whose IP originated in an SME. This can be performed by analyzing the chapters entitled "To market, to market" in the 10 years' editions of *Annual Reports in Medicinal Chemistry* from 1993 to 2002.[79–88]

The survey of the origins of new drugs over 10 years between 1993 and 2002 shows (**Table 5** and **Figure 1**) that the proportion of new drugs originating from the laboratories of big/traditional pharma has stayed steady at around 50–75% depending on individual company. The proportion of new drugs originating from the biotech sector – companies which are SMEs, or were SMEs when patents were filed – has grown substantially to 25–33%. Academic sources of drugs remain steady at 5–10%. The sector that has contracted is that of the small pharma organization, once a subsidiary of a larger conglomerate or pharmaceutical manufacturing or generics company. Many such Japanese companies were sources of new drugs in the early part of these 10 years. It should be noted that, over the years surveyed, the majority of the new drugs that were invented by SMEs were brought to the market by traditional pharma companies (after licenses were taken to third-party products or the originating SME companies were acquired).

Although the contribution of SMEs to drug discovery is clear, a similar analysis (not shown) over 1993–2002 reveals that few products invented by SMEs are actually marketed by them, particularly in the early years of a company, and when marketing has become viable, most such SMEs have become FIPCOs in their own right – Amgen, Genentech, Serono, etc. For most SMEs, the route to market of their precious candidate drugs still lies with the traditional pharma companies having licensed the product or acquired the company.

Table 5 NCEs/ new biological entities (NBEs) 1993–2002 as reported by *Annual Reports in Medicinal Chemistry 1994–2003*, showing the origin of each individual drug

Year	*Number originating in big/traditional pharma*	*Number originating in biotech-type SMEs*	*Number originating in academic laboratories*	*Number originating in other types of organization*[a]	*Total number of new drug launches*
1993	25	4	2	12	43
1994	29	5	1	9	44
1995	25	3	0	7	35
1996	29	7	0	2	38
1997	30	4	2	3	39
1998	23	3	0	1	27
1999	16	15	1	3	35
2000	19	10	3	3	35
2001	13	9	3	0	25
2002	26	5	2	1	34

[a] Typically this covers generic laboratories, pharmaceutical manufacturing companies that have had R&D laboratories attached, pharmaceutical subsidiaries of larger, broader organizations like Nippon Flour Mills, and small FIPCO traditional pharma companies, many of which have since been acquired or merged.

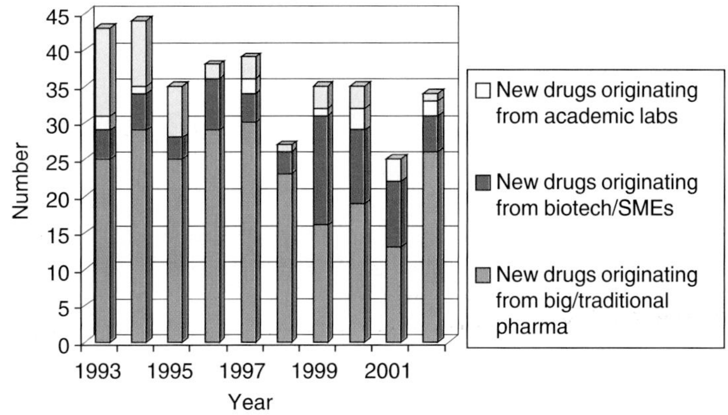

Figure 1 Graph of NCEs/NBEs 1993–2002 as reported by *Annual Reports in Medicinal Chemistry* 1994–2003, showing the origin of each individual drug.

At the time of writing (2005), the largest pharmaceutical company in the world was Pfizer. An analysis of its 2005 R&D pipeline in terms of compound origin (**Table 6**) allows a crude estimate that some 10% of the products that Pfizer might make available to patients in the future might have been invented in SME/biotech companies.

1.13.9.2 Summary

When the first edition of *Comprehensive Medicinal Chemistry* was being written, the majority of drug discovery was performed in the established large and small pharma, with no company having over 4% market size. Companies put into clinical trials drugs that were either discovered by the in-house teams of chemists, or cross-licensed between the large companies – little cross-licensing occurred before the clinical development stage. The companies were usually traditionally located close to their industrial roots, often relatively far apart geographically.

The situation in 2005 is vastly different. The impact of mergers since 1990 has created mammoth companies (SmithKline and Beecham, Wellcome, and Glaxo, all now part of GlaxoSmithKline; Warner Lambert, Pharmacia, and

Table 6 Product origins for 2005 Pfizer R&D projects by therapeutic area, analyzed from data reported by the Investigational Drug Database IDDB3

Therapeutic area	*Originated with Pfizer and acquired traditional pharma subsidiaries*	*Originated with traditional pharma*	*Originated with universities/ academic groups*	*Originated with biotech*	*Total*
Oncology	61	2	6	12	61
Cardiovascular	60	6	2	8	76
Neurology	50	8	0	5	63
Mental health	39	5	0	2	46
Musculoskeletal	35	4	1	0	40
Nonviral antiinfectives	31	4	0	4	39
Metabolic disorders	28	2	1	4	36
Respiratory	22	5	1	8	36
Gastrointestinal	22	5	1	4	32
Endocrine	18	7	1	4	30
Pain/anesthesia	24	6	0	1	31
Genitourinary and renal	24	1	0	4	29
Blood disorders	19	0	1	3	23
Viral infections	11	2	1	5	19
Dermatologicals	12	1	1	4	18
Immune disorder	13	2	0	2	17
Sensory organs	9	0	1	6	16
Other	3	0	0	2	5
Total projects	481	60	17	78	617
Total drugs at this date (many drugs overlap therapeutic areas)					387

http://www.iddb.com/

Pfizer all now being known as Pfizer). However, the combination of the major pharmaceutical companies has spawned a whole host of spinout companies, many in the field of drug discovery. In addition, many universities recognize the worth of the IP they generate and create spinout SME companies to maximize it. Drug discovery is such a multidisciplinary task that few SMEs ever become proficient in all disciplines, and the majority of SME companies have become intensely interdependent on each other. Such enforced sharing of objectives and ideals has formed the biotech communities of California, Massachusetts, and Medicon Valley, amongst others, and is being emulated in territories like Singapore. Each localized community has provided a forum for intellectual discussions that lead to innovation. This is now being recognized by big pharma which in some cases is moving its own R&D sites to such geographical clusters.

Of the many thousands of biotech/SMEs created in the past 30 years, only two have reached the exalted BIG PHARMA status, but another 10 or so are poised to join that exalted throng. It is salutary that, with the exception of a very few – Celltech (UK: now part of UCB) and Serono (Switzerland: an acquisition target in early 2006) – most are in the US.

In terms of drug discovery, in any one year, up to 10–33% (depending on company) of marketed products have their origins in biotech companies. This is because the vast development sums required to develop and market each potential medicine are usually too large for biotech to resource. As of 2005, big pharma appears to have maintained its traditional preponderance of invention of drugs. Nonetheless, it is likely that biotech will be a significant (if not ultimately the major) source of new drugs to big pharma and ultimately the patient population. The concept that big pharma may entirely abandon its own research groups and source its compounds entirely from SME/biotech organizations, however, still looks unjustified in 2006, and on the basis of the analysis of Pfizer's pipeline, remains some way from happening.

References

1. http://www.actetsme.org/usa/usa98.htm (accessed June 2006).
2. http://www.sbs.gov.uk/default.php?page = /statistics/default.php (accessed June 2006).
3. http://www.chusho.meti.go.jp/sme_english/outline/02/01.html (accessed June 2006).
4. http://commerce.ca.gov/ttca/pdfs/detail/biotechnology.pdf (accessed June 2006).
5. http://www1.imperial.ac.uk/medicine/ (accessed June 2006).
6. http://www.imperialinnovations.co.uk/ (accessed June 2006).
7. http://www.investinbiotech.com/pressroom_release.php?id = 1747 (accessed June 2006).
8. Stolis, A.; Goodman, D. E. *Improving the Lives of Millions – The Vital Role of Venture Capital in Life Science Innovation. An Assessment of the Impact of Venture-backed Life Sciences Companies on Patient Health in the United States and an Explanation of How Products Get from the Laboratory to Patients. A Report Commissioned by the National Venture Capital Association*; The Weinberg Group: London, 2004.
9. http://www.ardana.co.uk/news17022005.html (accessed June 2006).
10. http://www.biospace.com/news_story.cfm?StoryId = 19365820 (accessed June 2006).
11. http://www.biospace.com/company_region.cfm?RegionID = 11 (accessed June 2006).
12. http://www.biospace.com/hotbed/12/biotechbeach_map.cfm (accessed June 2006).
13. http://www.biocom.org/ (accessed June 2006).
14. http://www.erbi.co.uk/ (accessed June 2006).
16. http://www.mediconvalley.com(accessed June 2006).
17. http://www.sosei.com/ (accessed June 2006).
18. http://www.biomed-singapore.com/bms/sg/en_uk/index.html (accessed June 2006).
19. http://www.iddb.com/iddb3/iddb3_2/xsl_sheet.popSmartDevStatus?i_company_id = 19453&i_sortOrder = L0 (accessed June 2006).
20. http://www.iddb.com/iddb3/iddb3_2/xsl_sheet.popSmartDevStatus?i_company_id = 1005244&i_sortOrder = L0 (accessed June 2006).
21. http://www.iddb.com/iddb3/iddb3_2/xsl_sheet.popSmartDevStatus?i_company_id = 21991&i_sortOrder = L0 (accessed June 2006).
22. http://66.218.71.225/search/cache?p = cambridge + genetics + company&sm = Yahoo%21 + Search&toggle = 1&ei = UTF-8&xargs = 0&pstart = 1&b = 11&u = www.ukbusinesspark.co.uk/axs34322.htm&w = cambridge + genetics + company&d = 3DFED9ED41&icp = 1&.intl = us (accessed June 2006).
23. http://www.arakis.com/ (accessed June 2006).
24. http://www.idun.com/ (accessed June 2006).
25. http://www.oxtrust.org.uk/pooled/articles/BF_NEWSART/view.asp?Q = BF_NEWSART_67011 (accessed June 2006).
26. http://www.sosei.com/ (accessed June 2006).
27. http://www.athelas.com/ (accessed June 2006).
28. http://www.arrowt.co.uk/ (accessed June 2006).
29. http://www.iddb.com/ (accessed June 2006).
30. Chen, C.; Pontillo, J.; Flack, B. A.; Gao, Y.; Wren, J.; Tran, J. A.; Tucci, F. C.; Marinkovic, D.; Foster, A. C.; Saunders, J. *J. Med. Chem.* **2004**, *47*, 6821–6830.
31. http://www.memorypharma.com/ (accessed June 2006).
32. http://www.ardana.co.uk/ (accessed June 2006).
33. http://www.metabolex.com/news/mar032005.html (accessed June 2006).
34. http://www.arexis.se/ (accessed June 2006).
35. http://www.prostrakan.com/randd.html (accessed June 2006).
36. http://www.eirx.com (accessed June 2006).
37. http://www.kudospharma.co.uk/index.php (accessed June 2006).
38. http://www.sbio.com (accessed June 2006).
39. http://www.aerovance.com/index/index.htm (accessed June 2006).
40. http://www.theravance.com/ (accessed June 2006).
41. http://www.argentadiscovery.com (accessed June 2006).
42. http://www.avidex.com/ (accessed June 2006).
43. http://www.topigen.com/ (accessed June 2006).
44. http://www.cotherix.com/home.php (accessed June 2006).
45. http://www.arenapharma.com/ (accessed June 2006).
46. Ulven, T.; Kostenis, E. *J. Med. Chem.* **2005**, *48*, 897–900.
47. http://www.7tmpharma.com/ (accessed June 2006).
48. http://www.adenrx.com (accessed June 2006).
49. http://www.juvantia.com/ (accessed June 2006).
50. http://www.chemocentryx.com/product/overview.html (accessed June 2006).
51. http://www.xention.com/ (accessed June 2006).
52. http://www.activx.com (accessed June 2006).
53. http://www.gpc-biotech.com/en/about_the_company/index.html (accessed June 2006).
54. http://www.cellulargenomics.com/drug_disc/onc_prog.shtml (accessed June 2006).
55. http://www.piramed.com/overview.html (accessed June 2006).
56. www.serologicals.com/pdf/dundeeopen/pdf (accessed June 2006).
57. http://www.domantis.com/ (accessed June 2006).
58. http://www.medarex.com (accessed June 2006).

59. http://www.morphosys.com/ (accessed June 2006).
60. http://www.cambridgeantibody.com (accessed June 2006).
61. Allerson, C. P.; Sioufi, N.; Jarres, R.; Prakhsh, T. P.; Naik, N.; Berdeja, A.; Wanders, L. R. H.; Swayze, E. E.; Bhat, B. *J. Med. Chem.* **2005**, *48*, 901–904.
62. http://www.oxti.com/technology/primeboost.html (accessed June 2006).
63. http://www.oxti.com/products/immunotherapeutic_products.html (accessed June 2006).
64. http://www.paradigm-therapeutics.co.uk/pages/company/1/index.html (accessed June 2006).
65. http://www.pharmagene.com/ (accessed June 2006).
66. http://www.accelrys.com/ (accessed June 2006).
67. http://www.cresset-bmd.com/ (accessed June 2006).
68. http://www.denovopharma.com/ (accessed June 2006).
69. Kelly, M. D.; Mancera, R. *J. Med. Chem.* **2005**, *48*, 1069–1078.
70. http://www.inpharmatica.com/news/2001/290301.htm (accessed June 2006).
71. http://www.evotecoai.com/opencms/export/evotec/en/partners/casestudy_novartis.html (accessed June 2006).
72. http://phx.corporate-ir.net/phoenix.zhtml?c = 121941&p = irol-newsArticle&ID = 680109&highlight (accessed June 2006).
73. http://www.renasci.co.uk/ (accessed June 2006).
74. http://www.oncodesign.fr/ (accessed June 2006).
75. http://www.astex-technology.com/index.html (accessed June 2006).
76. http://www.crystalgenomics.com/ (accessed June 2006).
77. http://www.structuralgenomics.com/ (accessed June 2006).
78. http://www.plexxicon.com/ (accessed June 2006).
79. Cheng, X.-M. To Market, To Market 1993. In *Annual Reports in Medicinal Chemistry*; Bristol, J. A., Ed.; Academic Press: London, 1994; Vol. 29.
80. Cheng, X.-M. To Market, To Market, 1994. In *Annual Reports in Medicinal Chemistry*; Bristol, J. A., Ed.; Academic Press: London, 1995; Vol. 30.
81. Cheng, X.-M. To Market, To Market, 1995. In *Annual Reports in Medicinal Chemistry*; Bristol, J. A., Ed.; Academic Press: London, 1996; Vol. 31.
82. Galatsis, P. To Market, To Market, 1996. In *Annual Reports in Medicinal Chemistry*; Bristol, J. A., Ed.; Academic Press: London, 1997; Vol. 32.
83. Galatsis, P. To Market, To Market, 1997. In *Annual Reports in Medicinal Chemistry*; Bristol, J. A., Ed.; Academic Press: London, 1998; Vol. 33.
84. Gaudilliére, B. To Market, To Market, 1998. In *Annual Reports in Medicinal Chemistry*; Doherty, A. M., Ed.; Academic Press: London, 1999; Vol. 34.
85. Berna, P.; Gaudilliére, B. To Market, To Market, 1999. In *Annual Reports in Medicinal Chemistry*; Doherty, A. M., Ed.; Academic Press: London, 2000; Vol. 35.
86. Gaudilliére, B.; Bernardelli, P.; Vergne, F. To Market, To Market, 2000. In *Annual Reports in Medicinal Chemistry*; Doherty, A. M., Ed.; Academic Press: London, 2001; Vol. 36.
87. Bernardelli, P.; Gaudilliére, B.; Berna, P. To Market, To Market, 2001. In *Annual Reports in Medicinal Chemistry*; Doherty, A. M., Ed.; Academic Press: London, 2002; Vol. 37.
88. Boyer-Joubert, C.; Lorthiiois, E.; Moreau, F. To Market, To Market, 2002. In *Annual Reports in Medicinal Chemistry*; Doherty, A. M., Ed.; Elsevier Academic Press: London, 2003; Vol. 37.
89. http://www.finfacts.com/irelandbusinessnews/publish/printer_1000713.shtml (accessed June 2006).

Biography

Chris G Newton has been Senior Vice President at BioFocus DPI (a Galapagos company) located both in the Netherlands and in the UK since 2005. Prior to joining BioFocus DPI he was a board member, scientific founder and CSO at the UK service company Argenta Discovery from 2000 to 2005. Before moving to the biotech/service industry sector, Chris spent 21 years in Big Pharma, with the French group Aventis/Rhone-Poulenc Group of companies, holding senior management positions in Medicinal Chemistry, New Lead Generation, and Process Chemistry. He participated in many drug discovery and development programs, and between 1985 and 1987 was project leader for the anticancer agent temozolomide, now on the market with Schering Plough. He has also participated in the discovery of many potential medicines that reached clinical trial status in cardiovascular, respiratory, and inflammatory disorders. Chris received MSc and PhD degrees in Organic Chemistry (Turner Prize) from the University of Sheffield in 1978 and 1980 respectively, and was awarded BA and MA degrees in Natural Sciences (Emeleus Prize) by the University of Cambridge in 1976 and 1978, respectively. He is a member of the American Chemical Society, the Royal Society of Chemistry, the SCI, and the Society of Drug Research (UK) and has contributed to the organization of numerous scientific meetings.

Comprehensive Medicinal Chemistry II
ISBN (set): 0-08-044513-6

ISBN (Volume 1) 0-08-044514-4; pp. 489–525

1.14 The Role of the Pharmaceutical Industry

R Barker, Association of the British Pharmaceutical Industry (ABPI), London, UK
M Darnbrough, London, UK

1.14.1 Overview of the Work of the Pharmaceutical Industry and Two Examples of Its Products

It is unmet medical need that drives the pharmaceutical industry. There are patients with diseases who cannot be cured and for whom existing treatments are inadequate. Disease-causing organisms evolve and mutate; they move from one country to another, or even from animals to man. As mankind survives longer, the incidence of disease in each lifetime increases, and disease targets become more challenging as we tackle more complex degenerative disorders and the very processes of aging itself. But at the same time, the amount of fundamental research in the biological sciences is increasing and discoveries about the biochemistry and physiology of cells and organs, both in healthy people and in those with diseases or chronic conditions, are being made at an astonishing rate throughout the world.

Pharmaceutical companies harness these scientific developments that offer new clues about how a disease might be treated, and in many cases make basic discoveries themselves, and this starts the long and expensive process of turning them into medicines. Every aspect of this process is carefully monitored and controlled by regulatory agencies across the globe. Regulation covers: the way research is done in laboratories; the conduct of clinical trials and whether a new product can be authorized for manufacture and sale; the use of animals in studies to discover how potential new medicines work; the manufacturing processes; the labeling of products and the information given to patients; and how medicines may be promoted. Once a medicine has been authorized for marketing, companies and doctors are also obliged to report any adverse reactions observed in patients. The research and development of a new medicine is a high-risk business (less than 1 in 10 000 of the chemical compounds considered as potential new drugs emerge as a marketed medicine) and the process takes more than 10 years, costing many hundreds of millions of pounds, euros, or dollars. But without this process mankind would be at the mercy of infectious diseases, heart attacks, mental diseases, and cancer, not to mention literally hundreds of other disorders.

Let us start with two examples of medical advances that have saved the lives of many patients and which, in the process, have contributed substantially to national economies, by enabling treated patients to continue in productive employment, and through the business activity of the companies producing the medicines.

Ulcers in the stomach or duodenum affect 1 in 100 males in the UK. Although it was known from the 1900s that gastric acid was a causative agent, until the appearance of the first medicines in the 1970s most people who developed ulcers suffered severe bleeding, underwent abdominal surgery, and often had recurrence of the disease. Such treatment as there was consisted of eating 'a little and often' and a high intake of milk along with antacids. Research showed that histamine played a part in stimulating the production of gastric acid and the enzyme pepsin, both of which were overproduced in patients with ulcers. In the 1970s it was found that histamine receptors in the stomach (so-called H2 receptors) were different from those in other parts of the body, and could be selectively blocked, reducing the flow of acid. The first medicinal chemist to design a molecule to block the H2 receptors was Sir James Black of SmithKline. Cimetidine (Tagamet) reached the market in1976 and, as often happens, was closely followed by ranitidine (Zantac) in 1978 (developed by Glaxo), and subsequently by famotidine (Pepcid) from Merck Sharpe and Dohme, nizatidine (Axid) from Eli Lilly, and others. Researchers then discovered that an enzyme that causes acid to be pumped into the stomach could be inhibited and the so-called proton pump inhibitors were developed. These became the world's biggest selling medicines, the first being omeprazole (Losec) from Astra. However, scientists continued to explore the underlying mechanisms of the disease and discovered that infection by the bacterium *Helicobacter pylori* was the fundamental causative agent in most cases. Hence, through regimens of medicines combining antibiotics with the above agents, there has been a massive reduction in the severity of ulcers, the number of abdominal operations required, and chronic pain experienced by patients. In the UK, savings have been made because the number of consultations with general practitioners about ulcers has halved and medicines have reduced the number of surgical interventions (*see* 6.27 Gastric and Mucosal Ulceration).

The worldwide AIDS (acquired immune deficiency syndrome) epidemic was first recognized in the 1980s and soon stimulated international research. The retrovirus that causes the fundamental immunodeficiency, the human immunodeficiency virus HIV-1, was identified by Luc Montagnier and Robert Gallo in 1983,[1] building on work done by clinicians in Los Angeles. In 1997, the UN AIDS organization estimated that 18 million people throughout the world were infected with HIV and that about half of those were young people under the age of 25 years. In 2003, there were 38 million people living with HIV, 2 million of whom were children.[2] Since the virus attacks and destroys lymphocytes (T cells), which play a vital role in fighting infections, those with HIV infection become increasingly susceptible to infections, such as tuberculosis, *Candida*, viral retinitis, and herpes, and to some infections that are hardly ever found in those not infected with HIV, e.g., *Pneumocystis carinii* pneumonia. There is also an increased risk of some tumors and of non-Hodgkin's lymphoma. Death in AIDS patients usually results from overwhelming infection.

The discovery and sequencing of the virus opened up a clear target for medicinal chemistry, since its nucleic acid core codes for several enzymes capable of selective inhibition. Since the first medicine appeared in 1987, there have been a number of generations of drugs using different mechanisms of action at various points in the life cycle of the virus. Some drugs affect the entry or activation of the virus within the human cells, while others affect the process by which the virus reproduces itself. The latter drugs target the reverse transcriptase enzyme and a protease, which cuts proteins in a way needed to compose the viral structure. The first medicine to be introduced was azidothymidine (generic name zidovudine), which had been under development by Wellcome as a potential anticancer agent because of its action on T cell reverse transcriptase; it was marketed as AZT with the brand name Retrovir. By 1994, two other similar drugs were on the market: Didanosine from Bristol Myers Squibb and Zalcitabine from Roche, both acting on the reverse transcriptase. Drugs that inhibit protease activity include Saquinavir (Roche), Indinavir (Merck Sharpe and Dohme), and Ritonavir (Abbott). In 1995, three compounds had been approved and licensed in the UK and by 2002 17 had been authorized. These medicines, typically used in combination, do not eliminate the virus but they reduce the amount of virus in the body and improve the immune status of the infected person, thus slowing down the development of other infections. However, the HIV virus reproduces and mutates so quickly that it can rapidly develop resistance to the medicines, and this is the reason why trials have shown that combinations of several drugs used together, administered as soon as possible after infection is discovered, are particularly effective. The rapid mutation of the virus means that we cannot cease the search for new antivirals for HIV/AIDS. Research into development of an effective vaccine is also currently in progress. However, drug developments over the last two decades, assisted by fast-track approval processes, has changed AIDS in the developed world within 2–3 years from a virtual death sentence into a disease that can be 'lived with' over the long term. This has led to a reversal of the trend in AIDS-related deaths (see **Figure 1**). Of course, the situation in sub-Saharan Africa is very different, a subject to which we will return in Section 1.14.6.1.3 (*see* 7.12 Ribonucleic Acid Viruses: Antivirals for Human Immunodeficiency Virus).

The last century has seen major improvements in life expectancy in the developed countries (**Table 1**). While improvements in sanitation, the provision of clean water, and nutrition have probably been the most important factors, medicines have made a major contribution since the 1940s.[3] Infectious diseases such as tuberculosis (TB), pneumonia, meningitis, and polio once claimed millions of lives, but antibiotics and vaccines have dramatically reduced this toll. Insulin, first extracted from animal pancreas and used experimentally in the 1920s, is now made by recombinant DNA and extends the lives of diabetics significantly. And we are now beginning to assemble an armamentarium of anticancer drugs that not only extend life but also, in many cases, result in complete remission.

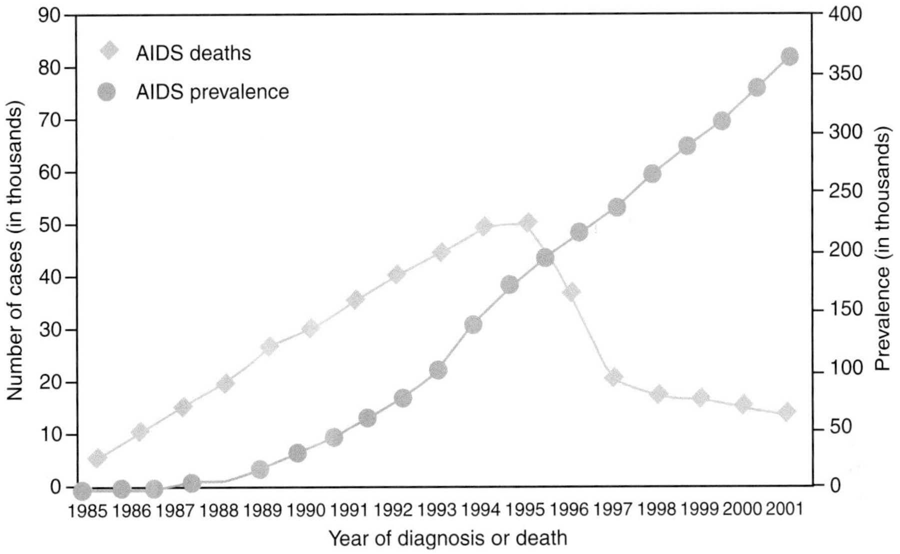

Figure 1 AIDS prevalence and deaths in the US, 1985–2001. From Center for Disease Control, USA.

Table 1 The fall in death rate for diseases treated with pharmaceuticals between 1965 and 1999

Disease	*Percentage drop in death rate*
Diseases of early infancy	80
Rheumatic fever and rheumatic heart disease	75
Atherosclerosis	68
Hypertensive heart disease	67
Ulcer of stomach or duodenum	61
Ischemic heart disease	41
Emphysema	31
Hypertension	22

Source: European Federation of Pharmaceutical Industries and Associations (EFPIA).

1.14.2 History of the Pharmaceutical Industry

The use of natural products as medicines predates medicinal chemistry by centuries. The effects of plant extracts and other natural products as treatments for various disorders and diseases have been documented in pharmacopoeias for centuries. The Chinese were among the first to explore and codify natural medicines; for example, use of wormwood, an extract from the qinghao plant, was recorded as having medicinal uses in 168 BC and it is now acknowledged by the World Health Organization (WHO) as being useful against malaria. The London Pharmacopoeia produced in 1618 included information on how to prescribe each substance. After similar documents had appeared in Edinburgh (1699) and Dublin (1807), a harmonized British Pharmacopoeia was eventually prepared in 1864. This reference work is regularly updated and it is this work that is the source of the abbreviation BP after the name of drugs. The Japanese Pharmacopoeia was first produced in 1886.

Systematic investigation of the physiological effects of natural products in the Western world began in the nineteenth century. French scientists, including Serturner, Pelletier, and Caventou, isolated alkaloids including quinine, which was already known to be useful against malaria, and morphine to counter pain.[4] The field of organic chemistry, which aimed to determine the structures of these products, synthesize them, or develop derivatives, was developed by numerous scientists in Europe and the US in the nineteenth and twentieth centuries. The first medicine to be produced synthetically on a large scale was acetylsalicylic acid (aspirin) back in 1898, when Felix Hoffmann started the process of turning the dye-making company Bayer into a biopharmaceutical company.

As the twentieth century progressed, medicinal chemistry added an increasingly rich list of compounds to the medicine chest, especially in the area of infectious disease. In 1910 Paul Ehrlich reported that an arsenic-based compound, arsphenamine, could be used successfully to treat syphilis – the compound was known by the tradename Salvarsan and was marketed by Hoechst.

Bayer's search for other physiologically active compounds showed that some of its azo dyes and acridines appeared to have useful effects in bacterial infectious diseases without harmful side effects. In the 1930s, Bayer's research director Gerhard Domagk instituted a systematic screening of a large number of synthetic dyes to look for curative effects, initially in mice. He found that the red dye known as Prontosil Rubrum was effective against streptococcal and, as later found, staphylococcal infections. The curative properties were shown to be from the sulfonamide moiety, and this Nobel Prize winning work (1939) led to the development of a new class of drugs – the sulfonamides. The screening approach used by Domagk was taken up by others. Howard Florey used the method in 1940 when he demonstrated that penicillin (whose powerful antibacterial properties had been discovered by Alexander Fleming's work on *Penicillium notatum* in 1928) cured infected mice.

The identification of the chemical entities responsible for observed physiological effects prompted chemists to synthesize related molecules and to screen them for desirable curative properties. It has been estimated that some 5000 compounds were synthesized in the late 1930s, including the biologically active compounds sulfapyridine, sulfathiazole, sulfadiazine, and sulfaguanidine. Sulfonamide antibacterial drugs saved thousands of lives during the Second World War (including that of Sir Winston Churchill) from pneumonia and other bacterial infections such as gangrene. US troops used them in the South Pacific campaign. Clinical trials in UK hospitals in the mid 1930s demonstrated their effectiveness in curing infections following childbirth, and in meningitis. Clinicians and the general public alike were enthusiastic during the 1930s and 1940s about the life-saving benefits of these new medicines – use spread rapidly and output in the US had exceeded $15 million by 1942. However, the antibacterial sulfonamides were soon superseded by penicillin. Other compounds in the sulfonamide series also proved clinically useful, e.g., bendrofluazide (a diuretic that lowers blood pressure), frusemide (a diuretic for use in kidney failure and heart disorders), acetazolamide (for treatment of glaucoma), and dapsone (for treatment of leprosy), and many are still in use today.

While the design of new molecules by medicinal chemists was becoming more strategic, applications were still often a matter of serendipity, as unexpected physiological effects were (and still often are) observed. Hence, screening of new molecules against a range of drug targets and animal disease models has become embedded in the industrial culture of drug discovery.

1.14.2.1 Other Advances from the Pharmaceutical Industry – Anesthetics and Vaccines

Although this chapter concentrates on the role of the pharmaceutical industry in researching, developing, and bringing to the market drugs to treat diseases and conditions, many pharmaceutical companies have also contributed to improvements in public health through the development of vaccines and in medicine through the development of anesthetics.

The idea of surgery taking place without anesthesia is now unthinkable. Initially, simple gases were used. In the 1840s Morton (Boston, MA) carried out a public demonstration of the anesthetic effects of nitrous oxide. Crawford W Long used ether; and chloroform was used in surgery from 1847. Today, there is a range of anesthetics for delivery intravenously as well as by inhalation. Many are opioids, a development pioneered by Abbott working with Loevenhart at Wisconsin University in the late 1920s.[5] One of the barbiturate anesthetic compounds to emerge was ethyl(1-methyl butyl)barbituric acid known later as Nembutal (Abbott's tradename) and marketed as pentobarbital by Lilly. Abbott also worked with other academic researchers and produced thiobarbital (Pentothal). Inhalation compounds such as halothane (Fluothan) were introduced in 1956.

Recent examples of intravenous anesthetics include propofol (Diprivan from AstraZeneca) available since 1989 and etomidate (Amidate from Boehringer Ingelheim) since 1972. There are specific anesthetics for use in epidural delivery such as levobupivacaine (Chirocaine, Abbott). However, the physiology of anesthesia is still not fully understood. Modern practice includes use of several anesthetics together with muscle relaxants (*see* 6.15 Local and Adjunct Anesthesia).

The story of vaccines begins with smallpox. In the seventeenth century in the UK, a practice known as variolation (which had been used for centuries in China) was used. It involved the injection of material from the pustules of a smallpox patient under the skin. This resulted in a (usually) mild bout of the illness followed by subsequent immune resistance. Edward Jenner (as early as 1798) improved on this process by demonstrating that the use of cowpox, which did not cause infection in humans, could confer protection against smallpox. It was not until 1880 that Louis Pasteur's theory that germs caused disease was demonstrated by administering attenuated cholera culture to chickens and showing resistance to cholera infection. He went on to develop vaccines against anthrax, cholera, and rabies. In 1890 von Behring and Kitasato showed that immunity was bestowed through cells in the blood serum and they induced immunity to tetanus and diphtheria by injecting serum from immune individuals. Von Behring's work led to the development of the use of serum from immunized horses to vaccinate people against diphtheria and for this he was awarded the Nobel Prize for medicine in 1901 – his invention was described as "a victorious weapon against illness and death." Robert Koch identified the bacillus that causes tuberculosis, and its culture enabled an attenuated strain to be developed for use as a vaccine – the Bacillus Calmette Guérin (BCG). This technique, which was first used in the 1930s, is still used today in the production of some vaccines, for example, the growth of human viruses in embryonated hens' eggs. A yellow fever vaccine was then produced using chick tissue extracts, followed by an antityphus vaccine, which was widely used in the Second World War. In the 1940s new methods of culture were developed and a poliomyelitis virus vaccine was produced by Enders using mammalian cells as the growth medium. Jonas Salk developed a killed vaccine for poliomyelitis in 1955, which was grown in monkey kidney cells. The Pasteur Institute in France made improvements to Salk's processes and Charles Merieux's institute obtained a license to manufacture a vaccine in Lyon. This work marked the first product from Merieux that was for use in humans (previous work had been for animal health). In the 1950s a vaccine against influenza was produced by growing the virus in chicken eggs and then killing the live vaccine making it an attenuated vaccine. The use of bird eggs meant that no animal viruses, which might affect humans, were contained in vaccine. Using such vaccines, a WHO campaign to eradicate polio throughout the world is underway, although the original target date of 2000 has been missed.[6]

Research to develop new vaccines continues. Rather than using killed or attenuated whole organisms, new techniques rely on subunits of the microorganisms against which vaccination is sought. This has enabled vaccines to be developed against bacterial pathogens, which have a tough outer coat of polysaccharides – the polysaccharides have been linked to more immunogenic proteins. The first genetically engineered vaccine was developed by Chiron and introduced in 1984 against hepatitis B. There is a particular focus on developing vaccines against HIV. However, the tendency of the virus to rapidly mutate makes this an extremely difficult task and there is no guarantee of early success.

1.14.3 Patterns of Pharmaceutical Innovation

When a completely novel type of medicine is discovered and developed and approved, it is known as the 'first in class.' However, it is unusual, and in fact undesirable, for there to be only one medicine in a drug category. Not only will the company that developed the original medicine seek related compounds with better efficacy and/or side effect profiles, but also other companies will often have embarked on similar research at around the same time. This is because the research programs of several companies are often triggered by publication of the basic discovery of the role of a protein in a disease process. The result is the so-called 'fast follower' products. They are important for a number of reasons. Firstly, the different profiles will result in a better match for some medicines than others for the specific genetic makeup of certain patients. Secondly, there is always a risk that the first-in-class medicine is withdrawn from the market because of unanticipated side effects. Thirdly, a variety of medicines give the doctor a choice and the health system the opportunity for price competition. However, the increasing speed at which these 'fast followers appear has also reduced the effective period of market exclusivity, prompting the original innovator to recover its investment over an ever-shorter period (**Table 2**).

Public criticism of 'me-toos' (the less complimentary name for multiple products) often misunderstands the timescale of innovation, the value of variety, and the fact that advances occur in the industry more frequently in small steps than by major breakthroughs. There are few other industries in which it would be thought desirable to have only one version of a product!

We can see the pattern clearly in the drug class, the statins, which lower lipids, including cholesterol, in the blood, by blocking a specific step in the biosynthesis of cholesterol. The first statin was Merck's product Mevacor (lovastatin), which was approved and marketed in 1987. Other statins have subsequently appeared: Lipitor (atorvastatin) from Pfizer/Warner Lambert in 1995, Lipobay (cerivastatin) from Bayer in 1997, Lescol (fluvastatin) from Novartis, and Lipostat (pravastatin) from Bristol Myers Squibb. The growth in the market, particularly stimulated by Lipitor, widened the use of these drugs and led to strong competition in the marketplace. Continuing research on the statins showed other beneficial effects and some have been submitted and approved by the regulatory authorities for additional uses. For example, fluvastatin is approved in the UK for slowing down the development of atherosclerosis in patients with congestive heart failure and moderate plaque in their blood vessels. Research is underway to explore potential use to reduce the risk of stroke or dementia.[7]

Even when a successful method of treating a disease or condition exists, companies will carry out research to try to find alternative mechanisms of action that may offer greater efficacy, less significant side-effects, or an improved drug–drug interaction profile. For example, in addition to beta blockers (which act by blocking the beta receptor sites for norepinephrine and so reduce the strength of the heart beat and dilation of blood vessels), there are now seven other classes of medicines to treat hypertension (**Table 3**): diuretics act by dilating arterial blood vessels, increasing sodium excretion and urine output and thus lowering the volume of the blood; angiotensin-converting enzyme (ACE) inhibitors prevent the formation of angiotensin II, which acts to constrict blood vessels; angiotensin II antagonists prevent it being bound properly to the site where it acts; calcium channel blockers inhibit the movement of calcium

Table 2 Period of exclusivity between first in class and follower medicine

First in class drug	*Year of introduction*	*Follower drug*	*Year of introduction*	*Exclusivity period (years)*
Inderal	1968	Lopressor	1978	10
Tagamet	1977	Zantac	1983	6
Retrovir	1987	Videx	1991	4
Diflucan	1990	Sporanox	1992	2
Invirase	1995	Norvir	1996	Less than 1 year

Table 3 Treatment of hypertension (raised blood pressure)

Action of drug	*Example of drug*
Diuretic	Frusemide, amiloride, chlorothiazide
Alpha 1 antagonist	Doxazosin, prazosin
Central acting	Clonidine, methyldopa
Beta blocker	Propanolol, atenolol
ACE inhibitor	Captopril, lisinopril, ramipril
Angiotensin II antagonist	Losartan, valsartan, candesartan, irbesartan
Calcium channel blocker	Nifedipine, amlodipine
Vasodilator	Hydralazine, glyceryl trinitrate

into the smooth muscles in blood vessel walls and in the heart and hence cause the muscles to relax; and alpha 1 antagonists block the nerve impulses that cause blood vessels to constrict. Combinations of the different classes of drugs are also in use (*see* 6.32 Hypertension).

1.14.3.1 Use of Animals by the Pharmaceutical Industry

The pharmaceutical industry conducts or pays for about a third of all the research using animals in the UK. Some of this is basic research into biological processes – often carried out in partnership with universities. Most are used in order to assess a potential new drug, and to compile data about the way it works, in order to obtain authorization from the regulatory authorities (Medicines and Healthcare Products Regulatory Agency (MHRA) in UK, Food and Drug Administration (FDA) in US, European Medicines Evaluation Agency (EMEA) for Europe) to put the drug onto the market.

The regulatory authorities demand evidence of the safety, quality, and efficacy of the drug. The regulations are set out in Council Directive 2001/83/EC for the EU, with guidance in CMPM/ICH 286/95 about conduct of nonclinical safety studies, and for the US there are extensive guidance notes from the FDA Center for Drug Evaluation and Research. The work must be carried out in accordance with Good Laboratory Practice (GLP) (87/18/EEC; 88/320/EEC).

The use of animals is required in assessing acute toxicity (the qualitative and quantitative study of toxic reactions which may result from a single administration of the active compound); repeated dose toxicity (to show any physiological and anatomo-pathological changes and to show the effects of different doses); mutagenic and carcinogenic potential; fetal/perinatal and teratological effects.

The regulations demand that the toxicology must be done on at least two different mammalian species. Rodents (mice and rats) are used but another nonrodent species is also needed. Dogs are frequently used, partly because a great deal of data already exists and many of their metabolic pathways are well understood and documented, and the data are readily accepted by the regulators. Small marmosets are often used for similar reasons and because only small amounts of a new compound are needed as the animals are tiny and the dose is given in proportion to body weight. In the UK, cats, dogs, nonhuman primates, and some other species are given special protection under the law and may only be used when no other species is suitable.

The way animals are treated is regulated very strictly in the UK under the Animals (Scientific Procedures) Act 1986 which brought up to date the earlier (1896) act. Similar (but slightly less strict) regulations exist for the EU (EU Directive 86/609/EEC). These provide protection to animals while recognizing the need to use them in research and safety testing. Under the 1986 act, a license is required from the Home Office for each individual researcher (this ensures that people are trained and are suitable) and for each project (this ensures that the number of animals used is minimized, that alternatives to animals are used where appropriate, and that any suffering is considered carefully against the potential benefits to both animals and humans). In addition, in the UK, the work may only be carried out at designated establishments which meet high standards of animal welfare and which have suitable veterinary and animal welfare personnel. These laboratories are inspected (without warning) by Home Office inspectors to ensure that high standards are maintained.

The numbers of animals used to assess a new chemical entity has been reduced as experimental designs have improved – the US Center for Drug Evaluation and Research (CDER) guidance for industry on acute toxicity testing says that studies should be designed so that the maximum amount of information is obtained from the smallest numbers of animals. However, the development of mice that have been genetically modified in order to provide better models for some diseases and also to show the effects of loss of a particular gene (so-called 'knock-out mice') has led to an increase at an early stage of basic research.

1.14.3.2 The Importance of Patent Protection

The ability of the innovator to protect a new medicine, and the processes that produce it, via patents is fundamental to the industry. Without patent protection the huge investments in developing and commercializing new products would not be made. This was not always recognized. Permission was refused by the American Medical Association to a delegate who wanted to register for the 1896 annual meeting because he had applied for a patent. The founder of the pharmaceutical company Squibb said that nothing should be patented by a physician or a pharmacist, and his company held only one patent by 1920. However, German companies had patented their dyes and medicines, and after the First World War the US Government abrogated German property, including pharmaceutical patents, selling the latter to US companies. Some of the patents originally owned by the German company Merck were obtained by their US branch, which was one of the first companies to set up their own research institute (in 1933).

In the 1940s, Merck investigated soil microorganisms as potential sources of antibiotics and identified streptomycin on which they sought patent protection. The US Patent Office decided that, because chemical modifications had to be made to the naturally occurring chemical substance, and that there was a therapeutic effect, a patent could be granted. Following that test case, other broad-spectrum antibiotics were subsequently patented (Chloromycetin to Parke Davis and Terramycin (oxytetracycline) to Pfizer, for example), and application for patent protection became the established practice of the industry. Typically, a 'composition of matter' patent is applied for as soon as a molecule is discovered to have useful biological activity, and often covers a family of molecules demonstrated to have similar activity. Subsequently, manufacturing and applications patents may also be applied for.

The patent system, which is accepted by most Governments throughout the world, encourages innovation through giving inventors a period of exclusivity. Governments recognize that new inventions may be of wide benefit to citizens and to the national economy. The patent system exists throughout the world but applications for patents have to be made in each country, and the rules differ, notably between Europe and the US. Once a patent has been granted (often several years after the application for patent protection has been made) the holder is permitted to have a monopoly on the patented product or process, typically for a period of 20 years. However, because of the very lengthy process of development, often only around 8 years of patent life remain at the launch of the product, and similar products may also come on the market within that period, further reducing market exclusivity. Even with the exclusivity provided by the patent system, only about a third of new drugs make sufficient profit to cover their development costs.

In some cases, extensions of the patent period (patent restorations) are granted in recognition of time lost in the processes of testing and by the review processes of the regulatory authorities. The US Drug Price Competition and Patent Term Restoration Act allows the US Patent and Trademark Office (PTO) to grant up to 5 years additional patent protection. In Japan, patents are granted for 20 years from the date of application and can be extended by up to 5 years taking into account the time elapsed from the start of clinical trials to the day before approval for marketing was received.

In Europe, supplementary protection certificates (SPCs) may be granted, which take effect on the expiry of the patent and are limited in scope, covering only the marketed pharmaceutical product for which regulatory approval has been obtained. The SPC, once granted, is in effect dormant until the normal expiry date of the patent and then has a duration equal to the time between patent filing and the grant of the first marketing authorization in the European Economic Area, reduced by 5 years, up to a maximum duration of 5 years. This has the effect that, if a marketing authorization is first granted between 5 and 10 years from the patent filing date, then the effective patent life will be 15 years. Patent protection can also be extended when research, carried out after a product has been put onto the market, shows that the medicine is effective in other disease indications beyond that for which it was originally developed and approved.

After the expiry of the patent, other companies may seek to use the information in the patent application to make the product and to compete with the original inventor. In the case of pharmaceuticals, products produced by companies other than the original inventor, or that company's licensee, are known as generic drugs. Generic manufacturers have to meet all the stringent requirements concerned with manufacturing but may make an abridged application not having to repeat the innovator's clinical trials that demonstrated the mode of action and efficacy of the medicine. In recognition of the innovator's investment, a period of data protection is often granted before the generic manufacturer can utilize such information, effectively extending the market exclusivity. In the European Union from October 2005 this period of data exclusivity will be 10 years, although generic companies may make their marketing authorization application after 8 years though the product will not be marketable until 10 years have elapsed.

The fact that each country has a different patent system means that it is costly and time consuming for companies to obtain worldwide patent protection. The United Nation's World Intellectual Property Organization (WIPO) encourages discussions that might one day lead to a unified global system or to acceptance by one country of the assessment of a patent application by another country.

Not surprisingly, intellectual property issues have become a significant factor in the World Trade Organization (WTO) rounds of trade negotiations, beginning in Uruguay in the 1980s. The WTO's agreement on Trade Related Aspects of Intellectual Property Rights (TRIPS) addressed the need for all countries to respect the system of patenting to encourage innovation and has set minimum standards in patents, copyright, and other aspects, to be gradually introduced in countries by 2010. At the Doha WTO meeting in 2001 the declaration on the TRIPS agreement said that TRIPS "does not and should not prevent WTO members from taking measures to protect public health," against a background of wanting to promote access to medicines for all. Under TRIPS, governments retain the right to over-ride the patent in exceptional circumstances such as national emergencies, and so issue a 'compulsory license' to someone other than the patent holder to make the patented product. There was a WTO Council decision in August 2003 on the implementation of paragraph 6 of the declaration on the TRIPS agreement and public health, which is to be implemented in Europe by a regulation, on compulsory licensing of patents relating to the manufacture of pharmaceutical products for export to countries with public health problems. By using WTO Members' obligations, under article 3f of the TRIPS agreement, this allows WTO members to grant compulsory licenses for the production and sale of patented pharmaceutical products intended for export to developing world countries that have insufficient or no manufacturing capacity. It includes substantial safeguards against trade diversion, and includes rules to ensure transparency and to provide for further replacement of the decision by an amendment to the TRIPS agreement (*see* 1.26 Intellectual Property Rights and Patents).

1.14.3.3 Naming of Pharmaceutical Products

Pharmaceutical products have more than one identifier. When a substance is undergoing preclinical and clinical trials it is often referred to by a company number. The product has a formal (and usually very complex) chemical name and a generic name (nonproprietary name), and it is also given a tradename (sometimes referred to as its brand name or proprietary name), which is normally registered with the intellectual property/patent offices as a trademark. For example one nucleoside reverse transcriptase inhibitor drug used in AIDS treatment was developed by Glaxo-Wellcome as 1592U89; its generic name is abacavir and its tradename is Ziagen.

1.14.4 The Process of Turning Discoveries into Marketed Products

Despite great advances in rational drug design and the introduction of high-throughput techniques, the process of identifying effective, safe drugs remains an unpredictable one. The unpredictability derives from the fact that we do not completely understand human biochemistry or physiology at the organ, cellular, or subcellular level, and complete understanding lies many decades ahead. Hence, the drug discovery process is a complex blend of theory and empiricism. So a potential drug still needs to be tested on animal models of disease and toxicology, assessing the effects on a wide range of tissues and body chemistry, but there are nevertheless nearly always some surprises when the drug is administered to healthy human volunteers and in patients. Frequently, these tests lead to drugs being used against different diseases from those originally envisaged (for example alpha-agonists were originally looked at as nasal decongestants but were found to be effective in reducing high blood pressure). Hence, so-called 'serendipity' remains a major feature of drug discovery in practice.

The modern pharmaceutical industry has research teams that assiduously keep up with the scientific literature and maintain consulting relationships with leading academics, and quickly incorporate new discoveries into their applied research projects. They also themselves engage in basic biomedical research (e.g., exploring the molecular basis of disease by looking at which genes are 'switched on' at each phase of the disease and looking at the proteins and systems that are affected by those genes). The sequencing of the human genome has enabled researchers to identify the genes that are 'switched on' at various stages of cellular activity, in particular organs of the body, in health and in disease, and hence to identify potential target proteins for drug therapy. The study of the physiological effects of the outputs from gene activity is known as functional genomics. However, there are approximately 30 000 human genes coding for perhaps 150 000 proteins, and these give rise to a complex web of protein–protein and protein–gene interactions. Hence, although this research may be helping to make the discovery and development of new medicines more predictable, it will be several decades before we have a functional genomic database. That database would include not only knowledge of every chemical subsystem that is active in different types of cells at different stages of development – in healthy organisms and those with conditions and diseases – but also knowledge of how the subsystems interact with one another. Understanding the complex interactions within the cell, within organs, and throughout the body, is known as 'systems biology' and that is a challenge of such a scale that it makes sequencing the genome pale in comparison.

Paul Ehrlich used the term 'magic bullet' in the early twentieth century to describe drugs that hit their 'target' without affecting other aspects of bodily functioning and hence act without causing undesirable side effects. However, it is still an elusive dream in most diseases. Only in antibiotics, perhaps, do we see a relatively simple transition from drug design to effective medicine. The process of discovering and developing a marketable new drug is illustrated in **Table 4**. As described above, there are many factors that can derail the process, but let us describe how it happens.

A disease-oriented, or mechanism-focused, multidisciplinary team within a company will decide to explore whether interfering with the action of a particular 'target' protein, enzyme, or cofactor will result in disease-modifying effects.

Table 4 The time table of pharmaceutical development

Years from start of research	*Activities*	*Specialists involved*
1	Select disease; identify possible biochemical targets; chemical synthesis including combinatorial chemistry; bio-informatics research; genomics	Researchers, bioinformatics, synthetic chemists
2–5	Target identification; development of assays; high-throughput screening; 3D molecular modeling; lead identification; assess lead compound for manufacturability, stability, solubility, etc.; preclinical in vitro and in vivo pharmacology tests to determine absorption, metabolism, sites of action, persistence, and elimination from organism; acute toxicity tests in animals; patent application for promising compounds; ideas and experimental design submitted to ethical committees in preparation for clinical studies; manufacture to cGMP standards of compound for clinical trials	Researchers, patent experts, contract research organizations, process plant designers, safety and toxicity specialists
5–6	Phase 1 clinical trial of lead compound in healthy human volunteers, market research, decide name of product	Clinicians, contract research organizations, commercial department
7–8	Phase II clinical trials. Pilot scale process development and patents. Long-term animal experiments continue. Reproductive studies done in animals	Toxicologists, manufacturing chemists, plant design
9–11	Phase III clinical trials. Decisions about how and where to manufacture, market analysis	
11–12	Submission of data for product approval and registration and marketing authorization country by country	Regulatory specialists
12–13	Negotiations with customer governments and healthcare agencies on price and reimbursement under healthcare systems	
14–20	Marketing and sales of drug under patent protection; pharmaco-vigilence to identify any emerging side effects and problems; application for supplementary protection certificate; discovery of additional uses for the compound and development of new formulations	Marketing. General practitioners
21–25	Sales under supplementary protection certificate	
25	Generic manufacturers seek authorization and market cheaper versions of the drug	

The biologists on the team identify a range of in vitro tests (assays) that indicate whether the target protein, and/or relevant metabolic pathways, are being affected. Structural biologists and biochemists, using x-ray crystallography and nuclear magnetic resonance (NMR) and other techniques, investigate the target protein structure (often with and without the natural cofactor) to identify the molecular features of the active site. The chemists then set about synthesizing molecules with the right characteristics (shape, charge distribution, hydrophilicity/hydrophobicity, and so on). In the absence of detailed molecular structure (many drug targets are membrane-bound and so pose a difficult challenge for structural biology) assays will focus on measurable downstream effects of successful binding (*see* 4.17 Chemogenomics in Drug Discovery – The Druggable Genome and Target Class Properties).

Large numbers of potential compounds are made (or sourced from the company's compound library) for this screening. So-called 'combinatorial chemistry' is used to ring the changes among families of compounds, relieving the chemists of the burden of individual synthesis. Compounds with binding constants in the micromolar range typically emerge from these large screens as starting points for the next stage, in which these 'leads' are optimized further in an attempt to improve desirable characteristics and produce molecules with binding constants at nanomolar levels. The goal is usually to produce several such candidate molecules, because many will prove to be unsuitable for further development as a result of their characteristics such as solubility, hydrophilicity, toxic breakdown products, and so on. Therefore, it is only a very small proportion of the synthesized compounds that go forward into preclinical testing. The preclinical phase is where the compound is tested in cells and tissues in vitro and then in whole animals (most frequently mice and rats) to assess the physiological effects and to follow the biochemical reactions taking place. Again, at this stage, many compounds fall out because of toxic or damaging side effects or inadequate efficacy, leaving a few

Table 5 Attrition rates of potential new drugs

Phase of testing	*Percentage of compounds failing*
Preclinical	90
Phase I clinical trials	82
Phase II clinical trials	72
Phase III clinical trials	34
Submission to regulators for market approval	10

Source: Association of the British Pharmaceutical Industry (ABPI).

true 'clinical candidates' for human clinical trial. The Center for Medicines Research International (CMR), based on data from 27 global companies in 2000, found that 90% of compounds selected for preclinical testing were rejected, mainly for reasons of bioavailability or toxicity (**Table 5**).[13]

1.14.4.1 Delivering the Drug to the Patient: Formulation of Medicines

In order to administer an active pharmaceutical compound to a patient, it has to be mixed with other materials to make a tablet or a capsule or put into a solution for injection. Other methods of drug delivery include aerosols, which deliver the compound through the bloodstream in the lungs, and topical (external) application. The making up of the physiologically active ingredient into its administered form is known as formulation; this is critical to ensuring that the required dose is administered and that the active ingredient is taken up at an appropriate rate into the bloodstream and into the organ at which its action is required. Expertise in developing formulations resides with the major pharmaceutical companies and is a very valuable and important part of the product development process. (Reference books known as formularies (different versions for different markets) give details of the different formulations and dose sizes of a product once it has been licensed for release on that particular market.)

As a promising compound emerges from the early stages of research, its solubility in water is looked at carefully (as an indicator of the extent to which the compound can be effectively administered orally). But, in order to get into cells where the majority of drug targets reside, a compound must be able to pass through the lipid-containing cell membrane. Water solubility and lipid solubility are of course in conflict to some extent, and selection of the 'best' drug is often a combination of various characteristics of the molecule (according to the so-called Lepinski rules). Most important, however, is a tight binding to the relevant receptor (being occupied at nanomolar concentrations). Particle engineering and nanotechnology are now being used in drug design. Use of liquid-filled soft or hard gel capsules has enabled drugs to be delivered to the stomach in a solubilized state ready for absorption. Some products are required to pass through the barrier that protects the brain (the blood–brain barrier) so that they can act on parts of the brain or central nervous system. A compound delivered in a tablet taken by mouth must withstand exposure to digestive enzymes in saliva, esophagus, and stomach where it is absorbed into the bloodstream and moved to the part of the body where its action is required. The bioavailability of a medicine, and the level of it in the blood or in a particular organ over time, are measured during clinical trials and are important aspects of the approval process. Coatings (enteric coatings) have also been developed to ensure that the product is not destroyed in the digestive system. In some cases, incidentally, the medicine administered is not the form in which biological activity occurs: it is a 'prodrug' converted by the body's metabolism into the required molecule.

When a medicine is administered in a single dose (e.g., a tablet) the concentration in the bloodstream spikes and poses problems either in excessive levels or in maintaining the therapeutic effect over 24 h. Hence, formulations offering sustained release of the active compound, evenly throughout a 24-h period, were introduced in the 1980s. In the 1990s steps were taken to try to mimic the bodily rhythms throughout the day, via delayed release profiles and pulsatile release mechanisms. Some medicines (contraceptives, antinausea, and antismoking products) are formulated in patches, which deliver the active ingredients over time through the skin.

1.14.4.2 Manufacture of a Novel Medicine

Another area for research, as a novel compound is investigated, is the development of a reliable, repeatable, well-documented method of manufacturing that will meet the regulatory requirements for that aspect of the pharmaceutical business. The standard for Good Manufacturing Practice (GMP) covers aspects of plant and facilities, management procedures, and quality assurance processes. The manufacture of a new medicine at commercial scale often requires a rethinking of the synthetic pathway to ensure reasonable costs, security of supply of raw materials, etc.

Typically, the manufacture of the active ingredient of a pharmaceutical product takes between 4 and 12 stages, sometimes in plants at different locations. In exceptional cases an even larger number of steps is required; for example, the new-generation HIV medicine Fuzeon was said to require 40 steps!. The active ingredient is then combined with excipients (chemicals or other materials that aid in formulating the medicine) and processed into tablets or capsules and packed and labeled in the appropriate language for each market. The manufacturing processes (both for the active ingredients and the downstream steps) are controlled under GMP, and inspectors from the regulatory agencies inspect the plant and interrogate the managers and operators about their procedures as well as examining written procedures and quality controls, before authorization to manufacture is granted to a product manufacturing site. The final processing into tablets, capsules, etc., and the packaging, is done in clean rooms maintained to international standard conditions. In order to ensure that packaged drugs arrive in good condition to each patient, each batch is bar coded and can be followed, and if necessary withdrawn, if a problem is subsequently discovered.

Trends in manufacturing technology include less use of batch processes and greater use of continuous processes, but also moves away from use of very large reaction vessels to smaller scale since that permits greater control and can reduce the proportion of impurities. Increasing use is being made of enzymes at various stages of manufacture of chemical ingredients, for example, to manufacture chiral compounds (since frequently it is only one of the enantiomers that has the required biological activity). Crystallization is an important part of purification and isolation of active ingredients, and new analytical methods are being used to refine and control the shape and size of crystals. While the use of natural products, and of fermentation processes to produce them, is now 50 years old, more recently we have seen the emergence of recombinant DNA-based processes for the manufacture of therapeutic proteins (including enzymes and monoclonal antibodies) at commercial scale. Such processes and facilities present unique problems of design and control: it is frequently found, for example, that the form of protein produced varies from one location to another even when the 'recipe' may not. The use of animal-derived raw materials in both fermentation and formulation is also an issue as a result of concerns about viral and prion contamination.

1.14.4.3 Clinical Trials

Before clinical trials can begin, the design has to be submitted to the ethical committees in all the hospitals or clinics that will take part. The protocol used for a clinical trial of a particular drug is often worked out in consultation with the regulatory authority that will first consider granting a marketing license. Dialog with the regulator is particularly important if there is something very novel about the compound or about the way it is likely to be delivered to the patient.

The first phase of clinical trials is conducted on healthy human volunteers. This phase has recently been the subject of regulatory changes in the EU making it necessary for the first time to submit details of the experimental design to the regulators. In this phase of human trials, volunteers are given a range of doses to make the first steps toward determining the lowest effective dose and the highest tolerated dose. This first phase provides data on the physiological effects of the compound, the sites and mode of action in the body, and extensive measurements are taken to determine the pharmacokinetic and pharmacodynamic properties. As with all aspects of laboratory analyses, done as part of clinical trials, the lab work must be done to GLP standards, which require the use of suitably qualified staff for each function, documented standard operating procedures and methods for the tests and assays, and a system of quality control. Many compounds fail at phase1 because they do not produce the desired effect in humans or because the effect is too short lived, localized, or generalized, or show indications of undesirable or dangerous side effects. For example, in the 1990s many potential compounds were found to affect heart rhythms (via so-called QT prolongation) and had to be rejected. Findings made at this stage are often fed back into the development cycle for other similar compounds, which may therefore be subjected to additional tests in mice and rats to rule out those with comparable unwanted effects (*see* 5.40 In Silico Models to Predict QT Prolongation). However, this too is not a precise science, and the numbers of volunteers involved (typically less than 100) cannot predict those side effects that will emerge for a small minority of patients.

The second stage of clinical trials is on patient volunteers and often involves between 150 and 500 individuals. Typically, 70% of compounds fail at this stage. For those compounds that still look promising at this phase of development, those responsible for designing the eventual manufacturing process will scale it up to pilot plant scale (for new chemical entities this means producing kilogram quantities rather than milligrams or grams).They will also devise analytical methods for quality control of the manufacturing process. Various ways of formulating the medicine to administer it to the patient will also be investigated – tablets, inhalers, injections, etc. Process patents may be applied for and plans made for design of manufacturing plant and for validation and certification to good manufacturing practice standards (cGMP) in anticipation that the company will commit to launch the product, subject to the next and longer phase of testing.

Phase 3 clinical trials (often referred to as 'pivotal trials') will subject several hundred, and often many thousands, of patients to the medicine at the dose and in the form intended for market. This is the most costly stage by far (typically tens of millions of dollars) and yet still some 35% of compounds fail, as a result of low incidence side effects or because of inadequate efficacy when a larger statistical sample is viewed.

Patients are recruited for clinical trials through their consultants and general practitioners, or in some cases through advertisement. The 'double blind' design used for trials means that those monitoring the effects on patients do not know whether the patient is receiving the drug, a comparator drug (in some cases), or an inactive placebo (in others). Placebos (where ethically justified) enable the statistical analysis of the study to separate the effects on the patient that arise simply from receiving regular attention and monitoring during the trial from the genuine effects of the drug. Trials must remain 'blinded' while they proceed, subject to the occasional 'unblinding' for review by an independent supervisory panel who can order the trial to stop. Reasons to stop the trial include findings that the medicine is clearly causing unacceptable side effects or that it offers such advantages to patients that it becomes unethical for there to be a control group who are not receiving it (*see* 1.22 Bioethical Issues in Medicinal Chemistry and Drug Treatment).

At the end of phase 3, companies have sufficient data to prepare documentation needed for product approval, registration, and market authorization. Extensive dossiers detailing the mode of action of the drug, its effects, its efficacy, safety and quality, and its manufacturing processes have to be submitted. No single regulatory agency authorizes a new medicine for worldwide marketing. The compound has to be assessed by the relevant authority for each country – for example: the FDA for the US; the EMEA for the most novel products and for biotechnology products throughout Europe; or by the regulatory authority in an individual European country (under a system called 'mutual recognition'); or the Pharmaceutical and Medical Devices Agency (PMDA, KIKO) of the Ministry of Health, Labor, and Welfare in Japan. As a result of steady addition of requirements the amount of documentation is now enormous and when printed on paper often fills a small truck!

While there may have been dialog between regulator and innovator during development (and companies are unlikely to submit dossiers on compounds for which they have no hope of getting approval) there is still an appreciable (10%) failure at this stage because compounds do not get through the authorization assessment.

Clinical research does not stop when the product has been approved and put onto the market. Phase 4 trials may include monitoring of responses to different doses, the long-term effects of taking the drug, and its use in new populations that were not included in earlier clinical trials. Some of these trials may be mandated by the regulatory authority when marketing approval was granted (so-called 'conditional approval'). Even if no such trials are required, the close monitoring of new drugs in use is a growing concern of authorities, via the 'yellow card' systems that report 'adverse events.' In light of recent major product withdrawals, there is increasing focus on making such systems more rigorous and involving the patients themselves more actively.

1.14.4.4 Patient Responses to Medicines

Individuals have different genetic make up and this means that some people will find a medicine effective while others, because of slight differences in their metabolism, may find that the medicine has little or no beneficial effect. Up to a third of individuals who are prescribed medicines are believed to receive no benefit from the medicine, and the result may be the prescription of a different medicine or discontinuation by the patient. Another sub-group of the population may experience side effects because of their genetic make up. Research is underway to map these individual differences now that the small differences in the sequences of DNA in different individuals can be identified (via so-called SNP or single nucleotide polymorphism mapping). This field of research is known as pharmacogenetics. This type of research can be used either to identify medicines with the maximum coverage of genetic subgroups or to target medicines more scientifically. In the latter case the medicine will be used in tandem with a genetic screen to identify whether a patient has the appropriate genetic make up. An example is Herceptin, an anticancer drug targeted toward aggressive breast tumors with high expression of her-2 receptors.

1.14.4.5 The Costs of Research and Development and Its Efficiency

On average it takes about 12 years to take a promising idea through research, testing, and regulatory approval to get a new drug on the market. The overall attrition rate during development is enormous – only one compound out of perhaps 10 000 undergoing preclinical screening reaches the market.[8] **Table 5** shows typical attrition rates at various

Table 6 New chemical and biological entities launched between 1999 and 2003

Area of world where launched	*Number of products launched*
Europe	62
US	73
Japan	28
Others	8

Source: Association of the British Pharmaceutical Industry (ABPI).

stages of drug development. The costs of development and testing are very high and they increase as the compound moves from preclinical trials into the increasingly expensive phases of clinical trials. Estimates of the total costs take into account the costs of developments of products that fail to reach the market and spread those costs over the small minority of successes. On this basis of calculation, data by DiMasi and co-workers[9] showed that the full costs of bringing a new drug to the market rose (in constant 2000 prices) from about €150 million in 1975, to €344 million in 1987 and €868 million in 2000. The Tufts Center for Study of Drug Development estimated, in May 2005, that the costs had risen to $897 million (*see* 1.18 The Apparent Declining Efficiency of Drug Discovery).[9,12]

1.14.4.6 Trends in New Product Launches

The number of new chemical and biological entities launched between 1999 and 2003 is given in **Table 6**. The launch of new products is under the control of company marketing departments, who start their work on their launch campaign well in advance (perhaps 2 years) of regulatory approval. They will be seeking to establish and communicate differentiation for their product from others in the same therapeutic area, and to use evidence on comparative efficacy or convenience to support pricing and reimbursement proposals. Most major new products are launched as simultaneously across the globe as possible, subject to successful regulatory approvals in the major countries. Central marketing groups craft the overall strategy, passing this to national marketing managements for local customization. Developing so-called 'opinion leader' support is a pivotal part of this process. This typically begins at international level, with the leading clinical investigators involved in the trials presenting their results to colleagues at major international conferences. Subsequently, a similar process will take place at national level, with local opinion leaders (usually consultants) introducing the new therapy to potential prescribers.

The launch phase is a further major investment for the innovator company, particularly when a new generation of product is involved for a disease with no or poor existing therapy.

1.14.5 The Pharmaceutical Industry and Its Markets

The world pharmaceutical market (**Table 7**) was estimated to have reached $550 billion in 2004 and has been growing at nearly 10% compound each year for the last decade.[10] The US is by far the largest market and is the quickest to take up new medicines – $236 billion in 2004 according to IMS (**Table 8**). Growth in the US market had been running at 10% per year but between 2003 and 2004 it had slowed to below 8%. Japan is the second biggest market, with Germany, France, and the UK (some 4% of the world market) next in the rankings. Individual European countries account for small percentages of the global market; there is not yet a single market for pharmaceuticals in the European Union because each country has a different healthcare system and each sets or negotiates product pricing and reimbursement directly with supplier companies. There were Government-imposed price cuts in pharmaceuticals in several countries in 2004 and for 2005; for example, a price cut of 4.2% in the Japanese market meant that it only grew by some 1.5% in 2004. The market in China is growing very fast – 28% in 2004 reaching some $9 billion and a rate of growth of around 20% per annum is forecast to continue.

In 2002 production of pharmaceuticals (ex factory prices) was almost equal in Europe and the US (37.5% in Europe and 35.5% in US) with 33% of the world's production taking place in Japan and 14% in the rest of the world.

Eurostat data[11] show the pharmaceutical industry to be the sector in Europe with the highest added value per person employed – €102 000 compared with €65 800 for all high-tech sectors and €51 400 for manufacturing industries. Given that the European Union leaders set targets at the Lisbon Summit in 2002 to develop high added value, science-based industry, the pharmaceutical industry represents one of Europe's best options.

Table 7 Total global pharmaceutical sales

Year	*Total world market US$ billion*	*Percentage increase over previous year*
1997	289	
1998	297	7
1999	332	11
2000	357	10
2001	387	12
2002	426	9
2003	493	10
2004	550	7

Source: IMS.

Table 8 Geographical markets, 2004

Market area	*Sales in 2004 US$billion*	*Percentage of world sales*
North America	248	48
European union	144	28
Rest of Europe	9	2
Japan	58	11
Asia/Africa/Australia	40	8
Latin America	19	4

Source: IMS Health MIDAS.

The R&D-based pharmaceutical industry has seen a succession of mergers and acquisitions, in search of global commercial scale. As a result the 'league table' of leaders has changed continuously over recent decades (**Table 9**). The current leader is Pfizer with some $50 billion in sales and a research expenditure of over $7 billion in 2003. Pfizer was originally set up in 1849 and has grown through a series of acquisitions over more than a century. Warner Lambert, with which Pfizer merged in 2000, was itself built up from its foundation in 1955 by acquisition including Parke-Davis (which itself had been set up in 1866 and which had been the world's largest drug maker). In 2003 Pfizer merged with Pharmacia and that company had been built up from Upjohn, Monsanto, and Searle, which came together with Pharmacia in 2000. GlaxoSmithKline (whose Anglo-US history is summarized in **Table 10**) is the next largest with sales of over $30 billion, followed by Sanofi-Aventis, which contains a wide range of largely European businesses (including Rhone Poulenc and Hoechst). All three market leaders also have some nonprescription businesses, but are dominated in sales and profits by prescription drugs. In contrast, the number four, Johnson and Johnson, has a wide variety of other businesses, each operating through a number of comparatively autonomous independent companies (of which Janssen and Ortho Biologicals are pharmaceutical businesses).

Also within the top 10 research-based pharmaceutical industries is Amgen – the only company founded as a biotechnology enterprise to join this group, and it has taken nearly 30 years to do so. Amgen has achieved this status on the basis of only 3–4 significant products, showing that it is possible to build a major profitable company based on a comparatively small number of products.

Young companies also evolve through mergers and acquisitions. For example, the Chiron Corporation, founded in 1981 in Emeryville, CA, has become a company with a revenue of $1.7 billion in 2004. Its history is as follows: in 1991, Chiron acquired Cetus (a neighboring biopharmaceutical company); in 1995, the company entered a partnership with Novartis (with 49% equity investment, in part paid for via transfer of revenue-earning business); in 1998, Chiron Vaccines was consolidated with Behring (Germany) and Schlavo (Italy); in 1998 and 1999, Chiron sold some businesses in order to concentrate on therapeutics, vaccines, and blood screening; in 2000, it acquired PathoGenesis (Seattle) to increase its anti-infective position; in 2002, it acquired Matrix Pharmaceutical Inc.; and in 2003 acquired Powderject

Table 9 Sales and R&D expenditure of the world's largest R&D-based pharmaceutical companies 2004

Company (country where registered)	*Sales ($ billion 2004)*	*R&D expenditure ($ billion 2003)*
Pfizer (US)	51	3.984
GlaxoSmithKline (UK)	33	2.791
Sanofi-Aventis (France)	27	2.988
Johnson & Johnson (US)	25	2.617
Merck Sharpe and Dohme (US)	24	1.775
Novartis (Switzerland)	23	2.098
AstraZeneca (UK)	22	1.927
Roche (Switzerland)	18	2.153
Bristol-Myers Squibb (US)	16	1.273
Wyeth (US)	14	1.170
Abbott laboratories (US)	14	0.968
Eli Lilly (US)	13	1.313
Schering-Plough (US)	7	0.821
Bayer (Germany)	6	1.701 (includes chemicals)

Source: IMS health; Thompson Datastream.

(UK), which contained a major UK-based vaccines business. Mergers have now begun to play a larger role in the traditionally quite separate Japanese industry, and the formation of Astellas in April 2005 from Japanese leaders Yamanouchi and Fujisawa is unlikely to be the last.

Despite the continuing process of mergers, most of the largest companies still have only single digit worldwide market shares. There is plenty of additional scope for consolidation, subject to antitrust limitations within specific therapeutic areas.

While the majority of the value of company stocks is accounted for by currently marketed products, a significant factor is also the state of the so-called pipeline of products, especially those at the later stages of development. The pipeline is usually described in a company's annual report, and is regularly presented to investment analysts. The pipeline shows the numbers of compounds under test and the diseases and conditions for which they are under development. Recent annual reports show, for example, that Sanofi-Aventis had 80 compounds in preclinical trials and 48 in phase II and III clinical trials; AstraZeneca had some 85 pipeline projects including completely new compounds at different phases of clinical trials and research into new uses for existing medicines. However, in the case of some biotechnology-based early stage companies a single product may account for the majority of its stock value and the failure of this product in trials can precipitate a crisis for the company.

Over the 1980s and 1990s, the pharmaceutical industry was one of the most favored by stock market investors. The combination of high margins and fast (10% or so) annual growth for some companies produced strong EPS (earnings per share) performance. Whilst there were some 'blips' (such as concerns about potential healthcare reforms in the US at the beginning of the Clinton presidency) the underlying belief was that the combination of high margins, remaining high unmet medical need, and aging populations would guarantee high long-term returns for industry investment. Over the last few years investor sentiment has shifted. There are concerns that increasing cost pressures on healthcare payers (particularly in the US) will force reductions in prices and margins, tipping the balance further on the profitability of new products. Also, the slowing of new product approvals has raised concerns about the innovative potential of the industry. The comparative stock market performance of the industry since 2000 has therefore been poor. However, many of the fundamentals of the industry remain extremely positive: unmet needs and aging populations are as evident as ever, and the willingness of populations to pay for pharmaceuticals from their own pockets is largely untested. Looking at potential innovation, there are more than 7000 potential new products in the industry pipeline, many in serious disease areas (such as cancer and heart disease) for which payers have demonstrated a willingness to reimburse. The industry has much potential still to capture.

Table 10 Companies that became GlaxoSmithKline

1830 J K Smith set up in Philadelphia			
	1842 Thomas Beecham set up laxative company in England		
			1873 J Nathan set up trading company in New Zealand
1875 company renamed Smith Kline and Co			
		1880 Henry Wellcome and Silas Burroughs set up Burroughs Wellcome and Co in London	
1891 acquired French Richard and Co			
		1924 Wellcome Foundation Ltd formed	
	1926 Macleans and Eno's acquired		
1929 renamed Smith Kline and French Laboratories			
	1945 Beecham group formed		
			1947 Glaxo absorbed Nathan and listed in London
			1958 acquired Allen and Hanbury's (founded in 1715)
		1959 acquired Cooper, McDougall & Robertson (animal health)	
1960 acquired Norden Animal Health			
1960s acquired RIT vaccines, 6 US labs and 1 Canadian lab			
	1972 unsuccessful bid for Glaxo		1972 unsuccessful bid for Boots
			1978 acquired Meyer Lab in the US
			1980 Glaxo Inc set up
1982 acquired Allergan; merged with Beckman Instruments Inc to become Smith Kline Beckman			
	1986 acquired Norcliff Thayer		
1988 acquired International Clinical Labs			
	1988 Beecham and Smith Kline Beckman merged to become Smith Kline Beecham		
	1994 animal health businesses sold		
		1995 Wellcome and Glaxo merged	
		1998 acquired Polfa Poznan	
		2000 Glaxo Wellcome and Smith Kline Beecham merged to form GlaxoSmithKline	

Table 9 shows sales of the largest companies together with their expenditure on R&D. They are among the largest private sector investors in R&D across all industries, both directly through their own research organizations and also via their investments in partnerships with smaller (biotech) companies. The scale of investment required to bring new products through to market launch means that very few, if any, of the biotechnology startups whose advances hit the headlines can hope to emerge as independent companies funding their own R&D. There is therefore an inevitable symbiosis between the large and small ends of the biopharmaceutical company spectrum. The distinction often drawn between so-called 'Big Pharma' and 'Biotech' is not a matter of technology. Both groups of companies use the same armory of technological tools and take similar approaches to drug discovery. The difference lies largely in scale, and in the organizational processes that change with scale.

Over the last 20 years there have also been major shifts in the geographical patterns of R&D investment, manufacturing, and ownership. At the end of the 1930s the German companies were the world's largest (see **Table 11**) and for a long period the European industry collectively led the world. However, more recently the US-based industry has grown, in both size and R&D investment, much faster than the European industry: in 1992 six of the 10 top selling medicines came from Europe but by 2002 eight of the 10 came from the US. This shift and the more recent emergence of India, China, and Singapore as centers for R&D and manufacturing, is causing significant concern among European governments about how best to encourage industry resurgence in Europe.

The economic impact of the industry in countries in which it is successful is much more significant than the scale of employment it directly generates. Because of high added value, it can be a major contributor to balance of trade: in the UK for example, the industry exports of more than £11 billion generate a positive trade balance of £3.5 billion. Direct employment is about 80 000, but perhaps three times as many jobs are indirectly created by the industry's presence. Less tangibly, relationships with universities and small companies help create clusters of bioscience activity and accompanying centers of excellence that in turn make the emergence of future bioscience-based (e.g., stem cell) enterprises more likely. The 'opportunity cost' of the loss of this strategic industry from major European economies would therefore be extremely high.

1.14.5.1 Generic Medicines

Once the period of patent protection expires, the details, which are in the public domain, can be used by companies other than the original researcher and developer company. Some of the blockbuster drugs of the 1990s with sales estimated at some $8 billion per year will become accessible to generics manufacturers from 2005. These include the antidepressant Zoloft developed by Pfizer ($3 billion sales), the cholesterol-lowering drug Zocor from Merck ($4 billion), Proscar for prostate treatment (Merck $370 million sales), and many more.

Generic medicines are pharmaceutical products that are no longer protected by patent and they are referred to, and prescribed by, their generic name rather than by the tradename used by the original company that researched and developed the product. (The exception is the so-called 'branded generics,' important in some markets: here the generic name is combined with, or replaced by, a tradename from the generic manufacturer.) Generic medicines contain the same active ingredient as the original product but they may have different formulations, provided that bioequivalence can be demonstrated. There is no limit to the number of different 'copies' of the original drug that can be put into the market but each manufacturer has to satisfy the regulatory authorities about the safety and quality of the product and the manufacturing process has to meet the same standards of GMP as the original manufacturer. When patents expire, the multiple generics market entries very rapidly erode the sales of the innovator's product. This is, of course, accelerated when prescriptions are already being written with the generic name of the drug.

Table 11 Shares of word pharmaceutical sales (percentage of total sales)

	1938	*1955*
Germany	39%	10%
US	13%	34%
UK	12%	16%
France	12%	12%
Switzerland	7%	14%

Reproduced from Cooper, M. H. *Prices and Profits in the Pharmaceutical Industry*; Pergamon Press: Oxford, 1966.

Table 12 Turnover of generics companies (2004)

Company (country where registered)	*Turnover (€ billions)*
Teva + Ivax (Israel/US)	6
Sandoz (Switzerland)	4
Ratiopharm (Germany)	1.5
Merck (Germany)	1.4
Alpharma (US)	1.3
Mylan (US)	1.1
Ranbaxy (India)	1
Perrigo (US)	1
ICN (US)	0.9
Stada (Germany)	0.8

Source: Le Figaro.

Generic manufacturers are able to offer medicines at much lower prices than the innovator, because the generics manufacturer did not have to incur the costs of research or of clinical trials. As a result of this cost reduction governments are keen to see doctors prescribing a medicine using its generic name rather than specifying a particular tradename. In England more than 55% of all prescriptions for National Health Service patients were written as generics by 2000 – an increase of 20% since the mid-1980s. In some cases, for example where a patient has a sensitivity or allergy to a particular ingredient in the formulated tablet, it is important that a specific brand name is prescribed. It is for that reason that the policy of generic substitution – the automatic decision by the dispensing pharmacist to provide a generic product rather than the more expensive brand name written on the prescription – cannot be implemented without some danger to patients.

While there are many generics companies around the world, especially in India, China, and Eastern Europe, some of the major innovator companies are also important generics suppliers, notably Novartis, whose Sandoz subsidiary is the second largest generics producer in the world. The cost structure of such companies is obviously completely different from the research-based innovators, and low-cost manufacturing and distribution are critical to success. They play an important role in the industry, since the low-cost medicines they supply after patent expiry allow 'headroom for innovation,' i.e., the ability of health systems to afford the more expensive new drugs. The turnover of companies producing generic drugs is shown in **Table 12** (*see* 1.21 Orphan Drugs and Generics).

Consolidation is also taking place in the generics sector, where worldwide economies of scale are still important (although not to the same extent as in the branded sector). The Swiss company Novartis bought generics manufacturers Hexal and Eon Labs in 2005 to become, briefly, the world's largest generics manufacturer. Its position was quickly overtaken by the merger of the Israeli generics company Teva and the US company Ivax.

1.14.6 Relationships between Pharmaceutical Industry and Stakeholders

The industry has many key stakeholders. Governments act as regulators and customers in almost all markets, while other private sector payers including employers, insurers, and health maintenance organizations (HMOs) are important in the US (and some other) markets. However, since many of the factors are similar for all payers, we will focus principally on governments. Medical professionals, in the form of opinion leaders and prescribers, are clearly critical, both in the development and marketing of medicines, as has been described above. Over recent years, patients and patient organizations have been of increasing significance for the industry. This section summarizes some of these trends.

1.14.6.1 Relationships with Governments

The relationships between the pharmaceutical industry and governments throughout the world are critical for the industry. Governments have a multifaceted interaction with pharmaceutical companies. Governments are keen to encourage the companies to invest in R&D and manufacturing facilities in their countries because of the high value added employment that they bring as well as the ready supply of medicines and vaccine. Hence, Government is acting as a sponsor of the industry. Governments are responsible for public health and for the health of their citizens and as such are a purchaser of medicines. They strictly regulate the pharmaceutical industry in order to ensure that medicines are effective, safe, and of reliable quality. They also fund basic research in biological sciences, chemistry, and the physical sciences, which underpin innovation in pharmaceuticals, and the academic research community provides the education needed by the specialists who are employed in the pharmaceutical industry. Governments also have programs of international aid for the developing world, which includes provision of medicines. Governments (in most cases) set the spending policies for their nation's health budgets and this determines the proportion of this spending devoted to medicines. Governments and their agents negotiate prices for novel pharmaceutical products. They decide whether to press physicians to prescribe generic drugs. Governments work together to develop global policies toward provision of drugs to the developing world. Their education policies affect the numbers and quality of skilled scientists and other professionals needed by the pharmaceutical industry. The relationship between the industry and each government affects the extent to which companies will carry out research and development and/or manufacture locally (*see* 1.24 The Role of Government in Health Research).

1.14.6.1.1 **Governments as regulator**

Governments establish the regulatory framework in which the industry operates, and regulatory agencies decide whether products can enter the market.

Government control of medicines is not new. In the UK for example, Henry the Eighth (1509–1547) gave the Royal College of Physicians of London the power to inspect apothecaries' products and to destroy defective stocks. Regulation to control the sale of poisons was introduced in the UK in the nineteenth century and operated through the Pharmaceutical Society of Great Britain. Licensing of medicinal products began with the 1925 Therapeutic Substances Act, which covered a wide range of products from toxins and antitoxins to vaccines, insulin, and surgical sutures. Licensing included inspection of manufacturing sites and demanded record keeping of the processes involved.

Up to the 1960s there was no regulation in the UK to give any organization the power to prevent the marketing or sale of a new medicinal product. Prompted by the thalidomide tragedy, a Committee on the Safety of Drugs (CSD) was set up in the UK with the aim of ensuring for each drug 'reasonable safety for its intended purpose.' The pharmaceutical industry agreed to submit safety (not efficacy) data to the committee and to abide by its decisions. In 1970 the Medicines Act came into operation in the UK bringing together existing legislation and establishing a comprehensive system of licensing controlling manufacturing, and sale, supply, and import of drugs in the UK. Similar evolution was taking place in other major national markets.

Across the European Union there has been gradual harmonization and in 1965 a Directive (EEC 65/65) on the control of medicines was passed. Each member State has an organization set up by the government to put the legislation into practice – in the UK it is the MHRA, formed by bringing together the Medicines Control Agency (MCA) and the Medical Devices Agency (MDA) in 2003. The role of the MHRA is to ensure that medicines, healthcare products, and medical equipment meet appropriate standards of safety, quality, performance, and effectiveness and are used safely. In 1995 the European Union set up the European Agency for the Evaluation of Medicinal Products (EMEA), with its headquarters in London. Its role is the protection and promotion of public (and animal) health through the evaluation and supervision of medicines. It coordinates the evaluation and supervision of products in the 25 member States through a network of 42 national authorities like the MHRA, assigning specific products for evaluation to national bodies with the relevant competence. A company can therefore make one single application for marketing authority for a new product, which is then given a single evaluation by the Committee for Medicinal Products for human use. If that committee is satisfied about the quality, safety, and efficacy of the product it gives a positive opinion to the Commission of the European Union, which issues single market authorization valid for the whole of the EU.

In the US drugs are regulated under the Federal Food Drug and Cosmetic Act. In order to get a drug approved for the US market an application has to be made to the FDA (even if the drug has been approved in the EU or elsewhere). The FDA has a CDER, which reviews New Drug Applications (NDAs) made by sponsors (usually companies) of a new drug. For its reviews, the FDA, like other authorities in other countries, uses physicians, statisticians, chemists, pharmacologists, and other specialists to assess the drug's safety and effectiveness and its mode of manufacture. CDER also looks at the data, which are proposed to be given to patients, and reviews the information on the proposed label. An NDA has to include all data from animal and human studies of the drug and analyses of that data: descriptions of the mode of action, the metabolism and excretion from the body, and also data about manufacture. The FDA has up to 60 days to decide whether an NDA is complete and ready for review. The CDER is then required to deal with 90% of all

NDAs within 10 months of receipt of the application (6 months for priority drugs). It is usual for the sponsor to meet the FDA at the end of phase II clinical trials to discuss the design of phase III trials, and another meeting often takes place before the NDA is prepared. Accelerated approval can be given for drugs that are going to treat serious life-threatening diseases or conditions for which no satisfactory treatment exists. This accelerated procedure uses surrogate endpoints to evaluate effectiveness and to predict likely benefits rather than waiting to collect data from patients in full phase III trials, but often then requires the company to conduct postapproval trials to demonstrate the benefits in use.

A separate system exists for Japan and for most other countries. In Japan, the Ministry of Health, Labor, and Welfare is responsible for pharmaceutical and food safety and it set up the Pharmaceutical and Medical Devices Agency (PMDA, KIKO) in 2004. The PMDA provides consultations with companies about clinical trials and it carries out approval reviews and surveys of the reliability of the data that have been submitted for product approvals. The underlying law and regulation is the Pharmaceutical Affairs Law, revised in 2002, as a result of which new approvals procedures came into effect in 2003 with further provisions, covering manufacturing and distribution and medical devices, coming into force in April 2005. Global pharmaceutical companies that manufacture products for the Japanese market outside Japan, must have a local 'in country caretaker' to interface with the regulatory bodies.

An international organization exists to work toward harmonization of the approvals processes for new drugs between the EU, Japan, and the US – the International Conference on Harmonization (ICH) of technical requirements for registration of pharmaceuticals for human use.

Regulation is constantly evolving in order to address public and government concerns about safety, introducing additional testing and submission of more data hence making product registration more complex and costly for companies. One example is the need to test pediatric medicines specifically on children. Every day, millions of children are prescribed medicines safely and effectively but most medicines do not have a specific license for use in children, as they have been through clinical trials on adults only. The unlicensed or off-label use of medicines in children is permitted under the discretion of the doctor, and any adverse reactions to them must be reported to the regulatory authorities in the same way as for licensed medicines. However, there are some specific biochemical differences at work when the systems in the body are still developing – the speed of metabolism of a drug may be different in a baby or a child and the dose required to be effective may need to be in a different proportion to body weight than in an adult. There have long been ethical concerns about conducting clinical trials on children since they cannot assess the risks and benefits of participation in trials themselves and anxious parents of ill children may be put in a difficult position when a new treatment is offered. The EU is preparing to introduce legislation that will require pediatric clinical trials for medicines that are likely to be used regularly in children. The ICH has already produced international guidelines on the development of medicines for children, and the UK Government has set up a Pediatric Research Network to bring together expert centers of excellence comprising specialists trained in pediatric pharmacology. The costs of pediatric trials will be high and there will need to be compensatory incentives for the industry – such as extensions of patent protection for the product. Also, where a new medicine is primarily designed for use in adults, industry and patients will be concerned that pediatric trials should not delay the licensing of the adult product.

Special regulations exist for drugs that are developed for rare diseases – so-called orphan drugs. There are some 4000–5000 diseases that have been described that fall into this category. Many countries, including the EU, have introduced laws and regulations that are designed to encourage the pharmaceutical and biotech industry to develop medicines to treat these diseases given that the costs will be high and the total market for sales will be small. Many of these products are proteins to treat genetic metabolic deficiency diseases that would otherwise cause premature death. The US passed its Orphan Drug Act in 1983 with related regulations in 1993, Japan passed regulations in 1993, and Australia in 1997. Each country has a slightly different definition of what constitutes orphan drug status – the incidence of the disease must be less than 0.75 per 1000 people in the US regulations, 0.4 per 1000 in Japan and 0.1 per 1000 in Australia. Europe set up a Committee on Orphan Medicinal Products (COMP) in 2001 to review applications from researchers and companies wanting to have drugs under development designated as orphan drugs (*see* 1.21 Orphan Drugs and Generics).

Despite the rigor of the approval process, the benefit/risk tradeoff for drugs can only be fully assessed when they have been taken by large numbers of people. Sometimes analysis of adverse events leads to restrictions on the patients who can receive the product; in other cases, where side effects are deemed serious and/or common enough, the drug may be removed from the market, either at the discretion of the company or the action of the regulator. These events are always controversial, since they obviously affect patients currently receiving benefit from the drug, and also the investors who may see significant value wiped off the company's stock literally overnight. While the trend for such product withdrawals has been a downward one in recent years recent events with the COX-2 inhibitor product class has highlighted this phenomenon in the world's media and led to demands for tighter regulation, greater transparency, and accountability. It is obviously important to respond to these concerns and industry has, for example, committed to much greater publication of clinical trial results. However, it is equally important not to expect the elimination of any risk, nor punitive legal remedies for side effects, or investment in novel (and by definition high-risk) products will cease.

1.14.6.1.2 Governments as purchasers

All governments carry responsibility for the health of their citizens and to that end they operate various systems of healthcare. In the UK the National Health Service provides medicines and surgery and other treatments at the point of need, free of charge – the costs being met from tax revenue. In the US medicines are provided through a wide variety of health insurance schemes, provided by employers via HMOs or by the Federal and State governments for the retired or uninsured (via Medicare and Medicaid). Changes are taking place in the US, with greater financial responsibility being passed to insured individuals, and legislation to provide at least part of the costs of drugs for the elderly and the poor.

The 'free market' price for a drug reflects its medical importance, differential effectiveness, and the price of competitive products. The US market is one of the few that allows this manufacturer-set price to prevail, since there has been little government-sponsored price control. As a result US prices have been the highest, although this has not prevented the market also being the fastest to take up new medicines. In most countries, governments carry out negotiations with the pharmaceutical industry to set a framework for the pricing of medicines. However, there are wide differences in how governments go about this process.

The UK scheme is called the Pharmaceutical Price Regulation Scheme (PPRS) and it is a voluntary agreement between the industry and the UK Department of Health that runs for 5 years – negotiations took place in 2004 to establish the PPRS for the years 2005–2010. The objectives of the UK PPRS are to secure the provision of safe and effective medicines for the National Health Service at reasonable prices, while promoting a strong pharmaceutical industry able to reinvest in R&D and encouraging the efficient and competitive development and supply of medicines. The scheme reflects the recognition by the Department of Health that the pharmaceutical industry makes a valuable contribution to the UK economy and also that continuous innovation is the key to competitive success in a research-based industry. For its part the industry recognizes that it is in the public interest that the prices of pharmaceutical products supplied under the National Health Service are fair and reasonable. The scheme achieves this balance by allowing companies to price new products at their discretion but determining reasonable limits to the overall profits (return on capital) to be made from their supply of medicines to the National Health Service. There is encouragement for R&D in the pricing arrangements through allowances for R&D costs (and smaller ones for costs of providing information and for marketing). One element of the 2005–2010 agreement was an overall cut in the prices of medicines covered by the PPRS of 7% from January 2005, but with stability thereafter for 5 years.

In some other countries (e.g., France) costs are controlled by negotiating the reimbursed price of each new product. In practice this process can delay new product introductions. In many countries, governments exercise control of the medicines bill simply by applying across-the-board price reductions, frequently unexpectedly. It is clear that such systems do not provide a stable investment environment for the industry, and greater effort needs to be made, particularly in continental Europe, to design systems that balance cost control with adequate 'reward for innovation.'

1.14.6.1.3 Governments and developing countries

Governments and agencies throughout the world recognize the close relationship between poverty and ill health. About 9 million children die each year in developing countries and about half of these deaths result from diarrhea, pneumonia, measles, malaria, HIV/AIDS, and malnutrition. Improving access to essential medicines and supporting health service infrastructure are key to tackling mortality in the developing world. Internationally agreed millennium development goals set in 2000 have a number of elements relating to disease and public health. The pharmaceutical companies are playing a major part in a wide range of programs that support the millennium development goal which is set out as 'in cooperation with pharmaceutical companies, provide access to affordable, essential drugs in developing countries.' The WHO has produced an essential drugs list and about 95% of those drugs are outside the period of patent protection so there are no problems about manufacturing licenses. For medicines covered by patents many companies are providing drugs at reduced prices or are donating them free of charge. GlaxoSmithKline has made a commitment to make its antiretroviral and antimalarial medicines available at not-for-profit prices to some 100 countries – and the products are carriage and insurance paid to their destination. That company, like many others, is involved in pilot projects in several African countries to assess the feasibility of offering a wide range of products at preferential prices. GlaxoSmithKline has developed R&D and voluntary license and technology transfer agreements with companies in South Africa, India, and with the Brazilian Government's Fiocruz. There are also many public–private partnerships involving companies and other agencies and governments. For example, as part of the partnership to eradicate onchocerciasis in Africa Merck is donating Mectizan (ivermectin). Pfizer and the Edna McConnell Clark Foundation have established the International Trachoma Initiative to eradicate blinding trachoma. Global alliances have been established to fight AIDS, TB, and malaria and there is also a global alliance program on vaccines and immunization.

1.14.6.2 The Increasing Importance of the Patient

Although patients are, of course, the ultimate consumers and beneficiaries of all the industry's efforts, it is only in recent years that they have had a prominent role in the process. This is for two main reasons. Firstly, until recently the majority of patients have been relatively passive recipients of their doctors' decisions, and have been poorly informed about the choices of treatment available. Secondly, patients have typically not paid for medicines out of their own pocket, so have not taken the kind of interest in treatment choices that they would in other areas of life.

While the second of these factors still applies in most markets, the involvement of patients in medical decision-making is increasing rapidly, as the 'baby boomer' generation seeks more information and a stronger role in the treatment of their parents, their children, and themselves. In the US, this has spawned the phenomenon of direct-to-consumer (DTC) marketing, via press and TV advertisements, which while controversial has increased patient awareness of treatment options and a willingness to approach doctors for specific medicines. While DTC marketing is unlikely to spread widely beyond the US, there is a broad recognition that the industry should provide much greater levels of information, involvement, and transparency to the patient. Public–private partnerships could be one route to supplying nonpromotional information to patients so that they can play their full part in decision-making. And public registers of clinical trials and their results should both dispel concerns about secrecy and bias and also enable those able to benefit from trials to identify those that are potentially relevant.

Patients are also gaining greater influence on government decisions about approval and reimbursement, typically via patient groups, some of whom are very effective lobbyists as well as valuable support organizations for their members.

1.14.7 Future Developments and Issues

The last 20 years have seen unprecedented success, scientific, clinical, and financial, from the global research-based pharmaceutical industry. With the explosion of knowledge in the life sciences and the growing numbers, greater health awareness and heightened health expectations of more wealthy aging consumers, the industry ought to be entering another golden age. But, as we have seen, the horizon has its dark clouds. Despite escalating investment, R&D productivity has been disappointing of late. Mounting criticism of drug prices in some markets, concern about industry spending and approaches in product promotion, and product safety issues lead to the threat of greater government intervention. Costs of launching new products into already crowded therapeutic areas continue to escalate. Even for 'first of kind' products, the time of marketplace exclusivity is shortening, as competitor drugs follow ever more closely behind. And, at the end of the product life cycle, generic competition is swift and brutal. Last, but by no means least, the huge complexity of managing globally fragmented mega-organizations is becoming ever more apparent.

The future success of the industry will depend on the answers it gives to some key questions now confronting it. In each case we give our own thoughts, but it is how the industry reacts that will set the direction:

1. Is there enough potential for innovation to sustain industry growth and valuation, and enable continued vigorous investment in innovation? Here we give an unqualified 'yes': the combination of need and scientific opportunity has never been greater (*see* 2.12 How and Why to Apply the Latest Technology).
2. Is there a new paradigm for drug discovery in the wings that takes full advantage of the explosion in biotech tools and information, and avoids miring projects in organizational complexity? This will be critical to turn the tide in R&D productivity. We see companies taking major new initiatives both in their own internal structuring of research, outsourcing activities to lower cost Asian centers and building productive long-term relationships with the smaller (biotech) companies that are inevitably nimbler and more innovative. So we are cautiously optimistic.
3. Will customized care or individualized medicine become the established standard? This issue cuts both ways. The ability to tailor therapy to the patient is obviously critical in both maximizing efficacy (remember the high proportion of drugs prescribed having little or no beneficial effect) and avoiding those side effects that can be predicted by patient biomarkers. However, genuinely customized care will be more costly: will prices for tailored therapies be allowed to rise? This is likely in (say) cancer, but doubtful elsewhere (*see* 1.05 Personalized Medicine).

4. Will healthcare socioeconomic forces permit continued high prices for innovative drugs? There is likely to be some leveling of the playing field here. US prices are unlikely to continue to rise as they have, while European and Japanese governments may begin to realize that greater reward for innovation is required to sustain their locally based industries (*see* 1.19 How Much is Enough or Too Little: Assessing Healthcare Demand in Developed Countries).
5. Can middle income and developing world markets be effectively served by mainstream companies? This is, of course, not just a matter of economics. Public opinion is demanding that pharmaceutical companies do not reserve their life-saving products for those who can pay premium prices. But the companies are rightly concerned when donated or at-cost products intended for developing world markets turn up back on their doorstep, and also when fast-growing middle-income countries, such as Brazil, China, or India, either expect bargain prices or set up in patent-breaking competition. This is certain to be one of the thorniest issues of the next decade (*see* 1.20 Health Demands in Developing Countries).
6. Can marketing expenditures be constrained or more selectively focused without loss of marketing momentum? There is likely to be continuing resistance to high-pressure marketing, whether targeted at prescribers or consumers, and reevaluation of the need for the very large sales forces built up in recent years. The industry has become accustomed to spending aggressively rather than scientifically: there is considerable scope to re-optimize.
7. How should the patient's role change? Will consumers (and their families) supplant the physician as the industry's most important customer? There are major changes underway worldwide, as 'baby boomer' patients demand a greater involvement in decision-making, have access to much more information on treatment options, and are more organized as political pressure groups. This must be one of the industry's highest priorities. Responsible engagement and information toward patients will not only shape demand for products, but will, still more importantly, create political support for an industry too often friendless when under attack (*see* 1.23 Ethical Issues and Challenges Facing the Pharmaceutical Industry).

References

1. http://www.diplomatie.gouv.fr/label_france/sciences/montagni/monta.html (accessed July 2006).
2. UNAIDS: 2004 Report on the Global HIV/AIDS Epidemic: 4th Global Report; ISBN 92 9173 3555. http://www.unaids.org/ (accessed July 2006).
3. Mackenbach, J. P. *J. Clin. Epidemiol.* **1996**, *49*, 1207.
4. Sykes, R. *New Medicines, The Practice of Medicine, and Public Policy*; Nuffield Trust/HMSO: London, UK, 2000.
5. Swann, J. P. *Academic Scientists and the Pharmaceutical Industry*; The Johns Hopkins University Press: Baltimore, London, 1988.
6. Seytre, B.; Shaffer, M. *A World without Polio*; Aventis Pasteur: Lyon, 2003.
7. Bartlett, S. *An A to Z of British Medicines Research*; Association of the British Pharmaceutical Industry: London, 2002.
8. European Federation of Pharmaceutical Industries and Associations (EFPIA). *The Pharmaceutical Industry in Figures*; EFPIA: Brussels, 2004
9. DiMasi, J. A.; Hansen, R. W.; Grabowski, H. G. *J. Health Econ.* **2003**, *22*, 151–185.
10. Aitken, M. Dynamics of the Global Pharmaceutical Market – 2004. In *Intelligence* 360: *Global Pharmaceutical Perspectives*; IMS: Fairfield, CT, USA, 2004
11. Eurostat: Statistics in Focus, 2003. http://epp.eurostat.cec.eu.int (accessed July 2006).
12. Kettler, H. E. *Updating the Cost of a New Chemical Entity*; Office of Health Economics: London, UK, 1999.
13. The Centre for Medicines Research International/SCRIP Pharmaceutical R&D Compendium 2000.

Biographies

Richard Barker is Director General of the Association of the British Pharmaceutical Industry, which represents companies researching, developing, and marketing medicines in the UK. In this capacity he is also a board member of the EFPIA (the European industry association) and Council member of the IFPMA (the International equivalent). His priorities include boosting the UK and Europe as a global leader in pharmaceutical innovation, strengthening the partnership between the industry and the health service, increasing patient access to new medicines in the UK and globally, and ensuring that the industry's external image reflects its major contribution to health and economic prosperity.

Dr Barker is also active in the biotechnology sector, as a board member of Adlyfe, a company developing early detection technology for diseases involving protein misfolding. He is also an advisory board member for Noxilizer, commercializing a novel approach to biological sterilization. Until recently he was Chairman of Molecular Staging, a biotechnology company whose market-leading products in whole genome amplification and protein biomarker development were acquired by Qiagen.

Prior to joining the ABPI, Dr Barker was president of New Medicine Partners, a firm focused on consulting and entrepreneurship in pharmaceuticals, biotechnology, molecular diagnostics, and biodefense. His past operating roles include CEO of iKnowMed, a clinical decision support and pharmaceutical services business in oncology, Chief Executive of Chiron Diagnostics, a global diagnostics company, and General Manager of IBM's Worldwide Healthcare Solutions business. He also led McKinsey's European pharmaceuticals and healthcare practice.

Dr Barker's academic research was in biological magnetic resonance at Oxford, Leeds and Munich.

Monica Anne Darnbrough, CBE, after studying psychology and physiology for a joint honors degree at Nottingham University, did a PhD on the effects of hormones on physical and behavioral development. She then joined the scientific civil service and worked in the Home Office (police scientific development branch), Department of Trade and Industry (DTI), and Foreign and Commonwealth Office (as Counselor for Science and Technology in Paris).

In the 1980s, she was seconded to the Cabinet Office and worked in the small team of Prime Minister Margaret Thatcher's Chief Scientific Adviser, and she also undertook a review of Government laboratories led by Sir Derek Rayner. From 1998 to 2005 Dr Darnbrough was responsible for developing policies to support the pharmaceutical and biotech sectors in the UK.

Comprehensive Medicinal Chemistry II
ISBN (set): 0-08-044513-6
ISBN (Volume 1) 0-08-044514-4; pp. 527–552

1.15 The Role of the Venture Capitalist

D J Bower, University of Dundee, Dundee, UK

1.15.1 Introduction

1.15.1.1 Young Firms and Large Projects

The emergence in the 1970s of the new techniques of DNA recombination[1] and hybridoma technology[2] whose industrial potential was perceived to be enormous, triggered the 'biotechnology revolution.'[3,4] Applications in healthcare, agriculture, waste disposal, and other industries were forecast to have markets worth many billions of dollars.[5,6] The early evolution of the industry took place in the San Francisco Bay Area of California.[3,4] However, although the industry has still tended to concentrate strongly in a relatively small number of geographic locations where scientists are actively contributing to the basic science,[4,7,8] interest and efforts to stimulate biotechnology innovation have been much more widespread.

The apparent potential of these new technologies captured the imagination of governments seeking to boost their economies, and of investors seeking rich returns on their investment. These high expectations encouraged policy-makers to invest increasing amounts in the science, and also to introduce many and diverse support measures to facilitate exploitation of the technology. These have included a range of direct support mechanisms. They have also included incentives to private sector involvement, such as fiscal incentives to invest in the commercialization of the stream of new life science technologies that has continued to flow from university and research institutes.

The US led the way in devising models for the commercialization of these technologies. The UK, which contributed substantially to the science, quickly followed the American lead, but with less success. Other countries followed and now there are few that do not have active government programs of support for the biotechnology industry.

The biotechnology revolution generated a technological discontinuity in the industries on which it impacted, of which the pharmaceutical industry has been the most affected thus far. It introduced major technological changes in the way drugs are discovered, developed, and manufactured relative to the traditional, chemical-based pharmaceutical framework.[9] Large firms that had the internal resources to undertake expensive R&D projects initially lacked the technical capabilities in these novel areas. The scientific capabilities were in the universities and research institutes. While there was a widely shared consensus among scientists, government, and technology investors that the new technologies could be the basis of new products and processes for many industries, the large firms that had traditionally developed products in these industries did not move immediately to acquire the technologies.

In the US, the perceived opportunity attracted the interest and involvement of independent investors who had experience of funding innovative firms. These independent providers of high-risk finance, known as venture capitalists (VCs), had become an important source of finance for young high-technology firms in the US by the 1970s. There was a strong concentration of VCs in Silicon Valley in California, which had been a key ingredient in the growth of the electronics industry.[3,10] VCs were willing to find ways of backing the academic scientists who had the ideas and the skills to invent biotechnology applications.[11–15] They provided venture funding, which underpinned much of the early stages of exploitation of the new technologies in biomedical, agricultural, and other industrial applications. It became possible to undertake major pharmaceutical R&D development projects in startup companies. As these new firms progressed their development projects, they were able to attract large pharmaceutical firms into strategic alliances with them. This provided the new entrants with additional financial support and access to the complementary skills of their partners.[16] At the same time it allowed the large firms to adapt to the new technologies, and eventually to build in-house competence.[17–19]

Taking candidate drugs and therapeutic molecules through all the phases of the highly regulated research and development process to the final stages of clinical approval and marketing is a slow and risky process. The associated costs are also high and escalating. Estimates of costs vary. Di Masi *et al.*'s[20] estimate for 68 new drugs developed by 10 large US pharmaceutical companies in 2001 was US$803 million per drug, up from US$231 million in 1990.[21] They took nearly 12 years on average to develop. Large pharmaceutical companies are set up to deal with such projects. They have large, well-funded R&D departments, extensive legal and regulatory expertise, and they have the marketing capabilities to generate revenues when the drugs are approved, which provides the financial underpinning for the enterprise.

Academic scientists, who had the technological expertise, lacked the organizational support that the pharmaceutical companies provided for their developments. However, the availability of VC cash to fund high-risk therapeutic developments outside of the large pharmaceutical company environment enabled entrepreneurial scientists to begin to initiate developments of novel candidate therapeutics in new companies set up for this purpose.[18,22–27]

1.15.1.2 Disaggregation of the Pharmaceutical Industry

In the years that followed, the competitive pressures that this new possibility created, compounded with the impact of other environmental factors, led to significant changes in the way that pharmaceutical companies operated. Previously, pharmaceutical companies had had a high degree of vertically integrated discovery, development, manufacturing, and marketing capability. As small firms founded by entrepreneurial scientists became increasingly important for discovery, large firms became willing to work with them to gain access to the novel candidate drugs. From these beginnings the 'biotechnology revolution' led to increasing disaggregation and outsourcing of functions by the large firms. This resulted in a 'distributed innovation system' in which it has become the norm for academic scientists, small technology spinouts, a variety of other specialist contract research organizations, and pharmaceutical companies to collaborate closely in bringing new therapeutic products to market.[28] There is now an active market for products and services for the development stages where formerly these activities took place inside the large firms of the industry. Specialist firms can now provide a wide range of services and tools for drug discovery and other functions, as well as candidate drugs.

Hence, there are now many opportunities for innovative small firms to make a business of addressing different parts of the pharmaceutical innovation process. In other industries as well, current trends toward outsourcing noncore functions have increased the range of opportunities for entrepreneurial, high-technology firms. These have included some opportunities for firms exploiting life science technologies.

1.15.1.3 From Ideas to Innovation

This does not imply that all new knowledge generated in the basic science laboratory should be expected to form the basis for the formation of a new firm.

It should be remembered that basic science laboratories are not primarily concerned with the creation of companies and products. For the most part, knowledge generated by life science research makes its contribution to society through increasing our understanding of living organisms. This knowledge underpins evidence-based policy-making and decision-making in healthcare, agriculture, environmental matters, etc. It rarely lends itself to industrial exploitation although through its informative function it can lead to improvements in practice in healthcare and other contexts.

From time to time, however, research generates new knowledge or novel research tools that are seen to have some potential for commercial application. If this potential is to be explored, a number of questions are raised:

1. Can the technology be sold or licensed, as it stands, to an external organization that has the resources to develop it?
2. If developed further, can the technology's commercial potential become sufficiently convincing that it can be sold or licensed to a third party?
3. Is the technology capable of forming the basis of an independent business that could develop and eventually sell products or services derived from the technology?

If the answer to the first question is positive, and the relevant stakeholders choose this route, then the problem of resourcing the development is transferred to the licensee. We will not consider this situation further here. If the answer to either question 2 or 3 is positive, this immediately raises the next question: where will the resources for the early development come from? The rest of this chapter considers the providers of financial resources, and the circumstances under which they are willing to invest in novel life science R&D projects. It discusses the key issues the founders need to address to harness the resources required to carry the project through the early stages, and to position it for success.

1.15.2 Resource Providers

Venture capital of various types has provided a significant share of the financial resource for the commercialization of life science technologies outside of large firms.

Large pharmaceutical firms have also been major providers of financial and other resource through partnership arrangements. As developments proceeded, firms have often been able to list on public stock markets, which has allowed them to access finance on a larger scale and from a wider range of investors.

The value of venture capital investment has not been confined to the cash investment. Venture capitalists have usually provided a range of nonfinancial support. They have assisted in finding suitably experienced management, they have helped guide the business through involvement as nonexecutive board members, their extensive business and industry networks have helped the firm to progress, and they have usually played an important role in securing follow-on finance or exit opportunities.

Large corporate partners have made significant contributions through their complementary technical and business expertise, and have often provided an exit through acquisition of the smaller firm.

1.15.2.1 What is Venture Capital?

1.15.2.1.1 US definition

The US definition of venture capital differs somewhat from that of the UK and much of Europe. In the US, venture capitalists are the providers of early stage and development capital for young, entrepreneurial businesses, whereas in the UK venture capital is synonymous with private equity, which is equity financing of unquoted companies ranging from seed and startup stage to large management buyouts.[29]

An additional distinction may be drawn between formal and informal venture capital. Informal providers of venture capital, known as 'business angels,' are wealthy private individuals who invest their own money directly in early stage firms. Providers of formal venture capital are intermediaries, which draw their funds from institutional investors such as pension funds, banks, and insurance companies, or from large industrial companies.[29] They usually provide larger amounts of investment than angels.

1.15.2.1.2 Biotechnology venture funding from the early days

Both formal and informal venture capital have played an important role in the creation and growth of many companies in the life science industries. However, the situation has changed from the early days of the biotechnology revolution. High technology startup companies offer high risks to the investor, but high rewards if they are successful. However, the risk/reward ratio must be seen to be favorable.[10,30,31] When investors believe that there is a good chance of success, and they think that they understand the risks, they are likely to invest. If they think success is not very likely and they feel that there are unknown or unquantifiable risks, they are reluctant to invest.

In 1985 Daly was able to write "Usually a biotechnology company is founded by venture capital in a number of rounds of financing."[11] At that time this was indeed the normal route, and there were a substantial number of companies in the US that had been founded in the 1970s and early 1980s following this model.[11,14,32] Daly's statement was made at a time when young companies such as Genentech,[78] Amgen,[79] and Biogen, now BiogenIdec,[80] had been financed through their early stages in this way and were showing considerable promise. These companies are now large biopharmaceutical companies with lucrative products, which they market worldwide. In the same era, however, many other companies in the US, the UK, and elsewhere, which were founded with generous venture capital funding, have now disappeared. Some were acquired, often for less than the investment that had been made in them. Others ran out of money and closed the doors. A very few have survived and become profitable, and even fewer have grown into large companies. This high failure rate is one of the reasons that venture capitalists today are only rarely willing to invest up front in early stage life science companies that will require large amounts of capital investment before there are any prospects of revenues.

The very long period required to get products to market in these industries also discouraged investors, especially when growing regulatory barriers increased the unpredictability and costs of the process. Being tied in for a lengthy period with uncertainty about how long it would take to get product approval was generally unattractive to investors.

In the early days of the biotechnology industry a sense of the vast commercial potential of the technologies was shared equally by the venture capital providers and the scientists, along with a degree of vagueness about the markets and industries they would address and precisely how. Kenney[12] has discussed the importance of venture capitalists' prior, successful experience in the electronics industry in Silicon Valley. In expectation of similar returns they were willing to provide the startup finance and contribute to the discussions about the novel strategies the new biotechnology companies would have to follow.

The pharmaceutical industry turned out to be very different from the electronics industry. The conventional drug development process had become a highly regulated process, requiring long years of testing for safety and some degree of efficacy. Developing novel therapeutics added other challenges as well, which were initially underestimated or overlooked. At the founding of Genentech in 1976, the founders and early investors believed there were only two new problems to be solved, both of which were technical.[32] The two challenges envisaged were: (1) to improve their gene isolation techniques, which would give them many potential products; and (2) to demonstrate that they could mass produce these novel materials. The task of building a pharmaceutical company was not seen as a major barrier. Great pharmaceutical companies had been built up before, mainly on the back of one or two great products. Why should they not be built up now? It turned out to be much more difficult than expected to develop biomanufacturing techniques. Genentech managed to deal with this, but this was one of the first indications that bringing novel therapeutics to market entailed more problems than had been predicted.

The business model established in its early stages by Genentech was to attract external resources of finance, technical, and marketing capability through a combination of venture finance and strategic alliances with pharmaceutical companies to complement Genentech's unique technical ability.[33] Further finance was raised through listing on public markets.[11,32] The plan as presented to investors from the start was to become a fully integrated pharmaceutical company. Apart from a few early hitches the model worked. Genentech's success appeared to confirm its validity. However, the first products, human growth hormone and human insulin, were not truly novel. These molecules had been used therapeutically for years, extracted from human or animal tissues. It was only the process of production, using cloned genes, and expressing them in animal cells that was novel. This is not to underestimate the considerable technical and business feats that Genentech had accomplished. However, the initial perception was that the new technologies would generate many new drugs. In fact, the number of successful and truly novel biopharmaceutical drug candidates has been low.

Many other companies were able to raise successive rounds of finance in this period, on the basis of business plans similar to Genentech's. These included companies targeting different industries, such as agrichemicals. Unfortunately, very few of them managed to imitate Genentech's technical and business success, and investors became less enthusiastic about the biotechnology sector. At the time of writing (2005), very few life science ventures are able to attract formal venture capital at the early stages of their development for unproven products.

1.15.2.2 Biotechnology Investment Trends

Fashions in investment have changed throughout the history of the biotechnology revolution, and they will no doubt continue to change. Any venture seeking investment is well advised to look carefully at the trends in the marketplace and consider the implications for the structuring of the venture, the business plan, and the likelihood of interest from the different categories of investors. In addition, absolute availability of investment varies significantly from year to year. Timing is critically important in seeking finance at any stage of development – ideally, sufficient investment should be taken on when it is available to last the venture through the lean period that inevitably follows, to the next period of relative abundance.

The profile of the companies receiving funding has changed considerably over the years. In the 1990s platform companies attracted most early stage finance. That is, companies with a generic technology that might form the basis of a family of drugs, but was still at an early, untested stage. After 2000, wary investors avoided early stage discovery projects and moved toward startups formed around later stage assets already in clinical trials, or further rounds of finance for companies with products close to market.[34] Profiles of US companies going public in 2004 showed that most had a phase 2 or later lead drug, indicating that the public markets were also unwilling to invest in early stage products.

In 2004, companies developing new drugs were back in fashion, receiving a near record US$4.09 billion out of the US$6.30 billion that venture capitalists invested in healthcare during the year. Seed/first round funding was about US$0.8 billion, up on 2003 but not as high as 2000 when about US$1.2 billion went into first round drug companies (Klausner[34] quotes VentureSource).

Considering global trends in biotechnology industry financing, venture capital funding totals have stayed fairly constant over the last year, but Europe has been claiming a progressively larger share, about a third of the total in Q1 of 2005 (the first quarter of 2005). Partnering has consistently accounted for more of the finance than venture capital globally over the last year. There was an upturn of initial public offering (IPO) money raised in Q1 of 2005, with seven companies in North America and three in Europe going public.[35]

1.15.3 Risk Factors in Biotechnology Developments

1.15.3.1 Technical Risk

As increasing numbers of companies were set up and developments proceeded, it became evident that there were many more risks associated with biotechnology companies than had been foreseen. The commonest problem was that the technology simply failed to live up to expectations, i.e., it did not work. It became increasingly evident that novel technologies that performed well in the laboratory rarely performed as well in the more complex and less controllable environments in which they were applied. This was discovered through empirical experience of testing the new applications in the clinic or in the field, depending on the industry. The many failures are widely documented and discussed in the industry and financial press of the 1990s. One well-analyzed example was the monoclonal antibody therapeutic Centoxin, developed by the US company Centocor, which was failed by the US regulator, the Federal Drug Authority, in 1993.[10] Centoxin had been widely hyped but failed to show efficacy in clinical trials.

1.15.3.2 Issues About Ethics/Values

It became apparent that the technical risks were considerable. However, they were not the only cause for concern that emerged. In addition, it became clear that bioprocesses and products raised ethical and other values-related issues that potentially affected their public acceptability. There were worries that scientists in changing the DNA of organisms were playing God, possibly creating monsters. Both the processes by which the changes were made and the products themselves led to feelings of unease in some quarters.[36–38] This affected their acceptability and hence became another source of risk. There could only be an adequate return on the high levels of investment required for these developments from products that were approved and widely accepted in wealthy societies.

1.15.3.2.1 The case of genetically modified foods

This issue of public acceptability became particularly problematic in the case of GM foods, where popular concerns created major hurdles, especially in Europe.[39] Polls had shown that public doubts about GM foods were widespread among the European public well before field trials were carried out.[40] When this stage was reached, action groups destroyed fields of trial crops.[41] European Union (EU) countries threatened to close their markets to GM foods.[42] A controversial study, which concluded that GM foods threatened the survival of the Monarch butterfly, an American icon, reawakened American concerns, which had been quiet since the 1980s.[43] One of the biggest US agri-trading businesses asked growers to segregate GM and non-GM crops, which led to a great increase in costs.[42] This led to loss of investor interest in this area of application. Small firms failed, for example, Axis Genetics, which used plants to produce human vaccine components, failed in September 1999 to raise the £10 million that was urgently required, and put itself up for sale.[43] Large firms in the sector experienced falls in their share prices, and in some cases divested the parts of their business associated with GM foods. The company Syngenta[81] came into existence in this way, divested by Swiss multinational Roche.

In spite of these doubts, GM crops are being planted in increasing areas each year. In 2003 about 167 million acres in 18 countries were planted with transgenic crops.[44] While 63% of GM crops in 2003 were grown in the US[44] much of the new development is in developing countries, where publicly funded organizations are playing an important role in creating new transgenics. A recent study of public research in 15 developing countries showed that GM crops derived from 201 genetic transformations of 45 different crops were under development.[45] However, the lucrative market of Europe has remained closed, although there are continuing efforts in the UK to change this.[46] It continues to be difficult to interest investors in this area.

The discussion of GM foods in the West has been dominated by a 'rhetoric of fear,'[47,48] which has demonized the technology, the products, and the companies. The tone of some of the rhetoric that has been used by activist groups and subsequently adopted by the media has shaped the trajectory of the ensuing discussion – Frankenfoods and related rhetoric has located the discussion about GM foods in a disturbing, unacceptable context, particularly for Europeans.[38,39,49]

The pharmaceutical industry and biotechnology companies targeting the healthcare markets with recombinant product have not (yet) been the subject of such violent public aversion. Biotechnology is producing acceptable products and corporate profits for a number of pharmaceutical and biotechnology companies with fairly strong market capitalizations in the healthcare industry. This indicates that the companies and products are perceived differently from the Ag-bio companies, in spite of using the same technologies. The implication is that it is not just the technology, but also contextual factors, including who controls the technology, and even the language which is first used to describe it in a lay context, that together engender public perceptions. Apparently technology is viewed differently when it is harnessed to a therapeutic application than when it changes our food crops. This negative public perception of GM foods is yet another risk factor that now impacts on investor interest.

1.15.3.2.2 The case of stem cell research

Medical applications are not automatically free of this type of risk. The acceptance of recombinant products when produced for therapeutic purposes should not be taken to imply that biomedical companies will all be automatically viewed favorably by the public. Nor can one assume that all societies will react in the same way. Different societies have responded differently. Some areas of biomedical innovation have aroused public controversy, which has placed barriers against development in some countries. For example, embryonic stem (ES) cell research has provoked much debate in Western societies, due to the destruction of human embryos in the process of making the ES cell lines.[50]

The application of stem cell technologies in regenerative medicine is currently believed to hold great promise in many disease areas, including cancer, diabetes, and heart disease.[51] Following pressures from Christian fundamentalist groups US President George Bush has severely curtailed the use of Federal funds for research in this area in spite of a high level of support for stem cell research. Following this, a number of states led by California, fearing that their cherished biotechnology industries would be damaged, have committed considerable amounts of public money to stem cell research.[50] Other countries have taken up a variety of positions. Far Eastern countries such as Japan and

South Korea, which have strong research programs in this area, have no cultural concerns with ES cell research. On the other hand, there has been considerable debate within the EU with countries such as Italy strongly opposed and the UK in favor of permitting ES cell research.

The favorable position in the UK was not reached easily. The debate about embryo research could have been lost. Indeed considerable efforts were required to win over many doubtful parties.[52] Mulkay's[47,48] analysis of the parliamentary embryo research debates underlines that these highly controversial and value-laden issues were recognized and effectively addressed by the UK Government. The Human Fertilization and Embryology Act (1990), which is administered by the Human Fertilization and Embryology Authority (HFEA), was passed after considerable debate. It addressed the ethical issues and provided a regulatory process designed to minimize ethical concerns and provide a route for medical research and treatment that would be generally acceptable. As Mulkay[47,48] has demonstrated, the debates that culminated in the Act created a favorable and acceptable discourse around the new capability, which positioned it as beneficial and health improving. Although the UK Department of Health's Report on Stem Cell Research[51] acknowledged that there was still a significant body of opinion that regarded the HFE Act as unethical, it was reached through a process that appeared to have achieved substantial public acceptance for the framework it provided, according to the UK Parliament's Select Committee on Science and Technology.[53]

1.15.3.2.3 Investors avoid 'ethical risk'

The cases of GM foods and stem cell research underline the issue of public acceptance, which is essential for the successful introduction of an innovation. Where there are doubts about public acceptance, investors fear to tread. Biotechnology developments are technically risky enough without adding a further dimension of risk due to negative public perceptions. Consequently, the question of ethics and values in relation to a particular development are factors that impact on the viability of a commercialization plan. Ethical issues are discussed further in Chapters 1.22 and 1.23.

1.15.3.3 The Problem of Long Development Times

The long development times for drugs have also created another problem for investors. In the prolonged periods when there were no clear measures of the progress by which to judge individual investee companies, investors used news events in the industry as a whole as a proxy. In other words, when company A's product X was failed in Phase II by the US Federal Drug Authority, confidence in the whole industry fell and investors withdrew from biotechnology. When Company's B's product Y did well in Phase III, confidence rose, all companies in the sector were viewed more favorably, and money flowed into the industry. A much-reported example of this was when Centoxin was failed by the Federal Drug Authority, after which the investment in the whole sector was reduced for some time.[10]

During the period when there were many new companies developing therapeutics, but very few which had actually made it to the market, this was a major factor in the availability of investment in biotechnology. Hence, in the 1990s investor interest fluctuated, with the flow of money increasing on good news stories and declining when products failed in clinical trials. Nevertheless, investment levels held up fairly well over the decade as a whole. During that period there was a wider confidence factor affecting all financial markets. The generally buoyant mood of the markets in the 1990s sustained a substantial level of investment, especially in high-technology stocks. However, the financial markets fell and venture capital investment in high technology collapsed in the wake of the dot.com bubble of 2000 and has not fully recovered.[54]

1.15.3.4 Failure of Expectations

Another factor today that affects investor interest is the failure of the biotechnology industry to live up to the wildly optimistic predictions of the 1980s.[55] This has dampened the enthusiasm of venture capitalists for the sector. Although there have been from the start a number of commentators who questioned the hype that surrounded biotechnology, they have become more visible recently.[55–57]

With a growing awareness of these risks investors have become increasingly selective. They are in general reluctant to invest in ventures until they appear to be near to bringing lucrative products to market. It is particularly difficult to assess risk at the early stages and this is a major reason for the reluctance of formal VCs to invest in life science ventures until they have reached a stage at which risks can be more reliably evaluated.

Today, it is difficult to find recently formed companies that have been able to follow Genentech's startup model. Venture capital is still playing a role in life science company development, but this has changed in line with changing perceptions of the associated risks. Professional VCs only rarely invest in the early stages now, preferring to invest later when there are products on or near the market. However, there are other sources of investment that are allowing therapeutic developments to take place outside the laboratories of the large firms in the sector. In the sections that follow, we consider the ways in which the development of therapeutics invented outside of established pharmaceutical companies are being resourced. We discuss the different types of resource providers and how their requirements fit with the needs of those who are directly involved in the development.

1.15.4 Resourcing the Early Innovation Process

The resources required, and the potential providers of the resources change over the course of the lengthy development process. If the option of licensing to an established pharmaceutical company is available and acceptable, then the inventors and the institution need not concern themselves with this. Otherwise it is important to be aware that taking an idea for a new therapeutic product and progressing it toward the marketplace requires expenditure on a range of technical and business skills, access to laboratories, supplies, and other costs. If the idea has been conceived in an academic setting it is likely that most of these skills will have to be sought elsewhere. The money to pay for them must be raised, usually from a combination of sources. Public funds, financial investors, and stock markets are all possible contributors to the early stages of an independent development. It may also be possible to generate some early commercial revenues through consulting, contract research, or other activities, which can be invested in the project. If this is possible it relieves some of the pressure to find external interest. It is easier to attract investment into a project if enough data has been generated to give some comfort that the technology will work as required in the application context.

However, in order to attract external investment for the commercial development from any source a number of steps must be taken. Each of the resource providers listed above has requirements which must be fulfilled. These differ in detail but they have some common elements. First, the intellectual property rights (IPR) to the novel technology must be protected through international patenting. In addition, a credible business plan must be prepared.

1.15.4.1 Intellectual Property Rights

The requirement for patenting is to give those who own the patented technology the legal right to control who uses the technology, and hence to prevent others who have not contributed to the investment from capturing the returns if the product is eventually marketed successfully. Financial investors and industrial organizations may be willing to take the risk of losing their money if the project fails, but only if they believe they will get their money back, with interest, if it succeeds. Hence they wish to be sure that they will have sufficient control over the IPR which underpins the project. Government funds also require IP protection, since they recognize that the necessary follow-on investment from private sources will not be forthcoming without it. IP law differs in important respects between different jurisdictions. For most purposes it is important to have protection at least in the countries of the EU, in the USA, Canada, and Japan. China and India are becoming important producers and markets for pharmaceuticals now, but there are questions about the possibility of enforcing IPR there. IPR is discussed in detail in Chapter 1.26. Decisions about patenting should be made with advice from experienced patent agents who have pharmaceutical industry specialists.

1.15.4.2 The Business Plan

A business plan is usually required by investors and even for publicly funded 'enterprise' grants or loans, since their objective is to support developments that are likely to become attractive to private investors when some of the uncertainties are removed.

1.15.4.2.1 Content

A business plan is essentially a narrative that explains what the technology does, how it can be applied in a specific context, who the customers would be for this application, how many would want it, why they would prefer it to the alternatives, and how much they would pay. As Gera[58] has pointed out, writing business plans for the development of novel, untested technologies poses special problems, since much of the information that is expected in a conventional business plan is often unavailable. For example, at very early stages the amount of work to establish proof of concept, i.e., that the technology is capable of the proposed application, is often impossible to estimate accurately. Hence, the plan must allow considerable flexibility. Academic technologies are often generic, in other words they have potential in a wide range of applications, often in more than one industry. At the outset, it may be difficult to identify the most promising application. It is normally impractical to seek to exploit all the possible applications, and so a decision must be made about which to pursue. Sometimes circumstances lead to an early change of opinion about this, and it is advisable to retain an option to explore other paths provided that this does not lead to a diffusion of resources. Nevertheless, potential investors, public or private, need to have some idea of how the original idea could lead to marketable products, and most limit their interest in investing to specific industries and types of products. While it is worth mentioning the alternatives, the business plan must address one specific application that is within the preferred investment area of the investor.

Drawing on all the information that is available, the business plan should also set out a route for developing the technology from its current state into a marketable product, approved by any regulators. This holds even if the plan contains the expectation that the technology will be sold on before there are marketable products. However, unless some organization can ultimately take it to the market, no one will want to fund the early stages. There will be no buyers of the technology if they do not see a way to overcome the barriers on the way to the market. Technical issues are only one of the problems to be overcome. Estimating accurately the time and cost to market of novel technologies for which there is no established regulatory route is not possible, especially if there are unresolved ethical issues. Investors will want to see a credible timeline for the development.

The business plan should be written with consideration for the roles and requirements of providers of equity finance and strategic partners. The plan should clearly present the estimated costs, the development schedule including critical points at which progress could be demonstrated, and the means by which investors would be able to reap their returns (the exit route). Investors in high-risk ventures usually expect to be able to exit from their investment within a few years.

It is important to demonstrate the capability and willingness of the key scientists to carry the project through to the point where their unique know-how and expertise is no longer required. Their professional standing and track record in demanding research projects should lend credibility to the plan.

1.15.4.2.2 The development vehicle

In most business plans the necessary rights to the technology are vested in a company, usually a vehicle set up specifically to take this project forward, and into which the investment also goes. In exchange for their money investors take a share in the ownership (known as the equity) of the company. From a practical perspective it is much easier to link the value created in the development if the IPR and the investment are packaged within an appropriate, separate legal entity. Equity holders then have a recognized legal route to claim, or to sell, their share of the value that is created. This claim can only be effectively exercised when the value is realized in some way.

1.15.4.2.3 Equity versus debt

Borrowing money to fund the development is not usually possible, since lenders like to have some security for a loan, either tradeable capital assets or a reliable source of revenue. Where there is no expectation of significant revenues for some years, and considerable investment will be required, the ability to even pay interest on a loan is absent. Debt is in any case an unsuitable way to finance such a project. Equity finance, where the investors share the risk of failure with the founders, and share in the rewards if the project is successful, is appropriate for risky projects of this type. If investors only get a return if the project is successful then they are motivated to give support to that end.

Investors may prefer complex financial instruments. The European Venture Capital Association publishes a helpful guide to this very specialist area.[59]

1.15.4.2.4 Project finance

Project finance is another possibility, most probably coming from a corporate strategic partner. This may not be available at the early stages but at later stages is increasingly desirable. This is discussed further below in the sections on strategic partners and corporate partnerships.

1.15.4.2.5 The exit

Investors (and founders too, very often) usually hope to realize the returns on their investment within a few years. For this to be achieved, the value that has been created in the company must be realized. This is usually achieved through selling the company that has the IPR to a large organization (trade sale) or through a listing on a public stock market (initial public offering (IPO)). Delafond[60] has written a helpful discussion of the exit process for lay readers.

1.15.4.2.6 Credibility of the plan

It is not difficult to write a business plan that promises great returns to investors. The difficulty lies in persuading any fairly sophisticated investor that its promises are likely to be realized. Potential investors who are easy to convince are not necessarily desirable. In most industrialized countries there are strict rules that prevent the solicitation of investment from private individuals unless they are wealthy and/or sophisticated investors. Consequently, it is advisable to seek investment from sources that are, on the one hand, knowledgeable and very exacting in their judgment of the general credibility of the business plan, while on the other, able to appreciate the need for flexibility in implementing it. The credibility of the plan will be interrogated by interested potential investors through their processes of due diligence before they commit to an investment.

1.15.4.3 Stakeholder Requirements

One of the factors that gives credibility to a business plan is attention to the requirements of the various stakeholders whose cooperation will be needed to drive forward the development.[61] These include:

- the institution where the technology was invented;
- the inventing scientists;
- the management team;
- the investors; and
- strategic partners.

The main requirements of these stakeholders are briefly summarized below.

1.15.4.3.1 The institution

The institution may contribute in a number of ways. It frequently owns the intellectual property (IP), and is responsible for the initial patent applications and the licensing or assignation of the technology to the company. It may take the lead in setting up the company, and is in any case likely to take a share in the founding equity. In the early stages, it is likely that the company will need to rent facilities in the institution, and it may well need help with R&D from employees of the institution.

In return for these contributions the institution will usually expect to have some equity in the company, unless it has a specific policy of not holding equity. If it licenses the technology to the company, it will expect a return in the form of royalties if and when the product is eventually marketed. The amount of equity expected, and the terms of any licenses vary widely between institutions and between deals. If the institution is too demanding it may fail to attract the interest of investors.

1.15.4.3.2 The inventing scientists

Technologies invented in the 'science base' are usually untested and little is known about their performance in the contexts in which they are to be applied.[62] Scientists in the team that invented them have the greatest familiarity with them and are best placed to carry them through the early stages until their characteristics are tested and their behavior is well codified. Consequently, their continuing involvement in the development, either through contract research or as employees of the company, is usually important for the success of the project. If their contribution is to be substantial, particularly if they are going into the company and taking on a high degree of career risk, their motivation should be aligned with those of the investors. This will be the case if their rewards are tied to the success of the project. This can be achieved if they have an appropriate share in the equity and/or future royalty streams.

1.15.4.3.3 The management team

Driving the development forward requires a combination of technical and business skills. Some of these will eventually be harnessed through employing staff with suitable experience within the company as it develops. Some can be provided by external sources of expertise. The nonexecutive members of the Board and the professional advisers can provide important support and advice to management, as well as valuable contacts, if they have relevant industry or professional backgrounds. Investors may be able to help recruit advisers and Board members with the skills and networks to help progress the business plan. The right to appoint one or more Board members is often a condition of investment.

1.15.4.3.4 The investors

Investors will expect a share of the equity that will be sufficient to meet their requirements for return on investment in a risky project if it is successful. They will also expect to see Management incentivized by the opportunity to hold an appropriate equity stake. They will want the company generally to be set up to support value creation. They will not normally wish the institution to play an active role in the business aspects of the company's development, since academic organizations can rarely offer relevant business expertise. They may wish the institution to provide scientific services and facilities, subject to mutual agreement. Whatever the level of involvement, investors will want the institution's role to be aligned with the roles of the other stakeholders in such a way that the project outcomes are facilitated.

Potential investors, in the course of carrying out their due diligence, will check on every aspect of the company's relationship with the Institution. For example, they will want to see strong patents and a licensing or assignation of IP that is clear and appropriately priced. If facilities and research support are needed the agreements should be put in place.

1.15.4.3.5 Strategic partners

Strategic alliances have played an important part in the successful development of life science products and processes. They have provided market-related assets such as marketing and distribution capacity, complementary technical skills, capital, and the value of an established reputation.[19,63–65] Except where the technology is sold at a fairly early stage, it will probably be important to forge one or more strategic alliances with partners if the project is to progress satisfactorily. The business plan should address this. Strategic partners will want some rights in the ultimate product in exchange for their contribution, but investors and founders will also want a return. The plan must offer a route that can give satisfaction to all.

1.15.4.3.6 The structure of the investment agreement

The terms of the investment agreement should formally clarify and support the roles and relationships of the stakeholders, and provide rules governing the process of their interaction as the development progresses. While legal documents cannot make relationships work, they can help set the expectations, and they can also provide remedies for dealing with unexpected and adverse situations, some of which will inevitably arise over time. The documentation that accompanies an investment is difficult to read for a nonlawyer. However, it is important for all parties to the agreement to be quite clear about what they are signing up to. Shori[66] has written a helpful guide for the layman to the legal structure of a private equity investment.

1.15.4.4 Categories of Investors

There are several categories of investors in growing life science companies. However, there is some overlap between them. The characteristics described in the following sections are generalizations. The precise characteristics and requirements of any potential investor can be checked through reading any information they publish. This will usually include lists of recent investments, which are a helpful indicator of the type of venture they like to invest in.

1.15.4.4.1 Preincorporation finance

Some publicly funded enterprise schemes, and some private funds, are willing to provide some early development finance before incorporation, in exchange for an agreement about how the resulting IP will be commercialized. Funds of this type are usually labeled 'proof of concept' or some such title. They are designed to provide data of the sort that an investor would require before making an early stage investment. The sums involved are usually quite small. As the development progresses beyond this stage a company or similar vehicle is formed. The objective of this is to ring fence the relevant intellectual property and the resources intended for its development within a suitable legal vehicle whose ownership is clearly specified and which is able to operate in such a way as to progress the development. It must also be a vehicle whose ownership can be transferred, provided there is a willing buyer or buyers, in order to allow the initial owners to harvest the value created by the development. Setting up this development vehicle is a critical stage, and it is important that it is set up in such a way as to facilitate the development. This means that the framework of legal agreements relating to the vehicle, usually a company, which identifies the founding ownership of the equity, the rights and obligations of the owners, assets of the company including intellectual property rights, and other relevant matters, is designed to encourage the development process and to permit additional rounds of investors to come in on acceptable terms where this is required.

1.15.4.4.2 Early stage finance

Providers of finance for early stage life science companies fall into a number of categories. The founding institution may contribute some of the earliest funding. If the academics who invented the technology plan to play a continuing role in the development they usually have an equity stake in the company. If they have savings they may themselves invest in the spinout. Academics often do make an investment in their own company, but the amounts available to them are usually insufficient for the needs of the company. There is commonly a need to seek external investment at an early stage. There are a number of categories of external investors who will invest at the early stages. All categories of investors vary considerably in their experience and expertise in dealing with academic spinouts. They also differ in their ability to bring complementary business and industrial skills and contacts to support the young venture, which is usually deficient in these.

1.15.4.4.2.1 Economic development funds

In the UK there are a number of small public funds linked to the universities and some public/private funds that are not primarily concerned with financial returns. These funds are mainly concerned with maximizing the number of viable starts. They have usually been set up to bridge a perceived funding gap and are really instruments of economic development policy. Some invest in R&D at both pre- and postincorporation stages. Investors in this category usually have a close relationship with technology transfer organizations (TTOs) in their area and provide modest amounts of seed funding in standard deals to ventures that fit their criteria. They do not usually provide follow-on funding. Depending on the individuals and organizations running them, they provide a variable amount of support, training, and advice to the academic entrepreneur.

1.15.4.4.2.2 Financial investors

On the other hand, there are investors who are looking for a substantial financial return and will only invest in ventures with high growth potential. This latter group may in some cases be able to draw down some public matching funds, but only where it does not compromise the financial objectives. This category includes business angels and VCs.

1.15.4.4.2.2.1 Business angels Business angels have been a major source of startup capital for biotech companies in the US and they are known to be active in this field in the UK as well.[67,68] Sohl[68] estimates that in the first half of 2004 angels put US$40–60 million into US biotech startups compared with US$10–20 million by VCs, although their appetite for biotech investment fell in the latter part of 2004.[69] However, the experience and advice they supply are valued even more than the capital they provide.[67,70]

Business angels are private individuals who invest individually or in syndicates. They are usually wealthy individuals who often have had a career in the industry in which they choose to invest.[29,67] However, they have very diverse backgrounds and range from entrepreneurs who have sold a business to financial services professionals who have accumulated substantial personal wealth. They are a growing force in early stage investment, although relatively few invest in academic spinouts. When they do they may invest alongside the economic development funds described above. Angels prefer to invest in ventures targeting industries with which they, or one of their syndicate, are familiar. They frequently choose to invest in ventures where they have industry or business expertise and contacts to contribute, as well as funds.[67,68]

It is usual for external investors to charge the costs they have incurred in making an investment to the investee company, although some angels do not follow this practice. When seeking investment it is important to check what the expectations of the investors are in this respect. The costs of due diligence on a company and the professional fees associated with drawing up the investment agreement and associated documentation can quickly mount up, and in a small, early stage investment this can eat up much of the cash that is needed to develop the business. Business angels usually aim to keep the transaction costs of an investment to a minimum in order to maximize the impact of their investment. As direct investors they can also avoid the regulatory costs incurred by professional investors, and they can usually take advantage of the tax breaks set up by governments anxious to encourage early stage venture finance. They often use standard agreements and try to simplify other aspects of the investment process.

Angels are organized in a variety of ways. Angel syndicates usually have websites that carry variable amounts of information about their investment preferences, criteria, processes and the names of their investee companies. A good example of this is the Archangel syndicate in Scotland.

Sohl[68] offers some advice to those who are looking for Angel investment. He warns that the process of finding a suitable source of angel finance, and going through the investment process is arduous. He notes that the route to a successful deal will involve extensive networking, presenting a sound business plan, and 3–4 months of due diligence. Angels look for investees within a half-day's journey in order to be able to keep a close eye on their investment. On the positive side, they expect to work closely in their mentoring role with the entrepreneur throughout the duration of the investment. They typically expect to stay with the investment for 5–7 years (longer than the average VC) and they are often described as 'patient capital.'

There are believed to be about 140 angel syndicates or 'angel groups' in the US at the time of writing.[68] The number in the UK is constantly growing. Many can be tracked down using a web search. However, by its nature angel investment is an informal and private process and even the syndicates can be difficult to track down. Individual angels are even harder to find. They may be sourced through networks, including contacts with lawyers, accountants, or fellow entrepreneurs.[68] Economic development agencies also may have contacts.

1.15.4.4.2.2.2 Venture capital Although formal VCs now rarely invest at the earliest stages of biotech developments, there are some exceptions, and in any case they still play an important role at later stages of development.

The British Venture Capital Association distinguishes among the different investment stages. Early stage funds include funds that fund concepts, early stage capital (firms less than 3 years old), and early development capital. Then come development funds (up to £2 million) followed by mid-size buy-in and buy-out (£2–10 million equity invested) then large buy-in and buy-out (>£10 million invested). Generalist funds invest across the different stages.

VC organizations can also be categorized as 'independent,' which raise their funds from external sources such as banks and insurance companies, and 'captive' funds (or corporate venturers), which obtain all or most of their funds from parent organizations.[29,54] The parents are often financial organizations. However, large industrial firms often have captive funds. In 2000, European corporate venture funds invested €49 million in biotechnology.[54]

VC funds that are independent or linked to financial services companies are run by professional investors. They usually have raised funds from groups of financial institutions and foundations to invest in a portfolio of ventures. Each fund has undertaken to aim for a high return on investment and has identified its investment philosophy in terms of the industries, risk profile of investee companies, and their stage of development. The fund has a contractual obligation to its own investors to meet its objectives by selling on its equity investments. It must return the profits to the group within a fixed period. Corporate venture funds are set up by individual large industrial companies, with variable remits – some invest in technologies which the parent company might adopt, while others have a wider mandate.[10,29]

As noted above, there are a small number of VC specialist funds that target early stage technology companies. However, most professional VCs do not invest in early stage life science ventures, especially since 2000. Some are prepared to do so, but are highly selective. However, this appears to be a growing trend. Mack[71] identifies a number of American funds that are specializing in early stage, very lean ventures, such as Angiosyn, founded by scientists at the Scripps Research Institute, which was sold to Pfizer within 2 years of the investment when it just had two scientist employees. This is a rather different model from that of Genentech. The aim with Angiosyn was to find a buyer at an early stage when the venture still just consisted of a product under development and the key scientists. With large drug companies currently needing to do deals to fill their development portfolio this is seen as a feasible route by some and may become more popular with VCs.

For most VCs, however, the level of risk at this stage is unattractive. A larger number of fund providers have some interest in entering later to follow-on when initial investments have given promising results, a good management team is in place, and customers are showing enough interest to give a clearer idea of the growth potential of the venture.

1.15.4.4.2.2.3 Corporate partnerships Project finance is sometimes available from industrial firms that are potential strategic partners. This most commonly comes from a large firm that has an interest in acquiring rights to the product under development provided it looks sufficiently promising. The finance may be provided through a captive corporate venture fund or may come directly from the department that has an interest in the technology that the venture is developing. It will sometimes provide equity investment, project finance, and even nonfinancial complementary development resources. Although the large firm's interest is most commonly to eventually acquire rights to the technology once proven, in some cases it may be as future customers or distributors. As noted above, these large firm partners can be a very good source of development finance and expertise. Their participation can enhance the interest of VCs and other coinvestors, since they lend credibility to the project and their nonfinancial contributions can reduce the risk of the venture. The extent to which such a partnership enhances the interest of other investors depends on the nature of the interest of the large company and the details of the deal. If it wants to eventually acquire the technology and the deal ties in the new venture too tightly, it may discourage other investors if it appears that they will not be able to realize sufficient returns. If the corporate investor has too much control it can eventually name its price or walk away, depending on the success of the technology. This is not a desirable situation for the other stakeholders including the founders.

If the large firm has no interest in acquisition, or is taking only limited rights in one or more products and leaving the founders more freedom to develop the business, it can be a very desirable source of investment. If its interest is primarily in having access to the technology as an outsourced service provided by the young firm, it can help finance the development of the venture through contracting for services that can also deliver valuable data on the performance of the technology in commercial contexts.

1.15.4.5 Selecting Investors

Since investors often bring nonfinancial resources, it is worth considering which nonfinancial resources are required, and how this impacts on the relative desirability of investors.

It is widely reported that angels and VCs can bring industry and business expertise, industry contacts and facilitate access to follow-on investment. However, the extent to which this is likely to be the case varies considerably depending on the knowledge base and network of contacts of the investor, and also with their desire to be 'hands-on.' In general, angels and VCs prefer to invest in nearby firms,[72] so that they can involve themselves closely with the firm. There is no fixed rule about how far this involvement extends. However, physical proximity is by no means essential to all VCs who invest in biotechnology. Zucker *et al.*[8] found that the key factor in the founding of biotechnology companies in an area was the presence of star scientists. In this US-based study, the number of VCs in an area was inversely proportional to the probability of foundings.

When looking for VC investment it is essential to consider what the firm needs apart from cash from its investors. The track record, experience, and willingness of potential investors to provide support can be investigated before moving toward any commitment. A number of indicators can provide some information. If the investor is an industry specialist, with strong industry links and a record of investment in the industry, they are more likely to understand industry requirements and to have good contacts in the industry. Information about current investments can usually be obtained, and the investee companies contacted to inquire about their experience with their investors.

Corporate investors can offer varied nonfinancial resources, usually to do with complementary technical, regulatory, legal, and marketing capabilities. The possibility of accessing these services should be explored before a deal is done. It is helpful to have written agreements in place that clarify the position of both parties.

Successful life science companies have often managed to obtain investment from a variety of corporate and VC sources, which together provide very substantial support to their developments. In these cases, the deals with the different stakeholders had to be mutually compatible, allowing each party to receive what it needed from the situation – financial returns for financial investors, products and services for corporate investors, and whatever mix of financial and business rewards the management were seeking.

1.15.4.6 Structuring the Deal

Raising early stage finance is usually far more difficult and takes far longer than inexperienced founders expect. When a decision is finally made to invest, the need may be so pressing that it is accepted regardless of the terms. However, the terms of the investment are likely to have a major impact on the eventual outcomes. A well-structured deal should incentivize the management team. It should not discourage future rounds of investment. Broadly speaking, it should facilitate the success of the venture. Brophy and Mourtada[73] consider the details of the transaction and how it affects the future of the venture. They point out that the key elements are:

- What is the venture worth before investment?
- How are gains and losses to be split?
- Who makes the main strategic and operating decisions?

Both the founders and the investors are interested in maximizing their profit and maintaining the maximum amount of control to protect their interests. On the other hand, most VCs realize that their objectives are most likely to be met when the incentives of the founder are aligned with their own.

1.15.4.6.1 Valuation

One of the key questions noted above is "How much is the venture worth?" The investors' assessment of this will be an important determinant of the amount of equity they demand in exchange for a given amount of investment.

As Bennett *et al.*[74] have remarked, the valuation of biotechnology companies has generally been a complicated business for investors. In their overview of methodologies used to value companies in the biotechnology and related life science arena, they note that the key metrics used for early stage companies that lack earnings are:

1. Comparable company analysis, in which the value is established by comparison with that of another that has been given a market valuation, e.g., by a third party equity investment, trade sale (acquisition of the company by another), or through the flotation of the company on a stock market.
2. Net present value (NPV) analysis – the value of a future cash, revenue, or earnings stream discounted back to today's value according to the investors' required rates of return. Biotechnology valuations usually calculate the NPV to be the sum of risk-adjusted cash flows arising from the sales of products of the company's development projects. In the case of a drug discovery company, the risk adjustment is conducted using clinical product attrition rates. Given the relatively higher rate of attrition of early stage projects (about 3% of preclinical versus 42% of phase III candidate drugs will make it to market),[74] valuations of companies whose projects are all at an early stage in the drug discovery process are very low due to the very high risk of failure. However, historical data show that therapeutic areas have different degrees of risk associated with them, for example, anti-infectives have a much higher success rate than CNS and respiratory drugs.[20]

Comparable company valuations are widely used by VCs in the US and Europe and are recommended by the Private Equity Guidelines Group of the USA where comparable companies with determinable fair values can be identified.[75]

1.15.4.6.2 Structuring partnership deals

Pharmaceutical companies have again become more interested in partnering or acquiring early stage (preclinical trial stage) biotechnology companies, after a period of low interest.[76] This increases the possibility of finance and also complementary expertise from a corporate partner. The objectives of corporate partners are somewhat different from those of financial investors, as noted above. They are likely to be able and willing to provide organizational resources to complement those of the new venture. Consequently, where investment comes from an industrial organization the deal should facilitate the formation of a relationship between the firms, which will facilitate the provision of complementary resources to support the development. It should aim for a 'win-win' situation in order to incentivize both partners. O'Donohue *et al.*[77] provide practical advice for the academic spinout on how to build and exploit corporate partnerships. They warn against becoming too focused on corporate requirements and thus losing viability in the wider marketplace. They also point out the need to avoid overdependence on the corporate for revenue, which overexposes the venture to the whim of the corporate.

1.15.5 Conclusion

Taking novel life science technologies into practical use is generally a very costly process. Time to market, especially for novel therapeutics and agri-food applications, may take many years. There are major regulatory constraints, which require exhaustive proof of safety and efficacy. Public concerns about the morality of deploying some technologies and applications have placed further regulatory constraints. In some cases, they have blocked substantial areas of development. GM foods are currently unsaleable in Europe and it appears unlikely that they will ever be accepted in this major market.

In spite of these hurdles, which grow ever greater, small firms have played a significant role in the commercialization of life science technologies during the last 30 years. In the 1970s, the possibility of developing new life science technologies invented in the science base within small firms set up as vehicles for the development project emerged as a realistic and attractive opportunity, and it shows no sign of losing popularity with founding scientists and governments.

From the earliest days of the biotechnology industry venture capital has played an important role in these developments. Through enlisting the support of venture capital providers and corporate strategic partners, it has become possible to carry risky projects through the early stages. In some cases, such as Genentech, Genzyme, Biogen and others, the early projects have formed the basis of a major independent business. Many others have been acquired by other organizations and gone on to contribute to successful innovation, as well as yielding a return, sometimes substantial, sometimes negative, on the initial investment.

The enthusiasm of venture capital providers, formal and informal, has waxed and waned with the perceived success of the biotechnology industry. Some sectors appear to have definitely fallen from favor, such as the agri-food sector, even though its GM crops are widely used. Where there is likely to be consumer resistance in the main markets of the wealthy industrialized world, profits are likely to be small and VCs cannot hope for the high returns they require to offset the risk of the projects.

Scientists who believe they have identified a potential commercial application of their technology need to be aware of a number of issues. Novel science-base technologies often have many possible applications in a number of industries. Some of these applications may raise ethical questions, while others may not face these particular problems. Some applications may have a clear, well-tested regulatory route to market, while others may lack this. These are important factors to consider before deciding on the specific application to pursue.

Equally important is the availability of finance for specific applications, from the earliest stages and through to the exit. Angel syndicates and venture funds usually state fairly clear investment preferences. This changes over time. Areas become fashionable when they appear to offer the prospects of good returns. They cease to be fashionable if they begin to look less promising, or if another area emerges that appears to be even more attractive.

Completely novel technologies or applications ideas can be difficult for investors to understand. If they cannot compare them with anything, they may be wary and reluctant to invest. In these cases a good business plan is particularly critical, and evidence underpinning its credibility may help attract the interest of investors.

The business plan and the credibility of its assumptions, of the scientists, and of the market are the key questions to address. If the prospects are sufficiently attractive and the risks are clear and reasonably assessable, the project may be able to win investment. At the earliest stages it may be impossible to make a strong enough case. Public grants and awards are very widely available and they are designed for just this eventuality.

Although the investment climate today is not as buoyant as in the 1990s, there are still many opportunities for scientists to attract funding into commercially promising projects. It usually takes much longer than expected to sign up investment, and the terms may be onerous. However, if the founders have made a strong and believable case in the business plan, if they believe in the potential of the project and are prepared to back their belief with a huge amount of effort, they have a reasonable chance of success.

References

1. Cohen, S. N.; Chang, A. C. Y.; Boyer, H. W.; Helling, R. B. *Proc. Natl. Acad. Sci. USA* **1973**, *70*, 3240–3252.
2. Köhler, G.; Milstein, C. *Nature* **1975**, *256*, 495–503.
3. Prevezer, M. *Small Bus. Econ.* **1997**, *9*, 255–271.
4. Audretsch, D. B. *Small Bus. Econ.* **2001**, *17*, 3–15.
5. US Congress. A Review of Past and Present Assessments of the Outlook for Biotechnology. In *Commercial Biotechnology: An International Analysis*; Office of Technology Assessment: Washington, DC, 1984, pp 6–18.
6. Walton, A. G.; Hammer, S. K. *Genetic Engineering and Biotechnology Yearbook*; Elsevier: Amsterdam, 1985.
7. Link, A. N.; Rees, J. *Small Bus. Econ.* **1990**, *3*, 1–38.
8. Zucker, L. G.; Darby, M. R.; Brewer, M. B. *National Bureau of Economic Research* (National Bureau of Economic Research (NBER), Cambridge, MA); NBER Working Paper No. 4653, 1994, 4653.
9. Tushman, M. L.; Anderson, P. C. *Admin. Sci. Quart.* **1986**, *31*, 439–465.
10. Gompers, P. A.; Lerner, J. *The Money of Invention: How Venture Capital Creates New Wealth*; Harvard Business School Press: Cambridge, MA, 2001.
11. Daly, P. *The Biotechnology Business: A Strategic Analysis*; Frances Pinter: London, 1985.
12. Kenney, M. *Biotechnology: The University/Industrial Complex*; Yale University Press: New Haven and London, 1986.
13. Teitelman, R. *Gene Dreams: Wall Street, Academia and the Rise of Biotechnology*; Basic Books (Harper Collins): New York, 1989.
14. Jones, S. *The Biotechnologists*; Macmillan: London, 1992.
15. Bower, D. J. *Company and Campus Partnership*; Routledge: London, 1992.
16. Shan, W. *Strat. Manag. J.* **1990**, *11*, 129–139.
17. Greis, N. P.; Dibner, M. D.; Bean, A. S. *Res. Pol.* **1995**, *24*, 609–630.
18. Zucker, G. L.; Darby, M. R. *Res. Pol.* **1997**, *26*, 429–446.
19. Rothaermel, F. T. *Res. Pol.* **2001**, *30*, 1235–1251.
20. Di Masi, J. A.; Hansen, R. W.; Grabowski, H. G. *J. Health Econ.* **2003**, *22*, 151–185.
21. DiMasi, J. A.; Hansen, R. W.; Grabowski, H. G. *J. Health Econ.* **1991**, *10*, 107–142.
22. Arora, A.; Gambardella, A. *J. Indust. Econ.* **1990**, *4*, 361–379.
23. Liebeskind, P. J.; Oliver, A. L.; Zucker, L. G.; Brewer, M. B. *Org. Sci.* **1996**, 7, 429–442.
24. Oliver, A. L.; Liebeskind, J. P. *Int. Stud. Manag. Org.* **1998**, *27*, 76–103.
25. Oliver, A. L.; Montgomery, K. *Hum. Relat.* **2000**, *53*, 33–55.
26. Powell, P. W.; Koput, K. W.; Smith-Doerr, L. *Admin. Sci. Quart.* **1996**, *41*, 116–145.
27. Zucker, G. L.; Darby, M. R. *Proc. Natl. Acad. Sci. USA* **1996**, *93*, 12709–12716.
28. Coombs, R.; Metcalfe, S. *Tech. Anal. Strat. Manag.* **2002**, *14*, 261–272.
29. Burgel, O. *UK Venture Capital and Private Equity as an Asset Class for Institutional Investors*; British Venture Capital Association: http://www.bvca.co.uk (accessed April 2006).
30. Von Burg, U.; Kenney, M. *Reg. Pol.* **2000**, *29*, 1135–1155.
31. Gompers, P. A.; Lerner, J. *The Venture Capital Cycle*; MIT Press: Cambridge, MA, 1999.
32. Bower, D. J. *R&D Manag.* **2003**, *33*, 97–106.
33. Lee, K. B., Jr.; Burrill, G. S. *Biotech '95 Reform, Restructure, Renewal*; Ernst & Young: Palo Alto, CA, 1994.
34. Klausner, A. *Nat. Biotechnol.* **2005**, *23*, 417–418.
35. Lawrence, S. *Nat. Biotechnol.* **2005**, *23*, 518.

36. Barnes, B. Biotechnology as Expertise. In *Nature, Risk and Responsibility: Discourses of Biotechnology*; O'Mahony, P., Ed.; Routledge: New York, 1999, pp 129–138.
37. Webster, A. *Curr. Sociol.* **2002**, *50*, 443–457.
38. Bower, D. J. *Technol. Anal. Strat. Manag.* **2005**, *17*, 183–204.
39. Krimsky, S. The Cultural and Symbolic Dimensions of Agricultural Biotechnology. In *Private Science: Biotechnology and the Rise of the Molecular Sciences*; Thackray, A., Ed.; University of Pennsylvania Press: Philadelphia, 1998, pp 67–92.
40. Martin, S.; Tait, J. Attitudes of Selected Public Groups in the UK to Biotechnology. In *Biotechnology in Public: A Review of Recent Research*; Durrant, J., Ed.; Science Museum for the European Federation of Biotechnology: London, UK, 1992, pp 28–41.
41. Miller, H. L. *Nat. Biotechnol.* **1999**, *17*, 730.
42. Tait, N. ADM calls for crop segregation. *Financial Times*, Sept 2, **1999**, p 32.
43. Fox, J. L. *Nat. Biotechnol.* **1999**, *17*, 1053–1054.
44. Human Genome Project Information. *What are Genetically Modified (GM) Foods?* **2005**. Available from: http://www.ornl.gov/sci/techresources/Human_genome/elsi/gmfoods.html (accessed April 2006).
45. Cohen, J. I. *Nat. Biotechnol.* **2005**, *23*, 27–33.
46. UK Cabinet Office. *Field Work: Weighing up the Costs and Benefits of GM Crops*; Cabinet Office Strategy Unit: London, 2003. Available from www.strategy.gov.uk (accessed April 2006).
47. Mulkay, M. *Soc. Stud. Sci.* **1993**, *23*, 721–742.
48. Mulkay, M. *The Embryo Research Debate: Science and the Politics of Reproduction*; Cambridge University Press: Cambridge, UK, 1997.
49. Tait, J. *More Faust than Frankenstein*; Paper 6; SUPRA: Edinburgh University, UK, **1999**.
50. Herrera, S. *Nat. Biotechnol.* **2005**, *23*, 775–777.
51. Department of Health. *Stem Cell Research: Medical Progress with Responsibility.* A Report from the Chief Medical Officer's Expert Group, June, 2000. Available from www.dh.gov.uk (accessed April 2006).
52. Parry, S. *New Genet. Soc.* **2003**, *22*, 183–201.
53. SCST *Developments in Human Genetics and Embryology.* 4th Report of the Science and Technology Committee of the UK Parliament, 2002. Available from www.publications.parliament.uk/pa/cm200102/cmselect/cmsctech/791/79103.htm (accessed April 2006).
54. European Venture Capital Association. *Corporate Venturing Activity Update*, 2001. Available from www.evca.com (accessed April 2006).
55. Nightingale, P.; Martin, P. *Biotechnology* **2004**, *22*, 564–569.
56. Horrobin, D. *Nat. Biotechnol.* **2001**, *19*, 1099–1100.
57. Williams, M. *Curr. Opin. Pharmacol.* **2003**, *3*, 571–577.
58. Gera, M. H. From Research to Spinout and Beyond: A Venture Capitalist's View. In *Taking Research to Market: How to Build and Invest in Successful University Spinouts*; Tang, K., Vohora, A., Freeman, R., Eds.; Euromone Books: London, UK, 2004, pp 47–55.
59. Stathopoulos, N. Financial Structure of a Private Equity Investment. In *European Venture Capital Association Entrepreneurship Education Course*; Module 4; EVCA, Zaventem: Belgium, 2002.
60. Delafond, S. IPOs, Trade Sales. In *European Venture Capital Association Entrepreneurship Education Course*, Module 8; EVCA, Zaventem: Belgium, 2002.
61. Bower, D. J.; Farmer, K. Components of a University Spinout Aligning Goals Drivers and Expectations. In *Taking Research to Market: How to Build and Invest in Successful University Spinouts*; Tang, K., Vohora, A., Freeman, R., Eds.; Euromone Books: London, UK, 2004, pp 9–20.
62. Howells, J.; Nedeva, M.; G eorghiu, L. *University of Manchester PREST,* December 1998; 98/70 Report.
63. Pisano, G. P. *Res. Pol.* **1991**, *20*, 237–249.
64. Stuart, T. E.; Hoang, H.; Hybels, R. C. *Admin. Sci. Quart.* **1999**, *44*, 315–349.
65. Baum, A. J.; Oliver, C. *Admin. Sci. Quart.* **1991**, *36*, 187–218.
66. Shori, S. Legal Structure of a Private Equity Investment. In *European Venture Capital Association Entrepreneurship Education Course*, Module 5; EVCA, Zaventem: Belgium, **2002**.
67. Nace, K.; Cotton, C. The Role of Business Angels. In *Taking Research to Market: How to Build and Invest in Successful University Spinouts*; Tang, K., Vohora, A., Freeman, R., Eds.; Euromone Books: London, UK, 2004, pp 114–121.
68. Sohl, J. *Nat. Biotechnol.* **2005**, *23*, 263–264.
69. Bouchie, A. *Nat. Biotechnol.* **2004**, *22*, 480–482.
70. Sohl, J. E. *Venture Capital: An International Journal of Entrepreneurial Finance* **1999**, *1*, 101–120.
71. Mack, G. *Nat. Biotechnol.* **2005**, *23*, 779.
72. Zook, M. A. *Int. J. Urban Reg. Res.* **2004**, *28*, 621–641.
73. Brophy, D. J.; Mourtada, W. R. Structuring the Transaction. In *Taking Research to Market: How to Build and Invest in Successful University Spinouts*; Tang, K., Vohora, A., Freeman, R., Eds.; Euromone Books: London, UK, 2004, pp 153–178.
74. Bennett, S.; Parkes, R.; Herrmann, M. Biotech Valuation: An Investor's Guide. In *ING Financial Markets*; ING Bank: London, UK, December 2004.
75. Private Equity Industry Guidelines Group. http://www.peigg.org (accessed April 2006).
76. Ratner, M. *Nat. Biotechnol.* **2005**, *23*, 509.
77. O'Donohue, J.; Winter, C.; Tang, K. Adding Value – Strategic Alliances and Corporate Partnerships. In *Taking Research to Market: How to Build and Invest in Successful University Spinouts*; Tang, K., Vohora, A., Freeman, R., Eds.; Euromone Books: London UK, 2004, pp 229–239.
78. Genentech Home Page. http://www.gene.com (accessed April 2006).
79. Amgen Home Page. http://www.amgen.com (accessed April 2006).
80. BiogenIdec Home Page. http://www.biogenidec.com (accessed April 2006).
81. Syngenta Home Page. http://www.syngenta.com (accessed April 2006).

Biography

D Jane Bower, PhD, FRSE is the Director of the Center for Enterprise Management at the University of Dundee, which delivers specialist management development programs for managers in high technology firms. She initially spent 15 years as a biomedical researcher at Edinburgh University, Stanford Medical School, and the MRC Human Genetics Unit. She became increasingly involved in issues surrounding the transfer of biotechnology and its commercialization in small and large firms, and managed a number of evaluation and commercialization projects for private and public sector organizations. These included the planning and financing of a joint venture with Massachusetts Institute of Technology (MIT) spinout, an investigation of pharmaceutical industry/university relationships in Japan, and the design and financing of a technology venture fund. She has published extensively on technology commercialization and managing innovation in networks of small and large organizations, both public and private. She was a member of the Scottish Higher Education Funding Council from 2001 to 2005, of the Scottish Science Advisory Committee from 2002 to 2004, and she is currently Chairman of the Scottish Stem Cell Network.

Comprehensive Medicinal Chemistry II
ISBN (set): 0-08-044513-6

ISBN (Volume 1) 0-08-044514-4; pp. 553–570

1.16 Industry–Academic Relationships

M Darnbrough, London, UK
M Skingle, GlaxoSmithKline, Stevenage, UK

1.16.1 Introduction

Fundamental research in biology and chemistry is the driving force behind innovation in pharmaceutical and biotechnology companies. Unmet medical need inspires today's researchers who work within companies to develop new medicines and treatments for patients by using the latest discoveries. However, the picture was not always like that. Many of today's pharmaceutical companies were set up in the 1880s as suppliers of natural products which had medicinal value. The companies did not have scientists on their staff because research was seen to be part of a university's function. Around that time the first academically educated scientists joined companies – but they were taken on to lead work on the quality control of products. The recruitment of scientists for a research function in companies first happened in the 1920s and 1930s – the period when Merck, Abbott, Lilly, and Squibb all set up in-house research programs. An early example of the evolution of company policy on research was from Carleton Palmer, the president of Squibb, who said: "We cannot afford to discontinue all research work, even though we connect with the Mellon Institute and use their data to solve our bigger problems." In the late 1920s, Abbott was spending $100 000 per year on research and Merck's research budget grew from $146 000 in 1931 to $906 000 in 1940 (4% of sales).

1.16.2 Why Do Companies and Academia Collaborate?

1.16.2.1 Who Invests in Basic Research?

Today, pharmaceutical companies themselves carry out a substantial proportion of the total research done throughout the world. The 10 largest pharmaceutical companies spent more than $40 billion on research and development (R&D) in 2004. Governments also fund a substantial amount of basic research which is carried out in universities, other higher education institutions (HEIs), and in government research laboratories. **Tables 1–3** show countries' expenditure on R&D. The US accounts for about 44% of the total expenditure on R&D from all Organization for Economic Co-operation and Development (OECD) countries. Of the total expenditure of $276 billion on R&D in the US in 2002, about 13% was carried out in universities and colleges (an estimated expenditure of $36 billion) while 70% was performed in industry ($194 billion). However, looking specifically at basic research, academia carried out the majority (54%) of basic research done in the US. The US federal government spent some $78 billion on R&D in 2002 – just over a quarter (28%) of the US total.[1]

Research is an international activity and most governments allocate funds to programs of basic research in all aspects of science – but increasingly in the life sciences – which are carried out in centers of excellence round the world. One current example, in an area of interest to pharmaceutical and biotech companies, is the field of stem-cell research. Expenditure on research in this field around the world is growing. For example, the state of California alone has allocated $3 billion over 10 years; a team leader in South Korea has been allocated $4 million per year, and in the UK some $80 million has been allocated to the field by the Research Councils. Companies will be watching developments in all the resultant centers and will collaborate with those whose expertise fits best with the company's interests.

By 2000, there were approximately 1 million scientific papers published each year throughout the world – in science, engineering, and medicine. About one-third come from the EU, one third from the US, and one-third from the rest of the world. Pharmaceutical and biotechnology companies are interested in research in a wide range of subjects and disciplines including chemistry, nanotechnology, genetics, pharmacology, biochemistry, animal physiology, molecular modeling, computer simulation, biomanufacturing, and large-scale chemical engineering.

Table 1 Research carried out in sample countries in 2003

	Total spend on research ($m)	*% paid for by industry*	*% paid for by government*
EU: 25 countries	211 194.7	54.5	34.8
US	284 584.3	63.1	31.2
Japan	114 009.1	74.5	17.7
China	84 618.3	60.1	29.9
All 33 OECD member countries (excludes China)	679 782.8	61.6	30.5

Source: OECD.[2]

Table 2 Organizations carrying out research (2003)

	% carried out in industry	*% carried out in higher education*	*% carried out in government labs*
EU: 25 countries	63.5	22.0	13.4
US	68.9	16.8	9.0
Japan	75.0	13.7	9.3
China	62.4	10.5	27.1
All 33 OECD member countries (excludes China)	67.3	18.7	10.9

Source: OECD.[2]

Table 3 Numbers of researchers in sample countries (2003)

	Number of researchers
EU: 25 countries	1 160 305
US	1 261 227
Japan	675 330
China	862 108

Source: OECD.[3]

International statistics show that interactions between universities and companies are increasing. In the G7 countries the proportion of R&D carried out in academia which is funded by industry grew from 2.6% in 1981 to 5.2% in 1990 and reached 6.0% in 1999.[2] Between 1985 and 2001, a total of 861 technology alliances were registered under the US National Cooperative Research and Production Act. About half of these were in information technology and communications but some were in life sciences. Fifteen percent involved a university and 12% included a federal laboratory. European Commission data[3] show that the proportion of academic-based research which is paid for by industry ranges from less than 4% in Portugal, Austria, Denmark, Japan, France, and Sweden; is around 5% in Finland, Italy, Greece, Netherlands; is over 6% in the US, Ireland, UK, Spain; and the proportion of research in academia, which is paid for by industry reaches over 10% in Belgium and Germany.

Researchers who are based in pharmaceutical and biotechnology companies keep up to date with research done in academia and in other research organizations through scientific journals. Companies' research groups extend their fields of investigation, building on the latest research reports, by setting up research collaborations with individuals or teams in universities and other HEIs. The subjects for study are those that the companies consider may offer potential for development of novel medicines. Companies sometimes commission specific work under contract with specialist researchers or with research teams in HEIs.

1.16.2.2 Research Databases

There are some fields of research that are generating substantial databases of information, which are of potential value in seeking novel medicines. Consequently, some companies have set up formal relationships with the research groups involved (see the case studies included in this chapter).

One of the major recent international collaborative projects was the sequencing of the human genome – the work was done by researchers in the UK, US, and Japan.[4] The sequencing data (produced by the teams funded by the governments and, in the UK, by the Wellcome Trust), were made publicly available for all researchers throughout the world, free of charge. (Sequencing data had also been produced by commercial companies – Celera and Human Genome Sciences and others.) Publicly available information about the gene sequences of many different species now exists. The sequence data have been built on – for example, through research about the location of genes on chromosomes – another international collaborative effort with teams in different countries focusing on specific human chromosomes. The basic sequencing information is only useful when the functions of the genes are known. Researchers are undertaking programs to discover at what stage in normal development and at what stages in the onset of a disease certain genes are active, and to study which enzymes and proteins are being produced as a result of the gene activity.

A major collaborative project between Canada and the UK – the Structural Genomics Consortium – is one example of work being done to build up knowledge about the human proteins that are made as a result of gene activity.[5] The consortium is a 3-year project, started in 2003, led by Canadian proteomics expert Dr Aled Edwards at the University of Toronto. The collaboration involves four Canadian funding partners (Ontario Research and Development Challenge Fund, Ontario Innovation Trust, Genome Canada, and the Canadian Institutes of Health Research) and the UK-based research charity the Wellcome Trust. The pharmaceutical company GlaxoSmithKline is also a partner. The project aims to determine the three-dimensional structure of more than 350 proteins and hopes that this knowledge will be used to develop novel medicines and treatments for diseases. In February 2005, a further collaboration between the Structural Genomics Consortium and the McLaughlin–Rotman Center for Global Health was announced. This further research partnership will focus on the structure of the proteins in parasites which cause malaria. The data from this project will be made freely available to the research community at large.

Within research-based pharmaceutical companies there are large libraries of compounds that have been synthesized and screened to look at their potential physiological and medical properties, but which have not been taken forward into development as drug targets. As part of the US National Institutes of Health (NIH) Roadmap Initiative, collaborative projects are being planned to enable academic research groups to have access to some of these compounds in order to perform further screening of their biological activity and this work may lead to a fundamental understanding of some cellular and membrane processes.[6]

Sharing of basic research data is one area of collaboration between researchers, regardless of whether they are working within companies, medical research charities, or in universities or government laboratories.

1.16.2.3 Relationships between Academia and Companies

There is considerable interdependence between pharmaceutical companies and HEIs. Firstly, the researchers and other specialists who are recruited and employed by companies are trained in universities and other HEIs. It is important that new recruits have been trained in the approaches used in industry and that they are familiar with the equipment and techniques used by innovative, leading-edge companies. Secondly, the areas of study in PhD research and in other projects in universities are sometimes related to disease or development issues which are of direct interest to companies – indeed the subjects are sometimes directly chosen because of the company interest.

Academic researchers also work with small science-based companies on specific projects that will enable the companies to develop innovative products or analytical services. Very importantly, some academic research is commercialized through the setting-up of a new small company specifically to develop the discoveries and inventions made in academia and take them to the market.

Some government-funded organizations, and some universities and hospitals, have unique facilities, equipment, and specialist expertise. For example, around the world, many synchrotrons (particle accelerators whose beam lines are used for imaging large biologically active molecules) have been built with funding from national and regional governments and the research programs that use them are also publicly funded. However, some synchrotrons and other large facilities (including specialist analytical or imaging equipment) are based within universities or in major teaching hospitals. Both company researchers and academic ones find access to some of these large facilities useful to monitor growth of tissues and cells within a living organism or to track the movement of radioactively labeled compounds as they move within the body and are metabolized and eliminated from the body.

Very close collaboration also takes place when new analytical techniques are emerging in HEI-based research teams and which offer the potential for development as novel laboratory analytical tools and other instruments. Collaborations to develop and commercialize analytical techniques often involve academic researchers, companies that design and manufacture analytical instruments, and also pharmaceutical and biotech companies who use the instruments in their drug development laboratories. Pharmaceutical companies sometimes become beta test sites for novel prototype analytical instruments to help the instrument makers finalize the equipment design before putting it on to the wider market. Access to the most precise, novel analytical devices can give the beta test site a competitive advantage if the new equipment enables analyses to be done to a new level of accuracy or with increased speed.

1.16.3 What Are the Types of Collaboration between Academia and Companies and What Are the Benefits to Each Partner in Collaboration?

Table 4 shows some of the different types of interaction that take place between companies and HEIs. These range from long-term commitment to fund a professorial chair to short-term contracts, for closely defined research, of only a few months' duration. When company-based researchers and academic researchers work together, to share knowledge

Table 4 Types of interaction between industry and academia

Interaction	*People involved*	*Type of agreement*
Exchange of information and ideas without financial commitment	Clinicians, lecturers, professors, and postgrad researchers	Confidential disclosure agreement
Funding of postgraduate award by company	Student selected by company with academic supervisor	PhD studentship agreement; freedom to publish; shared patent right
Funding of postgraduate award to include working in company	Student with academic and company supervisors	PhD studentship agreement; confidentiality agreement
Funding of a university lectureship by industry	Lecturer selected by university with representative from company on panel	Lectureship agreement
Endowment of a professorship by industry	Professor selected by university – company person on selection panel	Charitable donation; chair named as company requests
Collaborative research project	One or more postdoctoral researchers	Project grant award typically for 2 years with an option to extend to 3 years
Collaborative research program	Several postdoctoral researchers and technicians	Program grant award typically for 3 years with option to extend to 5 years
Research consortium	Team of researchers led by world-class professor with technical support and specialist facilities supported by several companies	Legal framework for access to precompetitive data and nonexclusive licensing
Provision of academic advice	World-class specialist with scarce expertise	Consultancy agreement
Contract research	Specific experts in academia	Closely defined program of work with specific deliverables set by company
Provision of biological or chemical materials		Material transfer agreement
Shared funding of facilities	Company, university, government or EU or regional grant	Contract setting out shares of costs and rights to knowledge

and to develop technical know-how, both benefit from the interaction. Also, many practical inventions – which have brought enormous benefit to patients – have come about as a result of such collaborations. It is not possible for a company to undertake research in all fields in-house and so it extends its portfolio of projects through collaborations with individual researchers or with groups in academia (and with small companies too). When new techniques are emerging, a company may want to be involved with more than one competing technology before it commits to one approach in its own laboratories – this watching can be done by participating in academic projects.

Most collaboration between industry and academia involves what is known as precompetitive research. That is fundamental research, which is not associated with a specific product (which a company would market in competition with another company). Precompetitive research findings can be shared by several companies and will provide an underlying basis of understanding on which company-based researchers will build specific and confidential projects which are specifically focused on identifying a potential novel medicine. The in-house company projects are competitive ones designed for product development and are therefore kept confidential.

During collaborative projects, the academic researcher often provides new ideas and causes company researchers to question their approaches. At the same time, the specific and focused concerns, which are the subject of the company projects may give a fresh orientation, impetus, and relevance to an academic project or program.

1.16.3.1 Collaborative Projects

There are many different types of interaction between pharmaceutical companies and academic institutions and with individuals based in academia. Interaction often involves projects of short-term and direct relevance to a company. For example, a pharmacologist or physiologist may include a compound that is under test by a company among a series of compounds, when researching into physiological, biochemical, or behavioral effects. At the other extreme, collaborative projects can involve novel fundamental research that may never be taken up by the company – for example, when the area of research is considered to offer potential insights that might lead to novel medicines or methods of treatment for patients (e.g., basic research on types of enzymes or proteins or determination of their three-dimensional structures or electrostatic characteristics; or research into the functioning of organelles within a cell or of cell membranes).

Some collaborations involve teams of researchers based in more than one academic institution. Sometimes these collaborations involve researchers based in more than one country.

There are a number of reasons why collaboration is mutually beneficial. The risks associated with new areas of research – which might lead to highly innovative approaches to drug development or lead directly to medicinal products – can be shared. A company can explore a new field of research without setting up its own in-house team and without buying specialist equipment or employing experts. By setting up a collaborative relationship with an academic team, a company can gain access to facilities, share the costs of researchers and technical support, and pay for specific studies. The academic partner benefits from additional funding, and from exposure to the different approach to problem-solving taken by the company sponsors. Risks are shared and so too are rewards – ideas, know-how, patents, and royalties. Networks of researchers are often established, which remain mutually supportive for many years and these are often across country boundaries.

Collaborations are often set up to support a long-term vision and to explore long-term research objectives. There are also some major programs within which a company may pay for the construction or installation of a facility – such as a scanner, or clinical beds equipped for metabolic measurement – and the facilities are located on a university or research hospital campus and can be used for other research which is nothing to do with the interests of the sponsoring company.

From a company point of view, one long-term opportunity which may arise from a collaboration is the eventual recruitment, into the company research team, of a researcher who has worked on the collaborative program – the collaboration gives the company a chance to assess the qualities of the potential employee.

1.16.3.2 What Criteria Are Used to Decide Whether to Go Ahead with a Collaboration?

There are many criteria used by a company to assess the merits of a possible collaboration. These include consideration of the extent to which the collaboration will fit with the long- or medium-term strategy of the company. This aspect will include decisions about which diseases are being worked on, which targets have been identified as the route to treatment, a requirement to explore emerging fields of knowledge (functional genomics, stem cells) or to explore the potential for novel techniques (in the past this included novel separation techniques like chromatography, use of gene chip arrays, and novel imaging). One of the most important criteria is the quality of the basic scientific research that is to be carried out. Consideration is also given to whether potentially useful discoveries or technologies are likely to emerge from the project or program. Other factors include availability of essential facilities and staff and the time period for deliverables.

The potential academic researcher looks carefully at the demands and expectations of the company in terms of the science, the tightness of the project specification, and the timescales. Many projects done for or with pharmaceutical and biotech companies involve interesting basic biology, but there may be unacceptable constraints on publication of the results or on the subsequent use of the knowledge gained in the project. Companies may be reluctant to pay full economic costs, expecting that some of the overheads and use of facilities will not be charged for. The academic will also consider the extent to which the collaboration will provide access to specialist equipment and to know-how that will add to the knowledge and experience of the academic group and enhance both research and teaching.

1.16.3.3 How Many Collaborations Are There in the UK?

In the UK, in 2003, UK-based pharmaceutical companies which had research activities located in the UK funded more than 1100 collaborations in 80 different institutions in the UK (**Table 5**). These included 685 postgraduate studentships, 418 postdoctoral researchers, and several research consortia. The majority of the studentships were co-funded with one of the Research Councils – particularly the Engineering and Physical Sciences Research Council (EPSRC) and the Biotechnology and Biological Sciences Research Council (BBSRC). Five of the universities in the UK accounted for a third of all these collaborations and were involved in more than 50 collaborations (University College London, Oxford, Manchester, Cambridge, and Imperial College London). Despite this concentration, the companies had collaborations of different types with a total of 60 different universities in the UK as well as with some of the Research Council Institutes, some National Health Service trusts, and many medical charities.

Table 5 Research collaborations and PhD studentships between companies and UK institutions (2003)

Company	*No. of studentships*	*No. of collaborative grants*
Astra Zeneca	124	120
GlaxoSmithKline	357	218
Celltech	21	5
Pfizer	108	58
Eli Lilly	27	0
Merck, Sharpe & Dohme	7	11
Novartis	14	0

Source: Association of British Pharmaceutical Industry academic liaison group.

1.16.3.4 What Are the Effects of Globalization and of Increasing Academic Costs?

Increasingly, as academic institutions throughout the world compete for collaborative projects with the world's leading research-based companies, the costs of academic research come into consideration. One issue facing UK academic researchers in 2005, who wish to work with companies, is the move toward charging full economic costs for all projects done in universities and Research Council Institutes. This may well make the UK uncompetitive and company collaborations may move to countries where the quality of the research teams and the facilities is equally high (or higher), but crucially the costs per researcher-hour are lower.

The pharmaceutical industry is a global one and companies seek to partner with academic groups that are the best in the world in their specialist field of work. This means that global companies have collaborative projects and programs with universities in many different countries. When there are several teams or individual experts of world class in a given field, companies can therefore compare the value for money provided in different countries. For example GlaxoSmithKline has, for many years, taken undergraduate chemistry students into the company's UK laboratories for a training period of 1 year. The quality of chemistry undergraduates from some French establishments and French universities has been found to be particularly high and **Table 6** shows that GlaxoSmithKline trained more French than British undergraduates in UK laboratories in 2004. Governments and university authorities include different items in their charges and take different approaches to charging for overheads. In countries that are keen to attract partnerships from pharmaceutical companies, some elements of costs may be covered by the state. The availability of instant electronic communication, teleconferencing, and video conferencing has also made it easy for company researchers from anywhere in the world to work easily with academic researchers based on another continent.

1.16.3.5 Collaborative Training Awards

Industrial sponsorship of good postgraduate students to study for PhD qualifications provides an ideal opportunity for initiating the interaction between research groups and industry. The major pharmaceutical companies typically support a few hundred such students every year – for example, GlaxoSmithKline supported more than 350 such students in the UK alone in 2002. GlaxoSmithKline would expect each student to carry out part of the project within the company's laboratories under the direct supervision of an industrial supervisor.

The advantages for the student from working in an innovative global company include the opportunity to use more modern equipment and different facilities from those in the academic department. The experience would also give the student the opportunity to see how research is directed and carried out within an industrial setting. This part of the project work would form part of the thesis submitted for the PhD or similar qualification. During the period of study, the academic supervisor and the industrial supervisor would exchange ideas, and long-lasting links between the company and the supervisor's academic department might become established. Sometimes at the end of a 3- or 4-year period of study the industrial supervisor might decide to offer employment to the student.

The purpose of a PhD training is to enable the graduate to be equipped to embark on a career as a research scientist requiring minimal supervision. It is not appropriate for a company to think that a student undertaking a PhD will be providing answers to specific industrial problems. Sometimes, however, the student's work will produce findings that are potentially commercializable and therefore it is important that a framework for ownership of intellectual property

Table 6 Chemistry undergraduates on industrial placements to GlaxoSmithKline in 2004

University where student is studying	*No. of sponsored students*
Lyon (France)	21
Sheffield (UK)	11
ESCOM (France)	10
Manchester (UK)	10
Bristol (UK)	8
York (UK)	8
Edinburgh (UK)	6
ENSCP (France)	5
Imperial College London (UK)	5
Strathclyde (UK)	5
ESPCI (France)	4
Montpellier (France)	4
Bath (UK)	4
Leeds (UK)	4
Glasgow (UK)	3
Loughborough (UK)	3
Newcastle (UK)	3
Exeter (UK)	2
Dublin (Ireland)	2

Source: GlaxoSmithKline, 2005.

and the eventual sharing of financial income is established. A formal agreement covering intellectual property is therefore a normal part of an industrially sponsored PhD.

In the UK, a scheme has been operating for 30 years to assist businesses to access the latest research thinking through inviting newly qualified graduates or postgraduates to work in the company for short periods. The scheme was known as the Teaching Company Scheme and has recently been renamed the Knowledge Transfer Partnership (KTP). The Department of Trade and Industry is the principal funder of the scheme but there are another 12 sponsors, including five of the Research Councils, the Departments of Health and of Food and Rural Affairs, the European Social Fund, and bodies in Northern Ireland, Scotland, and Wales. One of the English Regional Development Agencies (One North East) has recently become a sponsor of the scheme too. In 2004–2005, £32 million (US$60 million) was spent by the sponsoring organizations with a further £53 million (US$100 million) from the companies. In that year, 425 new projects were started and a total of 958 associates were working on projects.[7a,7b]

The aim of the KTPs is to strengthen the competitiveness, wealth creation, and economic performance of the UK by enhancing the knowledge, technology, and skills in businesses, and encouraging the stimulation of innovation, through collaborative projects involving the knowledge base in HEIs and other research organizations. The scheme operates by enabling one or more so-called associates (recently qualified people) to take part in a project of between 1 and 3 years' duration and thus transfer knowledge and know-how into the host business. The associate works in the business on a project that is core to the strategic development of the business. The associate is supervised jointly by a senior member of the business and by an academic or technical specialist from the partnering knowledge-base organization. As a result of the collaboration, the knowledge-base partner is provided with a relevant and improved understanding of the challenges that arise in companies and of their business requirements and operations. The academic institution sometimes develops business-related teaching materials and sometimes identifies new research themes for projects. Papers in academic journals are often published describing the results of the collaborative research.

This collaborative activity is also taken into account in the assessment of UK universities – the research assessment exercise. The associates themselves benefit because they gain experience of managing a challenging project that is central to a company's long-term growth, and they develop skills that enhance their employability. Fifty percent of the associates register for a PhD and 70% of them are offered employment by the host company at the completion of the project. KTP projects take place in many engineering and management and design disciplines as well as in biology (3% of the projects), chemistry (3%), and medicine (1%).

Some PhD students in the UK are funded jointly by a company and by a Research Council. CASE or CAST awards (Cooperative Awards in Science and Engineering or Science and Technology) provide the student with a higher grant than that provided for traditional PhD students and the HEI also receives an additional allowance from the company to support project costs. The student spends at least 3–4 months working within the company. Some of the Research Councils have allocated these awards to certain university departments and left it to the academic supervisor and student to approach a matching company. This means that companies sometimes received novel project ideas that they had not considered before. More recently, the Research Councils have allocated studentships to companies so that the companies can decide the area of research and can also choose the academic collaborator. In practice, the project takes shape through interactive discussion between the company and the academic partner.

1.16.3.6 Collaborative and Contract Research Compared

There are important differences between collaborative research and contract research. Collaborative research involves input from both the industrial and the academic partner, driven by a combination of shared interest in basic research and long-term interest to industry.

1.16.3.6.1 Collaborative research

Collaborative projects usually address fundamental scientific or medical issues. They may be designed to increase knowledge about the mechanisms that cause a disease; to explore a biochemical pathway; to elucidate the structure and functions of a little-understood group of enzymes, coenzymes, or hormones; to understand better the way molecules are transported within or between cells, for example. Some projects may also explore some fundamental aspects of biochemical engineering and process modeling. The results of such work are not immediately related to the identification or production of a new medicine, or its manufacture, but might contribute to future decisions about potential drug targets and ways they are made. The partners in a collaboration may bring different aspects to the work – one partner may have chemical reagents or particular experimental animals or cell lines; the other partner may have technical know-how (for instance, about methods to keep a cell line alive and healthy) or may bring analytical equipment or techniques to the project.

When collaborative projects are set up, the contribution and 'prior art' from each of the partners must be carefully assessed in order to structure an appropriate and fair legal agreement between the partners. Both academic and industry partners contribute to the work throughout the project and share the risk associated with it.

1.16.3.6.2 Contract research

Contracts for research are quite different. The project is always initiated by the industrial partner and usually has clearly defined objectives set by the company. The academic researchers would play the role of carrying out research according to the plan set by the company, and there would be little room for input of ideas from the academic team to affect the overall project plan. Timelines and deliverables would be set by the company. Examples of the type of contract research commissioned by a pharmaceutical company include screening of compounds in a specialist assay or work to develop a mechanism to deliver a drug to the patient. Usually the intellectual property would belong to the company, which would have all the rights. Although the results of the project might be published in the scientific literature or patent literature, the company would be able to reserve the right to set the timing for publication, and could decide to prevent publication altogether for reasons of commercial confidentiality. The results from research contracts are expected to be of immediate usefulness to the company. The academic partner also gains some benefit (in addition to the fee income) because contracts often provide access to novel compounds, which might elucidate a little-explored aspect of physiology or biochemistry and, although the findings might have to remain confidential, the tacit knowledge may be useful in shaping new hypotheses for other basic research projects.

Research contracts are given to companies as well as to academic groups – for example, to contract research organizations that carry out toxicological testing of potential new medicines using in vitro methods and animal studies.

Some contracts for synthesis of compounds, which a company believes might have potential as novel medicines are also increasingly placed with specialist companies as in-house research teams in major pharmaceutical companies limit the range of their exploratory studies.

1.16.3.7 Relationships between Academia and Start-Up Companies

A different relationship exists between academic researchers and start-up companies. Governments in most countries have been encouraging a spirit of entrepreneurship among publicly funded researchers over the last decade and this has led many researchers to consider building businesses on their discoveries and inventions. Many companies are initially set up within a higher education establishment and many of the founding researchers continue to work within their HEI at the same time as they are starting the company. However, if the company is to grow, there comes a time when the initiator has to decide whether to move full-time to the company, or whether to entrust the running of the new company to someone more experienced in the management of a small company while returning to full-time academic work. Sometimes the originator of the intellectual property remains as a scientific advisor to the company or maintains a relationship as provider of further novel ideas or discoveries to the company.

Some academic–company relationships have been set up with government funding. In the UK a scheme funded jointly by the Department of Trade and Industry and several of the Research Councils, known as LINK, helped to bring researchers in HEIs and Research Council Institutes together with companies. The objective was to accelerate the transfer of knowledge and to encourage the take-up and use of discoveries, particularly in small companies. The scheme provided up to 50% of project costs with the remainder being provided by the industrial participant. Sometimes more than one company took part in a project. While LINK covered all areas of science it proved particularly helpful in the life sciences and over more than a decade the UK government financed LINK programs in analytical biotechnology; applied biocatalysis and biotransformations; biochemical engineering; cell engineering; protein engineering; and applied genomics.[8] Some 130 projects were supported under these programs and two case studies are described later in this chapter.

1.16.4 What Legal Frameworks and Practical Arrangements and Incentives Exist to Encourage Collaboration?

1.16.4.1 Collaborative Research in the US

The US National Academy of Sciences set up the National Research Council in 1916 that encouraged collaborative research in the 1920s and 1930s. The National Research Fund, which was set up in 1925, attracted millions of dollars in funds from companies. In 1926, the Division of chemistry and chemical technology had a program to involve academic researchers with problems of interest to industry – but the academics were not told the name of the interested company until the work had been completed. Despite some success, funding from industry for these programs dried up in the depression in 1929.

In the early 1980s, reviews in the US showed there had been disappointingly few medical treatments or medicines, which had been commercialized from government-funded research programs. Laws were passed, and incentives put in place, to encourage companies to take a closer interest in commercializing academic research and to encourage more collaborative projects. One mechanism was the Cooperative Research and Development Agreements (CRADA) under which university and NIH researchers and biotech or pharmaceutical companies make a formal agreement to share personnel, equipment, and services and knowledge within a specific project. This mechanism was reviewed by the NIH in 1995 and found to have resulted in increased exchange of proprietary materials, experimental compounds, and research experience. The issues surrounding ownership of intellectual property were a matter of concern but the NIH work was usually early stage and far from the market. Dr Varmus – who was Head of the NIH – said that "there is an important distinction between having rights to a compound and having rights to a fully developed product. NIH does not license drugs that are ready for marketing. NIH biomedical technologies are early stage and, in almost all cases, require further research, development, and testing, usually in combination with other proprietary technologies, to bring a product to market."

The NIH RoadMap is designed to accelerate medical discovery and use it to improve health. The program recognizes that there is a need for more research collaboration among public and private sectors. One example is the work in the Osteoarthritis Initiative, which is achieving developments that neither the public nor the private sector could achieve alone – it is establishing a database of radiological images, physical examinations, and biomarkers to help develop ways to measure the progression of the disease in a patient. The project will take 7 years and involves several

NIH Institutes with the pharmaceutical companies Merck, Novartis, and Pfizer. The data will be made available to other researchers. A team has been set up to encourage public–private partnerships and it is based in the NIH Office of Science Policy which is part of the Office of the NIH Director. Their work will include advising on IP issues.

Like the US government, many governments in Europe, the EU itself, and many other countries have all made public statements designed to encourage academic–industry links and have set up many support mechanisms to encourage collaborations. One of the most successful EU initiatives worked because it enabled individuals to move and to transfer ideas and know-how as they worked in another country. This was the Marie Curie fellowships program that began under the fifth framework program in Europe. Within the programs available in 2005 there are Marie Curie host fellowships for the transfer of knowledge that are designed particularly to help small companies to develop new areas of competence by hosting experienced researchers. There is an aim to build up durable partnerships between academia and companies through exchange of experienced researchers. The program will integrate researchers in a research environment that is new to them and they will take their knowledge with them when they return to their original organization after a period of up to 2 years.

1.16.4.2 Collaborative Research in the UK

In the UK, the government-funded Research Councils that fund basic research in HEIs are now encouraging academics to form relationships with industrial companies. Following the 1993 White Paper *Realising Our Potential*, the Research Councils were restructured as part of the plan to enhance both the competitiveness of UK industry in global markets and quality of life for the people of the UK. The Councils introduced several schemes offering incentives to academics who collaborated with industry, in order to encourage the commercialization of emerging technology. The objective was to turn some of the basic research, which had been paid for by the taxpayer, into economic activity of value to the nation. Specific schemes in place at that time included the Teaching Company scheme, under which young researchers working for a PhD carried out their research partly in the company and partly in the university with supervision being done by someone from a company as well as an academic. (The Teaching Company scheme has now become the KTP scheme: *see* Section 1.16.3.5.) Another scheme was designed to enable established academic researchers to gain experience of the research environment within companies – the Senior Academics in Industry scheme.

The Medical Research Council (MRC) in the UK has set up a separate organization to commercialize research discoveries – MRC Technology. One of its initiatives is the drug discovery group. That group is looking for industrial partners who may wish to collaborate on projects or may want to take a license for assays, for lead compounds, for reagents, know-how, and other intellectual property. Potential projects are thoroughly assessed by a review committee before they are taken forward, and the assessment includes consideration of the novelty, characterization, and validation of the target; consideration of whether the target is 'druggable'; whether assays, reagents, or cell lines are available; whether a secondary/functional assay exists; and whether there is any competing intellectual property.

In 2005, the BBSRC in the UK announced a new pilot scheme – the industry interchange programme. This scheme is designed to encourage interchange, in both directions, between the science base and industry. The hope is that the interchanges will provide strategic advantage to the science base and to industry by providing reciprocal access to facilities, expertise, and knowledge and, most importantly, will stimulate long-lasting partnerships based on shared interests and concerns.

1.16.4.3 The Legal Framework for Academic–Industry Collaborations in the US

In the US, a law was passed in 1980 – the Bayh–Dole Act – to encourage use of the patent system to promote the commercialization of discoveries that had been funded from federal grants. The Act was introduced at a time when interaction between academia and industry was low. It provided a framework for the ownership of intellectual property arising from government-funded research in universities, small businesses, and in nonprofit organizations.

The Act had the effect of moving the ownership of intellectual property from the government (federal) funder of research to the research institution carrying out the work. As the owner of the intellectual property, the university has the right to grant licenses under the patents it holds. The Act affected research that was co-sponsored by requiring that the universities should own the intellectual property from their federally funded research. This means that if, say, 40% of total project costs come from federal funds and 60% is contributed by the partner company, the intellectual property automatically belongs to the university.

The Act also contained other provisions that give preference for manufacturing products arising from government-funded inventions and discoveries in the US and also a preference for granting licenses to small companies. It also provides uniform guidelines for the granting of licenses. Importantly, the US government retains the right to take title to any invention made by a government employee and to ensure that inventions are properly exploited – these rights are sometimes referred to as 'march-in rights.'

As a result of the Act, most universities set up technology transfer offices, many of which have become very effective in identifying intellectual property in their institutions, and there have been many successful examples of transfer of technology, particularly in establishing the flourishing US biotechnology sector.

The Bayh–Dole Act is seen by many in industry to act as a disincentive to companies to collaborate in fields where there will be high costs to the company to commercialize the results – such as medicines development. This is because the company is unable to own the intellectual property – even though it will be the company that invests the enormous sums to undertake all the research and development (clinical trials and formulation etc.), which is needed before the product can get to the market. Hence, sufficient future royalties and profits may not accrue to the company to offset their investment. There is a lack of flexibility in negotiating deals between companies and universities. The university may not assign ownership of the intellectual property to the company unless there was absolutely no government funding involved in the underlying research (and this is rare).

1.16.4.4 The Framework for Academic–Industry Collaboration in the UK

In 1999, the UK government introduced a specific stream of funding to support and encourage knowledge transfer from the universities. This funding was consolidated and reached £90 million (US$170 million) per year by 2005–2006. This funding has enabled universities to build up their activities to reach out to businesses of all sizes and to interact with regional agencies; has helped them market their research and teaching to businesses; has led to the establishment or strengthening of technology transfer and business liaison offices to provide advice and to negotiate consultancy, collaborative research, and licensing agreements as well as contracts for research; has assisted with the spinning out of start-up companies; provided entrepreneurship training for science and engineering graduates; and helped to find good-quality placements for students in industry.

In 2003, the UK government commissioned a report from Richard Lambert about business–university collaboration (the Lambert Review).[9] This found that British industry was not as R&D-intensive as industry in many other countries and this resulted in a comparatively weak demand side; however, the pharmaceutical and biotech industry sector was an exception. US universities started to commercialize their research earlier than UK universities and some were seen to have more mature relationships with industry. The review noted that US universities earn far more in license income, as a proportion of their research budgets, than UK universities. Nevertheless, the review reported a marked culture change in the UK's universities during the 1990s, with most of them seeking to play a more active part in the regional and national economies. Given that the quality of university research remained high by international standards, and that universities were paying more attention to management and governance issues, the review identified further opportunities. The review recognized that the best forms of knowledge transfer involve interaction between people, and so wanted to see more academics gain first-hand experience of working in companies and also to involve people from companies in taking part in university life.

The report noted that the employment contracts of many UK university researchers limited the number of days per year that the individual was permitted to spend on consultancy (as little as 20 days in some institutions) and contrasted this with the approach taken by the Massachusetts Institute of Technology (MIT) in the US. MIT contracts cover 9 months of the year and the rest may be spent on consultancy or other work for companies. Working with companies is positively encouraged at MIT by removing teaching responsibilities from those who bring income of $2 million or more into MIT, and administrative responsibilities are also removed for those who bring in $4 million. MIT has clear policies to avoid conflicts of interest and the review recommended that clear codes of conduct should be developed to avoid such conflicts that may arise.

The review commended collaborative research projects co-funded by the public and private sectors (citing, among others, those supported by large multinationals such as BAE Systems and GlaxoSmithKline). The review recognized the importance of establishing, at the outset, the ownership and exploitation rights for any intellectual property that might be generated. Both business and universities had informed the reviewer that negotiations over intellectual property were very time-consuming and costly. The use of model contracts drawn up by the UK's Association for University Research and Industry Links (AURIL)[11] and the Confederation of British Industry[12] were recommended as a starting point to help speed up the preparation of agreements.

Richard Lambert thought that the UK technology transfer offices could learn from the US Association of University Technology Managers and more formal professional training for technology transfer staff has been developed.

The review acknowledged that it is very difficult to quantify the economic returns of investment of this kind, since the path from research discovery to commercialization is long and convoluted.

Table 7 The Lambert Review model agreements

Lambert level	*Terms of agreement*	*Ownership of IP*
1	Sponsor has nonexclusive rights to use in specified field; no sublicenses	University
2	Right to an exclusive license to some or all IPR	University
3	Assignment (of specific results) to the sponsor	University
4	Sponsor owns, university has right to use for noncommercial purposes	Sponsor
5	Contract research: no publication without permission	Sponsor

Source: The Lambert Group on Intellectual Property 2005.[10]

The Lambert Review looked at the issue of intellectual property – the legal protection for inventions, brands, designs, and creative works, including patenting. Patent applications in the UK are low and have been falling relative to the US, France, and Germany. A large proportion of UK patent applications come from the life sciences and pharmaceutical sector. Intellectual property ownership is often contested when a research project has received funding from both the university and from a company. Universities want to retain the intellectual property in order to build on it and use it in further research; the company wants ownership in order to protect the further investment, which it will need to put in, in order to move toward commercialization. Companies told the reviewers that they considered that universities overvalued their intellectual property because so much further development work (and substantial costs) would need to be incurred before the research could be marketed.

The Lambert Review recommended that the starting point of negotiations should be that universities own any intellectual property that comes from a collaborative project but with industry free to negotiate license terms to exploit it. However, the review acknowledges that industry might own the intellectual property in projects to which the company has made a particularly significant contribution. The industry contribution often includes nonfinancial contributions such as providing equipment, technologies, and intellectual know-how, and these contributions need to be taken fully into account.

The UK government responded to the Lambert review and decided to task the regional development agencies in England, guided by their science and industry councils, to help a broader spectrum of business to develop more productive links with universities – including support for business-focused research. The extent of business–university collaboration will become one of the measures of the agencies' performance. Of greater importance to the pharmaceutical and biotechnology sectors is the government's decision to increase the Higher Education Innovation Fund (HEIF) to £110 million a year by 2007–08. A further study of intellectual property issues was also initiated and this led to the preparation of the model agreements shown in **Table 7**.

1.16.4.5 Intellectual Property Frameworks in Other Countries

In the UK, there has been no framework within which to help the two sides – industry and academia – to balance their competing interests. However, the UK's university patenting and licensing framework is rather better developed than that in many countries. France, Germany, and Denmark introduced legislation in the late 1990s to permit universities to claim ownership of intellectual property – this framework had been in place in the UK since the 1980s. Japan passed new laws in 2000 to encourage commercialization of university research and to allow university researchers to become involved in company activities.

1.16.5 Trends in Academic–Industry Relationships

Over the last decade, fierce competition in both industry and academia has radically changed the way in which academics and industrialists now interact. Pharmaceutical companies have faced tighter regulation by government authorities in most countries of the world and there has also been pressure to reduce prices. Mergers have also added to the pressure on companies to reduce their costs. As companies have merged, there has been streamlining and consolidation of their research programs, with the consequence that some in-house programs have been cut out. However, work in some fields has been continued through contracts with academic researchers. Some academic groups have realigned their research to coincide with the interests and research needs of industry, as HEIs have been under pressure to increase their income from companies and from collaborative projects.

Another result of mergers has been that some researchers have left companies and have set up new small companies to work on research or development projects, which are somewhere between the traditional work of academic researchers and that of the pharmaceutical companies.

There has been a dramatic change in culture within certain universities and many no longer consider funding from industrial companies to be unwelcome – rather, they positively welcome the income and the intellectual stimulus it can bring. Many academic researchers now consider the quality of research done in, with, and for, the pharmaceutical industry to be basic research of equal scientific interest to many projects funded through Research Councils. By working with industrial companies, academic researchers are able to gain access to equipment, know-how, resources, and novel technology that are not available in their university setting.

In the 1970s, as the pressure on UK universities to look for commercial applications for research increased, organizations were set up to support the growing number of technology transfer officers employed within HEIs. In 1995, a single organization was set up – AURIL. It is a network of professionals whose role is to develop partnerships between HEIs and industry in order to support innovation and competitiveness. Its members come from HEIs in the UK and in Ireland. AURIL runs professional theme groups to help its members to exchange best practice and to learn about policy and legal developments that affect their work. Their eight professional themes include intellectual property and commercialization, research, medicine, and health-related issues. It also has a group on ethics that was established following meeting with the Association of British Pharmaceutical Industries (ABPI) and Universities UK. Another highly respected organization, which has helped to improve the quality of technology transfer in the UK, is Praxis. Praxis is a training organization, which has involved 150 practitioners to train over 1000 people who are now leading their universities' and research organizations' links with industry.

There are many other bodies that play a role in technology and knowledge transfer in the UK – the UK Science Parks Association, which covers all the special locations that have been developed to enable start-up companies to get started, a field in which UK Business Incubation also plays an important part, as it includes managers of the specially built incubator facilities for small companies.[13,14]

1.16.6 Some History of Academic and Industry Relationships

As noted in the introduction, when companies were first set up to make and sell medicines, their staff did not include scientists who had had an academic research training. The initial use of trained scientists within companies was to bring about consistency in their products. May and Baker (which later became part of Rhône Poulenc and subsequently Aventis) won prizes in 1851 for the quality control of its products. Parke Davis was a pioneer in quality control and, by 1883, had 20 chemically assayed fluid extracts of plant-derived drugs, and they used academic chemists to standardize their products derived from ergot.

One of the earliest companies to recognize the value of science in the development of new products was H K Mulford in Philadelphia, whose president wanted to enact a policy of active product development through laboratory science. In 1894, he employed Joseph McFarland of the Medico-Chirurgical College and he, in turn, obtained help from academia in developing vaccines and antitoxins on a large scale. He used the University of Pennsylvania's veterinary school to produce antitoxin and that university's laboratory of hygiene to test it.[15]

At the turn of the nineteenth century in Germany, the synthetic chemistry industry, which was making innovative dyes and antibacterial sulfonamide drugs derived from dyes, relied on universities to do the research. The companies provided raw materials, and paid consultancy fees to the academics and paid patent royalties. The academic researchers supplied inventions and developed new production techniques. In addition, academic specialists became expert witnesses in patent cases. The curriculum in university chemistry courses reflected the interests of industry. In 1895, 500 students from the US went to study in Germany because of the industrial relevance and intellectual challenge, and because the cost of the courses was cheaper than in the US.

As early as 1907, the chemist Robert Kennedy Duncan set up industrial fellowships at the University of Kansas. He thought that industry was inefficient because it was not using research. Graduate students worked under Professor Duncan for a minimum of 2 years and did a few hours' teaching. Under his arrangements any discoveries made during the industrial fellowship became the property of the sponsor, but the firm usually paid a 10% royalty of the profits on any marketed discovery. The company had the right of approval over publications but the fellow was allowed to use the work in a thesis. The sponsor usually employed the fellow for 3 years; however, Duncan said: "it is positively necessary that the professor of industrial chemistry should, so far as any material sense is concerned, gain absolutely nothing otherwise his donors would not trust him with the details of their secret processes."

Duncan addressed issues that continue to concern those involved in academic and industrial collaborations today – questions about payment and trust, issues of ownership of discoveries, and sharing of income from resultant products. There were 10 fellowships at the University of Kansas supported by companies in 1909. When Duncan moved to the Mellon Institute of Industrial Research and School of Specific Industry at Pittsburgh University, the value of fellowships was $40 000 a year and it rose to $238 000 by 1918. By 1929, there were 135 fellows with $700 000 worth of contracts.

MIT set up a laboratory of applied chemistry in 1908 and by the 1920s this was receiving several thousands of dollars per year in research funding. By 1926–27 MIT's research contract income reached $172 000 but the institute closed in 1934 because companies were unable to continue to fund research during the great depression.[15]

In the 1840s, in the US, it was considered unethical for researchers in academia to hold patents and there was some antipathy there and in the UK toward academics who worked on applied research of interest to industry. Collaboration with industry was frowned upon by many professional bodies and learned societies and some expelled members who went as far as having patents for medical discoveries. As late as the 1960s, many academics considered research for industry to be second-class compared with pure research, for which no application or commercialization could be imagined. Some individuals demonstrated that it was possible to maintain respect as a researcher while also having links with industrial companies. One pioneer whose career spanned many different organizations was A N Richards, vice-president of medical affairs at the University of Pennsylvania, who, in the late 1930s became chairman of the wartime committee on medical research of the US Office of Science, Research and Development, President of the National Academy of Science, as well as being a longtime consultant to Merck, where he advised on the setting up of their research programs and was on the board of Merck for 10 years.

1.16.7 Case Studies of Types of Collaboration

1.16.7.1 Case Study 1: Insulin

In 1889 Mering and Minkowski in Strasbourg showed a link between the pancreas and diabetes mellitus and in 1901 Opie at Johns Hopkins showed that diabetes was linked to impairment in the islets of Langerhans within the pancreas. Bayliss and Starling at University College London published findings in 1902 about the way in which production of insulin is triggered, and Starling coined the term 'hormone' to describe the secreted stimulatory agent. Frederick Banting, a physician in London, Ontario, carried out experiments on dogs in the 1920s to investigate the effects of atrophy of the pancreas and he worked with John Macleod at University of Toronto, who had an interest in carbohydrate metabolism. Charles Best, a medical student, worked with Macleod and they were also joined by James Collip, a biochemist on sabbatical. Collip made significant improvements to the extracts of beef pancreas, which were the source of insulin and developed a bioassay from his observations that insulin lowered the blood glucose levels in normal healthy rabbits. In January 1922, that team showed that an extract of pancreas lowered the blood glucose levels in a human patient with diabetes.

The University of Toronto set up large-scale production facilities in the basement – the Connaught laboratories. But the team had recurrent problems in scaling up their production process and so they looked for outside help. They selected the company Lilly as a partner because Lilly had several therapeutic products derived from glands in its 1922 catalogue and had succeeded in making a physiologically stable extract of pituitary gland. The university disclosed all its findings to Lilly, including the methods used in production and standardization of the product.

In April 1922, a patent was applied for to give the University of Toronto – and not any one company – the monopoly with a view to subsequently making the process widely available. Bliss[16] quotes: "the patent would not be used for any other purpose than to prevent the taking out of a patent by other persons – when details of the method of preparation are published anyone would be free to prepare extract but no one could secure a profitable monopoly." Concerns about patenting were to dog the collaborative arrangement between the University of Toronto and Lilly. The original patent had been applied for before the first meeting had been held with Clowes, the Co-Vice-President of Lilly, at which a contract for collaborative research was agreed. That agreement prohibited Lilly from sharing information from the University of Toronto with others and requested Lilly to give information to the university about ways to improve the production, with the proviso that any patents arising should belong to the University of Toronto.

There was a time-limited exclusive license to Lilly, covering the experimental period, during which Lilly had the sole right to produce and sell the product. This right was granted for a year but could be extended by the University of Toronto. During this experimental period, Lilly was to sell the product at cost or supply it without charge to clinicians.

(About a quarter of the production at this time went to Toronto for further experiments.) The University of Toronto had planned to license the production to others after the first year and had guaranteed that Lilly would be granted a license throughout the patent life. Lilly and other licensees were to pay the university a 5% royalty and were to collaborate over how to advertise and over the use of the name Iletin.

The board of governors of the university set up an insulin committee to be responsible for licensing companies to produce insulin and to oversee the quality control for North America. They also screened Lilly's advertising.

The technical developments that made the production process efficient included reductions in the amount of solvent used in parts of the extraction process; carrying out the extraction under refrigeration (thus inhibiting the catalytic deterioration of the active ingredient and solving one of the biggest problems experienced by the university team); and improving the evaporation of the solvent to remove it. Problems over the loss of potency in large-scale production were solved by George Walden, using the novel isoelectric point approach, and by Joseph Loeb at Woods Hole, with whom Lilly had had other collaborations over a long period. Walden filed a patent on the isoelectric precipitation process, which included 30 innovative claims on aspects of the process.

In the 1920s, clinical trials were not compulsory but Clowes set up a clinical advisory group which looked at posology and at methods of administration of insulin to the patient. Clinicians set up 12 groups in eight states in the US and Canada to evaluate the effects of insulin. These groups received insulin free of charge, during the year of exclusivity, from centers set up by Lilly and so they became reliant on that source of supply.

The Insulin Committee rejected Lilly's application for an extension of the period of exclusivity because they had been approached by other companies. The UK MRC had been licensed by the University of Toronto to supervise production in the UK and the MRC had already granted licences to several companies. Lilly's 1923 license from the University of Toronto was nonexclusive and effectively the company was required to put its knowledge into a patent pool because it licensed the Toronto group – royalty-free –to use its production processes and Toronto licensed others, who paid 5% royalties.

Despite the nonexclusivity of its license, Lilly remained the sole producer in the US for another year because other potential manufacturers found it very difficult to standardize their product. In the US Stearns started to sell insulin in June 1924, Squibb in January 1925, and Sharpe and Dohme in April 1925.

The University of Toronto received some $8 million in royalties up to the 1950s when the patent expired – Lilly was paying $50 000 to the university a year by 1957. The university sold the Connaught laboratories in 1972 for $24 million. Lilly's sales of insulin reached $1.1 million in 1923 (14% of the company's total sales) and produced almost 50% of its profits that year.

1.16.7.2 Case Study 2: Integrative Pharmacology Fund in the UK

Given the importance of in vivo physiology, pharmacology, and toxicology in drug discovery and in basic functional genomics research, it was of great concern to UK-based pharmaceutical companies when surveys revealed a shortage of skills in these fields. These disciplines require knowledge and experience of working with live animals and living tissues. Many existing experts and trained animal technicians are due to retire in the next 5 years. In order to address this skills shortage, AstraZeneca, Pfizer, GlaxoSmithKline, and the British Pharmacological Society decided in 2004 to set up an initiative to revitalize the in vivo sciences research and training base in the UK and they hope that other pharmaceutical and biotechnology companies will join the activity. They have established a fund to be used over 4 years. The project is being funded by a wide range of partners – industrial companies are putting in £2 million, BBSRC £2.9 million, MRC £2 million, and the Higher Education Funding Council for England (HEFCE) £4 million, and the Scottish Funding Council (SFC) £1 million – a total of almost £12 million (US$22 million). The funds will be used to support in vivo research and training at the academic centers in the UK, which have the greatest expertise in these scarce skills and which have excellent facilities and very high standards of animal welfare.

This initiative fits in well with the strategic approach of the research councils and the funding councils – the BBSRC, which is looking to build skills in integrative or systems biology, and the MRC, which is also trying to build capacity in these fields. The BBSRC has set up industrial partnership awards and capacity-building awards to provide funds for established researchers. Some academic fellowships from the combined Research Councils for new lecturers and fellows are designed to build capacity. Some funds will be used to supply additional consumable items for postgraduate trainees so that they can use whole animals and have the equipment needed to keep them in high levels of welfare. The universities will be expected to provide organizational and physical infrastructure. The British Pharmacological Society will be supporting undergraduate training. These various awards will be taken up in more than 10 different universities around the UK. The universities will define the details of each award since they will be flexible. The universities will be expected to maintain the posts, after the initial pump-priming funding has ended, and the funders will seek a formal statement of long-term commitment by each university to continue in vivo work. The capacity-building awards will provide a greater pool of skilled researchers with whom companies can collaborate as well as providing individuals who can be recruited by companies for the essential skills they need. Young people trained in in vivo science will be able to work with companies and authorities to reduce and to refine the research methodologies involving living animals and also to develop alternatives to the use of animals.

1.16.7.3 Case Study 3: Dundee Kinase Consortium

Consortium projects are particularly appropriate when they involve big science or specialist facilities but also when there is a large amount of basic biology to be undertaken in a new or emerging field or on a class of compounds, which look particularly interesting in terms of potential applications in medicines. In a consortium, the costs and the risks can be shared.

Setting up a pioneering model for collaboration – which brings together the world's greatest concentration of expertise on kinases with the research strength of the MRC and the financial backing and drive of six of the world's major global pharmaceutical companies – was not easy. Two years of difficult negotiations (involving much discussion among lawyers about intellectual property issues) preceded the establishment of the consortium in 1998.

The consortium has been built around the University of Dundee Division of Signal Transduction Therapy. In 1998, a 5-year agreement was signed involving the University of Dundee (where the research is being carried out under Professor Sir Philip Cohen and a team of some 70 researchers) and the MRC (the funder of the initial research and facilities) and six major global pharmaceutical companies – Pfizer, AstraZeneca, GlaxoSmithKline, Boehringer Ingelheim, Merck Co. (New Jersey), and Merck KGaA (Darmstadt).

The research work is looking at a gene family, the kinases, which are involved in almost every physiological process in the body. The enzyme systems under study play a vital role in cell signaling and cell growth. It is possible that greater understanding of the structure and function of these families of compounds might lead to identification of potential drug targets as well as helping to understand the basic biology of processes within and between cells in the body. Targets in cancer, diabetes, cardiovascular diseases, inflammatory diseases, and allergies are emerging from the research.

The industry members of the consortium will get access to the know-how and research findings coming from more than 170 world-class scientists working on the projects; have a screening facility for a range of kinases; and be able to obtain proteins and biological reagents and also antibodies made through custom synthesis. In 1999, Pfizer described this consortium as its most important academic collaboration worldwide.

The university will benefit from access to the most selective chemical tools that industry can provide; benefit from increased efficiency in processes to synthesize compounds by using semi-industrialization available from the companies; and have access to industrial companies' know-how, best practice, and direction. The results from the consortium work will be published in top journals – with joint authorship.

In 2003, support for the consortium was renewed by the partners for a further 5 years and the team is being strengthened in bioinformatics and related areas.

1.16.7.4 Case Study 4: Bradford Particle Design

A group at the University of Bradford Pharmaceutical Technology Department received funding of £3 million (US$5 million) from Glaxo Wellcome in the early 1990s to develop reliable processes for producing pharmaceutical products as dry powders. Once progress had been made in this collaboration Bradford were able to apply for support from the UK's Department of Trade and Industry to optimize its prototype lab-scale equipment (1995–96) and then to build and optimize its pilot plant (1996–98). Further government funding – jointly from the Department of Trade and Industry and the BBSRC under the LINK scheme – in 1997–2000, enabled the group to develop processes for the production of proteins and biopharmaceuticals.

Bradford Particle Design had developed a technology for making particles using supercritical fluids. This technology enables small-molecule pharmaceuticals to be produced with small, highly crystalline particles of defined shape and size, in a reproducible and repeatable manner. The extension of this approach to the production of dry particulate biologicals was supported as a LINK project. The academic partners were the Centre for Biochemical Engineering at University College London and the School of Pharmacy and the Department of Biomedical Science at the University of Bradford.

Through this work 26 staff (with 10 PhDs and 10 graduates) collaborated with 14 different pharmaceutical companies. In December 2000, Inhale Therapeutic Systems offered $200 million for Bradford Particle Design. In 2003, a new corporate brand was established for the drug delivery company – Nektar – which then involved Inhale

Therapeutic Systems, Shearwater, and Bradford Particle Design. Nektar has worked with 25 biotech companies and 15 pharmaceutical companies. It has taken part in 50 partnered programs and about 20 of these have involved clinical testing of the compounds produced.

1.16.7.5 Case Study 5: Imperial College London/Hammersmith Hospital/GlaxoSmithKline Imaging Centre

One of the largest collaborations between industry and academia in the UK, in financial terms, is the establishment of an imaging center located at the Hammersmith Hospital site in London. The project will cost £76 million (US$143 million) and £28 million will be contributed by the industrial partner GlaxoSmithKline. Imperial College London has signed a 10-year research agreement with GlaxoSmithKline for a program of research on medical imaging. The center will have the latest magnetic resonance imaging equipment and positron emission tomography equipment.

The center will provide an opportunity to look in real time at the chemical processes taking place in cells in human organs and will be capable of revealing the immediate changes to biochemistry that occur when a medicine is administered.

1.16.7.6 Case Study 6: Nuclear Magnetic Resonance – Oxford University and Oxford Instruments

A long-standing collaboration exists between Oxford University and the company – Oxford Instruments – which develops, manufactures, and supplies equipment to generate high magnetic fields to scientific instrument makers.

In 1994, the world's first 4.2 K, 750 MHz magnet was operational in the Department of Biochemistry in Oxford. Support under the UK government's LINK scheme enabled the partners to design and develop the subsequent 800 and 900 MHz superconducting magnets needed for ultrahigh-field nuclear magnetic resonance systems. One area of work explored in collaboration included looking at the cost-effectiveness of moving from 600 to 750 MHz. The university researchers provided the biological specimens and worked on interpretation of the data. The academic department benefited by being able to attract key research staff and from the training that many received as part of the trials of the new equipment.

1.16.7.7 Case Study 7: Atracurium: A Patented Molecular Structure with Activity Against a Validated Target

During the mid-1970s, Professor John Stenlake and his team in the Department of Pharmaceutical Chemistry at the University of Strathclyde were working on a series of synthetic curare-like molecules that would be free of the side effects exhibited by the natural product. These muscle-relaxant compounds were designed to be used as an adjunct to anesthesia and the technical team was attempting to create molecules that were free of muscarinic (tachycardia) and sympathetic (hypotension) side effects.

Through a chance encounter between Professor Stenlake and a senior scientist within the Wellcome Foundation, an initial informal collaboration was set up between the university and the company. The arrangement was that the university group would continue to develop the chemistry associated with the novel molecules while the company would concentrate on testing the molecules against validated targets for muscle relaxation.

A patent was filed in 1976, the inventors of which were Professor Stenlake and a number of his group. The company assumed responsibility for the prosecution of the patent and negotiated an assignation of the rights to the company in return for a revenue-sharing agreement with the university. This was signed by the university in the late 1970s and, by the mid-1980s, the first royalty check was received by the university – the first payment in what was to result in a *c.* £26 million return to the university over 15 years. This financial return to the university was shared on a 50 : 50 basis with the research group.

The product, atracurium (marketed as Tracrium), was sold and marketed by the Wellcome Foundation for some 10 years and achieved a market value in the US and Europe of in the region of £100 million per year, making it the best-selling compound in its class in the US, while its competitor vecuronium (a compound marketed by Organon and co-developed by the pharmacology group at Strathclyde University) enjoyed this status in Europe. Despite achieving this financial success, the group at the university continued to develop new analogs of atracurium, as did the scientists at Wellcome. A further series of molecules was developed and patented by Wellcome, which led to the successor product, Nimbex. The university and the company (now GlaxoSmithKline) have reached a supplementary commercial agreement to cover Nimbex during the period that it was dominated by the original patent rights.

References

1. US National Science Foundation. *Science and Engineering Indicators – 2004 and Pocket Data Book 2000*; NSF: US.
2. OECD. *Main Science and Technology Indicators*; OECD: Paris, 2005.
3. OECD. *Science, Technology and Industry Outlook*; OECD: Paris, 2004.
4. Human Genome Project Information. http://www.ornl.gov/sci/techresources/Human_Genome/home.shtml (accessed Aug 2006).
5. Structural Genomics Consortium. http://www.sgc.utoronto.ca/ (accessed Aug 2006).
6. NIH Roadmap for Medical Research. http://nihroadmap.nih.gov/ (accessed Aug 2006).

7a. DTI Knowledge Transfer Partnerships. http://www.ktponline.org.uk/ (accessed Aug 2006).

7b. Department of Trade and Industry. http://www.dti.gov.uk/ (accessed Aug 2006).

8. DTI Innovation Site. http://www.innovation.gov.uk/ (accessed Aug 2006).
9. Lambert Review of Business–University Collaboration. *Final Report*. HMSO, 2003.
10. Science and Innovation Investment Framework 2004–2014. HM Treasury, Department of Trade and Industry (DTI), Department for Education and Skills; HMSO 2004. http://www.hmtreasury.gov.uk/.
11. Managing IP – A Guide to Decision Making in Universities; AURIL, Universities UK, Patent Office; 2002.
12. Partnerships for research and innovation between industry and universities; Confederation of British Industry (CBI), 2001.
13. The United Kingdom Science Park Association. http://www.ukspa.org.uk/ (accessed Aug 2006).
14. Cordis Business Incubator Site. http://www.cordis.lu/incubators/ (accessed Aug 2006).
15. Swann, J. P. Academic Scientists and the Pharmaceutical Industry: Cooperative Research in Twentieth Century America; Johns Hopkins University Press: Baltimore and London, 1988.
16. Bliss, M. *The Discovery of Insulin*. University of Chicago Press: Chicago; McClelland and Stewart: Toronto, 1982.

Biographies

Monica Anne Darnbrough, CBE, after studying psychology and physiology for a joint honours degree at Nottingham University, did research into the effects of hormones on physiological and behavioral development for her PhD. Having joined the scientific civil service, Monica worked in the UK Home Office (police scientific development branch), Department of Trade and Industry (DTI), and Foreign and Commonwealth Office (as Counselor for Science and Technology in Paris). In the 1980s, Monica was seconded to the Cabinet Office and worked in the small team of Prime Minister Margaret Thatcher's Chief Scientific Adviser and she also undertook a review of government laboratories led by Sir Derek Rayner. From 1998 until 2005, Monica was responsible for developing government policies to support the pharmaceutical and biotech sectors in the UK.

Malcolm Skingle has a BSc in pharmacology/biochemistry and a PhD in neuropharmacology. He has worked in the pharmaceutical industry for more than 30 years and has gained a wide breadth of experience in the management of research activities. Part of his former role as a research leader in a neuropharmacology department involved co-supervising collaborations with academics in the UK, continental Europe, and USA. He has more than 60 publications, including articles on the interface between industry and academia. For more than a decade he has managed Academic Liaison at GlaxoSmithKline and has staff in Stevenage, Research Triangle Park, and Philadelphia. This role involves close liaison with several groups outside the company, e.g., government departments, research and funding councils, small biotechnology companies, and other science-driven organizations. He sits on many external bodies, including the BBSRC Strategy Board, the East of England RDA Science and Industry Council, the CBI academic liaison group, and several UK university department advisory groups. He chairs several groups, including the BBSRC Bioscience for Industry Panel, the Diamond (Synchotron) Industrial Advisory Board, the Inner Core Lambert working group on boilerplate agreements, and the ABPI group working on academic liaison.

Comprehensive Medicinal Chemistry II
ISBN (set): 0-08-044513-6

ISBN (Volume 1) 0-08-044514-4; pp. 571–590

1.17 Is the Biotechnology Revolution a Myth?

M M Hopkins, University of Sussex, Brighton, UK
A Kraft and P A Martin, Nottingham University, Nottingham, UK
P Nightingale and S Mahdi, University of Sussex, Brighton, UK

1.17.1 Introduction

1.17.1.1 The Biotechnology Revolution and Policy

Over the last decade consultants, policy-makers, academics, and industrialists have promoted a model of technical change in which biotechnology in general, and genomics in particular, are seen as revolutionizing drug discovery and development.[1–7]

This conception of revolutionary technological change has engendered a range of widely shared expectations about the impact of medicinal biotechnology. In particular, it has helped generate an expectation that biotechnology has the potential to lead to a substantially increased number of more effective drugs, therapies, and diagnostics, and more effective ways of developing and prescribing drugs, all of which will lead to improvements in healthcare.[8–12]

The emergence and growth of biotechnology is also perceived to have the potential to induce structural change within the pharmaceutical industry by challenging the historical dominance of large drug companies. In particular, some see potential for the main locus of drug discovery to shift from large pharmaceutical firms to networks of biotechnology firms either agglomerated in regional clusters, or working together in virtual networks but as geographically separated firms.[13–15] In the 1990s, advocates of genomics-based approaches to drug innovation argued that the new genomics paradigm offers potential solutions to the pharmaceutical industry's productivity crisis. Taken together, all these envisaged changes have provided the basis for expectations of a fundamental transformation in healthcare and medicine, in which biotechnology is also cast as a source of increased wealth creation.[5,13,15–17]

As a result, agencies at the regional, national, and supranational levels are investing heavily in biotechnology and genomics in order to establish a foothold in what is seen as a key part of the New/Knowledge Economy.[6,17–19] Policies based on the revolutionary model of biotech take a number of forms, and include dedicated research funding programs, fostering knowledge/technology transfer, financial and technical support for start-up firms and regional clusters, R&D tax credits, and lower regulatory hurdles.[17–20] In the UK, a recent report by the Bioscience Innovation and Growth Team (BIGT) argued that, in order to realize the potential of the molecular biosciences, there is a need for significant changes in the relationship between the National Health Service and industry, to allow easier clinical trials, and earlier and cheaper access to new medicines.[5] Similar policies are being promoted to change the relationship between university research and industry.[17] The hopes and promises that attach to biotechnology are of far-reaching significance, not least because they underpin many national and regional economic development strategies[13] and changes in science and technology policy throughout the OECD[3,7] (for the USA,[9] and the EU[6]). In the private sector expectations about the impact of biotechnology have led to annual investments amounting to tens of billions of dollars in biotechnology firms by the investment community as well as substantial investments by leading large firms.[21,22]

Given that the expectation of a biotechnology revolution has exercised such powerful influence over government policy and private sector investment decisions there is a need for an empirical audit to evaluate the evidence for such a revolution, particularly as similar expectations built up around the dot.com industries less than a decade ago. It is important to bring any gap between expectations and reality to the attention of various stakeholder groups such as fledgling biotech start-ups, the wider pharmaceutical industry policy-makers, economists, and the investment community.

In this chapter we examine the evidence for a medicinal biotechnology revolution within healthcare, and specifically for drug innovation. In an extension of earlier work[23] we focus on the scale, scope, and speed of change, and the impact of biotechnology (particularly genomics) on the drug innovation process. For analytical purposes, drug innovation is divided into three distinct but connected stages: drug discovery, drug development, and clinical application postregulatory approval. Drug discovery refers to the early stages of finding a chemical lead (lead compound) or biological molecule that has promise as a potential drug candidate. Drug development includes the preclinical testing of lead compounds in vitro and in animals, as well as clinical testing in human studies of safety and efficacy. Clinical application refers to the extent to which new therapeutic and diagnostic products have been adopted into routine clinical practice within healthcare systems.

Since the field suffers generally from overuse of key terminology and ambiguities in their meaning, we begin by clarifying some of the most important terms.

1.17.1.2 What Do We Mean by Biotechnology?

Biotechnology can be broadly defined as the application of biological organisms, systems, and processes to manufacture products or provide services.[24] Three generations of biotechnology have been proposed, beginning with the use of whole organisms (initially, unknowingly) in fermentation – for example, in brewing.[25,26] The second generation exploited greater microbiological understanding and led to development of culture and extractive techniques in the first half of the twentieth century (e.g., for the production of antibiotics from fungi). The third generation, dating from the 1970s, is related to the isolation and application of restriction enzymes and monoclonal antibodies (e.g., recombinant production of insulin in bacteria, monoclonal drugs from mammalian cell hybridomas). The long history and breadth of biotechnology – in terms of, for example, activities, technologies, spheres of application – render precise and universal definitions of its meaning difficult. This chapter focuses on third-generation biotechnology, with specific reference to its application in the pharmaceutical innovation process. Thus we examine the use of recombinant DNA and monoclonal antibody-based technologies since the 1970s, the emergence of genomics as a subset of biotechnology in the 1990s, and wider developments in the postgenomics period.

1.17.1.3 What Do We Mean by Revolution?

The term 'revolution' is also problematic and its use and meanings have been the subject of debate amongst historians.[27–29] Within discussions of the impact of new technology, revolution is typically used to indicate radical change and draws on two distinct traditions of academic writing. The first addresses industrial revolutions, and is subject to wide variation in its use and the extent to which it is supported by evidence.[28] The second draws on the work of the Austrian economist Joseph Schumpeter (1883–1950), who argued that capitalist society is regularly subjected to revolutionary innovations that disrupt and destroy the existing economic order.[30] In both instances the term revolution is relative and relates to the time period under discussion. Used in this context, revolution metaphors not only imply a sense of speed, but also suggest that changes will be caused by the technology.

The historical literature on the impact of major technologies has highlighted the incremental, slow, and often indirect processes whereby new technology generates changes in organization, industrial structure, firm performance, and productivity.[31–37] Such studies suggest that new technologies can often take decades to produce an impact on productivity. When technologies are first introduced, typically as new processes for producing existing products, they are often in a very primitive form, focused on a narrow technological and sectoral use, and are subject to large and often increasing development costs.[34] Only after many years are the complementary technical and organizational innovations generated that allow technologies to be exploited fully.[31–34] The introduction of electricity, for example, required changes in factory organization as electricity allowed a more decentralized form of production than steam power which relied on shafts to distribute power.[38] This is in marked contrast to Schumpeter's position that took innovation as an event and conflated invention and the processes of innovation and diffusion,[39] thereby suggesting that the initial act of invention is rapidly followed by productive application. By contrast, Freeman and Louca[40] have proposed that it can take 40–60 years before the full impact of major innovations is felt. This timeframe for impacts has been endorsed by numerous studies, including von Tunzelmann[41] on steam power and David[31] on electricity. If biotechnology were demonstrated to be following this well-documented pattern of technological change, then this would give cause for scepticism about the notion of a biotechnology revolution in the currently accepted sense.

1.17.1.4 Structure of the Chapter

The aim of this chapter is to describe the contours of change associated with biotechnology in a range of contexts and measured by numerous criteria to assess the extent to which its impacts justify their description as revolutionary. In assessing the evidence for a biotechnology revolution we consider three areas in which discourse of revolutionary change has been prevalent. We begin in Section 1.17.2 with an analysis of changes in drug discovery, before turning in Section 1.17.3 to consider the impact of biotechnology in drug development. Section 1.17.4 then surveys changes brought about in the clinic as a result of the diffusion of biotechnology-derived drugs. In each of these sections we explore the empirical evidence for a biotechnology-driven revolution using both quantitative and qualitative indicators of change. For each of the three stages of the innovation process outlined in Sections 1.17.2–1.17.4 we examine the scale of change, the scope of change, and the speed of change, guided by the following questions:

- Scope: how different are the technologies of drug innovation and the products created from those produced before the advent of biotechnology (say 1975)?
- Scale: how widespread have these changes to the innovation process and the development of new products been since 1975? To what extent are improvements apparent in key metrics?
- Speed: how rapidly have new processes been adopted within the industry and new products applied in the clinic?

In Section 1.17.5, we critically assess the evidence and discuss the implications of our findings.

1.17.2 The Impact of Biotechnology on Drug Discovery

1.17.2.1 The Context of Change in Drug Discovery

Drug discovery can be defined as the process of creating chemical or biological compounds that have a desired biological effect in an appropriate testing or assay system, and have the potential to be developed as therapeutic agents. Drug discovery has changed markedly during the twentieth century. One of the most important shifts occurred during the 1970s with the move from the highly successful discovery paradigm of random screening toward what is referred to as rational drug design (RDD).[42] This shift was both target and theory driven, with greater emphasis on understanding the biology and physiology of the pathological process against which a drug was targeted and on the properties of drug candidate molecules.

RDD gave rise to many novel drugs, beginning with Captopril, and produced important changes in the practice and organization of drug discovery.[43] New classes of therapeutics were made possible by improved understanding of the physiology of pathology, exemplified by a new generation of highly profitable cardiovascular disease (CVD) medicines.[44] This changed the research portfolios of pharmaceutical companies and is associated with a shift away from antibiotics and the treatment of infectious diseases. New business strategies were adopted that focused on chronic diseases as highly profitable markets. Initially, this strategic shift paid off and was closely associated with the birth of the 'blockbuster culture' with Zantac (ranitidine), the antiulcer drug launched by Glaxo in 1982, becoming the first blockbuster drug (defined as generating sales in excess of $1 billion). By 1990 the portfolios of large pharmaceutical firms typically included multiple blockbuster drugs such as Tagamet (cimetidine), Vasotec (enalapril), and Mevacor (lovastatin).

1.17.2.2 The Scope of Change in Drug Discovery

The impact of biotechnology on drug discovery falls into three main areas. Firstly, the prospect of new therapeutic modalities where biotechnology provides the product itself (such as a protein, or nucleic acid). Secondly, the use of biotechnology as a set of research tools to guide the development of small-molecule drugs. Thirdly, biotechnology has impacted the industry more indirectly by helping to change its shape, particularly through the injections of risk capital and the founding of a diverse range of small to medium sized enterprises (SMEs) with novel business models. We examine each of these areas in turn below.

The first of these domains began to open up during the 1980s when great interest was expressed in the potential of molecular biology and biotechnology to create new classes of 'biological' products, such as therapeutic proteins, monoclonal antibodies, and vaccines. Whilst large companies were riding the wave of blockbuster successes based on RDD, this period also saw important advances in the nascent biotechnology sector. Based on recombinant DNA (rDNA) and hybridoma techniques, drug development within the biotechnology industry of the 1980s initially focused on recombinant protein or monoclonal antibody-based drugs, widely referred to as biopharmaceuticals. A significant number of the early biopharmaceuticals were based on established therapies, with innovation in manufacturing rather than in drug discovery, design or development. For example, during the 1980s 15% of clinical projects involved insulin or human growth hormone.[45] The first biopharmaceutical – recombinant insulin (Humulin) – was launched in 1982, the outcome of a joint project between Genentech and Eli Lilly. However, despite interest from a few leading companies, the majority of the pharmaceutical industry adopted a 'wait and see' approach in respect of these new (potential) sources of innovation. This mainly left the discovery and development of the first wave of biopharmaceutical products to first mover biotechnology companies, such as Genentech and Amgen. In the two decades following the discovery of the first biopharmaceuticals, this continued to be the case largely due to difficulties in establishing the therapeutic properties of protein drugs. There are a number of reasons for this: proteins are large molecules, susceptible to digestion and not suitable for oral delivery; they often cannot cross membranes and therefore have poor availability; they may be immunogenic, especially with long-term use; and may not be specific to the disease target.[46,47]

Only in more recent years have large companies started to invest more heavily in biopharmaceuticals in general and monoclonal antibodies in particular, although small firms continue to be the main source of new biologicals. This is a result of the increasingly urgent search for new products stimulated by dwindling pipelines and the recent commercial and clinical success of biological drugs. As a result, a number of large companies have established their own internal program to work on the discovery of protein drugs. Despite this, investment in protein drugs is far from universal and it remains difficult to discover novel proteins that are both safe and effective. As a consequence, a large number of new products in this class continue to be second or third generation derivatives of established protein therapies (as discussed further in Section 1.17.4).

In contrast, monoclonal antibodies (mAbs) are proving to be an increasingly successful class of therapeutics. The first generation of products was based on murine or partially humanized antibodies and suffered high immunogenicity as a consequence. However, the development of chimeric and fully humanized mAbs in the last decade has seen the discovery and development of a number of highly effective therapies. This has stimulated significant investment from the pharmaceutical industry in this technology, mainly through strategic alliances. The therapeutic progress of mAbs is discussed further in Sections 1.17.3 and 1.17.4.

The second promising domain offered by biotechnology began when molecular biology started to have an important impact on the discovery of new drugs from the mid-1980s onwards. By 1990, before the advent of genomics, expectations were already high that biotechnology could help transform drug discovery as illustrated by the quote below (from Knight, an industry commentator writing in the journal *Bio-Technology* at the time):

> It is perhaps a measure of how far biotechnology has come that its tools are now being applied to revolutionise traditional small molecule drugs. Recombinant protein receptors for crystallographic modelling and drug screening, engineered enzymes for stereospecific synthesis [reducing isoforms] and enhanced understanding of cellular biochemical processes will make major contributions to new drugs designed by a rational process.[48]

The overall integration of new biology-based technologies into the pharmaceutical innovation process since the 1980s is revealed by studying their pattern of publications in scientific literature. These can be used to produce a technological fingerprint of the largest companies. **Figure 1** describes the pattern of publications produced by GlaxoSmithKline from 1979 to 2003 and demonstrates the increasing accumulation of capabilities in biotechnology over time (e.g., rDNA, gene expression, etc.).

Initially, techniques such as molecular cloning, gene and protein sequencing, computer modeling, and an improved understanding of the molecular pathways involved in pathology were used to support the further development of RDD.[43] By the early 1990s these were widely used throughout the industry. However, it was only with the advent of genomics that the full potential of biotechnology in the drug discovery process became apparent. Ironically, this was achieved through a move away from RDD and back toward the directed screening of large compound libraries.[49]

The launch of the Human Genome Project and the availability of high-speed DNA sequencing equipment in the late 1980s and early 1990s marked an important shift in the application of biotechnology to the drug discovery process.[49] The powerful appeal of genomics lay in the hope that it might provide tools to generate and validate targets for a new generation of novel small-molecule, blockbuster drugs based on novel proteins coded for by recently discovered genes.

Competencies in DNA sequencing, gene expression, and bioinformatics technologies were rapidly developed in large companies throughout the industry through the creation of in-house research groups, as well as external collaborations with academia and the recently created genomics sector.[49] Critically, the increasing availability of these novel targets coincided with the introduction of important new platform technologies, most notably combinatorial

Subject	1979	1980	1981	1982	1983	1984	1985	1986	1987	1988	1989	1990	1991	1992	1993	1994	1995	1996	1997	1998	1999	2000	2001	2002	2003
Recombinant DNA					2	0	1	0	1	1	1	1	3	1	4	2	2	1	1	0	11	8	3	2	1
Sequencing DNA					1	0	3	1	1	5	4	10	7	12	9	15	13	25	20	21	14	41	12	13	15
Gene cloning						1	3	1	3	4	5	15	18	18	16	15	17	18	14	11	16	25	9	9	5
Protein sequence							3	0	4	6	4	13	20	21	20	27	27	42	29	38	28	39	11	25	20
Protein expression							1	0	0	2	0	0	1	3	3	2	2	3	6	10	6		2	11	9
Gene expression									1	2	1	2	8	6	6	11	17	9	25	30	28	28	21	26	21
Sequence homologies											1	2	6	1	4	2	0	4	5	5	7	6	0	7	4
Transgenic animals											1	2	0	1	1	2	1	5	5	6	5	7	4	2	2
DNA Database												6	3	5	5	5	6	8	7	12	6	7	5	3	1
Bioinformatics																1	0	2	3	4	4	11	12	10	9
Gene function																	1	1	3	1	5	2	7	3	1
Genome mapping																	1	0	1	1	2	4	2	2	1
Gene knockout																	1	0	1	2	1	2	1	1	0
Genotyping																		2	2	0	2	5	1	2	4
Population genetics																		1	1	1	1	5	0	5	0
Pharmacogenetics																		1	0	1	1	7	2	6	2
Microarray																			1	0	1	7	4	7	7
Proteomics																				1	2	9	6	6	6
SNP Analysis																					1	4	2	4	2
RNAi																								1	0

Figure 1 Accumulation of research interests related to exploitation of genomics of GlaxoSmithKline, 1979–2003. Years in which GlaxoSmithKline and its constituent predecessors (such as Wellcome) published in a field are highlighted in gray, with darker shades indicating higher activity. Where GlaxoSmithKline discontinues publishing in a field, that year is colored in yellow. The cream color indicates publishing activity outside the firm, while no shading indicates that the field had yet to emerge and thus there were no publications. The limitations of publication counts mean that they are an indication of internal interest in a field, rather than a measure. Counts are affected by a number of factors such as the time lag between research and publication, the changing propensity to publish over time within firms and even departments, and changes in the propensity to use the keywords used to classify the data (although these are minimized here by the use of sets of keywords to describe a field). (*Source*: Surya Mahdi, SPRU pharmaceutical industry database.)

chemistry and high-throughput screening (HTS). These enabled the generation of very large chemical libraries and their rapid screening against a particular target.[42]

However, the new targets flowing from genomics in the 1990s were often poorly characterized, with companies having little understanding of their role in disease pathology. The number of scientific papers associated with each target worked on by the industry fell from 100 in 1990 to eight in 1999.[50] Furthermore, the industry's failure rate of drugs against novel targets appears to be 50% greater than for drugs against clinically validated targets.[51] In other words, genomics vastly increased the potential for innovation within the industry, but shifted the bottleneck from the identification of targets to their biological characterization and functional validation through the elucidation of pathological processes for complex diseases at the molecular, cellular, and system levels. New techniques for validation have been developed in the last decade, including the use of more animal model systems, gene transfer techniques, gene expression profiling, gene knockout and knockdown tools, and large genetic databases.[52–54] The development of new genomics-based therapeutics therefore requires a vast new biological enterprise in order to establish and characterize disease correlates.

The third way in which biotechnology has impacted the pharmaceutical industry is that the promise of these new therapeutic and research approaches created potent expectations of technological change that resulted in structural change in the industry. Since the early 1980s a very large commercial sector has developed with between 4000 and 5000 businesses globally.[21] This industry is supported by a growing range of government policies[55] and billions of dollars in risk capital. For example in the five years up to 2004 the industry raised an estimated $80 billion, with $18 billion invested in biotechnology during 2003 in North America and the Europe alone.[21] Healthcare is by far the largest market focus for biotechnology companies, many of which survive the lean years awaiting maturation of their therapeutic pipeline by offering tools and services relating to drug discovery to the pharmaceutical industry. The scale of these activities is discussed below.

1.17.2.3 The Scale of Change in Drug Discovery

There has been a very significant increase in the number of small firms within the biotechnology industry seeking to discover new drugs since the 1980s. This is partly explained by the overall growth of the biotechnology sector as a whole, but also by the shift from the development of technology platforms and the provision of contract research services, to a focus on the discovery and early stage development of new therapeutic products.[56] Furthermore, in the early 1990s many biotech firms adopted a strategy of focusing on small-molecule drugs that are more easily absorbed, less likely to be immunogenic, and are orally available and therefore easier to distribute in the body.[48] This move is well illustrated within the genomics subsector, where the number of small genomics firms working on target identification and validation increased from fewer than 10 in 1990 to over 70 in 2002 (**Figure 2**).

At the same time as the rapid growth in the number of small firms working on drug discovery, large companies have been outsourcing increasing amounts of their research activities to the biotechnology sector. This has been achieved through the creation of a large number of strategic alliances with biotechnology SMEs. By the end of the 1990s a new form of heavily networked industrial structure was emerging[57] with large companies committing as much as 30% of

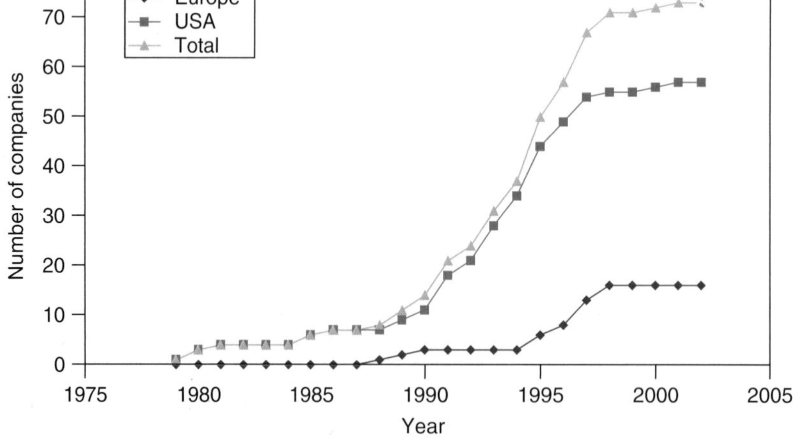

Figure 2 Genomics firms developing target identification and validation technology. (*Source*: IGBiS Biotechnology Database.)

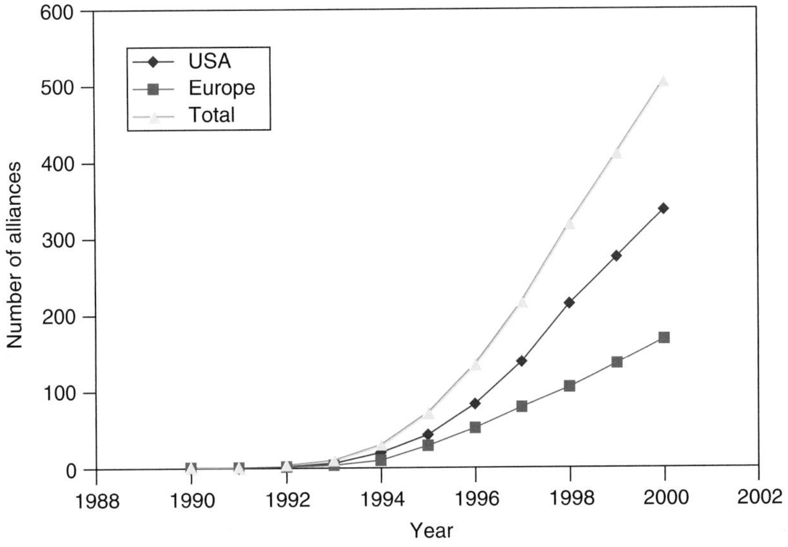

Figure 3 Formation of alliances in target identification and validation. (*Source*: Recap.com.)

their R&D budget to technology and product development collaborations with smaller companies. Part of this general trend has seen large companies increasingly outsourcing drug discovery activities. As a consequence, there has been a significant increase in the number of drug discovery collaborations in the last decade. For example, within the genomics sector the number of collaborations relating to the area of target identification and validation and recorded on the Recap alliances database had grown from zero in 1990 to over 500 by 2000 (**Figure 3**).

The increasing role of the biotechnology sector as a source of new drugs (both biological and small molecule) is illustrated by both a rise in the total number of biotech-derived products and the significant proportion of all new molecular entities submitted to the US Food and Drug Administration (FDA) (which we discuss in Section 1.17.4).

As **Figure 3** shows, target identification and validation has been a major focus for the sector and one where revolutionary change was expected over a short period of time. In the mid-1990s the number of molecular targets that all available therapeutics acted on was estimated to lie in the region of 417, and genomics was expected to raise these by an order of magnitude by providing between 3000 and 10 000 new drug targets over a time period of less than a decade.[58] More recent estimates of the number of 'druggable' targets (i.e. those proteins, encoded by genes, which small molecule drugs can bind to and moderate a disease process) are a more modest 600–1500, based on the estimated intersection of the set of small molecules with necessary pharmacological properties (such as oral availability) and the set of proteins with suitable sites for these to bind.[59] Nonetheless, a single target may be associated with multiple diseases and new biotechnology modalities expand the range of targets that drug developers can act on because their mode of action does not rely on the competitive binding to relatively small sites on proteins.[59] Thus even though the number of genes in the human genome is likely to be 20 000–25 000 rather than 100 000,[60] the number of new therapeutic targets available for exploration in early stage drug discovery has risen in an unprecedented manner in just a few years. However, many of the drug candidates acting on these targets remain undisclosed in early stage pipelines, whilst information in the public domain suggests few firms have taken drugs based on genomic targets to the clinic in recent years.[61]

With few specific data available, we must rely on relatively crude measures of productivity. Typically, these have been shown as total R&D investment divided by the number of new chemical entities (NCEs) approved in the same year.[62] However, the long time lag between investment and yield in pharmaceuticals make this a poor indicator of productivity in drug discovery. Nonetheless, by this widely used measure, overall productivity is falling.[62–64] **Figure 4** shows the number of drugs containing a novel active ingredient approved by the FDA each year and demonstrates that for example, in 2002, FDA approvals of NCEs (17 in total) were the lowest for eight years. This trend continues despite a rise in potential drug targets, change in the division of labor, and the very significant rise in industry and public sector R&D expenditure since the 1970s.

A more specific indicator of productivity in drug discovery is the number of new compounds discovered and patented. **Figure 5** shows the number of patents granted at the US Patent and Trademark Office (USPTO) in classes 424 and 514 over the period 1978–2002. These are the main patent classifications for therapeutically active compounds,

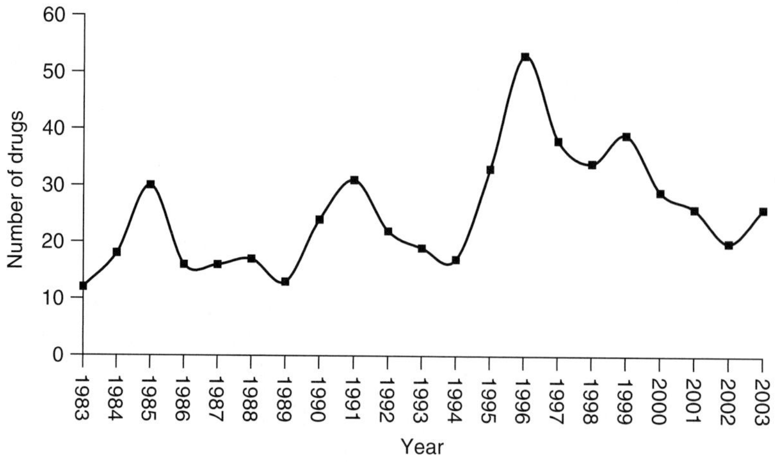

Figure 4 FDA approved prescriptive drugs containing novel active ingredients, 1983–2003. (Reproduced with permission from Nightingale, P.; Martin, M. *Trends Biotechnol.* **2004**, *22*, 564–569.) (*Source*: FDA *Orange Book*.)

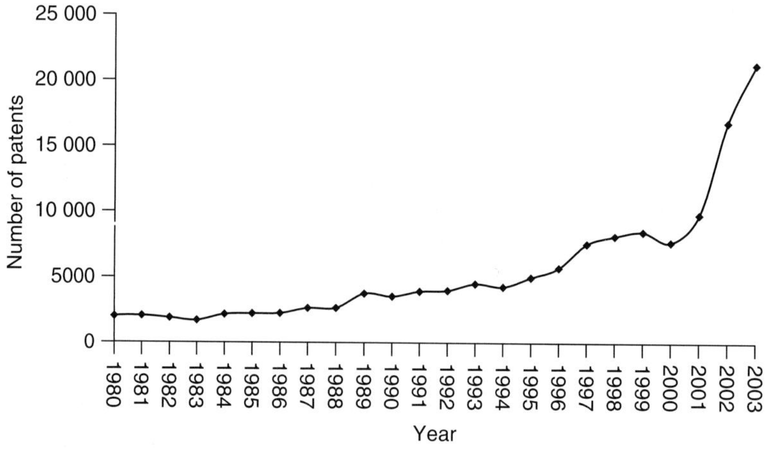

Figure 5 Patents on therapeutically active compounds at the USPTO, 1980–2003. (Reproduced with permission from Nightingale, P.; Martin, M. *Trends Biotechnol.* **2004**, *22*, 564–569.) (*Source*: USPTO.)

and can be used as an indicator of the number of molecules considered attractive enough to warrant patent protection, but not necessarily viable enough to enter development. This is a direct measure of the productivity of drug discovery and it is clear that this period saw a steady rise in the number of patented molecules. It should be noted that recent interviews with managers in large pharmaceutical companies suggest that firms are now patenting more heavily to protect each potential drug than previously. As a consequence, the growth in patents may overstate the productivity of discovery.

When this is set against the R&D investment data in **Figure 6** we see that R&D spending increased approximately tenfold in the period between 1978 and 2002, while patenting activity failed to keep this pace until after the year 2000, when there was a rapid rise in activity. Despite this recent rise the data do not suggest patenting activity has done much more than to keep pace with expenditure. The case for increasing R&D productivity in real terms is further weakened when more global trends such as the increasing propensity to patent in industry are taken into account.[65]

1.17.2.4 The Speed of Change in Drug Discovery

Although investment in the new therapeutic modalities provided by biotechnology began in the 1980s and developed in an incremental manner, the evidence above points to dramatic changes in the overall industry structure and application of biotechnology as a set of research tools. Particularly over the past decade this rapid pace of change in the industry structure has been indicated by the growth of alliances, as well as the relatively short space of time within which

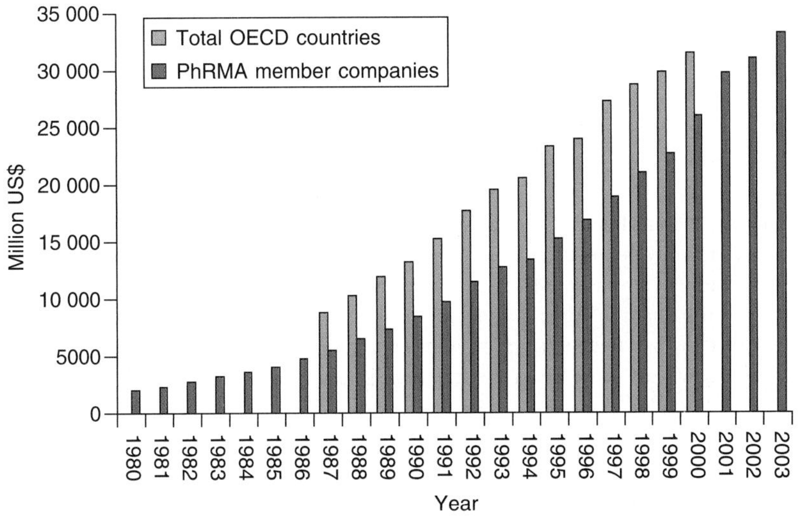

Figure 6 Increases in R&D expenditure for pharmaceutical industry in OECD countries and member countries of Pharmaceutical Research and Manufactures of America (PhRMA), 1980–2003. (Reproduced with permission from Nightingale, P.; Martin, M. *Trends Biotechnol.* **2004**, *22*, 564–569.) (*Source*: OECD and PhRMA.)

biotechnology has provided an increase in potential targets for therapeutics. The implications of this for drug innovation are profound. In effect, the key innovation bottleneck has shifted from the creation of novel small-molecule drugs against known targets (chemistry) to the functional characterization of large numbers of unknown drug targets (biology). Historically, publicly funded research has played a key role in establishing this important knowledge base, with a complete understanding of any particular biological pathway often taking decades rather than years. This raises the possibility of a significant lag between industry-wide investment in genomics, together with the associated heavy research costs, and gains in the productivity of discovery. Industry analysts have suggested that the upfront investment in genomics may not yield significant benefits for over a decade, whilst at the same time exacerbating the crisis in productivity in the short term,[66] which is consistent with the pattern of technical change identified by historians of technology.[34]

1.17.2.5 Conclusions on the Impact of Biotechnology on Drug Discovery

The impact of biotechnology on the process of drug discovery has to be set against a number of important trends, including the shift to RDD, the development of strategies to generate blockbuster products in the 1970s and 1980s, and an ongoing decline in R&D productivity. The initial application of molecular biology techniques lay in further developing RDD and the industry as a whole steadily integrated the biological sciences within discovery program during the 1980s. Yet it was only with the advent of genomics, combinatorial chemistry, and HTS in the 1990s that real gains in productivity were seen, as many more drug targets were identified against which leads could be rapidly generated. With progress in the validation of these new targets likely to be slow, the impact of genomics on small-molecule drug innovation may not be known for some time.

However, the development of biotechnology has led to the creation of new biopharmaceutical modalities, including recombinant therapeutic proteins and mAbs. During the 1980s a number of important biopharmaceuticals were produced by first-mover biotechnology firms, but innovation strategies based on these products were not widely adopted. Only in recent years has there been greater interest in the potential of biological drugs, with most attention being given to mAbs (as will be discussed further in Sections 1.17.3 and 1.17.4).

The rapid growth of the global biotechnology industry during the 1990s was accompanied by a broadening from biologicals to include small-molecule drugs. At the same time this coincided with moves to far greater outsourcing of R&D activities. As a result, by the end of the decade a new highly networked industrial structure had emerged based on a new division of labor, with biotechnology firms playing an increasing role in drug discovery. The significance of this shift is demonstrated by the growing number of biotechnology-derived products that are reaching the market (*see* Section 1.17.4).

Taken together, these factors make it clear that biotechnology has had a major impact of drug discovery by improving the efficiency of discovery of new small-molecule compounds, creating new therapeutic modalities and facilitating major changes in the industrial organization of discovery. However, it has not solved the deep-rooted R&D productivity crisis and the adoption of biological drug modalities as an integral part of the pipeline still remains patchy.

1.17.3 The Impact of Biotechnology in Drug Development

1.17.3.1 Drug Development and the Critical Path

The focus in this section is on change brought about by biotechnology within drug development (as opposed to discovery). We define development here in terms of the FDA's 'critical path' – which is the period of R&D from when a drug prototype enters preclinical testing to the time when it is approved by the regulatory authorities (**Table 1**).[64]

To assess the impact that biotechnology has had on drug development in terms of scope, scale, and speed we focus on several key metrics commonly discussed in the literature. These are the cost of drug development, which is linked to the attrition rate of products in development (generally measured as a percentage of phase I candidates that gain marketing approval), and the length of time drugs take to pass through clinical development, as well as the types of drugs being developed. Although biotechnology has promised advances along a broad front, here we focus mainly on the therapeutic modalities of small-molecule drugs, recombinant proteins and monoclonal antibodies.

1.17.3.2 The Scope of Change in Drug Development

The process of drug development shown in **Table 1** is often characterized as a process of attrition whereby 5000–10 000 potential drug candidates are narrowed down to generate a single molecule that has the appropriate characteristics of a drug.[22,67] Such a view implies drugs are truly discovered only after they survive clinical development.[68] While the fruits of the development process are reliant on the quality of candidates entering the clinical process, biotechnology also has the potential to improve the productivity of development.

Pharmacogenetics in particular holds out the promise of improving the efficiency of drug development in a number of ways: improving preclinical safety screening, enhancing the effectiveness of early stage trials, streamlining later stage trials, and even rescuing drugs that have failed late stage trials due to lack of efficacy in the whole population.

From the late 1990s pharmacogenetic profiling of trial populations was widely seen to offer opportunities to gain better clinical efficacy or safety in smaller trials by targeting subpopulations more likely to respond favorably to drugs, and by ensuring that all drug metabolizing enzyme alleles are accurately represented in phase I – so-called balanced trials. These are hoped to reduce the likelihood of failure in late stage trials.[63,69,70] Pharmacogenetics has been widely adopted by the pharmaceutical industry and the biotechnology sector. The overall range of strategies employed by these firms suggests pharmacogenetics data could feed into all stages of the R&D process.[71] However at present it has yet to yield widespread benefits.[72]

In the early 2000s gene expression studies were suggested as a useful toxicological tool for smarter candidate selection.[73] By removing unsuitable drugs from the pipeline in the preclinical testing stage, or even earlier, it was hoped that new toxicological and metabolic screens would reduce expensive failures in the clinical phases of development.[50,74] The majority of large pharmaceutical firms have attempted to integrate these technologies into their R&D efforts but these approaches are relatively new and it is too early to assess their effectiveness.[50] However, one major limitation at present is the lack of agreement on how to validate the gene, protein, or metabolic biomarkers used to distinguish

Table 1 The 'critical path' for drug development

Phase	*Rationale*	*Typical numbers of human subjects*
Preclinical	Synthesis/purification and animal testing in preparation for application to regulators for permission to test in humans	—
Phase I	Pharmacokinetic studies in patients/healthy volunteers to determine absorption, metabolism, and excretion	From 20 to around 100
Phase II	Effectiveness in patients/possible side effects	Several hundred
Phase III	More detailed study of effectiveness and side effects compared to placebo or other treatment, in preparation for submitting to regulators for marketing approval	Several hundred to several thousand
Marketing review and approval	Review of clinical evidence and resolution of remaining concerns	—

Source: Adapted from PhRMA and Tufts Centre for Study of Drug Development websites.

responses and separate patient groups.[73,75] This is a centrally important question for the assessment of biomarkers by regulatory authorities overseeing clinical trials and drug approval. Without the agreement of regulatory authorities and industry on such issues the potential benefits from these new approaches will be difficult to realize, for example in clinical trial design.

While biotech platform technologies have been seen as important for improving the efficiency of small-molecule drug development, the longer-established promise of biotechnology is the creation of entirely new classes of therapeutics, such as recombinant proteins and antibodies. These were initially seen as more likely to succeed in clinical trials because, unlike synthetic drugs, they are endogenous to the human body and were thought to be more 'natural,' more potent, and less toxic.[46,76,77] In fact this has not proved to be the case. Large molecular weight biological drugs have poor absorption characteristics and relatively short half-lives in the body, and they have often proved just as toxic as their small molecule predecessors and as prone to side effects.[46,77] Other modalities, such as gene therapy and antisense, have not been as successful in clinical trials as was hoped in the 1990s. Although one licensed antisense product (Vitravene) has been approved by the FDA for the treatment of cytomegalovirus retinitis, this is a niche product and few others are in late stage development. Similarly, no gene therapy has been licensed in Europe or North America, and only a handful of products may reach the market in the next few years.[78]

1.17.3.3 The Scale of Change in Drug Development

Of all NCEs that enter clinical trials only between 8% and 20% finally receive marketing approval.[64,79–81] These broad statistics hide a more nuanced picture. Of the NCEs entering phase I trials in the period 1981–1992, the anti-infectives enjoyed a survival rate of around 30%, while only 12% of respiratory drugs were successful. In oncology, around 16% of molecules passed successfully through trials at the end of the 1990s, although other estimates of 5% have been cited more recently.[64] When the price of failed drugs across the industry is taken into account, the cost of bringing a successful NCE to market has been estimated at between $800 million and $1.7 billion.[64] Although such figures are often disputed, the trend toward rising costs is not.[50,82] These rises disproportionately affect drug development where costs have increased from around 40% of total R&D to around 60%.[50,64] Some estimates suggest that in real terms development costs alone have grown from less than $500 million in the late 1990s to almost $1 billion in the early 2000s.[64] The increasing cost of pharmaceutical development is primarily being driven by a growing number of clinical failures, especially in phase II.[50,68] In a recent analysis of the industry's predicament, the FDA concluded that biotechnology had yet to make an impact on drug development:

> Today's revolution in biomedical science has raised new hope for the prevention, treatment, and cure of serious illnesses. However, there is growing concern that many of the new basic science discoveries that have been made in recent years may not quickly yield more effective, more affordable, and safe medical products for patients. This is because the current medical product development path is becoming increasingly challenging, inefficient, and costly. During the last several years, the number of new drug and biologic applications submitted to the FDA has declined significantly; the number of innovative medical device applications has also decreased. The costs of product development have soared over the last decade.[50]

Indeed, although the overall trend of declining R&D productivity predates the uptake of biotechnology by pharmaceutical firms, the application of biotechnology research tools, and in particular genomics, is blamed for making the situation worse.[50,83,84] This can be explained as follows: genomics has generated many new targets (as discussed in Section 1.17.2) that are more poorly characterized than those traditionally pursued. These targets are tested in experimental models where simulated conditions are further and further removed from those of the intended patient population (i.e., from patients, to animals, to cell cultures) and in models that are regarded with suspicion by those favouring more traditional deductive pharmacological approaches.[84] The molecules aimed at these novel targets then fail more often in clinical trials. Because lack of efficacy is generally only ascertained in the later stages of drug development, efficacy failures account for a significant share of the rising cost of drug R&D.[50] As such the financial risks to the pharmaceutical industry come particularly from the failure of first-in-class drugs.[68] For example, one study suggests candidates using novel approaches have a 5% survival rate in clinical trials compared to 8% for approaches where there is a precedent.[51] With the risk of novel therapies so high, the relative reward for developing so-called 'me-too' drugs has increased. This is illustrated by the fact that in recent years 'fast followers' as a group were more profitable than novel drugs collectively.[51] This is an industry-wide phenomenon affecting both biologicals and small molecule drug manufactures as has been noted by the FDA.[64]

It has been estimated that a 10% improvement in predicting failures before clinical trials could save $100 million in development costs per drug.[15] Hence there are big incentives for using biotechnology in drug development to

Table 2 Comparing the approval rates of different therapeutic modes

Drug type	*Cohort size*	*Success rate(%)*	*Dates of IND filing at FDA*	*References*[a]
New chemical entities	671	21	1982–1992	80
Recombinant proteins	91	26	1980–1989	45
Recombinant proteins	120	35	1990–1997	45
Chimeric monoclonal antibodies	20	29	1987–1997	79
Humanized monoclonal antibodies	46	27	1988–1997	79

[a] All data collected from Tufts studies.

compensate for the damage to productivity it has been responsible for in drug discovery to date. However, there is little objective evidence publicly available that this is being achieved.

As mentioned in Section 1.17.2, biotechnology promised to provide new modes of therapy that are more 'natural' and hence more likely to be effective and tolerated. Using data from three different studies from the same research group (**Table 2**), a comparison between the success rates of monoclonal antibodies, recombinant proteins, and new chemical entities appears to show that, overall, biological molecules that enter clinical trials are indeed less likely to fail to reach the market than small-molecule drugs. Hence there appears to be some support for the case that biological drugs are more efficient in terms of R&D costs than NCEs. However, in part this can be attributed to the smaller studies often undertaken to support these drugs' application. While small studies suit drug sponsors as long as they are adequate to prove efficacy, they may not be as effective in revealing side effects as large studies.[85] Recently European regulators have suggested that this has resulted in worrying deficiencies in biological marketing approval applications, which are not being evaluated on rigorous methodological criteria.[77]

Furthermore, this apparent high success rate is boosted by an increasing number of 'me-too' drugs that some suggest are little different from established products.[77] For example, it is interesting to note that in the two cohorts of protein therapeutics that filed for FDA investigational new drug (IND) in the periods 1980–89 and 1990–97 shown in **Table 2**, the second demonstrates a marked improvement in the rate of success. The increasing success rate in trials has been attributed to the reshaping of biotech product portfolios to mirror early successes. In particular this is illustrated by the focus on diabetes and endocrine products maintained by firms after initial success by pioneering biotechs.[45] These products focused on a narrow range of therapeutic areas (oncology, infection, inflammation/autoimmune, diabetes, and endocrinology). As a result, competition is expected to grow in coming years, as already witnessed in NCEs.[45,79] We might therefore expect failure rates, trial sizes, and overall costs to rise too, mirroring the pattern reported in NCE development, as the bar of commercial success is raised for drugs when they have to prove efficacy and safety improvements against benchmarks set by existing treatments.[50]

Table 2 shows that mAbs have also generated an impressive approval rate in recent years. However, these statistics do not show the poor performance of the initial wave of mAbs entering the clinic in the early 1980s that were derived from mouse cell lines and discontinued due to high immunogenicity and a low clinical success rate of 3%.[79] Since then monoclonal products have moved from wholly murine-derived antibodies, to chimeric antibodies, then humanized antibodies, and now toward smaller antibody fragment-based approaches. The recent success of monoclonal therapies is viewed as continuing in the future.[79] Perhaps then after a long gestation period and some false starts, mAbs are the best indication yet of biotechnology's revolutionary promise in the development of new products.

1.17.3.4 The Speed of Change in Drug Development

Biological drugs, notably mAbs and protein therapeutics, do offer some indication of more effective drug development speeds at least in the short to medium term as they spread into previously unoccupied therapeutic niches. However these benefits appear restricted to a small number of therapeutic areas. While a number of early protein therapeutics reached the clinic relatively quickly, mAbs have taken between 20 and 30 years to emerge as a viable and widely applicable modality, and other biological platforms such as gene therapy and antisense are yet to provide substantial success.

For NCEs, recent trends indicate that development times have fallen from 79.5 months in 1990–91 to 63.2 months in 2000–01 and approval times have also reduced from 31.3 months to 18.4 months over the same period.[85,86] As **Table 3** shows, overall drug development times (including approval times) are beginning to fall for the first time in

Table 3 Changes in drug approval times from the 1970s to the present (in years)

Decade	*Preclinical phase*	*Clinical phase*	*Approval phase*	*Total*
1970s	4.5 (41%)	4.3 (40%)	2.1 (19%)	10.9
1980s	5.3 (38%)	5.9 (42%)	2.8 (20%)	14.0
1990s	5.5 (39%)	6.7 (48%)	1.8 (13%)	14.0
2000s	5.5 (44%)	5.9 (45%)	1.5 (11%)	13.2

Reproduced with the permission from Tufts Center for the Study of Drug Development, Tufts University, Boston Unpublished Data, February 2005 (presentation: Christoper-Paul Milne, What Future for Big Pharma: From Challenge to Change? Innogen Conference, February 23–25, 2005, Edinburg)

several decades. It is important to emphasize at this stage that this is not likely to be early evidence of the impact of biotechnology on NCE development as such changes are the result of alterations in the regulatory landscape in the 1990s, such as accelerated approval and fast-tracking[87] and improved management practices, especially in large firms.[22]

Development times vary also for a number of specific reasons. The speed of patient enrolment may vary by condition.[50,81] Some diseases require longer treatment times (chronic diseases versus acute). For example anti-infectives need trials of 50.2 months versus 92.9 months for endocrine disorders. Some decreases in clinical trial times over the 1990s were dramatic – 41% for cancer drugs, 44% for respiratory drugs, but in other areas such as cardiovascular and pain they increased by 12% and 11%.[50,81]

It is notable that in the last decade the FDA has adopted new policies to enable it to become an active "partner in drug development rather than a hurdle."[50] A key change has been the FDA Modernization Act of 1997 which introduced the fast track mechanism allowing streamlined NDA submission, and a willingness to accept the use of surrogate end points to determine efficacy in clinical trials (for example, the use of CD4 cell counts and measurement of viral load rather than patient survival for anti-HIV drug approvals). This simplifies effectiveness studies and reduces development time and costs.[64]

Bolten and DeGregorio[81] suggest that the reductions in development time (as indicated in **Table 3**) resulted from these changes in FDA policy. This indicates that regulators and companies could, together and individually, reap further improvements through organizational changes.

1.17.3.5 Conclusions on the Impact of Biotechnology on Drug Development

Drugs fail to reach the market for one or more of three main reasons: they are not effective; they have serious side effects; or their markets are not profitable. Biotechnology may increase the effectiveness of NCE development in the future, but it has not achieved this to date. Where biotechnology has shown more immediate promise (e.g., mAbs and therapeutic proteins) there is a danger that in the longer term we will see the emergence of the same pressures on productivity seen in relation to NCEs. In particular, this may come about as the fast-track approval mechanisms and relatively lower regulatory hurdles in areas of unmet medical need give way to familiar increases in efficacy requirements and the commercial pressure to seek profits in more certain markets. Nonetheless, productivity improvements may be gained from better development process management and from enhanced modes of feedback from clinical development to drug discovery about which drugs work and which ones do not. Ultimately perhaps the greatest long-term hope of efficiency in development is reliant on changes in the discovery stage and preclinical stage to weed out poorer candidates, and a shrinking of trial size obtained through the targeting of responder subpopulations. However neither of these approaches is yet proven to provide the wide base of applicability for what might be termed a revolution in drug development.

1.17.4 The Impact of Biotechnology on Clinical Practice

1.17.4.1 Focus of Analysis

In this section, we explore the impact of biotechnology on clinical practice through an analysis of therapeutics approved by regulatory agencies and available on the market. Although we touch on other potential biotechnology applications in the clinic, such as diagnostics, we limit our analysis to therapeutics alone. The application of these drugs to clinical practice is assessed here in terms of the scope (namely the range of modalities made possible and the conditions they address), the scale of impact (in terms of the numbers of therapeutics and their perceived impact), and finally, the speed with which these therapies have become available (rather than the speed of their diffusion through communities of practice).

1.17.4.2 The Scope of Change in Clinical Practice

In Sections 1.17.2 and 1.17.3 we established that one of the main promises of biotechnology – a big increase in the number of drug targets – has yet to impact on the clinic, as potential new drugs based on genomics are still in the early stages of development.

A second important promise of biotechnology (in particular the Human Genome Project and Genomics) was the rapid application of genetic tests to predict and prevent disease by supporting early interventions, and more optimal or targeted use of therapies (pharmacogenetics).[89–93] Examination of developments in diagnostics is beyond the scope of this analysis, although by now it is already obvious that the timeframes suggested in many accounts have not been met. Recent analysis suggests many previous claims remain far from realization, and the future contribution of genomics in medicine remains hotly contested.[94,95] Certainly it is clear that for genomics to impact mainstream medicine it must offer solutions to prevalent complex disorders, where prediction of phenotypes (whether a disease or drug response) may rely on measurement of a number of genes and environmental factors.[85] Despite much effort, there is little evidence of any startling progress in this area.

The third and perhaps the oldest promise of biotechnology for the clinic remains the introduction of new therapeutic modalities – recombinant proteins, mAbs, gene therapy, stem cells, and gene silencing technologies such as RNA interference (RNAi). This is important because traditional pharmaceutical approaches generally generate antagonists that inhibit biological functions and have been less successful at generating agonists that simulate function.[85] For example, many of biotech's biggest successes, as indicated in **Table 4** have been protein replacement therapies (such as Factor VII for hemophilia) or agonists that promote biological processes (such as the increase in blood cell production generated by erythropoietin for treatment of anemia). As noted in Section 1.17.3 only recombinant proteins and mAbs have produced significant numbers of regulatory approved biological products and our discussion in the next section therefore focuses solely on these.

1.17.4.3 The Scale of Change in Clinical Practice

The number of therapeutics produced by biotechnology firms (alone or in partnership with pharmaceutical firms) has been growing, though not continuously, for over 20 years (**Figure 7**). However we need to differentiate between the apparent increasing productivity of biotechnology firms (or rather those firms that are not traditional pharmaceutical firms) and the yield of new technologies. On closer inspection the majority of these products (100 out of 192) are not recombinant protein or monoclonal antibody products, reflecting the fact that many firms classed as biotech firms are often producing small-molecule therapies. As such they are more like speciality pharmaceutical companies rather than true biotechnology firms. This may lead to overestimates of the impact of what we term third-generation biotechnology (for example compare Arundel and Mintzes[96] with Arnst[97]) as opposed to the combined impact of the new division of labor between small or medium sized enterprises and established large pharma, and the accompanying expansion of the drug discovery industry since the 1980s, as discussed in Section 1.17.2.

Indeed much of the apparent boom in clinically available treatments from biotechnology's new therapeutic modalities is based on their performance relative to small-molecule drugs, as a decline in the numbers of NCEs being approved makes the contribution of biological drugs look higher, even though there has not been an increase in the rate of biological drug approvals.[87] For example, the percentage of FDA drug approvals made up by biological drugs increased from 21% in 1998, to 24% in 2001, and 30% in 2002.[87] However, in absolute terms the number of biological drugs gaining FDA approval in recent years is still relatively low with only 26 biological drugs approved between 1998 and 2003 as compared with 144 NCEs.[98] Nonetheless, 33% of the NCEs originated outside traditional pharmaceutical laboratories.[98]

By 1993, biotechnology-based therapeutics accounted for nearly $3 billion in sales in the USA, capturing over 5% of the total US sales for therapeutics, and this has since followed an upward trend.[99] At present IMS Health estimate that $55 billion (10%) of global $550 billion pharmaceutical market revenues are derived from biopharmaceutical product sales. Around three-quarters of this is generated by the 15 therapeutic product classes in **Table 4**. However, the therapies listed in **Table 4** are mainly for rare diseases compared to the mass markets of many small-molecule drugs. For example in an analysis by Arundel and Mintzes[96] of the approximately 30 disease groups targeted by biopharmaceuticals, eight are very rare with prevalence rates of less than 1 in 10 000 in the USA and only four have prevalence rates close to or above 1 in 100 (diabetes, stroke, heart attacks, and rheumatoid arthritis). Of these four

Table 4 Top 15 biotech-derived therapeutics

Generic name	*Brands*	*Companies*	*Indications*	*2004 Global sales ($billion)*
Alpha and beta erythropoietin	Epogen, Epogin, Procrit, Eprex, NeoRecormon, Aranesp	Amgen, J&J, Roche, Kirin, Sankyo	Anemia	11.8
Alpha and beta interferon	PEG Intron, Pegasys Avonex, Rebif, Betasetron	Schering-Plough, Roche Biogen Idec, Serono, Schering AG, Chiron	Hepatitis C, multiple sclerosis	6.8
Human insulin	Novulin, Humalin, Humalog	Novo Nordisk, Lilly	Diabetes	5.6
Granulocyte colony stimulating factor G-CSF	Neupogen, Neulasta	Amgen, Roche, Schering	Granulocytes stimulator	3
Rituximab	Rituxan	Roche	Leukemia, lymphoma	2.8
Etanercept	Enbrel	Amgen, Wyeth	Rheumatoid arthritis	2.6
Infliximab	Remicade	J&J	Rheumatoid arthritis	2.1
Trastuzumab	Herceptin	Roche	Breast cancer	1.3
Human growth hormone	Serostim, Saizen, Humatrope, Protopin, Neutropin Synagis	Serono, Biogen Idec, Roche, Novo Nordisk, Akzo Nobel, Lilly	Dwarfism	1.8
Palivizumab	Synagis	Medlmmune	Respiratory symmetrical virms	0.95
Follicle stimulating hormone	Gonal F, Follistim	Serono, Akzo Nobel	Infertility	0.95
Glucocerebrosidase	Cerezyme, Ceradase	Genzyme	Gaucher's disease	0.88
Adalimuzad	Humira	Abbott	Rheumatoid arthritis	0.85
Facto VII	Novo Seven	Novo Nordisk	Hemophilia	0.76
Botulin toxin	Botox	Allergan	Wrinkles	0.7
Bevacizumad	Avastin	Genetech, Roche	Colon cancer	0.55

Source: Express Pharma Pulse (http://www.expresspharmapulse.com/20050428/edito2.shtml).

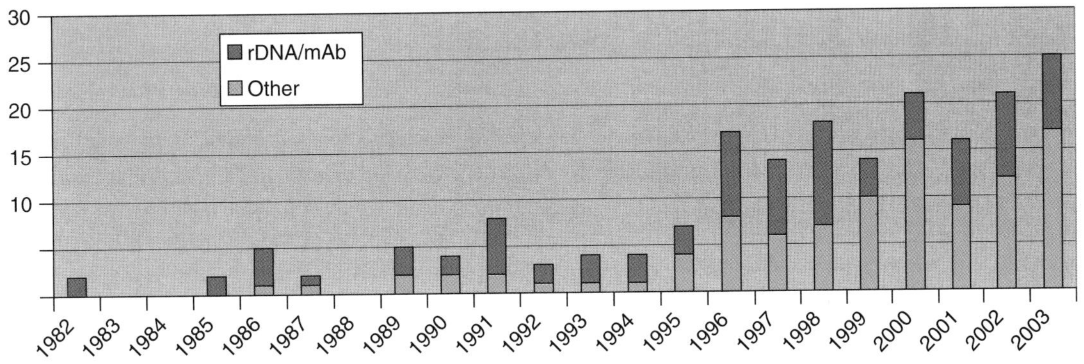

Figure 7 Annual FDA approvals of products from biotechnology firms, 1982–2003. (*Source*: Biotechnology Industry Organization.)

disease areas, in three the biopharmaceutical is only approved for use under specific circumstances or as a second-line drug when alternatives fail. Almost all of these therapeutics are used in specialist practice (secondary or tertiary care) rather than in general practice (primary care). The one exception is diabetes, although insulin was already available in the clinic before biotechnology. As such, biotechnology has had little impact on primary care medicine.

The majority of products in **Table 4** are proteins rather than mAbs, but this is likely to change, as 57 mAbs were launched by mid-2005[100] and around 200 are in clinical trials. Furthermore around 50% of these are predicted to be profitable, as compared to only 30% for traditional small-molecule drugs. However like the protein drugs before them, the monoclonal drugs developed so far are for a relatively narrow range of disease areas – albeit ones with large markets such as cancer and inflammatory diseases. They must be stored carefully, can only be injected, and tend to be delivered in specialist secondary or tertiary care settings. Therefore a small sales force and lower levels of marketing are needed than in the more competitive primary care market where most blockbuster small molecules compete. Secondly, the early mAbs faced low competition and targeted the most profitable markets. With oncology a focus for 40% of mAbs in clinical development at present, future drugs will face much greater competition.

There are few nonfinancial metrics available for a broad analysis of the scale of the impact that biotechnology has had on clinical practice. However, some indications are given by regulatory licensing documentation, and the views of physicians in surveys.

Joppi *et al.*'s review[77] of EMEA documentation found that only 15 out of 61 (25%) biologicals approved between 1995 and 2003 (including a number of vaccines) provided efficacy improvements on existing therapies or were targeted in therapeutic areas where effective treatment was not available. A further 22 (36%) provided nontherapeutic advantages, such as improved safety, or convenience (e.g., in administration). The largest share, 24 (39%) provided no clinical advance, and were merely 'me-too' applications.

These findings give further support to a study by Arundel and Mintzes[96] derived from data gathered from French physicians by *La Revue Prescrire*. (The *Prescrire* data set's collection is funded by journal subscriptions and as such gives the physician respondents' views of the medicinal value of treatments independently of industry or government influences.) The *Prescrire* data suggest that only 16 of the 48 (33%) biologicals (excluding vaccines) evaluated between January 1986 and June 2004 were better than 'minimal improvements' over pre-existing treatments.[96] However, this is better than traditional pharmaceuticals, only one in ten of which score at this level (i.e., rated as 4 or higher on a 7 point scale by prescribers). Furthermore, only 24% of biologicals evaluated were rated as offering no therapeutic advance, whereas the vast majority of other drugs (66%) achieved this level or lower. This can be explained by Joppi *et al.*'s[77] finding that many biological products have been aimed at markets with unmet medical needs. However, when the *Prescrire* data are separated into three time periods, Arundel and Mintzes demonstrate an increasing trend toward 'me-too' biological drugs (from 7% between 1986 and 1998 to 40% after 2000), which are less welcomed by physicians.[96] Furthermore the proportion of drugs viewed as offering some advance (6 on the scale) or a major advance (the top mark of 7) declined from 39% of the 28 treatments in the period 1986–98 to 13% of the 30 treatments evaluated after 2001.

The work of Joppi *et al.*[77] and Arundel and Mintzes[96] further supports the argument we made in Section 1.17.3, that the marked rate of advance apparently offered by biotechnology may not be sustained in the medium to long term. Biopharmaceuticals, like NCEs before them, are increasingly focused on securing economic benefits for developers rather than clinical benefits for patients. Furthermore biological drugs are notably more expensive than their traditional pharmaceutical equivalents.[77]

1.17.4.4 The Speed of Change in Clinical Practice

In this final part of our review of biotechnology's advance from research bench to clinic, it is appropriate to examine the speed with which biotechnology has generated its clinical advances and consider if this has been revolutionary in pace.

The initial biotechnology drugs reached the clinic in the first half of the 1980s, arriving only around 10 years after the development of third-generation biotechnology techniques. In the late 1980s these were rapidly followed by approvals of innovative protein therapies such as alteplase/tissue plasminogen activator for treatment of pulmonary occlusions following myocardial infarctions (1988), epoietin alfa for treatment of anemia (1989), and the first monoclonal antibody drug (muromomab-CD3) for treating kidney rejection (1989). These seemingly represented very swift advances, but the targets on which these drugs acted were known prior to the development of third-generation biotechnology techniques: insulin and human growth hormone were already in clinical use, alteplase had been isolated and purified in 1964, and epoietin in 1960. Once the genes for the molecules had been cloned, they moved from bench to approval in five to six years.[101] However, this is not dissimilar to the best-case examples of traditional pharmacological approaches. For example, at the height of the HIV/AIDS crises in the 1980s, azidothymidine moved from compound screening in cell cultures to approval on a named patient basis in less than three years. Biotechnology does not always yield such rapid results. Avastin, the first of the angiogenesis inhibitors, was approved in 2004, 35 years after the work on related angiogenesis research that led to its discovery began.[88] Yet this has been termed a 'low hanging fruit' by those in mAb firms.[100] By comparison some traditional pharmacological approaches have been more rapid, for example angiotensin-converting enzyme (ACE) inhibitors (discovered in 1968, approved for limited use in 1981, with license extended in 1985) and statins (from research hypothesis in 1971 to first approval in 1987).[101] Hence there does not appear to be much evidence that biotechnology provides a more rapid route from bench to clinic as yet.

1.17.4.5 Conclusions on the Impact of Biotechnology on Clinical Practice

The apparent scale of third-generation biotechnology's contribution to clinical practice is accentuated by the dwindling productivity of traditional pharmacological approaches and the increasing focus on small-molecule drugs by biotechnology firms. Nonetheless, biotechnology drugs do make up an increasing proportion of total therapeutic approvals. The evidence of clear clinical benefit from genomics is yet to emerge, although it is too early to expect any impact at this time.

Stories abound in the press about the impact biotechnology products are already having on the lives of a great many patients.[97,102] Certainly there is ample evidence that biotechnology has provided a number of highly effective and highly profitable therapies based on novel modalities in areas where no effective therapy previously existed. For example, mAbs such as Herceptin (Trastuzumab) provide significant improvements in the treatment of certain cancers. However, apart from these and other notable exceptions such as recombinant insulin, biological drugs are mainly used in the treatment of comparatively rare diseases by specialist centers. Furthermore, there is evidence to suggest that as early biotechnology modalities mature into routine clinical interventions they suffer from many of the same limitations as their small-molecule predecessors, such as unanticipated safety issues. As such, it is perhaps more prudent to regard biological drugs as novel and important rather than revolutionary.

1.17.5 Conclusions

1.17.5.1 Scope, Limitations, and Main Findings

To understand the complex ways in which biotechnology has given rise to promise and to critically assess progress to date, we have examined the impact of biotechnology in a range of contexts – drug discovery, drug development, and clinical practice. This provides powerful insights into where change has taken place, and on its extent and character. It is clear that the dynamics of change vary in terms of scale, scope, and speed from one location within the innovation process to another.

Our findings must be considered in context. Evidence presented in this chapter has largely focused on the impact of biotechnology on drug innovation. This approach means we have not addressed the impact of other changes in the industry as a whole, such as changes in managerial practices that may reduce development times and costs (for example see [50,103]). These include more efficient 'go/no go' decision-making processes, tighter controls on outsourced work undertaken at contract research organizations, and use of IT based clinical knowledge management systems. Firms at the forefront of development efficiencies in clinical trials attribute their success to these organizational approaches rather than to biotechnology.[104] As such some metrics presented here may overestimate the impact of biotechnology. If these were taken into account it would serve to further strengthen the central argument made here.

Similarly, we maintain that differences in definitions (of what constitutes a biotechnology company, for example), the changing nature of technology, and changing industry structure mean that the contribution of organizational improvements, new industrial structures, increased funding, and changes in the regulatory environment may have been attributed to biotechnology in accounts of biotechnology's promise made by others. Such factors may in part account for differences between the perspective provided here and those provided by proponents of the revolutionary model of biotechnology.

These caveats notwithstanding, strong evidence remains of widespread and important changes in drug discovery resulting from the emergence of biotechnology. However, this is yet to be accompanied by changes of a similar magnitude in either drug development or in clinical practice. Our findings suggest, therefore, that evidence for radical and widespread change rapidly diminishes as one moves along the innovation path from target discovery to target validation and into clinical development.

In terms of marketed drug products, biotechnology has already brought clear benefits to healthcare, and undoubtedly will continue to do so. However, the impact of biotechnology, including genomics, on clinical practice has been less striking and immediate than was envisaged at the time of each successive technological breakthrough or scientific advance arising since the inception of commercial biotechnology at the end of the 1970s. The output of marketed drugs from the biotech sector has proceeded steadily, but in general, products have provided modest benefits rather than being highly innovative. This is, in part, a reflection of the industry-wide tendency to pursue incremental innovation, which manifests itself in convergence around established drug targets and the predominance of so-called 'me-too' drugs. Tried and tested targets and drug classes are less expensive to develop and inherently less risky. This trend reflects the practical difficulties of developing novel drugs using new technologies and approaches. The low output rate from the biotech sector is also symptomatic of the difficulties of drug innovation per se, an issue that has become the subject of FDA concerns and led to a broadening dialog between industry and the regulatory authorities. Nevertheless, it is clear that translating biotechnology into new drugs is in fact proving more difficult, costly, and time-consuming than many policy-makers, scientists, investors, and other stakeholders envisaged. The development of mAb-based drugs exemplifies this pattern, as do revisions to the timescale within which genomics-based drugs might reach the market.

At the same time, some biotechnology-derived therapies have been received with great enthusiasm by patients and healthcare professionals – to whom they represent significant or even dramatic improvements over previous treatment options. Whilst acknowledging the important advances made by biotechnology in such areas, our review does not support sweeping claims of a biotechnology revolution overall. For those associated with the discovery, development, and use of novel treatments the biotechnology revolution may appear to have arrived, while for a great many others it remains a bitter disappointment.

1.17.5.2 The Biotechnology Revolution: Morals of the Myth

Overall, the empirical evidence does not support a revolutionary model of technological change in terms of either scale, scope, or speed. Rather, our findings suggest that biotechnology is following a well-established pattern of incremental technical change coupled with steady growth in applications seen in the emergence of other major technological paradigms, including steam power and electricity. Historical research suggests that major technological changes, such as those produced by the steam engine, the production line, or the electric motor, do not take place in a vacuum but typically require complementary technical and organizational innovations to facilitate their adoption. Studies of the adoption and diffusion of steel and electricity have shown that it may take between 40 and 60 years for major technologies to produce benefits discernible using broad measures such as productivity.

The integration of complex technologies such as biotechnology requires massive long-term investment to yield a return. The failure of biotechnology to generate proportionate returns to date cannot be taken as evidence that it will not do so in the future. However, longer timescales than expected may render many current business models unviable, especially in the small firm sector, and make some investments uneconomic when the future profits are properly discounted and opportunity costs taken into account. As such the myth of a biotechnology revolution might be viewed as a rhetorical device employed to generate the necessary political, social, and financial capital to allow perceived promise to emerge.[105] The promise of a biotechnology revolution holds powerful appeal in that it provides government policy-makers and industry strategists with a simple model for promoting costly and difficult programs of change. Nonetheless the reality, as our analysis confirms, is far more complex. This underlines the general and pressing need for more nuanced appraisals of technical change, together with careful assessments of the likely dynamics, impacts, and timescale. These would render decision-makers better placed to lend effective support to emerging technologies and industrial sectors in both the short and longer term. With the appropriate support the benefits of biotechnology will be realized, beyond the period of patent protection, and beyond timescales focused on by politicians, investors, and their intended markets. However those that would inspire continued support for emerging technologies are duty bound to carefully consider their claims, lest they unintentionally reduce the appetite to support further investment when they are not realized at first.

Acknowledgments

This paper has developed from a commentary by two of the authors (Nightingale and Martin) entitled 'The Myth of the Biotech Revolution' published in *Trends in Biotechnology*, *22*, 564–569 in November 2004. We are grateful to Elsevier Ltd for allowing the reproduction of some material from that paper here. We also thank our industry interviewees and anonymous referees, as well as the UK's Economic and Social Research Council, and Medical Research Council for supporting this research under the Innovative Health Technologies Programme (Grant L128 25 2087) and the joint ESRC/MRC fellowship initiative (Grant PTA-037-27-0029). This paper has greatly benefited from the incisive comments received from Anthony Arundel and Michael O'Neill as well as a number of readers of the original paper. Any remaining errors are our own.

References

1. Kevles, D.; Hood, L.; *The Code of Codes: Scientific and Social Issues in the Human Genome Project*; Harvard University Press: Cambridge, MA, 1992.
2. Organization for Economic Co-operation and Development. *Biotechnology and Medical Innovation: Socioeconomic Assessment of the Technology, the Potential and the Products*; Paris, 1997.
3. Organization for Economic Co-operation and Development. *Economic Aspects of Biotechnologies Related to Human Health Part II: Biotechnology, Medical Innovation and the Economy: The Key Relationships*: Paris, 1998.
4. Bell, J. I. *Nature* **2003**, *421*, 414–416.
5. Bioscience, Innovation and Growth Team (BIGT). *Bioscience 2015*; Department of Trade and Industry: London, 2003.
6. European Commission. *Life Sciences and Biotechnology: A Strategy for the European Community*; Brussels, 2002.
7. Organization for Economic Co-operation and Development. Committee for Scientific and Technological Policy at Ministerial Level "Science, Technology and Innovation for the 21st Century", Final Communique Meeting, 29–30 Jan, Paris, 2004.
8. Bell, J. *Br. Med. J.* **1998**, *316*, 618–620.
9. Collins, F. S.; Patrinos, A.; Jordan, E.; Chakravarti, A.; Gesteland, R.; Walters, L.; the members of the DOE and NIH planning groups. *Science* **1998**, 282, 682–689.
10. Lenaghan, J. *Brave New NHS?* Institute for Public Policy Research: London, 1998.
11. Lindpaintner, K. *Br. J. Clin. Pharmacol.* **2002**, *54*, 221–230.
12. Department of Health. *Our Inheritance, Our Future: Realizing the Potential of Genetics in the NHS*; HMSO: London, 2003.
13. Department of Trade and Industry. *Biotechnology Clusters*; HMSO: London, 1999.
14. Enriquez, J.; Goldberg, R. A. *Harvard Bus. Rev.* **2000**, *78*, 94–104.
15. Tollman, P.; Guy, P.; Altshuler, J.; Flanagan, A.; Steiner, M. *A Revolution in R&D: How Genomics and Genetics are Transforming the Biopharmaceutical Industry*; The Boston Consulting Group: Boston, MA, 2001.
16. House of Commons Science and Technology Committee. *Human Genetics: The Science and its Consequences*; HMSO: London, 1995.
17. Department of Trade and Industry. *Science and Innovation Strategy*; London, 2001.
18. Dohse, D. *Res. Policy* **2000**, *20*, 1111–1133.
19. Giesecke, S. *Res. Policy* **2000**, *29*, 205–223.
20. Reiss, T.; Hinze, S.; Dominguez Lacasa, I.; Mangematin, V.; Enzing, C.; van der Giessen, A.; Kern, S.; Senker, J.; Calvert, J.; Nesta, L. et al. *Efficiency of Innovation Policies in High Technology Sectors in Europe (EPOHITE)*; Office for Official Publications of the European Communities: Luxembourg, 2003.
21. Ernst and Young. *Refocus: The European Perspective Global Biotechnology Report*; London, 2004.
22. Gassmann, O.; Reepmeyer, G.; von Zedtwitz, M. *Leading Pharmaceutical Innovation: Trends and Drivers for Growth in the Pharmaceutical Industry*; Springer-Verlag: London, 2004.
23. Nightingale, P.; Martin, M. *Trends Biotechnol.* **2004**, *22*, 564–569.
24. Office of Technology Assessment. *Biotechnology in a Global Economy, OTA-BA-494*; US Government Printing Office: Washington, DC, 1991.
25. Sharp, M. *The New Biotechnology: European Governments in search of a strategy Industrial Adjustment and Policy: VI, Sussex European Papers No. 15*; Science Policy Research Unit, University of Sussex: Brighton, UK, 1985.
26. Bud, R. *The Uses of Life: A History of Biotechnology*; Cambridge University Press: Cambridge, UK, 1993.
27. Teich, M.; Porter, R.; *The Industrial Revolution in National Context: Europe and the USA*; Cambridge University Press: Cambridge, UK, 1996.
28. Cannadine, D. *Past Present* **1984**, *103*, 31–72.
29. Floud, R.; McCloskey, D. Eds.; *The Economic History of Britain since 1700, Vol. 1, 1700–1860*; Cambridge University Press: Cambridge, UK, 1984.
30. Schumpeter, J. A. *The Theory of Economic Development: An Inquiry into Profits, Capital, Credit, Interest, and the Business firm*; Harvard University Press: Cambridge, MA, 1934.
31. David, P. A. *Am. Econ. Rev.* **1990**, *80*, 355–361.
32. Rosenberg, N. *Technol. Culture* **1979**, *20*, 25–51.
33. Freeman, C. Technological Revolutions and Catching Up: Information and Communication Technologies and the Newly Industrializing Countries. In *The Dynamics of Technology, Trade and Growth*; Fagerberg, J., Verspagen, B., von Tunzelmann, N., Eds.; Edward Elgar: Aldershot, UK, 1993, pp 198–221.
34. von Tunzelman, G. N. Technological and Organizational Change in Industry during the Early Industrial Revolution. In *The Industrial Revolution and British Society*; O'Brien, P. K., Quinault, R., Eds.; Cambridge University Press: Cambridge, UK, 1993, pp 254–282.
35. Crafts, N. F. R. *J. Econ. Hist.* **1995**, *55*, 745–772.
36. Crafts, N. *Econ. J.* **2004**, *114*, 338–351.
37. Rosenberg, N.; Trajtenberg, M. *J. Econ. Hist.* **2001**, *64*, 61–99.
38. Devine, W. D., Jr. *J. Econ. Hist.* **1983**, *43*, 347–372.
39. Nightingale, P. *Res. Policy* **2004**, *33*, 1259–1284.
40. Freeman, C.; Louca, F. *As Time Goes By: From the Industrial Revolution to the Information Revolution*; Oxford University Press: Oxford, UK, 2002.
41. von Tunzelmann, G. N. *Steam Power and British Industrialization to 1860*; Clarendon Press: Oxford, UK, 1978.
42. Nightingale, P. *J. Indust. Corp. Change* **2000**, *9*, 315–359.
43. Bognor, W. C. *Drugs to Market*; Pergamon Press: Oxford, UK, 1996.
44. Vos, R. *Drugs Looking for Diseases*; Kluwer Academic: Amsterdam, Netherlands, 2001.
45. Pavlou, A. K.; Reichert, J. M. *Nat. Biotechnol.* **2004**, *22*, 1513–1519.
46. Edgington, S. M. *Bio/technology* **1992**, *10*, 1529–1534.
47. Brekke, O. H.; Sandlie, I. *Nat. Rev. Drug Disc.* **2003**, *2*, 52–62.
48. Knight, P. *Bio/technology* **1990**, 7, 105–107.
49. Nightingale, P.; Mahdi, S. The Evolution of the Pharmaceutical Industry. In *Knowledge Accumulation and Industry Evolution: The Case of Pharma-Biotech*; Mazzucato, M. Dosi, G. Eds.; Cambridge University Press: Cambridge, UK, 2005, Chapter 3.
50. Booth, B.; Zemmel, R. *Nat. Rev. Drug Disc.* **2004**, *3*, 451–456.
51. Ma, P.; Zemmel, R. *Nat. Rev. Drug Disc.* **2002**, *1*, 571–572.
52. Friedrich, G. A. *Nat. Biotechnol.* **1996**, *14*, 1234–1237.
53. Shriver, Z.; Raguram, S.; Sasisekharan, R. *Nat. Rev. Drug Disc.* **2004**, *3*, 863–873.
54. Kramer, R.; Cohen, D. *Nat. Rev. Drug Disc.* **2004**, *3*, 965–972.

55. Senker, J.; Enzing, C.; Joly, P.; Reiss, T. *Nat. Biotechnol.* **2000**, *18*, 605–609.
56. Rothman, H.; Kraft, A. *J. Commer. Biotechnol.* **2006**, *12*, 86–98.
57. Powell, W. W.; White, D. R.; Koput, K. W.; Owen-Smith, J. *Am. J. Sociol.* **2005**, *110*, 1132–1205.
58. Drews, J. *Nat. Biotechnol.* **1996**, *14*, 1516–1518.
59. Hopkins, A. L.; Groom, C. R. *Nat. Rev. Drug Disc.* **2002**, *1*, 727–730.
60. Stein, L. D.; Stein, L. D. *Nature* **2004**, *431*, 915–916.
61. Van Brunt, J. 2002. Genomics Drugs March to the Clinic. http://www.signalsmag.com (accessed May 2006).
62. Service, R. *Science* **2004**, *303*, 1796–1799.
63. Drews, J.; Ryser, S. *Nat. Biotechnol.* **1997**, *15*, 1318–1319.
64. Food and Drug Administration. *Innovation or Stagnation: Challenge and Opportunity on the Critical Path to New Medical Products*; Bethesda, MD, 2004.
65. Organization for Economic Co-operation and Development. *Compendium of Patent Statistics*; Paris, 2004.
66. Leheny, R. *The Fruits of Genomics: Drug Pipelines Face Indigestion Until the New Biotechnology Ripens*; Lehman Brothers and McKinsey & Co: New York, 2001.
67. Campbell, N. *Eur. Biopharma Rev.* **2001**, (Mar), 52–58.
68. Hopkins, A. *Drug Discov. Today* **2004**, *3*, 173–175.
69. Marshall, A. *Nat. Biotechnol.* **1997**, *15*, 1249–1252.
70. McCarthy, A. *New Genet. Society* **2000**, *19*, 135–144.
71. Webster, A.; Martin, P.; Lewis, G.; Smart, A. *Nat. Rev. Genet.* **2004**, *5*, 7–13.
72. Institute for Prospective Technological Studies. *Pharmacogenetics and Pharmacogenomics: State-of-the-Art and Potential Socio-Economic Impacts in the EU.* European Commission Joint Research Centre: Seville, Spain, 2006.
73. Hackett, J. L.; Lesko, L. J. *Nat. Biotechnol.* **2003**, *21*, 742–743.
74. Kola, I.; Landis, J. *Nat. Rev. Drug Disc.* **2004**, *3*, 711–715.
75. Van Brunt, J. **2004**. Biomarkers: The Pendulum Finally Swings. www.signalsmag.com (accessed May 2006).
76. Pollack, A. Rebellious Bodies Dim the Glow of "Natural Biotech Drugs." *New York Times*, July 30, 2002, p F5.
77. Joppi, R.; Bertele, V.; Garattini, S. *Br. Med. J.* **2005**, *331*, 895–897.
78. Martin, P. A.; Morrison, M. M. *Realizing the Promise of Genomic Medicine*; Royal Pharmaceutical Society: London, 2006.
79. Reichert, J. M.; Rosenweig, C. J.; Faden, L. B.; Dewtiz, M. C. *Nat. Biotechnol.* **2005**, *23*, 1073–1078.
80. DiMasi, J. A. *Clin. Pharmacol. Ther.* **2001**, *69*, 297–307.
81. Bolten, B. M.; DeGregorio, T. *Nat. Rev. Drug Disc.* **2002**, *1*, 335–336.
82. Dickson, M.; Gagnon, J. P. *Nat. Rev. Drug Disc.* **2004**, *3*, 417–429.
83. Horrobin, D. F. *Nat. Rev. Drug Disc.* **2003**, *2*, 151–154.
84. Higgs, G. *Drug Discov. Today* **2004**, *9*, 727–729.
85. Horrobin, D. F. *Nature* **2001**, *19*, 1099–1100.
86. Kaitin, K. I.; DiMasi, J. A. *Drug Inform. J.* **2000**, *34*, 673–680.
87. Reichert, J. M. *Nat. Rev. Drug Disc.* **2003**, *2*, 695–702.
88. Glassman, R. H.; Sun, A. Y. *Nat. Rev. Drug Disc.* **2004**, *3*, 177–183.
89. Cantor, C. The Challenges to Technology and Informatics. In *The Code of Codes: Scientific and Social Issues in the Human Genome Project*; Kevles, D., Hood, L., Eds.; Harvard University Press: Cambridge, MA, 1992, pp 98–111.
90. Gilbert, W. A Vision of the Grail. In *The Code of Codes: Scientific and Social Issues in the Human Genome Project*; Kevles, D., Hood, L., Eds.; Harvard University Press: Cambridge, MA, 1992, pp 83–97.
91. Hood, L. Biology and Medicine in the Twenty-First Century. In *The Code of Codes: Scientific and Social Issues in the Human Genome Project*; Kevles, D., Hood, L., Eds.; Harvard University Press: Cambridge, MA, 1992, pp 136–163.
92. Department of Health. *The Genetics of Common Diseases*; HMSO: London, 1995.
93. Roses, A. D. *Nature* **2000**, *405*, 857–865.
94. Royal Society. *Personalized Medicines: Hopes and Realities*; OML: London, 2005.
95. Cooper, R. S.; Psaty, B. M. *Ann. Intern. Med.* **2003**, *138*, 576–580.
96. Arundel, A.; Mintzes, B. *The Impact of Biotechnology on Health*; Merit-Innogen Working; Paper University of Edinburgh: Edinburgh, UK, 2004.
97. Arnst, C. Biotech, Finally. *Business Week*, June 13, 2005, pp 41–46.
98. Kneller, R. *Nat. Biotechnol.* **2005**, *23*, 529–530.
99. Dibner, M. *Trends Biomed. Res.* **1993**, Sept. 6, 553–558.
100. Mitchell, P. *Nat. Biotechnol.* **2005**, *23*, 906.
101. Sneader, W. *Drug Discovery: A History*; John Wiley: Chichester, UK, 2005.
102. Bergman, G. Today's Biotech Breakthroughs Are Saving the Lives of Patients. *The Financial Times*, Nov 18, **2004**, p 20.
103. Pisano, G. P. *The Development Factory: Unlocking the Potential of Process Innovation*; Harvard Business School Press: Boston, MA, 1997.
104. AstraZeneca 2001. *Annual Report and Form 20-F.* http://www2.astrazeneca.com/annualrep2001/default.asp (accessed May 2006).
105. Guice, J. *Res. Policy* **1999**, *28*, 81–98.

Biographies

Michael M Hopkins is a research fellow at SPRU: Science and Technology Policy Research, at the University of Sussex, UK. Originally trained as a biologist, he has a DPhil from the University of Sussex in Science and Technology Policy Studies. His research interests include the industrial dynamics of the biotechnology and pharmaceutical sector and the evolution of technical systems in healthcare. His recent work has focused on the implementation of genetic testing services, trends in the patenting of DNA, and the changing nature of pharmaceutical innovation processes. He has worked as a policy consultant for the European Commission, UK Department of Trade and Industry and Europa Bio, the European biotechnology trade association. He also lectures on "Biotechnology, Innovation and Science Policy" at the University of Sussex.

A biochemistry graduate, **Alison Kraft's** research interests have broadly centred on the history and commercialization of the life sciences. Her work is informed by a concern to understand the emergence of new medical technologies and how they, in turn, shape and impact clinical practice. Recent work has focused on the strategies of biotechnology firms attempting to exploit genomics. Currently, she is working on an ESRC-funded project looking at the development of hematopoietic stem cells (HSC) and the dynamics of expectations in HSC-based innovation.

Paul A Martin is Senior Lecturer and Deputy Director of the Institute for the Study of Genetics, Biorisks and Society (IGBiS) at the University of Nottingham, UK. He originally trained as a molecular biologist and subsequently worked as a health policy analyst before taking a DPhil in science and technology studies at SPRU: Science and Technology Policy Research at the University of Sussex. His research interests cover innovation in the biotechnology industry, the social and ethical issues raised by genetics and the regulation of new medical technologies. Dr Martin has written reports for or advised a number of national and international organizations including the Wellcome Trust's Medicine in Society Programme, UK Department of Trade and Industry, Genethon, the French Conseil d'Analyse Economique, and the European Parliament.

Paul Nightingale is a Senior Research Fellow at SPRU: Science and Technology Policy Research, at the University of Sussex. Originally an industrial R&D chemist, he has a DPhil in science and technology policy studies. His main research interests are innovation in the financial and pharmaceutical sectors. He has lectured and published widely on innovation theory, the relationship between science and technology, and the theory of the firm. He is currently joint UK editor of *Industrial and Corporate Change*.

Surya Mahdi is a Research Fellow at SPRU, University of Sussex, UK. Dr Mahdi is trained as a chemical engineer and has a DPhil in science and technology policy studies. He specializes in analyzing indicators of innovative activity in the pharmaceutical and biotechnology industry. His research interests are industrial and corporate science and technology policy analysis, patent analysis, and bibliometric analysis, particularly in agrochemicals and pharmaceuticals. He has extensive experience of patent and bilbiometric analysis using a range of public and private databases.

Comprehensive Medicinal Chemistry II
ISBN (set): 0-08-044513-6

ISBN (Volume 1) 0-08-044514-4; pp. 591–613

1.18 The Apparent Declining Efficiency of Drug Discovery

D Trist, E Ratti, and L Da Ros, GlaxoSmithKline, Verona, Italy

1.18.1 Background

The pharmaceutical industry owes its origins in the nineteenth century to the necessity that new chemicals that had been recently synthesized and shown to be clinically active, such as ether, chloroform, and amyl nitrate, needed to be made on a large scale.[1] At that time, a number of pharmaceutical wholesale traders such as E. Merck had started to develop skills of applied chemistry and were developing the manufacturing capacity that could produce large quantities of new agents. However, companies such as Agfa, Bayer, and Hoechst began making and selling new kinds of drugs, many based on by-products of dyestuffs. Much of this work was guided by the understanding of the structure of known medicines. For example, new chemicals with antipyretic, antiinflammatory, and analgesic activity came to light by synthesizing molecules from coal tar distillation that contained part of the structure of quinine. These included phenazone, acetanilide, phenetidin, phenacetin, and amidopyrine.

Aspirin (acetylsalicylic acid) itself was synthesized in 1853 by Charles Gerhardt, but its medicinal properties remained undiscovered until 1899, even though it bore resemblance to salicylic acid, a known antipyretic and analgesic derived from willow bark.

The new-found pharmaceutical industry continued to modify chemical structures, often with unexpected results. For example, in the 1950s when the new antituberculosis drug isoniazid caused a striking improvement in mood of patients suffering from tuberculosis, other related products such as iproniazid were found to be monoamine oxidase inhibitors and iproniazid also improved the mood of depressed patients.[2]

Also, in the 1950s, a new approach to finding tranquillizers is a good example of how chemists were moving away from exploiting existing structures and looking for novelty. Chemists in the Roche Laboratories in New Jersey, US, were pursuing molecules unrelated to barbiturates and other sedatives and one chemist, Leo Sternbach, applied his earlier work in searching for new dyestuffs to try benzodiazepines, leading to a revolution in treating anxiety and insomnia.[3]

In the 1960s and 1970s drug discovery moved toward a more rational approach. Now, instead of new molecules being found by modification of existing medicines or by serendipity (random screening), novel medicines were discovered by understanding the molecular determinants in hormones and neurotransmitters that convey agonism and affinity. The discovery of the histamine H_2-receptor antagonist cimetidine exemplifies this approach.[4] Starting with the selective H_2-agonist 4-methylhistamine, and through the identification of the selective partial agonist N^{α}-guanylhistamine, agonism could be removed. Changing the basic group to thiourea and lengthening the side chain gave rise to burimamide, the first H_2-receptor antagonist. Optimization of the burimamide structure produced cimetidine, a more potent, orally active medicine that revolutionized the treatment of excess gastric acid secretion and consequent ulceration.

There are now many examples of applying this method to develop selective antagonists and agonists that produced major medicines and it could be postulated that this tactic fueled the wave of new product registrations throughout the 1970s and 1980s.

The rational approach was possible as a result of the advances in pharmacology, in starting to classify receptor subtypes by the identification of selective chemical tools.

Unfortunately, this rational methodology is not easy to apply to receptors that bind peptides and proteins as it is more difficult to understand the determinants within their structures that give efficacy and affinity. In addition, closely related receptors are often indistinguishable by antagonists when starting from agonists that do not differentiate these receptors.

The advent of molecular biology offered a new approach to drug discovery. The first advantage is that human receptors can now be cloned, structured, and expressed in cultured cell systems. The use of animal tissues to develop selective drugs has not always worked. Unfortunately, activity on animal receptors does not always translate into efficacy in humans. Thus, by having assay systems that use human receptors, this pitfall has been removed. By combining human assays with advances in screening and synthetic chemistry, the knowledge of receptor structure and the screening of large chemical libraries offer a new approach for drug discovery that overcomes the disadvantages of rational drug design.

Most pharmaceutical companies have embraced high-throughput screening (HTS) as the method to discover novel ligands for specific receptors and enzymes. Initially, fermentation media were also used as a way to generate novel molecules. An early example of success in this area was the generation of specific cholecystokinin (CCK) antagonists. In 1985, scientists at Merck screening products from microbiological fermentation discovered a selective CCK_A receptor antagonist asperlicin. From the observation that within the asperlicin structure could be identified a 1,4-benzodiazepine, a whole new class of CCK_A antagonists was born.[5]

The question that remains is whether the newer approaches can meet the challenges that face the pharmaceutical industry.

Figure 1 shows an apparent decline in productivity within the pharmaceutical industry in terms of new chemical entities (NCEs) approved since 1996.[6] This conclusion would be in agreement with many other key opinions. However, the authors of this analysis, extending the data back to 1945 (using a 3-year rolling average), show that NCE approvals may be cyclic (**Figure 2**). Thus, the downward trend between 1959 and 1967 may be considered equivalent to a similar trend since 1996. Schmidt and Smith[6] noted that in both 1963 and 1998 there were significant changes in the regulatory climate with the introduction of more hurdles. This was almost certainly not the only motive as, for example, NCE approvals were already in decline before 1963 and 1998. It is highly likely that the two peaks each represent the pinnacle of a cycle of drug discovery. The first followed a major investment after the Second World War that led to a number of important discoveries, whilst the second was the culmination of a greater understanding of pharmacology linked to rational drug design. In both cases, the peaks will reflect initial first-in-class compounds followed by medicines with incremental improvements. In addition, the 1970s and 1980s saw the beginning of extending the utility of medicines for one indication to others. For example, beta-blockers were originally developed for angina pectoris to reduce the activity of adrenaline on the heart and thus have a sparing effect on oxygen consumption.

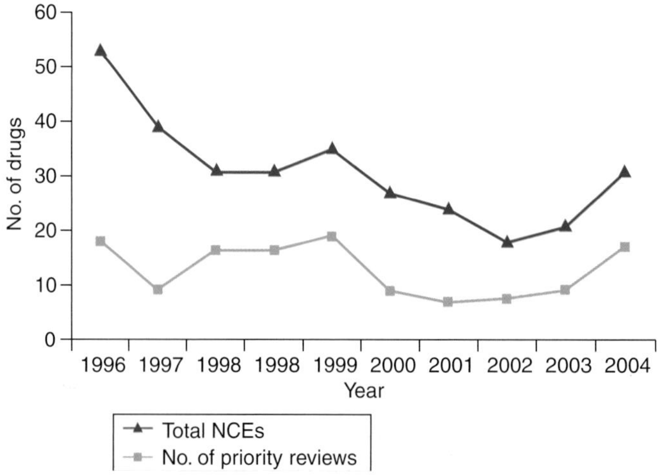

Figure 1 The number of total new chemical entities approved from 1996 to 2004. (Reproduced with permission from Schmidt, E. F.; Smith, D. A. *Drug Disc. Today* **2005**, *10*, 1031–1039 © Elsevier.)

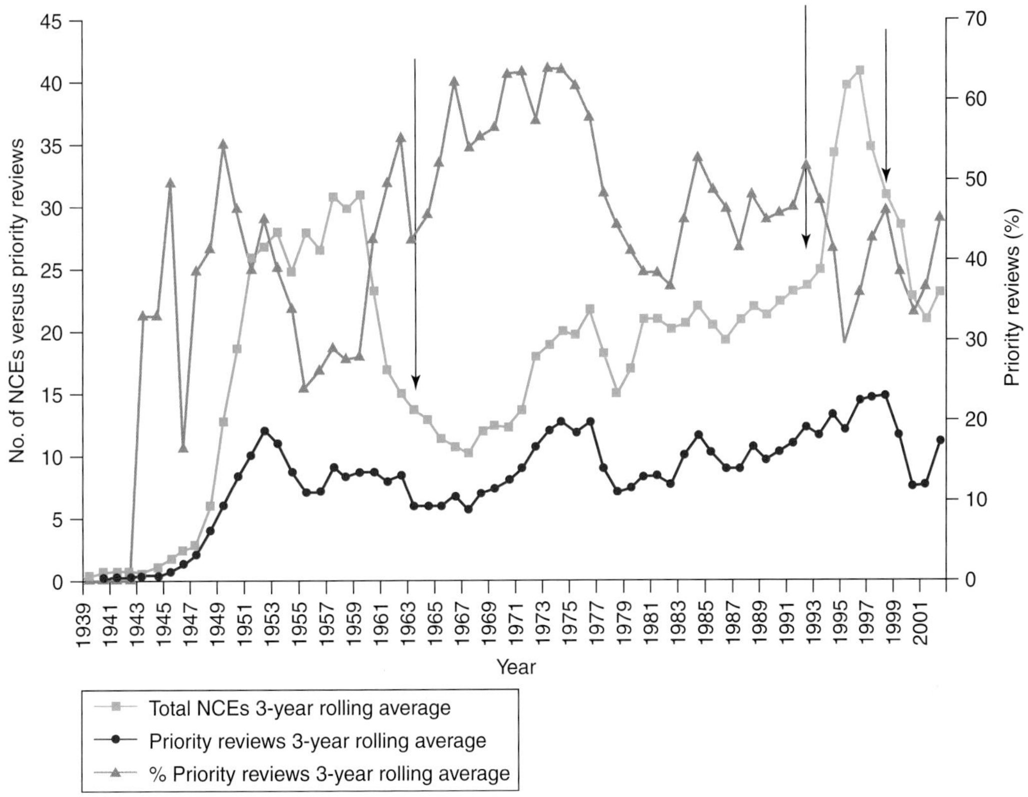

Figure 2 Analysis of NCE approvals since 1945, expressed as a 3-year rolling average. Arrows depict major changes in regulatory requirements for NCE registration. (Reproduced with permission from Schmidt, E. F.; Smith, D. A. *Drug Disc. Today* **2005**, *10*, 1031–1039 © Elsevier.)

However, subsequent investigations showed that these medicines, acting on different subtypes of beta-receptors could also affect blood pressure, cardiac arrhythmia, and glaucoma.

Discounting the two peaks, NCE approvals have been relatively constant at around 12–20 per year. However, pharmaceutical investment in research and development (R&D) has been anything but constant.

Since the 1960s the industry has developed more than 780 new medicines. However, during the last 25 years it has exponentially invested in obtaining NCE registrations (**Figure 3**). Thus, it could be said that there has been a declining return on investment in terms of registrations rather than an overall reduction in output. In 2004, in the US, R&D investment in the pharmaceutical industry reached more than US$ 30 billion.

Why has investment increased by more than 10-fold since 1980 to obtain, more or less, the same productivity? A number of causes can be singled out to explain the major increase in investment that the pharmaceutical industry has undertaken and these are discussed in more detail in the following sections.

1.18.2 Need for Greater Innovation

A constant driver for developing new medicines has always been the unmet medical need. However, there are now strong pressures to treat the underlying pathology rather than give solely symptomatic relief. This is leading the industry toward a more disease-based approach where understanding human pathology better should deliver targets that are involved in the causative processes of the disease.[7]

The pharmaceutical industry has always been a highly innovative enterprise. However, it could be said that much of the 'low-hanging' fruit has been harvested in a number of diseases and that the new challenge is to find more effective and better-tolerated medicines. In many cases, the challenge is to find disease-modifying agents rather than those that give symptomatic relief. Thus, more than ever, the identification and validation of novel mechanisms to human disease have become of paramount importance. This is especially important in therapeutic areas like the central nervous system where predictive animal models are few and far between. The complexity of diseases such as major depression, bipolar disorder, and schizophrenia have, so far, not been modeled in animals.[8] Therefore, in these diseases we are left

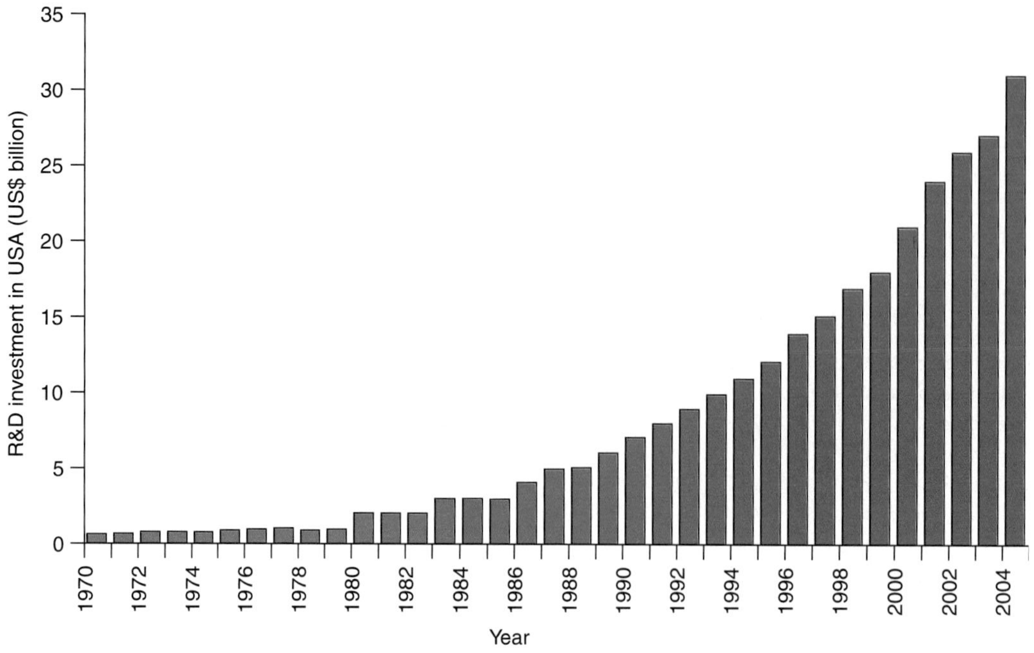

Figure 3 R&D investment in the US (US$ billion). (Reproduced with permission from Schmidt, E. F.; Smith, D. A. *Drug Disc. Today* **2005**, *10*, 1031–1039 © Elsevier.)

with targeting components such as receptor or biochemical systems, with no simple ways to validate these targets in the complex intact system.

In order to accomplish this, the investment that has already been initiated in technologies such as noninvasive imaging, clinical genetics, and genomics will increase. This is now assured with the publication of the human genome. It has been estimated that some 2000–3000 genes are druggable.[9] That is, these genes code for proteins that have structures that potentially could interact with small molecules. Interestingly, whilst this number of putative targets is not very different from previous estimates, there is a significant shift from the major target families, with, for example, fewer G protein-coupled receptors.

In these cases, where there are no simple ways of validating targets, the scientist is constrained to collecting a logical series of evidence that associates the target with the disease. This process will certainly be strengthened by the application of the technologies mentioned above. Already, imaging methods such as positron emission tomography and functional magnetic resonance imaging (MRI) are being used to understand the correlation between disease and specific receptors. Clinical genetics networks are being put into place to allow sufficient probands to be collected such that associations between particular gene(s) and disease can be made and ultimately lead to target validation and eventually identification. The discovery that the time to Alzheimer disease onset is significantly associated with having the homozygous allele of $APOE_4$ was the first robust finding of a disease-associated gene[10] and is pointing the way to more of such findings. Unfortunately, no drug has yet been developed based on this evidence.

The advent of the human genome's publication now offers a great opportunity for the understanding of the genetic make-up of disease and will furnish specific gene products and/or pathways as new targets that would not have been previously identified. Importantly, they will be born out of human data, so again adding to the level of confidence in the validity of the target.

The understanding of the human genome was thought to be a major strategy in understanding both target identification and obtaining evidence of target validation. The sequencing of the human genome has already given major insight into the numbers and types of genes that may be targets for drug discovery. However, to date knowledge of the human genome sequence has not yet delivered the benefits that were postulated at the inception of the initiative.

In addition, the application of pharmacogenetics to understand the genetic basis of efficacy and tolerability in patients when given specific medicines will feed back important information to develop more effective and better tolerated medicines. Thus, for example, patients who are high metabolizers may find less efficacy, whilst low metabolizers may have more side effects. This approach has been seen by some as the logical way forward to deliver the 'right medicine to the patient,' whilst others have been more skeptical, suggesting that such an approach will fragment the marketplace in an environment that is already hostile to high prices (*see* Section 1.18.5 below).

Innovation can be measured as NCEs launched into the marketplace. These are clearly more innovative than product line extensions, but are they all truly innovative? Using the central nervous system again as an example, there have been no really new mechanisms as antidepressants or as antipsychotics in more than 40 years. However, a large number of NCEs for both of these types of medicines have been launched, all showing incremental improvement. A novel mechanism that altered the course of the disease, that satisfied more than one disease domain, or acted within the prodromal phase of the disease might all be considered truly innovative medicines.

1.18.3 The Challenge of Attrition

A driver of change that is becoming increasingly important is the overall success rate of the drug discovery process. The probability of success of launching a candidate molecule into the market remains basically unchanged at around 10%[11] and the perception is that also time to market is not reducing, as might be expected with greater automation. In fact, it might even be rising!

As mentioned above, attrition has remained static despite the investment in many new technologies. This reflects the fact that good molecules need more than potency and selectivity to be successful, and it is in these areas where technology has been concentrating in the last few years. The challenges ahead lie in reducing the risk of not obtaining efficacy in humans, and in increasing the developability of the molecules.

1.18.3.1 Efficacy

Many new mechanisms fail when they get into humans through lack of efficacy. This is one of the risks that the industry takes when developing such molecules and has led to the strategy of many companies to have a pipeline that has a mix of novel and marketed mechanisms. One way to diminish risk is to get better validation in humans as soon as possible. The use of imaging, genetics, and genomics has already been discussed above as a way to help build early confidence in the target. Clinical readouts as early as possible are now being sought as part of making decisions as early as possible. It is now recognized that fast decision-making saves money and allows resources to be more effectively used. In addition, killing compounds in phase III is extremely costly. Thus, simple proof-of-concept (POC) studies are being sought in phase I or phase II, and this is discussed from an organizational concept in Section 1.18.4. The philosophy behind POC is that sufficient evidence can be generated in humans that allows the molecule to go forward. In some cases, the POC might be very simple. New antibacterials are known to work from extensive in vitro and in vivo studies. Thus, the issue in humans is to find a safe molecule with the right pharmacokinetic profile. This can be accomplished in phase I and could well constitute the POC, whereas until now a new stroke drug has only been shown to be active, or not, in phase III. The application of a smaller phase II study using MRI to follow structural damage progression postischaemic event is one example in which some evidence of efficacy in humans can be obtained to allow decisions on progression to be made.

The POC approach using small numbers of patients or volunteers will be actively followed in the future over a wide range of diseases. Unfortunately, in most cases small clinical trials using surrogates and/or novel endpoints have not yet been adequately validated and thus serve to improve confidence but do not substitute for traditional phase II trials. For this reason, the pharmaceutical industry has been looking outside its traditional country base to conduct such trials, both from an economic point of view and also from finding first-episode patients. Thus, investment in clinical studies in South America, India, the Pacific Rim, and parts of Africa are being seen. The cost of clinical trials is becoming ever more expensive. In a 10-year period the number of patients enrolled in clinical studies has increased by 135%.[12] In addition, in certain diseases the 'professional patient' is beginning to be seen. These are patients who participate in many trials, with the consequence that, based on their experience, they may try to second-guess the therapy that they are receiving.

Diagnostics will play a greater role in helping to choose patient populations, at least initially to show that the mechanism works. This will see greater and greater use of imaging, proteomics, and genetics in helping to identify the right patient group.

In the meantime, a better balance of novel molecules and those that are precedented will continue to be seen in the drug discovery portfolio. This will mean that a higher proportion of molecules will not fail for efficacy. However, this strategy creates its own problems in that to be successful in the marketplace the molecule will need to be differentiated from those already present. To do this in major clinical trials will add to the cost and to the overall cycle time (see below); thus, these problems will need to be addressed much earlier in the process.

1.18.3.2 Developability

A large proportion of all molecules that fail do so because of lack of developability – qualities of the molecule that give it poor safety, unacceptable side effects, and short time of action. Prentis *et al.*[13] suggest that this proportion is as high as 69%, broken down as toxicity (22%), poor biopharmaceutical properties (41%), and market reasons (6%). This is not a new revelation and efforts have been actively progressed throughout the industry to automate and miniaturize methods to measure solubility, stability, pK_a bioavailability, brain penetration, and hepatotoxicity, the main cause of toxicity. These methods (combinatorial lead optimization) are being applied to leads during optimization, but need to be developed further and applied even earlier to maximize their impact. This is particularly true for toxicity screens where it can be predicted that much effort will be done in the next few years. However, the goalposts are continuously moving. Over the years, regulators have added more and more criteria that individual molecules must pass before they can be registered. For example, these are being brought together in a number of guidelines under the International Conference on Harmonization of the European Medicines Agency. In recent times, cardiovascular safety (delayed ventricular repolarization (QT prolongation)), phototoxicity, genotoxicity of reactive metabolites, and immunotoxicity are just some of the items to which regulators have become more sensitive. The net effect is that industry attrition is likely to remain unchanged despite the efforts that the pharmaceutical industry is making to predict human toxicity of NCEs.

Much work is being done in the field of predictive algorithms and Pfizer has developed one that is known as the rule of five.[11] This is an awareness tool for medicinal chemists that suggests that there will be poor absorption if a molecule has two or more of the following: (1) more than 5 H-bond donors; (2) a molecular weight >500; (3) $\text{ClogP}>5$; and (4) the sum of Ns and Os (a rough measure of H-bond acceptors) >10.

Whilst it is inherently costly to try and fix poor developability by formulation, pharmaceutical development will become more actively engaged in alternative formulations and delivery systems during the lead optimization phase. The trend towards higher-potency compounds that reduces the cost of goods also allows, due to the smaller dose, alternative delivery systems such as inhalation, nasal, buccal, and sublingual absorption.

1.18.3.3 Cycle Times

Another driver of process evolution is to deliver molecules to the market quicker. A major reason for doing this is that the regulatory environment and the growing complexity of drug development have been potentially driving up times within phases. Screening automation and combichem initially reduced the time to candidate selection, but this may be increasing again with the need to find molecules that can overcome the greater developability hurdles. This will almost certainly decrease again by further application of techniques like chemoinformatics to aid library design, both for those to be used for random screening and those within the process of lead optimization.

As mentioned above, continual automation of developability criteria will also speed up the process by selecting out compounds with a high probability of not succeeding. This raises the concept that speed in each phase should not always be the major driver. A candidate for development goes forward with all of its associated baggage. Fixing problems becomes costly and may lead to a suboptimal product that cannot fulfill its medical and commercial potential. In addition, as discussed in Section 1.18.4, if the management organization is responsible for both the candidate selection and its POC, then poor candidates will eventually 'come home to roost.' Thus, spending time choosing the right candidate will have major benefits downstream, in terms of both speed and value. The same concept applies to development candidates in phase III. Differentiation may not be obvious if the mechanism is precedented with another marketed product. Thus, differentiation will become a challenge which potentially will increase the time in phase III. To aid in this process and help in choosing which differentiators to pursue, this problem will need to be addressed much earlier. This is beginning to stimulate automated assays for common side effects of drugs as part of the candidate selection criteria during the lead optimization stage.

1.18.4 Organization of R&D within Pharma

Over the years the pharmaceutical industry has been very successful at finding drugs using a variety of different structural organizations. Traditionally, organization has tended to follow the drug discovery process, being divided into distinct R&D organizations. Within research, the main functions were often organized on the disciplines of the day – biology and chemistry. Half a century ago candidates were found by serendipity and then 'tossed over the wall' to the development organization (**Figure 4**). As time progressed, and more systematic approaches were applied to finding chemical leads, research began to see new divisions such as biochemistry and pharmacology, medicinal chemistry, and physical chemistry. In addition, there was a tendency also to divide development into early and full development. However, candidates were still thrown from the research organization to the development organization.

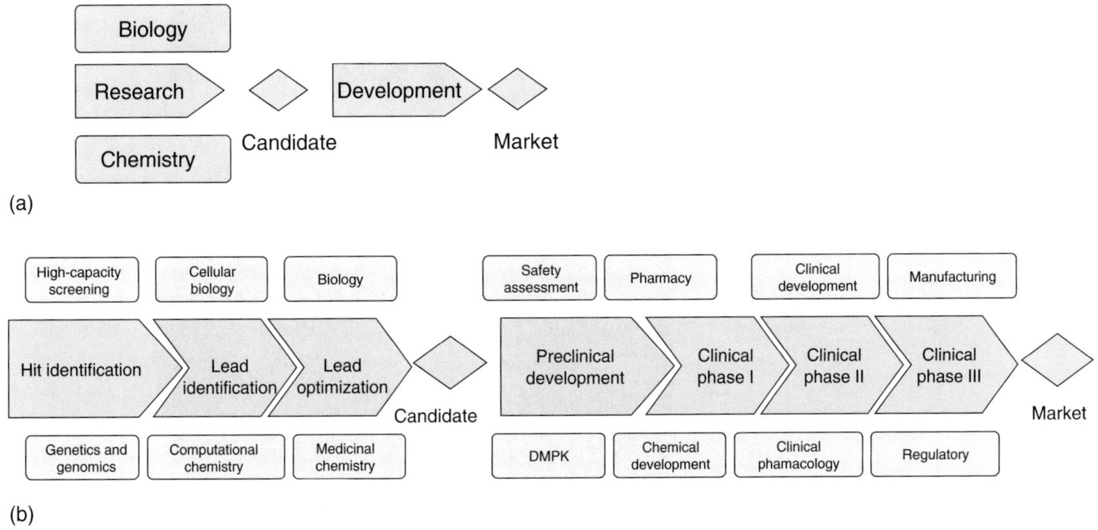

Figure 4 Evolution of the drug discovery process during the second half of the twentieth century. (a) The process as used in the 1950s and 1960s, where research was mainly driven by biological screening and serendipity. (b) The process has evolved, embracing novel technologies, but also in many cases greater organizational complexity. DMPK, drug metabolism and pharmacokinetics.

With the advent of consolidation within the industry throughout the 1980s and 1990s, it was beginning to be seen that the traditional organization modality was becoming managerially unwieldy, often crossing many sites and many countries. In addition, attrition was beginning to become an issue, with many candidates failing, often before they even reached humans. Initially, companies organized their research even more on technology, forming divisions of screening, genetics, molecular biology, and cell sciences. Research itself was often renamed discovery to describe better the principal activity required. However, the fundamental problem always remains the same – that of size. The bigger the organization was growing, the more its very size was impeding its output. Some companies moved to dedicated units based around therapeutic areas. This itself can simplify management structure, but can also create duplication to retain speed and interaction between functions. Thus, critical disciplines like safety assessment, pharmacy, chemical development, and clinical development have a tendency to be colocated with the therapeutic area discovery organization.

In 2001 the newly formed GlaxoSmithKline decided to organize itself in such a way that it could leverage the advantage that size brings to some important platforms like HTS, combinatorial chemistry, genetics research, and clinical development, but also put in place a drug discovery organization that combined the therapeutic focus of six small separate centers of excellence for drug discovery (CEDD), with the agility and innovation seen with biotech companies (**Figure 5**). In addition, the drug discovery process was elongated from lead identification to first clinical POC in humans. The wall between traditional R&D is now no longer there. The whole reason for introducing these changes was to provoke a step change in productivity. In 2001, GlaxoSmithKline had 25 NCEs in Phase I and 19 molecules in Phase II. By 2005 the number of NCEs within these phases had grown to 38 in Phase I and 46 within Phase II. It remains to be seen if this level of productivity is sustainable (**Figure 6**). The full development organization within GlaxoSmithKline has also been reorganized to form medical development centers (MDCs), again therapeutically aligned.

Other companies have implemented drug discovery organizations and have worked in focused therapeutic areas. However, it may be the sum of all of the changes that GlaxoSmithKline has implemented that has led to an apparent leap in NCEs entering the clinic.

Recently, companies such as Pfizer have announced organizational changes that share some of the elements of the GlaxoSmithKline structure. If successful, then productivity across the industry may well be heading up, simply due to removing major organizational 'road blocks.'

1.18.5 External Environment

The external environment in which pharma operates has significantly changed within the last half-century.

First of all, the environment has become highly regulated.[1] This has been discussed in part under developability (*see* Section 1.18.3.2). The regulatory requirements are a major factor in determining the development path an NCE

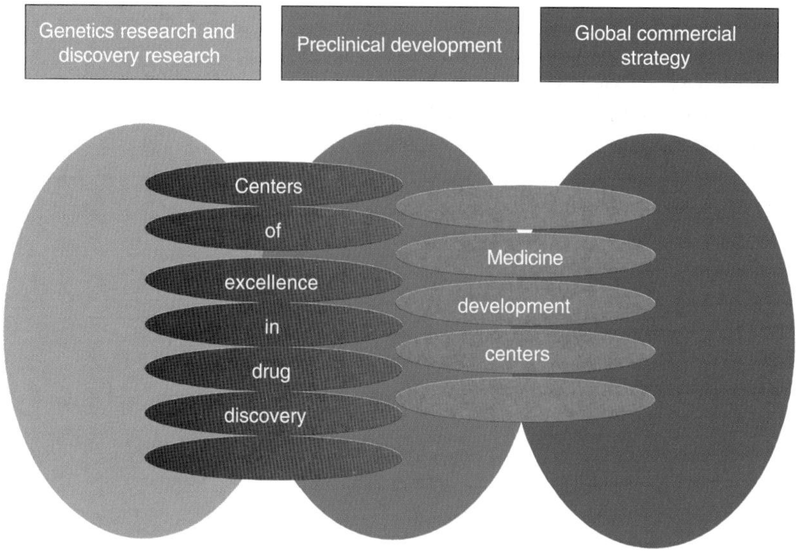

Figure 5 The R&D model in GSK. In this model, large technical platforms such as genetics research, discovery research, and preclinical development, support two groups of centers responsible for specific parts of the drug development process. Thus, the, CEDD are accountable for taking a tractable lead to first clinical evidence of effect (POC) within a specific therapeutic alignment. The asset is then transitioned to late-stage development process that is managed by medicine development centers that are also therapeutically aligned. These groups manage the asset to the market. From the GSK Annual Report 2004. GSK, GlaxoSmithKline.

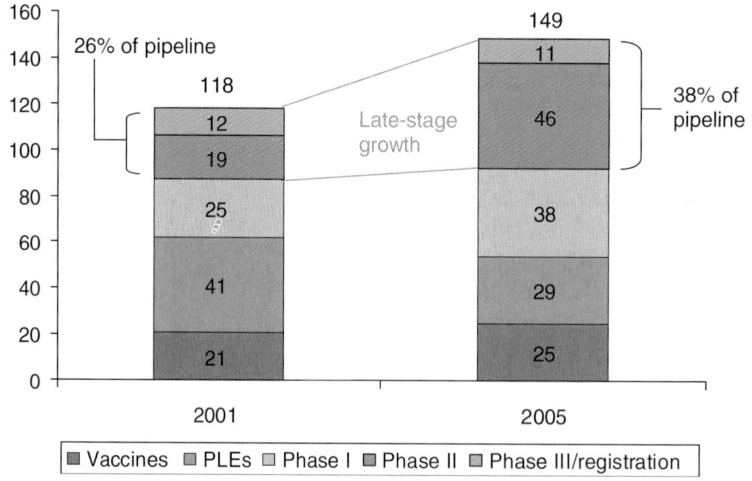

Figure 6 Productivity within GSK since 2001 as measured by Phase I, Phase II, and Phase III numbers (from the GSK Annual Report, 2005). It can be seen that since the inception of GSK there has been a sustained step change in the number of NCEs entering the late stage of the pipeline. GSK, GlaxoSmithKline.

may follow and the time and expense associated with this development.[14] Despite harmonization activities at a regional and global level, the regulatory environment is varied and changes rapidly. The Pharma industry has always been regulated, but changes in the way regulators think can be seen in subsequent launches during the next decade. Thus, Schmid and Smith[6] point out that following the Harris–Kefauver amendment introduced in the US in 1963 in response to the thalidomide crisis, the additional testing requirements introduced had a consequence of only 134 approvals in the following decade. The 10 years previously (1955–64) had seen 242 approvals.

The enabling legislation of the US Food and Drug Administration (FDA) requires that the agency approve a drug if found effective in well-conducted clinical trials. Recently, the FDA has shown that it can have a broader vision in the interests of better science.[14] For example, it will not allow a company to claim that its osteoporosis drug prevents fractures if the trial data only demonstrate an increase in bone mineral density. Thus, it should be within the purview of the FDA to decide what 'effective' really means.

The agency is also required to determine whether a drug is safe or not. This decision can only be reasonably made against the clinical usefulness of the potential medicine. This inevitably means consideration of how the drug compares with alternative therapies. Probably the agencies' (both the FDA and European equivalents) greater attention to safety issues in the past decade or so may have contributed to the lower number of NCE registrations.

The increasing attention on safety has been in part driven by increasing public concern about efficacy–risk–cost trade-offs that have been publicised of late. In the US this may well take center stage if Medicare becomes the US's largest drug purchaser.

Attention to cost of the health service of the new medicine is also driving down registrations. This is particularly relevant in Europe where protracted pricing discussions can delay the launch of medicines by a considerable time.

Adverse publicity of the industry has been more common of late, although the idea that the sick are being exploited for monetary gain has always been leveled at Pharma. With respect to the developing world, exploitation has been replaced by accusations of inaccessibility by pricing. Many companies within the industry now run initiatives to help the developing world that have increased low-cost (or sometimes free) sales to these countries and the prosecution of programs directed entirely at diseases of the Third World (e.g., malaria).

1.18.6 Toward the Future

The pharmaceutical industry must be ranked as one of the most successful over the last 50 years. It has helped to revolutionize how patients are treated and many diseases that would have been seen to be life-threatening do not now exist (e.g., smallpox) or might now considered in the main to be a nuisance (e.g., gastric ulcers). The industry has continually responded to the external environment, which has often been hostile. It has continued to be successful because it continues to invest in new technology and processes. However, the promise of screening mega libraries and that of the elucidation of the human genome are two areas that until now have not materialized as was originally expected, although delivery has probably only been delayed.

Biologicals are another area to which pharma has turned its attention. The finding that attrition is actually lower for this type of compound in development than the more conventional small molecule is an added plus in an area that is beginning to generate considerable sales across the industry. This may well be the instrument for regrowing the product registrations.

Costs will be restrained by outsourcing and offshoring activities. These activities may be taken further to reduce the process internally to its critical elements with all else being executed externally, i.e., virtualization of the process.

There will certainly be more and more business done with biotechs and academia, especially to access new mechanisms and to understand disease pathophysiology.

Delivering the right medicine to the right patient will eventually become a reality. As patients become more used to associating diagnostic tests with their disease, disciplines like pharmacogenetics will come into their own to help diagnose the type of disease from which the patient is suffering and, importantly, the probability that he/she will suffer from any specific side effect or safety issue associated with the medicine.

In conclusion, the pharmaceutical industry has probably never been so innovative in its history. The need to deliver better and better medicines is stimulating technology investment, smarter ways of working, and implementation of organizations that remove things that were simply getting in the way. The reduction in registrations in recent years probably simply reflects the cyclical nature of the business where delivery is sensitive to regulatory changes and the introduction of new waves of innovative medicines.

References

1. Weatherall, M. *In Search of a Cure*; Oxford University Press: New York, 1990.
2. Davis, W. A. *J. Clin. Exp. Psychopathol.* **1958**, *19* (Suppl.), 1–10.
3. Sternbach, L. H. *Prog. Drug Res.* **1978**, *22*, 229–266.
4. Black, J. W.; Duncan, W. A. M.; Durant, L. J.; Emmett, J. C.; Ganellin, C. R.; Parsons, M. E. *Nature* **1972**, *236*, 385–390.
5. Evans, B. E.; Rittle, K. E.; Bock, M. G.; DiPardo, R. M.; Freidinger, R. M.; Whitter, W. L.; Lundell, G. F.; Veber, D. F.; Anderson, P. S.; Chang, R. S. L. et al. *J. Med. Chem.* **1988**, *31*, 2235–2246.
6. Schmid, E. F.; Smith, D. A. *Drug Disc. Today* **2005**, *10*, 1031–1039.
7. Ratti, E.; Trist, D. *Pure Appl. Chem.* **2001**, *73*, 67–75.
8. Sams-Dodd, F. *Drug Disc. Today* **2005**, *10*, 139–147.
9. Russ, A. P.; Lampel, S. *Drug Disc. Today* **2005**, *10*, 1607–1610.

10. Strittmatter, W. J.; Roses, A. D. *Proc. Natl. Acad. Sci. USA* **1995**, *92*, 4725–4727.
11. Lipper, R. A. *Modern Drug Disc.* **1999**, *January/February*, 55–60.
12. Schmid, E. F.; Smith, D. A. *Drug Disc. Today* **2004**, *9*, 18–26.
13. Prentis, R. A.; Lis, Y.; Walker, S. R. *Br. J. Clin. Pharmacol.* **1988**, *25*, 387–396.
14. Avorn, J. *N. Engl. J. Med.* **2005**, *September 8*, 969–971.

Biographies

David Trist Vice President and Head of Strategy and Operations for the Psychiatry Centre of Excellence for Drug Discovery (CEDD), GlaxoSmithKline. Dr Trist obtained his first degree in biochemistry and his PhD in pharmacology. He has spent all of his working life within the pharmaceutical industry, mostly as an analytical pharmacologist. He was Director of Pharmacology in Verona, and presently is responsible for formulating the strategy for the Psychiatry CEDD.

Emiliangelo Ratti gained his doctorate in pharmacy at Milan University. He has specialized throughout his career in Neurosciences, particularly in amino acid research. Within Glaxo, he held positions of Project Leader, Departmental Head and Research Director. He was one of the three Drug Discovery Directors within Glaxo Wellcome before he was appointed Head of the Psychiatry Centre of Excellence for Drug Discovery at the formation of GlaxoSmithKline.

Lucio Da Ros Director, Portfolio and Performance Analysis – Psychiatry Centre of Excellence for Drug Discovery. Dr Da Ros is an engineer by training and has an MBA from the Bocconi University, Milan. He has many years of experience in clinical pharmacokinetics. He is a specialist in mathematical modeling of scientific and portfolio data. Dr Da Ros is interested in the quantitative analysis of drug discovery performance within the pharmaceutical industry.

Comprehensive Medicinal Chemistry II
ISBN (set): 0-08-044513-6

ISBN (Volume 1) 0-08-044514-4; pp. 615–625

1.19 How Much is Enough or Too Little: Assessing Healthcare Demand in Developed Countries

D Callahan, The Hastings Center, Garrison, NY, USA

1.19.1 Introduction

How much and what kind of healthcare do citizens of developed countries need?

Or should the question be: how much do they want or desire? Or, as an economist might put the question, what is the demand for healthcare, that is, how much are people willing to buy and at what price? And of course one might also ask, apart from people's expressed needs, desires, and demands, two further questions. One of them is: what ought people to want, desire, and demand – what will really do them the most good? The other is, from a government perspective: what kind and extent of healthcare should ideally be provided to citizens to maximize individual and population welfare (which are not quite the same)? How much healthcare is enough for individuals and for societies, and how much is too little?

Each one of these questions seems reasonable enough, but each requires a different approach when looking for an answer. How each is answered, however, will usually provide at least some illumination when trying to answer the other questions. My approach will be not to single any one of them out for special attention, but to treat them as an ensemble. If one is interested in developing plausible health policies, then each one of them will have to be considered; and how they are collectively answered will take us a long way in thinking well about effective policy.

In the nature of the case, some of the questions seem open to empirical answers (public opinion surveys can tell us what people want or desire), but others are normative in nature (what people ought to want); and still others will have to manage a mix of the empirical and the normative (what, given available economic resources, should government try to provide), often walking a fine line between expressed desires and its perceptions of what a country most needs? That older men may want Viagra does not prove it qualifies as an important national health need.

A first step is simply to recall some truisms; familiar enough on the one hand, but often neglected when thinking about health policy on the other:

- The more affluent and developed a country is, the more likely it is to spend on healthcare, both in terms of actual monies and as a percentage of its gross domestic product (GDP). But there is no perfect correlation between healthcare expenditures and health outcomes, even though money surely counts. The US spends more per capita on healthcare than any other developed country, but ranks no better than 18th on life expectancy.[1]
- There is no such thing as perfect health, either for individuals or populations. As the late distinguished biologist Rene Dubos reminded us some years ago in *The Mirage of Health* "Complete and lasting freedom from disease is but a dream remembered from imaginings of a Garden of Eden."[2] However healthy individuals are at one time or another

in their lives, they will eventually sicken and die. Every population in developed countries is in far better health now than 50 years ago, and most people are well rather than sick. No matter. We will all die sooner or later. The conquest of illness and disease can be likened to the exploration of outer space: however far one goes there is still further one could go. Western societies have what I call an 'infinity' model of medical progress and technological innovation at their core: no end result, other than open-ended progress, is envisioned, no final end point – just more.[3]

- The expectation and desire for good health, and the baseline of what counts as good health, will rise in step with medical progress and technological innovation. The healthier a country becomes the more it will spend on health. And there will be no correlation between progress and healthcare expenditures and people's evaluation of their health.[4] Becoming healthier, in fact, does not entail that we will demand less healthcare; we will just demand different and improved kinds of healthcare, often to treat the successor illnesses that have replaced those earlier cured or ameliorated illnesses.
- The main determinant of health status in both developing and developed countries is not medical or organized healthcare, but socioeconomic conditions.[5] Those conditions are estimated to account for 60% of a population's health status. The reason for that figure remains unclear, but there is no evidence that improvement in organized healthcare will ever become as important as such variables as education, income, and social status.

1.19.2 Balancing Individual and Community Needs

For convenience and clarity, I will not attempt in any systematic way to take up each of the earlier listed questions, but instead will return to them as necessary, aiming to touch on each point to show how they are related to one another. To this end, I will pose one overarching question: how ought government health policy balance the healthcare needs and desires of the individual with that of the community as a whole in a way that is economically affordable and sustainable over the long run? That question has three components, which are considered below: individual needs and desires, the needs of the community as a whole, and an economically affordable and sustainable policy.

1.19.2.1 Individual Needs and Desires

Imagine a continuum. It runs from one end of the healthcare spectrum to the other. On the left end of the spectrum is indisputable medical need: a broken leg to be set, a ruptured appendix to be removed, a growing malignant tumor that requires elimination, or a severe heart attack needing immediate treatment. On the right end of the spectrum might be Viagra for erectile dysfunction for the very old (over 80), cosmetic surgery for improved appearance, restoration of male hair growth in cases of balding, or sex selection of a male child. Yet even to suggest a list of items at the right end of the continuum represents desire rather than need is to enter into contested territory. For some people, their desires take on an obsessive flavor, which is part of their self-identity and hard to distinguish from real needs. Being bald makes some people as unhappy as if they had cancer. Yet no one's life will end for lack of fulfilling such desires and most people throughout history have adjusted to live good lives without such fulfillment. If everything at the right end of the continuum becomes increasingly open to argument as a putative need (increasingly being negotiated into something more important than desire) then everything between need and desire is in an ever deeper gray area, no less open to redefinition and negotiation.

Expensive new heart technologies and chemotherapies, often effective in ameliorating lethal diseases but not curing them, fall into this gray area. An implantable artificial heart is just over the horizon that will be able to save lives, even of those in their 80s and 90s. Measured by traditional standards of saving life, it will be medically acceptable, even conventional, and will meet a basic and indisputable need. But can the needs of the old be viewed in the same way as those of the young. If the artificial heart saves lives, should it be employed regardless of age? If so, then of course we are on an open frontier of progress. But do societies, in terms of the health of their population, need expensive new technologies to extend the life of the very old, who have a short life ahead of them? Does someone aged 100 need to have their life saved? Do societies need to save centenarians with the same zeal as they do children?

Few, if any, developed countries want to openly confront questions of that kind. They implicitly pit a conventional principle of healthcare, that of using whatever technology is necessary if it will clearly save lives, against the cost of that care. Rising costs increasingly make the honoring of such a principle difficult, even subject to anguish. The American Medicare program for the elderly has recently had to make some difficult decisions about new and expensive technologies.[6] So far, it has been able to solve the worst kinds of dilemmas by devising a set of medical standards to deal with them, but this will not always be possible. During the 1950s and up to at least the 1970s, the UK achieved a kind of notoriety in healthcare circles because of its unacknowledged practice of using age as a standard to deny dialysis to those over their mid-50s dying of end-stage renal disease.[7] Doctors told their patients that it was not 'medically indicated.' That kind of statement was not true at all (many other countries were dialyzing patients well into their 60s and 70s) but was much more acceptable to the public than openly admitting that the policy was nothing but age-based rationing. The public finally caught on and the practice was changed, but it is not difficult to imagine similar scenarios developing in the future in many countries. How much is enough? This is likely to be an unavoidable question with expensive treatments for the elderly.

1.19.2.2 Medical Necessity

Any effort to determine what counts as individual need in terms of current healthcare turns out to be a complex effort; easy enough with acute appendicitis or myocardial infarction, but increasingly difficult with many other conditions. The concept of 'medical necessity,' for instance, a term that can be found in much American healthcare legislation as a standard for the provision of care, has never been definitively defined. Its purpose has been to capture the notion of need in a clear and usable formula, but it has been remarkably elusive in practice. There does not appear to be any generally agreed upon, accepted, policy-relevant definition. The initial assumption was that the phrase 'medical necessity' was an empirical concept that was easily deployable: that which would save a life, cure a disease, or relieve pain. What could be more clear and obvious?

Just about anything it turns out. A variety of cultural, economic, medical, and economic realities obscure the issues. The cultural realities are many. Different national cultures, or regional differences within countries, can influence what people, including doctors, count as necessary treatment. Though there are now few medical obstacles to providing even the most advanced technological treatments to the elderly, there is considerable data to show significant variations from place to place in doing so, reflecting different values. Less than a decade ago it was accepted in the Czech Republic that elderly people become more forgetful as they get older. Alzheimer's disease and other dementias were simply not recognized, and thus there was no perceived 'need' for medical treatment of what was taken to be an ordinary feature of aging.[8] Different cultures have different attitudes toward pain, suffering, illness, and death.

In most developed countries the media is a variable influence on evaluations of medical need. By publicizing new treatments for previously neglected or untreatable conditions, it can change their perceived status, and doctors can expect to receive calls insisting on the new treatments.[9] This seems particularly true for the treatment of chronic conditions, such as arthritis, where no cure is available and many of its victims accept persistent pain as an unavoidable part of their daily life. As new treatments become available, and are trumpeted by the media, what was previously understood as a desire or want (for lack of any other choice) is gradually transformed into a need that should now be met. The history of medical progress is full of cases of conditions once tolerated being turned into medical conditions classified as needs. The conditions had not changed, but the possibility of doing something about them brought them into the domain of medicine and, often, into the more privileged domain of medical necessity.

The drug and device industries have also played a large role in transformations of this kind. The situation in the Czech Republic changed when the Pfizer Company provided some funding to create a professional association that aimed to increase physician interest in and education on dementia and, not incidentally, to help sell its Alzheimer's drug Aricept. Direct-to-consumer drug advertising (though allowed only in the US and New Zealand) has been most effective in prompting physicians to prescribe drugs that patients want but probably would not have been otherwise prescribed.[10] There is often a fine line between a drug that a patient simply desires and that same drug so insistently desired that it becomes a need in the patient's eyes.

American hospitals regularly advertise special services or available technologies designed to suggest to patients that if they really want to survive their illness, then their facilities are the best to meet that need. In those cases, there is undeniable need, and the aim of the advertising is to suggest that only this particular facility or service is likely to fully meet their needs. Furthermore, the media, either through its news stories or paid advertisements, is a major vehicle for helping to transform desires and wants into treatable needs.

1.19.2.3 Supply and Demand

The economist's understanding of 'demand' is relevant in this context. The term is typically defined as the amount of a good people are willing to buy at a certain price. Supply and demand are in balance when the price of a good matches the public's willingness to pay for it.[11] A common assumption of supply and demand analysis is that the higher the price of a good, the lower the quantity demanded. Competition among suppliers, it is thought, will ordinarily drive down costs. This may be true of the sale of cars, cell phones, and computers but not necessarily of healthcare, which regularly features cost increases in most developed countries that regularly exceed annual cost-of-living (COL) increases.

There is a long argument about the market and its role behind increases of that magnitude. One of them is that when third parties (insurance companies, for instance) are paying the costs of healthcare, there is little incentive for patients to worry about costs. While this is no doubt true, the highly effective marketing of drugs, aided by patent-protected monopolies, can effectively increase demand, even in the face of cost escalation well beyond COL increases. This may not only be attributed to the fact that many drugs meet well-established medical need, but also that the more expensive proprietary versions of drugs that are also available in their cheaper generic form are effectively advertised.[12] In addition, so-called 'me-too' drugs, i.e., those that vary only slightly from older drugs, are often wanted despite their marginal benefits. Once again, desire is transformed into need, and need commands higher prices than wants and desires (at least most of the time). By virtue of their willingness to control drug prices, European countries are more insulated from the force of the market than the US, which has been resistant to price controls for a long time. A refusal to allow direct-to-consumer advertising displays in most countries reflects an awareness of how powerful patient pressure on physicians can be.

It is safe to conclude that there are a variety of forces that work to turn wants and desires into needs. That has been the history of modern medicine, and it is now intensified by affluence, the market, and media attention. It seems to be a law of human nature that once we have moved beyond a stage of life or human development allowing us to control that which once seemed uncontrollable, it is hard to imagine how we put up with the earlier state and no less unimaginable to turn back the clock. If the history of the automobile, the telephone, and the airplane are not enough to convince one of this, then medical change – with all the attendant nonmedical pressures that force a response to it – should make it obvious enough. Yet to emphasize, as I have done, all the factors that shape the medical desires, wants, demands, and needs of individuals without taking account of the undeniable value of medical progress would be a mistake. That progress has made it possible for those in developed countries to put behind them a once dominant fatalism, and to take arms against a sea of bodily and mental miseries. If the progress has encouraged excessive expectations, it has also, and legitimately, allowed people to have a realistic hope that, with many earlier medical conditions now overcome, there will be many more triumphs in the future.

1.19.3 Community Needs and Desires

If it is the case that the nature of medical progress leads inexorably to a widening scope for the notion of medical need, it leaves unclear and ever-changing the meaning of 'medical necessity.' What is 'enough' medical care for individuals in developed society? The evidence would suggest that, with an infinity model of medical progress, the notion of 'enough' simply has no meaning. As long we continue to get sick, to get old, and to die – and to continue, as is our wont, to dislike those conditions – then there will never be enough healthcare for us. Yet a focus on individual needs and desires is not the only possible perspective. Many nations attempt to assess the health of their population as a whole, focusing on birth and death rates, life expectancy, quality of life, and access to healthcare, among other indicators.

A good example of such an effort is *Healthy People 2010*, the third in a series of studies (at 10-year intervals) of American health, which was published in 2000 by the US Department of Health and Human Services (DHHS).[13] Its aim is to establish goals for the future, and to do so in some detail, with the aim of "promoting health and preventing illness, disability, and premature death." The report explains that its focus on population health is important because "the health of the individual is almost inseparable from the health of the larger community."[14] For this edition of the study, two more specific goals are specified: "to increase the quality and years of healthy life," and to "eliminate health disparities." The report then lists 28 focus areas which includes a mix of healthcare services, behavioral change, and specific medical conditions, such as cancer, followed by 467 objectives (e.g., the subfocus goal for cancer is to "reduce the number of new cancer cases as well as the illness, disability, and death caused by cancer," involving 15 objectives, e.g., reducing lung cancer). It is an impressive collection of information and sets sensible goals.

Yet, in light of the aims of this paper, the 2010 report presents a number of problems. In some cases, the goals are open ended, with improvement of health status or services as the aim, as with cancer, childhood immunization, and access to care. In the instance of improving life expectancy, there is no suggestion that a longer average life expectancy per se would be appropriate, but mentions instead the fact that the US is well down on the list of life expectancy for developed countries (about 18th) and that there are significant disparities in life expectancy among subgroups of the American population. However, when the report discusses health disparities among the general population (race, gender and ethnicity, education and income), the aim is not just their improvement but also their elimination. At the same time, it recognizes the difficulty of achieving that goal: gender differences in life expectancy seem, in part, to have deep biological roots, and for no less obscure reasons women are at significantly greater risk for Alzheimer's disease and depression than men.

Nonetheless, the fact that 'elimination' is the goal for a number of disparities – even if this is unlikely to be achieved in the next decade – stands in contrast to the incremental aims of the overwhelming majority of medical conditions and health access problems that mark the rest of the study. Improvement, not victory, is the aim. How might one classify the 467 objectives using the categories of this paper: needs, wants, and demand? Most of them would seem to qualify as 'needs;' that is, they are thought to be significant enough to be listed in the report and each one of them points to an issue of importance.

There is no mention of cosmetic surgery, sex selection, or memory enhancement, suggesting that at least for the DHHS they do not qualify as needs. It speaks forcefully, however, about the public health needs: "Diabetes and other chronic conditions continue to present a serious obstacle to public health. Violence and abusive behavior continue to ravage homes and communities ... Mental disorders continue to go undiagnosed and untreated." There is no discussion of whether there is a need, for instance, to eliminate every case of diabetes, all abusive behavior, and each instance of mental disorder; all are approached in an incremental way. Nor is there any suggestion of priorities among the objectives.

1.19.4 Individual and Community Demand

There is endless room for speculation and analysis on the difference between individual need and desire, the way desires are transformed into needs, and, for that matter, the way new technologies can move people straight from utter disinterest in something to a powerfully felt need but with no intermediate desire stage. Where such an analysis begins to have some important policy implications is when governments (or, say, managed care organizations) are forced to live within limited budgets and thus have to decide on what to provide.

Policy formation may well be defined as finding a tolerable balance between what providers think their patients (or would-be patients) want and what the providers can afford to give them. Moreover, a balance must be struck between different levels of care, each of which will be important, and with different health implications, but which may have greater or less patient appeal. Among the different levels will be preventive care (aiming to keep people well), therapeutic care (aiming to treat them when they are sick), rehabilitative care (aiming to allow those with disabilities to function well), and palliative care (designed to relieve pain and suffering). While it is well known that preventive care can be a good economic investment, not to mention good for health, it does not have the kind of emotional force that the treatment of a lethal disease does.

1.19.4.1 Setting Priorities

A number of European countries in the 1990s, and the state of Oregon in the US, created commissions to set healthcare priorities. They were aimed at distinguishing different levels of need, on the one hand, and treatment for conditions that fell on the borderline between desire and need, on the other. A common classification, reflected in other commission reports, can be found in the Swedish findings.[15] Six levels of care were defined: (1) treatment of life-threatening acute diseases (some countries specified life-threatening diseases as those that would bring a premature death); (2) treatment of severe chronic diseases; (3) treatment of less severe acute and chronic diseases; (4) palliative terminal care; (5) treatment of diseases entailing reduced capacity for autonomy; and (6) habilitation/rehabilitation of disabled persons and provision of technical aids.

The Swedish commission then went on to distinguish between priorities at the political-administrative level (that "concern anonymous groups and are impersonal") and those in the clinic (that "affect individuals and are personal"). There were differences between these two categories, but the general idea remained the same: life-threatening conditions remained at the top of both lists, with lesser conditions lower down. The commission defined need as "something one feels deprived without." They then added "this means a hierarchy of needs determined by the gravity of the diseases or injuries."[16]

One of the categories that appears near the bottom of both the political and clinical priority lists is what the commission calls "Borderline cases – medical care for reasons other than disease or injury." The examples it provides are assisted conception, cosmetic plastic surgery, and treatment of shortness of stature. While it does not rule out treatment for those conditions, it places them low on the hierarchy, primarily because there is debate about how crippling such conditions are. Although it does not use that language, the implication is that they fall closer to desires than to needs, while at the same time conceding that with some individuals there can be a serious need; hence, its borderline designation.

In some countries, and many American states, legislatures have mandated coverage for fertility treatment, as has been the case with Viagra. In such instances, it was political pressure, not a medical judgment, that brought this about.

Since the Swedish commission distinguished between objective and subjective need, its judgment on the examples it provides of "medical care for reasons other than disease or injury," makes sense, although with the other categories it noted how the distinction can generate a range of other difficulties, notably when some people with lesser objective needs seem to subjectively suffer much more than their peers.

On paper, the work of the various priority-setting commissions has been impressive. In practice, there is little evidence (save for Oregon's Medicaid program) that politicians and healthcare administrators have made much use of their recommendations. Tidy, rationalistic schemes developed by academic and policy analysts seem to have little political bite, perhaps because politicians everywhere are used to, and prefer, the rough and tumble of politics, and because, as has happened in some places, political mandates are used to override priority schemes, as with Viagra and fertility services. While the Swedish commission gave a high priority to preventive healthcare services, in practice it would be hard to find a country that gives them a higher place than the provision of acute care medicine. Treating those who are already sick, even with conditions of lesser importance, has always had a larger following than the goal of keeping us well in the first place (and probably always will).

1.19.5 The Medicine Market Debate

Every developed country has for at least two decades had national debates on the role of the market. Canada and the western European countries have been resistant to the market, but it has been gingerly tried and experimented with in those countries. While there are many ways of describing the market debate, I want to single out one feature of it that seems particularly important: that of the appropriate role of government as the regulator of healthcare expenditures, determining in the end just what counts as important and necessary care. In Beveridge tax-based systems (e.g., Italy, Sweden, the UK), the government directly manages the system. The regulation of Bismarckian systems (e.g., France, the Netherlands, Switzerland), which are based on private insurance companies and mandated employer – employee contributions, is of a kind that sets some basic parameters for the system, such as price controls on drugs and regional placement of expensive diagnostic machines. The presumption behind a strong government role is that it is best placed to determine national health needs and to allocate resources accordingly.

Both the Beveridge and Bismarckian systems allow for some marketing practices, e.g., that of the 'internal markets' of the UK and competition among insurance companies in the Bismarckian countries. For both systems, however, the net result is that costs are better controlled than in the US. Western European countries have seen cost increases in the range of 4–5% in recent years, while in the US these have ranged from 9 to 13%. Not only are the costs of healthcare higher in the US than in western Europe, but also the healthcare system is much more dominated by market-oriented private insurers, health maintenance organizations, for-profit hospitals and clinics, and a pharmaceutical industry that has successfully fought off price controls.

At the heart of the market argument is a tension between two sets of values, with those in Canada and the western European countries being dominated by the values of solidarity and equity, bolstered by a commitment to government as the best source of both. Every citizen is to have access to good healthcare, and if the means of achieving this goal is a limitation on freedom, then that is a price worth paying. Canada has gone even farther than the western European countries in not allowing parallel private insurance for services available under its national Medicare program (though a recent Supreme Court ruling may change that situation).

In the US, by contrast, a market orientation is much more pronounced, with its advocates claiming that the market offers greater efficiency in managing healthcare than government, and that the value of choice is more important than that of equity. The failure of the once-promising Health Maintenance Organization (HMO) movement to control costs, temporarily achieved for a few years in the mid-1990s, can be traced to a revolt on the part of both American doctors and patients against various cost-controlling measures.[17] Patients wanted to be able to go directly to specialists without passing through gate-keeping primary care physicians, and doctors objected to the necessity of gaining approval from HMOs for costly procedures. It was clear that American doctors and patients would on the whole prefer higher costs and greater freedom than the opposite.

In the nature of the case, market thinking tends to erase the line between desire and need. Its emphasis on choice rests on a concept of choice that privileges individual preferences, and does so without passing a moral or social judgment on the content of the choices. Its most extreme manifestation would be various proposals to simply provide people with vouchers to buy whatever healthcare they desire, or even to spend the money on goods other than healthcare if that is their choice. Medical savings accounts are another means of fostering choice. The point is that people should be free to make the choices that are consistent with their values (not those fostered or imposed by government) and, at the same time, by virtue of being forced to make choices, to become personally responsible for their own health and healthcare. Government-controlled or -regulated healthcare systems, or even those paid for by employers, according to the pro-market argument, take choice away from people while at the same time allowing them to consume healthcare in a way that makes them indifferent to its actual cost.

A market-oriented system will not in principle, and logically cannot, encompass the setting of healthcare priorities, or the control of hospital, technology, or drug prices. Most importantly, it cannot determine what is an objective need, instead placing the emphasis heavily on the subjective element in determining need. If someone wants something to relieve their experienced pain and suffering, and they are willing to pay for it, then they should be free to get it. The reality, even in a country such as the US, is the recognition that many people, particularly the poor, simply do not have the economic resources to exercise much choice of any kind. Even the most ardent market advocates accept the need for some kind of government safety net, but one that should be kept to a minimum. Hostility to government control and regulation and (for the most part) welfare programs is a key part of the market advocacy agenda.

The American experience with the market is not encouraging if the emphasis is placed on health outcomes and public satisfaction with their healthcare. The fact that there are some 45 million uninsured, that 15% of GDP is spent on healthcare, and that steadily increasing costs are much greater than those experienced by other developed countries does not tell a happy story. Although Europeans generally have less access to high-technology medicine than Americans, and face waiting lists in some countries, they are on the whole satisfied with their healthcare. It may well be, of course, that the current economic problems of Europe will force inroads on its healthcare systems. Europe's welfare programs of all kinds are particularly targeted for reform, and reform will likely move in a market direction, but one less motivated by consumer choice than by getting the cost burdens of those programs off the back of government.

1.19.6 How Much Is Enough or Too Little?

This paper has so far skirted the issue of how much is enough healthcare and what is too little. As I have suggested, it is possible to ask that question of individual healthcare and population healthcare, and the answers will not necessarily be the same. However, I will specify some categories that might make the task a little easier. There are at least three possible ways of categorizing the adequacy of healthcare and the meeting of need. One of them would be to call adequate that state of healthcare that can be commanded by the most wealthy, who are able to move about anywhere in the world to find the best physicians, technologies, and hospitals. The wealthy can also set their own standard of need, and are able to bypass any limitations set by national healthcare systems in their own country. By that standard, 'too little' for individuals would be an inability to afford what the most wealthy can have; 'too little' for a population would be when most people fall below that standard. 'Enough' would be the best of everything, even though, as even the richest and most powerful would have to concede, the rich cannot be saved if even the best medicine does not have the capacity to do so.

A second category, more modest in its aims, would be so far as possible to enable everyone to have their needs met in a system that aimed to eliminate disparities in care, aiming to bring everyone up to decent level of care. The priority here would be equality of health outcome for the individual, requiring a bias toward meeting the needs of the worst off to give them a chance to draw even with the best off. The population aim would be to cluster most of the population at the average life expectancy level, aiming in particular at reducing premature deaths. Needs only would count, not wants and desires. A third category, depending far more heavily on market practices, would aim to maximize individual choice, making use of cash vouchers to allow people to buy whatever kind of healthcare pleased them, including the spending of the voucher benefits heavily to satisfy desires and wants, not needs, if that is what they want to do; the government's role would be minimal, serving only to facilitate choice. The health of the population would be of no formal interest to government, and only the satisfaction of individual desires would count.

The difficulty with each of these categories, even if superficially plausible, takes us back to the beginning observation of this paper, which can now be made more precise. It is that medical need, fixed in some ways, is increasingly an infinitely malleable concept. If medical progress can turn wants and desires into needs, it can no less stretch the notion of need into constantly new shapes.

A prime example comes to mind. The DHHS study cited above spoke of helping people to avoid a 'premature death'; according to the US government, 'premature death' is that before the age of 65. But that figure is not as meaningful as it was when officially adopted some 70 years ago, and much less when in Germany, Otto von Bismarck used it to mark the beginning of old age over a century ago. People live longer now, with those over 80 being the fastest growing age group as a result of declining mortality and disability rates for the elderly. Therefore, the 'need' to help people avoid a premature death no longer relates to a fixed age, but is a moving target. Even so, the body declines and will eventually die, and of course not everyone cares to settle for the minimal goal of avoiding a premature death. The need to live is a strong one, and what was once just a desire to live longer than the average life expectancy, has turned into a need, if only because medical progress makes that a desire whose fulfillment is now plausible. How to meet this need – increasingly an insistent one – will be one of the major challenges in health policy over the next few decades.

And yet if it is accurate to say that a longer life than can be had at present has passed beyond desire into need with individual elderly people, it is far less clear that it is an important population need. By that I mean that constantly lengthening life expectancies among the old may confer no population benefit, and in fact may constitute a social danger.[18] A growing proportion of elderly people and a declining proportion of young people may wreak havoc with procreative habits and traditions, job mobility, and age-based tradition, as well as weakening the kind of vitality that comes from young populations. In other words, it is by no means clear that a growing proportion of elderly people will make for better, more satisfying, and productive societies. We may already have a high enough average life expectancy in developed countries to sustain vital societies. It may be the case that however malleable and progress driven the notion of medical need may be, not nearly so malleable is the potential for societal improvement by the improvement of health status. There are few who would argue that the main problem facing developed societies is that of illness and death from disease, which is not the same as saying that there is a problem in coping with healthcare costs. But then that is another curiosity: the provision of healthcare is a growing economic problem and is near the very top of the list of problems for governments, and yet it is unlikely that illness and disease per se would appear anywhere near the top. Good healthcare is clearly something the public wants, which is why governments take it seriously. But it is far less clear that poor health, save for significant disparities, is a major objective problem in affluent, developed countries with the longest life expectancies in the world.

This point can be made clearer by the fact that AIDS, even more than poverty, is the main social problem facing many African nations, principally by virtue of its decimation of the infrastructure and social organization of those nations. Cancer and heart disease are the main killers in developed societies, but those societies can still function, even flourish, with those deaths in its midst in a way no African nation can with AIDS in its midst.

Enough healthcare for the individual, I conclude, would be that to enable one to live out, without significant pain or disability, the average life expectancy of developed countries, which is now almost 80 years. This life span is long enough to do most of the things that make human life fulfilling: to receive an education, to find meaningful work, to raise a family, to travel, and to enjoy culture. Too little healthcare would consist of those cases where gender, race, geography, or other factors limited either the chance to live out an average life expectancy or did so at a high price in terms of preventable pain and suffering.

Enough healthcare for a whole population would be that sufficient to run the crucial institutions of society: education, commerce, job creation, transportation, and information systems. It is obvious that a standard such as this does not require that everyone be in good health – only most people. However, it does require that no subgroup in the population have health (and healthcare) that keeps them from benefiting from the cultural, economic, and social benefits of society, i.e., 'too little' healthcare. Obviously, the standard I am proposing is consistent with many people suffering from many diseases, and particularly the chronic diseases of old age. My standard of enough is a functional one: whatever it takes to run a well-ordered successful society.

I quickly stipulate an important proviso. To say how much a society needs for healthcare is not to say that it is wrong to spend more than that. It can desire to meet individual needs over and above what is socially necessary; and it can aim to respond to some wants and desires as well. At that point the economics of healthcare, and healthcare demand, come into play. To what extent are people willing to pay higher taxes, or higher co-payments, or to take out private insurance, or to expect their employers to make a larger healthcare contribution (and how willing to do so are their employers)? In short, whatever the financing plan used by a country, decisions will have to be made about finding a balance between supply and demand.

Health demand will almost always exceed the financing necessary to supply that demand. Every government in every developed country now grapples with that problem. Those inclined toward government solutions look to increased taxation, or money taken from other budgets, or increased co-payments and supplementary private insurance to meet public demand. Those inclined toward market solutions look to medical savings accounts, even more private insurance and co-payments than might be tolerated in a government-controlled system, and in general a variety of measures to force patients to decide just how much of their healthcare demand they are prepared to meet out of their own pockets. The ambiguity and constantly changing concept of medical need or necessity, the increasing medicalization of wants and desires, and the steadily rising cost of healthcare everywhere guarantees that a determination of needs, desires, wants, and demands will remain the chronic disease of healthcare systems.

References

1. Organisation for Economic Co-Operation and Development (OECD). *Health At a Glance: OEDC Indicators, 2003*, OECD: Paris, 2003, p 11.
2. Dubos, R. *Mirage of Health: Utopias, Progress, and Biological Change*; HarperColophon Books: New York, 1979, p 2.
3. Callahan, D. *What Price Better Health: Hazards of the Research Imperative*; University of California Press: Berkeley, CA, 2003; Chapter 3, pp 57–84.

4. Barsky, R. *Worried Sick: Our Troubled Quest for Health*; Little-Brown: Boston, MA, 1988.
5. Wilkinson, R., Marmot, M., Eds; *Social Determinants of Health: The Solid Facts*, 2nd ed.; World Health Organization: Geneva, Switzerland, 2003.
6. Gillick, M. R. *N. Engl. J. Med.* **2004**, *350*, 2199–2203.
7. Aaron, H. J.; Schwartz, W. J. *The Painful Prescription: Rationing Health Care*; The Brookings Institution: Washington, DC, 1984.
8. Topinkova, E.; Callahan, D. *J. Appl. Gerontol.* **1994**, *18*, 411–422.
9. Lieberman, T. Covering Medical Technology: the seven deadly sins. *Columbia Journalism Review* **2001**, *25*, 23–27.
10. Henney, T. *JAMA* **2001**, *284*, 2242.
11. Rice, T. *The Economics of Health Reconsidered*; 2nd ed. Health Administration Press: Chicago, IL, 2003; Chapters 3 and 4.
12. Angell, M. *The Truth About Drug Companies: How They Deceive Us and What to do About it*; Random House: New York, 2004.
13. US Department of Health and Human Services. *Healthy People 2010: Understanding and Improving Health*; 2nd ed.; US Government Printing Office: Washington, DC, 2000.
14. US Department of Health and Human Services. *Healthy People 2010: Understanding and Improving Health*; 2nd ed.; US Government Printing Office: Washington, DC, 2000, p 3.
15. Health Care and Medical Priorities Commission. *No Easy Choices: The Difficult Priorities of Health Care;* The Ministry of Health and Social Affairs: Stockholm, Sweden, 1993.
16. Health Care and Medical Priorities Commission. *No Easy Choices: The Difficult Priorities of Health Care;* The Ministry of Health and Social Affairs: Stockholm, Sweden, 1993, p 79.
17. Ginzburg, E. *JAMA* **1997**, *272*, 1813.
18. Overall, C. *Aging, Death, and Human Longevity*; University of California Press: Berkeley, CA, 2003.

Biography

Daniel Callahan is Director of the International Program, The Hastings Center, Garrison, NY, USA. He has a BA from Yale and a PhD in philosophy from Harvard. He is a Senior Fellow of the Institute for Social and Policy Studies at Yale University as well as an honorary faculty member of Charles University Medical School, Prague, the Czech Republic. He is an elected member of the National Academy of Sciences and a Fellow of the American Association for the Advancement of Science. He is the author or editor of 39 books, the most recent being *Medicine and The Market: Equity v. Choice* (Johns Hopkins University Press, 2006).

Comprehensive Medicinal Chemistry II
ISBN (set): 0-08-044513-6

ISBN (Volume 1) 0-08-044514-4; pp. 627–635

1.20 Health Demands in Developing Countries

A A Wasunna, The Hastings Center, Garrison, NY, USA

1.20.1 Introduction: 'Developing Countries'

The term 'developing countries' is used to describe countries that have low average annual incomes, whose populations are mostly engaged in agriculture, and who live at or near the subsistence level. These countries generally do not have sophisticated industries, and they are heavily dependent on foreign capital and development aid. Having said that, there are important differences even within this collective category of 'developing countries'; for example, can China, considered to be a developing country, be reasonably compared to Malawi or Chile in terms of economic development? The answer is no. Therefore, even with this amorphous category referred to as 'developing countries,' there are several important distinctions among the countries.

The World Bank and other international financial institutions classify developing countries using gross national income per capita (GNI) – formerly gross national product, or GNP – as the main criterion for determining whether a country is low-income, middle-income (further subdivided into lower-middle and upper-middle), or high-income. Even then, classification by income does not necessarily reflect the development status of countries. Other indicators applied to classify countries include human resources, as measured by a composite index (augmented physical quality of life index), based on a variety of indicators such as life expectancy at birth, per capita calorie intake, combined primary and secondary school enrollment, and adult literacy. Other indicators include levels of economic diversification, as measured by a composite index (economic diversification index) based on the share of manufacturing in gross domestic product (GDP), the share of the labor force in industry, and annual per capita commercial energy consumption (this list was established by the United Nations Conference on Trade and Development (UNCTAD).[1] Because of these variables and lack of international consensus on what uniform measurement criteria to apply to determine whether a country should be considered developed or developing, there is no official United Nations listing of developed and developing countries. Having said that, this chapter intends to focus on a subcategory of developing countries generally referred to as 'least developed countries,' commonly referred to as LDCs or 'low-income countries,' the majority of which are in sub-Saharan Africa. Disease patterns in these countries differ greatly from those of industrialized countries. While noncommunicable diseases, including cardiovascular disease, mental illness, cancers, and injuries tend to be responsible for the majority of the disease burden in developed countries, communicable diseases and maternal-, perinatal-, and nutrition-related conditions account for the majority of ill health in LDCs.

For a quarter of the world's population, absolute poverty remains the principal determinant of their health status, as well as exposure to human immunodeficiency virus/acquired immune deficiency syndrome (HIV/AIDS) and high fertility levels. Average life expectancy at birth in most LDCs is less than 50 years; the developed world average is over 70 years.[2] In some developing countries, particularly those worst affected by HIV/AIDS, such as Botswana, Niger, Malawi, and Zambia, life expectancy at birth has fallen to less than 40 years. Three diseases, HIV/AIDS, tuberculosis, and malaria, kill around 5.4 million people a year in developing countries – more than a third of their annual death toll.

Many African countries achieved substantial reductions in morbidity and mortality when they achieved independence from colonial rule, but many of these gains are being reversed rapidly. It is important, therefore, to trace the history of health systems in these countries to understand better the poor state they find themselves in today.

1.20.2 A Brief History of Health Systems in Least Developed Countries

Many developing countries share similar colonial histories. One of the causes of the Scramble for Africa (1885–1910), which resulted in the colonization of all of Africa in just 25 years, was competition among European nations for colonies and their rich natural and human resources. No major European nation wanted to be without colonies. By the beginning of World War I in 1914, all of Africa, with the exception of Liberia and Ethiopia, had been colonized, and initial pockets of African resistance had been overcome by the colonial powers. Competition for colonies was particularly strong among the UK, France, and Germany – the strongest European nation-states in the late nineteenth century; these countries had the most colonies in Africa. Most of the current borders of African countries were arbitrarily imposed on Africans by these European nations. Often the people who drew these borders did not take into account the cultural, political, and ethnolinguistic diversity among groups, or existing political organizations at the time of colonialization. Sometimes they grouped together people who had never been united under the same government or administrative entity before. In other cases, they divided existing systems of government at the time of colonial conquest. The colonizing powers exploited these newly formed countries for their labor, agricultural produce, and minerals. Colonial administrators were able to rule the countries and suppress indigenous populations by exploiting systems of hierarchies, creating divide-and-rule policies, and entrenching the view, particularly among African elites, that the colonialists knew best. Most colonial countries established highly centralized, quasi-military bureaucracies that were paternalistic and authoritarian.[3]

During the colonial period in African countries, social services like healthcare and education were developed along segregated lines. Western healthcare services, for instance, were provided by the colonial government, to cater to their wounded soldiers; missionaries, to cater to their African converts; industries, to cater to their employees; and private practitioners, to cater to colonial administrators and the settler population.[4] Endemic diseases were the main concern of French and British regimes. The French tried to deal with vector-borne disease though programs organized on a paramilitary basis, while the British placed greater stress on controlling fecal contamination of food and water. Hospitals were built primarily for the benefit of colonial administrators and settlers, and rural clinics were usually the by-product of missionary activities. Public health programs like health education, mass immunizations, screening of populations for disease, and good nutrition played little or no role in colonial social policies.[5]

For the vast majority of black Africans, traditional healers were their primary health providers. Traditional healers were held in high regard politically, and they were to be found in almost all communities. Unfortunately, because traditional healers generally wielded much power in communities, colonial governments systematically undermined their professional abilities, and often referred to them as 'quacks,' or the more derogatory term of 'witchdoctors.' In an attempt to reduce their political power, the colonialists called upon educated Africans to denounce these healers and urged them to abandon their services.[6]

In the 1960s, several African countries gained independence from colonial governments. In most of these nations, independence came only after organized armed conflicts and violent uprisings against the brutal colonial regimes. These struggles for independence had a significant by-product – a greater appreciation of the concept of nationalism (oneness) among the different ethnic communities. Now, the different African groups found themselves with one common enemy – foreign colonialists – and one common goal – self-rule. Within their borders, African groups forged a unity that transcended local loyalties of clan, ethnic divisions, and region.[7]

Armed with this new-found idea of nationalism, African governments launched social and economic programs aimed at improving the lives of all their citizens, regardless of ethnic differences. The provision of free or highly subsidized healthcare services to the entire population was one of the most popular and ambitious policies adopted by these governments. Many of the countries establishing these sorts of health systems were inspired by the historic Alma-Ata conference in 1978.[8] At this conference, developing countries were urged to model their health systems using community-based approaches that emphasized preventive programs over curative care. Consequently, most African countries began to dismantle the highly fragmented and unjust systems they had inherited from the colonialists and sought to make the development of public healthcare programs an integral part of their long-term national health plans. The majority of countries ended up establishing centralized, hierarchical healthcare systems with the government as the main provider and financier of healthcare services (both primary care and curative care).

Kenya and Zambia provide interesting illustrations of this postindependence process. Kenya became independent from British rule in 1963, and the new government committed itself to providing free health services as part of its development strategy to alleviate poverty and improve the welfare and productivity of the nation.[9] This pledge was honored in 1964 with the discontinuation of the preindependence user fees in government-run hospitals, and the introduction of free outpatient services and hospitalization for all children in public health facilities. Services in public facilities remained free for all except those workers whose expenses were met by employers. Private health enterprises developed parallel to these government initiatives, with missionary hospitals charging subsidized fees for services offered, and private for-profit facilities moving toward full cost recovery.[10]

The Kenyan government also constructed new health facilities in rural areas and upgraded existing government health institutions. The state also tried to ensure that essential medical supplies and equipment were available throughout the country by constructing depots in strategically located administrative posts. Training was seen as a key element to improving healthcare, and, to this end, rural health training centers were established to provide practical field training to health officers. The government also expanded and promoted training opportunities and career development for health personnel through public continuing education courses, and assured graduates of employment in the public sector.[10] These efforts by the government and nongovernmental organizations (NGOs) to increase the population's access to health services translated into marked gains in the country's health status. From independence to 1993, the crude death rate dropped from 20 to 9 per 1000, the infant mortality rate declined from 120 to 60 per 1000 live births, and life expectancy increased from 40 to 60 years.[10]

Zambia became independent in 1964, and, like many other African countries, including Kenya, its government expanded health services throughout the country. Infant mortality was reduced from 123 per 1000 live births in 1965 to 85 in 1984. By 1981, the ratio of doctors to population had dropped from 1:11 300 in 1965 to 1:7110. Most of the doctors (76%) worked in urban areas where only 54% of the population resided.[11] The Zambian government, in its third national development plan (1979–1983), committed itself to providing primary care as the main national healthcare strategy, with the control and prevention of malaria, respiratory and diarrhoeal diseases, and measles as its focus. But Zambia's economy was devastated by the global economic crisis in the 1990s, which brought down the price of copper, Zambia's main export. By the mid-1990s, Zambia faced a negative growth of its economy, with the government budget for health falling from $29 per capita in 1970 to $2.6 per capita in 1998.[12]

For many LDCs, there were significant improvements in the health status of their populations in the years after independence. Infant mortality was cut by more than one-third, average life expectancy increased by more than 10 years, more people had access to safe drinking water, and, by the end of the 1980s, almost half of all Africans were able to travel to a healthcare facility within an hour.[13] Public spending for health (and all other social sectors) grew rapidly in the 1960s and 1970s, but slow economic growth and record budget deficits forced governments to reduce their public spending on health services. Rising incomes, aging populations, and urbanization increased demand for conventional hospital services and physicians, making the provision of primary and prevention services an inadequate option for the growing middle class in African countries.[14]

By the 1980s, governments in many LDCs were the main financiers of healthcare, and they also provided healthcare services. Under this centralized system, the government set budgets for healthcare facilities, and also paid salaries for healthcare personnel working in these healthcare facilities. The state generally funded these services from general taxes. Healthcare systems in these countries were, however, plagued with institutional problems, including corruption, overemployment of healthcare personnel in public health institutions, a lack of adequate funding to run these facilities, and a lack of qualified healthcare personnel, poor institutional management, and growing consumer dissatisfaction with the quality of healthcare services in government-run healthcare institutions.

There were also chronic shortages of drugs, poor equipment maintenance, inadequate logistical support for services, and a lack of clear procedures for monitoring and evaluating programs and services. For the citizens of African countries, this translated into a low confidence in the health system; for governments, this meant a large share of public expenditures in health were being wasted; and for private health service providers, this meant that they were often trying to deal with unmet needs with little guidance, assistance, or competition from the public sector.[15]

The global economic crisis of 1990s hurt the economies of LDCs greatly, and most of them saw the economic and social progress they made in the years after independence erode. Something had to be done, and for most LDCs, this inevitably meant experimenting with market-oriented reforms.

1.20.3 The Age of the Market – Reform of Health Systems

At the peak of the global economic crisis many LDCs, particularly those in Africa and Asia, sought financial assistance from the World Bank and International Monetary Fund (IMF) to help stabilize their economies. These institutions agreed to lend monies to these countries on the condition that they would engage in major reform of their economies. Although LDCs agreed to the terms and conditions of their loan arrangements with these financial institutions, many felt that they were coerced, and that they had no leverage to dispute most of the terms set by the lending agencies. If they did not agree to the conditions, they were simply denied access to credit and loans. Faced with overwhelming debt, shrinking national incomes, and no other funding avenues, most countries accepted the reform conditions set by the Bank and IMF. Between 1980 and 1990 70% of African countries (or 32 of 45 countries) began to engage in economic reform measures as a lending condition of the IMF and World Bank.[15]

For the health sector, reform meant limiting the role of the government in providing healthcare services, encouraging private sector growth, and placing emphasis on individual responsibility.[16] Initially, the World Bank proposed four major policy reforms that were seen as closely related and mutually reinforcing: (1) introducing user fees; (2) expanding health insurance; (3) decentralizing government health services; and (4) using nongovernmental resources effectively.

1.20.3.1 Introducing User Fees

This measure required government health institutes to charge users, particularly for drugs and curative care. This measure was supposed to increase resources available to the government health sector, allow more spending on underfunded programs, encourage better quality and more efficiency, and increase access for the poor.[17] The World Bank justified the implementation of user fees in three ways. First, it felt that payment for services would discourage frivolous use of health facilities. Second, the Bank presumed that patients would become more conscious of quality and demand it if they were paying for services; and third, it assumed that the greater availability of funds through user fees at the point of service would increase both the availability and quality of services.[18]

1.20.3.2 Expanding Health Insurance

The Bank encouraged LDCs to support well-designed health insurance programs to help mobilize resources for the health sector while simultaneously protecting households from large financial losses. The long-term goal was to relieve the government budget of the high costs of curative care.[19]

1.20.3.3 Decentralizing Government Health Services

This entailed decentralizing administrative tasks such as planning, budgeting, and purchasing for government health services. This measure was aimed at using market incentives where possible to improve motivation among staff, and to make allocation of resources more efficient. It was also supposed to allow revenues to be collected and retained as close as possible to the point of service delivery to improve fee collection and service effectiveness.[19]

1.20.3.4 Using Nongovernmental Resources Effectively

The Bank also asked countries seeking funding to encourage the nongovernmental sector to provide health services for which consumers were willing to pay. This move was aimed at allowing the government to focus its resources on programs that benefited whole communities (populations), rather than individuals. The Bank specifically wanted governments to create an environment to encourage the setting up of community-run and privately managed cooperative health plans.

LDCs implemented these measures with mixed success. For example, there is now evidence that user fees seems to have had a socially regressive impact in most African countries, including Zimbabwe, Zambia, Kenya, and Tanzania, partly because these countries did not have adequate mechanisms to protect vulnerable groups, particularly women, from its negative effects.[18] In Kenya, for example, after the imposition of user fees, attendance in public clinics dropped by about 50%. This decline prompted the government to suspend the fees for 7 months, during which attendance at government health centers increased by about 41%.[21]

The development of new health insurance programs has not taken root as hoped. For one thing, the main beneficiaries of such insurance in LDCs are the relatively small middle classes, and coverage extends to only about 10% of the population. Insurance coverage ranges from virtually nil in Uganda and Nigeria to between 1 and 6% in Zimbabwe (1997), 16% in South Africa (1997), and 11.4% in Kenya.[22] Another problem is that, in a largely unregulated environment, insurers in many LDCs are reluctant to cover high-risk individuals, and there have also been cases of excessive medicalization as healthcare providers seek to maximize their profits.

To make matters worse, some of the reform measures affected women and girls in LDCs negatively. They were, for instance, forced to take on greater caregiving roles in the family to compensate for the lack of access to publicly provided service.[23] Drastic government cuts in healthcare spending also affected the provision of reproductive and maternal health services. Under user fees arrangements, a Ugandan study found that women were likely either to forgo medical services, or to resort to home care treatment.[23]

A number of critics of healthcare reform programs argue that they have had a less than stellar effect on public health services and equitable access to these services. A 1993 study, for instance, found that structural adjustment had a negative impact on important health indicators such as infant and child mortality in African countries.[24] The study also found that the nutritional status of children in countries that engaged in market-oriented reform declined, and that structural adjustment policies induced a profound change in health policy, resulting in a widening gap between affected communities and policy-makers. There has been, however, some debate about whether the negative effects on health outcomes should be entirely attributed to structural adjustment, rather than to other more general factors such as the effects of the global economic crisis, high levels of corruption, and economic mismanagement.

Some policy-makers have argued that the Bank-led reforms did not take the local realities in the countries into account. The prescriptions were standard for all countries regardless of the historical, cultural, political, and social contexts, with too strong an emphasis on the content of the reforms, and too little attention given to the actors and processes involved.[25]

After the LDCs signed their loan agreements with the Bank and started to reform their health systems, it soon became clear that the potential effectiveness of the measures would be hampered by the lack of institutional capacity. Generally, for market ideas to function, health systems must possess a certain level of sophistication in policy analysis, research, information systems, management expertise, and logistics systems. These capacities were simply absent in many developing countries, particularly those in Africa.[26]

The World Bank and IMF have cited countries such as Mauritius and Thailand as successes. These countries, the Bank believes, had the right conditions for economic reform, including political stability – a feature that was absent in most other LDCs.[27] Having said that, there is generally, however, little evidence or data to support the success of reforms in achieving their broad objectives in LDCs. Though a number of proposals to evaluate health reform have been pitched, there remain difficulties in establishing an appropriate yardstick against which to measure change in the various countries.[28]

Of note, however, is the fact that not all developing countries (both middle- and low-income) experimented with market-oriented tools in the 1980s solely at the insistence of the World Bank and IMF. In Chile, for example, economic liberalization had already begun under the Pinochet regime, and the process was simply reinforced by the Bank and IMF. Likewise, consumer dissatisfaction drove the Vietnam government to embark on reforms, while in India; the use of market tools in health was not new – it has had a huge competitive and unregulated private sector providing important health services for several years.

Today, most LDCs have a heterogeneous mix of public, private, and nongovernmental sources of funding for their healthcare systems, and as they continue to experiment with a variety of market-oriented policies, they are encountering new challenges and demands.

1.20.4 Current Health Challenges and Demands in Least Developed Countries

1.20.4.1 Functional Public Health Systems

The greatest differences in health problems between LDCs and industrialized countries are not just in the kinds of disease that are prominent in one or the other, but in the quantitative issues; in LDCs, the demand for health services, together with serious shortages in health personnel, facilities, and materials.[29] An effective healthcare system must be designed to cater to immediate and urgent health needs; it must be able to reach into communities and homes, and it

Table 1 Mortality and morbidity statistics

Causes of death in least developed countries	*Number of deaths in 2001*
1. HIV/AIDS	2 678 000
2. Lower respiratory infections	2 643 000
3. Ischemic heart disease	2 484 000
4. Diarrhoeal diseases	1 793 000
5. Cerebrovascular disease	1 381 000
6. Childhood diseases	1 217 000
7. Malaria	1 103 000
8. Tuberculosis	1 021 000
9. Chronic obstructive pulmonary disease	748 000
10. Measles	674 000

Reproduced with permission from World Health Organization. *World Health Report: Reducing Risks, Promoting Healthy Life*; World Health Organization: Geneva, 2002.

has to play a role in influencing certain patterns of life. In an LDC, it is imperative that the healthcare system attend to public health issues such as public sanitation, protection of water, nutritional status of the population, as well as reproductive health services. Unfortunately, while the extent of poor health in LDCs is immense, the lack of effective health systems makes this problem worse.

Public health systems in LDCs have all but collapsed, and where they exist, they are in a terrible state, barely able to cope with demand (**Table 1**). Among 30 states in Nigeria, for example, the number of health facilities ranges from 1 per 200 people in Lagos state, to 1 per 129 000 in Benue state. Three-quarters of the country's public and private health facilities are located in urban areas, which contain just 30% of the population. In Kenya, there is one doctor on average per 500 people in the capital city of Nairobi, compared to just one doctor per 160 000 people in the rural Turkana district. In Angola, the supply of hospital beds ranges from 4 per 10 000 people in the province of Malage, to 42 per 10 000 in Luande Norte.[30]

These rural–urban imbalances in health service provision are also reflected by the pattern of government spending on health in LDCs. Hospitals in urban areas often receive half or more of the public funds spent on health, and commonly account for 50–80% of recurrent health sector expenditure for the government.[30] These major health establishments also employ the largest proportions of highly trained health personnel. In Kenya, 60% of all physicians and 80% of all nurses in the country are assigned to such establishments.[31] In terms of access to health services, rural Africa is therefore disproportionately affected, with more than 45% of the population living without any such access. Moreover, a large share of public funds for health – as much as 80% – goes to urban-based curative care rather than to rural primary care clinics.[32]

In 1985, parasitic and infectious diseases killed an estimated 4.25 million people in LDCs. Each year, there are about 100 million clinical cases of malaria, 1 million of which end in death. Tuberculosis kills almost 1 million people each year, and 1.7 million new cases are reported each year. In Africa, perinatal, infectious, and parasitic illnesses are responsible for 75% of infant deaths. Infectious and parasitic infections are also responsible for 71% of the deaths of children from age 1 to 4, and 62% of the deaths of children ages 4–14. Other causes of mortality and morbidity include diarrhea, measles, and respiratory infections, especially pneumonia.[33]

According to two leading British development agencies, Water Aid and Tearfund, deaths from preventable diseases are a silent tragedy made worse by a scandalous lack of political will among governments of LDCs to deal with the issue. Having said that, it should be noted that, although treatment and preventive strategies are available for many infectious diseases, they are, for the most part, too expensive for LDCs, and many poor countries argue that the cost of providing these strategies would consume all or most of their available health resources.[34] More donor funding needs to be channeled toward building health infrastructure for both primary and curative care in developing countries. Given the fact that the majority of death-causing diseases in developing countries are preventable, and that health and nutrition are essential to development, it is imperative that a better balance be achieved between preventing disease and merely treating its consequences.[35]

1.20.4.2 Access to Essential Drugs

If they are safe, available, and accessible, medicines or drugs offer a simple, cost-effective answer to many health problems in LDCs.[36] In 1975, the World Health Organization introduced the concept of 'essential drugs' as a pragmatic approach to accelerate the positive impacts of drugs on health status, particularly for developing countries. An essential drug is defined as one that is clinically proven to be safe and effective; available in a stable, easily managed form; made with only one active ingredient unless there is a good reason otherwise; designed to meet clearly defined healthcare needs; less expensive than comparable drugs; and appropriate for a wider range of local conditions.[37]

Under the World Health Organization's essential drugs approach, national governments are supposed to identify a list of drugs that satisfies the most urgent health needs of the majority of the population, and thereafter take steps to make the drugs available in the appropriate dosage forms in all government health facilities in the country.[38] The way in which certain drugs (and not others) make it to the list of essential drugs in countries is not just a scientific decision – it is a highly political process. Because of the potential for profit, various pharmaceutical companies lobby to ensure their drugs make the list. Some of these companies have been known to try to influence entry of their products into essential lists in a variety of ways, including "talking to right people in government," as well as producing branded generics, and justifying higher prices with the claim that brand-name generics guarantee quality.[39] Germany, Japan, Switzerland, and the US – home to many of the word's largest drug manufacturers – do not generally support the World Health Organization's essential drugs program.

After personnel costs, pharmaceuticals are generally the largest item of expenditure within public sector health budgets in most LDCs. Drug costs in developing countries represent between 10 and 40% of public health budgets, and between 20 and 50% of total healthcare expenditure, compared to an average of 12% in developed countries.[40] In most LDCs, drugs are mainly paid for out of pocket. Studies have shown that the poor, who do not have insurance to pay for the drugs, spend a disproportionate share of household income to purchase drugs – sometimes more than twice as much as the richest 10% of the population spends.[41]

Currently, most LDCs, particularly those in Africa and Latin America, have systems where drugs are supplied by government medical stores or state-owned wholesalers, and thereafter, dispensed by government health facilities, but paid for (in whole or in part) by user fees. Parallel to this system, all LDCs have private for-profit pharmaceutical sectors with patients paying the entire costs of drugs by purchasing from privately owned retail pharmacies.

The acquisition and use of drugs in LDCs suffer from several problems, including poor quality: surveys from several developing countries show that between 10 and 20% of sampled drugs fail quality control tests.[41] Fewer than one in three developing countries are estimated to have fully functioning drug regulatory authorities, and this has contributed to the growing problem of fake drugs entering LDCs. Other problems faced by developing countries include poor storage and distribution facilities.

Perhaps the most troubling challenge facing poor countries today is the increasing globalization of the international pharmaceutical markets. Today, multinational companies make decisions about pricing of drugs based on factors such as market size, protection of intellectual property and prices for competing products.[42] When prices for drugs are set at levels aimed at developed country markets, developing countries have no choice but either to forgo purchasing the drugs altogether, or buy them at cost and in so doing place great pressure on their already overstretched healthcare budgets.

1.20.4.2.1 Trade-related aspects of intellectual property rights and its impact on least developed countries

The patent system is profit-driven and has been long defended as a necessary protection of research investments. In 1995, the World Trade Organization (WTO) introduced intellectual property rules into the multitrading system for the first time by passing an agreement on trade-related aspects of intellectual property rights (TRIPS). This agreement covers copyright and related rights, trademarks, patents, including the protection of plants; the layout, designs of integrated circuits; and undisclosed information, including trade secrets and test data.

The TRIPS agreement is described as an attempt to strike a balance between the long-term objective of providing incentives for future inventions and creations, and the short-term objective of allowing people to use existing inventions and creations. Under TRIPS, WTO member countries have to provide patent protection for any invention, whether a product (such as medicine) or a process (such as a method of producing the chemical ingredients for a medicine), while allowing for certain exceptions.[43] Patent protection lasts for least 20 years from the date the patent application was filed.[44] In addition, members of the WTO cannot discriminate between different fields of technology in their patent regimes, nor can they discriminate between the place of invention, and whether the products are imported or locally produced.[45]

Under TRIPS, governments can refuse to grant patents for three reasons: (1) inventions whose commercial exploitation needs to be prevented to protect human, animal, or plant life or health; (2) diagnostic, therapeutic, and surgical methods for treating humans or animals; (3) certain plants and animal inventions.[44] However, these exceptions apply provided certain conditions are met; for example, the exceptions must not 'unreasonably' conflict with the 'normal' exploitation of the patent.[46] The TRIPS agreement also provides what is known as a 'Bolar' provision that allows researchers to use patented inventions for research, in order to understand the invention more fully. Some countries allow manufacturers of generic drugs to use the patented invention to obtain marketing approval – for example, from public health authorities – without the patent owner's permission and before the term of the patent expires. The generic producers can then market their versions as soon as the patent expires.[47]

TRIPS also allows governments to act, subject to certain conditions, to prevent patent owners and other holders of intellectual property rights from abusing intellectual property rights, 'unreasonably' restraining trade, or hampering the international transfer of technology.[47] The TRIPS agreement has been very controversial, and has spurred many debates about the potentially negative effects it will have on LDCs trying to implement domestic public health policies. LDCs have argued that conforming to the agreement by providing or strengthening the protection of pharmaceutical products with intellectual property rights poses a special challenge for them, worsening opportunities for access to medicines, particularly for the poor.[48] Specifically, LDCs argued that, under the agreement, they would be restricted from producing or importing generic drugs as a means of protecting public health in their own countries. After intense and persistent protests from LDCs and health development organizations, and upon recognition of the gravity of the health problems facing LDCs, not the least of which is HIV/AIDS, members made some concessions to cater to public health demands at a WTO ministerial meeting in Doha. The Doha Declaration, as it is now referred to, states, in part, that "intellectual property protection is important for the development of new medicines; however, the TRIPS agreement does not, and should not, prevent members from taking measures to protect public health."[49] Accordingly, the agreement can and should be interpreted and implemented in a manner supportive of WTO members' right to protect public health and, in particular to promote access to drugs for all.[49] This declaration was seen as a victory for poor countries, as they could now use this provision to circumvent patent protection and provide much-needed drugs for the citizens during public health emergencies.

Specifically, the Doha Declaration provided the following flexibilities under the TRIPS agreement:

1. The right to get compulsory licenses – compulsory licensing is when a government allows someone else to produce the patented product or process without the consent of the patent owner. Compulsory licensing must, however, meet certain requirements; for example, it cannot be granted exclusively to licensees – the patent-holder can continue to produce; it must be granted mainly to supply the domestic market; and compulsory licensing cannot be arbitrary.[50]
2. The right to determine what constitutes a national emergency or other circumstances of extreme urgency: it being understood that public health crises, including those relating to the HIV/AIDS, tuberculosis, malaria, and other epidemics, can represent a national emergency or circumstances of extreme urgency and trigger the granting of compulsory licensing.[51]
3. The right of members to determine their parallel import regimes – subject to the most favored nation (MFN), and national treatment provisions in articles 3 and 4.[52]
4. The right of LDC members to extend exemptions on pharmaceutical patent protection for LDCs until 2016.[53]

After the Doha Declaration, LDCs still had one issue to resolve – how to provide extra flexibility, so that countries unable to produce pharmaceuticals domestically could obtain generics or copies of patented drugs from other countries. Most LDCs have insufficient or no manufacturing capacities in the pharmaceutical sector – consequently, they could face difficulties in making effective use of the compulsory licensing clause under the TRIPS agreement. This issue, referred to as 'paragraph 6,' was resolved on August 30, 2003, when WTO members agreed on legal changes to make it easier for countries to import cheaper generics made under compulsory licensing if they are unable to manufacture the medicine themselves. The decision waives exporting countries' obligations under article 31 (f) – any member country can export generic pharmaceutical products made under compulsory licenses to meet the needs of importing countries, provided certain conditions are met.[54]

1.20.4.2.1.1 Case study

The Pharmaceutical Manufacturers' Association of South Africa *et al.* versus The President of the Republic of South Africa, the Honorable Mr. N. R. Mandela *et al.*[55]

At a time when over 4.5 million people were infected with HIV/AIDS, and thousands were dying each year from the disease in South Africa, a country with the highest rate of HIV prevalence in the world, the Pharmaceutical Manufacturers Association (PMA), representing 39 drug companies in South Africa, launched legal action against an amendment to an Act, passed by the South African parliament in 1997, that was intended to make essential medicines, particularly drugs for HIV/AIDS, more affordable.[56] In its pleadings, the PMA claimed that the amendment violated a range of its members' rights, citing specifically rights to property contained in the South African Constitution.[56]

Internationally, supporters of the pharmaceutical industry argued that the actions of the South African government threatened the international patent regime, encapsulated in the TRIPS agreement, and that the government's action made it a pariah state acting contrary to its obligations as a member of the WTO. (Under S 25 (1), "No one may be deprived of the right to property except in terms of law of general application, and no law may permit arbitrary deprivation of property.") Initially, this lobbying had some success in some countries, particularly in the US, which placed South Africa on a US Trade Representative 'watch list'.[57] A year later South Africa was removed from this list as a result of pressure from angry human rights activists.

This court case provoked both local and international demonstrations, petitions, and letters calling on the pharmaceutical companies to withdraw. The publicity surrounding the court case also intensified the price war on antiretroviral drugs, both between the pharmaceutical giants, and between them and the generic drug companies. Large companies such as Merck cut drug prices as they sought to recoup some public support, to blunt the offers from generic-drug companies, and to stave off growing public disquiet about patents on medicines.[57]

In April 2001, 3 years after the case was filed, the drug companies withdrew their case against the government in the Pretoria High Court. The case had become a disastrous public relations embarrassment for the pharmaceutical industry, and they decided to cut their losses, and bring the matter to a close by withdrawing their claim. This action by the companies cleared the way for the South African government to import cheaper anti-AIDS drugs, even if international companies hold the rights to manufacture and market the medicines in South Africa. As expressed at the time by the South African Minister of Health:

> We regard today's settlement as a victory in the sense that it restores to us the power to pursue policies that we believe are critical to securing medicines at affordable rates and exercising wise control over them. We have undertaken to include pharmaceutical manufacturers in such initiatives, where appropriate, and we fully intend to pursue this course of action.[58]

One of the issues that came to light a result of the South African case was that many developing countries lacked the technical skills to implement the provisions in the TRIPS agreement. Perhaps more importantly, as was the case in South Africa, they did not have the economic and political power to exercise the options available to them under the TRIPS provisions. Attempts to exercise these options have been met with pressure from pharmaceutical companies, and the threat of economic sanctions from powerful WTO members. In the end, LDCs have to adopt national legislation putting into effect compulsory licensing, parallel importing clauses, and other options in a manner most conducive for the individual health, economic, and development needs of their people.

1.20.4.2.2 Drug pricing

The South African court case also brought international practices on drug pricing for LDCs to the fore. Even though drug companies argued that they had huge stakes in maintaining their high prices for HIV/AIDS drugs, they began lowering their prices significantly in the face of fierce public criticism from people living with HIV/AIDS, civil society organizations, and international human rights organizations. Since then, prices for AIDS drugs have dropped significantly in African countries – from approximately \$10 000 to \$200 per patient, per year. Offers by generic drug producers such as Cipla and Ranbaxy in India to provide African countries with AIDS drugs at much lower prices have served as catalysts forcing drug companies to bring their prices down. In addition, some major pharmaceutical companies have entered into deals with the World Health Organization to offer substantial discounts for HIV/AIDS drugs in LDCs. In the long term, the key issue under TRIPS will be whether compulsory licensing will afford developing countries' indigenous pharmaceutical sectors enough flexibility to dilute the monopoly powers of foreign pharmaceutical and biotech companies. The fate of patents in the poorest countries is also unclear because compulsory licenses permit nations to import generic drugs if they cannot make them on their own, but concerns remain that the terms of this waiver are so complex that generic firms in exporting countries will be reluctant to enter foreign markets.

In the wake of the South African case, there has also been a shift in international policy regarding provision of treatment of people with HIV/AIDS: there is growing consensus that providing treatment to people living with HIV/AIDS is akin to a human right, and given the numbers of otherwise financially productive people in LDCs succumbing to the disease, it makes good economic sense to keep them alive (and productive). In the last 3 years, therefore, new programs to provide resources treatment have been launched. The Global Fund to Fight AIDS, the

US Presidential Emergency Plan for AIDS Relief (PEPFAR) to provide HIV/AIDS treatment to an initial 14 countries, and the World Health Organization's Three by Five Program, which aims to treat 3 million people in 5 years, are three such examples.

Global funding for HIV/AIDS in resource-constrained countries has, as a result, increased from just over $300 million in 1999 to an unprecedented $3 billion in 2002 and $4.7 billion in 2003, with additional funding promised by foreign governments and international donor agencies. Other institutions such as the William J Clinton Presidential Foundation, the World Bank, and United Nations International Children's Emergency Fund (UNICEF) are also assisting LDCs to gain access to inexpensive generic AIDS drugs by brokering deals with pharmaceutical companies and other manufacturers of HIV/AIDS diagnostic tools. These developments in funding for antiretroviral (ARV) treatments have also been accompanied by other positive steps that have made the administration of treatment in poor countries easier. The development of simpler treatment regimens, and consensus building around the setting-up of treatment protocols for resource-constrained settings, are some examples of such progress.[59]

Despite this development in making drugs available in poor countries, as we have alluded to in an earlier section, public health infrastructure in developing countries has all but collapsed. If these health systems do not operate efficiently and effectively, the proper delivery and administration of drugs will be difficult or impossible. Problems with health delivery systems arise mainly from a lack of resources, including the inability to manage complicated laboratory monitoring, a lack of trained personnel, and inadequate community education strategies to encourage compliance.[60] Most countries also have the problem of corruption at various levels as well as the absence of sufficient political will to improve health systems. To ensure fair access to essential drugs, health infrastructure in LDCs has to be scaled up.[61]

It is also important to note that HIV/AIDS represents only one of the major infectious diseases in the developing world (**Table 1**). Tuberculosis infects 8 million people a year, 77% of whom do not have any access to drugs. As a result, 2 million people die every year from this disease, which is much cheaper to treat than HIV/AIDS (although the threat of resistance is growing). Malaria is also a leading killer in the developing world. More than 300 million people are infected each year and between 1 and 2 million die from the disease. A more comprehensive approach is needed to prevent disease from occurring in the first place, but also to provide access to drugs needed to prevent death.

The patent debate, particularly on the question of how patent protection affects drug prices and access to drugs in LDCs, is by no means settled. A study published by proindustry analysts, the international firm of economists, National Economic Research Associates, found that improving intellectual property protection for pharmaceuticals did not raise drug prices in developing countries.[62] In 2001, the *Journal of the American Medical Association* published a study that caused much controversy, particularly since it reached conclusions that were viewed as favorable to the drug industry. The study found that only 22% of the 795 potential patents in Africa had been awarded, in most cases because the companies who owned them did not bother to apply for patent protection. The researchers came to the conclusion that patents generally did not appear to be a substantial barrier to antiretroviral treatment access in Africa. Instead, the researchers argued, the poverty of African countries, the high cost of antiretroviral therapy, national regulatory requirements for medicines, tariffs, and sales taxes and, above all, a lack of sufficient international financial aid to fund antiretroviral treatment posed greater barriers to access than patents per se.[63] This study was widely criticized on two facts: that the drugs most aggressively patented were drug cocktails made by the makers of generic drugs in India; and it was in the richer African countries, or those in which many of the Indian companies had offices, that most of the patents had been registered.[64]

1.20.4.3 Dearth of Relevant Research – The 90%:10% Gap

Although patent protection has increased over the last 20 years, the average innovation rate has fallen. In the last 25 years, almost 1400 new medicines were developed but only 1% were for tropical diseases which kill millions each year.[65] Less than 10% of the estimated US$56 billion spent annually on health research by the public and private sectors is devoted to diseases or conditions that account for 90% of the global burden of disease. This 90/10 divide shows the gross neglect of research for diseases afflicting the world's majority poor.[66]

According to the Global Forum for Health Research, the research priorities outlined for developing countries are child health and nutrition (including diarrhea, pneumonia, HIV, tuberculosis, malaria, other vaccine-preventable diseases, and malnutrition); maternal and reproductive health (including mortality, nutrition, sexually transmitted diseases, HIV, and family planning); noncommunicable diseases (including cardiovascular diseases, mental illness, and disorders of the nervous system); injuries; and health systems and health policy research.[67] Despite this need, there is virtually no research and development for new drugs for diseases like tuberculosis and malaria, even as resistance to drugs for these conditions increases. The main reason for this gap is that LDCs do not represent a profitable market for the international pharmaceutical industry.

Table 2 Building and sustaining health research capacity in LDCs

1. Building and sustaining research capacity within developing countries is an essential and effective means of accelerating research contributions to health and development. Nurturing individual scientific competence and leadership, strengthening institutions, establishing strong linkages between research and action agencies, and reinforcing national institutions through international networks are all important elements of capacity building
2. Capacity building for country-specific health research should be given top priority by every country because of its importance to policy and management decisions for the health sector. It is equally important to create demand for research results among those responsible for health policy and management through effective arrangements for communication and shared priority setting for research
3. National commitment is indispensable to secure the resources, and to create a positive environment for research capacity building
4. Bilateral and multilateral agencies and development banks should reduce their dependence on expatriate consultants and increase their investment in research capacity in developing countries. Special attention should be given to sub-Saharan African countries
5. Capacity building requires sustained support over an extended period. External agencies can assist more effectively by committing at the outset support for 10–15 years, subject only to demonstrating achievement in relation to agreed-upon milestones and normal agency legal and reporting requirements

Figure: Building and Sustaining Health Research Capacity in LDCs, from Health Research by Commission on Health Research for Development, copyright © 1990 by The Commission on Health Research for Development. Used by permission of Oxford University Press, Inc.

What can LDCs do to try to change this unacceptable state of affairs? Locally, they have to define clearly their health priorities and seek resources to fund research that addresses their needs. Priorities have to be based on sound scientific information about problems and possible solutions, and research should provide the necessary information and influence these choices. Ideally, national governments should increase their health research budgets to at least 2% of their national health expenditure. Internationally, research sponsors and researchers must adopt equity as the core value in setting priorities for research in developing countries if the research divide is to be bridged.[67] Ideally, international development agencies should invest 5% of their health budget in health research and capacity building. Clearly, incentives have to be developed to direct some of the existing technological capacity of developed countries into addressing the health problems of the poor, while also engaging developing countries in the process. Strengthening health research capacity is vital to achieving equitable access to healthcare services.[68]

Recent international debates on the ethical appropriateness of certain clinical research trials in developing countries signal the increasing role that poor countries are playing in international health research.[69] There is pressure on foreign research sponsors to conduct clinical trials expeditiously and inexpensively.[70] Therefore, researcher sponsors seem to be attracted to LDCs where a large population of people with high disease burden can serve as research volunteers, and where access to patients is generally easier due to fewer competing clinical trials and lax regulation, and where lower personnel cost may make the expense of conducting research much less than it would in a developed country.

The conduct of international biomedical research therefore raises many new ethical challenges. There is a need for accepted ethical principles that can be adapted to the unique cultural norms of each society, ethical principles related to the immeasurable health inequities between rich and poor countries, and procedural guidelines to ensure compliance with acceptable ethical standards. In the absence of such guidance, there is a danger that research participants in LDCs may be at a heightened risk of exploitation.[71] Potential research volunteers in resource-poor countries, who are poor, illiterate, and unfamiliar with their rights as research volunteers, may thus be vulnerable to exploitation by international medical researchers and research sponsors.

Table 2 lays out the main factors necessary to reverse the 90:10% research gap in developing countries.

1.20.4.4 Brain Drain: A Crisis in Health Personnel

LDCs are faced with the increasing migration of doctors, teachers, engineers, lawyers, and other professionals to developed countries. These professionals migrate from developing countries to wealthier countries in order to further their careers, or improve their economic or social situations. This south–north migration has created a brain drain in LDCs, and it is becoming a major obstacle in the socioeconomic development of these countries. According to the International Organization for Migration (IOM), an estimated 3 million health workers from the developing world, including poor African countries struggling under the burden of HIV/AIDS and malaria, now work in the west. Each would have cost the rich countries an estimated $184 000 to train, adding up to a staggering $552 billion – almost the same as poor countries' total debt to donor countries.[72] According to the IOM, a total of 23 000 African health workers emigrate to developed countries each year, "leaving their own stretched health service in dire straits."[73] One study has gone as far as to state that "the hemorrhage of health professionals from African countries is easily the single most serious human resource problem facing health ministries today."[73]

A report prepared by Joint Learning Initiative found that sub-Saharan Africa has a tenth the nurses and doctors for its population compared to Europe. Ethiopia has a fiftieth of the professionals for its population that Italy does.[74] The report further suggests that, in nearly all LDCs, there is a scarcity of public health specialists. Adding to this is the nagging problem of misdistribution of personnel – most of the health professionals are based in urban areas, leaving rural part of the countries without little professional health support.

This lack of health professionals has serious repercussions on LDCs with high rates of HIV/AIDS. Medical care of people living with HIV/AIDS requires the treatment of serious conditions such as tuberculosis (the leading cause of death in people with HIV), malaria, hepatitis, and sexually transmitted diseases. Yet most countries in Africa do not meet the World Health Organization's minimum standards for the number of physicians or nurses per 100 000 people, even if you take HIV/AIDS out of the equation.[75] In Lusaka, Zambia, for instance, the HIV prevalence among midwives is 39%, and among nurses, 44%. It has been projected that Botswana, a middle-income country in Africa, will require an additional 330 nurses to its current 4400 to support national ARV programs. In Tanzania, there are fewer than 100 physician specialists in the public sector and many regional district hospitals have no physicians or medical officers.[76] In Uganda, there is currently one nurse or midwife per 11 365 people, while Liberia and Haiti have 1 per 10 000 people. The whole continent of Africa graduates a mere 5000 doctors a year. This is a most vexing problem for the healthcare sector in developing countries in general, and Africa in particular, with no easy or immediate solutions.

The ethical issues arising from the migration of health professionals from poor countries to rich ones are complicated by the competition of legitimate interests, namely, each country's need for an adequate trained health workforce and an individual's right to travel and seek a better life. When health professionals go to industrialized countries for training, and then return to apply their skills, there are advantages to the home country. Additionally, emigrants of all social classes from poor countries typically send money home to relatives, although sub-Saharan African remittances, at less than US$5 billion, comprise the lowest dollar amounts remitted back when compared to those by other poor world regions.[77]

Country leadership and strategies should therefore target policies that provide the right incentives to retain health personnel in LDCs. There are several proposals being developed to deal with this brain-drain problem in LDCs. The Joint Learning Initiative, for instance, has recommended the creation of an Action Alliance – a temporary international body to bring together healthcare and human resource experts, to advocate for the importance of sustained attention to workforce issues, and to hold responsible actors accountable. Another proposal seeks compensation to LDCs for their loss of trained personnel. Calculation of loss would include the cost of training replacement personnel in the source country.[78] It has also been suggested that rich countries (or recipient countries) should directly invest in enhancing training and skills development in the exporting LDCs for health workers. There is, without question, a great and urgent need for some form of coordinated international investment in building healthcare human resource capacity in LDCs if they are to tackle their own problems adequately.

1.20.4.5 Human Immunodeficiency Virus/Acquired Immune Deficiency Syndrome: A Development Challenge

In the last two decades, over 30 million people have died of HIV/AIDS. In 2003, almost 5 million people became newly infected with HIV, the greatest number in any one year since the beginning of the epidemic. At the global level, the number of people living with HIV continues to grow – from 35 million in 2001 to 38 million in 2003. In the same year, almost 3 million were killed by AIDS; over 20 million have died since the first cases of AIDS were identified in 1981.[79] Approximately 28.5 million people are living with HIV/AIDS in sub-Saharan Africa, yet the continent is home to just over 10% of the world's population – and almost two-thirds of all people living with HIV.

In 2003, an estimated 3 million people became newly infected, and 2.2 million died (75% of the 3 million AIDS deaths globally that year). Outside Africa, the Caribbean (Haiti particularly), India, Russia, and China have the highest infection rates. Globally, just under half of all people living with HIV and AIDS are female. In sub-Saharan Africa, women and girls make up almost 57% of all people infected. In most other regions, women and girls represent an increasing proportion of people living with HIV, compared with 5 years ago.[80]

Differences in the spread of the epidemic can be accounted for by complex interactions of sexual behavior and biological factors. Sexual behavior patterns are generally determined by cultural and socioeconomic factors. In sub-Saharan

Africa, some traditions and socioeconomic factors have contributed to the extensive spread of AIDS, including the poor social status of women, widespread poverty, the collapse of public health systems, unemployment, rapid urbanization, wars, and population displacement.

Apart from HIV/AIDS being a challenge for health systems, it is also retarding years of development in LDCs due to the numbers of people succumbing to the disease. AIDS kills people in the prime of their lives; it is restructuring populations in LDCs, and destroying the social fabric of societies that are much needed for the development of these countries. In LDCs, HIV/AIDS threatens human welfare, socioeconomic advances, productivity, social cohesion, and even national security. The disease does not spare any segment of the community; it affects parents, children, and youth, teachers and health workers, rich and poor. It is, however, often the poorest that are the most vulnerable to HIV/AIDS, and on whom the consequences are most dire.

The impact of HIV/AIDS on the economies of African countries is difficult to measure. The economies of many of the worst-affected countries were already struggling with development challenges, debt, and declining trade before HIV/AIDS started to affect them. Together with other factors, HIV/AIDS is having a devastating effect on many countries' economies. The disease has a negative impact on labor supply, through increased mortality and morbidity. This is multiplied by the loss of skills in key sectors of the labor market. Long periods of AIDS-related illness also reduce labor productivity. Government income also declines, as tax revenues fall, and governments are pressured to increase their spending, to deal with the rising prevalence of AIDS. All of these factors are creating financial and development crisis in the most affected African countries. Current estimates suggest that HIV/AIDS has reduced the rate of growth of Africa's per capita income by 0.7 percentage points a year and that for those African countries affected by malaria, growth was further lowered by 0.3 percentage points per year.[80] Clearly then, not only is HIV/AIDS having a detrimental effect on the growth of African economies, it is reversing the modest gains made in recent times.

Many of the continent's economic development goals depend on Africa's ability to diversify its industrial base, expand exports, and attract foreign investment. By making labor more expensive and reducing profits, AIDS limits the ability of African countries to attract industries that depend on low-cost labor, and makes investments in African businesses less desirable. HIV/AIDS therefore threatens the very foundations of economic development in Africa.[81] HIV/AIDS affects all sectors of the economy, and the costs that are incurred as a consequence of the disease are not just financial in nature but fundamentally social and psychological.[82] Without a doubt, long-term structural policy reforms, aimed at combating gender inequality and the economic and social vulnerability of women, will be of paramount importance to turn this situation around.[83]

1.20.5 Reflections: The Way Forward

It is inexcusable that in the twenty-first century, even though we understand what causes most of the world's plagues and have already devised medical tools and sanitation strategies to fight them, infectious diseases still claim millions of lives each year. These diseases remain major killers largely because the tools and strategies do not reach the people who need them most.[84]

The other problem, of course, is a lack of real commitment to put an end to these epidemics, further exacerbated by a lack of resources to implement basic prevention, primary care and public health strategies in Africa. Many African countries faced with shrinking budgets for providing public healthcare services have been offered support by NGOs and other charitable organizations whose mission is to improve the health of Africans. NGOs have in effect begun to perform certain public health duties that traditionally belong to the state. This action on the part of NGOs has in some cases made for an uneasy relationship with governments, which feel that these organizations undermine public sector capacity. Despite these tensions, NGOs and charitable organizations remain a strong force in providing public health services in Africa.[85] However, the burden of providing these essential services cannot be placed squarely on the hands of NGOs.

African countries face several practical obstacles in boosting their public health systems; for example, there is currently little synergy among African countries to deal with public health problems collaboratively. Further, there is a poor understanding of the key principles of public health within the health sector, and preventive care services have not received much prominence because the achieved gains are generally invisible. There is also a lack of public health expertise within the health sector since health professionals in many African countries are generally trained to treat or cure, and not to prevent disease.[86]

To improve healthcare in Africa, a more ambitious, comprehensive public health plan is required, involving more players and stakeholders at different levels, including: reintroducing strong public health systems aiming to deal effectively with the most basic health needs, good sanitation, nutritious food, and at least a rudimentary primary care system.[87] As we have seen, in many countries that kind of system, or a version of it, did exist prior to the introduction of market experiments, and it is important for the successful elements of that system to be reintroduced. More public and political pressure should be placed on drug companies to reduce drug prices for African countries, which have a heavy dependence on drugs, and to support research for neglected disease.

The growing number of public–private partnerships (PPPs) being created to boost health research in developing countries is a positive one. PPPs for health are generally defined as arrangements that innovatively combine different skills and resources from institutions in the public and private sectors to address persistent global health problems.[88] If international financial institutions such as the World Bank and important private foundations such as the Gates and Ford Foundation work together with bilateral donors and research institutions in both developed and developing countries, creative programs and incentives can be developed to stimulate medical and scientific research for diseases that primarily affect the developing world.

Most important in the long run will be a continuing improvement in the overall economic and social welfare of those who live in poor countries. There are many signs of health improvement, most of which have little to do with healthcare, and it is becoming more evident that the concept of health in Africa should not be considered in isolation from other elements in the development process. African health policy-makers need to start to think outside formal 'health systems' if they are to improve the overall health of their populations. Good health is affected by socioeconomic factors such as income level and nutritional status. For example, children's ability to take full advantage of the education being offered to them depends on their mental and physical fitness. On the other hand, the extent to which health conditions can be improved depends on people's knowledge and attitude toward health practices. Standards of both health and education depend on the whole societal milieu, especially the prevailing attitudes and institutions.[89] It is imperative, therefore, that health problems should become fully integrated into the general development process of countries.

A 2005 report published by the World Health Organization's Commission on Social Determinants of Health found that the root causes of disease and health inequalities in some of the world's poorest and most vulnerable communities are the social conditions in which they live and work – social determinants.[90] The report also stated, as we have illustrated in this chapter, that, like other features of comprehensive primary healthcare programs in developing countries, concerted action on these social determinants was weakened by the neoliberal economic and political experiments that dominated the 1980s and 1990s, with their focus on privatization, deregulation, shrinking states, and freeing markets.[91] The idea that poor social and economic circumstances affect health throughout life is finally gaining recognition by health development agencies and policy-makers. People further down the social ladder usually run at least twice the risk of serious illness and premature death as those near the top.[91]

Recognizing the important correlation between social conditions and the concept of health, 189 countries (including most African nations) adopted in 2000, as part of the Millennium Development Goals, ambitious targets for reducing poverty and hunger, increasing access to education, improving child and maternal health programs, and controlling epidemic diseases. The goals also set important objectives for environmental protection, and the development of fair global trading systems, all of which are to be reached by the year 2015.[92] These goals, if reached, will go a long way toward tackling disease in African countries.

Improving the health of African citizens must involve reducing levels of educational failure, reducing insecurity and unemployment, and improving housing standards.[93] Societies that create an environment in which citizens can play a full and useful role in the social, economic, and cultural life of their society will be healthier than those where people face insecurity, exclusion, and deprivation. Integrating health into a comprehensive development policy is the correct way – the most sensible way – forward, in the pursuit of good health for all Africans.

References

1. United Nations Conference on Trade and Development (UNCTAD). http://www.unesco.org/ldc/list.htm (accessed Aug 2006).
2. World Health Organization. *The World Health Report, Shaping the Future*; World Health Organization: Geneva, 2003.
3. Sindima, H. *Religious and Political Ethics in Africa: A Moral Inquiry*; Westport: Greenwood, CT, 1998, p 34.
4. Turshen, M. *Privatizing Health Services in Africa*; Rutgers: New Brunswick, NJ, 1999, pp 15–18.
5. The World Bank. *Better Health in Africa: Lessons Learned*; World Bank: Washington, DC, 1994, p 46.
6. Owoahene-Acheampong, S. *Inculturation and African Religion: Indigenous and Western Approaches to Medical Practice*; Peter Lang: New York, 1998.
7. Sindima, H. *Religious and Political Ethics in Africa: A Moral Inquiry*; Greenwood: Westport, CT, 1998, p 22.
8. World Health Organization/UNICEF. *International Conference on Primary Health Care*; Alma-Ata: USSR, 1978.
9. Government of Kenya. *African Socialism and its Applications to Planning in Kenya, sessional paper number 10*; Government Printer: Nairobi, 1965.
10. Wasunna, O. *Delivery and Financing of Health Care Services in Kenya: Critical Issues and Research Gaps*; IPAR discussion papers series DP. no. 002; IPAR: Nairobi, 1997, p 2.
11. Turshen, M. *Privatizing Health Services in Africa*; Rutgers: New Brunswick, NJ, 1999, pp 15–18.
12. Ministry of Health. *National Health Core Financing Policy*; Lusaka, Zambia, 1998.
13. UNICEF and the OAU. *Africa's Children, Africa's Future*; UNICEF: New York, 1992.
14. World Bank. *Financing Health Care: An Agenda for Reform*; World Bank: Washington, DC, 1997, pp 10–13.

15. World Bank. *Adjustment in Africa: Reforms, Results, and the Road Ahead*; Oxford: New York, 1994, p 20.
16. Madeo, M.; Spinaci, S. *G. Ital. Med. Trop.* **2000**, *5*, 1–2.
17. World Bank. *Adjustment in Africa: Reforms, Results, and the Road Ahead*; Oxford: New York, 1994, p 3.
18. Griffin, G. C. *User Charges for Health Care in Principle and Practice in Health Care Financing, Proceedings of Regional Seminar on Health Care Financing*; Asian Development Bank: Philippines, 1987.
19. World Bank. *Adjustment in Africa: Reforms, Results, and the Road Ahead.*; Oxford: New York, 1994, p. 5.
21. Mwabu, G.; Mwanzia, J.; Liambila, W. *Hlth Policy Plann.* **1995**, *10*, 164–170.
22. World Bank. *World Development Report: The State in a Changing World*; Oxford: Oxford, 1997.
23. SAPRIN. *Structural Adjustment: The SAPRI Report. The Policy Roots of Economic Crisis, Poverty and Inequality*; Zed Books: London, 2004.
24. Loewenson, R. *Int. J. Hlth Serv.* **1993**, *23*, 717–730.
25. Gilson, L.; Mills, A. *Hlth Policy* **1995**, *32*, 215–243.
26. Brijal, V.; Gilson, L. *Understanding Capacity: Financial Management within the District Health System; CHP Monograph*; Center for Health Policy: Johannesburg, 1997.
27. Dollar, D.; Svennson, J. *What Explains the Success and Failure of Structural Adjustment Programs? Policy Research Working Paper 1938*; World Bank Macroeconomics and Growth Group: Washington, DC, 1998.
28. Hsiao, W. C. *J. Hlth Polit. Policy Law* **1992**, *17*, 613–636.
29. Bryant, J. *Health and the Developing World*; Cornell: Ithaca, NY, 1969, p 40.
30. World Bank. *Better Health in Africa: Lessons Learned*; World Bank: Washington, DC, 1994.
31. Bloom, G.; Segall, M.; Thube, C. *Expenditure and Financing of the Health Sector in Kenya: Abridged Report of a Study Performed for the Ministry of Health and the World Bank*; Institute of Development Studies (IDS): Sussex, 1993.
32. World Bank. *Better Health in Africa: Lessons Learned*; World Bank: Washington, DC, 1994.
33. Lopez, A. D. Causes of Death in Industrial and Developing Countries: Estimates for 1985–1990. In *Disease Control Priorities in Developing Countries*; Jamison, D. T., Mosley, E. H., Measham, A. R., Bobadilla, J. L., Eds.; Oxford: New York, 1993.
34. Bryant, J. *Health and the Developing World*; Cornell: Ithaca, NY, 1969.
35. World Health Organization. *World Health Report: Reducing Risks, Promoting Healthy Life*; World Health Organization: Geneva, 2002.
36. World Bank. *Better Health in Africa: Lessons Learned*; The World Bank: Washington, DC, 1994, p 67.
37. World Bank. *World Development Report, Investing in Health*; Oxford: Oxford, 1993.
38. Hartog, R. *Soc. Sci. Med.* **1993**, *37*, 897–904.39.
39. World Health Organization. *Communicable Disease Prevention and Control: New Emerging, and Re-emerging Infectious Diseases.* WHO document A48/15; World Health Organization: Geneva, 1995.
40. Turshen, M. *Privatizing Health Services in Africa*; Rutgers: New Brunswick, NJ, 1999.
41. Ramesh, G.; Chellaraj, G.; Murray, C. J. L. *Soc. Sci. Med.* **1997**, *44*, 2.
42. World Health Organization. *Framework for Containing Resistance to Antimicrobial Drugs*. WHO Presentation at Infectious Diseases Partners Initial Meeting, US Agency for International Development; Washington, DC, 1997.
43. Govindaraj, R.; Reich, M. R.; Cohen, J. C. *World Bank Pharmaceuticals*, HNP discussion paper; World Bank: Washington, DC, 2000, p 8.
44. Agreement on Trade-Related Aspects of Intellectual Property Rights, annex 1C to the Marrakesh Agreement Establishing the World Trade Organization, 15 April 1994 ("TRIPS"). Article 27.1 TRIPS.
45. Agreement on Trade-Related Aspects of Intellectual Property Rights, annex 1C to the Marrakesh Agreement Establishing the World Trade Organization, 15 April 1994 ("TRIPS"). Article 33. TRIPS.
46. Agreement on Trade-Related Aspects of Intellectual Property Rights, annex 1C to the Marrakesh Agreement Establishing the World Trade Organization, 15 April 1994 ("TRIPS"). Articles 27 (2), (3a) (b) TRIPS.
47. Agreement on Trade-Related Aspects of Intellectual Property Rights, annex 1C to the Marrakesh Agreement Establishing the World Trade Organization, 15 April 1994 ("TRIPS"). Article 30 TRIPS.
48. Agreement on Trade-Related Aspects of Intellectual Property Rights, annex 1C to the Marrakesh Agreement Establishing the World Trade Organization, 15 April 1994 ("TRIPS"). Articles 8 and 40 TRIPS.
49. Haochen, S. *J. World Trade* **2003**, *37* (1).
50. *Declaration on the TRIPS Agreement and Public Health*, WTO Ministerial Conference, fourth session Doha, WT/MIN (01)/DEC/2, para 3.
51. *Declaration on the TRIPS Agreement and Public Health*, WTO Ministerial Conference, fourth session Doha, WT/MIN (01)/DEC/2, para 5(b).
52. *Declaration on the TRIPS Agreement and Public Health*, WTO Ministerial Conference, fourth session Doha, WT/MIN (01)/DEC/2, para 5 (c).
53. *Declaration on the TRIPS Agreement and Public Health*, WTO Ministerial Conference, fourth session Doha, WT/MIN (01)/DEC/2, para 5 (d).
54. *Declaration on the TRIPS Agreement and Public Health*, WTO Ministerial Conference, fourth session Doha, WT/MIN (01)/DEC/2, para 7.
55. Abbott, F. M. Compulsory Licensing for Public Health Needs: The TRIPS Agenda at the WTO after the Doha Declaration on Public Health, occasional paper 9; Quaker United Nations Office: Geneva, p 17.
56. The Pharmaceutical Manufacturers Association and Others v. The President of the Republic of South Africa and Others, case no: 4183/98, High Court of South Africa (Transvaal Provincial Division).
57. Heywood, M. Debunking 'Conglomo-talk': a case study of the amicus curiae as an instrument for advocacy, investigation and mobilisation. Presented at Health, Law and Human Rights: Exploring the Connections, An International Cross-Disciplinary Conference Honoring Jonathan M. Mann, September 29 – October 1, 2001, Philadelphia, Pennsylvania.
58. Moore, O. *Globe Mail* **2001**, April 19.
59. Basu, P. *Nat. Biotechnol.* **2005**, *23*, 13–15.
60. Commission on HIV/AIDS and Governance in Africa (CHGA) Economic Commission for Africa. *Scaling Up AIDS Treatment in Africa: Issues and Challenges*. Background paper for CHGA Interactive: Gaborone, Botswana, July 2004.
61. Wasunna, A. Human Ethical Issues Arising in ARV Scale-Up in Resource Constrained Settings. In *Scaling Up Treatment for the Global AIDS Pandemic*; Curran, J., Debas, H., Arya, M., Kelley, P., Knobler, S., Pray, L., Eds.; The National Academies: Washington, DC, 2005, pp 352–369.
62. Rozek, R. P.; Berkowitz, R. *J. World Intellect. Prop.* **1998**, *1*, 179–243.
63. Attaran, A.; Gillespie-White, L. *JAMA* **2001**, *286*, 1886–1892.
64. Consumer Project on Technology, Essential Action, Oxfam, Treatment Access Campaign, Health Gap. *Comment on the Attaran/Gillespie-White and PhRMA Surveys of Patents on Antiretroviral Drugs in Africa*; October 16, 2001. http://www.cptech.org/ip/health/africa/dopatentsmatterinafrica.html (accessed Aug 2006).
65. Trouiller, P.; Olliaro, P.; Torreele, E; Orbinski, J.; Laing, R.; Ford, N. *Lancet* **2002**, *359*, 2188–2194.

66. Global Forum for Health Research. *The 10/90 Report on Health Research 2000*; World Health Organization: Geneva, 2000.
67. Siithi-amorn, C.; Somrongthong, R. *BMJ* **2000**, *321*, 775–776.
68. Kickbusch, I.; Payne, L. *Research Challenges for Global Health*; Global Forum Update on Research for Health: Pro-Brook: London, 2005.
69. Office of Inspector General, Department of Health and Human Services USA. *Recruiting Human Subjects: Pressures in Industry Sponsored Clinical Research.* OEI-01-97-00195. Department of Health and Human Services: Washington, DC, 2000.
70. Kass, N.; Hyder, A. A. Attitudes and experiences of US and developing country investigators regarding US human subjects regulations. Commissioned Paper in National Bioethics Advisory Commission, *Ethical and Policy Issues in International Research: Clinical Trials in Developing Countries;* US Dept of Commerce, Technology Administration, National Technical Information Service: Bethesda, 2001, Vol. 2.
71. Nuffield Council on Bioethics. The ethics of research related to health care in developing countries. Nuffield Council on Bioethics: London. http://www.nuffieldbioethics.org/publications (accessed Aug 2006).
72. Kimani, D. *East African*, Nairobi, February 7, 2005.
73. Sanders, D.; Dovlo, D.; Meeus, W.; Lehmann, U. Public Health in Africa. In *Global Public Health: A New Era*; Beaglehole, R., Ed.; Oxford: Oxford, 2003, pp 135–155.
74. Joint Learning Initiative. *Human Resources for Health: Overcoming the Crisis*; Harvard: Cambridge, MA, 2005, p 2.
75. Steinbrook, R. *N. Engl. J. Med.* **2004**, *351*, 739–741.
76. Attawell, K.; Mundy, J. *Scaling up the Provision of Antiretroviral therapy in Resource-Poor Countries: A Review of Experience and Lessons Learned*; DFID: London, 2003.
77. Stilwell, B.; Diallo, K.; Zurn, P.; Dal Poz, M. R.; Adams, O.; Buchan, J. *Hum. Res. Hlth* **2003**, *1*.
78. Hagopian, A.; Thompson, M. J.; Fordyce, M.; Johnson, K. E; Hart, L. G. *Hum. Res. Hlth* **2004**, *2*.
79. UNAIDS, 2004 *Report on the Global AIDS Epidemic 4th Global Report (UNAIDS/04.16E).* UNAIDS: Geneva, Switzerland.
80. Bonnel, R. *South African J. Econ.* **2000**, *68*, 820–855.
81. Rosen, S.; Vincent, J. R.; MacLeod, W.; Fox, M.; Thea, D. M.; Simon, J. L. *AIDS* **2004**, *18*, 317–324.
82. Ainsworth, M.; Over, A. M. The Economic Impact of AIDS in Africa. In *AIDS in Africa*; Essex, M., Ed.; Raven: New York, 1994, pp 559–588.
83. Seghal, J. M. *The Labour Implications of HIV/AIDS*; ILO: Geneva, 1999.
84. Ash, C.; Jasny, B. *Science* **2002**, *295*, 2035.
85. Wasunna, A. *Drug Dev. Res.* **2005**, *51*, 1–5.
86. Hunter, D. *Investing in Primary Health Care*. Paper presented at the National Association of Primary Care meeting held at the Royal College of Physicians, Regents Park, March 5, 2003.
87. Wasunna, A. *Drug Dev. Res.* **2005**, *51*, 1–5.
88. Widdus, R. *Bull. World Health Org.* **2001**, *79*, 713–720.
89. Bryant, J. *Health and The Developing World*; Cornell: Ithaca, New York, 1969, p 96.
90. Commission on Social Determinants of Health. *Action on the Social Determinants of Health: Learning from Previous Experience.* Background paper. World Health Organization: Copenhagen, Denmark, 2005, p 4.
91. Wilkinson, R.; Marmot, M. *Social Determinants of Health: the Solid Facts*; World Health Organization Regional Office for Europe: Copenhagen, Denmark, 1998, p 10.
92. United Nations Development Programme (UNDP). *Millennium Development Goals: A Compact among Nations to End Human Poverty.* Oxford University Press: New York, 2003.
93. Wilkinson, R.; Marmot, M. *Social Determinants of Health: The Solid Facts*; World Health Organization Regional Office for Europe: Copenhagen, Denmark, 1998, p 11.

Biography

Angela A Wasunna is a lawyer by training and an Advocate of the High Court of Kenya. Angela received a Bachelor of Laws degree from the University of Nairobi Kenya in 1996, and two Master of Laws degrees (with bioethics specializations) from McGill University and Harvard Law School in 2000. Angela's research and writing interests include international health law, health and human rights issues, reproductive law and policy, financing of health care in developing countries, ethical issues raised by international research and HIV/AIDS. Angela is an elected Board Member of the International Association of Bioethics, a member of the International Bar Association, the Pan African Bioethics Initiative, and the Law Society of Kenya. She is the author (with Daniel Callahan) of the book *Medicine and the Market: Equity v Choice*, Johns Hopkins University Press, 2006.

Comprehensive Medicinal Chemistry II
ISBN (set): 0-08-044513-6

ISBN (Volume 1) 0-08-044514-4; pp. 637–653

1.21 Orphan Drugs and Generics

C-P Milne and L A Cabanilla, Tufts University, Boston, MA, USA

1.21.1 Introduction

Illnesses that affect a large swath of the industrialized world's population, such as heart disease or common cancers, are often the focus of pharmaceutical and biotech R&D. However, there exists a wide array of diseases with relatively small prevalence that, when added up, affect a significant segment of the population. The number of orphan diseases has been estimated to be between 5000 and 8000 with 25 million people affected in the United States, another 25–30 million in Europe, and untold numbers throughout the rest of the world.[1,2] Orphan drug laws were designed to encourage research, development, and marketing of products for rare diseases and conditions. In the United States, a disease qualifies as rare if the prevalence is less than 200 000 persons.[3] The term 'orphan drug,' which first appeared in the 1970s, reflected the fact that no company would be interested in, or indeed could afford to commercialize, medicines for unique diseases that affected small patient populations because there was no reasonable expectation that a developer's R&D investment could be recouped from sales.[4]

The term 'orphan drug' has different meanings depending on the market, regulatory, or political context. To some, it means medicines for children who are referred to as 'therapeutic orphans' because they generally comprise too small a patient subpopulation to make testing of pediatric uses of drugs developed for adults feasible. To others, it evokes the commercial orphaning of millions of people who are affected by tropical diseases, yet live in impoverished economies and healthcare systems. Generally speaking, however, the term 'orphan drugs' refers to drugs that are both therapeutic and commercial orphans because they have too small a patient population to render their development commercially viable. These varied contemporary connotations for orphan drugs highlight the increasing diversity of groups interested in orphan drug development and signal that orphan drug issues will continue to be a hotbed of political debate at the national and international level.

1.21.2 The History of the Orphan Drug Act in the United States

Although a number of countries now have orphan drug laws, the United States has led the way for policy initiatives to encourage R&D for orphan drugs.[5] The early impetus of the initiative to develop an incentive program for R&D of rare diseases in the United States arose, in part, from civil defense concerns prevalent during the early Cold War era of the

1950s and 1960s when the USSR was stockpiling rare and very dangerous viruses and bacteria as potential weapons.[6] In 1964, a Public Health Service (PHS) Task Force was convened to determine if the needs for the treatment of rare and tropical diseases were being met at PHS hospitals. This information was largely unavailable, and the following year PHS undertook the first attempt to assess the actual availability of drugs for tropical and other rare diseases in the United States.[7]

In the 1970s, the Cold War threat of annihilation by weapons of mass destruction ameliorated somewhat, and the civil defense urgency waned, as did interest in many tropical and exotic diseases. In the interim, the 1963 Kefauver-Harris Amendments to the Federal Food, Drug, and Cosmetic Act (FFDCA) had greatly increased the difficulty for a drug to gain Food and Drug Administration (FDA) approval.[8] Thus, pharmaceutical companies of this era were far more focused on cost–benefit considerations than they had been in the past.[5] The net result was a shift in the pharmaceutical industry toward illnesses with large target populations, and thus a larger potential return on their R&D investment.

During this time period, however, rare disease patient organizations shifted from working individually to organizing under the umbrella of the National Organization of Rare Disorders (NORD), thus gaining political clout. As the focus of concern shifted away from protecting the United States against microbial invaders, NORD was able to direct some of the public health momentum toward providing drugs for indigenous rare diseases. Government recognition of this need took the form of a 1979 report by the Interagency Task Force on Significant Drugs of Limited Commercial Value, so-called orphan drugs, and shortly thereafter, congressional hearings began on the orphan drug problem.[5] The report identified several problem areas of drugs with high therapeutic but low economic value. These included drugs for rare diseases, single usage drugs (e.g., diagnostic and some vaccines), drugs that require lengthy development times (and thus are left with short effective patent lives), drugs with anticipated legal liability (e.g., birth control pills), drugs for use in diseases endemic to third world countries (i.e., countries with limited distribution capabilities and monetary resources), unpatentable drugs, and marketed drugs requiring FDA approval for use in new indications.[9] NORD and other proponents enlisted congressional sponsors and achieved passage of the Orphan Drug Act (ODA) by means of what was then a novel combination of patient advocacy and media hype to carry out healthcare policy. The actor Jack Klugman was featured in a story about orphan drugs on the show *Quincy* (a popular TV series about a medical examiner), and later rallied support for the orphan drug law by appearing at a hearing of the house on the ODA. The ODA, which amended both the FFDCA and the Public Health Service Act (PHS), was passed by an overwhelming majority soon afterwards in 1983.

1.21.3 The Challenges of Orphan Drug R&D

There are many challenges to overcome in orphan drug development: lack of knowledge about the pathophysiology of these diseases; longer clinical development time related to the difficulty in identifying and enrolling patients for illnesses, often with small and geographically dispersed affected populations; and the need to depart from traditional clinical study designs involving double-blind, parallel design with placebo controls in favor of less traditional (more problematic, less acceptable) designs such as open-label, historical controls, and crossover. There is also the difficulty of enrolling patients in controlled trials once the word is out that there is a potential treatment; however, there are FDA initiatives to help facilitate enrollment, active patient advocacy groups and close patient/doctor relationships with specialists.[10] Developers need guidance as to the choice of alternative trial designs when controlled trials are not feasible. Evaluating the risk–benefit balance is also more difficult since studies may not be able to detect even fairly common adverse events due to trial design and conduct limitations inherent with rare diseases. In addition, the clinical dossiers for orphan drugs are based on a limited amount of data. This presents the obvious difficulty in evaluating the safety profile of an orphan product with sufficient statistical confidence. On average, orphan drugs are associated with greater hazard than other products. While risks are high because the diseases are typically severe, heterogeneous, and little known, benefits are also high because often there are no alternative treatments available, and orphan drugs are generally regarded as having high therapeutic value (see **Figure 1**) and innovativeness.

Orphan products are not only more challenging in the premarket phase but also in the marketing phase. Orphan diseases are generally of interest to specialists, of whom there are far fewer than doctors in general practice, and require more time and resources by the sales force per patient than non-orphan drugs. Medical information departments expend much greater resources per information request for orphan drugs than for drugs intended for the general patient population. Orphan approvals are often based on just a few clinical studies and there is little supporting information to put in the prescribing information (often making it more difficult to develop adequate labeling) and promotional materials. Off-label use is common for orphan drugs due to the prolonged market life and because of product novelty and patient needs. Follow-up publications are typically based on case reports and uncontrolled experiments with these disorders and so questions on off-label uses are difficult to answer.

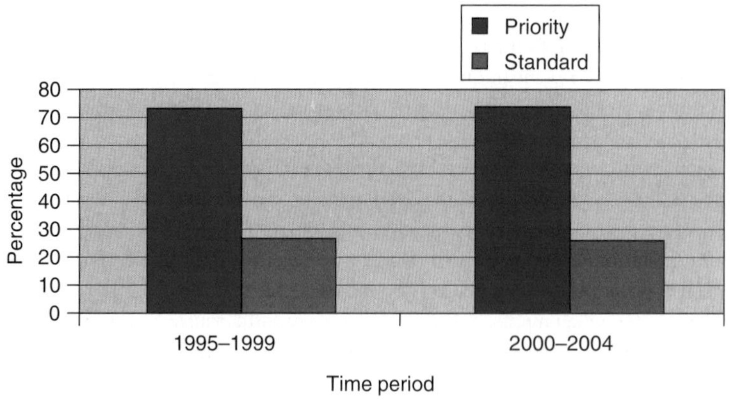

Figure 1 Comparison of priority (FDA review status for products that represent an improvement over currently marketed products) versus standard (all nonpriority products) status for orphan product approvals.

Moreover, the increased clinical complexity of new biotechnology approaches could further burden R&D for rare diseases and conditions. It will take longer to characterize therapeutic profiles of the products of such strategies, and to standardize their quality control, in part because this requires more complicated and less well-validated technologies, such as microarrays.[11]

Despite these challenges, the orphan drug law in the United States has been very successful. In the two decades following the passage of the ODA, there has been almost a fourfold increase in the number of products approved for orphan diseases, as compared to the two decades preceding its enactment. Moreover, the rate of orphan approvals has been increasing. Before the ODA was passed, two or three drugs that could have qualified as orphans were approved annually.[12] In the first 10 years after the ODA was passed in 1983, the average number approved increased to 8.5 per year. During the second decade of the ODA, the rate increased to 15.6 per year. Over 250 drugs and biologicals have been approved under the ODA, and close to 1500 investigational compounds are currently designated.

Dietary supplements are the latest trend in innovative orphan products. The FDA recently granted a designation for Tishcon's ubiquinone, called UbiQGel, which is a dietary supplement based on coenzyme Q10. Coenzyme Q10 is an essential factor in mitochondrial energy production and tends to be lacking in people with MELAS syndrome, Kearns-Sayre syndrome, and MERRF syndrome. The importance of this dietary supplement is that when and if it becomes approved, it will become eligible for insurance coverage just like prescription drugs.

Poisons have always been a focus of the ODA, by encouraging the development of antivenoms or antidotes to treat conditions caused by poisoning. Now the ODA is using a poison to treat disease. Arsenic trioxide has an orphan approval in the United States and an orphan designation in Europe to treat acute promyelocytic leukemia.

Orphan products are also expanding the frontiers of medicine by new drug delivery routes. Take for example, intravenously administered plasma-derived alpha1-antitrypsin (AAT), which has been used for more than 10 years to inhibit an excess of elastase enzyme that causes tissue damage in the lungs for individuals with congenital AAT deficiency and cystic fibrosis. PPL Therapeutics was recently awarded orphan designation for a formulation of AAT delivered directly to the lungs. The market for the 20 000 individuals affected worldwide with AAT deficiency is believed to be worth $400 million alone.

1.21.4 The Economics of Orphan R&D in the United States

In the year before the ODA was passed, annual sales for drugs addressing rare diseases were in the $3–5 million range. By 1991, average first year sales for orphan NCEs were $19.6 million, and the 'blockbuster orphan' phenomenon was beginning to emerge with four drugs achieving a mean first year sales of $116 million.[13] In 2001, average peak-year mean sales were estimated to be $34 million for orphan drugs; however, including a few blockbuster orphans in the cohort results in average peak-year sales of $80 million. Occasionally, what happens is that if the disease is severe and no other treatments exist, demand is insensitive to price and a few orphan drugs earn great returns despite small markets. Orphan drugs may be especially profitable in these circumstances because the variable costs of production, distribution, and marketing may be proportionately lower than for mainstream blockbuster drugs.

Some 11 million people in the United States are being treated with orphan drugs. Orphan products have been a success not only for patients but for sponsors as well. Orphan drugs were only 1/25 of the total market by sales in the year

2000. That total market consisted of 10 000 products, of which only approximately 125 products were brand-name products first approved for an orphan indication. So, by number of products, the orphan market was 1/80 of the overall market for prescription drugs. However, orphan drugs are more profitable as the average sales for each of the drugs in the general market was $14.5 million, while the orphan drugs have mean sales of approximately twice that at $34 million.

To put the size of the orphan sales market in perspective, it must be recognized that the above figures both underestimate and overestimate total sales from orphan drugs. It is an underestimate in that it will not represent sales of orphan products with the same brand name, for which the first approval was not an orphan indication. At the same time, it can overestimate because it does reflect sales from subsequent non-orphan indications (label and off-label) of a drug first approved as an orphan sold under the same brand name. Overall, it is more likely an underestimate. The value of the specialty drug market, described as the niche pharmaceutical market (but including drugs for AIDS patients and other niche patient markets, which may no longer be served by orphan drugs alone), provides an upper boundary for the potential value of the orphan drug market and was estimated to be $40 billion for 2005.[14]

1.21.5 Orphan Drug Market Factors

Factors affecting the predictability of the size of the sales market for orphans differ somewhat in scale and scope from those of non-orphan drugs. The first is chronicity of use, as the majority of orphan drugs can be considered chronic use drugs. A Tufts Center for the Study of Drug Development (CSDD) study showed that while 30% of orphan drugs required administration for less than 30 days, 31% needed to be used intermittently as symptoms recurred or called for specific periods of re-treatment, and 39% required administration over a patient's lifetime or continuously for a year or more. The second factor is off-label use. Off-label use of orphan drugs is believed to be greater than for non-orphans, but also more difficult to estimate due to the novel uses to which they are put where the circumstances of the patients are desperate. The third factor is foreign markets. Foreign markets for orphan drugs operate differently from those for general use. They tend to run to more extremes: at one end of the spectrum is Australia, which seeks to channel United States orphan drugs to its market and at the other is Japan, which now provides for many of its own needs in the same disease categories as those of the United States. Competition, the fourth factor, is also more difficult to predict, as erosion of the market does not happen routinely and gradually as in the market for general use drugs from the usual sources of 'me-toos' (i.e., therapeutically equivalent successor products) and generics (bioequivalent, but less expensive product). Orphans have little to fear from generics and 'me-toos' because they typically fall under the market size threshold of $50–100 million in sales, above which makes it worthwhile for competitors to enter the market. When competition arises in the orphan drug market, it can happen dramatically from competitor products divined by regulatory edict to be clinically superior (albeit however minimally) or off-label use. This can result in the sudden division of what the first sponsor had believed would be a captive market after winning the race to approval. Lastly, there is the staying power of orphan drugs – the typical peak-year sales curve for general-use drugs has its shoulder at 8–12 years, while that of the orphan drug extends from 7 to 15 years.[15]

In addition to these operational factors, some emerging market trends are impacting the orphan drug market. Despite the fact that one of the perceived benefits of orphan R&D is a captive market at launch, some of the blockbuster orphans are hedging their bets and using direct-to-consumer advertising. For example, Amgen is reportedly using cable TV ads to target new cancer patients for its anti-anemia product Neupogen, while also promoting a children's book about a 14-year-old boy with kidney disease in order to reach the 5000 pediatric patients with anemia from kidney dialysis.

The dawning of the age of e-commerce may have special significance for the orphan market. As of the beginning of the 2000s, two hundred million people are 'wired up' worldwide, 80 million of those being in the United States. Thirty percent of 50 000 drug stores in the United States have their own websites. While e-commerce was worth only $160 million in 1999, it was predicted to increase exponentially midway through the next decade. The orphan drug purchaser may be especially drawn to some of the benefits of e-commerce. Orphan purchasers are typically buying the drug for a long period of time and may be less mobile than your average consumer, and so need a convenient way to purchase their medicines, especially as drugs are often not available at their local drug store. They also may be more aggressive in seeking ways to save money on their drug purchases. Moreover, they are likely to know more about their disease than the average consumer and may not feel the need to talk to a pharmacist about it, and, in particular, may not want to go to a public place to buy it.

The utility of e-commerce for orphan drug distribution may have to be considered in the context of several other market trends. Mail order is capturing an increasing share of the distribution market, at the expense of the hospital share. While this channel of distribution currently occupies only about 10% of all purchases at this point in time, up to 70% of all prescription drugs could be delivered by mail, so the market has considerable room for growth. Along these lines, the specialty market itself is a growing area as the CVS Procare network is looking to cover the 32 largest urban markets this year in the United States. While the purpose is to offer a one-stop-shop for specialty drug buyers at convenient bricks-and-mortar locations, e-commerce would be a natural outgrowth of increasing customer familiarity with the specialty drugs market concept.

1.21.6 The Orphan Drug Law in the United States: Push and Pull Incentives

The economic rationale for the United States ODA was to provide a mechanism to make the orphan drug market more attractive to drug developers by making orphan drug development more like that of drugs for the general population. This was done by providing 'push' incentives that reduce the fixed costs of R&D, and 'pull' incentives that increase the expectation of profits, by providing monopoly market conditions.[16] The ODA contains five 'push-pull' incentives: technical and administrative assistance; research grants; waiver of user fees; tax credits; and market exclusivity. The first four are push measures that act to reduce fixed costs, while the last one is a pull mechanism that acts to enhance the likelihood that there will be sufficient return on investment by monopolizing profits from the small market. These measures affect the various segments of the industry in different ways. Tax credits appeal to all segments of the industry, but especially to larger companies, which carry the majority of orphan products through to approval. Grants and agency assistance are especially useful to smaller companies. Through these incentives, the FDA encourages smaller, less experienced pharmaceutical manufacturers to seek regulatory approval of orphan products.[17] In general, the incentives have worked well as orphan drugs now account for a significant portion of new drugs and biologicals approved each year (see **Figure 2**).

1.21.6.1 First Incentive

The first incentive is technical and administrative assistance with the identification and development of orphan products provided directly by the FDA. To this end, the Office of Orphan Product Development (OOPD) was created within the FDA in 1982. The OOPD monitors the review process and assists in resolving specific issues that may arise during the review process. The OOPD's mission also includes a directive to "assist and encourage the identification, development, and availability" of orphan drugs, as well as to facilitate the orphan status application.[3] Specific protocol assistance concerning what studies the sponsor needs to complete in order to obtain regulatory approval is available from the FDA reviewing divisions. Protocol assistance helps the sponsor decide what tests and experiments the sponsors need to complete to get approval. Open protocols, for example, allow the drug sponsors to sell the drug to patients while the trials are ongoing, enabling sponsors to recoup some of their costs, while accumulating additional test data.

1.21.6.2 Second Incentive

The second incentive is the availability of FDA grants to cover clinical trial expenses. Whereas the ODA pertains primarily to drug and biological products, the orphan grants program has been expanded to include eligibility for

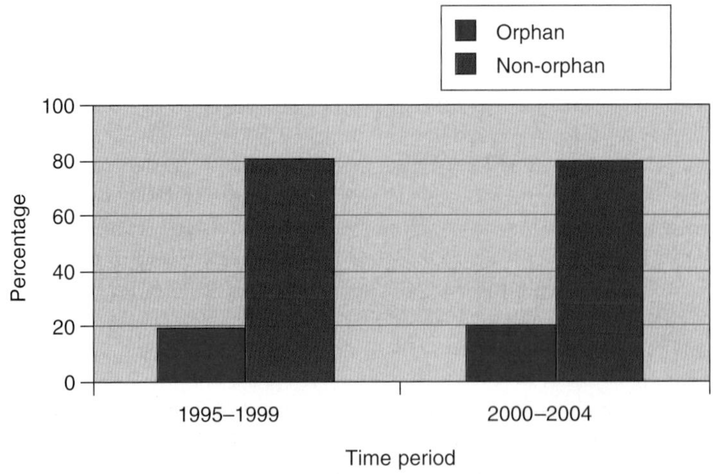

Figure 2 New molecular entity and significant biologic orphan versus non-orphan approvals for 1995–99 and 2000–04.

medical foods and devices that meet the 'orphan' criteria.[3] The FDA's FY2006 budget has \$13.2 million available for these grants, with \$9.2 million for noncompetitive continuation of previous awards, and \$4 million for 10–12 new awards annually.[18] The grants are awarded for periods of 1–3 years, but multiple sequential grant awards have been made. For FY2006, grant limits are based on total costs (including direct and indirect costs), and have a maximum value of \$200 000 for a phase I trial, and \$350 000 for a phase II or III trial.[18] Currently, the ODA is funding a total of 94 such projects.[19] While the grants may help companies, particularly smaller companies, at the aggregate level the grants represent a small percentage of the total cost of drug development.

1.21.6.3 Third Incentive

The third incentive is the automatic waiver of user fees, which are registration fees paid to the FDA by a sponsor for review of a marketing application. These are automatically waived unless the application also includes an indication for other than a rare disease or condition.[20] For FY2005, the full New Drug Application (NDA) or Biological License Application (BLA) fee is \$672 000, and the fee for approval applications without clinical data is \$336 000.[21] The FDA may also waive, upon request by the product sponsor, establishment and product maintenance fees, if proper justification is provided.[20] This incentive again encourages small and medium size firms more than the large ones, and represents only a fraction of the total cost of development.

1.21.6.4 Fourth Incentive

The fourth incentive is the availability of tax credits. These credits can apply to as much as 50% of clinical development costs. The tax credits were established in 1997, and can be carried forward for 15 years or backward for 3 years, thus allowing smaller companies more time to recoup large R&D costs. This tax benefit covers the cost of studies conducted overseas, if the company can justify the need to do testing outside the United States.[22] To some critics of indirect government subsidies, tax credits are considered a more palatable incentive than market exclusivity.[23] However, the amounts saved are still small, relative to the fixed costs of R&D, and are not considered by some to be a strong incentive for the industry.[16]

1.21.6.5 The Fifth Incentive

While the preceding four incentives help to reduce the R&D costs of product development, the fifth incentive, market exclusivity, is consistently cited by pharmaceutical manufacturers and analysts as the most important incentive.[16,17,24–26] Genentech, a leading United States biotechnology firm, stated that the company's orphan drug successes were "... due to the certainty that stems from the [orphan drug] law's promise of market exclusivity."[27] Sponsors value orphan exclusivity, especially when their products cannot be patented as products of nature, are known synthetic compounds, are products with expired patents, or a patent would provide them with only a weak form of intellectual property protection (e.g., a process patent). During this 7-year period of market exclusivity, the FDA cannot approve another marketing application for the 'same' drug treating the 'same' orphan indication.

Small molecule products are the 'same' if they share the same active moiety; for large molecule products, the rules are necessarily more complex and involve separate definitions for various classes of products. If products are the same, they can still be distinguished if one is clinically superior to the other. To establish a drug as clinically superior, a sponsor must demonstrate that its product has a 'significant therapeutic advantage' over the approved version. This can be accomplished through evidence of either greater effectiveness (preferably on the basis of direct comparative trials), or a better safety profile in a substantial portion of the target population, or in 'unusual cases,' by showing that the drug otherwise "makes a major contribution to patient care." The FDA has indicated that in making determinations under the last category, the following factors may be considered, particularly when severe or life-threatening diseases are involved: treatment convenience; duration of treatment; patient comfort; improvements in drug efficiency; advances in the ease and comfort of drug administration; longer periods between doses; and potential for self-administration.

There are two ways in which the exclusivity can be withdrawn. If a sponsor cannot produce enough of the drug to meet patient needs, FDA may revoke exclusivity (this action was threatened by FDA recently against Bayer regarding its hemophilia A drug, Kogenate). Or a sponsor can voluntarily consent in writing to approval of another sponsor's application for the same orphan indication/drug. However, orphan drugs are not protected from competition through off-label uses, the approval of clinically superior versions, and the approval of drugs with a different mode of action for an already approved orphan indication.

1.21.7 Orphan Drug Laws Worldwide

The number of orphan diseases has been estimated to be between 15% and 30% of the total universe of known diseases, with around 50 million people affected in the United States and Europe, and a much larger number existing throughout the rest of the world. Orphan product programs exist in the United States, Europe, Japan, Australia, and Singapore. The South Korean and Israeli ministries of health are creating orphan drug offices in preparation for the anticipated passage of orphan drug legislation in those countries. Canada has regulatory mechanisms that provide some of the same incentives as orphan drug laws.

1.21.7.1 Orphan Drugs in Europe

Legislation on orphan drug products in the European Union (EU) grew out of the European Commission (EC) Directive 91/507/EEC, a 1992 amendment to the French Public Health Code (article L601-2), and a 1994 National Institute of Health and Medical Research (INSERM) report on orphan drugs, which aimed to facilitate the registration process of drugs for rare conditions. Regulation (EC) No. 141/2000 of the European Parliament and the Council on Orphan Medicinal Products was published in January 2000, and enacted legislation to establish EC procedures for designation of orphan products and to establish incentives for orphan product research, development, and marketing. The EC's regulation on orphan medicinal products for drugs and biologicals was published in the *Official Journal of the European Community* in January 2000 followed by implementing regulations in April 2000.

The basic criterion for orphan eligibility is that the product is intended for the diagnosis, prevention, or treatment of a life-threatening or chronically debilitating condition affecting not more than 5 in 10 000 persons in the EC. The European Commission grants designations, but decisions on applications are made by an orphan drugs committee hosted by the European Medicines Agency (EMEA).

Many of the incentives closely mirror those created in the United States, primarily market exclusivity. In the EU, a 10-year period of market exclusivity, along with protocol assistance and the possibility of fee waivers, is issued to orphan drugs. The waivers, tax incentives, and protocol assistance are relatively small, but significant for smaller companies. Tax credits, grants for R&D costs, and fee reductions for marketing approval through the mutual recognition procedure are being considered by the individual EC member states. As the EU is the second largest region for pharmaceutical sales, 10 years of market exclusivity is a very lucrative incentive for companies. However, the EU reviews every orphan drug after 5 years to determine if it has been 'sufficiently profitable' to no longer warrant protection. In cases where this is determined to be the case, market exclusivity ends after the sixth year.[28] This policy is due to be reviewed in early 2006. While many legislators like the idea, a recent study commissioned by the European Commission found it was exceedingly difficult to determine whether a drug meets the criteria.[29]

The EU has the same provisions allowing companies to form marketing and research partnerships to mollify the risk of investing in R&D for orphan drugs. Also, the European Rare Diseases Therapeutic Initiative (ERDITI),[30] sponsored by the European Science Foundation, functions in much the same way as the OOPD does in the United States. ERDITI recognizes the need for support of rare diseases research and seeks to:

- provide academic teams with facilitated access to available compounds developed by companies;
- provide a streamlined facilitated process of collaboration between public and private partners; and
- guarantee continuity all the way from preclinical research to development and commercialization of the drug.

Legislation aside, due to the vagaries of European healthcare systems, approval is not the only hurdle for orphan drugs. A study of the first five EMEA-approved orphans found much variability across EU countries with regard to availability, market availability delays, and prices.[31] While some countries such as France and the Netherlands have been quick to accept orphan drugs, other countries have lagged behind, even making orphans available in some areas of a country, but not in others.[29] The number of orphan drugs available seems to be correlated to the mean orphan drug prices, with countries with more orphan drugs available also having lower prices. According to the authors of the study, this could reflect national policy priority on patients with rare diseases.[31]

1.21.7.2 Orphan Drugs in Japan

Although two-thirds of orphan drugs originate in the United States or EU, Japan also has a well-regarded orphan drug program. This began with a Ministry of Health and Welfare (MHW) Pharmaceutical Affairs Bureau Notification in 1985. This Bureau allows for submission, or notification, of stability validations later and for flexibility in acceptance of foreign clinical data. Japan's orphan products law was the result of an amendment to the Pharmaceutical Affairs Law in April 1993 and required the following criteria to be fulfilled:

- the patient population should be less than 50 000 (about 4 in 10 000 inhabitants);
- no alternative drugs should be available, or the drug must offer higher safety or efficacy compared with existing drugs; and
- there should be a theoretical basis for the indication and a feasible plan to develop the drug.

The Organization for Pharmaceutical Safety and Research (OPSR) has responsibilities similar to the OOPD in that it provides services relating to the promotion of orphan drug development. The MHW is responsible for issuing orphan drug designations, based on advice from the Central Pharmaceutical Affairs Council (CPAC).[32]

Support measures are very similar to those in the United States and EU in that they include grants to subsidize research expenses, tax deductions for the cost of R&D, free or at cost preapproval consultation and instruction from the OPSR, fast track designation, a deduction for user-fees, priority in review, and a 10-year guarantee before re-examination (most new drugs are given 6-year guarantees). The laws also allow an increase by 10% over the formula for the National Health Insurance price for reimbursement.[33] The re-examination period dictates when postmarketing re-evaluation of the drug will occur to determine if a product should remain on the market. Because abbreviated applications are not allowed before re-evaluation, the length of the re-examination period is effectively market exclusivity.[32] After introduction of support measures, the number of orphan approvals doubled in Japan as compared to the number approved when only the Bureau was in place.[32] Currently, 167 orphan drug designations are listed on the website of Japan's Organization for Pharmaceutical Safety & Research (OPSR or KIKO), which jointly administers with the Pharmaceuticals and Medical Devices Evaluation Center and the Ministry of Health & Welfare,[10] and over 100 approvals.

1.21.7.3 Orphan Drugs in Australia

Australia set up the Orphan Drug Unit in the Drug Safety and Evaluation Branch (DSEB) of the Therapeutic Goods Administration in 1998.[33] Australia's orphan drugs program has several features that could bode favorably for a future role as a source of drugs and vaccines for neglected diseases, and as a model program for smaller countries. The DSEB has an interagency agreement with the FDA in the United States for expedited approval of United States orphan drugs. If the FDA approves the orphan indication and there are acceptable FDA evaluation reports available, and the approval of the application is also proposed by the delegate, then the delegate may proceed to decision without referral to the Australian Drug Evaluation Committee (ADEC). However, an unedited version of the FDA evaluation report is generally not available online and must be requested via a three-step process.[33]

To qualify for orphan status in Australia, a drug must be intended to treat, prevent, or diagnose a rare disease; and must not be commercially viable.[32] A drug cannot be an orphan if the FDA, the UK Medicines Control Agency, the Bureau of Pharmaceutical Assessment of Canada, the Medical Products Agency of Sweden, the Medicines Evaluation Board of the Netherlands, or the European Medicines Evaluations Agency (EMEA) has refused to approve its use for the disease for a reason related to the medicine's safety. Eligible conditions are those with a prevalence less than 11 per 100 000 and not affecting more than 2000 at a given time. Drugs that qualify receive many of the same incentives offered in the United States and EU.

As part and parcel of the US–Australia collaboration, orphan products from the United States are approved here unless there is some overriding public health concern unique to Australia. Up until mid-2005 there were over 100 orphan drug designations.[34] Not surprisingly, orphan products from the United States do quite well in Australia: by the year 2000, 35% of United States marketed orphans were also marketed in Australia.

Australia, a country of over 20 million people, has very distinctive goals for its orphan product program, based on unique patient needs related to its indigenous poisonous plants and animals, aboriginal inhabitants, and tropical diseases. The program is administered by the Therapeutic Goods Agency.

1.21.7.4 Orphan Drugs in Singapore

Singapore is a good example of a country whose orphan drug program is smaller in extent than those of the major pharmaceutical markets, such as the United States and EU. In 1991, Singapore was the second country after the United States to establish guidelines for orphan drugs. The focus of these laws pertains to ensuring an adequate supply of drugs for orphan diseases through the regulation of importation of orphan drugs from other countries rather than through R&D. Singapore uses the same definition of an orphan drug as Japan does, and allows drugs approved in certain countries with well-recognized orphan programs to be imported as such.

1.21.8 Drivers of Orphan R&D

1.21.8.1 Biotechnology

When the original version of the ODA was passed, the biotech industry was in its nascent stages; however, the industry grew quickly in the late 1990s, especially in the field of orphan drugs. Today, the biotechnology industry commands the lion's share of orphan designations while big pharma still dominates approvals (see **Figures 3** and **4**). This illustrates the symbiosis that occurs among the public sector (universities and governments), biotech start-ups, and big pharmaceutical and biotechnology companies. Basic research emanating from the public sector is integrated into the biotech sector, which interacts with the public sector both through academic and business activities, which in turn seeks business collaborations with larger pharmaceutical and biotechnology companies.

Orphan drug exclusivity has been especially important for biotechnology-derived medicines. Until recently, products of nature were not considered patentable, except for process patents, which are a weaker form of intellectual property protection than so-called compound patents. For example, several types of processes may be utilized to achieve a particular end product. The life-blood of start-up biotechnology companies is venture capital. The quid pro quo for speculative investment is a guarantee that the fruits of the labor will be protected, otherwise there is a double-ended risk – from both success and failure. Strengthened protection for process patents and clarification of the utility requirements for biopharmaceutical patents have made the intellectual property environment at least more predictable, but have certainly not eclipsed the need for the orphan drug exclusivity. Patents are also vulnerable to legal challenges and must be enforced by the patentee, while the FDA enforces market exclusivity. Despite the fact that

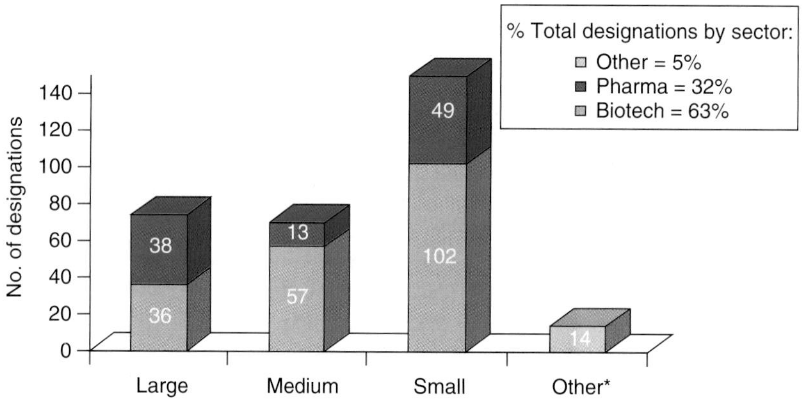

Figure 3 Orphan designations by company size and sponsor type 2000–04. Large: 24% of industry; medium: 22% of industry; small: 49% of industry; other: 5% of industry. * Includes hospitals, universities, government, and sole investigators.

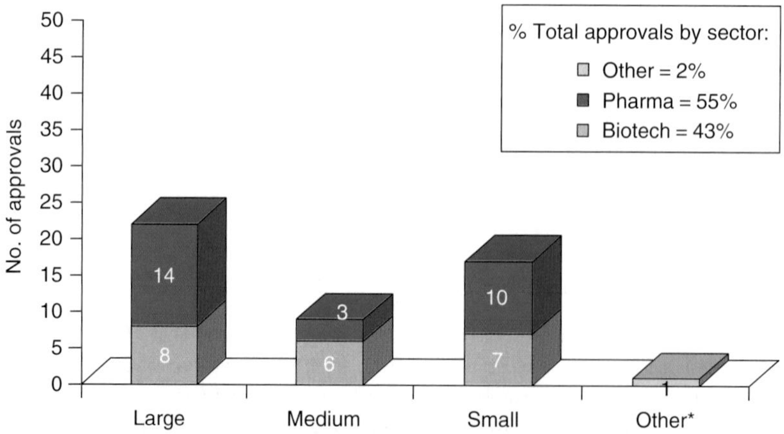

Figure 4 Orphan approvals by company size and sponsor type 2000–04. Large: 45% of industry; medium: 18% of industry; small: 35% of industry; other: 2% of industry. * Includes hospitals, universities, government, and sole investigators.

exclusivity protection is not as encompassing as patent protection, getting orphan drug designation and approval is less resource-intensive than preparing and pursuing a patent application.[27] Patent prosecution requires a lengthy process of drafting and filing an application that explains the invention, and then convincing the Patent and Trademark Office that the invention meets the statutory requirements for patentability. In contrast, requests for orphan drug designation are straightforward, involving only the compilation and submission of information to the FDA, as provided by regulations.

In absolute terms, an orphan drug's exclusivity period of 7 years is obviously shorter than a 20-year patent term. However, a biotechnology product's patent term begins when the patent application is filed, typically early in the development process. The time taken in patent prosecution and the time required to conduct preclinical and clinical studies to obtain regulatory approval eliminate a substantial portion of the 20-year term, leaving a relatively short period of effective patent life in which the sponsor can enjoy monopoly status for its product. This is exacerbated due to the complexities of both orphan diseases and biotech development. Even though orphan drug exclusivity does not have the versatility of a patent, and the uncertain status of biotechnology patents is slowly being resolved in the courts, orphan exclusivity can serve as a powerful substitute for a patent.[35]

1.21.8.2 Pharmacogenomics

Pharmacogenomics studies the effect of genotype on individual drug responses in order to improve the safety and efficacy of drug therapy.[36] This field is similar to orphan drugs in that it is targeted at a small affected patient population, who have no existing effective treatments for their illnesses. The field represents a potentially large percentage of the orphan drug market in the future. The Director of the OOPD, Marlene Haffner, believes that pharmacogenomic advances portend that many more diseases will be recognized as orphan disorders. At present, pharmacogenomic products, such as Gleevec for chronic myeloid leukemia, are just beginning to appear in the marketplace. The field is highly specialized, and expensive, and is dependent on mechanisms, such as orphan drug status, to improve chances for profitability. However, many of the pioneer drugs in this field have had problems receiving orphan designation, most commonly because of disagreements over how a target population is defined.[37]

With advances in research, such as the sequencing of the human genome, many experts believe that the future of pharmaceuticals will become more specialized than ever before. As such the future for orphan diseases, especially the mechanisms underlying many rare genetic disorders, may be unraveled in the next few years, thus opening the door for the possibility of many new drugs for their treatment.[30] Beyond what is being discovered, the thousands of compounds that have been developed by pharmaceutical companies for more common diseases, but that were abandoned or failed to achieve registration for various reasons, represent a treasure trove of compounds for treatment of rare disorders, especially for pharmacogenomic companies willing to invest in the research necessary to make these connections.[30]

Orphan product development is better positioned to take advantage of pharmacogenomics because orphan diseases often have genetic causes or predisposing factors, which may involve more discrete genetic targets as opposed to the multifactorial genetic influences of more common diseases and metabolic pathways more amenable to enzyme interventions. Also, orphan product development is becoming increasingly dominated by biotechnology research, which is better suited to incorporating new discovery sciences such as proteomics.

Financial analysts predict that pharmacogenomics will radically alter the traditional mass-market paradigm, while medical researchers assert that 'personalized medicines' will be available in a few years time with significant impacts within another 5 years. One recent example is a genetic test developed by a team of researchers at Johns Hopkins, which will assist in targeting patients affected with gliomas for whom such therapeutic agents as the orphan drug carmustine will be more effective.

1.21.8.3 Patient Advocacy Power and Private–Public Partnerships

The role of patient advocacy has been well recognized in the United States, particularly in the area of AIDS and cancer. The United States' ODA has a legacy of responsiveness to advocacy. The influence of the NORD on passage of the ODA has been officially recognized in the legislative history of the ODA. When the ODA was passed, however, rare cancers, AIDS, and diseases affecting the elderly and other special populations were not contemplated.[38] Yet, the influence of the AIDS Action Council (acknowledged in the legislative record of the ODA Amendments of 1992) on orphan designations and approvals for AIDS and AIDS-related conditions has been remarkable: 100 designations and 24 approvals. The National Cancer Society and other cancer advocacy groups have had a similar impact. From 1983 to 1991, only 10 orphan drugs were approved for rare cancers, but from 1992 to 1999, that number nearly quadrupled, to

37.[39] In the last decade, advocacy efforts on behalf of special subpopulations have had demonstrable effects on orphan designation trends, as well as on orphan program goals for the coming decade.[19,38]

Those responsible for drafting the European regulation also recognized the emergent role of patient advocacy when they included three patient representatives on the Committee for Orphan Medicinal Products, "underlining the increasing contribution of patients to the health care debate."[40] The European advocacy groups can play an important intermediary role by fostering the needs of the developing countries, especially regarding R&D of neglected diseases such as malaria, tuberculosis, schistosomiasis, and sleeping sickness.

The trend in orphan designations for neglected diseases highlights the importance of advocacy for specific orphan illnesses. For example, there were 11 designations for neglected diseases (out of 213, or 5%) in the first full 5 years of the orphan drug program (1984–88), but only 8 designations (out of 778, or 1%) over the next 12 years (1989–2000). In the 1980s and early 1990s, the FDA and the OOPD espoused tacit support of the use of the ODA to encourage R&D for tropical disease medicines.[41,42] Lack of industry interest, the competing R&D needs of other emerging diseases, and the lack of sustained advocacy efforts for neglected diseases may have been responsible for the downturn in R&D for these diseases.

Similarly, there are problems with orphan incentives for vaccine R&D. These products can be for patient populations that, although quite numerous and medically needy, may not be capable of supporting a sufficiently profitable market or sales profile for a drug. Moreover, patent exclusivity may be poorly adhered to, further decreasing incentives. These realities have acted as barriers to the participation of vaccine manufacturers in the orphan drug program. In addition, there were concerns early on related to the inadequacy of orphan grant amounts and the capacity of vaccines to meet the eligibility requirements for orphan designation.[43] It should come as no surprise that as of 2000, there had been no approved orphan vaccines, and only 8 were designated by the FDA, 7 of which are therapeutic vaccines (such as for cancer or sickle cell anemia), leaving only one designated for prevention.[44]

Orphan drug laws can be used to encourage firms to focus on these problems as a means of global health improvement, to avoid potential re-emergence of such diseases in their own countries, and to treat immigrants who arrive with infections, or citizens who may be infected while vacationing or working in endemic areas. In Canada, for example, there were on average 538 malaria infections per year in the 1990–2004 period.[45] The ODA prevalence limitation is generally not relevant as an obstacle to orphan status. An AIDS vaccine, for example, intended for use in developing countries, would not be thwarted by the fact that AIDS is no longer an orphan disease in the United States, and the vaccine could be targeted at a strain other than that most prevalent in the United States. As long as one case of the strain being developed under an orphan designation exists in the United States, the research will qualify for orphan status.

However, such neglected disease R&D programs require considerable resources, while prospects for return on investment are limited, thus necessitating simultaneous inputs from both the public and private sectors to spread the risk. To this end, the emergence of public–private partnerships (PPPs) has been important in coordinating collaborative efforts between the two sectors. A PPP is a third party, which facilitates coordination between a public sector entity, such as governments, universities or foundations, and a private company. In some cases, a PPP can work with a nonprofit research group and a pharmaceutical company to bring an orphan drug to the market. This can create the necessary resources and scale through coordination, which would probably not occur otherwise.[46] Thus, many PPPs focus on disease-specific indications that, while funded by philanthropic sectors, fund projects that involve for-profit pharmaceutical companies.[47] PPPs now conduct three-quarters of all identified neglected disease drug development, sometimes in partnership with these industry initiatives.

1.21.8.4 Going Global

On a worldwide stage, the orphan drug development paradigm is the wave of the future. It focuses on subpopulations, such as children and women, which are now the focus of legislative and societal initiatives to promote investigation of indications for drugs already marketed as well as new drugs. Also, orphan drug legislation has traditionally been considered a vehicle to incentivize the study of tropical diseases, now referred to as neglected diseases such as malaria, tuberculosis, leprosy, and sleeping sickness. There are already about 20 approvals for these diseases in the United States. While the sales potential for these are not very significant in the United States, the worldwide market is potentially lucrative based on volume and the need for long-term or preventative use for these drugs. The global interest in orphan drug incentives as well as harmonization through the Common Technical Document (CTD) and other International Conference on Harmonization (ICH) initiatives will only enhance the potential for worldwide marketing of orphan drugs.

Patient advocacy also could play a role at the international level. NORD has analogous organizations developing in Europe such as the European Alliance of Genetic Support Groups (EAGS), European Rare Disorders Organization (EURORDIS), and the International Alliance of Patient Organizations (IAPO) – an umbrella group that also hopes to enlist member organizations worldwide. Moreover, global patient networks, the Internet, and global outsourcing services make global recruitment of patients a reality. International harmonization of orphan drug regulations may facilitate global registration and work synergistically with global availability and utilization of resources. What this means at the practical level is that orphan product developers here in the United States can begin to plan for the possibility of expansion to global markets. For example, Protherics recently received approval in the United States for its rattlesnake antivenom product CroFab, which at $6000–9000 per treatment will have an estimated market worth of $40 million per year. But there is potential to expand to the South American market because there are a number of rattlesnakes common to South America and the United States.

1.21.9 The Future

By most standards, the ODA has been an unqualified success as an incentive program for R&D on treatments for rare diseases. Multiple sclerosis is an example of where the planes of pharmacogenomics and patient advocacy all intersect. Medically plausible subsets will increasingly be discovered with patients in those subsets wanting tailored drugs, a movement driven by the multiple sclerosis society through their professional and consumer education campaigns. Lower R&D entry barriers, protected markets, expanding patient populations, and long duration of use bode well for the future of orphan product development. More recently, the emergence of AIDS, the persistence of endemic tuberculosis and malaria as well as bioterrorism concerns worldwide have reinvigorated research in anti-infective R&D as well. Since then, new laws and policy initiatives have created favorable markets for products outside the mainstream patient populations, while the emergence of the biotechnology industry has expanded the capacity for orphan drug R&D. Likewise, orphan laws in the EU, Australia, Japan, and other countries have further increased interest and investment in this field of drug R&D.

1.21.10 Generic Drugs

1.21.10.1 The Generic Drugs Market

If you were to go to the pharmacy with a prescription for a drug, such as omeprazole, there are many relatively cheap options available. In contrast, if you need a drug that has just been released, such as pramlintide, there is only one brand-name option. This is because a patent protects the latter from competition, while the former has several generic versions that largely compete by lowering their price. A generic drug is identical, or bioequivalent, to a brand-name drug in terms of intended use, dosage, safety, strength, quality, performance characteristics, and method of ingestion.[48] The size of the generic drug market is considerable. In the United States, 52% of all prescriptions dispensed are for generic drugs.[49] Of the top five United States pharmaceutical companies, based on the number of prescriptions dispensed, four are generic companies. Worldwide generic sales represent 23% of the global volume of the pharmaceuticals market.[50] In middle and lower income countries, such as Central and Eastern Europe, generic drugs account for as much as 70% of the market.[51] The list of essential drugs from the World Health Organization (WHO), which is designed to assist such countries in deciding which drugs to make available, is almost entirely comprised of drugs that are available in generic form.[52]

In the United Sates and elsewhere, a government will issue a patent to a company or individual as a reward for innovation. A patent allows the innovator to sell their product at monopoly prices and recoup their investment in research and development (R&D) costs. Generic drugs are lower cost bioequivalent copies of brand-name drugs that are available after a patent expires. In the drug market this is particularly important, given the cost of bringing a drug to market has an average capitalized pre-approval cost of $1.1 billion.[53] However, after the patent period expires, the entry of generic products is beneficial to society in that it lowers the cost of vital medicines, and encourages the brand-name drug industry to continue to innovate and search for breakthrough drugs to earn future patents.

1.21.10.2 The Origins of Generic Drugs

Generic drugs, and modern generic firms, emerged in the mid-1960s with regulatory changes resulting from the tragedy of women who used thalidomide during pregnancy.[54] The resulting legislation from this era created clear procedures for generic drugs to gain access to markets through an abbreviated approval process.[55] In 1984, the Hatch-Waxman Act created a balance to ensure market availability for generic drugs, and in so doing provided the seminal moment in the development of the generics industry in the United States and, ultimately, the rest of the world.[56] Although generics markets have evolved at different times, in slightly different ways, throughout the world, they all share the same rationale: to provide an equally effective, cheaper alternative to a brand-name drug.

Countries that allowed pharmaceuticals the highest degree of freedom over pricing, such as the United States, United Kingdom, Germany, and the Netherlands, were the first to reach a substantial market share of generic drugs. Whereas countries, such as France, which controls market pricing, have been slow to incorporate generics into their market.[56] At present, however, every country has some market for generic drugs. The changes have been the most profound in less developed countries with extremely limited budgets. Although generics are sold at a lower cost than their brand-name counterparts, the industry is financially robust, and dominated by multinational firms, some of which, such as Novartis, are actually brand-name manufacturers as well.[56]

1.21.10.3 Market Economics

Although generic drugs are chemically identical, they are typically sold at substantial discounts from the branded price.[48] A recent research article found that, with antiviral medications sold in the United States, the entry of a single generic reduced the price from \$60 per prescription to \$30, and the price was further reduced to \$4 with substantial generic competition.[57] In 2004, the average price of a prescription generic drug was \$28.74, while the average price of a brand-name prescription drug was \$96.01.[58] This represents potentially tens of billions of dollars in yearly prescription savings for consumers and insurance providers in the United States and around the world.

The reason generic companies can sell their products at substantially lower prices than brand-name manufacturers is that their costs for R&D are substantially lower. The generic drug development process can take 3–5 years,[48] as opposed to 8–12 years for a pioneer drug.[59] Generic product development starts with the purchase or production of the active ingredient of a brand-name drug and developing a similar formulation. This generally requires 6–18 months. Bioequivalence testing can take another 6–12 months, and FDA approval takes approximately 18–30 months.[60] The up-front investment for the typical generic development process was estimated to be on the order of \$1–2 million in the mid-1990s,[60,61] while out-of-pocket expenses for R&D of 'new' drugs (i.e., containing an active ingredient never before tested in humans) by pioneer drug makers have been calculated to have an out-of-pocket cost of \$400 million in the 1990s.[53]

1.21.11 Generic Drugs in the United States

1.21.11.1 Early Laws and Regulation in the United States

In 1938, the 'modern' era of drug regulation began when Congress enacted the Federal Food, Drug and Cosmetic Act (FDCA), which required premarket notification for new drugs. The FDA then created regulatory procedures requiring new drugs to be tested on humans for safety, in accordance with the Investigative New Drug Application (IND) regulations, and also that a New Drug Application (NDA) be submitted. This allowed the FDA to certify that any drug sold would meet certain safety criteria. The FDCA defined a 'new drug' as any drug not, 'generally recognized to be safe' (GRAS). The FDCA provided the FDA with the authority to disapprove a new drug, but not to affirmatively approve the drug. The degree of regulation for new drugs during this era was lax, and more so for copies of known drugs. Many drugs gained market access by simply claiming not to be new, or bypassed the FDA all together.[54]

1.21.11.2 Policy Shifts and Public Concern

Although not approved in the United States, the teratogenic drug thalidomide caused birth defects in other countries and increased public interest in drug regulations in the early 1960s. In 1962, the Congress amended the FDCA to require the FDA to approve new drugs. Consequently, FDA approval for NDAs required evidence of safety and effectiveness for all drugs activated between 1938 and 1962.[62] Moreover, the FDA concluded that generic drugs would require their own approved NDAs.[63] This greatly increased the time and money required for a company to introduce a generic drug to market, and was the first real barrier to the entry of generic substitutes.

In 1970, the FDA instituted Abbreviated NDAs (ANDAs) for the approval of drugs instituted prior to 1962. The ANDA required all information for an NDA, but no data regarding safety and efficacy testing.[63] The ANDA approval process relied on information from the National Academy of Sciences' Drug Efficacy Study Implementation (DESI) program for the approval process.[50] While generics of pre-1962 pioneer drugs benefited from the ANDA process, the FDA had no such process for post-1962 drugs. Beyond this limitation, all information for IND and NDA applications were regarded as confidential proprietary business information, and could not be disclosed by the FDA to the public or any competitor. As a result, post-1962 generics required independent animal and human tests to demonstrate safety and effectiveness.[62] This created significant time and cost investments for new generic drugs entering the market.

In 1981, the FDA began to allow some generic versions of post-1962 drugs to be marketed under a 'paper NDA' policy whereby FDA approval could be sought based on evidence derived primarily from published reports of well-controlled studies.[63] However, the FDA requirements to meet this definition were strict, as they estimated this information to be unavailable for 85% of all post-1962 drugs. Furthermore, the lack of clear specific guidelines, and excessive FDA discretionary power prevented any significant number of approvals from paper NDAs.[63–65] The legislation and litigation of the period from 1962 to 1984 made generic market entry difficult, and as a result only one in ten drugs had a generic equivalent.[66]

1.21.11.3 Defining Generics

Bringing a generic drug to market is not as simple as replicating the chemical compound of a pioneer drug and marketing it. Although generic drugs are identical in structure, or bioequivalent, they are still subject to regulations and approval processes that parallel their brand-name counterparts. Scientific problems such as polymorphs and bioequivalence require that generics undergo a fairly rigorous approval process. The FDA criteria for a generic drug have a significant impact on the approval process as well as the issuance of patents. To qualify as a generic, a drug must:

- Contain the same active ingredient(s) as the innovator drug. Inactive ingredients can, and often do, vary from those of the pioneer drug. As such, generics can vary in size, shape, and color from the pioneer.
- Be identical in strength, dosage form, and route of administration.
- Have the same use indications.
- Be bioequivalent.
- Meet the same batch requirements for identity, strength, purity, and quality.
- Be manufactured under the same strict standards of the FDA's good manufacturing practice regulations required for innovator products.[67,68]

How FDA Defines Equivalence for Generics

Pharmaceutical equivalence – Drug products are considered pharmaceutical equivalents if they have the same active ingredients, the same dosage form and are identical in strength, quality, purity, and identity as the brand-name product, but they may differ in characteristics such as shape, packaging, and excipients (e.g., colors, flavors, and preservatives).

Bioequivalence – A generic drug is considered bioequivalent if the rate and extent of absorption does not show a significant difference from the brand-name drug.

Therapeutic equivalence – A generic drug is considered therapeutically equivalent to the comparable brand when the FDA determines the generic is safe and effective, pharmaceutically equivalent, bioequivalent, adequately labeled, and manufactured in compliance with current regulations. (NB The concept of therapeutic equivalence, as used in the Orange Book (*see* Section 1.21.15), applies only to drug products containing the same active ingredient(s) and does not encompass a comparison of different therapeutic agents used for the same condition).[69]

The amount of active drug that a product provides to the site of drug action is known as bioavailability, and equivalence in bioavailability is referred to as bioequivalence. The critical concept for determining bioequivalence is absorption, which has two important pharmacokinetic features: the extent of absorption and the rate of absorption. The former is measured by comparing the area under the plasma concentration–time curve (AUC) after administration of the formulation with the AUC of an intravenous injection. Intravenous injection provides an extent of absorption of 1, since the entire dose reaches the blood or systemic circulation. The rate of absorption may be measured by determining the maximum (peak) plasma concentration (C_{max}) and the time taken to reach C_{max} (t_{max}).[68] Bioequivalence was defined in the Act as no significant difference between the rate and extent of absorption of the generic drug and the brand-name drug, later established by FDA to be a range of concentration of the active ingredient in the blood from -20% to $+25\%$.[70] Thus, for a generic copy, safety and effectiveness is established by showing that it will provide the same amount of the active ingredient at the site of action in the body at the same rate as the pioneer drug.[71]

1.21.11.4 Polymorphs

Polymorphs are different crystalline forms of the same pure substance according to the International Conference on Harmonization (under ICH Q6A). However, a drug substance can include crystalline forms, solvates, hydrates, and amorphous forms. Understanding that not every polymorph of an active ingredient will result in bioequivalence is important in understanding the need for scientific testing of generic drugs, even when they are identical in substance. Polymorphs can differ in terms of quality, efficacy, and safety due to differences in their physical properties, such as solubility and dissolution (bioavailability), or their chemical properties, such as reactivity. When characterizing polymorphs of active ingredients, researchers generally rely on x-ray powder diffraction, differential scanning calorimetry (DSC)/thermo-analysis, or microscopy and/or spectroscopy. To ensure drug safety, researchers must also contend with the problem of conversion once a drug is produced. The most stable polymorphic form should be used. This is easier to characterize when the form has been commercialized previously. When research is needed, dissolution testing is typically used to detect the potential for conversion, but in rare cases, solid characterization methods have been used.[72]

1.21.12 Generic Drugs in the European Union

1.21.12.1 European Generics

Generic markets in Europe have been largely ignored by consumers as direct-to-consumer advertising has been prohibited, and social insurance structures traditionally offered little to no pricing incentives for consumers to utilize generics. Furthermore, as many countries in the EU control the pricing of pharmaceuticals, the potential savings from switching to generic drugs was limited. For this reason the generics market in European countries did not develop as early as in the United States. For example, Italy introduced generics drugs in its market for the first time in 1996.[73] In recent years, however, from a government standpoint, generics have increasingly become an important measure in healthcare cost containment. Many European countries implemented regulations, such as reference-based pricing, or requiring pharmacies to offer customers generic versions of drugs, as a means to increase the use of generic drugs in their own markets. In 2004, generic drug sales accounted for 41% of the total volume of drug sales in Germany, 49% in the United Kingdom,[74] and in Eastern and Central European countries the percentage was as high as 70%.[51] As the pressure on national budgets in Europe encourages generic drug substitution, and less-developed nations seek out low-cost treatments for their citizens, the trend is predicted to continue in the immediate future.[75]

1.21.12.2 European Union Regulation

Historically, the EU has had a more complex structure with regard to generic drugs regulation than did the United States. Whereas the FDA in the United States supervises the generic market, the system was decentralized in the EU. More recently, the European Medicines Agency (EMEA) has adopted a central evaluation process for generic and innovator drugs involving a single evaluation that is carried out through the Committee for Medicinal Products for Human Use (CHMP). If the relevant Committee concludes that quality, safety, and efficacy of the medicinal product is sufficiently proven, a single market authorization is given, which can be used as a basis for market authorization in any EU country. The passage in 2005 of the New Medicines Legislation (NML) provided for the first time a clear scientific definition of generic drugs. NML also provides generic drugs with a 'Bolar exemption,' which allows them to begin R&D testing before patent expiry (based on the US court ruling discussed in Section 1.21.14.1). Many of the new laws, such as the Bolar exemption, closely mirror the evolution of the issues (and resolutions) the generic industry faced in the United States. Indeed, in recent decades, the regulatory structure of the EU and the US has begun to converge.

1.21.12.3 Reference-Based Pricing and Generic Market Creation

An important practice for the success of the generic drug market has been the utilization of reference pricing. Reference pricing is a practice whereby a new drug is priced based on its quality and characteristics relative to comparable or interchangeable drugs for a therapeutic class. This might entail increasing the co-payment for a patient who would rather use a brand-name drug than a generic equivalent. Although the practice officially began in Germany in 1989, the essential concept has existed in the UK and the United States Medicare program for many years.[76] Such programs shifted some of the cost difference between generic and brand-name drugs to the consumer, thus providing them with a financial incentive to use cheaper generic drugs. This process has been instrumental in the generic industry gaining an initial market foothold in countries where third-party reimbursement traditionally negated the incentive to utilize generics. It is now common for public and private insurers to use reference pricing as a means of encouraging generic substitution for their beneficiaries, although many may not call it reference pricing per se.[76]

1.21.13 Generic Drugs in Emerging Economies and Less Developed Countries (LDCs)

1.21.13.1 Essential Medicines and Less Developed Countries

In countries with very limited budgets for public and private healthcare spending, generics are of the utmost importance. To help LDCs accomplish the most for their limited budgets, the WHO maintains an 'Essential Medicines List (EML)', comprised of 319 drugs, which are the most efficient drugs designated for the treatment of priority illnesses from the public health perspective as well as the possibility of safe and effective treatment.[77] The vast majority of drugs on the EML are off patent, and available in generic form. More importantly, a study of valid patents on the EML found that, even when legally available, pharmaceutical companies usually did not seek, or enforce patents in LDCs. As such, patents did not create a financial impediment to access of 'Essential Medicines' for 98.6% of the listed drugs. The study suggests that this is the case because many of the drugs are older and not eligible for patent protection, as well as the fact that in many instances there is not a large enough market to warrant a patent. Moreover, in poorer countries where brand names are available, they are often sold at deep discounts.[52]

When dealing with the poorest of countries, the availability of generic drugs still has only a limited impact on public health. People subject to poverty, defined by the WHO as living on $2 per day or less (3 billion people), are incapable of purchasing drugs at almost any price. Even if a drug or vaccine were made available for $10 per treatment, it would be beyond the budget of many people.[79] Interventions to improve health in such situations tend to focus on improvements in the basic quality of living, such as water quality. However, in situations where a country has some funds to address epidemic or endemic illness, the approach taken often must be limited to interventions of a preventative or basic public health nature. For example, some health authorities confronting AIDS in Africa consider it most cost-effective to focus limited resources on prevention in the form of condom distribution, and disrupting mother-to-child transmission by single dose nevirapine and short-course zidovudine.[80]

1.21.13.2 Emerging Economies and Trade-Related Aspects of Intellectual Property Rights

In countries with large populations where there is a sizeable middle class, such as India, China, and Brazil, the generic drug markets have different characteristics than those of the poorest countries. In India, for example, pharmaceutical industries have been created that produce large volumes of generic drugs. India creates 20% of the world's drugs by volume, many of which are used within India or exported to poor countries that have no domestic pharmaceutical manufacturing capabilities.[81] About two-thirds of India's exports go to developing countries.[82] For the past 35 years, India did not recognize patent exclusivity; as such India would not only produce generic drugs, but would also reverse engineer new drugs and make them available at deeply discounted prices within India and other countries.[81] While this practice was a violation of international agreements for Intellectual Property Rights (IPR), many governments looked the other way because it greatly lowered the price of potentially lifesaving drugs. In the case of an antiretroviral triple AIDS therapy, an Indian company can make a drug costing $10 000 per year in the West available for $200.[83]

1.21.13.3 Trade-Related Aspects of Intellectual Property Rights and the Future of Generic Availability in Less Developed Countries

As a WTO member state, India is required to follow the Agreement on Trade-Related Aspects of Intellectual Property Rights (TRIPS). The agreement was passed in 1994, and went into effect January 1st 2005, and requires member countries to recognize patent rights for new drugs. The patents apply to new drugs, and as such, Indian production of cheap generics as detailed on the WHO's essential drug list are not affected by TRIPS. However, TRIPS is important for new drugs, or drugs which have yet to be invented, as well as incentives for drug manufacturers to sell to underserved markets. In India, a third of the population, over 300 million people, can afford to pay for drugs. This creates an incentive for drug companies to patent their drugs and focus on sales to this portion of the population.[81] A provision known as the 'August 30th decision' requires licensing between countries and stipulates that a country cannot be primarily focused on exporting drugs.[83] This may increase the difficulty a poor country has for importing drugs, and may dissuade India from serving these markets. Aside from TRIPS, it is possible that Indian drug companies will shift their focus from low-cost high-quality generic production for poor countries to production for export to richer markets such as the United States and EU.[81]

At present, it appears that Indian drug manufacturers are focusing on Western markets, creating partnerships, or attempting to sell the company itself to Western companies.[73] If there is a major shift in India's production focus away from poorer countries, which cannot produce their own drugs, they may face a shortage of reliable drug imports.

1.21.14 Current Regulatory Environment in the United States

1.21.14.1 The Bolar Decision and Government Policies

A final hurdle for a generic drug's entry into the United States market was the law against the use of patent-protected active ingredients by generics for premarket testing. The 1984 case of Roche Products Inc. versus Bolar Pharmaceutical Co., Inc. in the United States Court of Appeals for the Federal Circuit ruled that generic drug testing of pioneer drug formulations prior to patent expiration infringed on the patent.[64] Subsequent to this case, laws prohibiting preclinical testing before a patent expires have been called Bolar Laws. In countries with Bolar Laws, generic drug producers are required to wait for the end of a patent to begin clinical testing for their generic drug, thus effectively increasing the length of a patent by several years, and greatly decreasing economic incentives to produce a low-cost generic. For example, following the Hatch-Waxman Act of 1984, there was an average period of 3–4 years between the end of a patent and generic entry, versus 1–3 months after the implementation of the Bolar exemption.[66] During this era, healthcare and pharmaceutical costs rose steadily, and the United States government, through social insurance programs, as well as employee and military healthcare programs, largely shouldered these increasing costs. In Europe, where price controls were in place for pharmaceuticals, the Bolar Laws persisted until 2005 when the EU unanimously abolished them.[75]

1.21.14.2 The Hatch-Waxman Act

In 1984, the Hatch-Waxman Act established the current regulatory environment for generic drugs markets in the United States. The act was intended to strike a balance between the need for competition, and the need for incentives to match the substantial costs and risks associated with innovation in the drug industry. The Act amended FDCA to revise the procedures for NDAs, as well as Title 35 of the United States Code to authorize the extension of patents. This authorized a process for submitting ANDAs for generic copies of drugs that were deemed bioequivalent to the original approved drug. As an incentive for generic firms to challenge brand-name patents, Title I granted the first company to submit an application for a generic in a new market a 180-day exclusivity period during which the company would be the only generic sold.[62] As part of the dual purpose of lowering entry barriers for generics while increasing incentives for innovation, Congress stipulated a period of market protection of 3–5 years (depending on whether it was a new drug or new indication) following FDA approval, during which the FDA could approve no generic, regardless of the pioneer drug's patent status. Furthermore, in Title II Congress directed the restoration of a patent life for a pioneer drug for up to 5 years to compensate for the time lost to FDA regulatory review.[84] As previously discussed, Hatch-Waxman negated the *Bolar* court decision by adding Section 271 (e)(1) to the Patent Act.[71] Lastly, section III.e. of the Act required the FDA to update and make publicly available a listing of approved drugs, patents, and equivalence ratings on a monthly basis in what is known as the Orange Book.[70]

1.21.15 Food and Drug Administration: The Orange Book

Once approved by the FDA, all products – including prescription and over-the-counter drugs, both pioneer drugs as well as their generic versions – are listed in the FDA's *Approved Drug Products with Therapeutic Equivalence Evaluations*, also known as the Orange Book. Of the 10 375 drugs listed in the FDA's Orange Book, 7602 have generic counterparts, which account for over half of all prescriptions dispensed in 2004.[86] The listing of therapeutic equivalence evaluations designates a single reference listed drug to serve as the standard for comparisons of patented, and unpatented drug compounds, which generic versions must show to be bioequivalent, along with the basis for FDA's ratings as indicated by code letters (see Box below).[70]

The Orange Book also lists patents broadly defined as covering drug substances, drug products, and methods of use. Process patents are not permitted, unless they can be considered intermediate patents that fall within FDA's definition of an active ingredient. However, for a single drug there are typically multiple patents for various aspects of the drug, such as polymorphs, metabolites, pro-drugs, isomers, anhydrates and hydrates, as well as various salt forms.[55] If a generic applicant disputes the relevancy or accuracy of such a listed patent, the applicant can notify FDA, stating the grounds for this belief. Unless the NDA holder voluntarily agrees to withdraw or amend the information, the FDA will not remove the patent. At this point the FDA cannot act further, and the generic company must proceed through legal channels.

All generic drugs are rated 'A' or 'B.' 'A' drugs are considered bioequivalent to the brand-name original. They either have been demonstrated to be so by human bioavailability study (AB), or are considered inherently unlikely to have bioavailability problems (AA), typically oral solutions or oral drugs that dissolve in water. Other 'A' designations (AN, AO, AP, AT) refer to nonoral formulations considered bioequivalent by the FDA. Only 'A'-rated drugs are interchangeable with their brand-name equivalents. 'B' drugs have not been demonstrated to be bioequivalent by an in vivo test. These are generally older drugs that were approved by the FDA on the basis of chemistry, manufacturing controls, and in vitro dissolution tests. Less than 3% of marketed generic drugs have a 'B' rating.[70,85]

1.21.16 Patents and Market Exclusivity: Provisions of Hatch-Waxman

1.21.16.1 Section 505 and Paragraphs I–IV

Applications made under sections 505(j) and 505(b)(2) must include patent information and must notify FDA and patent holders of the application. Under Hatch-Waxman, generic drug applicants must make one of the four following certifications; essentially patent status notifications, regarding the pioneer drug when filing an ANDA:

- Paragraph I – the drug has not been patented
- Paragraph II – the patent has already expired
- Paragraph III – the generic drug will not go on the market until the patent expiration date passes
- Paragraph IV – the patent is not infringed upon, or is invalid.[67]

These clauses are commonly referred to by the paragraph, although they are technically subclauses. The common term, 'Paragraph IV' refers to the fourth, and most common subclause submission for a generic company entering a new market.

1.21.17 Generic Drug Regulation and Patents

1.21.17.1 New Drug Application and Abbreviated New Drug Application Regulation

An ANDA may be approved by FDA for a generic version of any 'listed drug,' i.e., a drug approved for safety and efficacy that is listed in the Orange Book, as long as all relevant patents and periods of market exclusivity for the pioneer drug have expired, or when existing patents are invalid, or not infringed upon. There are two basic situations in which an ANDA may be submitted. The first situation is where the generic version is the same as the pioneer version in all-material respects (i.e., a duplicate). The second is where the generic version is significantly different from the pioneer drug (e.g., a different route of administration, dosage form, or strength). In the second circumstance, the generic applicant must first submit to FDA a 'suitability petition' to justify why the difference between the drugs is not significant enough to preclude an ANDA, thus rendering new studies for safety and efficacy unnecessary.

In all other respects the regulations and requirements for an ANDA are the same as those for a full NDA.[49,52] The gray zone and an area of debate for the drug industry is 505(b)(2) NDAs, which pertains to modifications to existing drugs such that they are substantially different from the original drug, or intended to demonstrate new uses or indications for an existing drug. In the latter instance, safety and efficacy from previous data may be combined with other data. This gray zone is a contentious issue for pioneer and generic drug makers as the extent to which a drug is 'substantially different' or intended for a 'new application' can determine the degree of market protection awarded and increase profits for the pioneer drug company.

1.21.17.2 Paragraph IV and Evergreening

The drug market is a multibillion dollar industry and, as was mentioned in previous sections, the prices charged, and profits realized between an on-patent and off-patent drug are substantial. Competition can result in as much as a 94% decrease in price.[67] As such, pioneer firms have an incentive to evergreen, or protect the patent status of their drugs for as long as possible by repeatedly making new NDAs for a single drug. The original patent denies generic firms the chance to enter the market for a drug under a Paragraph I clause. If a generic drug company wants to compete in a new drug market under the provisions of Paragraph II or III, they will be forced to wait until all relevant patents expire. Evergreening can significantly extend the effective time of patent protection. Paragraph IV provision and, as such, most litigation and interest center around this clause.

An important provision of Hatch-Waxman related to Paragraph IV and patent abuse is the 30-month litigation period afforded to pioneer firms who sue generic firms entering the market under a Paragraph IV provision.

The 30-month Stay Period – To begin the FDA approval process, a generic applicant must (a) make a Paragraph IV certification, and (b) notify the patent holder of their submission of an ANDA. If the patent holder files an infringement suit against the generic applicant within 45 days of the ANDA notification, FDA approval to market the generic version is automatically postponed for 30 months. This postponement is negated if, before that time, the patent expires, or a court ruling determines the objection to be invalid.[67]

This 30-month postponement is intended to allow time for a court to decide the legality of a patent holder's claim against generic competition. Brand-name firms often use the process as a means of extending their patent, and accrue profits.[87,88] In fact, the brand-name sues the first-to-file generic company 72% of the time to gain this injunction, although the court finds in their favor 15% of the time, and out of court settlements are reached in 38% of cases.[84] In the past, these firms would 'stack' multiple postponements for various NDA filings they held to substantially increase their effective patent protection period while courts decided the viability of their claims. However, in August 2003, changes to the FDA review process for ANDA/505 (b)(2) filings, as part of the Medicare Modernization Act (MMA) blocked multiple filings, while still allowing a single 30-month postponement.[89] In the year following the enactment of the MMA (August 2003 to August 2004), the FDA recorded a 36% drop in Orange Book patent listings.[90]

1.21.17.3 Paragraph IV, Generic Exclusivity, and Authorized Generics

The 180-day exclusivity period is granted to the first company to submit an ANDA with Paragraph IV certification to the FDA. This gives exclusive rights to the company to sell their generic drug for 180 days. The start date of this period is calculated from the earliest date of either the first commercial marketing of the generic drug, or the date of the court decision declaring the patent on the pioneer drug invalid, unenforceable, or not infringed.[67] This is known as a triggering event.

This exclusivity period was intended to provide generic companies with a sort of 'mini-patent' to encourage them to invest in the required product testing and to mitigate expense of legal challenges to innovator products. The period accounts for a substantial portion of a generic drug's profit potential, as firms generally charge 70% of the brand-name, versus 30% when the exclusivity ends.[84]

The MMA closed several loopholes which the brand-name industry utilized, yet did not prohibit 'authorized generics' – the practice of having a brand-name company's subsidiary to sell a generic version of the drug during the exclusivity period. Thus the brand-name industry increasingly utilized authorized-generics in the wake of the MMA. The brand-name manufacturer produces the actual drug, although it is packaged differently, thus avoiding the requirement for an FDA ANDA approval. This practice allows the name-brand manufacturer to split the profits earned by a generic company during the exclusivity period, and capture part of the profits from generic sales. This practice decreases the financial incentive for generics to enter a new market, and in the long term will deter generic companies attempting a Paragraph IV entrance into a market. The generic drug industry has focussed on the issue and brought a lawsuit which the Supreme Court refused to hear in June of 2006, and the FDA has thus far refused to take action.[91,92]

1.21.18 Biogenerics and the Hatch-Waxman Act

In the mid-1980s, when Hatch-Waxman was implemented, biotech was still an industry in its infancy, and as such overlooked by the Act. The need for an approval process for generic versions was non-existent given the dearth of pioneer biological products. However, since the late 1990s the industry has experienced tremendous growth; by 2004, biotech products accounted for 27% of the active R&D pipeline, and 10% of global sales.[93] In 2004, the United States market for biologics reached $27.5 billion in revenues, representing nearly a three-fold increase over a 5-year period.[94] This is almost certain to increase as in 2003 there were 270 new biopharmaceuticals in the pipeline.[95] Worldwide, the percent of New Active Substances considered biologics rose from the mid-20th percentile throughout the 1990s to the low-30th percentile by 2004.

1.21.19 Public Health Services Act and Biogenerics

A provision in the Public Health Services Act (PSHA)[96] makes clear that the FDA has the ability to approve biologic products under the FDCA, as is the case with insulin and human growth hormones such as EPO.[56] As a means of dealing with biologics and their unique attributes the FDA requires a biologic license application (BLA), which is issued when an establishment and product meet FDA guidelines.[56] Clinical trials for biologics follow the same IND regulations, and in this respect are similar to other drug trials. However, the PSHA does not provide provisions or a process by which to approve a generic biologic.[56] In this respect, generic biologics are currently in a limbo similar to generic drugs prior to the Hatch-Waxman Act. As of yet, the FDA has no mechanism to determine the equivalency of generic biotech-based drug products and has taken few steps toward formalizing any such policy.[64,97] The complex nature of biologics, and their dissimilarity to traditional drug chemical structures has prevented the implementation of an ANDA process for therapeutic biologics. This is evidenced by the fact that in 2006 the FDA approved a therapeutic equivalent biologic, Omnitrope. However, the FDA explicitly stated that Omnitrope is not a 'biologic generic' but rather a 'follow-on protein product.' Both lawmakers and the FDA will continue to define a regulatory framework in the coming months and years.[84,98,99]

1.21.20 European Biotechnology Regulation

Although generic biologics are available in Europe and Asia, these governments and regulatory agencies are also debating the approval process for biogenerics, and the debate and degree of understanding of the biogeneric industry remains in the nascent stages. In Europe, some generics have been approved at the national level, prior to the creation of the European Medicines Agency (EMEA), which is the European Union's oversight organization for biogeneric regulation. The agency has gone well beyond the FDA in setting regulations, standards, and definitions for the industry.[100] The framework for regulations began in 2003 with the European Pharma Review. In Article 10, paragraph 4 (Directive), they began to form a structure for their policy. The article stated:

> "4. Where a biological medicinal product which is similar to a reference biological product does not meet the conditions in the definition of generic medicinal products, owing to, in particular, differences relating to raw materials or in manufacturing processes of the biological medicinal product and the reference biological medicinal product, the results of appropriate pre-clinical tests or clinical trials relating to these conditions must be provided ... The results of other tests and trials from the reference medicinal product's dossier shall not be provided."[95]

While this directive begins to form the regulatory environment it was not created with sufficient legislation or guidelines to avoid ambiguities for generic marketing authorization applications (MMA), or to define the extent of testing needed for biogenerics at the pre-clinical, and more importantly, clinical level.

1.21.21 Arguments For and Against Biogenerics

The Biotechnology Industry Organization (BIO), which represents innovation firms in the biotech industry argues that a generic biotech should undergo the same application process as for any biogeneric. This would mean that biotech would have no ANDA process for generic approval, and as such the industry would represent a significant barrier for generic competition. The argument is valid in several respects given the nature of the biotech industry. Primarily, substances created by biotech firms are much larger compounds, which are more complex and not as well studied as those in the pharmaceutical industry. The basis for biotech products is often organic tissue from humans or animals, which undergo an extensive process to reform the shape of the molecules. The process by which this is done is a trade secret which manufacturers are not required to make publicly available.[84] Moreover, procedures to test the efficacy of a biogeneric are not as well-defined, and animal testing is often ineffective.[84]

However, changes in technology, such as increased sensitivity of mass spectrometers, have to some extent helped to increase the accuracy of biotech comparisons. Thus far the EMEA has adopted a case-by-case attitude for deciding the extent of research necessary to establish bioequivalence.[99] For a recent biogeneric study of erythropoietin, a human growth hormone, the EMEA required a 600 patient clinical trial to be conducted for 6 months, a much less stringent approval process than was required for the pioneer drug. This study as well as others suggests that the biogeneric approval process will be far more stringent than the generic drug approval process, yet less demanding than the approval project for a new biologic. Even if there is a clear resolution on the approval process, the debate promises to continue on the issues of prescription guidelines and other more detailed aspects of the market.

The FDA's opinion on biogenerics has evolved over recent years, as has the public and scientific perception of the safety of biogenerics. Given the relatively new nature of the industry there is no traditional course of action, and no strong basis for future decision. At present, the FDA's public stance on biogenerics is guarded because of unresolved scientific issues regarding the safety of biogenerics.[100] In 2004, the FDA informed Sandoz International that while the company's application for the follow-on protein Omnitrope had no deficiencies, the FDA did not plan to approve the drug. Sandoz subsequently filed a lawsuit, which helped the drug ultimately gain approval.[101]

1.21.22 The Economic Question: Will Biogenerics Save Money?

As the current Congress has declined to take action on the issue of biogenerics, it appears as though any significant legislation will not come for several years. However, the actual amount of savings realized from biogenerics is uncertain. The increased costs associated with the research and production for biogenerics suggests they will not be as deeply discounted as traditional generic drugs. Whereas small molecule generics are discounted 40–80% observers predict that biogenerics will be discounted 10–20%.[100] The varying degrees of complexity for the approval of biogenerics will create a high level of variation in actual cost, and the approval mechanisms, which will eventually be in place, will certainly affect this. Recently, proponents of biogenerics have emphasized the return to taxpayers for public investments in biomedical research and the benefit of having multiple manufacturers for biodefense purposes as circumstances favoring biogenerics.[66]

1.21.23 Judicial, Regulatory, and Legislative Directions in the United States

The FDA Commissioner, the Pharmaceutical and Research Manufacturers of America (PhRMA), and the Generics Pharmaceuticals Association (GPA) have all voiced opinions that FDA regulation of biogenerics should not be undertaken without enabling legislation,[66,102,103] which leaves it in the purview of the United States Congress. There were two bills that addressed the issue legislatively, in the summer of 2002, and both failed.[104,105] In 2005, Representative Waxman stated that defining guidelines for testing generic biologics will be a high priority,[100] and Senator Hatch is also reputed to be working on legislation to permit approval of biogenerics.[98,100] The contours of what shape a compromise legislative proposal might take are beginning to emerge. Both advocates and opponents seem to agree that inherent differences between biologics and small molecule drugs make it impossible to apply the traditional generic model. What may be possible is to create categories of biologics that conceivably could be approved based on relatively less clinical data and those that would require much more extensive human testing prior to approval.[104] Much of this remains up in the air as the FDA has yet to finalize a guidance paper on biotech and biogeneric definitions and regulations, which is based on the International Conference on Harmonization (ICH-Q5E).[80]

The wild card in all the machinations surrounding biogenerics is what happens if (or as some believe, 'when') the various issues surrounding the topic go before the courts. Diverging court views indicate that the court is inclined to defer to FDA on 'technical' determinations such as safety and effectiveness, but strictly interprets the language of the statutes regarding periods of exclusivity.[63] As the case of Sandoz versus the FDA is set to begin, it will be interesting to observe how the courts handle the issue of biogenerics, since it involves decisions related to both bioequivalence determinations as well as interpretation of whether Congress intended or even contemplated extending Hatch-Waxman incentives to therapeutic biologics.

1.21.24 Innovation and Cost Containment: Striking a Balance

Although the new stream of drugs arriving on the market can be superior to previous drugs, new drugs are not always better than the old generic drugs they are competing against. A recent large-scale study of drugs designed to treat schizophrenia found older generic medicines were as effective as most newer and more expensive brand names.[106] This fact is underappreciated, as the approval process for drugs in most countries is not designed to compare the available treatments, or to highlight the respective strength and weaknesses for drug treatment options.[106] As cost containment pressures increase, it will become increasingly necessary for new drugs to study these variables to warrant inclusion in formularies, or even to receive approval by national regulatory authorities.

Generic drugs have been the focus of the quest for greater affordability, but not without equal concern that weakened patent protection could debilitate the financial constructs supporting future innovation.[71] Generics now represents a sizeable portion of the drug market, while private insurers and government officials continue to push for increased utilization, they must appreciate the tradeoff made, and the need for stability and predictability in the market. Likewise, the brand-name companies must adapt to the new market realities of limited patent protection, and increased cost competition from generic firms in the United States, and, increasingly internationally, from India and China.

In the American pharmaceuticals market the pendulum at one time strongly favored the brand-name industry. However, since 1984 and the Hatch-Waxman Act, this has changed quite drastically, and as a result more affordable generic versions of drugs are available for nearly any malady. Internationally the trend ran contrary, for decades drug producers in India did not recognize patents, and produced generic versions of drugs without regard for intellectual property. However, with TRIPS, India has mended its ways and now recognizes drug patents, and as a result innovative and imitative R&D have increased, and Indian generics are now available in the United States and Europe. In the biotech industry, generics are just beginning to struggle to gain a foothold, as patents expire and governments begin to define regulations. Despite the promise of an arduous legal process the market potential bodes well.

Competition between generic and brand-name drugs is desirable; however, creating the correct environment to ensure that one market does not dominate and destroy the other is an evolving and complex process. Balance can only be accomplished with clear and consistent healthcare policy, well conceived economic rational, and a stable, fair, and predictable regulatory environment. This must occur at the national and international level, as the drug industry has become truly global in scale. It will also necessitate government policies that do not sacrifice long-term goals in medicine and science for short-term gains in healthcare cost containment.

References

1. Brochure, Genetic and Rare Diseases Information Center, NIH publication number 02-5212. National Institutes of Health and Human Services, September, 2002.
2. Milne, C. P. *Pharmaceutical Industry Dynamics Decision Resources Report*; Spectrum Life Sciences: Walthan, MA, 2001.
3. Office of *Orphan Products Development*. http://www.fda.gov/orphan/faq/index.htm (accessed July 2006).
4. Milne, C. P. *Nature Biotechnol.* **2002**, *20*, 780–784.
5. Meyers, A. *Understanding the History of the Orphan Drug Act*. Proceedings of the Orphan Drug Development Conference, September 2000. http://www.rarediseases.org/news/speeches/orp_drug (accessed July 2006).
6. Alibek, K. *Perspective* **1998**, *9*. http://www.bu.edu/iscip/vol9/Alibek.html (accessed July 2006).
7. Sarett, L. H. *Clin. Res. Pract. Drug Regul. Affairs* **1984**, *2*, 97–107.
8. Haffner, M. E. Office of Orphan Products, FDA. Personal Communication, 2001.
9. Asbury, C. *Int. J. Health Serv.* **1981**, *11*, 451–462.
10. KIKO. *Status of Approval, Development, and Revocation by Year of Designation*. http://www.mhlw.go.jp/english/ (accessed July 2006).
11. Ashton, G. *Nature Biotechnol.* **2001**, *19*, 307–311.
12. US Food and Drug Administration. *Orphan Drug Regulations: Final Rule*, Dec 29, 1992, *57*. Fed. Reg62076.
13. Tufts CSDD. Impact Report, Tufts Center for the Study of Drug Development, 2001, *3*.
14. Knowledge Source. http://knowsource.ecnext.com/coms2/summary_0233-203_ITM (accessed July 2006).
15. Grabowski, G. G.; Vernon, J. *Pharmacoeconomics* **2000**, *18*, 21–32.
16. Peabody, J. W.; Ruby, A.; Connon, P. *Pharmacoeconomics* **1995**, *8*, 374–384.
17. Rohde, D. D. *Food Drug Law J* **2000**, *55*, 125–143.
18. Department of Health and Human Services (HHS), Public Health Service (PHS), Food and Drug Administration (FDA) Federal Register, Clinical Studies of Safety and Effectiveness of Orphan Products; Availability of Grants; Request for Applications. Notice, 1999 July 23; *64*, 40012–40016.
19. FDA. Grants Page. http://www.fda.gov/orphan/grants/awarded.htm (accessed July 2006).
20. User Fee Waiver. Federal Food, Drug and Cosmetic Act, Chapter VII, Subchapter C § 736 (739 h).
21. FDA. http://www.hhs.gov/news/press/2004pres/20040516a.html (accessed July 2006).
22. Mathieu, M. *New Drug Development: A Regulatory Overview*, 5th ed.; PAREXEL International Corporation: Waltham, MA, 2000, pp 377–378.
23. Love J. Unpublished work. http://www.cptech.org/ip/health/rnd/carrotsnsticks.html (accessed July 2006).
24. Arno, P. S.; Bonuck, K.; Davis, M. *Milbank Quart.* **1995**, *73*, 231–252.
25. Benzi, G.; Ceci, A. *Pharm. Res.* **1997**, *35*, 89–93.
26. Shulman, S. R.; Manocchia, M. *Pharmacoeconomics* **1997**, *12*, 312–326.
27. Rich, T. *Pharmacoeconomics* **1996**, *10*, 191–192.
28. Sheridan, C. *Nat. Biotechnol.* **2004**, *22*, 1203–1205.
29. *Scrip Magazine* 2004, 3016. SCRIP, Dec 24, 2004, 4–5, 3016.
30. Fischer, A.; Borensztein, P.; Roussel, C. *PLoS Medicine* **2005**, *2*, 845–848.
31. European Organisation for Rare Diseases (EURODIS), 2003–2004 Annual Report. http://www.eurodis.org/IMG/pdf/eurodis_activity_report 03–04.pdf (accessed Sept 2006).
32. Shiragami, M.; Kiyohito, N. *Drug Inform. J.* **2000**, *34*, 829–837.
33. Department of Health and Aged Care. http://www.tga.gov.au/docs/pdf/orphrev.pdf (accessed July 2006).
34. Therapeutic Goods Agency. http://www.tga.gov.au/ (accessed July 2006).
35. Pulsinelli, G. A. *Santa Clara Comp. High Technol. Law J.* **1999**, *299*, 221–240.
36. Shah, J. *Biotechnology* **2003**, *21*, 747–753.
37. Fogarty, M. HMS Beagle, 1998, *44*. http://news.bmn.com/hmsbeagle/ (accessed July 2006).
38. Haffner, M. E. *Drug Inform. J.* **1994**, *28*, 495–503.
39. Milne, C. P. Commission on Macroeconomics and Health. CMH Working Paper Series. Paper No. WG2: 9. 2001. http://www.cmhealth.org/docs/wg2_paper9.pdf (accessed July 2006).
40. Tambuyzer, E. T. *J. Biolaw Business* **2000**, *4*, 57–59.
41. Norris, J. A. *Orphan Drug Development in the United States*. Proceedings of Meeting on Drugs for Human Life, Rome, Italy, Oct 1986, pp 131–136.
42. Haffner, M. E. *Food and Drug Law Journal* **1994**, *49*, 593–602.
43. Lasagna, L. *Impediments to Vaccine Research*; Medicine in the Public Interest, Inc.: Boston, MA, 1984.
44. Lang, J.; Wood, S. *Emerg. Infect. Dis.* **1999**, *5*, 1195–1201.
45. MacLean, J. D.; Demers, A. M.; Ndao, M.; Kokoskin, E.; Ward, B. J.; Gyorkos, T. W. *Emerg. Infect. Dis.* **2004**. http://www.cdc.gov/ncidod/EID/vol10no7/03-0826.htm (accessed July 2006).
46. Evans, T. United Nations Development Programme – *Global Public Goods Financing: New Tools for Challenges*. A Policy Dialogue, 2000. http://www.undp.org/ods/monterrey-papers/evans.pdf (accessed July 2006).
47. Moran, M. A. *PLoS Medicine* **2005**, *2*, 828–832.
48. Office of Generic Drugs. Center for Drug Evaluation, Food and Drug Administration (Online). Available from URL: http://www.fda.gov/cder/ogd/index.htm (accessed July 2006).

49. IMS Health Incorporated. IMS Reports. 2005 (Online). Available from URL: http://www.imshealth.com/ims/portal/front/articleC/0,2777,6599_3665_71496463,00.html (accessed July 2006).
50. *Scrip Magazine* **2003**, *2824*, 18–19.
51. EMEA Overview (Online). Available from URL: http://www.emea.eu.int/htms/aboutus/emeaoverview.htm (accessed July 2006).
52. Attaran, A. *Health Affair.* **2004**, *3*, 155–166.
53. DiMasi, J. A.; Hansen, R. W.; Grabowski, H. G. *J. Health Econ.* **2003**, *2*, 151–185.
54. Meyer, G. F. *Transplant P.* **1999**, *31*, 10S–12S.
55. Rheinstein, P. H. *Am Fam. Physician* **1993**, 48, 1357–1360.
56. Lofgren, H. *Industry Project Equity,* Working Paper Series. 2002. Working Paper No. 10; *21 C.F.R.* § 601.2 (2004), p 21.
57. Wiggins, S.; Maness, R. *Econ. Inq.* **2004**, *2*, 247–263.
58. Gross, D.; Schondelmeyer, S. W.; Raetzman, S. *AARP Policy Institute* 2005 (Online). Available from URL: http://www.aarp.org/research/health/drugs/aresearch-import-869-2004-06–IB69.html (accessed July 2006).
59. Reichert, J. M.; Milne, C.-P. *Am. J. Ther.* **2002**, *12*, 543–555.
60. Findlay, R. J. *Food Drug Law J.* **1999**, *2*, 227–232.
61. Grabowski, H. Duke University. 2002. Unpublished. Available from URL: http://www.econ.duke.edu/Papers/Other/Grabowski/Patents.pdf (accessed July 2006).
62. Hutt, P. B. *Textbook of Pharmaceutical Medicine*, 4th ed.; BMJ Books: London, 2002, pp 653–698.
63. Abbreviated new drug application regulations (final rule). *Fed. Regist.* **1992**, *82*, 17950.
64. Behrendt, K. E. *Food Drug Law J.* **2002**, *2*, 247–271.
65. *Statement of the Pharmaceutical Manufacturers Association before the Subcommittee on Health and Environment of the Committee on Energy and Commerce. House of Representatives on H.R. 3605.* Washington, DC: Pharmaceutical Manufacturers Association, 1983, July 29.
66. Grabowski, H. Patents and Generic Competition. Duke University. Presentation March 28, 2001, pp 17968–17969.
67. *Drug Price Competition and Patent Term Restoration Act 1984. Public Law 98-417. In: United States Statutes at Large containing laws and concurrent resolutions enacted during the 2nd session of the 98th Congress of the United States of America: 98 STAT. 1585.*
68. Halbert, G. *Textbook of Pharmaceutical Medicine*, 14th ed.; BMJ Books, London, 2002, Chapter 2, pp 118–120. Codified at 35 U.S.C.§§ 1-376
69. Dickinson, E. H. *Food Drug Law J.* **1999**, *2*, 195–203.
70. Center for Drug Evaluation, Food and Drug Administration. *Approved Drug Products with Therapeutic Equivalents (The Orange Book)*, 22nd ed.; Rockville, MD: FDA, 2003.
71. Strongin, R. J. *George Washington University National Health Policy Forum Background Paper*, 2002.
72. Yu, L. X. Proceedings of the GphA 2002 Fall Technical Workshop, Bethesda, MD, October 15–16, 2002. FDA Perspectives on Polymorphism in ANDAs.
73. Ghislandi, S.; Krulichova, I.; Garattini, L. *Health Policy* **2005**, *72*, 53–63.
74. The European Generic Medicines Association. 2005 (Online). Available from URL: http://www.egagenerics.com/doc/fac-GxMktEur_2004.pdf (accessed July 2006).
75. In Pharm. Executive Summary (Online). Available from URL: http://www.inpharm.com/static/intelligence/pdf/European_Exec_Summ_(PDF).pdf (accessed July 2006).
76. Danzon, P. M. *Coleccion de Economia de la salud y Gestion Sanitaria.* Verlag: Barcelona, 2001; Chapter 5.
77. WHO List of Essential Medicines, 14th ed.; March 2005.
78. World Bank (Online). Available from URboL: http://web.worldbank.org/WBSITE/EXTERNAL/TOPICS/EXTPOVERTY/0,contentMDK:20153855~menuPK:373757~pagePK:148956~piPK:216618~theSitePK:336992,00.html (accessed July 2006).
79. WHO. 2005 (Online). Available from URL: http://www.wpro.who.int/NR/rdonlyres/588F3CB4-22F9-4543-A65E-BB447D8425AA/0/Regional_Strategy_for_Improving_Access.pdf (accessed July 2006).
80. Creese, A.; Floyd, K.; Alban, A.; Guinness, L. *The Lancet* **2002**, *359*, 2059–2064.
81. Department of International Development (UK), Health Systems Resource Centre. *Issue Paper* 2004.
82. Russell, A. Fact sheet. 2004 (Online). Available from URL: http://www.healthgap.org/tools/papers.html (accessed July 2006).
83. Médecins Sans Frontières. Will the lifeline of affordable medicines for poor countries be cut? 2005. External briefing document.
84. D. W.; Vukhac, K.-L.; Friedrich, A. M.; Lohman, D. M. *The Future of the Generic Drug Industry*; Banc of America Securities, New York, 2003; Chapters 1–3, pp 1–49.
85. Generic Drugs. *Med. Lett.* **2002**, *1141*, 89–90.
86. Generic Pharmaceutical Association (Online). Available from URL: http://www.gphaonline.org/aboutgenerics/factsabout.html (accessed July 2006).
87. Murphy, T.; Montana, S. *J. Biolaw & Business* **2003**, *3*, 73–78.
88. Brown, D. G. *J. Biotechnol.* **2003**, *3*, 199–208.
89. US Code Title 35, Part II, Chapter 14.
90. The Pink Sheet. **2005**, *14*, 16–17.
91. Abboud, L. *Wall Street Journal.* **2004**, *July 6*, C3.
92. *MEDADNEWS* **2005**, 7.
93. IMS Health. Press Room. 2005 (Online). Available from URL: http://www.imshealth.com/ims/portal/front/articleC/0,2777,6599_3665_71496463,00.html (accessed July 2006).
94. Mathews, A. W. *Wall Street Journal* **2005**, *September 14*, D4.
95. *Main Outcomes of the Pharma Review: After the Compromise between the Council and the European Parliament.* The European Generic Medicines Association, 2003.
96. *21 C.F.R. § 601.2 (2004).*
97. Beers, D. O. Generic and Innovator Drugs: A Guide to FDA Approval Requirements: 2003 Supplement. New York: Aspen, 2003, pp 74–76.
98. *The Pink Sheet*, **2005**, May 16.
99. Wess, L. *BioCentury, The Bernstein Report on BioBusiness* **2005**, *32*, 1–5.
100. Ainsworth, S. *Chem. Eng. News* **2005**, *23*, 21–29.
101. Media Release. **2005**, *September 13* (Online). Available from URL: http://dominoext.novartis.com/NC/NCPrRe01.nsf/0/a6e89e5832ca9-ce3c125707c0023d651/$FILE/FDA_Omnitrope_%20News%20release.pdf (accessed July 2006).
102. Biotechnology Industry Organization News (BIO News). *BIO Bull.* **2002**, *November 6*, 1–2.

103. Public Health Service Act 42. United States Code (USC) §§ 201 et seq.
104. Biotechnology Industry Organization. BIO letter to Rep. Michael Bilirakis, Chairman, Energy & Commerce Committee, Subcommittee on Health, 2001.
105. *The Pink Sheet*, **2002**, *41*, 7–8.
106. Abboud, L. *Wall Street Journal* **2005**, *September 20*, D1–D6.

Biographies

After graduation from veterinary school, **C-P Milne** practiced veterinary medicine in New Jersey and Maryland for several years. In 1987–88, he attended Johns Hopkins University in Baltimore, Maryland, where he earned a master's degree in public health with a concentration in epidemiology and health statistics. Upon graduation, he spent the next six years working for the New Jersey Department of Health, initially as a researcher in environmental health risk assessment, then as manager of a legislative and regulatory review program, and finally as Emergency Response Coordinator. In 1997, Dr Milne graduated from law school at the Franklin Pierce Law Center in Concord, New Hampshire, where he studied environmental and health law.

In 1998, Dr Milne became a Senior Fellow at Tufts Center for the Study of Drug Development (Tufts CSDD), which is located at the biomedical sciences campus of Tufts University in downtown Boston. Dr Milne's role at Tufts CSDD is to address legal and regulatory issues that affect the research and development of new drugs and biologicals. His current research interests include: challenges to the R&D of new medicines; incentive programs for pediatric studies; issues related to the R&D of treatments for serious and life-threatening illnesses (fast track diseases), rare disorders and conditions (orphan diseases), and neglected diseases of the developing world; and, the role of the clinical trials outsourcing industry. Dr Milne has lectured and published widely on these topics. Dr Milne is currently Associate Director of the Tufts CSDD and a licensed attorney in New Hampshire.

Louis A Cabanilla is currently working with Dr Milne to characterize national and international drug policy and regulation, and maintains the CSDD's databases on orphan and fast track drugs. He has also assisted in characterizing the WHO Essential Medicines List. He is currently researching drug development efficiency issues.

Comprehensive Medicinal Chemistry II
ISBN (set): 0-08-044513-6

ISBN (Volume 1) 0-08-044514-4; pp. 655–680

1.22 Bioethical Issues in Medicinal Chemistry and Drug Treatment

P R Wolpe, University of Pennsylvania, Philadelphia, PA, USA
M Sahl, MJSahl Consulting, Philadelphia, PA, USA
J R Howard, University of Pennsylvania, Philadelphia, PA, USA

1.22.1 Introduction

1.22.1.1 Bioethics

The discipline known as bioethics is the modern manifestation of the venerable field of medical ethics. Bioethics includes the categories traditionally known as medical ethics – the proper way to treat patients, ethical principles around death and dying, abortion, euthanasia, confidentiality, and so on. One characteristic that distinguishes bioethics from its forbearers, however, is the attention it pays to biotechnological solutions for health problems. From genetic medicine, stem cells, and biologics to brain imaging, artificial hearts, and other biomechanical treatments, bioethics grapples with the impact of our extraordinary technological virtuosity on the human body.

The field of bioethics is intrinsically interdisciplinary. Philosophers, social scientists, theologians, historians, and other disciplinary academics interact with lawyers, physicians, biologists, chemists, and other clinicians and scientists to try to understand the implications of biotechnological advances and to establish ethical guidelines that will inform treatment. Given the diversity of values in our pluralistic society, the ethical complexity of the issues, and the very different religious approaches to medicine and the body that are represented in Western Society, it is not surprising that there are many bioethical challenges where there is sharp disagreement over the proper ethical course.

The degree of disagreement should not be overestimated, however. Despite media and professional attention to the disputes – that is where the action is, after all – there is overall consensus on a surprisingly large number of ethical principles. The right of individuals to have autonomy over their bodies except under specific circumstances, such as incompetence or gross self-mutilation, the importance of informed consent, the fiduciary responsibility of clinicians to their patients, the corrupting effects of monetary inducements on clinicians to promote particular treatments, the nature of individual and institutional conflicts of interest, pursuit of equity and justice in access to treatment, and many other principles have been agreed upon, and their specifics have generally been worked out, even if they are not always actualized in practice.

Of course, the use of drugs in treating illness has long been an activity with ethical implications. Even the earliest medical codes of ethics discussed the dispensing drugs as a primary topic for ethical guidance: in the Hippocratic Oath, for example, the clinician vows never to give a person a deadly drug, even if requested, and not to give a woman an abortive remedy. On the other hand, the way drugs were tested and administered before the twentieth century seems almost casual to our eyes today. Prior to the Pure Food and Drug Act of 1906, for example, there were no consumer regulations about drug development, few research subject protections, and no review bodies such as the Food and Drug Administration (FDA) or institutional review boards. The latter half of the twentieth century, in contrast, has been a time of rapid development of regulations and regulatory bodies to ensure that both clinical research and clinical care conform to ethical and safety standards. Some aspects of that development are surprisingly late: for example, the Common Rule, the set of regulations that covers ethical standards for using human beings as research subjects in all federally funded research, was not finally codified until 1991.

In this chapter, we will look at some of the important ethical implications of drug development and treatment.

1.22.1.2 The Ethics of Drug Treatment

Attempts to prevent illness, treat illness, or otherwise modify the functioning or malfunctioning of the human body through ingestion are as old as humanity. In fact, it is almost certainly older than humanity; animals regularly use targeted medicinal ingestion to treat or prevent symptoms.[1] Some scientists have even postulated a continuity between animal and human medicine.[2] Traditional human societies have sometimes astonishing sophistication in their knowledge of local plants, and have in many cases worked out multistep processing to prepare plants and herbs for ingestion to treat illnesses.[3] Medicinal preparation and ingestion is an ancient and integral part of human existence.

The fundamental ethical acceptability of treating human conditions with drugs has itself rarely been questioned. Insofar as opponents object to the use of drugs to treat disease it is usually from within a broader religious belief in the exclusive province of Divine healing (e.g., Christian Science) or it is targeted at a certain type or class of drugs (e.g., psychotropics). Some also question the increasing emphasis on drug use, especially in Western societies, over less costly, more benign, or more socially acceptable techniques or technologies for treatment. However, the prudent use of ingestible, bioactive substances is itself almost universally accepted as an ethical means to alleviate human disease and suffering.

While the use of drugs to treat the human condition is itself ethically acceptable, a variety of ethical challenges arise as we try to treat real people with our imperfect remedies. Drugs are powerful substances with a great capacity to cause harm if misused or overused. Even when used correctly, pharmaceuticals can elicit disputes about their relative harms and benefits in specific situations. The establishment of modern intermediaries – physicians or pharmacists – in the allocation of the most powerful drugs increases the potential for conflict between those who control the resource and those who desire access to it. The size and influence of the pharmaceutical–industrial complex places disproportionate power in those whose interests lie in promoting and expanding pharmaceutical use in society. The expense of certain drugs complicates equitable allocation, and the concentration of pharmaceutical power in Western industrialized countries promotes research and drug discovery disproportionately for diseases that are prevalent in the wealthier nations. The increasing sophistication of drug action challenges the traditional model of using drugs as a means to treat pathological conditions and processes, and raises the specter of life style and enhancement uses of pharmaceuticals. New means of drug discovery, such as the use of stem cells, have elicited debate about the relative values placed on the status of the embryo and the potential treatments for intractable conditions that could result from stem cell research. Finally, the overall emphasis on drugs as the first line defense against what ails us has provoked some societal soul-searching: What messages do we want to send to ourselves and our children about how to solve problems and face difficulties in our lives?

Despite these concerns over drug use, Western industrialized nations are consuming pharmaceuticals at an unprecedented rate. The National Health and Nutrition Examination Survey found a 13% increase between 1988–1994 and 1999–2000 in the proportion of Americans taking at least one drug, and a 40% jump in the proportion taking three or more medicines. Forty-four percent reported taking at least one drug in the past month and 17% were taking three or more in the 2000 survey.[4] So, therein lies the paradox about ethical reflection on drug use: modern societies are taking more and more drugs even as they agonize over drug proliferation.

In this chapter we will explore a number of the more important ethical questions related to drug discovery, production, and distribution. Many of these topics are the subject of heated debate and disagreement. Thus, the goal is not so much to solve the ethical dilemmas they represent, or to provide a handbook of correct ethical action, but rather to clarify and illuminate the nature of these ethical concerns so that the reader can better reflect on his or her own values and judgments.

1.22.2 **Safety**

Drugs work by altering bodily function, which can lead to undesired results if they are taken inappropriately. A large number of drugs are toxic in nature, requiring careful dosing and monitoring when used therapeutically. Others have undesired effects if taken in the absence of a particular pathology or antigen. Safety issues include not only the side effects and toxicity of drugs, but adverse drug interactions, contraindications, off-label use, and black market trade in pharmaceuticals. It is also important to remember that in virtually every medical encounter, a value judgment is made about whether the side effects or toxicity of the treatment is outweighed by its positive therapeutic effects. Often, clinicians and patients differ about that value equation.

It is an ethical mandate that drugs be safe within the medical meaning of that term (i.e., if taken appropriately and with knowledge of possible side effects and toxicities). However, the standards for drug safety, the evaluation of drug safety, and the regulation of safe pharmaceutical development do not inherently raise ethical questions unless they violate some reasonable standard of ethical competence. A clear ethical issue does arise, however, in the attempt to avoid or minimize safety standards, especially in the pursuit of profits or market share. In a broader sense, there is also an ethical discussion to be had about taking the risk of ingesting drugs for life style or enhancement purposes, or under duress. Is it ethical to prescribe psychopharmaceuticals for people without diagnosed mental illness simply because they seek to alter their personality or sense of self?[5] Is it acceptable to make psychopharmaceutical treatment a condition of employment or release from jail? Is it ethical, for that matter, to mandate treatment for those facing trial or on death row in order to try them or to execute them?[6] Balancing the risk of life style pharmaceuticals with a growing consumer demand is one of the biggest ethical challenges of drug development in the coming decades (*see* Section 1.22.6.1).

Determining drug safety is complicated by the widespread off-label use of pharmaceuticals. Drugs developed and approved for a primary purpose are commonly used to treat other, sometimes unrelated, conditions.[7] For example, the American Pain Society, in their 1994 letter to the FDA, cited a number of drugs used for pain management in clinical practice that were not originally approved for that purpose such as antidepressants, anticonvulsants, corticosteroids, beta blockers, and amphetamines. While the FDA acknowledges the prevalence of such practices, their approach is generally to encourage the pharmaceutical industry to go through the proper channels of drug approval for new uses of drugs already approved for another purpose. It rarely happens that way, however, given the fact that these secondary applications of previously approved agents rarely have the 'critical mass' of potential patient volume to justify the time and expense for an additional FDA approval process.[8]

Recently, a number of high-profile cases have resulted in the removal of drugs from the market. Four examples are offered below, illustrating four different avenues whereby unsafe drugs come to be widely prescribed, but other examples could be offered. In these four cases, safety became an ethical issue when: (1) marketing or hype overrode safety concerns, (2) studies that should have been conducted were resisted or deemed proprietary and, therefore, concealed from scrutiny, (3) drug combinations were used without proper precautions, and (4) faddish treatments were prescribed without proper empirical support. A fifth case is then offered, that of the use of antidepressant selective serotonin reuptake inhibitors (SSRIs) in adolescents, to illustrate the difficulties of determining when the safety concerns of a drug outweigh its benefits. The purpose of discussing these cases is not to condemn the pharmaceutical companies involved or to impugn their motives, but rather to illustrate that the significant investment put into producing and marketing a drug, the complex nature of verifying its safety, and the vagaries of prescribing drugs in a large medical system can lead to undesirable outcomes if the highest standards of safety are not upheld. Error should always be on the side of public safety.

1.22.2.1 Cyclooxygenase-2 (COX-2) Inhibitors and Drug Safety Regulation

For nearly 40 years, chronic pain was treated with nonsteroidal anti-inflammatory drugs (NSAIDs). However, this class of drugs caused gastrointestinal side effects. In 1987, the idea emerged that selective inhibitors of COX-2 could relieve inflammatory pain without gastrointestinal side effects. Two large studies of the concept were published in 2000, the Celecoxib Long-Term Arthritis Safety Study (CLASS) and the Vioxx Gastrointestinal Outcomes Research (VIGOR) study. CLASS did not show COX-2 inhibitors to have a gastrointestinal protective effect and researchers believed this was due to the continued use of low-dose aspirin in tested patients. VIGOR prohibited the use of low-dose aspirin, which did indeed reduce the incidence of gastrointestinal lesions, but a serious problem emerged. There was a fivefold higher incidence of myocardial infarction in the rofecoxib (Vioxx, Merck) group than in the group treated with naproxen,[9] an NSAID.

Researchers did not test a placebo group, so it was unclear to them whether this adverse cardiovascular side effect was due to an actual increase in cardiovascular risk associated with rofecoxib or a protective effect of naproxen. While scientists were not completely sure, evidence suggested early on that COX-2 inhibitors could induce adverse cardiovascular events by disrupting the balance between two fatty acids, prostacyclin and thromboxane, that work together to control blood clotting. Despite these early warning signs, no randomized, controlled trials were implemented to follow-up on the proposed question of cardiovascular toxicity.[9]

Concerns about Vioxx's cardiovascular effects were serious enough for the FDA to warrant labeling changes in 2002. Merck voluntarily withdrew Vioxx from the market in September 2004 after their adenomatous polyp prevention on VIOXX (APPROVe) trial revealed that groups assigned to 25 mg of Vioxx daily for more than 18 months had a fourfold greater incidence of serious cardiovascular events than the placebo group.[10] In September 2004, Merck sponsored a trial examining whether the pill could prevent precancerous colon tumors and instead found that Vioxx doubled the risk of heart attacks and stroke in patients using it for more than 18 months.[11]

Similarly, the National Cancer Institute halted a trial of celecoxib (Celebrex, Pfizer) when an independent panel of cardiovascular experts reviewed the data and found a greater risk of cardiovascular events in treated patients. In 2005, Pfizer published a study of cardiac surgery patients stating that participants taking valdecoxib (Bextra) reported nearly three times the rate of cardiovascular events compared with those on placebo.[12] Parecoxib also showed an increased incidence of cardiovascular events after 30 days among patients who had received a total of only 10 days of COX-2 inhibition.[9] Ultimately, the increased cardiovascular risk appeared to be a class effect. The concerns raised by these trials echoed the problems that arose over 5 years earlier.

An epidemiologist in the FDA's Office of Drug Safety, David Graham, was the first in the agency to indicate the possible cardiovascular risk of Vioxx. He conducted a study using the database of Kaiser Permanente and warned his bosses that high doses of Vioxx significantly increased the risk of heart attack and sudden death. In November 2004, Graham testified before a US Senate hearing that Vioxx had caused between 88 000 and 140 000 excess cases of serious coronary heart disease and at least 26 000 deaths from heart attacks in the US alone in the 5 years before Merck withdrew it from the market. By the time Vioxx was withdrawn, though, an estimated 80 million people had taken the drug worldwide.[10]

The problems associated with COX-2 inhibitors emphasized the difficulty in associating a drug directly with the increased risk of common health disorders. While it is much easier to track rare adverse drug reactions after a new drug is introduced into the market, identifying an increased incidence of common events such as heart attack and stroke proves to be more problematic, especially when an adverse event could stem from an underlying disease. Epidemiologic studies of cardiovascular events are often not clear, making it difficult to determine the true risk associated with a particular treatment.[9] In the end, it seems that harms are extremely difficult to anticipate or measure in extent until many patients have been exposed. The FDA relies largely on a reactive reporting system, where doctors report possible cases of side effects that they believe to be associated with a drug. Thus, it is extremely difficult to detect dangerous side effects that already commonly occur in a population such as the heart attacks and strokes caused by COX-2 inhibitors.[11]

Monitoring the side effects of many types of drugs relies largely on pharmaco-epidemiological research like that performed by David Graham at the FDA. Some believe that there is a need to develop clinical databases that pharmaco-epidemiologists can use to investigate the possibility of adverse drug events. The UK General Practice Research Database is the largest database that incorporates full records of patient medical history. With over 2.5 million patient records, it proved successful in determining the risk of suicidal behavior in patients on antidepressants. The hope is that these types of studies could be extended to the US, where drugs are approved earlier and taken by larger numbers of people.[13] The FDA's Office of Drug Safety is responsible for monitoring and assessing the safety of existing drugs, but it has very limited funding for its own epidemiological studies. George W. Bush's proposed 2006 budget designates only 6% of the FDA's Center for Drug Evaluation and Research's (CDER's) budget for postmarket surveillance while new-drug review takes up about 80%.[11]

Many have criticized the FDA, saying the agency should have identified the problem and pulled COX-2 inhibitors from the market. Graham testified that the FDA's organizational structure and corporate culture were biased toward the approval of new drugs and that safety monitoring of drugs on the market was only a second priority. He further noted that reviewers at the Office of New Drugs who approve therapeutics have a vested interest in the drugs' success, often ignoring or overruling post-market safety concerns raised by those in the Office of Drug Safety. However, Janet Woodcock, former director of CDER, argues "risk and benefit are inextricably linked for any drug." She points out that drugs with risky side effects should not necessarily be denied approval, but rather their risks and benefits must be weighed up by the regulatory authorities who approve them and by the physicians and patients who use them.[11]

Many have argued that the FDA lacks the legal influence necessary for proper regulation of therapeutic drugs. After the FDA approves a medication, the agency cannot require a company to pay for and conduct postmarket safety studies. Nor can it limit the use of a drug to certain medical subspecialties as with medical devices. It does not have the explicit authority to change warnings on drug labels. Rather, it can negotiate label changes with manufacturers. Ultimately, the FDA does not aim to systematically monitor dangerous side effects.

The hype surrounding COX-2 inhibitors emphasized its advantage over NSAIDs. However, clinical trials have shown NSAIDs, aspirin, and acetaminophen to be just as effective in relieving pain as the COX-2 inhibitors. The moderate improvements on gastrotoxicity hardly seem worth the now proven increased risk of cardiovascular events.

1.22.2.2 Propulsid

The popular medicine Propulsid (cisapride) for the treatment of nighttime heartburn has experienced trouble ever since it hit the market in 1993. More than 100 patients taking Propulsid had already suffered serious heart problems by March 1998. The FDA ordered strengthened warnings for the drug in June 1998 after numerous reports of heart rhythm abnormalities and nearly 38 deaths. While the FDA could not prove the drug caused any death, it noted that Propulsid was already known to cause irregular heartbeat rhythm, or arrhythmias, when taken in conjunction with particular medicines. Warnings about the drug said that Propulsid should not be used in patients taking certain antibiotics, antidepressants, antifungals, protease inhibitors, or various other drugs. The drug was also contraindicated for patients with certain disorders such as congestive heart failure, multiple organ failure, kidney failure, and chronic obstructive pulmonary disease, which causes serious respiratory problems, and advanced cancer.[14]

Propulsid quickly became a popular drug for the treatment of acid reflux in infants, even though they are particularly at risk for adverse side effects and the drug was not systematically tested for infants. In 1995, Johnson & Johnson failed to receive approval for pediatric sales for Propulsid and the label did not recommend it for use in children. Doctors are free to prescribe medicines beyond the confines of labels and they insisted on having access to the drug, despite the side effects, because the label was confusing. By 1996, FDA officials warned the company that reports suggested pediatric patients were at greater risk for cardiac problems. Despite this, the company continued to finance programs that encouraged the drug's pediatric use. When federal officials warned Johnson & Johnson that the drug may have to be banned for children or withdrawn, the government and the company simply negotiated new warnings for the drug's label. Following this label change, Johnson & Johnson continued to promote Propulsid's use in children, resulting in 20% of babies in neonatal intensive care units being given the drug in that year.[15]

By 2000, Johnson & Johnson was forced to pull Propulsid off the market when it was linked to nearly 80 deaths and hundreds of heart attacks and the government threatened to publicize the drug's history of trouble. Janssen Pharmaceutical (a unit of Johnson & Johnson) agreed to pay up to $90 million to settle lawsuits involving claims that 300 people died and nearly 16 000 were injured as a result of taking their drug.[15]

Documents show that Johnson & Johnson failed to conduct safety studies urged by federal regulators and that company consultants could have easily revealed the dangers associated with the drug early on. However, the FDA also failed to disclose company research that placed serious doubt on Propulsid's effectiveness against digestive disorder, arguing that the studies were trade secrets. While Johnson & Johnson could not directly promote Propulsid for children without approval for pediatric use, FDA rules allowed the company to support educational efforts among doctors, and those programs tacitly endorsed pediatric usage. Physicians were ultimately never made fully aware of the FDA's concerns with Propulsid. In the end, the opinion of the FDA was that Propulsid was only minimally efficacious and therefore no safety risk was acceptable.[15]

1.22.2.3 Fen-Phen

In 1992, several published articles suggested that the combined use of phentermine and fenfluramine could result in significant weight loss when used over an extended period of time.[16] While fenfluramine and phentermine were individually approved by the FDA as appetite suppressants for the treatment of obesity, the FDA had not approved the use of the combination. Nevertheless, physicians began prescribing this drug combination, named fen-phen, for use in weight loss programs, and fen-phen became popular at commercial diet clinics. By 1996, the total number of prescriptions for fenfluramine and phentermine in the US exceeded 18 million.[17]

Following a study conducted at the Mayo Clinic reporting cases of cardiac valvulopathy in persons taking fenfluramine or dexfenfluramine, the FDA issued a public health advisory on 8 July 1997. The *New England Journal of Medicine* also reported that the two drugs increase the risk of pulmonary hypertension, particularly when patients receive high doses for more than 3 months. Valvular disease had been reported after exposure to serotonin-like drugs.[17]

This led to the voluntary withdrawal of the two drugs from the US markets on 15 September 1997. Several months later, the US Department of Health and Human Services recommended that all users of these pills undergo a medical history and cardiovascular examination. Practitioners were also encouraged to obtain an echocardiogram for all patients who had used these drugs before undergoing any invasive procedure for antimicrobial prophylaxis where endocarditis is indicated. This applied to most users because such prophylaxis is recommended even for common dental and oral procedures like teeth cleaning and fillings.[16] In 1999, American Home Products Corporation, makers of the fen-phen combination, agreed to pay $3.75 billion in compensation to thousands who used the drug before it was taken off the market.

1.22.2.4 Hormone Replacement Therapy in Menopausal and Postmenopausal Women

By the 1980s, the US population of postmenopausal women began to increase dramatically. With this increase also came a higher incidence of cardiovascular, neoplastic, and neurologic diseases among older persons. Estrogen replacement was seen as a panacea for postmenopausal women, thought to prevent coronary artery disease and delay the onset of Alzheimer's disease.[18] However, later studies began to severely question these claims.

Not a single controlled trial had ever shown that hormone replacement therapy prevented cardiovascular disease, stroke, Alzheimer's disease, or wrinkles, or that it was an effective treatment for depression or incontinence. The Heart and Estrogen/Progestin Replacement Study (HERS) failed to show a benefit of hormone replacement therapy (HRT) to women with heart disease. Despite this, pharmaceutical companies persuaded physicians that HERS actually showed a benefit. They argued that women with cardiovascular disease had dysfunctional cardiovascular systems that were unable to respond to hormonal assistance, but healthy women could still benefit.[19]

A study by the National Cancer Institute found that women who used estrogen only replacement therapy, particularly for 10 or more years, were at significantly increased risk of ovarian cancer.[20] Nonetheless, the risk of HRT is being minimized by rhetoric, with the hormone promoters trivializing HRT-caused breast cancers as 'better behaved' breast cancers. Furthermore, a trial conducted by the Women's Health Initiative (WHI) found that the risk of pulmonary embolism began to rise shortly after the trial began. Stroke risk increased after the first year, and invasive breast cancer rates rose after the fourth year. In fact, the WHI was stopped early due to treated women experiencing higher rates of breast cancer, cardiovascular disease, and overall harm.[21] One randomized controlled trial found that HRT significantly increased daily depression compared with pretreatment levels.[19]

The Estrogen Replacement and Atherosclerosis Study showed no benefit of either HRT or estrogen alone in preventing progression of atherosclerosis. While many clinicians are still convinced that HRT prevents Alzheimer's disease and improves mood, incontinence, and general well-being, randomized controlled trials have proved each of these claims to be false.[19]

1.22.2.5 Antidepressants and Suicidal Behavior in Youth

Antidepressants have been criticized for a lack of efficacy and poor side-effect profiles. The advent of SSRIs changed depression treatment strategy and they have emerged as the dominant pharmacological therapy for depression, in large part because they have been marketed as safe and effective.[22]

However, concern has been raised about whether SSRIs increases suicidality, especially early in treatment. Even small incremental risks can have serious implications for countries like the US, where there were 24.5 million patient visits for depression in 2001, a 70% increase over 15 years. Sixty-nine percent of patient visits for depression result in prescriptions for SSRIs.[23] Studies evaluating the existence of increased risk of suicide attempts often contradict one another.[24] More recently, however, the controversy has heated up over the use of SSRIs and the increased risk of suicidality in adolescents.

According to reports by the American College of Neuropsychopharmacology (ACNP), suicide is the third leading cause of death among 15- to 24-year-olds in the US. Depression and other psychiatric disorders are the major causes behind these suicide cases. Depression occurs in 10% of youth, with a majority of these cases going untreated and undiagnosed. Controversy over the use of SSRIs in youth arose when case reports began to arise in the 1990s indicating a chance of increased suicidal tendencies in patients (mostly adults) undergoing SSRI treatment. In 2003, regulatory agencies in both the US and the UK expressed concern over the treatment of depression in children and adolescents through the use of SSRI treatment and its possible link to increased risk of suicidal thinking or behavior for depressed youth under the age of 18.[22] The concerns culminated in a 2004 FDA Public Health Advisory warning of the possibility of increased suicidal ideation and behavior in children and adolescents being treated with antidepressant medications. The FDA directed manufacturers of all antidepressant drugs to revise the labeling for their products to include a boxed warning and expanded warning statement.[25]

Regulatory agencies in both the US and UK found data on the safety and efficacy of SSRIs in children problematic. Following analysis of 24 trials involving over 4400 patients, they found that greater risk of suicidality exists during the first few months of treatment. The average risk of these events on the drug was 4%, twice the placebo risk of 2%.[25] The UK Committee on Safety of Medicines (CSM) also found that the risk of suicidal thoughts and behavior was between 1.5 and 3.2 times more likely in young patients taking paroxetine (Paxil) compared to placebo.[26] Further studies on Paxil not only found increased rates of suicidal ideation and suicide attempts, but an expert working group also concluded that Paxil did not demonstrate efficacy in depressive illness in patients under 18.[27]

Findings that suggest serious risks associated with SSRI treatment in youth are largely based on unpublished studies, raising questions over the appropriate level of information a drug manufacturer should be required to disclose before an antidepressant is widely available on the market. A study conducted by researchers at the Center for Outcomes Research and Effectiveness at the University College London, the Royal College of Psychiatrists' Research Unit, and other organizations reevaluated the risk–benefit profile of five SSRIs used in the treatment of depression in youth. The study was based on both published and unpublished trial data on antidepressants including Prozac, Zoloft, and Paxil and revealed that increased risk of serious adverse effects and suicide-related events greatly outweighed the benefits of treatment for all the SSRIs except Prozac. The results of this study and several others flew in the face of studies of efficacy and safety based solely on published data, which indicate favorable risk–benefit profiles. In many instances, negative trials were simply never reported.[28]

However, it becomes difficult to determine if attempts at suicide are a manifestation of depression or a side effect of the drug. As Christopher Varley writes in *JAMA*, "Suicide attempts may occur as depression is lifting and an individual is energized enough to act on thoughts of self-harm. Since suicide is rare in children and adolescents, ascertaining whether there is a meaningful increased risk of suicidal ideation, suicide attempts, or suicide completion associated with any medication used to treat depression will require review of large numbers of patients."[27] The ACNP Task Force on SSRIs and Suicidal Behavior in Youth is quick to point out that suicidal behavior captured in adverse events reports did not include trials that set out specifically to determine whether medications led to suicidal behavior. Similarly, recent studies reported in *JAMA* note that the epidemiology of suicidal behavior in persons taking antidepressant drugs is not well documented by formal observational studies.[29]

The question remains whether the efficacy of newer antidepressants in childhood depression have exaggerated their benefits, while the adverse effects have been downplayed. Some researchers believe that the fact that serious adverse effects with newer antidepressants are common enough to be detected in randomized controlled trials already raises serious concerns about their potential for harm. As Jon N. Jureidini and others write in the *British Medical Journal*, "The magnitude of benefit is unlikely to be sufficient to justify risking those harms, so confidently recommending these drugs as a treatment option, let alone as a first line of treatment, would be inappropriate."[30]

1.22.2.6 Conclusion

Determining safety in drugs that have rare side effects is difficult, and it cannot always be done before a drug is brought to market. For that reason, reports of 'phase IV' adverse events – that is, adverse events reported postmarketing – must be followed-up vigorously. The exact moment that a rare adverse event becomes compelling enough to require withdrawal of a drug from the market is a judgment call, and well-intentioned actors can differ on when that moment has come. In many of the cases above, however, extenuating circumstances delayed a reasonable response. It is when self-interest obscures or trumps safety that an ethical breach has occurred.

1.22.3 Economics

The ethics of drug use in the modern world cannot be considered in isolation from its means of production. The modern drug culture is enormously expensive. The US pharmaceutical industry hit the quarter trillion mark in annual sales for the first time in 2004,[31] a figure that is growing at about 12% a year (down from a high of 18% in 1999). The economic impact of pharmaceuticals on healthcare policy such as Medicare and Medicaid, employee health benefits, and corporate profitability are profound. In addition, individual decisions about such things as marriage, employment, elder care, child bearing and rearing, retirement, and so on can be influenced by the need to access pharmaceuticals.

The cost of drugs, the institutional bodies created to manufacture them, and the strategies used to distribute them have both implicit and explicit ethical repercussions. In this section, we will look at drug profitability, diseases of affluence, and equitable drug distribution.

1.22.3.1 Pharmaceutical Profitability and Drug Discovery

Drug costs are the fastest growing part of the USA health-care bill, a fact that has brought much hand-wringing by those attacking and those defending the pharmaceutical industry. The elderly are particularly hard-hit; in 2002, the average price of the 50 drugs most commonly used by the elderly in the US was $1500 for a year's supply.[32] As health costs become a larger part of corporate and governmental expenditure, the pressure on big Pharma is increasing. Pharmaceutical companies are constantly faced with the conflicting pressures of research and drug discovery, public expectation, marketing and sales, and shareholder demand. For over two decades, as medicine expanded its scope, the drug industry has been the most profitable industry in the US, losing that status only in the last few years.[32] The recognition that brand-name drugs are more expensive in the US than in other markets has increased public resentment of pharmaceutical profits and power.

The pharmaceutical industry relies on a variety of strategies to maintain targeted levels of profitability. Industry consolidation coupled with increased shareholder demand for financial performance has created a competitive marketplace that takes full advantage of the industry's capabilities and resources. The industry has come to rely increasingly on blockbuster drugs, but the development of such drugs can be elusive. The industry has traditionally kept its profits robust by continually pursuing new molecules and searching for new indications and formulations for existing drugs. The process of discovering, testing, and releasing new therapeutic compounds is referred to as the pipeline. In general, the pipeline for new drug introduction is drying up; according to trade publication news, for example, 56 new drugs were introduced in 1996 compared with just 21 new drugs in 2003. In addition, many blockbusters of the past are about to go off patent, further draining the industry's potential for profit-making.[33]

In the US, drug prioritization falls to market forces, and the major pharmaceutical companies, as is the case for most industries, try to find the highest profit margins as an obligation to their shareholders. The pharmaceutical industry's concern over declines in profit is complicated by nonpharma competition looming on the horizon. The growth in fields like biotechnology, genomics, and even nanotechnology threatens to surpass the pharmaceutical industry in uncovering new approaches to therapy or disease management.[34] In response to these kinds of pressures, the pharmaceutical industry has changed its tactics. For example, there has been a significant decline in the development of fully innovative agents in the pharmaceutical industry. In its place, the industry has increasingly relied on stopgap drugs that offer only limited therapeutic advantage but do so for a large market niche, such as, minor improvements for Alzheimer patients, life-enhancement drugs, such as medicinal remedies to relieve dry mouth in patients with Sjogren's syndrome, life style drugs for relatively healthy people looking to enhance normal functioning (*see* Section 1.22.6.1), and me-too drugs, which closely mimic existing successful products.[35]

1.22.3.2 Me-Too Drugs

FDA approvals for uniquely new drugs, or new molecular entities (NMEs), which provide unequivocal advances in disease treatment compared to existing pharmaceutical remedies, have been declining. As mentioned above, the NME approvals numbered only 21 in 2003 – a startling contrast to the 35 NMEs receiving FDA approval in 1999.[36] The FDA classified three-quarters of the drugs it approved in 2004 as me-too drugs.[37] Me-too drugs can be highly profitable: Nexium, a me-too drug for stomach acid, has earned 4 billion dollars for AstraZeneca.

The increased emphasis on me-too drugs has ethical implications. Critics contend that the focus on me-too drugs detracts from drug discovery, privileges profits over therapeutic need, and duplicates effort rather than pursuing new cures or less well-served diseases. Pharmaceutical firms, on the other hand, justify the development of such drugs by arguing that me-too drugs improve on existing drugs by lowering side effect profiles or targeting drug-resistant patients, can increase compliance, and are usually cheaper than the prototype.[38,39] While new drugs that offer improved side-effect profiles and lower prices are clearly beneficial, the pursuit of me-too drugs has clearly contributed to the diminished number of NMEs. If me-too drugs are being pursued to increase industry profits at the expense of new drug development it clearly poses an ethical problem, particularly as pharmaceutical companies use the development of new drugs as a justification for high drug prices.

1.22.3.3 Antibiotics: An Example

There is a major global need for new antimicrobial agents, or new antibiotics. Development of new antibiotics has lagged, and there are few new antibiotics in the pipeline. In fact, of the 89 new drugs approved by the FDA in 2002, none were new antibiotics.[40]

New antibiotics are needed to address the increasing incidence of antibiotic resistance, especially in relation to the prevalence of infectious disease in developing countries.[41] Antibiotic needs can be classified according to the three groups of infectious diseases they target. The first tier of infectious diseases accounts for nearly 4 out of every 10 deaths from communicable infection, which include diseases such as tuberculosis, AIDS, and malaria. The second-level category represents neglected diseases (e.g., schistosomiasis, onchocerciasis, and leprosy), which leave their victims with lifetime burdens of illness such as blindness, mental retardation, and extreme physical deformities. Finally, the third group of infectious diseases seems to appear without warning, often in epidemic fashion, similar to the recent outbreak of severe acute respiratory syndrome (SARS).

The pharmaceutical industry has had enormous success in tackling many of these diseases in the past. More recently, big pharmaceutical companies have dedicated relatively few dollars to developing new antibiotics despite the growing need. Malaria, for example, has become a reemergent threat in Africa, Latin America, Southeast Asia, and parts of Eastern Europe as the result of gradual resistance to antimicrobial agents, which had been adequate for decades. In the case of tuberculosis, the industry has not developed any new compounds in more than three decades. This lack of on-going responsiveness has resulted in a failure to address the needs of new subgroups disease. It has exposed the failure to address populations that require shorter or fewer supervised doses for compliance conformity, that have built resistance to existing protocols, or that harbor latent strains of the infection that could initiate a new round of the epidemic.[42] The reemergence of tuberculosis in inner city populations in the US has emphasized the potential for crisis when immunity to older drugs develops in the absence of newly developed agents to take their place.[43]

There are a number of reasons for industry neglect of antibiotic development. Certainly, the industry's search for blockbuster drugs is one key reason for the decline in new antibiotic research and development. In addition, the antibiotics that have been developed are broad-spectrum drugs rather than targeted antibiotics, designed for the widest application and thus the largest potential market. However, by avoiding the additional cost and development time necessary to create bacteria- or condition-specific antibiotics, the industry has promoted drug resistance by encouraging the overuse of a small number of generalized agents.

Congress has made little effort to correct the situation, despite a long-standing history of targeted legislative rulings mandated to specifically encourage or discourage various pharmaceutical industry trends and behavior.[44] Innovative incentives will be needed to encourage the pharmaceutical industry to target communicable diseases, thereby supporting a more ethical balance in public health around the world. As the development of new antimicrobial agents has been waning, so has the intellectual pool of potential scientists interested in and trained with the necessary professional and technical expertise needed in this field of study.

1.22.3.4 Drug Pricing

The US stands alone among developed nations in its absence of a national policy to control escalating drug prices. The cost of drugs in this country can be twice that in Europe, and three times the cost of equivalent drugs in Japan.[43]

The pharmaceutical industry asserts that their research and development costs account for their need to recoup expenses through pricing margins. It is true that the costs of new drug development are substantial. However, drug companies are often not the ones incurring these R&D costs. The majority of new drug development is based on bench research funded by the federal government through grants to university research centers that the industry can then negotiate with over intellectual property rights.[43] Federal legislation, in fact, serves to underwrite significant dollars for pharmaceutical drug development and market distribution through tax subsidies and partial industry exemptions for patents, licenses, and intellectual property rights. The higher costs to wealthier nations are a de facto subsidy to poorer nations. While wealthier nations should burden more of the costs than poor nations, the present system accomplishes this without transparency and through arbitrary pricing determined by each nation's healthcare policy. It would be far better to set pricing policy in a rational, coordinated manner.

While economics can keep even lower price pharmaceutical agents from being utilized in developing countries, often the reverse is true of wealthier nations. When prescription medication expenses are covered by health plans, the true cost of drugs becomes less transparent to the consumer, giving way to the potential for overutilization.[45] The result is overutilization by wealthier countries who buy the drugs at the highest prices, and underutilization by poor countries who have subsidized prices. The pharmaceutical companies have addressed these issues through a variety of programs in specific countries and specific disease areas, and deserve credit for those programs. But there is still an enormous amount of randomness and profiteering in the system.

1.22.3.5 Regulation and Recommendations

The FDA is the oversight body responsible in the US for "advancing the public health by helping to speed innovations that make medicines and foods more effective, safer, and more affordable; and helping the public get the accurate, science-based information they need to use medicines and foods to improve their health."[46] Over the last decade, however, questions have been raised about the FDA's impartiality. The FDA has been accused of ignoring scientists' concerns about the dangers of some approved products on the market (e.g., Vioxx) and avoiding internal debates about drug safety factors. Critical discussions have been discouraged under the combined pressures of recent approval process acceleration mandates funded by the very (pharmaceutical) corporations whose efforts the agency is judging (i.e., the 1992 Prescription Drug User Fee Act).[47] Similarly, the UK's National Health Service has recently been accused of lapses in their drug approval and monitoring role, and in allowing the pharmaceutical industry's influence to sway providers' and consumers' increased reliance on medication.[48]

The example regarding the lack of new antibiotic development efforts is indicative of a broader issue of ethical concern underlying the motives and rationale often driving the direction of pharmaceutical R&D. In the case of antibiotics, the Infectious Diseases Society of America (IDSA), a professional organization dedicated to policy, advocacy, education, and practice guidelines addressing the health impact of infectious diseases, has put forth a number of recommendations that could be applied to R&D for any number of otherwise narrowly focused agendas. Other leaders in this field have suggested measures such as better business models as well as different regulatory approaches. The recommendations include, but are not limited to[40,41]:

1. congressional re-casting of industry-based statutory incentives and/or disincentives for drug development prioritization, such as R&D tax credits, patent extensions, antitrust exemptions, and liability protections;
2. Congressional and National Institutes of Health (NIH) accommodations for targeted intellectual property rights;
3. expanding the NIH's role in sponsoring research to generate otherwise neglected agent needs;
4. expanding the NIH's role in encouraging the translation of bench-to-marketplace research;
5. creating an empowered independent national commission charged with setting drug discovery priorities;
6. creating a not-for-profit drug company that could work in tandem with the pharmaceutical industry and provide a noncompetitive setting for industry scientists on sabbatical to work on needed global projects. The home pharmaceutical company would receive tax incentives for their temporary leave;
7. FDA's acceleration of published guidelines for less market-popular clinical trials and drug review status;
8. restructuring the FDA review process and decision-making efforts to ensure greater public transparency;
9. requiring FDA approval of combinations of drugs (i.e., combination therapy) that target a smaller patient population, rather than approval of singular broad-spectrum agents for combating infectious diseases. Institute approval of individual pathogen-specific antibiotics, in an effort to delay the emergence of antibiotic resistance;
10. re-organizing the antibiotic approval process to evaluate applicants not only for drug safety and efficacy in relation to existing drugs on the market, but for resistance to immunity – a more difficult and costly process, but one with better long-term outcomes;
11. increasing funding and grant support directed at less profitable drug discovery;
12. supporting joint public–private ventures aimed at more widely focused needs;

13. banning new antibiotics from widespread use in healthy animals; and
14. encouraging multinational drug firms to contract their manufacturing to lower- and middle-income countries that could then benefit from both the medical and economic advances available on their own soil.

1.22.4 Allocation

1.22.4.1 Equitable Distribution of Drug Resources

At the start of the twenty-first century, one-third of the international community lacked access to critical pharmaceutical drugs. Within the world community, particularly in the impoverished regions of Africa and Southeast Asia, populations are without life-saving pharmaceuticals that are readily available to wealthier neighbors.[49] Communicable diseases continue to devastate Third World countries, further widening the health gap and increasing the tremendous inequities for poor people in developing nations. About 14 million children under the age of 5 in these regions die each year. Ninety-five percent of childhood deaths occur in developing countries. Seventy percent of those fatalities are caused by infectious diseases for which vaccines are elsewhere available.[50] Unfortunately, the pharmaceutical agents needed to treat these diseases are either unaffordable or unavailable to the populations in developed countries.[51]

The global expenditure on drugs in 2004 was $550 billion. Yet 88% of that half-trillion dollars was spent by North America (248 billion), Europe (144 billion), and Japan (58 billion).[52] The burden of disease coupled with disparities in access to pharmaceuticals makes for a stark picture:

- In 1998, 6.1 million people in tropical Third World nations died of malaria, tuberculosis, or acute lower respiratory tract infections. That figure does not include the additionally high number of AIDS deaths in those regions for lack of access to and/or funding for AIDS drugs, which could run upwards of $15 000 or more per AIDS victim per year.[43]
- By 2004, that annual number grew to more than 10 million infectious disease-related deaths estimated to have occurred in developing countries for lack of safe, inexpensive essential drugs.[53]
- Acute lower respiratory tract infections alone claim over 3.5 million people a year, representing the third highest cause of death in underdeveloped countries. The youngest citizens of these nations, i.e., the children, are most vulnerable.[43]
- From 1975 to 1997, only 1% (13 of 1223) of new drugs developed by multinational pharmaceutical corporations were targeted to treat communicable diseases that devastate tropical regions. The balance of new pharmaceuticals primarily targeted 'diseases of affluence' or conditions resulting from overconsumption.[43]
- 50–90% of drugs in developing and transitional economies are paid for out-of-pocket.[53]
- Less than one in three developing countries has fully functioning drug-regulatory authorities.[53]
- Antimicrobial resistance is increasing for many major infectious diseases including, but not limited to, bacterial diarrhea, gonorrhea, malaria, pneumonia, and tuberculosis.[53]
- $1 billion is the World Health Organization's estimate of the cost to reduce by half the 1.1 million annual deaths caused by malaria. That amount is also estimated as Pfizer's 1999 revenue from sales of Viagra.[54]

1.22.4.2 Diseases of Affluence

The term 'diseases of affluence' has been defined as "diseases which are thought to be the result of increasing wealth in a society."[55] It refers to the association of a malady with a particular level of social class. Examples of such conditions include obesity, heart disease, type 2 diabetes, some cancers, high blood pressure, allergies, autoimmune diseases, asthma, alcoholism, depression, and other psychiatric disorders. These illnesses stand in sharp contrast to the infectious communicable diseases more common to Third World countries.[55] Different disease patterns require alternative forms of treatment, which results in different patterns of drug consumption in various regions of the world.

The relation of disease trends to class is not a new concept. During the early years of the twentieth century, health professionals were faced with the growing epidemic of poliomyelitis, and found that it did not follow the traditional path where infectious diseases were most prevalent in unsanitary environments. In fact, children living in unclean, poverty-ravaged slums at that time had a better chance of developing a certain protective level of immunity to polio, which struck at least as heavily in middle-class neighborhoods where cleanliness was the norm.[56] Today, chronic conditions tend to be more common in developed, industrialized nations that have a surplus of assets and resources, which encourage high-fat diets of processed foods, discourage adequate levels of health-benefiting physical activity, and accommodate for significant rates of tobacco, alcohol, and substance abuse. The dietary trends of developed nations also contribute to diseases of affluence in their use of refined foods that lack the necessary nutritional content needed to maximize overall population health.[57]

The pharmaceutical industry has focused its research and development overwhelmingly on diseases of affluence, such as chronic conditions and life style enhancements.[40,58] Worldwide, money invested in health-care research has grown dramatically over the past two decades. In 1986, the Commission on Health Research for Development estimated an annual global health research expenditure of approximately $30 billion per year. Ten years later, the WHO's Ad Hoc Committee on Health Research Relating to Future Intervention Options put that figure at $55 billion for 1992. The most recent estimates, from the Global Forum for Health Research for the year 2001, suggest a global health research amount of nearly $106 billion.[59,60]

Though significant in dollar values, the distribution is highly skewed. Over the past 20 years, one concept has remained fairly stable – that of the 10/90 gap. The 10/90 gap describes the imbalance between research expenditures and disease prevalence, with less than 10% of the world's health research dollars applied to diseases that represent more than 90% of the world's health challenges – most of which are concentrated in developing nations.[59] Evidence of the imbalance has been recently documented in an evaluation of published research in the professional literature. The analysis of eight selected disease categories, prevalent in either industrialized or developing regions, indicates an enormous range of variation in research investment. At the top of the range, an estimated $9.4 billion is spent per year on cardiovascular research, yet only $0.3 billion annually on research for malaria. This divergence demonstrates the emphasis currently placed on diseases of affluence to the detriment of research for communicable diseases.[61]

1.22.4.3 Global Burden of Disease

A Global Burden of Disease Study, conducted in the early 1990s, arrived at three broad categories of cause-of-death diseases[62]:

- Category 1: communicable/perinatal/maternal/nutritional,
- Category 2: noncommunicable diseases (NCDs), and
- Category 3: injuries.

In developing nations, mortality rates for Category 1 diseases outweighed that of the developed nations by almost 17-fold. The study found, however, that poorer countries were catching up with wealthier nations that in 1990 with NCD rates rising to just under half that of developed countries. Thus, along with economic development comes a shift in disease emphasis.

Forty percent of the 2001 global burden of disease was represented by four communicable diseases among the top 10 leading causes of death worldwide (**Table 1**). Those diseases included lower respiratory tract infections, HIV/AIDS, diarrheal disease, and tuberculosis. Another 40% of the top killers were NCDs, such as ischemic heart disease, cerebrovascular disease, chronic obstructive pulmonary disease, and pulmonary cancers. Developed countries are significantly more burdened by those conditions falling under the label of diseases of affluence. Poorer nations, while plagued by infections to a much greater degree, are also beginning to experience the burden of western life style illnesses.[62–64] It is now becoming apparent that what had once been considered diseases of affluence, concentrated in industrialized nations, are now impacting the morbidity and mortality rates in developing countries, in addition to the hardships of infectious communicable diseases.[65] Contrary to public opinion, NCDs are now responsible for more global deaths annually than the so-called infectious diseases of Third World nations.[62]

1.22.4.4 Epidemiological Transition

The term epidemiological transition, which reflects the parallels between evolving economies and disease patterns, now suggests that chronic diseases, specifically cardiovascular disease, represent emerging threats in the less developed regions of the world. As such conditions increase in prevalence, the corresponding workforces and economies will feel the impact. Stemming the tide of this impact requires acknowledgment of the basis for the change as well as innovative proposals to minimize its toll.[66]

The same may be true of mental illness. For example, depression, while not in the top 10 leading causes of death, is being diagnosed and treated at increasing rates in more developed countries, as are drug dependence, anxiety, and compulsive behaviors. As a result, antidepressants are one of the key target areas for the pharmaceutical industry's profitability.[67] Mental illness also represents a significant 11% of the worldwide burden of disease, and approximately

Table 1 Top 10 leading causes of death: global, developed, and developing regions (version 2) estimates, 2000 (from Mathers, 2002; adapted from WHO sources)[168]

Rank	*Cause of death: global*	*% Total deaths*	*Rank*	*Cause of death: developed countries*	*% Total deaths*	*Rank*	*Cause of death: developing countries*	*% Total deaths*
1	Ischemic heart disease	12.6	1	Ischemic heart disease	23.3	1	Ischemic heart disease	9.2
2	Lower respiratory tract infections	11.1	2	Cerebrovascular disease	13.4	2	Cerebrovascular disease	8.4
3	Cerebrovascular disease	9.6	3	Trachea, bronchus, lung cancers	4.4	3	Lower respiratory tract infections	7.9
4	Chronic obstructive pulmonary disease (COPD)	4.7	4	Lower respiratory tract infections	3.6	4	Perinatal conditions	6.0
5	HIV/AIDS	4.6	5	COPD	3.2	5	HIV/AIDS	6.0
6	Perinatal conditions	4.5	6	Colon and rectum cancers	2.3	6	COPD	5.2
7	Diarrheal diseases	3.6	7	Self-inflicted injuries	1.8	7	Diarrheal disease	4.6
8	Tuberculosis	2.9	8	Diabetes mellitus	1.7	8	Tuberculosis	3.6
9	Road traffic accidents	2.2	9	Stomach cancer	1.7	9	Malaria	2.7
10	Trachea/bronchus/lung cancers	2.1	10	Hypertensive heart disease	1.7	10	Road traffic accidents	2.4

1% of annual deaths around the globe. Although it is generally more prevalent – or better recognized – in developed societies, it is becoming increasingly significant in developing nations as well.[68]

On the other hand, some conditions are still disproportionately characteristic of developed nations. Obesity has not historically been categorized as a disease; therefore, treatment options covered by health insurance plans have been limited. In 2004, the US Centers for Medicare and Medicaid Services (CMS) reversed its tradition of denying Medicare coverage for obesity-related clinical care in light of the growing epidemic of overweight seniors and the impact that it has, in turn, on further escalating the cost of chronic care. According to the National Obesity Forum in Nottingham, UK, obesity itself can be the fundamental cause of numerous other illnesses and conditions, including cardiovascular disease, hypertension, stroke, hyperlipidemia, diabetes, cancers, osteoarthritis, respiratory disorders, sleep apnea, fertility problems, Alzheimer's disease, depression, and other psychological disorders.[69,70]

Many risk factors that serve as precursors to chronic diseases may not differ markedly from those present in developed countries. A combination of genetics, life style behaviors, and comorbidities represent the lion's share of determinants. For the poorer nations, however, additional factors may come into play. The ability to correct life style behaviors (e.g., smoking, sedentary existence, etc.) represents a clinical challenge for the medical community. Prevention is the preferred approach through avenues such as primary carebased patient education, population-based health promotion programs to manage otherwise wrongly directed societal trends, and political and economic policies that cross over more than just health-care boundaries to underscore positive behavioral priorities for individuals, local governments, and corporations.[66]

The People's Republic of China provides one clear example of the epidemiological transition model. Generally classified as a developing country, infectious diseases took the lives of many Chinese citizens before reaching old age. Over the last 30 years, however, China's economic improvements and urbanization have opened the way for greater longevity and the growth of an aging population. In turn, morbidity rates now reflect the prevalence of more chronic and degenerative diseases found in Western societies, or diseases of affluence.[71] Currently (Category 2), NCDs account for about two-thirds of China's burden of disease.[68] Nevertheless, both older and new forms of infectious diseases are also beginning to plague that nation.[71]

Finally, evolving population pyramids reflect the changes in a developing population's longevity and fertility trends. Third World children who manage to survive their early experiences with serious infectious diseases could be at increased risk and vulnerability to NCDs in adulthood.[62]

1.22.4.5 International Policy and Regulatory Efforts

The uneven worldwide distribution of health-care assets, resources, and critical pharmaceuticals results in poor health for large populations of certain countries and regions. This, in turn, practically assures these regions dire states of poverty. Good health, in other words, is a fundamental requirement for a society to be productive economically and otherwise. Conversely, poverty supports the exacerbation of morbidity and mortality in undeveloped countries, whereby populations can either afford medicines that may be available, or are at a loss when proper agents are nonexistent.[70,72] The devastating cycle of poverty and poor health feeds upon itself.

Options are available to tighten the reins of pharmaceutical pricing growth and development focus. Federal negotiations with the industry for intellectual property rights and reasonable pricing can consider the public policy advantages of quid pro quo agreements to help strengthen the drug industry's social conscience in lieu of shareholders' profits.[43] In addition to mandated parameters for patented drug pricing, proposals for the industry's embrace of voluntary, albeit limited, licensing flexibility to aid affordability for poor nations would demonstrate good will.[73]

Over the past several years, the UK has established a Commission on Intellectual Property Rights to inform the government and citizenry regarding the pharmaceutical industry's role in preserving the fundamental human right to health through socially responsible development of essential drugs as a global public good.[74] In fact, embracing a stronger position of corporate social responsibility may serve the pharmaceutical industry's bottom line as well. As the spotlight on the industry's behavior shines ever more brightly, benchmarks are being applied, such as applicable international codes and guidelines developed by World Health Organization (WHO) and Organization for Economic Co-operation and Development (OECD), to measure their public 'goodness.' Being shown in a poor light could be undesirably costly for industry firms and their shareholders.[73]

Pharmaceutical companies need to anticipate the world's shifting patterns of diseases in order to protect the public from unlikely new strains of old plagues. Concern has been expressed, for example, that remaining unmonitored samples of the poliovirus previously employed in the development of the vaccine may find their way into the mainstream population and quickly spread.[75] Other potential threats to global health include a pandemic reemergence of pharmaceutical-resistant tuberculosis, as well as an outbreak of yellow fever in highly urbanized areas. Cholera has become pandemic for seven different outbreaks, and meningitis has reemerged in a new strain not treatable with existing vaccines. Finally, the influenza epidemic of the 2004–05 flu season hit the US in a climate of unpreparedness that should have been predictable.[76]

International cooperation is needed to ensure the production, availability, and adequate distribution of essential drugs determined to be crucial to public health in a variety of international settings. The term essential drugs refers to a compendium of pharmaceutical agents first compiled by the WHO in 1977, and regularly updated thereafter, to represent the basics for optimal treatment of a variety of conditions found in a population. The WHO's 1977 list included 208 drugs believed to provide "safe, effective treatment for the majority of communicable and noncommunicable diseases."[77]

In 2003, the Trade Related aspects of Intellectual Property Rights (TRIPS) and Public Health Decision initiatives laid groundwork for developing nations to avail themselves of necessary essential medicines more economically. Licenses issued to regulate importing and exporting mandates will increase the supply of patented pharmaceuticals released after 2004–05 (depending on the country of origin) at reduced pricing for poor nations. Market forces will still apply in an arena of negotiated prices that would encourage pharmaceutical manufacturers to reach agreements with developing countries at more economically feasible levels for the poor.[73] The Hastings Center is currently conducting a 2-year (2004–06) review of the relationship between TRIPS and access to beneficial biological materials, with the goal of developing recommendations for international pharmaceutical policy.[49]

1.22.4.6 Ethical Implications

According to a WHO Working Group on medicinal R&D priorities, adequate levels of pharmaceutical agent research are currently ongoing for communicable diseases such as HIV/AIDS and other sexually transmitted diseases. Perhaps not surprisingly, these are diseases that are prevalent in Western societies as well as in developing nations, though their toll on health and life is far greater in developing regions such as Africa. Infectious disease in general receives attention in the developed world, as there is general recognition today that modern transportation renders all infectious diseases as global threats. Still, many infectious conditions no longer generally plague industrialized nations. Diseases that are prevalent predominantly in developing countries, such as tuberculosis and malaria, tend to be neglected by the research community. In addition to these more well-known illnesses, others such as trypanosomiasis, Chagas' disease, and leishmaniasis are foreign to economically established countries and receive scant attention in pharmaceutical

R&D.[78] A 2004 study of the relationship between published research on randomized clinical trials (RCTs) of pharmaceutical agents and the global burden of disease found that nearly half of the top 40 leading causes of death were not among any of the published RCT research articles.[79]

Clearly there needs to be more attention paid to the global burden of disease. Pharmaceutical companies are, of course, only one player in a complex set of institutional and political dynamics that create barriers to solving the problems of disease in the developing world. Solutions must be pursued at local, regional, and international forums, and should include nongovernmental organizations (NGOs), international relief agencies, and charitable foundations, for example, as well as national and international political entities.

Still, the level of investment in drugs that offer marginal benefits or for which there are already existing, efficacious drugs, while millions in developing nations suffer from diseases that could benefit from those funds, is ethically unsupportable. The economic factors are vexing, but ultimately unconvincing; an effective vaccine or affordable treatment for malaria or tuberculosis would be profitable, and there are avenues to subsidize R&D into less common maladies. For example, newer approaches to funding the development of necessary drug agents have partnered private and public organizations. In one instance, the pharmaceutical giant Novartis has joined with the government of Singapore to establish the Institute of Tropical Diseases. The Institute's initial goals will entail tackling remedies for dengue fever and drug-resistant tuberculosis.[80] Direct donor gifts of philanthropy are also making a difference. The Bill and Melinda Gates Foundation has provided $168 million in grant funding for treatment and research targeting malaria. Overall, the Gates Foundation has contributed over $3 billion to global health research.[81]

Today's international health inequities can be minimized through the equitable development and distribution of crucial pharmaceutical research agents. The increasing influence of the global economy coupled with the growth of population migration and international travel have created a climate of multinational interdependence that is difficult to ignore.[50] As a consequence, the health and productivity of all communities represents the vested interest of each nation in one another. As the pharmaceutical industry has strengthened its multinational influence, its presence – or absence – is keenly felt on all continents of the world. Encouraging signs such as the partnering of pharmaceutical companies and other interested parties are the proper first steps in battling the devastating toll disease takes on the developing world.

1.22.5 Mental Illness

The proper role of drugs in treating mental illness has been a topic of ethical concern at least since the second half of the nineteenth century and the issue periodically captures public attention. According to some medical historians, the use and distribution of psychotropic medication can be divided into three basic periods.[82] The first period was marked by the introduction of morphine in the late 1800s and concluded in the synthesis of barbital in 1903. The development of barbiturates characterized the second period, which took place during the first half of the twentieth century, through the 1949 discovery of the therapeutic effect of lithium for the treatment of mania. The third period commenced with the development of the first set of modern psychotropic drugs in the 1950s.

Corresponding to the development of more powerful and specific drugs in the second half of the twentieth century was a marked decrease in the number of inpatients at mental hospitals. Deinstitutionalization resulted in an increase in outpatients whose medical management consisted primarily of medication.[83] As a result of deinstitutionalization, effective medication, and effective education efforts on the part of psychiatry, public fears and misunderstanding of mental illness began to dispel to some degree.[84] By the 1970s, as pharmacotherapy became the primary treatment modality in psychiatry, the expectations for psychotropics and neuropharmacology were quite high. The emerging field of biological psychiatry was predicted to enable psychiatrists to understand the pathophysiology of mental illness. In addition, psychopharmacology was expected to generate the feedback needed to develop more effective and selective pharmacological treatments.[85] In the late 1980s, fluoxetine (Prozac), an SSRI, hit the market, and within 3 years it became the most highly prescribed antidepressant in the world.

1.22.5.1 Treatment

In terms of prescription practices, there are five principal symptoms for which psychoactive drugs are most commonly prescribed: inability to cope, depression, anxiety, sleeplessness, and pain.[86] However, these common complaints may be secondary to other problems. The point at which these symptoms become severe enough to warrant medical treatment is somewhat arbitrary and varies among patients and doctors. The rate and type of psychotropic drug usage also varies greatly across the globe. While the US and Canada are the most prominent consumers of amphetamine-type stimulants, primarily methylphendate, amphetamine, and various anorectics, many European countries consume a notably high amount of benzodiazpine-type hypnotics, sedatives, and anxiolytics.[87] The blurry line between prescription of psychotropics for diagnosed mental illness and for complaints that do not fulfill criteria for mental illness makes it difficult to separate out medical and life style usages of psychotropics from treatment for recognized mental disorders.

WHO estimates that 450 million people suffer from a form of mental or brain disorder, including alcohol and substance abuse disorders.[88] The proportion of those suffering from mental and brain disorders on a global level is projected to rise to 15% by 2020.[89] In most industrialized countries, psychopharmaceuticals are prescribed for most mental illnesses, with or without additional or ancillary therapies.

There are significant ethical concerns regarding the use of many psychotropic medications. First, mental illnesses can be chronic, requiring many years or a lifetime of treatment. Yet, the side effects of psychotropics can be quite significant over the long term. Physicians are responsible for assessing the risks of, for example, long-acting neuroleptic medications, which include recovery time from adverse effects, lag time in building therapeutic dose, psychological disturbances due to injection, increased risk of neuropleptic malignant syndrome, tardive dyskinesia, and other extrapyramidal symptoms. Clearly, the introduction of atypical antipsychotic agents such as clozapine, risperadone, and ziprasidone (among others) has produced notable improvement in the treatment of mood disorders, some behavioral disorders, and other forms of mental illness.[90,91] While these drugs are less likely to cause the involuntary movements that are so problematic with the older antipsychotic drugs, they still have numerous side effects, including chest pain, high blood pressure, agitation, confusion and memory loss, sleep disturbances, and others. Mental illness is debilitating and involves significant suffering, yet clinical care requires a careful consideration of the long-term consequences of psychotropic drugs.

In general, therapeutic decisions should be based on the clinician–patient relationship, accurate assessment and diagnosis by the clinician, and careful consideration of the various treatment options, including expected benefits and risks. Longer acting (or depot) drugs, which release slowly into the bloodstream after being injected, have recently been introduced. They can help to solve the problem of adherence to regimens in those with mental illness. According to one study, serious mental illness adherence rates are at about 50%.[92] The consequences of not adhering to medication include exacerbation, rehospitalization, major disruptions in relationships, loss of employment, and even loss of housing or involvement in the criminal justice system.[93] However, depot drugs can also complicate medical decision-making. As there is a large difference in the half-life of some depot medications compared to older medications, it takes longer to eliminate side effects. Administration of long-acting medication changes a patient's timeframe for reversing a treatment decision. Further, long-acting medication can sometimes be used to treat an individual with a mental illness against his or her will. These ethical concerns, as well as the societal value of individual choice, are responsible for the lower than expected use of long-acting antipsychotic medications in the US.[94] Similarly, the introduction of surgically implantable medication delivery systems can provide psychiatric patients with uninterrupted access to medication for up to 14 months. Implanted under the skin, these systems can be removed, allowing reversibility. While some patients find such technologies more desirable because they simplify drug-taking, others report feeling controlled by such technologies.[95]

With the increased use of psychotropics, has also come a more lax attitude toward them among the general public as well as some clinicians. Administration of psychotropic drugs has become more casual, and their distribution more common. Increased use has been criticized by some who credit it to such things as "uninformed prescribing, inconsistent or lax prescribing; willful and consistent misprescribing for misuse; self-prescribing and self-administration" caused by "inadequate training; shortage of information; lenient or lax attitudes; lack of a sense of professional responsibility; unethical behaviour; personal drug addiction; criminal behaviour or direct financial interest."[86] However, the increased use of psychotropics is also part of an expansion of the definition of mental illness and the greater receptivity of both the public and practitioners for the general management of mood and cognitive states with drugs.

1.22.5.2 Decision-Making, Competence, Informed Consent

Informed consent has become the legal and philosophical cornerstone of physician–patient relationships. It is a key factor for all drug treatment, though here we will discuss it in relation to mental illness. The legal basis of the doctrine of informed consent is also a philosophical one, in that persons have a right to privacy and bodily integrity, and that, generally, only the person can decide whether to be treated and in what manner. The issue of competence revolves around an assessment of whether an individual has the mental capacity to make an autonomous decision, to understand the facts, to appreciate the consequences of making a particular decision, and to be consistent in their decision-making about it. Competence, however, is a legal term, and is binary; one is either competent or incompetent. Since the ability to perform a certain action, such as making a decision about medical care, is task-specific, the notion of capacity is preferable to the more global notion of competence. In this sense, even an individual who has been declared legally incompetent with respect to his or her financial affairs, for example, is not necessarily so incapacitated that he or she cannot make a decision about how much medication can comfortably be tolerated. Even when it has been determined that an individual is incapable of making a particular decision, protestations against continuing even indicated, standard treatment (let alone innovative or investigational therapy) should be taken seriously and weighed carefully against the expected benefits and the amount of time required to manifest them. Alternative approaches should be considered under these circumstances, considering likely benefits to the patient and the inherent cost of forced treatment to the patient's well-being. The values underlying informed consent, especially patient self-determination, have also become reference points for identifying acceptable alternatives when a patient lacks the capacity to agree to a treatment or test.

In the mental illness consumer-movement, self-determination has become the guiding principle. Practitioners should respect patient autonomy unless there is a compelling reason against doing so.[96] Some even argue that if the patient is capable of expressing their wishes regarding treatment, those wishes must always be respected regardless of seeming irrationality.[97] Mental capacity is the main issue in balancing patient autonomy and practitioner responsibility to protect the patient from harm,[98] but capacity remains an extremely vague and controversial standard.

Some suggestions to mitigate the capacity problem have included such things as 'Ulysses contracts' or 'psychiatric living wills.' These are documents that are written when a person with mental illness is stable or in remission, and empowers another person to act on their behalf when they become mentally incapacitated, even if they are legally incompetent. For example, a person with bipolar disorder may empower a loved one to coerce them to take drugs when they are first entering a manic phase, a point at which people are generally feeling very good and will reject the idea of taking medication. Such documents are problematic, however, if they require action while a person is still technically competent; if a person is legally competent right now and refusing treatment, by what authority does a document written before have precedence over the individual's decision now? Such a document is legally unenforceable. However, once the individual becomes impaired enough to require a surrogate decision-maker, a psychiatric living will can be a useful expression of their desires.

The Interaction Model of Client Health Behavior (IMCHB) assumes that: "(1) clients are capable of making informed, independent, and competent choices, (2) those choices are affected by client singularity and by client-provider interactions, and (3) clients should be given the maximum amount of control feasible."[99] There is a noted link between the medication adherence of persons with mental illness to their relationship with their provider. Including patients with mental illness in decision-making also helps in the decision-making process, increases self-perceived competencies, promotes the team-player role, teaches specific skills, empowers the client, and recognizes, addresses and even sometimes overcomes system-based barriers.[100] Patient involvement in health-care decision-making has led health providers to shift the focus from simply alleviating symptoms to helping patients adapt to a life with chronic mental illness. Effective health-care requires informed, active, and independent clients who participate in determining the treatment goals, monitoring symptoms, evaluating the regimen, and in revising regimens.[99]

1.22.5.3 Competence and Consent in Research

Informed consent is not only important in clinical care, but in a subject's competence to make a decision about research participation. Berg and Appelbaum[101] have outlined four main standards for determining decision-making competence based on a framework developed by Appelbaum and Grisso.[102] The first standard, most widely used by courts and legislatures, is the ability to communicate a choice. Many potential subjects fail to reach this standard because they are unable to coherently communicate, whether due to chronic schizophrenia, various levels of consciousness due to psychotic episodes, or other disorders. The ability to communicate choice does not necessarily translate into the capacity to make a choice autonomously. Comatose, mute, catatonic, or severely depressed persons, individuals with manic or catatonic excitements, and persons with severe psychotic thought disorders or severe dementia will fall under this category.

A second standard used is the ability to understand relevant information. This understanding means that the potential subjects have the ability to comprehend the concepts involved in the informed consent disclosure. Understanding itself is not enough; if a person understands the information, but is not able to retain the information long enough to make a decision, they are not competent enough to consent. Impairments of intelligence, attention, and memory can all affect this ability. A third standard is the ability to appreciate the nature of the situation and its likely consequences. The subject must be able to apply the information to his or her own situation. Denial, delusions, and psychotic levels of distortion can all impair this ability. The final standard is the ability to manipulate information rationally. It is necessary that a potential subject possess reasoning capacity and the ability to employ logic to compare the risks and benefits of the treatment options.

Recently, psychiatric research has been scrutinized out of fear that individuals with mental disorders are at a higher risk of being exploited due to the effect of mental illness on decision-making ability. Patients with schizophrenia, for example, experience delusions, apathy, lack of insight, and impaired memory and mental flexibility, which can impede informed consent to research.[103] Subjects with psychiatric disorders have often failed to appreciate the nature of research and its possible impact on their treatment. Two multinational studies found that a diminished ability to appreciate that one is ill is prevalent among those with schizophrenia. Patients with schizophrenia or severe depression also had an impaired ability to appreciate the potential value of treatment.[101] The MacArthur Treatment Competence Study[104] found that subjects with schizophrenia were more likely to completely deny the presence of illness than were depressed patients (35% versus 4%). Most dementia patients lack the ability to recognize that they have a memory problem. This lack of ability to recognize their mental state could indicate an impaired ability to relate the research to their own situation.

Alzheimer's disease can also affect decision-making.[105] While many cognitive functions account for impaired decision-making abilities in Alzheimer's disease, neuropsychological measures of executive dysfunction were the best indicators of impaired decisional ability in Alzheimer's patients.[106] The issue of competence in elderly patients suffering from dementia has become increasingly significant as the clinical research on Alzheimer's disease is rapidly accelerating and aiming to develop methods of early detection and prevention. Dr Scott Kim and his colleagues write in the *American Journal of Psychiatry* that more subjects with relatively mild illness will begin to be invited to participate in therapeutic and nontherapeutic Alzheimer's research. Thus, there will likely be a large range in the ability of individuals to give consent, with some "not capable even while they maintain their 'social graces' and their expressive abilities."[107] In assessing the competency of patients with Alzheimer's disease under various legal standards, researchers have found that even mild to moderate Alzheimer's disease has a significant impact on treatment consent capacity.[108] Researchers do, however, reinforce the ethical principle that diagnosis of dementia does not necessarily imply incapacity.[109] Therefore, distinguishing capable from incapable Alzheimer's patients remains a considerable challenge.

Unlike research involving other populations considered vulnerable, such as children, prisoners, and fetuses, no additional federal regulations specifically govern research involving potential subjects who are cognitively impaired. The recommendations of the National Commission for the Protection of Human Subjects are similar to those made with respect to children.[110] Many psychiatric researchers, and a number of patient advocacy groups, have argued that many patients with mental disorders still possess substantial decision-making capacity. They believe that providing additional regulations for the mentally ill is paternalistic and that it would reinforce a social stigma about the disease and further impede research.

1.22.6 Emerging Issues

1.22.6.1 Enhancement and Life Style Drugs

Human use of ingestibles to achieve mind-altering effect predates recorded history. Paleoethnobotanical evidence suggests that the Middle Paleolithic Shandinar Neanderthals may have used *ephedra altissima* to obtain amphetamine-like effects as early as 50 000 BC, in what is now modern day Iraq.[111] Around 3000 BC the Sumerians in the southern Mesopotamia planted poppies and extracted its juice, which they called 'lucky' or 'happy,' an indication that they utilized it for its mood-brightening properties.[112] The use of fermented grapes was not only an early human discovery, but its importance is written into early records of human societies and embodied in religious rituals that persist to this day. Both Western and Eastern medicines traditionally recommended eating particular foods to induce proper cognitive as well as physical health. Nineteenth century America was particularly enamored of developing nutritional philosophies of health, from the botanical medicine of Samuel Thompson to the bland diets developed by Will Kellogg (corn flakes were invented as a bland breakfast to avoid stirring up adolescent passions) or Sylvester Graham (whose now-famous cracker was designed toward the same ends as corn flakes). If the use of ingestibles to try and change the human mind is ancient, so undoubtedly is the moral debate about the degree to which the activity is acceptable.

Though the attempt to change human functioning with food and drugs is ancient, the power to do so was rather limited until recent pharmaceutical advancements brought enhancement and life style drugs to the forefront of drug development and public debate. Enhancement and life style drugs will, for our purposes, refer to pharmaceuticals that change a human function in a desirable way in the absence of pathological processes. However simple that definition seems, it is fraught with problems – problems with differentiating pathological processes from natural ones, differentiating food from drugs, and differentiating activities we tend to think of as medical from those we categorize differently.

Life style drugs have therefore been defined in a number of ways, including drugs that: (1) alleviate or enhance life style problems or conditions, regardless of the cause,[113] (2) address health problems for which the underlying cause is in the realm of personal responsibility,[114] (3) address nonhealth problems,[115] or (4) improve general well-being.[116] It is how a drug is used, rather than its medicinal chemistry, that classifies it as a life style drug.[114] Though the parameters of the definition may be fuzzy, the increasingly selective alteration of our cognitive and affective states through neurochemical alteration promises more frequent, and specific, use of drugs by those who desire to improve a function that is already within the normal range.

The demand for life style drugs has been fueled by advances in neuroscience and neuropharmacology. For example, better understanding of the pathophysiology of depressive disorders and the discovery of SSRIs has led to safer and more effective treatments for mood disorders. New drugs with ever more selective actions on the neurochemistry of mood, anxiety, attention, and memory are under development. New agents acting on entirely new pathways and targets are in the research pipeline and offer immense potential in the treatment of neuropsychiatric disease.[115]

In the wake of consumer demand, pharmaceutical companies have increasingly begun to focus on lucrative life style drugs, such as remedies for impotence (Pfizer's Viagra), baldness (Merck's Propecia), weight reduction (Abbott's Meridia), and facial wrinkles (Allergan's Botox).[117] The 2003 life style drug market was estimated at $23 billion.[118] In 2003, the top selling drugs in the US, by therapeutic class, were cholesterol-lowering drugs, followed by drugs for treating heartburn, anemia, thyroid conditions, elevated blood pressure, and depression.[36] SSRI drugs are now the second largest selling class of drugs in the US, with over 146 million total dispensed prescriptions written in 2005.[119] The 2004 world sales for erectile dysfunction drugs rose to $2.7 billion, with US sales for the first 11 months topping $406 million (only $22 million less than US sales for Coca Cola).[120] The increasing focus on life style drugs raises ethical questions not only about proper use of these drugs, but issues of drug company priorities and third party payer responsibilities in responding to consumer demand.

Psychological faculties such as memory, appetite, mood, libido, and sleep, and executive functions such as attention, working memory, and inhibition, represent the most attractive targets for pharmacological enhancement.[121] Until recently, psychotropic medications for the treatment of diseases associated with these functions carried undesired side effects and high risks that made them attractive only for the amelioration of severe mental illness. Increasing knowledge of chemical neurotransmission, however, has enabled the formulation of drugs that affect their intended targets more specifically with fewer and less severe side effects. Not only are these drugs more effective in treating diseases, but they also present unique and increased abilities to heighten normal cognition, emotional and executive functions.

1.22.6.1.1 Enhancing human functions

The debate over human enhancement centers primarily on the attempt to bypass mechanisms such as learning or behavioral reinforcement and directly moderate brain electrochemistry or structure.[122] Drawing on the body's own resources, or manipulating the external environment to effect change, does not raise the same ethical challenges. The myriad ethical and social challenges posed by enhancement pharmaceuticals, as well as other emerging neurotechnologies, have led to the emergence of the field of neuroethics. Neuroethics is defined as the analysis of, and remedial recommendations for, ethical challenges posed by chemical, organic, and electromechanical intervention in the brain.[123] Rather than base itself on a specific philosophical model, neuroethics is characterized by the particular technologies it examines. These technologies include psychopharmacology as well as brain imaging, brain–computer interfaces, cell transplants, and external and internal stimulation of the brain. Neuroethical inquiry into pharmaceuticals asks under which conditions chemical intervention in mental processes is ethical. What are the implications of using a drug that is developed for the treatment of disease to alter personality or to improve normal human abilities or characteristics? What standards should exist? Will advances in psychopharmacology be used as forms of social control? Might this potentially contribute to a widening gap of social inequality? Or, at the other extreme, might it encourage conformity of personality – are those a bit more irascible going to be encouraged, or coerced, to conform to a chemically induced standard of effect?

The use of life style or enhancement drugs poses thorny questions. The more philosophical questions are about categorization: what do terms such as average or normal functioning mean if we can improve functioning across the entire range of human capability? If Prozac can lift everyone's mood, what, then, is a normal or typical affect?[5] Will sadness or inner struggle be pathologized? If we can all be happy and well adjusted through Prozac, should insurance companies pay to reach that state of bliss? Should physicians be the vehicles for prescribing life style mood-altering technologies to their patients? What are the implications of using drugs or other neurotechnologies to micromanage mood, improve memory, maintain attentiveness, or improve sexuality?[122] Is there anything wrong with the emerging field of 'cosmetic neurology,' i.e., using pharmaceuticals to achieve the same ends as cosmetic surgery?[124] The questions are both medical, in terms of the proper role of healthcare professionals, and social, in terms of whether broader society should encourage or discourage people to ingest pharmaceuticals to enhance behaviors, skills, and traits.

The specificity of modern neuropharmaceuticals raises concerns that are distinct from the generalized effects of previous life style drugs such as alcohol or nicotine. Let us take as an example the effort to develop drugs targeted at improving memory in humans.[125] The improvement of memory sounds attractive in the abstract, and certainly the development of drugs to boost or enhance memory function is desirable for those suffering from Alzheimer's or other conditions that affect memory. But there are many unknowns in the use of such drugs in the cognitively intact. The assumption is that memory drugs will simply increase the amount of memory we have available, leaving all other cognitive and affective processes unaffected. But in fact, memory is a selective, delicate process. There are experiences and data that the brain filters out, choosing specific kinds of data to remember, while specifically forgetting others. Who needs to remember the hour waiting in a long line at the bank, staring at the ceiling tiles, or to recall the amnesia induced directly after a personal trauma? Will memory enhancement drugs impair our selectivity process? Or might they target and enhance only certain kinds of memories, such as the traumatic or emotional memories, positive and negative, that the brain tends to retain? Might we end up awash in memories that are troubling to us, unable to forget a painful past? And how might a memory drug affect associated mental processes – such as mood (which is closely connected to memory) or attentiveness (daydreaming is often fueled by a sudden recollection)? Perhaps evolution has stabilized at a particular level of memory capacity because any additional capacity would sacrifice a certain cognitive flexibility, a plastic brain may have advantages over one crammed with memory.[122]

The concern is not only speculative: in 1999, scientists reported in *Nature* that they had genetically engineered mice with increased ability to perform learning tasks.[126] The scientists inserted a gene in mouse zygotes that increased the production of the protein subunit NR2B, part of the NMDA receptor. The mice also displayed physiological changes in the hippocampus (associated with learning) when compared to nontransgenic mice. However, subsequent research demonstrated that the mice with enhanced NR2B seemed to have a greater sensitivity to pain.[127] The original creators of the mice argued the mice may not feel the pain more acutely, but simply learn about pain more readily and thus seem to react more strongly. Still, it is troubling that even the most preliminary research on memory enhancement has already raised the question of whether enhancing memory might have unexpected collateral effects. Perhaps there is a link we do not understand between memory and pain, either at the structural or behavioral level. What other unexpected linkages might be discovered in attempts to change cognitive functions through induced physiological modification?

While most of the cognitive enhancement discussed in the literature focuses on memory or attentiveness, the range of cognitive abilities, of course, exceeds just these two traits. Learning, language, skilled motor behaviors, and executive functions (e.g., decision-making, goal setting, planning, and judgment) are all part of general cognition, and a drug that manages to enhance a greater range of function (especially executive function) may be more desirable than one that narrowly enhances memory alone.[128] But if memory drugs alone have collateral effects, how much more so might a drug that influences a greater range of cognitive functioning?

As a variety of such drugs begin appearing on the market, each individual will be challenged to explicitly consider the kind of self he or she wants to be. The debate is already engaged, with one group arguing that our astounding ability to manipulate our own biology is an integral part of who we are as human beings.[129,130] Opposed to them are those who believe that new technologies are an affront to our humanity, that they diminish what is essentially human nature.[131,132] The argument has no fundamentally right or wrong answers, emerging as it does from two philosophically different visions of human life. It will, however, have a profound impact on the reception of life style drugs in the coming decades.

An illustrative example of the use of a drug for a range of symptoms ranging from the clearly pathological through to enhancement is the debate over Ritalin and its analogs in children diagnosed (or not) with attention deficit disorder (ADD) and attention deficit/hyperactive disorder (ADHD). An examination of the case reveals a broader set of concerns about enhancement technologies.

1.22.6.1.2 Ritalin in attention deficit disorder/attention deficit/hyperactive disorder

Attention deficit/hyperactivity disorder (ADHD) is a commonly diagnosed neurobehavioral disorder in children. Prevalence rates are estimated to be anywhere from 1.7% to 17.8% depending on diagnostic criteria and population studies,[133] though the rate is more commonly cited as being between 3% and 10%.[134,135] The Center for Disease Control reported that among the almost 4 million children 3–17 years old in US, about 6% of that group were diagnosed with ADHD in 2003.[136] Although diagnostic criteria have evolved over time, the condition remains characterized by above normal levels of impulsivity, inattention, and hyperactivity. The most obvious manifestations, as well as the most common reason children are referred to physicians for diagnosis and treatment, are unacceptable classroom behavior and poor academic achievement. Currently, boys are more commonly diagnosed than girls, though that may be changing.[137]

There are no clinical tests for ADHD, so the diagnosis is subjective and situational. According to the Diagnostic and Statistical Manual of Mental Disorders (DSM-IV), a manual published by the American Psychiatric Association that provides standardized criteria for diagnosis of psychiatric conditions, symptoms must begin before the age of 7 years and be present for at least 6 months and cause a significant impairment of function in more than one setting (e.g., social, academic, familial). However, determining the appropriateness of a preschooler's impulsivity in various setting is highly subjective. Children are prescribed stimulants in the absence of DSM criteria for ADHD, often at the behest of teachers and particularly when the children tend to demonstrate oppositional defiant disorder.[138] Even the youngest children are being diagnosed and treated, pharmacoepidemiological studies have documented a rise in overall psychoactive medication treatment for preschool children in the last decade, with a threefold rise in prescription rates specifically for stimulants in this age group since 1990.[139]

The treatment of choice for ADHD is amphetamines. Ritalin (methylphenidate) is a stimulant medication originally developed to treat ADHD, and newer stimulants have since entered the market, including Adderall, Strattera, and Concerta. Stimulants improve "disruptive behavioral inhibition, impulse control, selective attention, active working memory and executive functioning."[140] Significantly, Ritalin confers these benefits on normal people as well as those with ADHD.[141]

Controversy has surrounded the diagnosis of ADD and ADHD in children. While the use of stimulants to improve children's behavior was used as far back as the 1930s,[142] alarm has been raised at the rapid increase in diagnosis prescription rates. From 1990 to 1995, the annual US production of Ritalin increased fivefold, far exceeding that of any other country. By the late 1990s, the US was producing and consuming 90% of the world's Ritalin.[143] Concern over the use of drugs in small children led to widespread media and medical attention. A general perception arose that ADHD was overdiagnosed and stimulants overprescribed. Others have argued, in contrast, that the trends reflect better diagnosis, more effective treatment, and increased education and recognition of the syndrome.[144]

Part of the controversy lies in how children tend to be diagnosed. Teachers are often the first to bring up the possibility of ADHD when the child does not conform to classroom behavioral standards.[145] As children with ADHD are fidgety and seek out sources of stimulation, it has been suggested that the modern classroom setting, where children are required to sit still in a chair facing a teacher for long periods of time, is precisely the wrong kind of learning environment for these children.[146] Classrooms that permit more physical activity and interactive learning and are more developmentally appropriate are more likely to have fewer referrals for ADHD diagnosis.[147]

Pharmaceutical amphetamines have become an enhancement technology used by thousands of otherwise healthy people.[148] College students freely admit to using each others' stimulant pills as study aids, and students with prescriptions can do a brisk business in the dormitories, prescription amphetamines are among the most used and abused drugs among young people.[143] In younger children, it is difficult to determine the degree to which pressure from parents and teachers to put unruly children on stimulants can be untangled from more objective diagnoses. It is clear that some schools have pressured parents of difficult-to-manage children to administer Ritalin, some even by using threats of expulsion.[147] In wealthier school districts where competitive performance pressure is high, however, it tends to be parents who push the use of stimulants. In the absence of clear physiological pathologies or discrete functional identifiers, the constellation of traits that characterize ADHD are applicable to some degree or another in a large percentage of children.

Stimulant use for ADHD is a perfect example of how the line between medical treatment for recognized disorders and the use of drugs for enhancement is becoming blurred. There is little doubt that ADHD and ADD are seriously impeding the ability of certain children to perform well in school, get along with peers, and cooperate in family units. However, the disproportionate diagnosis of ADHD in the US suggests a strong cultural component. It has been said that American culture, with its emphasis on speed and constant sensory stimulation, is particularly ADD-ogenic.[149] However, the distinction may also stem from differences in cultural tolerances for specific child behaviors.[150]

1.22.6.1.3 Conclusion: paying for life style drugs

The difference between a life style drug and a clinically therapeutic medicine can come down to the simple matter of who pays. US health plans deny coverage for pharmaceuticals not prescribed for a specific medical diagnosis. For the US federally funded Medicare program, as well as for national health plans in other developed countries such as the UK, pharmaceuticals that some could construe as primarily life style drugs are often covered by insurance. Examples include drugs for erectile dysfunction, menopausal symptoms, smoking cessation, and birth control. It is not the intention here to trivialize these conditions by giving them a life style classification, nevertheless, it can be argued that they are outside the realm of pathology as classically defined. They do, however, bring into question the challenge of allocating scarce national (economic) healthcare resources intended for medical necessity. While rationing may not be the preferred method of allocating healthcare dollars, funds that support research, development, and distribution of pharmaceutical agents for anything less than medical need increase the challenges for government and employer-based funding.[116] A number of states (Washington, New Jersey, Illinois) in the US are currently considering imposing a vanity tax on cosmetic surgery and Botox injections – another indication that such pharmaceutical agents are viewed as not only elective in nature, but also unrelated to health matters requiring curative or medical care.

Clearly, we are currently witnessing only the leading edge of a wave of pharmaceuticals that will be used, and may even be intended and designed to be used, to enhance functions that would otherwise be considered in the normal range. The market will be lucrative and pharmaceutical companies may feel the pressure to dedicate more and more resources to developing markets that include large populations of nonmedical consumers. The danger is that drugs for specific pathological conditions may be neglected.

1.22.6.2 Pharmacogenomics/Pharmacogenetics

Pharmacogenomics and pharmacogenetics are both terms referring to the science of how genetics influences an individual's response to medication. The terms are often used interchangeably, although technically pharmacogenetics refers to the study of inherited differences in drug metabolism (pharmacokinetics) and response (i.e., receptors, pharmacodynamics), while pharmacogenomics refers to the study of the overall array of different genes that determine drug behavior. The field in general is being touted as a significant advance in improving how drugs will be developed and prescribed in the future. Despite major advances in drug therapy for many diseases, treatment remains suboptimal for a significant proportion of individuals because of unpredictable side effects or lack of response. Understanding how small differences (called single nucleotide polymorphisms (SNPs)) in genetic make-up can predict drug response may allow clinicians to tailor drug therapy to individual patients and avoid the morbidity and costs associated with adverse drug reactions or lack of effectiveness.

Right now companies are engaged in cataloging as many SNPs as possible, hoping to capitalize on their usefulness once the field matures. The development of pharmacogenomic research has opened up the potential for widespread changes in the way we approach the discovery and development of medical therapies and drugs. The ability to use genomic technology to understand how individuals with particular genotypes will respond to various drugs may allow manufacturers to streamline clinical testing and drug development. Pharmacogenetic drugs will allow targeted therapies, and may thus save money by avoiding less than optimal treatment strategies. They will improve dosing by helping predict optimal doses in a particular patient, and may decrease drug interactions. Collective pharmacogenomic statistics may aid national medication programs and formularies to choose the best overall drugs for a population. Finally, increased pharmacogenomic knowledge promises advanced screening for genetic disease, better preventive medicine, and even perhaps targeted genetic vaccines. At the same time, pharmacogenomics also holds the potential to drastically alter how we view the relationship between genes and disease, while posing ethical questions concerning the treatment of patients and data in pharmacogenomic clinical trials.

Current clinical trials aim to determine the efficacy of medical therapies and procedures through the use of pharmacogenetic profiling, attempting to correlate genotypes with disease or drug responsiveness. In order to reduce pharmacokinetic variability or the incidence of adverse events, these studies typically stratify research subjects, including or excluding subjects from the trial based on genotype.[151] It is precisely this type of genotypic stratification that poses several ethical as well as scientific challenges.

First, genotyping as either an inclusion or exclusion criteria could lead to a loss of benefits in research participation, unfair representation in clinical trials, or subject selection biases.[151] The categorization of patients into responders and nonresponders of drugs could lead to the development of new drugs licensed only for the specific genetic group of good responders.[152] Also, there has historically been underrepresentation of groups such as women, children, and the elderly in trials. Could the same underrepresentation occur with genetic subgroups such as race and ethnicity?[153] This is problematic because studies involving subject selection biases will not accurately reflect the response and adverse event profiles of the general population.

Some have argued that pharmacogenomics and the dawn of tailored medicine has the potential to create new therapeutic orphan populations. These would be genetically defined groups with limited access to new and more effective therapy, justified either by their small numbers, making drug development uneconomical, or by the nature of their genotype whereby no effective therapy can be discovered for them from existing technology.[152] It is precisely these genetically and socially marginalized groups that may be the ones most in need of genetically tailored treatment.

There could be several economic ramifications to the tailoring of drugs to specific subpopulations. Will the market incentive to develop drugs for particularly small groups be lessened or eliminated? Orphan populations may also be defined as too small in size to be economically advantageous; drugs for these groups would be very expensive if developed. The US Orphan Drug Law (in effect since 1983) provides financial incentives for companies to direct their efforts toward the development of pharmaceutical therapies for the 5000 or so orphan conditions.[154,155]

Pharmaceutical companies can also use pharmacogenomic studies largely to their advantage. Experts have suggested that pharmaceutical companies may design trials that are focused toward favoring certain drugs, using pharmacogenomic profiling to position their particular drug advantageously in the market.[156] Others suggest that the development of pharmacogenomics-based drugs targeted to specific subpopulations will lead to a narrowing of the markets for drugs.[157] Companies could create demand for drugs by offering tests to identify people who respond to that drug, while entire populations might be ignored in this market-driven style of drug development.[151] With this in mind, regulatory guidelines over pharmacogenomic trial design and conduct need to prevent companies from luring certain patients or avoiding particular genotypes that could adversely affect the positioning of their drug in the market.

Additionally, many are concerned that pharmacogenomic technology may exacerbate current global inequities in medical care. If these drugs and therapies are premium priced, will these innovations be restricted to the wealthy? What do we make of less developed countries who will have limited access to improved therapeutic treatments and expensive pharmacogenomic testing technology?[152]

Pharmacogenomic profiling and patient stratification could lead to the creation of new disease classifications or categories of conditions that are largely social in origin. Individuals with no serious health problems who are informed that they possess a drug-associated genetic polymorphism may now label themselves as ill. Genotyping could result in individuals being categorized as difficult to treat, less profitable to treat, or more expensive to treat.[158] Not only does this contribute to the social construction of disease, but introduces implications for how insurance is determined and how medical care is rationed, especially in the age of managed care. Fear of stigmatization may also affect willingness to participate in clinical trials.

How closely are differences in drug response related to genetic differences that arise from race and ethnicity? There exists an ongoing discussion over this precise question.[159] The recent development of ethnically targeted therapies has reopened the controversial debate over pharmacogenomics and race. BiDil, a new drug treatment for heart failure tested solely in one racial group, recently became the first drug to be approved by the FDA to treat heart failure in African-Americans only.[160–162] NitroMed began the African-American Heart Failure Trial (A-HeFT) in 2001, the first ever heart failure trial to be comprised solely of African-American patients. The study claimed that "observed racial disparities in mortality and therapeutic response rates in Black heart failure patients may be due in part to ethnic differences in the underlying pathophysiology of heart failure."[163,164] This raises some important questions. Are there significant genetic differences between ethnic and racial groups? How good are current racial labels as indicators of genetics differences? Should these classifications even be used in this manner? It is possible that developments like these can breed discrimination and racism, reinforcing "discredited crude biological notions of race,"[165] with whole population groups becoming stigmatized.[166] Historically, clinical decisions based on ethnic or racial classification often leads to poor or ineffective care.[167]

1.22.7 Conclusion

In this chapter we have attempted to profile some of the ethical issues in drug development and delivery. The use of pharmaceuticals to cure a variety of ills is one of the great success stories of human technology, and has resulted in symptom relief and cures that were scarcely imagined by our forebears. In order to continue developing and using drugs as they increase in strength and specificity, it is important to clearly define the ethical basis of proper drug use and the pitfalls of our current means of creating and distributing them. To do so is to ensure continued development for the health and well-being of future generations.

References

1. Engel, C. *Wild Health: How Animals Keep Themselves Well and What We Can Learn from Them*; Weidenfeld & Nicolson: London, 2002.
2. Fabrega, H., Jr. *Evolution of Sickness and Healing*; University of California Press: Berkeley, CA, 1997.
3. Martin, G. J. *Ethnobotany: A Methods Manual*; Earthscan: London, Sterling, 2004.
4. National Center for Health Statistics. *Health, United States, With Chartbook on Trends in the Health of Americans*. National Center for Health Statistics: Hyattsville, MD, 2004; p 1232.
5. Kramer, P. D. *Listening to Prozac*; Viking: New York, 1993.
6. Zonana, H. V. *J. Am. Acad. Psychiatry Law* **2003**, *31*, 372–376.

7. Conroy, S.; Choonara, I.; Impicciatore, P.; Mohn, A.; Arnell, H.; Rane, A.; Knoeppel, C.; Seyberth, H.; Pandolfini, C.; Raffaelli, M. P. et al. *Br. Med. J.* **2000**, *320*, 79–82.
8. Angarola, R. T.; Joranson, D. E. *Am. Pain Society Bull.* **1995**, *5*, 14–15.
9. Drazen, J. M. *N. Engl. J. Med.* **2005**, *352*, 1131–1132.
10. Maxwell, S. R.; Webb, D. J. *Lancet* **2005**, *365*, 449–451.
11. Wadman, M. *Nature* **2005**, *434*, 554–556.
12. Couzin, J. *Science* **2005**, *307*, 1183–1185.
13. Frantz, S. *Nature* **2005**, *434*, 557–558.
14. New Warning for Heartburn Drug Propulsid. CNN Interactive: Washington DC, June 30, 1998. http://www.cnn.com/HEALTH/9806/30/heartburn.drug/ (accessed March 2006).
15. Harris, G.; Koli, E. Lucrative Drug, Danger Signals and the F.D.A. *New York Times*, June 10, 2005; p A1.
16. Blanck, H. M.; Khan, L. K; Serdula, M. K. *Prev. Med.* **2004**, *39*, 1243–1248.
17. Connelly, H. M.; Crary, J. L.; McGoon, M. D.; Hensrud, D. D.; Edwards, B. S.; Edwards, W. D.; Schaff, H. V. *N. Engl. J. Med.* **1997**, *337*, 581–588.
18. Noller, K. L. *JAMA* **2002**, *288*, 368–369.
19. Fugh-Berman, A.; Pearson, C. *Pharmacotherapy* **2004**, *22*, 1205–1208.
20. Lacey, J., Jr.; Mink, P. J.; Lubin, J. H.; Sherman, M. E.; Troisi, R.; Hartge, P.; Schatzkin, A.; Schairer, C. *JAMA* **2002**, *288*, 334–341.
21. Writing Group for the Women's Health Initiative Investigators. *JAMA* **2002**, *288*, 321–333.
22. American College of Neuropsychopharmacology (ACNP). *Executive Summary: Preliminary Report of the Task Force on SSRIs and Suicidal Behavior in Youth*, Nashville, TN, Jan 21, 2004.
23. Fergusson, D.; Doucette, S.; Glass, K. C.; Shapiro, S.; Healy, D.; Hebert, P.; Hutton, B. *Br. Med. J.* **2005**, *330*, 396–403.
24. Jick, H.; Kaye, J. A.; Jick, S. S. *JAMA* **2004**, *292*, 338–343.
25. U.S. Food and Drug Administration (USFDA). *Public Health Advisory: Suicidality in Children and Adolescents Being Treated with Antidepressant Medications*, Oct 15, 2004. http://www.fda.gov/ (accessed March 2006).
26. Meek, C. *Can. Med. Assoc. J.* **2004**, *170*, 455.
27. Varley, C. K. *JAMA* **2003**, *290*, 1091–1093.
28. Whittington, C. J.; Kendall, T.; Fonagy, P.; Cottrell, D.; Cotgrove, A.; Boddington, E. *Lancet* **2004**, *363*, 1341–1345.
29. Jick, H.; Kaye, J. A.; Jick, S. S. *JAMA* **2004**, *292*, 338–343.
30. Jureidini, J. N.; Doecke, C. J.; Mansfield, P. R.; Haby, M. M.; Menkes, D. B.; Tonkin, A. L. *Br. Med. J.* **2004**, *328*, 879–883.
31. NDC Health Corporation. NDCHealth Reviews 2004 Pharmaceutical Market Information. http://www.prnewswire.com/cgi-bin/micro_stories.pl?ACCT = 603950&TICK = NDC&STORY = /www/story/05-02-2005/0003537082&EDATE = May + 2, + 2005 (accessed March 2006).
32. Angell, M. The Truth About the Drug Companies. *New York Review of Books* [Online] **2004**, *51*, July 15, 2004. http://www.nybooks.com/articles/17244 (accessed March 2006).
33. Viswanathan, S. *Pharmaceutical Formulation and Quality* [Online] **2004**, March/April. http://www.pharmaquality.com/Cover%20Story5.htm (accessed March 2006).
34. Burke, M. A.; de Francisco, A. *Managing Financial Flows for Health Research*; Global Forum for Health Research: Geneva, Switzerland, 2004.
35. L. Goshman, *J. Pharmacy Soc. Wisconsin* 2001, July/August.
36. Peters, C. P. *Fundamentals of the Prescription Drug Market*; National Health Policy Forum: Washington, DC, 2004.
37. Spector, R. *Stanford Med. Magazine* [Online] **2005**, *22*. http://mednews.stanford.edu/stanmed/2005summer/drugs-metoo.html (accessed March 2006).
38. Garattini, S. *J. Nephrol.* **1997**, *10*, 283–294.
39. Lee, T. *N. Engl. J. Med.* **2004**, *350*, 211–212.
40. Infectious Diseases Society of America. *Bad Bugs, No Drugs*; Alexandria, VA, 2004.
41. Nathan, C. *Nature* **2004**, *431*, 899–902.
42. Global Alliance for TB Drug Development. *The Economics of TB Drug Development*; New York, 2001.
43. Silverstein, K. Millions for Viagra, Pennies for Diseases of the Poor. *The Nation*. July 19, 1999. http://www.thenation.com/doc.mhtml%3Fi = 19990719&s = silverstein (accessed March 2006).
44. US Food and Drug Administration. *Milestones in U.S. Food and Drug Law History.* http://www.fda.gov/opacom/backgrounders/miles.html (accessed March 2006).
45. Doran, E.; Robertson, J.; Henry, D. *Social Sci. Med.* **2004**, *60*, 1437–1443.
46. US Food and Drug Administration. *FDA's Mission Statement*. http://www.fda.gov/opacom/morechoices/mission.html (accessed March 2006).
47. Okie, S. *N. Engl. J. Med.* **2005**, *352*, 1063–1066.
48. Kmietowicz, Z. *Br. Med. J.* **2005**, *330*, 805.
49. Wasunna, A.; Johnston, J. *Intellectual Property Rights in Pharmaceuticals and Biological Materials: Ensuring Innovation or Barring Access?* July 2002–March 2006. http://www.thehastingscenter.org (accessed March 2006).
50. Obaro, S. K.; Palmer, A. *Vaccine* **2003**, *21*, 1423–1431.
51. Mayor, S. *Br. Med. J.* **2005**, *330*, 748.
52. IMS Health. *IMS Reports 2004 Global Pharmaceutical Sales Grew 7 percent to $550 Billion*; Press Release; Fairfield, CT, March 9, 2005. http://www.imshealth.com/ims/portal/front/articleC/0,2777,6599_3665_71496463,00.html (accessed March 2006).
53. World Health Organization. *What are Essential Medicines? Access, Quality and Rational Use of Medicines and Essential Medicines.* http://www.who.int/ (accessed March 2006).
54. World Watch Institute. Viagra, Malaria, and the Future of Health Care. *World Watch: Matters of Scale* [Online] **2000**, May/June. http://www.worldwatch.org/pubs/mag/2000/133/mos/ (accessed March 2006).
55. Wikipedia. *Diseases of Affluence*. http://en.wikipedia.org/wiki/Diseases_of_affluence (accessed March 2006).
56. Tomes, N. J. *N. Engl. J. Med.* **1993**, *328*, 670.
57. Ely, J. J. *Rev. Environ. Health* **2003**, *18*, 111–129.
58. Leeb, M. *Nature* **2004**, *431*, 892–893.
59. The Global Forum for Health Research. *The 10/90 Gap Now 2005.* http://www.globalforumhealth.org/site/003__The%2010%2090%20gap/001__Now.php (accessed March 2006).

60. White, C. *Br. Med. J.* **2004**, *329*, 1064–1065.
61. Lewison, G.; Rippon, I.; de Francisco, A.; Lipworth, S. *Outputs, Expenditures on Health Research in Eight Disease Areas, 1996–2001*, Global Forum for Health Research, Forum 8, Mexico City, Nov 2004.
62. Murray, C. J.; Lopez, A. D. *Lancet* **1997**, *349*, 1269–1276.
63. World Health Organization. *The World Health Report 2002: Reducing Risks, Promoting Healthy Life*, Geneva, Switzerland, 2002.
64. World Health Organization. *Epidemiology and Burden of Disease*, Geneva, Switzerland, 2003.
65. Swafford, S. *Br. Med. J.* **1997**, *314*, 1367.
66. Greenberg, H.; Raymond, S.U.; Leeder, S. R. *Health Affairs* [Online] 2005, DOI: 10.1377/hlthaff.w5.31.
67. Hamilton, C. Diseases of affluence and other paradoxes. *The Australian Financial Review*, Oct 15, 2004, p 8.
68. Medical College of Wisconsin. *Global Health in the 21st Century*. http://www.healthlink.mcw.edu/article/977858884.html (accessed March 2006).
69. Roche UK. *Weight Management*. http://www.rocheuk.com/html/health/WeightManagement.asp (accessed March 2006).
70. Hilton, I. A bitter pill for the world's poor. *The Guardian*, London, Jan 5, 2000.
71. Cook, I. G.; Dummer, A.; Trevor, J. B. *Health Policy* **2004**, *67*, 329–343.
72. Triggle, D. J. *Drug Devel. Res.* **2003**, *59*, 269–291.
73. Granville, B. The Right Response for the Pharmaceutical Industry. In *Delivering Essential Medicines: The Way Forward*; Attaran, A., Ed.; Royal Institute of International Affairs: London, 2003.
74. Foster, J. W. North-South Institute; Social Watch 2004 [cited March 5, 2005]. http://www.socwatch.org/en/informesTematicos/81.html (accessed March 2006).
75. Bulter, D. *Nature* **2003**, *424*, 604.
76. World Health Organization. Global Defense against the Infectious Disease Threat. In *Communicable Diseases 2002*; Kindhauser, M. K., Ed.; World Health Organization: Geneva, Switzerland, 2003, pp 56–103.
77. World Health Organization. What are Essential Medicines? Access, Quality and Rational Use of Medicines and Essential Medicines. http://www.who.int/ (accessed March 2006).
78. World Health Organization/Industry Drug Development Working Group. Working Paper on Priority Infectious Diseases Requiring Additional R & D. WHO-IFPMA Round Table: Geneva, Switzerland, July 2001.
79. Rochon, P. A.; Mashari, A.; Cohen, A.; Misra, A.; Laxer, D.; Streiner, D. L.; Dergal, J. M.; Clark, J. P.; Gold, J.; Binns, M. A. *Can. Med. Assoc. J.* **2004**, *170*, 1673–1677.
80. Butler, D. *Nature* **2003**, *426*, 754.
81. Butler, D. *Nature* **2003**, *425*, 331.
82. Ban, T. A. *Prog. Neuro-Psychopharmacol. Biol. Psych.* **2001**, *25*, 709–727.
83. *The Triumph of Psychopharmacology and The Story of CINP*; Ban, T. A.; Healy, D.; Shorter, E., Eds.; Animula Publishing House: Budapest, 2000, pp 9–10.
84. Ghodse, H.; Khan, I. *Psychoactive Drugs: Improving Prescribing Practices*; World Health Organization: Geneva, Switzerland, 1988.
85. Ban, T. A. They Used to Call it Psychiatry. In *The Psychopharmacologists*; Healy, D., Ed.; Chapman and Hall: London, 1996, pp 587–620.
86. Ghodse, H. *Br. J. Psychiatry* **2003**, *183*, 15–21.
87. International Narcotics Control Board. *Report of the International Narcotics Control Board for 2000 (E/INCB/2000/1)*. United Nations: New York, 2001.
88. World Health Organization. *The World Health Report 2001. Mental Health: New Understanding, New Hope*; Geneva, Switzerland, 2001.
89. World Health Assembly. *Mental Health: Responding to the Call for Action.* Report by the Secretariat A55/18. World Health Organization: Geneva, Switzerland, 2002.
90. Buckley, P. F. *Biol. Psychiatry* **2001**, *50*, 912–924.
91. Dawkins, K.; Lieberman, J. A.; Lebowitz, B. D.; Hsiao, J. K. *Schizoph. Bull.* **1999**, *25*, 395–405.
92. Fenton, W. S.; Byler, C. R.; Heinssen, R. K. *Schizoph. Bull.* **1997**, *3*, 637–651.
93. Pyne, J.; Bean, D.; Sullivan, G. *J. Nervous Mental Dis.* **2001**, *89*, 146–153.
94. Roberts, L. W.; Geppert, C. M. *Compreh. Psychiatry* **2004**, *45*, 161–167.
95. Irani, F.; Dankert, M.; Brensinger, C.; Bilker, W. B.; Nair, S. R.; Kohler, C. G; Kanes, S. J.; Turetsky, B. I.; Moberg, P. J.; Ragland, J. D.; et al. *Neuropsychopharmacology* **2004**, *29*, 960–968.
96. Rand, C. S.; Sevick, M. A. *Control. Clin. Trials* **2000**, *21*, 241S–247S.
97. Chamberlin, J. *Psychiatr. Rehab. J.* **1998**, *21*, 405–408.
98. Wong, J. G.; Clare, I. C. H.; Gnn, M. J.; Holland, A. J. *Psychol. Med.* **1999**, *29*, 437–446.
99. Mahone, I. H. *Arch. Psychiatr. Nursing* **2004**, *18*, 126–134.
100. Barrett, K. E.; Taylor, D. W.; Pullo, R. E.; Dunlap, D. A. *Psychiatr. Rehab. J.* **1999**, *21*, 241–249.
101. Berg, J. W.; Appelbaum, P. S. Subject's Capacity to Consent to Neurobiological Research. In *Ethics in Psychiatric Research*; Pincus, H. A., Lieberman, J. A., Ferris, S., Eds.; American Psychiatric Association: Washington DC, 1999, pp 81–106.
102. Appelbaum, P. S.; Grisso, T. *N. Engl. J. Med.* **1988**, *319*, 1635–1638.
103. Cohen, B. J.; McGarvey, E. L.; Pinkerton, R. C.; Kryzhanivska, L. *J. Am. Acad. Psychiatry Law* **2004**, *32*, 134–143.
104. Appelbaum, P. S.; Grisso, T. *Law Human Behav.* **1995**, *19*, 105–126.
105. National Bioethics Advisory Commission. *Research Involving Persons With Mental Disorders That May Affect Decision-making Capacity.* NBAC: Rockville, MD, 1998.
106. Marson, D. C.; Annis, S. M.; McInturff, B.; Bartolucci, A.; Harrell, L. E. *Neurology* **1999**, *53*, 1983–1992.
107. Kim, S. Y. H.; Caine, E. D.; Currier, G. W.; Leibovici, A.; Rayn, J. M. *Am. J. Psychiatry* **2001**, *158*, 712–717.
108. Marson, D. C.; Ingram, K. K.; Cody, H. A.; Harrell, L. E. *Arch. Neurol.* **1995**, *52*, 949–954.
109. Kim, S. Y. H.; Karlawish, J. H. T.; Caine, E. D. *Am. J. Geriatr. Psychiatry* **2002**, *10*, 151–165.
110. Penslar, R. L. *Institutional Review Board Guidebook. Chapter VI: Special Classes of Subjects*. http://www.hhs.gov/ohrp/irb/irb_chapter6.htm#g5 (accessed March 2006).
111. Merlin, M. D. *Econ. Bot.* **2003**, *57*, 295–323.
112. Spiegel,, R. *Psychopharmacology: An Introduction*; Wiley: West Sussex, England, 2003.
113. Mitrany, D. *PharmacoEconomics* **2001**, *19*, 441–448.
114. Farah, M. J.; Wolpe, P. R. *Hastings Cent. Rep.* **2004**, *34*, 35–45.
115. Gilbert, D. *Br. Med. J.* **2000**, *321*, 1341–1344.

116. Carlsson, A. *Brain Res. Bull.* **1999**, *50*, 363.
117. *Physician's Desk Reference* (web version). Thomson Healthcare: Montvale, NJ, 2005. http://www.pdr.net (accessed March 2006).
118. Coe, J. *The Lifestyle Drug Outlook to 2008: Unlocking New Value in Well-Being*; Business Insights: Dallas, 2003.
119. IMS. *Leading 20 Therapeutic Classes by Total U.S. Dispensed Prescriptions, MAT (Moving Annual Total).* http://www.imshealth.com/ims/portal/front/articleC/0,2777,6599_73914140_73916014,00.html (accessed March 2006).
120. Shmit, J. Impotence drugs selling slowly. *USA Today* [Online] February 2, 2005.
121. Farah, M. J.; Illes, J.; Cook-Deegan, R.; Garnder, H.; Kandel, E.; King, P.; Parens, E.; Sahakian, B.; Wolpe, P. R. *Nat. Rev. Neurosci.* **2004**, *5*, 421–425.
122. Wolpe, P. R. *Brain Cogn.* **2002**, *50*, 387–395.
123. Wolpe, P. R. Neuroethics. In *Encyclopedia of Bioethics*, 3rd ed.; Post, S. G., Ed.; Macmillan Reference: New York, 2004; Vol. 4, pp 1894–1898.
124. Chatterjee, A. *Neurology* **2004**, *63*, 874–968.
125. Furey, M. L.; Pietrini, P.; Haxby, J. V. *Science* **2000**, *290*, 2315–2319.
126. Tang, Y. P.; Shimizu, E.; Dube, G. R.; Rampon, C.; Kerchner, G. A.; Zhuo, M.; Liu, G.; Tsien, J. Z. *Nature* **1999**, *401*, 63–69.
127. Wei, F.; Wang, G. D.; Kerchner, G. A.; Kim, S. J.; Xu, H. M.; Chen, Z. F.; Zhuo, M. *Nat. Neurosci.* **2001**, *4*, 164–169.
128. Whitehouse, P. J.; Juengst, E.; Mehlman, M.; Murray, T. H. *Hastings Center Report* **1997**, *27*, 14–22.
129. Hughes, J. H. *Citizen Cyborg*; Westview Press: Boulder, CO, 2004.
130. Stock, G. *Redesigning Humans*; Houghton Mifflin: Boston, MA, 2003.
131. Fukuyama, F. *Our Posthuman Future*; Farrar, Straus and Giroux: New York, 2002.
132. McKibben, B. *Enough: Staying Human in an Engineered Age*; Times Books: New York, 2003.
133. Elia, J.; Ambrosini, P. J.; Rapport, J. L. *N. Engl. J. Med.* **1999**, *340*, 780–788.
134. Spencer, T. J.; Biederman, J.; Wilens, T. E.; Faraone, S. V. *J. Clin. Psychiatry* **2002**, *63*, 3–9.
135. Swanson, J. M.; Seargent, J. A.; Taylor, E.; Sonuga-Barke, E. J.; Jensen, P. S.; Cantwell, D. P. *Lancet* **1998**, *351*, 429–433.
136. Dey, A. N.; Bloom, B. *Vital & Health Statistics*; Series 10: Data From the National Health Survey, **2005**, *223*, 1–78.
137. Robison, L. M.; Skaer, T. L.; Sclar, D. A.; Galin, R. S. *CNS Drugs* **2002**, *16*, 129–137.
138. Angold, A.; Erkanli, A.; Egger, H. L.; Costello, E. J. *J. Am. Acad. Child Adolesc. Psychiatry* **2000**, *39*, 975–984.
139. Connor, D. F. *J. Dev. Behav. Pediatr.* **2002**, *23*, S1–S9.
140. National Health and Medical Research Council (Australia). *Attention Deficit Hyperactivity Disorder – Part II Management, Section 4.2 Stimulant Medication*, 1997. Commonwealth of Australia. http://www.nhmrc.gov.au/publications/adhd/contents.htm (accessed March 2006).
141. Kuhn, C.; Swartzwelder, S.; Wilson, W. *Buzzed: The Straight Facts About the Most Used and Abused Drugs from Alcohol to Ecstasy*, 2nd ed.; W.W. Norton & Company: New York, 2003.
142. Bradley, C. *Am. J. Psychiatry* **1937**, *94*, 577–588.
143. Elliott, C. *Better than Well*; Norton: New York, 2003.
144. Accardo, P.; Blondis, T. A. *J. Pediatr.* **2001**, *138*, 6–9.
145. Powers, T. Race for perfection: children's rights and enhancement drugs. *J. Law Health* **1998**, *13*, 141–169.
146. Yaroshevsky, F.; Bakiaris, V. Deficit of Attention Disorder. In *Family Therapy as an Alternative to Medication*; Prosky, P. S.; Keith, D., Eds.; Brunner-Routledge: New York, 2003.
147. Kohn, A. *What to Look for in a Classroom: and Other Essays*; Jossey-Bass: San Francisco, 1998.
148. Farah, M. *Nat. Neurosci.* **2002**, *5*, 1123–1129.
149. DeGrandpre, R. *Ritalin Nation*; W.W. Norton: New York, 1999.
150. Cohan, J. A. Psychiatric ethics and emerging issues of psychopharmacology in the treatment of Depression. *J. Contemp. Health Law Policy* **2003**, *20*, 115–172.
151. Issa, A. M. *Nat. Rev.: Drug Disc.* **2002**, *1*, 300–304.
152. Smart, A.; Martin, P.; Parker, M. *Bioethics* **2004**, *18*, 322–343.
153. Wolpe, P. R. *Pharm. Therapeut.* **2001**, 16–17.
154. National Organization for Rare Diseases (NORD). *Understanding the Orphan Drug Act.* http://www.rarediseases.org (accessed March 2006).
155. Food and Drug Administration. *Miscellaneous Provisions Relating to the Orphan Drug Act.* http://www.fda.gov/ (accessed March 2006).
156. Bodenheimer, T. *N. Engl. J. Med.* **2000**, *342*, 1539–1544.
157. Regaldo, A. *Am. J. Health Sys. Pharm.* **1999**, *56*, 40–50.
158. Rothstein, M. A.; Epps, P. G. *Nat. Rev. Genet.* **2001**, *2*, 228–231.
159. Burchard, E. G. *N. Engl. J. Med.* **2003**, *384*, 1170–1175.
160. Rahemtulla, T.; Bhopal, R. *Br. Med. J.* **2005**, *330*, 1036–1037.
161. Khan, J. D. *Yale J. Health Policy Law Ethics* **2004**, *4*, 1–46.
162. Food and Drug Administration. FDA Approves BiDil Heart Failure Drug for Black Patients. *FDA News* July 23, 2005.
163. NitroMed, Inc. *NitroMed Initiates Confirmatory BiDil Trial in African American Heart Failure Patients*; Press Release March 17, 2001. http://www.nitromed.com/press/03-17-01.html (accessed March 2006).
164. Taylor, A. L.; Ziesche, S.; Yancy, C.; Carson, P.; D'Agostino, R.; Ferdinand, K.; Taylor, M.; Adams, K.; Sabolinski, M.; Worcel, M. et al. *N. Engl. J. Med.* **2004**, *351*, 2049–2057.
165. Witzig, R. *Ann. Intern. Med.* **1996**, *125*, 675–679.
166. Dolgin, J. L. *Studies in the History and Philosophy of Biological and Biomedical Sciences* **2001**, *32*, 705–721.
167. Bhopal, R. *Br. Med. J.* **1998**, *316*, 1239–1242.
168. Mathers, C. D.; Stein, C.; Fat, D. M.; Rao, C.; Inoue, M.; Tomijima, N.; Bernard, C.; Lopez, A.D.; Murray, C. J. L. Global Burden of Disease 2000: Version 2 Methods and Results. In *Global Programme on Evidence for Health Policy*, WHO: Geneva, Switzerland, 2002, pp 32–37.

Biographies

Paul Root Wolpe, PhD, is a professor in the Department of Psychiatry at the University of Pennsylvania, where he also holds appointments in the Department of Medical Ethics and the Department of Sociology. He is a Senior Fellow of Penn's Center for Bioethics, is the Director of the Program in Psychiatry and Ethics at the School of Medicine, and is a Senior Fellow of the Leonard Davis Institute for Health Economics. Dr Wolpe serves as the first Chief of Bioethics for the National Aeronautics and Space Administration (NASA), where he is responsible for formulating policy on space ethics and safeguarding research subjects. Dr Wolpe is also the first National Bioethics Advisor to the Planned Parenthood Federation of America, advising them on the reproductive implications of emerging reproductive technologies and changing social dynamics.

Dr Wolpe is the author of numerous articles and book chapters in sociology, medicine, and bioethics, and has contributed to a variety of encyclopedias on bioethical issues. His research examines the role of ideology, religion, and culture in medical thought, especially in relation to emerging biotechnologies, including nanotechnology, reproductive and genetic technology, and neurotechnology. Dr Wolpe is a founder of the field of neuroethics, which specifically focuses on the social, legal, and ethical issues related to new brain technologies and psychopharmaceuticals, and is considered an expert on the impact of biotechnology and psychopharmaceuticals on the human body. He is the author of the textbook *Sexuality and Gender in Society* and the end-of-life guide *In the Winter of Life*, and is currently writing a book on emerging technologies.

Dr Wolpe sits on a number of national and international nonprofit organizational boards and working groups, and is a consultant to the biomedical industry. He is Associate Editor of the *American Journal of Bioethics*, and is a member of the National Board of the American Society for Bioethics and Humanities. He serves on the Ethical Advisory Board of the State of Pennsylvania, and is bioethics advisor to the Philadelphia Department of Human Services, Children and Youth Division. A dynamic speaker who has won a number of teaching and writing awards, Dr Wolpe has been chosen by The Teaching Company as a 'Superstar Teacher of America,' and two of his courses are nationally distributed on audio and videotape. Dr Wolpe is a regular columnist on biotechnology for the *Philadelphia Inquirer*, and appears frequently in the broadcast and print media.

Michelle Sahl is principal and owner of MJSahl Consulting, a Philadelphia healthcare management and consultancy firm specializing in strategic planning, market assessment, economic feasibility studies, and public policy analyses for healthcare, governmental, and nonprofit organizations. For over 20 years, Ms Sahl's healthcare industry experience has ranged from mental health research and program administration in major academic medical centers, to university-level teaching engagements. Prior to establishing her own consultancy business, Ms Sahl held positions in various organizations, including New England Medical Center, Harvard University, Laventhol & Horwath, and the University of Pennsylvania Health System. Recently, Ms Sahl has focused her interests in the areas of urban health, medical ethics, and population-based healthcare analyses. She holds degrees from Northwestern University (BA), Boston University (MEd), and Northeastern University (MBA). Currently, Ms. Sahl is completing dual programs for a Master's degree in Bioethics (University of Pennsylvania) and a doctoral degree in Health Policy (The University of the Sciences in Philadelphia). More information on Ms Sahl's experience is available on her website at http://home.earthlink.net/~healthypublicpolicy/index.html

James R Howard is from Concord, New Hampshire. He received his Master of Bioethics from the University of Pennsylvania in 2005. He served as a Clinical Research Project Manager for the University of Utah Department of Neurology, and performed research on diencephalic amnesia while an undergraduate at the University of New Hampshire. He is currently the Administrator for the University of Pennsylvania Institutional Review Board #1. He has worked with the American Association for the Advancement of Science, the Pharmaceutical Research and Manufacturers of America, the Biotechnology Industry Organization, and DuPont/Pioneer Hi-Bred on projects related to bioethics, health policy, and regulatory affairs.

Comprehensive Medicinal Chemistry II
ISBN (set): 0-08-044513-6

ISBN (Volume 1) 0-08-044514-4; pp. 681–708

1.23 Ethical Issues and Challenges Facing the Pharmaceutical Industry

M L Eaton, Stanford University, Graduate School of Business, Stanford, CA, USA

1.23.1 Introduction

Rarely is any drug or other medical product used in a patient without first having been commercialized by a corporate entity, either a pharmaceutical or biotechnology company, or some related company. And rarely does such a company commercialize a drug without having to face some ethical challenge along the way. The reason for this phenomenon is that, among the products produced by all industries, drugs have special moral importance. People depend on them for their lives and health, the products have a propensity to harm, they are often used in sick and therefore vulnerable people, and, since they are often costly, they create access disparities. In addition, many new products developed by biotechnology companies (drugs derived from fetal tissue or embryo cloning, genetic diagnostics, gene therapies, etc.) challenge people's notions about the acceptable limits of medical technologies. A second reason that ethics has special relevance to the pharmaceutical industry is that its primary customers, physicians, have since the time of Hippocrates operated within an ethical framework. Physicians are therefore trained to place patient and human subject interests above their own. Physicians are therefore sensitized to spot pharmaceutical corporate actions that do not adhere to their ethical frameworks. Since the 1960s, with the advent of the discipline of biomedical ethics, physicians' perspective on ethics has grown to encompass the applied biotechnologies, and ethics debates on these topics is strong in the profession. The vital importance of the products and the sensitivities of the physicians who prescribe them also ensures that company research, development, and marketing will be closely scrutinized by governments, physicians, patients, and the press. Such scrutiny means that any lapse in corporate ethical conduct is increasingly detected and engenders a loss of public trust in an industry that needs it the most. The combination, therefore, of the ethical significance of drug products, the ethical viewpoint of the customers, and the attention paid to the companies indicates that managers in these companies need to incorporate an ethical perspective into their corporate decision making.

Avoiding the harm that flows from unethical conduct is always a solid justification for adopting corporate ethics practices. Failing to behave ethically can result in legal or regulatory sanctions, avoidance by the capital markets, or retaliation from business partners, trade groups, and activists, all of which can influence the financial value (stock price) of a company and a company's ability to do business. Corporations can also be harmed by negative reactions from regulators, customers, the media, and medical organizations. Employees also suffer and can rebel when they feel compelled to ignore their moral compass when they walk through the workplace door. Issues that have generated such negative reactions for this industry include aggressive intellectual property practices, animal and human research ethical lapses, lack of research data integrity, product safety problems, selective publication practices, product pricing, lack of access to products, misleading advertising, self-serving legislative politics, and failure to address stakeholder concerns. The pharmaceutical industry is not alone in this regard. Increasingly, societies have come to doubt the integrity of business managers and, as a result, demand better behavior. A recurrent international public opinion poll has found that over two-thirds of 20 000 consumers surveyed across 20 countries believe that large companies should do more than "focus on making a profit, paying taxes, providing employment and obeying all laws."[1] History has repeatedly shown that when the expectations of society for business are not met, penalties are sure to follow.

Avoiding harm is one reason for companies to adopt ethical practices but it is not the only reason. Companies often benefit when they gain a reputation for ethical conduct and exhibit good corporate citizenship.[2,3] Ethical practices can have strategic value by improving investor, regulatory, consumer, and media relations. Brand image also improves. In some countries like the US, criminal penalties are reduced for white collar crimes committed by managers of companies with an ethics code and practice. Companies also usually prefer to do business with other companies they can trust and chambers of commerce welcome such companies into their communities. Having a reputation for high ethical standards also attracts employees who will work harder and better when they believe they serve a good company.

Others cite the intrinsic value of ethical standards – that ethical norms appeal to our common sense and our rationality – such as when we say that we should treat others fairly (justice as fairness), or we attempt to make choices that maximize social benefits and minimize harm (utilitarianism), or we conclude that there are rules such as not using people as a means to an end (rights and duties). It is also rational to say that ethical conduct is ethical conduct regardless of whether the actor is a person or a company. Therefore, the basic rules of ethics should apply to business. Another branch of ethics called virtue ethics, derived from the philosophy of Aristotle, focuses on the question of what makes someone a good person. This approach to ethics is also relevant to business. Those who believe that business ethics is intrinsically valuable often make analogies between personal and corporate integrity. If it is wrong for a person to behave in a certain way, the analogy goes, it is also wrong for the company. This approach to keeping an ethical focus has resulted in some companies adopting some version of an ethics quick test where managers ask certain questions of themselves before implementing decisions: If I do it, will I feel badly about myself?; How will it look in the newspaper? Commonly, the intrinsic value of ethical behavior is championed by those who recognize that the corporate environment can alter the way a person justifies decisions or behaves and by maintaining a personal focus it is easier to identify when those changes result in ethically questionable conduct.

Experienced managers support business ethics programs that avoid legal and financial harm. However, there is less (or even no) support for ethics when the appeal is to the actual or intrinsic corporate benefits that can be obtained. These seem to many managers to be more tenuously related to the primary traditional duty of a business manager, which is to maximize shareholder value. Anything that detracts from that focus is often shunned. But, for those who believe that business ethics is a modern intrusion on the responsibilities of business managers, consider the writings of Adam Smith, the economist and philosopher who is credited with establishing the economic doctrine of free enterprise. According to Adam Smith in his 1776 treatise *The Wealth of Nations*, unrestrained self-interested business activity guides the most efficient use of resources in a nation's economy, which in turn produces maximum public welfare. However, Smith qualified this view when he wrote that:

> Every man, *as long as he does not violate the laws of justice* [emphasis added], is left perfectly free to purse his own interest his own way, and to bring both his industry and capital into competition with those of any other man, or order of men.

Milton Friedman, the 1976 Nobel laureate economist and an ardent capitalist of the Adam Smith stripe, is also known best for his view that society benefits most from the incentives that result from operating businesses. Friedman wrote that the responsibility of business is to "conduct the business in accordance with [owners] desires, which generally will be to make as much money as possible." Yet, Friedman's quote went on to say that this responsibility must be exercised "while conforming to the basic rules of society, both those embodied in law and those embodied in ethical custom." The fact that both Adam Smith and Milton Friedman believed that neither justice nor other ethical customs should be violated in the pursuit of business reinforces the view that ethics should play a role in the operation of a company.

If a business manager accepts this proposition, he or she can seek education or training in business ethics. But for managers with more constraints on their time and attention, two valuable substitutes are recommended – one is to gain awareness of the common situations in the business that raise ethical issues and the second is to know how ethical advice can be obtained when needed. This chapter addresses both of these approaches. Awareness of where the ethical pitfalls lie and where to obtain help in resolving them can break a common practice of addressing these issues only in times of emergency and allow managers to anticipate potential problems and address them proactively.

The remainder of this chapter discusses ethical dilemmas commonly encountered in the pharmaceutical and biotechnology industry. Following is a discussion of the ways in which companies can obtain ethics advice and the pros and cons associated with these various approaches.

1.23.2 Ethical Issues Confronting Companies in the Early Twenty-First Century

1.23.2.1 Conflicts of Interest with Academia

Frequently, pharmaceutical companies seek the research expertise of academics to conduct the studies that lead to product approval. Universities welcome these research contracts since they offset ever-dwindling public research support, generate publications, and create training experiences for students. Despite the mutual benefits, however, conflicts of interest are common in this setting since the two entities engaged in this activity are very different from one another. Companies do not want competitors to learn of their early drug research activity and often seek to keep this information confidential. Confidentiality is also important to protect intellectual property and trade secrets. The company sponsor also needs to design the research to achieve a commercial goal. In contrast, academics operate under conditions of openness and free discussion, and faculties are accustomed to the freedom to research any matter regardless of its short-term practical application. Universities also pride themselves as sources of objective knowledge and some faculty members suspect that corporate funding influences their research focus or scientific judgment about a particular technology. Too much corporate research support, it is argued, can undermine the credibility of faculty members and the public's trust in the university. In addition, academics have other campus obligations (teaching, governance) that can be compromised if a faculty member spends too much time on corporate sponsored projects. Sometimes, also, industry sponsored research can be too mundane to advance science and learning. As a consequence, even though faculty may agree to conduct corporate research, universities often refuse to abide by corporate requests to maintain confidentiality, delay publication, or alter other basic academic practices.[4,5]

Health policy researcher David Blumenthal and colleagues at Harvard University study these conflicts of interest and commitment and have demonstrated that a certain level of industry-sponsored academic research is beneficial to both parties.[6,7] Beyond a certain point, however, the risks seem to outweigh the benefits. When academic researchers obtain more than two-thirds of their research funding from commercial sources, they tend to be less open, lose academic credibility, and become less productive academic citizens. Other problems include intellectual property disputes, disagreements about study design integrity and data interpretation, and conflicts over company-mandated publication delays. A certain amount of publication delay to submit patent applications is usually tolerated but not when delay is to preserve competitive advantages such as by restricting the dissemination of unfavorable data. On the other side of the deal, companies can be unhappy with university researchers when, for instance, they assign the work to junior faculty or students, cause delays, breach confidentiality, fail to follow the study protocol, and commit research misconduct.

An example of a research collaboration that created these kinds of conflict occurred when Sandoz collaborated with academic physicians to study the cardiac effects of one of its blood pressure drugs compared to another common drug. Some of the faculty researchers alleged that Sandoz had wielded undue influence in controlling how the research results were reported. For this reason, these faculty researchers made public their refusal to have their names included as authors on the publication.[8] At other times, the complaint is about the magnitude of corporate research funding. Research collaborations such as those between Sandoz and the Scripps Research Institute in 1992 and Novartis and the University of California at Berkeley in 1998 were so large that protestors alleged that companies were taking over and subverting the missions of premier American biomedical research and academic institutions. In the Scripps case, National Institutes of Health objections and congressional investigations forced a drastic reduction in scope.[9] Faculty objections at Berkeley delayed the Novartis deal for over 6 months.[10] Another well known case involved the company Immune Response, which sued researchers at Harvard University and University of California at San Francisco (UCSF) over publication of data on the company's anti-HIV drug Remune.[11] The academic researchers found a lack of effectiveness. Corporate scientists analyzed the data differently, found a clinical benefit, and withheld data from the academic researchers in an attempt to prevent their publication of negative study results. When the academics eventually published their data, the company sued claiming damages of US$7 to US$10 million.

Problems like this led many universities to adopt conflict of interest and conflict of commitment policies. These policies often prevent faculty from committing too much time to corporate sponsored work, dictate the permissible financial ties to industry, require control over study design, mandate disclosures of corporate funding sources, and prevent certain controls over and delays in publishing.[12] Becoming aware of the sources of potential conflict between independent scientists and corporations and dealing with them proactively is often the best means to avoid conflict in this endeavor.[13] Several resources can assist the researcher in this regard, including the reports and policies of the American Association of Medical Colleges, the National Health and Medical Research Council of Australia, and guidelines generated by the International Committee of Medical Journal Editors. Careful attention to the avoidance and management of these conflicts serve the goal of generating mutually constructive and medically beneficial research collaborations.

1.23.2.2 Biased Study Design and Data Interpretation

A related ethical question relates to a concern that corporate research protocol designs are manipulated or the data interpreted to achieve a result favorable to the commercialization of a drug.[14] This is an ethical issue faced by both academic researchers and those employed by industry. The dilemma arises because companies need to design trials to meet their marketing needs. For instance, if the company plans to market a drug for a severe form of a particular disease, the studies will exclude animal and human subjects with mild disease. Whether such a study design is considered biased hinges on intent, which is often revealed by the manner in which the company interprets and markets the data. Using the data, for instance, to claim that the drug is useful in the disease per se is obviously misleading. The study may also be considered irresponsible if the company knows or strongly suspects that the drug is ineffective or toxic in mild disease and that unsuspecting physicians will tend to do what they always do – use the drug in all forms of the disease. The company may respond to this situation with a range of actions:

- claim that it is someone else's responsibility to produce study data on how the drug behaves when used 'off-label' (the term off-label refers to the use of a drug for other than approved indications), in this case, for patients with the mild form of the disease;
- state in the labeling that the drug is useful in severe forms of disease;
- state that no data exists on how the drug behaves in mild disease;
- disclose the suspicion that the drug is not useful or safe in the wider spectrum of diseased patients; and
- act on the suspicion and conduct studies to show how the drug behaves when prescribed in the real world, in this case, for patients with the mild form of the disease.

Disagreements are increasingly common about whether drug study designs intentionally enhance efficacy outcomes or conceal side effects. Merck experienced criticism of this nature after it withdrew its anti-inflammatory drug Vioxx in 2004, a drug that was generating US$2.5 billion per year. After 5 years on the market, the drug was shown to enhance the risk for heart attacks and strokes prompting questions about whether the company had delayed the discovery of these risks. Leaks of internal documents revealed comments by the company's vice president for clinical research that suggested she knew that Vioxx could cause cardiovascular side effects and was contemplating designing an important clinical trial to obscure this finding.[15] In another widely publicized case, the lead researcher responsible for ImClone's anticancer drug was called to testify at a congressional investigation hearing and asked to address allegations that his lucrative company ties had led him to manipulate drug study data.[16] Whether Merck or ImClone scientists had designed studies in order to heighten efficacy or mask side effects is unknown. Yet, commentary in the scientific literature indicates a long-held uneasiness about whether company-sponsored drug trials are skewed to produce positive data.[14] Others say that this conclusion is too facile – that companies sponsor only those trials where prior data support a reasonable expectation that outcomes will be positive.

Another questioned practice is covert reporting of the same data. For instance, researchers found that meta-analysis of the side effects of the antipsychotic risperidone was skewed because what appeared to be 20 separate studies was actually only 9 studies reported multiple times, often with different lead authors.[17] Re-publication of favorable studies artificially skews the balance of opinion in favor of a drug and some researchers suspect that drug companies do this deliberately. Another criticized study design practice deals with how companies define and code adverse drug reactions (ADRs). Obviously, the number and kind of ADRs reported can be influenced by company decisions, for instance, to characterize an event as disease rather than drug related. This issue arose when regulators were attempting to determine if the antidepressants selective serotonin reuptake inhibitors (SSRIs) caused an increase in suicidal behavior. Some critics accused companies of wrongly attributing all suicidal behavior to the depression the drugs were designed to treat. Subsequently, research showed that these drugs could increase suicidal behavior and black box warnings to this effect were added to the US drug labels almost 16 years after the first of these drugs was marketed.[18]

Another related issue involves whether companies or company ties influence data interpretation or opinions about drug products.[19] Recent studies have given credence to an association if not causation in this regard. For instance, in 1998, a debate existed about the safety of calcium-channel antagonists. One study at the time showed that 96% of medical journal authors writing on the topic who supported the use of the drugs had a financial relationship with the manufacturers of the drugs compared with 37% who were critical of them.[20] This study and others have been cited as evidence that an association with industry lessens objectivity about the benefits and risks of drugs. Another possibility, of course, is that scientists and physicians form a favorable opinion about a product, which then leads to a research affiliation with a company.

Again, the possibility that conflicts of interest can lead to biased studies and opinions has led to university and journal policies requiring disclosure of industry ties. University policies, based on the belief that the amount of money determines the amount of influence, also limit the extent to which faculty can financially benefit from commercial affiliations. These policies cannot, however, eliminate the potential for bias. Sensitivity to the problem should prompt researchers to remain vigilant – to carefully review all protocols to eliminate biasing design elements and to maintain critical and objective data interpretation.

1.23.2.3 Selecting Drugs to Develop

Several ethical and social questions exist about the choice of drugs to develop. The first involves whether the result is worth the effort and cost. Researchers often wish to devote their time and expertise to developing drugs that will make a major difference in the lives of sick patients. The socially beneficial aspect of their work is what drives them and keeps them in the lab. In contrast, they chafe when asked to develop a chemically related drug that offers only minimal improvement over existing therapies, so-called 'me-too' drugs. The term me-too comes from the fact that as soon as a large market prototype drug becomes available, several other similarly active compounds immediately follow from competing companies. This kind of drug is attractive for companies to develop since they can capitalize on the research and approval of the successful prototype, develop follow-ons more efficiently, and take an easier share of a large established market. Me-too drugs are also developed to preserve a market for a blockbuster drug, the patent for which is about to expire. Patenting a similar drug can allow the company to maintain premium pricing rather than watch revenues decline when the drug loses its patent and becomes generic.

Most often, the follow-ons offer some benefit over the prototype, such as lesser toxicity, improved pharmacokinetic profile, more favorable compatibility with comorbid conditions, or easier use. Patients benefit when the me-too approach offers significant medical benefit and generates price competition.[21] The reason that many researchers dislike developing me-toos, however, is that too many offer only a marginal benefit, so small, claim some, that the intellectual and financial development resources expended far outweigh any medical or social benefit. Some claim that the situation worsens when aggressive marketing obscures the small differences between a highly priced new drug and a cheaper generic. Scientists and physicians prefer intrinsic medical value rather than marketing to determine which drugs get prescribed. The fact that these drugs are considered duplicative and wasteful of resources angers consumer groups as well. Criticism of me-too drugs is relevant even in the situation where the existence of the me-too is the result of a development race rather than post hoc imitation. Once a strong class breakthrough drug has been approved, some favor a system that will allow only those follow-ons that demonstrate significant improvements over the prototype. Currently, most regulatory systems approve drugs based on their intrinsic merits rather than whether they offer significant benefits over existing drugs.

An example of the me-too phenomena is captopril, the prototype of the angiotensin-converting enzyme (ACE) inhibitors, a class of drugs used to control hypertension. The approval of captopril was followed by at least 15 me-too ACE inhibitors, most of which have equivalent efficacy and side effect profiles. The similarities are so close that one pharmacologist suggested that physicians should routinely prescribe whatever brand was cheaper.[22] An example of a drug developed to preserve revenues from a patented blockbuster is the acid reflux drug Nexium by Astra Zeneca. This drug is the *S*-isomer of the company's blockbuster drug Prilosec, which had been a major revenue producer for the company until its US patent expired in 2001. Astra Zeneca took advantage of the fact that isomeric forms are considered by the Food and Drug Administration (FDA) to be different compounds and can be separately patented. The company studies showed the *S*-isomer to be somewhat more effective than the racemic mix and, while other companies can experience a loss of up to 85% of a drug's revenues within a year of patent loss, Astra Zeneca was able to preserve its franchise for Prilosec by marketing Nexium, the sales for which were US$3.3 billion in 2003. However, this move created a storm of criticism when patients understood that they paid up to eight times as much for the patented Nexium as for the generic and very similar (and now over-the-counter) Prilosec. Company reputation suffered as a result and a class action lawsuit was filed alleging that deceptive advertising of a medically significant benefit over Prilosec led to the demand for Nexium.[23] Whether prescribing a me-too based on price or prescribing an equally effective generic drug, the cost saving can be substantial and, at a time when financial resources for health are limited, economic considerations are not a minor point. The profusion of these follow-on drugs has led commentators to ask whether this conventional form of pharmaceutical business competition serves society as well as it should.

This me-too value issue relates to a larger question about the role of the pharmaceutical industry in society and whether the industry has an obligation to focus their efforts on innovative therapies and those for neglected illnesses. Some dislike the question since it implies a social obligation that is more appropriate for governments than business. The criticism seems aimed more at the existence of the capitalist system that would become seriously eroded if companies were required to compromise their focus on the bottom line to benefit public health. In contrast, the focus on corporate profits contributes to serious social injustice since drugs are not developed for diseases that exist in poor countries.[24] Because of increasing pressure from underserved groups, this issue will continue to be debated by the industry and policy makers.

1.23.2.4 Human Research Ethics

Because of a long-standing history of abuse, the ethical issues associated with human research have received widespread attention. Multiple national and international guidelines and regulations were developed to ensure that human subjects are treated ethically and responsibly.[25] Because these resources are readily available, this section will deal only with the general categories of research topics that generate the most ethical concern.

1.23.2.4.1 Risk versus benefit

The first consideration for any researcher is to assess whether the potential benefit of the proposed research outweighs the potential harm, which can include physical, psychological, social, or economic harm. In addressing this issue, most research can be justified if benefit and risk is applied broadly in a utilitarian sense, i.e., any harm to the relatively few subjects is justified either because the research leads to new medical products that can benefit the many or, with failed trials, there is benefit from gaining medical knowledge and sparing of many patients from a useless or unsafe drug. However, ethical norms require that the risk-benefit calculus consider primarily the effects of the research on the human subjects and, additionally, on the potential to advance medical knowledge. For instance, in the US, IRBs (the institutional review boards that approve human research) are instructed to assess risk and benefit as follows:

> Risks to subjects [must be] reasonable in relation to anticipated benefits, if any, to subjects, and the importance of the knowledge that may reasonably be expected to result. In evaluating risks and benefits, the IRB should consider only those risks and benefits that may result from the research (as distinguished from risks and benefits of therapies subjects would receive even if not participating in the research). The IRB should not consider possible long-range effects of applying knowledge gained in the research (for example, the possible effects of the research on public policy) as among those research risks that fall within the purview of its responsibility.[26]

The Declaration of Helsinki (Section A.5) is more narrow and states that the well being of the human subject should take precedence over the interests of science and society. Other aspects of minimizing risks to subjects require that sufficient prior research should be performed to understand the potential risks as much as possible. Researchers also must be qualified and adhere to standards of good research practice so as to preserve the benefits of the research as much as possible, thus not exposing subjects to risk for no good purpose. Minimizing the risks and burdens to subjects in these ways adheres to the ethical norm of non-malfeasance or doing no harm, which is the principal obligation of any research physician.

1.23.2.4.2 Informed consent

Over the years, lack of informed consent has been at the heart of most unethical studies. Failure to fully inform subjects means that people are used as a means to an end (to produce data), disrespects the right to personal autonomy, and offends our sense of fairness. A commonly cited example occurred in a San Antonio, Texas study in the 1970s that enrolled mostly poor Mexican-American women who had had multiple pregnancies and were seeking contraceptive advice or medication. The double-blind, randomized, placebo-controlled study was intended to evaluate the cause of oral contraceptive side effects. Subjects enrolled in the study were not told that they could receive a placebo and, as expected, there were a high number of unplanned pregnancies in the placebo group.[27] Ethical lapses such as this are the reason for mandatory requirements that subjects be provided with a clear and accurate explanation of risks, benefits, and alternatives before enrolling in any clinical study.[28] In order for consent to be meaningful, the potential subject must be given understandable information, must be informed of all of the expected consequences that a reasonable person would want to know about, and must be given an opportunity to ask questions and an adequate time to reflect before committing. Special attention to the consent process is needed when studies are complex or particularly risky and when subjects' ability to understand is compromised. Some researchers in these situations have required subjects to pass a comprehension test before enrolling them in studies.

1.23.2.4.3 Free consent

Informed consent is not the same thing as free consent. Respect for persons requires that consent to research be truly voluntary. For various reasons, some patients are more vulnerable than others and are therefore more susceptible to the influence of a trusted physician who asks for consent to join a drug study. Especially when patients are desperate for any chance of treatment, physician researchers need to be careful to minimize the possibility of coercion or undue influence in the recruitment process. Vulnerable patients also include those who are economically and medically disadvantaged and who may volunteer simply because money is offered or because research participation offers the only access to medical care. Those who lack mental capacity and cannot competently consent to research need a proxy who represents the patient's best interest. In addition, many believe that patients with mental deficits should be included in drug research only if the mental condition is a necessary characteristic of the research population. Most countries also have provisions forbidding or controlling the use of prisoners in research because of the difficulty in dissuading the incarcerated from the belief that cooperation with researchers will result in favorable prison treatment.

In general, little regulation exists regarding the methods or safeguards that should be employed to ensure that vulnerable or incapable subjects consent freely. As a result, individual researchers need to tailor their recruitment methods to minimize any coercive influences.

1.23.2.4.4 Privacy

The information collected in drug research is medical in nature and, as such, must be kept confidential. Also, the fact that a subject has entered a clinical study often reveals his or her diagnosis, so enrollment is also private information. Every precaution should be taken to respect the privacy of the subject and the confidentiality of the patient's information.

1.23.2.4.5 Research in children

Pediatric drug research has been generally neglected because of the reluctance to subject children to unknown risks given that they cannot protect their own interests through informed consent. In addition, the protective instincts of parents and guardians make them unwilling to enroll their children in research. The resultant lack of data on how children react to drugs means that most pediatric treatment proceeds through trial and error. Regulatory agencies are trying to correct this disadvantage by offering drug companies inducements to conduct pediatric research and by specifying the ethical requirements necessary to enroll these subjects in research. For instance, the FDA regulations list the safeguards that must be satisfied depending on the risk of the research, the potential benefit to the child, the potential to produce generalizable knowledge about the child's disease, and the opportunity to understand, prevent, or alleviate a serious problem affecting the health or welfare of children. The regulations also specify the qualifications of parents or other consent providers and require the assent of the child if feasible.[29]

1.23.2.4.6 Placebo controls

The ethical debate about the use of placebo controls centers on the scientific benefits of eliminating study variables versus the risks to subjects of forgoing standard treatment. The use of placebo controls, according to most researchers, best satisfies the fundamental research goal of producing scientifically accurate and dependable data that can distinguish the effect of the drug from other influences, such as spontaneous change in the course of the disease, placebo effect, biased observation, or chance. Regulatory agencies will not approve a drug without sufficiently controlled data and the FDA especially adheres to the belief that no control is better than placebo.[30] Thus, drug companies often rely on the use of placebo controls because the resulting study data are considered trustworthy by the regulatory agencies and by physicians. Placebo controls are also desirable since, by producing fewer confounding variables, studies can be conducted more rapidly with smaller populations than clinical trials using active controls. Shorter clinical trials with fewer subjects can save time and money and put effective products on the market more quickly.[31]

The problem with placebos is that human subjects must agree to random and blind assignment to a therapeutically inert substance masked as an active treatment. Obviously, the sicker the patient, the greater the potential harm from placebo administration. The duplicity (even though consented to) and the fact that diseased subjects receive no active treatment are considered unethical by some researchers and standards bodies. For instance, the World Medical Association (in the Declaration of Helsinki, C. 29) states that placebo controls should be used only in the absence of existing proven therapy. With few provisos, the Declaration requires that, if other therapies are available, experimental drugs should be tested against the best of them. The World Medical Association's stance on the use of placebos was promulgated after some controversial AIDS research in Africa and was intended to reflect a widespread interest by Association members to place subject safety over concerns about data certainty.

In the African studies, the drug AZT was tested to prevent the transmission of HIV infection from a pregnant woman to her newborn. The studies were carried out in 11 impoverished developing countries. Women in these countries were given a shorter and less expensive AZT treatment regimen than was used effectively in Western countries. The studies, which were double-blind and placebo-controlled, ultimately determined that the new treatment protocol was better than no treatment but was not as effective in preventing infant infection as the full Western treatment. According to the researchers, the use of placebos was justified because the study women had no access to prenatal care and, if they did, could not afford (nor could their countries afford) the drug as it was prescribed in the West. Thus, no study women were denied access to any locally available treatment. Plus, the studies offered at least some chance that these women would receive active treatment.[32] Heated debates among medical professionals and medical ethicists followed publication of the study results.[33] The major objections included the exploitative nature of the research and doubts about whether there was full informed consent (always an issue since many people do not understand the concept of placebo control even after an explanation). But the primary ethical objections focused on the fact that the HIV infection rate in the babies born of the study subjects would have been significantly less if the study had used an active treatment control with the effective Western treatment rather than placebo.

Given the differences of opinion on this topic between regulatory agencies and medical organizations, debates about placebo controls are expected to continue. In this environment, it is important for researchers to explicitly justify the use of placebos based on compelling medical and scientific criteria.

1.23.2.4.7 Adherence to protocol

Research protocols need to be carefully designed in order to produce scientifically valid data and to protect the rights and interests of human subjects. Failure to adhere to the protocol undermines both. A widely publicized example of this kind of failure occurred in 1999 when an 18-year-old human subject Jesse Gelsinger died in a gene therapy trial conducted at the University of Pennsylvania. Gelsinger's death resulted in investigations that disclosed multiple failures to follow protocol requirements, including failure to stop the study if subjects developed certain levels of toxicity, enrolling subjects who failed to meet eligibility criteria, failure to report side effects, and failure to notify the IRB of protocol changes. In Gelsinger's case it was likely that these failures contributed to his death and the family sued as a result. In addition, the FDA ordered the University of Pennsylvania to halt all gene therapy experiments at its Institute for Human Gene Therapy and brought debarment proceedings against its director.[34] The tragedies associated with this incident were many and, in addition to the death, included a major setback in the progress of gene therapy research and the career injury suffered by one of the country's best gene therapy scientists. Most of the time, protocol violations do not produce such widespread destruction. But at a minimum, the failure to adhere to protocol requirements compromises the integrity of the resultant study data and thus undermines all subsequent research and clinical decisions based on that data. Consequently, the importance of this research requirement should not be underestimated.

1.23.2.4.8 Monitoring

Vigilant study monitoring is required so that studies can be stopped as soon as the balance of harms and benefits becomes unfavorable. Monitoring has become more difficult as companies have been conducting multicenter studies often managed by independent contractors (clinical research organizations). Hoechst Marion Roussel (this company later merged with France's Rhone-Poulenc SA to become Aventis SA) experienced a problem in this regard when it conducted trials on Cariporide, a drug expected to reduce cardiac tissue damage after a heart attack. The Naval Hospital in Argentina was one of 26 Argentinean study sites and one of 200 medical centers worldwide that participated in the 11 500-subject trials. In 1998, Hoechst and its US-based clinical research organization, Quintiles Transnational, noticed some irregularities with the Naval Hospital trial and notified the local authorities. Criminal investigations led prosecutors to allege that 137 patients had not consented to the study treatment and signatures on at least 80 consent documents had been forged. Prosecutors also collected evidence that subjects' records contained duplicated electrocardiograms with the characteristic findings needed to justify entering the patient in the trial. At least 13 subjects had died, and prosecutors claimed that some of these deaths were attributable to the experimental drug.[35] While no one thought that the drug company or the clinical research organization had been responsible for these problems, the multitude of offshore studies and the fact that both the company and the study monitor were in different countries most likely contributed to the problem.

1.23.2.4.9 Compensation for injury

Most countries have requirements that research subjects be told whether they will receive medical care or compensation for injuries experienced in clinical trials. However, there is no uniform practice about the provision of treatment or compensation. Some researchers provide neither and others provide both. And there is a range of practice between these two options. Companies attempt to obtain insurance to underwrite the risk but it is often not available because of the unknown risks inherent in research. Injured human subjects who are not treated or compensated for their expenses and losses are left to their own devices and they sometimes sue the study sponsor. Many commentators see the unfairness in such situations and have lobbied for some uniform treatment and compensation practices. During the 1970s, a consensus began to evolve among medical ethicists that compensatory justice principles required that subjects who were injured as a result of participating in clinical research be entitled to full compensation for both medical- and nonmedical-related costs. Policy reasoning that supported this conclusion stemmed from the view that society as a whole has a direct stake in the conduct of scientific research, including the knowledge benefits that come from experimental failure. If society shares the benefits, so should it share the financial burden of the research activity by paying for the costs of research-related injuries. Legal and ethical commentators argue that even if these injuries occurred through no fault of the researcher or sponsor, accountable public policy required that responsibility be fixed wherever it will most effectively reduce the hazards to life and health inherent in the research endeavor. If research sponsors were committed in advance to pay for injuries, the tendency to engage in excessively dangerous research would also be minimized and research sponsors would take more care in designing safe trials, monitoring for injury, and stopping trials as soon as the data revealed excessive risk. Also, companies are most often in a better financial position to bear the costs of research injury by including the cost of injury compensation in the price of the product. Finally, people would more likely consent to participate in research if they knew they would not be responsible for the costs of any injuries. Many proponents of this view also argue that since the primary goal of compensatory justice is to restore the injured person as much as possible to his or her original condition, injury compensation should include medical and nonmedical costs of injury, the latter including such things as lost wages or services provided while the injured subject was disabled.[36]

Over the years, these arguments have been presented to various national and medical commissions,[37] but consistent policy has failed to result primarily because of disagreements about who should bear the expense. Consequently, researchers need to consider in advance how they will manage this aspect of their research activity.

1.23.2.4.10 Poststudy responsibilities to subjects

After a clinical drug study ends, subjects become patients again and are most often referred back to their primary care physician for any needed follow-up care. There are times, however, when subjects will expect something more from the company. Subjects who experience long-term side effects from study drugs can expect help from the company despite any contrary provisions in the consent form. In other cases, when subjects benefit from the experimental drug, it is natural that they would want to continue taking it. Prohibitions on selling experimental drugs prevent such access and, especially when the drug was an overall failure and will not be marketed, subjects can feel abandoned and unfairly denied effective therapy. The acuteness of the situation is heightened when the drug treated cancer or some other life-threatening illness. For these reasons, the Declaration of Helsinki contains a provision that effective drugs are provided to all study subjects at the conclusion of research. Not all researchers abide by this provision and it remains a matter of debate.[38] Envisioning all possible study outcomes in advance will help researchers devise strategies to address questions about poststudy support.

1.23.2.5 Publication of Negative Data and Data Access

Drug companies are not required to publish study results. Except under rare circumstances, regulatory agencies keep company data confidential and only require disclosure of selected summaries of study data in the drug labeling. As a result, negative drug study data often remains unpublished. Nothing stimulated an interest in changing this status quo as much as the revelation about unpublished data on the efficacy and safety of the use of SSRIs in depressed children.

All SSRI manufacturers had obtained approval to market the drugs for use in depressed adults but for years physicians had been prescribing the drugs for children off-label. Eli Lilly & Company then performed studies and on the basis of the results obtained approval to market its SSRI Prozac for use in depressed children. When GlaxoSmithKline applied for similar approval in the UK for its drug Paxil (Seroxat in the UK), the resulting investigation shed light on the phenomenon of selective publication. The UK's Medicines Control Agency (MCA) discovered that GlaxoSmithKline had conducted nine studies of its drug in depressed children but had published only one of them. When the MCA and then the FDA conducted analyses of the individual and combined data in all of the nine submitted studies, the combined data showed that the drug was not effective in pediatric depression. Furthermore, in total, these studies showed that 3.4% of children who were taking or had recently stopped Paxil had attempted suicide or thought more about it, compared with 1.2% of the children taking a placebo, a statistically significant difference. In contrast, the one study that GlaxoSmithKline had published showed that the drug was better than placebo in children and was not associated with any suicidal attempts or ideation. Further investigations revealed that other companies had also sponsored but not published negative pediatric SSRI studies. GlaxoSmithKline had not been marketing the drug for childhood depression. However, critics, regulators, and prosecutors claimed that the company knew that the drug was widely prescribed for children and that, by publishing only the positive study, the company misled physicians and patients about the efficacy and safety of the drug. The press reporting on the story convinced the public that this kind of publication bias was common, and the ethical, clinical, and scientific integrity of the industry was called into question. The resultant furor involved the drug regulatory agencies (they started to reconsider their practice of keeping study data confidential), drug companies (they began to face lawsuits and criminal charges were brought against GlaxoSmithKline), medical journal editors (they sought wider disclosure standards), the US Congress (which convened investigations), and academics and patient advocates (who demanded full access to all drug company research data).[39–43] An article in the *Washington Post* stated that the FDA had not approved Paxil for use in children, but, since physicians can prescribe drugs for nonapproved uses, more than 2.1 million Paxil prescriptions for children had been written in 2002, according to the legal complaint.[40]

The SSRI situation stimulated a renewal by national and international groups to encourage registration of all drug studies and disclosure of all drug study data.[44] Registries had existed for this purpose but had not been widely used by companies. Surveys had shown that existing registries had not produced the openness and transparency that would have prevented the problem seen with pediatric SSRI data. Even companies that had attempted to address publication bias had not been successful in avoiding problems. For instance, 6 years before the pediatric SSRI problem, GlaxoSmithKline's predecessor Glaxo Wellcome announced on its website that it would register, make accessible, and publish all of its clinical trials.[45]

Openness and access, while well intentioned, is operationally difficult. The problem with publishing all studies is that some are merely small trial balloons and small sample sizes tend to magnify or diminish influences and findings. In addition, some studies are recognized after the fact as being flawed in some important way. Publication of such studies would constitute misrepresentation of the behavior of a particular drug. Publication of negative studies could also lead physicians to prematurely reject a medicine. Researchers learn from failed trials and many times are able to change the study conditions to improve a drug's performance. Companies are also reluctant to publish if it undermines their ability to patent or keep sensitive business information confidential. Others wonder if the mountain of data that will become available will be digestible and practically useful for physicians and ask why physicians should not continue to rely on the regulatory agencies for the vetting of drug data. Despite these difficulties, the SSRI case stimulated renewed and wider efforts by companies, journals, and medical and scientific organizations to eliminate publication bias, allow for greater transparency and access to trial data, and improve the ability to evaluate data, flawed or not.

1.23.2.6 Product Pricing and Access to Drugs

New drugs are expensive because manufacturers need to recoup the considerable expense of getting a new drug to market (upwards of $800 million in the US at the time of writing).[46] Included in the cost of developing a successful drug are the costs of developing and testing drugs that never reached the market. The price paid by patients is sometimes covered by insurance or public health programs but, frequently, uninsured patients in developed countries and most patients in undeveloped countries cannot afford to pay for new drugs.[47] This disparity has prompted members of society to ask whether the industry is going to address the problem. Many large pharmaceutical companies have programs that provide needy patients with drugs and/or assist patients in obtaining public or private insurance to cover the cost of its drugs.[48] Yet, these programs help only a relatively few patients prompting countries and patient groups to continue lobbying the industry to make medicines more affordable. In the US, Congress frequently considers bills that would mandate drug price controls. And in Germany, where price controls exist, companies went so far as to contribute €200 million (about US$189 million in 2002) to the national health plan to prevent the country from imposing a further 4% price cut on prescription drugs.[49] Some states in the US have even sued drug companies for allegedly inflating prices.[50] Patients in the US, where drug prices are often the highest, hire tour companies to bus them to Canada where drugs can cost as much as 50% less.[51] Activities such as this prompt debates about allowing importation of cheaper drugs,[52] or controlling patent monopolies.[53]

Desperately needed drugs create extra pressure on companies such as when the AIDS activist group Act-Up was created to harass Burroughs Wellcome to lower the price of the first anti-AIDS drug AZT. Act-Up's tactics were very disruptive and included convincing the National Institutes of Health to adopt a 'reasonable pricing rule' imposed on companies who develop drugs with the assistance of publicly funded research[54] and prompting Congress to investigate Burroughs Wellcome's drug pricing policy. The company did lower AZT prices and other companies producing AIDS drugs were made to follow suit.[55] Most of the pressure, however, comes from third world countries where drugs are not affordable, even for the generics used to treat common diseases such as tuberculosis and malaria. The global AIDS crisis has added fuel to the protests against drug prices and spawned multiple efforts to develop generic drugs in violation of drug company patents.[56] All of this activity has put pressure on the pharmaceutical industry to disclose its price setting practices, lower prices, and/or relinquish patent rights to allow the manufacture of cheaper copies of patented drugs.

The pharmaceutical industry has been hard pressed to convince the public that forcing these changes, while producing short-term relief, would dampen incentives for industry to engage in expensive R&D and would ultimately deprive society of valuable drugs. The US Pharmaceutical Research and Manufacturers of America reported that, in 2003, its member companies invested an estimated US$33.2 billion in research to develop new disease treatments. This figure amounted to an estimated 17.7% of domestic sales spent on R&D – a higher R&D to sales ratio than any other US industry.[57] Without the ability to freely price, this vigorous activity will inevitably diminish. Given the high cost of development and the crucial need for prescription drugs, this cost versus value versus access debate is sure to continue for many years.

1.23.2.7 Marketing and Advertising

There are three main ethical issues associated with drug company marketing and advertising. The first is whether the material is truthful and fairly balances information about benefit and risk. The second issue has to do with whether companies medicalize conditions to generate a market for drugs or otherwise stimulate off-label use. The third issue concerns the amount of money companies spend on advertising.

Since prescription drug regulations have been on the books, company marketing material has been required to be truthful, contain all material information, and fairly balance risk and benefit information.[58] The industry has also adopted voluntary codes to promote and support ethical marketing practices.[59] These regulations and codes are a result of continuous negotiation between regulatory agencies, drug companies, and physician groups about the boundaries of legal and ethical prescription drug marketing and advertising practices. Nothing has tested these requirements more, however, than when companies in the US were allowed to market drugs directly to patients. Until recently, it was widely believed that direct-to-consumer ads (DTC ads) for prescription drugs were inappropriate because lay people lacked the education to understand the technical medical information that was conveyed. In addition, drug ads can be misleading since they promote one drug only and fail to educate patients about other treatment options. Another primary objection concerns the fear that manufacturers, in their effort to persuade, attempt to downplay the risks and give patients the false impression that the drugs are totally safe. The FDA struggles with DTC ads because it is understaffed to police them and because drug companies always seem to be pushing the envelope of tolerability, especially for televised ads where the time to present risk information is limited. Physicians generally dislike DTC drug ads because they prompt patients to self diagnose and put pressure on the doctor to prescribe, which often forces doctors to spend valuable time explaining why a certain advertised drug is inappropriate.[60] For all of these reasons, most countries do not allow DTC drug advertising. In contrast, this activity has been widespread in the US since the late 1980s. The drug companies claim a right to commercial free speech and the FDA, after conducting some surveys, decided that patients had a legitimate right and need to learn about newly available prescription drugs and these ads can be educational and empowering and can prompt patients to seek needed medical attention. Despite the expected benefits, however, debates continue about whether DTC prescription drug advertising is good or bad for the medical profession and the public.[61,62]

DTC drug advertising is an effective way to increase prescribing. Research has shown that drugs that are heavily advertised, whether to physicians or patients, are prescribed more. In one study, researchers at Harvard University and the Massachusetts Institute of Technology looked at the effect of DTC drug advertising on spending for prescription drugs. The study found that every US$1 the industry spent on DTC advertising in 2000 yielded an additional US$4.20 in drug sales. In addition, DTC advertising was responsible for 12% of the increase in prescription drugs sales, or an additional US$2.6 billion, in that same year.[63] Numbers like this are the reason that the pharmaceutical industry in the US is likely to continue this activity.

The so-called medicalization issue involves the question of whether marketing departments in pharmaceutical companies intentionally (and irresponsibly) create an expansion in what is considered a disease condition. For example, one reason that attention deficit hyperactivity disorder (ADHD) became a legitimate diagnosis was the discovery that Ritalin could control it. A similar change occurred when drugs were found to control hair loss, male erectile dysfunction, short stature, and menopausal and premenstrual symptoms. This kind of expansion has been given the name diagnostic bracket creep by critics and has come to mean that what was once considered within the range of normal experience and behavior is now classed as illness. The introduction of new drugs is the primary reason that diagnostic bracket creep occurs. Debates exist about whether this phenomenon generates better and more enlightened treatment for patients or an irresponsible medicalization of conditions that should not be considered diseases. This debate is currently robust within the psychiatry and psychology professions in the case of approval and marketing of one company's SSRI antidepressant for a new disease the company has called social anxiety disorder (SAD). Since many people are often uncomfortable in new social situations, critics accused the company of using its ads to expand the market for its drug by blurring the distinction between a legitimate psychiatric disorder and normal personality variation.[64]

These criticisms of drug marketing practices have led to the question of cost versus benefit. If the advertising (especially the DTC kind) does not improve health awareness or provide other benefits and, worse, if the ads are harmful, the cost of advertising becomes an issue. In 2003, US companies spent US$3.22 billion in DTC advertising[65] and it is reasonable to assume that this amount is incorporated, at least to some extent, into the price of the products. Industry statistics indicate that the primary effect of DTC ads is to increase drug spending by increasing the number of prescriptions written, not the prices of the advertised drugs.[62] Supporters of DTC advertising believe that it can have a positive effect, such as when the ads prompt people to seek treatment with cholesterol-lowering drugs. Advertising lipid-lowering drugs, for instance, can decrease healthcare costs from the decrease in cardiovascular morbidity in treated patients. Critics, however, deplore the spending and wonder if the increase in the number of prescriptions written for advertised drugs is contributing significantly to patient health. In the midst of this debate, some companies, such as Johnson & Johnson, are skewing their patient ads to highlight the educational and safety aspects. According to the company, the move was to reduce the animosity directed at these ads and preserve the right to market directly to patients by educating and counseling consumers to improve their health.[66]

1.23.2.8 Postmarketing Safety Monitoring

Drug companies have increasingly understood that the commitment they must make to a new drug is often more than the usually quoted statistic of 10 years and US$800 million to get the drug to market. The postmarket research and surveillance phase of a drug's life is now often as active as the premarket phase. Companies are either obliged by the regulatory agencies or voluntarily undertake these postmarket programs. Reasons for doing so are usually related to the fact that clinical research can never predict precisely how a drug will behave once prescribed in the open market. Plenty of examples exist of drugs approved for marketing only to be withdrawn within a short period because of unexpected side effects.[67] Historically, the rate of drug recalls in the US is 2–3%. Drug companies also know that physicians will prescribe a drug for many patients who were not included in the premarket studies – children, pregnant patients, the elderly, patients with comorbid conditions, or even patients with other diseases. These circumstances give rise to two questions that have ethical relevance. How much postmarket safety monitoring should the company perform? And, to what extent is it the company's responsibility (as opposed to the physician's) to collect data on or study off-label uses? The question about postmarket safety monitoring depends on the extent of premarket research, the utility of the drug, and the severity of potential side effects. For instance, if there was extensive premarket research, the drug is for an unmet medical need, and the potential for side effects low, then the need for safety monitoring or further study is probably on the low side of the spectrum. The question about off-label use is more difficult to resolve. On the one hand, physicians need to take responsibility for their prescribing. If there is little data to indicate that a drug is safe for a particular use, this suggests that the physician should use the drug under research rather than clinical conditions and the company should not be responsible if the physician is reckless in this regard. On the other hand, if the company knows that the drug has a high potential for toxicity, doing nothing seems irresponsible, especially since the company benefits from the expansion in the market attributable to off-label use. This is exactly the view taken by critics in the Vioxx debate (described above). The significant legal, financial, and reputational harm that resulted in that case indicates that companies need to devote considerable attention to their responsibilities for postmarket drug safety.

1.23.3 Obtaining Ethics Advice

Companies have several options when seeking ethics advice to manage the issues described above. Large companies tend to have in-house ethics officers who are either primarily responsible for legal compliance and report to the general counsel or, even better, are more narrowly focused on ethics and liaise with the legal department. Resources to support such people include the Ethics Officers Association[68] and Business for Social Responsibility,[69] both of which exist to educate and enhance communication about business ethics. Some companies also empanel ethics advisory committees to address ethics issues as they arise or seek outside individual consultants who work on discreet issues. Several aspects of using outsiders need to be addressed when pursuing this course of action. Ground rules and agreements must be established about confidentiality of discussions, whether the company discloses its use of ethics advice (this can seem self-serving), compensation (too much seems like buying an opinion, too little will deter participation), and how to manage ethics advice that conflicts with management decisions. The advisors themselves also need to be carefully selected. Although a company will not be well served by retaining a committed industry critic, neither is it helpful to hire a 'yes man' who will justify and endorse all company decisions. Companies will benefit maximally from learning about all sides of a debate, and should retain an advisor who will provide objective advice while retaining a willingness to support the overall corporate effort. Expertise is also a factor. A good company advisor will be educable about the technology, knowledgeable about general pharmaceutical company business and legal affairs and about business ethics, and able to provide practical advice. Because it is rare for one person to possess all of these skills, companies may want to convene a panel of advisors with a variety of expertise.[70]

Regardless of the method used, an awareness and incorporation of an ethical perspective into business decision-making will serve pharmaceutical company interests. Both economic and social goals are advanced when business ethics programs assist in responsible development and marketing of medical products, support physicians in their treatment of patients, and advance the important social benefits of pharmaceutical and biotechnologies.

1.23.4 Resources

Readers should consult the references given below for further information on business ethics.[71–78]

References

1. Corporate Social Responsibility Monitor. *Global Public Opinion on the Changing Role of Companies*; 2001. Available at: http://www.bsdglobal.com/issues/sr_csrm.asp (accessed April 2006).
2. Business for Social Responsibility. http://www.bsr.org (accessed April 2006).
3. Longest, B. B.; Lin, C. J. *J. Pharm. Fin. Econom. Policy* **2004**, *13*, 33–43.
4. Witt, M. D.; Gostin, L. O. *JAMA* **1994**, *271*, 547–551.
5. Bodenheimer, T. *N. Engl. J. Med.* **2000**, *342*, 1539–1544.
6. Blumenthal, D.; Campbell, E.; Causino, N.; Louis, K. *N. Engl. J. Med.* **1996**, *335*, 1734–1739.
7. Blumenthal, D.; Causino, N.; Campbell, E.; Louis, K. *N. Engl. J. Med.* **1996**, *334*, 368–373.
8. Applegate, W.; Furberg, K.; Byungton, R.; Grimm, R. *JAMA* **1996**, *277*, 297–298.
9. Rose, C. D. Scripps Deal with Drug Firm Approved; Sandoz is Partner in Scaled-Back Alliance. *San Diego Union-Tribune*, May 17, 1994, p A1.
10. Rodarmor, W. Dangerous Liaison? *California Monthly* December, 1998.
11. Niiler, E. *Nat. Biotechnol.* **2000**, *18*, 1235.
12. See, for example, Stanford University Faculty Policy on Conflict of Commitment and Interest. http://www.stanford.edu/dept/DoR/rph/4-1.html (accessed April 2006).

13. Moses, H., III; Braunwald, E.; Martin, J. B.; Their, S. O. *N. Engl. J. Med.* **2002**, *347*, 1371–1375.
14. Bodenheimer, T. *N. Engl. J. Med.* **2000**, *342*, 1539–1544.
15. Mathews, A. W.; Martinez, B. Warning Signs, e-mails Suggest Merck knew Vioxx's Dangers at Early Stage. *Wall Street Journal* November 1, 2004, p A1.
16. Uraneck, K. Balancing Business and Science at ImClone: Researchers in Business Struggle to Manage Conflicts of Interest. *The Scientist* December 9, 2002.
17. Rennie, D. *JAMA* **1999**, *282*, 1766–1788.
18. FDA Launches a Multi-Pronged Strategy to Strengthen Safeguards for Children Treated with Antidepressant Medications. *FDA News*, October 15, 2004.
19. Zuckerman, D. *Int. J. Health Serv.* **2003**, *33*, 383–389.
20. Stelfox, H. T.; Chua, G.; O'Rourke, K.; Detsky, A. S. *N. Engl. J. Med.* **1998**, *338*, 101–106.
21. Dimasi, J. A.; Paquette, C. The Economics of Follow-On Drug Research and Development: Trends in Entry Rates and the Timing of Development. *Pharmacoeconomics* **2004**, *22*, 1–14.
22. Garattini, S. *J. Nephrol.* **1997**, *10*, 283–294.
23. Prescription Access Litigation Project. http://www.prescriptionaccesslitigation.org (accessed April 2006).
24. Hartog, R. *Soc. Sci. Med.* **1993**, *37*, 897–904.
25. See generally, the Nuremberg Code (available at http://ohsr.od.nih.gov/guidelines/nuremberg.html.), The Declaration of Helsinki (available at http://www.wma.net/e/policy/b3.htm (accessed April 2006)), The Belmont Report: Ethical Principles and Guidelines for the Protection of Human Subjects of Research, Report of the National Commission for the Protection of Human Subjects of Biomedical and Behavioral Research, 44 Fed. Reg. 23,192, Department of Health, Education & Welfare, April 18, 1979.
26. 21 *Code of Federal Regulations*, Section 111(a)(2).
27. Veatch, R. M. *Hastings Center Rep.* **1971**, *1*, 2–3.
28. The FDA regulations on informed consent are at 21 *Code of Federal Regulations*, Section 50.20 *et seq*.
29. See 21 *Code of Federal Regulations*, Section 50.50–50.56.
30. In the United States, the Regulations that Specify the Requirement for Reliable Study Data are found at 21*Code of Federal Regulations*, Section 314.126.
31. Harrington, A., Ed. *The Placebo Effect: An Interdisciplinary Exploration*; Harvard University Press: Cambridge, MA, 1999; Jost, T. S. *Am. J. Law Med.* **2000**, *26*, 175–186; Temple, R.; Ellenberg, S. *Annals Int. Med.* **2000**, *133*, 455–463.
32. Bayer, R. *Am. J. Pub. Health* **1998**, *88*, 567–570.
33. Angell, M. *N. Engl. J. Med.* **1997**, *337*, 847–849; Resnik, D. B. *Bioethics*, **1998**, *12*, 286–300; Perinatal HIV Intervention Research in Developing Countries Workshop Participants. *Lancet* **1999**, *353*, 832–835.
34. FDA Notice of Opportunity for Hearing, James Wilson MD PhD, Institute for Gene Therapy, University of Pennsylvania, February 8, 2002.
35. Borger, J. Dying for drugs: volunteers or victims? Concern grows over control of drug trials. *The Guardian* February 14, 2001, p 4.
36. Levine, R. J. Compensation for research-induced injury. In *Ethics and Regulation of Clinical Research*; Yale University Press: New Haven, CT, 1988, pp 155–161.
37. See, for example, US President's Commission for the Study of Ethical Problems in Medicine and Biomedical and Behavioral Research. *Compensating for Research Injuries: the Ethical and Legal Implications of Programs to Redress Injured Subjects*; Department of Health and Human Services, Washington, DC, 1982; Vol. 1.
38. Lie, R. K.; Emanuel, E.; Grady, C.; Wendler, D. *J. Med. Ethics* **2004**, *30*, 190–193.
39. Steinbrook, R. *N. Engl. J. Med.* **2002**, *347*, 1462–1470.
40. Masters, B. A. New York sues Paxil Maker over Studies on Children. *Washington Post* June 3, 2004, p E1.
41. Couzin, J. *Science* **2004**, *305*, 468–470.
42. Letter to Bernard J. Poussot, President, Wyeth Pharmaceuticals, from Hon. James C. Greenwood, Chairman, Subcommittee on Oversight and Investigations, US House of Representatives, February 3, 2004. Available from URL: http://energycommerce.house.gov/108/Letters/02032004_1202.htm (accessed April 2006).
43. Kennedy, D. *Science* **2004**, *305*, 451.
44. Arzberger, P.; Schroeber, P.; Beaulieu, A.; Bowker, P.; Casey, K.; Laaksonen, L.; Moorman, D.; Uhlir, P.; Wouters, P. *Science* **2004**, *303*, 1777–1778.
45. For an example of corporate trial registry, see GlaxoSmithKline at http://ctr.gsk.co.uk/welcome.asp (accessed April 2006). For a collaborative industry proposal, see the Joint Position on the Disclosure of Clinical Trial Information via Clinical Trials Registries and Databases at http://www.efpia.org (accessed April 2006). See also the Good Publishing Practice (GPP) for Pharmaceutical Companies at http://www.gpp-guidelines.org (accessed April 2006). The International Committee of Medical Journal Editors (ICMJE) uniform requirements for manuscripts is at http://www.icmje.org (accessed April 2006). See also, the Consolidated Standards of Reporting Trials (CONSORT) statement at http://www.consort-statement.org (accessed April 2006).
46. DiMasi, J. A. *Clin. Pharmacol. Ther.* **2001**, *5*, 286–296.
47. Kucukarslan, S.; Hakim, Z.; Sullivan, D.; Taylor, S.; Grauer, D.; Haugtvedt, C.; Zgarrick, D. *Clin. Ther.* **1993**, *15*, 726–738.
48. Merck's Patient Assistance Program. http://www.merck.com/pap/pap/consumer/index.jsp (accessed April 2006).
49. Fuhrmans, V.; Naik, G. In Europe, Drug Makers Fight Against Mandatory Price Cuts. *Wall Street Journal* June 7, 2002, p A1.
50. Gold, R. Minnesota Joins List of States Suing Firms over Drug Prices. *Wall Street Journal* June 19, 2002, p D3.
51. Weil, E. Grumpy old drug smugglers. *New York Times* May 30, 2004, p 42.
52. Kennedy, D. *Science* **2003**, *301*, 895.
53. Brown, D. Group says U.S. should Claim AIDS Drug Patents. *Washington Post*, May 26, 2004, p A4.
54. Adams, C.; Harris, G. When NIH Helps Discover Drugs, should Taxpayers Share Wealth? *Wall Street Journal*, June 5, 2000, p B1.
55. Cimons, M.; Zonana, V. F. Manufacturer Reduces Price of AZT by 20%. *Los Angeles Times*, September 19, 1989, p 1.
56. Gellman, B. An Unequal Calculus of Life and Death; As Millions Perished in Pandemic, Firms Debated Access to Drugs; Players in the Debate Over Drug Availability and Pricing. *Washington Post* 2000, December 27, 2000, p A1.
57. The Pharmaceutical Research and Manufacturers of America (PhRMA) Home Page. http://www.phrma.org (accessed April 2006).
58. The U.S. Regulations for Drug Marketing and Advertising are Found at 21 *Code of Federal Regulations*, Part 202.
59. See, for example, Code of Pharmaceutical Marketing Practices of the International Federation of Pharmaceutical Manufacturers Associations, available from http://www.ifpma.org/ (accessed April 2006).

60. Pines, W. L. *Food Drug Law J* **1999**, *54*, 489; Terzian, T. V. *Am. J. Law Med.* **1999**, *25*, 149; Hollon, M. F. *JAMA* **1999**, *281*, 382–384; Wilkes, M. S.; Bell, R. A.; Kravtiz, R. L. Direct-to-Consumer Prescription Drug Advertising: Trends, Impact, and Implications. *Health Affairs* **2000**, *19*, 110–128.
61. t'Hoen, E. *Am. J. Health Syst. Pharm.* **1998**, *55*, 594–597.
62. United States General Accounting Office. *FDA Oversight of Direct-to-Consumer Advertising has Limitations*. Report to Congressional Requesters GAO-03-177, October, 2002.
63. Kaiser Family Foundation. Impact of direct-to-consumer advertising on prescription drugs pending, **2003**, June. http://www.kff.org/rxdrugs/6084-index.cfm (accessed April 2006).
64. Vedantam, S. Drug ads Hyping Anxiety Make Some Uneasy. *Washington Post* July 16, 2001, p A1.
65. Hensley, S. As Drug ad Spending Rises, a Look at Four Campaigns. *Wall Street Journal* February 9, 2004, p R9.
66. Hensley, S. In switch, J&J gives straight talk on drug risks in new ads. *Wall Street Journal* March 21, 2005, p B1.
67. Kleinke, J. D.; Gottlieb, S. *Brit. Med. J.* **1998**, *317*, 899.
68. European Association for Osseointegration (EAO) Home Page. www.eao.org (accessed April 2006).
69. Brotherhood of Railroad Signalmen Home Page. www.brs.org (accessed April 2006).
70. Eaton, M. Affymetrix, Inc., Using Corporate Ethics Advice. *BioIndustry Ethics*; Elsevier Inc.: New York, 2004.
71. Barber, N. *Encyclopedia of Ethics in Science and Technology*; Facts on File, Inc.: New York, 2002.
72. Benowitz, S. *The Scientist* **1996**, *10*, 1.
73. Brower, V. Biotechs Embrace Ethics. *BioSpace*, 1999, June 14.
74. Chadwick, R. *The Concise Encyclopedia of the Ethics of New Technologies*; Academic Press: San Diego, CA, 2001.
75. Dhanda, R. K. *Guiding Icarus, Merging Bioethics with Corporate Interests*; Wiley-Liss: New York, 2002.
76. Eaton, M. *Ethics and the Business of Bioscience*; Stanford University Press: Stanford, CA, 2004.
77. Levine, R. J. *Ethics and the Regulation of Clinical Research*, 2nd ed.; Yale University Press: New Haven, CT, 1988.
78. Velasquez, M. G. *Business Ethics, Concepts and Cases*. Prentice-Hall: Upper Saddle River, NJ, 1998.

Biography

Margaret L Eaton currently teaches at the Stanford University Graduate School of Business and prior to that she was a Senior Research Scholar in the Center for Biomedical Ethics at the Stanford University School of Medicine. She has undergraduate and graduate degrees in pharmacy, clinical pharmacy, and law and has held faculty positions at the University of Minnesota School of Pharmacy and Stanford University. Her non-academic work has included positions in the pharmaceutical and biotechnology industry and as a lawyer representing pharmaceutical companies in products liability litigation. At Stanford University, she has also served as a medical attorney and was a member and chair of the hospital ethics committee. Dr Eaton's current teaching includes courses in biotechnology and pharmaceutical business ethics, medical law, and biomedical ethics. Her primary academic work focuses on the ethical and social issues that impact the biotechnology, pharmaceutical, medical device, and related industries. Among her publications are 12 business school case studies and two books on this subject: Ethics and the Business of Bioscience, and (with co-authors) BioIndustry Ethics. She has co-authored another book with Donald Kennedy, Innovation in Medical Technology; Ethical Issues and Challenges, which is in press.

Comprehensive Medicinal Chemistry II
ISBN (set): 0-08-044513-6
ISBN (Volume 1) 0-08-044514-4; pp. 709–723

1.24 The Role of Government in Health Research

J K Ozawa, SRI International, Arlington, VA, USA
Q C Franco, StratEdge, Washington, DC, USA

1.24.1 Introduction

Health innovations are scientific and technological advances that improve human health by increasing lifespans, reducing the incidence of disease, and improving the quality of life of individuals suffering from chronic or other diseases. This definition spans the spectrum of biomedical advances from new drug and vaccine discovery to better drug delivery systems with reduced side effects. It includes more efficient production of therapeutics, better diagnostic tools, more effective public health initiatives, and improved data management for health research and healthcare. Driving the pace of health innovation is health research.

Heath innovation generates both economic and socially desirable outcomes, and this is at the heart of why governments are active participants, to a greater or lesser extent, in health-related industries. The market for health products and services is sizeable, employing large numbers of people and generating significant revenues for companies and governments (through corporate taxes). The global pharmaceutical market alone is approximately US$550 billion annually.[1] With an eye on the graying of populations in North America, Europe, East Asia (Japan, China, Korea), and South America (Brazil, Argentina), policy-makers view health-related industries as important, high-growth sectors both domestically and globally. According to the UN Population Division, in 2050 the median age will exceed 40 years in countries such as the UK, US, China, and Brazil. In the same year, the median age will exceed 50 years in Japan, Italy, and South Korea. Health innovations in particular industry segments in a country can provide that country's industry with a competitive advantage that can sustain industry growth for decades to come. In addition, developed country social safety nets that provide healthcare for senior citizens will be severely strained in the coming years, and governments are seeking more cost-effective remedies for illnesses and disabilities associated with aging. For all these reasons, governments the world over are making large investments enacting policies to promote health innovations.

In 2003, the US government allocated PPP$31.5 billion of its nondefense research and development (R&D) budget to health and environment programs; the EU15 governments appropriated a total PPP$10.1 billion; and the Japanese government appropriated PPP$1.9 billion.[3] (PPP$ refers to national currencies that have been converted to purchasing power parity (PPP) dollars for international comparison purposes. Although both PPP$ and US$ conversions have pros and cons, PPP is generally used in Organization of Economic Co-operation and Development (OECD) data, because PPP normalizes for what a similar basket of goods and services actually costs residents in each country.) These are sizeable annual expenditures when one considers that the entire multicountry, 15-year Human Genome Project (HGP) was funded at less than US$300 million per year. In developed countries, as well as many industrialized Asian and South American countries, health research is a priority area within public R&D budgets. In addition, governments the world over are struggling with such public policy issues as intellectual property (IP) rights regimes, prescription drug pricing, and ways to enhance university–industry interactions, all designed to increase the impact and efficiency of the health innovation system.

Given the significant expenditures on health research around the world and the numerous other ways in which governments intervene in health-related markets, governments are undeniably influencing the pace, direction, and scope of health research. It is, therefore, important to consider the public policy rationale for what the role of government should be in health innovation and to investigate what the actual role of government has been across countries. This chapter first examines the theoretical underpinnings of the role of government in spurring innovation and scientific advancement in general, including a discussion of different policy levers governments have used to promote innovation in all sectors. It then moves to the specific case of health innovation, presenting the major reasons why governments actively intervene in this sector and how the health sector differs from other innovation sectors. A significant portion of this chapter focuses on the conduct of health-related R&D around the world because of the sizeable impact R&D has on health innovation. However, we also review how government influences health innovation through its role as market regulator and by its effect on the end-use of innovations. Finally, outcomes of health innovation and government's role in fostering these outcomes are reviewed.

1.24.2 Role of Government in Spurring Innovation

Before exploring the role of government in the particular case of health innovation, it is useful to summarize current thinking regarding innovation and to lay out the generic case for government intervention in innovation markets. This section begins by discussing models of innovation, outlines the causes of market failure in innovation markets, and presents a short review of the government role in innovation markets in general.

1.24.2.1 Models of Innovation

To many, innovation primarily evokes images of scientists and engineers working in laboratories. These researchers make discoveries that ultimately find their way on to market shelves (e.g., Post-it notes) or influence behavior (the understanding that disease is transmitted by microbes leads people to wash their hands more often). While vastly oversimplified, the Schumpeterian model of innovation divides the innovation process into three sequential stages, and it is a useful construct. Invention, the first stage, refers to the generation of new ideas and insights – what is often referred to today as basic research. In the second stage, innovation – or applied R&D – the invention is developed into a final product or process which, at the third stage, is introduced into the marketplace.

While often presented as a linear pipeline – research to products to market acceptance – this process is, in reality, a complex, risky system with numerous feedbacks and interactions between each phase and between different actors. Sometimes research leads to new products. But often, these new products stimulate follow-on research or require further development to realize their market potential. A few breakthrough products will generate enormous profits for their inventors, encouraging others to attempt to imitate or to build upon the original invention. Most often, of course, basic research leads to very little in the near term. Research often simply increases our knowledge of how the universe works, which may, at some later time, be combined with other insights to generate new products or activities. Furthermore, developing a better product does not automatically lead to market success. Sound management, efficient production, talented marketing, access to financing, and effective delivery systems are all essential.

As these considerations highlight, the process of innovation is much more complex than the linear model suggests. At each stage, the broader environment, or system, plays a critical role in shaping how innovation takes place. The concept of innovation systems has gained considerable attention throughout the 1990s as a framework for thinking of innovation in the real world.[2] While there is no widely accepted standard definition, an innovation system generally refers to the local, regional, or national environment for innovative activity, encompassing a wide range of actors, their relationships, and the regulatory and policy environment in which they operate. Innovation requires ideas and knowledge that, in turn, build upon an existing knowledge base that may come from research conducted by private firms, but also from universities, government laboratories, and even individual inventors. Individual researchers' activities are influenced by their social networks – peer groups, professional societies, partners in different organizations. Industrial incentives and resources, in turn, are affected by the market structure of the industry, government regulation, financial arrangements, and access to risk capital. A regional or industry innovation system refers to these and other components and to how they interact to affect the pace and direction of innovation.

1.24.2.2 The Generic Case for a Government Role in Innovation

Analysts generally see two broad justifications for a role of government in research and innovation. One is based on neoclassical economic market failure arguments first applied in the 1950s to government support of basic research. This argument holds that if the private market fails to operate in an economically efficient manner, then there exists a case for government action intended to correct the source of the inefficiency. As innovation markets are characterized by numerous market failures, government action can be justified in addressing those particular failures.

The second broad justification is based on evolutionary or new growth theories of development, which draw attention away from immediate market failures to the needs of a dynamic market system in the pursuit of socially desirable outcomes (competitiveness, structural change, and improved health outcomes).[3] While experts debate the fine points, economists generally agree that innovation plays a crucial role in economic growth. Evolutionary growth theories emphasize that a country's rate of economic growth is increasingly driven by the stock of knowledge and innovative ideas that are held by its people, firms, universities, and governments. This stock of ideas and knowledge is, in certain key respects, a public good that is drawn on by innovators, entrepreneurs, and society in general to generate innovations and productivity improvements. Therefore, public attention and resources are warranted to ensure sufficient investments in education and scientific R&D to enhance economic growth prospects and to achieve long-term social objectives.

Regardless of their theoretically different underpinnings, the implication of both of these justifications is that public resources and efforts should be focused on those areas (market failures and/or high-impact activities) that will generate the highest societal payoff. While the specific underlying sources of private underinvestment in innovation are different in each sector and across countries and regions, we first present a general outline of why companies may fail to invest sufficiently in innovation before turning to the specific case of health innovation and research.

Economists have expended much effort analyzing and assessing markets and identifying under what conditions they fail to generate socially optimal outcomes. There are three basic types of market failures, all of which are present in innovation markets: (1) indivisibility; (2) uncertainty; and (3) externalities.[4] Indivisibility refers to a situation where a product's costs are shared among units and cannot be easily divided among them. For example, developing a word-processing software package requires a costly upfront investment. However, once the software program has been developed, the production and distribution costs for each individual unit of the software are relatively low. At the extreme, an online download costs the software company next to nothing. In contrast, the material and labor costs of a traditional product are incurred for each unit produced. Every meal in a restaurant requires the restaurant to purchase inputs (pasta, vegetables, spices), allocate a certain amount of the chef's and a waiter's time, and then all the plates must be washed. These costs are relatively easily allocated to each individual meal whereas developing the copy–paste function in the software program is shared among all users.

Market efficiency requires that prices be set at marginal costs (the cost of producing one more meal or one more software CD, which does not include sunk costs such as the R&D investment that went into software development). Marginal cost pricing is required for market efficiency under classical economic models because, at other prices, resources are misspent and there are welfare-improving trades that are left unrealized. At prices below marginal costs, firms spend more to produce goods than those goods are worth to consumers. Pricing above marginal costs means some potential customers are priced out of the market even though they would be willing to pay more for an additional unit than it would cost companies to produce. However, in technology- or knowledge-intensive markets, if a firm charged its marginal costs (which might be close to zero, for example when a piece of software is sold over the internet), it would never be able to recover its initial R&D investment expenses (the sunk costs). The conundrum is that marginal cost pricing is required for economic efficiency, yet marginal cost is not a true reflection of production cost in technology- or knowledge-intensive markets. Consequently, these markets may fail.

A second source of market failure is uncertainty, and innovation markets are highly risky and uncertain. Research investments are not guaranteed to succeed, and even if successful are not guaranteed to lead to market success. Furthermore, R&D is a long-term endeavor. Basic scientific breakthroughs today may not generate marketable products until decades down the line. Because of this, firms with short-term profit horizons may be unwilling to undertake long-term R&D efforts even though it would be beneficial from a social perspective.

Externalities, where there are spillover benefits from investments, are also frequently observed in innovation markets. In order for a market to work efficiently, the returns from one's investment should be fully appropriable. For example, when a home-owner replaces her leaky ceiling, she is able to benefit immediately from her investment, and is rewarded when she sells her home with a higher sales price. Innovations, however, typically have profound benefits that cannot be captured by the innovator. First, innovations can more or less easily be copied or adopted by others, such that successful innovators soon find themselves with hearty competition (though patent and copyright systems have been enacted to control this). Second, innovations provide significant value to users of the innovation and, by contributing to the generally available stock of knowledge, enable further innovations in the future. These socially beneficial but nonprivate returns to innovation can be substantial.

Taken together, these and other failures of innovation markets imply the need and usefulness of government action in helping to resolve the market failures, and to encourage sufficient investment in innovation for improving future standards of living.[5] It is important to note that, while the market failure and public good characteristics of innovation and research overall are widely accepted, there are many aspects of the innovation process and system that suffer from few market failures. For example, there are high costs associated with the transfer and exploitation of technological knowledge. The fact that much of the technical knowledge needed to produce a flu vaccine is publicly available does not mean that people just make their own. Considerable technological skills, infrastructure, and other resources are required to produce and successfully market technological products. The further one moves away from basic science towards product development, the weaker the case for government intervention. Technology and innovation markets, therefore, are most appropriately characterized as including a mix of market failures and functioning markets.

Of course, government failures also exist. Public policies to provide incentives for innovation, while well intentioned, may turn into general subsidies with little impact on the national level of innovative activity. IP protections designed to help innovators capture more of the returns to their investments also exclude potential users who are unable to pay the higher costs and may impede useful follow-on innovations by other firms. The appropriate role of government in innovation, therefore, depends on the specific circumstances of the innovation system, varying from industry to industry and from country to country.

1.24.2.3 Short Overview of Government's Role in Fostering Innovation

In response to the market failures apparent in innovation markets, and in light of the importance innovation plays in increasing standards of living, governments have enacted a number of policy, regulatory, and direct support activities to promote innovation. Government activities, or interventions, take many forms and can be thought of as falling into three broad categories or roles: (1) supply-side support for innovation; (2) market regulation; and (3) demand-side or adoption policies. The remainder of this section presents a brief overview of these government roles.

1.24.2.3.1 Supply-side support for innovation

One of the most prevalent and obvious ways that governments have promoted innovation is through supply-side policies. These interventions are fundamentally designed to increase the availability or supply of new ideas, techniques, knowledge, and technologies that subsequently lead to the production of new products and their market acceptance. The most prominent example of this is direct public funding of R&D, which takes three basic forms: (1) basic research carried out in academic settings or in government laboratories; (2) basic research and technology development supporting public agency missions (defense, energy, health); and (3) private-sector R&D support (direct funding of commercial technology development projects or broader R&D tax subsidies).

This direct support of research and technology development affects innovative outcomes not only by increasing the supply of knowledge and ideas, but, perhaps as importantly, through the training of scientists and engineers. Some studies find that industrial research managers place a higher value on publicly funded R&D's training of scientists and engineers in experimental methodologies and science over the traditional research outputs of R&D.[6] Public R&D also often leads to the development of new research instruments and techniques that reduce the costs of conducting industrial R&D and improving its effectiveness.

Governments pursue many other supply-side policies that aim to improve the performance of the R&D system broadly. These types of policies include those directed more specifically at improving the supply of technical personnel at all levels, such as general support of the educational system, scholarships and fellowships, subsidized loans, tuition reimbursements, support for industry–academic exchange programs, and continuing education. Because of the importance that university–industry interactions play in the innovation process, governments have implemented varying approaches to improve these relationships. In Germany, the Fraunhofer Institutes conduct R&D that acts as a bridge between industry and university activities.[7] In the US, the National Science Foundation's (NSF's) Engineering Research Centers and the University–Industry Cooperative Research Centers programs are designed to serve as centers of research excellence able to tap the skills and resources of both universities and industry. Similar programs exist around the world (*see* 1.16 Industry–Academic Relationships).

Recognizing that innovation and technology development are particularly risky and difficult at their early-stages, and that it is small firms or independent innovators who are often the sources for radical innovations, governments have also endeavored to support small innovative entrepreneurs. Governments have contributed to the establishment of seed and venture capital (VC) financing that is crucial to early and mid-stages of technology development by a combination of direct funding of venture funds and enacting favorable tax and regulatory treatments of such funds (see below). Universities, cities, and national extension programs have also developed technology incubator programs that aid the early-stage growth of technology-based companies by providing shared facilities such as space, equipment, business consulting assistance, access to financing, and access to technology expertise.[8]

While governments have identified certain innovation support gaps and have enacted policies to address these, governments operate in a dynamic environment. Consequently, the actual outcomes of even well-intended policies are uncertain, given the feedbacks and interplay between government and industry and other external factors. An example of public-sector action not necessarily achieving a desired end is recent US public-sector involvement in early-stage VC funds. About 35 US states run some type of VC funds. Example of these include the Maryland Venture Fund (started in 1994), the public–private Pennsylvania Early-stage Partners (started in 1997) and Massachusetts Technology Development Corporation (started in 1978). Identifying a gap in funding for early-stage, high-risk projects, states across the US have initiated programs to provide directly (through state-run VC funds) or indirectly (through public–private or private VC funds) candidate companies with US$100 000–US$500 000 of early-stage funds. Some states have gone so far as to allocate a small percentage of state pension funds (1–3%) for early-stage VC investment through private VC funds. A common goal of these programs is to use public VC funds as a catalyst for attracting additional early-stage financing from the private sector by helping to demonstrate the feasibility of new technologies and companies (*see* 1.15 The Role of the Venture Capitalist).

It is interesting to see, then, that despite the public sector stepping in to provide more early-stage financing, there has been a significant decline in US VC financing of early-stage projects over the past 25 years (**Figure 1**). Early-stage disbursements (e.g., seed, startup, and other capital) have declined from 49.9% of total disbursements in 1980 to 20.8% in 2002. Later-stage disbursements have risen from 50.1% to 79.9% of total disbursements over the same period.[9] While more than one factor is likely to be influencing the divergence in early state and later-stage financing, this divergence generates some interesting policy questions. Has the public sector's provision of early-stage financing created a disincentive for the private sector to provide early-stage financing? Are these risky and administratively expensive projects so high-cost that it is not efficient for the private sector to undertake them? If this is the case, should tax dollars be used for early-stage VC investment?

1.24.2.3.2 Market regulation

The second major role of governments, establishing and enforcing the rules that govern economic activity, has an indirect though very important role in fostering innovation. Industry regulation aims to ensure that private activities achieve socially desirable outcomes that markets would not otherwise meet. Governments, therefore, regulate such

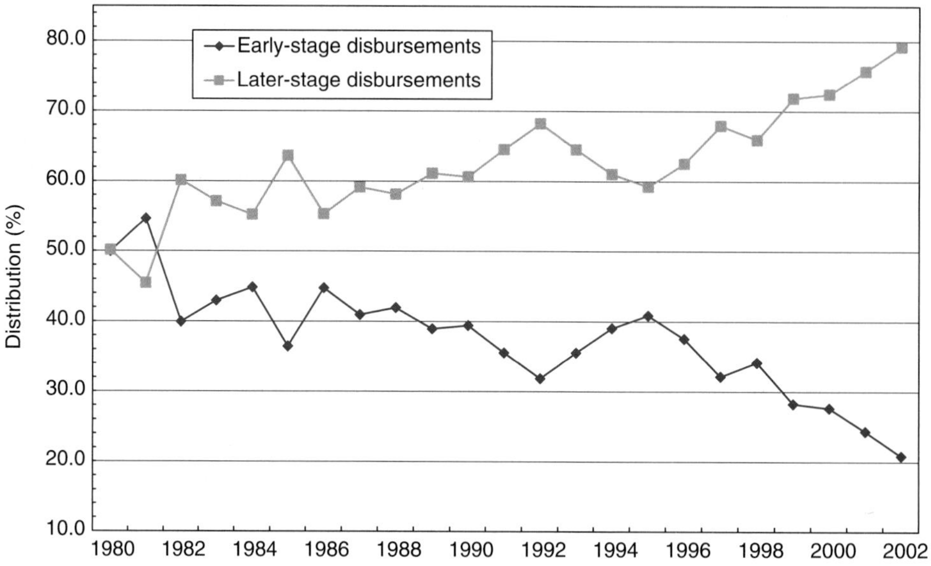

Figure 1 US venture capital disbursements by financing state, 1980–2002. (*Source*: National Science Board.[9])

things as environmental and health safety standards and promote competition. Of particular importance to innovation, governments protect the IP rights of inventors through copyright and patent law, enabling inventors to control and benefit from the fruits of their inventions.

The first modern patent law protecting IP was established in Venice in 1474. It gave city inventors a 10-year exclusive right to their novel inventions. IP systems work largely the same today. While national systems differ in their details, IP systems generally confer an exclusive right for producing, selling, and using inventions for a limited period of time. IP protections encourage firms and inventors to bring their innovations to the marketplace by providing some assurance that their innovations will not be copied and sold by competitors. By so doing, the inventor is able to appropriate more of the potential returns to the innovation as users are required to purchase or otherwise license the invention from the inventor. This ensures that inventors have adequate incentives to innovate.

There is a certain tradeoff between protecting IP to provide incentives for innovation, and competition policy aimed at ensuring open access and market efficiency. Moreover, providing exclusive rights to ideas would seem to run counter to the view that innovation requires an open, free exchange of ideas. For these reasons, patents are only granted to innovations that are novel, nonobvious, synthetic, and useful. Patents are also limited in duration (usually 20 years), after which the inventor relinquishes exclusive ownership of the invention. Similarly, patent disclosure rules require inventors to make information about their innovations public, contributing to the stock of knowledge upon which other inventors can build. In these ways, governments balance the benefits of providing incentives to invest in innovation with the costs of granting exclusive rights to the inventor.

Similar challenges also exist in industry regulation and competition policy as they relate to innovation. Regulators around the world are struggling to weigh the near-term benefits of price competition (the traditional focus of antitrust regulation) and health and safety regulation (the focus of industry regulation) with the long-term impacts on innovation. For example, the high-risk and long-term nature of some technology development implies that it may be beneficial to relax antitrust rules to allow firms to cooperate – the long-term innovation benefits may outweigh short-term fears of increased collusion and reduced competition. For these reasons, the US National Cooperative Research and Production Act (1984) provides some protection against antitrust liability for qualifying industry R&D and production joint ventures. Similarly, industry regulators have increasingly incorporated innovation considerations in their rule-making. For example, a central characteristic of environmental regulation in the US has been industry's need to develop and adopt new technologies to meet and surpass these regulatory goals (*see* 1.26 Intellectual Property Rights and Patents).

1.24.2.3.3 **Support for technology adoption**

Inventions only have an impact when people actually use them. Since the process of industrial and consumer uptake of innovations is not smooth and suffers from its own share of market failures and barriers, governments have stepped in to promote and encourage the adoption process. This is often done through agricultural extension and energy efficiency programs around the world, as well as through technology extension programs, such as the US Manufacturing Extension Program. These programs provide information and other technology adoption support services to local producers. Government support also takes the form of tax credits or subsidies for the purchase of beneficial technologies (such as the purchase of hybrid or electric vehicles). To combat informational problems, governments often carry out technology demonstration projects and serve as (or support the formation of) respected clearinghouses of information on new technologies and techniques.

In addition, the public sector is itself, in many instances, a user of innovations. This is especially the case in sectors such as defense, space exploration, and health, where governments can dominate the end-use markets. In these situations, government procurement policies and strategies directly guide innovation.

1.24.3 Role of Government in Promoting Health Innovation

The previous section discussed why and how governments intervene in markets to foster innovation that supports economic growth and competitiveness and enhances societal welfare. Many types of government interventions are common throughout industry sectors. For example, governments support basic research in a variety of scientific fields that are relevant to many industries, from the pharmaceutical industry, to aerospace, to electronics and telecommunications. However, important differences specific to each sector shape the distinct role that governments have played in the industrial and technological development of each industry.

This section discusses the role of government in promoting innovation in the health sector. It examines the economic and social objectives of governments that might induce them to support health innovation. It also describes the key characteristics specific to the health sector that differentiate government support for health innovation from other innovation sectors. Finally, we analyze the government's role in health research in particular, because supporting health R&D is an innovation role that has received significant attention by governments in recent years.

1.24.3.1 Motivations for Government Involvement in Health Innovation

In assessing government's role in health innovation, it is useful to distinguish between the health industry and the healthcare system. As a major industry involved with the generation and development of health solutions that are in high demand around the world, health is a particularly attractive industrial sector offering relatively high average wages and significant growth prospects. A government's role in promoting health innovation is, in this respect, comparable to its role in promoting the competitiveness and growth of other industries – the government supports health innovation in order to generate new jobs, stimulate economic growth, and increase standards of living.

At the same time, a country's healthcare system supports and promotes the physical and mental well-being of its citizens. As such, the level and diffusion of innovation in a healthcare system have a direct bearing on a country's overall health outcomes (e.g., morbidity, longevity, disability), as well as affecting the costs related to providing healthcare. In this respect, health innovation refers to a country's capacity to enact evidence-based clinical care (e.g., prenatal care for pregnant women), to carry out successful public health initiatives (e.g., childhood vaccination campaigns or bird flu containment) and to provide access to medical treatment. Innovation contributes to the pursuit of both economic and societal health goals (*see* 1.19 How Much is Enough or Too Little: Assessing Healthcare Demand in Developed Countries).

1.24.3.2 Specific Characteristics of the Health Innovation System

The health innovation system is a large, complex, international system with several key characteristics relevant to evaluating the role of government. First, to a greater extent than most industries, innovation in healthcare in the absence of strong IP protections is a public good. The provision of healthcare necessitates that extensive information about health products, procedures, and clinical practices be diffused throughout the system. Characteristics of new drugs, for example, are made very public in a way that the performance and other characteristics of a new refrigerator are not. Second, the introduction of new products and the process of developing and bringing these products to market affect human health directly. Consequently, there is a role for government in regulating the biosafety of health products, as well as in regulating the clinical trials that rely on human subjects for the determination of product efficacy, dosage, and other critical aspects. Thirdly, the active ingredients and processes used in the production of many pharmaceuticals (and increasingly biologics) can be replicated relatively easily by other chemical and pharmaceutical

Figure 2 Sources of funding for preventive HIV vaccine R&D, 2004. (*Source*: HIV Vaccines and Microbicides Resource Tracking Working Group. Tracking Funding for Preventative HIV Vaccine Research & Development: Estimates of Annual Investments and Expenditures 2000 to 2005.)

companies. This is evidenced by a large and growing global generics drug market. Generics accounted for 53% of prescriptions filled in the US, and US$18.1 billion out of a total US$217.4 billion in US pharmaceutical sales of brand products.[10] These characteristics mean that IP protections are crucial if the health industry is to continue to invest in the creation and development of novel products (*see* 1.21 Orphan Drugs and Generics).

Even with today's strong system of IP protections (e.g., drug patents average 20 years), research suggests that health innovators are able to appropriate only a small portion of the benefits of their inventions. For example, looking specifically at new human immunodeficiency virus (HIV)/acquired immune deficiency syndrome (AIDS) therapies introduced since the late 1980s, researchers have estimated that innovators appropriated only 5.0% of the total gains from the new therapies.[11] This suggests that the private sector may invest less than the socially optimal level of R&D required for health innovation, because of these weakened incentives. For example, in 2004 only 9.9% (US$68 million) of the world's US$683 million total investment in preventive HIV vaccine R&D came from pharmaceutical and biotech companies. Governments assumed critical funding role, providing the largest share of HIV vaccine R&D investment – 88.0% of the total (**Figure 2**).

Another key characteristic of health innovation is the central role that basic science plays in health innovation. While our understanding of how the human body works has progressed enormously, fundamental discoveries are still being made and technological advances are being achieved that have enormous implications for the development of health products and solutions. For example, advances in genomics, combinatorial chemistry, high-throughput screening, x-ray crystallography, and computer-aided drug design have made a rational approach to drug design a greater possibility than ever before.[12] Governments played a lead role in funding the basic research that supported the aforementioned advances, and governments will continue to play a strong role in the future because of the public-good characteristics of health research.

Finally, private industry plays a quintessential role in health innovation. The vast majority of health products in use today have come through a process of product development and commercialization financed by private companies. Even in the US, which has by far the largest government expenditures on health research in the world, the private sector has outspent or equaled health research spending by the public sector.

1.24.3.3 Appropriate Role of Government in Health Research

In recent years, the US government has assumed a very large role in one aspect of health innovation through its doubling of public funding for health R&D (from US$11.3 billion in 1998 to US$25.3 billion in 2003). US government spending on health R&D in 2003 was triple the health R&D spending by the EU15 and 15 times that of Japan. (In 2003, OECD data indicate the US government spent PPP$31.5 billion on health and environment R&D programs, compared to PPP$10.1 billion by the EU15 and PPP$1.9 billion by Japan.)[13] In addition, the Gates Foundation, the Wellcome Trust, and other foundations have entered the global health innovation arena with significant funding for health research. This emphasis on health R&D as a means of achieving economic and societal health objectives has put tremendous pressure on the governments of other developed and developing countries to follow suit. Rightly or wrongly, the magnitude of US public investments in health R&D will provide economists and science and technology (S&T) policy analysts with fertile ground for analysis in future years. Given the significant attention paid to health research worldwide, this section looks specifically at why governments support health research and what the appropriate role of government is in this area.

Government supports health R&D because it is a significant source of health innovation. Health R&D yields new knowledge that: (1) reduces the costs of developing new technologies and products; (2) trains researchers and personnel for a variety of occupations in health-related areas; (3) leads to enhancements in the overall healthcare system; and (4) creates entirely new industries and markets (e.g., automated DNA sequencers and microarrays).

In their support of health research, governments face two major challenges. First, because of its nonrival and nonexcludable characteristics, basic research is generally considered to be purely a public good. However, there are many valuable public goods, e.g., K-12 education, information and communications infrastructure, prenatal care. Consequently, it is difficult for governments to know how much of which particular public good it should fund vis-à-vis others. Second, any action that the government takes affects the incentives of private-sector participants. The most efficient outcome for society may be one in which the government enhances incentives for the private provision of public goods instead of attempting to supplant the private sector altogether. For example, the entrance of Celera Genomics as a private-sector competitor in the race to sequence the human genome reduced the time to completion by 2 years, saving hundreds of millions of dollars. On the other hand, if Celera had been allowed to patent all the new genes it identified, this would end up costing drug development companies and, hence, consumers billions of dollars.

In tackling the challenges presented by the public-good nature of basic health research, policy-makers would want to know several things in deciding upon an appropriate government role: (1) if there is a positive relationship between government interventions and economic growth and improved health outcomes; (2) what policy tools government can use to increase research intensity and effectiveness; and (3) how government policy affects private-sector incentives to invest in research. The literature related to these issues is sizeable, and contain some interesting considerations and empirical findings.

Regarding the economic impact of government interventions, Salter and Martin[14] compare different theoretical models for evaluating the economic impact of research and review various methodological approaches to measuring the economic payoff from publicly funded research. They find a limited number of empirical studies that have examined the social rates of return on publicly funded R&D or basic research, and several weaknesses make the numerical results unreliable. Nevertheless, the empirical evidence suggests a large and positive social rate of return on government-funded R&D, but one which is less than the social rate of return on private R&D.

Garber and Romer[15] tackle the second issue of what policy tools the government can use to promote health research in both the public and private sectors. They acknowledge a wide variety of legitimate government roles: government-funded basic research, government-performed research, government subsidies to private research, regulation of product markets, and IP protections. In general, however, they suggest that the most appropriate role of the government is to subsidize researcher training. This policy has the fewest distortions and supports public and private research through the distribution of tacit knowledge embodied in these researchers.

The fundamental problem confronting policy-makers is that because basic health research is a nonrival good (i.e., the resulting knowledge can be taken or duplicated and used by others at no cost), there is no efficient market outcome. That is to say, governments are left with second-best policy choices such as deciding to provide IP protections to innovators in order to make innovation markets approximate private markets. However, this entails denying the public free access to the results and cannot help but create some inefficient and market-distorting outcomes.

Factoring in the costs to society of publicly funded research brings us to the third issue of how government policy affects private-sector incentives to invest in and perform research. Empirically, the social rate of return on private-sector R&D typically exceeds the social rate of return on publicly funded basic research. (In part, these empirical results could be due to the very nature of publicly funded basic research which has only indirect market commercialization objectives and may have diffuse impacts over broad markets that are not easy to measure.) (For summaries of such empirical studies, see Griliches[16] and Office of Technology Assessment report.[17]) Consequently, policy-makers should strive to fund high-risk, long-term and high-reward basic research that private companies are unwilling to perform, and to enact policies that will encourage and direct private-sector investment in R&D. These would include fiscal and IP policies (e.g., R&D tax credits, tech transfer legislation), as well as support for researcher training, university research, and small business innovation research awards that support private-sector health research and innovation. However, government funding of health research is not costless. There are direct costs (actual grant amount) and indirect costs (administrative overhead, selection of poor research projects) associated with each type of research subsidy. These costs are transferred directly to the private sector in the form of higher taxes, and less obviously in the form of distorted incentives.

Therefore, while it may be clear that there is a role for government in supporting basic health research, it is less clear what the magnitude of government support should be, how this affects the private sector, and how important this role is relative to others. Supporting health research is just one of the many critical roles the government plays in promoting health innovation. As mentioned earlier, others include market regulation, biosafety regulation, and end-user of new health products and services. A more thorough examination of these roles is the focus of Section 1.24.5.

1.24.3.4 The Human Genome Project

The first serious debates on the desirability and feasibility of sequencing the 3 billion basepair human genome began in the US, UK, France, Germany, and Japan in the mid-1980s. The HGP was launched in 1990 with an initial 5 years of funding. At the time, sponsors thought the task would take 15 years at a total cost of US$3 billion. The HGP was led by the US Department of Energy (DOE) and the NIH. The UK Medical Research Council and Wellcome Trust assumed a significant role in the project with the opening of the Wellcome Trust Sanger Institute in 1993. The project involved several other noted research institutes in Japan, France, Germany, and China.

In 1998, former NIH biologist Craig Venter caused a stir when he founded Celera Genomics, a for-profit company, with the goal of sequencing the human genome within 3 years for a total US$300 million. Venter brought a novel sequencing method and the latest high-throughput DNA sequencers to the task, and built upon publicly available data. This direct challenge to the HGP resulted in the DOE and NIH pushing up their completion date to 2003 (from 2005). The Wellcome Trust committed an additional US$200 million (£110 million) to the project in 1998. On February 12, 2001, Celera Genomics and the public HGP published initial working drafts of the human genome in the journals of *Science*[45] and *Nature*.[46] On April 14, 2003, the two groups celebrated the completion of the HGP.

The HGP exemplifies many of the policy issues and challenges of government support for health innovation. This was a 'big science' project intended to lay the foundation for future investigations into the molecular basis for disease predisposition and resistance in humans. (For example, in 1990, at the project start, scientists knew of fewer than 100 genes linked to human diseases. In 2003, following the completion of the project, 1400 such genes were known.) It was an expensive project that required funding for research not only to support the actual sequencing, but also to support related research and technology development to hasten the pace of discovery and reduce costs. The program therefore funded such research as sequencing instrumentation, computational methods for gene mapping, information management systems, and technologies for the analysis of gene expression. Finally, the HGP raised many ethical, social, and IP issues. The undertaking of the HGP heralded the biotech age, and has created new commercial opportunities in numerous industries from software and drug development to reagents and imaging equipment. (The first private company founded on the concept of patenting information derived from the HGP was Genome Corporation in 1987.)

While private-sector companies may have hastened the completion of the HGP project and were vital in commercializing related R&D results, public support for the project was instrumental. Particularly, public funding contributed by:

- providing initial-stage support to demonstrate feasibility
- strengthening the US national laboratories' genomics expertise and experience undertaking large, multidisciplinary projects
- developing the global DNA repository, GenBank, where HGP data is stored
- funding early research on capillary-based DNA sequencing instruments that resulted in the dramatic increase in throughput in the late 1990s.

1.24.4 Government Support of Health Research

In this section we move from theoretical considerations of the role of government in fostering health innovation to review public-sector support for health research over time and across countries. This section first examines research priorities and how these priorities are set. It then examines historical trends in R&D funding and the health research share of this funding. The section closes with a discussion of the direction of health research policies.

1.24.4.1 Determining Research Priorities

In the broadest terms, the national determination of research priorities is a function of the relative importance policy-makers place upon: (1) national security interests; (2) economic competitiveness considerations; and (3) social objectives (**Table 1**). These emphases change over time with a country's level of development, economic success in science-based industries, access to resources, participation in the global economy, and development of events that impact the country (e.g., disease, war, natural disasters). For example, India's government might stress nuclear energy research more than Thailand's because of India's need to generate large amounts of electricity (with much lower carbon dioxide emissions) to meet its rapidly growing energy demands, as well as for national security reasons. Similarly, US funding of health research is motivated by both public health and economic concerns.

The weight placed on each of these three broad considerations helps explain the cross-country differences in the distribution of government research funding across scientific fields. For example, a 2004 study[18] of US and Japanese R&D funding priorities found shared public investments in Japan's four R&D priority areas specified in its Second S&T Basic Plan (2001–2005): (1) life sciences; (2) information and communications technologies (ICT); (3) environment; and (4) nanotechnology. Although the US does not formally articulate national research priorities or produce research or S&T plans, the study found that total US R&D funding in these four areas exceeded Japan's level of funding. Other findings included a greater emphasis by the Japanese government on energy, manufacturing, and social infrastructure research and greater emphasis by the US government on frontiers (space and frontiers of the oceans, atmosphere, and the poles) research. Shifting research emphases in the US are illustrated in **Figure 3**.

Table 1 Examples of research priority setting by national objectives

National security	*Economic competitiveness*	*Social*
Biodefense	Biotechnology (agricultural, medical, industrial/environmental)	Biotechnology (agricultural, medical, industrial/environmental)
Information and communications technology	Information and communications technology	Information and communications technology
Aerospace	Electronics	Energy
Advanced materials	Nanotechnology	Environment
Nuclear energy	Energy	
	Advanced materials	

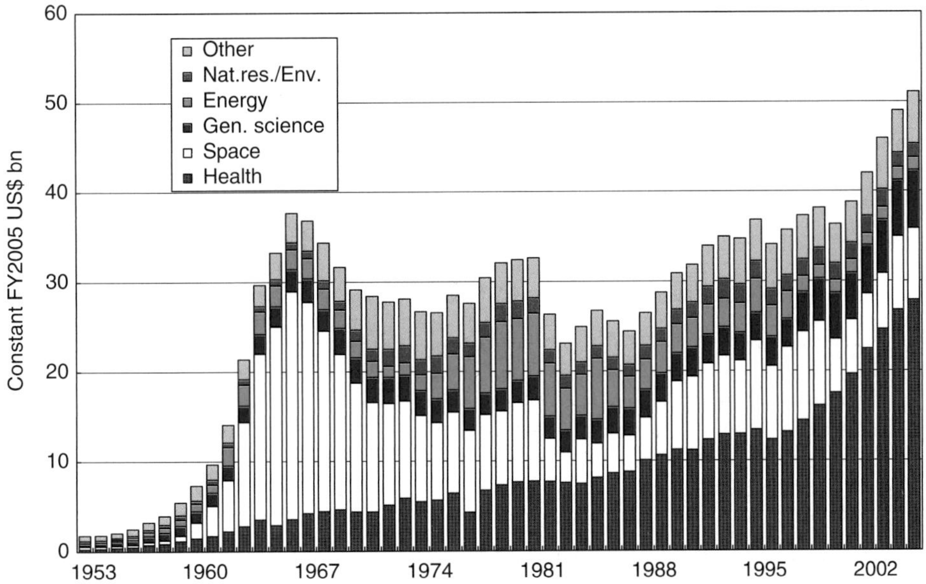

Figure 3 Trends in US nondefense R&D, 1953–2005. FY, financial year. (*Source*: American Association for the Advancement of Science (AAAS), based on US Office of Management and Budget (OMB) data, February 2005.)

The setting of research priorities is fundamentally a political process. It first involves decision making at the national/ministerial level where national budget appropriations are made. Advocates for different research agendas compete for resources among themselves and must compete against other demands on public resources. Once significant priority areas are decided, there is a second tier of decision making within institutions regarding the distribution of funds by type of research (basic, applied, research training), focus (scientific field or disease), and funding mechanisms (e.g., competitive or programmatic, centers-based or investigator-initiated).

One of the quintessential problems that vexes priority-setting at all levels is that scientific discovery is inherently unpredictable and, therefore, difficult to plan for. As the US NIH points out in an article on priority setting: "The emergence of new diseases (AIDS or West Nile Virus), the rise of importance of others as our society changes (Alzheimer's disease), and the resurgence of old ones (tuberculosis, malaria) all require urgent attention. The expense of supporting new and unforeseen research, however, does not displace the need to continue investigation into heart disease, muscular dystrophy, arthritis, diabetes or asthma."[19] Furthermore, it is often the unplanned and untargeted basic research that leads to the most important scientific breakthroughs and advances. Nevertheless, government S&T agencies around the world strive to collect data, forecast trends, and evaluate research outputs and outcomes to allocate public R&D funds better.

1.24.4.2 Priority-Setting at the World's Largest Health R&D Agency

Health research has been a priority sector in the US's nondefense R&D budget for some time. Recently, annual funding for the NIH more than doubled, increasing from US$11.3 billion in 1998 to US$25.3 billion in 2003.[20] As the world's largest funder of health R&D, it is interesting to examine how such an influential government agency distributes its money.

NIH is composed of 27 institutes and centers, each with a statutory mission, e.g., the investigation of a particular class of diseases, certain aspect of human life, or technology area. By law, each of these institutes must be funded. Each year, Congress appropriates funds based on the overall budget situation, national priorities, and presidential requests, among other considerations. As part of this process, each institute submits a request for funds outlining how it intends to allocate this money among different activities based on the institute's research objectives. This request includes each institute's breakdown of current obligations (multiyear grants to investigators, previously established research centers) and desired level of funding for investigator-initiated grants, intramural research programs, and research training.

Each NIH institute can set research topic emphases within its overall mission and select which specific research grants to fund among those submitted by researchers at universities, research centers, foundations, and even companies. Grants are awarded for an average of 4 years. It is the net effect of the myriad of these decisions – rather than a top-down directive – that determines how much of the entire NIH budget is devoted to research on a particular disease or focused on a particular scientific discipline.

The NIH specifies the following criteria as influencing its overall R&D budget allocation:

- incidence, severity, and cost of specific public health needs
- scientific quality of research proposals
- understanding that not all problems are equally approachable at a given point in time
- goal of maintaining a large and diverse portfolio – unforeseen discoveries may lead to significant advances
- view that investment in research training and physical infrastructure and instruments are critical to enabling research programs

1.24.4.3 Gross Expenditures on Health R&D

Spending on health research is influenced by overall levels of R&D spending. Therefore, this section first looks at trends in total R&D expenditures across countries. We then narrow our focus to expenditures on health R&D, which is broken down by government funding of health R&D and business expenditures on health R&D. It is important to note that, although these data are useful for illustrative purposes, these indicators are far from perfect measures. There are several difficulties related to collecting and using this type of data, including the fact that they are often generated for other purposes (e.g., R&D tax credits in the case of businesses) rather than careful definitions and data tailored to these definitions that are similar across countries.

1.24.4.3.1 Overall R&D trends

Increasing R&D expenditures has been a common goal and defining characteristic of developed countries and many developing countries over the last 20–25 years. This objective has been driven by the growing recognition of the relationship between the level of R&D activity and the pace of innovation-based economic growth. One common measure used to examine this growth is gross expenditures on R&D (GERD) as a percentage of GDP. (This is often referred to as R&D intensity.) GERD includes both public- and private-sector spending on R&D and, when expressed as a percentage of GDP, facilitates cross-country comparisons by normalizing for the varying sizes of economies.

If one looks at GERD levels and changes in R&D intensity over the last 25 years (**Figures 4** and **5**), several things stand out. First, developed countries have R&D intensities on the order of 2–3% of GDP. In 2003, Japan led this spending with a R&D intensity of 3.2% of GDP; the US followed with 2.6%; and the EU15 had 2.0%. Secondly, Asia's

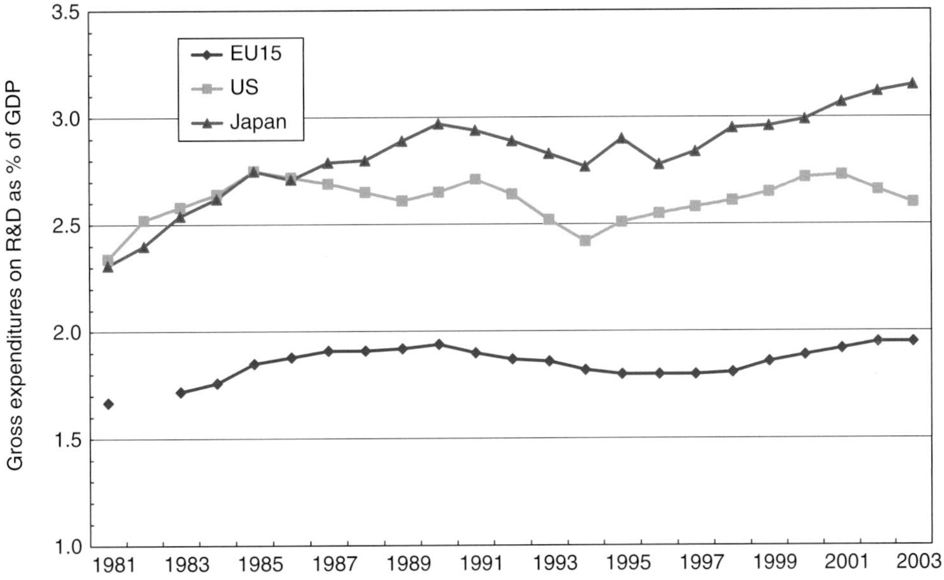

Figure 4 Trends in R&D intensity in the EU15, Japan, and US, 1981–2003.[13]

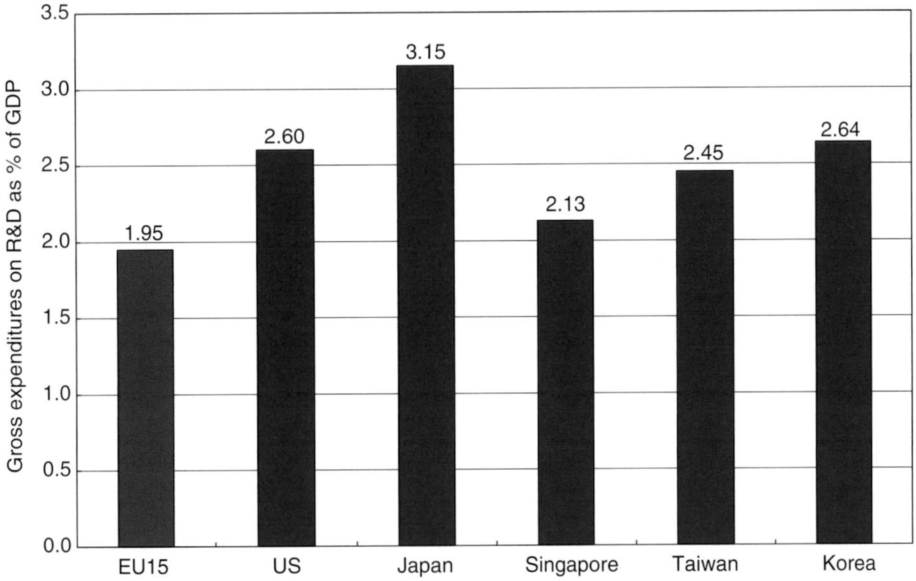

Figure 5 R&D intensity for select developed countries and Asian NICs, 2003.[13]

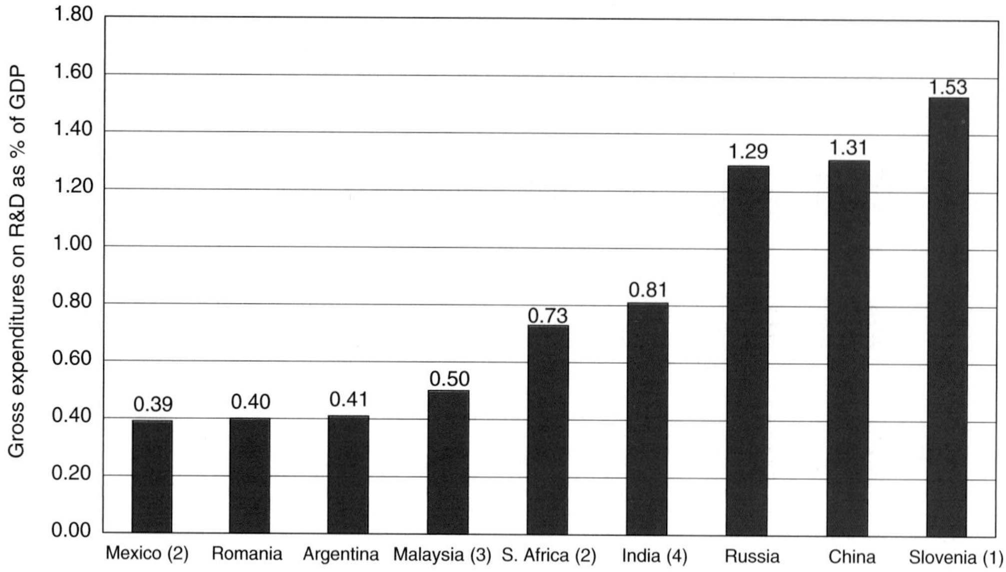

Figure 6 R&D intensity for select developing countries, 2003. Note: data for some countries are latest available: (1) 2002; (2) 2001; (3) 2000; and (4) 2000. Data from OECD, Main Science and Technology Indicators 2005-1[13]; India National Science and Technology Management Information System (NSTMIS); and Malaysian Science and Technology Information Centre (MASTIC).

newly industrialized countries (NICs) – Singapore (2.1%), Taiwan (2.5%), and Korea (2.6%) – had R&D intensities equal to or greater than those in the EU and the US.

Turning to developing countries, there is less R&D data, and these data cover a more limited time period (**Figure 6**). Nevertheless, available data indicate that an R&D intensity of 0.3–1.5% is more common in developing countries. For these countries, R&D intensity has tended to remain stable or has increased over the past two decades. The exceptions are the former Eastern-bloc countries that have experienced declines in R&D intensity: Russia (2.0% in 1990 to 1.3% in 2003), Romania (0.8% in 1991 to 0.4% in 2003), and Slovenia (1.6% in 1993 to 1.5% in 2002). Looking at the two largest developing countries, China has made a concerted effort to double its R&D intensity (0.7% in 1991 to 1.3% in 2003), while India's R&D intensity (0.8% in 1986 and 0.8% in 1999) has remained relatively flat. South Africa's R&D intensity has fluctuated in a range between 0.6% and 0.8% from 1983 to 2000.

R&D intensity is important, but so, too, are actual levels of R&D spending (**Figures 7** and **8**). Even though China's R&D intensity is much lower than that of the Asian NICs (Korea, Singapore, Taiwan), its absolute R&D expenditures (PPP$84.6 billion) are 3.5 times greater than Korea's (PPP$24.4 billion), six times greater than Taiwan's (PPP$13.7 billion), and nearly 40 times greater than Singapore's (PPP$2.2 billion). When one removes China and India, developed countries' annual expenditures dwarf R&D expenditures by higher-income developing countries. For example, Japan's and China's total R&D expenditures are somewhat comparable (PPP$114.0 billion compared to PPP$84.8 billion), but Russia's (PPP$16.9 billion) or Mexico's (PPP$3.6 billion) are significantly less (*see* 1.20 Health Demands in Developing Countries).

1.24.4.3.2 Trends in government funding for health R&D

What is the health sector's share of this R&D spending (**Figure 9**)? In developed countries, as well as many industrialized Asian and South American countries, health research is a priority area within public R&D. Health research appropriations have risen steadily over the past 15–20 years, and the majority of countries for which data are available dedicate 10–20% or more of public R&D budgets to health research. (Japan is an outlier among developed countries, with less than 10% of its civilian government budget outlays on R&D allocated to health research.) Fueling increases in health research funding are concerns over the rapid aging of populations, the social and economic costs of healthcare and disease, and the excitement surrounding the commercial promise of biotechnology.

The US government is the largest funder of health research in the world. In 2003, health and environment sectors received PPP$31.5 billion, or two-thirds of total US civilian R&D appropriations. This is triple the health and environment R&D budget of EU15 countries (PPP$10.1 billion) and 15 times that of Japan (PPP$1.9 billion).

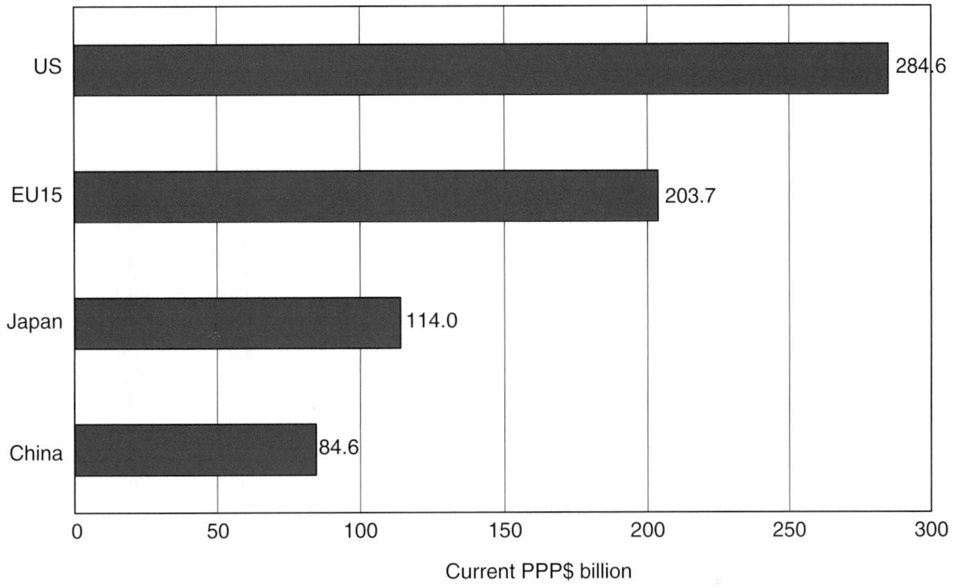

Figure 7 GERD levels for select countries, 2003.[13]

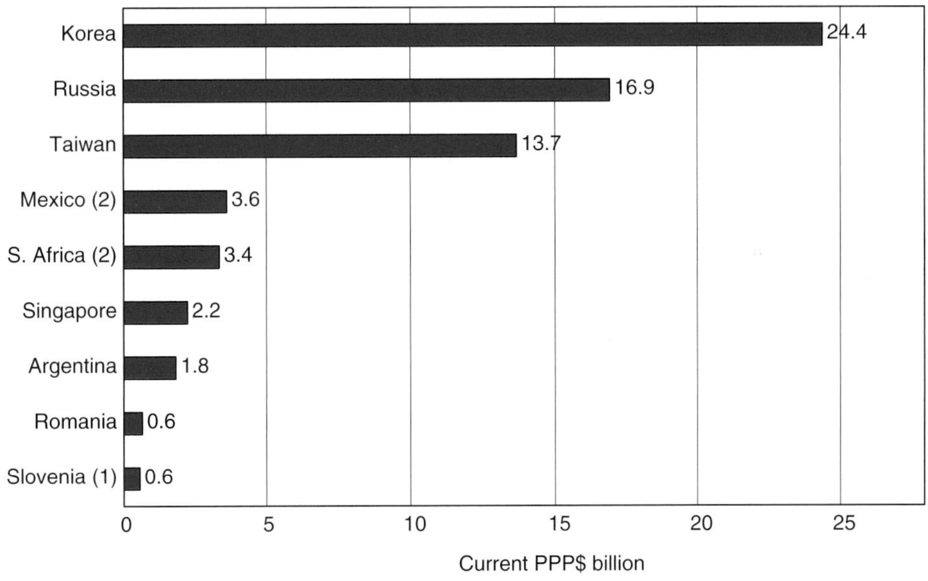

Figure 8 GERD levels for select developing countries and Asian NICs, 2003. Note: data for some countries are latest available: (1) 2002; and (2) 2001.[13]

Prior to 1998, the difference between US and EU government funding of health R&D was less: US funding levels were 1.8 times larger than EU levels. However, from 1998 to 2003 the US government pursued and met a 5-year doubling plan for government health research spending through the NIH (**Figure 10**). This occurred during a time when spending on most other US R&D programs remained flat or kept pace with inflation, explaining the increase in the share of US nondefense health and environment R&D in the total nondefense R&D budget from 20.1% in 1997 to 66.3% in 2003.[21] By comparison, EU15 government spending on health and environment R&D programs for this same 5-year period grew by an average 3.6% per year and Japan's by 9.9% year. Despite relatively rapid average annual growth of 10.2% from 1988 to 2003, Japan's health and environment R&D program funding is more comparable to that of Korea

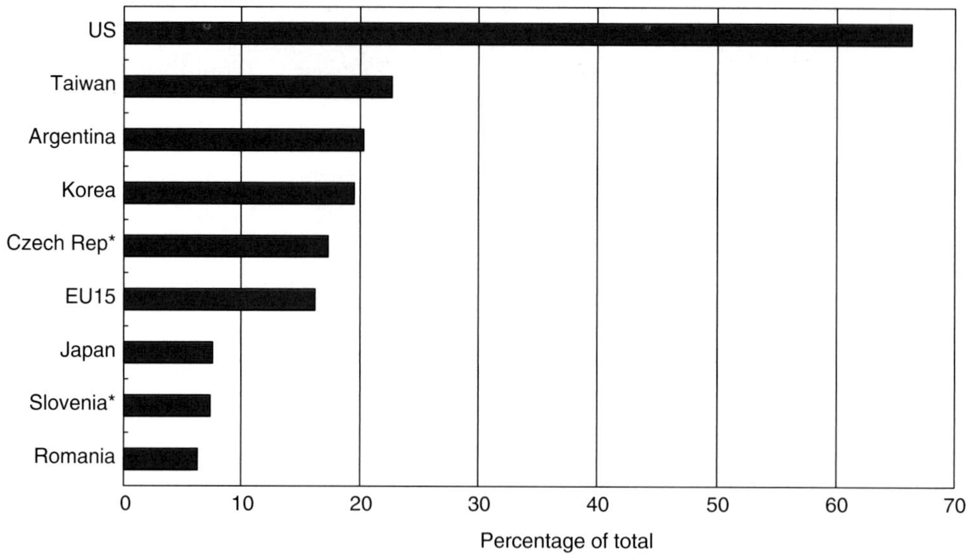

Figure 9 Share of government nondefense R&D budget going to health and environment R&D programs in select countries, 2003. *Data for the Czech Republic and Slovenia are for 2002.[13]

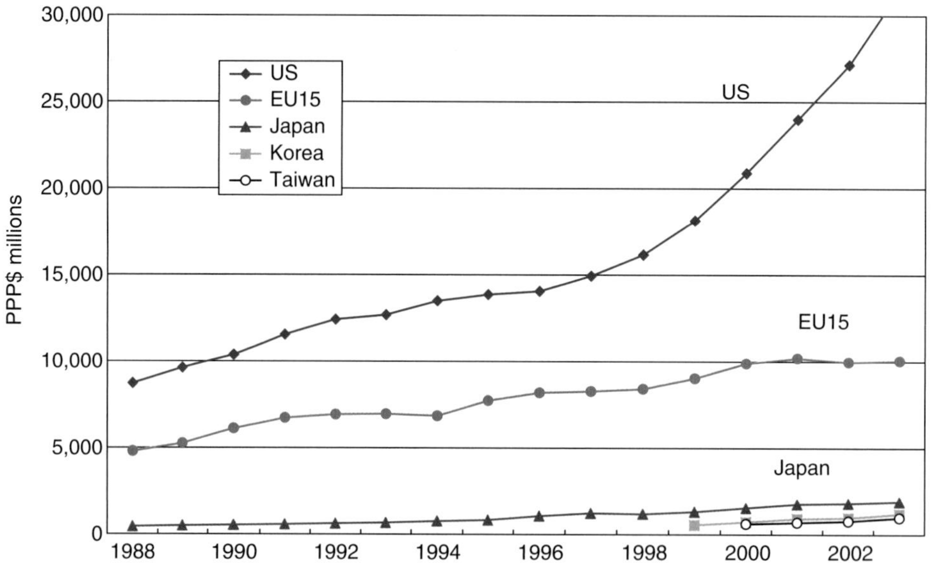

Figure 10 Government spending on civilian health and environment R&D programs in select developed countries, 1988–2003.[13]

(PPP$ 1.2 billion) or Taiwan (PPP$ 1.0 billion), although Korea and Taiwan are significantly smaller economies. (In 2003, OECD data indicate the size of Japan's economy was PPP$3.6 trillion, Korea's was PPP$922 billion, and Taiwan's was PPP$559 billion. The size of an economy is a measurement of the country's annual GDP, i.e., the sum of value added by all resident producers in the economy.)

In developing countries, nondefense health and environment R&D programs are considerably smaller (**Figure 11**). OECD data are not available for China and India, but the annual budget appropriations for health research in other developing countries are well below PPP$500 million (compared to over PPP$1 billion for small Asian NICs like Taiwan and Korea). Again, we see a sharp decline in health and environment R&D appropriations for former Eastern-bloc countries (Russia and Romania) from the beginning of the 1990s, and cyclical fluctuations in health and environment R&D budget appropriations for Mexico and Argentina.

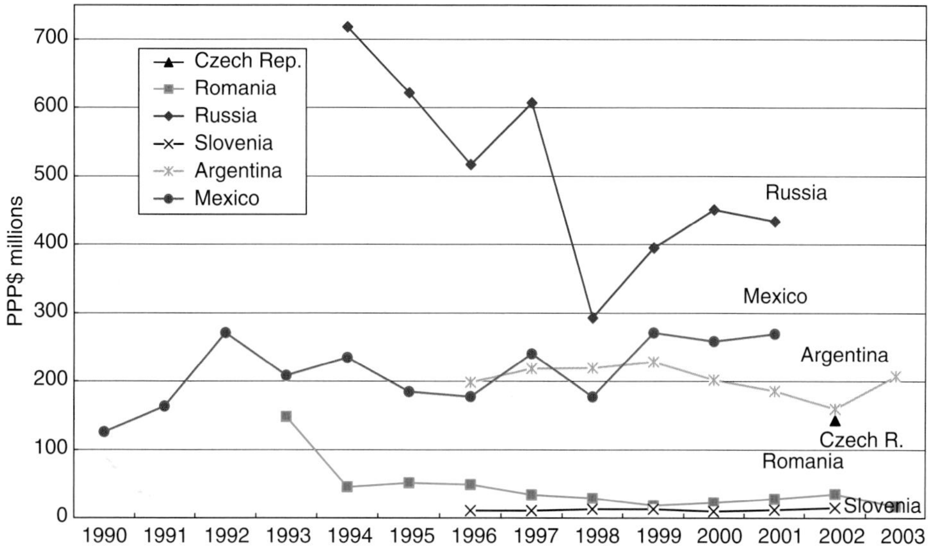

Figure 11 Trends in government spending on civilian health and environment R&D programs in select developing countries, 1990–2003.[13]

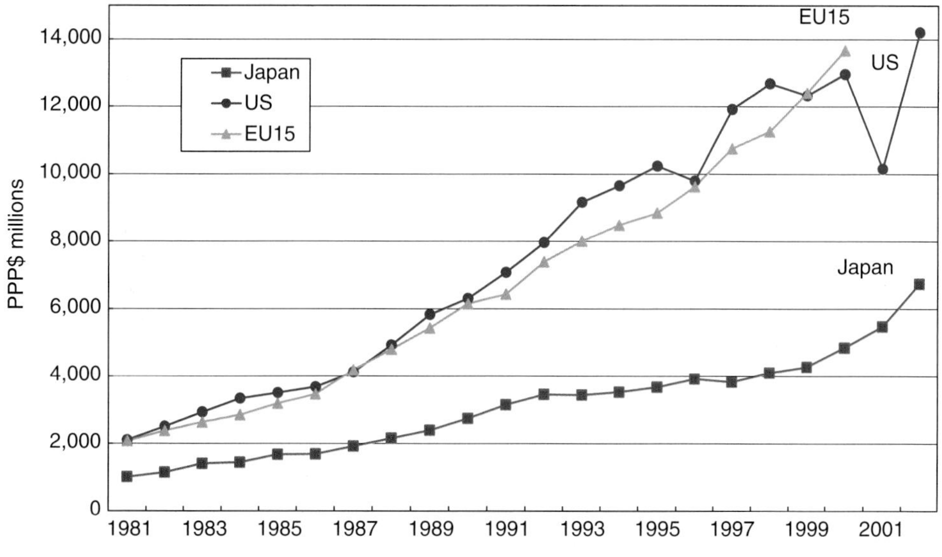

Figure 12 Trends in European, Japanese, and US business expenditures on R&D performed in the pharmaceutical industry, 1981–2002.[13]

1.24.4.3.3 Trends in private-sector funding and performance of health R&D

In addition to looking at government funding for health R&D, it is also important to look at private-sector funding and performance of health R&D (**Figure 12**). In most developed countries, the private sector is the largest funder and performer of health R&D. Health R&D covers a wide range of industrial R&D from pharmaceutical and biological products to research tools and medical devices. Here, we focus on a narrow definition of health R&D because of the limited availability of cross-country data on this topic: the OECD's data on business expenditures on R&D (BERD) performed in the pharmaceutical industry.

The data show large increases in business-performed pharmaceutical R&D from 1981 to 2002: annual average growth of 9.6% in the case of the US, 10.5% in the EU15 (EU15 growth rate is for the latest available data for 1981–2000, whereas the US and Japan data span 1981–2002), and 9.5% in Japan. Data for the latest available year indicate that US businesses performed PPP$14.2 billion of pharmaceutical R&D in 2002; EU15 businesses spent PPP$13.6 billion in 2000; and Japanese businesses performed PPP$6.7 billion in 2002.

It is likely that the US pharmaceutical industry BERD is significantly underreported in OECD statistics. The Pharmaceutical Research and Manufacturers of America (PhRMA), an industry association, reports aggregated R&D expenditures and sales from an annual survey of PhRMA members. In recent years, PhRMA's R&D survey estimates have increasingly diverged from the OECD's estimates of US pharmaceutical BERD, which come from the NSF *Survey of Industrial Research and Development*. For example, in 2002 OECD/NSF estimated US pharmaceutical industry BERD to be US$14.2 billion compared to PhRMA's estimated US$25.6 billion.

Analysis conducted by NSF found differences in survey methodology and definitions to be the source of these dissimilar R&D estimates. By including the R&D expenditures of one additional industry category (US$6.8 billion), NSF found its 2002 estimate came much closer (82.0%) to the 2002 PhRMA estimate: US$21.0 billion versus US$25.6 billion. NSF found that some pharmaceutical companies are migrating to the Wholesale Trade: Drugs and Druggists' Sundries (4222) NAICS industry classification from the Chemicals–Pharmaceuticals and Medicines (3254) NAICS classification (NAICS, North American Industry Classification System). As these large companies are increasingly marketing their own products, the bulk of payrolls, which are used to assign NAICS codes, are going to employees engaged in trade activities. Because of this migration between industry classifications, the NSF data may not sufficiently capture true pharmaceutical industry R&D. NSF is investigating this matter. The majority of the remaining difference owes to PhRMA's inclusion of company spending on: (1) clinical evaluations: phase IV and (2) process development for manufacturing and quality and control accounted, which NSF does not include as R&D. Similarly, it is possible that EU and Japanese pharmaceutical industry BERD could also be underreported due to problems with industrial classifications of companies and the R&D categories included for data reporting.

Nevertheless, US pharmaceutical industry BERD is likely to have been closer to PPP$25.6 billion in 2002, compared to the EU15's PPP$13.6 billion in 2000; and Japan's PPP$6.7 billion in 2002 (**Table 2**). This would make US business-performed R&D in the pharmaceutical industry nearly equal to 2002 government appropriations for the nondefense health and environment program R&D (PPP$27.2 billion).

By contrast, in developing countries, BERD in the pharmaceutical sector are much smaller than government funding for health research (**Table 3**). For example, in 2001, Mexico's government appropriated PPP$268.5 million for health and environmental program R&D, whereas pharmaceutical industry BERD for the same year was PPP$34.0 million. Russia's government appropriated PPP$433.2 million, whereas Russian businesses spent PPP$11.4 million. However, the data point to rapidly growing pharmaceutical industry BERD in many developing countries: Mexico's pharmaceutical BERD grew by an average 69.1% per year from 1992 to 2001; Slovenia's grew by 13.9% per year from 1993 to 2002; and the Czech Republic's by 11.5% per year from 1992 to 2003.

The Asian NICs follow the same developing-country pattern where government is the larger investor in health research: government's health R&D of PPP$1.2 billion versus the pharmaceutical industry's PPP$0.3 billion in 2003 in Korea, and government's PPP$1.0 billion versus industry's PPP$0.1 billion in 2003 in Taiwan. However, government funding for health research in these countries is growing more rapidly than the private sector's. The limited data available suggest that Korean government funding for health research grew by 21.4% per year from 1999 to 2003, while business expenditures on pharmaceutical R&D grew by a much slower 8.8% per year over the same period. Taiwan's government funding for health research grew by an average 16.0% per year from 2000 to 2004, while pharmaceutical industry BERD grew by 4.9% per year from 2000 to 2003 (and 9.2% per year from 1998 to 2003) (*see* 1.14 The Role of the Pharmaceutical Industry).

1.24.4.4 Major Policies and Programs Influencing Level and Scope of Health Research

What does the future of public funding for health research hold? If one examines the S&T strategy and planning documents of developed and developing countries around the world (see, for example, Japan's S&T Basic Plan (2001–2005)[47]; the European Community's Sixth Framework Programme for Research, Development and Technological Development (2002–2006)[48]; Korea's Biotech 2000 Plan (1993) and Bio-Star Project (2005) for biotech commercialization[49,50]; India's National Biotechnology Development Strategy (2005)[51]; and South Africa's A National Biotechnology Strategy for South Africa (2004–2007)[52] one sees a similar emphasis on the medical applications of biotechnology among the top five thematic areas. (Other important thematic areas identified in these plans include energy, nanotechnology/materials science, ICT, the environment and manufacturing technologies.) Health research, referred to in these national plans as life sciences or biotechnology, is being emphasized as much for economic reasons (economic growth and employment) as for public health reasons. Biotechnology is viewed as a key economic driver, in part due to excitement over sequencing the human genome, the promise of new approaches to drug design, and the commercialization of many new molecular-based research tools, therapeutics, diagnostics, and drug delivery systems.

Table 2 Comparison of Japanese, European, and US business expenditures on pharmaceutical R&D and government appropriations for nondefense health and environment R&D, 1981–2005

Year	*Japan*		*EU15*		*US*	
	Govt (PPP$m)	*Business (PPP$m)*	*Govt (PPP$m)*	*Business (PPP$m)*	*Govt (PPP$m)*	*Business (PPP$m)*
1981	–	991.7	3247.8	2057.2	5340.0	2084.8
1982	–	1130.9	3516.0	2357.8	5172.0	2491.7
1983	–	1388.3	3302.2	2615.2	5581.0	2913.5
1984	–	1421.8	3703.8	2831.6	6182.0	3319.3
1985	–	1658.4	3866.5	3166.7	6952.0	3484.5
1986	–	1668.0	4047.8	3444.0	7100.0	3658.0
1987	–	1902.7	4387.2	4160.9	8209.0	4100.0
1988	447.0	2135.3	4814.9	4777.0	8734.0	4906.2
1989	489.9	2373.1	5,275.1	5408.0	9635.0	5807.7
1990	530.8	2724.3	6131.3	6131.8	10 398.0	6287.4
1991	578.1	3132.1	6748.6	6413.3	11 561.0	7060.8
1992	627.4	3437.6	6943.9	7368.2	12 436.0	7944.0
1993	677.4	3420.7	6971.8	7985.5	12 703.0	9146.0
1994	756.7	3509.7	663.0	8462.6	13 513.0	9633.0
1995	828.2	3653.1	7745.6	8823.9	13 874.0	10 215.0
1996	1064.9	3897.0	8207.6	9604.8	14 079.0	9773.0
1997	1222.7	3809.5	8293.2	10 729.2	14 953.5	11 898.7
1998	1190.9	4081.9	8,428.5	11 238.5	16 186.5	12 666.9
1999	1317.2	4254.9	9043.3	12 389.7	18 160.7	12 304.5
2000	1546.1	4827.2	9913.8	13 647.6	20 912.4	12 945.2
2001	1759.5	5455.5	10 220.0	–	24 007.7	10 137.0
2002	1811.5	6722.0	9988.3	–	27 179.4	14 186.0
2003	1907.9	–	10 071.7	–	31 530.9	–
2004	–	–	–	–	32 332.2	–
2005	–	–	–	–	33 086.9	–

Data from OECD, Main Science and Technology Indicators 2005-1.[13]

Consequently, governments are looking for ways to support the development of the biotech sector through a combination of increased funding for biotech research and cross-cutting policies that improve the innovation environment. A summary of these policies includes:

- improved infrastructure (facilities, equipment, computing power) at universities, government labs, and science parks
- development of human resources, e.g., increased competitive grant support for researchers and more incentives and programs for workforce training
- enhanced technology diffusion through industry–university–government lab research collaboration support
- competitive funding to high-performing clusters or universities (centers of excellence)

Table 3 Comparison of select developing countries' business expenditures on pharmaceutical R&D and government appropriations for nondefense health and environment R&D, 1981–2005

Year	*Russia*		*Mexico*		*Czech Republic*		*Slovenia*	
	Govt (PPP$ m)	*Business (PPP$ m)*	*Govt (PPP$ m)*	*Business (PPP$ m)*	*Govt (PPP$ m)*	*Business (PPP$ m)*	*Govt (PPP$ m)*	*Business (PPP$ m)*
1990	–	–	125.0	–	–	–	–	–
1991	–	–	162.5	–	–	–	–	–
1992	–	–	270.1	0.3	–	12.5	–	–
1993	–	–	207.6	13.5	–	9.8	–	32.2
1994	717.4	–	233.7	8.7	–	25.8	–	36.0
1995	620.9	16.2	183.7	13.9	–	15.8	–	37.2
1996	516.2	20.1	176.5	14.0	–	21.2	10.7	46.4
1997	606.4	10.7	239.2	17.1	–	13.7	10.8	48.5
1998	292.1	–	176.3	26.0	–	15.7	13.1	52.5
1999	394.3	13.6	270.0	28.3	–	21.8	13.1	60.1
2000	450.3	–	257.6	29.4	–	31.9	9.8	68.5
2001	433.2	11.4	268.5	34.0	–	31.5	12.1	88.1
2002	–	–	–	–	143.1	45.1	15.0	103.7
2003	–	6.7	–	–	–	41.2	–	–
2004	–	–	–	–	–	–	–	–
2005	–	–	–	–	–	–	–	–

Data from OECD, Main Science and Technology Indicators 2005-1.[13]

- IP and business development assistance and improved access to capital for biotech commercialization and business startup
- streamlined regulatory procedures

Governments are looking for results – vibrant private-sector growth in health-related industries and health innovations that improve the quality of life of their citizens. Consequently, much of the increased research funding across countries is taking the form of competitive grants to researchers, universities, and companies. One problem some regions and countries have encountered is an insufficient number of competitive proposals. Thus, governments are likely to struggle to balance efforts to improve the capacity of national health innovation systems (e.g., physical infrastructure, the quality of higher education) on the one hand, and direct competitive research funding on the other.

1.24.5 Regulation and Health Innovation

Government regulation of IP, industry practices (biosafety), and competition play an indirect but crucial role in health innovation by shaping the incentives faced by the private sector to conduct health research and product and market development. There is a need for government to play these regulatory roles in most markets (see discussion in Section 1.24.2.3.2 above). However, in the health sector, the need for government regulation is particularly acute given the knowledge- and research-intensive nature of the industry, the direct impact of products on human health, and the market concentration of industry segments. This section explores the specific ways governments address these issues.

1.24.5.1 Intellectual Property

The knowledge- and research-intensive nature of health innovation makes IP protections vital to protecting market incentives for innovation. Recent estimates put the cost of bringing a new medicine to market at between US$800 million and US$1.7 billion.[24a,b] The process typically takes over 8 years from preclinical toxicology testing through drug approval. In order to make those investments, companies need some security that a competitor will not simply copy their successful products. Of course, large product development investments are required in many industries – it takes around 4 years to bring a new car model into the market from concept to production. What is unique, however, is the degree to which many drugs can be copied and produced by competitors at very low costs. In other words, the production and distribution costs of a drug represent a very small portion of the final price of a drug compared to their discovery and development costs. For this reason, IP protections are particularly important for health markets.

For the most part, health innovations are governed by general national and international IP protection systems – patents are awarded for a limited time (20 years) for novel, nonobvious, synthetic, useful inventions. Patent disclosure rules ensure that the nature of the innovation is made public and is available to other innovators to build upon. While IP systems are being harmonized internationally, it is still largely a national process and IP protections for pharmaceutical and other health innovations have varied considerably across countries. For example, Japan (until 1976) and many European countries did not confer patents for pharmaceutical products, but only for the processes used to produce them.[25] As a result, companies in those countries tended to concentrate their research efforts on process improvements rather than discovering and developing new medicines.

Today, there is a heated debate over the appropriateness of the IP system, particularly in health and biotechnology. The system is criticized first for maintaining high prices that exclude many consumers from the benefits of new medicines, and second for conferring exclusive rights to ideas that critics contend should be freely available. Many life-saving drugs, for example antiretrovirals and other drugs for treating HIV/AIDS, are very expensive, reflecting the enormous investments that have gone into their development and the high willingness to pay for drugs that have such a direct life-sustaining impact. HIV/AIDS affects 40.3 million people worldwide,[26] so the life-saving potential of these drugs is enormous. Unfortunately, many of those affected are the very poor in developing countries (25.8 million live in sub-Saharan Africa) who are unable to afford these drugs.

Some countries, such as Brazil, that have significant pharmaceutical production capacity, have either begun producing cheap generic versions of patented drugs or have threatened to do so and, thereby, been able to negotiate cheaper prices with patent holders. Brazil was able to reduce costs dramatically and to provide patented antiretroviral drugs free of charge to its citizens, reducing the share of its population dying from HIV/AIDS. Such arguments for circumvention of IP rights (IPR) for pressing national needs have even been recognized, to some degree, by representatives at the World Trade Organization. A ministerial declaration during the Doha round of trade talks specifically recognizes the principle that initiatives to promote widely available medicines should not be unduly hindered by IPR regimes.[27]

Such efforts to relax stringent IPR are not restricted to poor countries attempting to acquire more affordable drugs. Faced with an anthrax scare after the terrorist attacks of September 11, 2001, the US Congress considered legislation that would have forced the German manufacturer of an antianthrax drug to license it to American producers. Likewise, a debate is raging in the US over the legality of allowing prescription drug imports from neighboring countries with considerably lower drug prices (e.g., Canada and Mexico) over the protests of patent-holders. Some states have begun to direct residents to prescreened foreign pharmacy websites from which they can purchase cheaper imported drugs, a move strongly opposed by the federal government.[28]

Another difficult area for governments is the scope of IP protections in fundamentally new scientific areas. Especially in biotechnology, where new insights and tools are being brought to bear on human health issues, there have been concerns over how to apply the statutes of patent law. In the past, the discoverer of a new chemical element was not able to patent that element or its use. Similarly, with the harnessing of electricity, early inventors were allowed to patent machinery and technologies for the production and transmission of electricity, but could not patent the fundamental knowledge of how electricity works. Today, patent law in the US, Europe, and Japan has moved towards conferring patent protections on more basic discoveries (genes and gene sequences) if a useful, specific application is noted and if it meets the additional criteria of novelty, nonobviousness, and enablement. For more information on the history of human DNA sequence patenting, see.[29]

While the modern IP system has been credited with creating a conducive environment for health innovations – the US' lead in biotechnology has partly been attributed to the strength of its IP system – that system has struggled in recent years to find the right balance between current needs (for more widely available drugs and technologies) and incentives for innovation.

1.24.5.2 Biosafety Regulation

Health innovations, particularly pharmaceuticals, are strictly regulated as they directly impact human life. A complex regulation web governs the use and introduction of new health products in many countries, and one of the most salient aspects of this system is how a new product is approved for use. The approval process typically requires that a company submit to the national regulatory body (e.g., the Food and Drug Administration in the US, the National Sanitary Surveillance Agency, referred to as ANVISA, in Brazil, etc.) proof of efficacy of the new drug based on adequate, documented clinical trials. (A process is required for all types of medical products, including pharmaceuticals, biologics, devices, and diagnostic tools. Here we focus on pharmaceutical and biologic agents.) The regulatory body also sets standards for the conduct of the clinical trials. The drug approval process is governed by national rules, and approval must be sought in every country where a drug is to be marketed. Therefore, even if a drug were already in use in Europe, it must also undergo clinical testing in Japan if it is to be released in Japan.

Both the substance of the drug approval process and the resources dedicated to the public agencies that manage that process vary widely from country to country. For example, analysts estimate that it takes, on average, a third longer to introduce a new drug in Europe than in the US.[30] New drugs therefore can get to market earlier, and affect health outcomes much more quickly in the US than in Europe, all else equal. With the slow loss in competitiveness of the European pharmaceutical industry, Europe has begun to place more emphasis on improving its business environment for drug companies by taking steps to speed up the drug approval process. Meanwhile, in the US, the Food and Drug Administration (FDA) and NIH have also undertaken initiatives to accelerate the drug approval system, which has been faced with an enormous backlog of pending drugs. Reforms in the US have reduced average new drug application review times at the FDA by more than 40%.[31] Of course, speed can also have its costs, as exemplified by recent high-profile withdrawals of drugs such as Vioxx over safety concerns. Governments must constantly balance the pressures to speed beneficial drugs to market while protecting the public from harm.

1.24.5.3 Competition

There is vigorous overall competition in the health industry. In pharmaceuticals, the largest company, Pfizer, has less than a 10% share of the global market. However, for any particular health problem in a given country, there are often only a few producers at most, each with government-conferred patent rights to a product. Traditional competition concerns, therefore, are restricted for IP protection reasons. For this reason, competition policy with respect to health innovation focuses not so much on competition in the current market for drugs (where standard competition policy applies, but whose scope is restricted for patent reasons), but on ensuring competition that drives future health innovation. (This is not to say that traditional competition concerns are not important. Regulatory and court activity surrounding agreements between patent-holders and generic competitors in the US attests to the importance of competition regulation in the market. Rather, such competition concerns are common across most industries.)[32]

These concerns were first applied in merger policy in the US in the early 1990s. (See review of these and other cases in Katz and Shelanski.[33]) The US Federal Trade Commission (FTC) challenged the merger of Genentech and Roche Holdings on the grounds that the merger would reduce potential competition in the market for human growth hormone-deficiency treatments. Furthermore, the FTC was concerned that Roche and Genentech were both working hard to develop HIV/AIDS treatments, an area where only a handful of groups were working at the time. A merger of the two companies would concentrate R&D efforts in this important field, and so the FTC blocked the merger.

The FTC had similar concerns in 1996 with the proposed merger of Ciba-Geigy and Sandoz, two Swiss firms with significant operations in the US. Here, the FTC had traditional concerns with increased concentration of several product lines where the two firms were competitors. But the Commission was further concerned with competition in the innovation market for the development of gene therapy products (even though no such products were then on the market). Looking ahead, however, the FTC found that the two firms were among a small group that had the IP assets and technological capacity to develop gene therapy products. A combined firm would have controlled critical patents and been able to inhibit commercial development of the entire industry. As a result, the FTC negotiated an agreement whereby Ciba and Sandoz agreed to license critical IP which would allow other firms to develop competing gene therapy products.

As these cases illustrate, competition regulators examine not just the current market activities of firms, but also their R&D activities in order to preserve competition in the markets for future products.

1.24.6 Government as End-User

A final way government impacts the scope and pace of health innovation is through its purchases of health products and services, so-called demand-side policies and practices. In a free-market economy, inventors invest in discovering and developing new health products with the expectation that, if successful, they will be able to sell their solution, generating returns that will pay for their initial investment and some degree of profit. The types of products and services demanded by the market (consumers and the health system overall), and the prices that end-users are willing to pay for them affect the type of health research carried out. Demand completes a critical feedback loop in the innovation system by providing the incentives and signals that affect research-funding decisions by governments, companies, and nongovernmental organizations.

Government's role in influencing the demand for health innovations varies around the world with the public sector's role in the health system. At one extreme, where there is public provision of healthcare, like the UK's National Health Service, the government is the end-user and therefore has an immediate and direct impact on demand. Its willingness to purchase new innovative health products will directly influence the incentive of companies to develop and market those products in the country. At the other extreme, where the private provision of healthcare dominates, insurers and individuals are the end-users and it is their willingness to provide coverage for certain treatments, or directly pay for them out of pocket that matters. Even in this case, governments influence demand through tax and other policies that affect the behavior of insurers and citizens.

Among developed countries, the major differences are between pricing regimes for pharmaceuticals. Central government negotiations with drug companies result in considerably lower prices in Europe and Canada than are typically charged for the same product in the US' free-market approach. Europe spends about 60% less per capita on prescription drugs, a large portion of which is due to these lower prices.[34] Americans, therefore, pay a larger portion of the costs of developing new treatments because of these pricing decisions.

In today's global economy, innovators focus on the global market for a prospective health innovation. This means that it is the global aggregate willingness to pay for innovations that influences the direction and depth of private innovative activity. The US, with its relatively lavish spending on health, represents a large share of that market. Its decisions, e.g., not to negotiate lower prices with drug companies and not to allow drug importation from countries that do, directly affect the incentives facing health innovators. Studies have shown, for example, that drugs are introduced sooner in the US than they are in Europe, partly for these reasons.

1.24.7 Health R&D Outcomes and Conclusions

Given the large investments in health R&D and other government policies aimed at improving the health innovation system, what has been the impact of governments' support for health innovation? Research that increases our fundamental understanding of human biological processes and disease transition mechanisms is necessary and critical. However, policy-makers must make difficult tradeoffs in prioritizing and funding public policies. They, therefore, need to know (or must make an educated guess as to) what the impact of an additional dollar spent on health research is vis-à-vis the impact of that dollar being spent somewhere else (e.g., expanding early-childhood education, physics research, etc.).

Measuring the impact of research is a complex undertaking. Only in the past couple of decades have governments begun to require more rigorous impact analyses of these investments. Research evaluation as a field has lagged behind the rapid increases in research investments and growth of the research enterprise in many countries. There are no tried and true evaluation techniques for assessing how much an additional euro spent on health research will generate.

More fundamentally, there is little consensus among policy-makers as to what research outcomes are most desirable (**Table 4**). As discussed previously, research motivations run the gamut from research enterprise/knowledge outcomes (articles published, students trained) which generate societal health outcomes (increased longevity) and economic development outcomes (exports, jobs, new industries). While these categories of impacts are not mutually exclusive, there is considerable debate over what the goals of health R&D should be and which should be emphasized.

1.24.7.1 Research Enterprise/Knowledge Outcomes

Evaluations of research outputs and outcomes have traditionally focused on assessing the contribution to knowledge through citation and patent analysis (the number of publications, their quality, and impact). This is the science-enabling aspect of research – the creation of new knowledge contributes to new and better approaches to tackling important scientific questions. A recent study by Paraje *et al.*[35] finds that scientific publications on health topics from 1992 to 2001 were disproportionately distributed and highly concentrated among the world's wealthiest countries. The top five health-related publication producers in 2001 included: the US (34.7%), UK (8.2%), Japan (8.4%), Germany (7.1%), and France (4.8%). The top 20 producers accounted for 90% of global health-related publications. While shares

Table 4 Examples of different research outcomes

Research outcome	*Evaluation topics*
Intermediate research enterprise/ knowledge outcomes	• Is this level of investment resulting in a sufficient pipeline of new scientists and engineers? • What is country A's share of global biomedical publications and patents?
Societal health outcomes	• Is there a correlation between increasing health research investments and longer lifespans, reduced disability, or reduced incidence of disease? • Are increasing health research investments benefiting people equitably?
Economic development/competitiveness outcomes	• How many higher-than-average-wage jobs have been generated? • What is country A's global market share of this industry? • Is country A actively commercializing discoveries and process innovations?

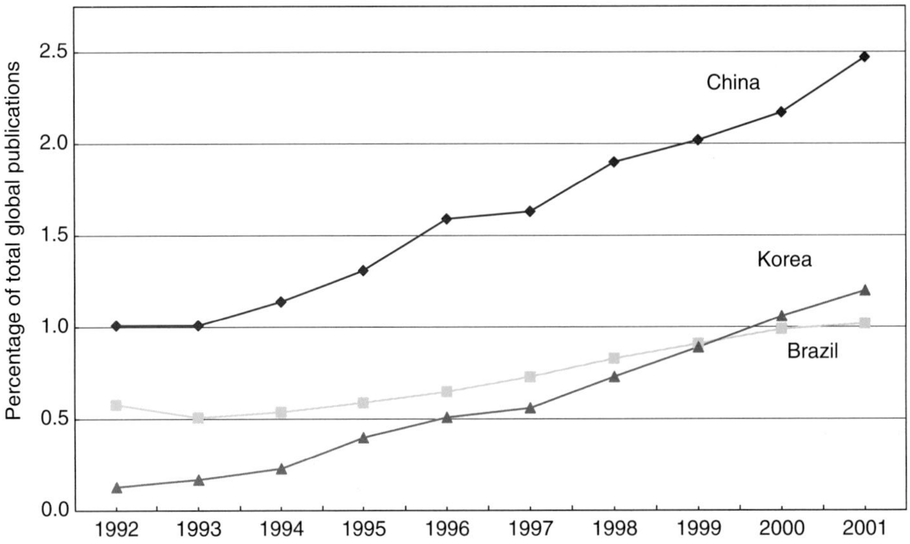

Figure 13 Trends in Chinese, Brazilian, and Korean share of worldwide health-related publications, 1992–2001.[35]

have remained relatively constant over this time period, the study notes a clear upward trend in the share of worldwide publications by China (2.5%), Brazil (1.0%), and Korea (1.2%) (**Figure 13**).

A slightly different aspect used to evaluate investments in R&D is how it enhances a country's absorptive capacity, or ability to adopt and adapt innovations from elsewhere. The learning by doing resulting from carrying out research builds capacity for adapting global research to local needs. This is a primary rationale for developing country funding of R&D. A commonly used indicator with regard to absorptive capacity is the number of scientists and engineers employed in research occupations in a country or the number of people holding science and engineering degrees. Data on the number of full-time equivalent researchers per 1000 workforce population indicate much greater capacity on the part of developed countries. Japan and the US had 9.7 and 9.0 full-time equivalent researchers per 1000 workers, respectively, in 1999.[36] The EU15, Taiwan, Singapore, and Russia had full-time equivalent researcher per 1000 workers numbers that ranged from 5.3 to 6.8 (**Figure 14**). Mexico and China, the only developing countries represented, had fewer than 1.0 full-time equivalent researchers per 1000 workers. However, in absolute numbers this translates to 531 million full-time equivalent researchers in China's workforce compared to 22 000 full-time equivalent researchers in Mexico.

1.24.7.2 Societal Health Outcomes

An important goal of government support for health innovation, even support for basic research, is the improvement of health outcomes. New medicines today offer treatments that improve the quality and length of life for people with

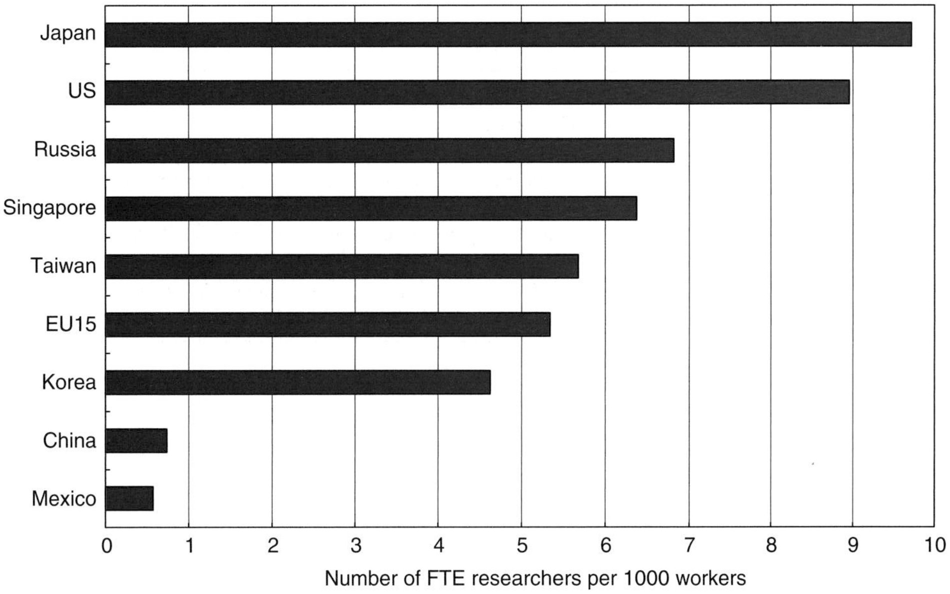

Figure 14 FTE researchers per 1000 workers in select countries, 1999.[13] FTE, full-time equivalent.

conditions for which there were previously few treatment options. For example, in the early 1980s, being diagnosed with HIV/AIDS was a fatal, short-term prognosis. The development of effective diagnostic tests, changes in the management of blood banks, increased public education about the disease, and the development of new drugs to slow the advance of the disease have both slowed the rates of new infections in many parts of the world and improved life prospects of many of those infected. Statins, angiotensin-converting enzyme inhibitors, and other medicines are being used to prolong life and reduce risks for people suffering from cardiovascular diseases or predisposed to developing these diseases. Drugs are now used to treat depression, taking the place of sometimes severe and less effective treatments – or no-treatment – of the past. In the US, life expectancy at birth has increased from 47 years in 1900 to 77 years in 2002.[37] Health innovations have contributed significantly to these improvements.

Despite the very clear contributions that specific health innovations have had, it has been extremely difficult to link health innovations empirically to improvement in a country's health outcomes. Ultimate improvements in health outcomes are determined by public and private investments in healthcare, behavioral and environmental factors, as well as by available treatment options. For example, the US Centers for Disease Control and Prevention[38] attributes 25 years of the 30-year gain in US life expectancy from 1900 to 1999 to improvements in public health. Many of those improvements, however, are not due to particular health innovations, but rather to environmental quality improvements (safer drinking water), behavioral modifications (wearing seatbelts, smoking cessation), and systemic changes (safer workplaces). Nevertheless, health innovations have contributed to the control of infectious diseases,[39,40] to a decline in death rates from heart disease and stroke of 56% since 1950,[41] to healthier mothers and babies from improved maternal and neonatal medicines,[42] and so on. Looking broadly at improvements in health outcomes, Garfinkel *et al.*[43] point out that most of the improvements in US health outcomes over the last 60 years are attributable not to basic health research since World War II, but to technological development and systems changes.

At the global level, health research critics have stronger evidence of a mismatch between research and public health outcomes. A startling statistic is that less than 10% of global investment in health research is devoted to the diseases that affect 90% of the world's population.[44] The 2003 launch of the Grant Challenges in Global Health (**Table 5**), a US$436 million medical research initiative, announced at the World Economic Forum in Davos, Switzerland, by Bill Gates, is one response to the growing concern over the poor connection between actual research funding directions and global societal health priorities. (The Grand Challenge in Global Health is managed and administered by the Bill & Melinda Gates Foundation and supported by the Bill & Melinda Gates Foundation, Canadian Institutes of Health Research, Foundation for the NIH, and Wellcome Trust. The initiative was launched with an initial Gates Foundation grant of US$200 million to the Foundation for the NIH. In 2005, the initiative was supporting 43 separate research projects related to these challenges.) The initiative describes a grand challenge as a call for a specific scientific or

Table 5 Grand challenges (GC) in global health

Improve childhood vaccines
GC 1 Create effective single-dose vaccines
GC 2 Prepare vaccines that do not require refrigeration
GC 3 Develop needle-free delivery systems for vaccines
Create new vaccines
GC 4 Devise reliable tests in model systems to evaluate live attenuated vaccines
GC 5 Design antigens for protective immunity
GC 6 Learn which immunological responses provide protective immunity
Control insects that transmit agents of disease
GC 7 Develop a genetic strategy to control insects
GC 8 Develop a chemical strategy to control insects
Improve nutrition to promote health
GC 9 Create a nutrient-rich staple plant species
Improve drug treatment of infectious diseases
GC 10 Find drugs and delivery systems to limit drug resistance
Cure latent and chronic infection
GC 11 Create therapies that can cure latent infection
GC 12 Create immunological methods to cure latent infection
Measure health status accurately and economically in developing countries
GC 13 Develop technologies to assess population health
GC 14 Develop versatile diagnostic tools

Source: Grand Challenges in Global Health, http://www.gcgh.org.

technological innovation that would remove a critical barrier to solving an important health problem in the developing world. In this way, the funded research is anticipated to produce tangible, actionable results with large health benefits in a short timeframe. Another example is the UN Global Fund to Fight AIDS, Tuberculosis, and Malaria, which takes a results-based approach to grant-making. The Fund disperses research grant awards incrementally based on measures of progress.

1.24.7.3 Economic Development/Competitiveness Outcomes

Over the past few decades, developed countries have increasingly shed manufacturing jobs and increased their economy-wide share of service sector jobs. This transition has been difficult for many countries as manufacturing jobs generally pay higher relative wages than some of the service sector jobs replacing them (e.g., in the retail, food, and accommodation industries). As structural economic changes have occurred, it has become clear that maintaining and increasing standards of living in the future will be directly related to a country's ability to promote the development of more knowledge-intensive industries. Research is a cornerstone of knowledge-intensive industry development, especially in the health sector: evidence-based healthcare, discovery of new drugs and therapeutics, medical devices, diagnostic tests, drug delivery systems, health information management, and research tools.

In this regard, patents are used as one research output indicator of potential economic impact. For example, one can examine the technologically active countries in emerging areas of importance, such as DNA patenting. The number of active assignees refers to the number of unique organizations filing patent applications and includes companies, universities, nonprofit organizations, government agencies, and individuals. A study of active assignees for DNA patents in the US, Europe, and Japan found that the US has had the most organizations (not including individual inventors) filing patent applications for human DNA sequences nearly every year from 1980 to 1999, followed by Japan and then Europe depending on the year (**Figure 15**).

With regard to the type of organization that has been most active, the study found that companies dominate human DNA patenting overall. However, there are striking differences among countries. For example, the US and UK had the largest number of universities seeking patents for human DNA sequences (163 and 27 respectively from 1995 to 1999), although companies still dominated in both countries. By comparison, Germany, Japan, France, and Israel had very few universities, nonprofits or government agencies that sought human DNA sequence patents compared to the number of companies.

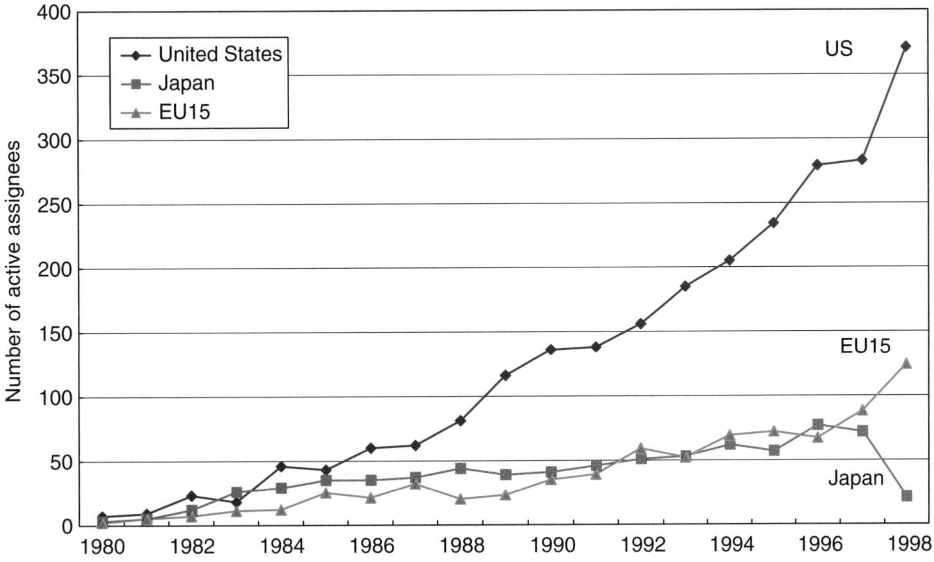

Figure 15 Trends in the number of active assignees for DNA patents in the US, EU15, and Japan, 1980–1998. (Reproduced from National Science Board.[9])

1.24.7.4 Conclusions

Governments are intimately involved in markets for health innovation. The precise role governments play varies from country to country, depending in part on the particular characteristics of the country's level of development, its healthcare industry, and its broader health innovation system. In all cases, though to different extents, governments support the generation of innovations, set and enforce market rules, and oversee and encourage the use of health innovations. Policy-makers must make difficult decisions in all of these roles – for example, on budget levels for funding health research, on IP rules, and on prescription drug-pricing systems. They must weigh competing interests in maximizing long-term public welfare. Tradeoffs must be made between the near-term needs of reducing drug costs and funding the delivery of current health solutions versus long-term investments in R&D and the need to provide strong incentives for private-sector investment in innovation. These decisions are particularly difficult given the dynamism of innovation in the health industry. Governments will be struggling with these issues for many years to come. In the meanwhile, health innovations have already contributed enormously to health improvements around the world, and governments have undoubtedly played a major role in the process.

References

1. El Feki, S. *Prescription for Change: A Survey of Pharmaceuticals* (Reprint from *Economist* 18 June print edition); The Economist: London, England, 2005, p 1.
2. Nelson, R. R., Ed. *National Innovation Systems: A Comparative Analysis*; Oxford University Press: New York, 1993.
3. Metcalfe, J. S. The Economic Foundations of Technology Policy. In *Handbook of Industrial Innovation*; Stoneman, P., Ed.; Blackwell: Oxford, UK, 1995, pp 409–512.
4. Arrow, K. J. Economic Welfare and the Allocation of Resources for Invention. In *The Rate and Direction of Innovative Activity*; Nelson, R. R., Ed.; National Bureau of Economic Research, Princeton University Press: Princeton, NJ, 1962, pp 609–625.
5. Teece, D. J. *Res. Policy* **1986**, *15*, 285–305.
6. Nelson, R. *Am. Econ. Rev.* **1986**, *76*, 186–189.
7. Keck, O. The National System for Innovation in Germany. In *National Innovation Systems: A Comparative Study*; Nelson, R. R., Ed.; Oxford University Press: New York, 1993, pp 115–157.
8. Kumar, U.; Kumar, V. *Incubating Technology: Best Practices. Prepared for the Federal Partners in Technology Transfer*; Logitech Systems Management Consultants: Ottawa, ON, 1997.
9. National Science Board (NSB) *Science and Engineering Indicators 2004*; National Science Board: Arlington, VA, 2004, Vol. 2, Table 6-16.
10. Generic Pharmaceutical Association *Generic Pharmaceutical Facts at a Glance*; Generic Pharmaceutical Association: Arligton, VA, 2005.
11. Philipson, T. J.; Anupam B. J. *NBER Working Papers* **2005**, 11810.
12. Kubinyi, H. *J. Recep. Signal Trans. Res.* **1999**, *19*, 15–39.
13. OECD *Main Science and Technology Indicators (MSTI), 2005-1*; OECD: Paris, France, 2005.
14. Salter, A. J.; Martin, B. R. *Res. Pol.* **2001**, *30*, 509–532.
15. Garber, A. M.; Romer, P. M. *Proc. Natl. Acad. Sci.* **1996**, *93*, 12717–12724.

16. Griliches, Z. R&D and Productivity: Econometric Results and Measurement Issues. In *Handbook of Industrial Innovation*; Stoneman, P., Ed.; Blackwell: Oxford, UK, 1995, pp 52–89.
17. Office of Technology Assessment (OTA) *Research Funding as an Investment: Can we Measure the Returns? A Technical Memorandum*; Government Printing Office: Washington, DC, 1986.
18. Koizumi, K. *U.S. Funding of Japanese Priority Areas and U.S. Priorities in R&D*; American Association for the Advancement of Science (AAAS): Washington, DC, 2004.
19. National Institutes of Health (NIH) *Setting Research Priorities at the National Institutes of Health*; NIH: Bethesda, MD, 2005.
20. American Association for the Advancement of Science (AAAS) *AAAS Report XXIX: Research and Development FY2005*; AAAS: Washington, DC, 2005; reports I–XXIX.
21. Koizumi, K. *U.S. Funding of Japanese Priority Areas and U.S. Priorities in R&D*; American Association for the Advancement of Science (AAAS): Washington, DC, 2004.
24a. DiMasi, J. A.; Hansen, R. W.; Grabowski, H. G. *J. Health Econ.* **2003**, *22*, 151–186.
24b. Gilbert, J.; Henske, P.; Singh, A. *In Vivo: The Business and Medicine Report* **2003**, 21, 1–4, 6–10.
25. Malerba, F.; Orsenigo, L. *Ind. Corp. Change* **2002**, *11*, 667–703.
26. UNAIDS/WHO *AIDS Epidemic Update: December 2005*; UNAIDS: Geneva, Switzerland, 2005, p 1.
27. World Trade Organization; *Declaration on the TRIPS Agreement and Public Health*, WT/MIN(01)/DEC/2; World Trade Organization: Geneva, Switzerland, 2001.
28. Barry, P. *AARP Bulletin* **2004**, October.
29. National Science Board (NSB) *Science and Engineering Indicators 2004*; National Science Board: Arlington, VA, 2004, Vol. 2, pp 6–28.
30. *The Economist* **2004**, 29 Jan.
31. Berndt, E. R.; Gottschalk, A. H. B.; Philipson, T. J.; Strobeck, M. W. *Nat. Rev. Drug Disc.* **2005**, *4*, 545–554.
32. Bulow, J. The Gaming of Pharmaceutical Patents. In *Innovation Policy and the Economy*; Jaffe, A. B., Lerner, J., Stern, S., Eds.; National Bureau of Economic Research, MIT Press: Cambridge, MA, 2004; Vol. 4, pp 145–187.
33. Katz, M. L.; Shelanski, H. A. *NBER Working Paper Ser.* **2004**, 10710.
34. *The Economist* **2004**, 29 Jan.
35. Paraje, G.; Ritu, S.; Karam, G. *Science* **2005**, *308*, 959–960.
36. OECD *Main Science and Technology Indicators (MSTI), 2005-1*; OECD: Paris, France, 2005.
37. Arias, E. *Nat. Vital Stat. Rep.* **2004**, *53*, 1–40.
38. US Centers for Disease Control and Prevention (CDC). *MMWR Week* **1999**, *48*, 241–243.
39. CDC. *MMWR Week* **1999**, *48*, 243–248.
40. CDC. *MMWR Week* **1999**, *48*, 621–629.
41. CDC. *MMWR Week* **1999**, *48*, 649–656.
42. CDC. *MMWR Week* **1999**, *48*, 849–858.
43. Garfinkel, M. S.; Sarewitz, D.; Porter, A. L. *Am. J. Public Health* **2006**, *96*, 441–446.
44. Specter, M. *The New Yorker* **2005**, *24*, 56–71.
45. http://www.sciencemag.org/content/vol291/issue5507/ (accessed Aug 2006).
46. http://www.nature.com/ (accessed Aug 2006).
47. Japan's Science and Technology Basic Plan (2001–2005). http://www8.cao.go.jp/cstp/english/basic/2nd-BasicPlan_01-05.pdf (accessed Aug 2006).
48. European Community's Sixth Framework Programme. http://ec.europa.eu/research/fp6/pdf/fp6-in-brief_en.pdf (accessed Aug 2006).
49. Wong, J.; Quach, U.; Thorsteinsdóttir, H.; Singer, P. A.; Daar, A. S. *Nat. Biotechnol.* **2004**, *22*, DC42–DC47.
50. Sung-jin, K. *The Korea Times* **2005** 21 Feb. http://times.hankooki.com/ (accessed Aug 2006).
51. India's National Biotechnology Development Strategy (2005). http://dbtindia.nic.in/biotechstrategy/BiotechStrategy.pdf (accessed Aug 2006).
52. A National Biotechnology Strategy for South Africa (2004–2007). http://www.pub.ac.za/resources/ (accessed Aug 2006).

Biographies

Jennifer K Ozawa is a Senior Economist at SRI International, specializing in technology-based economic development. Ms Ozawa has crafted policy options for public and private sector clients who want to create an environment conducive to nurturing dynamic, high-growth sectors of the economy and enhancing firm attraction efforts, with special emphasis on the biotech industry. Her past biotech industry experience in developing countries includes a project to identify and assess viable biotech clusters in Brazil, India, South Africa, and China, involving field work, and the development of investment strategies to bring financing to these sectors. Ms Ozawa evaluated life sciences technologies over many years for an international award. For public and private sector clients, she has analyzed high-tech industry trends and regional capacity to stimulate technology-based economic growth. Ms Ozawa has an MSc in Development Economics from the School of Oriental and African Studies (SOAS), University of London, and a BA in International Relations and Economics from the College of William & Mary, Virginia.

Quindi C Franco is a director at StratEdge who specializes in developing and applying innovative analytic tools to address complex problems. He has led and participated in a wide range of engagements for public and private sector clients around the world in areas including innovation policy, economic development, and impact assessment. Mr Franco worked for Harvard University on a long-term competitiveness and economic growth project in Latin America. He was an Economist at the US White House Council of Economic Advisors conducting economic analysis of policy issues, including science and technology, energy, and the environment. He also served in the Chief Economist's office of the Federal Communications Commission where he covered innovation and market structure issues. He has a Masters in Public Policy from Harvard University and a BA from Pomona College, California.

Comprehensive Medicinal Chemistry II
ISBN (set): 0-08-044513-6

ISBN (Volume 1) 0-08-044514-4; pp. 725–753

1.25 Postmarketing Surveillance

A Li Wan Po, Centre for Evidence-Based Pharmacotherapy, Nottingham, UK

1.25.1 Introduction

Adverse drug reactions are the cause of considerable mortality and morbidity in both primary-care and secondary-care settings. While the frequency of adverse drug reactions varies from country to country, the general picture shows that it is consistently high. Prospective studies suggest an incidence of 2–4% of hospital admissions being related to adverse drug reactions.[1–4] Quite aside from the ethical imperative to minimize such iatrogenic effects, there is a strong economic incentive to do so.[5–7] For example, a 1997 estimate suggests that the cost of adverse drug events (ADEs) for a 700–bed hospital in the US was about $5.6 million. Of this, about $2.8 m was attributable to preventable ADEs.[6] It was suggested that this was an underestimate as harm to patients and malpractice costs were not included.

ADEs are therefore an important aspect of healthcare and indeed, in most countries, the drug regulatory frameworks in place today have been shaped by various drug disasters. In the US, the death of 10 children caused by diphtheria antitoxin, contaminated with live tetanus bacilli, led to the passing of the Biologics Control Act in 1902. The US Food and Drug Administration (FDA) was set up as a regulatory body in 1906, with the passing of the Food and Drugs Act. The Federal Food, Drug and Cosmetics Act (1937) followed the 105 deaths from ingestion of elixir sulfonilamide, formulated with diethylene glycol (antifreeze) the previous year.

In the UK, the thalidomide tragedy led to the Medicines Act (1962), and the establishment of formal regulatory systems for the licensing of medicines. The Committee on Safety of Drugs (CSD), set up in 1963, evolved into the Committee on Safety of Medicines, and today's Commission on Human Medicines, operating under the umbrella of the Medicines and Healthcare Products Regulatory Agency (MHRA). The MHRA, a governmental agency funded through the licensing fees, operates within the supranational European agency, the European Medicines Evaluation Agency (EMEA).

Although the vigilance of one of its FDA assessors meant that thalidomide was never licensed in the US, the disaster led to the Kefauver–Harris Amendments. These required all clinical testing of drugs to be reviewed by the FDA for both efficacy and safety, prior to marketing. Today, the US FDA provides the most extensive publicly accessible database of drug evaluations.

Table 1 highlights some of the major historical events in the development of postmarketing surveillance as part of drug regulation.

Table 1 Historical events in the development of regulatory postmarketing surveillance

Event	*Consequence (country, year)*
Deaths due to contamination of diphtheria antitoxin by live tetanus bacilli (1902)	Biologics Control Act passed (US, 1902)
Deaths due to diethylene glycol present in elixir sulfanilamide[8]	Passing of the Federal Food, Drug and Cosmetics Act (US, 1938)
Thalidomide disaster[9,10] (McBride, 1961)	Kefauver–Harris Amendments (US, 1962) requiring drugs to be shown to be both safe and effective before marketing. Vetting to be done by the FDA
	Setting up of the Committee on Safety of Drugs (UK, 1962) and passing of the Medicines Act (UK, 1962)
First report of torsade de pointes[11]	Subsequent reports were to lead to evaluation of possible QTc prolongation as an important aspect of drug safety evaluation during the premarketing stage
Diethylstilbestrol, given for threatened abortion to mothers, shown to be associated with vaginal carcinoma in their daughters one or two decades later[12]	Despite some concerns[90] about the quality of the incriminating evidence, the association is generally accepted (AFSS, 2003).[82] Tightening of premarketing screening of drugs in general and contraindication of use of diethylstilbestrol in pregnant women. Subsequent studies also suggest a higher risk of urogenital abnormalities in sons
European Committee for Proprietary Medicinal Products (CPMP) set up, 1975	Pan-European evaluation of drugs
Claimed association of MMR vaccination and autism[13]	Widely regarded as one of the most high-profile false alarms in the history of pharmacovigilance[18]
Establishment of the European Medicines Evaluation Agency, 1995	Central procedures for drug regulatory approval implemented

1.25.1.1 International and Multi-Institutional Collaborations

International and multi-institutional collaboration on drug licensing occurs through: (1) the International Conference on Harmonization (ICH), which brings together the US FDA, the EMEA, the Japanese drugs regulatory agency, and experts from the pharmaceutical industry, and (2) the Council for International Organization of Medical Sciences (CIOMS). The aim of the ICH is to harmonize requirements for product registration and hence minimize duplication. The ICH is sponsored by the European Commission, the European Federation of Pharmaceutical Industries and Associations, the Japanese Ministry of Health, Labour and Welfare, the US FDA, and the Pharmaceutical Research and Manufacturers of American. CIOMS was established by the World Health Organization and the United Nations Educational, Scientific and Cultural Organization (UNESCO) in 1949. Its aim was primarily to act as a forum for views on developments in biology and medicine and to explore their social, ethical, moral, administrative, and legal implications. A CIOMS I working group was established in 1986 with the aim of internationally standardizing ADR reporting. Subsequent CIOMS groups have addressed other safety issues, including benefit–risk balance for marketed drugs, pragmatic approaches in pharmacovigilance, and safety monitoring and evaluation during clinical trials. Some of the subjects dealt with in the ICH drug safety documents include definitions and standards for expedited reporting of adverse events, and data elements for transmission of individual case safety reports.[98]

Major drugs therefore currently go through a rigorous process of formal evaluation prior to marketing and, in the major trade blocks, the US, EU, and Japan, the standard of evaluation is similar. These markets account for close to 90% of the drugs market by sales. The drug regulators hope that, through their oversight, only safe and effective drugs reach patients. Unfortunately, postmarketing drug withdrawals are not uncommon, as the examples in **Table 2** demonstrate. Drug regulators hail these withdrawals as successes for their postmarketing surveillance systems. Critics say that the withdrawals demonstrate clear failure of the regulatory systems. The truth, as is often the case with sharply opposing positions, is probably somewhere in the middle.

Regulatory authorities monitor specific drugs more intensely and the UK MHRA, for example, uses the following criteria for selecting such drugs:

1. a new active substance
2. a new combination of active substances
3. administration via a new route, which is significantly different from existing routes

Table 2 Recent postmarketing withdrawal of pharmaceuticals

Drug	*Year licensed*	*Year withdrawn*	*Reason*
Natalizumab (Tysabri)	2004	2005	Progressive multifocal leukoencephalopathy, caused by reactivation of clinically latent JC polyomavirus infection
Rofecoxib	1999	2004	Myocardial infarction/stroke
Valdecoxib	2001	2005	Cutaneous adverse effects
Etretinate	1985	1999	Birth defects
Rapacuronium	1998	2001	Severe respiratory distress
Oral rotavirus vaccine (Rotashield)	1998	1999	Possible association between the use of Rotashield and the development of intussusception
Cerivastatin	1996	2001	Rhabdomyolysis
Troglitazone	1996	2000	Hepatotoxicity
Cisapride	1991	2000	Torsade de pointes, arrhythmias
Grepafloxacin	1997	1999	Torsade de pointes, arrhythmias
Mibefradil	1996	1998	Arrhythmias
Bromfenac	1995	1998	Hepatotoxicity

4. a novel drug delivery system
5. a significant new indication, where this is likely to result in a significantly different population being exposed to the drug, or where there are potential safety concerns associated with the new indication.

1.25.1.2 Drug Withdrawals

One of the aims of postmarketing surveillance is to identify adverse effects that make drugs unsafe for continued marketing. However, as discussed earlier, risk–benefit assessment is ultimately judgmental. Therefore, sometimes drugs are withdrawn from some markets but not others. In a survey of 121 drugs withdrawn from the worldwide market during the period 1960–1999, Fung *et al.* [14] reported that 42% were withdrawn from European markets only, 5% from North America only and 50% from multiple continents. Hepatic (26%), hematologic (11%), cardiovascular (9%), dermatological (6%), and carcinogenic adverse effects were the most common reasons for the withdrawals. Nonsteroidal anti-inflammatory drugs (13%), nonnarcotic analgesics (8%), antidepressants (7%), and vasodilators (6%) were the most common drug classes involved. Of the 87 products for which the timings of marketing and withdrawals were both available, the median time on the market was 5.4 years, with about one-third withdrawn within 2 years of first introduction. Given the short intervals, the pattern of drug withdrawals therefore reflects markedly the cohort of drugs introduced. With the emergence of the biopharmaceuticals era, an increase in the number of withdrawals of such products can be expected (**Table 2**).

1.25.2 What Then Is Postmarketing Surveillance?

Postmarketing surveillance is the monitoring of drug performance in clinical practice and the taking of appropriate action to improve patient safety. The action taken can range from changes in product labeling (e.g., dose regimen alterations, drug interaction alerts, warnings about previously unknown adverse effects) to product withdrawal from the market (**Tables 2** and **3**). The changes can be instituted either voluntarily by the drug companies concerned or enforced through regulatory action. Actors within the postmarketing surveillance arena are drug companies, drug regulators, patients, specialist drug surveillance organizations, and, increasingly, consumer advocacy groups, and specialist commentators such as those represented by the International Society of Drug Bulletins.

Table 3 Label changes and serious safety warnings issued by the FDA in 2005

Drug	*Label change*
Rosiglitazone (Avandia)	Warning of worsening of diabetic macular edema and concurrent peripheral edema
Paroxetine	Women warned of increased risk of congenital malformations if taken during the first trimester of pregnancy
Darbepoetin alfa (Aranesp)	Warning about pure red-cell aplasia and severe anemia, with or without other cytopenias, associated with neutralizing antibodies to erythropoietin
Epoetin alfa	Warning about pure red-cell aplasia and severe anemia, with or without other cytopenias, associated with neutralizing antibodies to erythropoietin
Coagulation factor VIIa recombinant (NovoSeven)	Increased thromboembolic events
Tamsulosin (Flomax)	Intraoperative floppy iris syndrome during phacoemulsification cataract surgery
Long-acting β2-agonists (salmeterol, formoterol)	Severe asthma episodes and resultant increased risk of death
Ethinylestradiol (an estrogen) and norelgestromin (a progestin) contraceptive patch (Evra)	Exposure to greater amounts of estrogen than conventional contraceptive pills and hence increased risk of thromboembolic events
Alefacept (Amevive)	Reduction of CD4+ T-lymphocyte count and hence contraindicated in HIV patients
Morphine sulfate extended-release capsules (Avinza)	Warning that patients should not consume alcohol while taking Avinza. Additionally, patients must not use prescription or nonprescription medications containing alcohol while on Avinza therapy
Ibritumomab tiuxetan (Zevalin)	Warning that patients experiencing a severe cutaneous or mucocutaneous reaction should not receive any further components of Zevalin

An important point to note from the lists of drugs withdrawn and of new adverse reaction alerts issued (**Tables 2** and **3**) is that they include both new and old drugs and new formulations. This illustrates well that in pharmacovigilance constant alert is necessary. It is doubtful whether the risk of congenital malformations associated with paroxetine would have emerged without the close scrutiny which it received because of its other association with suicidal ideation in children. The FDA has since October 2004 required manufacturers of most antidepressants to warn that "antidepressants increase the risk of suicidal thinking and behavior (suicidality) in children and adolescents with major depressive disease (MDD) and other psychiatric disorders."

1.25.3 Rationale for Postmarketing Surveillance

Modern drug development is a long and expensive journey, as illustrated in **Figure 1**. Recent estimates suggest that bringing a drug to the market costs close to $900 million,[15] although the assumptions used in such calculations have been disputed.[16] While the randomized controlled trial is the gold standard for estimating both efficacy and adverse effects, such trials are very expensive and difficult to conduct (**Table 4**). Therefore, at the design stage a cost-effectiveness trade-off has to be made when deciding how big a trial ought to be.

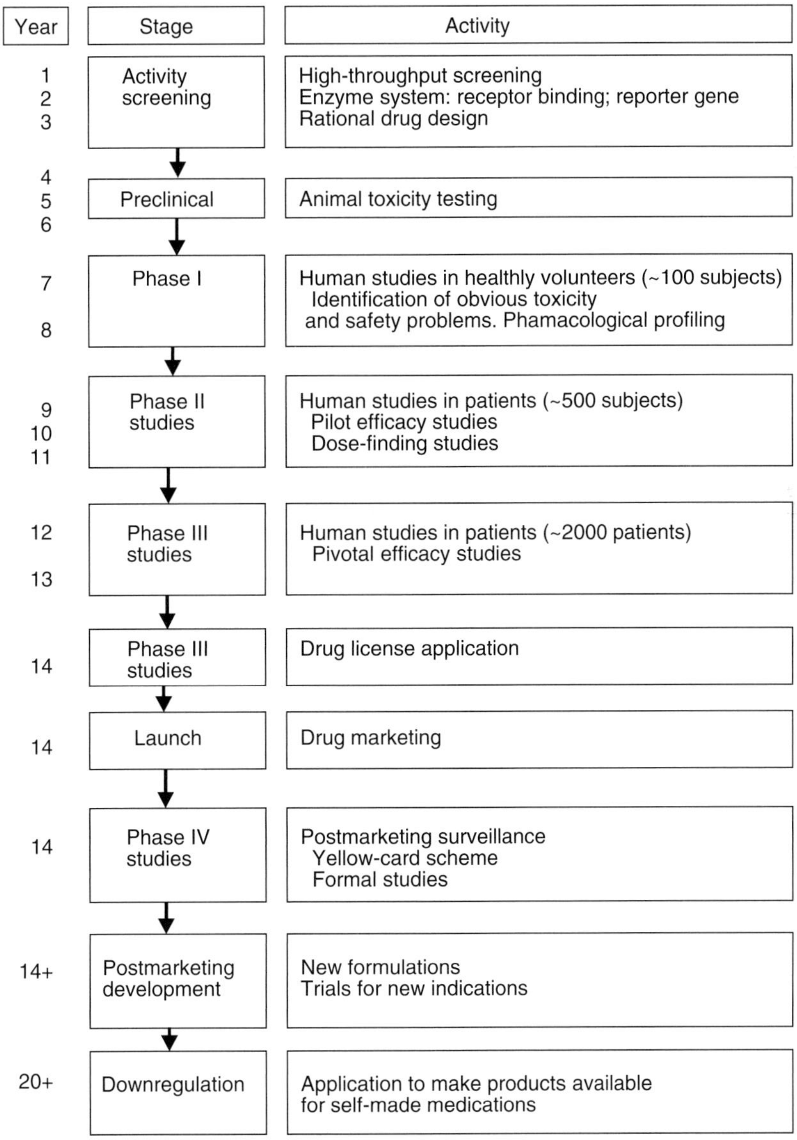

Figure 1 Stages in the drug development process.

Table 4 A widely accepted hierarchy of study design for generating clinical evidence in descending order of quality

Randomized controlled trial
Nonrandomized trials with contemporaneous controls
Nonrandomized trials with historical controls
Cohort study
Case-control study
Cross-sectional study
Spontaneously reported surveillance data
Case reports

Table 5 Drugs fast-tracked by the FDA and licensed in 2005

Brand name (generic)	*Manufacturer*	*Date of license approval*	*Indications*
Remicade (infliximab)	Centocor	9/15/2005	Treatment of moderately to severely active ulcerative colitis in patients who had an inadequate response to conventional therapy
Aryplase (galsulfase)	BioMarin	5/31/2005	Treatment of patients with mucopolysaccharidosis type VI (MPS-VI)
Retisert (fluocinolone acetonide)	Bausch & Lomb	4/8/2005	Treatment of chronic noninfectious uveitis affecting the posterior segment of the eye
Pegasys (peginterferon alfa-2a)	Hoffmann-La Roche	2/25/2005	Alone or in combination with ribavirin (Copegus) for treatment of chronic hepatitis C in patients coinfected with HIV, who have clinically stable HIV disease
Lamivudine 150 mg; zidovudine 300 mg tablets co-packaged with nevirapine 200 mg tablets	Aspen Pharmacare	1/24/2005	Treatment of HIV
Abraxane (paclitaxel protein-bound particles for injectable suspension)	American BioScience	1/7/2005	Treatment of breast cancer after failure of combination chemotherapy for metastatic disease or relapse within 6 months of adjuvant chemotherapy

In practice, the general consensus within the regulatory bodies is that trials ought to be sufficiently powered to detect clinically meaningful efficacy but need not be large enough to detect rare adverse events. This consensus is based on the realization that to identify rare adverse reactions reliably, impossibly large trials are often required. Consequently, in practice, the rare adverse reaction profile of a new drug is usually predicted from preclinical animal studies and validated in phase IV postmarketing studies.

Formal postmarketing studies are sometimes requested from manufacturers at the time of marketing authorization . In the US, there are now mandatory postmarketing studies for two types of products: (1) fast-track products, approved on an accelerated basis; and (2) products for which safe use in children needs to be determined or more clearly defined. Fast-track products are those intended for the treatment of serious and life-threatening illnesses and likely to provide meaningful therapeutic benefit over existing therapies. **Table 5** lists some of the drugs recently fast-tracked by the FDA.

The Food and Drug Administration Modernization Act (FDAMA) passed by the US Congress in 1997, as an amendment of the Federal Food, Drug, and Cosmetic Act, provided the FDA with the authority to monitor the progress of postmarketing studies that drug manufacturers have agreed to conduct during the licensing process. This clause (section 506B (21 U.S.C. 356b) of the FDAMA) was enacted in response to concerns about the timeliness of completion of postmarketing studies by manufacturers, and the need to update product labeling with any new dangers identified. The new provision requires: (1) drug sponsors to submit an annual report on the progress of any postmarketing studies to which they have committed themselves; and (2) the FDA to publish summaries of these reports and an evaluation of sponsor performance with respect to the postmarketing studies. The FDA provides this in its annual report to Congress[99] and has issued draft guidance for industry on how to fulfill their commitments.[100]

Examination of the full postmarketing commitment profiles of a single new drug is informative, as shown in **Table 6**. Bevacizumab (Avastin), a monoclonal antibody against vascular endothelial growth factor (VEGF), was licensed in February 2004 as first-line treatment for metastatic colorectal cancer. VEGF is overexpressed in many human tumors and interruption of VEGF signaling leads to inhibition of growth of the epithelial cells of the tumor vasculature and inhibition of angiogenesis. The only phase III trial submitted as part of the licensing dossier randomized only 402 patients to bevacizumab.[17] Therefore only very frequent adverse effects could possibly have been identified from such a meager prelicensing patient database. However, given that the drug, when combined with chemotherapy, was at least as good as the best treatments then available for metastatic colorectal cancer, marketing authorization was granted to Genentech but with the requirement that the company committed itself to a number of postmarketing studies.

Another major reason why reported trials are often too small to detect the rarer adverse effects reliably is that the disease itself is rare and there are simply not enough patients to be recruited. Sometimes, the disease is of such severity and the available treatments so ineffective that expediency is required. Smaller trials than desirable are therefore conducted. This applies to most orphan diseases. In some cases patients themselves may not be willing to wait until the risk–benefit trade-off is better defined. For example, there has been considerable pressure put on health authorities by breast cancer patients to be put on trastazumab (Herceptin) despite the fact that its cardiovascular toxicity is ill defined.[18] In the early days of the development of drugs for the treatment of the acquired immune deficiency syndrome (AIDS), patients were also not willing to wait for extended safety data on drugs against the human immunodeficiency virus (HIV). In the words of the philosopher Jean-Louis Laya (1731–1833), "When one has nothing to lose, everything can be put at risk."

In addition to sample size, other aspects of the design of randomized controlled trials also bias them against identifying adverse effects. The main reasons are due to narrow inclusion criteria, making patients not representative of those met in routine clinical practice. Their carefully controlled use also excludes exposure to other factors, which additively or synergistically predispose to the adverse effects (**Table 7**). In other words, while the trials have good internal validity, they do not generalize well to the real-world patients and actual in-use conditions.[42]

1.25.4 Data Acquisition in Postmarketing Surveillance

The sources of information and systems in place for drug safety data are made up of:

1. case reports published in journals
2. spontaneous reports received by the industry and regulators and other health organizations (Medwatch, ADROIT)
3. databases, often with record linkage, of various organizations with a focus on pharmacovigilance (e.g., Drugs Safety Research Unit, PHARMO, General Practice Research Database)
4. formal postmarketing studies required by regulatory agencies
5. postmarketing randomized controlled trials aiming to extend the indications for the drugs concerned (e.g., extension of use of cyclooxygenase-2 (COX-2) inhibitors for prophylaxis of colorectal cancer and various diseases of the elderly)

1.25.4.1 Spontaneous Reporting Systems

The yellow-card system was established in 1964 by the CSD. To ensure a high response rate, the chairman of the CSD provided reassurance that all reports would be treated with complete professional confidence by the CSD and its staff. Moreover, the ministers gave an undertaking that the information supplied would never be used for disciplinary purposes or for enquiries about prescribing costs. Initial enthusiasm was high and more than 14 000 reports were made by 1968. By 1991, this had increased to some 20 000 reports annually. Sadly, since then, despite numerous efforts made by the Committee on Safety of Medicines to increase the reporting rate, this has remained essentially static. In 2004, there were 20 142 reports. During the same period, 1991–2004, the number of prescriptions issued in England soared from 407 to 686 million. One interpretation is that drugs have become safer. However, most observers interpret this as a real fall in reporting rate. If anything, as drugs have become more potent, their adverse effects have become more serious. Therefore, while spontaneous reporting systems are relatively inexpensive to operate, they are sparse and often incomplete. In addition to low reporting rates, various biases make interpretation of data in such databases difficult to interpret. For example, reporting rates for different drugs vary markedly.

The main reasons, identified in various surveys,[19–21] for the low reporting of ADEs are listed in **Table 8**.

Table 6 Postmarketing studies of bevacizumab (Avastin)

Aim of study as described by the FDA	*Proposed completion date (status)*
1. Prospectively collect and analyze data characterizing the incidence and clinical course (including duration and medical management) of hypertension in patients during treatment and following the discontinuation of bevacizumab and in concurrent control patients	06/30/2008
2. Collect data and conduct analyses within study NO16966 that will characterize the clinical consequences of both full-dose and low-dose anticoagulation therapy and assess the role of the international normalization ratio (INR) as a predictor of subsequent hemorrhage and/or thrombosis in patients treated with bevacizumab	09/28/2007
3. Conduct analyses to characterize the comparative incidence of proteinuria, risk factors associated with proteinuria, and the clinical course of proteinuria (including time to resolution) using available data from ongoing trials AVF2107 g, AVF2192 g, and AVF2119g	12/30/2005 (submitted)
4. Assess for risk factors associated with proteinuria by prospectively collecting and analyzing data to characterize the incidence and clinical course (including duration) of proteinuria in patients during treatment with bevacizumab and following the discontinuation of bevacizumab and in concurrent control patients	06/30/2008
5. Explore patient factors associated with the risk of development of proteinuria, characterize the clinical course of proteinuria, and assess screening strategies that more accurately identify patients at increased risk of high-grade proteinuria and nephrotic syndrome in 100 patients treated with bevacizumab alone or in combination with [...] in study AVF2938g	06/30/2005 (submitted)
6. Conduct analyses to characterize the comparative incidence of hypertension in patients treated with bevacizumab to those not receiving bevacizumab, risk factors associated with hypertension, and the clinical course of hypertension (including time to resolution), using available data from studies AVF2107g, AVF2192g, and AVF2119g	12/30/2005 (submitted)
7. Prospectively collect and analyze data characterizing the incidence and clinical course (including duration and medical management) of hypertension in patients during treatment and following the discontinuation of bevacizumab and in concurrent control patients	06/30/2008
8. Provide narrative descriptions of each vascular adverse event (myocardial infarction, cerebrovascular accident, peripheral arterial event, vascular aneurysm, or other vessel wall abnormalities, and venous thromboembolic events) for patients enrolled in study AVF2540g and provide descriptive statistics of the incidence of vascular events (overall and each subtype)	06/30/2005 (submitted)
9. Collect data and conduct analyses of the comparative incidence of delayed vascular events (myocardial infarction, cerebrovascular accident, peripheral arterial event, vascular aneurysm, or other vessel wall abnormalities, and venous thromboembolic events) in bevacizumab-treated patients following the discontinuation of bevacizumab (from 12 to 24 months after initiation of treatment) and in concurrently enrolled control patients (over the same time interval – 12–24 months after initiation of treatment) in NSABP study C-08	06/30/2008
10. Assess the relative impact on fertility and gonadal function of bevacizumab in combination with chemotherapy, as compared to patients receiving chemotherapy alone	06/30/2008
11. Examine the long-term impact of bevacizumab on pregnancy outcome. This will be evaluated through inclusion of a special section in the periodic adverse experience report (PAER) containing a thorough and cumulative evaluation of pregnancy, spontaneous abortion, and fetal malformation	02/28/2007
12. Directly assess the pharmacokinetic interactions between irinotecan and bevacizumab in a single-arm, cross-over study in approximately 32 evaluable subjects	09/29/2006
13. Assess the pharmacokinetic profile of bevacizumab in a rodent model of hepatic dysfunction	12/31/2004 (submitted)
14. Perform additional analyses of clinical pharmacokinetic data from studies AVF0780g and AVF2107g in order to provide a comparison of clearance in patients with hepatic dysfunction	06/30/2004 (submitted)
15. Obtain further information on the pharmacokinetics of bevacizumab by assessing bevacizumab drug levels at 3 and 6 months posttreatment in NSABP study C-08	12/31/2008

Table 6 Continued

Aim of study as described by the FDA	*Proposed completion date (status)*
16. Develop a standardized approach to the collection of data and generation of narrative descriptions of selected adverse events (gastrointestinal perforation, intraabdominal abscess, fistula, wound dehiscence) that will include description of the event, surgical operative and pathology reports, and outcome/resolution information for all such patients enrolled in study NO16966 and NSABP study C-08	06/30/2008
17. Provide the final study report for study E3200, examining the comparative safety and effectiveness of single agent bevacizumab, bevacizumab in combination with the [...] regimen, and [...] alone	03/31/2006
18. Provide the study report for study AVF2192g examining the comparative efficacy and safety of 5-fluorouracil and leucovorin with and without bevacizumab in patients with newly diagnosed metastatic colorectal cancer who are unable to tolerate irinotecan-based therapy	09/30/2004 (submitted)
19. Develop a validated, sensitive, and accurate assay for the detection of an immune response (binding and neutralizing antibodies) to bevacizumab, including procedures for accurate detection of antibodies to bevacizumab in the presence of serum containing bevacizumab and vascular endothelial growth factor	09/30/2004 (submitted)
20. More accurately characterize the immune response to bevacizumab in NSABP study C-08 using the more sensitive, validated assay described above	12/31/2008

Table 7 Main reasons for randomized controlled trials, particularly at the premarketing stage, missing important drug adverse effects

1. Inadequate statistical power (rare adverse outcomes missed)
2. Narrow selection criteria for trial patients (e.g., disease severity and age group)
3. Deliberate exclusion of some patient groups (e.g., pregnant women, infants, and the very old)
4. Close control of patients (e.g., diet and concomitant medications), and hence important drug–drug and drug–diet interactions not tested for
5. Close monitoring of patients (e.g., numerous in-trial biochemical tests identify liver and renal abnormalities before adverse effects are seen)
6. Noncompliance controlled
7. Off-label prescribing or use disallowed
8. Dose used not optimized and closely controlled. Effects of overdosing ill defined
9. Signals for unusual adverse effects may be missed (e.g., mitochondrial toxicity)
10. Usual criteria softened in response to societal demands for more rapid access to potentially innovative drugs
11. Inadequate follow-up time to pick up adverse effects with long latency or due to chronic dosing
12. Some exposures are unethical in premarketing trials (e.g., infectious agents)

Table 8 Main reasons for low reporting of adverse reactions

1. Uncertainty about causal relationship
2. Adverse reaction thought to be trivial
3. Adverse reaction perceived to be well known already
4. Time constraint
5. Did not know how to report
6. Worried about legal implications

1.25.4.1.1 The Weber effect

The Weber effect refers to the common observation that the number of reported adverse effects of a drug rises during the initial 1 or 2 years of marketing and then declines despite increasing use. Whether this effect, first reported for nonsteroidal anti-inflammatory drugs,[22] is a general effect that applies to all new drugs is still controversial.[23] The possible existence of the Weber effect makes interpretation of spontaneous reporting data very difficult as valid signals

Table 9 The Bradford–Hill criteria for causality assessment

1. Strength of association
2. Consistency of association
3. Specificity of effect
4. Temporality – exposure precedes effect
5. Biological gradient – the higher the exposure, the larger the effect
6. Plausibility
7. Coherence
8. Experimental supporting evidence
9. Analogy

Table 10 Items scored in the Naranjo causality algorithm

1. Previous conclusive reports on reaction concerned
2. Drug administration preceding the adverse event
3. Adverse reaction improving on drug discontinuation or administration of a specific antagonist
4. Reaction reappeared on rechallenge
5. Possible alternative explanations for the adverse reaction
6. Reaction not reappearing with administration of a placebo
7. Drug detected in body fluids at concentrations known to be toxic
8. Dose response with respect to the drug and reaction
9. Similar reaction to a prior exposure to the drug or similar drugs
10. Confirmation of adverse event by objective evidence

cannot be dissociated from false alarms early in the life of a new drug. That is precisely the period when concern about the safety of a new drug is highest and when false alarms can be most damaging to the marketing potential of a new drug. Given such potential biases in spontaneous adverse reporting systems, there is a need for protocols to assess causality of reported associations between drug and adverse effect.

1.25.5 Causality Assessment in Postmarketing Surveillance

1.25.5.1 The Bradford–Hill Criteria for Causal Association

Observational studies are subject to a wide variety of biases, which make causality assessment difficult except for highly specific and unusual adverse effects. Thus, it was relatively easy to assign causality to the association between phacomelia and thalidomide because of the highly specific and unusual nature of the drug's adverse effect. In contrast, the association of cervical cancer with hormonal replacement therapy was not thought to be causal for many years. From their groundbreaking work on establishing the causal link between cigarette smoking and lung cancer, Sir Richard Doll and Bradford Hill[24,25] developed what are now widely known as the Bradford–Hill[26] criteria for causality assessment (**Table 9**).

1.25.5.2 Causality Assessment Algorithms

While the Bradford–Hill criteria are used for causality assessment of a body of information, causality assessment algorithms for assessing individual case reports are available. One of the best known, the Naranjo algorithm,[27] perhaps better referred to as a scorecard, will be used for illustrative purposes. The score-card has 10 items, with some items having both negative and positive scores (**Table 10**). The higher the score, the higher the likelihood that the association between the drug and the adverse effect is causal.

1.25.6 Decision-Making in Postmarketing Surveillance

1.25.6.1 Acceptable Risk and Risk–Benefit Ratio

> "A thing is safe if its risks are judged to be acceptable.... This definition emphasizes the relativity, and judgmental nature of the concept of risk... Two very different activities are required for determining how safe

things are: measuring risk, an objective and probabilistic pursuit; and judging the acceptability of that risk (judging safety), a matter of personal and social value judgment."[78]

"risk assessments generally are not viewed by their protagonists as science in the usual sense, but rather involve decision making in the face of uncertainty."[76]

Given that all drugs can produce some adverse effects, the decision to license a drug takes on board the concept of what is an acceptable risk to patients and society at large.[28,29] The drug regulatory perspective is therefore a societal one. For the individual patient, the risk–benefit trade-off is a personal one and some patients are highly risk-averse and others risk-takers. This conflict between the average trade-off and the individual trade-off applies not only to drugs but to most activities, including the siting of nuclear power plants and the riding of motor cycles. It is therefore not surprising that there are frequent debates about whether particular drugs have been irresponsibly licensed or not. The recent debates about the link between autism and the triple vaccine, mumps, measles, and rubella (MMR),[13] illustrate the complexities well. There is little doubt that the MMR vaccine is cost-effective at the societal level. An important minority of the population, notably parents of children requiring vaccination, however, finds the risk of autism unacceptable, no matter how unlikely the association is between MMR vaccination and autism. In fact, many parents have shown that they preferred not to vaccinate their children rather than accept the perceived risk. Whoever does the trade-off, five factors need to be taken into account:

1. the seriousness of the risk
2. the likelihood of the risk
3. the benefit the drug is expected to produce
4. the likelihood of obtaining the benefit
5. alternative treatments that are available and their risk–benefit profile

In the discussion above, terms such as risk and likelihood are used in the colloquial sense, with their meanings to be inferred from the context. The statistical definitions of relevant terms are discussed below.

Decision-making about whether a drug should be withdrawn is based on its risks relative to its effectiveness. The concept of risk has two quantitative dimensions: (1) the harmfulness of an adverse effect in terms of its impact on quality of life (e.g., liver toxicity is more harmful than a transient headache, as adverse effects); and (2) a frequency dimension (how often a particular adverse effect occurs; a drug causing more frequent adverse effects is less unacceptable). For example, the health state associated with a given type of liver toxicity can be valued in terms of utility and hence the effect of liver toxicity on this utility scale can be estimated. This is the approach used in pharmacoeconomic evaluations. The benefit to be derived from drug treatment can also be similarly quantified in terms of its effect on utility. If one were able to measure the incidence of the adverse and beneficial effects, then the utilities can be used to estimate the number of quality-adjusted life-years (QALY) lost or gained. One is always interested in the incremental change in that whether a drug is acceptable or not depends on the next best available option. A simple decision rule can then follow. If the net effect on QALY is positive then the drug is deemed useful. Otherwise, it is not. However, as we have seen earlier, incidence rates of adverse effects cannot be reliably estimated from observational studies as we do not have the denominators and we cannot adequately control sources of bias in such studies. Therefore when regulators consider the risk–benefit of a drug in regulatory decisions, they are not considering a proper dimensionless ratio, as a ratio of two rates would be. Instead, they are dealing with a fuzzy, at best semiquantitative assessment of overall harm relative to the overall benefit. Therefore, it has been suggested that the term 'benefit–harm assessment' be used instead of 'risk–benefit ratio.' However, it is important to note that the new term is no more quantitative than the latter. Decision rules are therefore never quantitatively and comprehensively explicit with drug withdrawals.

Figure 2 shows the risk–benefit trade-off plane. If one assumes rational decision-making, then the better choices offered by quadrants A and D are clear. The choices represented by quadrants B and C are less clear and application of personal preferences is required.

1.25.7 Estimating Frequencies of Adverse Drug Events

When case reports of an association between a drug and an adverse effect suggest that the association is causal, an estimate of relative frequency is then sought to enable a risk–benefit assessment to be made. The most reliable estimate is obtained from an appropriately designed randomized controlled trial. However, as indicated earlier, such a trial is often not possible. Moreover, in other cases only an approximate estimate is necessary. Prospective cohort

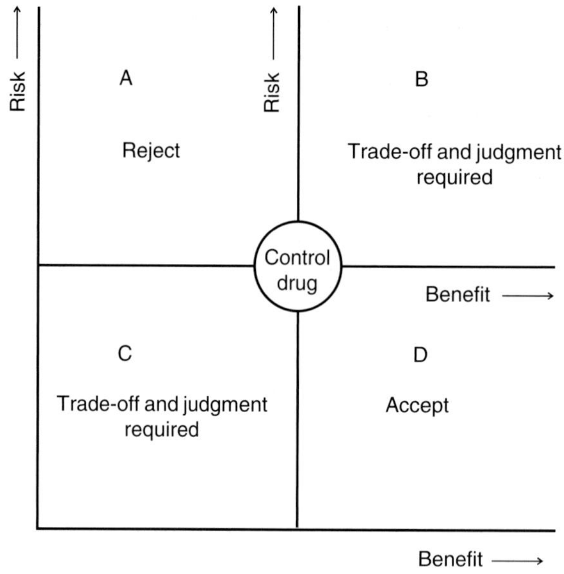

Figure 2 Risk–benefit plane for rational decision-making.

studies and case-control studies may give the necessary risk ratio and odds ratio respectively. For signal generation, an even cruder estimate of relative rates may be adequate.

While in the absence of background occurrence of an adverse effect, a sample size of some 5000 would be adequate to pick it out reliably if it were to occur at a frequency of 1 in 1000, in general a much larger sample size is necessary. Thus with a control group similar in size to a treatment group and with a background incidence rate of 1 in 1000, to pick out an additional incidence of 1 in 10 000 would require over 2 million subjects in each group.[30] As discussed, phacomelia caused by thalidomide was easy to pick out because it was a rare condition in the population not exposed to the drug and it was easy to diagnose. Hypertension, on the other hand, is an extremely difficult adverse effect to pick out, particularly in older patients.

1.25.7.1 Rule of Three[31]

The rule of three is an approximate method for estimating the maximum long-term risk associated with an observation of no event in a given sample of subjects. It states that, if none of n subjects (patients) suffers an event (e.g., ADE), we can be 95% confident that the chance of this event is at most 3 in n ($3/n$). For example, if none of 50 subjects given clozapine develops a blood dyscrasia, then the point estimate is 0 and the upper 95% confidence limit is approximately 3/50. The approximation improves with increasing n but even with n as low as 10, the approximation is adequate (30% versus 29%).

1.25.7.2 Signal Detection

Data-mining approaches are now widely used in postmarketing surveillance and all share the same approach but often use different adverse reactions databases, statistical algorithms for detecting departure from predicted rates and decision rules. The main components of such a system are:

1. a database of spontaneously reported adverse drug reactions (e.g., yellow-card database (ADROIT), the World Health Organization adverse reaction database, the General Practice Research Database, and the FDA MedWatch database) using standard codes (e.g., those in the Medical Dictionary for Regulatory Activities (MedDRA))
2. a statistical algorithm for detecting departure from predicted rates (e.g., gamma Poisson shrinker[32] and Bayesian confidence neural network[33])
3. a numerical and graphical summary of the data
4. appropriate decision rules, which are usually built into the statistical algorithm

1.25.7.2.1 Numerical algorithms for signaling departure from predicted frequencies

In order to generate a departure from normality signal, there is a need to define a statistical model for calculating the expected frequencies and then a cut-off point. This is characteristic of any hypothesis test. The null hypothesis states that there is no difference in frequency of a particular adverse effect between the drug being monitored and the comparator. In classical hypothesis-testing a significance level of 5% is employed but, to reduce the risk of false alarms, lower significance levels may be used. The comparator may be a different drug or a number of drugs within a therapeutic class or indeed all other drugs in a database.

Common statistical distributions used are the binomial distribution, the Poisson distribution, and the gamma distribution. Their typical application in pharmacovigilance and their probability functions are given below.

1.25.7.2.1.1 Binomial distribution

This is used to model the number of successes (x) in a set of n Bernoulli trials (i.e., trials each of which has two outcomes only; e.g., adverse effect or no adverse effect), with success occurring with probability p:

$$B(n,p) \sim \binom{n}{x} p^x q^{n-x}$$

The distribution has a range of [0, 1, 2, ... n], a mean of np, and a variance of npq.

1.25.7.2.1.2 Poisson distribution

This is used to model the number of events in a Poisson process during a fixed time:

$$\text{Poisson}(\mu) \sim \frac{e^{-\mu}\mu^x}{x!}$$

The distribution has a range of [0, 1, 2, ...] and both its mean and variance are equal to μ.

Note that Poisson distribution is an approximation of the binomial distribution $B(n, \frac{\mu}{n})$ with large n and small p, as is often the case in pharmacovigilance (large number of trials (patients) and small number of events (number of patients with a particular adverse effect)).

1.25.7.2.1.3 Gamma distribution

This gives the distribution for time to observe n events in Poisson process.

$$\text{Gamma} \sim \frac{\lambda^n x^{n-1} e^{-\lambda x}}{(n-1)!}$$

The distribution has a range of $x \geqslant 0$, a mean of $\frac{n}{\lambda}$ and a variance of $\frac{n}{\lambda^2}$.

Further details on the application of those distributions in signal generation are given by Hauben and Zhou.[80]

1.25.7.2.2 The cumulative sum approach

This generic approach is based on the fact that, if any observed variability is random then, when the deviations are summed, the negative deviations should balance out the positive deviations and hence the sum of deviations should center around zero. Therefore if one were to set tolerance limits of, say, three standard deviations from the expected value, then departure from this window would trigger an alarm. This technique, with its origin in the engineering industry,[34] is used with great success in the quality control of manufactured products (e.g., tablet weights, width of screws, drug content in capsules). A graphical plot of the cumulative sum with subgroup highlights any departure from predicted trends.

1.25.7.2.2.1 Proportional rate ratio

Two approaches may be used for comparing whether the number of cases of an adverse reaction spontaneously reported for a drug is above random variation and background incidence.[35] The first estimates the rate of occurrence of the adverse reaction and compares it to the background or comparator rate, using denominator data. The denominator is itself usually estimated from prescription data. Thus the estimated reporting rate (RR) would be given by:

$$\text{RR} = \frac{\text{Number of Reports}}{\text{Number of Prescriptions}}$$

Table 11 Contingency table for calculating the reporting rate ratio

	Drug of interest	*All other drugs*
Reaction of interest	*a*	*b*
All other reactions	*c*	*d*

Table 12 Example of calculation using hypothetical data for thioridazine

	Thioridazine	*All other drugs*
QTc prolongation	52	350
All other reactions	500	620 980
Total	552	621 330

This RR can be compared with the corresponding rate for a comparator as rate differences (RRD) or rate ratios (RRR).

$$\mathrm{RRR} = \frac{\mathrm{RR_{Drug}}}{\mathrm{RR_{Comparator}}}$$

Although this statistic can be then modeled to take account of the effect of covariates using techniques such as logistic regression, estimates of the covariates themselves are highly unreliable. For example, the Weber effect is probably highly variable and is likely to depend on calendar time, adverse publicity, and how many other drugs there are.

In order to overcome some of these biases, Evans *et al.*[36] proposed the proportional reporting rate ratio (PRR). The PRR uses the total number of reports for the drug itself as denominator. **Tables 11** and **12** illustrate how the data are arranged into contingency tables (tables with each cell giving the counts) and the equations show the necessary algebraic manipulations.

$$\mathrm{PRR} = \frac{\left(\frac{52}{552}\right)}{\left(\frac{350}{621\,330}\right)} = 167$$

$$\mathrm{PPR} = \text{Chi-squared} \quad (1\ \mathrm{df}) = 47\,978$$

The PRR method essentially adopts an internal standardization approach akin to that used in chemical analysis. The analytical conditions may be perturbed by the environment (Weber effect or adverse publicity, in the case of drug surveillance; temperature, fluid flow, and pH in the case of chemical analysis), but all components should hopefully be affected similarly. For example, the Weber effect is assumed not to single out any particular adverse effect. Of course, this is a strong assumption, which clearly is not valid under at least some circumstances. For example, if a new adverse effect receives popular media attention, then that particular effect might be preferentially reported and hence bias the reporting rates even more. As the authors themselves further acknowledge, a very strong signal for a particular drug will reduce the magnitude of the PRR for other signals with that drug. Class effects, such as gastrointestinal bleeding, will also similarly bias PRR for other drugs. A strong signal for one effect may drown the signals for other effects. That perhaps contributed to regulators missing the cardiovascular adverse effects of the COX-2 inhibitors. The authors suggest removing the drugs generating the striking signals from the 'all other drugs' part of the equation prior to calculating the PRR. However, decision rules need to be established from empirical data before this can be done.

1.25.7.2.3 **Multi-item gamma Poisson shrinker**[32]

Another method for detecting overrepresented associations within spontaneously reported adverse reaction data is the MGPS, used by the FDA. The algorithm uses an empirical Bayes updating to derive signal scores (adjusted observed/expected counts) exclusively from the numerator data.

1.25.7.2.4 False alarms: when the statistics say 'yes'

False alarms about ADEs help no one. Patients are deprived of useful drugs and society loses at least part of its investments in drug development and research. In the worst cases, as exemplified by the MMR episode, spread of disease may be the result. Algorithms for causality assessment have already been discussed. However, when the statistical tests suggest a significant association, people often find it difficult to believe that the association could be fortuitous and not causal. Therefore it is essential that maximum effort be put into minimizing confounders or alternative causes of any observed differences in the frequency of drug effects in those receiving the test drug and the control drug. This needs to be considered both at the postmarketing study design stage and at the statistical analysis and interpretation stage.

Confounding by indication is perhaps the most common source of erroneous inferred association between drug and adverse effect. This confounding effect, also called the channeling effect, describes the fact that doctors, quite appropriately, exercise judgment when prescribing for the individual patient. Thus a prescriber would be expected to avoid aspirin in those with a history of gastrointestinal bleed and may prescribe acetaminophen (paracetamol) instead. Similarly, a prescriber may avoid a tricyclic antidepressant for a person who may be particularly susceptible to its anticholinergic effect and prescribe a selective serotonin reuptake inhibitor instead. Therefore, unless adjustment is made for baseline differences in risk factors for a particular adverse effect, the safer drug may paradoxically emerge as the less safe alternative. Without randomization and blinding, the treatment groups being compared may be highly imbalanced. Therefore alternative strategies are necessary to minimize confounding, particularly in observational studies.

1.25.8 Adjustment for Covariates

1.25.8.1 Propensity Scores

The classical approach for taking account of possible confounders is to include the covariate in a regression model. This usually takes the form of logistic modeling within the generalized linear modeling framework.[37] For example, age group, disease severity status, and age group could be used as categorical predictor variables and logit of the response as the y variable. An extension of this method, which has gained considerable attention recently, is the use of propensity scores, developed by Rosenbaum and Rubin.[38] The propensity score is the conditional probability of assignment to a particular treatment given a vector of observed covariates. Therefore, what a propensity score does is to derive a weighting factor, on a probability scale, from a set of measured covariates.

1.25.8.2 Instrumental Variables[39,40]

An alternative and indeed also complementary approach to probing the existence of and accounting for bias is the use of instrumental variables. An instrumental variable is an observable characteristic that is related to choice of treatment but independent of both outcomes and characteristics of patients. For example, in a study of whether more intensive treatment of myocardial infarction in the elderly reduces mortalilty, the authors used patients' differential distances to alternative types of hospitals as the instrumental variable. While this variable was highly correlated with how intensively a patient was treated, it was apparently uncorrelated with health status, and hence presumably outcome. Use of an instrumental variable serves as a quasirandomization method but the problem is, of course, not knowing to what extent there is a lack of correlation between the instrumental variable and the outcome of interest.

1.25.8.3 Sensitivity Analysis[40]

Sensitivity analysis, like instrumental variable analysis, is also a technique borrowed from econometrics. Whenever there is uncertainty about a parameter estimate (e.g., probability of death as an adverse event), sensitivity analysis can be used to assess the extent to which a hypothetical confounder would have to be related to both mortality and use of the apparently higher risk agent, to make any observed increased risk spurious, if none existed. Sensitivity analysis is now widely used in pharmacoeconomic analysis. In any sensitivity analysis, the range of values to use for a parameter is always judgmental but may also be derived from other studies, including meta-analyses. Often the 95% confidence interval from a prior study is used as the range.

An interesting recent application of propensity score, instrumental variable, and sensitivity analysis in the analysis of risk of death in elderly users of conventional versus atypical antipsychotics agents is described by Wang *et al.*[39] They firstly derived the propensity scores from predicted probabilities in logistic regression modeling of the use of the two types of antipsychotic agents. The resulting scores were stratified into deciles for Cox modeling. Analysis, using the prescribing physician's preference with respect to conventional and atypical antipsychotic agents as the instrumental variable, was then undertaken to account for any important confounding variables that were not measured. The doctor's preference was determined by her choice of medication for her most recent patient. In their sensitivity analysis, they suggested that a large relative risk of 7 or more would be necessary between any hypothetical confounder and both the use of conventional antipsychotic agents and death, to reverse the observed increased risk associated with the conventional agents, if no increased risk truly existed.

1.25.8.3.1 Use of aggregate data

The rationale for use of aggregated data is that confounding at the individual-patient level would be averaged out at the aggregate level. With such an analysis it may also be possible to include confounders, such as deprivation, which are available at an aggregate level only. Epidemiological studies undertaken with this level of data are referred to as ecological studies. An obvious extension of this approach is to combine both individual-level covariate data with aggregate covariate data within a hierarchical or multilevel statistical model, as demonstrated by a recent study for an efficacy study.[41]

1.25.9 Lessons from Clinical Practice: Illustrative Case Vignettes

When one has nothing to lose, everything can be put at risk Jean-Louis Laya, 1731–1833.

To illustrate the complexity of factors involved in generation of adverse reactions, pharmacovigilance and associated decision-making, a few illustrative case vignettes of both older and newer drugs are presented.

1.25.9.1 Case 1: NeutroSpec (Technetium (99mTc) Fanolesomab)[42,43]

NeutroSpec is a diagnostic agent made up of a murine (mouse) immunoglobulin M monoclonal antibody (fanolesomab) and radioactive technetium. Fanesolomab targets CD-15, found primarily on the surface of white blood cells, which concentrate in infected tissue foci. Such concentrations are identified by the radiation emitted by the technetium payload (gamma scintigraphy).

The intravenous product was licensed in June 2004 for the diagnosis of equivocal appendicitis, and marketed in September that same year. Appendicitis, usually characterized by the presence of fever, an elevated white cell count, and pain in the right lower portion of the abdomen, needs to be accurately diagnosed in order to avoid unnecessary major surgery. When one or more of the classical symptoms are absent, diagnosis is equivocal and alternative diagnostic methods such as ultrasound and computed tomography are required. More invasive laparoscopy is reserved for the more difficult cases. NutroSpec is scientifically an innovative alternative, and premarketing trials suggest that it correctly classified 90% of patients with equivocal appendicitis.

The premarketing trials included 523 subjects, about half of whom had clinically equivocal appendicitis. The other subjects exposed were healthy volunteers, or had other infections. By December 2005, some 11 000 patients had the product, mostly for off-label use. Seventeen serious adverse events were reported to the FDA. In addition to two deaths, three patients required treatment in intensive care units, two were given cardiopulmonary resuscitation, and five oxygen therapy. Forty-six other patients developed milder hypoxia, hypotension, and dyspnea. Cardiopulmonary adverse events were detected in the premarketing studies but were all of the milder type.

The severe adverse events developed within minutes of the patients receiving the diagnostic agent. Both of the deaths were of diabetic subjects. Indeed, most, but not all, of the patients with the serious adverse events had serious medical conditions. Given the severity of some of the adverse effects and the availability of alternative methods for diagnosing appendicitis, even in equivocal cases, the FDA recommended withdrawal of the product. The manufacturers and distributors withdrew the product on 19 December 2005.

Important features of this case are:

1. the small number of subjects exposed to the product during the premarketing studies.
2. the inability of picking up the serious adverse events during the premarketing period despite their apparent high frequency during the postmarketing period
3. the extensive off-label use of the innovative product (Mesmer effect)
4. the rapid onset of symptoms suggests an anaphylactic reaction. However, given that the patients were unlikely to have received the product previously, any sensitivity reaction must have arisen from previous exposure to foreign proteins, sufficiently similar to the mouse monoclonal antibodies. The increasing use of genetically engineered proteins, including insulin, might have been be a factor

Table 13 Drugs contraindicated with mibefradil at the time

Amiodarone	Flecainide	Quinidine
Astemizole	Flutamide	Simvastatin
Bepridil	Halofantrine	Tacrolimus
Cisapride	Ifosfamide	Tamoxifen
Ciclosporin	Imipramine	Terfenadine
Cyclophosphamide	Lovastatin	Thioridazine
Desipramine	Mexiletine	Vinblastine
Erythromycin	Pimozide	Vincristine
Etoposide	Propafenone	

1.25.9.2 Case 2: Mibefradil Withdrawal

Mibefradil was a novel drug with selective T-channel calcium blocker activity. There was therefore much excitement for the drug, which was vigorously promoted as first-line therapy in hypertension or angina. However, almost exactly 1 year after its introduction, the drug was withdrawn, based on case and spontaneous reports of serious interactions with other drugs (**Table 2**). The interactions were predictable from the pharmacology of the drug, which inhibits cytochrome P450 3A4. Indeed, the original labeling specifically listed three drugs (astemizole, cisapride, and terfenadine) that could be expected to accumulate to dangerous levels. All three are associated with prolongation of the QT interval and torsade de pointes. Three months after introduction of the drug the labeling of the drug was strengthened and two statins (simvastatin and lovastatin) were added to the list of contrandicated drugs. Both the public and doctors were warned of the potentially fatal interactions. At the time of mibefradil withdrawal, no fewer than 26 drugs were on the 'to be avoided' drugs list (**Table 13**). Given the fact that, despite its interesting pharmacology, no special benefits could be demonstrated for the drug relative to better-validated antihypertensives and antianginal agents, the risk was judged too high and the drug was withdrawn. The drug sponsors explained that "In principle drug interactions can be addressed by appropriate labeling; however, with respect to Posicor (mibefradil), Roche believes that the complexity of such prescribing information would be too difficult to implement."

What lessons can one learn from this withdrawal? Pharmacological plausibility plays a large part in the licensing of drugs, as is evidenced by the wide use of surrogate outcomes. Perhaps it is time to be less enamored by pharmacological rationale and to accept that the true value of a drug and whether it should be a first-line agent can only be assessed through properly conducted randomized controlled trials with hard outcomes.[44] However, randomized controlled trials do not generalize well to the real-world patients, who are likely to harbor numerous other risk factors. In this case one of the major interactants, the statins, are precisely the drugs that patients on mibefradil are likely to take concomitantly. Noninteracting statins are available, but for many busy practitioners, the information overload is simply too much – a phenomenon we have referred to as the bollards and flashing lights syndrome.[45] Drug–drug interactions also contributed significantly to the more recent withdrawal withdrawal of cerivastatin.

1.25.9.3 Case 3: Withdrawal of Bromfenac[46,47]

Bromfenac (Duract), a nonsteroidal anti-inflammatory drug, was also withdrawn about a year after its introduction. Despite the fact that there were no fewer than 21 other nonsteroidal anti-inflammatory drugs at the time, bromfenac was licensed in 1997 for the treatment of acute pain. Furthermore, there was evidence of liver enzyme elevations serious enough for the FDA to recommend a maximum duration of treatment of 10 days. Unfortunately, in the real world, not all prescribers and patients adhered to this recommendation and cases of severe liver toxicity started to appear. This led, 6 months later, to labeling changes to reemphasize the 10-day limit. However, within 3 months of this warning, after a year of licensing, 12 deaths and liver transplants had been reported and the company concluded that "further steps to limit use of a potent analgesic such as Duract to just 10 days would not be feasible or effective" and withdrew the product. The major lesson to be drawn in this case is that if a drug can be abused, it will be abused. Cautionary labeling provides little protection.

Liver toxicity is notoriously difficult to predict as liver enzyme elevations do not necessarily lead to clinical effects. For example, diclofenac is often associated with liver enzyme elevations but it rarely causes clinical hepatoxicity.

1.25.9.4 Case 4: Withdrawal of Natalizumab (Tysabri)[48–51]

Natalizumab (Tysabri) received accelerated FDA approval in November 2004 for reducing the frequency of exacerbations in patients with remitting–relapsing multiple sclerosis. It is a monoclonal antibody that selectively neutralizes α_4 integrins. These adhesion molecules provide mechanical stability in cell–cell and cell interactions. Multiple sclerosis is a disabling condition that is usually treated with beta-interferon. In the premarketing trials natalizumab, when added to beta-interferon therapy, reduced the risk of exacerbations by 54% compared to interferon alone. This, together with its clear efficacy relative to placebo, persuaded the regulators that the drug brought 'important and meaningful benefit' to multiple sclerosis patients. However, at the time of approval only about 1100 patients with multiple sclerosis had received the drug for at least 1 year. The company was therefore asked to conduct confirmatory studies to demonstrate continued benefit of the drug after 2 years of treatment. During these studies, two cases of progressive multifocal leukoencephalopathy (PML) were observed. A further case was observed in another trial of the drug for Crohn's disease, not an approved indication at the time. PML is a is a lethal opportunistic infection of the central nervous system. It is caused by reactivation of a clinically latent JC polyomavirus infection and there is no effective treatment against the virus, which leads to multifocal demyelination and hence severe neurologic dysfunction. PML usually occurs in the severely immunocompromised patient (AIDS, renal transplant, or leukemic patients). Therefore its occurrence is totally unexpected in multiple sclerosis patients, thereby suggesting a likely causal association with the drug. Only three cases of PML were sufficient to trigger withdrawal of the drug in February 2005. Pharmacovigilance in this instance involved close monitoring of follow-up clinical trials.

1.25.9.5 Case 5: Vaccine Safety

The cost-effectiveness of most of the widely used vaccines is generally accepted. The success of smallpox eradication and the near conquest of infections such as rubella virus and poliomyelitis are impressive. Despite the recalcitrance of infections such as HIV, hepatitis in its many forms, and influenza in its periodic reincarnations, successes in trials of vaccines against viruses such as the papillomavirus, associated with cervical cancer, show that research persistence will eventually be rewarded.[52] However, as the recent controversy over the MMR vaccine demonstrates, parents are highly risk-averse when it comes to prophylactic treatments, particularly when infants and children are involved. Therefore it is imperative that good pharmacovigilance is in place for new vaccines so as not to damage the fragile confidence which many people have in vaccines. Without high vaccine uptake, the success of any vaccination campaign is unlikely. The Vaccine Adverse Event Reporting System (VAERS) was set up to identify problems with vaccines as rapidly as possible.[53] However, as a spontaneous reporting system it shares the same weaknesses as those discussed for standard drugs. Active vaccine safety-monitoring programs have been suggested.[54] Nonetheless, the VAERS still represents an additional useful safety barrier. It was instrumental in identifying the association between intussusception and the rhesus rotavirus vaccine (Rotashield).[53] There have been several interesting reports of postmarketing surveillance of vaccines recently.[55–57] Izurieta *et al.*[57] noted that some 2.5 million people had received the influenza vaccine by the time of their review and 450 adverse event reports were recorded in the VAERS. These included seven reports of anaphylaxis, two reports of Guillain–Barré syndrome, and one of Bell's palsy. This low frequency of adverse events has been welcomed[58] but whether this vaccine will retain its acceptance, only time will tell. The recent withdrawal of another intranasal influenza vaccine calls for continued vigilance.[54]

1.25.10 Who Should Conduct Postmarketing Surveillance?

Recent high-profile drug withdrawals, notably that of rofecoxib (Vioxx), has led to calls for a complete reexamination of current practices in pharmacovigilance.[59,60] Neither the drug manufacturers nor the regulators, who derive much of their funding from the former, are judged to be impartial enough, by some critics. Even academic, supposedly independent, researchers have raised expressions of concern about how faithfully they report trial data.[61,62] Indeed, some editors go further and now ask for independent analysis of clinical trial data.[63] Given that in the end, someone has to pay for the studies and therefore by definition, there is a risk of bias creeping in, the best safeguard against misinterpretation is full data transparency.

1.25.11 Limitations of Formal Postmarketing Surveillance Studies

While formal marketing surveillance studies are useful for identifying common adverse events, they are less useful for identifying rare adverse events. Their well-documented failure in identifying iatrogenic prolongation of the QT interval and associated torsade de pointes, which may lead to sudden death, provides a useful illustrative example.[64]

Table 14 Drugs withdrawn due to torsade de pointes or proarrhythmic effect from at least one major market

Drug	*Year withdrawn*	*Therapeutic class*
Prenylamine	1988	Antianginal
Lidoflazine	1989	Antianginal
Terodiline	1991	Initially antianginal, then for urinary frequency and incontinence
Encainide	1991	Antiarrhythmic
Terfenadine	1998	H_1 antihistamine
Sertindole	1998	Antipsychotic
Mibefradil	1998	Antihypertensive and antianginal
Astemizole	1999	H_1 antihistamine
Grepafloxacin	1999	Antibiotic
Cisapride	2000	Prokinetic
Droperidol	2001	Antipsychotic
Levacetylmethadol	2001	Opioid analgesic
Thioridazine	2005 (Canada)	Antipsychotic

Prolongation of the QT interval can usually be predicted from the pharmacological mode of action of drugs. Drugs which block potassium channels are expected to prolong the QT interval. However, under what circumstances such prolongation of the QT interval becomes proarrhythmic is not well understood. Indeed, class III antiarrhythmic agents are used to prevent the development of arrhythmia by prolonging the QT interval. When drugs prolong the QT interval unnecessarily, they may lead to ventricular tachyarrhthmias, torsade de pointes, and sudden death. For this reason drugs that prolong the QT interval as a side effect are undesirable. Unfortunately, a number of both cardiovascular and noncardiovascular drugs have this side effect. Drugs which have been withdrawn from the market, mainly or partly because of their propensity to induce torsade de pointes, are listed in **Table 14**.

Formal postmarketing surveillance studies prior to withdrawal of the drugs concerned failed to identify any risk of torsade de pointes sufficiently well to activate regulatory intervention. For example, four formal postmarketing studies using different approaches spanning over 8 years examining cardiac events associated with cisapride failed to show any worrisome risk.[65–68]

1.25.11.1 Prescription Event Monitoring Study[65]

A cohort of 13 234 cisapride users and a control cohort of 319 168 from a general practice setting in England and Wales were followed. All patients who were dispensed cisapride over a 6-month period were included and followed up from the date of dispensing to the stop date or end of survey date. Children from age 3 months and adults were included and events selected were disorders of heart rate (tachycardia and bradycardia) and disorders of rhythm (arrhythmias and palpitation). The crude event rate for disorders of cardiac rhythm rate were 2.9/1000 cisapride users compared to 3.3/1000 in the control group.

1.25.11.2 Open Prospective Cohort Study[66]

In this study, undertaken within the German general practice setting, data from a cohort of 37 925 adult patients, aged 19–70 years, prescribed cisapride were followed prospectively from date of prescription to end of treatment (mean 26.6 days). Some 1830 patients had at least one adverse event but none were cardiovascular events.

1.25.11.3 Retrospective Cohort and Nested Case-Control Study Using General Practice Research Database[67]

This was a two-country retrospective cohort study using the General Practice Research database and the Saskachewan Health database of medical claims. The UK cohort consisted of 18 571 cisapride users and 16 909 nonusers, while the

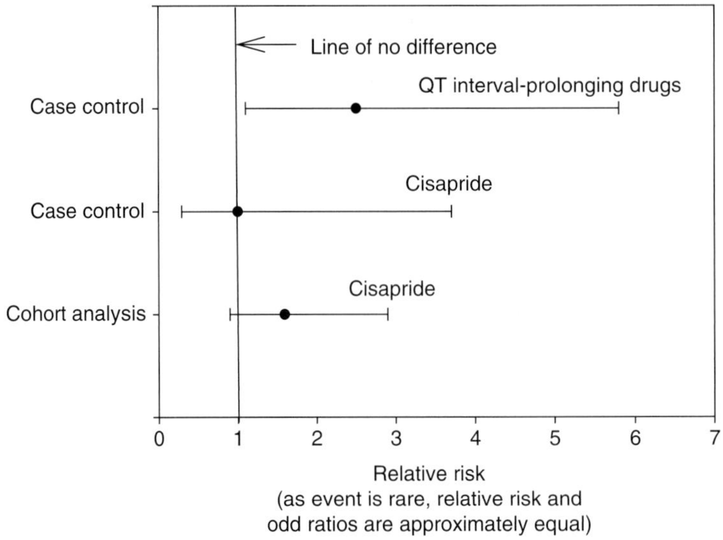

Figure 3 Estimate of relative risk of arrhythmic events with cisapride compared to controls. (Data from Walker, A. M.; Szneke, P.; Weatherby, L. B.; Dicker, L. W.; Lanza, L. L.; Longhlin, J. E.; Yee, C. L.; Dreyer, N. A. *Am. J. Med.* **1999**, *107*, 356–362.)

Canadian cohort consisted of 18 171 users and 16 058 nonusers. A case-control study nested within the cohorts was also undertaken. In the cohort analysis, the incidence of arrhythmic events was 1.6 times higher but not significantly so (95% confidence interval (CI) 0.9–2.9) in periods of recent (prior 8 weeks) cisapride use. With adjustments for clinical history, use of cytochrome 3A4 (CYP3A4A) inhibitors and use of drugs known to prolong QT interval, the odds ratio for cisapride and cardiac events was 1.0 (95% CI 0.3–3.7) relative to controls. Concomitant dispensing of QT interval-prolonging drugs and CYP3A4A did not increase the risk of cardiovascular events. The authors concluded that serious rhythm disorders were not associated with cisapride use and cautioned that "the upper confidence bounds do not rule out an increase in risk." However, what was of concern was the nonsymmetric distribution of the confidence intervals around the point of no difference (odds ratio of 1; **Figure 3**). They found that their positive controls (QT interval-prolonging drugs) were associated with an increase in the risk of arrhythmic events (odds ratio 2.5 (95% CI 1.1–5.8). At the time this would have provided false reassurance about the safety of cisapride.

1.25.11.4 Retrospective Cohort Study[68]

This was also a general practice study but using the Mediplus database. A total of 25 320 prescriptions were issued and the authors concluded that there was no evidence to associate cisapride with cardiac arrhythmias.

Similarly, formal postmarketing surveillance studies of terolidine failed to identify its cardiovascular adverse effects, even when spontaneous adverse reaction reports have generated a significant signal.[69]

To overcome the shortcomings of the more traditional formal postmarketing surveillance studies, it has been suggested that prospective active surveillance programs, registries of rare events, and pharmacogenetic studies may be useful.[70] However, what is even more important is recognition that for rare events much larger databases or patient cohorts than currently used are necessary for risk assessments.

1.25.11.5 Pharmacogenetics

The importance of genetics in determining a person's susceptibility to adverse reactions is being increasingly recognized. Increased susceptibility to the adverse effects of drugs may be due to polymorphism of metabolizing enzymes and drug receptors and transporters. Therefore, the adverse reaction profile of a drug in one country may be quite different from that seen in another because of differences in distribution of particular genetic polymorphisms. For example, there are considerable differences in the distributions of CYP2D6, CYP2C9, and CYP2C19, which are frequently involved in the metabolism of commonly used drugs.[71–74] However, there is often a complex interplay between genetic and nongenetic factors in the process leading to adverse reactions and genetic testing on its own is unlikely to be sufficiently predictive to guide clinical practice and make postmarketing surveillance unnecessary for a few years yet. Polymorphism of the thiopurine-metabolizing enzyme thiopurine methyltransferase (TPMT) is often put forward as a classical example of the value of genetic testing for predicting adverse drug reactions. Yet clinical studies suggest that close to 80% of adverse reactions to thiopurine drugs were not associated with TPMT polymorphisms.[75]

1.25.12 Conclusions

Drug regulatory agencies derive an important proportion, if not all, of their income from fees. They are therefore under considerable pressure to speed up the licensing process and indeed drug licensing is sometimes seen to be a race. Thus FDA officials, including one commissioner, once proclaimed that "the United States outpaces both Germany and Japan in approving important new drugs" and "the United States has a number of therapies with significant health benefits that are not yet available in the United Kingdom."[75] Critics claim that this haste may put patients at increased risk.[77] Irrespective of the validity of the claims and counterclaims, premarketing trials as currently undertaken will always mean that rare adverse effects and those with long latency will be missed prior to licensing and postmarketing surveillance will remain essential.[79]

> The dominant and analytic difficulty is pervasive uncertainty. Risk assessment draws extensively on science ... However, data may be incomplete, and there is often great uncertainty in the estimates of the types, probability, and magnitude of health effects associated with a chemical agent, of the economic effects of a proposed regulatory action, and of the extent of current and future human exposures ... To make judgments amid such uncertainty, risk assessors must rely on a series of assumptions.[81]

For further reading on this subject, readers are advised to consult references.[82–97]

References

1. Pouyanne, P.; Haramburu, F.; Imbs, J. L.; Begaud, B. *Br. Med. J.* **2000**, *320*, 1036.
2. Lazarou, J.; Pomeranz, B. H.; Corey, P. N. *JAMA* **1998**, *279*, 1200–1205.
3. Roughead, E. E.; Gilbert, A. L.; Primrose, J. G.; Sansom, L. N. *Med. J. Aust.* **1998**, *168*, 405–408.
4. Pirmohamed, M.; James, S.; Meakin, S.; Green, C.; Scokk, A. K.; Walley, T. J.; Farrar, K.; Park, K.; Breckenridge, A. M. *Br. Med. J.* **2004**, *329*, 15–19.
5. Classen, D. C.; Pestotnik, S. L.; Evans, R. S.; Lloyd, J. F.; Burke, J. P. *JAMA* **1997**, *277*, 301–306.
6. Bates, D. W.; Spell, N.; Cullen, D. J.; Burdick, E.; Laird, N.; Petersen, L. A.; Small, S. D.; Sweitzer, B. J.; Leape, L. L. *JAMA* **1997**, *277*, 307–311.
7. Johnson, J. A.; Bootman, J. L. *Arch. Intern. Med.* **1995**, *155*, 1949–1956.
8. Geiling, E.; Cannon, P. *JAMA* **1938**, *111*, 919–926.
9. Lenz, W.; Knapp, K. *Dtsch Med. Wochenschr.* **1962**, *87*, 1232–1242.
10. Lenz, W.; Knapp, K. *Arch. Environ. Health* **1962**, *5*, 100–105.
11. Dessertenne, F. *Arch. Mal. Coeur Vaiss.* **1966**, *59*, 263–272.
12. Herbst, A. L.; Ulfelder, H.; Poskanzer, D. C. *N. Engl. J. Med.* **1971**, *284*, 878–881.
13. Wakefield, A.; Murch, S.; Anthony, A. et al. *Lancet* **1998**, *351*, 637–641.
14. Fung, M.; Thornton, A.; Mybeck, K.; Wu, J.-H.; Hornbuckle, K.; Muniz, E. *Drug Inf. J.* **2001**, *35*, 293–317.
15. DiMasi, J. A.; Hansen, R. W.; Grabowski, H. G. *J. Health Econ.* **2003**, *22*, 151–185.
16. Public Citizen. Rx R&D myths. The case againsts the drug industry's R&D 'scare card'. Washington 2001.
17. Hurwitz, H.; Fehrenbacher, L.; Novotny, W.; Cartwright, T.; Hainsworth, J.; Heim, W.; Berlin, J.; Baron, A.; Griffing, S.; Holmgren, E. et al. *N. Engl. J. Med.* **2004**, *350*, 2335–2342.
18. Horton, R. *Lancet* **2004**, *363*, 747–749.
19. Hasford, J.; Goettler, M.; Munter, K. H.; Muller-Oerlinghausen, B. *J. Clin. Epidemiol.* **2002**, *55*, 945–950.
20. Tubert-Bitter, P.; Haramburu, F.; Begaud, B.; Chaslerie, A.; Abraham, E.; Hagry, C. *Pharmacoepidemiol. Drug Safe.* **1998**, *7*, 323–329.
21. Key, C.; Layton, D.; Shakir, S. A. *Pharmacoepidemiol. Drug Safe.* **2002**, *11*, 143–148.
22. Weber, J. Epidemiology of Adverse Reactions to Nonsteroidal Antiinflammatory Drugs. In *Advances in Inflammation Research*; Rainsford, K. D., Ed.; Raven Press: New York, 1984, pp 1–6.
23. Hartnell, N. R.; Wilson, J. P. *Pharmacotherapy* **2004**, *24*, 743–749.
24. Doll, S. R. *Am. J. Respir. Crit. Care Med.* **2000**, *162*, 4–6.
25. Doll, R.; Hill, A. B. *Br. Med. J.* **1956**, *12*, 1071–1081.
26. Li Wan Po, A. *Statistics for Pharmacists*; Blackwell Science: Oxford, 1998.
27. Naranjo, C. A.; Busto, U.; Sellers, E. M.; Sandor, P.; Ruiz, I.; Roberts, E. A.; Janecek, E.; Domecq, C.; Greenblatt, D. J. *Clin. Pharmacol. Ther.* **1981**, *30*, 239–245.
28. Lane, D. A.; Hutchinson, T. A. *J. Chron. Dis.* **1987**, *40*, 621–625.
29. Fischhoff, B.; Lichtenstein, S.; Slovic, P.; Derby, S. L.; Keeney, R. *Acceptable Risk: A Critical Guide*; Cambridge University Press: Cambridge, 1981.
30. Lewis, J. A. *Trends Pharmacol. Sci.* **1981**, *2*, 93–94.
31. Hanley, J. A.; Lippman-Hand, A. *JAMA* **1983**, *249*, 1743–1745.
32. Szarfman, A.; Machado, S. G.; O'Neill, R. T. *Drug Safe.* **2002**, *25*, 381–392.
33. Bate, A.; Lindquist, M.; Edwards, I. R.; Orre, R. *Drug Safe.* **2002**, *25*, 393–397.
34. Shewhart, W. A. *The Economic Control of the Quality Control of Manufactured Product*; Macmillan: New York, 1931.

35. Egberts, A.; Meyboom, R.; van Puijenbroek, E. *Drug Safe.* **2002**, *25*, 453–458.
36. Evans, S. J. W.; Waller, P. C.; Davis, S. *Pharmacoepidemiol. Drug Safe.* **2001**, *10*, 453–486.
37. McCullagh, P.; Nelder, J. A. *Generalized Linear Models*, 2nd ed.; Chapman and Hall: London, 1989.
38. Rosenbaum, P.; Rubin, D. *Biometrika* **1983**, *70*, 41–55.
39. Wang, P. S.; Schneeweiss, S.; Avorn, J.; Fischer, M. A.; Mogun, H.; Solomon, D. H.; Brookhart, M. A. *N. Engl. J. Med.* **2005**, *353*, 2335–2241.
40. McMahon, A. D. *Pharmacoepidemiol. Drug Safe.* **2003**, *12*, 551–558.
41. Johnston, S. C. *J. Clin. Epidemiol.* **2000**, *53*, 1236–1241.
42. NeutroSpec. Letter from the manufacturers. http://www.fda.gov/medwatch/safety/2005/NeutroSpec_DHPL11-30-2005_signatures.pdf.
43. NeutroSpec. FDA Public health advisory. http://www.fda.gov/cder/drug/advisory/technetium99.htm.
44. Li Wan Po, A.; Zhang, W. Y. *Lancet* **1998**, *351*, 1829–1830.
45. Li Wan Po, A. *J. Clin. Pharm. Ther.* **2005**, *30*, 97–99.
46. Duract (Bromfenac) http://www.fda.gov/bbs/topics/ANSWERS/ANS00879.html.
47. Duract (Bromfenac) http://www.fda.gov/medwatch/SAFETY/1998/duract3.htm.
48. Berger, T.; Deisenhammer, F. *N. Engl. J. Med.* **2005**, *353*, 1744–1746; author reply 1744–1746.
49. Alvarez-Cermeno, J. C.; Masjuan, J.; Villar, L. M. *N. Engl. J. Med.* **2005**, *353*, 1744–1746; author reply 1744–1746.
50. Tysabri (Natalizumab) http://www.fda.gov/medwatch/safety/2005/tysabri_dearhcp2_4-1-05.pdf.
51. Tysabri (Natalizumab) http://www.fda.gov/medwatch/safety/2005/safety05.htm#Tysabri.
52. Cohen, J. *Science* **2005**, *308*, 618–621.
53. Zhou, W.; Pool, V.; Iskander, J. K.; English-Bullard, R.; Ball, R.; Wise, R. P.; Haber, P.; Pless, R. P.; Mootrey, G.; Ellenberg, S. S. et al. *MMWR Surveill. Summ.* **2003**, *52*, 1–24.
54. Li Wan Po, A. *Br. Med. J.* **2004**, *329*, 62–63.
55. Black, S.; Shinefield, H.; Baxter, R.; Austrian, R.; Bracken, L,; Hansen, J.; Lewis, E.; Fireman, B. *Pediatr. Infect. Dis. J.* **2004**, *23*, 485–489.
56. Casey, C. G.; Iskander, J. K.; Roper, M. H.; Mast, E. E.; Wen, X. J.; Torok, T. J.; Chapman, L. E.; Swerdlow, D. L.; Morgan, J.; Heffelfinger, J. D. et al. *JAMA* **2005**, *294*, 2734–2743.
57. Izurieta, H. S.; Haber, P.; Wise, R. P.; Iskander, J.; Pratt, D.; Mink, C.; Chang, S.; Braun, M. M.; Ball, R. *JAMA* **2005**, *294*, 2720–2725.
58. Neuzil, K. M.; Griffin, M. R. *JAMA* **2005**, *294*, 2763–2765.
59. Juni, P.; Nartey, L.; Reichenbach, S.; Sterchi, R.; Dieppe, P. A.; Egger, M. *Lancet* **2004**, *364*, 2021–2029.
60. Bresalier, R. S.; Sandler, R. S.; Quan, H.; Bolognese, J. A.; Oxenius, B.; Horgan, K.; Lines, C.; Riddell, R.; Morton, D.; Lanas, A. *N. Engl. J. Med.* **2005**, *352*, 1092–1102.
61. Stelfox, H. T.; Chua, G.; O'Rourke, K.; Detsky, A. S. *N. Engl. J. Med.* **1998**, *338*, 101–106.
62. Curfman, G. D.; Morrissey, S.; Drazen, J. M. *N. Engl. J. Med.* **2005**, *353*, 2813–2814.
63. Fontanarosa, P. B.; Flanagin, A.; DeAngelis, C. D. *JAMA* **2005**, *294*, 110–111.
64. Shah, R. R. *Drug Safe.* **2004**, *27*, 145–172.
65. Inman, W.; Kubota, K. *Br. Med. J.* **1992**, *305*, 1019.
66. Liehr, H.; Schmidt, R. *Scand. J. Gastroenterol.* **1993**, *195*, 54–58, discussion 8–9.
67. Inman, W.; Kubota, K. *Br. Med. J.* **1992**, *305*, 1019.
68. Walker, A. M.; Szneke, P.; Weatherby, L. B.; Dicker, L. W.; Lanza, L. L.; Longhlin, J. E.; Yee, C. L.; Dreyer, N. A. *Am. J. Med.* **1999**, *107*, 356–362.
69. Tooley, P. J.; Vervaet, P.; Wager, E. *Pharmacoepidemiol. Drug Safe.* **1999**, *8*, 57–58.
70. Shah, R. R. *Withdrawal of Terolidine: A Tale of Two Toxicities* 2006 (in press).
71. Layton, D.; Key, C.; Shakir, S. A. *Pharmacoepidemiol. Drug Safe.* **2003**, *12*, 31–40.
72. Kim, K.; Johnson, J.; Derendorf, H. *J. Clin. Pharmacol.* **2004**, *44*, 1083–1105.
73. Xie, H. G.; Kim, R. B.; Wood, A. J.; Stein, C. M. *Annu. Rev. Pharmacol. Toxicol.* **2001**, *41*, 815–850.
74. Abraham, B.; Adithan, A. *Ind. J. Pharmacol.* **2001**, *33*, 147–169.
75. van Aken, J.; Schmedders, M.; Feuerstein, G.; Kollek, R. *Am. J. Pharmacogenomics* **2003**, *3*, 149–155.
76. Kessler, D. A.; Hass, A. E.; Feiden, K. L.; Lumpkin, M.; Temple, R. *JAMA* **1996**, *276*, 1826–1831.
77. House of Commons Health Select Committee. The influence of the pharmaceutical industry. Fourth Report of Session 2004–05. March 2005. Stationery Office: London.
78. Lowrance, W. W. *Of Acceptable Risk*; William Kaufman: Los Altos, CA, 1976.
79. Breslow, N. *Statist. Sci.* **1988**, *3*, 3–56.
80. Hauben, M.; Zhou, X. *Drug Safe.* **2003**, *26*, 159–186.
81. National Academy of Sciences. *Managing the Process: Risk Assessment in the Federal Government*; National Academy of Sciences: Washington, 1983.
82. AFSS (Agence Francaise de Sécurite Sanitaire). Mise au point sur le diéthylstilbestrol (D.E.S.) (Distilbène®, Stilboestrol-Borne®) et le risque de complications génitales et obstétricales. http://agmed.sante.gouv.fr/htm/10/des/des1.htm
83. Almenoff, J.; Tonning, J. M.; Gould, A. L.; Szarfman, A.; Hauben, M.; Ouellet-Hellstrom, R.; Ball, R.; Hornbuckle, K.; Walsh, L.; Yee, C. et al. *Drug Safe.* **2005**, *28*, 981–1007.
84. Angrist, J.; Imbens, G.; Ruben, D. *J. Am. Stat. Assoc.* **1996**, *91*, 444–455.
85. Aronson, J. K.; Ferner, R. E. *Drug Safe.* **2005**, *28*, 851–870.
86. Levine, J.; Tonning, J. M.; Szarfman, A. *Br. J. Clin. Pharmacol.* **2005**, *61*, 105–113.
87. Li Wan Po, A.; Kendall, M. J. *Drug Safe.* **2001**, *24*, 793–799.
88. Mann, R. D. *Br. J. Clin. Pharmacol.* **1998**, *46*, 195–201.
89. McClellan, M.; McNeil, B. J.; Newhouse, J. P. *JAMA* **1994**, *272*, 859–866.
90. McFarlane, M. J.; Feinstein, A. R.; Horwitz, R. I. *Am. J. Med.* **1986**, *81*, 855–863.
91. Page, E. S. *Technometrics* **1961**, *3*, 1–9.
92. Shah, B. R.; Laupacis, A.; Hux, J. E.; Austin, P. C. *J. Clin. Epidemiol.* **2005**, *58*, 550–559.

93. Shah, R. R. *Adv. Drug React. Toxicol. Rev.* **2001**, *20*, 199–225.
94. Stillman, R. J. *Am. J. Obstet. Gynecol.* **1982**, *142*, 905–921.
95. Strom, B. *Pharmacoepidemiology*, 4th ed.; John Wiley: Chichester, 2005.
96. Talbot, J.; Waller, P. *Stephen's Detection of New Adverse Drug Reactions*; John Wiley: Chichester, 2004.
97. Wood, L.; Martinez, C. *Drug Safe.* **2004**, *27*, 871–881.
98. International Conference on Harmonization. www.ich.org.
99. http://www.fda.gov/CbER/fdama/pstmrktfdama130.htm (accessed Sept 2006).
100. http://www.fda.gov/cber/gdlns/post040401.htm (accessed Sept 2006).

Comprehensive Medicinal Chemistry II
ISBN (set): 0-08-044513-6

ISBN (Volume 1) 0-08-044514-4; pp. 755–777

1.26 Intellectual Property Rights and Patents

J Cockbain, Frank B Dehn & Co, Oxford, UK

> Early Scots invented the wheel first… The earliest evidence for wheeled transport… is a wooden disc wheel from Blair Drummond, near Stirling, which the Oxford radiocarbon accelerator unit has dated to between 12000 BC and 8000 BC… *The Times* (2005)[1]

> The [2001] IgNobel Technology Prize was… awarded jointly to John Keogh of Hawthorn, Victoria, Australia, for patenting the wheel in the year 2001, and to the Australian Patent Office for granting Innovation Patent # 2001100012" Abrahams (2002: 172–175)[2]

> Lord Mustill once observed that 'some of the most penetrating legal minds, both on and off the bench, have directed themselves to the evolution of patent law' [Genentech[2a] (1989: 258)]… He put this down to the fact that "the industrial revolution happened where and when it did" or, putting the matter another way, the tendency for brains to follow money. Lord Hoffmann (2004: xi) in Vaver *et al.* (2004)[3]

1.26.1 Introduction

It has been estimated that the cost of bringing a new drug onto the market is in the region of US$800 million.[4] While this precise figure is somewhat controversial,[5,6] it sets the scene for this chapter.

Unless drug development is funded by the state or by charities, the costs must be born by companies. While a proposal is currently before the US Congress that would effectively replace patents for pharmaceuticals by a reward for bringing a new pharmaceutical to the market – H.R. 417 The Medical Innovation Prize Act of 2005[7] – this is as likely to fly as a lead balloon. For a company to make an investment of that magnitude there must be a reasonable expectation that the sales of the drug will pay back the investment; in other words the drug must be sold at a price significantly higher than the production and distribution costs. If other companies are free to sell the same drug, then the profit component of the sales price may be driven so low that the investment cannot be recouped or can be recouped only so slowly that it would make economic sense to invest in something else instead.

One system that has developed to encourage the production of new things is a set of laws referred to collectively as intellectual property (IP) law by virtue of which creators (or their assignees) may be granted property rights in their creations that are enforceable by the Courts. Several different types of intellectual property right (IPR) exist, but in the field of medicinal chemistry the most important are patent rights since patents may be used to prevent not just the sales of identical drugs but also of similar drugs.

IP rights hinder competition and limit access to the products or processes they cover. They are not universally accepted as being justified socially, morally, or economically.[8–12]

Traditionally, the range of IPR available has varied from country to country, and many countries have strengthened or weakened the scope of such rights as social, economic, and political conditions have changed. Thus, many countries at an early stage of industrial development granted no or only limited IPR, for example refusing to grant patents for pharmaceuticals. The countries in question are not simply the smaller countries nowadays considered to be net technology importers. Thus, Maskus (2000: 34)[8] mentions some of the United States' own history in this regard:

> International variations in IPRs have been the subject of trade conflict for a long time. For example, the US Copyright Act adopted by the first American Congress actively sought to encourage the development of the publishing industry by awarding rights to print, reprint, publish and sell literary works only to domestic citizens and residents… Foreigners were not allowed to obtain copyrights and the law explicitly permitted parallel importation of works copyrighted abroad. In consequence, American publishers were able to publish and sell foreign literary creations cheaply; this attracted sharp criticism, especially from British authors.

Japan's history has likewise been described recently by Fisher.[13] For the UK, see for example MacLeod.[14]

However, following the Uruguay Round of negotiations on the General Agreement on Tariffs and Trade (GATT), on 15 April 1994 the Marrakesh agreement Establishing the World Trade Organization (WTO) was signed and included, as

Annex 1C, the Agreement on Trade-Related Aspects of Intellectual Property Rights which is now generally referred to as TRIPS.[15] Article 27 of TRIPS essentially requires all the member states of the WTO to adapt their criteria for patentability to become no less favorable than those that existed in Europe and, for example, to grant patents for pharmaceuticals. For most countries, the requirement to harmonize their patent laws had a deadline of 1 January 2005; for the least developed countries, this has been extended to 1 January 2016.[16] The rights and obligations deriving from TRIPS are explained at length in UNCTAD-ICTSD.[17]

How IP entered into the GATT negotiations is an interesting story in itself and the reader is referred to the books by Drahos *et al.*[18] and Sell.[19]

1.26.2 Viewpoint

With the advent of TRIPS, national and regional patent laws are becoming increasingly similar in terms of what can be patented and what the requirements for patentability are, and thus this chapter will focus on European and US patent law as representative of patent laws in general in this regard. European and US patent laws do differ significantly, at least partially because of the fundamental question as to who is entitled to a patent, the first to file his patent application (Europe) or the first to make her invention (US), and the European law is closer to those applicable in the rest of the world.

1.26.3 What Is a Patent?

> The king's grants are also a matter of public record... These grants, whether of lands, honors, liberties, franchises, or ought besides are contained in charters, or letters patent that is, open letters, *litterae patentae*, so called because they are not sealed up, but exposed to open view, with the great seal pending at the bottom: and are usually directed by the king to all his subjects at large. Blackstone (1765–69)[20]

A patent is a document issued by or for a country that describes an invention in sufficient detail that it could be put into effect and which (in its 'claims') defines the technical scope of the property right that that country has granted the patent owner in that country for as long as the patent is in force. For most countries, unless revoked by the Courts or other authorities, a patent may be kept in force for 20 years from the date it was applied for. In certain countries, for example in the US and Europe, and for certain inventions, for example pharmaceuticals, it is possible for the patent term to be extended by some years (*see* Section 1.26.37).

The text of a patent, its 'specification,' generally follows a standard pattern: introduction/background; summary of invention; fallbacks; detailed description; claims; abstract; and, optionally, drawings. This standard pattern makes reading patents relatively straightforward. The introduction/background section sets the scene for the invention, usually explaining to the reader the purpose or importance of the invention and often describing the problem that is solved by the invention. Particularly with patents written in the US, this section will often review the prior art in order to distinguish it from the invention. The summary section will generally set out the broad definitions of the invention, which are equivalent to the independent claims in the claims section. Since objections to the scope of the broad definitions may be made during examination of the patent application, e.g., objections of lack of novelty over an item of prior art not previously encountered by the applicant, there will usually be a 'fall-backs' section, which provides the basis for possible limitations to the claims, for example, to define a chemical genus more narrowly, or to specify temperature or pressure ranges for a process. The detailed description section is intended to provide sufficient information for the reader to perform the invention, for example details of how particular compounds may be made or of how one embodiment of a claimed apparatus may be constructed. Often this section corresponds essentially to the experimental section of a scientific paper. One word of warning however to scientist readers: some of the Examples in a patent may be in the present tense and may be 'predictive,' i.e., the experimental work may not actually have been done before the patent application was filed. The claims section contains one or more numbered sentences defining the subject matter for which patent protection is sought (in the case of an application) or has been granted (in the case of a granted patent). The abstract is usually quite short and often corresponds simply to the broadest claim. Finally, if Figures are present, e.g., showing embodiments of a claimed apparatus, graphs, or images, these are separate from the written text and appear at the front (US) or back (Europe) of the printed text.

1.26.4 What Is an Invention?

That is a little more difficult to explain, but I will attempt to do so in Section 1.26.29.

1.26.5 What Rights Does a Patent Give?

A patent does not give its owner, the 'patentee,' the positive right to carry out the invention it covers. What it does is to give the patentee the right to ask the Courts in the country for which it is granted to order another party not to carry out the invention in that country, for example, by making, distributing, or importing an infringing object or by performing an infringing process, and, if appropriate, to pay damages for past infringements. The patent does not free its owner from the possibility of being sued for infringement of another party's patent, and a single product may easily be within the scope of many patents owned by many different owners. 'Trading' in patent rights is therefore often an important part of bringing a new product to the market. Thus, as with other property, for example a house or a car, a patent may be sold or rented (i.e., licensed) or used as collateral.

1.26.6 What Can Be Patented?

Patents are granted for things or ways of doing things (i.e., products or processes) that are new, nonobvious, useful, and not otherwise excluded from patentability. The precise requirements vary from country to country and the previous sentence is an approximation of the generally applicable requirements.

Article 27 TRIPS requires the member countries to grant patents for any invention in any field of technology but permits exclusion from patentability of: diagnostic, therapeutic and surgical methods for the treatment of humans or animals; plants and animals other than microorganisms; and essentially biological processes for the production of plants or animals (other than nonbiological and microbiological processes).

By contrast, the US Supreme Court, in Diamond vs Chakrabarty,[21] confirmed the earlier statements, made in the run up to the current US Patent Law being brought into force, that US Patent Law permits patenting of "anything under the sun that is made by man" (this is an allusion to Ecclesiastes, Chapter 1, Verse 9, which stated that "there is nothing new under the sun" (Bible, 1995: 626)).[22] The US Patent and Trademark Office (USPTO), however, has since adopted the limitation that humans may not be patented and in its recent rejection of a patent application directed to a human/animal chimera has argued that it should be up to the legislature to decide whether that limitation should be removed (see Newman *et al.*, 2003, pages 18–24 of the nonfinal rejection of March 5 2003 relating to USSN 10/308,135, a patent application published as US-A-2003-0079240, details accessible through the USPTO application file contents (see **Table 1**)).

The European Patent Convention or EPC,[23] in Article 52 EPC, goes beyond Article 27 TRIPS to exclude from patentability: discoveries, scientific theories, and mathematical methods; aesthetic creations; schemes, rules, and methods for performing mental acts, playing games or doing business; programs for computers; and presentations of information. However these exclusions are made on the basis that these are not inventions, and in any event the European Patent Office (EPO) has nonetheless found it possible to grant patents for both business methods and computer programs (see Singer *et al.*, 2003: 71–76[24] and Pila, 2005[25]). Moreover, despite plant or animal 'varieties' being excluded from patentability by Article 53(b) EPC, the EPO will issue patents for animals and plants (see Singer *et al.*, (2003: 93–97)[24]), and indirect protection for methods of medical treatment is likewise available through the medium of 'use' claims (see Singer *et al.*, 2003: 78–84[24]).

For the field of medicinal chemistry, perhaps the most important categories of patentable inventions include: pharmaceutical substances; compositions containing such substances; processes for making such substances; intermediates in such processes; catalysts for such processes; methods of treatment using such substances; plants or microorganisms capable of making such substances; genetic material capable of causing such substances to be made; drug discovery methods; drug screening methods; drug design methods; drug delivery systems; animal models; assay methods; diagnostic methods; prognostic methods; reagents (including microorganisms or genetic materials) usable in such assay, diagnostic and prognostic methods; and research tools.

Table 1

EPO status: http://register.epoline.org/espacenet/ep/en/srch-reg.htm
EPO file contents: http://ofi.epoline.org/view/GetDossier
USPTO application file contents: http://portal.uspto.gov/external/portal/pair
UKPO status: http://webdb4.patent.gov.uk/patents/
USPTO granted patents: http://www.uspto.gov/patft/index.html
JPO status (in English): http://www4.ipdl.ncipi.go.jp/Tokujitu/tjsogodben.ipdl?N0000 = 115
WIPO file contents: www.wipo.int/patentscope/en/database/search-adv.jsp

The general message thus is that virtually any medicinal chemistry related invention, which meets the requirements of novelty, nonobviousness and usefulness, can be patented in some way, in at least some countries of commercial interest.

In practice therefore, when an invention is made in the medicinal chemistry field, the questions to be addressed are should it be patented and, if so, where?

1.26.7 Should a Company or Individual Decide to Patent an Invention?

Patents can serve many purposes. At one extreme, a patent may simply be used to provide a period during which there is no commercial competition (or where such competition may be stopped via recourse to the Courts) so that the invention may be developed, brought to the market, and begin to turn a profit for its owner. At another extreme are the 'vanity' patents, simply expensive bits of paper providing the owner proof of his or her status as an 'inventor.'[26] Filing patent applications can be internal public relations for a company or institution, reassuring the inventors that their contributions are deemed valuable, or external public relations, letting customers, competitors, and investors know that the company or institution is actively engaged in 'cutting edge' research. Patent applications may be filed in order to be sold or licensed to create a source of income. Patent applications and the inventions to which they relate may be the basis for generating investment in a start-up company. Patent applications may also be used as leverage for access to another company's technology.

Vanity and public relations cases aside, the key question is whether or not the patent could give a commercial advantage: if a product or process that is at least comparable in performance can nonetheless be produced or performed in a commercially competitive manner without infringing, then the expense of patenting might not be justified. However, even then, if the new product or process is to be commercialized, a patent gives extra protection against a competitor launching the identical product or process, and then branding, packaging, advertising, marketing, pricing, and service may create and maintain a market position.

In the pharmaceutical field, a new, improved formulation or a novel end use or an easier-to-use package for an existing drug can help maintain sales and profits even when the basic patent for the drug expires; in the chemical field in general, a new synthesis route can improve yield, reduce waste product disposal expenses, improve purity, or use less expensive reagents. All of these have the potential to give the patentee, or a licensee, a commercial advantage, and patenting should be considered.

1.26.8 Where to Patent?

The answer to this question is essentially the same as that in the previous section: wherever the commercial advantage potentially outweighs the cost (or wherever the desired vanity or public relations effect can be achieved at an acceptable cost).

In practice, the countries to choose are primarily those in which potential sales profits are high, and those where competitors operate, in each case bearing in mind that the choice has to be made relatively early in the patenting process and that profits and competition will be greatest toward the end of the patent term and thus that countries that seem uninteresting now may be crucial in 15 or so years time.

A recent article by Silverman[26a] focused on patent filing strategies for pharmaceutical products and listed the top 10 countries/regions to proceed with nationally or at the 30-month stage of an international patent application as US, EPO, Canada, Japan, India, Australia, Brazil, Mexico, China, and Venezuela, and the top 10 countries to proceed with at the grant stage of a European patent application as UK, Germany, France, Belgium, Italy, Switzerland, Spain, Ireland, Turkey, and Greece. These listings take into account translation costs and are from an American perspective, but nonetheless are instructive.

Finally, in this section, it should be pointed out that various countries, the US in particular, have requirements that where an invention is made in such a country, either a patent application must be filed first in that country or official clearance must be obtained before a patent application is filed outside that country.

1.26.9 The Requirements for Patentability

As mentioned above, loosely stated, an invention must be new, nonobvious, useful, and not otherwise excluded from patentability for it to be patentable. Unfortunately, these requirements are not as simple as they might seem. So, what is 'new' to a research scientist may not be 'new' for patent purposes in many countries, and what appears to the research scientist not to be 'new' may likewise be 'new' in patent terms. Moreover, the requirements might appear to be the same, but may be defined or interpreted differently in different countries.

1.26.10 Novelty

Within Europe, or at least the countries covered by the EPC, something is new if, as of the filing date of the patent application (or the priority date where priority of an earlier application is claimed – *see* Section 1.26.16 below), it does not "form part of the state of the art" (Article 54 EPC). The 'state of the art' in this context means everything that has been made available to the public anywhere in written or oral form or in any other way, e.g., by use in public, as well as anything disclosed in an as yet unpublished European patent application of earlier date. This definition of the state of the art in the EPC as including anything made available any way and anywhere is sometimes referred to as an 'absolute novelty' requirement. In other places and at other times, 'local novelty' sufficed, i.e., the state of the art only included things made public in the country in question. Made available to the public in this context means one or more persons could have received the information and could have used it without breaching confidentiality.

However, for a piece of prior art to have a novelty-destructive effect, it must not only unambiguously and explicitly or implicitly disclose something falling within the definition of the invention (the claims), but it must also be an enabling disclosure, i.e., enough for a person skilled in the relevant field to be able to create something falling within the definition of the invention. A'black box' disclosure is not novelty-destructive. A disclosure of a range, e.g., 1 to 10, is only novelty-destructive for the range limits, i.e., 1 and 10. A disclosure of a composition containing two components, one selected from A, B, C, D, and E and the other selected from F, G, H, I, and J, is not of itself novelty-destructive for any of the particular possible combinations, e.g., A+H or C+F. An English Appeals Court Judge, Sachs LJ, once said on this topic:

> To anticipate the patentee's claim the prior publication must contain clear and unmistakable directions to do what the patentee claims to have invented... A signpost, however clear, upon the road to the patentee's invention will not suffice. The prior... [discloser] must be clearly shown to have planted his flag at the precise destination before the patentee. General Tire vs. Firestone (1972: 486)[27]

There is one limitation to what forms part of the state of the art under the EPC. As long as the European patent application itself is filed within 6 months of a disclosure of the invention taking place, that disclosure is not treated as forming part of the state of the art if it was 'an evident abuse in relation to the applicant or his legal predecessor,' e.g., if it was made in breach of confidence, or if it was a display of the invention by the applicant or predecessor at an officially recognized international exhibition falling within the terms of the Convention on international exhibitions signed in Paris in 1928 (see Article 55 EPC).

It is therefore of critical importance that inventors do not make nonconfidential disclosures of their inventions before a patent application is filed – this includes posters, seminars, and conversations in a bar.

As it usually is with patent law, in the matter of novelty, the US is a special case. Section 102 of the US Patent Law sets out what is or is not prior art for novelty purposes. (US Federal laws are referred to as 'Titles' in the 'United States Code' (U.S.C). The patent law is Title 35,[28] and Section 102 in this law is referred to as 35 U.S.C. 102.) In view of the importance of the US where IPR and medicinal chemistry is concerned, it is worth setting out 35 U.S.C. 102 in full:

> A person shall be entitled to a patent unless –
>
> (a) the invention was known or used by others in this country, or patented or described in a printed publication in this or a foreign country, before the invention thereof by the applicant for patent, or
> (b) the invention was patented or described in a printed publication in this or a foreign country or in public use or on sale in this country, more than one year prior to the date of the application for patent in the United States, or
> (c) he has abandoned the invention, or
> (d) the invention was first patented or caused to be patented, or was the subject of an inventor's certificate, by the applicant or his legal representatives or assigns in a foreign country prior to the date of the application for patent in this country on an application for patent or inventor's certificate filed more than twelve months before the filing of the application in the United States, or
> (e) the invention was described in – (1) an application for patent, published under Section 122(b), by another filed in the United States before the invention by the applicant for patent or (2) a patent granted on an application for patent by another filed in the United States before the invention by the applicant for patent, except that an international application filed under the treaty defined in Section 351(a) shall have the effects for the purposes of this subsection of an application filed in the United States only if the international application designated the United States and was published under Article 21(2) of such treaty in the English language, or
> (f) he did not himself invent the subject matter sought to be patented, or

(g) (1) during the course of an interference conducted under Section 135 or Section 291, another inventor involved therein establishes, to the extent permitted in Section 104, that before such person's invention thereof the invention was made by such other inventor and not abandoned, suppressed, or concealed, or (2) before such person's invention thereof, the invention was made in this country by another inventor who had not abandoned, suppressed, or concealed it. In determining priority of invention under this subsection, there shall be considered not only the respective dates of conception and reduction to practice of the invention, but also the reasonable diligence of one who was first to conceive and last to reduce to practice, from a time prior to conception by the other.

While US patent lawyers may cringe at this, the most salient points may perhaps be summarized a bit simplistically as follows:

1. The disclosures in an earlier US patent application, once published, retroactively form part of the state of the art as if published on the date of filing of that application.
2. An offer for sale of the invention by the inventor or their assignee forms part of the state of the art.
3. Use or oral disclosure in the US, but not elsewhere, forms part of the state of the art.
4. Printed publications form part of the state of the art.
5. An item of prior art may be ignored if the US patent application was filed within 12 months of its date of publication and the invention was made before that date of publication.

Where the item mentioned in point (5) is a US patent or an accepted US patent application, it may still pose problems if it claims the same or essentially the same invention (*see* Section 1.26.27 below). Equally, a later-filed US patent application may cause problems if it claims the same or essentially the same invention. This is due to the 'first-to-invent' basis for US patent law.

As in Europe, an item of prior art must contain an enabling disclosure to be novelty-destructive in the US.

In general, when considering whether or not a patent claim is or is not novel, each item of prior art should be considered separately, if necessary in conjunction with a further item of prior art to which it refers explicitly.

1.26.11 Nonobviousness

The requirement of nonobviousness in the European Patent Convention is set out in Articles 52(1) and 56 EPC. Thus, to be patentable, an invention must 'involve an inventive step' and is considered to do so 'if, having regard to the state of the art, it is not obvious to a person skilled in the art.' The 'state of the art' for this purpose is the same as for novelty, but excluding earlier patent applications unpublished at the priority date of the claim in question.

Unlike with novelty, however, for the determination of (non)obviousness, the notional 'person skilled in the art' need not consider the prior art items separately, so for example one document saying that "there is a problem in process X because reagent Y precipitates" can be combined with another document saying that "precipitation of reagent Y may be prevented by modification Z" to establish obviousness for a claim to "process X wherein reagent Y is subject to modification Z."

As can be expected in patent law, the requirement for an 'inventive step' leads directly to the question 'What is an inventive step?' and the definition of this requirement simply leads on to the further questions 'What is meant by obvious?' and 'What mental capability can be expected of the person skilled in the art?'

The decision as to whether or not something involves an inventive step (or is or is not obvious) is not taken by a real person active in the relevant technological field. Patents are not peer-reviewed publications. The decision is taken by a patent office examiner or a Court.

The 'person skilled in the art' is purely notional and is considered to be someone, or a group of people, incapable of inventive thought. In the field of medicinal chemistry, it might be useful to visualize this reference point for the determination of inventiveness as the postgraduate who needs to be told what to do, but once told knows how to do it. If such a person would not have arrived at the invention in the light of the prior art under consideration, then the likelihood is that a patent will be granted.

Another generally accepted 'rule of thumb' for estimating whether or not something should be considered inventive, is the question: 'Was it obvious to try with a reasonable expectation of success?'

The degree of inventiveness required to justify the grant of a property right that can be used to prevent others from doing something that they would otherwise have been free to do is subject to much debate, especially in any newly emerging area of technology. It has been a particularly prominent topic in recent years in the field of biotechnology, especially with regard to claims to proteins, gene sequences, single nucleotide polymorphisms (SNPs), and other naturally occurring products. If the barrier to patentability is too high, the argument goes, then the investment necessary to bring the new technology to the market will not be forthcoming. Conversely, if the barrier is too low, the development of any new product will require licenses under so many patents that the product will either never reach the market or will be insufficiently profitable if it does, and as a result research is stifled by the existence of the patents. In practice, the barrier does change as a technological field matures: what once might have been considered to be inventive, may later be judged to be obvious.

James[29] recently proposed an interesting approach to the determination of whether or not an invention was obvious. At page 79, James states that:

> The test for obviousness in patent law is equivalent, in the psychological model, to asking: *could* the invention have been made by the hypothetical worker? That is, is the association of concepts represented by the invention already part of the prior art base? Or, does the 'invention' represent, relative to the prior art base, an historically novel association between previously disparate concepts?

In the US, the nonobviousness requirement is set out in Section 103(a) of the US Patent Law, i.e., 35 U.S.C. 103(a). This reads as follows:

> A patent may not be obtained though the invention is not identically disclosed or described as set forth in section 102 of this title, if the differences between the subject matter sought to be patented and the prior art are such that the subject matter as a whole would have been obvious at the time the invention was made to a person having ordinary skill in the art to which said subject matter pertains. Patentability shall not be negatived by the manner in which the invention was made.

The relevant prior art for the determination of inventiveness in the US is essentially the same as that for the determination of novelty discussed above, with the exception of Section 102(e), (f), or (g) prior art (*see* Section 1.26.10 above for a recitation of 35 U.S.C. 102 in full) "where the [prior art] subject matter and the claimed invention were, at the time the invention was made, owned by the same person or subject to an obligation of assignment to the same person."

A major difference in practice, however, is the greater likelihood of multiple items of prior art being combined by a USPTO examiner to support an obviousness objection to a patent application.

1.26.12 **Usefulness**

In the European Patent Convention, the 'usefulness' requirement for patentability is expressed in Article 52(1) EPC as the need for the invention to be "susceptible of industrial application" and this is later defined in Article 57 EPC as meaning that the invention must be capable of being "made or used in any kind of industry, including agriculture." Virtually any commercializable invention in medicinal chemistry would meet this requirement; however, in Article 52(4) EPC, the European Patent Convention explicitly defines methods of treatment of the human or animal body by surgery or therapy and diagnostic methods practised on the human or animal body as not being susceptible of industrial application.

In the US, the requirement is instead expressed as one for 'utility.' This is set out in Section 101 of the US Patent Law as follows

> Whoever invents or discovers any new and useful process, machine, manufacture, or composition of matter, or any new and useful improvement thereof, may obtain a patent therefor, subject to the conditions and requirements of this title.

In the US, three types of utility are referred to: general utility, meaning that the invention can be put into effect; specific utility, in that the invention can be performed to serve the functions and secure the specific result intended; and beneficial utility, in that the invention should not be frivolous or injurious to the well-being, good policy, or sound morals of society (see Chisum, 1997, Vol. I, Section 4.01).[30]

The question occasionally does arise as to whether or not a new chemical does indeed have the usefulness required for patentability. This is particularly the case with compounds produced by academics interested in novel synthetic techniques and physicochemical properties. Long ago, as a chemistry student, I was told of the compound cubane (my family name is pronounced 'co-bane'), a cube-shaped C_8H_8 molecule reported by Eaton *et al.*,[31] which seemed to be a clear example of a compound without utility. Nevertheless, Hook[32] later stated that:

> Until recently, cubane was considered as just an academic curiosity, a novelty with no actual use within modern day chemistry. Yet further research found that several possible applications in industry, materials science and medicine could exist, giving cubane a new future...

More recently, the question of usefulness has been of great significance in relation to patent applications filed for gene sequences and SNPs of unknown function. A general statement that such sequences 'may be useful in assays' is not enough. *The Manual of Patent Examining Procedure*[33] (a helpful but immense tome), states on pages 2100–29 that:

> A claimed invention must have a specific and substantial utility. This requirement excludes 'throw-away,' 'insubstantial,' or 'nonspecific' utilities, such as the use of a complex invention as landfill, as a way of satisfying the utility requirement of 35 U.S.C. 101.

The EPO Guidelines[34] likewise state on page 7, Part C, that:

> ...in relation to certain biotechnological inventions, i.e. sequences and partial sequences of genes, the industrial application is not self-evident. The industrial application of such sequences must be disclosed in the patent application.

and on page 54, Part C, that:

> A mere nucleic acid sequence without indication of a function is not a patentable invention (EU Dir. 98/44/EC rec. 23 (i.e., the EU Biotechnology Directive – *see* Section 1.26.14 below)). In cases where a sequence or partial sequence of a gene is used to produce a protein or part of a protein, it is necessary to specify which protein or part of a protein is produced and what function this protein or part of a protein performs. Alternatively, when a nucleotide sequence is not used to produce a protein or part of a protein, the function to be indicated could e.g. be that the sequence exhibits a certain transcription promoter activity.

Thus, to fulfill the usefulness requirement, it can generally be stated that a specific, realistic, and moral end use for the invention should be identified in the patent application.

1.26.13 Excluded from Patentability

The European Patent Convention defines certain subject matter as not being inventions, e.g., discoveries, aesthetic creations, etc, as mentioned above. It defines certain subject matter as being not susceptible of industrial application and hence unpatentable, e.g., methods of medical treatment. However, it also states explicitly in Article 53 EPC that certain subject matter may not be patented. This effectively acknowledges that this excluded subject matter may be an otherwise patentable invention. The subject matter in question is "inventions the publication or exploitation of which would be contrary to 'ordre public' or morality," "plant or animal varieties," and "essentially biological processes for the production of plants or animals."

The EPO Guidelines in relation to 'ordre public' and morality state at page 12, Part C, that:

> It should be noted that the omission, from the publication of the [European patent] application, is mandatory for the first category. Examples of the kind of matter coming within this category are: incitement to riots or acts of disorder; incitement to criminal acts; racial, religious or similar discriminatory propaganda; and grossly obscene matter.

A commonly given example of subject matter that should be excluded is an improved design for a letter bomb.

However, the legal fiction of declaring certain subject matter as not being an invention or as not meeting the novelty, inventiveness, and usefulness requirements for patentability, i.e., as is done in the EPC in relation to methods of medical treatment, has been extended in 2005 with the amendment of the Belgian patent law in relation to inventions that are based on naturally occurring biological material. Thus, under Belgian Patent Law, if an invention is developed from biological material taken or exported in violation of certain Articles of the Convention on Biological Diversity[35] (*see* Section 1.26.30 below), then that invention shall be considered to be contrary to 'ordre public' or morality, and hence unpatentable.

Within Europe, more particularly within the European Union and the European Patent Office, the range of patentable subject matter in the field of biotechnological inventions has been 'clarified' by 'Directive 98/44 of the

European Parliament and of the Council of July 6, 1998 on the Legal Protection of Biotechnological Inventions,' the 'EU Biotechnology Directive,' or 'Directive 98/44/EC.'[36,37] This is discussed in Section 1.26.14 below.

The question of the possible exclusion of subject matter from patentability for moral reasons has recently received much attention, in particular in relation to the patentability of humans and other animals, of genetically modified plants, of body parts, and of naturally occurring gene sequences. This will be returned to in Section 1.26.41 below.

1.26.14 The European Union (EU) Biotechnology Directive

As mentioned above, the EU Biotechnology Directive serves to 'clarify' what is and is not patentable in the field of biotechnology. In Articles 4 and 6 of the EU Biotechnology Directive, the exclusion from patentability of plant and animal varieties, essentially biological processes and inventions contrary to ordre public or morality earlier required by Article 53 EPC and available under Article 27 TRIPS, is restated, while Article 5 of the EU Biotechnology Directive provides a blanket exclusion from the field of patentable inventions of the human body and 'the simple discovery of one of its elements.' More interestingly, these Articles of the EU Biotechnology Directive have effectively been used as a vehicle to interpret and limit these exclusions. It is instructive therefore to read these Articles in full:

Article 4

1. The following shall not be patentable:

   (a) plant and animal varieties;
   (b) essentially biological processes for the production of plants or animals.

2. Inventions which concern plants or animals shall be patentable if the technical feasibility of the invention is not confined to a particular plant or animal variety.
3. Paragraph 1(b) shall be without prejudice to the patentability of inventions which concern a microbiological or other technical process or a product obtained by means of such a process.

Article 5

1. The human body, at the various stages of its formation and development, and the simple discovery of one of its elements, including the sequence or partial sequence of a gene, cannot constitute patentable inventions.
2. An element isolated from the human body or otherwise produced by means of a technical process, including the sequence or partial sequence of a gene, may constitute a patentable invention, even if the structure of that element is identical to that of a natural element.
3. The industrial application of a sequence or a partial sequence of a gene must be disclosed in the patent application.

Article 6

1. Inventions shall be considered unpatentable where their commercial exploitation would be contrary to ordre public or morality however, exploitation shall not be deemed to be so contrary merely because it is prohibited by law or regulation.
2. On the basis of paragraph 1, the following, in particular, shall be considered unpatentable:

   (a) processes for cloning human beings;
   (b) processes for modifying the germ line genetic identity of human beings;
   (c) uses of human embryos for industrial or commercial purposes;
   (d) processes for modifying the genetic identity of animals which are likely to cause them suffering without any substantial medical benefit to man or animal, and also animals resulting from such processes.

Article 4(2) of the EU Biotechnology Directive thus provides a way of justifying the grant of patents covering plants or animals as long as the invention is not claimed at the variety/species level even though a claim to any one individual plant or animal, i.e., any member of the claimed group of plants or animals, might not be patentable. Thus 'a mouse [i.e., an animal of the genus Mus], having characteristic X' is a valid claim format while 'a mouse of subspecies Mus musculus domesticus, having characteristic X' may not be. The question of what taxonomic level is implied for animals by the word 'variety' and the question of the morality of patenting genetically modified animals has been explored at length in the EPO in proceedings relating to a patent directed to mice genetically engineered to increase susceptibility to cancers[38] (*see* Section 1.26.40 below) and is reflected by Article 6(2) (d) of the EU Biotechnology Directive.

Likewise, Article 5(2) of the EU Biotechnology Directive confirms that the exclusion of body parts from patentability is only absolute when such a part is in its naturally occurring state, i.e., within the body in which it was found and in the form in which it was found. This likewise reflects the current patent office attitude to the patentability of 'discoveries' in general (*see* Section 1.26.15 below).

The European Commission has recently reported on the progress made in implementing the EU Biotechnology Directive in "Report from the Commission to the Council and the European Parliament: Development and implications of patent law in the field of biotechnology and genetic engineering."[39] According to this report, as of June 2005, 21 member states of the European Union had notified the European Commission "of instruments implementing the Directive" and proceedings had been started against the noncomplying states. This report also referred to efforts to revisit the questions as to what scope of patent coverage should be permissible for elements isolated from the human body (e.g., gene sequences) and as to whether human pluripotent embryonic stem cells and stem cell lines obtained from them should be patentable, in particular whether, on ethical grounds, human gene sequences should receive different treatment to that given to chemical substances and whether the granting of broad claims should be constrained so as to avoid inhibiting further research and innovation. The report reached the following interesting conclusions (page 6):

1. There is "no ambiguity regarding the patentability of material isolated from the human body."
2. "While it can be argued ... that there are no objective grounds for limiting the traditional protection granted by patent law to inventions relating to sequences or partial sequences of genes isolated from the human body, other issues have also been raised relating to ethics, to research and to economics."
3. "It appears that totipotent stem cells should not be patentable, on grounds of human dignity."
4. "There is no immediate answer to the question of the patentability of embryonic pluripotent stem cells and indeed at this stage it would appear to be premature to come to a definitive conclusion."

These comments deserve careful consideration – the issues they relate to are perhaps not so settled.

On a separate subject, the EU Biotechnology Directive also introduces the concept of informed consent. Thus Recital 26 states that:

> Whereas if an invention is based on biological material of human origin or if it uses such material, where a patent application is filed, the person from whose body the material is taken must have had an opportunity of expressing free and informed consent thereto, in accordance with national law.

Thus, if human biological material is removed to be used for scientific research that may lead to an invention that may be patented, care should be taken to obtain the patient's consent to such use, and that informed consent should if possible be documented.

Recital 27 of the EU Biotechnology Directive further states that where an invention is based on or uses biological material of plant or animal origin "the patent application should, where appropriate, include information on the geographical origin of such material, if known." While the recital continues by indicating that failure to include such information would not prevent a valid patent from being granted, it should be born in mind that the UN Convention on Biological Diversity[35] recognizes national sovereignty over biological resources and that the country of origin may in due course wish to take a share of the benefits flowing from the invention.

1.26.15 **Discoveries**

The exclusions from patentability in the European Patent Convention can be seen to fall into four categories: moral, artificial, nontechnological, and natural. By moral, I mean that it has been felt that it is somehow ethically wrong for certain inventions to be the subject of patent monopolies. By artificial, I mean that something that would otherwise be seen to be a technological invention, e.g., a computer program, is simply excluded from coverage. By nontechnological, I mean that things that solve no technical problem, e.g., aesthetic creations, are excluded. The final category is perhaps more interesting. Things that exist naturally, even if we are currently unaware of their existence, e.g., discoveries, theories, and mathematical laws, are excluded. (The exclusion of 'discoveries' in the European Patent Convention has as an amusing counterpoint the very basis for US patent law, the statement in the US Constitution Article I, Section 8, Clause 8, that "[The Congress shall have the power] To promote the Progress of Science and useful Arts, by securing for limited Times to Authors and Inventors the exclusive Right to their respective Writings and Discoveries."[40]) This exclusion of the natural, however, is interpreted extremely narrowly by the EPO, which takes the stance that if something natural is discovered, then a technical use of it may be considered to be an invention even if that use would be obvious once the discovery was made.

1.26.16 How Are Patents Applied for?

Patents are granted by or for individual countries. For a patent to be granted in country X, an application must be filed with the patent office of country X or with a regional patent office empowered by country X to issue patents on its behalf. If an applicant wishes to have a patent in more than one country, it is usual to file a first patent application at one patent office, often but not always the national patent office in the applicant's home country, and then within 1 year to file one or more national, regional, or international patent applications to cover the set of countries they are interested in. As long as such subsequent applications are filed within 1 year, then in accordance with the provisions of the Paris Convention of 1883[41] any claims in those subsequent applications that are supported by the initial filing, the 'priority application,' are treated as having the filing date of the initial application. There is of course the requirement that the Paris Convention binds the initial and subsequent countries; however, this is the case for most countries of commercial interest.

The history of the adoption of the Paris Convention has interesting parallels with that of TRIPS, and these are at least in part memorialized by the inclusion in the Convention's title of the words 'Industrial Property,' a term that, unlike 'Intellectual Property,' excludes copyright. This history was admirably reported by Penrose in her difficult-to-find book *The Economics of the International Patent System*.[42]

However, for an analysis of the Paris Convention, the reader is referred to Bodenhausen.[43]

There are four regional patent conventions: one covering most of Europe (EP), one covering much of the former Soviet Union (EA), one covering much of French-speaking Africa (OAPI), and one covering much of English-speaking Africa (ARIPO). Under these conventions, a regional patent application, when granted, results in the grant of patents in one or more of the member states.

There is also an international patent treaty, the Patent Cooperation Treaty or PCT,[44] which is administered by the World Intellectual Property Organization (WIPO) in Geneva, using which an applicant can file an international patent application covering some or all of the countries party to the treaty. WIPO, however, cannot grant patents and the applicant must convert his or her PCT application into national or regional patent applications by a deadline, which is usually 30 months after the filing date of the priority application (or after the filing date of the PCT application if no priority is claimed). The PCT thus allows the applicant to delay taking the decision as to which countries to apply for patents in, as well as to delay translation and other costs. As of August 2005, 128 countries are party to the PCT and PCT applications may be converted into regional patent applications at all four regional patent offices. The countries are listed in WIPO[45]; the larger countries not party to the PCT tend to be in South America, the Middle East, and South East Asia and include, for example, Argentina, Bangladesh, Chile, Iran, Iraq, Jordan, Malaysia, Pakistan, Peru, Saudi Arabia, Taiwan, Thailand, and Venezuela.

It is occasionally the case that the applicant is not ready to proceed with foreign patent applications by the end of the year from the initial patent application filing. This might be due to lack of funds, lack of immediate commercial interest in the invention or lack of sufficient data to support an adequately broad patent claim. It is possible to 'restart the clock' for claiming priority by withdrawing and refiling the initial patent application or an improved version of it. This, however, requires that the initial application must not have been used as the basis for a priority claim and that the refiling must take place in the same country as the initial filing (see Article 4c(4) Paris Convention).

1.26.17 European Union Patents

A proposal for an EU Patent, the Community Patent Convention or CPC,[46] has been waiting to be brought into force for 20 years. This would make it possible for patents to be found invalid or infringed in relation to one EU country (more precisely any or all EU countries) either by a central court or by a national court in another EU country. Despite valiant efforts, the CPC has not been brought into effect as of August 2005. SCRIP[47] stated that:

> The European Commission is to try again to push through plans for a single EU patent, although in view of continuing resistance from some member states the signs are not good. Charlie McCreevy, the EU's internal market commissioner told members of the European Parliament he was prepared to make "one more attempt" to

> secure a Community Patent (CP) legally enforceable across all 25 EU member states.... The Community Patent concept has had a rocky ride since it was first mooted some 30 years ago... and its very survival is now at stake. The crunch came last year [2004] when it was formally put on hold after the commission claimed progress was being stalled by "vested interests" in certain member states. Spain, for one, fears its lawyers will lose business from Latin America if the single patent is based on just one or two key EU languages. Spain also wants patents to be translated into all EU languages and for all of them to carry some legal weight in the event of disputes, a suggestion contested by the Germans, among others. German lawyers are also concerned about losing business from the European Patent Office (EPO), which is a non-EU body based in Munich.

We, who are patent attorneys in Europe, continue to hold our breath.

1.26.18 United States (US) Law Changes

A bill, H.R. 2795 the Patent Reform Act of 2005,[48,49] was placed before the US Congress on 7 June 2005 that, if passed, would result in the US adopting the 'first-to-file' system generally accepted by the rest of the world. Prior art, for novelty purposes (cf. Section 1.26.10 above) would become anything: (1) publicly known more than 1 year before the filing or priority date; (2) publicly known before the filing or priority date other than as a result of the inventor's own disclosures; and (3) in a US patent or patent application having an earlier filing or priority date.

The Patent Reform Act of 2005 would also abolish the requirement to disclose the best mode known to the inventor for putting the invention into effect.

Given the reference to inventors in the US Constitution mentioned in Section 1.26.15 above, and the lucrative nature of interference practice referred to in Section 1.26.27 below, rapid acceptance of the Patent Reform Act of 2005 is perhaps unlikely.

1.26.19 Drafting the Patent Application

Section 1.26.3 above explained briefly the usual general structure of a patent specification. Two components are essential: a description of how to perform the invention that is sufficiently detailed that one skilled in the relevant technological field can put the invention into effect without having to make a further invention in order to do so; and a definition of the scope of protection sought.

In practice, drafting the patent application is usually a collaborative effort involving the inventor and a patent attorney (see Myers,[50] which provides a description of the drafting of two patent applications, one by a US patent attorney and the other by a UK patent attorney). The aim is to produce a text that will result in the grant of a patent that will be commercially unfeasible to 'design around' for most if not all of the lifetime of the patent or of the sales period of the product or process it covers. The specification will thus differ in two crucial respects from a scientific paper that might be written based on the same ideas and results: first, it will extrapolate beyond those ideas and results; and second, it will not use references to show that those ideas and results followed logically from earlier publications.

On the first point, the extrapolation is required in order to make the patent difficult or impossible to design around. For this point, it might be appropriate to refer to the invention of alumoxane-activated metallocene catalysts for olefin polymerization. The initial invention was by Kaminsky and the resulting European Patent, EP-B-35242, defined the invention at its broadest as involving the combination of an alumoxane and a cyclopentadienyl group containing zirconocene. As research over the last 20 years has shown, the definition of the metallocene was too narrow – a claim to "a process for the catalysed polymerisation of olefins, characterized in that the catalyst used is an eta-bonded metal ion activated by an alumoxane" would have covered most if not all of the catalysts developed from Kaminsky's initial invention. However, it is easy to be wise with hindsight.

On the second point, the avoidance of a trail of references is in order to present the invention as something that was not obvious. This second point was helpfully commented on by McSherry (2001: 174)[51]:

> [Attorneys] prefer that you make every invention by accident... What the patent attorney's trying to do is establish that there's no mechanism, [that] you couldn't have foreseen this. Which is the exact opposite of the faculty inventor who's trying to establish that their understanding of the mechanism and predictability led to this discovery... That scares patent attorneys to death. People could say "Wait a minute, you mean anyone could have formed this hypothesis based on what Professor Joe Schmoe said in this paper and all you did was test [that idea]?"

Indeed, references in a patent specification are included to perform four functions: (1) to demonstrate that the invention was necessary, i.e., there was a problem waiting to be solved; (2) to reduce the length of the text and hence the drafting and translation costs, thus, for example, a reference to a publication describing the production of a starting material obviates the need to describe its production in detail in the patent specification; (3) to distinguish over prior art likely to be encountered during the examination of the patent application and so reduce the expense and duration of the examination; and (4) in US patent applications, to meet the applicant's duty to inform the USPTO examiner of all relevant prior art.

As far as references in patent specifications are concerned, there are two more minor points to note. The details given must be adequate for the reference to be identified – but any more than the minimum increases translation costs, so expect to find 'Smith *et al. JACS* **111**: 1212–1234 (1999)' (this reference is fictitious) rather than a full list of the authors, the paper's title, and the journal's full name. Where the reference is a patent publication, a two-letter code is used to identify the country, and a one letter code is used to identify whether the publication is a patent application (usually 'A') or a granted patent (usually 'B'). In most cases, patent applications are published about 18 months after their priority or filing dates, whichever is earlier. One exception is US patent applications where there are no foreign equivalents, which can remain unpublished until the US patent is granted. The major country codes the medicinal chemist is likely to encounter are as follows: WO, an international patent application; US, the United States; EP, a European patent or patent application; JP, Japan; DE, Germany; GB, the United Kingdom; CA, Canada; CN, China; and FR, France. GB-B-2012345 is thus the accepted 'shorthand' for granted UK Patent No. 2012345. Most of the country codes are readily recognized; however, a few are quite confusing, e.g., DZ for Algeria and HR for Croatia. A listing by WIPO is available (2005: 24).[45]

Two other requirements may have to be met when the patent specification is drafted. First, for the US, the text must advise the reader of the best way of performing the invention known to the inventor at the time the application was filed. Second, where a biological material, e.g., a microorganism, is required for the performance of the invention and it is not possible to describe in words how to make this material, a deposit of such material must have been made at a recognized culture collection and must be identified in the specification.[34,52,53]

A patent claim can be considered to be a circle, the boundary of which is set by the words used to define the invention – everything within the circle is covered by the claim. The starting point for drafting the claim is usually a point or set of points (e.g., the experimental results relating to the invention), or a smaller or larger circle (e.g., the inventors' concept as they perceive it). The patent attorney's job involves working from that starting point to a definition that will meet the needs of the patent applicant, e.g., one of commercially adequate scope that has sufficient likelihood of being accepted by a patent office, etc. This involves determining which features of the invention are critical, and which are simply preferred.

Thus, the inventor may have found that compound A, previously produced by reacting compounds B and C together, may be produced more efficiently by reacting compounds D and E, at temperature X, in solvent Y, using catalyst Z.

The patent attorney will want to learn whether the conditions used are critical or depend on other parameters such as the reagents, solvent, and catalyst, whether alternative solvents and catalysts may be used, whether useful analogs of compound A may be produced using analogs of compounds D and E, whether the form in which compound A is produced is different from its conventional form, e.g., in purity, molecular weight (e.g., for a polymer), optical properties, etc., and whether the use of catalyst Z to produce materials like compound A was known.

Thus, from this simple example, the broad definitions of the invention could perhaps turn out to be:

1. a process for producing A or A-like compounds by reacting together (a) D or D-like compounds and (b) E or E-like compounds;
2. the use of Z or a Z-like material as a catalyst for the preparation of A or A-like compounds; or
3. A or an A-like material, in the novel form attainable by synthesis from D/D-like and E/E-like reagents.

The features that merely resulted in particularly good results, e.g., the use of temperature X, or solvent Y, will generally be consigned to the dependent claims, i.e., the claims containing narrower definitions of the invention.

In Europe, the EPO prefers the independent claims to be set out in a two-part format, with the features known from the closest prior art in the first part, and the new feature in the second part, the 'characterizing' clause. It must be stressed that this is really only for convenience when considering novelty: when (non)obviousness is considered, the question is whether the claim as a whole represents something inventive and not whether the feature in the characterizing clause is inventive on its own.

In most jurisdictions, where a patent application contains several independent claims, these will only be accepted if the patent office examiner can see that they all contain the same unifying inventive feature, e.g., an independent claim to a set of novel compounds and an independent claim to compositions containing those novel compounds will generally be allowed in the same patent. As a result, particularly where the patent attorney is European, the independent claims will routinely define the core of the invention in the same way. However, in the US, it can be more appropriate to include several overlapping but not coextensive core definitions.

Examples of the claim formats that have been accepted for biotechnology-related inventions may be found in Jaenichen *et al.* (2002).[54]

The wording of the claims should be clear and concise, making it possible for the reader to work out whether something does or does not fall within the definition of the invention. For this reason, 'relative' language is generally to be avoided, e.g., if the claim specifies that a process step is 'performed quickly,' how is the reader to know where the borderline between 'quickly' and 'not quickly' occurs? Where a term is somewhat unclear, this may be acceptable if the rest of the text of the patent application makes it clear. Nevertheless, although it has been said that a patent may serve as its own dictionary, where a term has a clear and well-accepted meaning, this may not necessarily be overridden by an alternative definition in the text.

The patent attorney, who works with the inventor in preparing the draft, will be someone with at least a degree in a science subject, often followed by years of training to become qualified as a patent attorney. In my experience, it usually takes many years before a patent attorney is fully able to draft and prosecute patent applications. This has the downside that the patent attorney's scientific expertise is getting a bit dated by the time he or she is fully competent on the patent side. As a result, the inventor must understand that the attorney will not be as conversant as they are with a new technological field. The inventor must bring the patent attorney up to the level required – this is normal and part of the enjoyment of the job for the patent attorney. Where the patent attorney is already conversant with a new technological field, it can be because he or she has worked in that field for another client and so is 'conflicted' and cannot work for you, i.e., there is a 'conflict of interest,' which due to their professional rules of engagement means they cannot handle your case.

1.26.20 Discovery Methods and Reach-Through Claims

One type of invention for which it is particularly difficult to obtain adequate patent protection is the drug discovery method, i.e., a process by which a potential drug candidate or a model for such a candidate can be identified, e.g., high-throughput screening, panning of phage display libraries, computer simulation of chemical structures and interactions, computer aided molecular design, multivariate analysis, etc. Typically, such methods provide either the identity of the drug candidate or a set of properties for such a drug candidate that may be used to select or design it. The final candidate may turn out to be a known compound and its identity will depend on the specific target used in the method, e.g., a known receptor or enzyme the activity of which is to be stimulated or suppressed. Where the method may be applied to several targets, the inventor may only know the details of the method and not the drug candidates when the patent application is filed; however, after the patent application is published, the application of the method to particular targets may be considered obvious to a patent office examiner and there is then a risk that no valid patent can subsequently be obtained for the drug candidates once they are identified using the method. Since the product obtained directly by the operation of the method is the identity of a drug candidate and not the drug itself, claims to the method may be of little practical value since interested parties can always choose to operate the method outside the countries in which it is patented. The patent applicant will thus wish to obtain grant of a patent covering the final, as yet unknown, commercial product, e.g., using a product claim such as "a compound identified by the new method" or less directly using a process claim such as "a process involving identifying a candidate compound using the new method, optionally from that identifying a functionally analogous compound, manufacturing the candidate or analogous compound, and formulating it as a pharmaceutical composition." Such claims are termed 'reach-through' claims and are highly unpopular with the patent offices, even though the occasional one has been granted.[55,56]

The optimum approach to protecting a drug discovery method invention (and analogous discovery inventions) will depend on the nature of the owner and the nature and location of his or her potential competitors and licensees, and professional advice is strongly recommended. For an academic inventor, where neither she nor the institution intends to operate the method commercially, the approach might be to sell the invention to an existing company or to patent and offer relatively inexpensive nonexclusive licenses. For a company, the decision might instead be to rely on secrecy or to offer licenses for noncompeting targets. Secrecy has the advantage that the chances of obtaining valid patent cover for compounds identified by the method are higher; however, secrecy can be lost and the method may be independently developed and patented by other companies (*see* Section 1.26.40 below).

1.26.21 Patent Attorneys

In many countries there are professional associations of qualified patent attorneys that will be prepared to provide the names and addresses of their members. Such details may also be found in telephone directories and via the internet. For representation before the European Patent Office, the professional organization is the Institute of Professional Representatives before the European Patent Office; in the UK the organization is the Chartered Institute of Patent Agents; in the US a comparable organization is the American Intellectual Property Law Association.

1.26.22 Patent Examination

The world has moved on since Charles Dickens' time when obtaining a patent meant an almost never-ending sequence of paying fees to register one's patent.[57]

In most commercially important countries nowadays, it is not a simple act of registration to obtain a patent. The national or regional patent office will examine a patent application to see whether: (1) formal requirements such as forms, paper size, line spacing and assignment, and authorization are met; (2) the application relates to a single invention; (3) the specification provides an enabling disclosure of the invention; (4) the claims are concise and clear in scope and supported by the rest of the specification; and (5) the invention claimed is new, nonobvious, useful, and not otherwise excluded from patentability.

This involves a three-part examination by such patent offices: formalities; prior art search; and patentability opinion. The formalities bit can be left to your patent attorney. The search reveals to the patent applicant the prior art that the patent office examiner will take into account when deciding whether or not to grant a patent. The patentability opinion will be based on the examiner's decision as to whether the subject matter claimed is new, is nonobvious, is useful, has unity, is sufficiently described in the specification, and is claimed in clear and unambiguous language.

Documents cited by the patent office search as prior art in a PCT or European patent application will generally be categorized as X (potentially novelty-destructive), Y (relevant to the question as to whether the invention is obvious or not), or A (pertaining only to the general technological background for the invention).

On the basis of the search, the patent applicant will generally decide whether or not to pursue his or her claim for a patent – if the prior art wholly anticipates the claimed invention, there is little point in wasting any more money and it may be desirable to abandon the application in such a way that it does not get published. However, it is important for the inventor to be able to assess the citations properly. To my mind, this is best done by having the inventor concentrate on the core features of his or her invention as he or she understands it (rather than as it has been defined in the claims of the application), then having him or her read through each of the citations to see what features of that core are or are not disclosed, and to see whether any citation teaches away from any key features or their combination. With such information in hand, the patent attorney should then be able to say whether or not there is a reasonable chance of convincing a patent office to grant a patent on the application, which covers the essential core of the invention. If the likelihood of a patent being granted is nil, it may be the best option simply to withdraw the patent application so that it (and the inventor's or his or her company's interest in the field) does not get published.

The third stage of examination is for patentability: is the invention claimed new; is it nonobvious; does it have utility; is the language of the claims clear; do the claims relate to a single invention or a group of inventions linked by a common inventive concept; and is the scope of the claim supported by the disclosure of the specification? At this stage, it may be necessary to offer limitations to the claim scope, e.g., to exclude subject matter that is not new or is otherwise excluded from patentability. Any limitations must be based on the text of the patent application as filed (which is why the text is drafted to include potential fallbacks) and the limitations will generally be selected to ensure that the limited claim remains difficult to design around to produce a commercially competitive product or process or to ensure that cover of a product or process sold or under development by a competitor or licensee is maintained. To some extent the selection of candidate limitations is based on an understanding of the relevant technology; to a large extent, however, it is based on an understanding of the activities and capabilities of competitors and licensees. At this stage it may also be necessary to submit arguments and possibly evidence to demonstrate that a piece of prior art is not novelty-destructive or does not point toward the claimed invention, to demonstrate that the advantages claimed for the invention are real and significant, or to demonstrate that the text does provide an adequate teaching of how the invention can be performed. Here, input from the inventors, or others skilled in the technical field, is of primary importance, and such scientists may be required to submit Declarations, or to appear as experts or witnesses at hearings.

The order in which the different stages of examination occur varies from country to country.

Once a patent office has accepted that a patent application meets all the formal requirements and that the claims are acceptable in language and scope, the application generally is accepted and a patent is granted. As usual, the formalities and costs involved in grant vary depending on the patent office responsible. Currently, for a European patent application, the grant stage is a relatively expensive exercise with the patent owner being required to submit translations of the patent in local official languages to virtually all of the national patent offices in which the European Patent is to enter into force. There have been attempts to remove the local translation requirement, in particular the London Agreement on the Application of Article 65 EPC of 17 October 2000[58]; however, at the time of writing, this has not yet occurred. For the London Agreement to enter into force, eight countries, including the UK, France, and Germany, must ratify or accede to it; by August 2005, six countries, not including France, had done so. Removal of the translation requirement is a politically sensitive area: small enterprises might find themselves infringing a patent in their home territory when the legal document that defines the protected monopoly is not available in a local language; national courts and lawyers may be asked to enforce a property right that is not defined in a local language; and national patent attorney professions may lose a major source of their income (patent translation) and hence the maintenance of a viable local patent attorney profession may be put into jeopardy.

After grant, patents may be open to attack by interested parties. This may be through the national Courts or the national or regional patent offices. Thus, for example, in the European Patent Office, granted European Patents may be opposed by such interested parties in the nine months following grant, while in the US interested parties may request that a granted US patent be re-examined by the USPTO. In the US, there is a further possible hurdle flowing from the 'first-to-invent' concept that underpins US patent law: thus, where an accepted US patent application is considered to claim the same or closely similar invention to that of a granted US patent, the USPTO may declare an 'interference' in order to determine which party is entitled to a patent for the invention. This is discussed further briefly in Section 1.26.27 below.

1.26.23 Divisional Applications

Where a patent office has argued that a patent application covers more than one distinct invention, and where the applicant has been unable to persuade the patent office that this is incorrect, then the claims of the application relating to all but one such invention may be deleted by the applicant and patent protection for the deleted subject matter may be sought by filing a new patent application, a 'divisional' application. The text and the effective dates of the divisional are usually the same as for the parent application and a divisional is usually filed before the parent proceeds to grant.

1.26.24 Continuation Applications and Submarines

Among the delights of the US patent system is the 'continuation' application (covered by 35 U.S.C. 120) and its brother in-law the 'continuation-in-part' (covered by 35 U.S.C. 132(b)) in which new material may be introduced into the patent application text. Such applications can be filed at any time during the pendency of the parent application and can be used to wear down a USPTO examiner, to gain further patent cover, or simply to buy time for the applicant. Until June 1995, they also were a technique by which the term of a patent could be shifted forward so as to cover a more economically beneficial period, i.e., when the patentee's product or a competitor's product was actually on the market, since the term of a US patent was 17 years from grant rather than the current 20 years from filing. Lemley *et al.* (2003: 1–2)[59] introduce the topic of continuation applications rather nicely:

> One of the oddest things to an outsider about the United States patent system is that it is impossible for the U.S. Patent and Trademark Office (PTO) ever to finally reject a patent application. While patent examiners can refuse to allow an applicant's claims to ownership of a particular invention, and can even issue what are misleadingly called "Final Rejections," the patent applicant always gets another chance to persuade the patent examiner to change her mind. Even stranger, perhaps, is that the PTO can't even finally grant a patent. Even when the examiner concludes that an invention is patentable and issues a "notice of allowance," the patent applicant always retains the right to abandon the application that was deemed patentable and start the process over again. Alternatively, an applicant can take the patent the PTO has awarded and at the same time seek additional or broader claims arising out of the same application. In all three cases, the culprit is what is known as the "continuation" application. Applicants dissatisfied with the course of patent prosecution can abandon an application and file a continuation. Alternatively, a patentee can prosecute one or more patents to issue and also keep a continuation application on file, hoping to win a better patent from the PTO in the future.

Since, until the mid 1990s, US patent applications were not open to public inspection until the patent was granted, and indeed are still kept secret where no corresponding application is being filed outside the US, a granted US patent could appear 'out of the blue' with claims covering products and processes already in commercial use in the US by others. These patents, which could emerge from nowhere and sink your business, were known as 'submarine' patents. Again to quote from Lemley *et al.* (2003: 13–14)[59]:

> While some applicants file continuation applications in order to have a further opportunity to persuade the PTO to issue the claims they originally sought, others file continuation applications in order to have an opportunity to modify their claims. Some are innocuous... Other explanations are more problematic... Inventors can keep an application pending in the PTO for years, all the while monitoring developments in the marketplace. They can then draft claims that they can be sure will cover those developments. In the most extreme cases, patent applicants add claims during the continuation process to cover ideas they never thought of themselves, but instead learned from a competitor. The most egregious example is Jerome Lemelson, who regularly rewrote claims over the decades his patents were in prosecution in order to cover technologies developed long after he first filed his applications. Lemelson filed eight of the ten continuation patents with the longest delays in prosecution in our study. These Lemelson patents spent anywhere from thirty-eight to more than forty-four years in the PTO.

Lemley *et al.* studied the US patents issued between 1976 and 2000 and found that, while the average time from application to grant was about 2 years, over 20 000 took more than 10 years and one took 52 years (see Lemley *et al.*, 2003: 59–60).[59]

Lemley *et al.* (2003: 39)[59] were scathing in their comments on continuations, "In short, there are not many good reasons for society to allow continuation applications. The most that can be said in their defense is that patent prosecutors will have a somewhat harder job if continuations are abolished"; however, they are an extremely useful way of obtaining broader patent cover.

Lemelson's patent tactics are described by Morando *et al.* [60] and examples of his patents include US-A-5249045, US-A-5281079, US-A-5283641, US-A-5351078, US-A-5491591, US-A-5570992, and US-A-5966457.

1.26.25 Deadlines

The world of patents is bedeviled with deadlines. With the 'first-to-file' system in countries other than the US, the first deadline is the date of filing of the initial or priority application. Any delay increases the risk that another party may file first and so reduce the chances that the applicant will obtain as broad patent cover as he or she wishes. Once the initial patent application is filed, this triggers the 1-year deadline for filing corresponding patent applications for other countries party to the Paris Convention.

In some cases, the patent attorney may be so pressed for time, for example because the inventor is disclosing the invention at a conference the same day, that he or she will choose to file the priority application in a country in a different time zone.

Deadlines are set in days, weeks, months, or years. If the priority application was filed on 5 August 2005, the 1-year deadline is 5 August 2006, unless this is a holiday in the country in which the subsequent application is filed (a '*dies non*'), in which case the deadline is the following working day. This calculation of deadlines is the norm for patent applications, but it should be noted that a 4-month deadline from 31 May is 30 September, and that for responses to objections set by letters from the European Patent Office (rather than to meet deadlines set by the European Patent Convention itself) the deadline is calculated 10 days from the date on the letter. This may seem a trivial matter; however, in practice any extra time to meet a deadline is gratefully received by both the applicant and their patent attorney. In some, but not all cases, particularly where a patent office has set a deadline for responding to objections to the patentability of the subject matter claimed, it is possible for an extension of a month or so to be obtained; whether this is feasible, whether it incurs a fine, and how long an extension is possible varies among patent offices. In some cases, where a deadline is missed, it is possible to avoid the patent application lapsing. This is possible, for example, where a deadline for responding to objections regarding patentability raised by the European Patent Office is missed, e.g., by invoking the procedure known as 'further processing'; however, as a general rule, deadlines must be accepted as being final unless you have been advised otherwise by the patent attorney.

1.26.26 Status

Increasingly, patent offices are making it possible for interested parties to check the status of pending patent applications and granted patents on-line, and even in some cases to view on-line the documents on the patent office file for published patent applications and granted patents. The relevant websites for the European Patent Office, the USPTO, the UK Patent Office, the Japanese Patent Office, and WIPO are included in **Table 1**.

The USPTO also provides a website where the texts of granted US patents and published US patent applications may be searched (see **Table 1**). This is particularly useful for 'quick-and-dirty' searches to see whether an idea has already been covered by a US patent or application.

1.26.27 Interference and Notebook Policy

It is not that unusual for scientists, working independently, to make the same invention at about the same time and for each to file patent applications directed to that invention. Since patent applications are generally only published about 18 months after their priority dates, the potential conflict between patent applications for the same invention does not always become immediately clear.

Take the example where inventor A conceives of his or her invention on day A1, tries it out on day A2, files their initial patent application on day A3, files their application in the country in which the conflict occurs on day A4 (for the sake of simplicity here assume they use the same text as on A3), and first publishes their invention on day A5; again for the sake of simplicity assume that this is as a result of one of A's patent applications being published at about A3 plus 18 months. For inventor B, code the equivalent days B1, B2, B3, and so on but assume that A3 precedes B3, i.e., A has the earlier priority date. In either case, A3 and A4 (or B3 and B4) could be the same day where the initial application is the application in the country of conflict, and the invention may not have been tried out so day A2 (or B2) may not exist.

For Europe, where the 'first-to-file' system operates and as long as A's application in the country of conflict is published, then if A3 precedes B3, B cannot validly obtain grant of a patent in the conflict country for claims that lack novelty over the disclosure of A's application. If A files no application in the conflict country, or if their application in the conflict country is withdrawn before publication, then B can still obtain a valid patent in the conflict country as long as B3 precedes A5.

Where the texts used by A and B are different, then B may still be able to craft a novel claim if B3 precedes A5; however, if A5 precedes B3 then A's text is prior art for the determination of both novelty and nonobviousness.

For the US, as long as B3 (or B4 if B3 and B4 are the same) is no more than 1 year after A5, then if B1 precedes A1, B potentially has the right to the US patent to the invention since the US operates the 'first-to-invent' system. B may be unable to benefit from their earlier conception date, however, if he or she has abandoned their invention between B1 and B2/B3/B4 or if he or she has not been diligent in their development of the invention between those dates.

Thus, if it is found, for example by the USPTO, that a granted US patent and an accepted US patent application seem to claim the same invention, an 'interference' is declared to determine which of the inventors conceived of the invention first and whether the inventors were diligent in 'reducing their inventions to practice' either by actually performing the invention or by filing a patent application directed to it.

It will be appreciated that this determination involves submission of evidence as to the activities of the inventors, and for this reason it is important for scientists in the US and elsewhere to document their activities in a manner that will make such evidence credible.

One way to achieve this is to require researchers to note down their ideas and experiments, both intended and performed, on a regular basis, preferably daily, in a format that is not susceptible to later amendment and also to provide a separate potential witness besides the scientist himself. One of the simplest ways of achieving this is to use bound notebooks, with numbered pages, which are completed leaving no blank spaces into which material could later be inserted, and which are read and 'signed off' on a regular basis by a supervisor or colleague.

Such a notebook policy was not essential outside the US until the mid 1990s, as the US previously recognized only activities within the US in determining when conception occurred and whether diligence was sufficient. In that respect, it is possible that the US was in breach of its 'national treatment' obligations under the Paris Convention (see Article 2 Paris Convention).

1.26.28 Maintenance Fees

Before certain national and regional patent offices it is necessary to pay fees to maintain a patent application before it is granted. If these, which are usually due annually, are not paid, then the application lapses or is deemed withdrawn. Likewise, most national patent offices require maintenance fees to be paid annually to maintain a granted patent in force. The US requires maintenance fees for granted patents, but not on an annual basis.

These maintenance fees represent a significant fraction of the income of the patent offices and may, in certain cases and to some extent, be seen as a way of subsidizing the costs of examination for patentability. Moreover, they serve to limit the number of patents and patent applications a company has to consider when determining whether or not activities it is or may become involved with might infringe the patent rights of others.

In some countries, for example the UK, a patentee may reduce the size of the maintenance fees by agreeing in advance to grant licenses to any interested party.

1.26.29 So, What Is an Invention?

The question as to what is an invention for is to some extent clarified by the legal definitions of what is a patentable invention, which clearly accept that the terms invention and patentable are not synonymous. Thus, for example, Article 54 EPC, in reciting the conditions under which an invention is considered to be new, implicitly confirms that something may not be new but may still be an invention. Likewise, US interference practice does not always operate to grant the patent rights to the first to invent, implicitly accepting that the same invention may be made at different times by different people. By stating in Article 52(2) EPC that "The following [e.g., discoveries, aesthetic creations, etc.] ... shall not be *regarded* as inventions" for which patents may be granted, it is again tacitly acknowledged that these excluded objects or actions may be regarded as inventions for other purposes. The question as to whether or not something is an invention thus appears to be a subjective rather than an objective one; an invention may simply be something the creator or someone else believes to be new. The child going to their mother with a contraption made of cardboard, sticky tape, and empty washing-up bottles and saying "Mummy, look at my invention" may be correct in concluding that they have made an invention, albeit not necessarily a patentable one.

The research scientist should not therefore be too concerned as to whether her creation is an invention or not – the relevant questions are simply "is it patentable?" and "is it worth patenting?"

Moreover, since two of the most common justifications for granting patents are that the act of inventing should be rewarded and that, in accordance with natural law, by virtue of his or her labors the creator should own their creations, it is perhaps therefore not unjust that patents may be granted to an inventor (e.g., the first to file their patent application) rather than only to the first inventor.

1.26.30 One-Size-Fits-All

Starting in the 1970s, the US began to use trade-related laws to address the 'problem' of 'free-riding' or 'intellectual piracy,' i.e., where other countries allowed the local use of US-derived intellectual property without payment to the creator. Thus, in 1984, the US designated inadequate protection of patents, trademarks, and copyrights as an unfair trade practice that could invoke retaliation under Section 301 of the Trade Act of 1974 (Maskus, 2000: 1).[8] This led to bilateral agreements that allowed stronger protection of US-deriving intellectual property, to the inclusion in the North American Free Trade Association Treaty[61] of required minimum levels of protection for intellectual property rights, and in due course to the 'one-size-fits-all' requirement for minimal levels of IPR protection at a first world level for all those countries that wanted to enjoy the trade relationships mandated by GATT (now WTO). The relevant provisions of NAFTA appear in Article 1709, paragraphs 1 to 3:

1. ...each Party shall make patents available for any inventions, whether products or processes, in all fields of technology, provided that such inventions are new, result from an inventive step and are capable of industrial application. For the purposes of this Article, a Party may deem the terms "inventive step" and "capable of industrial application" to be synonymous with the terms "non-obvious" and "useful," respectively.
2. A Party may exclude from patentability inventions if preventing in its territory the commercial exploitation of the inventions is necessary to protect ordre public or morality, including to protect human, animal or plant life or health or to avoid serious prejudice to nature or the environment, provided that the exclusion is not based solely on the ground that the Party prohibits commercial exploitation in its territory of the subject matter of the patent.
3. A Party may also exclude from patentability:

   (a) diagnostic, therapeutic and surgical methods for the treatment of humans or animals;
   (b) plants and animals other than microorganisms; and
   (c) essentially biological processes for the production of plants or animals, other than non-biological and microbiological processes for such production.

4. Notwithstanding subparagraph (b), each Party shall provide for the protection of plant varieties through patents, an effective scheme of sui generis protection, or both.

Up to the 1970s, the levels of IPR protection now required under TRIPS have been rejected by developed countries such as the US, Japan, Germany, the UK, and so on, in view of their then level of industrial development or due to political, economic, or other pressures. Economically, without appropriate safeguards, such levels of IPR protection have been seen to be of doubtful value. Socially, the development is potentially catastrophic in terms of the possibility of affordable healthcare being removed from availability, as cheap generic drugs become a thing of the past.

IP laws, as with most other laws, are usually enacted on a national basis on the assumption that their enactment is in the present and future interest of the citizens of the state in question. They are property laws, i.e, laws that restrict access of the citizens to something they might otherwise have had access to. Thus, unless the decision to bring in laws lies only with the powerful and wealthy, in order for such laws to be beneficial they must benefit the citizens, now or in the future.

The possible benefit from IP laws (to the technology importing countries), unless accepted as negative in a trade-off against some other benefit (as perhaps was thought to be the case with TRIPS), lies in accessible knowledge, the increase in creativity in general, the increase in technology transfer, and the increase in foreign direct investment in such countries. In contrast, the dangers include the increased cost or reduced availability (for example, due to unmeetable prices) of technology, in particular life-saving or standard of living-improving technologies, the setting in stone of creativity, and the crystallization of technology importer/exporter status.

Historically, countries have used a variety of techniques to balance the potential benefits of permitting personal property rights in abstract objects (i.e., IPR) with the potential damages. These have included: the ability to grant compulsory licenses; the ability to require inward technology transfer or the local production or performance of the invention; the ability to grant relatively short-term or narrow scope IPR; the ability to exclude from protection or enforceability industries or activities seen as critical (e.g., the food or drug industries or the actions of the medical profession); or the ability to require a share in the resultant profits, nationally or internationally.

Of these, to my mind, the key counterbalances are the ability to grant compulsory licenses and the ability to require inward technology transfer. In this way, the availability of the protected invention to the citizenry may be guaranteed and the development of local industry may be promoted so avoiding de facto 'permanent' status as an industrially backward country.

Compulsory licenses, where the state forces a patentee to grant a licence to another party, are a means by which unduly negative effects of granting a monopoly may be tempered, e.g., where the patentee is not working the invention in the state, where it is supplying the invention at a level less than the demand, or where it is charging more for the use of the invention than the citizens can afford. For luxury goods or goods and services that are pleasant but inessential, such negative effects are perhaps not that important; however, where the invention has a life-saving or productivity maintaining effect as is the case for many drugs, being told that the invention will be available in a cheap generic form when the patent expires is no comfort to a dying patient or their family. TRIPS severely constrained the conditions under which WTO member states could grant compulsory licenses, stating in Article 31 TRIPS:

> Where the law of a Member allows for other use of the subject matter of a patent without the authorization of the right holder..., the following provisions shall be respected:... (b) such use may only be permitted if, prior to such use, the proposed user has made efforts to obtain authorization from the right holder... This requirement may be waived by a Member in the case of a national emergency or other circumstances of extreme urgency or in cases of public, non-commercial use.... [and] (f) any such use shall be authorized predominantly for the supply of the domestic market of the Member authorizing such use...

Is a life-threatening or productivity-threatening disease (e.g., AIDS) that has been widespread within a country for a prolonged period a national 'emergency' or a circumstance of extreme 'urgency'?

The developing countries that signed up to TRIPS apparently in order to achieve better access to first world markets for their exports, belatedly came to appreciate that agreeing to patent protection for pharmaceuticals and limiting access to compulsory licenses as provided for in Article 31 TRIPS threatened the availability to their citizens of affordable, life-saving drugs. In paragraph 6 of the Doha Declaration on the TRIPS Agreement and Public Health of 14 November 2001 (see Pires de Carvalho, 2005: 507–508),[16] relating to the subsequent round of negotiations relating to GATT/WTO, it was stated that:

> We recognize that WTO Members with insufficient or no manufacturing capacities in the pharmaceutical sector could face difficulties in making effective use of compulsory licensing under the TRIPS Agreement. We instruct the Council for TRIPS to find an expeditious solution to this problem and to report to the General Council before the end of 2002.

This is discussed further in Section 1.26.31 below.

The history and operation of compulsory patent license systems is reviewed by Salamolard[62]; however, compulsory licenses played an interesting part in the discussions leading up to the adoption of the Paris Convention in the nineteenth century. Thus, Penrose (1951: 51–52)[42] states:

> On the principle of expropriation in the public interest, the [Paris Conference of 1878]... admitted the right of the state to revoke a patent if manufacture was not undertaken in the country in a specified time, at least if the patentee could not justify his inaction. In other words, the conference approved what is now known as compulsory working – revocation of the patent if it was not worked. Even compulsory licensing as the sanction for failure to work was rejected.... It seems curious, however, that in the 1878 conference, where the rights of inventors were upheld in the most extreme forms, the principle of compulsory licensing for failure to work was rejected in favor of outright revocation of the patent. This position was the direct result of the natural property theory of patents adopted by the conference. M. Charles Lyon-Caën, a prominent French lawyer, argued heatedly that compulsory licensing was a derogation of the right of property and compared the inventor subject to such licensing to a man who owned his house but was required to allow all who requested it, to live with him on the payment of a rent. Compulsory licensing only aided other private groups... The conference accepted M. Lyon-Caën's point of view and approved the resolution that: Patents ought to assure, during their entire duration, to inventors or their assigns, the exclusive right to the exploitation of the invention and not a mere right to a royalty paid to them by third parties.

The other fundamental counterbalance mentioned above was a requirement for technology transfer. Historically, this has been set out as a requirement that a patentee must work their invention in the patented country within a certain period of time or risk the patent being revoked or compulsorarily licensed. In this way, a set of the population of the country learned how to effect the invention, a skill they would be able to use competitively once the patent expired. However, a local working requirement is inherently economically unattractive to multinational companies (making your product in one country and exporting to one hundred is likely to be cheaper than making it in each country merely to meet the local demand), and a 'compromise' is to require technology to be transferred to the patent-granting country rather than to require local manufacture of the specific patented product. TRIPS pays lip service to this in Article 7 TRIPS where it is stated that:

> The protection and enforcement of intellectual property rights should contribute to the promotion of technological innovation and to the transfer and dissemination of technology, to the mutual advantage of producers and users of technological knowledge and in a manner conducive to social and economic welfare, and to a balance of rights and obligations.

However, there are no teeth to this provision, a provision that Pires de Carvalho (2005: 130)[16] points out relates to the objectives of IPR rather than an objective of TRIPS. According to Pires de Carvalho:

> Patent protection does promote the dissemination of technology... technology dissemination takes place in an automatic fashion just by making the contents of patent documents available to the public. Today, millions of patent documents are available on the Internet, and comprehensive searches can be performed at a cost close to zero by accessing the websites of the major patent offices around the world. Dissemination of technology may also occur in consequence of the spreading of the knowledge that results from the general availability of patented products, the use of which leads to the technical education of consumers.

If this is indeed the correct interpretation of Article 7 TRIPS, then the Article was surely somewhat redundant.

Moreover, WIPO even seems to think that enabling a country to promulgate first-world style patent laws meets the requirement to promote technology transfer.

As mentioned above, one national defense to the effects of IPR being granted which could, at the relevant time, be politically, socially or economically undesirable, has been to limit the range of subject matter that can be patented, e.g., by excluding from patentability drugs, food, or methods of medical treatment. While TRIPS has not broadened the range of patentable subject matter to equal that in the US, it has required member countries to allow subject matter of a range comparable to that accepted under the European Patent Convention to be patented (*see* Section 1.26.6 above). Thus, for most countries, complying with their obligations under TRIPS means giving up this line of defense. An alternative facet of this line of defense has been to limit the range of actions that can be the subject of patent infringement actions, i.e., actions that may be covered by valid patent claims but in relation to which the patentee cannot ask the Courts to prevent continuing infringement or to award damages for past infringement. One example of this is the exclusion from infringement of the actions of a pharmacist in the extemporaneous production of a pharmaceutical composition that is found in Section 60(5) (c) of the UK Patents Act 1977,[63] another is the exclusion from actionable infringement of the action of a medic in the treatment of his or her patients that found its way into US patent law in 1996 (see 35 U.S.C. 287(c)). Of particular relevance to the research chemist, whether in industry or academia, however, is the extent to which the use of patented inventions in research is permitted. This is discussed in Section 39.

As discussed in Section 1.26.36, the optimum economic return might theoretically be achieved were the scope and duration of patent claims to be tailored to the nature of the invention covered and the field of technology in which it is useful, e.g., allowing only short-term cover in a new and rapidly developing field. This has, to some extent albeit indirectly, been an approach historically adopted as a way to avoid nationally undesired effects of strong IPR, with longer patent terms and broader claims or claims to a broader range of technology becoming more acceptable as the relative level of industrial development of a country has increased. The ability to limit patent term downwards, however, has disappeared with TRIPS, which in Article 33 mandates a minimum patent term of 20 years and the ability to discriminate among technological fields has also been excluded by TRIPS, which states in Article 27(1) that, subject to certain limitations, "patents shall be available and patent rights enjoyable without discrimination as to the place of invention, the field of technology and whether products are imported or locally produced." (This Article in TRIPS also disposed of the possibility of a local working requirement discussed earlier in this section.)

The final category of defensive measures against the possible use of the patent system against national interests that I listed at the start of this section was the ability to require a share in the profits made from an invention, either locally or internationally.

One development in this area relates to the UN Convention on Biological Diversity[35] and its provisions relating to traditional knowledge. Thus, Article 8 CBD states that:

> Each Contracting Party shall, as far as possible and as appropriate:...
> (j) Subject to its national legislation, respect, preserve and maintain knowledge, innovations and practices of indigenous and local communities embodying traditional lifestyles relevant for the conservation and sustainable use of biological diversity and promote their wider application with the approval and involvement of the holders of such knowledge, innovations and practices and encourage the equitable sharing of the benefits arising from the utilization of such knowledge, innovations and practices;...

The 'equitable sharing of benefits,' is achievable by the recognition under the CBD that countries have 'sovereign rights' over biological resources and traditional knowledge, in the same way that they have long been acknowledged to have rights over physical property such as hydrocarbons and minerals. This is set out in Article 15 CBD as follows:

1. Recognizing the sovereign rights of States over their natural resources, the authority to determine access to genetic resources rests with the national governments and is subject to national legislation....
2. Access, where granted, shall be on mutually agreed terms and subject to the provisions of this Article.
3. Access to genetic resources shall be subject to prior informed consent of the Contracting Party providing such resources, unless otherwise determined by that Party.

The benefit sharing is also intended to include inward technology transfer (see Article 16 CBD).

1.26.31 Doha and Agreement on Trade-Related Aspects of Intellectual Property Rights (TRIPS)-Plus

The Uruguay round of multilateral trade negotiations led to TRIPS and the establishment of WTO and of a system that authorizes trade sanctions against countries that do not meet their treaty obligations, e.g., the obligations relating to IPR set out in TRIPS. The current negotiation round is the 'Doha round' and in this many countries have realized that they should address the question of affordable access to medicines, in particular medicines for the treatment of diseases prevalent in developing countries, such as AIDS, malaria, and various tropical diseases.

Prior to TRIPS, many developing countries could obtain pharmaceuticals at relatively low prices since there were established manufacturers of generics in countries such as India, which did not provide patent protection for pharmaceuticals as such, and, where they too did not permit pharmaceutical per se protection, the products could be imported without infringing patents. The activities of such manufacturers of generics drove the prices of the products down to a level where significantly more of the affected population could be treated.

With the current version of TRIPS, however, such developing countries may in theory be able to authorize compulsory licenses to permit local manufacture of drugs covered by local patents, but this is of little value to a country with no manufacturing capability and with a population too small or too poor to justify establishment of such capacity. One goal is therefore to confirm that compulsory licenses can be issued to companies to manufacture in other countries for export to the affected country. The General Council of the WTO approved this aim on 30 August 2003 (see Pires de Carvalho, 2005: 513–517),[16] and in May 2004 Canada became the first country to amend its patent law to permit compulsory licenses for export of pharmaceuticals.[64] The change in Canadian law took effect in May 2005.

Efforts still continue, however, to persuade countries to adopt stronger IPR protection than absolutely required by TRIPS. One example is in the field of data exclusivity where countries are being pressed to legislate a monopoly period of protection for the data used in obtaining regulatory approval for pharmaceuticals, in effect a de facto extension of patent term since a generic manufacturer is forced to undergo the expense of clinical trials or wait until the period of exclusivity terminates. Moreover, such trials are questionable ethically since they involve exposing volunteers to potential harm merely to generate information that is already in the hands of the regulatory authorities.

Data exclusivity is mentioned in Article 39(3) TRIPS as follows:

> Members, when requiring, as a condition of approving the marketing of pharmaceutical or of agricultural chemical products which utilize new chemical entities, the submission of undisclosed test or other data, the origination of which involves considerable effort, shall protect such data against unfair commercial use. In addition, Members shall protect such data against disclosure, except where necessary to protect the public, or unless steps are taken to ensure that the data are protected against unfair commercial use.

The use to be protected against, however, is 'commercial' use and national drug regulatory authorities are not commercial operations. Thus, Article 39(3) TRIPS does not seem to prevent a national drug regulatory authority from using the fact that it or another recognized regulatory authority has received sufficient data to justify issuing a product license as the basis for requiring that a further product licence request from another party, e.g., a generic manufacturer, need only be supported by data showing bioequivalence. In this way, the further party would not be in possession of or able to use the undisclosed data, rapid appearance onto the market of competing versions of the same drugs (once the patent has expired) can be assured hence increasing the availability of affordable drugs, and unnecessary human or animal experiments may be avoided.

1.26.32 Piracy and Biopiracy

> In 1740 George [later Lord] Anson set out from England under orders from King George to harass and distress the Spanish coast of South America. Starting with a fleet of eight old, rotting ships and a crew composed mostly of inexperienced farm boys pressed into service, lunatics and invalids from Navy hospitals, he returned to England nearly four years later – seven ships lost, decimated by scurvy, but having circumnavigated the globe and captured one of the fabulously wealthy Spanish galleons on the way. Walter (1974: rear cover).[65]

Piracy has been an acceptable part of the past of some of today's most developed countries. Nonetheless, much of the rhetoric behind the recent drive toward globally uniform strong IPR protection has been based on an analogy between the copying of knowledge and piracy, i.e., the theft of real objects. This analogy of course is flawed: theft of a real object deprives its owner of the use or enjoyment of that object, while what is lost when knowledge is copied is the ability to sell goods or services at profit in places where such a profit was not feasible due to the freedom of others to compete openly.

Advocates of national sovereignty over biological resources and traditional knowledge have copied this line of argument to justify international recognition of a property right in such subject matter, referring to its use without payment as theft or biopiracy. They have argued that those countries in which, over generations, traditional knowledge has been developed and handed on or in which biodiversity has been fostered have a right to a share in the profits made when profitable inventions are made based on the conserved biological resources or the preserved traditional knowledge. This debate has used emotive language and has not been helped either by the narrow interpretation of the exclusion from patentability of discoveries by patent offices such as the EPO or the exclusion from the prior art to be considered in the US of unpublished knowledge or use outside the US. Particular examples of acts considered to be 'biopiracy' have included the development of an appetite suppressant based on a chemical found in the plant Hoodia, used by San tribes people in Namibia, by the UK firm Phytopharm, and the patenting of fungicidal effects of neem oil, from the tree *Azadirachta indica* from India, e.g., in EP-B-436257 and US Patent 5124349.[66]

Nonetheless, CBD exists and we can expect to see developments requiring patentees to identify the source of any biological materials or traditional knowledge used in the development of their inventions, and possibly also to share profits with the country of origin. Scientists are thus advised to document any such use of biological resources and traditional knowledge so as to be able to meet these requirements if they arise.

1.26.33 Witch-Hunting

The nature of the world today is that we can't send the marines into a country that we may have an intellectual property dispute with... US Commissioner of Patents and Trademarks, Bruce Lehman (Lehman, 1995)[67]

As mentioned above, and as discussed in the literature,[8,18,19] the recent explosion in global upwards-harmonization, i.e., imposition of more IPR standards that are favorable to the rights owner, has largely been driven by unilateral, bilateral, and multilateral actions and discussions in relation to international trade. One of the most powerful tools in this regard has been Section 301 of the US Trade Act 1974, now Chapter 12 of Title 19 of the United States Code. Under 19 USC 2242, the US Trade Representative annually lists those countries that, in the opinion of the US, deny "adequate and effective intellectual property rights protection." In such annual listings, one of the points taken into account is whether the country is "making sufficient progress in bilateral or multilateral negotiations to provide" such IPR protection. In accordance with 19 USC 2411, a country might be found to be guilty of providing inadequate or ineffective IPR protection even if it is complying with its obligations under TRIPS, i.e., signing up to TRIPS has not removed the threat of trade sanctions. The sanctions that may be imposed include, for example, suspension of trade agreement concessions, imposition of increased import dues or of import quotas, and limitation or removal of duty free status for imports.[68] For many countries, the US is a critical export market and the threat of these sanctions is a matter to be taken seriously. However, it may interest the reader to learn that the countries listed by the US Trade Representative in 2005 included Israel and Russia (on the priority watch list) and Canada, Italy, and the EU as a whole (on the watch list).[69]

1.26.34 Ownership of Inventions

In general, a creation (e.g., a literary or artistic work or an invention) belongs to the creator unless ownership is overridden by law or by an agreement, e.g., an assignment or a contract of employment.

In the UK, the relevant provision of the UK Patents Act is Section 39, which states that:

1. Notwithstanding anything in any rule of law, an invention made by an employee shall, as between him and his employer, be taken to belong to his employer for the purposes of this Act and all other purposes if –

   (a) it was made in the course of the normal duties of the employee or in the course of duties falling outside his normal duties, but specifically assigned to him, and the circumstances in either case were such that an invention might reasonably be expected to result from the carrying out of his duties; or
   (b) the invention was made in the course of the duties of the employee and, at the time of making the invention, because of the nature of his duties and the particular responsibilities arising from the nature of his duties he had a special obligation to further the interests of the employer's undertaking.

2. Any other invention made by an employee shall, as between him and his employer, be taken for those purposes to belong to the employee.

However, the UK Patents Act goes on to confirm in Section 42 that contracts of employment cannot act to vest rights in an invention in the employer where the making of the invention was not part of the duties of the employee referred to in Section 39(1). Thus, for example, in the UK if a cleaner in a factory making drugs invents a new drug, the ownership will stay with the cleaner.

Statutory provisions transferring rights from employed researchers in industry generally seem reasonable. Problems do arise where academics and other external collaborators make inventions, e.g., consultants and employees of other companies involved in a joint venture, and especially with academic medics who may simultaneously hold positions as industry consultants, as university lecturers, and as healthcare organization employees. Since negotiations regarding ownership and remuneration are significantly more tense once a commercially viable invention has been made, it is therefore highly desirable that the position concerning ownership and remuneration is agreed clearly and unambiguously in advance where external collaborators are involved.[70] This may involve negotiations with more than one of the organizations with which the external collaborator is involved and may require the collaborator's various different employment or consultation agreements to be carefully reviewed.

The position relating to inventions made by academics is especially fraught, in view of the academic's need to publish in order to forward their professional career and of the tradition in many countries that academics owned their own results/inventions and were free to sell or license them or donate them to the public as they saw fit. This tradition, in recent years, has been coming to an end as countries have sought to derive revenue from or promote commercial exploitation of academic inventions. Thus, in recent years, Denmark, Norway, and Germany have enacted laws vesting ownership rights in academic inventions in the universities, and the US has passed the Bayh-Dole Act[71] encouraging out-licensing of academic inventions. Indeed, it appears that the only major European Union countries in which academics still own their inventions may be Sweden and Italy.

Nevertheless, the position is still complicated, with some universities apparently claiming a broader ownership of rights to academic inventions than the statute laws justify. Thus, for example, the Statutes and Regulations of the University of Oxford[72] state, in Section 5 of Statute XVI, that:

> The University claims ownership of all intellectual property specified in Section 6 of this statute which is devised, made, or created: (a) by persons employed by the University in the course of their employment; (b) by student members in the course of or incidentally to their studies; (c) by other persons engaged in study or research in the University who, as a condition of their being granted access to the University's premises or facilities, have agreed in writing that this Part shall apply to them; and (d) by persons engaged by the University under contracts for services during the course of or incidentally to that engagement.

The Section 6 referred to lists patentable (and nonpatentable) inventions, among other intellectual property. Thus, ownership of the rights to students' inventions is allegedly automatically vested in the university, a position that is clearly in conflict with the provisions of the UK Patents Act as discussed above and one which is contradicted by statements made on the website of the UK Patent Office.[73]

Similarly, the relevant statute law in Norway, the "Lov om retten til oppfinnelser som er gjort av arbeidstakere,"[74] allows employers, e.g., a university, to claim ownership of inventions that are patentable in Norway, thus apparently leaving the inventor with the ownership of inventions that are unpatentable in Norway (e.g., due to lack of novelty in accordance with the provisions of Norwegian law, or due to explicit exclusion from patentability) but still patentable elsewhere, i.e., in the US in particular.

It should also be borne in mind that many university researchers, e.g., undergraduates carrying out research projects, are not employees of the university and that rights to inventions that they may make will not be transferred to the university by provisions relating to employees. Parties to any joint or collaborative research and development project should therefore take care to sort out at the beginning of the project who will own any IPR resulting from the project.

Although Universities are normally considered to be relatively 'benevolent' institutions, where inventions have been made that are covered by contracts entered into by the Universities or that are potentially major sources of revenue, a benevolent attitude can not always be guaranteed. The story of the University of South Florida and Taborsky is mentioned in McSherry (2001: 231–232)[51]:

> In a 1992 University of South Florida case, a student was sent to a work camp for patenting in his own name research conducted under industrial contract. Peter Taborsky, an undergraduate in biology and chemistry... allegedly "disappeared" with several research notebooks and used information in them to file and win two patents... USF was able to have him sentenced to three and a half years of work camp for his "theft" of the notebooks.

The story is told at greater length by Shulman.[75] Taborsky was released early, after about 1 year.

In general, in the US, ownership is governed by contract law and contracts may contain terms more or less favorable to the inventor/employee/consultant than might be the case in the UK.

The position regarding academic inventions is discussed further by McSherry (supra), Monotti *et al.*,[76] and Janssens[77] while examples of articles discussing the recent change in German law and the effects of the Bayh-Dole Act in the US include those by Leistner [78] and Nelson.[79]

1.26.35 Property Rights in the Products of Labor

In terms of how property rights can or should come into being in relation to products of physical or mental labor, e.g., ideas or objects, it is clear that most products of labor build on existing knowledge or physical material available to all or to a subset of the population, the 'commons.' Isaac Newton is quoted as saying in a letter of 5 February 1676 to Robert Hooke that if he had seen further, it was "by standing on the shoulders of giants." The commons include all knowledge that is publicly accessible; however, it could also be seen to include 'discoveries,' i.e., knowledge that is simply waiting to be discovered such as gene sequences, the identity of a bioactive compound in a plant, or a mathematical law.

Where access is open to all, those commons are inclusive; where access is only to a limited group, e.g., members of a particular tribe or citizens of a particular country, they are exclusive. Where the property right requires the consent of the others having access to the commons it is a positive right, but where no consent is required it is a negative right (i.e., as under Natural Law as proposed by John Locke).[80]

Drahos[81] thus divided the possible systems for deriving property from the use of the commons into four (exclusive negative, exclusive positive, inclusive negative, and inclusive positive) and considered which system provided the most efficient way of encouraging the development of new knowledge. He concluded that this was the inclusive positive system, which is indeed the system underpinning most patent laws. The acknowledgement in the UN Convention on Biological Diversity of national sovereignty in relation to biological resources and Traditional Knowledge, however, seems more to be founded on the exclusive positive system.

Philosophical analysis of the basis for rights to intellectual property has identified three candidates: Natural Law, i.e., by their labor the creator owns their creation; IPR as a reward for creating; and Utilitarianism, i.e., a system that operates to produce the maximum good for the maximum number.

Natural Law is quite simply not the basis of patent law as it is. While lip service has been paid to it in the past, especially in continental Europe, it is inconsistent with a requirement for patents to be applied for and for their terms to be limited. The reward basis has its echoes in the US patent law and perhaps also in the system of inventors' certificates in the old Soviet Union, but the value of the reward is not matched to the value of the invention. However, while the utilitarianism pattern of thought seems to be the one that dominates, for nationally applicable laws it should surely match national conditions and needs – something that a globally uniform strong IPR system, applicable irrespective of the wealth or developmental stage of a country, does not appear to do. As Sterckx[12] signals in the title of her article, patenting has uneasy ethical justifications.

1.26.36 Economics

In his Foreword to Penrose[42] (1951: vii), Machlup stated that "Judging from the share which the subject of patents has had in the literary output of economists of the last fifty years... one may say that economists have virtually relinquished the field. Patent lawyers were probably glad to see them go." He continued, "One may safely predict that many members of the American patent bar... will intensely dislike some of the views, and perhaps all of the suggestions contained in this book."

Writing in 2000, Maskus[8] (2000, 81) suggests that the situation has not yet changed radically:

> Despite the obvious practical importance of the question, economists did not attempt until the 1990s to assess empirically the effects of international variations in the strengths of IPRs. Without such evidence, the field lay open to strong claims on both sides of the debate – a situation that remains largely true today. ... The situation was even more stark than the earlier acute shortage of information on how patent systems affect innovation and welfare, which had prompted Priest[82] to lament, "The ratio of empirical demonstration to assumption in this literature must be very close to zero."

The most prevalent justification for the granting of patents is that it is in the public interest to stimulate technological progress and development and the addition to the store of technological knowledge and that this can be achieved by offering the carrot of wealth; this has been referred to as a 'consequentialist' or 'utilitarian' justification.

This, or equivalent lines of justification or argument, may be found in Venetian law from 1474, in the US Constitution,[40] in proposals from the European Commission[83], in statements by patent commentators (e.g., Chisum, 1997: Vol. I, Chapt. III, Section 3.01, pp 3–3),[30] and in comments by people such as Jeremy Bentham and John Stuart Mill, the founding fathers of utilitarianism, and Abraham Lincoln, respectively:

> ...he who has no hope that he shall reap will not take the trouble to sow. Bentham (1785)[84]

> ...if all were at once allowed to avail themselves of [the inventor's] ingenuity, without having shared the labours and expenses [incurred]..., such labours and expenses would be undergone by nobody, except very opulent and very public-spirited persons ... Mill (1866)[85]

The patent system ... added the fuel of interest to the fire of genius in discovery and production of new and useful things. Lincoln (1953: 363)[86]

The argument is that the incentive of a property right is required for the timely development of new knowledge and that the price paid by the populace by the grant of that right, e.g., the constraints placed on the populace by the grant of patents, are outweighed by the benefit flowing from the new knowledge. The new knowledge can be viewed as a public good and three options for encouraging its production were offered by David (1993: 27–29)[87] who stated that:

One [mechanism] is that society should give independent producers of public goods publically financed subsidies [e.g., research grants or prizes] and require that the goods be made available to the public freely or at a nominal charge. A second mechanism would have the state levy general taxes to finance its direct participation in production and distribution of the good, furnish and manage the requisite facilities, and contract when necessary with private agents to carry out the work... again with the objective... to supply the good without having to charge prices for it [e.g., by funding state-owned research and development organizations]. The third solution is to create a publicly regulated private monopoly authorized to charge consumers prices that will secure a "normal" rate of profit [i.e., an IPR system].

This leaves four questions to be answered in relation to the third solution, a patent system:

1. is the development of new knowledge sufficiently desirable to allow access to it to involve paying prices higher than would be the case if it were immediately open to copying?;
2. would the new knowledge not have developed in a sufficiently timely fashion in the absence of the incentive of a property right?;
3. is the extent of the right granted commensurate with the public value of the new knowledge, i.e., is the cost too high?; and
4. can the grant of the property right cause undue undesired effects, either currently, such as duplication of investment and hence waste of resources, or in the future, such as disincentives to the generation of yet further new knowledge?

Regarding the first question Maskus (2000: 28–29)[8] explains:

Because intellectual property is based on information, it bears the traits of a public good in two separate but important ways. First, it is *nonrivalrous*: one person's use of it does not diminish another's use. Consider a new means of production, a musical composition, a brand name, or a computer program. All may be used or enjoyed by multiple individuals. In this context, it is optimal in a static sense to permit wide access to intellectual property. Indeed, the public interest is extreme in that the marginal cost of providing another blueprint, diskette, or videotape to an additional user may be low or zero. Unlike the case of physical property, a multiplicity of users does not raise congestion costs in the exploitation of intellectual property. The second characteristic is that intellectual property may be *non-excludable* through private means: it may not be possible to prevent others from using the information without authorization. If an intellectual effort is potentially valuable but easily copied or used by others, there will be free riding by second comers. In turn, there may be no incentive to incur the costs of creating intellectual property. Society has a dynamic interest in avoiding this outcome by providing defined property rights in information. ... The fundamental trade-off in setting IPRs is inescapable. On the one hand static efficiency requires wide access to users at marginal social cost, which may be quite low. On the other hand, dynamic efficiency requires incentives to invest in new information for which social value exceeds development costs. These are both legitimate public goals, yet there is a clear conflict between them.

In other words, the answer to the first question seems to be yes, as long as the extent of protection afforded (the negative in the static sense) reflects the benefit the public receives (the positive in the dynamic sense).

The chemical, pharmaceutical, and biotechnological industries appear to be virtually the only ones for which empirical evidence suggests that the answer to the second question given above is an unqualified 'no,' and even then the answer results from a survey of companies operating in fields where virtually all patentable inventions derive from developed countries with strong patent systems and ferocious regulatory systems. Thus, Maskus (2000: 43)[8] states that:

Mansfield[88] sampled 100 firms in 12 US manufacturing industries on their views of whether patents are important to their decisions about investment in innovation. His results suggested that only in the pharmaceutical and chemical industries were patents considered essential; here more than 30 percent of the inventions would not have been developed without potential protection. In these sectors, fixed costs of R&D are high and imitation is fairly easy. In three industries…, patents were seen as important in the development of 10 to 20 percent of inventions; in the other seven industries, patents were viewed as unimportant or only marginally significant in inducing R&D.

More specifically, the study by Mansfield[88] stated that in the pharmaceutical sector 65% of the inventions would not have been brought to the market had patent protection not been available.

Before turning to the third and fourth questions, both concerned with costs and benefits, it is perhaps worthwhile taking a jaundiced look at what the potential benefits are, and to whom they accrue, both historically and currently.

There seem to be three categories of potential beneficiaries: individuals in their own right; individuals viewed collectively as the society as a whole; and the state. Companies may be beneficiaries, but the benefits obtained flow through to one or other of the three categories, e.g., owners and employees benefit as individuals in their own right. Thus, the first category may perhaps be subdivided into owners and employees. Of course, an individual may be a small-scale owner by virtue of a pension fund or he or she may be both an owner and an employee.

Put bluntly: owners benefit through income, current or potential; employees (including state employees paid from increased national taxes) benefit through income (salary); all individuals, i.e., the individuals viewed collectively as a society, benefit through access (although possibly delayed access or access at an unaffordable price) to new technology and the potential to derive income from the other categories or to reduce income lost through taxation to support the other categories; and the state benefits through increased tax income or reduced tax spending to support unemployed, poor, or retired citizens, e.g., by virtue of increased employment, improved health, better personal pension provisions, and increased income, sales, or import taxes.

Since the large-scale owners are few in number, the cost–benefit balance is clearly skewed in the average citizen's favor where the benefits from the access to the new technology, or its derivatives, are high (e.g., life saving) and it is affordable, and where the technology is highly revenue-generating for his or her country.

The cost–benefit balance was relatively clearly weighted on the benefit side in the days before large-scale international trade and currently remains so for net technology exporting states; however, for net technology importing states the position is less clear; and even for net technology exporting states, the balance may not remain favorable if manufacturing and services relocate to countries where costs are lower.

Returning to the third and fourth questions above, these both relate to whether or not the cost of granting a property right exceeds the benefit. This point is discussed by Dutfield *et al.* (2004: 379–380)[89] who comment that:

> The intellectual property systems of Europe and the United States suffer from an identity crisis and lack of consistency in the sense that much of the rhetoric behind strong protection does not match the reality of what industrial and cultural innovation are really about…. The contradiction lies in the fact that by granting [the proponents of strong protection].. such strong protection in the form of, *inter alia*, stringent compulsory licence provisions and narrowly defined exemptions, we actually hinder follow-on innovation of both cumulative and breakthrough varieties. To put it another way, we face the "innovation dilemma," which is that to protect cumulative innovation, we require a low threshold. But this low threshold, which benefits the innovators of today, may hinder the innovation of tomorrow.

More specifically, Dutfield *et al.*[89] identify three problems that have arisen with the patenting of biotechnology: first, the enormous numbers of patents being granted, e.g., on genes and gene fragments, raises the cost of doing research because of the need to take licenses; second, the scope of the patent claims may be so broad as to cover applications undreamed of by the applicant, so providing a disincentive to further research; and third, there is a concentration of patent ownership in the hands of a small number of players, so hindering market access by new players. One aspect of the first of these problems has been called the "tragedy of the anticommons,"[90] a situation where increasing patenting of 'upstream research' or research tools may stifle the development of life-saving innovations further downstream in research and development (see Dutfield *et al.*, 2004: 404).[89] In relation to the second problem, Dutfield *et al.* (2004: 405)[89] exemplify this with reference to "US patent 5,159,135 awarded in 1992 to Agracetus for *all* transgenic cotton [where t]he patent claims covered any variety of cotton produced by any gene transfer technology."

1.26.37 Patent Term Extension

In certain countries it is possible to extend the term of various IPR.[91] Thus, it is generally possible to renew trademark registrations indefinitely and copyright in a few books has been extended beyond its normal term in some countries (e.g., J. M. Barrie's copyright in Peter Pan, which was given to Great Ormond Street Hospital in 1929 and should have expired in 1987), and currently patent terms for pharmaceuticals or agrochemicals may be extended, e.g., in the EU and the US to compensate for the time taken in obtaining regulatory approval before the product can be marketed and the resultant reduction in the period during which the manufacturer can use the existence of the patent to recoup investment costs before generic competition begins. There is currently a proposal[92] for a special patent term extension in the EU for pediatric formulations for drugs in order to encourage development of such formulations.

In the US, patent term extension is governed by 35 U.S.C. 155–156.

In the EU, patent term extension is achieved by virtue of a supplementary protection certificate (SPC), the requirements for which are set out in Regulations Nos. 1768/92 and 1610/96 (see appendices in Vitoria *et al.*[63]). SPCs serve to extend the protection conferred by the patent in relation to the product for which regulatory approval has been obtained and must be applied for nationally within 6 months of patent grant or local regulatory approval, whichever is later. The SPC extends the patent protection by 5 years from the date of first regulatory approval within the EU. A 'hiccup' in the system recently came to light when it was realized that first regulatory approval in the EU could be the approval from Switzerland, despite the fact that Switzerland is not an EU member state, since Swiss approval extended to Liechtenstein, which is an EU member state. This has been dealt with by amendment of Swiss regulatory approval so that it will not extend to Liechtenstein.[93]

1.26.38 **Infringement**

What Actions may be actionable infringements of patents, i.e., actions that the patentee may ask the Courts to have stopped, are governed by national laws and vary from country to country. However, it cannot be assumed that because the action is not done for profit, e.g., where it forms part of an academic's research or is performed by a not-for-profit organization, it is not an actionable infringement. The topic of what is permitted in the course of research is addressed in Section 1.26.39 below. It is also important here to correct two common fallacies relating to patent infringement: thus, it is not the case that because something is patented it cannot infringe another patent and it is not the case that because something was invented later than the patent then it cannot infringe.

In an infringement action, the relief the patentee may seek is typically a court order (an injunction) requiring the infringer to stop or not to start the infringing activity, to pay damages, and in some countries also to pay the patentee's legal costs.

As mentioned earlier, patents cover either things (products) or ways of doing things (processes). Infringement of a product claim might typically involve importing, making, keeping, using or disposing of (e.g., exporting, selling, giving away, loaning, or leasing) the product, while infringement of a process claim typically involves performing the process. Beyond this, infringement may involve offering to do any of these actions.

In the case of process claims, it may also be an infringement to import, keep, use, or dispose of the product of the process.

Actionable infringement (or the equivalent) may also occur where there is no actual involvement with the subject matter as claimed. This typically arises in three different circumstances: where the action relates to something used to put the invention into effect, e.g., a reagent for a process or a component for a product; where the action involves encouraging or conspiring with others to cause an infringement; and where the alleged infringement is substantially similar to the claimed product or process but doesn't exactly fall within the definition provided by the claims.

The first of these is termed "contributory infringement" and is covered, for example, by Section 60(2) and (3) of the UK Patents Act which state:

> (2) Subject to the following provisions of this section, a person (other than the proprietor of the patent) also infringes a patent for an invention if, while the patent is in force and without the consent of the proprietor, he supplies or offers to supply in the United Kingdom a person other than a licensee or other person entitled to work the invention with any of the means, relating to an essential element of the invention, for putting the invention into effect when he knows, or it is obvious to a reasonable person in the circumstances, that those means are suitable for putting, and are intended to put, the invention into effect in the United Kingdom.
>
> (3) Subsection (2) above shall not apply to the supply or offer of a staple commercial product unless the supply or the offer is made for the purpose of inducing the person supplied, or as the case may be, the person to whom the offer is made to do an act which constitutes infringement of the patent...

and by Section 271(c) of the US Patent Law which states:

> Whoever offers to sell or sells within the United States or imports into the United States a component of a patented machine, manufacture, combination, or composition, or a material or apparatus for use in practicing a patented process, constituting a material part of the invention, knowing the same to be especially made or

especially adapted for use in an infringement of such patent, and not a staple article or commodity of commerce suitable for substantial non-infringing use, shall be liable as a contributory infringer.

The third of these categories is frequently referred to by the term used in US patent law in relation to the construction of claim language to determine whether or not such a product or process infringes, the 'Doctrine of Equivalents.' The key UK and US Court decisions on this subject are Improver (1990)[94] and Festo (2002),[95] respectively; however perhaps the easiest way to illustrate this system by which a claim may be read to cover something that on a literal reading it does not is by reference to the UK Court (House of Lords) decision in Catnic (1982)[96] where a claim to a box-shaped lintel in which one component was required to be "extending vertically" was held to be infringed by an equivalent lintel in which the component was a few degrees off vertical yet still functioned in the same way.

In other words, around the periphery of a patent claim is a 'gray area' in which activity may still be considered to infringe. This vagueness to the periphery of the claims is explicitly sanctioned by the 'Protocol on the Interpretation of Article 69 of the [European Patent] Convention,' which requires that the scope of protection conferred by a European Patent should not be determined either by a strict literal interpretation of the words used in the claims nor by relying only on the claims as a guideline. This Protocol was necessary in order to facilitate a common approach to determining the scope of protection across the various countries covered by the EPC, countries which had adopted very different approaches.

An infringement action may even relate to a product or component deriving from the patentee or its licensee. Where that product was supplied with conditions attached, e.g., that it must not be resold, or where it was placed on the market in a different country, then reselling or importing may constitute infringement. Similarly, buying the reagents required for a patented process from the patentee may not give the purchaser a license to perform the patented process unless such a license was explicit or implicit. In various countries or groups of countries a principle of 'exhaustion of rights' does exist, whereby importation or resale of a product placed on the market by the patentee or with their consent prevents the patentee from claiming patent infringement. This is the case for products sold in one country of the EU and imported into and resold in another EU state and is currently of concern in particular to pharmaceutical companies that sell at different prices in different EU states. Certain companies are reported to have adopted or to be threatening to adopt policies of refusing to supply wholesalers in certain EU countries in order to reduce the profit losses arising from the resultant 'parallel importation.'[97] The supply of products with conditions attached relating to resale or reuse will have been experienced by most users of computer software but is also of concern to purchasers of seeds, semen, or animals covered by patents as such conditions frequently forbid practices traditionally practiced by farmers, e.g., planting some of the seed from the crop or selling or exchanging such seed to or with neighboring farmers. Shulman[75] tells the story of a farmer who encountered problems with such patents.

Normally, patent infringement is dealt with under civil law, i.e., the defendant generally faces a financial penalty or a court order to cease infringing; however, in June 2005 the European Commission proposed to bring certain IPR infringements under criminal law, with the possibility of a prison sentence of up to 4 years.[98]

1.26.39 Research and Development Exemptions

Where research, either academic or commercial, involves use of a patented invention for the purposes for which it was developed, e.g., the use of PCR to produce multiple copies of polynucleic acids or the use of a novel form of mass spectrometer to analyze molecular masses, the fact that the work being done is 'research' will not generally provide a defense to a claim of patent infringement. Where the product of such work is sold, licensed, or given away to others when it might have been sold to them by the patentee or his or her licensees makes the defense even less likely to succeed.

Where research is being carried out only to determine whether or not an alleged invention actually works, and the product is not used in any way, the likelihood of infringement being found is less.

Where the work is being carried out in order to obtain regulatory approval for a patented product, where sales are to take place once the patent has expired, the situation varies from country to country and the position should be checked with a patent attorney before such trials are started. In the US, 35 U.S.C. 271(e) (1) permits data generation for a submission for regulatory approval, while 35 U.S.C. 271(e) (2) makes the act of submission for regulatory approval an infringement. The question of generating data for regulatory approval is discussed by Gilat[99] and O'Gorman.[91]

1.26.40 The Secrecy Trap

Where a company invents a process and uses it to produce products that it sells but keeps the details of the process secret, then unless the details can be worked out by analysis of the product, the process is still novel for patent purposes and could be patented by the company or by a competitor that independently invents the process. The original company, however, will not be able to patent the process in the US on an application filed more than 1 year after it first started selling or offering to sell the product in the US.

If the competitor patents the process, then the original company will infringe the competitor's patents if it uses the process in the patent country or exports the products into the patent country unless the laws of that country provide that continuing to carry out an action begun before the patent was applied for is not an actionable infringement. Not all patent laws provide such 'prior user rights' and it can be the case that the original company, because it chose secrecy, will have to stop making or exporting the product, even if it has been doing so for years.

1.26.41 Morality and Other Reasons for Denying Patentability

> Last month the [US] government granted its first patent on something that can look you in the eye. Is this a small step for a mouse or a giant leap backward for mankind? New Republic (1988)[100]

The reasons, good or bad, for denying patentability of certain inventions can perhaps be summarized as follows:

1. It is not in the national interest, e.g., for economic reasons, to grant patents.
2. The invention can be encouraged better by another system, e.g., a different IPR, for example for plant varieties or nontechnological inventions such as literary or artistic works, or a reward or state-funding system.
3. The invention is an obvious danger to society, e.g., a better letter bomb.
4. It is in the national interest not to place barriers in the way of those looking after its members' health and safety and thus a medic (or fireman, etc.) should not be hindered from carrying out his or her duty to look after the welfare of members of society.
5. Natural phenomena, whether or not yet recognized, are part of our common heritage and should remain in the intellectual commons.
6. In some way it seems morally wrong to a significant portion of the public to grant a property right in the invention, for example, of a human or a process of genetic manipulation (i.e., such activities have a flavor of 'playing God').

Categories 1–5 seem clear, or at least clearly definable and thus excludable from patentability if desired.

However, category 6 is not clear – a patent does not grant ownership of all the things falling within its claims to the patentee, so where is the problem? Are there things that quite simply should not be patented, i.e., made the subject of a property right? At one extreme the answer is that a patent does not give ownership of the subject matter it covers to the patentee and does not give the patentee the right to put the invention into practice. The logical result of this is the Diamond vs. Chakrabarty ruling of the US Supreme Court[21] to the effect that anything new, nonobvious, and useful should be patentable. Even the US Patent and Trademark Office has shied away from this extreme, refusing to grant patent claims that cover humans on the basis that this would be contrary to the 13th amendment to the US Constitution, which, in Section 1, reads "Neither slavery nor involuntary servitude, except as a punishment for crime whereof the party shall have been duly convicted, shall exist within the United States, or any place subject to their jurisdiction."

A better cause for supporting some ethical limitation to the range of subject matter that should be patentable was expressed by the UK barrister Daniel Alexander QC (Alexander, 2000: 310),[101] where he said that he "believes that there are many people... who may think that it is one thing to engage in this kind of research [i.e., research involving animal testing], but it is quite another thing for a public body [i.e., a patent office] to go about putting its imprimatur on it, unless it can be shown that the relevant research will not go on without a patent." In other words, since a patent is a document produced by or on behalf of the state, issuing a patent for a particular technology carries a strong implication that that technology is ethical. Similar references to the official "imprimatur" given by patents are made and referred to by Chisum (1997, Vol. 2, Section 4.04).[30] Alexander also said that meeting the ethical requirements of the EPC has resulted in it being said by the patent office, time and time again, that "we are not qualified to do that, why do you place the burden upon us?" While expressing sympathy, Alexander said that while they, the examiners, may not be "well qualified to make these decisions... [t]he question may then be raised 'who else should?' [and that the]... importance for this particular purpose is that it is only in that kind of institution [i.e., a patent office] that these issues may be addressed on a sufficiently case-by-case basis [and that]... an institution such as the EPO is in fact not unsuited, but might be particularly suited to making decisions of this kind and ought to do so" (see Alexander, 2000: 255–256).[101]

Article 6(2) (d) of the EU Biotech Directive and the Transgenic Animals/HARVARD decision,[38] explicitly instruct the EPO to consider a cost–benefit approach to the ethical question of patentability when it comes to genetically modified animals. Thus, the EPO now seems to be required to perform some ethical analysis. However, how should a particular technology be determined to be ethically acceptable? Are the views of the scientists in the field all that count, should the views of ethicists be canvassed, and are the views of lay people of any relevance? Or should it be left to patent office examiners? The ferocity of the public debate regarding genetically modified crops, animal testing, stem cell research, and abortion makes it clear that the views of scientific 'experts' alone, even if they are unanimous, may for good or bad reasons not be acceptable to large swathes of the public; the reactions in the past to Gallileo and Darwin make it clear that this is nothing new and that what is acceptable may change with time. Recent research papers have strongly suggested that on ethical decisions relating to the implementation, and hence perhaps also the patenting, of divisive technologies, the opinions of the lay public should be canvassed and considered seriously, rather than simply rejected as 'uninformed.' After all, scientific experts have no particular authority when it comes to making *moral* judgments.[102,103]

One approach could be to require the examiner to submit any ethically doubtful cases to an internal review board, and for such a review board to refer any new questions, or any old questions where the old answer is challenged, to a broader review by a panel of ethicists. However, how should such a board or panel address the question?

The final decision is a simple yes (patentable) or no (not patentable) one. The submissions made on either side will generally be a combination of facts (proved, provable, or simply alleged), judgments (i.e., interpretations of the facts or attempts to influence the decision makers), ideology (after all, patents are property rights, and property rights limit the freedom of others), and principles. The first task of a review board would seem to be to separate facts from (pre)judgment, in order to identify the extent to which the 'facts' are proven, or simply based on belief, e.g., as a result of 'scientific/technological optimism,' and to identify the principles at issue and which will be violated by the decision going one way or another.

The assessment would then have to split into two separate approaches, which represent the two main approaches in the field of ethics or moral philosophy. First, a moral cost – benefit analysis, essentially a 'utilitarian' approach – which option renders the greater good? Second, a principle-matching exercise involving prioritizing the principles at issue which will or will not be violated by the decision to grant or refuse to grant a patent in terms of their importance (e.g., it is bad to steal food, but it is worse not to steal from someone with an excess if by failing to do so you or your family will starve), essentially a 'deontological' approach.

If both approaches result in the same answer, the decision is clear. If one points toward a yes, and the other toward a no, then the decision will have to be taken as to which is the 'superior' approach. It is quite likely that in practice the utilitarian approach will triumph over the deontological unless an issue with particular religious significance is involved.

One example of the type of principle that may be invoked is that 'man should not play God' response to genetic manipulation. This is nicely discussed in more secular terms by Myskja (2006)[102] when he refers to the Greek concept of hubris:

> The crux of this argument [against transgenic modification]... is not primarily avoidance of unpredictable harm, but respect for the "otherness" of nature.... The fundamental lesson of the Greek myths about hubris is found in this lack of respect. Man believes he can be equal to the gods, but lacks understanding of the rules set by them. He transgresses the limits set by these rules and is severely punished for his lack of respect for that which is beyond his understanding.

If we revisit the Transgenic Animals/HARVARD case then, what questions might have been posed and what principles taken into account?

Before looking at those questions, it may be helpful to set out the independent claims of the patent (European Patent No. 169672) as finally maintained by the EPO in its Appeal Board decision T0315/03 Transgenic Animals/ HARVARD of 6 July 2004[38]:

> 1. A method for producing a transgenic mouse having an increased probability of developing neoplasms, said method comprising chromosomally incorporating an activated oncogene sequence into the genome of a mouse.
>
> 19. A transgenic mouse whose germ cells and somatic cells contain an activated oncogene sequence as a result of chromosomal incorporation into the animal genome, or into the genome of an ancestor of the animal...

It may also be useful to mention the criteria adopted by the EPO, namely those set out in the Appeal Board decision T0019/90[104] issued earlier in respect of the same case, and in Rule 23d(d) EPC. These require three matters to be taken into account: the suffering of the animal covered by the patent claim; the medical benefit claimed to result from the invention; and the correspondence between the two. In T0315/03, the following conclusions were reached:

1. Any patent should only extend to those animals whose suffering is balanced by a medical benefit.
2. The level of proof required in relation to the suffering and the benefit is the same – a likelihood.
3. The degree of suffering does not need to be taken into account.
4. The decision must be made in relation to the position at the filing (or priority) date of the patent application, i.e., while "[e]vidence need not be limited to that available at the filing or priority date.... evidence becoming available thereafter must be restricted to the position at that date."

The requirement in Article 53(a) EPC that patents should not be granted for inventions where their exploitation is contrary to morality, implies that performing an invention that has been patented in Europe is viewed by the state as morally acceptable. This is in agreement with the intuition among members of the public that the act of granting a patent in some way confirms that the relevant technology is officially sanctioned, it has the state's 'imprimatur' as Alexander (2000: 310)[101] and others have said. However, where the invention is a genetically modified animal, performance of the invention implicitly requires animal experimentation. Since animals capable of striving for survival or avoiding suffering also have interests (though some may deny this), such experimentation, even seen from a utilitarian standpoint, is moral only if the interests of society (e.g., the provision of safe drugs) outweigh the interests of the animals in question. That animals have the right not to be mistreated is already acknowledged by existing laws forbidding cruelty to animals, etc. As early as 1831, Marshall Hall, an English physician, outlined five principles that should be considered before animal experiments are carried out:

1. An experiment should never be performed if the necessary information could be obtained by observations.
2. No experiment should be performed without a clearly defined and obtainable objective.
3. Scientists should be well-informed about the work of their predecessors and peers in order to avoid unnecessary repetition of an experiment.
4. Justifiable experiments should be carried out with the least possible infliction of suffering (often through the use of lower, less sentient animals).
5. Every experiment should be performed under circumstances that would provide the clearest possible results, thereby diminishing the need for repetition of experiments.

Principles 3 and 5 are essentially the same – if the information is clear, there is no moral justification for replicating animal experiments. Principle 1 perhaps could be brought up to date by stating instead that the experiment should not be performed if the information can be obtained by an experiment that does not involve animals, e.g., an in vitro test, or which involves animals of a species less capable of experiencing suffering. Principle 2 could perhaps be reformulated as a requirement that the possibility of a beneficial result and the magnitude of the potential benefit together must be weighed against the likelihood and severity of the suffering or death of the animal, i.e., unnecessary, trivial, or scientifically dubious experiments should be avoided. Here, any patent office examiner or review board, unless they have access to independent advice, would have to take into account that in a patent application the applicant will tend to be optimistic as to the likelihood and magnitude of the possible benefits of an invention.

The questions to be posed by the patent office examiner thus seem to be:

1. Is there an alternative technique that could be used that does not require animal experimentation or that involves animals of a species less capable of experiencing suffering?
2. What is the extent and likelihood of suffering that will occur?
3. What is the nature of the potential benefit (e.g., relief from disease, pain, or other suffering, or improved attractiveness, pleasure, or lifespan)?
4. How likely is it that the potential benefit will be achieved?

The principles used to reach an answer on the question as to whether exploitation of the invention is or is not moral might be the modified innocence principle or the principle of utility outlined by Jamieson.[105]

The task of reviewing whether an invention meets the requirements of morality, and hence of patentability thus does not seem to be beyond the capabilities of a competent patent office examiner, as long as he or she is entitled to refer more problematic cases to a patent office review board.

1.26.42 Further Forms of Intellectual Property Right

While patents serve to protect concepts, other IPRs exist that serve to protect presentation or reputation, i.e., the form in which an object or process is presented or the manner in which it is linked to its source. Examples of the first category include copyright, design, and, to some extent, plant variety protection; examples of the second category include trademarks, unfair competition/passing off, and geographical indications.

Clearly, these other IPR are more relevant to objects and processes that have already been developed, and indeed they can serve to strengthen or extend the protection that might be provided by a patent. Thus, by registering as a trademark the combination of colors used on a pill or capsule or on an inhaler, a pharmaceutical company can use the good will generated amongst the users of its products during the period of patent protection to encourage continuing purchase of those products once the patent has expired. Likewise, by registering the design of a user-friendly or eye-catching container, competitors unable to use that design for a generic equivalent are placed at a market disadvantage.

Such other forms of IPR are beyond the scope of this chapter; however, they are discussed in texts such as those by Cornish *et al.* [106] and Jacob *et al.*[107]

Plant variety protection is normally achieved under the International Convention for the Protection of New Varieties of Plants, the UPOV Convention[108] and UPOV.[109] In the US, which is a member of UPOV, plants may additionally be protected by plant patents (see 35 U.S.C. 161–164). The European position is discussed by Van der Kooij.[110]

References

1. *The Times (International Edition)*, Aug 8, 2005, p 45.
2. Abrahams, M. *IgNobel Prizes*; Orion Books: London, UK, 2002.
2a. Genentech Inc's Patent [1972] RPC 147.
3. Vaver, D., Ed. *Intellectual Property in the New Millennium*; Cambridge University Press: Cambridge, UK, 2004.
4. Tufts Center for the Study of Drug Development. *Tufts Center for the Study of Drug Development Pegs Cost of New Prescription Medicine at $802 Million*, Press Release, 30 November, 2001.
5. Goozner, M. *The $800 Million Pill*; University of California Press, 2004.
6. Global Alliance for TB Drug Development. The Economics of TB Drug Development, 2001. http://www.tballiance.org (accessed April 2006).
7. US. H.R. 417 The Medical Innovation Prize Act of 2005. http://www.govtrack.us/congress/billtext.xpd?bill = h109-417 (accessed April 2006).
8. Maskus, K. E. *Intellectual Property Rights in the Global Economy*; Institute for International Economics, 2000.
9. Granstrand, O., Ed. *Economics, Law and Intellectual Property*; Kluwer, 2003.
10. Landes, W. M. et al. *The Economic Structure of Intellectual Property Law*; The Belknap Press of Harvard University Press: Cambridge, MA, 2003.
11. Scotchmer, S. *Innovation and Incentives*; MIT Press, 2004.
12. Sterckx, S. In *Death of Patents*; Drahos, ed.; Lawtext Publishing: Witney, UK, 2005; Chapter 6.
13. Fisher, M. *I.P.Q.* **2004**, 85–113.
14. MacLeod, C. *Inventing the Industrial Revolution. The English Patent System, 1660–1800*; Cambridge University Press, 1988.
15. Gervais, D. *The TRIPS Agreement: Drafting History and Analysis*, 2nd ed.; Sweet & Maxwell, 2003.
16. Pires de Carvalho, N. *The TRIPS Regime of Patent Rights*, 2nd ed.; Kluwer, 2005.
17. UNCTAD-ICTSD. *Resource Book on TRIPS Development*; Cambridge University Press: Cambridge, UK, 2005.
18. Drahos, P. et al. *Information Feudalism*; Earthscan, 2002.
19. Sell, S. K. *Private Power, Public Law*; Cambridge University Press, 2003.
20. Blackstone, W. *Commentaries on the Laws of England*; University of Chicago Press: Chicago, USA, 1765–1769; Vol. 2, p 346.
21. Diamond vs. Chakrabarty. 447 US 303, 1980. http://caselaw.lp.findlaw.com/scripts/getcase.pl?court = us&vol = 447&invol = 303 (accessed April 2006).
22. The Holy Bible – New Revised Standard Version; Anglicized Edition; Oxford University Press, Oxford, UK, 1995.
23. EPO. *Convention on the Grant of European Patents (European Patent Convention)*, 11th ed.; EPO: Munich, Germany, 2002.
24. Singer, M. et al. *European Patent Convention*, 3rd ed.; Carl Heymanns Verlag: Cologne, Germany, 2003.
25. Pila, J. 36 IIC 173-191 **2005**.
26. Kingston, W. *I.P.Q.* **2004**, 369–378.
26a. Silverman, R. *AIPLA Q. J.* **2005**, *3*, 153–187.
27. General Tire vs. Firestone [1972]. RPC 457.
28. US. Title 35 of the United States Code. http://www.law.cornell.edu/uscode/html/uscode35/usc_sup_01_35.html (accessed April 2006).
29. James, L. S. In *Death of Patents*; Drahos, P. Ed.; Lawtext Publishing: Witney, UK, 2005; Chapter 3, 67–109.
30. Chisum, D. S. *Chisum on Patents. A Treatise on the Law of Patentability*; Matthew Bender: New York, 1997.
31. Eaton, P. E. et al. *J. Am. Chem. Soc.* **1964**, *86*, 3157.
32. Hook, E. 2004. http://www.chemsoc.org/exemplarchem/entries/2004a/Bristol_hook/applications/index.html (accessed April 2006).
33. USPTO. *Manual of Patent Examining Procedure*, 8th Edn.; USPTO: Washington, DC, 2001.
34. EPO. *Guidelines for Examination in the European Patent Office*; EPO: Munich, Germany, 2003 (the 2005 edition of the EPO Guidelines is available on-line at http://www.european-patent-office.org/legal/gui_lines/index.htm (accessed April 2006)).
35. CBD. Convention on Biological Diversity, 1992. http://www.biodiv.org/convention/default.shtml (accessed April 2006).
36. EU. *Directive 98/44 of the European Parliament and of the Council of July 6, 1998 on the Legal Protection of Biotechnological Inventions*. OJ L 213, 30.7.1998, pp 13–21.
37. Kamstra, G. et al. *Patents on Biotechnological Inventions: The EC Directive*; Sweet & Maxwell, 2002.
38. Harvard. EPO Technical Board of Appeal Decision T0315/03 Transgenic Animals/HARVARD, 2005.

39. EU. *Report from the Commission to the Council and the European Parliament: Development and Implications of Patent Law in the Field of Biotechnology and Genetic Engineering*, Brussels, 14 July, 2005; COM(2005) 312 final; Commission of the European Communities: Brussels, Belgium.
40. US. US Constitution Article I, Section 8, Clause 8' 2005. http://www.house.gov/ (accessed April 2006).
41. WIPO. *Paris Convention for the Protection of Industrial Property*; WIPO: Geneva, Switzerland, 1991.
42. Penrose, E. T. *The Economics of the International Patent System*; The Johns Hopkins Press, 1951.
43. Bodenhausen, G. H. C. *Guide to the Application of the Paris Convention for the Protection of Industrial Property*; WIPO, 1969.
44. WIPO. *Patent Cooperation Treaty (PCT)*; WIPO: Geneva, Switzerland, 2004.
45. WIPO. *PCT Newsletter* **2005**, August, pp 11–24.
46. EU. European Communities No. 18; *Convention for the European Patent for the Common Market (Community Patent Convention)*, Luxembourg, 15 December, 1975; HMSO: London, UK, 1976, Cmnd.6553.
47. *SCRIP* 15 June, 2005, p 2.
48. US. H.R. 2795 The Patent Reform Act of 2005. http://www.govtrack.us/congress/billtext.xpd?bill = h109-2795 (accessed April 2006).
49. Geier, P. Bill in Congress to Overhaul Patent Law Seeks to Quell Suits. *The National Law Journal*, Aug 19, 2005.
50. Myers, G. *Social Stud. Sci.* **1995**, *25*, 57–105.
51. McSherry, C. *Who Owns Academic Work?*; Harvard University Press, 2001.
52. Bostyn, S. J. R. *Enabling Biotechnological Inventions in Europe and the United States*; EPO, 2001.
53. Wegner, H. G. *Patent Law in Biotechnology, Chemicals & Pharmaceuticals*; Stockton Press, 1992.
54. Jaenichen, H. R. et al. *From Clones to Claims*, 3rd ed.; Carl Heymanns Verlag: Cologne, Germany, 2002; Chapters 22 and 23, pp 493–517.
55. Lim, A. et al. *I.P.Q.* **2005**, 236–266.
56. UKPO. *The CIPA Journal* **2004**, 125–129.
57. Phillips, J. *Charles Dickens and the 'Poor Man's Tale of a Patent'*; ESC Publishing: Oxford, UK, 1984, pp 15–21.
58. EPO. London Agreement on the Application of Article 65 EPC of 17 October 2000. *OJ EPO* **2001**, 549–553.
59. Lemley, M. A. et al. *Ending Abuse of Patent Continuations*; George Mason University Law and Economics Working Paper No. 03-52, UC Berkeley Public Law Research Paper No. 140, Virginia, US, 2003.
60. Morando, J. W. et al. Silent Enemies. *Recorder*, 4 May 1994, p 10.
61. NAFTA – North American Free Trade Association. http://www-tech.mit.edu/Bulletins/nafta.html (also http://www.nafta-sec-alena.org/DefaultSite/index_e.aspx?DetailID = 168#A1709) (accessed April 2006).
62. Salamolard, J. M. *La Licence Obligatoire en Matière de Brevets d'Invention*; Droz, 1978.
63. Vitoria, M. et al. *Encyclopedia of United Kingdom and European Patent Law*. Sweet, & Maxwell: London, UK, 1977–2006.
64. Ogilvy Renault. *The CIPA Journal* **2004**, 474–475.
65. Walter, R. *Anson's Voyage Round the World in the Years 1740–44*; Dover Publications, 1974.
66. Dutfield, G. et al. *I.P.Q.* **2004**; 379–421.
67. Lehman, B. A. Hearings before the Subcommittee on Courts, Intellectual Property of the Committee on the Judiciary, House of Representatives, 104th Congress, 1st Session, 8 June and 1 November, 1995; US Government Printing House: Washington, DC, 1996, p 47.
68. Thompson, G. et al. *Intellectual Property Rights and United States International Trade Laws*; Oceana Publications: Dobbs Ferry, NY, 2002, chapter XLVI and appendix C, pp 229–235 and 409–430.
69. *SCRIP* 13 May, 2005, p 18.
70. Cockbain, J. *Global Outsourcing Rev.* **2004**, *6*, 51–52.
71. US. Bayh-Dole Act, 37 C.F.R. 401, 2005. http://access.gpo.gov/nara/cfr/waisidx_02/37cfr401_02.html.
72. Oxford. Statutes and Regulations of the University of Oxford, Statute XVI, Section 5. http://www.admin.ox.ac.uk/statutes/790-121.shtml (accessed April 2006).
73. UKPO. http://www.patent.gov.uk/patent/indetail/whomay.htm (accessed April 2006).
74. Norway. Lov om retten til oppfinnelser som er gjort av arbeidstakere 1970-04-17 nr 0021, as amended by law 73 of 6 December, 2002, http://patentstyret.gazette.no/templates/Page_295.aspx
75. Shulman, S. *Owning the Future*; Houghton Mifflin: Boston, MA, 1999, Chapters 6 and 7.
76. Monotti, A. L. et al. *Universities and Intellectual Property*; OUP, 2003.
77. Janssens, M. C. *Uitvindingen in Dienstverband met bijzondere Aandacht voor Uitvindingen aan Universiteiten*; Bruylant, 1996.
78. Leistner, M. 35 IIC 859–872, 2004.
79. Nelson, R. R. *I.P.Q.* **2001**, 1–9.
80. Locke, J. *Two Treatises of Government and a Letter Concerning Toleration*, 2nd Treatise; Shapiro, I. ed.; Yale University Press: New Haven, CT, 2003, Chapter V.
81. Drahos, P. *A Philosophy of Intellectual Property*; Ashgate, 1996.
82. Priest, G. L. In *Research in Law and Economics: Vol. 8: The Economics of Patents and Copyrights*; Palmer, J., Ed.; JAI Press, 1986, p 19.
83. EU. *Proposal for a European Parliament, Council Directive on the Legal Protection of Biotechnological Inventions*, COM(88)496 final – SYN 159 of 17 October 1988, section 11, p 6.
84. Bentham, J. *A Manual of Political Economy*; Works of Jeremy Bentham; Bowring, J., Ed.; Tait: Edinburgh, UK, 1838–1843, p 1785.
85. Mill, J. S. *Principles of Political Economy With Some of Their Applications to Social Philosophy*; Longman Green: London, UK, 1871.
86. Lincoln, A. *The Collected Works of Abraham Lincoln*; Basler, R. P., Ed.; Rutgers University Press: New Brunswick, NJ, USA, 1953.
87. David, P. A. In *Global Dimensions of Intellectual Property Rightsin Science, Technology*; Wallerstein, M., Ed.; National Academy Press, 1993, pp 19–61.
88. Mansfield, E. *Manag. Sci.* **1986**, *32*, 173–181.
89. Dutfield, G. *Intellectual Property, Biogenetic Resources and Traditional Knowledge*; Earthscan: London, UK, 2004.
90. Heller, M. A. et al. *Science* **1998**, *280*, 698–701.
91. O'Gorman, E. *The CIPA Journal* **2004**, 337–343.
92. *SCRIP* 20 July, 2005, p 2.
93. *SCRIP* 29 April, 2005, pp 2–3.
94. Improver. Improver vs. Remington (1990). *FSR* 181.
95. Festo. Festo Corp. vs. Shoketsu Kinzoku Kogyo Kabushiki Co., 535 U.S., 28 May, 2002.
96. Catnic. *Catnic Components v. Hill & Smith* [1982] RPC 183.
97. *SCRIP* 16 March, 2005, p 4.

98. EU. *Proposal for a European Parliament and Council Directive on Criminal Measures Aimed at Ensuring the Enforcement of Intellectual Property Rights. Proposal for a Council Framework Decision to Strengthen the Criminal Law Framework to Combat Intellectual Property Offences*, Brussels, 12 July, 2005; COM(2005)276 final; Commission of the European Communities: Brussels, Belgium.
99. Gilat, D. *Experimental Use and Patents*; IIC Studies in Industrial Property and Copyright Law; VCH: Weinheim, Germany, 1995, Vol. 16.
100. New Republic. *The New Republic*, 23 May, 1988.
101. Alexander, D. In *Biotechnology, Patents and Morality*, 2nd ed.; Sterckx, S., Ed.; Ashgate: Aldershot, UK, 2000; pp 255–256 and 310.
102. Myskja, B. K. Themoral difference between in tragenic, transgenic modification of plants. *J. Agri. Environ. Ethics* **2006**, in press.
103. Stirling, A. In *Science and Citizenship*; Leach, M. et al., ed.; Zed Books: London, UK, 2005.
104. Harvard. EPO Technical Board of Appeal Decision T0019/90 Oncomouse/HARVARD. *OJ EPO* **1990**, 476.
105. Jamieson, D. *Morality's Progress*; Clarendon Press: Oxford, UK, 2002, pp 118–134.
106. Cornish, W. R. et al. *Intellectual Property*, 5th ed.; Sweet & Maxwell, 2003.
107. Jacob, R. et al. *A Guidebook to Intellectual Property*, 5th ed.; Sweet & Maxwell, 2004.
108. UPOV. International Convention for the Protection of New Varieties of Plants, 2005. http://www.upov.int/ (accessed April 2006).
109. UPOV. *International Union for the Protection of New Varieties of Plants: What It Is, What It Does*; UPOV Publication No. 437(E); UPOV: Geneva, Switzerland, 2005.
110. Van der Kooij, P. *Introduction to the EC Regulation on Plant Variety Protection*; Kluwer, 1997.

Biography

J Cockbain has a degree (1976) and a doctorate (1980) in Chemistry from the University of Oxford, UK, is qualified as a British Patent Attorney (1983) and as a European Patent Attorney (1984), and is registered as a Trade Mark Attorney in Britain and in the European Union. He has been a partner in the firm of Frank B Dehn & Co since 1985 and specializes in obtaining and defending patents as well as in opposing and designing around patents belonging to his clients' competitors, particularly in fields such as pharmaceuticals, medical imaging, medical diagnostics, hydrocarbons, and animal feedstuffs. His clients range from individuals to large companies, and include Universities and academics. He is not a specialist in patent litigation and he is not a US patent attorney, and this chapter is written from the perspective of a patent attorney working in Europe and focuses on issues relevant to medicinal chemistry. Cockbain is a firm believer in the usefulness of the patent system, particularly to investment-intensive industries, especially the pharmaceutical industry. The views expressed in this chapter, a standard statement in articles written by attorneys, are his own and not those of his firm or his clients.

Comprehensive Medicinal Chemistry II
ISBN (set): 0-08-044513-6
ISBN (Volume 1) 0-08-044514-4; pp. 779–815

INDEX FOR VOLUME 1

Notes

Abbreviations

ADME – Absorption, Distribution, Metabolism, Excretion
ADMET – Absorption, Distribution, Metabolism, Excretion, Toxicity
HTS – high-throughput screening
SAR – structure–activity relationships
SME – small/medium-sized enterprise

Cross-reference terms in italics are general cross-references, or refer to subentry terms within the main entry (the main entry is not repeated to save space). Readers are also advised to refer to the end of each article for additional cross-references – not all of these cross-references have been included in the index cross-references.

The index is arranged in set-out style with a maximum of three levels of heading. Major discussion of a subject is indicated by bold page numbers. Page numbers suffixed by T and F refer to Tables and Figures respectively. *vs.* indicates a comparison.

Names of scientists included in subentries refer to their development role, unless otherwise specified.

This index is in letter-by-letter order, whereby hyphens and spaces within index headings are ignored in the alphabetization. Prefixes and terms in parentheses are excluded from the initial alphabetization.

Any method, model or other subject, associated with the name of the developer (e.g. name's model) does NOT imply that Elsevier, nor the indexers, have assumed the right to name models/methods after the authors of the papers in which they are described. This is merely a succinct phrase to refer to a model/method developed/described by the relevant author, so that the subentry could be alphabetized under the most pertinent name.

A

B

C

D

E

F

G

H

I

M

N

O

P

S

U

V